STUDENT'S SOLUTIONS MANUAL

NORMA F. JAMES
New Mexico State University

A GRAPHICAL APPROACH TO PRECALCULUS WITH LIMITS

THIRD EDITION

John Hornsby
University of New Orleans

Margaret L. Lial
American River College

Gary K. Rockswold
Minnesota State University–Mankato

Boston San Francisco New York
London Toronto Sydney Tokyo Singapore Madrid
Mexico City Munich Paris Cape Town Hong Kong Montreal

Reproduced by Addison-Wesley from electronic files supplied by the author.

ISBN 0-201-79264-8

1 2 3 4 5 6 7 8 9 10 VG 05 04 03 02

PREFACE

This book provides complete solutions for the following exercises in A Graphical Approach To Precalculus with Limits, third edition, by E. John Hornsby, Jr., Margaret L. Lial, and Gary Rockswold.

All of the odd-numbered exercises
All of the exercises in *Relating Concepts*
All of the exercises in *Reviewing Basic Concepts*
All of the *Chapter Review* and *Chapter Test* exercises

This book should be used as an aid as you work to master your course work. Try to solve the assigned exercises before you refer to the solutions in the book. Then, if you have any difficulty, read these solutions to guide you in solving the exercises. The solutions have been written so that they are consistent with the methods used in the textbook.

The graphics presented in this manual are hand drawn, TI generated, and provided by the Maple Program. You may notice two zeros at the intersection of the axes, and the x and the y labels on those axes may be placed under, over, or at the side of the axes. The graphs may also *look* differently than those you obtain on your graphics calculator.

You may find that some of the solutions are presented in greater detail than others. Thus, if you cannot find an explanation for a difficulty that you encountered in one exercise, you may find the explanation in a solution to a previous exercise.

I would like to express my appreciation to Professor Janis M. Cimperman, St. Cloud State University of Minnesota, for her suggestions and her excellent work in proof- reading these solutions. I feel fortunate to have been able to work with authors of such high caliber as John Hornsby, Margaret Lial, and Gary Rockswold.

If you have any suggestions or corrections, I would appreciate hearing from you. Please write to *Addison Wesley, attention Christine O'Brien, Editor*, 75 *Arlington Street, Suite* 300, *Boston, MA* 02116.

Good luck with your study of Precalculus With Limits with graphing calculators. It is an exciting approach.

CONTENTS

CHAPTER 1

Section 1.1

1. (a) 10

(b) 0, 10

(c) $-6,\ -\frac{12}{4}\ (-3),\ 0,\ 10$

(d) $-6,\ -\frac{12}{4}\ (-3),\ -\frac{5}{8},\ 0,\ .31,\ .\overline{3},\ 10$

(e) $-\sqrt{3},\ 2\pi,\ \sqrt{17}$

(f) All are real numbers.

3. (a) None are natural numbers.

(b) None are whole numbers.

(c) $-\sqrt{100},\ -1$

(d) $-\sqrt{100},\ -\frac{13}{6},\ -1,\ 5.23,\ 9.\overline{14},\ 3.14,\ \frac{22}{7}$

(e) None are irrational numbers.

(f) All are real numbers.

5.

−4 −3 −2 −1 0 1

7.

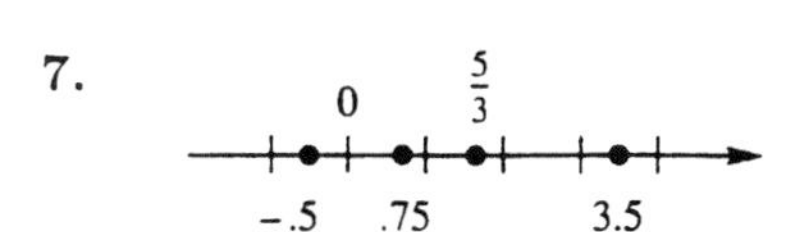

9. A Rational number can be written as a fraction, $\frac{p}{q}, q \neq 0$, where p and q are integers. An irrational number cannot.

Graph for 11. – 19:

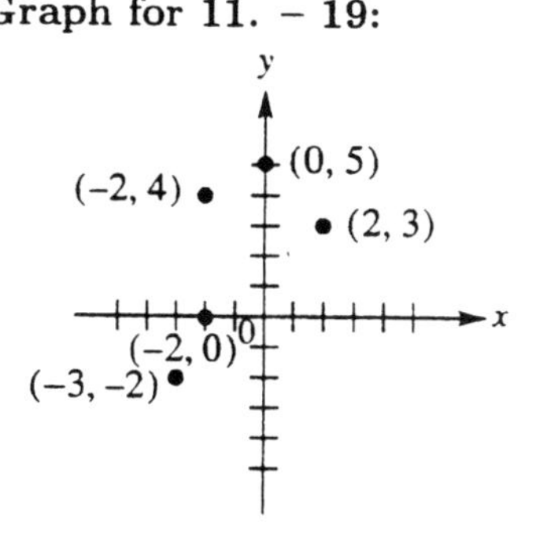

11. $(2,\ 3):$ I

13. $(-3, -2):$ III

15. $(0,\ 5):$ No quadrant

17. $(-2,\ 4):$ II

19. $(-2,\ 0):$ No quadrant

21. $xy > 0$

$x > 0$ and $y > 0$; **Quad I**

or **or**

$x < 0$ and $y < 0$; **Quad III**

23. $\frac{x}{y} < 0$

$x > 0$ and $y < 0$; **Quad IV**

or **or**

$x < 0$ and $y > 0$; **Quad II**

25. $(0,\ b)$ is on the y-axis.

27. Answers will vary.

29. $[-5,\ 5]$ by $[-25,\ 25]$

31. $[-60,\ 60]$ by $[-100,\ 100]$

33. $[-500,\ 300]$ by $[-300,\ 500]$

35. $[-10,\ 10]$ by $[-10,\ 10]$

See *Answers to Selected Exercises* at the back of your textbook.

37. [−5, 10] by [−5, 10]
See *Answers to Selected Exercises* at the back of your textbook.

39. [−100, 100] by [−50, 50]
See *Answers to Selected Exercises* at the back of your textbook.

41. There are no tick marks, which is a result of setting the x-scale and the y-scale $= 0$.

43. $\sqrt{58} \approx 7.615773106 \approx 7.616$

45. $\sqrt[3]{33} \approx 3.20753433 \approx 3.208$

47. $\sqrt[4]{86} \approx 3.045261646 \approx 3.045$

49. $19^{1/2} \approx 4.358898944 \approx 4.359$

51. $\dfrac{5.6-3.1}{8.9+1.3}$
Calculator: $(5.6-3.1)\,/\,(8.9+1.3) \approx .25$

53. $\sqrt{\pi^3+1}$
Calculator: $\sqrt{\ }(\pi\hat{\ }3+1) \approx 5.66$

55. $3\,(5.9)^2 - 2\,(5.9) + 6 = 98.63$

57. $\sqrt{(4-6)^2+(7+1)^2}$
Calculator:
$\sqrt{\ }\left((4-6)^2+(7+1)^2\right) \approx 8.25$

59. $\dfrac{\sqrt{\pi-1}}{\sqrt{1+\pi}}$
Calculator: $\sqrt{\ }(\pi-1)\,/\sqrt{\ }(1+\pi) \approx .72$

61. $\dfrac{2}{1-\sqrt[3]{5}}$
Calculator: $2/\left(1-\sqrt[3]{5}\right) \approx -2.82$

63.
$$\begin{aligned} a^2+b^2 &= c^2 \\ 8^2+15^2 &= c^2 \\ 64+225 &= c^2 \\ 289 &= c^2 \\ c &= 17 \end{aligned}$$

65.
$$\begin{aligned} a^2+b^2 &= c^2 \\ 13^2+b^2 &= 85^2 \\ 169+b^2 &= 7225 \\ b^2 &= 7056 \\ b &= 84 \end{aligned}$$

67.
$$\begin{aligned} a^2+b^2 &= c^2 \\ 5^2+8^2 &= c^2 \\ 25+64 &= c^2 \\ c^2 &= 89 \\ c &= \sqrt{89} \end{aligned}$$

69.
$$\begin{aligned} a^2+b^2 &= c^2 \\ a^2+\left(\sqrt{13}\right)^2 &= \left(\sqrt{29}\right)^2 \\ a^2 &= 29-13 \\ a^2 &= 16 \\ a &= 4 \end{aligned}$$

71. $P(-4,\ 3)$, $Q(2,\ 5)$

(a) $d = \sqrt{(x_2-x_1)^2+(y_2-y_1)^2}$
$d = \sqrt{(2-(-4))^2+(5-3)^2}$
$d = \sqrt{(6)^2+(2)^2} = \sqrt{36+4}$
$\boxed{d = \sqrt{40} = 2\sqrt{10}}$

M.P. $= \left(\dfrac{x_1+x_2}{2},\ \dfrac{y_1+y_2}{2}\right)$

(b) M.P. $= \left(\dfrac{-4+2}{2},\ \dfrac{3+5}{2}\right) = \left(\dfrac{-2}{2},\ \dfrac{8}{2}\right)$
$\boxed{\text{M.P.} = (-1,\ 4)}$

73. $P(5,\ 7)$, $Q(13,\ -1)$
(a) $d = \sqrt{(5-13)^2+(7-(-1))^2}$
$d = \sqrt{(-8)^2+8^2}$
$d = \sqrt{128}$ $\boxed{d = 8\sqrt{2}}$

(b) $\boxed{\text{M.P.} = \left(\dfrac{5+13}{2},\ \dfrac{7-1}{2}\right) = (9,\ 3)}$

75. $P(-8, -2)$, $Q(-3, -5)$

(a) $d = \sqrt{(-8-(-3))^2 + (-2-(-5))^2}$

$d = \sqrt{(-5)^2 + 3^2}$

$\boxed{d = \sqrt{34}}$

(b) $\text{M.P.} = \left(\frac{-8-3}{2}, \frac{-2-5}{2}\right)$

$\boxed{\text{M.P.} = (-\frac{11}{2}, -\frac{7}{2}) \text{ or } (-5.5, -3.5)}$

77. (1991, 10017), (1998, 15380)

$\text{M.P.} = \left(\frac{1991+1998}{2}, \frac{10017+15380}{2}\right)$

$= (1995, 12698.5)$

$\boxed{\$12{,}698.50 \text{ is the cost.}}$

79. 1965: $M = \left(\frac{1960+1970}{2}, \frac{3022+3968}{2}\right)$

$= (1965, 3495): \boxed{\$3495}$

1985: $M = \left(\frac{1980+1990}{2}, \frac{8414+13359}{2}\right)$

$= (1985, 10886.5): \boxed{\$10{,}886.50}$

81. distance from (0, 0) to (3, 4):

$d_1 = \sqrt{(3-0)^2 + (4-0)^2}$

$d_1 = \sqrt{9+16} = \sqrt{25} = 5$

distance from (3, 4) to (7, 1):

$d_2 = \sqrt{(7-3)^2 + (1-4)^2}$

$d_2 = \sqrt{16+9} = \sqrt{25} = 5$

distance from (0, 0) to (7, 1):

$d_3 = \sqrt{(7-0)^2 + (1-0)^2}$

$d_3 = \sqrt{49+1} = \sqrt{50} = 5\sqrt{2}$

Since $d_1 = d_2$, the triangle is isosceles.

83. (a) Distance between cars:

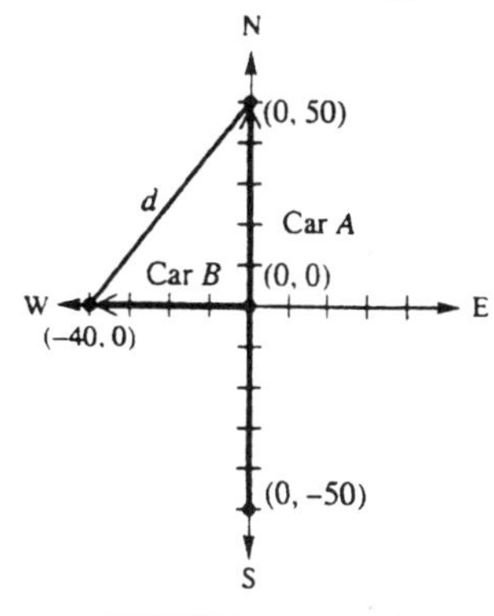

(b) $d = \sqrt{(50-0)^2 + (0-40)^2}$

$d = \sqrt{2500 + 1600}$

$d = \sqrt{4100} \approx 64.0312$ $\boxed{\approx 64.0 \text{ miles}}$

85. (a) The length of a side of the large square is $\underline{a+b}$, so its area is $\underline{(a+b)^2}$ or $\underline{a^2 + 2ab + b^2}$.

(b) The area of the large square may also be found by obtaining the sum of the areas of the four right triangles and the smaller square. The area of each right triangle is $\underline{\frac{1}{2}ab}$, so the sum of the areas of the four right triangles is $\underline{2ab}$. The area of the smaller square is $\underline{c^2}$.

(c) The sum of the four right triangles and the smaller square is $\underline{2ab + c^2}$.

(d) Since the areas in (a) and (c) represent the area of the same figure, the expressions there must be equal. Setting them equal to each other we obtain $\underline{a^2 + 2ab + b^2} = \underline{2ab + c^2}$.

(e) Subtract $2ab$ from each side of the equation in (d) to obtain the desired result $\underline{a^2 + b^2} = \underline{c^2}$.

Section 1.2

1. $(-1, 4)$

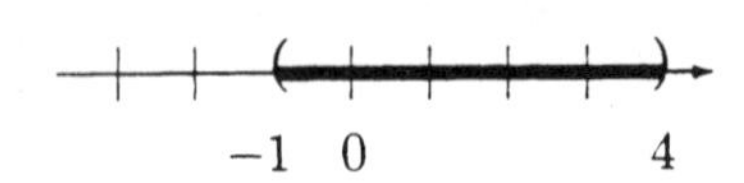

3. $(-\infty, 0)$

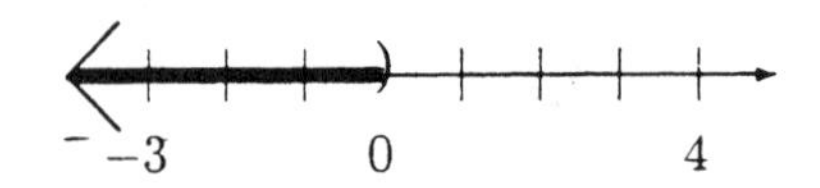

5. $[1, 2)$

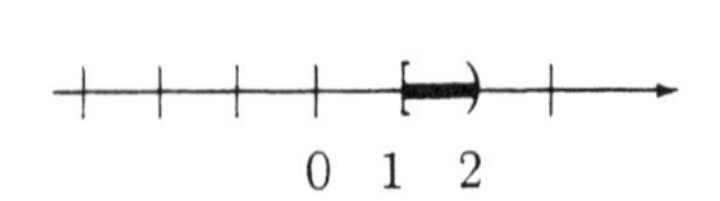

7. $(-4, 3) \rightarrow \{x \mid -4 < x < 3\}$

9. $(-\infty, -1] \rightarrow \{x \mid x \leq -1\}$

11. $\{x \mid -2 \leq x < 6\}$

13. $\{x \mid x \leq -4\}$

15. Parenthesis: $<$ or $>$
Square bracket: $\leq$ or $\geq$

17. Function;
Domain: $\{3, 4, 5, 7\}$
Range: $\{1, 2, 6, 9\}$

19. Not a Function;
Domain: $\{4, 3, -2\}$; Range: $\{1, -5, 3, 7\}$

21. Function;
Domain: $\{1, 2, 3, 4\}$; Range: $\{-5\}$

23. Function;
Domain: $(-\infty, \infty)$
Range: $(-\infty, \infty)$

25. Not a function;
Domain: $[-4, 4]$
Range: $[-3, 3]$

27. Function;
Domain: $[2, \infty)$
Range: $[0, \infty)$

29. Not a function;
Domain: $[-9, \infty)$
Range: $(-\infty, \infty)$

31. Function;
Domain: $\{-5, -2, -1, -.5, 0, 1.75, 3.5\}$
Range: $\{-1, 2, 3, 3.5, 4, 5.75, 7.5\}$

33. Function;
Domain: $\{2, 3, 5, 11, 17\}$
Range: $\{1, 7, 20\}$

35. $f(x) = -2x + 9$
$f(3) = -2(3) + 9 = 3$

37. $f(11) = 7$

39. $f(x) = 2x$
$f(-3) = 2(-3) = -6$

41. $f(x) = 5$
$f(6) = 5$

43 $f(x) = x^2$
$f(-6) = (-6)^2 = 36$

45. $f(x) = \sqrt{x}$
$f(100) = \sqrt{100} = 10$

47. $f(-2) = 3 \rightarrow (-2,\ 3)$ is a point on f.

49. $(7,\ 8) \rightarrow f(7) = 8$

51. (a) $f(-2) = 0$ (b) $f(0) = 4$
(c) $f(1) = 2$ (d) $f(4) = 4$

53. (a) $f(-2) = -3$ (b) $f(0) = -2$
(c) $f(1) = 0$ (d) $f(4) = 2$

55. (a) – (d) Answers will vary;
See the definitions.

57. $f(x) = x - 3$

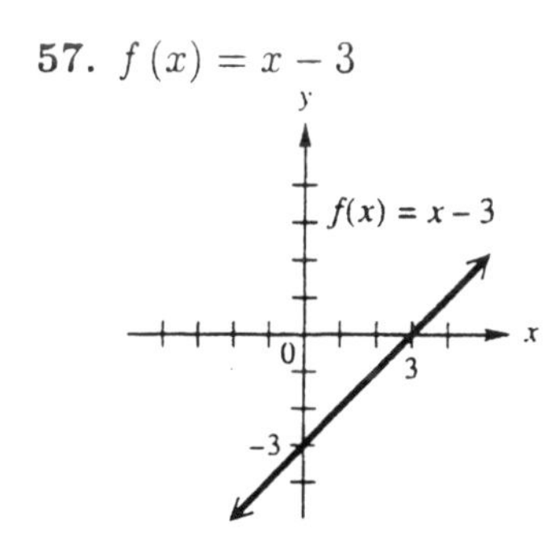

59. $f(x) = 3$

61. $f(x) = x^2$

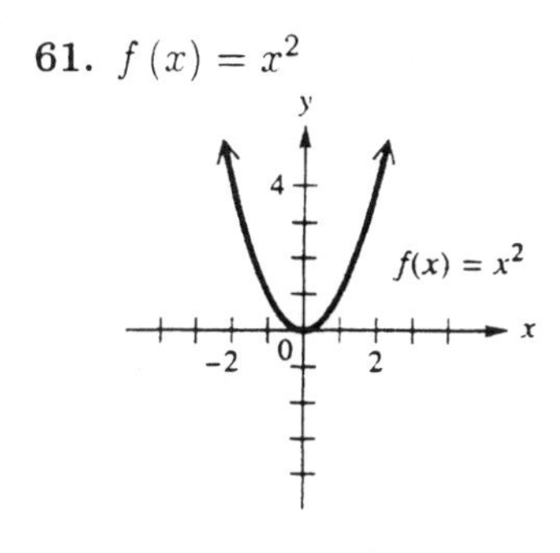

63. $f(x) = \dfrac{x}{5}$

(a) $f(15) = \dfrac{15}{5} = 3$
Approximately 3 miles between observer and lightning when delay is 15 seconds.

(b) $y = f(x)$:

65. $f(x) = .075x$
$f(86) = .075(86) = 6.45$
The sales tax is \$6.45.

67. $f(x) = 92x + 75$
$f(11) = 92(11) + 75 = 1087$
11 credits cost \$1087.

Reviewing Basic Concepts (Sections 1.1 and 1.2):

1. Plotting points:

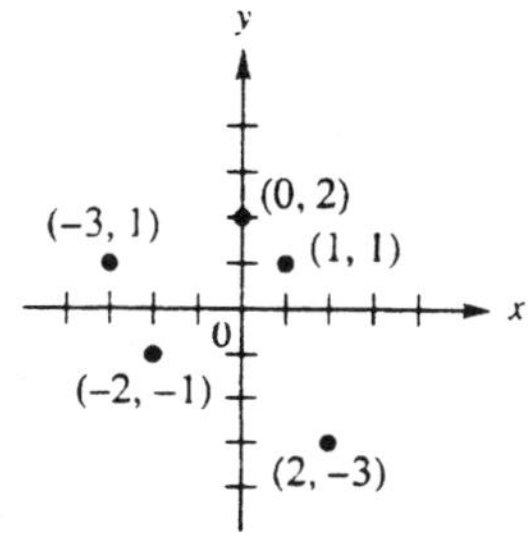

2. $P(-4, 5)$, $Q(6, -2)$:

$$d = \sqrt{(6-(-4))^2 + (-2-5)^2}$$

$$d = \sqrt{100+49} = \sqrt{149}$$

$$\text{M} = \left(\frac{-4+6}{2}, \frac{5-2}{2}\right) = \left(1, \frac{3}{2}\right)$$

3. $\dfrac{\sqrt{5+\pi}}{\sqrt[3]{3}+1} \approx 1.168$

4. $a^2 + b^2 = c^2$

$11^2 + b^2 = 61^2$

$b^2 = 61^2 - 11^2 = 3600$

$b = 60$ inches

5. $\{x \mid -2 < x \leq 5\}$: $(-2, 5]$

$\{x \mid x \geq 4\}$: $[4, \infty)$

6. Not a function;

domain: $[-2, 2]$

range: $[-3, 3]$

7. $f(x) = 3 - 4x$

$f(-5) = 3 - 4(-5) = 23$

8. $f(2) = 3$; $f(-1) = -3$

9. $f(x) = \frac{1}{2}x - 1$;

Section 1.3

1. $f(x) = x + 2$

(a) $f(-2) = 0$; $f(4) = 6$

(b) $(-2, 0) \Longrightarrow$ zero is -2

(c) The x-intercept is -2 and corresponds to the zero of f.

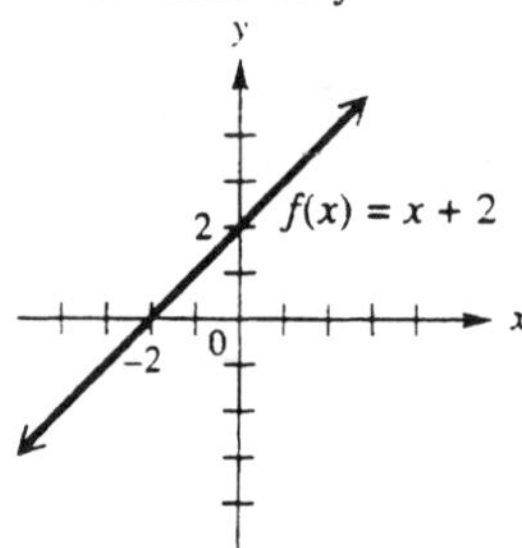

3. $f(x) = -3x + 2$

(a) $f(-2) = 8$; $f(4) = -10$

(b) $\left(\frac{2}{3}, 0\right) \Longrightarrow$ zero is $\frac{2}{3}$

(c) The x-intercept is $\frac{2}{3}$ and corresponds to the zero of f.

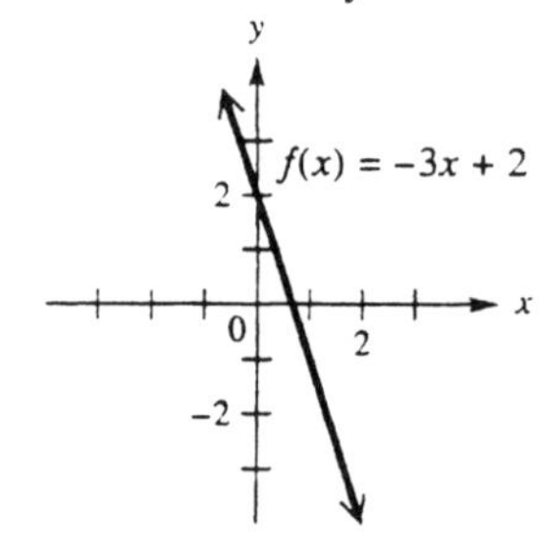

5. $f(x) = \frac{1}{3}x$

(a) $f(-2) = -\frac{2}{3}$; $f(4) = \frac{4}{3}$

(b) $(0, 0) \Longrightarrow$ zero is 0

(c) The x-intercept is 0 and corresponds to the zero of f.

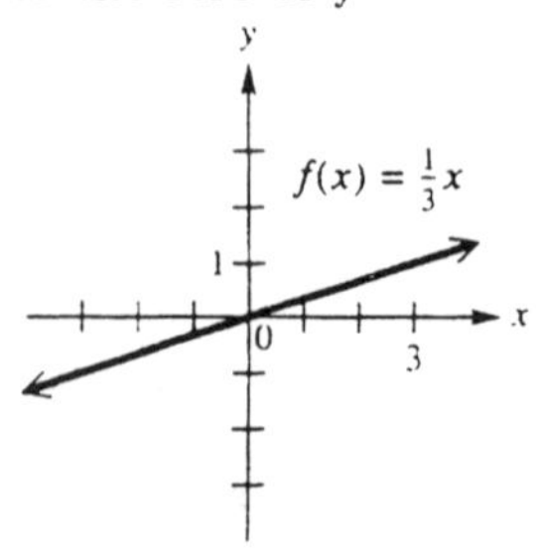

7. $f(x) = .4x + .15$

(a) $f(-2) = -.65$; $f(4) = 1.75$

(b) $(-.375, 0) \Longrightarrow$ zero is $-.375$

(c) The x-intercept is $-.375$ and corresponds to the zero of f.

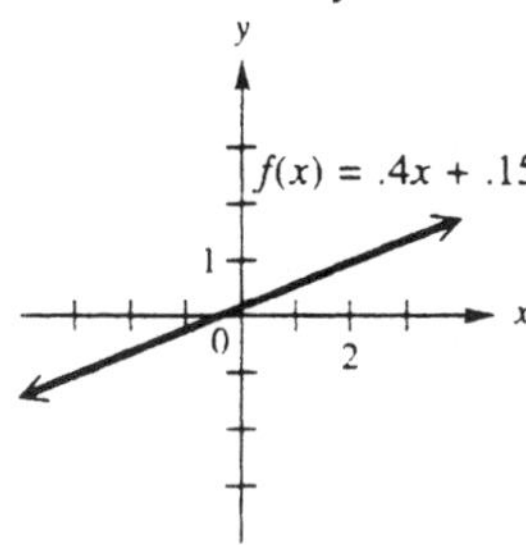

13. $f(x) = 3x - 6$

(a) x-intercept $= 2$

(b) y-intercept $= -6$

(c) Domain: $(-\infty, \infty)$

(d) Range: $(-\infty, \infty)$

(e) Slope $= 3$

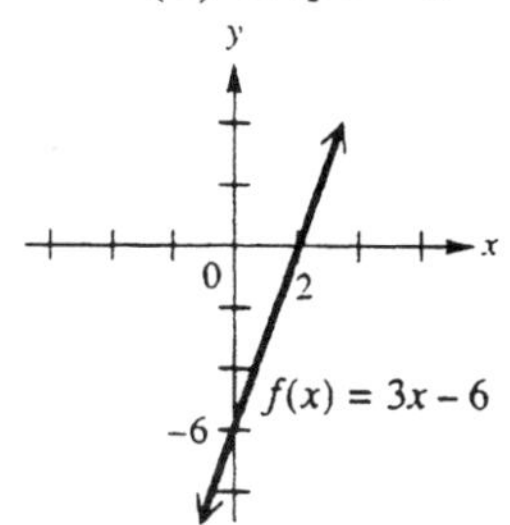

9. $f(x) = \dfrac{2-x}{4}$

(a) $f(-2) = 1$; $f(4) = -\dfrac{1}{2}$

(b) $(2, 0) \Longrightarrow$ zero is 2

(c) The x-intercept is 2 and corresponds to the zero of f.

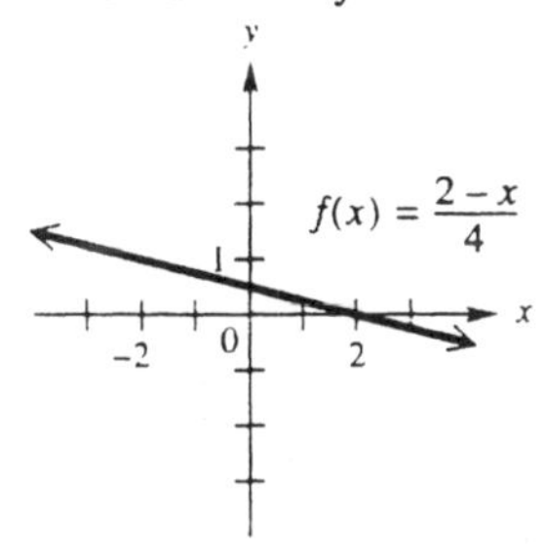

15. $f(x) = -\dfrac{2}{5}x + 2$

(a) x-intercept $= 5$

(b) y-intercept $= 2$

(c) Domain: $(-\infty, \infty)$

(d) Range: $(-\infty, \infty)$

(e) Slope $= -\frac{2}{5}$

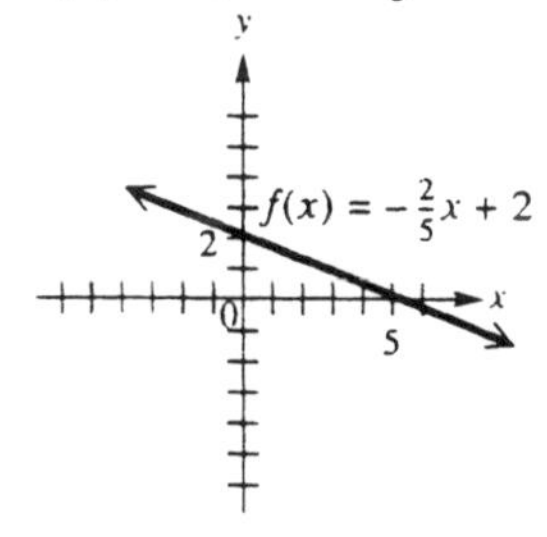

11. $f(x) = x - 4$

(a) x-intercept $= 4$

(b) y-intercept $= -4$

(c) Domain: $(-\infty, \infty)$

(d) Range: $(-\infty, \infty)$

(e) Slope $= 1$

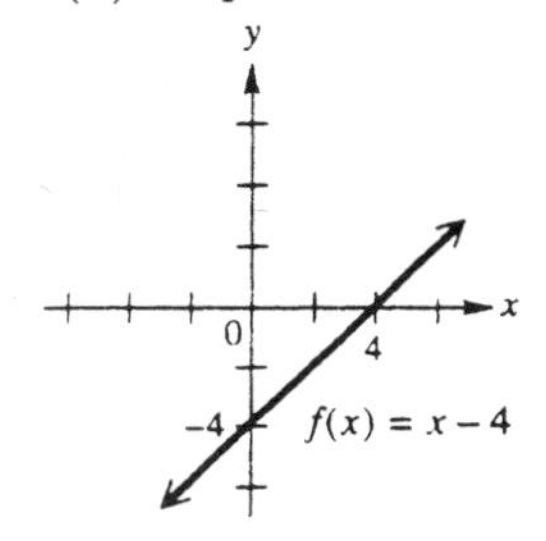

17. $f(x) = 3x$

(a) x-intercept $= 0$

(b) y-intercept $= 0$

(c) Domain: $(-\infty, \infty)$

(d) Range: $(-\infty, \infty)$

(e) Slope $= 3$

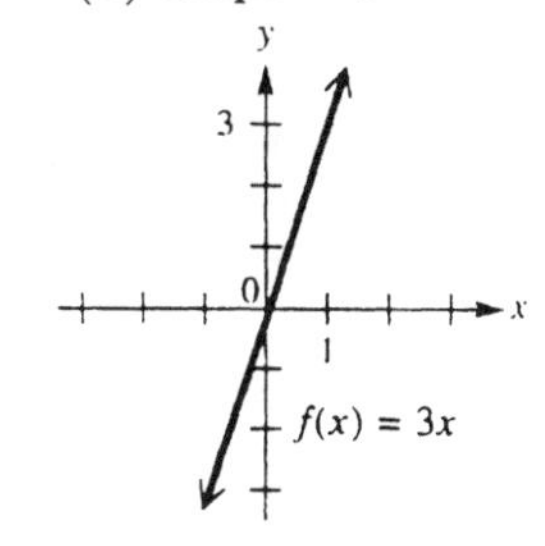

19. $y = ax$ always goes through $(0, 0)$.

21. $f(x) = -3$

(a) x-intercept: none
(b) y-intercept $= -3$
(c) Domain: $(-\infty, \infty)$
(d) Range: $\{-3\}$
(e) Slope $= 0$

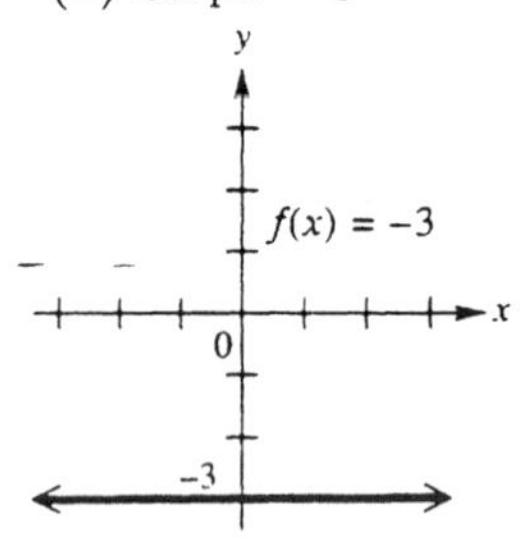

23. $x = -1.5$

(a) x-intercept $= -1.5$
(b) y-intercept: none
(c) Domain: $\{-1.5\}$
(d) Range: $(-\infty, \infty)$
(e) Slope is undefined

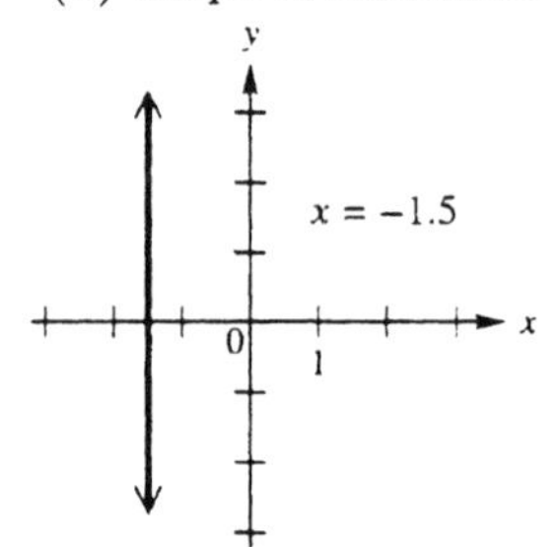

25. $x = 2$

(a) x-intercept $= 2$
(b) y-intercept: none
(c) Domain: $\{2\}$
(d) Range: $(-\infty, \infty)$
(e) Slope is undefined

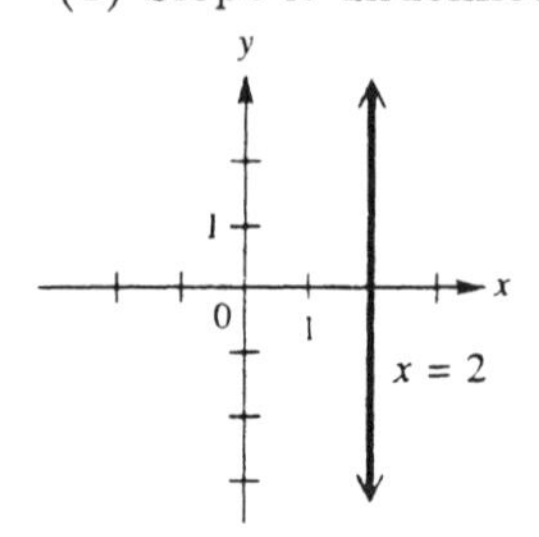

27. $f(x) = a$ is a constant function.

29. x-axis: $y = 0$

31. $f(x) = 4x + 20$: Window B

33. $f(x) = 3x + 10$: Window B

35. $m = \dfrac{2 - 1}{3 - (-2)} = \dfrac{1}{5}$

37. $m = \dfrac{4 - (-3)}{8 - (-1)} = \dfrac{7}{9}$

39. $m = \dfrac{5 - 5}{-6 - 12} = \dfrac{0}{-18} = 0$

41. line through $(0, -1)$ and $(1, 1)$:

$$m = \frac{y_2 - y_1}{x_2 - x_1} = \frac{1 - (-1)}{1 - 0} = \frac{2}{1} = \boxed{2}$$

43. line through $(0, 1)$ and $(3, -1)$:

$$m = \frac{y_2 - y_1}{x_2 - x_1} = \frac{-1 - 1}{3 - 0} = \frac{-2}{3} = \boxed{-\frac{2}{3}}$$

45. line through $(0, -3)$ and $(3, -3)$:

$$m = \frac{y_2 - y_1}{x_2 - x_1} = \frac{-3 - (-3)}{3 - 0} = \boxed{0}$$

horizontal line

47. $(-2, -6)$, $(0, 2)$

Any two points can be used.

$$m = \frac{-6 - 2}{-2 - 0} = \frac{-8}{-2} = 4$$

y-intercept is $(0, 2)$, so $b = 2$.

EQ: $y = mx + b$
$y = 4x + 2$

49. (a) $y = k$: A
(b) $y = -k$: C
(c) $x = k$: D
(d) $x = -k$: B

51. $(-1, 3)$, $m = \frac{3}{2}$ goes through $(1, 6)$

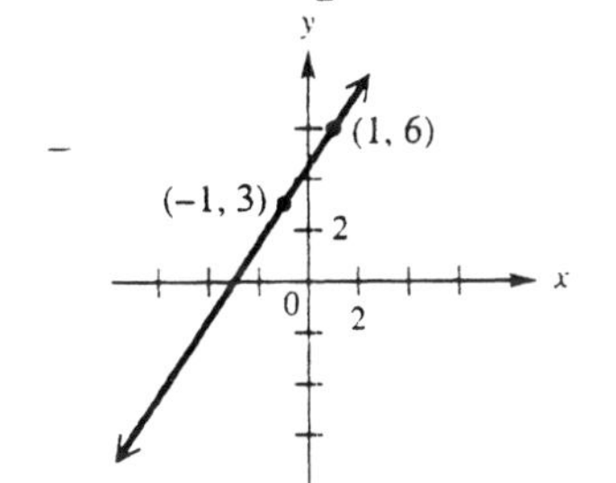

53. $(3, -4)$, $m = -\frac{1}{3}$ goes through $(6, -5)$

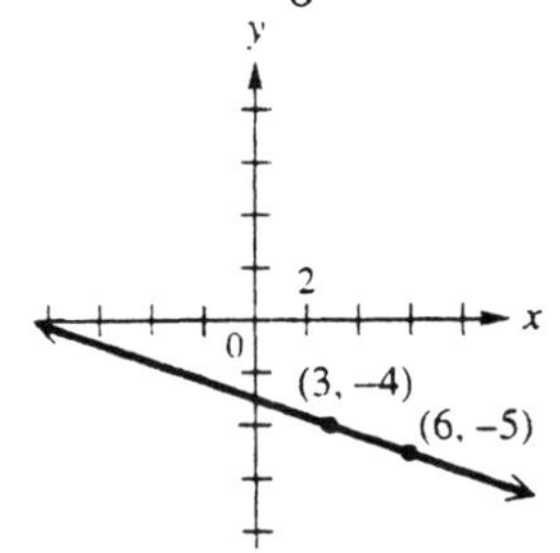

55. $(-1, 4)$, $m = 0$ goes through $(2, 4)$

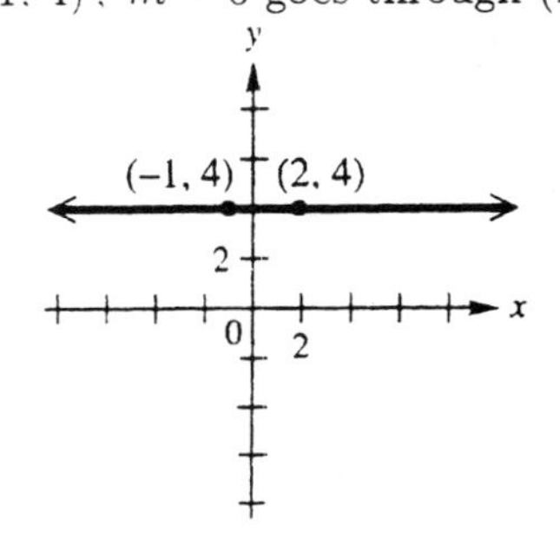

57. $(0, -4)$, $m = \frac{3}{4}$ goes through $(4, -1)$

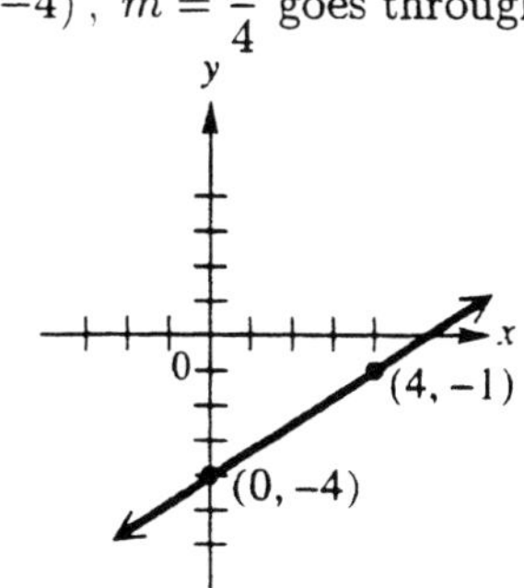

59. $m = \frac{3}{4}$, through $(0, -4)$

$$y = mx + b$$
$$y = \frac{3}{4}x - 4$$

61. $f(x) = \frac{1}{4}x + 3$

at 2:30pm $\rightarrow x = 2.5$ (2.5 hrs past noon)
$f(2.5) = 3.625$ inches

63. (a) line through $(0, 2000)$, $(4, 4000)$

$$m = \frac{y_2 - y_1}{x_2 - x_1} = \frac{4000 - 2000}{4 - 0} = 500$$
$$b = 2000$$
$$f(x) = 500x + 2000$$

(b) Water is entering the pool at 500 gallons per hour; the pool has 2000 gal initially.

(c) After 7 hrs, approximately 5500 gallons.

Section 1.4

1. through $(1, 3)$, $m = -2$:

$$y - 3 = -2(x - 1)$$
$$y - 3 = -2x + 2$$
$$\boxed{y = -2x + 5}$$

3. through $(-5, 4)$, $m = -1.5$:

$$y - 4 = -1.5\,(x - (-5))$$
$$-y - 4 = -1.5x - 7.5$$
$$y = -1.5x - 7.5 + 4$$
$$\boxed{y = -1.5x - 3.5}$$

5. through $(-8, 1)$, $m = -.5$:

$$y - 1 = -.5(x - (-8))$$
$$y - 1 = -.5x - 4$$
$$\boxed{y = -.5x - 3}$$

7. through $(-4, -6)$ and $(6, 2)$:

$$m = \frac{2 - (-6)}{6 - (-4)} = \frac{8}{10} = \tfrac{4}{5} = .8$$
$$y - 2 = .8(x - 6)$$
$$y - 2 = .8x - 4.8$$
$$\boxed{y = .8x - 2.8}$$

9. through $(2, 3.5)$ and $(6, -2.5)$:

$$m = \frac{3.5 - (-2.5)}{2 - 6} = \frac{6}{-4} = -1.5$$
$$y - 3.5 = -1.5(x - 2)$$
$$y - 3.5 = -1.5x + 3$$
$$\boxed{y = -1.5x + 6.5}$$

11. through $(-2, 2)$ and $(-4, -4)$:

$$m = \frac{2 - (-4)}{-2 - (-4)} = \frac{6}{2} = 3$$
$$y - 2 = 3(x - (-2))$$
$$y - 2 = 3x + 6$$
$$\boxed{y = 3x + 8}$$

13. through $(-2, -4)$ and $(-1, 4)$:

$$m = \frac{-4 - 4}{-2 - (-1)} = \frac{-8}{-1} = 8$$
$$y - 4 = 8(x - (-1))$$
$$y - 4 = 8x + 8$$
$$\boxed{y = 8x + 12}$$

15. $x - y = 4$

Let $y = 0 : x = 4$ $\boxed{x\text{-intercept: } (4, 0)}$

Let $x = 0 : -y = 4$

$y = -4$ $\boxed{y\text{-intercept: } (0, -4)}$

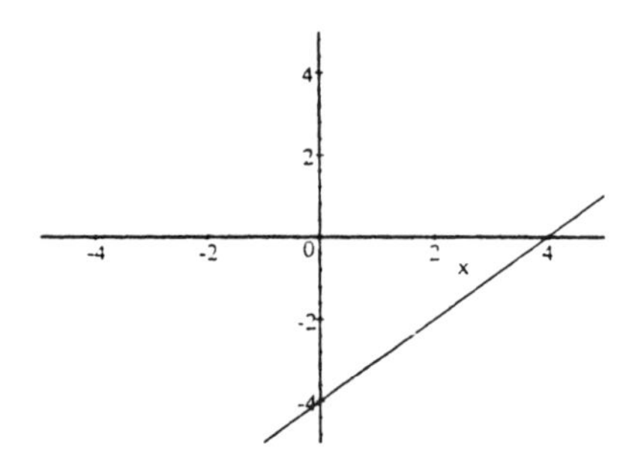

17. $3x - y = 6$

Let $y = 0 : 3x = 6$

$x = 2$ $\boxed{x\text{-intercept: } (2, 0)}$

Let $x = 0 : -y = 6$

$y = -6$ $\boxed{y\text{-intercept: } (0, -6)}$

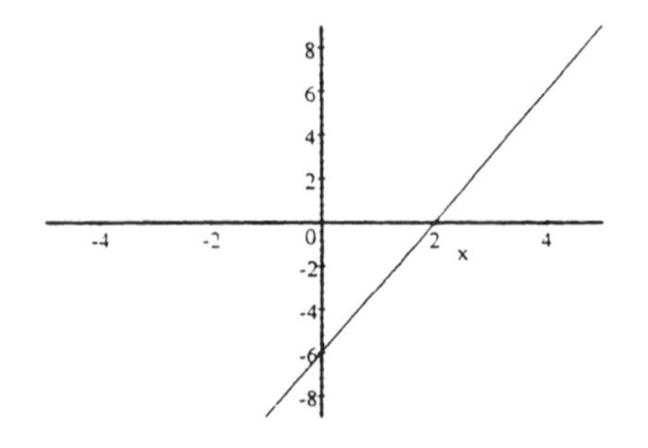

19. $2x + 5y = 10$

Let $y = 0: 2x = 10$

$x = 5$ $\boxed{x\text{-intercept: } (5, 0)}$

Let $x = 0: 5y = 10$

$y = 2$ $\boxed{y\text{-intercept: } (0, 2)}$

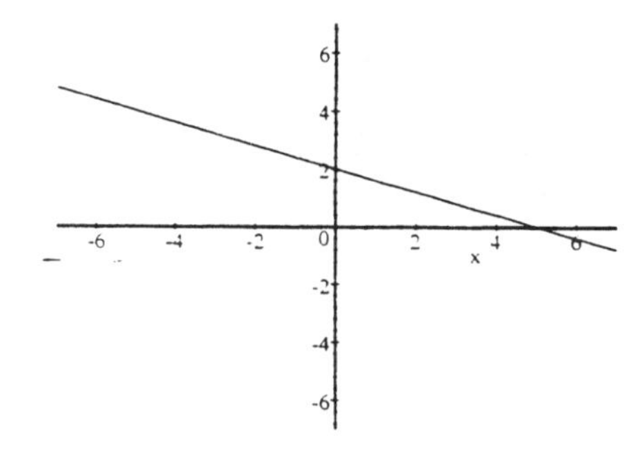

21. $y = 3x$

Let $x = 1$, then $y = 3$

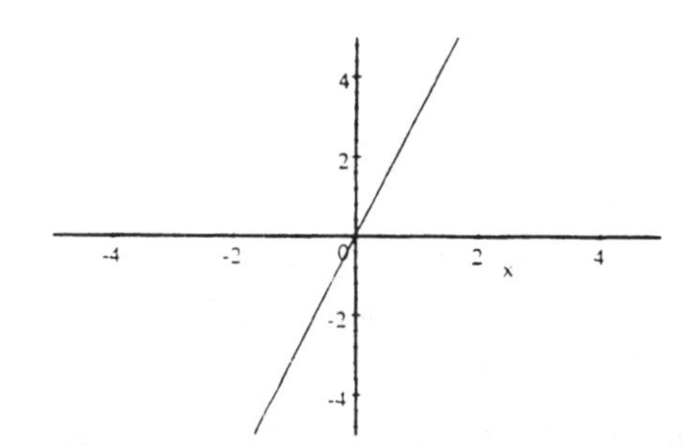

23. $y = -.75x$

Let $x = 4$, then $y = -3$

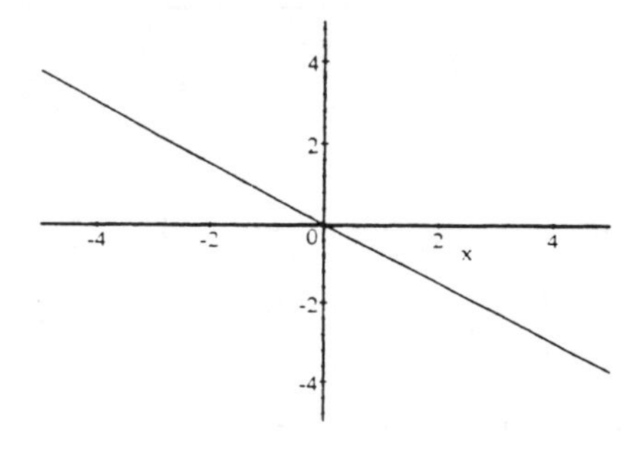

25. $5x + 3y = 15$

$3y = -5x + 15 \Longrightarrow y = -\dfrac{5}{3}x + 5$

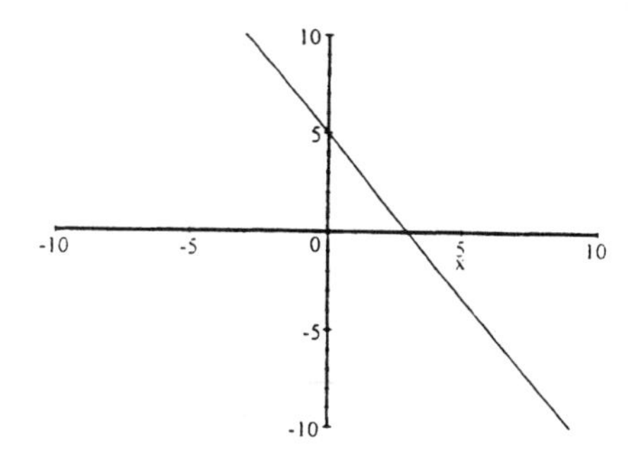

27. $-2x + 7y = 4$

$7y = 2x + 4 \Longrightarrow y = \dfrac{2}{7}x + \dfrac{4}{7}$

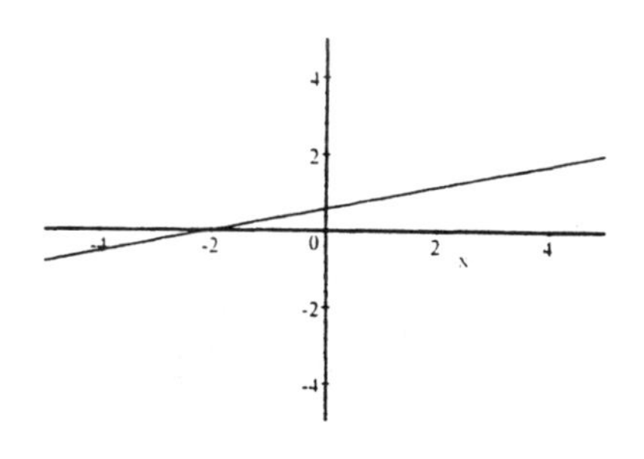

29. $1.2x + 1.6y = 5.0$

$12x + 16y = 50$

$16y = -12x + 50$

$y = -\dfrac{12}{16}x + \dfrac{50}{16} \Longrightarrow y = -\dfrac{3}{4}x + \dfrac{25}{8}$

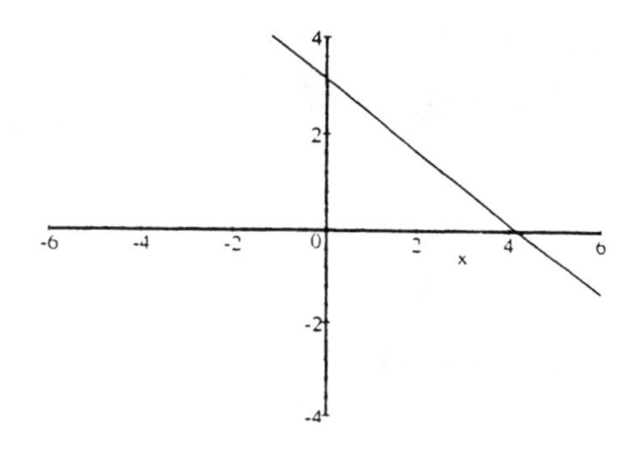

31. $x + 3y = 5$

$3y = -x + 5$

$y = -\frac{1}{3}x + \frac{5}{3} \quad \left(m = -\frac{1}{3}\right)$

Parallel lines have equal slopes,

so $m = -\frac{1}{3}$, $(-1, 4)$:

$y - 4 = -\frac{1}{3}(x - (-1))$

$y - 4 = -\frac{1}{3}x - \frac{1}{3}$

$y = -\frac{1}{3}x - \frac{1}{3} + \frac{12}{3} \Longrightarrow \boxed{y = -\frac{1}{3}x + \frac{11}{3}}$

33. $3x + 5y = 1$

$5y = -3x + 1$

$y = -\frac{3}{5}x + \frac{1}{5} \quad \left(m = -\frac{3}{5}\right)$

$\perp$ lines: negative reciprocal slopes,

so $m = \frac{5}{3}$, $(1, 6)$:

$y - 6 = \frac{5}{3}(x - 1)$

$y - 6 = \frac{5}{3}x - \frac{5}{3}$

$y = \frac{5}{3}x - \frac{5}{3} + \frac{18}{3} \Longrightarrow \boxed{y = \frac{5}{3}x + \frac{13}{3}}$

35. $y = -2$ has a slope $= 0$;

Undefined $\perp$ slope; equation of the form $x = a$.

$x = a$ through $(-5, 7)$ gives $\boxed{x = -5}$

37. $y = -.2x + 6$ has slope $= -.2$;

Parallel slope would also $= -.2$, through $(-5, 8)$:

$y - 8 = -.2(x - (-5))$

$y - 8 = -.2x - 1 \Longrightarrow \boxed{y = -.2x + 7}$

39. $2x + y = 6$

$y = -2x + 6 \ (m = -2)$

Perpendicular slope would be $\frac{1}{2}$ $(.5)$,

through $(0, 0)$: $\boxed{y = \frac{1}{2}x}$

41. Slope of $y_1 = 2.3$,
slope of $y_2 = 2.3001$.

Since non-vertical parallel lines have exactly equal slopes, these lines are not parallel.

43. **(a)** The Pythagorean Theorem and its converse.

(b) $(0, 0)$ to (x_1, m_1x_1)

$d(0, P) = \sqrt{(x_1)^2 + (m_1x_1)^2}$

(c) $(0, 0)$ to (x_2, m_2x_2)

$d(0, Q) = \sqrt{(x_2)^2 + (m_2x_2)^2}$

(d) (x_1, m_1x_1) to (x_2, m_2x_2)

$d(P, Q) = \sqrt{(x_1 - x_2)^2 + (m_1x_1 - m_2x_2)^2}$

(e) $[d(0, P)]^2 + [d(0, Q)]^2 = [d(P, Q)]^2$

$(x_1)^2 + (m_1x_1)^2 + (x_2)^2 + (m_2x_2)^2 = (x_1 - x_2)^2 + (m_1x_1 - m_2x_2)^2$

$(x_1)^2 + (m_1x_1)^2 + (x_2)^2 + (m_2x_2)^2 = (x_1)^2 - 2x_1x_2 + (x_2)^2 + (m_1x_1)^2 - 2m_1m_2x_1x_2 + (m_2x_2)^2$

$0 = -2x_1x_2 - 2m_1m_2x_1x_2$

(f) $0 = -2x_1x_2(1 + m_1m_2)$

(g) Product $= 0$, so each factor $= 0$:

$-2x_1x_2 = 0$, which is not possible;

$1 + m_1m_2 = 0 \rightarrow m_1m_2 = -1$

(h) The product of the slopes of two perpendicular, lines, neither parallel to an axis, is -1.

Relating Concepts (45 – 54):

45. $(0, -6)$, $(1, -3)$: $m = \frac{-6 - (-3)}{0 - 1} = \frac{-3}{-1} = 3$

46. $(1, -3)$, $(2, 0)$: $m = \frac{-3 - 0}{1 - 2} = \frac{-3}{-1} = 3$

47. If we use any two points on a line to find its slope, we find that the slope is <u>equal</u> in all cases.

48. $(0, -6), (1, -3)$

$$d = \sqrt{(0-1)^2 + (-6-(-3))^2} = \sqrt{1+9} = \sqrt{10}$$

49. $(1, -3), (3, 3)$

$$d = \sqrt{(1-3)^2 + (-3-3)^2}$$
$$= \sqrt{(-2)^2 + (-6)^2} = \sqrt{4+36} = 2\sqrt{10}$$

50. $(0, -6), (3, 3)$

$$d = \sqrt{(0-3)^2 + (-6-3)^2}$$
$$= \sqrt{(-3)^2 + (-9)^2} = \sqrt{9+81} = 3\sqrt{10}$$

51. $\sqrt{10} + 2\sqrt{10} = 3\sqrt{10}$, which is equal to the answer in Exercise 76.

52. If points A, B, and C lie on a line in that order, then the distance between A and B added to the distance between $\underline{B}$ and $\underline{C}$ is equal to the distance between $\underline{A}$ and $\underline{C}$.

53. $(0, -6), (6, 12)$

$$\text{M.P.} = \left(\frac{0+6}{2}, \frac{-6+12}{2}\right) = (3, 3)$$

The midpoint equals the middle entry.

54. $(4, 6), (5, 9)$

$$\text{M.P.} = \left(\frac{4+5}{2}, \frac{6+9}{2}\right) = \boxed{(4.5, 7.5)}$$

55. **(a)** $m = \dfrac{161-128}{4-1} = \dfrac{33}{3} = 11$

$y - 128 = 11(x-1)$
$y - 128 = 11x - 11$
$y = 11x + 117$

(b) Speed is 11 mph (slope)

(c) At $x = 0$, $y = 117$ miles

(d) At $x = 1.25$, $y = 11(1.25) + 117$
$y = 130.75$ miles

57. **(a)** Let $x = 0$ correspond to 1979.

Then the line through $(0, 1480)$, $m = 280$:

$y = 280x + 1480$

(b) $1988 - 1979 = 9$

At $x = 9$, $y = 280(9) + 1480$
$= 4000$ pollution incidents

59. **(a)** Temperature Scales:

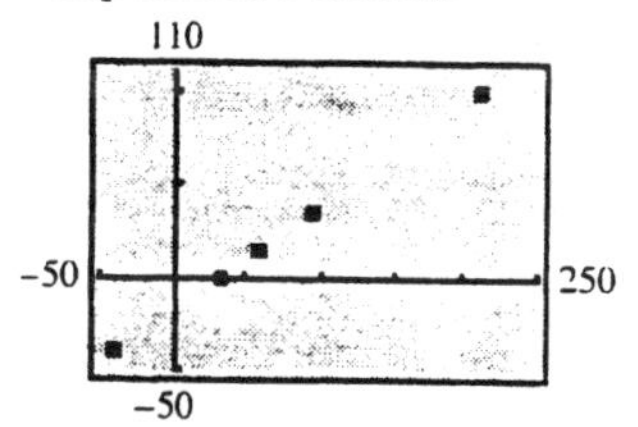

(b) $(-40, -40)$, $(32, 0)$

$$m = \frac{0-(-40)}{32-(-40)} = \frac{40}{72} = \frac{5}{9}$$

$$C(x) = \frac{5}{9}(x - 32)$$

Slope of $\frac{5}{9}$ means that Celsius temperature changes $5°$ for every $9°$ change in Fahrenheit temperature.

(c) $C(83) = \frac{5}{9}(83 - 32) = 28\frac{1}{3}°C$

61. **(a)** $m = \dfrac{60.0 - 37.7}{1999 - 1960} = \dfrac{22.3}{39} = \dfrac{223}{390}$

$$y - 37.7 = \frac{223}{390}(x - 1960)$$
$$y = \frac{223}{390}(x - 1960) + 37.7$$

(b) 1970: $y = \frac{223}{390}(1970 - 1960) + 37.7 = 43.4\%$

1980: $y = \frac{223}{390}(1980 - 1960) + 37.7 = 49.1\%$

1990: $y = \frac{223}{390}(1990 - 1960) + 37.7 = 54.9\%$

The value for 1970 is slightly higher, while the other two are slightly lower.

63. **(a)** $y \approx 635.042x - 1,254,358.2$

(b)

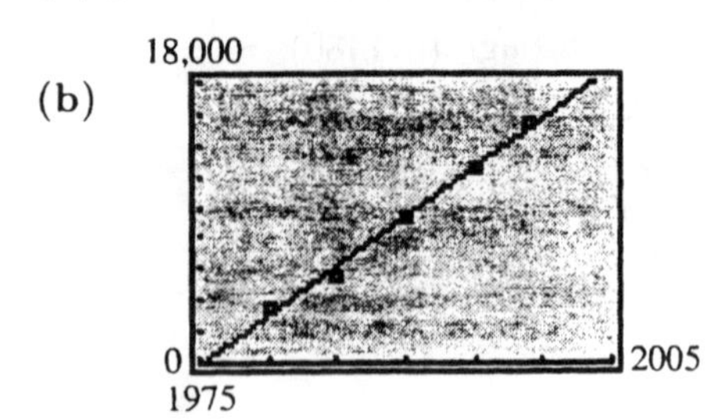

(c) At $x = 1992$, $y \approx \$10,645$,
which is close to the actual value of \$10,498.

65. **(a)** $y \approx 14.68x + 277.82$

(b) Graph y_1 (from (a)) and $y_2 = 37,000$;
intersection: $(2501.5,\ 37,000)$;
approximately 2500 light-years.

67. $\{x\} = \{63, 63.28, \ldots, 151\} \rightarrow L_1$
$\{y\} = \{26, 24.15, \ldots, 16\} \rightarrow L_2$

$y = .101x + 11.6; \quad r \approx .909$

There is a fairly strong positive correlation, because .909 is close to +1.

Reviewing Basic Concepts (Sections 1.3, 1.4)

1. $m = 1.4,\ b = -3.1$

$f(x) = 1.4x - 3.1$
$f(1.3) = 1.4(1.3) - 3.1 = -1.28$

2. $f(x) = -2x + 1$

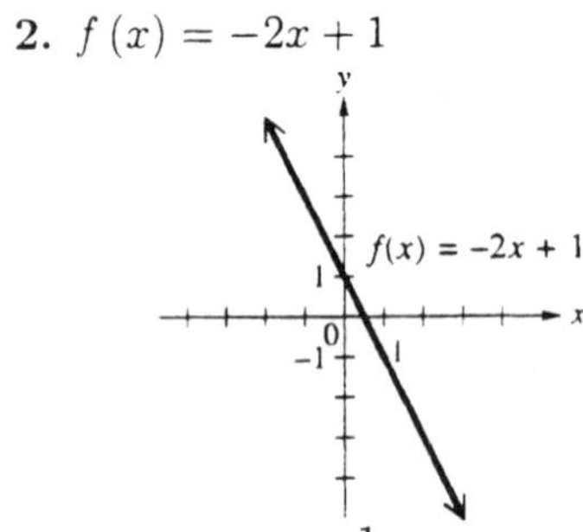

x-intercept: $\frac{1}{2}$, y-intercept: 1, slope $= -2$

domain: $(-\infty, \infty)$; range: $(-\infty, \infty)$

3. $(-2, 4)$ and $(5, 6)$: $m = \dfrac{6-4}{5-(-2)} = \dfrac{2}{7}$

4. Through $(-2,\ 10)$:
vertical: $x = -2$; horizontal: $y = 10$

5. $f(x) = .5x - 1.4$

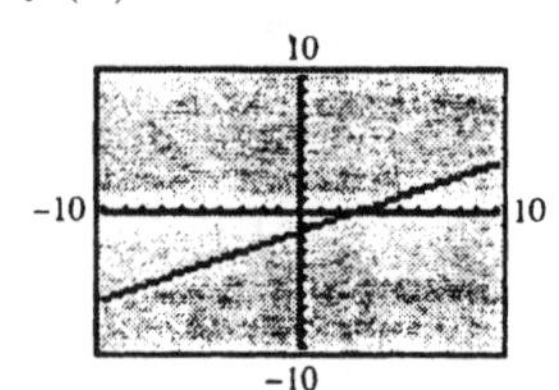

X	Y1
-3	-2.9
-2	-2.4
-1	-1.9
0	-1.4
1	-.9
2	-.4
3	.1

Y1=.5X−1.4

6. line through $(0, -3)$ and $(2,\ 1)$:

$$m = \frac{-3-1}{0-2} = \frac{-4}{-2} = 2; \quad b = -3 \Longrightarrow y = 2x - 3$$

7. $m = \dfrac{4-2}{-2-5} = \dfrac{2}{-7} = -\dfrac{2}{7}$

$$y - 4 = -\frac{2}{7}(x+2)$$
$$y - 4 = -\frac{2}{7}x - \frac{4}{7}$$
$$y = -\frac{2}{7}x - \frac{4}{7} + \frac{28}{7} \Longrightarrow y = -\frac{2}{7}x + \frac{24}{7}$$

8. $3x - 2y = 5$

$$-2y = -3x + 5$$
$$y = \frac{3}{2}x - \frac{5}{2}; \quad m = \frac{3}{2}$$

$\perp m = -\dfrac{2}{3}$ through $(-1,\ 3)$:

$$y - 3 = -\frac{2}{3}(x+1)$$
$$y - 3 = -\frac{2}{3}x - \frac{2}{3}$$
$$y = -\frac{2}{3}x - \frac{2}{3} + \frac{9}{3} \Longrightarrow y = -\frac{2}{3}x + \frac{7}{3}$$

9. **(a)**

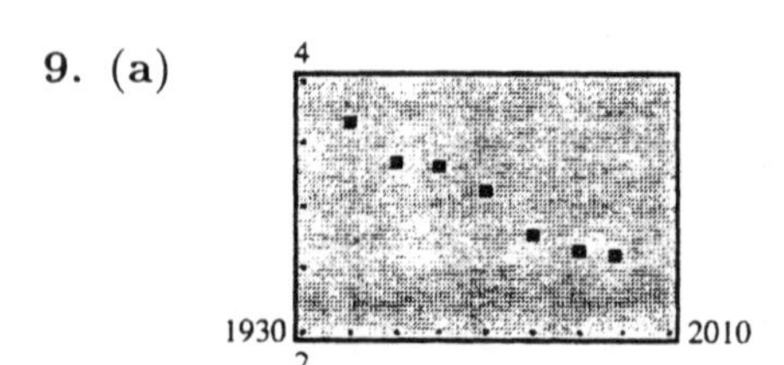

(b) Negative correlation coefficient

(c) $y \approx -.019x + 40.577$
$r \approx -.9792$

(d) 3.05; the estimate is close to the actual value of 2.94.

Section 1.5

1. Solve $f(x) = 0$:

$$-3x - 12 = 0$$
$$-3x = 12 \qquad x = \boxed{-4}$$

3. Solve $f(x) = 0$:

$$5x = 0 \qquad x = \boxed{0}$$

5. Solve $f(x) = 0$:

$$2(3x - 5) + 8(4x + 7) = 0$$
$$6x - 10 + 32x + 56 = 0$$
$$38x + 46 = 0$$
$$38x = -46 \qquad x = -\frac{46}{38} = \boxed{-\frac{23}{19}}$$

7. Solve $f(x) = 0$:

$$3x + 6(x - 4) = 0$$
$$3x + 6x - 24 = 0$$
$$9x = 24 \qquad x = \frac{24}{9} = \boxed{\frac{8}{3}}$$

9. Solve $f(x) = 0$:

$$1.5x + 2(x - 3) + 5.5(x + 9) = 0$$
$$1.5x + 2x - 6 + 5.5x + 49.5 = 0$$
$$9x + 43.5 = 0$$
$$9x = -43.5$$
$$x = -\frac{43.5}{9} = -\frac{435}{90}$$
$$x = \boxed{-\frac{29}{6}}$$

11. $y_1 = y_2 : x = \{10\}$

13. When $y = 0$, $x = \{-.8\}$.

15. When $x = 10$ is substituted in y_1 or y_2, the result is 20.

17. There is no real solution if $y_1 - y_2$ yields a contradiction, $y = b$, $b \neq 0$. The special name of the equation is a **contradiction**.

19. $2x - 5 = x + 7$

$$2x - x - 5 = x - x + 7$$
$$x - 5 + 5 = 7 + 5 \qquad \boxed{x = \{12\}}$$

Check: $2(12) - 5 = 12 + 7$

$$19 = 19$$

21. $.01x + 3.1 = 2.03x - 2.96$

$$100[.01x + 3.1 = 2.03x - 2.96]$$
$$1x + 310 = 203x - 296$$
$$x - 203x + 310 = 203x - 203x - 296$$
$$-202x + 310 - 310 = -296 - 310$$
$$-202x = -606 \qquad \boxed{x = \{3\}}$$

Check:

$$.01(3) + 3.1 = 2.03(3) - 2.96$$
$$.03 + 3.1 = 6.09 - 2.96$$
$$3.13 = 3.13$$

23. $-(x + 5) - (2 + 5x) + 8x = 3x - 5$

$$-x - 5 - 2 - 5x + 8x = 3x - 5$$
$$2x - 7 = 3x - 5$$
$$2x - 3x - 7 = 3x - 3x - 5$$
$$-x - 7 + 7 = -5 + 7$$
$$-x = 2 \qquad \boxed{x = \{-2\}}$$

Check:

$$-(-2 + 5) - (2 + 5(-2)) + 8(-2) = 3(-2) - 5$$
$$-3 - (-8) - 16 = -6 - 5$$
$$-11 = -11$$

25. $\dfrac{2x + 1}{3} + \dfrac{x - 1}{4} = \dfrac{13}{2}$

$$12\left(\frac{2x + 1}{3} + \frac{x - 1}{4}\right) = 12\left(\frac{13}{2}\right)$$
$$4(2x + 1) + 3(x - 1) = 6(13)$$
$$8x + 4 + 3x - 3 = 78$$
$$11x + 1 = 78$$
$$11x = 77 \qquad \boxed{x = \{7\}}$$

Check:

$$\frac{2 \cdot 7 + 1}{3} + \frac{7 - 1}{4} = \frac{13}{2}$$
$$\frac{15}{3} + \frac{6}{4} = \frac{13}{2}$$
$$\frac{60}{12} + \frac{18}{12} = \frac{13}{2}$$
$$\frac{78}{12} = \frac{13}{2} \implies \frac{13}{2} = \frac{13}{2}$$

27. $.40x + .60(100 - x) = .45(100)$

$100\,[.40x + .60(100 - x) = .45(100)]$

$40x + 60(100 - x) = 45(100)$

$40x + 6000 - 60x = 4500$

$-20x + 6000 = 4500$

$-20x = -1500$ $\boxed{x = \{75\}}$

Check:

$.40(75) + .60(100 - 75) = .45(100)$

$30 + .60(25) = 45$

$30 + 15 = 45$

$45 = 45$

29. $2\,[x - (4 + 2x) + 3] = 2x + 2$

$2\,[x - 4 - 2x + 3] = 2x + 2$

$2[-x - 1] = 2x + 2$

$-2x - 2 = 2x + 2$

$-4x - 2 = 2$

$-4x = 4$ $\boxed{x = \{-1\}}$

Check:

$2\,[-1 - (4 + 2(-1)) + 3] = 2(-1) + 2$

$2\,[-1 - 4 + 2 + 3] = -2 + 2$

$2[0] = 0 \Longrightarrow 0 = 0$

31. $\frac{5}{6}x - 2x + \frac{1}{3} = \frac{1}{3}$

$6\left(\frac{5}{6}x - 2x + \frac{1}{3} = \frac{1}{3}\right)$

$5x - 12x + 2 = 2$

$-7x + 2 = 2$

$-7x = 0$ $\boxed{x = \{0\}}$

Check:

$\frac{5}{6}(0) - 2(0) + \frac{1}{3} = \frac{1}{3}$

$0 - 0 + \frac{1}{3} = \frac{1}{3} \implies \frac{1}{3} = \frac{1}{3}$

33. $5x - (8 - x) = 2\,[-4 - (3 + 5x - 13)]$

$5x - 8 + x = 2\,[-4 - 3 - 5x + 13]$

$6x - 8 = 2[-5x + 6]$

$6x - 8 = -10x + 12$

$6x + 10x = 12 + 8$

$16x = 20$

$x = \frac{20}{16}$ $\boxed{x = \left\{\frac{5}{4}\right\}}$

Check:

$5 \cdot \frac{5}{4} - (8 - \frac{5}{4}) = 2\left[-4 - (3 + 5 \cdot \frac{5}{4} - 13)\right]$

$\frac{25}{4} - \frac{32}{4} + \frac{5}{4} = 2\left[-4 - 3 - \frac{25}{4} + 13\right]$

$-\frac{2}{4} = 2\left[6 - \frac{25}{4}\right]$

$-\frac{1}{2} = 2\left[-\frac{1}{4}\right]$

$-\frac{1}{2} = -\frac{1}{2}$

35. When $x = \{4\}$, both Y_1 and Y_2 have a value of 8.

37. When $x = \{1.5\}$, both Y_1 and Y_2 have a value of 4.5, so

$Y_1 - Y_2 = 4.5 - 4.5 = 0.$

39. $y_1 = 4(.23x + \sqrt{5})$ (y-intercept ≈ 8.9))

$y_2 = \sqrt{2}x + 1$ (y-intercept $=1$)

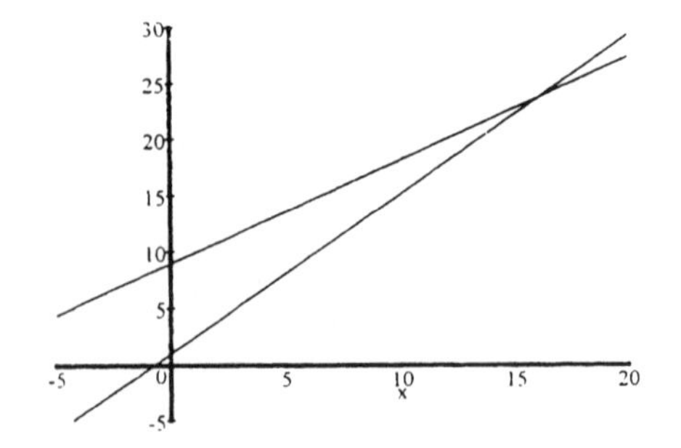

Solution set: $\boxed{\approx \{16.07\}}$

41. $y_1 = 2\pi x + \sqrt[3]{4}$ (y-intercept ≈ 1.6)
$y_2 = .5\pi x - \sqrt{28}$ (y-intercept ≈ -5.3))

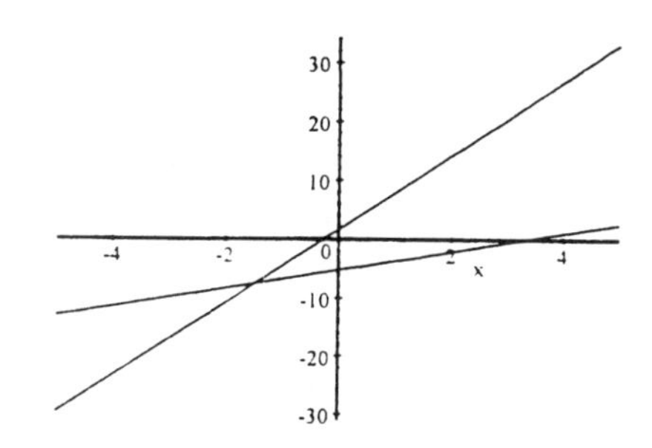

Solution set: $\boxed{\approx \{-1.46\}}$

43. $y_1 = .23(\sqrt{3} + 4x) - .82(\pi x + 2.3)$ $(m < 0)$
$y_2 = 5$ (horizontal line)

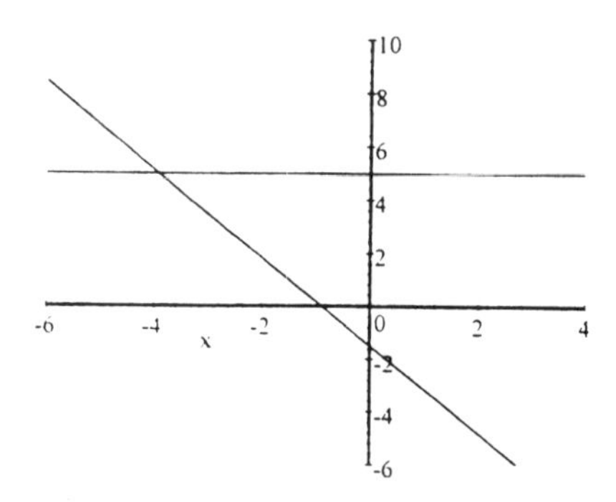

Solution set: $\boxed{\approx \{-3.92\}}$

45. $5x + 5 = 5(x + 3) - 10$
$5x + 5 = 5x + 15 - 10$
$5x + 5 = 5x + 5$
$5 = 5 \Longrightarrow$ Identity

Table: $y_1 = 5x + 5$; $y_2 = 5(x + 3) - 10$
start at $x = -2$. $\triangle = 1$.

47. $6(2x + 1) = 4x + 8\left(x + \dfrac{3}{4}\right)$
$12x + 6 = 4x + 8x + 6$
$12x + 6 = 12x + 6$
$6 = 6 \Longrightarrow$ Identity

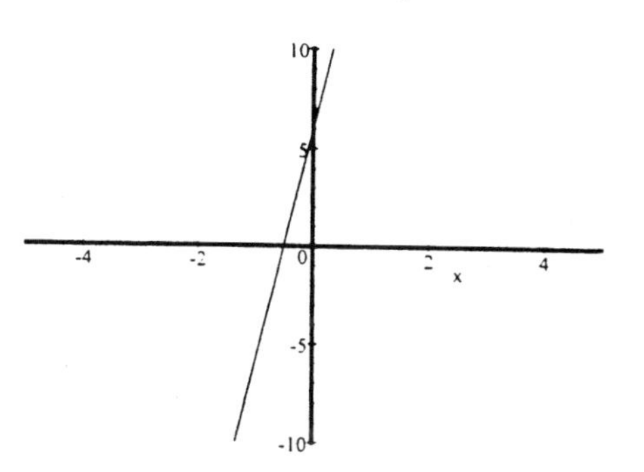

49. $-4[6 - (-2 + 3x)] = 21 + 12x$
$-4[6 + 2 - 3x] = 21 + 12x$
$-4[8 - 3x] = 21 + 12x$
$-32 + 12x = 21 + 12x$
$-32 = 21 \Longrightarrow$ Contradiction

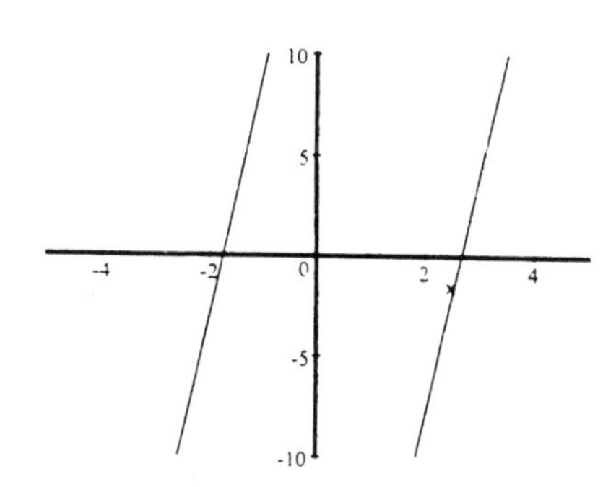

51. $f(x) = g(x)$ at $x = 3 : \{3\}$

53. $f(x) < g(x) : (3, \infty)$

55. $y_1 - y_2 \geq 0 \Longrightarrow y_1 \geq y_2 : (-\infty, 3]$

57. **(a)** $f(x) > 0 :$ $(20, \infty)$
(b) $f(x) < 0 :$ $(-\infty, 20)$
(c) $f(x) \geq 0 :$ $[20, \infty)$
(d) $f(x) \leq 0 :$ $(-\infty, 20]$

59. If $f(x) \geq g(x)$ at $[4, \infty)$

(a) $f(x) = g(x): \quad \{4\}$

(b) $f(x) > g(x): \quad (4, \infty)$

(c) $f(x) < g(x): \quad (-\infty, 4)$

61. (a) $9 - (x+1) < 0$

$9 - x - 1 < 0$

$-x + 8 < 0$

$-x < -8 \Rightarrow x > 8: \quad (8, \infty)$

(b) $9 - (x+1) \geq 0: \quad (-\infty, 8]$

63. (a) $2x - 3 > x + 2$

$x > 5: \quad (5, \infty)$

(b) $2x - 3 \leq x + 2: \quad (-\infty, 5]$

65. (a) $10x + 5 - 7x \geq 8(x+2) + 4$

$3x + 5 \geq 8x + 16 + 4$

$3x - 8x \geq 20 - 5$

$-5x \geq 15 \Rightarrow x \leq -3 \quad \boxed{(-\infty, -3]}$

(b) $10x + 5 - 7x < 8(x+2) + 4$

$\boxed{(-3, \infty)}$

67. (a) $x + 2(-x+4) - 3(x+5) < -4$

$x - 2x + 8 - 3x - 15 < -4$

$-4x - 7 < -4$

$-4x < 3 \Rightarrow x > -\frac{3}{4} \quad \boxed{\left(-\frac{3}{4}, \infty\right)}$

(b) $x + 2(-x+4) - 3(x+5) \geq -4$

$\boxed{\left(-\infty, -\frac{3}{4}\right]}$

69. $\frac{1}{3}x - \frac{1}{5}x \leq 2$

$15\left[\frac{1}{3}x - \frac{1}{5}x \leq 2\right]$

$5x - 3x \leq 30$

$2x \leq 30$

$x \leq 15 \quad \boxed{(-\infty, 15]}$

71. $\frac{x-2}{2} - \frac{x+6}{3} > -4$

$6\left[\frac{x-2}{2} - \frac{x+6}{3} > -4\right]$

$3x - 6 - 2x - 12 > -24$

$x - 18 > -24$

$x > -6 \quad \boxed{(-6, \infty)}$

73. $.6x - 2(.5x + .2) \leq .4 - .3x$

$.6x - 1x - .4 \leq .4 - .3x$

$10[.6x - 1x - .4 \leq .4 - .3x]$

$-4x - 4 \leq 4 - 3x$

$-4x + 3x \leq 4 + 4$

$-x \leq 8$

$x \geq -8 \quad \boxed{[-8, \infty)}$

75. $-\frac{1}{2}x + .7x - 5 > 0$

$-\frac{1}{2}x + .7x > 5$

$10\left[-\frac{1}{2}x + .7x > 5\right]$

$-5x + 7x > 50$

$2x > 50$

$x > 25 \quad \boxed{(25, \infty)}$

77. (a) Moving away from Omaha; distance is increasing

(b) 100 miles at $x = 1$ hour
200 miles at $x = 3$ hours

(c) $[1, 3]$

(d) $(1, 6]$

79. $4 \leq 2x + 2 \leq 10$

$4 - 2 \leq 2x + 2 - 2 \leq 10 - 2$

$2 \leq 2x \leq 8$

$1 \leq x \leq 4 \quad \boxed{[1, 4]}$

Graph $y_1 = 4$, $y_2 = 2x + 2$, $y_3 = 10$

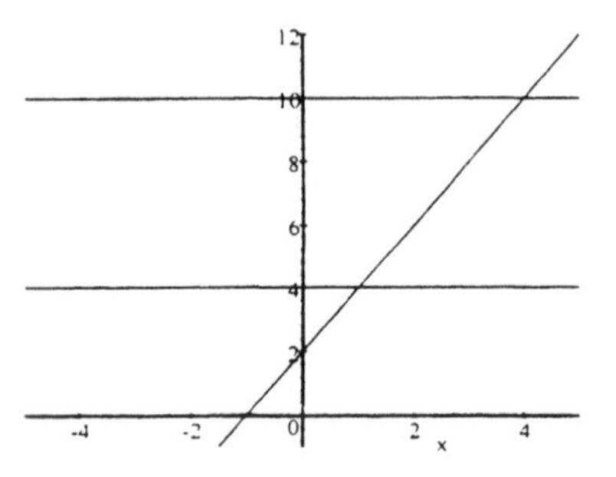

81. $-10 > 3x + 2 > -16$

$-10 - 2 > 3x + 2 - 2 > -16 - 2$

$-12 > 3x > -18$

$-4 > x > -6$

$-6 < x < -4$ $\boxed{(-6, -4)}$

Graph $y_1 = -10$, $y_2 = 3x + 2$, $y_3 = -16$

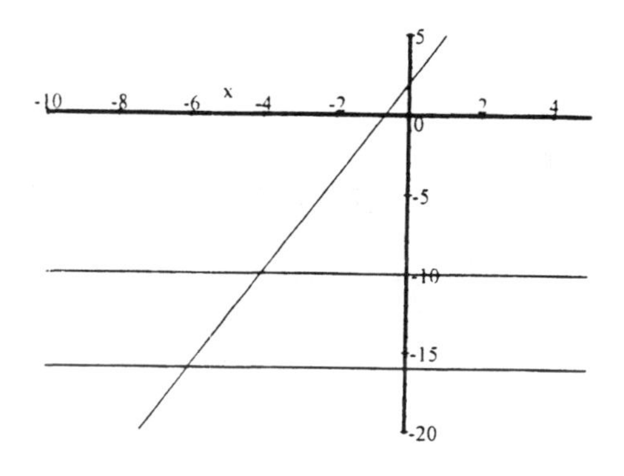

83. $-3 \le \dfrac{x-4}{-5} < 4$

$15 \ge x - 4 > -20$

$15 + 4 \ge x - 4 + 4 > -20 + 4$

$19 \ge x > -16$

$-16 < x \le 19$ $\boxed{(-16, 19]}$

Graph $y_1 = -3$, $y_2 = \dfrac{x-4}{-5}$, $y_3 = 4$

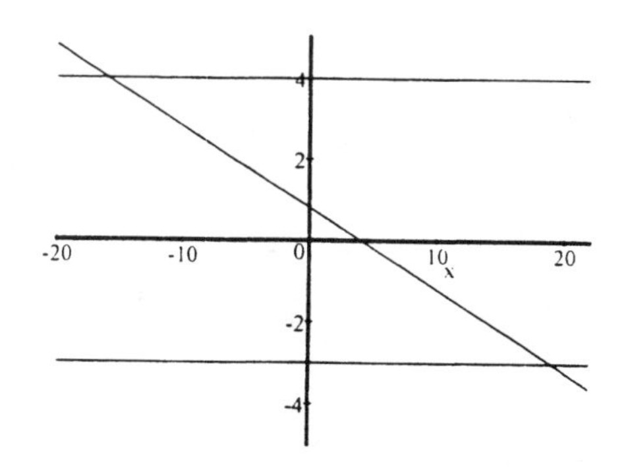

85. $\sqrt{2} \le \dfrac{2x+1}{3} \le \sqrt{5}$

$3\sqrt{2} \le 2x + 1 \le 3\sqrt{5}$

$3\sqrt{2} - 1 \le 2x \le 3\sqrt{5} - 1$

$\dfrac{3\sqrt{2}-1}{2} \le x \le \dfrac{3\sqrt{5}-1}{2}$:

$$\left[\frac{3\sqrt{2}-1}{2}, \frac{3\sqrt{5}-1}{2}\right]$$

Graph $y_1 = \sqrt{2}$, $y_2 = \dfrac{2x+1}{3}$, $y_3 = \sqrt{5}$:

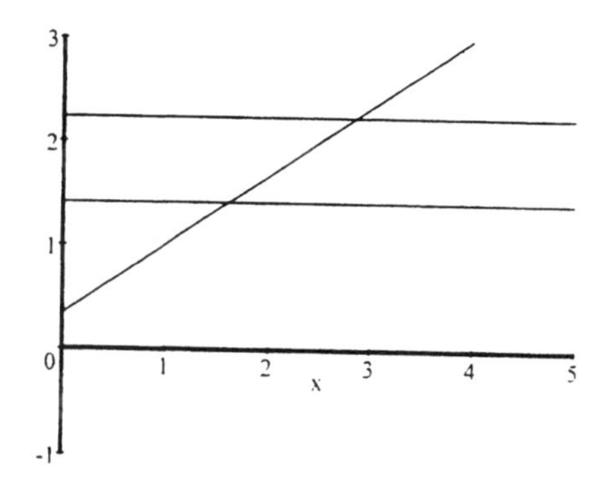

Relating Concepts (Exercises 87 - 90)

87. $3.7x - 11.1 = 0$; $\{3\}$

There is one solution to this equation.
There is one solution to any linear equation.

88. $3.7x - 11.1 < 0$: $(-\infty, 3)$

$3.7x - 11.1 > 0$: $(3, \infty)$

The value of $x = 3$ represents the boundary between the sets of real numbers given by $(-\infty, 3)$ and $(3, \infty)$.

89. $-4x + 6 = 0$: $\{1.5\}$

$-4x + 6 < 0$: $(1.5, \infty)$

$-4x + 6 > 0$: $(-\infty, 1.5)$

90. **(a)** $ax + b = 0,\ a \neq 0$

$ax = -b \Rightarrow x = -\dfrac{b}{a}: \quad \left\{-\dfrac{b}{a}\right\}$

(b) $a > 0$ (positive slopes)

$ax + b < 0: \quad \left(-\infty, -\dfrac{b}{a}\right)$

$ax + b > 0: \quad \left(-\dfrac{b}{a}, \infty\right)$

(c) $a < 0$ (negative slope)

$ax + b < 0: \quad \left(-\dfrac{b}{a}, \infty\right)$

$ax + b > 0: \quad \left(-\infty, -\dfrac{b}{a}\right)$

91. **(a)** $T(x) = 65 - 29x;\quad D(x) = 50 - 5.8x$

The graphs intersect at $\approx (.6466,\ 46.25)$, where $x =$ altitude. Clouds will not form below .65 miles: $[0,\ .65)$.

(b) $T(x) > D(x)$

$65 - 29x > 50 - 5.8x$

$-23.2x > -15 \Rightarrow x < \dfrac{15}{23.2}: \quad \left[0, \dfrac{15}{23.2}\right)$

93. $C = 2\pi r$

$1.99 \leq r \leq 2.01$

$2\pi(1.99) \leq 2\pi r \leq 2\pi(2.01)$

$3.98\pi \leq C \leq 4.02\pi$

95. **(a)** $(0, -1.5),\ (2,\ 4.5)$

$m = \dfrac{4.5 + 1.5}{2 - 0} = \dfrac{6}{2} = 3$

$y + 1.5 = 3(x - 0)$

$y = 3x - 1.5$

$f(x) = 3x - 1.5$

(b) $f(x) > 2.25$

$3x - 1.5 > 2.25$

$3x > 3.75$

$x > 1.25$

Section 1.6

1. $I = PRT$ for P

$\dfrac{I}{RT} = \dfrac{PRT}{RT} \qquad \boxed{\dfrac{I}{RT} = P}$

3. $P = 2L + 2W$ for W

$P - 2L = 2W$

$\dfrac{P - 2L}{2} = \dfrac{2W}{2} \qquad \boxed{W = \dfrac{P - 2L}{2} \text{ or } W = \dfrac{P}{2} - L}$

5. $A = \frac{1}{2}h(b_1 + b_2)$ for h

$2A = h(b_1 + b_2)$

$\dfrac{2A}{(b_1 + b_2)} = \dfrac{h(b_1 + b_2)}{(b_1 + b_2)}$

$\boxed{h = \dfrac{2A}{b_1 + b_2}}$

7. $S = 2\pi rh + 2\pi r^2$ for h

$S - 2\pi r^2 = 2\pi rh$

$\dfrac{S - 2\pi r^2}{2\pi r} = \dfrac{2\pi rh}{2\pi r}$

$\boxed{h = \dfrac{S - 2\pi r^2}{2\pi r} \text{ or } h = \dfrac{S}{2\pi r} - r}$

9. $F = \dfrac{9}{5}C + 32$ for C

$5F = 9C + 160$

$5F - 160 = 9C$

$\dfrac{5F - 160}{9} = C$ or $\boxed{C = \dfrac{5}{9}(F - 32)}$

11. Let x represent the width; then the length, $L = 2x - 3$ and the perimeter, $P = 2L + 2W = 54$

$2L + 2W = 54$

$2(2x - 3) + 2(x) = 54$

$4x - 6 + 2x = 54$

$6x = 60$

$x = 10$

$\boxed{\text{Width} = 10\,\text{cm},\quad \text{length} = 17\,\text{cm}}$

13. Let P represent the perimeter, l represent the length, and w represent the width.

Given $P = 28803.2$ in, $l = 11757.6 + w$
$w = l - 11757.6$

$$P = 2l + 2w$$
$$28803.2 = 2l + 2(l - 11757.6)$$
$$28803.2 = 4l - 23515.2$$
$$4l = 52318.4$$
$$l = 13079.6 \text{ inches}$$
$$= \frac{13079.6}{36} = \boxed{363\tfrac{29}{90} \text{ yards}}$$

15. Let x represent the number of gallons of 94-octane gasoline:

$$.94(x) + .99(400) = .97(x + 400)$$
$$94(x) + 99(400) = 97(x + 400)$$
$$94x + 39600 = 97x + 38800$$
$$-3x = -800$$
$$x = 266\tfrac{2}{3}$$

$\boxed{\text{Mix } 266\tfrac{2}{3} \text{ gallons of 94-octane gas.}}$

17. Let x represent the number of liters of pure alcohol:

$$.10(7) + 1.00(x) = .30(x + 7)$$
$$10(7) + 100x = 30(x + 7)$$
$$70 + 100x = 30x + 210$$
$$70x = 140 \rightarrow x = 2$$

$\boxed{\text{Add 2 liters of pure alcohol.}}$

19. Let x represent the number of liters of water (0%):

$$0(x) + .18(20) = .15(x + 20)$$
$$0x + 18(20) = 15(x + 20)$$
$$360 = 15x + 300$$
$$15x = 60$$
$$x = 4$$

$\boxed{\text{Add 4 liters of water.}}$

21. Let $y =$ pressure, $x =$ depth:
direct proportion: $y = kx$.

Given $y = 13 \text{ lbs/ft}^2$ when $x = 30$ ft:

$$y = kx$$
$$13 = k(30)$$
$$k = \frac{13}{30},$$
$$\text{so } y = \frac{13}{30}x$$

Given $x = 70$ ft, $y = \frac{13}{30}(70)$

$\boxed{30\tfrac{1}{3} \text{ lbs/ft}^2}$

Show that $(70, 30\tfrac{1}{3})$ is on the graph of $y_1 = \frac{13}{30}x$.

23. Let $y =$ height, $x =$ length of shadow; given $y = 2'$, $x = 1.75'$; direct proportion: $y = kx$

$$2 = k(1.75)$$
$$k = \frac{2}{1.75} = \frac{200}{175} = \frac{8}{7} \text{ so}$$
$$y = \frac{8}{7}x; \quad \text{given } x = 45 \text{ ft},$$
$$y = \frac{8}{7}(45) = \frac{360}{7} \quad \boxed{51\tfrac{3}{7} \text{ ft}}$$

Show that $\left(45, \frac{360}{7}\right)$ is on the graph of $y_1 = \frac{8}{7}x$.

25. With direct proportion,

$y_1 = kx_1$ and $y_2 = kx_2$,

$$k = \frac{y_1}{x_1} = \frac{y_2}{x_2}$$

$y_1 = 250$ tagged fish, $y_2 = 7$ tagged fish,
$x_2 = 350$ sample fish; find x_1, total fish.

$$\frac{250}{x_1} = \frac{7}{350}$$
$$7x_1 = 87500$$
$$x_1 = 12500 \quad \boxed{\approx 12,500 \text{ fish in the lake}}$$

27. $y = kx$

$(3, 7.5)$: $7.5 = 3k \Longrightarrow \boxed{k = 2.5}$

$y = 2.5x$

$(8, y)$: $y = 2.5(8) = 20 \Longrightarrow \boxed{(8, 20)}$

29. $y = kx$

$(25, 1.50): \quad 1.50 = 25k \Longrightarrow \boxed{k = .06}$

$y = .06x$

$(x, 5.10): \quad 5.10 = .06x$

$x = 85 \Longrightarrow \boxed{(\$85, \$5.10)}$

31. $t = kc$; t = tuition, c = credits

$(11, 720.50): \quad 720.50 = 11k \Longrightarrow k = 65.5$

$t = 65.5c$

$t = 65.5(16) = \boxed{\$1048;\ k = 65.5}$

33. $w = kd$; w = weight, d = distance

$3 = 2.5k \Longrightarrow k = \frac{6}{5} = 1.2$

$w = 1.2d$

$17 = 1.2d \Longrightarrow d = \boxed{14\tfrac{1}{6} \text{ inches}}$

35. **(a)** Fixed cost = \$200; variable cost = \$.02

$C(x) = .02x + 200$

(b) $R(x) = .04x$

(c) $C(x) = R(x)$ for which x?

$.02x + 200 = .04x$

$200 = .02x$

$\boxed{x = 10{,}000 \text{ envelopes}}$

(d) Graph $C(x)$ and $R(x)$:

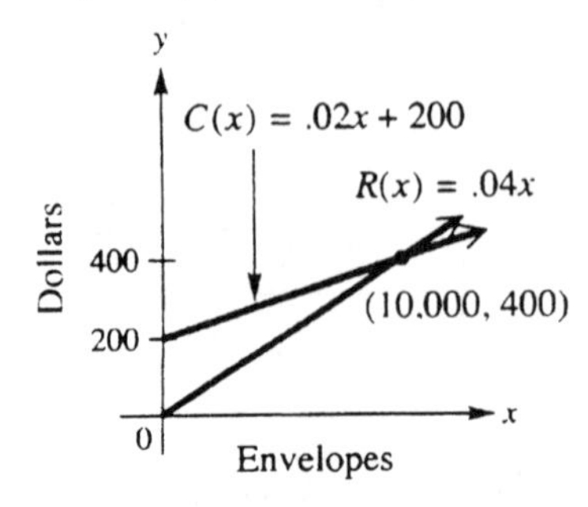

Loss at $x < 10{,}000$; Profit at $x > 10{,}000$

37. **(a)** Fixed cost = \$2300; variable cost = \$3.00

$C(x) = 3.00x + 2300$

(b) $R(x) = 5.50x$

(c) $C(x) = R(x)$ for which x?

$3.00x + 2300 = 5.50x$

$2300 = 2.5x$

$\boxed{x = 920 \text{ deliveries}}$

(d) Graph $C(x)$ and $R(x)$:

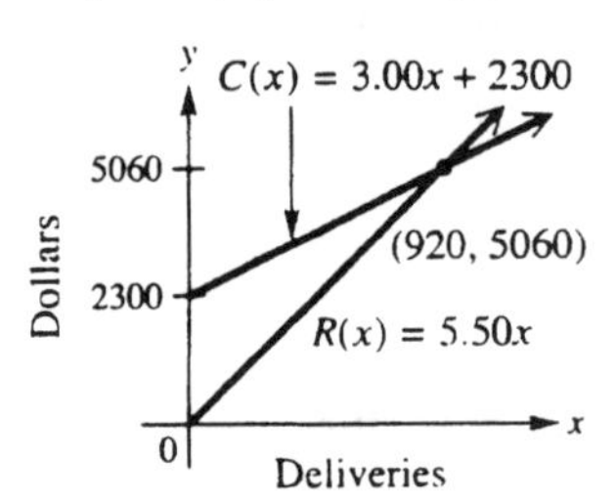

Loss at $x < 920$; Profit at $x > 920$.

39. Let x = Hien's grade on the 4th test:

$$\text{Average} = \frac{84 + 88 + 92 + x}{4}$$

$$\frac{84 + 88 + 92 + x}{4} = 90$$

$$84 + 88 + 92 + x = 360$$

$$264 + x = 360$$

$$x = 96$$

$\boxed{\text{Hien must score 96.}}$

41. $y = 10(x - 65) + 50; \quad x = 76$

$y = 10(76 - 65) + 50$

$y = 10(11) + 50 = 160$

$\boxed{\text{Jose's fine was \$160.}}$

43. Since the restriction, $x \geq 65$, is given, tickets are issued at speeds of 65 mph or more.

45. (10, 7500) and (20, 13900)
[(#heaters, cost)]

(a) $m = \dfrac{7500 - 13900}{10 - 20} = \dfrac{-6400}{-10} = 640$

$y - 7500 = 640(x - 10)$
$y - 7500 = 640x - 6400$
$y = 640x + 1100$

(b) $y = 640(25) + 1100$
$y = 16000 + 1100 = \$17,100$

(c) Show (25, 17100) is on the graph of $y_1 = 640x + 1100$.

47. [(years since '90, cost)]
(0, 120,000) and (10, 146,000)

(a) $m = \dfrac{120,000 - 146,000}{0 - 10} = 2600$

$y - 120000 = 2600(x - 0)$
$y - 120000 = 2600x$
$y = 2600x + 120,000$

(b) $y = 2600(14) + 120,000$
$y = 36400 + 120,000$
$y = \$156,400$ in 2004

(c) Slope is the rate of change of the value of the house per year (the annual appreciation of $2600).

49. (a) Since each student needs 15 ft^3/min, and there are 60 min/hr, the ventilation required by x students per hour would be:

$V(x) = 60(15x) = 900x$

(b) The number of air exchanges/hr:

$A(x) = \dfrac{900x}{15,000} = \dfrac{3}{50}x$

(c) If $x = 40$, then

$A(40) = \dfrac{3}{50}(40) = 2.4$ ach

(d) It should be increased by $\dfrac{50}{15} = 3\frac{1}{3}$ times. Smoking areas require more than triple the ventilation.

51. (a) The risk for one year would be

$\dfrac{R}{72} = \dfrac{1.5 \times 10^{-3}}{72} \approx .000021/$ individual

(b) $C(x) = .000021x$

(c) $C(100000) = .000021(100,000) = 2.1$
There are approximately 2.1 cancer cases for every 100,000 passive smokers.

(d) $C = \dfrac{.44(281,000,000)(.26)}{72} \approx 446,478$
There are approximately 446,000 excess deaths caused by smoking each year.

53. $y = \dfrac{5}{3}x + 455$

(a) $y = \dfrac{5}{3}(27) + 455$
$y = 45 + 455$ $\boxed{500 \text{ cm}^3}$

(b) $605 = \dfrac{5}{3}x + 455$
$150 = \dfrac{5}{3}x$ $\boxed{90° \text{ Celsius}}$

(c) $0 = \dfrac{5}{3}x + 455$
$-455 = \dfrac{5}{3}x$ $\boxed{-273° \text{ Celsius}}$

Reviewing Basic Concepts (Sections 1.5 and 1.6)

1. $3(x - 5) + 2 = 1 - (4 + 2x)$
$3x - 15 + 2 = 1 - 4 - 2x$
$3x - 13 = -3 - 2x$
$5x = 10 \Rightarrow x = 2$ $\boxed{\{2\}}$

Graph $y_1 = 3(x - 5) + 2$ and $y_2 = 1 - (4 + 2x)$; the intersection point is $(2, -7)$.

2. $\pi(1 - x) = .6(3x - 1)$
Graph $y_1 = \pi(1 - x)$ and $y_2 = .6(3x - 1)$; the intersection point is $\approx (.75716, .76289)$.
The solution is the x-value: $\boxed{\{.757\}}$

3. $f(x) = \dfrac{1}{3}(4x - 2) + 1$

Solve $\dfrac{1}{3}(4x - 2) + 1 = 0$
$4x - 2 - 3 = 0$
$4x + 1 = 0$
$4x = -1 \Longrightarrow x = -\frac{1}{4}$ $\boxed{\{-.25\}}$

Graph $y_1 = \dfrac{1}{3}(4x - 2) + 1$; the x-intercept is $-\dfrac{1}{4}$.

4. (a) $4x - 5 = -2(3 - 2x) + 3$
$4x - 5 = -6 + 4x + 3$
$4x - 5 = 4x - 3$
$-5 = -3$: Contradiction

(b) $5x - 9 = 5(-2 + x) + 1$
$5x - 9 = -10 + 5x + 1$
$5x - 9 = 5x - 9$
$-9 = -9$: Identity

(c) $5x - 4 = 3(6 - x)$
$5x - 4 = 18 - 3x$
$8x = 22$
$x = \frac{22}{8} = \frac{11}{4}$: Conditional

5. $2x + 3(x + 2) < 1 - 2x$
$2x + 3x + 6 < 1 - 2x$
$5x + 6 < 1 - 2x$
$7x < -5$
$x < -\frac{5}{7}$: $\left(-\infty, -\frac{5}{7}\right)$

Graph $y_1 = 2x + 3(x + 2)$ and $y_2 = 1 - 2x$;
$y_1 < y_2$: $\left(-\infty, -\frac{5}{7}\right)$.

6. $-5 \le 1 - 2x < 6$
$-6 \le -2x < 5$
$3 \ge x > -\frac{5}{2}$
$-\frac{5}{2} < x \le 3$: $\left(-\frac{5}{2}, 3\right]$

7. $f(x)$ and $g(x)$ intersect at $(2, 1)$
$f(x) = g(x)$: $\{2\}$
$f(x) \le g(x)$: $[2, \infty)$

8. $\frac{T}{27} = \frac{6}{4}$ where T = height of tree
$4T = 162 \implies T = 40.5$
The tree is 40.5 ft tall.

9. $Rev = 5.50x$; $Cost = 1.50x + 2500$
(a) $R(x) = 5.5x$
(b) $C(x) = 1.5x + 2500$
(d) Break-even: $R(x) = C(x)$
$5.5x = 1.5x + 2500$
$4x = 2500$
$x = 625$
Sell 625 discs to break-even.

10. Solve $V = \pi r^2 h$ for h:
Divide both sides by πr^2: $h = \frac{V}{\pi r^2}$

Chapter One Review

1. $(-1, 16)$, $(5, -8)$
$d = \sqrt{(-1 - 5)^2 + (16 - (-8))^2}$
$d = \sqrt{(-6)^2 + (24)^2}$
$d = \sqrt{36 + 576}$
$d = \sqrt{612}$
$\boxed{\text{Distance} = 6\sqrt{17}}$

2. $(-1, 16)$, $(5, -8)$
Midpoint $= \left(\frac{-1 + 5}{2}, \frac{16 - 8}{2}\right)$
$\boxed{(2, 4)}$

3. $(-1, 16)$, $(5, -8)$
$m = \frac{\Delta y}{\Delta x} = \frac{16 - (-8)}{-1 - 5} = \frac{24}{-6}$
$\boxed{\text{Slope} = -4}$

4. $(-1, 16)$, $(5, -8)$, $m = -4$
$y - 16 = -4(x - (-1))$
$y - 16 = -4x - 4$
$\boxed{y = -4x + 12}$

5. $3x + 4y = 144$
$4y = -3x + 144$
$y = -\frac{3}{4}x + 36$
$\boxed{\text{Slope} = -\frac{3}{4}}$

6. x-intercept: let $y = 0$
$3x + 4(0) = 144$
$3x = 144$
$x = 48$
$\boxed{x\text{-intercept is } 48}$

7. y-intercept: let $x = 0$
[#5]: $y = -\frac{3}{4}(0) + 36$
$\boxed{y\text{-intercept is } 36}$

8. One such window is
$[-10, 50]$ and $[-40, 40]$.

9. $f(3) = 6;\ f(-2) = 1$

Line through $(3, 6)$, $(-2, 1)$

$$m = \frac{6 - 1}{3 - (-2)} = \frac{5}{5} = 1$$

$$y - 6 = 1(x - 3)$$
$$y = x - 3 + 6;\ y = x + 3$$

$$f(x) = x + 3$$
$$f(8) = 8 + 3 = \boxed{11}$$

10. $y = -4x + 3$, through $(-2, 4)$:

Slope $= -4$, $\perp$ slope $= \frac{1}{4}$

$$y - 4 = \frac{1}{4}(x - (-2))$$

$$y - 4 = \frac{1}{4}x + \frac{1}{2} \Longrightarrow \boxed{y = \frac{1}{4}x + \frac{9}{2}}$$

11. **(a)** $(2, -4)$, $(-1, 5)$

$$m = \frac{-4 - 5}{2 - (-1)} = \frac{-9}{3} = -3$$

(b) $y - (-4) = -3(x - 2)$
$$y + 4 = -3x + 6$$
$$y = -3x + 2$$

(c) M.P.$= \left(\frac{2 - 1}{2}, \frac{-4 + 5}{2}\right) = (.5, .5)$

(d) $d = \sqrt{(2 + 1)^2 + (-4 - 5)^2}$
$$= \sqrt{3^2 + (-9)^2} = \sqrt{90} = 3\sqrt{10}$$

12. **(a)** $(-1, 1.5)$, $(-3, -3.5)$

$$m = \frac{1.5 - (-3.5)}{-1 - (-3)} = \frac{5}{2} = 2.5$$

(b) $y - 1.5 = 2.5(x + 1)$
$$y - 1.5 = 2.5x + 2.5$$
$$y = 2.5x + 4$$

(c) M.P. $= \left(\frac{-1 + (-3)}{2}, \frac{1.5 + (-3.5)}{2}\right)$
$$= \left(-\frac{4}{2}, -\frac{2}{2}\right) = (-2, -1)$$

(d) $d = \sqrt{(-1 - (-3))^2 + (1.5 - (-3.5))^2}$
$$= \sqrt{2^2 + 5^2} = \sqrt{29}$$

13. C **14.** F **15.** A

16. B **17.** E **18.** D

19. Domain: $[-6, 6]$; Range: $[-6, 6]$

20. False: Although the difference of the slopes is very small, the lines are not parallel. In fact, they intersect at $(9000, 45006)$.

21. I **22.** K

23. B **24.** A

25. I **26.** M

27. O **28.** K

29. $5[3 + 2(x - 6)] = 3x + 1$
$$5[3 + 2x - 12] = 3x + 1$$
$$5[2x - 9] = 3x + 1$$
$$10x - 45 = 3x + 1$$
$$7x = 46$$
$$x = \boxed{\left\{\frac{46}{7}\right\}}$$

30. $\frac{x}{4} - \frac{x + 4}{3} = -2$
$$12\left[\frac{x}{4} - \frac{x + 4}{3} = -2\right]$$
$$3x - 4(x + 4) = -24$$
$$3x - 4x - 16 = -24$$
$$-x = -8 \Longrightarrow x = \boxed{\{8\}}$$

31. $-3x - (4x + 2) = 3$
$$-3x - 4x - 2 = 3$$
$$-7x = 5$$
$$x = -\frac{5}{7}:\ \boxed{\left\{-\frac{5}{7}\right\}}$$

32. $-2x+9+4x=2(x-5)-3$
$2x+9=2x-10-3$
$9=-13$
Contradiction; No solution

33. $.5x+.7(4-3x)=.4x$
$.5x+2.8-2.1x=.4x$
$-1.6x+2.8=.4x$
$-2x=-2.8$
$x=1.4$ or $\left\{\frac{7}{5}\right\}$

34. $\frac{x}{4}-\frac{5x-3}{6}=2$
$12\left(\frac{x}{4}-\frac{5x-3}{6}\right)=12(2)$
$3x-2(5x-3)=24$
$3x-10x+6=24$
$-7x=18 \quad x=\left\{-\frac{18}{7}\right\}$

35. $x-8<1-2x$
$3x<9$
$x<3: \quad (-\infty, 3)$

36. $\frac{4x-1}{3}\geq\frac{x}{5}-1$ (multiply by 15)
$5(4x-1)\geq 3x-15$
$20x-5\geq 3x-15$
$17x\geq -10$
$x\geq -\frac{10}{17}: \quad \left[-\frac{10}{17}, \infty\right)$

37. $-6\leq\frac{4-3x}{7}<2$
$-42\leq 4-3x<14$
$-46\leq -3x<10$
$\frac{46}{3}\geq x>-\frac{10}{3}$
$-\frac{10}{3}<x\leq\frac{46}{3} \quad \boxed{\left(-\frac{10}{3}, \frac{46}{3}\right]}$

Graph for **38**:

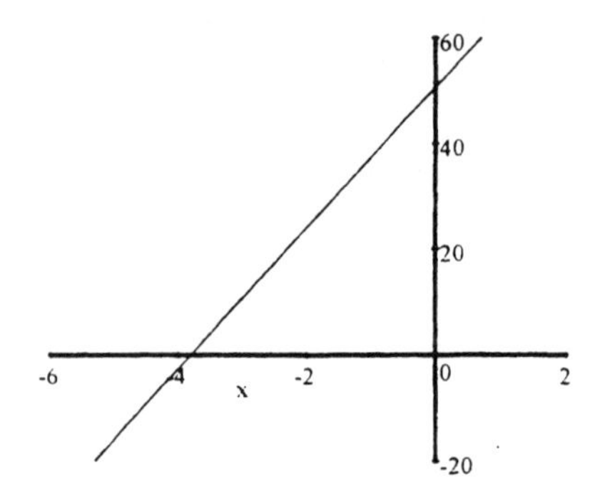

38. (a) Let $y_1=5\pi x+(\sqrt{3})x$
$-6.24(x-8.1)+\sqrt[3]{9}x$;
find the x-intercept: $\boxed{x\approx\{-3.81\}}$
(b) $f(x)<0: \quad (-\infty, -3.81)$
(c) $f(x)\geq 0: \quad [-3.81, \infty)$

39. Fixed cost $=\$150$; variable cost $=\$30$
$C(x)=30x+150$

40. $R(x)=37.50x$

41. $R(x)=C(x)$ for which x?
$30x+150=37.5x$
$150=7.5x \quad \boxed{x=20}$

42. When $C(x)>R(x)$, the cost exceeds the revenue (loss); at the intersection point ($x=20$); $C(x)=R(x)$ [break-even]; when $C(x)<R(x)$, the revenue exceeds the cost (profit).

43. $A=\frac{24f}{B(p+1)}$; Solve for f
$AB(p+1)=24f$
$f=\frac{AB(p+1)}{24}$

44. $A = \dfrac{24f}{B(p+1)}$; Solve for B

$AB(p+1) = 24f$

$B = \dfrac{24f}{A(p+1)}$

45. **(a)** $f(x) = -3.52x + 58.6$

$f(5) = -3.52(5) + 58.6 = \boxed{41° \text{ F}}$

(b) $-15 = -3.52x + 58.6$

$3.52x = 73.6$

$x \approx 20.9091 \quad (\times 1000) \approx \boxed{21{,}000 \text{ feet}}$

(c) Graph $y_1 = -3.52x + 58.6$. Then find the coordinates of the point where $x = 5$ to support the answer in **(a)**. Finally, find the coordinates of the point where $y = -15$ to support the answer in **(b)**.

46. $P = 1.06F + 7.18$

$50 = 1.06F + 7.18$

$5000 = 106F + 718$

$4282 = 106F$

$F \approx 40.3962 \quad \boxed{\approx 40 \text{ liters/sec}}$

47. **(a)** $(0, .25), (53, 4.25)$

$m = \dfrac{4.25 - .25}{53} = \dfrac{4}{53} \approx .075$

$b = .25 \quad \boxed{y = .075x + .25}$

(b) The data fit the linear model fairly well.

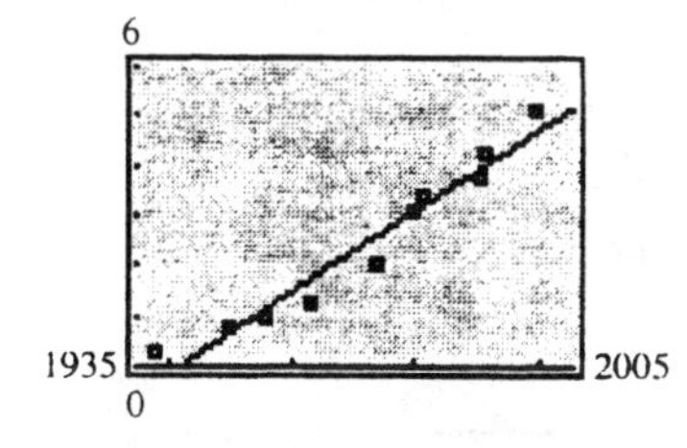

(c) $\{1938, \ldots, 1999\} \to L_1$

$\{.25, \ldots, 5.15\} \to L_2$

Linear Regression model:

$y = ax + b$

$y \approx .08191x - 159.05$

$r \approx .974$

(d) $1997 \to x = 59$, find y:

$y = .0819(59) - .311$

$y = 4.521 \approx \boxed{\$4.52}$

\$4.52 is quite a bit lower than the actual amount of \$5.15.

48. Let $1980 = 0$, $1981 = 1$, etc.

$\{0, 1, \ldots, 12\} \to L_1$

$\{192.256, \ldots, 232.482\} \to L_2$

$y = ax + b$

$y \approx 2.764x + 197.8$

Find x when $y = 250$:

$250 = 2.764x + 197.8$

$52.2 = 2.764x$

$x \approx 18.886 \approx 19 \quad \boxed{\text{About 1999}}$

49. $(50, 320), (80, 440)$

$[(x \text{ mph}, \text{ distance})]$

$m = \dfrac{440 - 320}{80 - 50} = \dfrac{120}{30} = 4$

$y - 320 = 4(x - 50)$

$y - 320 = 4x - 200$

$y = 4x + 120$

1 mph increase in x: $x + 1$

$y_1 = 4(x+1) + 120$

$= 4x + 4 + 120 = 4x + 124$

$y_1 - y = (4x + 124) - (4x + 120) = \boxed{4 \text{ feet}}$

[Or a slope of 4 indicates an increase of 4 feet for each mile per hour.]

50. Let h = the height of the box. Use the formula for the surface area of a rectangular box; $S = 496 \text{ ft}^2$, $l = 18$ ft, $w = 8$ ft

$S = 2lw + 2wh + 2hl$

$496 = 2(18)(8) + 2(8)h + 2h(18)$

$496 = 288 + 16h + 36h$

$208 = 52h$

$4 = h \quad \boxed{\text{The height of the box is 4 feet.}}$

51. Convert 100 meters to feet:

$$100 \text{ m}\left(\frac{3.281 \text{ ft}}{\text{m}}\right) = 328.1 \text{ feet}$$

Convert 26 miles to feet:

$$26 \text{ mi}\left(\frac{5280 \text{ ft}}{\text{mi}}\right) = 137,280 \text{ feet}$$

Equation: $\dfrac{9.85 \text{ sec}}{328.1 \text{ ft}} = \dfrac{x \text{ sec}}{137280 \text{ ft}}$

$x \approx 4121.32886315$ sec

Convert time to hr, min, sec:

$$\begin{aligned} &4121.32886315 \text{ sec}\left(\frac{1 \text{ hr}}{3600 \text{ sec}}\right) \\ &\approx 1.1448135731 \text{ hr} \\ &\approx 1 \text{ hr} + .148135731(60) \text{ min} \\ &\approx 1 \text{ hr} + 8.688814386 \text{ min} \\ &\approx 1 \text{ hr} + 8 \text{ min} + .688814386(60) \text{ sec} \\ &\approx 1 \text{ hr} + 8 \text{ min} + 41.32886 \text{ sec} \end{aligned}$$

$\boxed{1 \text{ hr, } 8 \text{ min. } 41.3 \text{ sec.}}$

52. $C = \dfrac{5}{9}(F - 32) = \dfrac{5}{9}(864 - 32)$

$= \dfrac{5}{9}(832) = \boxed{462\tfrac{2}{9}\,^\circ\text{C}}$

53. Let p = pressure, d = distance.
Direct variation: $p = kd$

$p = 3000 \text{ kg/m}^2$ when $d = 4$ m.

Find k: $3000 = k(4)$

$750 = k$

Equation: $p = 750d$

Find p when $d = 10$ m:

$p = 750(10)$ $\boxed{7500 \text{ kg/m}^2}$

54. Take-home pay= (100% − 26%) of weekly pay.

$592 = .74x$ $\boxed{x = \$800}$

55. Answers may vary.

(a) Using (1994. 504.), (2000. 514):

$$m = \frac{514 - 504}{2000 - 1994} = \frac{10}{6} = \frac{5}{3}$$

$$y - 504 = \frac{5}{3}(x - 1994)$$

or $y = \dfrac{5}{3}x - 2819\tfrac{1}{3}$

Regression Equation: $y \approx 1.707x - 2899$

(b) $y(1997) = 509$

56. **(a)** $(-3, 6.6), (-2, 5.4)$

$$m = \frac{6.6 - 5.4}{-3 + 2} = \frac{1.2}{-1} = -1.2$$

$$\begin{aligned} y - 6.6 &= -1.2(x + 3) \\ y - 6.6 &= -1.2x - 3.6 \\ y &= -1.2x + 3 \end{aligned}$$

(b) at $x = -1.5$: $y = 4.8$
at $x = 3.5$: $y = -1.2$

57. Let x = the amount of pure alcohol (100%) to be added:

$$\begin{aligned} .10(12) + 1.00x &= .30(x + 12) \\ 10(12) + 100x &= 30(x + 12) \\ 120 + 100x &= 30x + 360 \\ 70x &= 240 \\ x &= \frac{24}{7} \end{aligned}$$

$x = \boxed{3\tfrac{3}{7} \text{ liters}}$

58. Let x = the amount of 5% acid:

$$\begin{aligned} .20(120) + .05x &= .10(x + 120) \\ 20(120) + 5x &= 10(x + 120) \\ 2400 + 5x &= 10x + 1200 \\ -5x &= -1200 \Longrightarrow x = \boxed{240 \text{ ml}} \end{aligned}$$

59. Break-even: $R(x) = C(x)$.

The company will at least break even when $R(x) \geq C(x)$:

$$\begin{aligned} 8x &\geq 3x + 1500 \\ 5x &\geq 1500 \\ x &\geq 300 \end{aligned}$$

$\boxed{[300, \infty)}$

60. Let m = mental age, c = chronological age.

$$IQ = \frac{100m}{c}$$

(a) $130 = \dfrac{100m}{7}$

$910 = 100m \Longrightarrow m = \boxed{9.1 \text{ years}}$

(b) $IQ = \dfrac{100(20)}{16} = \boxed{125}$

Chapter 1 Test

1. (a) Domain: $(-\infty, \infty)$; Range: $[2, \infty)$
   No x-intercepts; y-intercept: 3

   (b) Domain: $(-\infty, \infty)$; Range: $(-\infty, 0]$
   x-intercept: 3; y-intercept: -3

   (c) Domain: $[-4, \infty)$; Range: $[0, \infty)$
   x-intercept: -4; y-intercept: 2

2. (a) $f(x) = g(x)$: $\{-4\}$
   (b) $f(x) < g(x)$: $(-\infty, -4)$
   (c) $f(x) \geq g(x)$: $[-4, \infty)$
   (d) $y_2 - y_1 = 0$: $\{-4\}$

3. (a) $y_1 = 0$: $\{5.5\}$
   (b) $y_1 < 0$: $(-\infty, 5.5)$
   (c) $y_1 > 0$: $(5.5, \infty)$
   (d) $y_1 \leq 0$: $(-\infty, 5.5]$

4. (a) Solve $f(x) = g(x)$:

$$3(x-4) - 2(x-5) = -2(x+1) - 3$$
$$3x - 12 - 2x + 10 = -2x - 2 - 3$$
$$x - 2 = -2x - 5$$
$$3x = -3 \qquad \boxed{x = \{-1\}}$$

Check:

$$3(-1-4) - 2(-1-5) = -2(-1+1) - 3$$
$$3(-5) - 2(-6) = 2(0) - 3$$
$$-15 + 12 = -3$$
$$-3 = -3$$

(b) y_1, y_2 :

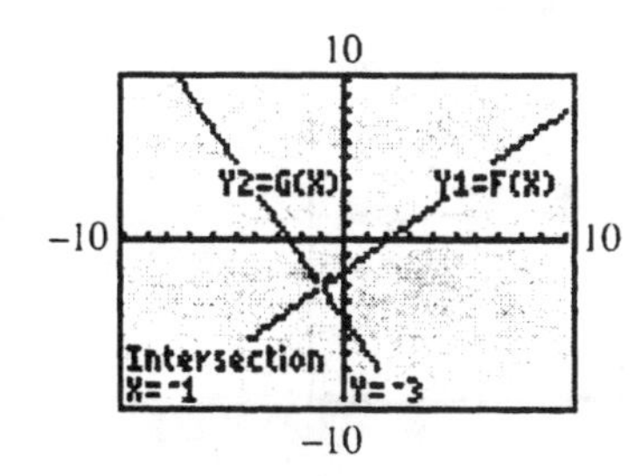

$f(x) > g(x)$: $(-1, \infty)$

The graph of $y_1 = f(x)$ is above the graph of $y_2 = g(x)$ for domain values greater than -1.

(c) $f(x) < g(x)$: $(-\infty, -1)$

The graph of $y_1 = f(x)$ is below the graph of $y_2 = g(x)$ for domain values less than -1.

5. (a) $-\frac{1}{2}(8x+4) + 3(x-2) = 0$

$$-4x - 2 + 3x - 6 = 0$$
$$-x - 8 = 0$$
$$-x = 8 \Longrightarrow x = \boxed{\{-8\}}$$

(b) $-\frac{1}{2}(8x+4) + 3(x-2) \leq 0$

$$-4x - 2 + 3x - 6 \leq 0$$
$$-x - 8 \leq 0$$
$$-x \leq 8 \Longrightarrow x \geq -8 \quad \boxed{[-8, \infty)}$$

(c) $y = f(x)$:

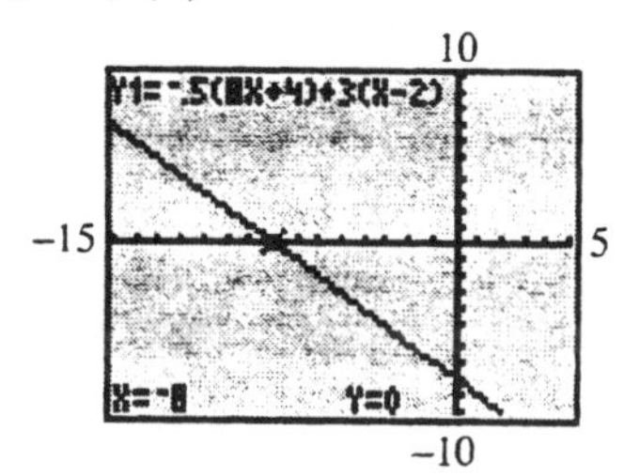

The x-intercept is -8, supporting the result of part (a). The graph of the linear function lies below or on the x-axis for domain values greater than or equal to -8, supporting the result of part (b).

6. (1980, 7.85), (2000, 30.36)

   (a) 1990 : M.P. $= \left(\frac{1980 + 2000}{2}, \frac{7.85 + 30.36}{2}\right)$

$$= (1990, 19.11) \qquad \boxed{\$19.11}$$

(b) $m = \frac{30.36 - 7.85}{2000 - 1980} = \frac{22.51}{20} = 1.1255$

The monthly rate increased on the average \$1.13 per year during the years 1980 – 2000.

7. (a) $(-3, 5)$ parallel to $y = -2x + 4$:

Slope of the line is -2; since parallel lines have equal slopes, $m = -2$.

$$y - 5 = -2(x - (-3))$$
$$y - 5 = -2x - 6 \qquad \boxed{y = -2x - 1}$$

(b) $(-3, 5) \perp$ to $-2x + y = 0$ $(y = 2x)$:

Slope of the line is 2; since perpendicular lines have slopes whose product equals -1, $m = -\frac{1}{2}$ $\left[2\left(-\frac{1}{2}\right) = -1\right]$.

$$y - 5 = -\frac{1}{2}(x - (-3))$$
$$y - 5 = -\frac{1}{2}x - \frac{3}{2}$$
$$y = -\frac{1}{2}x - \frac{3}{2} + \frac{10}{2} \qquad \boxed{y = -\frac{1}{2}x + \frac{7}{2}}$$

8. $3x - 4y = 6$

x-intercept: 2

$3x - 4(0) = 6$

y-intercept: $-\dfrac{3}{2}$

$3(0) - 4y = 6$

slope $= \dfrac{3}{4}$

$y = \dfrac{3}{4}x - \dfrac{3}{2}$

9. Horizontal line through $(-3, 7) : y = 7$
Vertical line through $(-3, 7) : x = -3$

10. **(a)** Least Squares Line:

$\{10, \ldots, 35\} \to L_1$

$\{28, \ldots, 11\} \to L_2$

$a \approx -.65,\ b \approx 32.7$

$y = ax + b$ | $y = -.65x + 32.7$, $r \approx -.97826 \approx -.98$ |

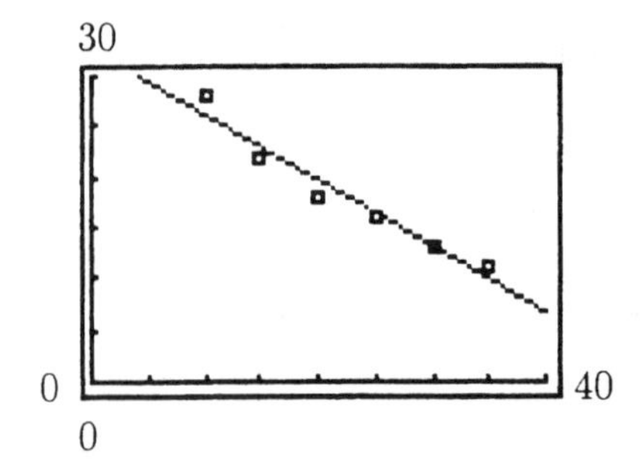

(b) Find y when $x = 40$:

$y = -.65(40) + 32.7 =$ $\boxed{6.7\,^\circ\text{F}}$

11. **(a)** $V = \pi r^2 h$; 20 feet = 240 inches

$V = \pi (240)^2 (1)$

$V = 57600\pi \approx 180,956 \text{ in}^3$

(b) 1 gallon = 231 cubic inches

$g(x) = \dfrac{180956}{231}x$

(c) $g(2.5) = \dfrac{180956}{231}(2.5) \approx 1958.398$

≈ 1958 gallons

(d) No; it will handle $2.5(400) = 1000$ gallons. There should be 2 downspouts.

Chapter 1 Project

Predicting the Height and Weight of Athletes

1. $\{74, 74, \ldots, 73\} \longrightarrow L_1$

$\{169, 181, \ldots, 194\} \longrightarrow L_2$

LinReg: $y \approx 4.51x - 154.4$, $r \approx .86$, indicates that taller female athletes tend to weigh more, but the correlation is not perfect.

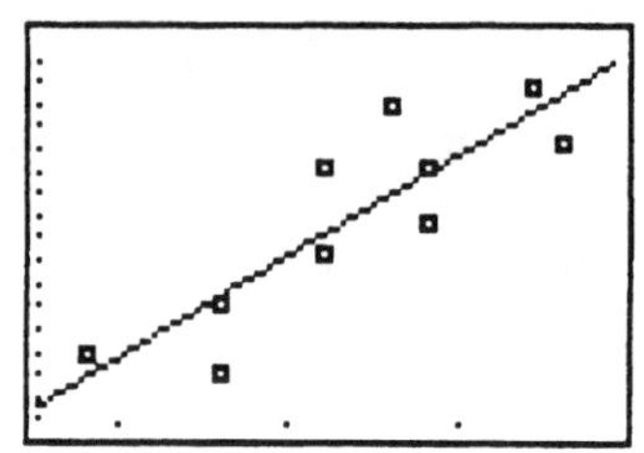

2. $4.51(75) - 154.4 = 183.85$
75 inches tall $\approx$ 184 pounds

3. $m = 4.51 \longrightarrow$ for each 1 inch increase in height, the corresponding increase in weight is ≈ 4.5 pounds.

4. $\{74, 75, \ldots, 85\} \longrightarrow L_1$

$\{197, 202, \ldots, 246\} \longrightarrow L_2$

LinReg: $y \approx 4.465x - 133.3$, $r \approx .90$, indicates that taller male athletes tend to weigh more, but the correlation is not perfect.

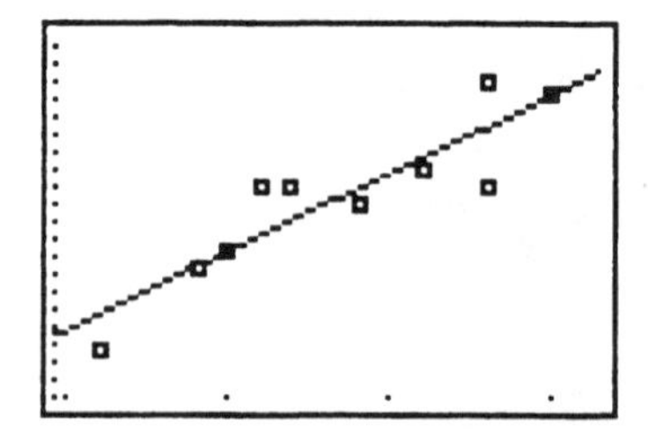

5. $4.465(80) - 133.3 = 223.9$
80 inches tall $\approx$ 224 pounds

6. $m = 4.46 \longrightarrow$ for each 1 inch increase in height, the corresponding increase in weight is ≈ 4.5 pounds.

CHAPTER 2

Section 2.1

1. The domain and the range of the identity function are both $(-\infty, \infty)$.

3. The graph of the cubing function changes from *opening downward* to *opening upward* at the point $(0, 0)$.

5. The cube root function increases on its entire domain.

7. The graph of the relation $x = y^2$ is symmetric with respect to the x-axis.

9. The function defined by $f(x) = x^3 + x$ is an odd function.

11. $(-\infty, \infty)$

13. $[0, \infty)$

15. $(-\infty, -3); (-3, \infty)$

17. (a) Increasing: $[3, \infty)$
    (b) Decreasing: $(-\infty, 3]$
    (c) Constant: none
    (d) Domain: $(-\infty, \infty)$
    (e) Range: $[0, \infty)$

19. (a) Increasing: $(-\infty, 1]$
    (b) Decreasing: $[4, \infty)$
    (c) Constant: $[1, 4]$
    (d) Domain: $(-\infty, \infty)$
    (e) Range: $(-\infty, 3]$

21. (a) Increasing: none
    (b) Decreasing: $(-\infty, -2] \cup [3, \infty)$
    (c) Constant: $(-2, 3)$
    (d) Domain: $(-\infty, \infty)$
    (e) Range: $(-\infty, \infty)$

23. $f(x) = x^5$: Increasing: $(-\infty, \infty)$

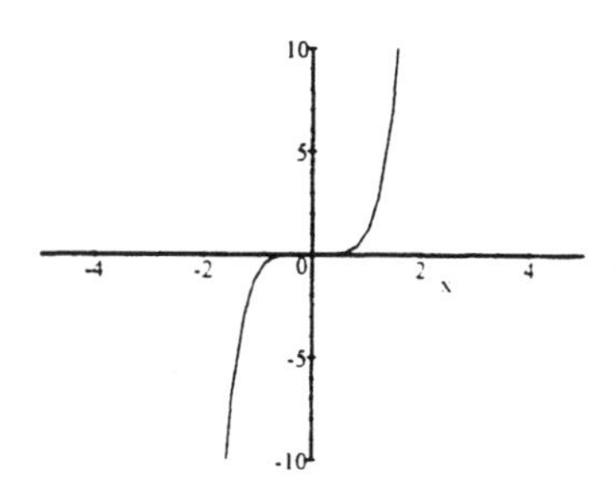

25. $f(x) = x^4$ Decreasing: $(-\infty, 0]$

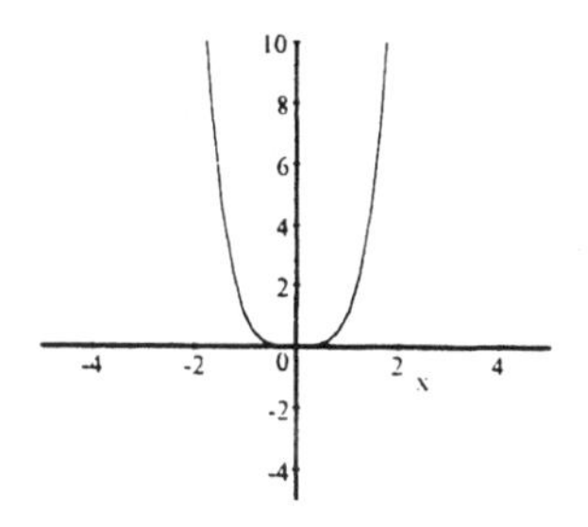

27. $f(x) = -|x|$ Increasing: $(-\infty, 0]$

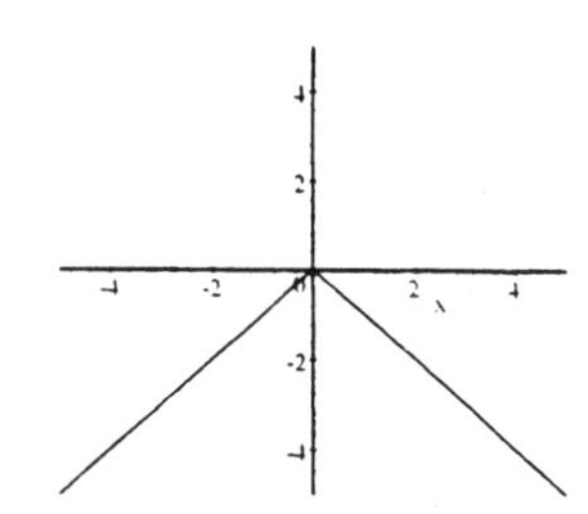

29. (a) x-axis: No
    (b) y-axis: Yes
    (c) origin: No

31. (a) x-axis: Yes
    (b) y-axis: No
    (c) origin: No

33. (a) x-axis: Yes
    (b) y-axis: Yes
    (c) origin: Yes

35. (a) x-axis: No
(b) y-axis: No
(c) origin: Yes

37. Symmetric with respect to the y-axis: $f(x) = f(-x)$;
if $(1.625,\ 2.0352051)$
then $(-1.625,\ 2.0352051)$

39. Symmetric with respect to the x-axis: replace y with $-y$;
$(.5,\ 0.84089642)$
and $(.5,\ -0.84089642)$

41. Symmetric with respect to the origin:
$f(-x) = -f(x)$;
replace x with $-x$ <u>and</u> replace y with $-y$:
$(5.092687, -0.9285541)$ and
$(-5.092687,\ 0.9285541)$.

43. f is an even function:

x	$f(x)$
-3	21
-2	-12
-1	-25
0	0
1	-25
2	-12
3	21

45. (a) $f(-x) = f(x)$: (b) $f(-x) = -f(x)$:

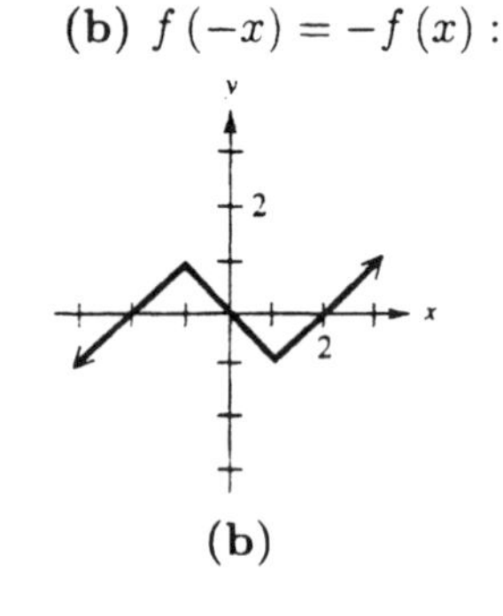

(a) (b)

47. Since $f(x) = f(-x)$: **Even**

49. $f(x) = x^4 - 7x^2 + 6$

$$f(-x) = (-x)^4 - 7(-x)^2 + 6 = x^4 - 7x^2 + 6$$

Since $f(x) = f(-x)$: **Even**

51. $f(x) = 3x^3 - x$

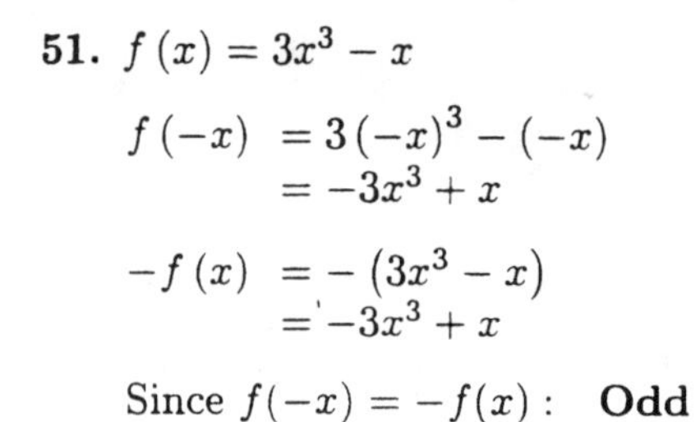

$$f(-x) = 3(-x)^3 - (-x) = -3x^3 + x$$

$$-f(x) = -(3x^3 - x) = -3x^3 + x$$

Since $f(-x) = -f(x)$: **Odd**

53. $f(x) = -x^3 + 2x$

$$f(-x) = -(-x)^3 + 2(-x) = x^3 - 2x$$

$$-f(x) = -(-x^3 + 2x) = x^3 - 2x$$

$f(-x) = -f(x)$; symmetric with respect to the origin.

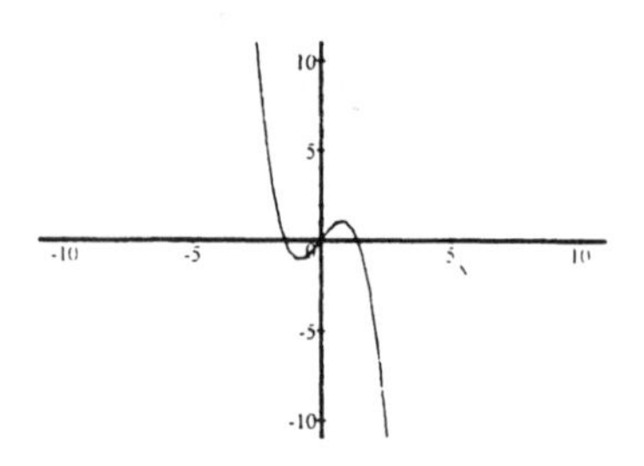

55. $f(x) = .5x^4 - 2x^2 + 1$

$$f(-x) = .5(-x)^4 - 2(-x)^2 + 1 = .5x^4 - 2x^2 + 1$$

$$-f(x) = -(.5x^4 - 2x^2 + 1) = -.5x^4 + 2x^2 - 1$$

$f(x) = f(-x)$; symmetric with respect to the y-axis.

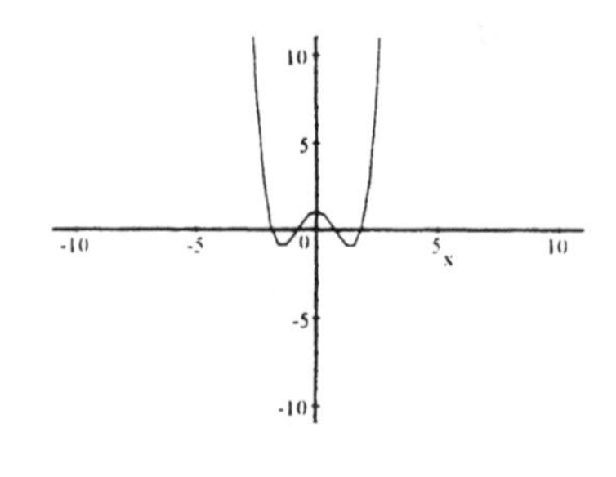

57. $f(x) = x^3 - x + 3$

$f(-x) = (-x)^3 - (-x) + 3 = -x^3 + x + 3$

$-f(x) = -\left(x^3 - x + 3\right) = -x^3 + x - 3$

$f(x) \neq f(-x) \neq -f(x)$; no symmetry

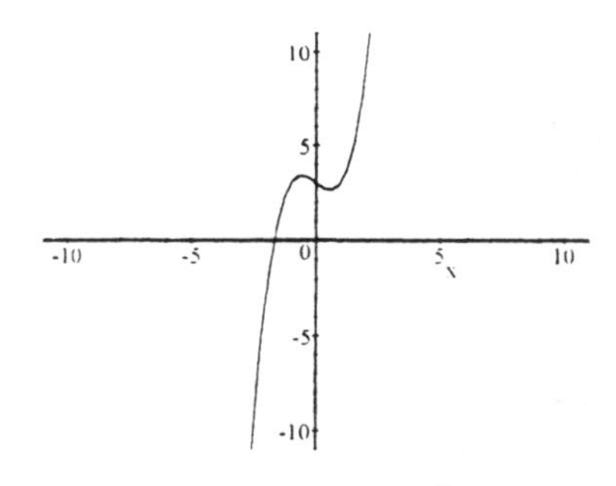

59. (a) even: Exercises 55, 56
(b) odd: Exercises 53, 54
(c) neither even nor odd: Exercises 57, 58

61. Symmetric with respect to the y-axis.

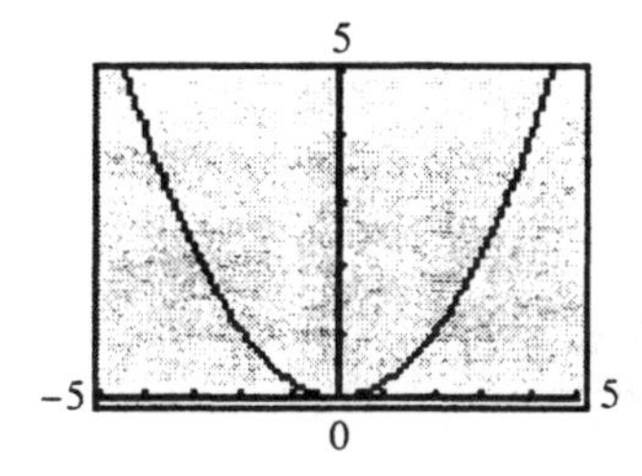

Relating concepts (63 – 66):

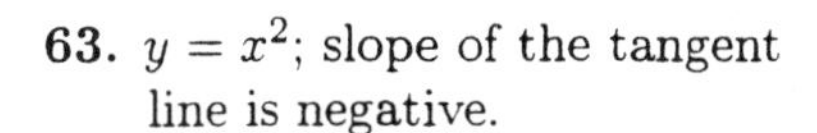

63. $y = x^2$; slope of the tangent line is negative.

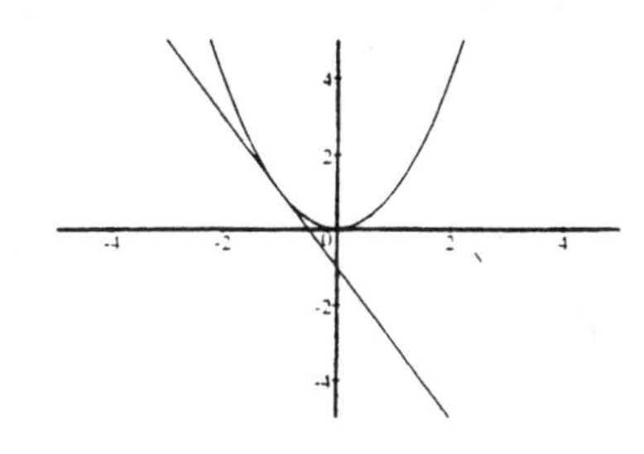

64. $y = x^2$; slope of the tangent line is positive.

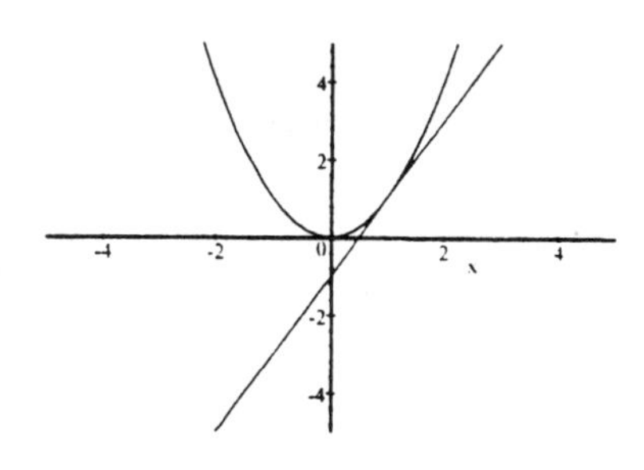

65. $y = x^2$; slope of the tangent line zero.

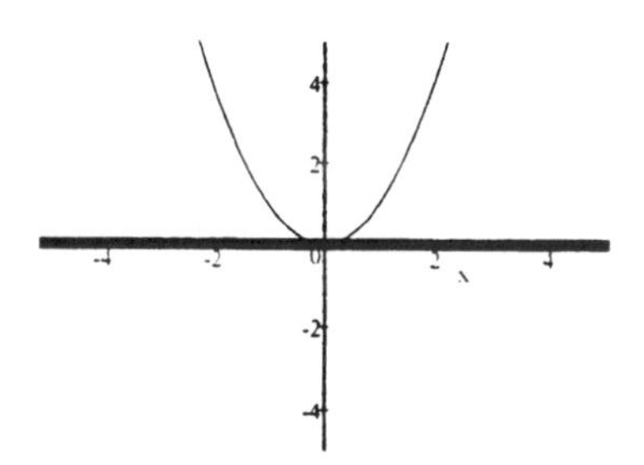

66. If a tangent line is drawn at a point where the function is decreasing, the slope is negative. If a tangent line is drawn at a point where the function is increasing, the slope is positive. If a tangent line is drawn at a point where the function changes from decreasing to increasing, the slope is 0.

67. $y = \dfrac{x^2 - 9}{x + 3} = \dfrac{(x-3)(x+3)}{x+3}$

$y = x - 3,\ x \neq -3$

While y seems to be continuous at first glance, there is a discontinuity at $x = -3$, as indicated by the calculator not giving a corresponding value of y when we trace to $x = -3$. This happens because -3 causes the denominator to $= 0$, which is undefined.

Section 2.2

1. Start with $y = x^2$
shifted 3 units units upward: $y = x^2 + 3$

3. Start with $y = \sqrt{x}$
shifted 4 units units downward: $y = \sqrt{x} - 4$

5. Start with $y = |x|$
shifted 4 units to the right: $y = |x - 4|$

7. Start wtih $y = x^3$
shifted 7 units to the left: $y = (x + 7)^3$

9. Shift the graph of f 4 units upward to obtain the graph of g.

11. $y = x^2 - 3$: **B**

13. $y = (x + 3)^2$: **A**

15. $y = |x + 4| - 3$: **B**

17. $y = (x - 3)^3$: **C**

19. $y = (x + 2)^3 - 4$: **B**

21. $y = x^2 - 3$
domain: $(-\infty, \infty)$; range: $[-3, \infty)$

23. $y = |x + 4| - 3$
   (a) Domain: $(-\infty, \infty)$
   (b) Range: $[-3, \infty)$

25. $Y_2 = Y_1 + k$ (Choose any x)
At $x = 0$:
$19 = 15 + k$ $\boxed{k = 4}$

27. $Y_2 = Y_1 + k$
At $x = 6$:
$-1 = 2 + k$ $\boxed{k = -3}$

29. $y = (x - h)^2 - k$
shift right h units, down k units: **B**

31. $y = (x + h)^2 + k$
shift left h units, up k units: **A**

33. $y = f(x) + 2$: add 2 to each y-coordinate:

$(-3, -2) \Rightarrow (-3, 0)$
$(-1, 4) \Rightarrow (-1, 6)$
$(5, 0) \Rightarrow (5, 2)$

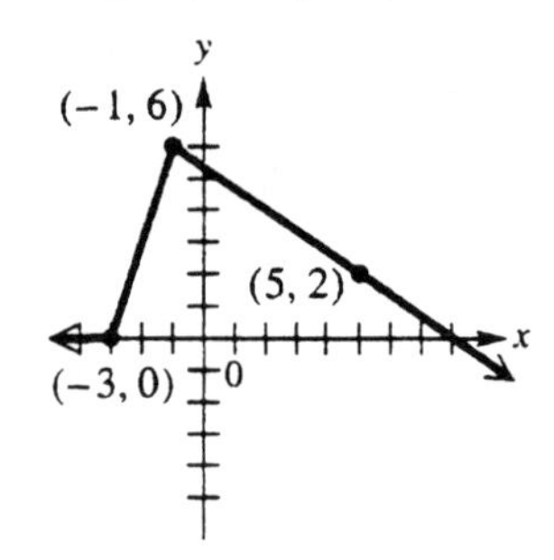

35. $y = f(x + 2)$: subtract 2 from each x-coordinate:

$(-3, -2) \Rightarrow (-5, -2)$
$(-1, 4) \Rightarrow (-3, 4)$
$(5, 0) \Rightarrow (3, 0)$

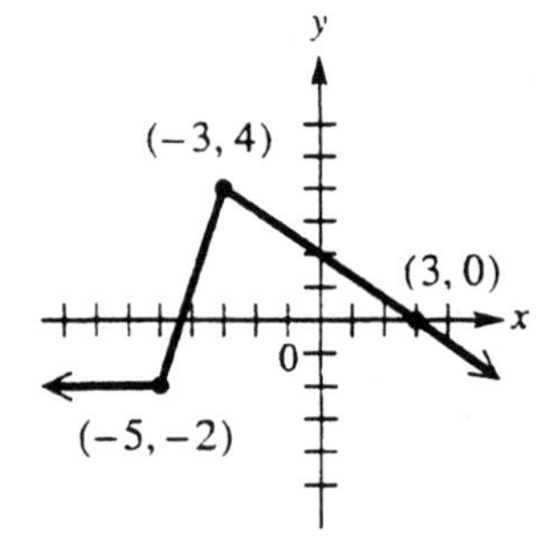

37. Basic function: $y = x^2$
Shift left 4, up 3: $y = (x + 4)^2 + 3$
   (a) Increasing: $[-4, \infty)$
   (b) Decreasing: $(-\infty, -4]$

39. Basic function: $y = x^3$
Shift down 5: $y = x^3 - 5$
   (a) Increasing: $(-\infty, \infty)$
   (b) Decreasing: none

Relating Concepts (41 – 44):

41. (a) $f(x) = 0: \{3, 4\}$
(b) $f(x) > 0: (-\infty, 3) \cup (4, \infty)$
(c) $f(x) < 0: (3, 4)$

42. (a) $f(x) = 0: \{\sqrt{2}\}$
(b) $f(x) > 0: (\sqrt{2}, \infty)$
(c) $f(x) < 0: (-\infty, \sqrt{2})$

43. (a) $f(x) = 0: \{-4, 5\}$
(b) $f(x) \geq 0: (-\infty, -4] \cup [5, \infty)$
(c) $f(x) \leq 0: [-4, 5]$

44. (a) $f(x) = 0: \emptyset$
(b) $f(x) \geq 0: [1, \infty)$
(c) $f(x) \leq 0: \emptyset$

45. $y = |x| = |x + 0| + 0$

D: $(-\infty, \infty)$; R: $[0, \infty)$

$y = |x - h| + k$ shifted left 3, up 1 :

$y = |x - (-3)| + 1$
$= |x + 3| + 1$

$\boxed{h = -3,\ k = 1}$

47. $y = 661.4x + 5459.6$
If $x = 1 \Longrightarrow 1995$
$x = 2 \Longrightarrow 1996$, then
$y = 661.4(x - 1994) + 5459.6$

49. (a) $y \approx 153.351x + 1101.565$
$y \approx 153.4x + 1101.6$
(b) $y \approx 153.4(x - 1984) + 1101.6$

Relating Concepts (51 – 58):

51. Through $(1, -2)$, $(3, 2)$:

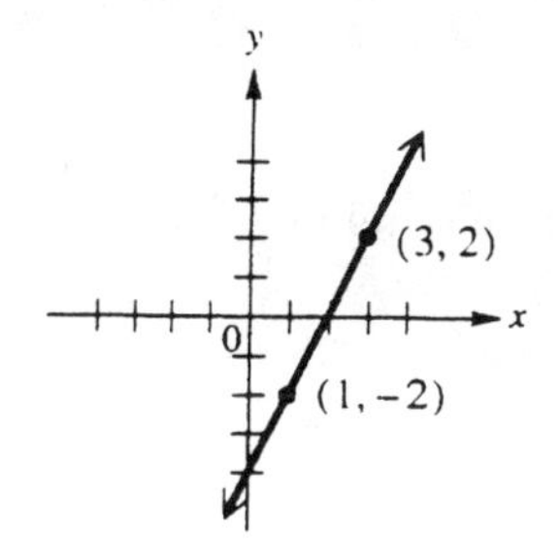

52. $m = \dfrac{\Delta y}{\Delta x}$

$= \dfrac{2 - (-2)}{3 - 1} = \dfrac{4}{2} = \boxed{2}$

53. $m = 2$, through $(1, -2)$:

$y_1 - (-2) = 2(x - 1)$
$y_1 + 2 = 2x - 2$
$\boxed{y_1 = 2x - 4}$

54. $(1, -2) \Rightarrow (1, 4)$
$(3, 2) \Rightarrow (3, 8)$

55. $m = \dfrac{\Delta y}{\Delta x}$

$= \dfrac{4 - 8}{1 - 3} = \dfrac{-4}{-2} = \boxed{2}$

56. $m = 2$ through $(1, 4)$:

$y_2 - 4 = 2(x - 1)$
$y_2 - 4 = 2x - 2$
$\boxed{y_2 = 2x + 2}$

57. $y_2 = y_1 + 6$; $y_1 = 2x - 4$
$= 2x - 4 + 6$
$= 2x + 2$

Translate vertically upward 6 units.

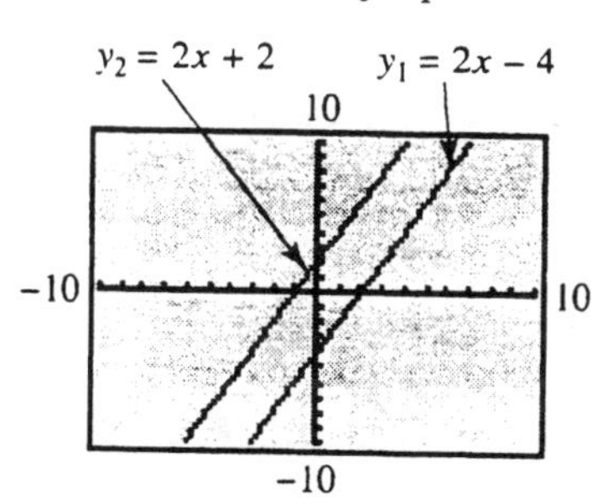

58. If the points (x_1, y_1) and (x_2, y_2) lie on a line, then when we add the positive constant c to each y-value, we obtain the points $(x_1,\ y_1 + \underline{c})$ and $(x_2,\ y_2 + \underline{c})$. The slope of the new line is <u>the same as</u> the slope of the original line. The graph of the new line can be obtained by shifting the graph of the original line $\underline{c}$ units in the <u>upward</u> direction.

Section 2.3

1. Squaring function: $y = x^2$
   Vertical stretch by a factor of 2: $y = 2x^2$

3. Square root function: $y = \sqrt{x}$
   Reflected across the y-axis: $y = \sqrt{-x}$

5. Absolute value function: $y = |x|$
   Vertical stretch by a factor of 3 $y = 3|x|$
   Reflected across the x-axis: $y = -3|x|$

7. Cubing function: $y = x^3$
   Vertical shrink by a factor of .25: $y = .25x^3$
   Reflected across the y-axis: $y = .25(-x)^3$
   or $y = -.25x^3$ since $(-x)^3 = -x^3$

9. $y_1 = x$
   $y_2 = x + 3$
   $y_3 = x - 3$

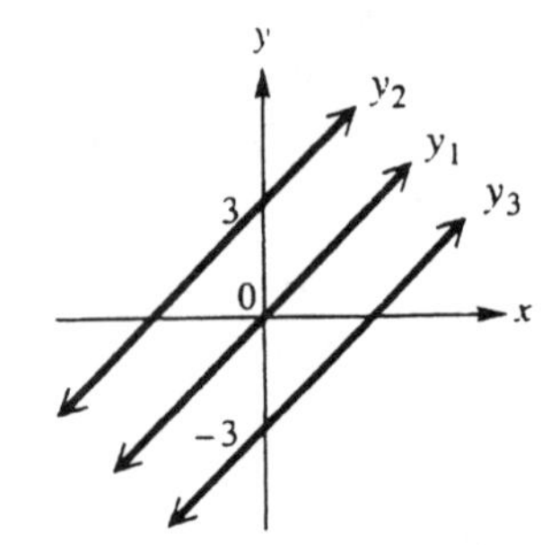

11. $y_1 = |x|$
    $y_2 = |x - 3|$
    $y_3 = |x + 3|$

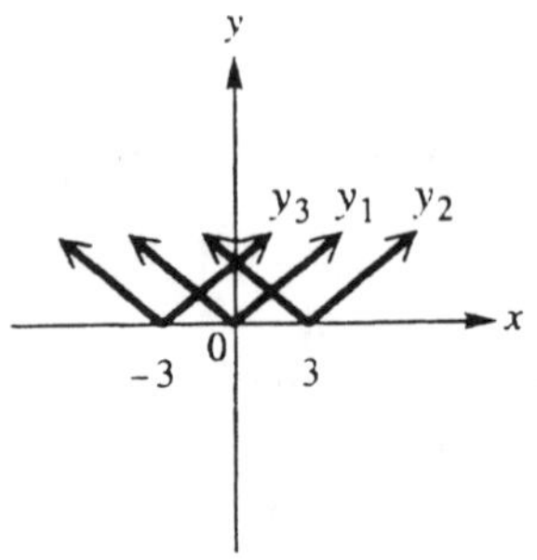

13. $y_1 = \sqrt{x}$
    $y_2 = \sqrt{x + 6}$
    $y_3 = \sqrt{x - 6}$

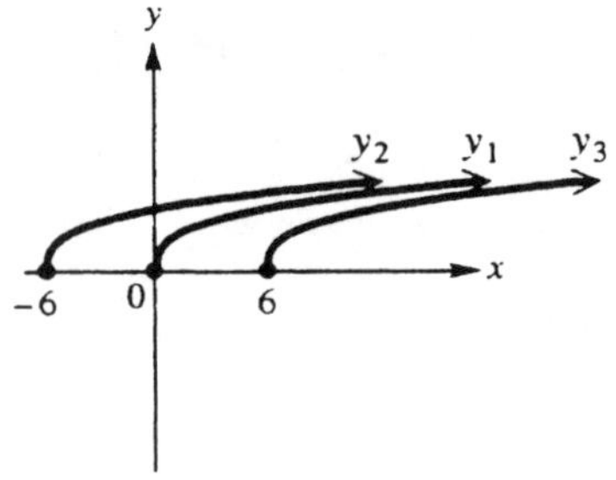

15. $y_1 = \sqrt[3]{x}$
    $y_2 = -\sqrt[3]{x}$
    $y_3 = -2\sqrt[3]{x}$

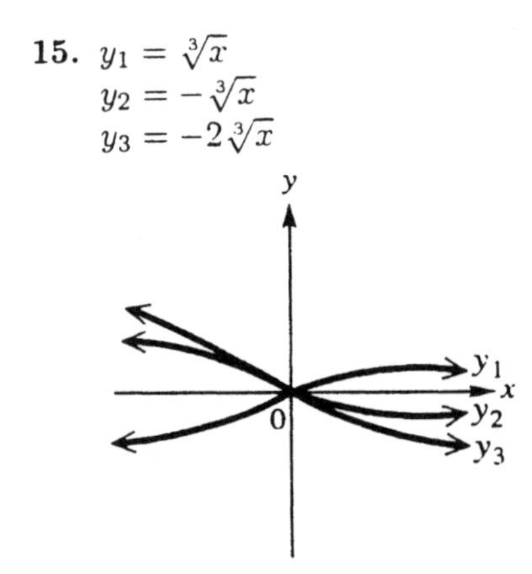

17. $y_1 = |x|$
    $y_2 = -2|x - 1| + 1$
    $y_3 = -\frac{1}{2}|x| - 4$

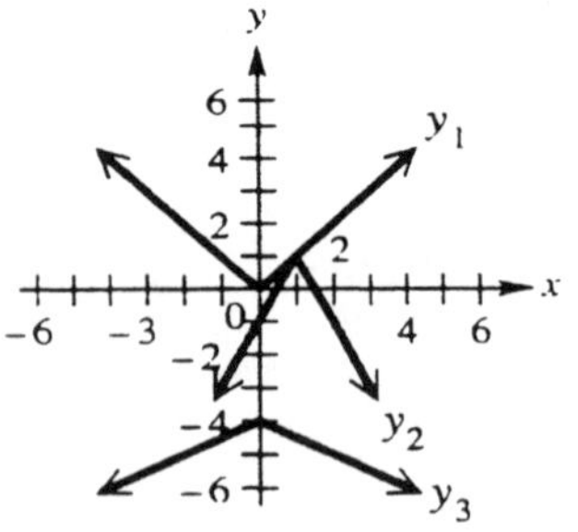

19. The graph of $y = -4x^2$ can be obtained from the graph of $y = x^2$ by vertically stretching by a factor of <u>4</u> and reflecting across the <u>x</u>-axis.

21. The graph of $y = -\frac{1}{4}|x+2| - 3$ can be obtained from the graph of $y = |x|$ by shifting horizontally <u>2</u> units to the <u>left</u>, vertically shrinking by a factor of <u>$\frac{1}{4}$</u>, reflecting across the <u>x</u>-axis, and shifting vertically <u>3</u> units in the <u>negative (downward)</u> direction.

23. The graph of $y = 6\sqrt[3]{x-3}$ can be obtained from the graph of $y = \sqrt[3]{x}$ by shifting horizontally <u>3</u> units to the <u>right</u>, and vertically stretching by a factor of <u>6</u>.

25. $y = x^2$

Vertical shrinking by a factor of $\frac{1}{2}$: $y = \frac{1}{2}x^2$

Shift down 7 units: $\boxed{y = \frac{1}{2}x^2 - 7}$

27. $y = \sqrt{x}$

Shift right 3 units: $y = \sqrt{x-3}$

Vertical stretch by factor of 4.5: $y = 4.5\sqrt{x-3}$

Shift down 6 units:

$\boxed{y = 4.5\sqrt{x-3} - 6}$

29. Y_2 : **F**

31. Y_4 : **D**

33. Y_6 : **B**

35. $y = x^2$

Shift 5 units to right: $y = (x-5)^2$

Reflect across the x-axis:

$y = -(x-5)^2$

Shift down 2 units:

$\boxed{g(x) = -(x-5)^2 - 2}$

37. **(a)** $y = -f(x)$

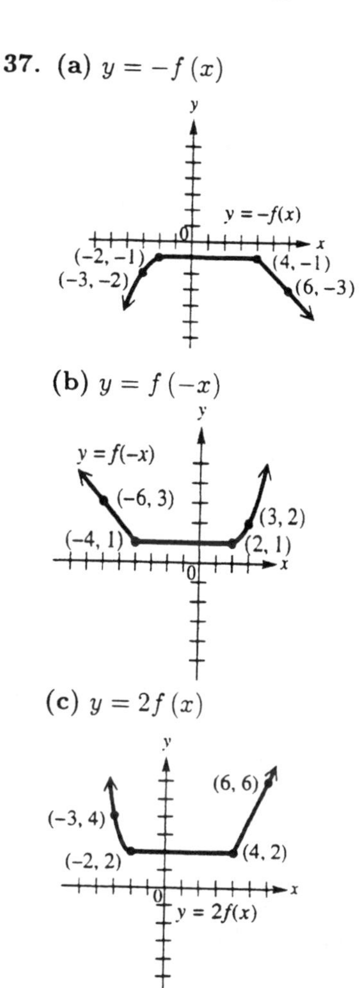

(d) $f(0) = 1$

39. **(a)** $y = -f(x)$

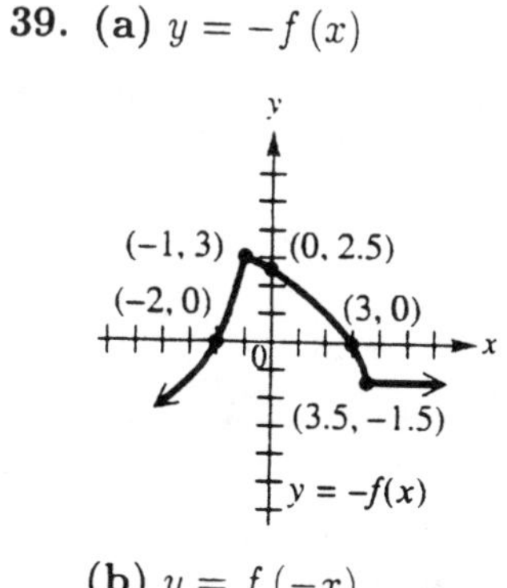

(b) $y = f(-x)$

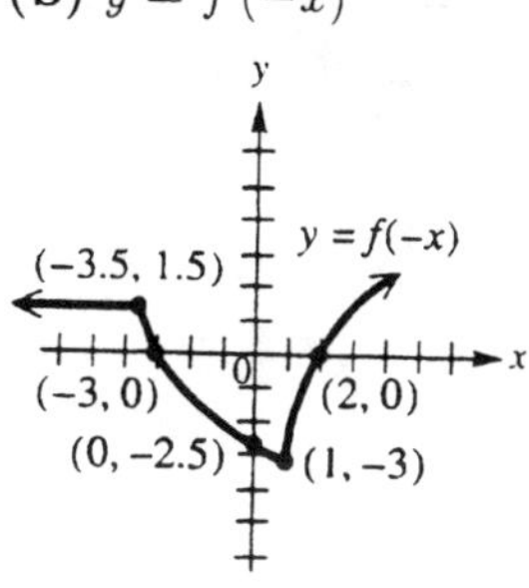

continued

39. continued

(c) $y = f(x+1)$

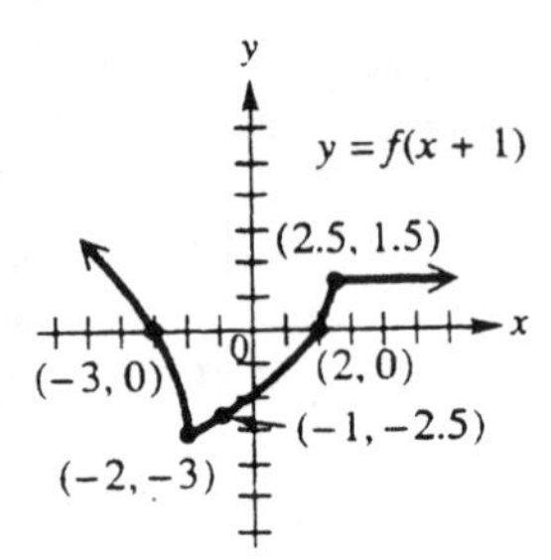

(d) x-intercepts of $y = f(x-1)$ are -1 and 4.

41. (a) $y = -f(x)$

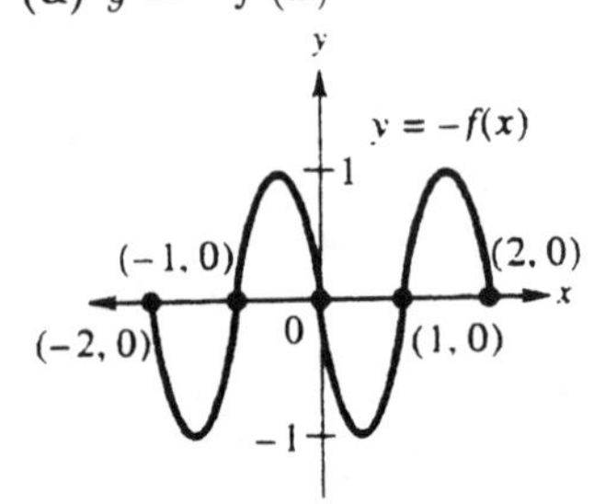

(b) $y = f(-x)$
This graph is the same as $y = -f(x)$, shown in part (a).

(c) $y = .5f(x)$

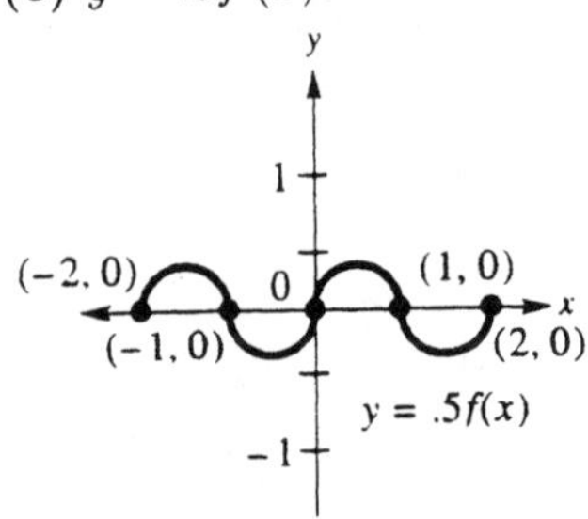

(d) Symmetric with respect to the origin.

43. (a) $y = -f(x)$; r is an x-intercept
(b) $y = f(-x)$; $-r$ is an x-intercept
(c) $y = -f(-x)$; $-r$ is an x-intercept

45. $y = f(x)$ increases $[a, b]$;
$y = -f(x)$ <u>decreases</u> $[a, b]$.

47. $y = f(x)$ increases $[a, b]$;
$= -f(-x)$ <u>increases</u> $[-b, -a]$.

49. D: $(-\infty, \infty)$; R: $[-3, \infty)$
(a) Increasing: $[-1, 2]$
(b) Decreasing: $(-\infty, -1]$
(c) Constant: $[2, \infty)$

51. D: $(-\infty, \infty)$; R: $(-3, \infty)$
(a) Increasing: $[1, \infty)$
(b) Decreasing: $[-2, 1]$
(c) Constant: $(-\infty, -2]$

53. $y = \sqrt{x}$: (endpoint $(0, 0)$)
$y = \sqrt{x-20}$: (endpoint $(20, 0)$)
$y = 10\sqrt{x-20}$: (endpoint $(20, 0)$)
$y = 10\sqrt{x-20} + 5$:
(endpoint $(20, 5)$)

$\boxed{\text{D: } [20, \infty);\ \text{R: } [5, \infty)}$

55. $y = \sqrt{x}$: (endpoint $(0, 0)$)
$y = \sqrt{x+10}$: (endpoint $(-10, 0)$)
$y = -.5\sqrt{x+10}$: (endpoint $(-10, 0)$)
$y = -.5\sqrt{x+10} + 5$:
(endpoint $(-10, 5)$)

$\boxed{\text{D: } [-10, \infty);\ \text{R: } (-\infty, 5]}$

57. $y_1 = \sqrt[3]{x}$, $y_2 = 5\sqrt[3]{x}$
$(x_1, y_1) \Rightarrow (x_1, 5y_1)$
$(8, 2) \Rightarrow \boxed{(8, 10)}$

Reviewing Basic Concepts (Sections 2.1 - 2.3)

1. Given $f(3) = 6 \Longrightarrow (3, 6)$
(a) symmetric with respect to the origin:
$(-3, -6) \Longrightarrow f(-3) = -6$
(b) symmetric with respect to the y-axis:
$(-3, 6) \Longrightarrow f(-3) = 6$
(c) $f(-x) = -f(x)$ (odd function)
$(-3, -6) \Longrightarrow f(-3) = -6$
(d) $f(-x) = f(x)$ (even function)
$(-3, 6) \Longrightarrow f(-3) = 6$

2. (a) B (b) D (c) E (d) A (e) C

3. (a) $y = x^2 + 2$: B
   (b) $y = x^2 - 2$: A
   (c) $y = (x + 2)^2$: G
   (d) $y = (x - 2)^2$: C
   (e) $y = 2x^2$: F
   (f) $y = -x^2$: D
   (g) $y = (x - 2)^2 + 1$: H
   (h) $y = (x + 2)^2 + 1$: E

4. (a) $y = \sqrt{x} + 6$: B
   (b) $y = \sqrt{x + 6}$: E
   (c) $y = \sqrt{x - 6}$: C
   (d) $y = \sqrt{x + 2} - 4$: D
   (e) $y = \sqrt{x - 2} - 4$: A

5. (a) $f(x) = |x|$
   $f(x) = |x + 1|$; 1 unit left
   $f(x) = -|x + 1|$: reflected across x-axis
   $f(x) = -|x + 1| + 3$; 3 units upward

   (b) $f(x) = \sqrt{x}$
   $f(x) = \sqrt{x + 4}$; 4 units left
   $f(x) = -\sqrt{x + 4}$: reflected across x-axis
   $f(x) = -\sqrt{x + 4} + 2$: 2 units upward

   (c) $f(x) = \sqrt{x}$
   $f(x) = \sqrt{x + 4}$; 4 units left
   $f(x) = 2\sqrt{x + 4}$; vertical stretch by a factor of 2
   $f(x) = 2\sqrt{x + 4} - 4$; 4 units downward

   (d) $f(x) = |x|$
   $f(x) = |x - 2|$; 2 units right
   $f(x) = \frac{1}{2}|x - 2|$; vertical shrink by a factor of $\frac{1}{2}$
   $f(x) = \frac{1}{2}|x - 2| - 1$: 1 unit downward

6. (a) $g(x) = f(x) + c$
   since $f(x)$ is moved up 2 units, $c = 2$.
   (b) $g(x) = f(x + c)$
   since $f(x)$ is moved left 4 units, $c = 4$

7. The graph of $y = F(x + h)$ is a horizontal translation of the graph of $y = F(x)$. The graph of $F(x) + h$ is not the same as the graph of $y = F(x + h)$, because $y = F(x) + h$ is a vertical translation of the graph of $y = F(x)$.

8. The effect is either a vertical stretch or a shrink, and perhaps a reflection across the x-axis. If $c > 0$, there is a stretch/shrink by a factor of c. If $c < 0$, there is a stretch/shrink by a factor of $|c|$. If $|c| > 1$, a stretch occurs, while when $|c| < 1$, a shrink occurs.

9.

x	(a) $f(x)$ even	(b) $f(x)$ odd
−3	4	4
−2	−6	−6
−1	5	5
0	0	0
1	5	−5
2	−6	6
3	4	−4

10.
$$\begin{aligned} g(x) &= f(x - 1982) \\ &= -.012053(x - 1982 + 1.93342)^2 + 9.07994 \\ &= -.012053(x - 1980.06658)^2 + 9.07994 \end{aligned}$$

Section 2.4

1. $f(a) = -5$, then $|f(a)| = |-5| = 5$

3. The range of $f(x) = -x^2$ is $(-\infty, 0]$. Since $y = |f(x)| = |-x^2| = x^2$, the range is $[0, \infty)$.

5. If the range of $y = f(x)$ is $(-\infty, -2]$, the range of $y = |f(x)|$ is $[2, \infty)$ since all negative values of y are reflected across the x-axis.

7. $y = |f(x)|$

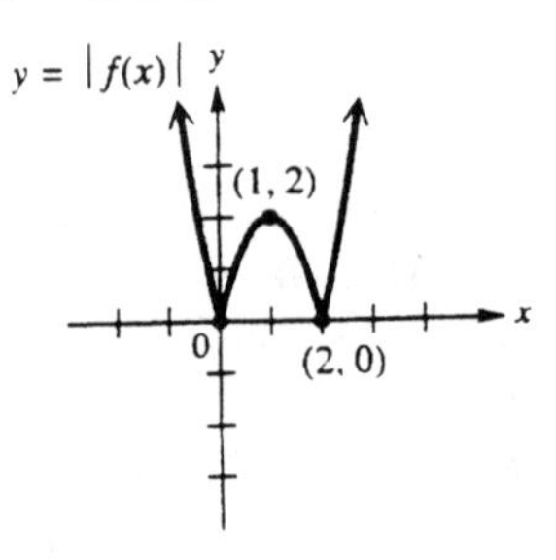

9. $y = |f(x)|$

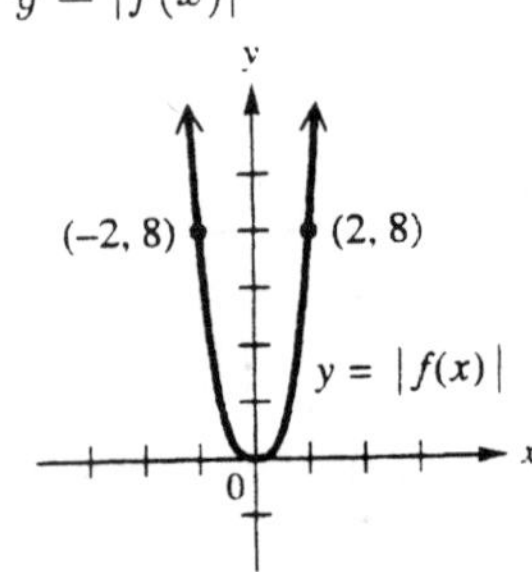

11. $y = |f(x)|$ (Same as $f(x)$)

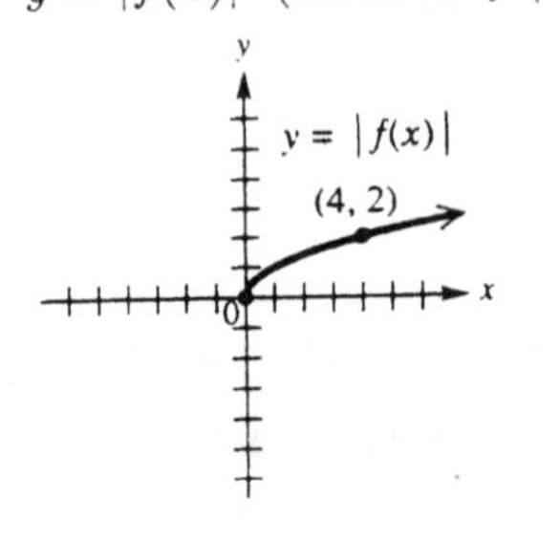

13. $y = |f(x)|$

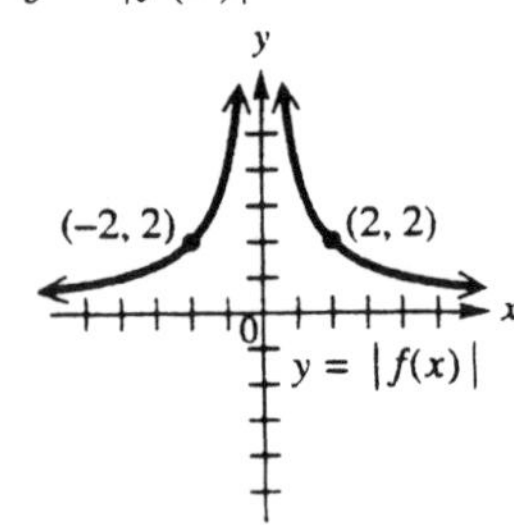

15. $y = |f(x)|$

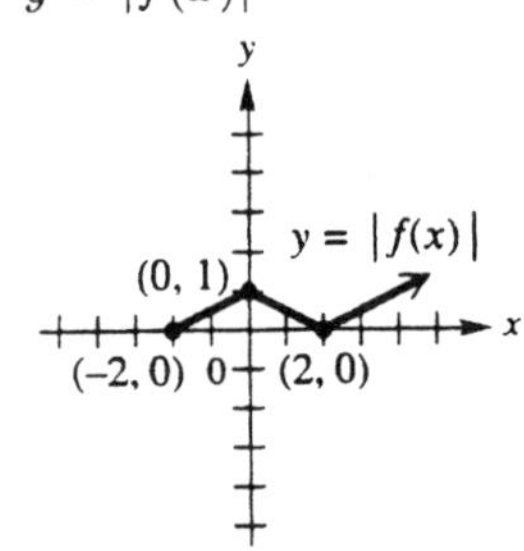

17. $y = |f(x)|$: D

19. $y = |f(x)|$: C

21. (a) $y = f(-x)$

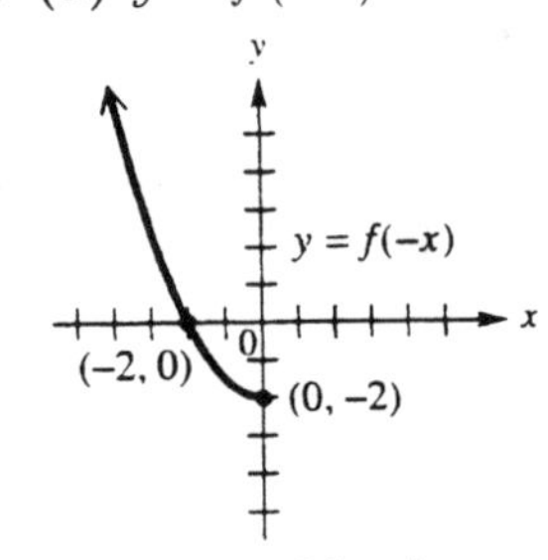

(b) $y = -f(-x)$

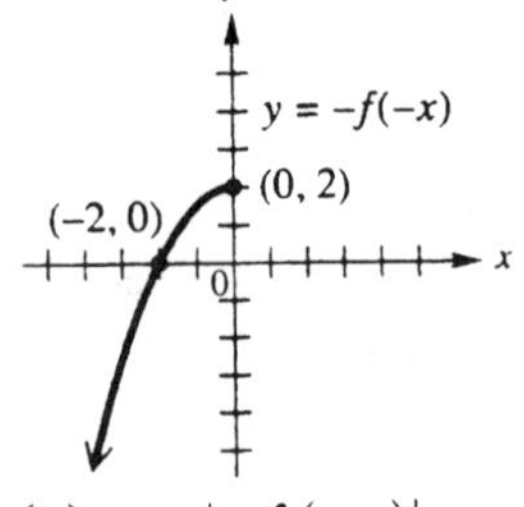

(c) $y = |-f(-x)|$

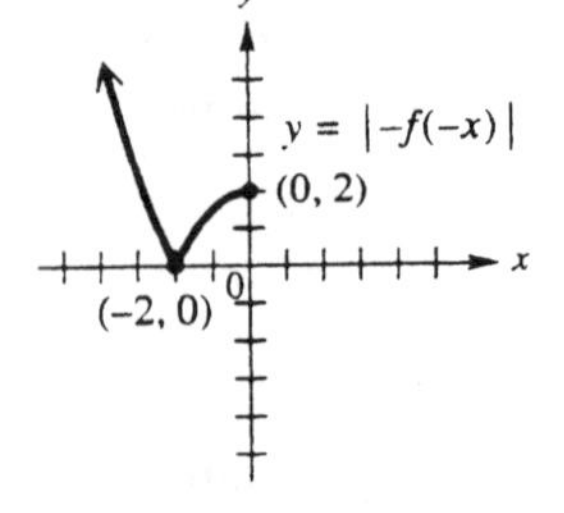

23. $y = f(x)$: **A** ; $y = |f(x)|$: **B**

25. (a) $\{-1, 6\}$
(b) $y_1 < y_2 : (-1, 6)$
(c) $y_1 > y_2 : (-\infty, -1) \cup (6, \infty)$

27. (a) $\{4\}$
(b) $y_1 < y_2 : \emptyset$
(c) $y_1 > y_2 : (-\infty, 4) \cup (4, \infty)$

Relating Concepts (29 – 34)

29. Basic function $P(x) = |x|$ is a V-shaped graph; $f(x)$ is $P(x)$ after shifting and vertically shrinking.

30. Basic function $Q(x) = ax + b$ is a straight line graph; $g(x)$ is $Q(x)$ after shifting and vertically stretching.

31. The graphs intersect at $(8, 10)$, so the solution is $\{8\}$.

32. $f(x) > g(x) : (-\infty, 8)$

33. $f(x) < g(x) : (8, \infty)$

34. The solution is $\{8\}$.

35. (a) $|x + 4| = 9$

$$x + 4 = 9 \quad \textbf{or} \quad x + 4 = -9$$
$$x = 5 \quad \textbf{or} \quad x = -13$$

Solution set: $\boxed{\{-13, 5\}}$

(b) $|x + 4| > 9$

$$x + 4 > 9 \quad \textbf{or} \quad x + 4 < -9$$
$$x > 5 \quad \textbf{or} \quad x < -13$$

Solution set: $\boxed{(-\infty, -13) \cup (5, \infty)}$

(c) $|x + 4| < 9$

$$-9 < x + 4 < 9$$
$$-13 < x < 5$$ Solution set: $\boxed{(-13, 5)}$

37. (a) $|-2x + 7| = 3$

$$-2x + 7 = 3 \quad \textbf{or} \quad -2x + 7 = -3$$
$$-2x = -4 \quad \textbf{or} \quad -2x = -10$$
$$x = 2 \quad \textbf{or} \quad x = 5$$

Solution set: $\boxed{\{2, 5\}}$

(b) $|-2x + 7| \geq 3$

$$-2x + 7 \geq 3 \quad \textbf{or} \quad -2x + 7 \leq -3$$
$$-2x \geq -4 \quad \textbf{or} \quad -2x \leq -10$$
$$x \leq 2 \quad \textbf{or} \quad x \geq 5$$

Solution set: $\boxed{(-\infty, 2] \cup [5, \infty)}$

(c) $|-2x + 7| \leq 3$

$$-3 \leq -2x + 7 \leq 3$$
$$-10 \leq -2x \leq -4$$
$$5 \geq x \geq 2$$
$$2 \leq x \leq 5$$

Solution set: $\boxed{[2, 5]}$

39. (a) $|2x + 1| + 3 = 5$

$$|2x + 1| = 2$$
$$2x + 1 = 2 \quad \textbf{or} \quad 2x + 1 = -2$$
$$2x = 1 \quad \textbf{or} \quad 2x = -3$$
$$x = \frac{1}{2} \quad \textbf{or} \quad x = -\frac{3}{2}$$

Solution set: $\boxed{\left\{-\frac{3}{2}, \frac{1}{2}\right\}}$

(b) $|2x + 1| + 3 \leq 5$

$$|2x + 1| \leq 2$$
$$-2 \leq 2x + 1 \leq 2$$
$$-3 \leq 2x \leq 1$$
$$-\frac{3}{2} \leq x \leq \frac{1}{2}$$ Solution set: $\boxed{\left[-\frac{3}{2}, \frac{1}{2}\right]}$

(c) $|2x + 1| + 3 \geq 5$

$$|2x + 1| \geq 2$$
$$2x + 1 \geq 2 \quad \textbf{or} \quad 2x + 1 \leq -2$$
$$2x \geq 1 \quad \textbf{or} \quad 2x \leq -3$$
$$x \geq \frac{1}{2} \quad \textbf{or} \quad x \leq -\frac{3}{2}$$

Solution set: $\boxed{\left(-\infty, -\frac{3}{2}\right] \cup \left[\frac{1}{2}, \infty\right)}$

41. **(a)** $|7x-5|=0$

$$7x-5=0$$
$$7x=5$$
$$x=\frac{5}{7}$$

Solution set: $\boxed{\left\{\frac{5}{7}\right\}}$

(b) $|7x-5|\geq 0$

$7x-5\geq 0$ **or** $7x-5\leq 0$

$7x\geq 5$ **or** $7x\leq 5$

$x\geq\frac{5}{7}$ **or** $x\leq\frac{5}{7}$

Solution set: $\boxed{(-\infty,\infty)}$

(c) $|7x-5|\leq 0$

$$0\leq 7x-5\leq 0$$
$$5\leq 7x\leq 5$$
$$\frac{5}{7}\leq x\leq\frac{5}{7}$$

Solution set: $\boxed{\left\{\frac{5}{7}\right\}}$

43. **(a)** $|\sqrt{2}x-3.6|=-1$

Absolute value is always positive.

Solution set: $\boxed{\emptyset}$

(b) $|\sqrt{2}x-3.6|\leq -1$

Absolute value cannot be less than or equal to a negative number.

Solution set: $\boxed{\emptyset}$

(c) $|\sqrt{2}x-3.6|\geq -1$

Absolute value is always greater than a negative number.

Solution set: $\boxed{(-\infty,\infty)}$

45. $|3x+1|=|2x-7|$

(a) $3x+1=2x-7$ **or** $3x+1=-(2x-7)$

$x=-8$ $\qquad 3x+1=-2x+7$

$5x=6$

$x=\frac{6}{5}=1.2$

Solution set: $\boxed{\left\{-8,\frac{6}{5}\right\}}$

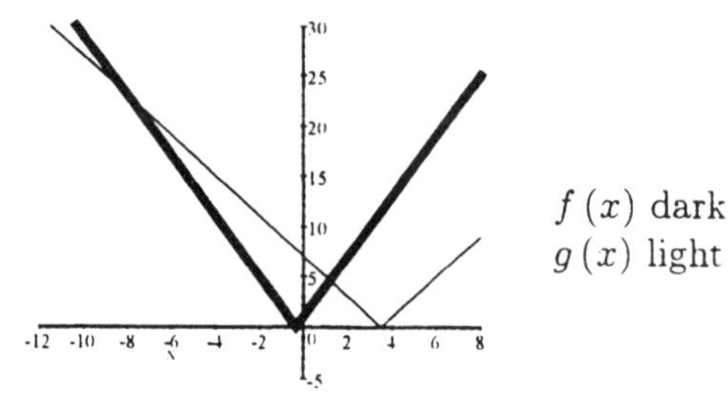

$f(x)$ dark
$g(x)$ light

(b) $|f(x)|>|g(x)|$: $\boxed{(-\infty,-8)\cup\left(\frac{6}{5},\infty\right)}$

(c) $|f(x)|<|g(x)|$: $\boxed{\left(-8,\frac{6}{5}\right)}$

47. **(a)** $|-2x+5|=|x+3|$

$-2x+5 = x+3$ **or** $-2x+5=-(x+3)$

$-3x=-2$ $\qquad -2x+5=-x-3$

$x=\frac{2}{3}$ $\qquad -x=-8$

$x=8$

Solution set: $\boxed{\left\{\frac{2}{3},8\right\}}$

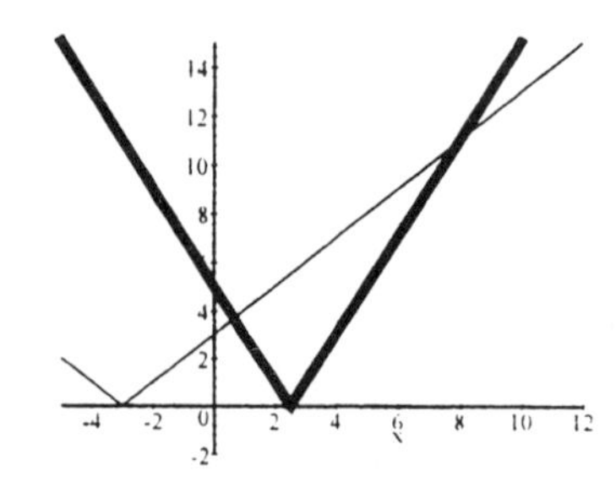

$f(x)$ dark
$g(x)$ light

(b) $|f(x)|>|g(x)|$: $\boxed{\left(-\infty,\frac{2}{3}\right)\cup(8,\infty)}$

(c) $|f(x)|<|g(x)|$: $\boxed{\left(\frac{2}{3},8\right)}$

49. (a) $\left|x-\frac{1}{2}\right| = \left|\frac{1}{2}x-2\right|$

$$x-\frac{1}{2}=\frac{1}{2}x-2 \quad \textbf{or} \quad x-\frac{1}{2}=-\left(\frac{1}{2}x-2\right)$$

$$\frac{1}{2}x=-\frac{3}{2} \qquad x-\frac{1}{2}=-\frac{1}{2}x+2$$

$$x=-3 \qquad \frac{3}{2}x=\frac{5}{2} \Longrightarrow x=\frac{5}{3}$$

Solution set: $\boxed{\left\{-3,\ \frac{5}{3}\right\}}$

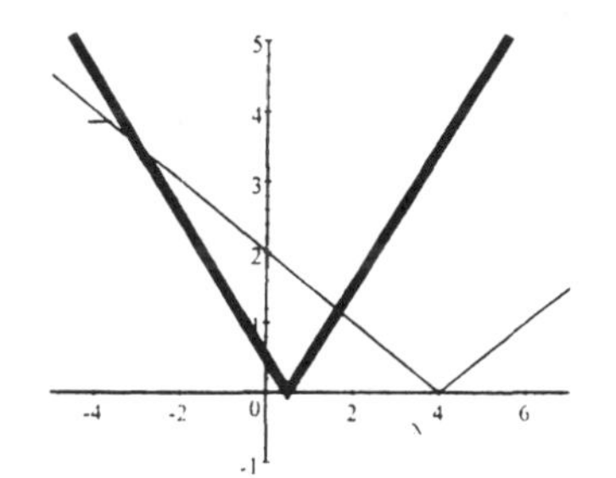

$f(x)$ dark
$g(x)$ light

(b) $|f(x)| > |g(x)|$: $\boxed{(-\infty,-3) \cup \left(\frac{5}{3},\infty\right)}$

(c) $|f(x)| < |g(x)|$: $\boxed{\left(-3,\ \frac{5}{3}\right)}$

51. (a) $|4x+1| = |4x+6|$

$$4x+1=4x+6 \quad \textbf{or} \quad 4x+1=-(4x+6)$$

$$1=6 \qquad 4x+1=-4x-6$$

$$\emptyset \qquad 8x=-7$$

$$x=-\frac{7}{8}$$

Solution set: $\boxed{\left\{-\frac{7}{8}\right\}}$

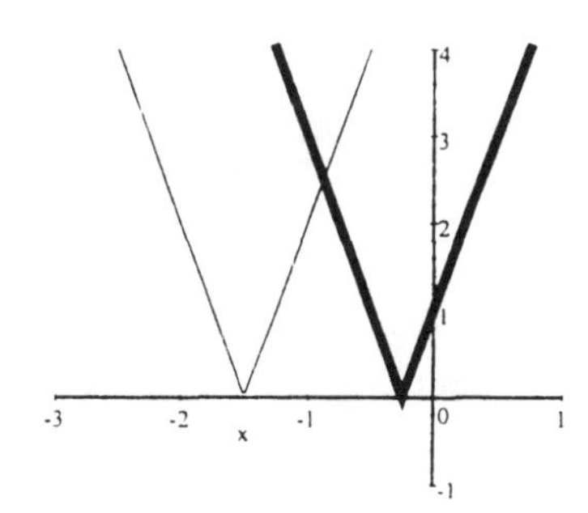

$f(x)$ dark
$g(x)$ light

(b) $|f(x)| > |g(x)|$: $\boxed{\left(-\infty,\ -\frac{7}{8}\right)}$

(c) $|f(x)| < |g(x)|$: $\boxed{\left(-\frac{7}{8},\ \infty\right)}$

53. (a) $|.25x+1| = |.75x-3|$

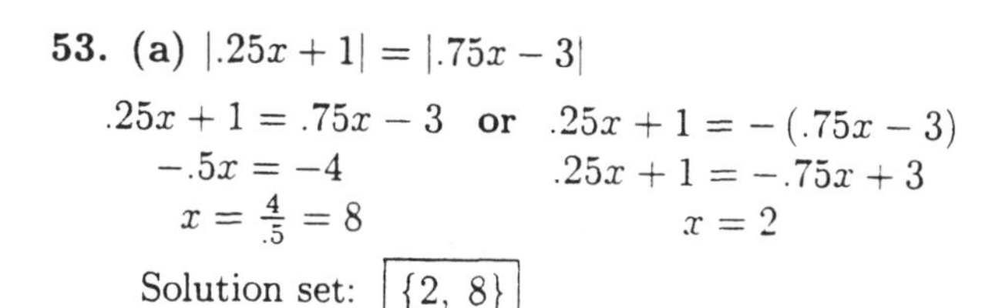

$$.25x+1=.75x-3 \quad \textbf{or} \quad .25x+1=-(.75x-3)$$

$$-.5x=-4 \qquad .25x+1=-.75x+3$$

$$x=\frac{4}{.5}=8 \qquad x=2$$

Solution set: $\boxed{\{2,\ 8\}}$

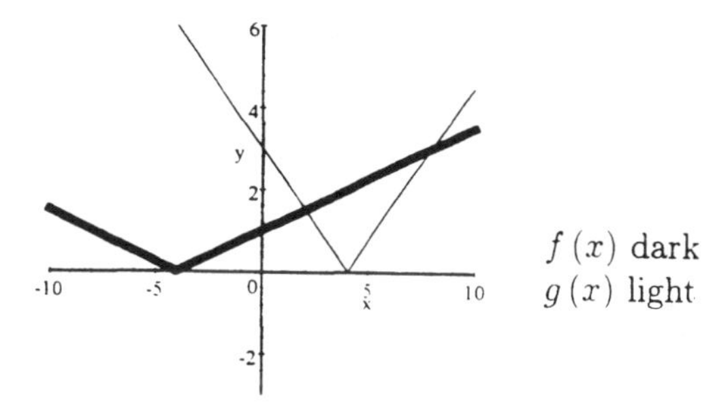

$f(x)$ dark
$g(x)$ light

(b) $|f(x)| > |g(x)|$: $\boxed{(2,\ 8)}$

(c) $|f(x)| < |g(x)|$: $\boxed{(-\infty,\ 2) \cup (8,\ \infty)}$

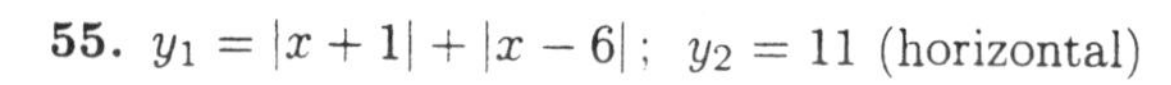

55. $y_1 = |x+1| + |x-6|$; $y_2 = 11$ (horizontal)

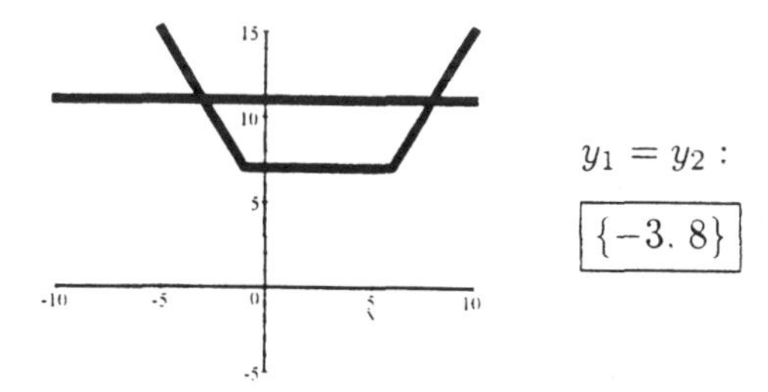

$y_1 = y_2$:

$\boxed{\{-3,\ 8\}}$

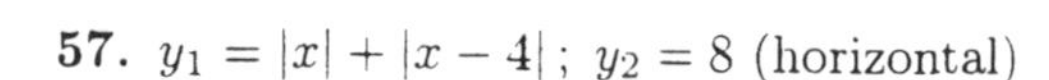

57. $y_1 = |x| + |x-4|$; $y_2 = 8$ (horizontal)

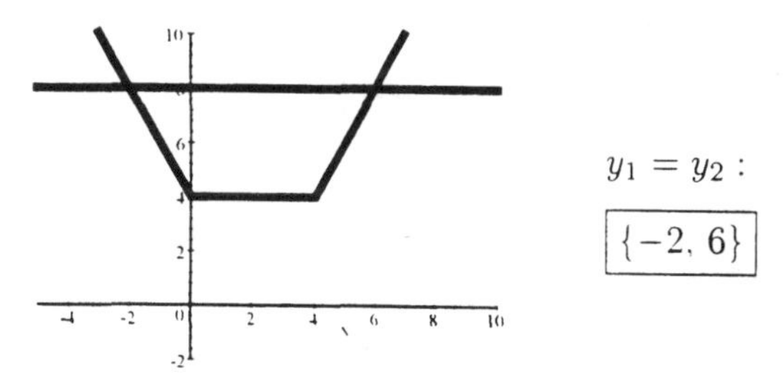

$y_1 = y_2$:

$\boxed{\{-2,\ 6\}}$

59. (a) $|T-43| \le 24$

$$-24 \le T-43 \le 24$$

$$19 \le T \le 67$$

(b) The average monthly temperatures in Marquette vary between a low of 19°F and a high of 67°F. The monthly averages are always within 24° of 43°F.

61. (a) $|T - 50| \leq 22$

$-22 \leq T - 50 \leq 22$

$28 \leq T \leq 72$

(b) The average monthly temperatures in Boston vary between a low of 28°F and a high of 72°F. The monthly averages are always within 22° of 50°F.

63. (a) $|T - 61.5| \leq 12.5$

$-12.5 \leq T - 61.5 \leq 12.5$

$49 \leq T \leq 74$

(b) The average monthly temperatures in Buenos Aires vary between a low of 49°F and a high of 74°F. The monthly averages are always within 12.5° of 61.5°F.

65. $P_d = |P - 125|$

(a) $P_d = |116 - 125| = |-9| = \boxed{9}$

(b) $17 = |P - 130|$

$P - 130 = -17$ or $P - 130 = 17$

$P = 113 \qquad P = 147$

Systolic blood pressure: 113 or 147

67. $|x - 8.0| \leq 1.5$

$-1.5 \leq x - 8.0 \leq 1.5$

$6.5\,\text{lbs} \leq x \leq 9.5\,\text{lbs} \quad \boxed{[6.5,\ 9.5]}$

69. 15 mph at 30°F: 9

10 mph at −10°F: −33

$|9 - (-33)| = |42| = \boxed{42°\text{F}}$

71. 30 mph at −30°F: −94

15 mph at −20°F: −58

$|-94 - (-58)| = |-36| = \boxed{36°\text{F}}$

73. $y = 2x + 1$;

difference between y and $1 < .1$.

$|y - 1| < .1$

$|2x + 1 - 1| < .1$

$|2x| < .1$

$-.1 < 2x < .1$

$-.05 < x < .05 \quad \boxed{(-.05,\ .05)}$

75. $y = 4x - 8$;

difference between y and $3 < .001$

$|y - 3| < .001$

$|4x - 8 - 3| < .001$

$|4x - 11| < .001$

$-.001 < 4x - 11 < .001$

$10.999 < 4x < 11.001$

$2.74975 < x < 2.75025$

$\boxed{(2.74975,\ 2.75025)}$

77. $y_1 = |2x + 7| - (6x - 1)$

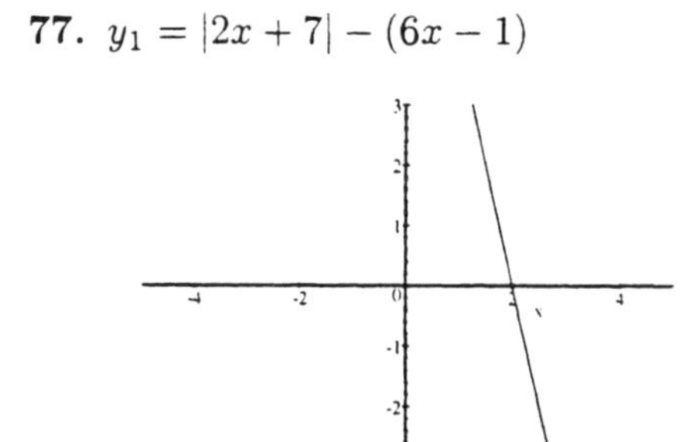

Solution set: $\boxed{\{2\}}$

79. $y_1 = |x - 4| - .5x + 6$

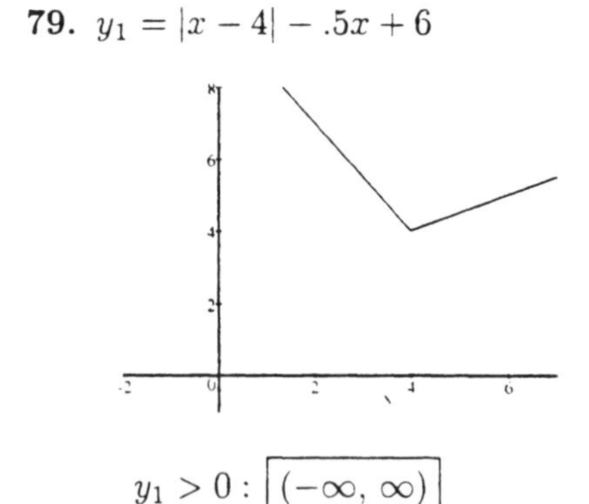

$y_1 > 0 : \boxed{(-\infty,\ \infty)}$

81. $y_1 = |3x + 4| + 3x + 14$

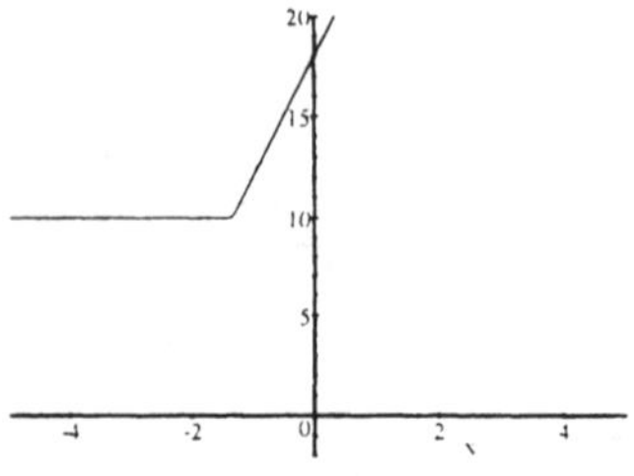

$y_1 < 0 : \boxed{\emptyset}$

Section 2.5

1. (a) maximum: 55 mph; minimum: 30 mph
   (b) $4 + 4 + 4 = 12$ miles
   (c) $f(4) = 40$. $f(12) = 30$. $f(18) = 55$
   (d) discontinuous at $x = 4, 6, 8, 12$, and 16.
   The speed limit changes at each discontinuity.

3. (a) initial: 50.000 gallons; final 30,000 gallons
   (b) constant: $[0, 1]$ and $[3, 4]$ days
   (c) $f(2) \approx 45000$, $f(4) = 40000$
   (d) between 1 and 3: $\dfrac{40000 - 50000}{2}$; 5000 gal/day

5. $f(x) = \begin{cases} 2x & \text{if } x \le -1 \\ x - 1 & \text{if } x > -1 \end{cases}$
   (a) $f(-5) = 2(-5) = -10$
   (b) $f(-1) = 2(-1) = -2$
   (c) $f(0) = (0) - 1 = -1$
   (d) $f(3) = (3) - 1 = 2$

7. $f(x) = \begin{cases} 2 + x & \text{if } x < -4 \\ -x & \text{if } -4 \le x \le 2 \\ 3x & \text{if } x > 2 \end{cases}$
   (a) $f(-5) = 2 + (-5) = -3$
   (b) $f(-1) = -(-1) = 1$
   (c) $f(0) = -(0) = 0$
   (d) $f(3) = 3(3) = 9$

9.

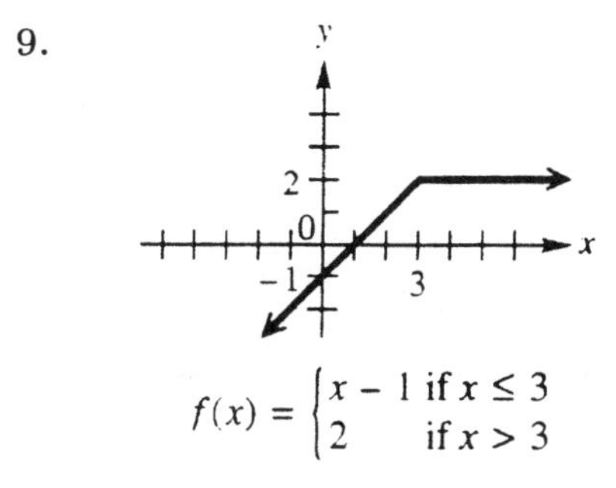

$f(x) = \begin{cases} x - 1 & \text{if } x \le 3 \\ 2 & \text{if } x > 3 \end{cases}$

11.

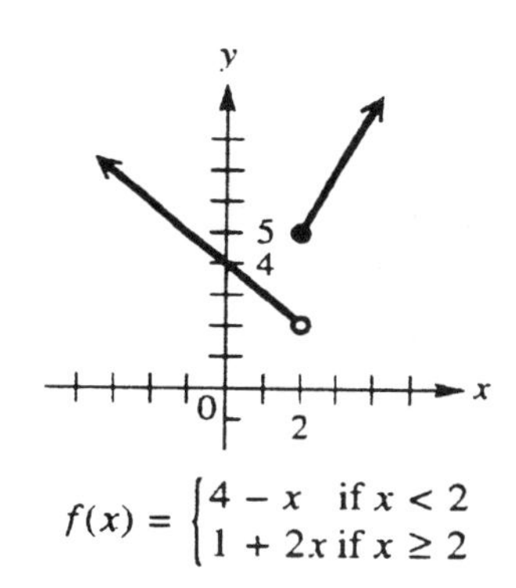

$f(x) = \begin{cases} 4 - x & \text{if } x < 2 \\ 1 + 2x & \text{if } x \ge 2 \end{cases}$

13.

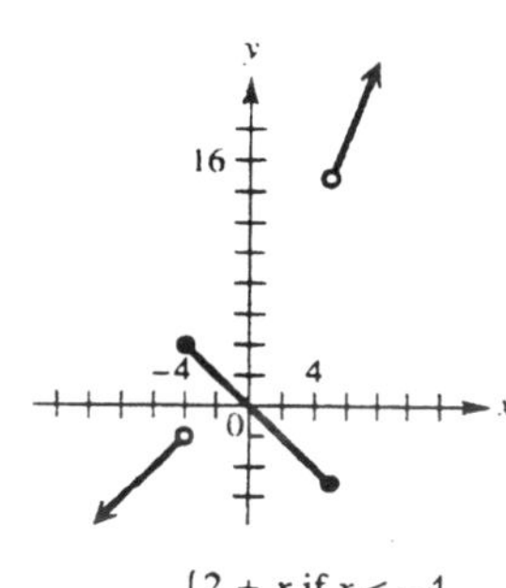

$f(x) = \begin{cases} 2 + x & \text{if } x < -4 \\ -x & \text{if } -4 \le x \le 5 \\ 3x & \text{if } x > 5 \end{cases}$

15.

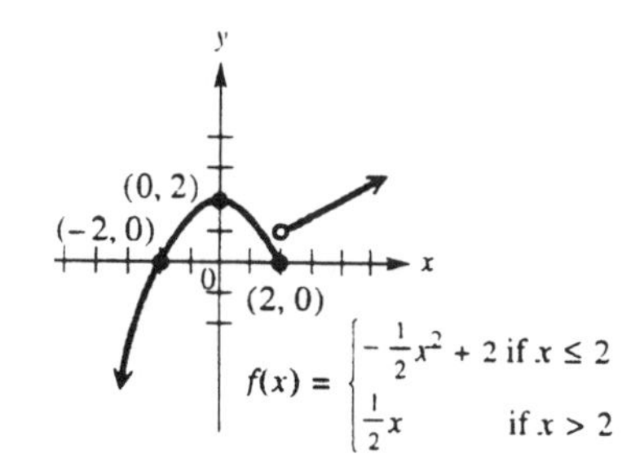

$f(x) = \begin{cases} -\frac{1}{2}x^2 + 2 & \text{if } x \le 2 \\ \frac{1}{2}x & \text{if } x > 2 \end{cases}$

17. B

19. D

21.

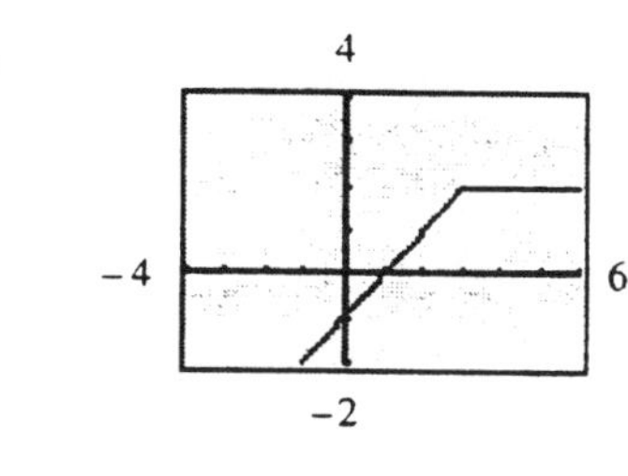

$f(x) = \begin{cases} x - 1 & \text{if } x \le 3 \\ 2 & \text{if } x > 3 \end{cases}$

23. 

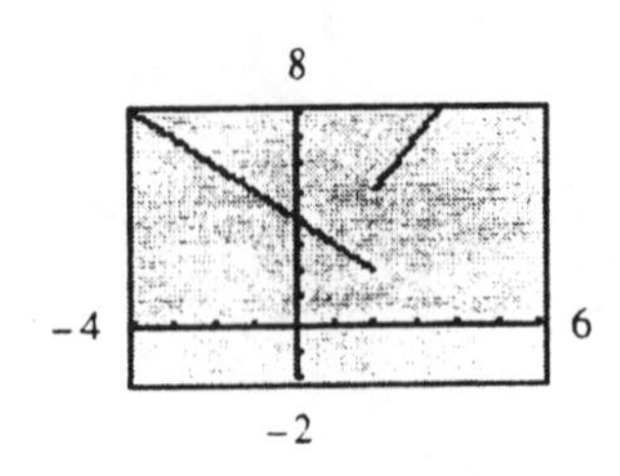

$$f(x) = \begin{cases} 4 - x & \text{if } x < 2 \\ 1 + 2x & \text{if } x \geq 2 \end{cases}$$

25.

20

-12 12

-6

$$f(x) = \begin{cases} 2 + x & \text{if } x < -4 \\ -x & \text{if } -4 \leq x \leq 5 \\ 3x & \text{if } x > 5 \end{cases}$$

27.

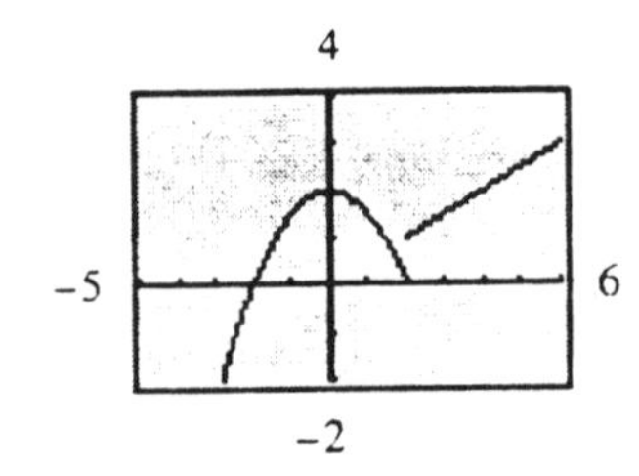

$$f(x) = \begin{cases} -\frac{1}{2}x^2 + 2 & \text{if } x \leq 2 \\ \frac{1}{2}x & \text{if } x > 2 \end{cases}$$

29. $f(x) = \begin{cases} 2 & \text{if } x \leq 0 \\ -1 & \text{if } x > 1 \end{cases}$

Domain: $(-\infty, 0] \cup (1, \infty)$

Range: $\{-1, 2\}$

31. Line through $(-1, -1)$ and $(0, 0)$:

$m = \dfrac{-1}{-1} = 1,\ y = x,\ x \leq 0$

Line through $(0, 2)$ and $(2, 2)$:

$y = 2,\ x > 0$

$$\boxed{f(x) = \begin{cases} x & \text{if } x \leq 0 \\ 2 & \text{if } x > 0 \end{cases}}$$

$D: (-\infty, \infty)$ $R: (-\infty, 0] \cup \{2\}$

33. Curve through $(-8, -2), (-1, -1), (0, 0), (1, 1)$;

$y = \sqrt[3]{x},\ x < 1$

Line through $(1, 2)$ and $(2, 3)$;

$m = \dfrac{1}{1} = 1,\ y - 2 = 1(x - 1)$

$y = x + 1,\ x \geq 1$

$$\boxed{f(x) = \begin{cases} \sqrt[3]{x} & \text{if } x < 1 \\ x + 1 & \text{if } x \geq 1 \end{cases}}$$

$D: (-\infty, \infty)$ $R: (-\infty, 1) \cup [2, \infty)$

35. $f(x) = \begin{cases} x + 7 & \text{if } x \leq 4 \\ x^2 & \text{if } x \geq 4 \end{cases}$

There is an *overlap* of intervals, since the number 4 satisfies both conditions. To truly be a function, every x-value is used only once.

37. Start with $y = [[x]]$; shift the graph downward 1.5 units to obtain $y = [[x]] - 1.5$.

39. Start with $y = [[x]]$; reflect the graph across the x-axis to obtain $y = -[[x]]$.

41. $y = [[x]] - 1.5$

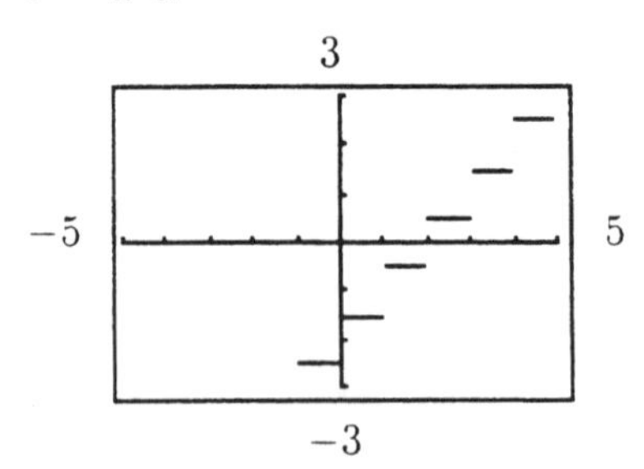

43. $y = -[[x]]$

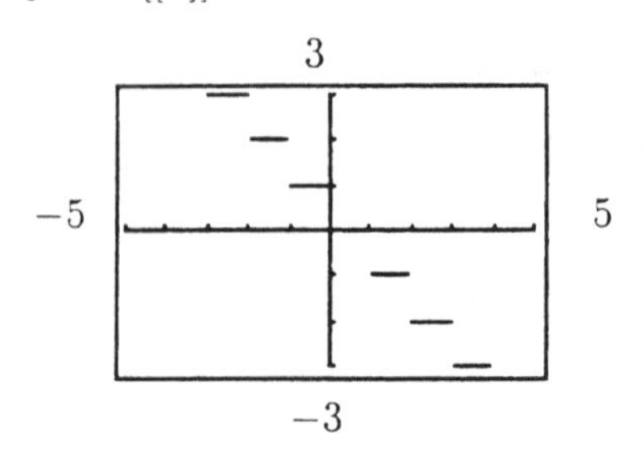

45. When $0 \le x \le 3$, the slope is 5, which means the inlet pipe is open and the outlet pipe is closed; when $3 < x \le 5$, the slope is 2, which means both pipes are open; when $5 < x \le 8$, the slope is 0, which means both pipes are closed; when $8 < x \le 10$, the slope is -3, which means the inlet pipe is closed and the outlet pipe is open.

47. $f(t) = \begin{cases} 40t + 100 & \text{if} \quad 0 \le t \le 3 \\ 220 & \text{if} \quad 3 < t \le 8 \\ -80t + 860 & \text{if} \quad 8 < t \le 10 \\ 60 & \text{if} \quad 10 < t \le 24 \end{cases}$

Let 6 AM = 0

(a) 7 AM: $f(1) = 40(1) + 100 = 140$

(b) 9 AM: $f(3) = 40(3) + 100 = 220$

(c) 10 AM: $f(4) = 220$

(d) noon: $f(6) = 220$

(e) 2 PM: $f(8) = 220$

(f) 5 PM: $f(11) = 60$

(g) midnight: $f(18) = 60$

(h) $f(t)$:

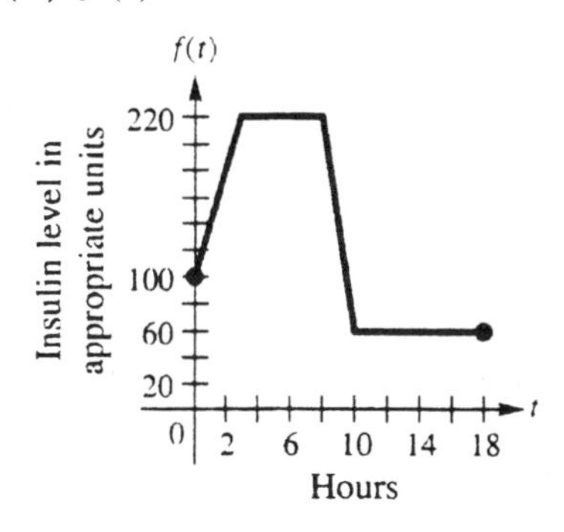

49. **(a)** In each case, the number of cases increases. From 1988 to 1990, there is an increase of 100 cases per year $(m = 100)$. From 1990 to 1992, there is an increase of 1700 cases per year $(m = 1700)$.

continued

49. continued

(b) first line: $(0, 4800)$, $m = 100$

$y = 100x + 4800,\ 0 \le x \le 2$

second line: $(2, 5000)$, $m = 1700$

$y - 5000 = 1700(x - 2)$

$y - 5000 = 1700x - 3400$

$y = 1700x + 1600,\ 2 < x \le 4$

so $f(x) = \begin{cases} 100x + 4800 & \text{if} \quad 0 \le x \le 2 \\ 1700x + 1600 & \text{if} \quad 2 < x \le 4 \end{cases}$

51. **(a)**

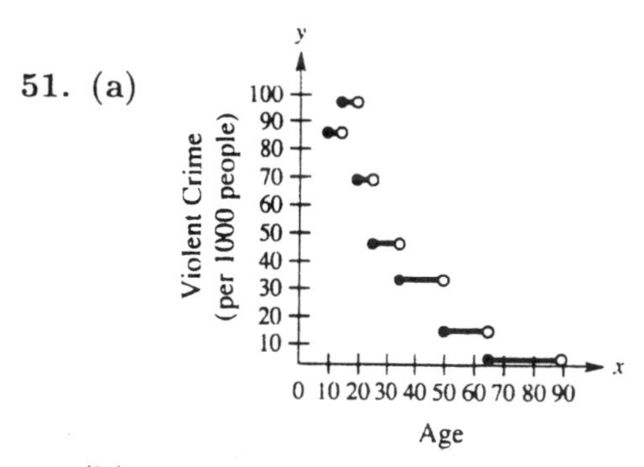

(b) The likelihood of being a victim of crime peaks from age 16 up to age 19, then decreases.

53. **(a)** 3.5 min = \$.50 first minute + \$.25/min:

$.50 + .25\,[|3.5|] = \$1.25$

(b) $f(x) = \begin{cases} .50 & \text{if} \quad 0 < x \le 1 \\ .75 & \text{if} \quad 1 < x \le 2 \\ 1.00 & \quad 2 < x \le 3 \\ 1.25 & \quad 3 < x \le 4 \\ 1.50 & \quad 4 < x \le 5 \end{cases}$

(c) It is not convenient because it rounds fractional values down rather than up.

55. 5 gal/min = 20 min to fill 100 gallons: graph from $(0, 0)$ to $(20, 100)$.
2 gal/min = 50 min to drain; graph from $(20, 100)$ to $(70, 0)$.

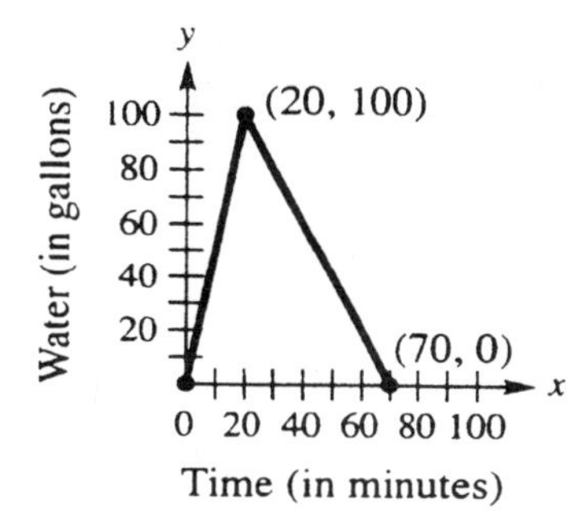

Section 2.6

For problems 1 - 6:

$f(x) = x^2;\ g(x) = 2x - 5$

1. $(f+g)(x) = x^2 + 2x - 5:$ **E**

3. $(fg)(x) = 2x^3 - 5x^2:$ **F**

5. $(f \circ g)(x) = f(2x-5) = (2x-5)^2 = 4x^2 - 20x + 25:$ **A**

7. $f(x) = 4x^2 - 2x,\ g(x) = 8x + 1$

$$\begin{aligned}(f \circ g)(x) &= 256x^2 + 48x + 2 \quad [\#9]\\ (f \circ g)(3) &= 256(3)^2 + 48(3) + 2\\ &= 2304 + 144 + 2 = \boxed{2450}\end{aligned}$$

9. $f(x) = 4x^2 - 2x,\ g(x) = 8x + 1$

$$\begin{aligned}(f \circ g)(x) &= f[g(x)] = f(8x+1)\\ &= 4(8x+1)^2 - 2(8x+1)\\ &= 4(64x^2 + 16x + 1) - 16x - 2\\ &= 256x^2 + 64x + 4 - 16x - 2\\ &= \boxed{256x^2 + 48x + 2}\end{aligned}$$

11. $f(x) = 4x^2 - 2x,\ g(x) = 8x + 1$

$$\begin{aligned}(f+g)(x) &= 4x^2 - 2x + 8x + 1\\ &= 4x^2 + 6x + 1\\ (f+g)(3) &= 4(3)^2 + 6(3) + 1\\ &= 36 - 18 + 1 = \boxed{55}\end{aligned}$$

13. $f(x) = 4x^2 - 2x,\ g(x) = 8x + 1$

$$\begin{aligned}(fg)(x) &= (4x^2 - 2x)(8x+1)\\ (fg)(4) &= \left(4(4)^2 - 2(4)\right)(8(4)+1)\\ &= (64 - 8)(32 + 1) = \boxed{1848}\end{aligned}$$

15. $f(x) = 4x^2 - 2x,\ g(x) = 8x + 1$

$$\begin{aligned}\left(\frac{f}{g}\right)(x) &= \frac{4x^2 - 2x}{8x + 1}\\ \left(\frac{f}{g}\right)(-1) &= \frac{4(-1)^2 - 2(-1)}{8(-1) + 1} = \frac{4+2}{-8+1} = \boxed{-\frac{6}{7}}\end{aligned}$$

17. $f(x) = 4x^2 - 2x,\ g(x) = 8x + 1$

$$\begin{aligned}(f \circ g)(x) &= 256x^2 + 48x + 2 \quad [\#9]\\ (f \circ g)(2) &= 256(2)^2 + 48(2) + 2\\ &= 1024 + 96 + 2 = \boxed{1122}\end{aligned}$$

19. $f(x) = 4x^2 - 2x,\ g(x) = 8x + 1$

$$\begin{aligned}(g \circ f)(x) &= 32x^2 - 16x + 1 \quad [\#10]\\ (g \circ f)(2) &= 32(2)^2 - 16(2) + 1\\ &= 128 - 32 + 1 = \boxed{97}\end{aligned}$$

21. $f(x) = 4x - 1,\ g(x) = 6x + 3$

(a)
$$\begin{aligned}f + g &= 4x - 1 + 6x + 3 = 10x + 2\\ f - g &= 4x - 1 - (6x + 3) = -2x - 4\\ fg &= (4x - 1)(6x + 3) = 24x^2 + 6x - 3\end{aligned}$$

(b) Domain $f + g: (-\infty, \infty)$
Domain $f - g: (-\infty, \infty)$
Domain $fg: (-\infty, \infty)$

(c) $\dfrac{f}{g} = \dfrac{4x - 1}{6x + 3},\ x \neq -\dfrac{1}{2}$

Domain : $\left(-\infty, -\frac{1}{2}\right) \cup \left(-\frac{1}{2}, \infty\right)$

or: $\left\{x \mid x \neq -\frac{1}{2}\right\}$

(d)
$$\begin{aligned}f \circ g &= f[g(x)] = f(6x + 3)\\ &= 4(6x + 3) - 1\\ &= 24x + 12 - 1\\ &= 24x + 11\end{aligned}$$

Domain $f \circ g: (-\infty, \infty)$

(e)
$$\begin{aligned}g \circ f &= g[f(x)] = g(4x - 1)\\ &= 6(4x - 1) + 3\\ &= 24x - 6 + 3\\ &= 24x - 3\end{aligned}$$

Domain $g \circ f: (-\infty, \infty)$

23. $f(x) = |x+3|$, $g(x) = 2x$

(a) $f + g = |x+3| + 2x$

$f - g = |x+3| - 2x$

$fg = 2x\,|x+3|$

(b) Domain $f+g : (-\infty, \infty)$

Domain $f-g : (-\infty, \infty)$

Domain $fg : (-\infty, \infty)$

(c) $\dfrac{f}{g} = \dfrac{|x+3|}{2x}$, $x \neq 0$

Domain $\frac{f}{g} : (-\infty, 0) \cup (0, \infty)$

or $\{x \mid x \neq 0\}$

(d) $f \circ g = f[g(x)] = f(2x) = |2x+3|$

Domain $f \circ g : (-\infty, \infty)$

(e) $g \circ f = g[f(x)] = g(|x+3|) = 2\,|x+3|$

Domain $g \circ f : (-\infty, \infty)$

25. $f(x) = \sqrt[3]{x+4}$, $g(x) = x^3 + 5$

(a) $f + g = \sqrt[3]{x+4} + x^3 + 5$

$f - g = \sqrt[3]{x+4} - x^3 - 5$

$fg = \left(\sqrt[3]{x+4}\right)\left(x^3+5\right)$

(b) Domain $f+g : (-\infty, \infty)$

Domain $f-g : (-\infty, \infty)$

Domain $fg : (-\infty, \infty)$

(c) $\dfrac{f}{g} = \dfrac{\sqrt[3]{x+4}}{x^3+5}$, $x \neq \sqrt[3]{-5}$

Domain : $\left(-\infty, \sqrt[3]{-5}\right) \cup \left(\sqrt[3]{-5}, \infty\right)$

or $\left\{x \mid x \neq \sqrt[3]{-5}\right\}$

(d) $f \circ g = f[g(x)] = f\left(x^3+5\right)$

$= \sqrt[3]{(x^3+5)+4}$

$= \sqrt[3]{x^3+9}$

Domain $f \circ g : (-\infty, \infty)$

(e) $g \circ f = g[f(x)] = g\left(\sqrt[3]{x+4}\right)$

$= \left(\sqrt[3]{x+4}\right)^3 + 5$

$= x + 4 + 5 = x + 9$

Domain $g \circ f : (-\infty, \infty)$

27. $f(x) = \sqrt{x^2+3}$, $g(x) = x + 1$

(a) $f + g = \sqrt{x^2+3} + x + 1$

$f - g = \sqrt{x^2+3} - x - 1$

$fg = \left(\sqrt{x^2+3}\right)(x+1)$

(b) Domain $f+g : (-\infty, \infty)$

Domain $f-g : (-\infty, \infty)$

Domain $fg : (-\infty, \infty)$

(c) $\dfrac{f}{g} = \dfrac{\sqrt{x^2+3}}{x+1}$, $x \neq -1$

Domain : $(-\infty, -1) \cup (-1, \infty)$

or $\{x \mid x \neq -1\}$

(d) $f \circ g = f[g(x)] = f(x+1)$

$= \sqrt{(x+1)^2 + 3}$

$= \sqrt{x^2 + 2x + 4}$

Domain $f \circ g : (-\infty, \infty)$

(e) $g \circ f = g[f(x)] = g\left(\sqrt{x^2+3}\right)$

$= \sqrt{x^2+3} + 1$

Domain $g \circ f : (-\infty, \infty)$

29. **(a)** $(f+g)(2) = f(2) + g(2) = 4 + (-2) = \boxed{2}$

(b) $(f-g)(1) = f(1) - g(1) = 1 - (-3) = \boxed{4}$

(c) $(fg)(0) = f(0) \cdot g(0) = 0 \cdot (-4) = \boxed{0}$

(d) $\left(\dfrac{f}{g}\right)(1) = \dfrac{f(1)}{g(1)} = \dfrac{1}{-3} = \boxed{-\dfrac{1}{3}}$

31. **(a)** $(f+g)(-1) = f(-1) + g(-1) = 0 + 3 = \boxed{3}$

(b) $(f-g)(-2) = f(-2) - g(-2) = -1 - 4 = \boxed{-5}$

(c) $(fg)(0) = f(0) \cdot g(0) = 1 \cdot 2 = \boxed{2}$

(d) $\left(\dfrac{f}{g}\right)(2) = \dfrac{f(2)}{g(2)} = \dfrac{3}{0}$: $\boxed{\text{undefined}}$

33. **(a)** $(f+g)(2) = f(2) + g(2) = 7 + (-2) = \boxed{5}$

(b) $(f-g)(4) = f(4) - g(4) = 10 - 5 = \boxed{5}$

(c) $(fg)(-2) = f(-2) \cdot g(-2) = (0)(6) = \boxed{0}$

(d) $\left(\dfrac{f}{g}\right)(0) = \dfrac{f(0)}{g(0)} = \dfrac{5}{0}$: $\boxed{\text{undefined}}$

35.

x	$(f+g)(x)$	$(f-g)(x)$	$(fg)(x)$	$\left(\frac{f}{g}\right)(x)$
-2	6	-6	0	0
0	5	5	0	$\emptyset$
2	5	9	-14	$-\frac{7}{2}$
4	15	5	50	2

37. (a) $(f \circ g)(4) = f[g(4)] = f(0) = \boxed{-4}$

(b) $(g \circ f)(3) = g[f(3)] = g(2) = \boxed{2}$

(c) $(f \circ f)(2) = f[f(2)] = f(0) = \boxed{-4}$

39. (a) $(f \circ g)(1) = f[g(1)] = f(2) = \boxed{-3}$

(b) $(g \circ f)(-2) = g[f(-2)] = g(-3) = \boxed{-2}$

(c) $(g \circ g)(-2) = g[g(-2)] = g(-1) = \boxed{0}$

41. (a) $(g \circ f)(1) = g[f(1)] = g(4) = \boxed{5}$

(b) $(f \circ g)(4) = f[g(4)] = f(5) : \boxed{\text{undefined}}$

(c) $(f \circ f)(3) = f[f(3)] = f(1) = \boxed{4}$

43. $g(3) = 4; \quad f(4) = 2$

45. $x = -1;\ Y_3 = (Y_1 \circ Y_2)(-1)$

$$= Y_1(Y_2(-1))$$
$$= Y_1(-1)^2 = Y_1(1)$$
$$= 2(1) - 5 = \boxed{-3}$$

47. $x = 7;\ Y_3 = (Y_1 \circ Y_2)(7)$

$$= Y_1(Y_2(7))$$
$$= Y_1(7^2) = Y_1(49)$$
$$= 2(49) - 5 = \boxed{93}$$

49. $f(x) = 4x + 2,\ g(x) = \frac{1}{4}(x-2)$

$$\begin{aligned}(f \circ g)(x) &= f\left(\frac{1}{4}(x-2)\right)\\ &= 4\left(\frac{1}{4}(x-2)\right) + 2\\ &= x - 2 + 2 = \mathbf{x}\end{aligned}$$

$$\begin{aligned}(g \circ f)(x) &= g(4x+2)\\ &= \frac{1}{4}((4x+2)-2)\\ &= \frac{1}{4}(4x) = \mathbf{x}\end{aligned}$$

51. $f(x) = \sqrt[3]{5x+4},\ g(x) = \frac{1}{5}x^3 - \frac{4}{5}$

$$\begin{aligned}(f \circ g)(x) &= f\left(\frac{1}{5}x^3 - \frac{4}{5}\right)\\ &= \sqrt[3]{5\left(\frac{1}{5}x^3 - \frac{4}{5}\right) + 4}\\ &= \sqrt[3]{x^3 - 4 + 4}\\ &= \sqrt[3]{x^3} = \mathbf{x}\end{aligned}$$

$$\begin{aligned}(g \circ f)(x) &= g\left(\sqrt[3]{5x+4}\right)\\ &= \frac{1}{5}\left(\sqrt[3]{5x+4}\right)^3 - \frac{4}{5}\\ &= \frac{1}{5}(5x+4) - \frac{4}{5}\\ &= x + \frac{4}{5} - \frac{4}{5} = \mathbf{x}\end{aligned}$$

53. The graph of y_2 can be obtained by reflecting the graph of y_1 across the line $y_3 = x$.

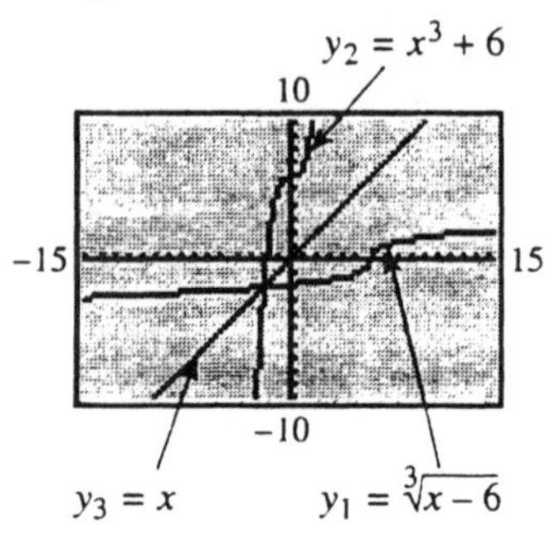

55. $f(x) = 4x + 3$

$$\frac{f(x+h) - f(x)}{h} = \frac{4(x+h) + 3 - (4x+3)}{h} = \frac{4x + 4h + 3 - 4x - 3}{h} = \frac{4h}{h} = \boxed{4}$$

57. $f(x) = -6x^2 - x + 4$

$$\frac{f(x+h) - f(x)}{h}$$
$$= \frac{-6(x+h)^2 - (x+h) + 4 - (-6x^2 - x + 4)}{h}$$
$$= \frac{-6(x^2 + 2xh + h^2) - x - h + 4 + 6x^2 + x - 4}{h}$$
$$= \frac{-6x^2 - 12xh - 6h^2 - x - h + 4 + 6x^2 + x - 4}{h}$$
$$= \frac{-12xh - 6h^2 - h}{h}$$
$$= \boxed{-12x - 6h - 1}$$

59. $f(x) = x^3$

$$\frac{f(x+h) - f(x)}{h} = \frac{(x+h)^3 - x^3}{h} = \frac{x^3 + 3x^2h + 3xh^2 + h^3 - x^3}{h} = \frac{3x^2h + 3xh^2 + h^3}{h} = \boxed{3x^2 + 3xh + h^2}$$

61. $h(x) = (6x - 2)^2$

For example:

Let $f(x) = x^2$, $g(x) = 6x - 2$,

then $(f \circ g)(x) = f(6x - 2) = (6x - 2)^2$

63. $h(x) = \sqrt{x^2 - 1}$

For example:

Let $f(x) = \sqrt{x}$, $g(x) = x^2 - 1$,

then $(f \circ g)(x) = f(x^2 - 1) = \sqrt{x^2 - 1}$

65. $h(x) = \sqrt{6x} + 12$

For example:

Let $f(x) = \sqrt{x} + 12$, $g(x) = 6x$

then $(f \circ g)(x) = f(6x) = \sqrt{6x} + 12$

67. Fixed cost = \$500
Variable cost = $\$10x$
Selling price = $\$35x$

(a) $C(x) = 10x + 500$

(b) $R(x) = 35x$

(c) $P(x) = R(x) - C(x) = 35x - (10x + 500) = 25x - 500$

(d) $25x - 500 > 0$
$25x > 500$
$x > 20$ $\boxed{\text{Produce 21 items}}$

(e) Show that if $P(x) > 0$, then $x > 20$.

69. Fixed cost $= \$2700$

Variable cost $= \$100x$

Selling price $= \$280x$

(a) $C(x) = 100x + 2700$

(b) $R(x) = 280x$

(c) $P(x) = R(x) - C(x)$
$= 280x - (100x + 2700)$
$= 180x - 2700$

(d) $180x - 2700 > 0$
$180x > 2700$
$x > 15$ [Sell 16 items]

(e) Show that if $P(x) > 0$, then $x > 15$.

71. (a) $V(r) = \frac{4}{3}\pi r^3$

Increase by 3 inches:

$$V(r+3) = \frac{4}{3}\pi(r+3)^3$$

Volume gained:

$$\boxed{V(r) = \frac{4}{3}\pi(r+3)^3 - \frac{4}{3}\pi r^3}$$

(b) The graph appears to be part of a parabola; when simplifying $V(r)$, the cubic terms drop out, leaving a squaring function.

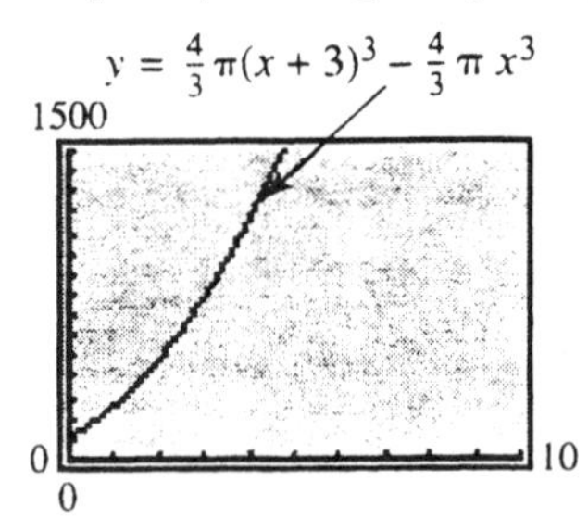

(c) Trace to an x-value of 4;
y-value ≈ 1168.67 in^3

(d)
$$\begin{aligned} V(r+3) &= \frac{4}{3}\pi(r+3)^3 - \frac{4}{3}\pi r^3 \\ V(4+3) &= \frac{4}{3}\pi(7)^3 - \frac{4}{3}\pi(4)^3 \\ &= \frac{4}{3}\pi(7^3 - 4^3) \\ &= \frac{4}{3}\pi(279) \approx \mathbf{1168.67} \text{ in}^3 \end{aligned}$$

73. x : width; $2x$: length

(a)
$$\begin{aligned} P &= 2l + 2w \\ P(x) &= 2(2x) + 2x = 6x \end{aligned}$$
This is a linear function.

(b) x : width; y : perimeter;
4 represents the width of the rectangle and 24 represents the perimeter.

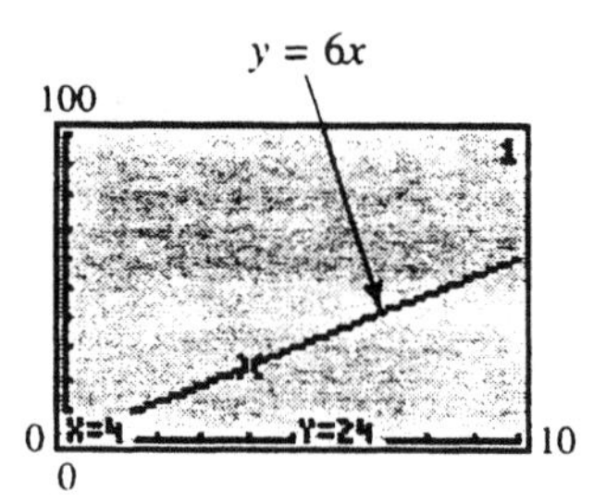

(c) See graph: If x is 4, $2x = 8$, and $P = 6(4) = 24$ which is the y-value.

(d) The point (6, 36) means that a rectangle with a perimeter of 36 will have a width of 6.

75. $A(x) = \frac{\sqrt{3}}{4}x$

(a) $A(2x) = \frac{\sqrt{3}}{4}(2x)^2 = \frac{\sqrt{3}}{4}(4x^2) = \sqrt{3}x^2$

(b) $A(16) = \frac{\sqrt{3}}{4}(16)^2 = 64\sqrt{3} \approx 110.85$

(c) Show that the point (16, 110.85) is on the graph of $y = \frac{\sqrt{3}}{4}x^2$.

77. (a)

x	1991	1992	1993	1994	1995	1996
h(x)	38	42	44	50	54	61

(b) $h(x) = f(x) + g(x)$

79. **(a)** $(f+g)(1970) = f(1970) + g(1970)$
$= 32.4 + 17.6 = 50$

(b) $(f+g)(x)$ computes the total emissions of SO_2 from coal and oil during year x.

(c)

x	1860	1900	1940	1970	2000
(f + g) (x)	2.4	12.8	26.5	50.0	78.0

81. **(a)** $h(x) = g(x) - f(x)$

(b) $h(1995) = g(1995) - f(1995)$
$= 800 - 600 = 200$

$h(2000) = g(2000) - f(2000)$
$= 950 - 700 = 250$

(c) For $h(x)$: $(1995, 200)$, $(2000, 250)$

$$m = \frac{250-200}{2000-1995} = \frac{50}{5} = 10$$

$y - 200 = 10(x - 1995)$
$y - 200 = 10x - 19950$
$y = 10x - 19750$

$\boxed{h(x) = 10x - 19750}$

Reviewing Basic Concepts (Sections 2.4 - 2.6)

1. **(a)** $\left|\frac{1}{2}x + 2\right| = 4$

$\frac{1}{2}x + 2 = 4$ **or** $\frac{1}{2}x + 2 = -4$
$x + 4 = 8$ **or** $x + 4 = -8$
$x = 4$ **or** $x = -12$ $\boxed{\{-12, 4\}}$

(b) $\left|\frac{1}{2}x + 2\right| > 4$ $\boxed{(-\infty, -12) \cup (4, \infty)}$

(c) $\left|\frac{1}{2}x + 2\right| \le 4$ $\boxed{[-12, 4]}$

2. $y = |f(x)|$

3. **(a)** $|R_L - 26.75| \le 1.42$
$-1.42 \le R_L - 26.75 \le 1.42$
$25.33 \le R_L \le 28.17$

$|R_E - 38.75| \le 2.17$
$-2.17 \le R_E - 38.75 \le 2.17$
$36.58 \le R_E \le 40.92$

(b) Multiply by 225:

$5699.25 \le T_L \le 6338.25$
$8230.5 \le T_E \le 9207$

4. $f(x) = \begin{cases} 2x+3 & \text{if } -3 \le x < 0 \\ x^2+4 & \text{if } x \ge 0 \end{cases}$

(a) $f(-3) = 2(-3) + 3 = - -3$
(b) $f(0) = 0^2 + 4 = 4$
(c) $f(2) = 2^2 + 4 = 8$

5. **(a)** $f(x) = \begin{cases} -x^2 & \text{if } x \le 0 \\ x-4 & \text{if } x > 0 \end{cases}$

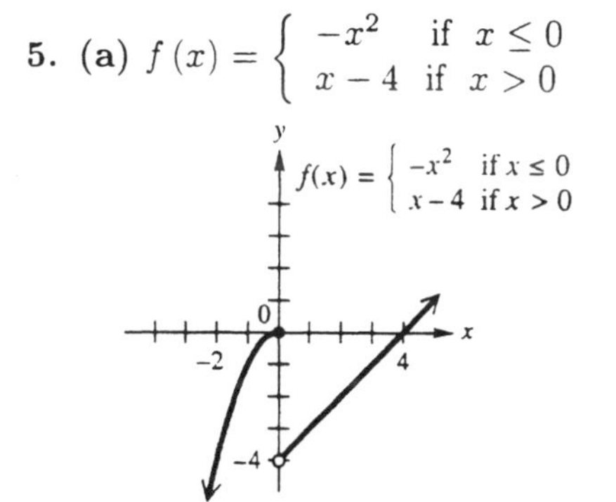

(b)

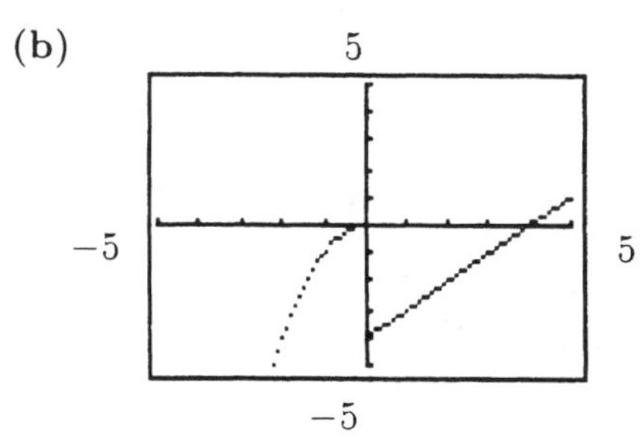

6. $f(x) = -3x - 4$, $g(x) = x^2$

(a) $(f+g)(1) = f(1) + g(1) = -7 + 1 = \boxed{-6}$
(b) $(f-g)(3) = f(3) - g(3) = -13 - 9 = \boxed{-22}$
(c) $(fg)(-2) = f(-2) \cdot g(-2) = 2 \cdot 4 = \boxed{8}$
(d) $\left(\frac{f}{g}\right)(-3) = \frac{f(-3)}{g(-3)} = \boxed{\frac{5}{9}}$
(e) $(f \circ g)(x) = f(g(x)) = f(x^2) = \boxed{-3x^2 - 4}$
(f) $(g \circ f)(x) = g(f(x)) = g(-3x - 4)$
$= (-3x-4)^2 = \boxed{9x^2 + 24x + 16}$

7. Answers may vary. One example:
$f(x) = x^4, \quad g(x) = x + 2$

8. $f(x) = -2x^2 + 3x - 5$

$$\frac{f(x+h) - f(x)}{h}$$
$$= \frac{-2(x+h)^2 + 3(x+h) - 5 - (-2x^2 + 3x - 5)}{h}$$
$$= \frac{-2(x^2 + 2xh + h^2) + 3x + 3h - 5 + 2x^2 - 3x + 5}{h}$$
$$= \frac{-2x^2 - 4xh - 2h^2 + 3x + 3h - 5 + 2x^2 - 3x + 5}{h}$$
$$= \frac{-4xh - 2h^2 + 3h}{h} = \boxed{-4x - 2h + 3}$$

9. **(a)** $r = 4\%;\ I = prt;\ P = x$

$y_1(x) = x(.04)(1)$ $\boxed{y_1(x) = .04x}$

(b) $r = 2.5\%;\ p = x + 500$
$y_2(x) = (x + 500)(.025)\,1$

$\boxed{y_2(x) = .025x + 12.5}$

(c) $y_1 + y_2$ represents the total interest earned in both accounts for 1 year.

(d) $y_1 + y_2 = .04x + .025x + 12.5$

100, 0, 1000, −20

X=250 Y=28.75

\$250 invested yields \$28.75 interest.

(e) $(y_1 + y_2)(x) = .04x + .025x + 12.5$
$= .065x + 12.5$

$(y_1 + y_2)(250) = .065(250) + 12.5$
$= 16.25 + 12.5 = \boxed{\$28.75}$

10. $S = \pi r\sqrt{r^2 + h^2};\quad h = 2r$

$S = \pi r\sqrt{r^2 + (2r)^2}$

$S = \pi r\sqrt{5r^2} = \pi r^2\sqrt{5}$

Chapter 2 Review

Graphs for problems 1 - 10:

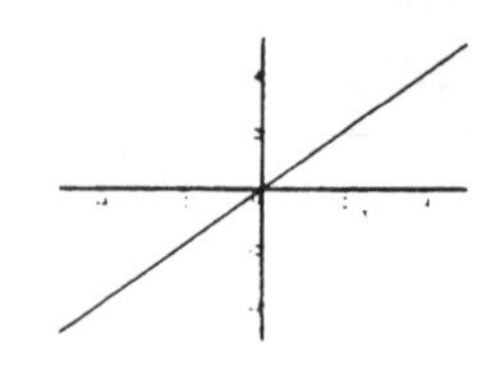

$y = x$

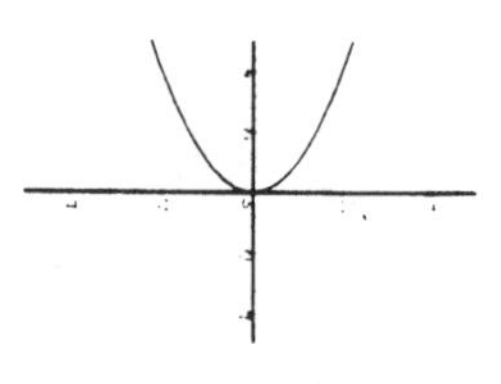

$y = x^2$

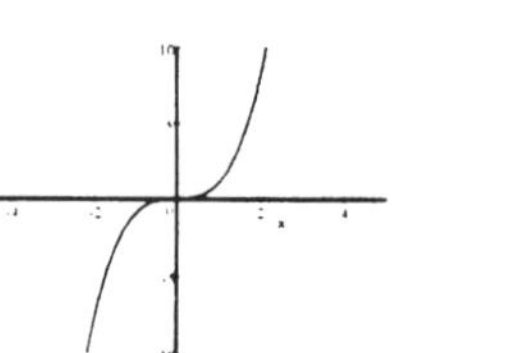

$y = x^3$

$y = \sqrt{x}$

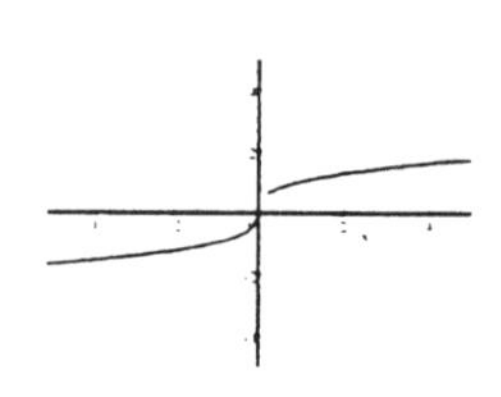

$y = \sqrt[3]{x}$

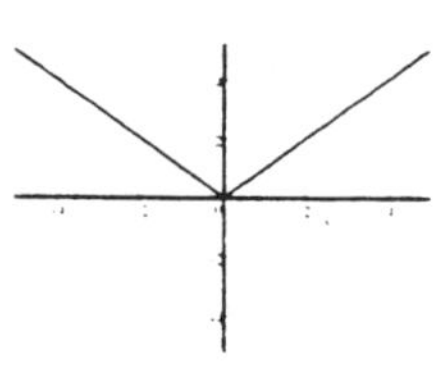

$y = |x|$

1. True: range of $f(x) = x^2$ is $[0, \infty)$ and range of $f(x) = |x|$ is $[0, \infty)$.

2. True; $f(x) = x^2$ increases on $[0, \infty)$ and $f(x) = |x|$ increases on $[0, \infty)$.

3. False; $f(x) = \sqrt{x}$ has domain $[0, \infty)$, while $f(x) = \sqrt[3]{x}$ has domain $(-\infty, \infty)$.

4. False; $f(x) = \sqrt[3]{x}$ never decreases; it is always increasing.

5. True; $f(x) = x$ has domain and range of $(-\infty, \infty)$

6. False; $f(x) = \sqrt{x}$ is continuous on $[0, \infty)$; it is not defined on $(-\infty, 0)$.

7. True; all functions are increasing on $[0, \infty)$.

8. True; $f(x) = x$ and $f(x) = x^3$ are both symmetrical with respect to the origin.

9. True; $f(x) = x^2$ and $f(x) = |x|$ are both symmetrical with respect to the y-axis.

10. True; no graphs are symmetrical with respect to the $x-$axis.

11. $f(x) = \sqrt{x}$
Domain: $[0, \infty)$

12. $f(x) = |x|$
Range: $[0, \infty)$

13. $f(x) = \sqrt[3]{x}$
Range: $(-\infty, \infty)$

14. $f(x) = x^2$
Domain: $(-\infty, \infty)$

15. $f(x) = \sqrt[3]{x}$
Increasing: $(-\infty, \infty)$

16. $f(x) = |x|$
Increasing: $[0, \infty)$

17. Domain of $x = y^2 : [0, \infty)$

18. Range of $x = y^2 : (-\infty, \infty)$

19. (a) Continuous: $(-\infty, -2)$, $[-2, 1]$, $(1, \infty)$
(b) Increases: $[-2, 1]$
(c) Decreases: $(-\infty, -2)$
(d) Constant: $(1, \infty)$
(e) Domain: $(-\infty, \infty)$
(f) Range: $\{-2\} \cup [-1, 1] \cup (2, \infty)$

20. $x = y^2 - 4$
$y^2 = x + 4$
$y = \pm\sqrt{x+4}$
Graph $y_1 = \sqrt{x+4}$, $y_2 = -\sqrt{x+4}$

21. Symmetric with respect to the x-axis, y-axis, and the origin; not a function.

22. No symmetry;
$F(x) = x^3 - 6$
$F(-x) = -x^3 - 6$
$-F(x) = -x^3 + 6$
Neither even nor odd

23. Symmetric with respect to the y-axis.
$f(x) = y = |x| + 4$
$f(-x) = |-x| + 4 = |x| + 4$
$-f(x) = -|x| - 4$
$f(x) = f(-x)$: Even function

24. No symmetry;
$f(x) = \sqrt{x-5}$
$f(-x) = \sqrt{-x-5}$
$-f(x) = -\sqrt{x-5}$
Neither even nor odd

25. Symmetric with respect to the x-axis; not a function.

26. Symmetric with respect to the y-axis;
$f(x) = 3x^4 + 2x^2 + 1$
$f(-x) = 3x^4 + 2x^2 + 1$
$-f(x) = -3x^4 - 2x^2 - 1$
$f(x) = f(-x)$: Even function

27. true: a graph that is symmetrical with respect to the x-axis means that for every (x, y) there is also $(x, -y)$, which is *not* a function.

28. true; since an even function and one that is symmetric with respect to the y-axis both contain the points (x, y) and $(-x, y)$.

29. true; since an odd function and one that is symmetric with respect to the origin both contain the points (x, y) and $(-x, -y)$.

30. false: for an even function, if (a, b) is on the graph, then $(-a, b)$ is on the graph.

31. false; for an odd function, if (a, b) is on the graph, then $(-a, -b)$ is on the graph.

32. true; if $(x, 0)$ is on the graph of $f(x) = 0$, then $(-x, 0)$ is on the graph.

33. $y = x^2$

$y = (x+4)^2$: Shift 4 units left.

$y = 3(x+4)^2$:
Stretch vertically by a factor of 3.

$y = -3(x+4)^2$: Reflect across the x-axis.

$y = -3(x+4)^2 - 8$:
Shift 8 units downward.

34. $y = \sqrt{x}$

$y = \sqrt{-x}$: Reflect across y-axis.

$y = -\sqrt{-x}$: Reflect across x-axis.

$y = -\frac{2}{3}\sqrt{-x}$: Shrink vertically by a factor of $\frac{2}{3}$.

$y = -\frac{2}{3}\sqrt{-x} + 4$: Shift 4 units up

35. $y = f(x) + 3$; shift upward 3 units

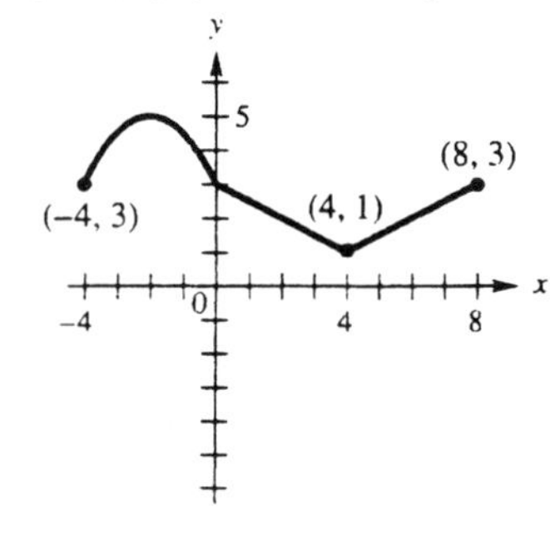

36. $y = f(x-2)$; shift right 2 units

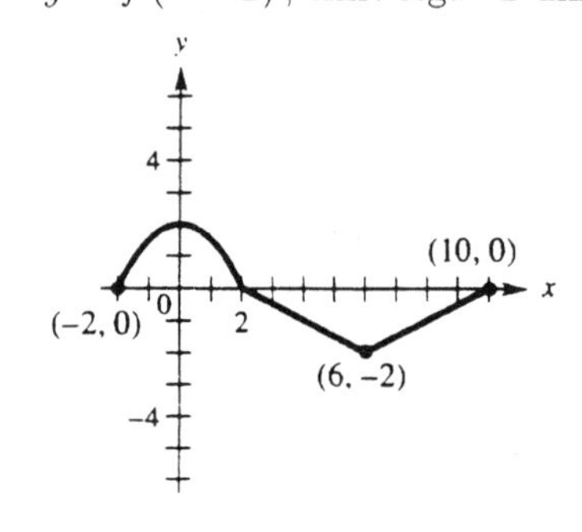

37. $y = f(x+3) - 2$; shift left 3 units then shift downward 2 units

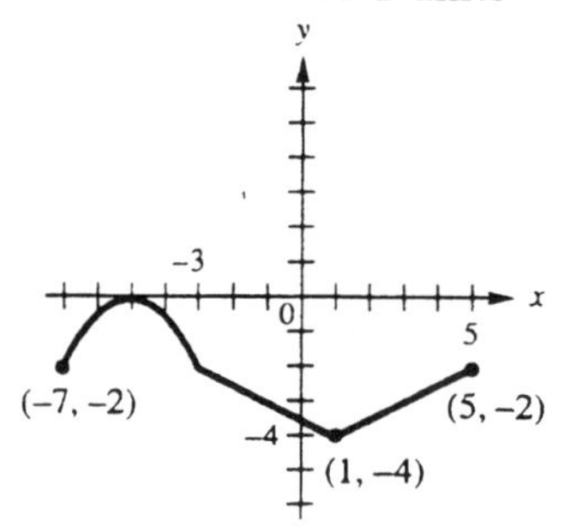

38. $y = |f(x)|$; reflect all $f(x) < 0$ across the x-axis

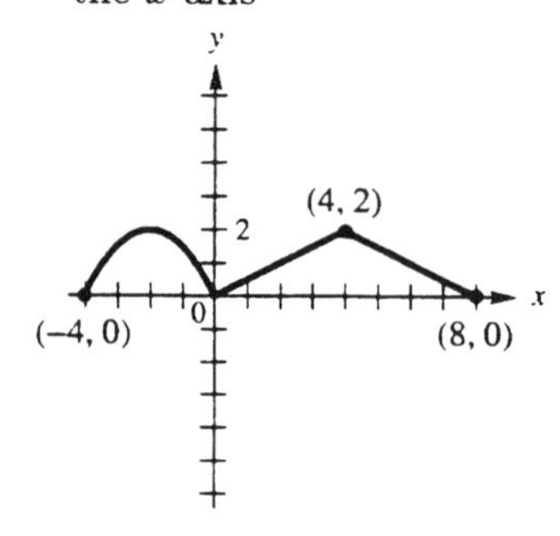

39. $y = |f(x)|$

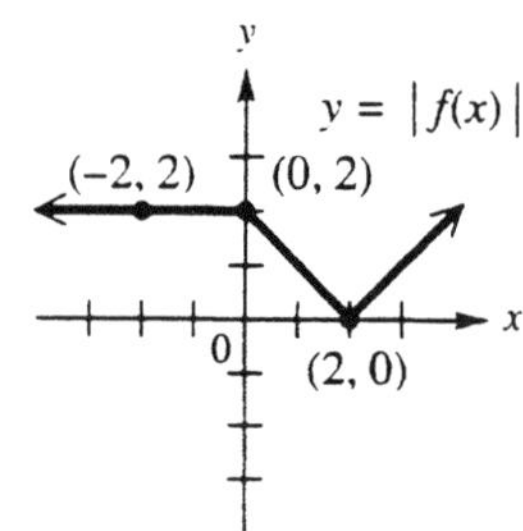

40. $y = |f(x)| = |x^2 - 2|$

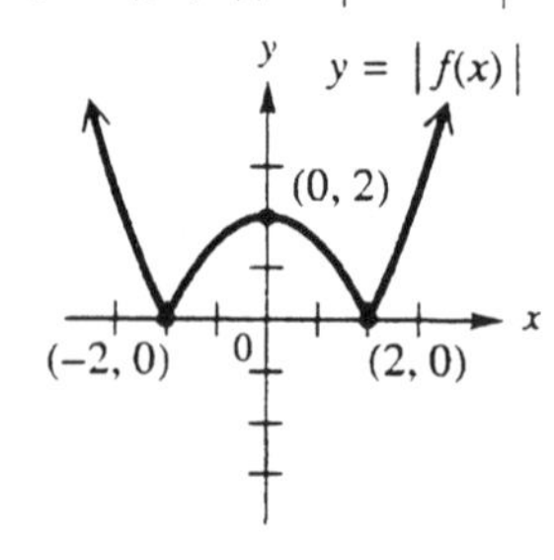

41. $y = |f(x)|$ and $y = f(x)$ are the same graph.

42. $y = |f(x)|$

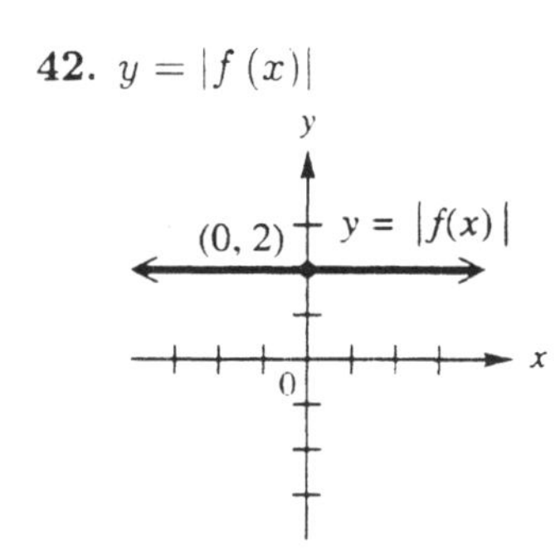

43. $|4x + 3| = 12$

$$\begin{aligned} 4x + 3 &= 12 \quad \text{or} \quad & 4x + 3 &= -12 \\ 4x &= 9 \quad \text{or} \quad & 4x &= -15 \\ x &= \frac{9}{4} \quad \text{or} \quad & x &= -\frac{15}{4} \end{aligned}$$

Solution set: $\boxed{\left\{-\frac{15}{4}, \frac{9}{4}\right\}}$

44. $|-2x - 6| + 4 = 1$

$|-2x - 6| = -3$ $\quad \boxed{\emptyset}$

There is no number whose absolute value is < 0.

45. $|5x + 3| = |x + 11|$

$$\begin{aligned} 5x + 3 &= x + 11 \quad \text{or} \quad & 5x + 3 &= -(x + 11) \\ 4x &= 8 & 5x + 3 &= -x - 11 \\ x &= 2 & 6x &= -14 \\ & & x &= -\frac{14}{6} = -\frac{7}{3} \end{aligned}$$

Solution set: $\boxed{\left\{-\frac{7}{3}, 2\right\}}$

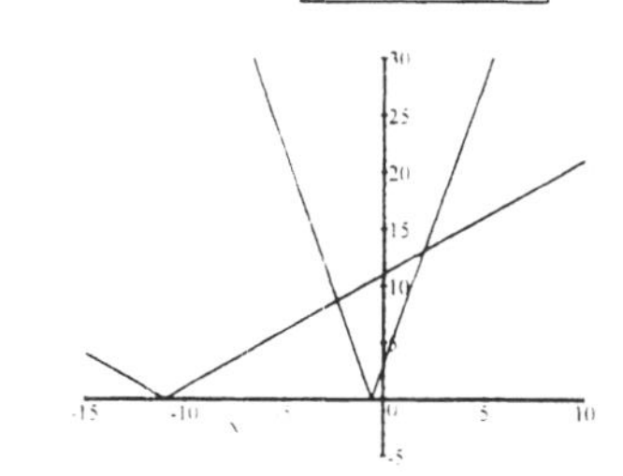

46. $|2x + 5| = 7$

$$\begin{aligned} 2x + 5 &= 7 \quad \text{or} \quad & 2x + 5 &= -7 \\ 2x &= 2 \quad \text{or} \quad & 2x &= -12 \\ x &= 1 \quad \text{or} \quad & x &= -6 \end{aligned}$$

Solution set: $\boxed{\{-6, 1\}}$

47. $|2x + 5| \le 7$

$-7 \le 2x + 5 \le 7$
$-12 \le 2x \le 2$
$-6 \le x \le 1$

Solution set: $\boxed{[-6, 1]}$

48. $|2x + 5| \ge 7$

$$\begin{aligned} 2x + 5 &\ge 7 \quad \text{or} \quad & 2x + 5 &\le -7 \\ 2x &\ge 2 \quad \text{or} \quad & 2x &\le -12 \\ x &\ge 1 \quad \text{or} \quad & x &\le -6 \end{aligned}$$

Solution set: $\boxed{(-\infty, -6] \cup [1, \infty)}$

49. The graph of $y_1 = |2x + 5|$ intersects the graph of $y_2 = 7$ at x-coordinates of -6 and 1. The solution sets are

$y_1 = y_2$: $\{-6, 1\}$
$y_1 \le y_2$: $[-6, 1]$
$y_1 \ge y_2$: $(-\infty, -6] \cup [1, \infty)$.

50. (a) Graph $y_1 = |x + 1| + |x - 3| - 8$; the x-intercepts are -3 and 5.

Solution set: $\boxed{\{-3, 5\}}$

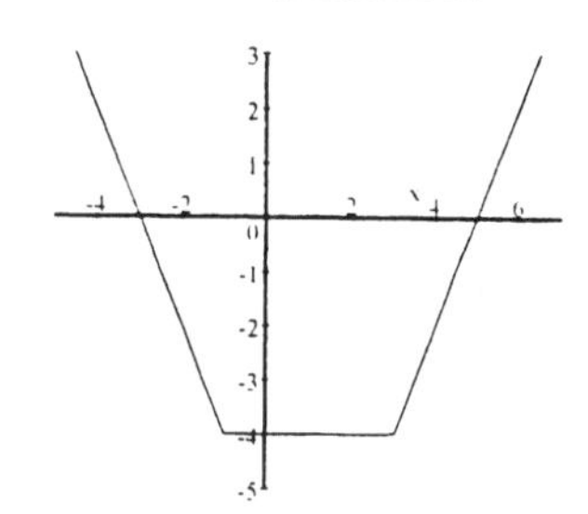

Check: $|x + 1| + |x - 3| = 8$

$\{-3\} : |-3 + 1| + |-3 - 3|$
$= |-2| + |-6|$
$= 2 + 6 = 8$

$\{5\} : |5 + 1| + |5 - 3|$
$= |6| + |2|$
$= 6 + 2 = 8$

51. Initially, the car is at home. After traveling 30 mph for 1 hr, the car is 30 mi away from home. During the 2nd hr the car travels 20 mph until it is 50 miles away. During the 3rd hr the car travels toward home at 30 mph until it is 20 mi away from home. During the 4th hr the car travels away from home at 40 mph until it is 60 mi away from home. During the last hr, the car travels 60 mi at 60 mph until it arrives home.

52.

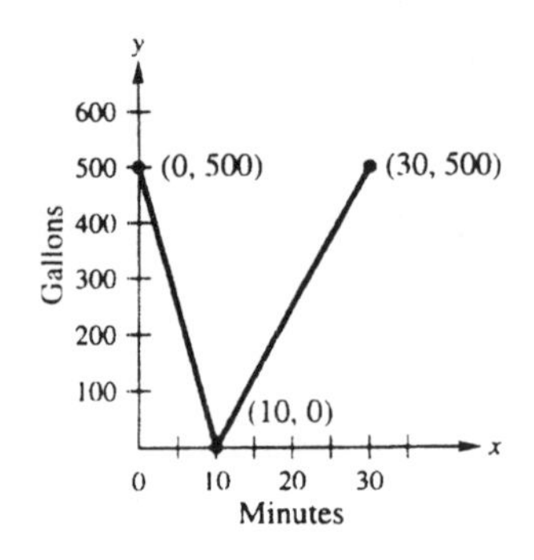

53.

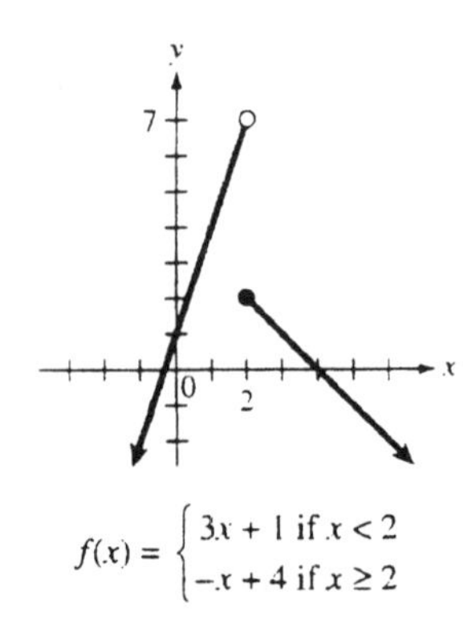

54.

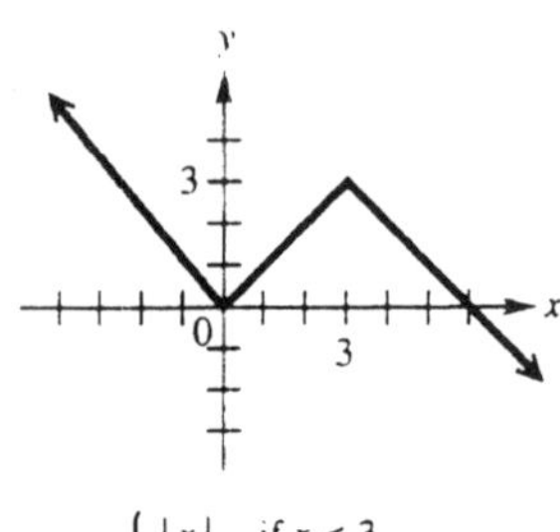

55. $f(x) = \begin{cases} 3x+1 & \text{if} \quad x<2 \\ -x+4 & \text{if} \quad x \geq 2 \end{cases}$

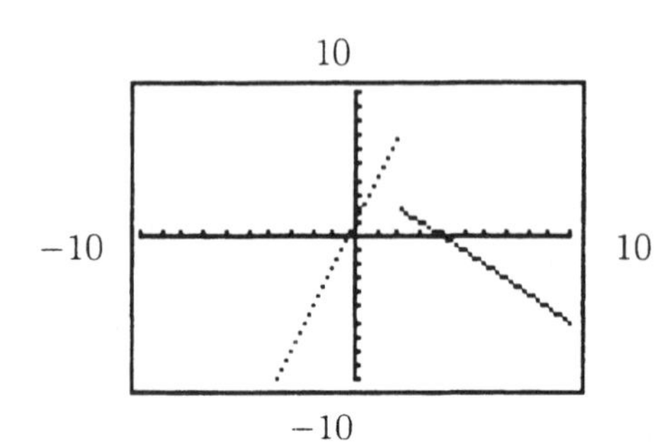

56. $f(x) = [[x-3]]$

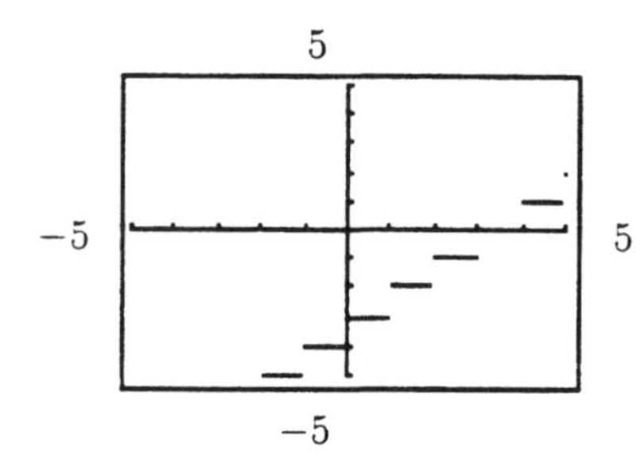

Closed circle on left part of line segment; open circle on right part of line segment.

57. $(f+g)(1) = f(1) + g(1) = 2 + 3 = 5$

58. $(f-g)(0) = f(0) - g(0) = 1 - 4 = -3$

59. $(fg)(-1) = f(-1) \cdot g(-1) = 0 \cdot 3 = 0$

60. $\left(\frac{f}{g}\right)(2) = \frac{f(2)}{g(2)} = \frac{3}{2}$

61. $(f \circ g)(2) = f[g(2)] = f(2) = 3$

62. $(g \circ f)(2) = g[f(2)] = g(3) = 2$

63. $(g \circ f)(-4) = g[f(-4)] = g(2) = 2$

64. $(f \circ g)(-2) = f[g(-2)] = f(2) = 3$

65. $(f+g)(1) = f(1) + g(1) = 7 + 1 = 8$

66. $(f-g)(3) = f(3) - g(3) = 9 - 9 = 0$

67. $(fg)(-1) = f(-1) \cdot g(-1) = (3)(-2) = -6$

68. $\left(\frac{f}{g}\right)(0) = \frac{f(0)}{g(0)} = \frac{5}{0}$: undefined

69. $(g \circ f)(-2) = g[f(-2)] = g(1) = 2$

70. $(f \circ g)(3) = f[g(3)] = f(-2) = 1$

71. $f(x) = 2x + 9$

$$\frac{f(x+h) - f(x)}{h} = \frac{[2(x+h)+9] - (2x+9)}{h} = \frac{2x + 2h + 9 - 2x - 9}{h} = \frac{2h}{h} = \boxed{2}$$

72. $f(x) = x^2 - 5x + 3$

$$\frac{f(x+h) - f(x)}{h} = \frac{\left[(x+h)^2 - 5(x+h) + 3\right] - (x^2 - 5x + 3)}{h}$$

$$= \frac{x^2 + 2xh + h^2 - 5x - 5h + 3 - x^2 + 5x - 3}{h} = \frac{2xh + h^2 - 5h}{h} = \boxed{2x + h - 5}$$

73. $h(x) = (x^3 - 3x)^2$

Let $f(x) = x^2$ and $g(x) = x^3 - 3x$;

then $(f \circ g)(x) = h(x)$.

Only one of many possible choices for f and g is given.

74. $h(x) = \dfrac{1}{x - 5}$

Let $f(x) = \dfrac{1}{x}$ and $g(x) = x - 5$;

then $(f \circ g)(x) = h(x)$.

Only one of many possible choices for f and g is given.

75. $V(r) = \dfrac{4}{3}\pi r^3$; increase r by 4 inches:

$$V(r+4) = \frac{4}{3}\pi (r+4)^3$$

Volume gained $= V(r+4) - V(r)$

$$\boxed{V(r) = \frac{4}{3}\pi (r+4)^3 - \frac{4}{3}\pi r^3}$$

76. **(a)** $V = \pi r^2 h$; $h = d = 2r$, $r = \dfrac{d}{2}$

$$V(d) = \pi \left(\frac{d}{2}\right)^2 d = \pi d \left(\frac{d^2}{4}\right)$$

$$\boxed{V(d) = \frac{\pi d^3}{4}}$$

(b) Surface area = Side area + 2(top area)

$$S = \pi d h + 2\pi r^2; \quad r = \frac{d}{2}, \quad h = d$$

$$S(d) = \pi d (d) + 2\pi \left(\frac{d}{2}\right)^2 = \pi d^2 + \frac{\pi d^2}{2}$$

$$\boxed{S(d) = \frac{3\pi d^2}{2}}$$

77. $f(x) = 36x$, $g(x) = 1760x$

$(f \circ g)(x) = (g \circ f)(x) = 63,360x$

78. x : width; $2x$: length

$$P = 2l + 2w$$
$$P(x) = 2(2x) + 2x$$
$$P(x) = 6x$$

This is a linear function.

Chapter 2 Test

1. (a) $f(x) = \sqrt{x} + 3$
   Domain: $[0, \infty)$, **D**
   (b) $f(x) = \sqrt{x-3}$
   Range: $[0, \infty)$. **D**
   (c) $f(x) = x^2 - 3$
   Domain: $(-\infty, \infty)$, **C**
   (d) $f(x) = x^2 - 3$
   Range: $[3, \infty)$. **B**
   (e) $f(x) = \sqrt[3]{x-3}$
   Domain: $(-\infty, \infty)$. **C**
   (f) $f(x) = \sqrt[3]{x} + 3$
   Range: $(-\infty, \infty)$. **C**
   (g) $f(x) = |x| - 3$
   Domain: $(-\infty, \infty)$, **C**
   (h) $f(x) = |x + 3|$
   Range: $[0, \infty)$. **D**
   (i) $x = y^2$
   Domain: $[0, \infty)$, **D**
   (j) $x = y^2$
   Range: $(-\infty, \infty)$, **C**

2. (a) $y = f(x) + 2$

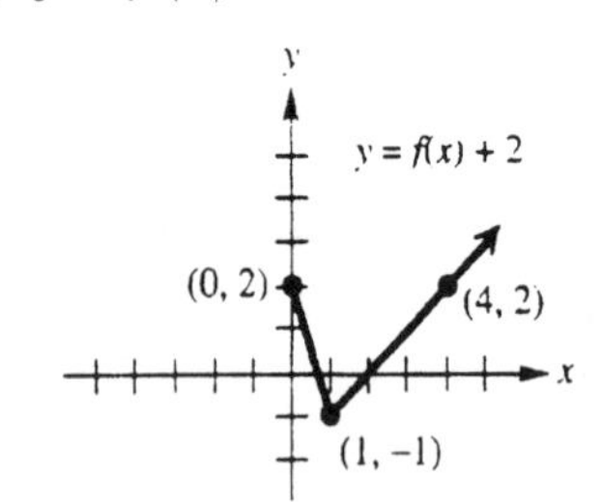

(b) $y = f(x + 2)$

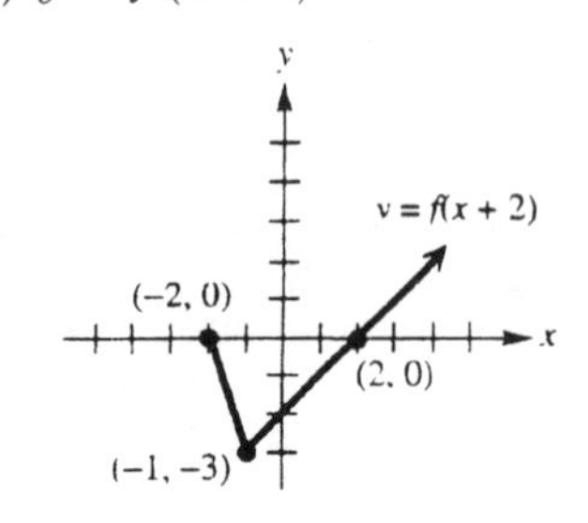

2. (c) continued

$y = -f(x)$

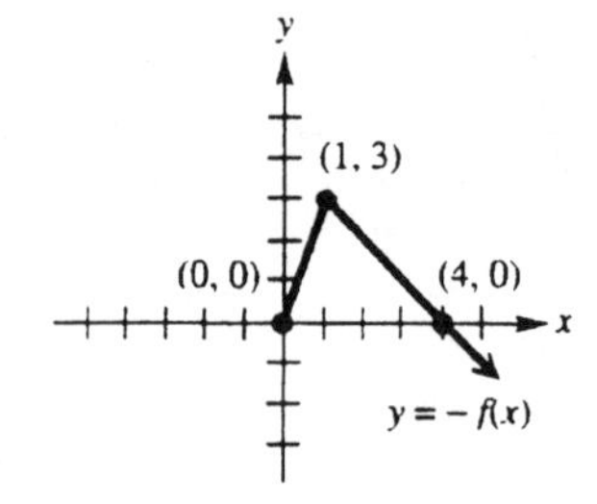

(d) $y = f(-x)$

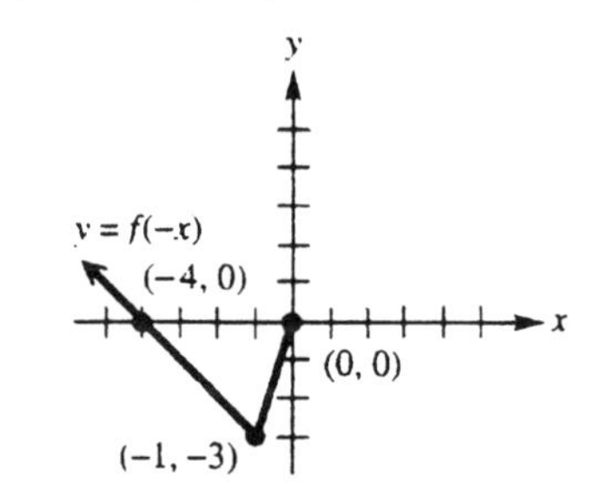

(e) $y = 2f(x)$

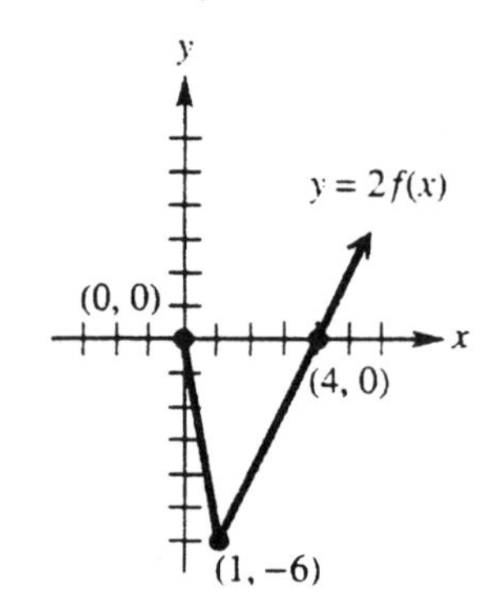

(f) $y = |f(x)|$

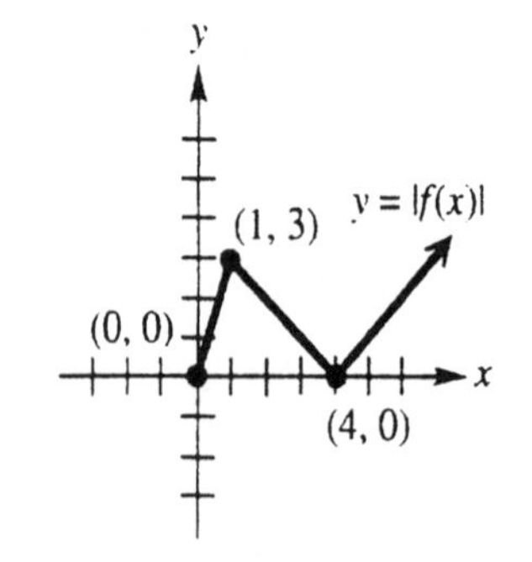

3. (a) Symmetric with respect to the y-axis:

If $(x, y) \Rightarrow (-x, y)$, so
$(3, 6)$ and $(-3, 6)$

(b) Symmetric with respect to the origin:

If $(x, y) \Rightarrow (-x, -y)$, so
$(3, 6)$ and $(-3, -6)$

(c) Symmetric with respect to the y-axis:

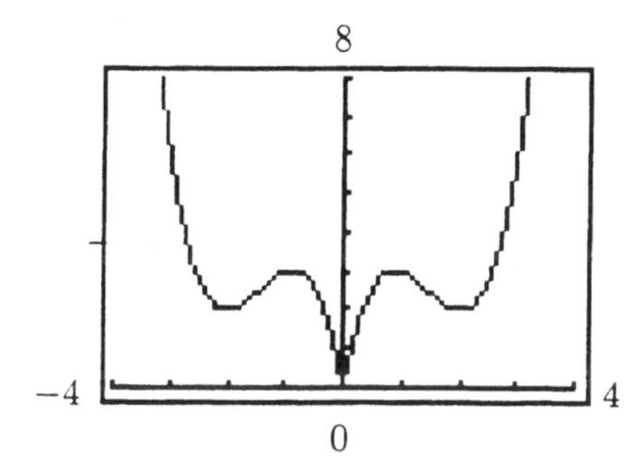

4. (a) $y = \sqrt[3]{x}$

Shift two units to the left: $y = \sqrt[3]{x+2}$

Stretch by a factor of 4: $y = 4\sqrt[3]{x+2}$

Shift five units down: $y = \sqrt[3]{x+2} - 5$

(b) $y = -\frac{1}{2}|x-3| + 2$

Domain: $(-\infty, \infty)$; Range: $(-\infty, 2]$

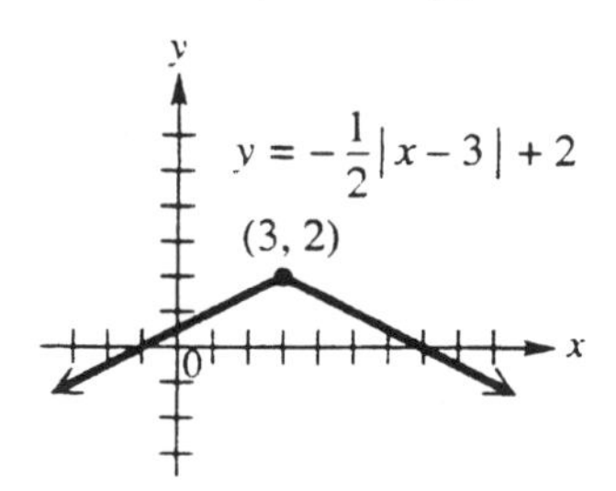

5. (a) Increasing: $(-\infty, -3)$

(b) Decreasing: $(4, \infty)$

(c) Constant: $[-3, 4]$

(d) Continuous: $(-\infty, -3)$, $[-3, 4]$, $(4, \infty)$

(e) Domain: $(-\infty, \infty)$

(f) Range: $(-\infty, 2)$

6. $y_1 = |4x+8|$; $y_2 = 4$

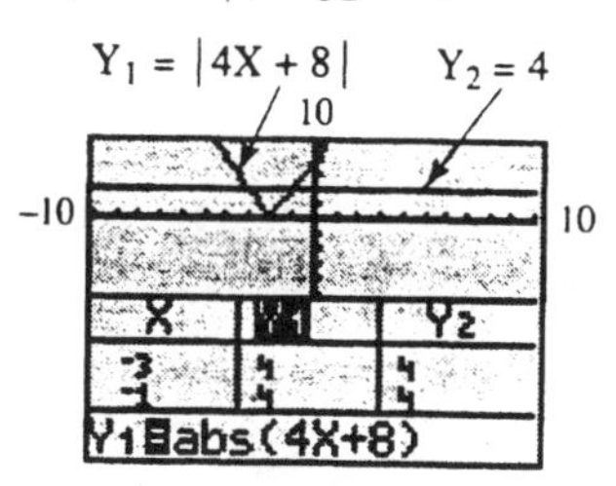

(a) $|4x+8| = 4$

$$4x+8=4 \quad \text{or} \quad 4x+8=-4$$
$$4x=-4 \qquad 4x=-12$$
$$x=-1 \qquad x=-3 \qquad \boxed{\{-3, -1\}}$$

The x-coordinates of the points of intersection of the graphs of y_1 and y_2 are -3 and -1.

(b) $|4x+8| < 4$

$$-4 < 4x+8 < 4$$
$$-12 < 4x < -4$$
$$-3 < x < -1 \qquad \boxed{(-3, -1)}$$

The graph of y_1 lies below the graph of y_2 for x-values between -3 and -1.

(c) $|4x+8| > 4$

$$4x+8>4 \quad \text{or} \quad 4x+8<-4$$
$$4x>-4 \qquad 4x<-12$$
$$x>-1 \qquad x<-3$$
$$\boxed{(-\infty, -3) \cup (-1, \infty)}$$

The graph of y_1 lies above the graph of y_2 for x-values less than -3 or for x-values greater than -1.

7. $f(x) = 2x^2 - 3x + 2$, $g(x) = -2x + 1$

(a) $(f-g)(x)$

$$= (2x^2 - 3x + 2) - (-2x + 1)$$
$$= 2x^2 - 3x + 2 + 2x - 1$$
$$= \boxed{2x^2 - x + 1}$$

(b) $\frac{f}{g}(x) = \boxed{\dfrac{2x^2 - 3x + 2}{-2x + 1}}$

(c) Domain of $\frac{f}{g}$: $-2x + 1 \neq 0$

$$-2x \neq -1$$
$$x \neq \frac{1}{2}$$
$$\boxed{\left(-\infty, \frac{1}{2}\right) \cup \left(\frac{1}{2}, \infty\right)}$$

continued

7. (d) continued

$$\begin{aligned}(f \circ g)(x) &= f(g(x)) \\ &= f(-2x+1) \\ &= 2(-2x+1)^2 - 3(-2x+1) + 2 \\ &= 2(4x^2 - 4x + 1) + 6x - 3 + 2 \\ &= 8x^2 - 8x + 2 + 6x - 1 \\ &= \boxed{8x^2 - 2x + 1}\end{aligned}$$

(e) $\dfrac{f(x+h) - f(x)}{h}$

$$\begin{aligned}&= \frac{\left[2(x+h)^2 - 3(x+h) + 2\right] - \left[2x^2 - 3x + 2\right]}{h} \\ &= \frac{2(x^2 + 2xh + h^2) - 3x - 3h + 2 - 2x^2 + 3x - 2}{h} \\ &= \frac{2x^2 + 4xh + 2h^2 - 3h - 2x^2}{h} \\ &= \frac{4xh + 2h^2 - 3h}{h} \\ &= \boxed{4x + 2h - 3}\end{aligned}$$

8. (a) $f(x) = \begin{cases} -x^2 + 3 & \text{if } x \le 1 \\ \sqrt[3]{x} + 2 & \text{if } x > 1 \end{cases}$

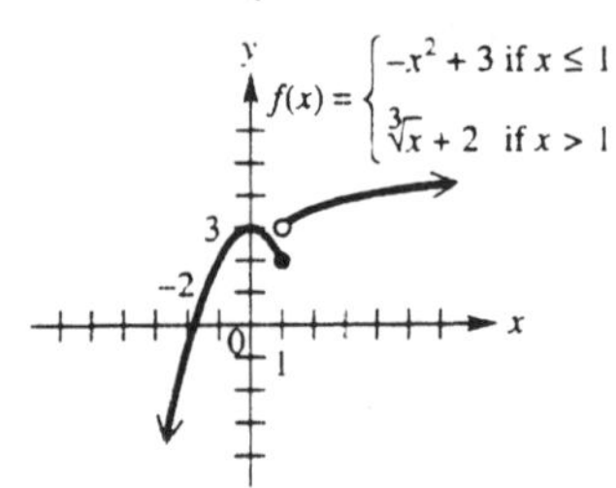

(b) $y_1 = (-x^2 + 3) / (x \le 1)$

$y_2 = (\sqrt[3]{x} + 2) / (x > 1)$

(May not need / on certain calculators.)

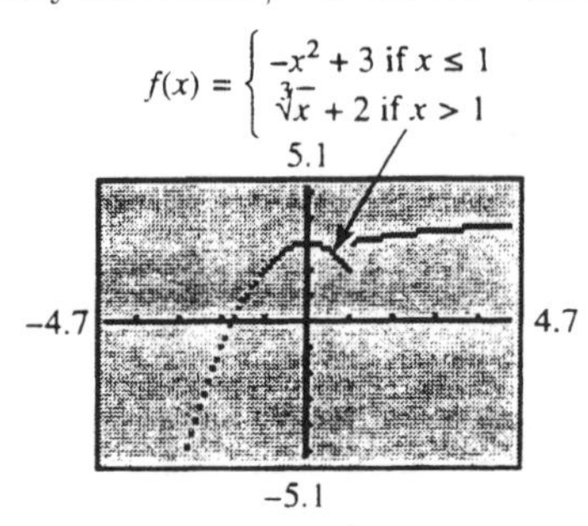

9. (a) $f(x) = .40\,[[x]] + .75$

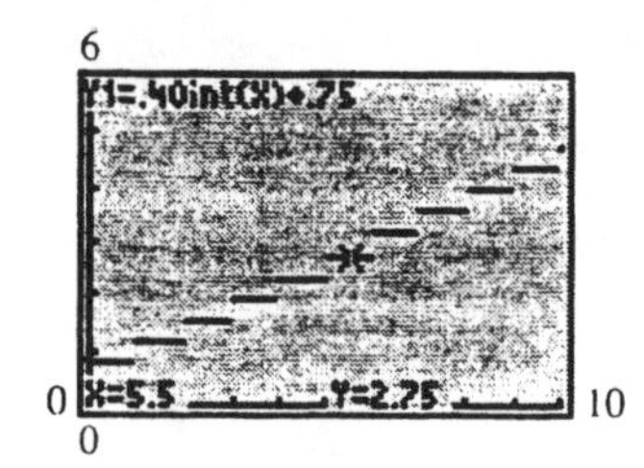

(b) $\begin{aligned} f(5.5) &= .40\,[[5.5]] + .75 \\ &= .40(5) + .75 \\ &= \boxed{\$2.75} \end{aligned}$

The point (5.5, 2.75) is on the graph.

10. (a) Cost = initial cost + 4.5 (x)

$\boxed{C(x) = 3300 + 4.50x}$

(b) $\boxed{R(x) = 10.50x}$

(c) Profit = Revenue − Cost

$P(x) = 10.50x - (3300 + 4.50x)$

$\boxed{P(x) = 6.00x - 3300}$

(d) $P(x) > 0$

$6.00x - 3300 > 0$

$6.00x > 3300$

$x > 550$ $\boxed{551}$

(e) The first integer x-value for which $P(x) > 0$ is 551.

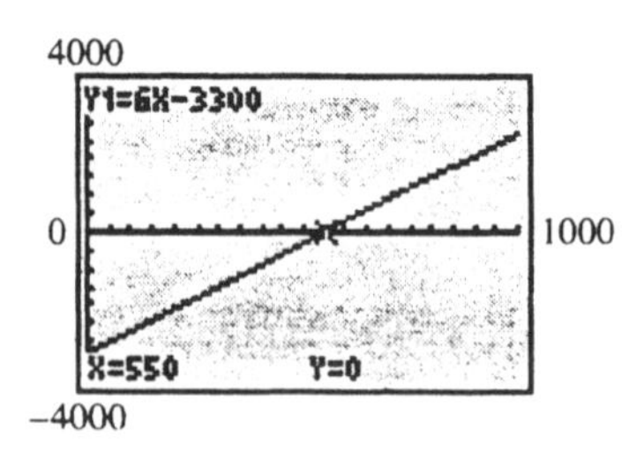

Chapter 2 Project

Modeling The Movement of a Cold Front

1. In 4 hours, the cold front has moved $4(40) = 160$ miles, which is $\frac{160}{100} = 1.6$ units downward on the graph. To show this movement, shift the graph 1.6 units downward:

$$y = \frac{1}{20}x^2 - 1.6$$

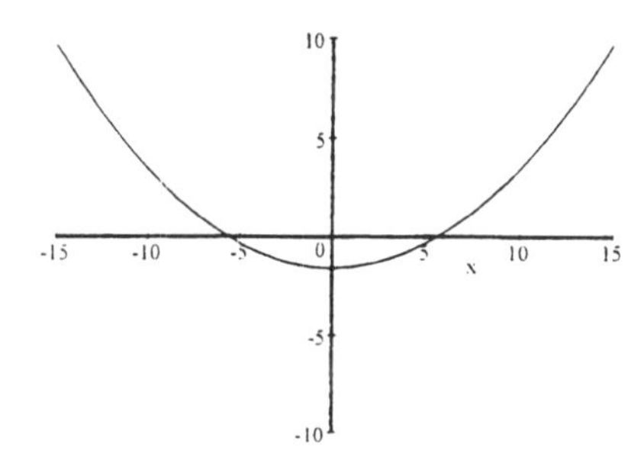

2. (a) The graph of the cold front has shifted $\frac{250}{100} = 2.5$ units downward and $\frac{210}{100} = 2.1$ units to the right. The new equation would be:

$$y = \frac{1}{20}(x - 2.1)^2 - 2.5$$

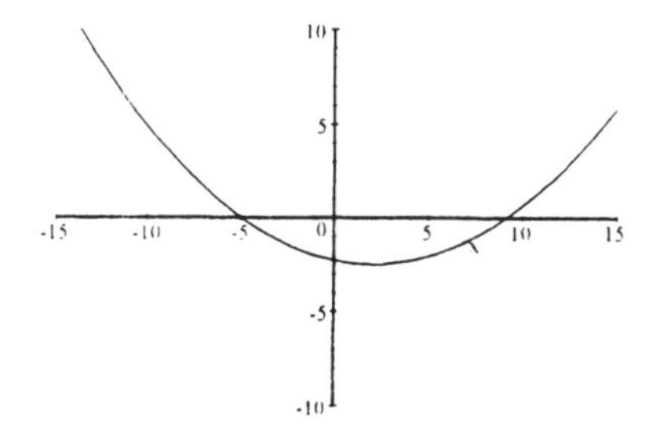

The location of Columbus, Ohio is at $(5.5, -.8)$ relative to the locations of Des Moines at $(0, 0)$. Since the point $(5.5, -1.922)$ is on the graph, $(5.5, -.8)$ is within the parabola representing the cold front at midnight; the front reaches Columbus by midnight.

(b) The location of Memphis, Tennessee is at $(1.9, -4.3)$ relative to the locations of Des Moines at $(0, 0)$. Since the point $(1.9, -2.498)$ is on the graph, $(1.9, -4.3)$ is not within the parabola representing the cold front at midnight; the front does not reach Memphis by midnight.

(c) The location of Louisville, Kentucky is at $(4.2, -2.3)$ relative to the locations of Des Moines at $(0, 0)$. Since the point $(4.2, -2.2795)$ is on the graph, $(4.2, -2.3)$ is not within the parabola representing the cold front at midnight; the front does not reach reach Louisville by midnight..

CHAPTER 3

Section 3.1

1. $-9i$: (a) Real: 0
(b) Imaginary: -9
(c) Imaginary

3. π : (a) Real: π
(b) Imaginary: 0
(c) Real

5. $i\sqrt{6}$: (a) Real: 0
(b) Imaginary: $\sqrt{6}$
(c) Imaginary

7. $2+5i$: (a) Real: 2
(b) Imaginary: 5
(c) Imaginary

9. $\sqrt{-100} = i\sqrt{100} = \boxed{10i}$

11. $-\sqrt{-400} = -i\sqrt{400} = \boxed{-20i}$

13. $-\sqrt{-39} = \boxed{-i\sqrt{39}}$

15. $5+\sqrt{-4} = 5+i\sqrt{4} = \boxed{5+2i}$

17. $9-\sqrt{-50} = 9-i\sqrt{25\cdot 2} = \boxed{9-5i\sqrt{2}}$

19. $\sqrt{-5}\cdot\sqrt{-5} = i\sqrt{5}\cdot i\sqrt{5} = 5i^2 = \boxed{-5}$

21. $\dfrac{\sqrt{-40}}{\sqrt{-10}} = \dfrac{i\sqrt{4\cdot 10}}{i\sqrt{10}} = \dfrac{2i\sqrt{10}}{i\sqrt{10}} = \boxed{2}$

23. A real number a, may be written in the form $a+0i$, a complex number; however, $a+bi$, $b \neq 0$, cannot be written without the imaginary term, so it is not a real number.

25. $(3+2i)+(4-3i)$
$= (3+4)+(2-3)\,i = \boxed{7-i}$

27. $(-2+3i)-(-4+3i)$
$= (-2-(-4))+(3-3)\,i = \boxed{2}$

29. $(2-5i)-(3+4i)-(-2+i)$
$= (2-3-(-2))+(-5-4-1)\,i$
$= \boxed{1-10i}$

31. $(2+4i)(-1+3i)$
$= -2+6i-4i+12i^2$
$= -2+2i+12(-1) = \boxed{-14+2i}$

33. $(3+2i)^2$
$= 9-12i+4i^2$
$= 9-12i+4(-1) = \boxed{5-12i}$

35. $(2+3i)(2-3i)$
$= 4-9i^2$
$= 4-9(-1) = \boxed{13}$

37. $(\sqrt{6}+i)(\sqrt{6}-i)$
$= 6-i^2$
$= 6-(-1) = \boxed{7}$

39. $i(3-4i)(3+4i) = i(9-16i^2)$
$= i(9-16(-1)) = \boxed{25i}$

41. $(6-i)(1+i)^2$: $\boxed{2+12i}$

43. $i^5 = (i^4)^1 \times i = 1^1 \times i = \boxed{i}$

45. $i^{15} = (i^4)^3 i^3 = (1^3)(-1)\,i = \boxed{-i}$

47. $i^{64} = (i^4)^{16} = 1^{16} = \boxed{1}$

49. $i^{-6} = \dfrac{1}{i^6} = \dfrac{1}{i^4 \times i^2} = \dfrac{1}{1(-1)} = \boxed{-1}$

51. $\dfrac{1}{i^9} = \dfrac{1}{(i^4)^2 \times i} = \dfrac{1}{1^2(i)}\cdot\dfrac{i}{i} = \dfrac{i}{-1} = \boxed{-i}$

53. $\left(\frac{\sqrt{2}}{2}+\frac{\sqrt{2}}{2}i\right)^2$

$$= \left(\frac{\sqrt{2}}{2}\right)^2 + 2\left(\frac{\sqrt{2}}{2}\right)\left(\frac{\sqrt{2}}{2}i\right) + \left(\frac{\sqrt{2}}{2}i\right)^2$$
$$= \frac{2}{4} + 2\left(\frac{2}{4}i\right) + \frac{2}{4}i^2$$
$$= \frac{1}{2} + 1i - \frac{1}{2} = i$$

55. conjugate of $5-3i$ is $5+3i$

57. conjugate of $-18i$ is $18i$

59. $$\frac{-19-9i}{4+i} = \frac{-19-9i}{4+i}\cdot\frac{4-i}{4-i}$$
$$= \frac{-76+19i-36i+9i^2}{16-i^2}$$
$$= \frac{-76-17i-9}{16-(-1)}$$
$$= \frac{-85-17i}{17} = \boxed{-5-i}$$

61. $$\frac{1-3i}{1+i} = \frac{1-3i}{1+i}\cdot\frac{1-i}{1-i}$$
$$= \frac{1-i-3i+3i^2}{1-i^2}$$
$$= \frac{1-4i-3}{1-(-1)}$$
$$= \frac{-2-4i}{2} = \boxed{-1-2i}$$

63. $$\frac{-6+8i}{4+3i} = \frac{-6+8i}{4+3i}\cdot\frac{4-3i}{4-3i}$$
$$= \frac{-24+18i+32i-24i^2}{16-9i^2}$$
$$= \frac{-24+50i+24}{16+9}$$
$$= \frac{50i}{25} = \boxed{2i}$$

65. $$\frac{4-3i}{4+3i} = \frac{4-3i}{4+3i}\cdot\frac{4-3i}{4-3i}$$
$$= \frac{16-12i-12i+9i^2}{16-9i^2}$$
$$= \frac{16-24i-9}{16-(-9)}$$
$$= \frac{7-24i}{25} = \boxed{\frac{7}{25}-\frac{24}{25}i}$$

67. $$\frac{-7}{3i} = \frac{-7}{3i}\cdot\frac{-3i}{-3i} = \frac{21i}{-9i^2} = \frac{21i}{9} = \boxed{\frac{7}{3}i}$$

69. We are multiplying by 1, the multiplicative identity.

Relating Concepts (71 - 74)

71. $x^3-x^2-7x+15=0;\ x=-3:$
$$(-3)^3-(-3)^2-7(-3)+15$$
$$= -27-9+21+15 = 0$$

72. $x^3-x^2-7x+15=0;\ x=2-i:$
$$(2-i)^3-(2-i)^2-7(2-i)+15$$
$$= 8-12i+6i^2-i^3$$
$$-\left(4-4i+i^2\right)-14+7i+15$$
$$= (8-12i+6(-1)-(-1)i)-4$$
$$+4i-(-1)+1+7i$$
$$= 8-12i-6+i-4+4i$$
$$+1+1+7i$$
$$= 0+0i = 0$$

73. $x^3-x^2-7x+15=0;\ x=2+i:$
$$(2+i)^3-(2+i)^2-7(2+i)+15$$
$$= \left(8+12i+6i^2+i^3\right)$$
$$-\left(4+4i+i^2\right)-14-7i+15$$
$$= (8+12i+6(-1)+(-1)i)-4$$
$$-4i-(-1)+1-7i$$
$$= 8+12i-6-i-4-4i+1+1-7i$$
$$= 0+0i = 0$$

74. Solutions (**72**, **73**): complex conjugates.

Section 3.2

1. $y = (x-4)^2 - 3$: **B**

3. $y = (x+4)^2 - 3$: **D**

5. $P(x) = 2x^2 - 2x - 24$

(a) $\dfrac{P(x)}{2} = x^2 - x - 12$

$\dfrac{P(x)}{2} + 12 = x^2 - x$

$\dfrac{P(x)}{2} + 12 + \dfrac{1}{4} = x^2 - x + \dfrac{1}{4}$

$\dfrac{P(x)}{2} + \dfrac{49}{4} = \left(x - \dfrac{1}{2}\right)^2$

$\dfrac{P(x)}{2} = \left(x - \dfrac{1}{2}\right)^2 - \dfrac{49}{4}$

$\boxed{P(x) = 2\left(x - \dfrac{1}{2}\right)^2 - \dfrac{49}{2}}$

(b) $\boxed{\text{V: } \left(\dfrac{1}{2}, -\dfrac{49}{2}\right) \text{ or } (.5, -24.5)}$

(c) Graph $P(x)$; the minimum point is $(.5, -24.5)$.

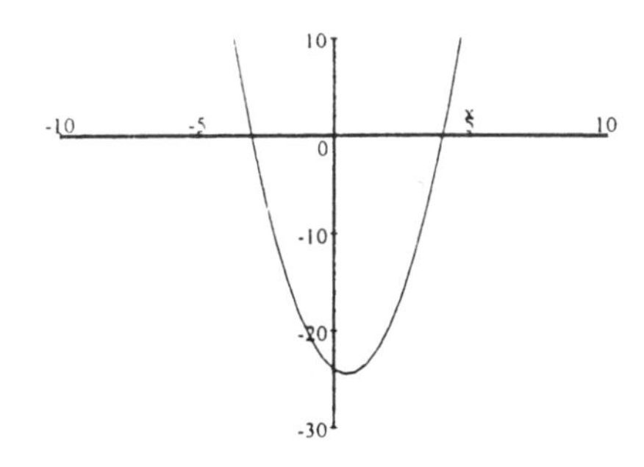

7. $y = x^2 - 2x - 15$

(a) $y + 15 = x^2 - 2x$

$y + 15 + 1 = x^2 - 2x + 1$

$y + 16 = (x-1)^2$

$\boxed{y = (x-1)^2 - 16}$

(b) $\boxed{\text{V: } (1, -16)}$

continued

7. continued

(c) Graph y; the minimum point is $(1, -16)$.

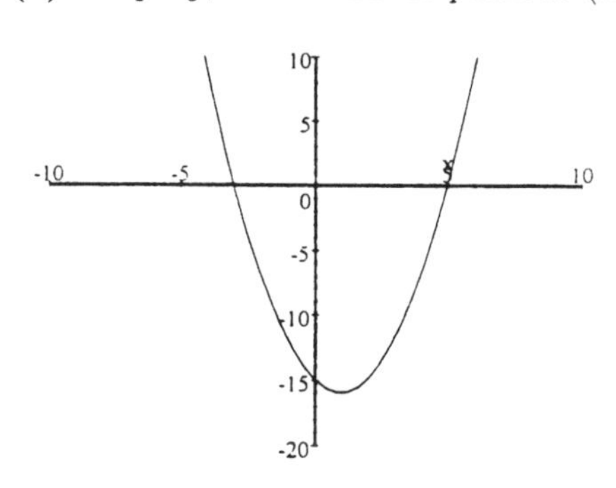

9. $f(x) = -2x^2 + 6x$

(a) $\dfrac{f(x)}{-2} = x^2 - 3x$

$-\dfrac{f(x)}{2} + \dfrac{9}{4} = x^2 - 3x + \dfrac{9}{4}$

$-\dfrac{f(x)}{2} + \dfrac{9}{4} = \left(x - \dfrac{3}{2}\right)^2$

$-\dfrac{f(x)}{2} = \left(x - \dfrac{3}{2}\right)^2 - \dfrac{9}{4}$

$\boxed{f(x) = -2\left(x - \dfrac{3}{2}\right)^2 + \dfrac{9}{2}}$

(b) $\boxed{\text{V: } \left(\dfrac{3}{2}, \dfrac{9}{2}\right) \text{ or } (1.5, 4.5)}$

(c) Graph $f(x)$; the maximum point is $\left(\dfrac{3}{2}, \dfrac{9}{2}\right)$.

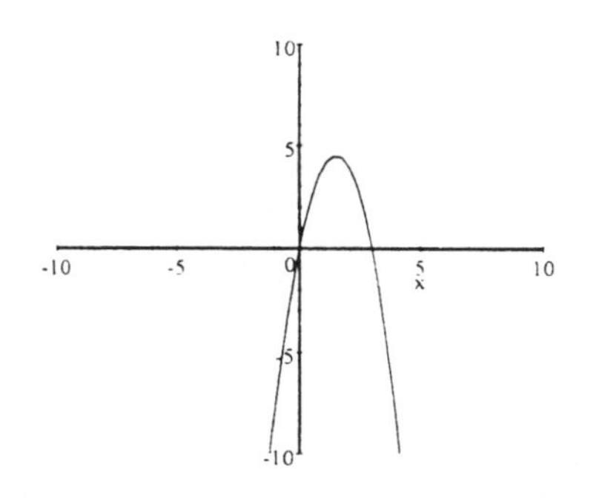

11. $P(x) = 4x^2 - 22x - 12$

(a) $\dfrac{P(x)}{4} = x^2 - \dfrac{11}{2}x - 3$

$\dfrac{P(x)}{4} + 3 = x^2 - \dfrac{11}{2}x$

$\dfrac{P(x)}{4} + 3 + \dfrac{121}{16} = x^2 - \dfrac{11}{2}x + \dfrac{121}{16}$

$\dfrac{P(x)}{4} + \dfrac{169}{16} = \left(x - \dfrac{11}{4}\right)^2$

$\dfrac{P(x)}{4} = \left(x - \dfrac{11}{4}\right)^2 - \dfrac{169}{16}$

$\boxed{P(x) = 4\left(x - \dfrac{11}{4}\right)^2 - \dfrac{169}{4}}$

(b) $\boxed{\text{V: } \left(\dfrac{11}{4}, -\dfrac{169}{4}\right) \text{ or } (2.75, -42.25)}$

(c) Graph $P(x)$; the minimum point is $\left(\dfrac{11}{4}, -\dfrac{169}{4}\right)$.

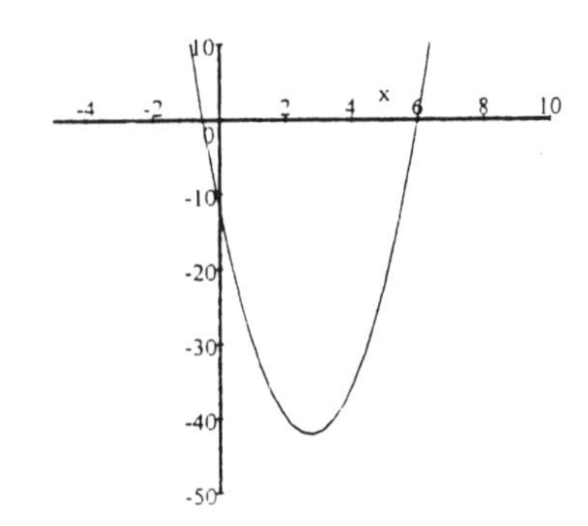

13. (a) $y = (x-4)^2 - 2$: **D**
(b) $y = (x-2)^2 - 4$: **B**
(c) $y = -(x-4)^2 - 2$: **C**
(d) $y = -(x-2)^2 - 4$: **A**

15. $P(x) = (x-2)^2$

(a) Vertex: $(2, 0)$

(b) D: $(-\infty, \infty)$ R: $[0, \infty)$

(c) Axis of Symmetry: $x = 2$

(d) Increasing: $[2, \infty)$

(e) Decreasing: $(-\infty, 2]$

(f) Minimum: $P(2) = 0$

17. $y = (x+3)^2 - 4$

(a) Vertex: $(-3, -4)$

(b) D: $(-\infty, \infty)$ R: $[-4, \infty)$

(c) Axis of Symmetry: $x = -3$

(d) Increasing: $[-3, \infty)$

(e) Decreasing: $(-\infty, -3]$

(f) Minimum; $f(-3) = -4$

19. $f(x) = -2(x+3)^2 + 2$

(a) Vertex: $(-3, 2)$

(b) D: $(-\infty, \infty)$ R: $(-\infty, 2]$

(c) Axis of Symmetry: $x = -3$

(d) Increasing: $(-\infty, -3]$

(e) Decreasing: $[-3, \infty)$

(f) Maximum: $f(-3) = 2$

21. $P(x) = x^2 - 10x + 21$

(a) $\dfrac{-b}{2a} = \dfrac{-(-10)}{2(1)} = \dfrac{10}{2} = 5$

$P(5) = 5^2 - 10(5) + 21$
$= 25 - 50 + 21 = -4$

$\boxed{\text{Vertex: } (5, -4)}$

(b)

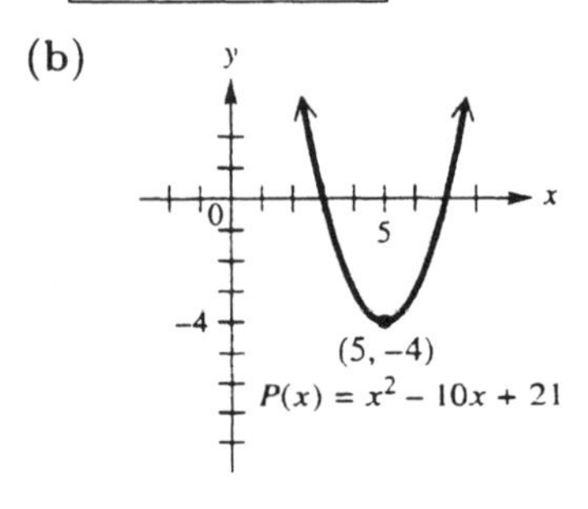

23. $y = -x^2 + 4x - 2$

(a) $\dfrac{-b}{2a} = \dfrac{-4}{2(-1)} = \dfrac{-4}{-2} = 2$

$(2, y): \ y = -(2)^2 + 4(2) - 2$
$= -4 + 8 - 2 = 2$

Vertex: (2, 2)

(b)

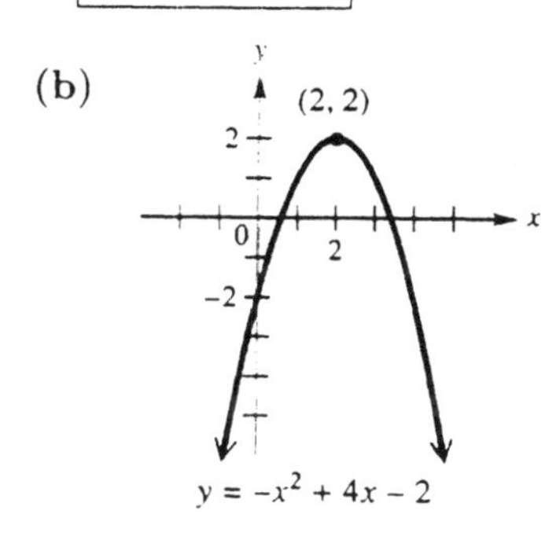

25. $f(x) = 2x^2 - 4x + 5$

(a) $\dfrac{-b}{2a} = \dfrac{-(-4)}{2(2)} = \dfrac{4}{4} = 1$

$f(1) = 2(1)^2 - 4(1) + 5$
$= 2 - 4 + 5 = 3$

Vertex: (1, 3)

(b)

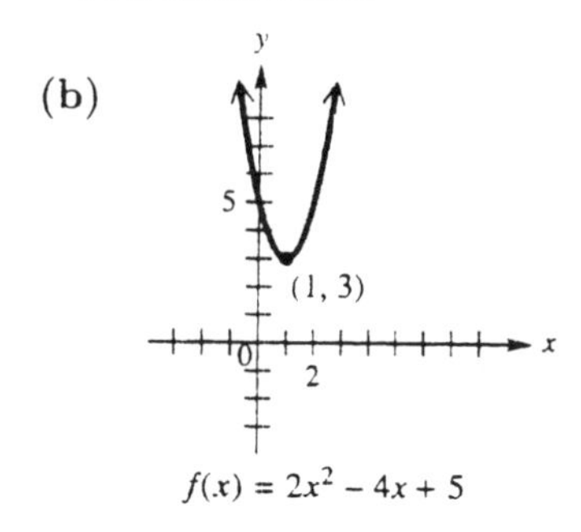

27. $P(x) = -.32x^2 + \sqrt{3}\,x + 2.86$

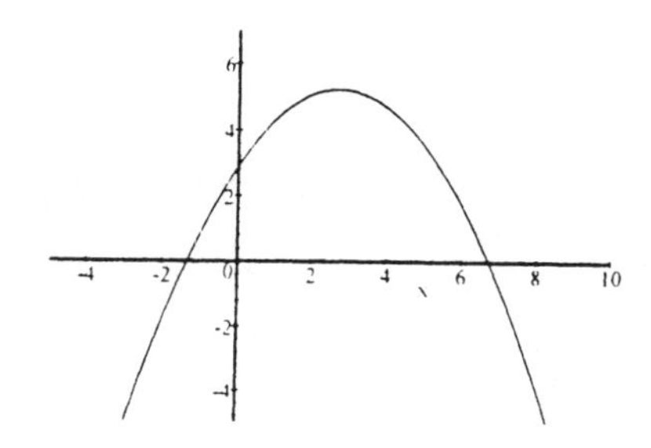

(a) Vertex: (2.71, 5.20)

(b) x-intercepts: $\{-1.33, 6.74\}$

29. $y = 1.34x^2 - 3x + \sqrt{5}$

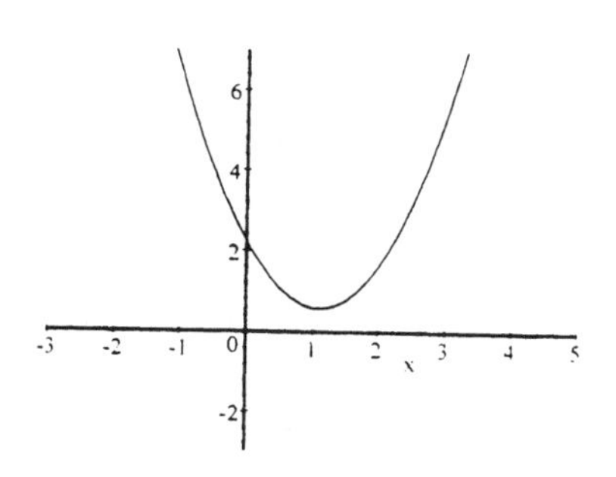

(a) Vertex: (1.12, .56)

(b) There are no x-intercepts.

31. $f(x)$ is smallest for $x = -3$

33. (a) Vertex: $(4, -12)$
[Halfway between $x = 3$ and 5]

(b) Minimum point
[-12 is lowest y-value]

(c) -12 is the minimum value.

(d) Range: $[-12, \infty)$

35. (a) Vertex: $(1.5, 2)$
[Halfway between $x = 1$ and 2]

(b) Maximum point
[2 is highest y-value]

(c) 2 is the maximum value.

(d) Range: $(-\infty, 2]$

37. quadratic; $a < 0$ **39.** quadratic; $a > 0$

41. linear, positive slope

43. (a) Scatter plot; Quadratic; $a > 0$

(b) $f(x) = 2.797x^2 - 22.18x + 117.7$

$\dfrac{-b}{2a} = \dfrac{-(-22.18)}{2(2.797)} \approx 3.965 \approx 4$

$f(4) = 2.797(4)^2 - 22.18(4) + 117.7 \approx 74$

Vertex: (4, 74)

(c) In 1994, company bankruptcy filings will reach the minimum number of 74. The number 74 is higher than the table value of 70.

45. (a) y-value will be maximum

(b) $f(x) = -.566x^2 + 5.08x + 29.2$

$$\frac{-b}{2a} = \frac{-5.08}{2(-.566)} \approx 4.4876 \approx 4.5$$

$$1990 + 4.5 = 1994$$

$$f(4.5) = -.566(4.5)^2 + 5.08(4.5) + 29.2 = 40.5985 \approx 41$$

The maximum occurred in 1994 and the oyster catch was 41 million.

47. (a) The value of t cannot be negative since t represents time elapsed from the throw.

(b) Since the rock was thrown from ground level, s_o, the original height of the rock is 0.

(c) $s(t) = -16t^2 + 90t$

(d) $s(1.5) = -16(1.5)^2 + 90(1.5) = 99$ feet

(e) $\frac{-b}{2a} = \frac{-90}{2(-16)} = 2.8125$

$s(2.8125) = 126.5625$

After 2.8125 seconds, the maximum height, 126.5625 feet, is attained. Graph $y_1 = -16x^2 + 90x$; the vertex is at $(2.8125, 126.5625)$.

(f) Find the zeros of y_1 : $(0, 0)$ and $(5.625, 0)$. The rock will hit the ground after 5.625 seconds.

49. $s(t) = -16t^2 + 150t$

(a) $y_1 = -16x^2 + 150x$ and $y_2 = 355$

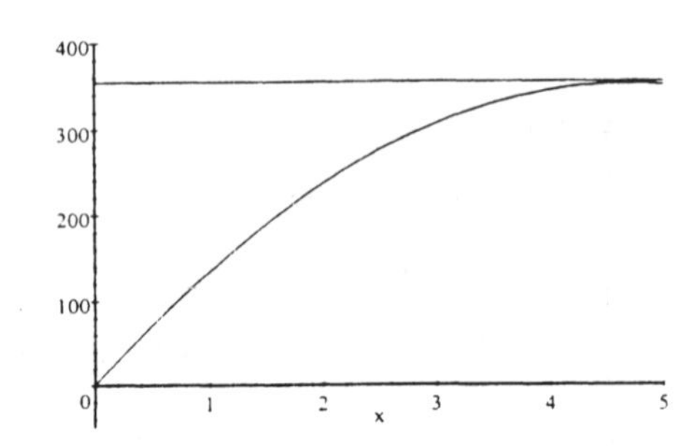

The ball will not reach 355 feet because the graphs do not intersect.

continued

49. continued

(b) $s(t) = -16t^2 + 250t + 30$

$y_1 = -16x^2 + 250x + 30$ and $y_2 = 355$

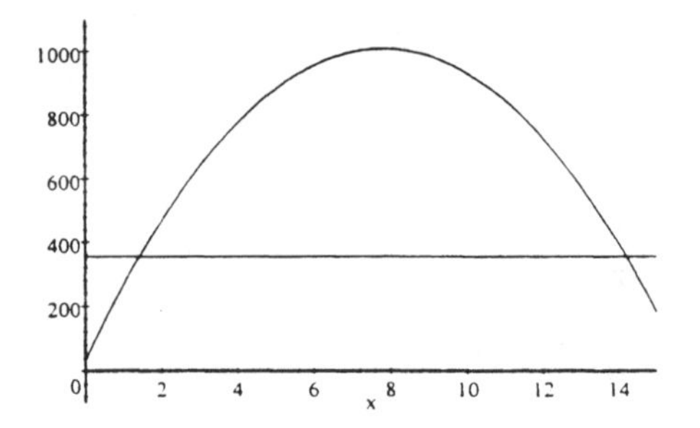

The graphs intersect at $x \approx 1.43, 14.19$. The times are at $t \approx 1.4$ seconds and $t \approx 14.2$ seconds.

51. Vertex is $(-1, -4)$ through $(5, 104)$:

$$P(x) = a(x-h)^2 + k$$
$$P(x) = a(x+1)^2 - 4$$

Let $x = 5$, $y = 104$:

$$104 = a(5+1)^2 - 4$$
$$108 = a(36) \qquad \mathbf{a = 3}$$

$$P(x) = 3(x+1)^2 - 4$$
$$P(x) = 3(x^2 + 2x + 1) - 4$$
$$\boxed{P(x) = 3x^2 + 6x - 1}$$

53. Vertex is $(8, 3)$ through $(10, 5)$:

$$P(x) = a(x-h)^2 + k$$
$$P(x) = a(x-8)^2 + 3$$

Let $x = 10$, $y = 5$:

$$5 = a(10-8)^2 + 3$$
$$2 = a(4) \qquad \mathbf{a = .5}$$

$$P(x) = .5(x-8)^2 + 3$$
$$P(x) = .5(x^2 - 16x + 64) + 3$$
$$\boxed{P(x) = .5x^2 - 8x + 35}$$

55. Vertex is $(-4,-2)$ through $(2,-26)$:

$$P(x)=a(x-h)^2+k$$
$$P(x)=a(x+4)^2-2$$

Let $x=2,\ y=-26$:

$$-26=a(2+4)^2-2$$
$$-24=a(36) \qquad \mathbf{a}=-\frac{2}{3}$$
$$P(x)=-\frac{2}{3}(x+4)^2-2$$
$$P(x)=-\frac{2}{3}(x^2+8x+16)-2$$
$$\boxed{P(x)=-\frac{2}{3}x^2-\frac{16}{3}x-\frac{38}{3}}$$

57. Vertex $(-2,-3)$; through $(1,4)$:

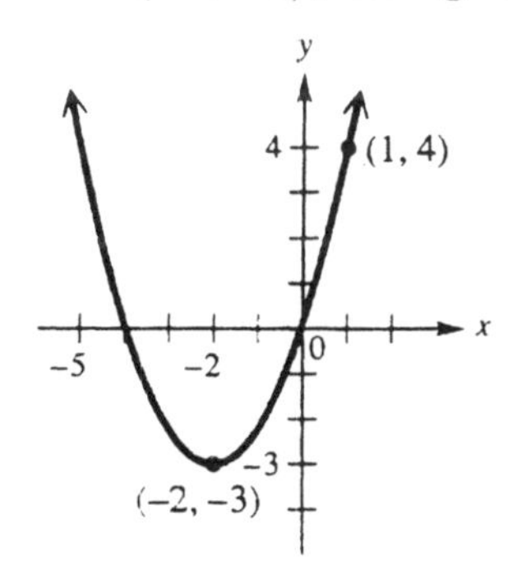

59. maximum value of 1 at $x=3$;
y-intercept is -4 :

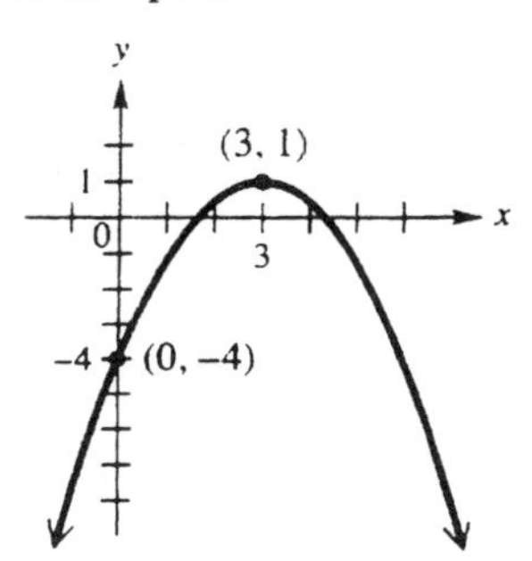

Section 3.3

1. (**d**): $(3x+1)(x-7)=0$

$$x=\boxed{\left\{-\frac{1}{3},\ 7\right\}}$$

3. (**c**): $x^2+x=12$

$$x^2+x+\frac{1}{4}=12+\frac{1}{4}$$
$$\left(x+\frac{1}{2}\right)^2=\frac{49}{4}$$
$$x+\frac{1}{2}=\pm\frac{7}{2}$$
$$x=-\frac{1}{2}\pm\frac{7}{2} \qquad x=\boxed{\{-4,\ 3\}}$$

5. $x^2=16$ $\quad\boxed{x=\{\pm 4\}}$

Show that the graphs of $y_1=x^2$ and $y_2=16$ intersect at $x=-4,\ 4$.

7. $3x^2=27$

$x^2=9$ $\quad\boxed{x=\{\pm 3\}}$

Show that the graphs of $y_1=3x^2$ and $y_2=27$ intersect at $x=-3,\ 3$.

9. $x^2=-16$

$x=\pm\sqrt{-16}$ $\quad\boxed{x=\{\pm 4i\}}$

11. $x^2=-18$

$x=\pm\sqrt{-18}$ $\quad\boxed{x=\{\pm 3i\sqrt{2}\}}$

13. $(3x-1)^2=12$

$$3x-1=\pm\sqrt{12}=\pm 2\sqrt{3}$$
$$3x=1\pm 2\sqrt{3} \qquad \boxed{x=\frac{1\pm 2\sqrt{3}}{3}}$$

15. $(5x-3)^2=-3$

$$5x-3=\pm\sqrt{-3}=\pm i\sqrt{3}$$
$$5x=3\pm i\sqrt{3} \qquad \boxed{x=\frac{3}{5}\pm\frac{\sqrt{3}}{5}i}$$

17. $x^2 - 2x - 4 = 0; \quad a = 1, b = -2, c = -4$

$$x = \frac{-(-2) \pm \sqrt{(-2)^2 - 4(1)(-4)}}{2(1)}$$

$$x = \frac{2 \pm \sqrt{4 + 16}}{2} = \frac{2 \pm \sqrt{20}}{2}$$

$$x = \frac{2 \pm 2\sqrt{5}}{2} = \frac{2(1 \pm \sqrt{5})}{2}$$

$$\boxed{x = \{1 \pm \sqrt{5}\}}$$

Show that the x-intercepts of $y_1 = x^2 - 2x - 4$ are $1 \pm \sqrt{5} \approx (-1.24,\ 3.24)$.

19. $2x^2 + 2x = -1$

$2x^2 + 2x + 1 = 0; \quad a = 2, b = 2, c = 1$

$$x = \frac{-(2) \pm \sqrt{(2)^2 - 4(2)(1)}}{2(2)}$$

$$x = \frac{-2 \pm \sqrt{4 - 8}}{4}$$

$$x = \frac{-2 \pm \sqrt{-4}}{4} = \frac{-2 \pm 2i}{4}$$

$$x = \frac{2(-1 \pm i)}{4}$$

$$\boxed{x = \left\{-\frac{1}{2} \pm \frac{1}{2}i\right\}}$$

21. $x(x - 1) = 1$

$x^2 - x - 1 = 0; \quad a = 1, b = -1, c = -1$

$$x = \frac{-(-1) \pm \sqrt{(-1)^2 - 4(1)(-1)}}{2(1)}$$

$$x = \frac{1 \pm \sqrt{1 + 4}}{2}$$

$$\boxed{x = \left\{\frac{1 \pm \sqrt{5}}{2}\right\}}$$

Show that the x-intercepts of $y_1 = x(x - 1)$ are $\frac{1 + \sqrt{5}}{2}$ and $\frac{1 - \sqrt{5}}{2} \approx (-.62,\ 1.62)$.

23. $x^2 - 5x = x - 7$

$x^2 - 6x + 7 = 0; \quad a = 1, b = -6, c = 7$

$$x = \frac{-(-6) \pm \sqrt{(-6)^2 - 4(1)(7)}}{2(1)}$$

$$x = \frac{6 \pm \sqrt{36 - 28}}{2} = \frac{6 \pm \sqrt{8}}{2}$$

$$x = \frac{6 \pm 2\sqrt{2}}{2} = \frac{2(3 \pm \sqrt{2})}{2}$$

$$\boxed{x = \{3 \pm \sqrt{2}\}}$$

Show that the x-intercepts of $y_1 = x^2 - 5x - (x - 7)$ are $3 \pm \sqrt{2} \approx (1.59,\ 4.41)$.

25. $4x^2 - 12x = -11$

$4x^2 - 12x + 11 = 0; a = 4, b = -12, c = 11$

$$x = \frac{-(-12) \pm \sqrt{(-12)^2 - 4(4)(11)}}{2(4)}$$

$$x = \frac{12 \pm \sqrt{144 - 176}}{8}$$

$$x = \frac{12 \pm \sqrt{-32}}{8} = \frac{12 \pm 4i\sqrt{2}}{8}$$

$$x = \frac{4(3 \pm i\sqrt{2})}{8}$$

$$\boxed{x = \frac{3}{2} \pm \frac{\sqrt{2}}{2}i}$$

27. $\frac{1}{3}t^2 + \frac{1}{4}t - 3 = 0$

$12\left(\frac{1}{3}t^2 + \frac{1}{4}t - 3 = 0\right)$

$4t^2 + 3t - 36 = 0; \ a = 4, \ b = 3, \ c = -36$

$$t = \frac{-3 \pm \sqrt{(3)^2 - 4(4)(-36)}}{2(4)}$$

$$= \frac{-3 \pm \sqrt{9 + 576}}{8} = \frac{-3 \pm \sqrt{585}}{8}$$

$$= \boxed{\frac{-3 \pm 3\sqrt{65}}{8}}$$

29. $x^2 + 8x + 16 = 0; \quad a = 1, b = 8, c = 16$

discriminate: $b^2 - 4ac = 64 - 4(16) = \boxed{0}$

$\boxed{\text{1 real solution: rational}}$

31. $4x^2 = 6x + 3$
$4x^2 - 6x - 3 = 0; a = 4, b = -6, c = -3$

discriminate: $b^2 - 4ac = (-6)^2 - 4(4)(-3)$
$= 36 + 48 = \boxed{84}$

$\boxed{\text{2 real solutions; irrational}}$

33. $9x^2 + 11x + 4 = 0; a = 9, b = 11, c = 4$

discriminate: $b^2 - 4ac = (11)^2 - 4(9)(4)$
$= 121 - 144 = \boxed{-23}$

$\boxed{\text{No real solutions}}$

35. $x = 4, x = 5:$

$(x - 4)(x - 5) = 0$
$x^2 - 9x + 20 = 0$
$\boxed{a = 1, b = -9, c = 20}$

37. $x = 1 + \sqrt{2}, x = 1 - \sqrt{2}:$

$(x - 1 - \sqrt{2})(x - 1 + \sqrt{2}) = 0$
$x^2 - x + x\sqrt{2} - x + 1 - \sqrt{2} - x\sqrt{2} + \sqrt{2} - 2 = 0$
$x^2 - 2x - 1 = 0$
$\boxed{a = 1, b = -2, c = -1}$

In exercises 39 - 43, answers may vary.

39. $a < 0, b^2 - 4ac = 0$

$a < 0$, opens downward
$b^2 - 4ac = 0 \Rightarrow$ 1 real solution

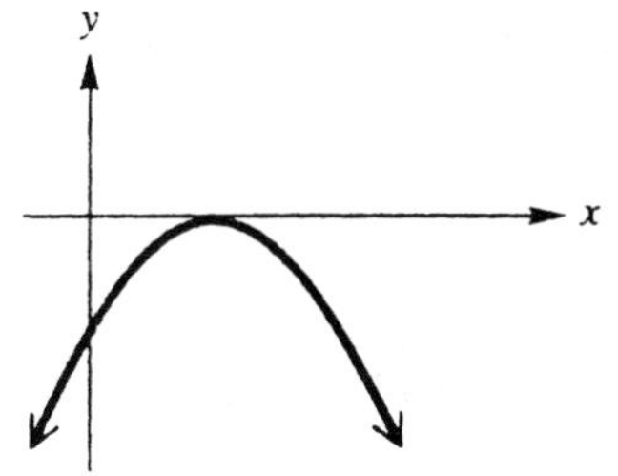

41. $a < 0, b^2 - 4ac < 0$

$a < 0$, opens downward
$b^2 - 4ac < 0 \Rightarrow$ no real solution

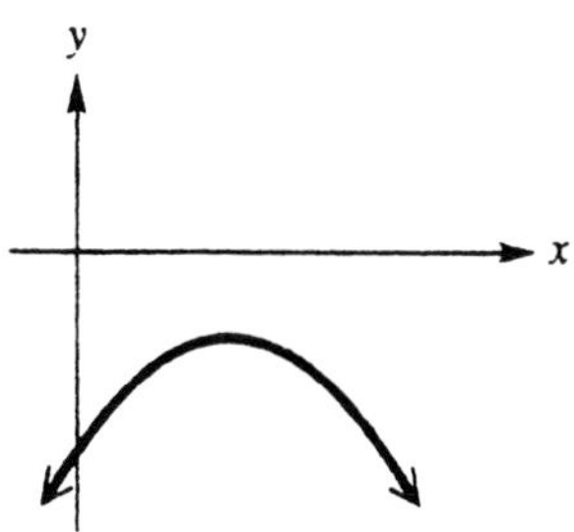

43. $a > 0, b^2 - 4ac > 0$

$a > 0$, opens upward
$b^2 - 4ac > 0 \Rightarrow$ 2 real solutions

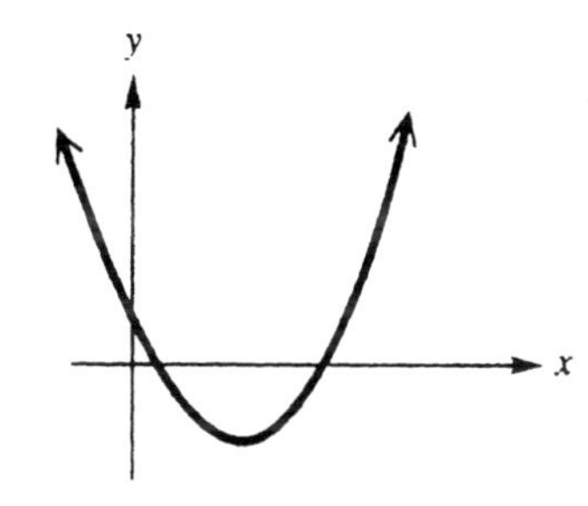

45. $f(x) = 0 : \{2, 4\}$

47. $f(x) > 0 : (-\infty, 2) \cup (4, \infty)$

49. $g(x) < 0 : (-\infty, 3) \cup (3, \infty)$

51. $h(x) > 0 : (-\infty, \infty)$

53. $h(x) = 0 :$ No real solutions;
2 imaginary complex solutions.

55. The axis of symmetry would be $x = 3$, half-way between 2 and 4; thus, the x-coordinate would be 3.

57. Yes, $g(x)$ does have a y-intercept; it would be negative.

59. $x^2 + 4x + 3 = 0$ $\quad x = -3,\ -1$
$(x+3)(x+1) = 0$

$(x+3)$	−	+	+
$(x+1)$	−	−	+
(*prod*)	+	−	+
		−3	−1

(a) $x^2 + 4x + 3 \geq 0$

$(-\infty, -3] \cup [-1, \infty)$

(b) $x^2 + 4x + 3 < 0$

the interval $(-3, -1)$

61. $2x^2 - 9x + 4 = 0$ $\quad x = \frac{1}{2},\ 4$
$(2x-1)(x-4) = 0$

$(2x-1)$	−	+	+
$(x-4)$	−	−	+
(*prod*)	+	−	+
		3/2	4

(a) $2x^2 - 9x + 4 > 0$ $\quad \left(-\infty, \frac{1}{2}\right) \cup (4, \infty)$

(b) $2x^2 - 9x + 4 \leq 0$ $\quad \left[\frac{1}{2}, 4\right]$

63. $-x^2 + 2x + 1 = 0$
$-\left(x^2 - 2x - 1\right) = 0$

$$x = \frac{2 \pm \sqrt{4+4}}{2}$$

$$x = \frac{2 \pm 2\sqrt{2}}{2}$$

$x = 1 \pm \sqrt{2}$ $\quad (\approx -0.4, 2.4)$

(-1)	−	−	−
$(x - 1 - \sqrt{2})$	−	−	+
$(x - 1 + \sqrt{2})$	−	+	+
(*prod*)	−	+	−
		−.4	2.4

continued

63. continued

(a) $-x^2 + 2x + 1 \geq 0$

$\left[1 - \sqrt{2},\ 1 + \sqrt{2}\right]$

(b) $-x^2 + 2x + 1 < 0$

$\left(-\infty, 1 - \sqrt{2}\right) \cup \left(1 + \sqrt{2}, \infty\right)$

Relating Concepts (65 – 70)

65. $y_1 = x^2 + 2x - 8$
$y_1 = (x+4)(x-2)$

x-intercepts at $-4,\ 2$

66. Graph is concave up, x-intercepts: $\{-4,\ 2\}$

$x^2 + 2x - 8 < 0:$ the interval $(-4, 2)$

67. $y_2 = -x^2 - 2x + 8$
$= -1\left(x^2 + 2x - 8\right)$
$y_2 = -y_1$;
Reflection of y_1 across the x-axis.

68. Graph is concave down, x-intercepts: $\{-4,\ 2\}$

$x^2 + 2x - 8 > 0:$ the interval $(-4, 2)$

69. The solutions are the same for **66** and **68**.

70. Because multiplying a function by -1 causes a reflection across the x-axis, the original solution set of $f(x) < 0$ is the same as that of $-f(x) > 0$. The same holds true for the solution sets of $f(x) > 0$ and $-f(x) < 0$.

71. $s = \frac{1}{2}gt^2$ for t:

$$2s = gt^2$$

$$\frac{2s}{g} = t^2 \quad \rightarrow \quad t = \pm\sqrt{\frac{2s}{g}} \cdot \frac{\sqrt{g}}{\sqrt{g}}$$

$$\boxed{t = \frac{\pm\sqrt{2sg}}{g}}$$

73. $F = \frac{-kMv^4}{r}$ for v:

$$Fr = kMv^4$$

$$\frac{Fr}{kM} = v^4 \quad \rightarrow \quad v = \pm\sqrt[4]{\frac{Fr}{kM}} \cdot \frac{\sqrt[4]{k^3M^3}}{\sqrt[4]{k^3M^3}}$$

$$\boxed{v = \frac{\pm\sqrt[4]{Frk^3M^3}}{kM}}$$

75. $P = \frac{E^2R}{(r+R)^2}$ for R:

$$P(r+R)^2 = E^2R$$

$$P(r^2 + 2rR + R^2) = E^2R$$

$$Pr^2 + 2PrR + PR^2 - E^2R = 0$$

$$PR^2 + R(2Pr - E^2) + Pr^2 = 0$$

$a = P.\ b = 2Pr - E^2,\ C = Pr^2$

$$R = \frac{(E^2 - 2Pr) \pm \sqrt{(2Pr - E^2)^2 - 4(P)(Pr^2)}}{2P}$$

$$= \frac{E^2 - 2Pr \pm \sqrt{E^4 - 4PrE^2 + 4P^2r^2 - 4P^2r^2}}{2P}$$

$$= \frac{E^2 - 2Pr \pm \sqrt{E^4 - 4PrE^2}}{2P}$$

$$= \frac{E^2 - 2Pr \pm \sqrt{(E^2 - 4Pr)E^2}}{2P}$$

$$= \frac{E^2 - 2Pr \pm |E|\sqrt{E^2 - 4Pr}}{2P}$$

$$= \boxed{\frac{E^2 - 2Pr \pm E\sqrt{E^2 - 4Pr}}{2P}}$$

77. $4x^2 - 2xy + 3y^2 = 2$

For x: $4x^2 - 2xy + (3y^2 - 2) = 0$

$$x = \frac{2y \pm \sqrt{(-2y)^2 - 4(4)(3y^2 - 2)}}{2(4)}$$

$$= \frac{2y \pm \sqrt{4y^2 - 48y^2 + 32}}{8}$$

$$= \frac{2y \pm \sqrt{-44y^2 + 32}}{8}$$

$$= \frac{2y \pm \sqrt{4(-11y^2 + 8)}}{8}$$

$$= \frac{2y \pm 2\sqrt{8 - 11y^2}}{8} = \frac{y \pm \sqrt{8 - 11y^2}}{4}$$

For y: $3y^2 - 2xy + (4x^2 - 2) = 0$

$$y = \frac{2x \pm \sqrt{(-2x)^2 - 4(3)(4x^2 - 2)}}{2(3)}$$

$$= \frac{2x \pm \sqrt{4x^2 - 48x^2 + 24}}{6}$$

$$= \frac{2x \pm \sqrt{-44x^2 + 24}}{6}$$

$$= \frac{2x \pm \sqrt{4(-11x^2 + 6)}}{6}$$

$$= \frac{2x \pm 2\sqrt{6 - 11x^2}}{6} = \frac{x \pm \sqrt{6 - 11x^2}}{3}$$

79. $S(x) = 0$: $-.076x^2 - .058x + 1 = 0$

$$x = \frac{.058 \pm \sqrt{(-.058)^2 - 4(-.076)(1)}}{2(-.076)}$$

$\approx -4.029,\ 3.266$

$x \approx 3.3$; the probability that anyone will live 33 years past 65 is 0. (There are, of course, exceptions).

Reviewing Basic Concepts (Sections 3.1 - 3.3)

1. $(5+6i)-(2-4i)-3i$
$= 5+6i-2+4i-3i$
$= 3+7i$

2. $i(5+i)(5-i) = i(25-i^2) = i(25+1) = 26i$

3. $\dfrac{-10-10i}{2+3i}\cdot\dfrac{2-3i}{2-3i} = \dfrac{-20+30i-20i+30i^2}{4-9i^2}$
$= \dfrac{-20+10i-30}{4+9}$
$= \dfrac{-50+10i}{13} = -\dfrac{50}{13}+\dfrac{10}{13}i$

4. $P(x) = 2x^2+8x+5$

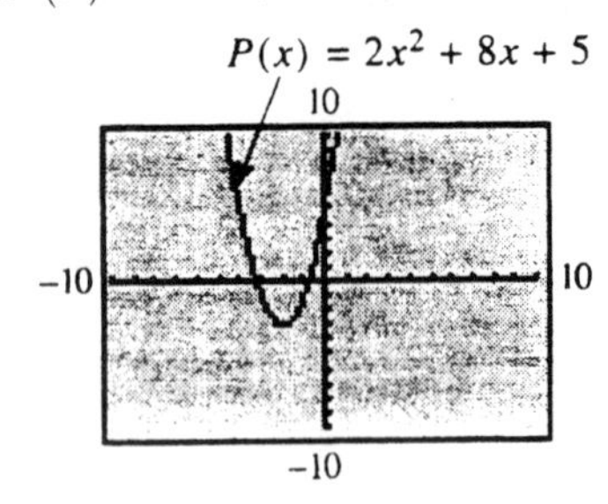

5. $\dfrac{-b}{2a} = \dfrac{-8}{4} = -2$
$f(-2) = 2(-2)^2+8(-2)+5 = -3$
Vertex: $(-2, -3)$; minimum

6. Axis of symmetry: $x = -2$

7. domain: $(-\infty, \infty)$; range: $[-3, \infty)$

8. $9x^2 = 25$
$x^2 = \dfrac{25}{9}$
$x = \pm\sqrt{\dfrac{25}{9}} = \left\{\pm\dfrac{5}{3}\right\}$

9. $3x^2-5x = 2$
$3x^2-5x-2 = 0$
$(3x+1)(x-2) = 0$
$x = \left\{-\dfrac{1}{3}, 2\right\}$

10. $-x^2+x+3 = 0$
$x = \dfrac{-1\pm\sqrt{1-4(-3)}}{2(-1)}$
$= \dfrac{-1\pm\sqrt{13}}{-2} = \dfrac{1\pm\sqrt{13}}{2}$
$\approx \{-1.3028, 2.3028\}$

11. $3x^2-5x-2 \le 0$
$(3x+1)(x-2) \le 0$
$x = -\dfrac{1}{3}, 2$

$(3x+1)$	$-$	$+$	$+$
$(x-2)$	$-$	$-$	$+$
$(prod)$	$+$	$-$	$+$
		$-1/3$	2

$\left[-\dfrac{1}{3}, 2\right]$

12. $x^2-x-3 > 0$ (See #10) $x = \dfrac{1\pm\sqrt{13}}{2}$

$\left(x-\frac{1+\sqrt{13}}{2}\right)$	$-$	$-$	$+$
$\left(x-\frac{1-\sqrt{13}}{2}\right)$	$-$	$+$	$+$
$(prod)$	$+$	$-$	$+$
		-1.3	2.3

$\left(-\infty, \dfrac{1-\sqrt{13}}{2}\right)\cup\left(\dfrac{1+\sqrt{13}}{2}, \infty\right)$

Section 3.4

1. $A = l \cdot w = 40,000 \text{ yd}^2$
$(2x + 200)(x) = 40,000$: **(A)**

3. Let x be the first number; sum $= 30$.

(a) Second number: $\boxed{30 - x}$

(b) Restrictions on x :
$x > 0$ and $x < 30$ $\boxed{0 < x < 30}$

(c) $P(x) = x(30 - x)$ $\boxed{P(x) = -x^2 + 30x}$

(d) Maximum occurs at vertex:

$$x = \frac{-b}{2a} = \frac{-30}{-2} = 15$$

$30 - x = 15$
$P(15) = 15(15) = 225$
Vertex: $(15, 225)$

$\boxed{1^{st}\ \# : 15,\ 2^{nd}\ \# : 15,\ \text{Product: } 225}$

Locate $(15, 225)$ on the graph of $P(x)$.

5. Total length $= 640$ ft; x represents the length of the 2 parallel sides.

(a) $2x + y = 640$ $y = \boxed{640 - 2x}$

(b) Restrictions on x :
$x > 0$ and $2x < 640$ $\boxed{0 < x < 320}$

(c) $A(x) = x(640 - 2x)$
$\boxed{A(x) = -2x^2 + 640x}$

(d) For y : $(30,000,\ 40,000)$

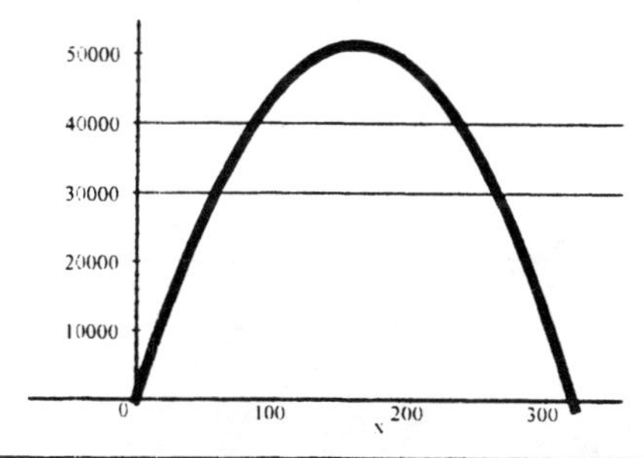

$\boxed{x : (57.04\text{ ft},\ 85.17\text{ ft}) \cup (234.83\text{ ft},\ 262.96\text{ ft})}$

(e) $x = \dfrac{-b}{2a} = \dfrac{-640}{-4} = 160$
Vertex: $(160,\ 51,200)$
$y = 640 - 2(160) = 320$
$\boxed{160\text{ ft} \times 320\text{ ft} = 51,200\text{ ft}^2}$

Locate $(160,\ 51,200)$ on the graph of $A(x)$.

7. (a) $w = x$, $\boxed{l = 2x}$

(b) $\boxed{w = x - 4,\ l = 2x - 4}$

$x > 0$, $\quad x - 4 > 0$, $\quad 2x - 4 > 0$
$\quad\quad\quad x > 4 \quad\quad\quad x > 2$

So $\boxed{x > 4}$

(c $\quad V = l \cdot w \cdot h$
$V(x) = (2x - 4)(x - 4)(2)$
$V(x) = 2(2x^2 - 12x + 16)$
$\boxed{V(x) = 4x^2 - 24x + 32}$

(d) $320 = 4x^2 - 24x + 32$
$4x^2 - 24x + 32 - 320 = 0$
$4x^2 - 24x - 288 = 0$
$x^2 - 6x - 72 = 0$
$(x - 12)(x + 6) = 0$
$x = 12,\ -6$
$w = x - 4 = 12 - 4 = \boxed{8\text{ in}}$
$l = 2x - 4 = 24 - 4 = \boxed{20\text{ in}}$

Locate $(12,\ 320)$ on the graph of $V(x)$.

(e) $400 < \text{Vol} < 500$; x : $\boxed{13.0\text{ in to } 14.2\text{ in}}$

9. $S.A. = 600\text{ in}^2$, $h = 4.25$ in

$S.A. = 2\pi r^2 + 2\pi rh$
$600 = 2\pi r^2 + 2\pi r(4.25)$
$2\pi r^2 + 8.5\pi r - 600 = 0$

$$r = \frac{-8.5\pi \pm \sqrt{(8.5\pi)^2 - 4(2\pi)(-600)}}{4\pi}$$

≈ 7.875 inches

11. $A = s^2$: $800 = s^2$
$s = \sqrt{800} = 20\sqrt{2}$

diagonal: $d^2 = (20\sqrt{2})^2 + (20\sqrt{2})^2$
$d^2 = 1600$
$d = 40$

Since the diagonal is $2 \times$ radius, $\boxed{r = 20\text{ feet}}$

13. $h^2 + 12^2 = (2h + 3)^2$
$h^2 + 144 = 4h^2 + 12h + 9$
$3h^2 + 12h - 135 = 0$
$h^2 + 4h - 45 = 0$
$(h - 5)(h + 9) = 0$

$\boxed{h = 5\text{ feet}}$

15. $10^2 + 13^2 = l^2$, where l = length of ladder

$100 + 169 = l^2$

$l^2 = 269$

$l = \sqrt{269} \approx 16.4$

A 16 foot ladder will not be long enough, so he will need a 17 foot ladder.

17. (a) $N(x) = \boxed{80 - x}$

(b) $r(x) = \boxed{400 + 20x}$

(c) $R(x) = \#$ rented $\cdot$ rent

$R(x) = (80 - x)(400 + 20x)$

$\boxed{R(x) = -20x^2 + 1200x + 32000}$

(d) $37500 = -20x^2 + 1200x + 32000$

$20x^2 - 1200x + 5500 = 0$

$x^2 - 60x + 275 = 0$

$(x - 5)(x - 55) = 0$

$\boxed{x = \{5,\ 55\}}$

(e) $x = \dfrac{-b}{2a} = \dfrac{-1200}{-40} = 30$

$$\begin{aligned} r(x) &= 400 + 20x \\ r(30) &= 400 + 20(30) \\ &= 400 + 600 = \boxed{\$1000} \end{aligned}$$

19. $y = \dfrac{-16x^2}{.434v^2} + 1.15x + 8$

(a) $10 = \dfrac{-16(15)^2}{.434v^2} + 1.15(15) + 8$

$10 = \dfrac{-3600}{.434v^2} + 25.25$

$-15.25 = \dfrac{-3600}{.434v^2}$

$-6.6185v^2 = -3600$

$v^2 = \dfrac{3600}{6.6185}$

$v \approx 23.32$ ft/sec

(b) Yes. graph $y_1 = \dfrac{-16x^2}{.434(23.32)^2} + 1.15x + 8$

(c) maximum height is ≈ 12.88 feet

21. (a)
$$\begin{aligned} h(x) &= -0.5x^2 + 1.25x + 3 \\ h(2) &= -0.5(4) + 1.25(2) + 3 \\ &= -2 + 2.5 + 3 = \boxed{3.5 \text{ feet}} \end{aligned}$$

(b)
$$\begin{aligned} h(x) &= -0.5x^2 + 1.25x + 3 \\ 3.25 &= -0.5x^2 + 1.25x + 3 \\ 0 &= -0.5x^2 + 1.25x + .25 \\ 0 &= 50x^2 - 125x - 25 \\ 0 &= 2x^2 - 5x + 1 \end{aligned}$$

$$x = \frac{5 \pm \sqrt{25 - 4(2)}}{4} = \frac{5 \pm \sqrt{17}}{4}$$

$x \approx 2.281,\ 0.219$ $\boxed{\approx 2.3 \text{ feet and } .2 \text{ feet}}$

(c) $h(x) = -0.5x^2 + 1.25x + 3$

$x = \dfrac{-b}{2a} = \dfrac{-1.25}{2(-0.5)} = 1.25$ $\boxed{1.25 \text{ feet}}$

(d) $h(x) = -0.5x^2 + 1.25x + 3$

$$\begin{aligned} h(1.25) &= -0.5(1.25)^2 + 1.25(1.25) + 3 \\ &= 3.78125 \quad \boxed{\approx 3.78 \text{ feet}} \end{aligned}$$

23. $D = .1s^2 - 3s + 22$

$800 = .1s^2 - 3s + 22$

$0 = .1s^2 - 3s - 778$

$$\begin{aligned} s &= \frac{3 \pm \sqrt{9 - 4(.1)(-778)}}{2(.1)} \\ &= \frac{3 \pm \sqrt{320.2}}{.2} \approx \boxed{104.5 \text{ feet/second}} \end{aligned}$$

25. $T(x) = .00787x^2 - 1.528x + 75.89$

(a)
$$\begin{aligned} T(50) &= .00787(50)^2 - 1.528(50) + 75.89 \\ &= 19.165 \quad \boxed{\approx 19.2 \text{ hours}} \end{aligned}$$

(b) $3 = .00787x^2 - 1.528x + 75.89$

$.00787x^2 - 1.528x + 72.89 = 0$

$x = 84.338225317 \approx 84.3$

Graph $y = .00787x^2 - 1.528x + 72.89$; the intercepts are $x \approx 84.3,\ 109.817$.

$\boxed{84.3 \text{ ppm}}$ $(109.817 \notin [50,\ 100])$

27. (a) Americans expected to be over 100 years old:

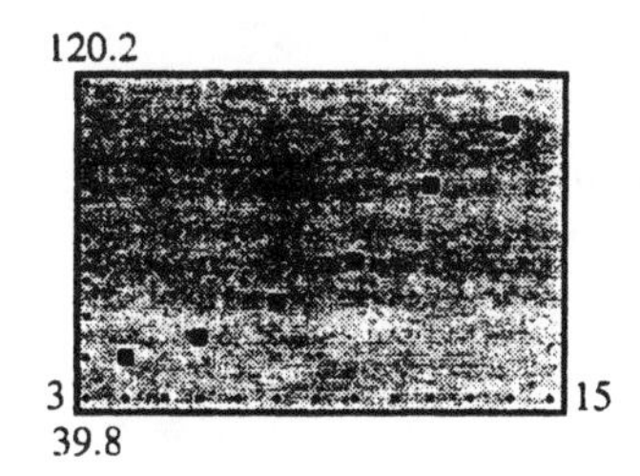

(b) $f(x) = a(x-h)^2 + k$

$f(x) = a(x-4)^2 + 50$

$(14,\ 110):\quad 110 = a(14-4)^2 + 50$

$60 = 100a$

$a = .6$

$$\boxed{f(x) = .6(x-4)^2 + 50}$$

(c) $f(x)$ and data; there is a very good fit.

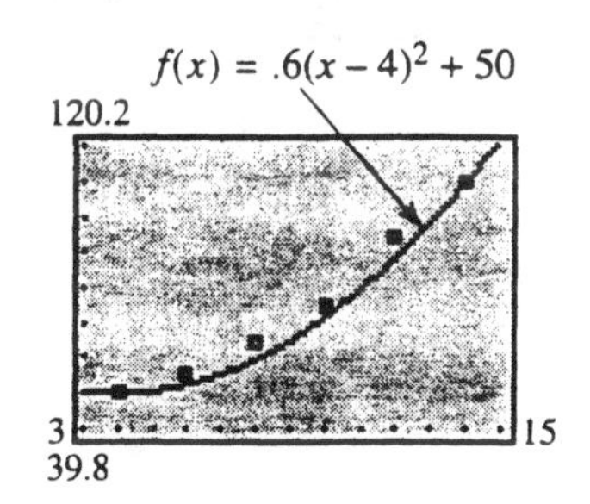

(d) $\{4, 6, 8, \ldots, 14\} \to L_1$

$\{50, 56, 65, \ldots, 110\} \to L_2$

STAT / CALC / QuadReg

$y = ax^2 + bx + c$

$$\boxed{g(x) = .402x^2 - 1.175x + 48.343}$$

(e) 2006, $x = 16$

$f(16) = 136.4 \approx 136$

$g(16) \approx 132.455 \approx 132$

29. (a) Automobile Stopping Distances

$f(x) = .056057x^2 + 1.06657x$

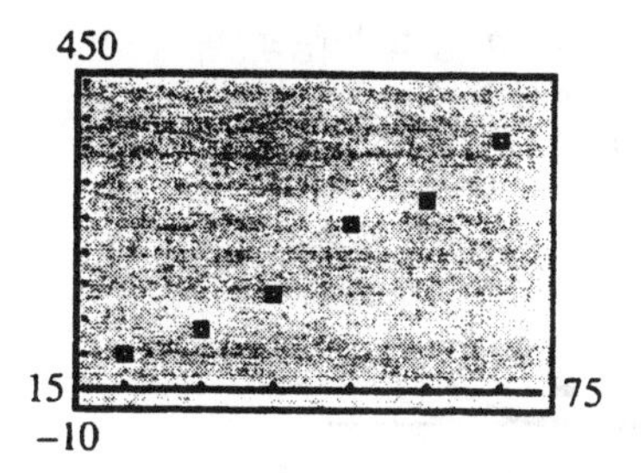

(b) $f(45) = .056057(45)^2 + 1.06657(45)$

≈ 161.5

When the speed is 45 mph, the stopping distance is 161.5 feet.

(c) See the graph above. The model is quite good, although the stopping distances are a little low for the higher speeds.

Section 3.5

1. As the odd exponent 'n' gets larger, the graph *flattens out* in the window $[-1, 1]$ by $[-1, 1]$.

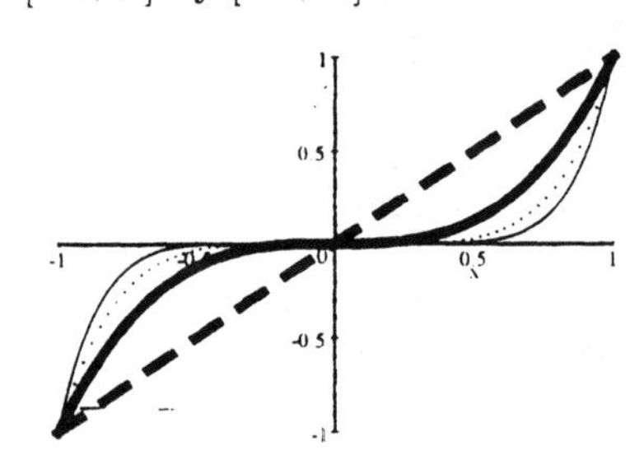

The graph of $y = x^7$ will be between $y = x^5$ and the x-axis in this window.

3. $f(x) = x^n$ for $n \in \{\text{positive odd integers}\}$ will take the shape of $f(x) = x^3$, but gets steeper as n gets larger.

Relating Concepts (5 – 8)

5. $y = x^4$

Shift 3 units left:
$y = (x + 3)^4$

Stretch vertically by a factor of 2:
$y = 2(x + 3)^4$

Shift 7 units downward:
$y = 2(x + 3)^4 - 7$

$y = 2(x^4 + 12x^3 + 54x^2 + 108x + 81) - 7$
$y = 2x^4 + 24x^3 + 108x^2 + 216x + 155$

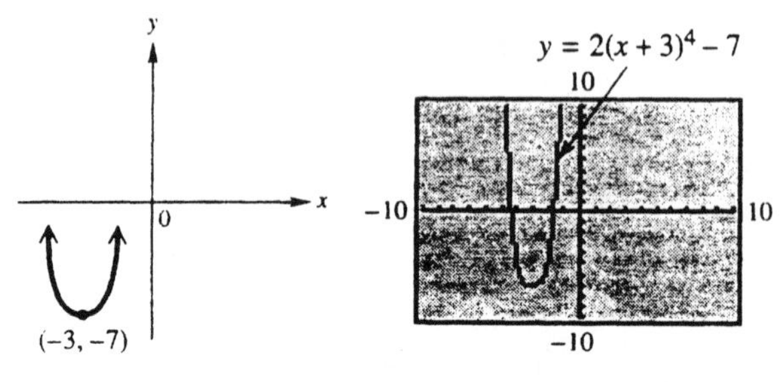

6. $y = x^4$

Shift 1 unit left:
$y = (x + 1)^4$

Stretch vertically by a factor of 3:
$y = 3(x + 1)^4$

Reflect across the x-axis:
$y = -3(x + 1)^4$

Shift 12 units upward:
$y = -3(x + 1)^4 + 12$

$y = -3(x^4 + 4x^3 + 6x^2 + 4x + 1) + 12$
$y = -3x^4 - 12x^3 - 18x^2 - 12x + 9$

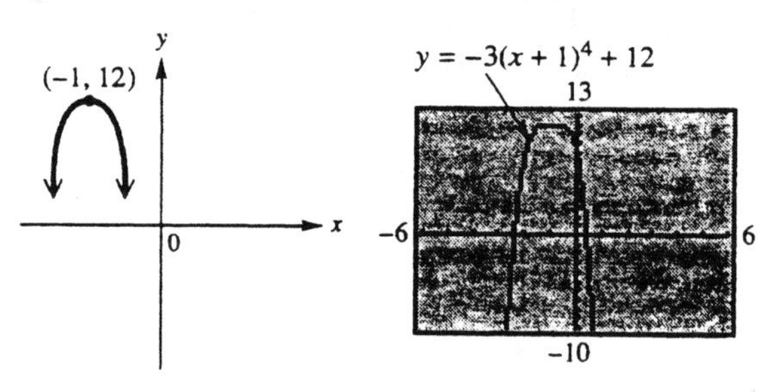

7. $y = x^3$

Shift 1 unit right:
$y = (x - 1)^3$

Stretch vertically by a factor of 3:
$y = 3(x - 1)^3$

Reflect across the x-axis:
$y = -3(x - 1)^3$

Shift 12 units upward:
$y = -3(x - 1)^3 + 12$

$y = -3(x^3 - 3x^2 + 3x - 1) + 12$
$y = -3x^3 + 9x^2 - 9x + 15$

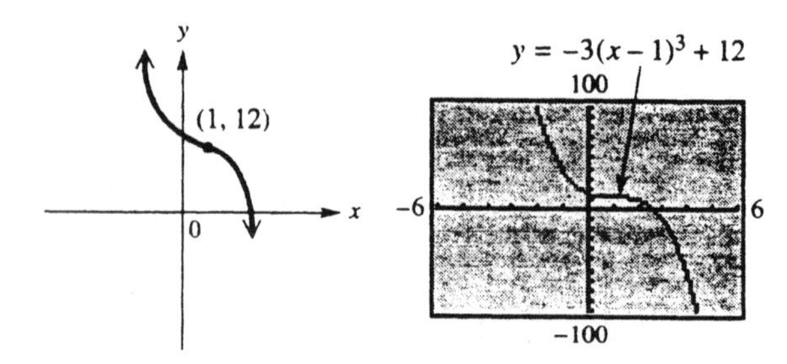

8. $y = x^5$

Shift 1 unit right:

$y = (x-1)^5$

Shrink vertically by a factor of .5:

$y = .5(x-1)^5$

Shift 13 units upward:

$y = .5(x-1)^5 + 13$

$y = .5(x^5 - 5x^4 + 10x^3 - 10x^2 + 5x - 1) + 13$

$y = .5x^5 - 2.5x^4 + 5x^3 - 5x^2 + 2.5x + 12.5$

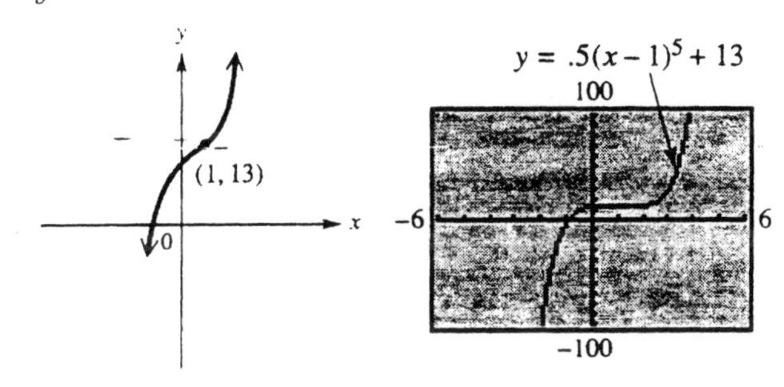

9. 3 turning points ⟶ at least degree 4

11. (a, b) : local maximum
(e, t) : local minimum
(c, d) : local maximum

13. (a, b) : absolute minimum

15. Local maximum values at b and d;
local minimum value at t;
absolute maximum value at b.

17. $y = \frac{1}{3}x^3 - \frac{5}{2}x^2 + 6x - 1$

local maximum at $(2, 3.67)$;
local minimum at $(3, 3.5)$.

19. $y = -x^3 - 11x^2 - 40x - 50$

local maximum at $(-3.33, -1.85)$;
local minimum at $(-4, -2)$.

21. $y = x^2 - 4.25x + 4.515$

two: 2.10 and 2.15

23. $y = -x^2 + 6.5x - 10.60$

window: $[2, 5]$ by $[-2, 0]$; none

25. $P(x) = \sqrt{5}\,x^3 + 2x^2 - 3x + 4$: ↙ ↗

27. $P(x) = -\pi x^5 + 3x^2 - 1$: ↖ ↘

29. $P(x) = 2.74x^4 - 3x^2 + x - 2$: ↖ ↗

31. $P(x) = -\pi x^6 + x^5 - x^4 - x + 3$: ↙ ↘

33. $f(x) = 2x^3 + x^2 - x + 3$: D

35. $h(x) = -2x^4 + x^3 - 2x^2 + x + 3$: B

37. $P(x) = x^n$, n is even, is always concave up.

39. $y = x^3 + x^2 + x$

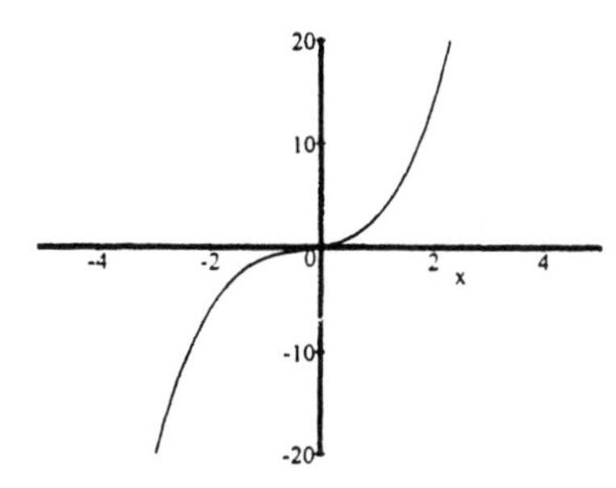

is concave down for $x < 0$,
and concave up for $x > 0$.

41. False; $f(x) = x^3 + 2x^2 - 4x + 3$ has at most 3 real zeros.

43. True; A polynomial function of even degree with $a < 0$ has endpoint behavior of ↗↘; with a positive y-intercept, it must cross the x-axis at least two times.

45. True; $f(x) = -3x^4 + 5$ has a positive y-intercept and opens downward with the shape of a parabola, so it must cross the x-axis twice.

47. False; a fifth degree polynomial has endpoint behavior of ↙↗ or ↖↘; it is possible that it will have 2 pairs of conjugate complex roots, thus crossing the x-axis only one time.

49. $y = x^3 - 3x^2 - 6x + 8$: **A**

51. The graph in **C** has <u>one</u> real zero.

53. **B** and **D** are not graphs of cubic polynomial functions.

55. The function graphed in **A** has <u>one</u> negative real zero.

57. The graph in **B** is that of a function whose range is not $(-\infty, \infty)$.

59. $y = -2x^3 - 14x^2 + 2x + 84$

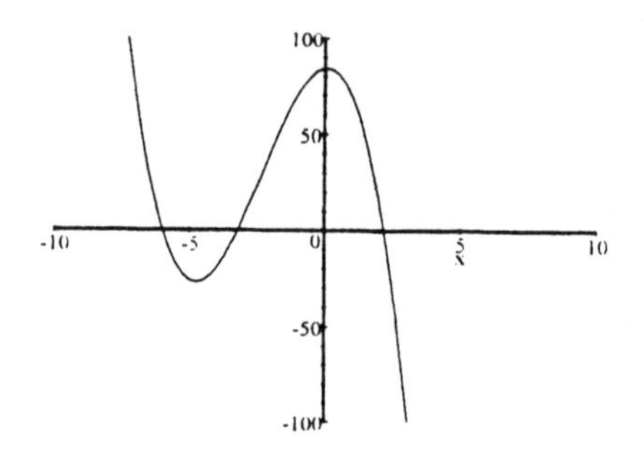

(a) Domain: $(-\infty, \infty)$

(b) Local minimum: $(-4.74, -27.03)$; not an absolute minimum

(c) Local maximum: $(.07, 84.07)$; not an absolute maximum

(d) Range: $(-\infty, \infty)$

(e) x-intercepts: -6, -3.19, 2.19
y-intercept: 84

(f) Increasing: $[-4.74, .07]$

(g) Decreasing: $(-\infty, -4.74]$; $[.07, \infty)$

61. $y = x^5 + 4x^4 - 3x^3 - 17x^2 + 6x + 9$

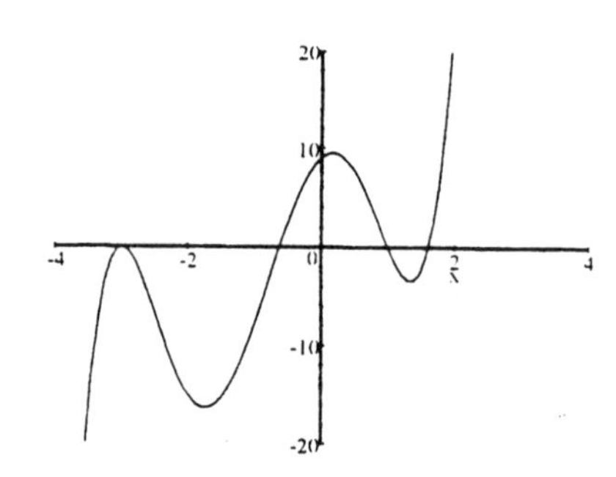

(a) Domain: $(-\infty, \infty)$

(b) Local minimums: $(-1.73, -16.39)$, $(1.35, -3.49)$; neither is an absolute minimum

(c) Local maximums: $(-3, 0)$, $(.17, 9.52)$; neither is an absolute maximum

(d) Range: $(-\infty, \infty)$

(e) x-intercepts: -3, $-.62$, 1, 1.62
y-intercept: 9

(f) Increasing: $(-\infty, -3]$; $[-1.73, .17]$; $[1.35, \infty)$

(g) Decreasing: $[-3, -1.73]$; $[.17, 1.35]$

63. $y = 2x^4 + 3x^3 - 17x^2 - 6x - 72$

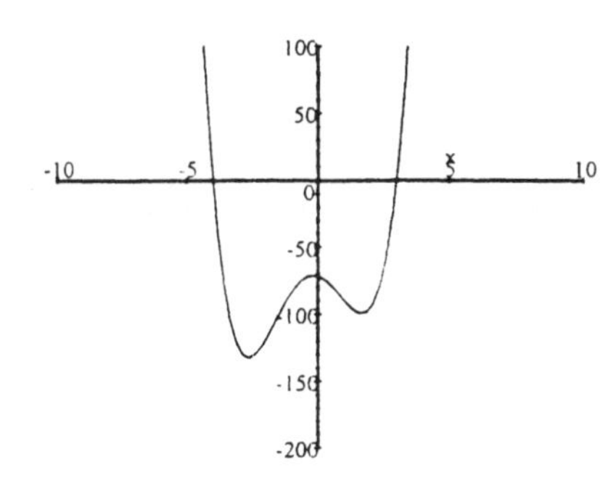

(a) Domain: $(-\infty, \infty)$

(b) Local minimum: $(-2.63, -132.69)$ is the absolute minimum; and $(1.68, -99.90)$ (not absolute)

(c) Local maximum: $(-.17, -71.48)$; not an absolute maximum

(d) Range: $[-132.69, \infty)$

(e) x-intercepts: -4, 3; y-intercept: -72

(f) Increasing: $[-2.63, -.17]$; $[1.68, \infty)$

(g) Decreasing: $(-\infty, -2.63]$; $[-.17, 1.68]$

65. $y = -x^6 + 24x^4 - 144x^2 + 256$

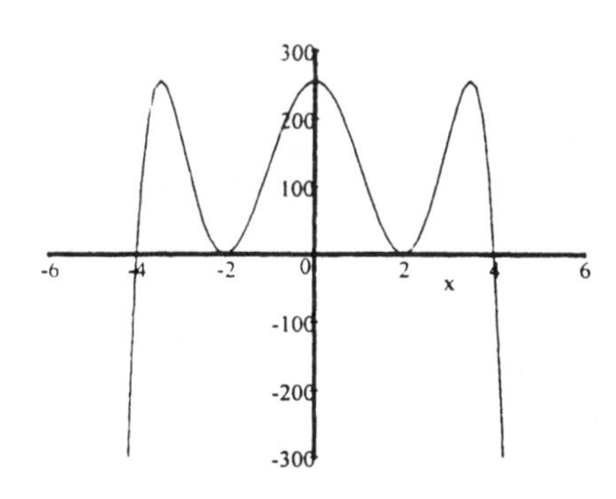

(a) Domain: $(-\infty, \infty)$

(b) Local minimum: $(-2, 0)$; $(2, 0)$ neither is an absolute minimum

(c) Local maximum: $(-3.46, 256)$; $(0, 256)$; $(3.46, 256)$; all are absolute maximums

(d) Range: $(-\infty, 256]$

(e) x-intercepts: ± 2, ± 4; y-intercept: 256

(f) Increasing: $(-\infty, -3.46]$; $[-2, 0]$; $[2, 3.46]$

(g) Decreasing: $[-3.46, -2]$; $[0, 2]$; $[3.46, \infty)$

For questions 67 - 71, answers may vary.

67. $y = 4x^5 - x^3 + x^2 + 3x - 16$
$[-4, 4]$ by $[-40, 20]$, among others.

69. $y = 2.9x^3 - 37x^2 + 28x - 143$
$[-20, 30]$ by $[-1000, 100]$, among others.

71. $y = \pi x^4 - 13x^2 + 84$
$[-5, 5]$ by $[-100, 300]$, among others.

73. (a) (i) $f(x) = .2(x - 1990)^2 + 12.6$

(ii) $g(x) = .55(x - 1990) + 12.6$

(iii) $h(x) = .0075(x - 1990)^3 + 14.12$

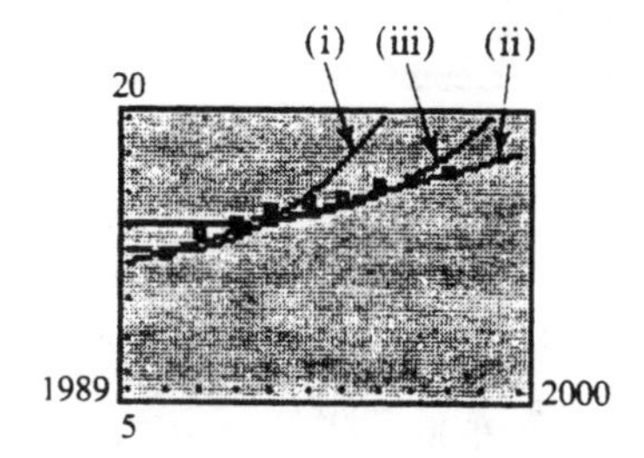

(b) All 3 models approximate the data up to 1994, but only $g(x)$ approximates the data for 1995-2000.

Reviewing Basic Concepts (Sections 3.4 and 3.5)

1. (a) Total length: $2L + 2x = 300 \Longrightarrow L = 150 - x$

(b) $l = 150 - x, \quad w = x$
$A(x) = x(150 - x)$

(c) $0 < x < 150$ since $L = 150 - x$

(d) $5000 = -x^2 + 150x$
$x^2 - 150x + 5000 = 0$
$(x - 100)(x - 50) = 0$
$x = 100, \quad x = 50$
$l = 50, \quad l = 100$
dimensions: 50 meters by 100 meters

2. (a) NSF Research funding:

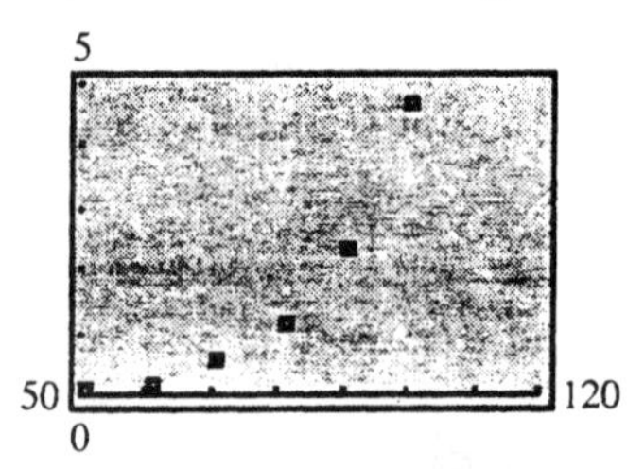

(b) $f(x) = a(x - 51)^2 + .1$
$(101, 4.7): \quad 4.7 = a(101 - 51)^2 + .1$
$4.6 = 50^2 a$
$a = \dfrac{4.6}{2500} = .0018$
$f(x) = .0018(x - 51)^2 + .1$

(c) $g(x) \approx .0026x^2 - .3139x + 9.426$

(d) f, g, and data:

$f(x) = .0018(x - 51)^2 + .1$

5

50 120

0

$g(x) = .0026x^2 - .3139x + 9.426$

The regressions function fits slightly better because it is closer to or through more data points. Neither function would fit the data for $x < 51$.

3. $P(x) = 2x^3 - 9x^2 + 4x + 15 \quad \swarrow \quad \nearrow$

4. maximum number of extrema: 2
maximum number of zeros: 3

5. $P(x) = x^4 + 4x^3 - 20 \quad \nwarrow \quad \nearrow$

6. $P(x) = x^4 + 4x^3 - 20$

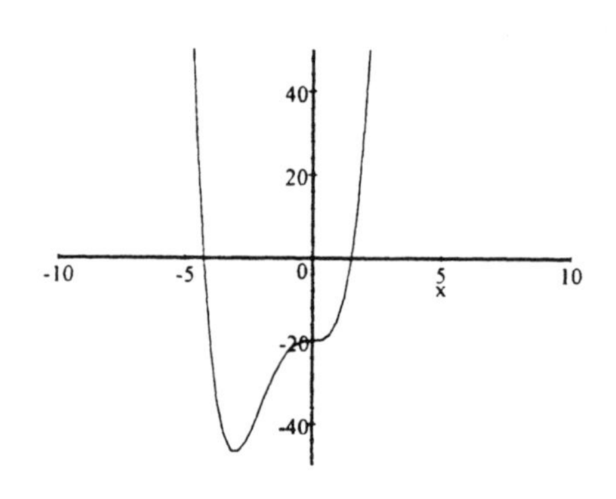

7. $(-3, -47)$ is the absolute minimum.

8. x-intercepts: -4.26, 1.53;
y-intercept: -20

Section 3.6

1. $P(x) = 3x^2 - 2x - 6$

$P(1) = 3(1)^2 - 2(1) - 6 = -5$

$P(2) = 3(4) - 2(2) - 6 = 2$

$\boxed{\text{Zero} \approx 1.79}$

3. $P(x) = 2x^3 - 8x^2 + x + 16$

$P(2) = 2(8) - 8(4) + 2 + 16 = 2$

$P(2.5) = 2(15.625) - 8(6.25) + 2.5 + 16 = -0.25$

$\boxed{\text{Zero} \approx 2.39}$

5. $P(x) = 2x^4 - 4x^2 + 3x - 6$

$P(2) = 2(16) - 4(4) + 3(2) - 6 = 16$

$P(1.5) = 2(5.0625) - 4(2.25) + 3(1.5) - 6 = -.375$

$\boxed{\text{Zero} \approx 1.52}$

7. $P(2) = -4, \quad P(2.5) = 2$

There is at least one zero between 2 and 2.5.

9. $P(x) = x^3 + 2x^2 - 17x - 10$; $\div$ by $x + 5$

−5)	1	2	−17	−10
		−5	15	10
	1	−3	−2	0

$\boxed{Q(x) = x^2 - 3x - 2}$

11. $P(x) = 3x^3 - 11x^2 - 20x + 3$; $\div$ by $x - 5$

5)	3	−11	−20	3
		15	20	0
	3	4	0	3

$\boxed{Q(x) = 3x^2 + 4x + \frac{3}{x - 5}}$

13. $P(x) = x^4 - 3x^3 - 4x^2 + 12x$;
$\div$ by $x - 2$

2)	1	−3	−4	12	0
		2	−2	−12	0
	1	−1	−6	0	0

$\boxed{Q(x) = x^3 - x^2 - 6x}$

15. $P(x) = x^2 - 4x + 5$; $k = 3$

3)	1	−4	5
		3	−3
	1	−1	$\boxed{2}$

$\boxed{P(3) = 2}$

17. $P(x) = 5x^3 + 2x^2 - x + 5$; $k = -2$

−2)	5	2	−1	5
		−10	16	−30
	5	−8	15	$\boxed{-25}$

$\boxed{P(-2) = -25}$

19. $P(x) = x^2 - 5x + 1$; $k = 2$

2)	1	−5	1
		2	−6
	1	−3	$\boxed{-5}$

$\boxed{P(2) = -5}$

21. $P(x) = x^2 + 2x - 8$; $k = 2$

2)	1	2	−8
		2	8
	1	4	$\boxed{0}$

$\boxed{\text{Yes; since } P(2) = 0, \text{ 2 is a zero.}}$

23. $P(x) = 2x^3 - 6x^2 - 9x + 6$; $k = 4$

4)	2	−6	−9	6
		8	8	−4
	2	2	−1	$\boxed{2}$

$\boxed{\text{No; since } P(4) \neq 0, \text{ 4 is not a zero.}}$

25. $P(x) = 4x^3 + 12x^2 + 7x + 1;\ k = -.5$

$$\begin{array}{r|rrrr} -.5) & 4 & 12 & 7 & 1 \\ & & -2 & -5 & -1 \\ \hline & 4 & 10 & 2 & \boxed{0} \end{array}$$

$\boxed{\text{Yes; since } P(-.5) = 0,\ -.5 \text{ is a zero.}}$

Relating Concepts (27 – 32)

27. Linear factors:

$(x - (-3)),\ (x - 1),\ (x - 4)$

$\boxed{(x + 3),\ (x - 1),\ (x - 4)}$

28. $P(x) = 0: \quad x = \{-3,\ 1,\ 4\}$

29. Zeros: $-3,\ 1,\ 4$

30. $P(x) = x^3 - 2x^2 - 11x + 12$

$$\begin{array}{r|rrrr} 2) & 1 & -2 & -11 & 12 \\ & & 2 & 0 & -22 \\ \hline & 1 & 0 & -11 & -10 \end{array}$$

$\boxed{\begin{array}{c} R(2) = -10 \\ P(2) = -10 \end{array}}$

31. $P(x) > 0 : (-3,\ 1) \cup (4,\ \infty)$

32. $P(x) < 0 : (-\infty, -3) \cup (1,\ 4)$

33. $P(x) = x^3 - 2x + 1;\ x = 1$

$$\begin{array}{r|rrrr} 1) & 1 & 0 & -2 & 1 \\ & & 1 & 1 & -1 \\ \hline & 1 & 1 & -1 & 0 \end{array}$$

$x^2 + x - 1$

Solve $x^2 + x - 1 = 0$:

$$\begin{aligned} x &= \frac{-1 \pm \sqrt{1 - 4(-1)}}{2} \\ &= \frac{-1 \pm \sqrt{5}}{2} \\ &= \boxed{\frac{-1 - \sqrt{5}}{2},\ \frac{-1 + \sqrt{5}}{2}} \end{aligned}$$

35. $P(x) = 3x^3 + 5x^2 - 3x - 2;\ x = -2$

$$\begin{array}{r|rrrr} -2) & 3 & 5 & -3 & -2 \\ & & -6 & 2 & 2 \\ \hline & 3 & -1 & -1 & 0 \end{array}$$

$3x^2 - x - 1$

Solve $3x^2 - x - 1 = 0$:

$$\begin{aligned} x &= \frac{1 \pm \sqrt{1 - 4(-3)}}{6} = \frac{1 \pm \sqrt{13}}{6} \\ &= \boxed{\frac{1 - \sqrt{13}}{6},\ \frac{1 + \sqrt{13}}{6}} \end{aligned}$$

37. $P(x) = 2x^3 - 3x^2 - 17x + 30;\ k = 2$

$$\begin{array}{r|rrrr} 2) & 2 & -3 & -17 & 30 \\ & & 4 & 2 & -30 \\ \hline & 2 & 1 & -15 & 0 \end{array}$$

$2x^2 + x - 15$

$(2x - 5)(x + 3)$

$\boxed{P(x) = (x - 2)(2x - 5)(x + 3)}$

39. $P(x) = 6x^3 + 25x^2 + 3x - 4;\ k = -4$

$$\begin{array}{r|rrrr} -4) & 6 & 25 & 3 & -4 \\ & & -24 & -4 & 4 \\ \hline & 6 & 1 & -1 & 0 \end{array}$$

$6x^2 + x - 1$

$(3x - 1)(2x + 1)$

$\boxed{P(x) = (x + 4)(3x - 1)(2x + 1)}$

41. $P(x) = 2x^3 - 4x^2 + 2x + 7$

$+ \; - \; + \; +$

2 sign changes

2 or 0 possible positive real zeros

$P(-x) = -2x^3 - 4x^2 - 2x + 7$

$- \; - \; - \; +$

1 sign change

1 negative real zero

actual positive zeros: 0

43. $P(x) = 5x^4 + 3x^2 + 2x - 9$

$+ \; + \; + \; -$

1 sign change

1 positive real zero

$P(-x) = 5x^4 + 3x^2 - 2x - 9$

$+ \; + \; - \; -$

1 sign change

1 negative real zero

45. $P(x) = x^5 + 3x^4 - x^3 + 2x + 3$

$+ \; + \; - \; + \; +$

2 sign changes

2 or 0 possible positive real zeros

$P(-x) = -x^5 + 3x^4 + x^3 - 2x + 3$

$- \; + \; + \; - \; +$

3 sign changes

3 or 1 possible negative real zeros

actual: positive zeros : 0
negative zeros : 1

Section 3.7

1. Zeros: 4, $2+i$
The conjugate of $2+i = 2-i$ is also a zero.

$$\begin{aligned} P(x) &= (x-4)(x-(2+i))(x-(2-i)) \\ &= (x-4)(x-2-i)(x-2+i) \\ &= (x-4)(x^2-4x+5) \\ &= x^3-4x^2+5x-4x^2+16x-20 \end{aligned}$$

$$P(x) = x^3 - 8x^2 + 21x - 20$$

3. Zeros: 5, i. The conjugate of $i = -i$ is also a zero.

$$\begin{aligned} P(x) &= (x-5)(x-i)(x-(-i)) \\ &= (x-5)(x^2+1) \end{aligned}$$

$$P(x) = x^3 - 5x^2 + x - 5$$

5. Zeros: 0, $3+i$
The conjugate of $3+i = 3-i$ is also a zero.

$$\begin{aligned} P(x) &= (x-0)(x-(3+i))(x-(3-i)) \\ &= (x)(x-3-i)(x-3+i) \\ &= (x)(x^2-6x+10) \end{aligned}$$

$$P(x) = x^3 - 6x^2 + 10x$$

7. Zeros: -3, -1, 4; $P(2) = 5$

$$\begin{aligned} P(x) &= a(x-(-3))(x-(-1))(x-4) \\ &= a(x+3)(x+1)(x-4) \\ &= a(x+3)(x^2-3x-4) \\ &= a(x^3-13x-12) \end{aligned}$$

$$\begin{aligned} P(2): \quad & a(8-26-12) = 5 \\ & -30a = 5 \\ & a = -\frac{5}{30} = -\frac{1}{6} \end{aligned}$$

$$P(x) = -\frac{1}{6}(x^3 - 13x - 12)$$

$$P(x) = -\frac{1}{6}x^3 + \frac{13}{6}x + 2$$

9. Zeros: $-2,\ 1,\ 0;\ P(-1) = -1$

$$\begin{aligned} P(x) &= a(x-(-2))(x-1)(x-0) \\ &= a(x+2)(x-1)(x) \\ &= ax\left(x^2+x-2\right) \end{aligned}$$

$$\begin{aligned} P(-1):\ \ a(-1)(1-1-2) &= -1 \\ -a(-2) &= -1 \\ 2a = -1 \ \rightarrow\ a &= -\frac{1}{2} \end{aligned}$$

$$P(x) = -\frac{1}{2}x\left(x^2+x-2\right)$$

$$\boxed{P(x) = -\frac{1}{2}x^3 - \frac{1}{2}x^2 + x}$$

11. Zeros of $4,\ 1+i;\ P(2) = 4$
Find $P(x) = ax^3 + bx^2 + cx + d$

$$\begin{aligned} Q(x) &= (x-4)(x-(1+i))(x-(1-i)) \\ &= (x-4)(x-1-i)(x-1+i) \\ &= (x-4)\left(x^2-2x+2\right) \\ &= x^3-2x^2+2x-4x^2+8x-8 \\ &= x^3-6x^2+10x-8 \end{aligned}$$

$$\begin{aligned} P(x) &= a\left(x^3-6x^2+10x-8\right) \\ 4 &= a\left(2^3-6(4)+10(2)-8\right) \\ 4 &= a(8-24+20-8) \\ 4 &= a(-4) \rightarrow a = -1 \end{aligned}$$

$$P(x) = -1\left(x^3-6x^2+10x-8\right)$$

$$\boxed{P(x) = -x^3+6x^2-10x+8}$$

13. $P(x) = x^3 - x^2 - 4x - 6$; 3 is a zero

$$\begin{array}{r|rrrr} 3) & 1 & -1 & -4 & -6 \\ & & 3 & 6 & 6 \\ \hline & 1 & 2 & 2 & 0 \end{array}$$

$$x^2+2x+2 = 0$$

$$\begin{aligned} x &= \frac{-2 \pm \sqrt{4-4(2)}}{2} = \frac{-2 \pm \sqrt{-4}}{2} \\ &= \frac{-2 \pm 2i}{2} = -1 \pm i \end{aligned}$$

$$\boxed{\text{Zeros: } -1+i,\ -1-i}$$

15. $P(x) = x^4 + 2x^3 - 10x^2 - 18x + 9$;
zeros: ± 3

$$\begin{array}{r|rrrrr} -3) & 1 & 2 & -10 & -18 & 9 \\ & & -3 & 3 & 21 & -9 \\ \hline & 1 & -1 & -7 & 3 & 0 \end{array}$$

$$x^3 - x^2 - 7x + 3 = 0$$

$$\begin{array}{r|rrrr} 3) & 1 & -1 & -7 & 3 \\ & & 3 & 6 & -3 \\ \hline & 1 & 2 & -1 & 0 \end{array}$$

$$x^2 + 2x - 1 = 0$$

$$\begin{aligned} x &= \frac{-2 \pm \sqrt{4-4(-1)}}{2} \\ &= \frac{-2 \pm 2\sqrt{2}}{2} = -1 \pm \sqrt{2} \end{aligned}$$

$$\boxed{\text{Zeros: } -1+\sqrt{2}\ \ -1-\sqrt{2}}$$

17. $P(x) = x^4 - x^3 + 10x^2 - 9x + 9$;
Zero: $3i$ (also $-3i$)

$$\begin{array}{r|rrrrr} 3i) & 1 & -1 & 10 & -9 & 9 \\ & & 3i & -9-3i & 9+3i & -9 \\ \hline & 1 & -1+3i & 1-3i & 3i & 0 \end{array}$$

$$\begin{array}{r|rrrr} -3i) & 1 & -1+3i & 1-3i & 3i \\ & & -3i & 3i & -3i \\ \hline & 1 & -1 & 1 & 0 \end{array}$$

$$x^2 - x + 1 = 0$$

$$x = \frac{1 \pm \sqrt{1-4(1)}}{2} = \frac{1 \pm \sqrt{-3}}{2}$$

$$\boxed{\text{Zeros: } -3i,\ \ \frac{1}{2}+\frac{\sqrt{3}}{2}i,\ \ \frac{1}{2}-\frac{\sqrt{3}}{2}i}$$

#19 – #29: Answers may vary

19. Zeros: $5,\ -4$

$$\begin{aligned} P(x) &= a(x-5)(x-(-4)) \\ &= a(x-5)(x+4) \\ &= a\left(x^2-x-20\right) \end{aligned}$$

If $a = 1$: $\boxed{P(x) = x^2 - x - 20}$

21. Zeros: $-3,\ 2,\ i$; (also $-i$)

$$\begin{aligned} P(x) &= a(x-(-3))(x-2)(x-i)(x-(-i)) \\ &= a(x+3)(x-2)(x-i)(x+i) \\ &= a\left(x^2+x-6\right)\left(x^2+1\right) \\ &= a\left(x^4+x^3-5x^2+x-6\right) \end{aligned}$$

If $a=1$: $\boxed{P(x)=x^4+x^3-5x^2+x-6}$

23. zeros: $1+\sqrt{3},\ 1-\sqrt{3},\ 1$

$$\begin{aligned} P(x) &= a\left(x-1-\sqrt{3}\right)\left(x-1+\sqrt{3}\right)(x-1) \\ &= a\left(x^2-x+\sqrt{3}x-x+1-\sqrt{3}-\sqrt{3}x+\sqrt{3}-3\right) \\ &\qquad \times (x-1) \\ &= a\left(x^2-2x-2\right)(x-1) \\ &= a\left(x^3-x^2-2x^2+2x-2x+2\right) \\ &= a\left(x^3-3x^2+2\right) \end{aligned}$$

If $a=1$: $\boxed{P(x)=x^3-3x^2+2}$

25. zeros: $3+2i,\ -1,\ 2$

Since $3+2i$ is a zero, $3-2i$ is a zero.

$$\begin{aligned} P(x) &= a(x-3-2i)(x-3+2i)(x+1)(x-2) \\ &= a\left(x^2-3x+2ix-3x+9-6i-2ix+6i-4i^2\right) \\ &\qquad \times \left(x^2-x-2\right) \\ &= a\left(x^2-6x+13\right)\left(x^2-x-2\right) \\ &= a\left(x^4-x^3-2x^2-6x^3+6x^2+12x+13x^2-13x-26\right) \\ &= a\left(x^4-7x^3+17x^2-x-26\right) \end{aligned}$$

If $a=1$: $\boxed{P(x)=x^4-7x^3+17x^2-x-26}$

27. zeros: $6-3i,\ -1$

Since $6-3i$ is a zero, $6+3i$ is a zero.

$$\begin{aligned} P(x) &= a(x-6+3i)(x-6-3i)(x+1) \\ &= a\left(x^2-6x-3ix-6x+36+18i+3ix-18i-9i^2\right) \\ &\qquad \times (x+1) \\ &= a\left(x^2-12x+45\right)(x+1) \\ &= a\left(x^3-12x^2+45x+x^2-12x+45\right) \\ &= a\left(x^3-11x^2+33x+45\right) \end{aligned}$$

If $a=1$: $\boxed{P(x)=x^3-11x^2+33x+45}$

29. zeros: $2+i,\ -3$ (multiplicity 2)

Since $2+i$ is a zero, $2-i$ is a zero.

$$\begin{aligned} P(x) &= a(x-2-i)(x-2+i)(x+3)^2 \\ &= a\left(x^2-2x+ix-2x+4-2i-ix+2i-i^2\right) \\ &\qquad \times \left(x^2+6x+9\right) \\ &= a\left(x^2-4x+5\right)\left(x^2+6x+9\right) \\ &= a\left(x^4+6x^3+9x^2-4x^3-24x^2-36x+5x^2+30x+45\right) \\ &= a\left(x^4+2x^3-10x^2-6x+45\right) \end{aligned}$$

If $a=1$: $\boxed{P(x)=x^4+2x^3-10x^2-6x+45}$

31. $P(x)=x^4+2x^3-7x^2-20x-12$;
$x=-2$, mult. 2

$-2\,)$	1	2	-7	-20	-12
		-2	0	14	12
	1	0	-7	-6	0

$-2\,)$	1	0	-7	-6
		-2	4	6
	1	-2	-3	0

$$x^2-2x-3=0$$
$$(x-3)(x+1)=0$$

Zeros: $-2,\ 3,\ -1$

$\boxed{P(x)=(x+2)^2(x-3)(x+1)}$

33. $P(x)$ of degree 5 can have **5**, **3**, or **1** real zeros (complex numbers occur in conjugate pairs).

35. Zeros of 1, 2, and $1+i$ for a polynomial with real coefficients means that $1-i$ is also a zero. The polynomial, therefore, must be at least degree 4.

37. $P(x) = 2x^3 - 5x^2 - x + 6$
$= (x+1)(2x-3)(x-2)$

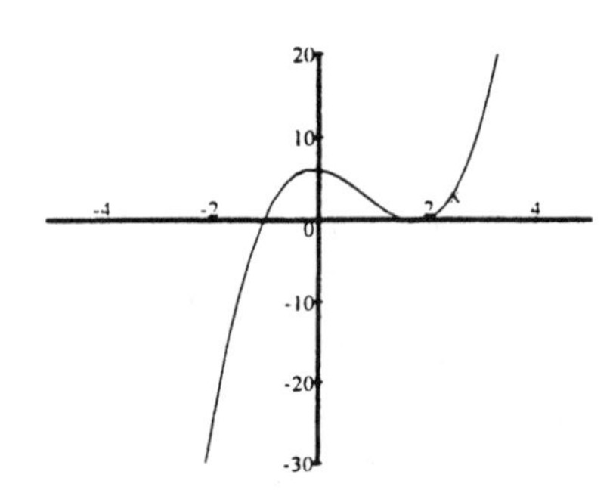

39. $P(x) = x^4 - 18x^2 + 81 = (x-3)^2(x+3)^2$

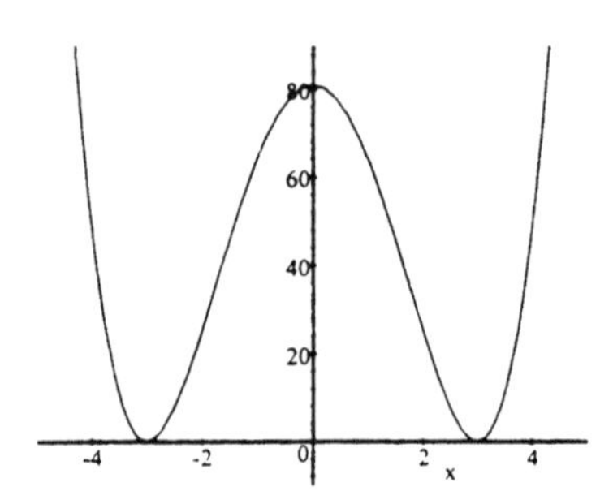

41. $P(x) = 2x^4 + x^3 - 6x^2 - 7x - 2$
$= (2x+1)(x-2)(x+1)^2$

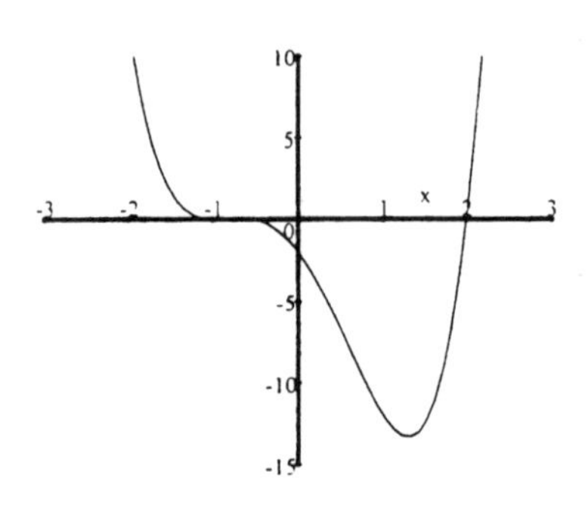

43. $P(x) = x^3 - 2x^2 - 13x - 10$

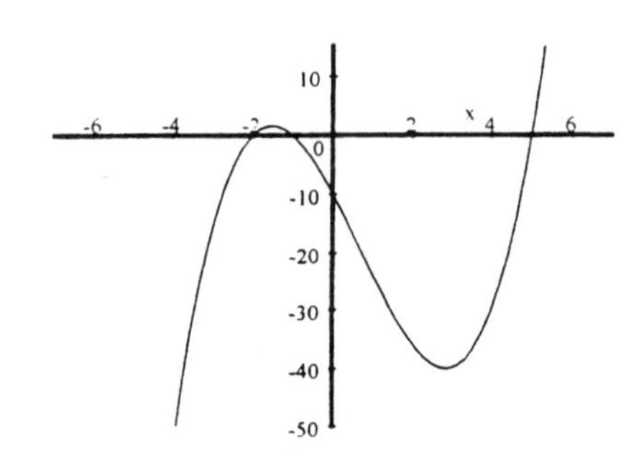

continued

43. continued

(a) $\frac{p}{q}$: $\pm 1, \pm 2, \pm 5, \pm 10$

(b) From the above graph, eliminate values less than -2 or greater than 5.

(c) Rational zeros:

$$\begin{array}{r|rrrr} -2 & 1 & -2 & -13 & -10 \\ & & -2 & 8 & 10 \\ \hline & 1 & -4 & -5 & \boxed{0} \end{array}$$

$$\begin{array}{r|rrr} -1 & 1 & -4 & -5 \\ & & -1 & 5 \\ \hline & 1 & -5 & \boxed{0} \end{array}$$

$x - 5 = 0 \rightarrow x = 5$

$\boxed{x = -2, -1, 5}$

(d) $\boxed{P(x) = (x+2)(x+1)(x-5)}$

45. **(a)** $P(x) = x^3 + 6x^2 - x - 30$

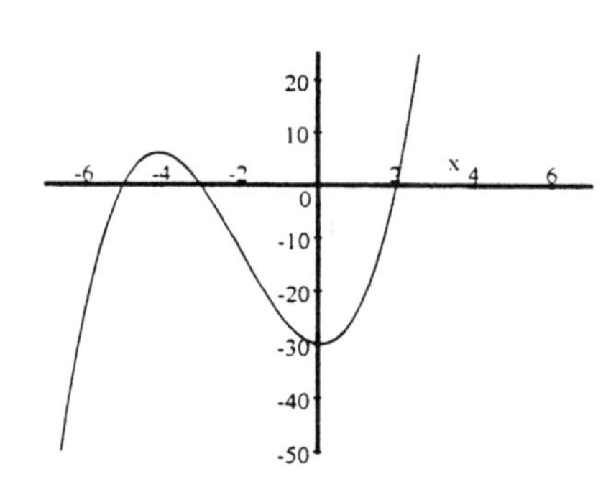

$\frac{p}{q}$: $\pm 1, \pm 2, \pm 3, \pm 5, \pm 6, \pm 10, \pm 15, \pm 30$

(b) From the above graph, $-5 \le x \le 2$.

(c) Rational zeros:

$$\begin{array}{r|rrrr} -5 & 1 & 6 & -1 & -30 \\ & & -5 & -5 & 30 \\ \hline & 1 & 1 & -6 & \boxed{0} \end{array}$$

$$\begin{array}{r|rrr} -3 & 1 & 1 & -6 \\ & & -3 & 6 \\ \hline & 1 & -2 & \boxed{0} \end{array}$$

$x - 2 = 0 \rightarrow x = 2$

$\boxed{x = -5, -3, 2}$

(d) $\boxed{P(x) = (x+5)(x+3)(x-2)}$

47. (a) $P(x) = 6x^3 + 17x^2 - 31x - 12$

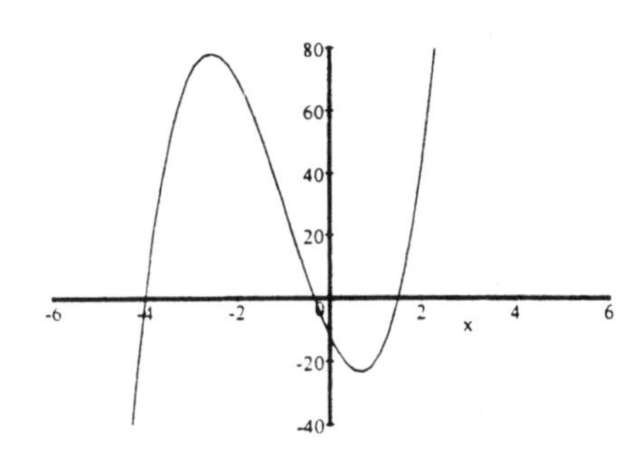

$\frac{p}{q}$: $\pm\frac{1}{2}, \pm\frac{3}{2}, \pm\frac{1}{3}, \pm\frac{2}{3}, \pm\frac{4}{3}, \pm\frac{1}{6},$
$\pm 1, \pm 2, \pm 3, \pm 4, \pm 6, \pm 12$

(b) From the above graph, $-4 \le x \le \frac{3}{2}$.

(c) Rational zeros:

$$\begin{array}{r|rrrr} -4) & 6 & 17 & -31 & -12 \\ & & -24 & 28 & 12 \\ \hline & 6 & -7 & -3 & \boxed{0} \end{array}$$

$$\begin{array}{r|rrr} -\frac{1}{3}) & 6 & -7 & -3 \\ & & -2 & 3 \\ \hline & 6 & -9 & \boxed{0} \end{array}$$

$6x - 9 = 0 \rightarrow x = \frac{3}{2}$

$\boxed{x = -4, -\frac{1}{3}, \frac{3}{2}}$

(d) $\boxed{P(x) = (x+4)(3x+1)(2x-3)}$

49. (a) $P(x) = 12x^3 + 20x^2 - x - 6$

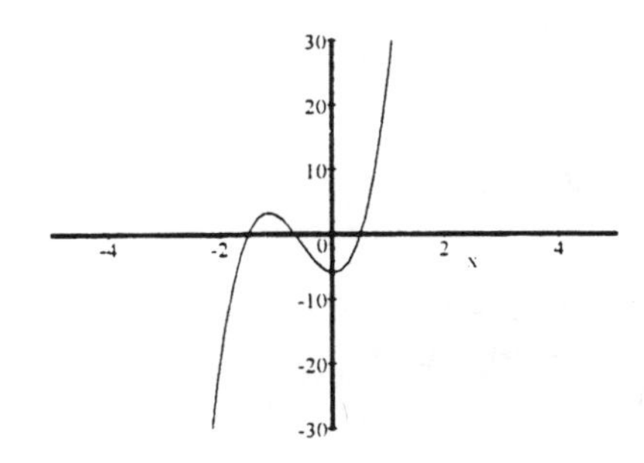

$\frac{p}{q}$: $\pm 1, \pm 2, \pm 3, \pm 6, \pm\frac{1}{2}, \pm\frac{3}{2},$
$\pm\frac{1}{3}, \pm\frac{2}{3}, \pm\frac{1}{6}, \pm\frac{1}{12}, \pm\frac{1}{4}, \pm\frac{3}{4}$

(b) From the above graph, $-\frac{3}{2} \le x \le \frac{1}{2}$.

continued

49. continued

(c) Rational zeros:

$$\begin{array}{r|rrrr} -\frac{3}{2}) & 12 & 20 & -1 & -6 \\ & & -18 & -3 & 6 \\ \hline & 12 & 2 & -4 & \boxed{0} \end{array}$$

$$\begin{array}{r|rrr} -\frac{2}{3}) & 12 & 2 & -4 \\ & & -8 & 4 \\ \hline & 12 & -6 & \boxed{0} \end{array}$$

$12x - 6 = 0 \rightarrow x = \frac{1}{2}$

$\boxed{x = -\frac{3}{2}, -\frac{2}{3}, \frac{1}{2}}$

(d) $\boxed{P(x) = (3x+2)(2x+3)(2x-1)}$

51. $P(x) = x^3 + \frac{1}{2}x^2 - \frac{11}{2}x - 5$
$Q(x) = 2x^3 + x^2 - 11x - 10$
$\frac{p}{q}$: $\pm 10, \pm 5, \pm 1, \pm\frac{1}{2}, \pm\frac{5}{2}$
From graphing, check -1, -2:

$$\begin{array}{r|rrrr} -1) & 2 & 1 & -11 & -10 \\ & & -2 & 1 & 10 \\ \hline & 2 & -1 & -10 & 0 \end{array}$$

$2x^2 - x - 10 = 0$
$(2x-5)(x+2) = 0$

$\boxed{\text{Zeros: } -2, -1, \frac{5}{2}}$

53. $P(x) = \frac{1}{6}x^4 - \frac{11}{12}x^3 + \frac{7}{6}x^2 - \frac{11}{12}x + 1$
$Q(x) = 2x^4 - 11x^3 + 14x^2 - 11x + 12$
$\frac{p}{q}$: $\pm 1, \pm 2, \pm 3, \pm 4, \pm 6, \pm 12, \pm\frac{1}{2}, \pm\frac{3}{2}$
From graphing, check $\frac{3}{2}$, 4:

$$\begin{array}{r|rrrrr} \frac{3}{2}) & 2 & -11 & 14 & -11 & 12 \\ & & 3 & -12 & 3 & -12 \\ \hline & 2 & -8 & 2 & -8 & 0 \end{array}$$

$$\begin{array}{r|rrrr} 4) & 2 & -8 & 2 & -8 \\ & & 8 & 0 & 8 \\ \hline & 2 & 0 & 2 & 0 \end{array}$$

$2x^2 + 2 = 0 \rightarrow x^2 = -1$; not real

$\boxed{\text{Zeros: } \frac{3}{2}, 4}$

55. (a)
$$\begin{aligned}\bar{c}+\bar{d} &= \overline{a+bi}+\overline{m+ni}\\ &= a-bi+m-ni\\ &= (a+m)-(b+n)i\\ &= \overline{(a+m)+(b+n)i}\\ &= \overline{c+d}\end{aligned}$$

(b)
$$\begin{aligned}\bar{c}\cdot\bar{d} &= \left(\overline{a+bi}\right)\left(\overline{m+ni}\right)\\ &= (a-bi)(m-ni)\\ &= am-ani-bmi+bni^2\\ &= (am-bn)-(an+bm)i\\ &= \overline{(am-bn)+(an+bm)i}\\ &= \overline{cd}\end{aligned}$$

(c)
$$\begin{aligned}x &= a+0i\\ &= a-0i=\overline{x}\end{aligned}$$

(d) $\overline{(c^n)}=(\bar{c})^n$

(i) Let $n=3$:

$$\begin{aligned}\overline{(a+bi)^3} &= \overline{a^3+3a^2bi+3a(bi)^2+(bi)^3}\\ &= \overline{a^3+3a^2bi-3ab-b^3i}\\ &= \overline{(a^3-3ab)+(3a^2b-b^3)i}\\ &= \left(a^3-3ab\right)-\left(3a^2b-b^3\right)i\end{aligned}$$

$$\begin{aligned}(a-bi)^3 &= a^3+3a^2(-bi)+3a(-bi)^2-bi^3\\ &= a^3-3a^2bi-3ab+b^3i\\ &= \left(a^3-3ab\right)-\left(3a^2b-b^3\right)i\end{aligned}$$

(ii) Let $n=4$:

$$\begin{aligned}&\overline{(a+bi)^4}\\ &= \overline{a^4+4a^3bi+6a^2b^2i^2+4ab^3i^3+b^4i^4}\\ &= \overline{a^4+4a^3bi-6a^2b^2-4ab^3i+b^4}\\ &= \overline{(a^4-6a^2b^2+b^4)+(4a^3b-4ab^3)i}\\ &= \left(a^4-6a^2b^2+b^4\right)-\left(4a^3b-4ab^3\right)i\end{aligned}$$

$$\begin{aligned}&(a-bi)^4\\ &= a^4+4a^3(-bi)+6a^2(-bi)^2\\ &\qquad +4a(-bi)^3+(-bi)^4\\ &= a^4-4a^3bi-6a^2b^2+4ab^3i+b^4\\ &= \left(a^4-6a^2b^2+b^4\right)-\left(4a^3b-4ab^3\right)i\end{aligned}$$

This pattern will continue, and can be proved using mathematical induction.

57. The conjugate zeros theorem states that all coefficients are real numbers. The coefficient $i+1$ is not real.

Section 3.8

1. $P(x)=7x^3+x$

$$\begin{aligned}0 &= x\left(7x^2+1\right)\\ x=0 \quad 7x^2+1 &= 0\\ 7x^2 &= -1\\ x^2 &= -\tfrac{1}{7}\\ x &= \pm\frac{\sqrt{-1}}{\sqrt{7}}\\ x &= \pm\frac{i}{\sqrt{7}}\cdot\frac{\sqrt{7}}{\sqrt{7}}=\pm\frac{i\sqrt{7}}{7}\end{aligned}$$

$$\left\{0,\ \pm\frac{i\sqrt{7}}{7}\right\}$$

3. $P(x)=3x^3+2x^2-3x-2$

$$\begin{aligned}0 &= x^2(3x+2)-1(3x+2)\\ 0 &= (3x+2)\left(x^2-1\right)\\ 0 &= (3x+2)(x-1)(x+1)\end{aligned}$$

$3x+2=0 \quad x-1=0 \quad x+1=0$

$x=-\frac{2}{3} \quad x=1 \quad x=-1$

$$\left\{-\frac{2}{3},\ 1,\ -1\right\}$$

5. $P(x)=x^4-11x^2+10$

$$0=\left(x^2-10\right)\left(x^2-1\right)$$

$x^2-10=0 \quad x^2-1=0$

$x^2=10 \quad x^2=1$

$x=\pm\sqrt{10} \quad x=\pm\sqrt{1}=\pm1$

$\{\pm1,\ \pm\sqrt{10}\}$

7. $4x^4-25x^2+36=0$

Let $t=x^2$:

$$\begin{aligned}4t^2-25t+36 &= 0\\ (4t-9)(t-4) &= 0\end{aligned}$$

$t=\frac{9}{4} \quad t=4$

$x^2=\frac{9}{4} \quad x^2=4$

$x=\pm\frac{3}{2} \quad x=\pm2$

$\boxed{\{-2,\ -1.5,\ 1.5,\ 2\}}$

Show that -2, -1.5, 1.5 and 2 are the x-intercepts of $y=4x^4-25x^2+36$.

9. $x^4 - 15x^2 - 16 = 0$
Let $t = x^2$:

$t^2 - 15t - 16 = 0$
$(t - 16)(t + 1) = 0$

$t = 16$	$t = -1$
$x^2 = 16$	$x^2 = -1$
$x = \pm 4$	$x = \pm i$

$\boxed{\{-4,\ 4,\ -i,\ i\}}$

Show that -4, and 4 are the only x-intercepts of $y = x^4 - 15x^2 - 16$.

11. $x^3 - x^2 - 64x + 64 = 0$
$x^2(x - 1) - 64(x - 1) = 0$
$(x - 1)(x^2 - 64) = 0$
$(x - 1)(x - 8)(x + 8) = 0$

$\boxed{\{-8,\ 1,\ 8\}}$

Show that -8, 1 and 8 are the x-intercepts of $y = x^3 - x^2 - 64x + 64$.

13. $-2x^3 - x^2 + 3x = 0$
$-x(2x^2 + x - 3) = 0$
$-x(2x + 3)(x - 1) = 0$

$\boxed{\{-1.5,\ 0,\ 1\}}$

Show that -1.5, 0, and 1 are the x-intercepts of $y = -2x^3 - x^2 + 3x$.

15. $x^3 + x^2 - 7x - 7 = 0$
$x^2(x + 1) - 7(x + 1) = 0$
$(x + 1)(x^2 - 7) = 0$
$(x + 1)(x - \sqrt{7})(x + \sqrt{7}) = 0$

$\boxed{\{-1,\ \pm\sqrt{7}\ (\approx \pm 2.65)\}}$

Show that -2.65, -1, and 2.65 are the x-intercepts of $y = x^3 + x^2 - 7x - 7$.

17. $3x^3 + x^2 - 6x = 0$
$x\underbrace{(3x^2 + x - 6)} = 0$

$$x = \frac{-1 \pm \sqrt{1 - 4(-18)}}{6} = \frac{-1 \pm \sqrt{73}}{6}$$

$$\boxed{\begin{array}{c}\left\{0,\ \dfrac{-1 + \sqrt{73}}{6},\ \dfrac{-1 - \sqrt{73}}{6}\right\} \\ \{0,\ \approx 1.26,\ \approx -1.59\}\end{array}}$$

Show that -1.59, 0, and 1.26 are the x-intercepts of $y = 3x^3 + x^2 - 6x$.

19. $3x^3 + 3x^2 + 3x = 0$
$3x\underbrace{(x^2 + x + 1)} = 0$

$$x = \frac{-1 \pm \sqrt{1 - 4(1)}}{2} = \frac{-1 \pm i\sqrt{3}}{2}$$

$$\boxed{0,\ -\frac{1}{2} - \frac{\sqrt{3}}{2}i,\ -\frac{1}{2} + \frac{\sqrt{3}}{2}i}$$

Show that 0 is the only x-intercept of $y = 3x^3 + 3x^2 + 3x$.

21. $x^4 + 17x^2 + 16 = 0$
Let $t = x^2$:

$t^2 + 17t + 16 = 0$
$(t + 16)(t + 1) = 0$

$t = -16$	$t = -1$
$x^2 = -16$	$x^2 = -1$
$x = \pm 4i$	$x = \pm i$

$\boxed{\{-4i,\ -i,\ i,\ 4i\}}$

Show that there are no x-intercepts of $y = x^4 + 17x^2 + 16$.

23. $x^6 + 19x^3 - 216 = 0$

Let $t = x^3$: $t^2 + 19t - 216 = 0$

$(t-8)(t+27) = 0$

(i) $t = 8 \Longrightarrow x^3 = 8$

$x^3 - 8 = 0$

$(x-2)(x^2+2x+4) = 0$

$$x = 2 \quad x = \frac{-2 \pm \sqrt{4-16}}{2} = \frac{-2 \pm 2i\sqrt{3}}{2} = -1 \pm i\sqrt{3}$$

(ii) $t = -27 \Longrightarrow x^3 = -27$

$x^3 + 27 = 0$

$(x+3)(x^2-3x+9) = 0$

$$x = -3 \quad x = \frac{3 \pm \sqrt{9-36}}{2} = \frac{3 \pm 3i\sqrt{3}}{2} = \frac{3}{2} \pm \frac{3\sqrt{3}}{2}i$$

$-3,\ 2,\ -1 - i\sqrt{3},\ -1 + i\sqrt{3},\ \frac{3}{2} + \frac{3\sqrt{3}}{2}i,\ \frac{3}{2} - \frac{3\sqrt{3}}{2}i$

Show that -3 and 2 are the only x-intercepts of $y = x^6 + 19x^3 - 216$.

Relating Concepts (25 – 28):

25. $y = x^4 - 28x^2 + 75$. This is an even function $[f(x) = f(-x)]$, so it is symmetric with respect to the y-axis. The graph is incomplete.

26. Symmetry with respect to the y-axis.

27. $x = \{-5, -\sqrt{3}, \sqrt{3}, 5\}$

28. There are no imaginary solutions; y is of degree 4 and there are at most 4 solutions, all of which are real.

29. $P(x) = x^3 - 3x^2 - 6x + 8$

$= (x-4)(x-1)(x+2)$

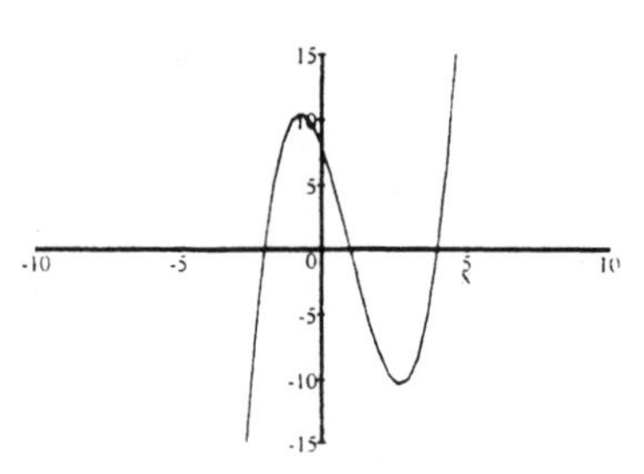

(a) $P(x) = 0: \{-2, 1, 4\}$

(b) $P(x) < 0: (-\infty, -2) \cup (1, 4)$

(c) $P(x) > 0: (-2, 1) \cup (4, \infty)$

31. $P(x) = 2x^4 - 9x^3 - 5x^2 + 57x - 45$

$= (x-3)^2(2x+5)(x-1)$

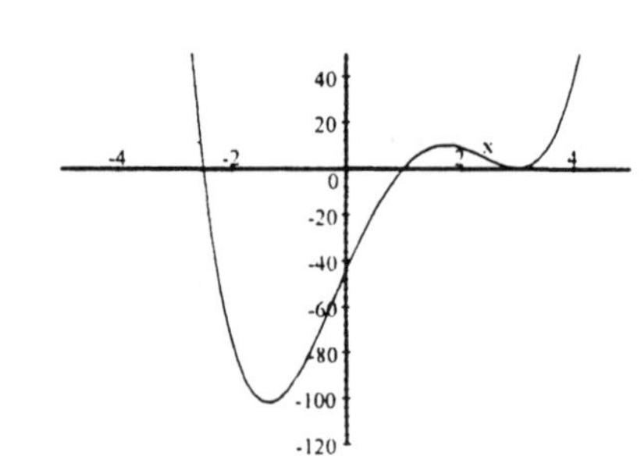

(a) $P(x) = 0: \{-2.5, 1, 3 \text{ (multiplicity 2)}\}$

(b) $P(x) < 0: (-2.5, 1)$

(c) $P(x) > 0: (-\infty, -2.5) \cup (1, 3) \cup (3, \infty)$

33. $P(x) = -x^4 - 4x^3 + 3x^2 + 18x$

$= x(2-x)(x+3)^2$

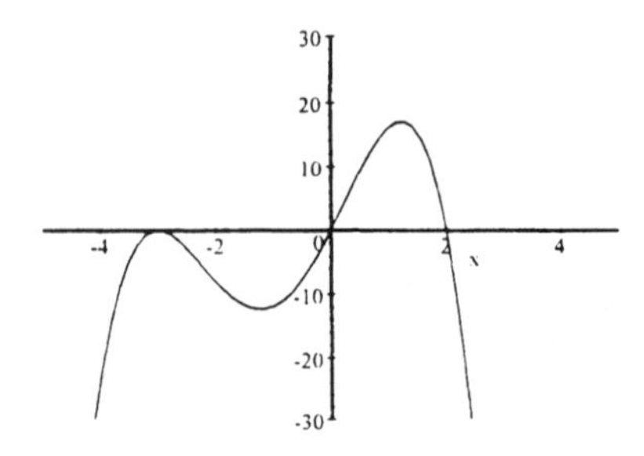

(a) $P(x) = 0: \{-3 \text{ (multiplicity 2)}, 0, 2\}$

(b) $P(x) \geq 0: \{-3\} \cup [0, 2]$

(c) $P(x) \leq 0: (-\infty, 0] \cup [2, \infty)$

35. $.86x^3 - 5.24x^2 + 3.55x + 7.84 = 0$

$y_1 = .86x^3 - 5.24x^2 + 3.55x + 7.84.$

x-int: $-.88,\ 2.12,\ 4.86$

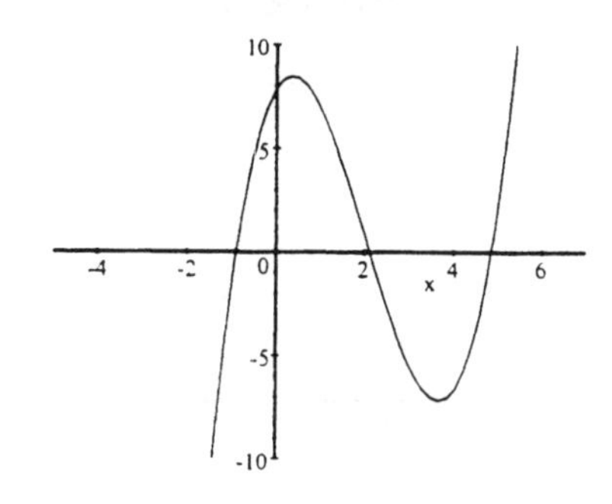

37. $-\sqrt{7}\,x^3 + \sqrt{5}\,x^2 + \sqrt{17} = 0$

$y_1 = -\sqrt{7}\,x^3 + \sqrt{5}\,x^2 + \sqrt{17}$

x-intercept: 1.52

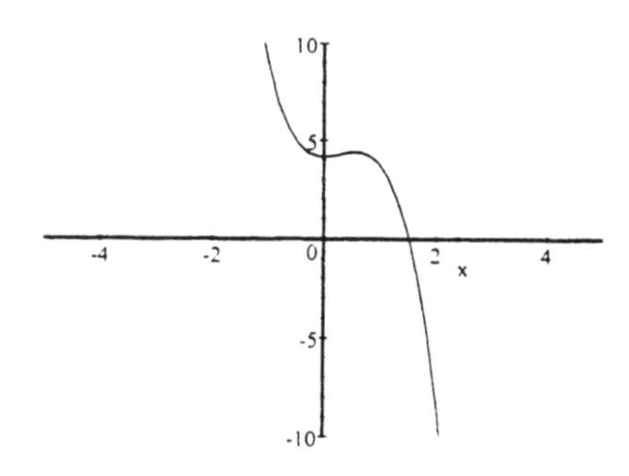

39. $2.45x^4 - 3.22x^3 = -.47x^2 + 6.54x + 3$

$y_1 = 2.45x^4 - 3.22x^3 + .47x^2 - 6.54x - 3$

x-intercepts: $-.40,\ 2.02$

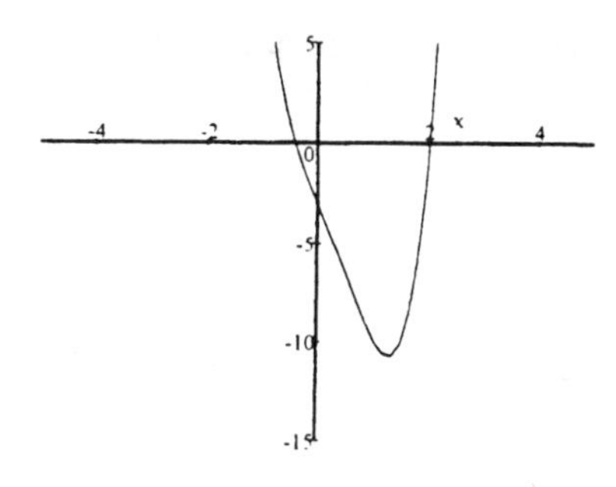

41. $x^2 = -1$

$x = \pm\sqrt{-1}$ $\{-i,\ i\}$

43. $x^3 = -1$

$x^3 + 1 = 0$

$(x+1)\left(x^2 - x + 1\right) = 0$

(i) $x + 1 = 0 \rightarrow x = -1$

(ii) $x^2 - x + 1 = 0$

$$x = \frac{1 \pm \sqrt{1-4}}{2} = \frac{1 \pm i\sqrt{3}}{2}$$

$$x = \left\{-1,\ \frac{1}{2} + \frac{\sqrt{3}}{2}i,\ \frac{1}{2} - \frac{\sqrt{3}}{2}i\right\}$$

45. $x^3 = 27$

$x^3 - 27 = 0$

$(x-3)\left(x^2 + 3x + 9\right) = 0$

(i) $x - 3 = 0 \rightarrow x = 3$

(ii) $x^2 + 3x + 9 = 0$

$$x = \frac{-3 \pm \sqrt{9 - 4(9)}}{2}$$

$$= \frac{-3 \pm \sqrt{-27}}{2} = \frac{-3 \pm 3i\sqrt{3}}{2}$$

$$x = \left\{3,\ -\frac{3}{2} - \frac{3\sqrt{3}}{2}i,\ -\frac{3}{2} + \frac{3\sqrt{3}}{2}i\right\}$$

47. $x^4 = 16$

$x^4 - 16 = 0$

$\left(x^2 - 4\right)\left(x^2 + 4\right) = 0$

$(x-2)(x+2)\left(x^2 + 4\right) = 0$

(i) $x - 2 = 0 \rightarrow x = 2$

(ii) $x + 2 = 0 \rightarrow x = -2$

(iii) $x^2 + 4 = 0$

$x^2 = -4$

$x = \pm\sqrt{-4} = \pm 2i$

$x = \{-2,\ 2,\ -2i,\ 2i\}$

49. $f(x) = \dfrac{\pi}{3}x^3 - 5\pi x^2 + \dfrac{500\,\pi\,d}{3}$

(a) $d = .8:\ f(x) = \dfrac{\pi}{3}x^3 - 5\pi x^2 + \dfrac{500\,\pi\,(.8)}{3}$

$f(x) = \dfrac{\pi}{3}x^3 - 5\pi x^2 + \dfrac{400\,\pi}{3}$

The smallest positive zero is at $x \approx 7.1286$;

≈ 7.13 cm; the ball floats partly above the surface.

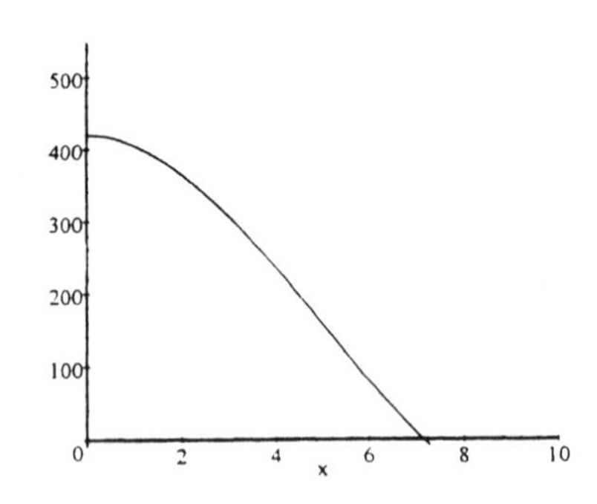

(b) $d = 2.7:\ f(x) = \dfrac{\pi}{3}x^3 - 5\pi x^2 + \dfrac{500\,\pi\,(2.7)}{3}$

$f(x) = \dfrac{\pi}{3}x^3 - 5\pi x^2 + \dfrac{1350\,\pi}{3}$

The graph has no zeros; the sphere sinks below the surface because it is more dense than water.

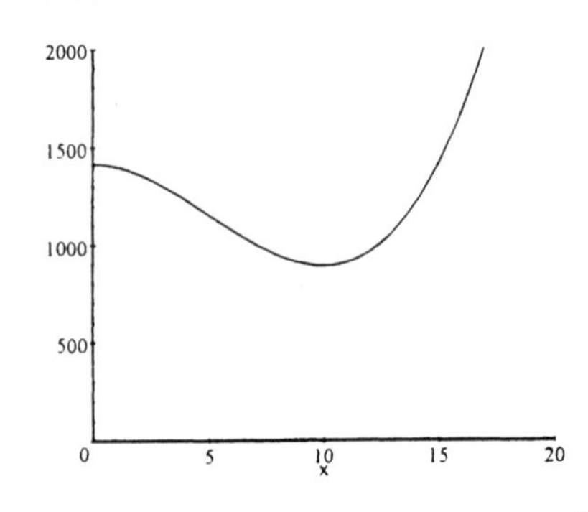

(c) $d = 1:\ f(x) = \dfrac{\pi}{3}x^3 - 5\pi x^2 + \dfrac{500\,\pi\,(1)}{3}$

$f(x) = \dfrac{\pi}{3}x^3 - 5\pi x^2 + \dfrac{500\,\pi}{3}$

The zero is 10 cm; the ball floats even with the surface.

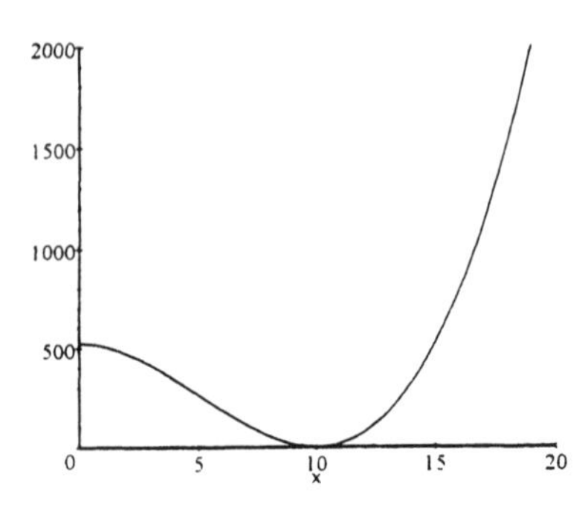

51. $f(x) = \dfrac{\pi}{3}x^3 - 10\pi x^2 + \dfrac{4000\,\pi\,d}{3}$

$d = .6:\ f(x) = \dfrac{\pi}{3}x^3 - 10\pi x^2 + \dfrac{4000\,\pi\,(.6)}{3}$

$f(x) = \dfrac{\pi}{3}x^3 - 10\pi x^2 + 800\pi$

The smallest positive zero is at $x \approx 11.341378 \approx 11.34$ cm.

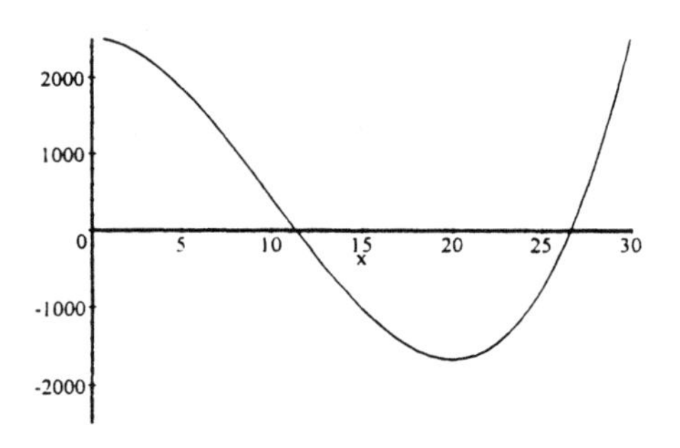

53. (a) $12 - 2x > 0$ and $18 - 2x > 0$

$-2x > -12 \qquad -2x > -18$

$x < 6 \qquad x < 9$

$\boxed{0 < x < 6}$

(b) $V = lwh$

$V(x) = (18 - 2x)(12 - 2x)(x)$

$V(x) = x\left(4x^2 - 60x + 216\right)$

$\boxed{V(x) = 4x^3 - 60x^2 + 216x}$

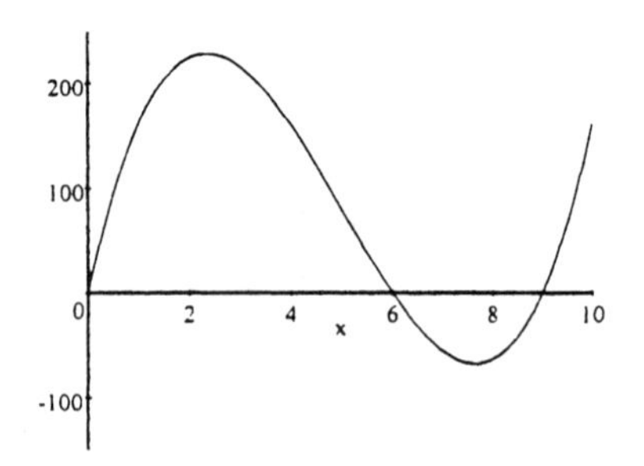

(c) $x \approx 2.35$, maximum volume $\approx 228.16\,\text{in}^3$

(d) $.42 < x < 5$

55. $x^3 - 12x^2 + 64 = 0$

If $y = 0,\ x = 2.61$ $\boxed{\approx 2.61\text{ inches}}$

57. **(a)** Leg length $= x - 1$

(b) $a^2 + b^2 = c^2$

$$(x-1)^2 + b^2 = x^2$$
$$b^2 = x^2 - (x-1)^2$$
$$b^2 = x^2 - x^2 + 2x - 1$$
$$b^2 = 2x - 1$$

$$\boxed{b = \sqrt{2x-1}}$$

$$\left[\text{or } b = \sqrt{x^2 - (x-1)^2}\right]$$

(c) $A = \frac{1}{2}bh = 84$

$$\frac{1}{2}\left(\sqrt{2x-1}\right)(x-1) = 84$$
$$\left(\sqrt{2x-1}\right)(x-1) = 168$$
$$\left[\left(\sqrt{2x-1}\right)(x-1)\right]^2 = 168^2$$
$$(2x-1)\left(x^2-2x+1\right) = 28224$$

$$\boxed{2x^3 - 5x^2 + 4x - 28225 = 0}$$

(d) x-intercept: 25

hypotenuse: $x = 25$ in

1 side: $x - 1 = 24$ in

1 side: $\sqrt{49} = 7$ in

$\boxed{7 \text{ in} \times 24 \text{ in} \times 25 \text{ in}}$

59. Let x represent the side of each square;

$l = 8.5 - 2x,\ w = 11 - 2x,\ h = x.$

(a)
$$\begin{aligned} V &= lwh \\ &= (8.5-2x)(11-2x)(x) \\ &= \left(93.5 - 17x - 22x + 4x^2\right)(x) \\ &= 4x^3 - 39x^2 + 93.5x \end{aligned}$$

All values must be positive: $x > 0$,

$8.5 - 2x > 0$ $\quad$ $11 - 2x > 0$

$x < 4.25$ $\quad$ $x < 5.5$

so $0 < x < 4.25$

Using the graphing calculator:

Maximum $\approx (1.59, 66.15)$

The maximum volume is $\approx 66.15 \text{ in}^3$ when $x \approx 1.59$ inches.

X	Y1
1.55	66.123
1.56	66.135
1.57	66.143
1.58	66.148
1.59	66.148
1.6	66.144
1.61	66.136

Y1=4X^3-39X²+93...

(b) $V = 4x^3 - 39x^2 + 93.5x$

Find x when $V > 40$:

X	Y1
.5	37.5
.51	38.072
.52	38.637
.53	39.195
.54	39.747
.55	40.293
.56	40.832

Y1=4X^3-39X²+93...

X	Y1
2.89	41.033
2.9	40.716
2.91	40.398
2.92	40.079
2.93	39.759
2.94	39.438
2.95	39.117

Y1=4X^3-39X²+93...

Since $x < 5.5$, $\boxed{.54 \text{ in} < x < 2.92 \text{ in}}$

61. **(a)** Water pollution

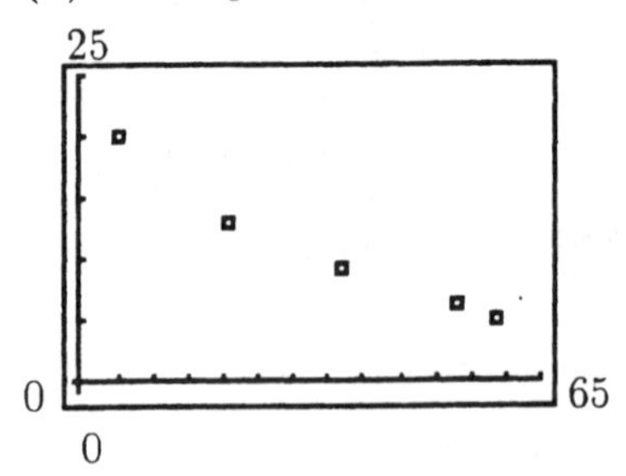

(b) $y = .0035x^2 - .49x + 22$

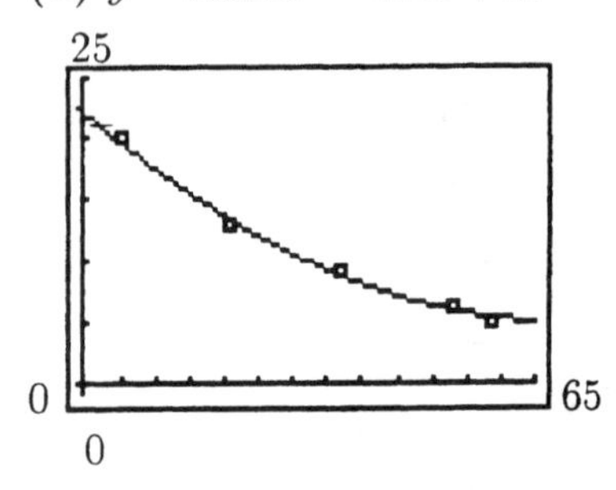

(c) $y = -.000068x^3 + .00987x^2 - .653x + 23$

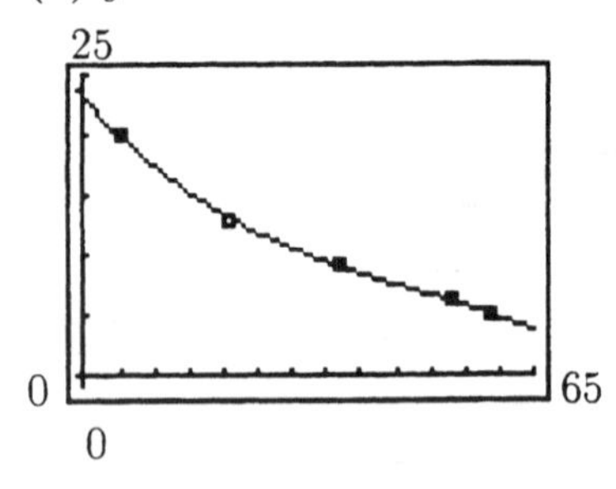

(d) The cubic function is a slightly better fit.

(e) Graph $y_1 = \mathbf{c}$ and $y_2 = 10$. The intersection point is $\approx (31.92,\ 10) : 0 \le x < 31.92$.
Answers may vary depending upon rounding.

63. $g(x) = -.006x^4 + .140x^3 - .053x^2 + 1.79x$

Graph $y_1 = g(x)$; $y_2 = 10$

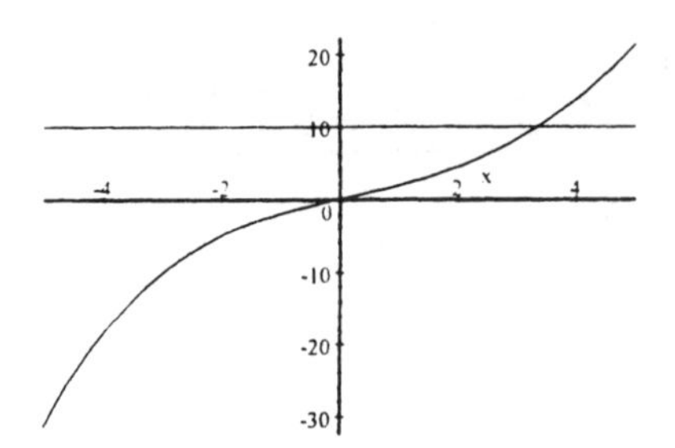

The graphs intersect at $\approx (3.367,\ 10)$.
At approximately 3.4 seconds.

Reviewing Basic Concepts (Sections 3.6 - 3.8)

1. $P(x) = 2x^4 - 7x^3 + 29x - 30$

$P(3) = 2(3)^4 - 7(3)^3 + 29(3) - 30$

$P(3) = \boxed{30}$

2. $P(x) = 2x^4 - 7x^3 + 29x - 30$

$P(2) = 2(2)^4 - 7(2)^3 + 29(2) - 30$

$P(2) = 4$, so $P(2)$ is not a zero.

3. Since $2 + i$ is a zero, $2 - i$ is a zero:

$2+i)$	2	-7	0	29	-30
		$4+2i$	$-8+i$	$-17-6i$	30
	2	$-3+2i$	$-8+i$	$12-6i$	0

$2-i)$	2	$-3+2i$	$-8+i$	$12-6i$
		$4-2i$	$2-i$	$-12+6i$
	2	1	-6	0

$2x^2 + x - 6 = (2x-3)(x+2)$

$\boxed{P(x) = (2x-3)(x+2)(x-2-i)(x-2+i)}$

4. zeros: $\frac{3}{2}$, i (also $-i$); $P(3) = 15$

$P(x) = a\left(x - \frac{3}{2}\right)(x-i)(x+i)$

$P(x) = a\left(x - \frac{3}{2}\right)(x^2+1)$

$P(x) = a\left(x^3 - \frac{3}{2}x^2 + x - \frac{3}{2}\right)$

$(3,\ 15): \quad 15 = a\left(3^3 - \frac{3}{2}(3)^2 + 3 - \frac{3}{2}\right)$

$15 = a\left(27 - \frac{27}{2} + 3 - \frac{3}{2}\right)$

$15 = a\left(30 - \frac{30}{2}\right)$

$15 = 15a \Longrightarrow a = 1$

$\boxed{P(x) = x^3 - \frac{3}{2}x^2 + x - \frac{3}{2}}$

5. zeros: -4 (multiplicity 2), $1+2i$, $1-2i$:

$$P(x) = a(x+4)^2(x-1-2i)(x-1+2i)$$
$$P(x) = a(x^2+8x+16)(x^2-2x+5)$$
$$P(x) = a(x^4+6x^3+5x^2+8x+80)$$

If $a=1$: $\boxed{P(x) = x^4+6x^3+5x^2+8x+80}$

6. $P(x) = 2x^3+x^2-11x-10$

possible zeros: $\pm1, \pm2, \pm5, \pm10, \pm\frac{1}{2}, \pm\frac{5}{2}$

from the graph: zeros at -2, -1, 2.5

$$\begin{array}{r|rrrr} -2) & 2 & 1 & -11 & -10 \\ & & -4 & 6 & 10 \\ \hline & 2 & -3 & -5 & \boxed{0} \end{array}$$

$$\begin{array}{r|rrr} -1) & 2 & -3 & -5 \\ & & -2 & 5 \\ \hline & 2 & -5 & \boxed{0} \end{array}$$

$$\begin{array}{r|rr} \frac{5}{2}) & 2 & -5 \\ & & 5 \\ \hline & 2 & \boxed{0} \end{array} \qquad \left\{-2,\ -1,\ \frac{5}{2}\right\}$$

7. $3x^4-12x^2+1=0$

$$x^2 = \frac{12\pm\sqrt{144-12}}{6} = \frac{12\pm\sqrt{132}}{6}$$
$$= \frac{12\pm2\sqrt{33}}{6} = \frac{6\pm\sqrt{33}}{3}$$
$$x = \left\{\pm\sqrt{\frac{6\pm\sqrt{33}}{3}}\right\}$$

8. $\{0,5,8,10\} \longrightarrow L_1$; $\{9,40,100,126\} \longrightarrow L_2$

STAT / CALC / CubicReg

$y = -.3125x^3+5.7875x^2-14.925x+9$

$P(x) \approx -.31x^3+5.8x^2-15x+9$

$P(9) \approx -.31(9)^3+5.8(9)^2-15(9)+9 \approx 117.81$

In 1999, about 118 debit cards were issued.

Answers may vary due to rounding.

Chapter 3 Review

1. $w=17-i$, $z=1-3i$

$$w+z = (17-i)+(1-3i)$$
$$\boxed{w+z = 18-4i}$$

2. $w=17-i$, $z=1-3i$

$$w-z = (17-i)-(1-3i)$$
$$w-z = (17-1)+(-i-(-3i))$$
$$\boxed{w-z = 16+2i}$$

3. $w=17-i$, $z=1-3i$

$$wz = (17-i)(1-3i)$$
$$wz = 17-51i-i+3i^2$$
$$wz = 17-52i-3$$
$$\boxed{wz = 14-52i}$$

4. $w=17-i$

$$w^2 = (17-i)^2$$
$$w^2 = 289-34i+i^2$$
$$w^2 = 289-34i-1$$
$$\boxed{w^2 = 288-34i}$$

5. $z=1-3i$

$$\frac{1}{z} = \frac{1}{1-3i}$$
$$\frac{1}{z} = \frac{1}{1-3i}\cdot\frac{1+3i}{1+3i}$$
$$\frac{1}{z} = \frac{1+3i}{1-9i^2} = \frac{1+3i}{10}$$
$$\boxed{\frac{1}{z} = \frac{1}{10}+\frac{3}{10}i}$$

6. $w=17-i$, $z=1-3i$

$$\frac{w}{z} = \frac{17-i}{1-3i}$$
$$\frac{w}{z} = \frac{17-i}{1-3i}\cdot\frac{1+3i}{1+3i}$$
$$\frac{w}{z} = \frac{17+51i-i-3i^2}{1-9i^2}$$
$$\frac{w}{z} = \frac{17+50i+3}{1+9} = \frac{20+50i}{10}$$
$$\boxed{\frac{w}{z} = 2+5i}$$

7. $P(x) = 2x^2 - 6x - 8$; Domain: $(-\infty, \infty)$

8. $P(x) = 2x^2 - 6x - 8$

$$x = \frac{-b}{2a} = \frac{-(-6)}{4} = \frac{3}{2}$$

$$P\left(\frac{3}{2}\right) = 2\left(\frac{9}{4}\right) - 6\left(\frac{3}{2}\right) - 8$$

$$= \frac{9}{2} - \frac{18}{2} - \frac{16}{2} = -\frac{25}{2}$$

$\boxed{\text{Vertex: } \left(\frac{3}{2}, -\frac{25}{2}\right)}$

9. $P(x) = 2x^2 - 6x - 8$: $\nwarrow \nearrow$

10. $P(x) = 2x^2 - 6x - 8$

$2x^2 - 6x - 8 = 0$
$x^2 - 3x - 4 = 0$
$(x - 4)(x + 1) = 0$
$x = 4, -1$

$\boxed{x\text{-intercepts: } -1, 4}$

11. $P(x) = 2x^2 - 6x - 8$

$P(0) = 2(0)^2 - 6(0) - 8 = -8$

$\boxed{y\text{-intercept: } -8}$

12. $P(x) = 2x^2 - 6x - 8$

Range: $\left[-\frac{25}{2}, \infty\right)$

13. $P(x) = 2x^2 - 6x - 8$

Increasing: $\left[\frac{3}{2}, \infty\right)$; Decreasing: $\left(-\infty, \frac{3}{2}\right]$

14. $P(x) = 2x^2 - 6x - 8$

(a) $P(x) = 0 : \{-1, 4\}$

(b) $P(x) > 0 : (-\infty, -1) \cup (4, \infty)$

(c) $P(x) \leq 0 : [-1, 4]$

15. $P(x) = 2x^2 - 6x - 8$

The graph intersects the x-axis at -1 and 4 [ans **(a)**]; it lies above the x-axis for $(-\infty, -1) \cup (4, \infty)$ [ans **(b)**]; it lies below or on the x-axis for $[-1, 4]$ [ans **(c)**].

16. $P(x) = 2x^2 - 6x - 8$

Axis of symmetry; through the vertex: $x = \frac{3}{2}$

17. $P(x) = -2.64x^2 + 5.47x + 3.54$

$$b^2 - 4ac = (5.47)^2 - 4(-2.64)(3.54)$$
$$= 29.9209 + 37.3824$$
$$= 67.3033$$

Since the discriminant is > 0, there are 2 x-intercepts of the graph.

18. $P(x) = -2.64x^2 + 5.47x + 3.54$

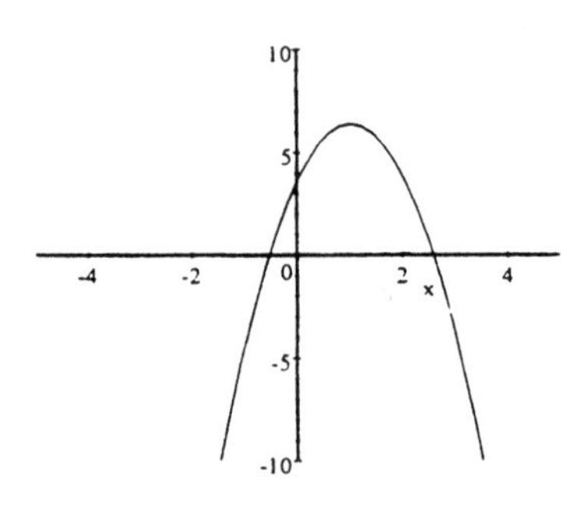

x-intercepts of graph of $P(x)$: -0.51777 and 2.5897

$\boxed{\{-0.52, 2.59\}}$

19. $P(x) = -2.64x^2 + 5.47x + 3.54$

See graph in #19:

Vertex: $(1.03599, 6.37342)$

Vertex $\approx$ $\boxed{(1.04, 6.37)}$

20. $P(x) = -2.64x^2 + 5.47x + 3.54$

$$\frac{-b}{2a} = \frac{-5.47}{2(-2.64)} = 1.03598 \approx 1.04$$

$P(1.04)$

$$= -2.64(1.04)^2 + 5.47(1.04) + 3.54$$
$$= -2.855424 + 5.6888 + 3.54$$
$$= 6.373376$$

$\boxed{\text{Vertex: } (1.04, 6.37)}$

21. $y = f(x)$

(a) Maximum value: 4
(b) Maximum when $x = 1$
(c) $f(x) = 2$: two
(d) $f(x) = 6$: none

22. $s(t) = -16t^2 + 800t + 600$
$s(0) = 600$

Height is 600 ft

23. $s(t) = -16t^2 + 800t + 600$
Vertex: (25, 10600)

Maximum height after 25 sec

24. $s(t) = -16t^2 + 800t + 600$
Vertex: (25, 10600)

Maximum height: 10,600 ft

25. $s(t) = -16t^2 + 800t + 600$
$s(t) > 5000$

(6.3 sec, 43.7 sec)

26. $s(t) = -16t^2 + 800t + 600$
x-intercept: 50.739

It will be in the air $\approx$ 50.7 sec.

27. Let x = width of cardboard
$3x$ = length. The base of the box is $(x - 8)$ and $(3x - 8)$:

(a) $Volume = lwh$

$V(x) = (3x - 8)(x - 8)(4)$
$V(x) = 12x^2 - 128x + 256$

(b) $2496 = 12x^2 - 128x + 256$

$0 = 12x^2 - 128x - 2240$
$0 = 3x^2 - 32x - 560$

$$x = \frac{32 \pm \sqrt{1024 + 6720}}{6}$$

$= 20, \ -9\frac{1}{3}$ (discard neg. #)

$x = 20, \quad 3x = 60$

20 in × 60 in

(c) Graph $y_1 = V(x)$, $y_2 = 2496$; show that they intersect $x = 20$.

28. (a) Atmospheric CO_2:

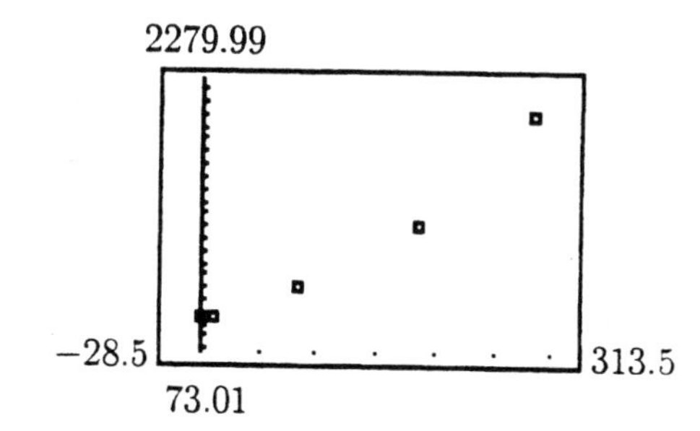

(b) $f(x) = a(x - h)^2 + k$
$f(x) = a(x - 0)^2 + 353$

(285, 2000):

$2000 = a(285)^2 + 353$
$1647 = 81225a$
$a = .020277$

$f(x) = .0203x^2 + 353$

(c) $\{0, 10, 85, 185, 285\} \rightarrow L_1$
$\{353, 375, \ldots, 2000\} \rightarrow L_2$
STAT / CALC / QuadReg

$g(x) \approx .0167x^2 + .968x + 363$

29. $\dfrac{x^3 + x^2 - 11x - 10}{x - 3}$

$$\begin{array}{r|rrrr} 3) & 1 & 1 & -11 & -10 \\ & & 3 & 12 & 3 \\ \hline & 1 & 4 & 1 & -7 \end{array}$$

$Q(x) = x^2 + 4x + 1; \quad R = -7$

30. $\dfrac{3x^3 + 8x^2 + 5x + 10}{x + 2}$

$$\begin{array}{r|rrrr} -2) & 3 & 8 & 5 & 10 \\ & & -6 & -4 & -2 \\ \hline & 3 & 2 & 1 & 8 \end{array}$$

$Q(x) = 3x^2 + 2x + 1; \quad R = 8$

31. $P(x) = -x^3 - 5x^2 - 7x + 1$; find $P(2)$:

$$\begin{array}{r|rrrr} 2) & -1 & 5 & -7 & 1 \\ & & -2 & 6 & -2 \\ \hline & -1 & 3 & -1 & -1 \end{array}$$

$P(2) = -1$

32. $P(x) = 2x^3 - 3x^2 + 7x - 12$; find $P(2)$:

2)	2	−3	7	−12
		4	2	18
	2	1	9	6

$\boxed{P(2) = 6}$

33. $P(x) = 5x^4 - 12x^2 + 2x - 8$; find $P(2)$:

2)	5	0	−12	2	−8
		10	20	16	36
	5	10	8	18	28

$\boxed{P(2) = 28}$

34. $P(x) = x^5 + 4x^2 - 2x - 4$; find $P(2)$:

2)	1	0	0	4	−2	−4
		2	4	8	24	44
	1	2	4	12	22	40

$\boxed{P(2) = 40}$

35. If $7 + 2i$ is a root of a polynomial with real coefficients, its conjugate, $7 - 2i$ must also be a root.

In 36 – 39, other answers are possible.

36. Zeros: $-1, 4, 7$:

$$\begin{aligned} P(x) &= (x+1)(x-4)(x-7) \\ &= (x+1)\left(x^2 - 11x + 28\right) \\ &= x^3 - 11x^2 + 28x + x^2 - 11x + 28 \end{aligned}$$

$\boxed{P(x) = x^3 - 10x^2 + 17x + 28}$

37. Zeros: $8, 2, 3$:

$$\begin{aligned} P(x) &= (x-8)(x-2)(x-3) \\ &= (x-8)\left(x^2 - 5x + 6\right) \\ &= x^3 - 5x^2 + 6x - 8x^2 + 40x - 48 \end{aligned}$$

$\boxed{P(x) = x^3 - 13x^2 + 46x - 48}$

38. Zeros: $\sqrt{3}, -\sqrt{3}, 2, 3$:

$$\begin{aligned} P(x) &= \left(x - \sqrt{3}\right)\left(x + \sqrt{3}\right)(x-2)(x-3) \\ &= \left(x^2 - 3\right)\left(x^2 - 5x + 6\right) \\ &= x^4 - 5x^3 + 6x^2 - 3x^2 + 15x - 18 \end{aligned}$$

$\boxed{P(x) = x^4 - 5x^3 + 3x^2 + 15x - 18}$

39. Zeros: $-2 + \sqrt{5}, -2 - \sqrt{5}, -2, 1$:

$$\begin{aligned} P(x) &= \left(x - \left(-2 + \sqrt{5}\right)\right)\left(x - \left(-2 - \sqrt{5}\right)\right) \\ &\qquad \times (x+2)(x-1) \\ &= \left(x^2 + 4x - 1\right)\left(x^2 + x - 2\right) \\ &= x^4 + x^3 - 2x^2 + 4x^3 + 4x^2 - 8x \\ &\qquad -x^2 - x + 2 \end{aligned}$$

$\boxed{P(x) = x^4 + 5x^3 + x^2 - 9x + 2}$

40. $P(x) = 2x^4 + x^3 - 4x^2 + 3x + 1$; is -1 a zero?

−1)	2	1	−4	3	1
		−2	1	3	−6
	2	−1	−3	6	−5

$\boxed{\text{Since } P(-1) = -5 \neq 0,\ -1 \text{ is not a zero.}}$

41. $P(x) = x^3 + 2x^2 + 3x + 2$; Is $x + 1$ a factor?

−1)	1	2	3	2
		−1	−1	−2
	1	1	2	0

$\boxed{\text{Since } P(-1) = 0,\ (x+1) \text{ is a factor.}}$

42. Zeros: $3, 1, -1 - 3i$; find a polynomial of degree 4 where $P(2) = -36$.

Since $-1 - 3i$ is a zero, $-1 + 3i$ is also.

$$\begin{aligned} Q(x) &= (x-3)(x-1)(x+1+3i)(x+1-3i) \\ &= \left(x^2 - 4x + 3\right)\left(x^2 + 2x + 10\right) \\ &= x^4 + 2x^3 + 10x^2 - 4x^3 - 8x^2 - 40x \\ &\qquad +3x^2 + 6x + 30 \\ &= x^4 - 2x^3 + 5x^2 - 34x + 30 \end{aligned}$$

$P(x) = a\left(x^4 - 2x^3 + 5x^2 - 34x + 30\right)$

$P(2) = -36$:

$-36 = a\left(2^4 - 2\left(2^3\right) + 5\left(2^2\right) - 34(2) + 30\right)$

$-36 = a(-18)$

$a = 2$

$\boxed{P(x) = 2x^4 - 4x^3 + 10x^2 - 68x + 60}$

43. A fourth degree polynomial having exactly two distinct real zeros is any function of the form:

$P(x) = a(x-b)^2(x-c)^2$

One example is:

$P(x) = 2(x-1)^2(x-3)^2$

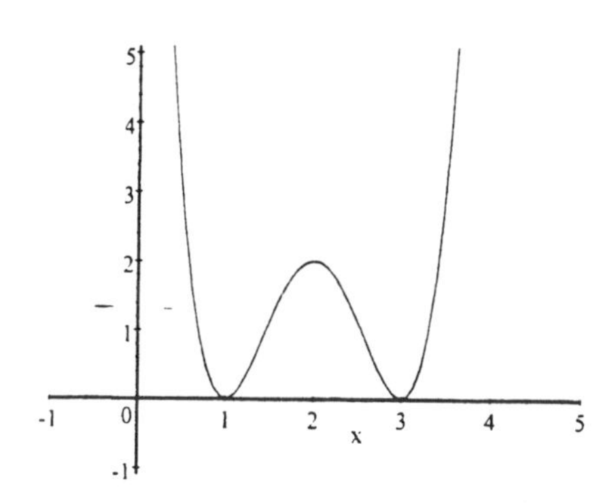

44. A third degree polynomial having exactly one real zero is any function of the form: $P(x) = a(x-b)^3$.

One example is: $P(x) = 2(x-1)^3$

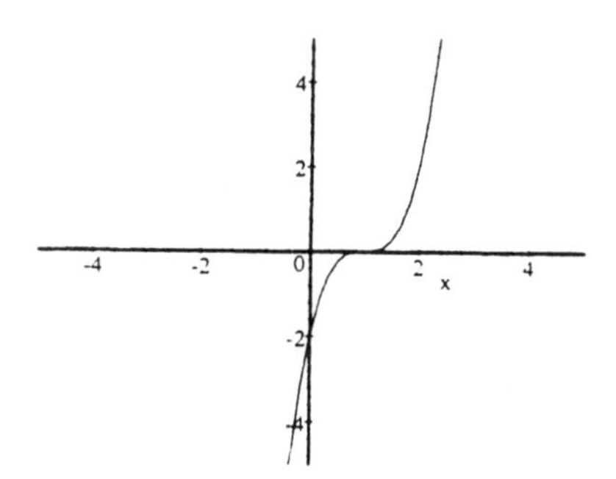

45. $P(x) = x^4 - 3x^3 - 8x^2 + 22x - 24$

Zeros: $1-i$, $1+i$

$1-i$)	1	-3	-8	22	-24
		$1-i$	$-3+i$	$-10+12i$	24
	1	$-2-i$	$-11+i$	$12+12i$	0

$1+i$)	1	$-2-i$	$-11+i$	$12+12i$
		$1+i$	$-1-i$	$-12-12i$
	1	-1	-12	0

$x^2 - x - 12$

Solve $x^2 - x - 12 = 0$:

$(x-4)(x+3) = 0$

$x = 4, -3$

$\boxed{-3, 4, 1-i, 1+i}$

46. $P(x) = 2x^4 - x^3 + 7x^2 - 4x - 4$

Zeros: $1, 2i, -2i$

1)	2	-1	7	-4	-4
		2	1	8	4
	2	1	8	4	0

$2i$)	2	1	8	4
		$4i$	$-8+2i$	-4
	2	$1+4i$	$2i$	0

$-2i$)	2	$1+4i$	$2i$
		$-4i$	$-2i$
	2	1	0

$2x+1$

Solve $2x + 1 = 0$:

$x = -\frac{1}{2}$

$\boxed{\left\{-\frac{1}{2}, 1, 2i, -2i\right\}}$

47. $P(x) = 3x^5 - 4x^4 - 26x^3 - 21x^2 - 14x + 8$

Possible rational zeros:

$\pm1, \pm2, \pm4, \pm8, \pm\frac{1}{3}, \pm\frac{2}{3}, \pm\frac{4}{3}, \pm\frac{8}{3}$

From graphing, check -2, 4, $\frac{1}{3}$

-2)	3	-4	-26	-21	-14	8
		-6	20	12	18	-8
	3	-10	-6	-9	4	0

4)	3	-10	-6	-9	4
		12	8	8	-4
	3	2	2	-1	0

$\frac{1}{3}$)	3	2	2	-1
		1	1	1
	3	3	3	0

Solve $3x^2 + 3x + 3 = 0$:

$x^2 + x + 1 = 0$

$x = \frac{-1 \pm \sqrt{-3}}{2}$; not real

$\boxed{\left\{-2, \frac{1}{3}, 4\right\}}$

48. $P(x) = x^3 - 2x^2 - 4x + 3$

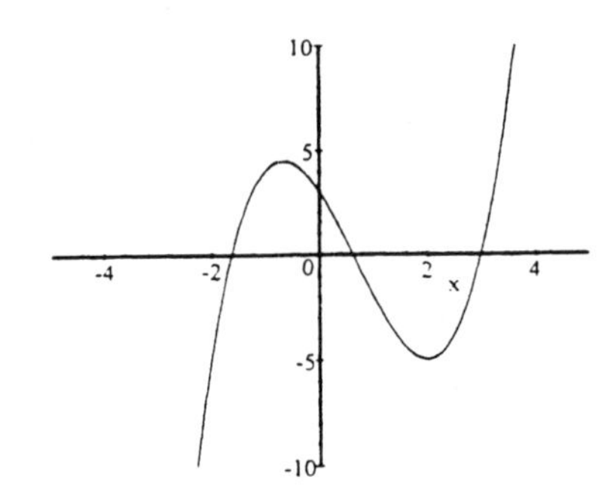

x-intercepts: $-1.62,\ 0.62,\ 3$

3 real solutions; $\boxed{\text{integer root: } \{3\}}$

49.
$$\begin{array}{r|rrrr} 3) & 1 & -2 & -4 & 3 \\ & & 3 & 3 & -3 \\ \hline & 1 & 1 & -1 & 0 \end{array}$$

$x^2 + x - 1$

$\boxed{(x-3)\left(x^2+x-1\right)}$

50. $(x-3)\left(x^2+x-1\right)$

$x^2 + x - 1 = 0$

$$x = \frac{-1 \pm \sqrt{1-4(-1)}}{2} = \frac{-1 \pm \sqrt{5}}{2}$$

$$\boxed{\left\{\frac{-1-\sqrt{5}}{2},\ \frac{-1+\sqrt{5}}{2}\right\}}$$

51. x-intercepts: $-1.62,\ 0.62,\ 3$

$$\frac{-1-\sqrt{5}}{2} \approx -1.62$$

$$\frac{-1+\sqrt{5}}{2} \approx 0.62$$

52. **(a)** $P(x) > 0$:

$$\left(\frac{-1-\sqrt{5}}{2},\ \frac{-1+\sqrt{5}}{2}\right) \cup (3,\ \infty)$$

(b) $P(x) \le 0$:

$$\left(-\infty,\ \frac{-1-\sqrt{5}}{2}\right] \cup \left[\frac{-1+\sqrt{5}}{2},\ 3\right]$$

53. $x^3 + 2x^2 + 5x = 0$

$x\left(x^2 + 2x + 5\right) = 0$

(i) $x = 0$

(ii) $x^2 + 2x + 5 = 0$

$$x = \frac{-2 \pm \sqrt{4-20}}{2} = \frac{-2 \pm 4i}{2} = -1 \pm 2i$$

$\boxed{x = \{0,\ -1 \pm 2i\}}$

$\boxed{\text{The only } x\text{-intercept: } 0}$

54. $P(x) = x^4 - 5x^3 + x^2 + 21x - 18$

x-intercepts: $-2,\ 1,\ 3$ (mult. 2)

$\boxed{P(x) = (x+2)(x-1)(x-3)^2}$

55. Degree of $g(x)$: even

56. Degree of $f(x)$: odd

57. Leading coefficient of $f(x)$ is positive.

58. $g(x)$ has no real solutions.

59. $f(x) < 0 : (-\infty,\ a) \cup (b,\ c)$

60. $f(x) > g(x)$: open interval $(d,\ h)$

61. $f(x) - g(x) = 0 : \{d, h\}$

62. If $r + pi$ is a solution, then $r - pi$ is a solution.

63. Since $f(x)$ has 3 real zeros, and a polynomial of degree 3 can have at most 3 zeros, there can be no other zeros, real or imaginary.

64. $y_1 = x^3 - x^2 - 19x + 4$; since one point is $(-4, 0)$, one factor is $(x+4)$:

$$\begin{array}{r|rrrr} -4) & 1 & -1 & -19 & 4 \\ & & -4 & 20 & -4 \\ \hline & 1 & -5 & 1 & 0 \end{array}$$

$$x^2 - 5x + 1 = 0$$

$$x = \frac{5 \pm \sqrt{25-4}}{2} = \frac{5 \pm \sqrt{21}}{2}$$

$$\boxed{(x+4)\left(x - \left(\frac{5+\sqrt{21}}{2}\right)\right)\left(x - \left(\frac{5-\sqrt{21}}{2}\right)\right)}$$

65. Because $y_1 = -x^3 - 5x + 4$ is a polynomial function, the intermediate value theorem indicates that there is a real zero between .7 and .8, as there is a sign change: $y > 0$ for $x = .7$, $y < 0$ for $x = .8$.

66. False; a 7th degree function can have at most 7 x-intercepts.

67. True; a 7th degree function can have at most 6 local extrema.

68. True; when $x = 0$ for $f(x)$, $y = 500$.

69. True; the end behavior is ↙ ↗; with a positive y-intercept, it must cross the negative x-axis at least one time.

70. True; an even degree function where $a > 0$ with a negative y-intercept will have end behavior ↖ ↗; it must cross the x-axis at least 2 times.

71. False; the conjugate of $-\frac{1}{2} + i\frac{\sqrt{3}}{2}$ is $-\frac{1}{2} - i\frac{\sqrt{3}}{2}$ not $\frac{1}{2} + i\frac{\sqrt{3}}{2}$.

Graph for #72 – #76:

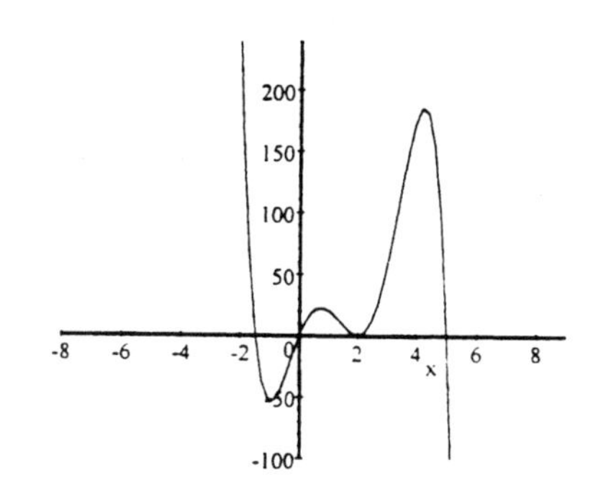

$$P(x) = -2x^5 + 15x^4 - 21x^3 - 32x^2 + 60x$$

72. Two local maxima

73. Local minimum: $(2, 0)$

74. Synthetic Division:

$$\begin{array}{r|rrrrrr} 5) & -2 & 15 & -21 & -32 & 60 & 0 \\ & & -10 & 25 & 20 & -60 & 0 \\ \hline & -2 & 5 & 4 & -12 & 0 & 0 \end{array}$$

$$\boxed{Q(x) = -2x^4 + 5x^3 + 4x^2 - 12x}$$

75. Range: $(-\infty, \infty)$

76. Local minimum: $(-0.97, -54.15)$

77. $3x^3 + 2x^2 - 21x - 14 = 0$
$x^2(3x+2) - 7(3x+2) = 0$
$(3x+2)(x^2-7) = 0$

$$x = -\frac{2}{3},\ \pm\sqrt{7}$$

$$\boxed{\left\{-\sqrt{7},\ -\frac{2}{3},\ \sqrt{7}\right\}}$$

Show that the x-intercepts are $\approx \pm 2.65,\ -0.67$.

78. $P(x) = -x^4 + 3x^3 + 3x^2 + 17x - 6$
$= (-x+2)(x-3)(x+1)^2$

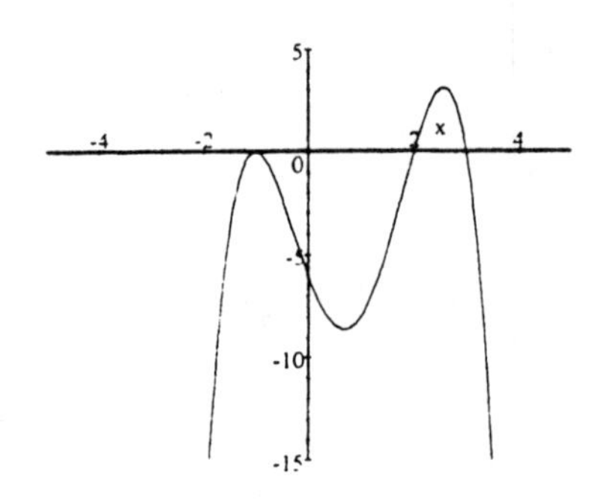

(a) $P(x) = 0 : \{-1, 2, 3\}$
(b) $P(x) > 0$: open interval $(2, 3)$
(c) $P(x) < 0 : (-\infty, -1) \cup (-1, 2) \cup (3, \infty)$

79. $V = s^3$: Let $x =$ the length of a side of the cube.
$V = (x)(x)(x-2) = 32$
$x^3 - 2x^2 - 32 = 0$

Graph to check for possible roots.

4)	1	−2	0	−32
		4	8	32
	1	2	8	0

$x^2 + 2x + 8 = 0$

No real solutions: 4 is the only zero.

4 inches by 4 inches by 4 inches

80. (a) Since the zero is at ≈ 9.26, the restrictions on t are: $0 \le t \le 9.26$.

(b) $s(t) = -4.9t^2 + 40t + 50$

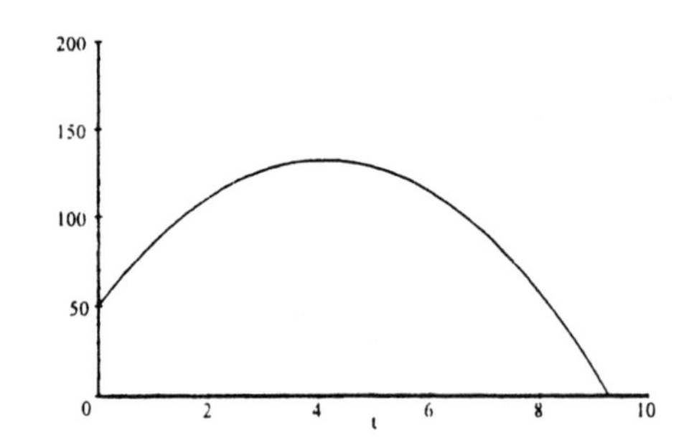

(c) Maximum point $\approx (4.08, 131.63)$
The object reaches its highest point at ≈ 4.08 seconds.
(d) The maximum height is ≈ 131.63 meters.
(e) $s(t) = 0 : -4.9t^2 + 40t + 50 = 0$

$$t = \frac{-40 \pm \sqrt{1600 - 4(-4.9)(50)}}{2(-4.9)}$$
$$= \frac{-40 \pm \sqrt{2580}}{-9.8} \approx -1.10, 9.26$$

Approximately 9.26 seconds.

81. $y = .0045x^3 - .072x^2 + .19x + 3.7$

at $x = 10$:
$y = .0045(10)^3 - .072(10)^2 + .19(10) + 3.7$
$= 2.9$ 2.9 million

82. (a) $\{8, 10, 15, 20, 25, 30, 35\} \longrightarrow L_1$
$\{18.6, 19.3, 21.7, 24.7, 27.5, 28.3, 28.6\} \longrightarrow L_2$

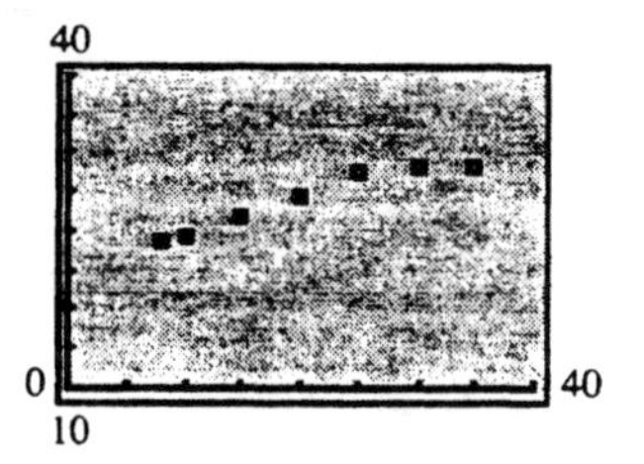

(b) STAT / CALC / QuadReg
$f(x) \approx -.011x^2 + .869x + 11.9$; $r = \sqrt{.986}$

(c) STAT / CALC / CubicReg
$f(x) \approx -.00087x^3 + .0456x^2 - .219x + 17.8$;
$r = \sqrt{.997}$

(d)

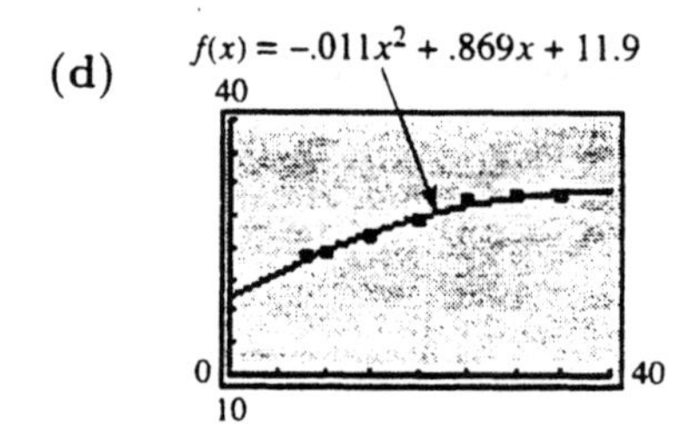

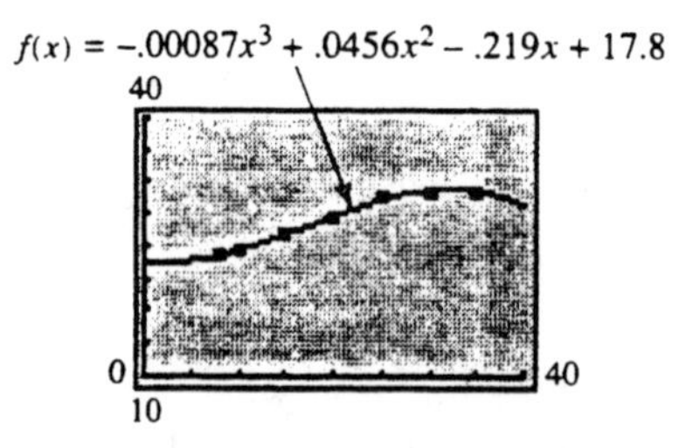

(e) Both functions approximate the data well. The quadratic function is probably better for prediction since it is unlikely that the percent of out-of-pocket spending would decrease after 2025 unless changes were made in Medicare law.

(f) Since the data points lie in a curved pattern, a linear model would not be appropriate.

Chapter 3 Test

1. (a) $(8-7i)-(-12+2i)$
$$= 8-7i+12-2i$$
$$= (8+12)+(-7-2)i$$
$$= \boxed{20-9i}$$

(b) $$\frac{11+10i}{2+3i}\cdot\frac{2-3i}{2-3i}$$
$$= \frac{22-33i+20i-30i^2}{4-9i^2}$$
$$= \frac{22-13i+30}{4+9}$$
$$= \frac{52-13i}{13}$$
$$= \boxed{4-i}$$

(c) $i^{65} = (i^4)^{16}(i)$
$$= 1^{16}(i) = \boxed{i}$$

(d) $2i(3-i)^2 = 2i(9-6i+i^2)$
$$= 2i(9-6i-1)$$
$$= 2i(8-6i)$$
$$= 16i-12i^2$$
$$= \boxed{12+16i}$$

2. $f(x) = -2x^2-4x+6$

(a) $x = -\dfrac{b}{2a} = -\dfrac{-4}{-4} = -1$
$$f(-1) = -2(-1)^2-4(-1)+6$$
$$= -2+4+6 = 8$$
$\boxed{\text{Vertex: } (-1,\ 8)}$

(b) Find the maximum on the graph:

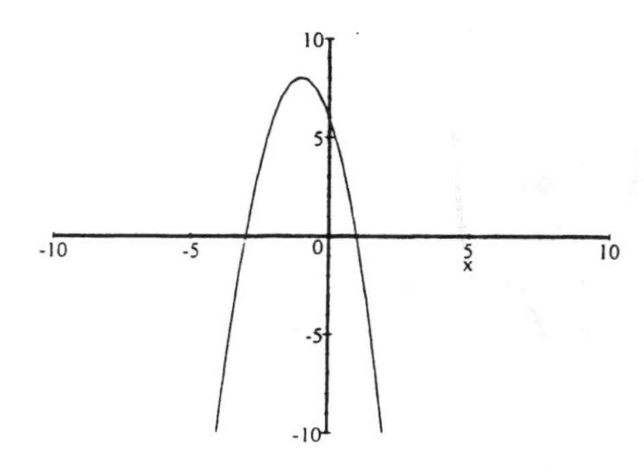

continued

2. continued

(c) $f(x) = -2(x^2+2x-3)$
$$= -2(x+3)(x-1)$$
$x = -3,\ 1$

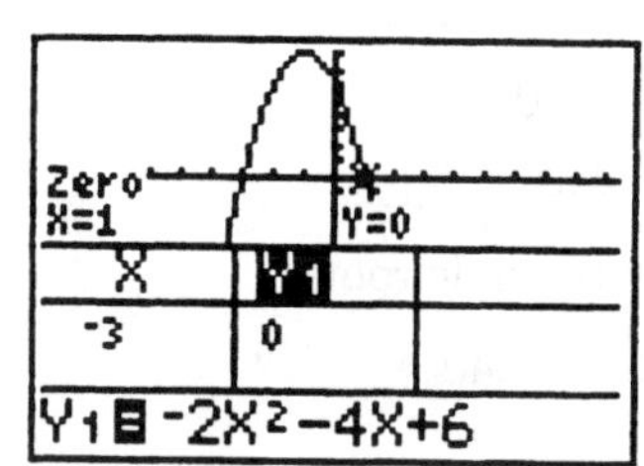

(d) y-intercept; let $x = 0$:
$f(0) = -2(0)-4(0)+6 = \boxed{6}$

(e) Domain: $(-\infty,\ \infty)$; Range: $(-\infty,\ 8]$

(f) Increasing: $(-\infty,\ -1]$
Decreasing: $[-1,\ \infty)$

3. (a) $3x^2+3x-2=0$
$$x = \frac{-3\pm\sqrt{3^2-4(3)(-2)}}{2(3)}$$
$$= \frac{-3\pm\sqrt{9+24}}{6}$$
$$= \boxed{\left\{\frac{-3-\sqrt{33}}{6},\ \frac{-3+\sqrt{33}}{6}\right\}}$$

(b) $f(x) = 3x^2+3x-2$

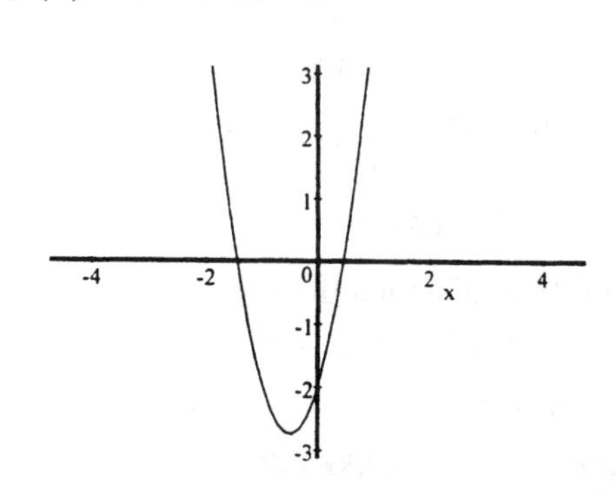

(i) $f(x)<0$: $\left(\dfrac{-3-\sqrt{33}}{6},\ \dfrac{-3+\sqrt{33}}{6}\right)$

(ii) $f(x)\geq 0$:
$$\left(-\infty,\ \frac{-3-\sqrt{33}}{6}\right]\cup\left[\frac{-3+\sqrt{33}}{6},\ \infty\right)$$

4. Let $x =$ the height; then
$w = 3x$, $l = 11 + x$, Vol is 720 in^3.

$$V = lwh$$
$$720 = (11 + x)(3x)(x)$$
$$720 = 3x^3 + 33x^2$$
$$0 = 3x^3 + 33x^2 - 720$$
$$0 = x^3 + 11x^2 - 240$$

Graph to check for possible roots.

$$\begin{array}{r|rrrr} 4) & 1 & 11 & 0 & -240 \\ & & 4 & 60 & 240 \\ \hline & 1 & 15 & 60 & 0 \end{array}$$

$x^2 + 15x + 60 = 0$; No real solutions

4 is the only zero.

$h = x = 4$
$w = 3x = 12$
$l = 11 + x = 15$

$\boxed{\text{15 in by 12 in by 4 in}}$

5. **(a)** $\{0, 2, 4, 6, 8, 10\} \longrightarrow L_1$
$\{470, 475, 530, 580, 620, 670\} \longrightarrow L_2$

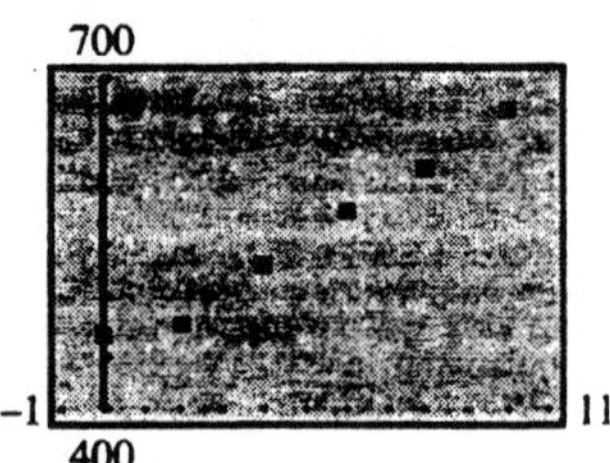

(b) $f(x) = a(x - h)^2 + k$
$f(x) = a(x - 0)^2 + 470$
$(10, 670): \ 670 = a(10)^2 + 470$
$200 = 100a \Longrightarrow a = 2$
$f(x) = 2x^2 + 470$

(c) STAT / CALC / QuadReg
$g(x) = .737x^2 + 13.8x + 461$
The regression function is the best fit.

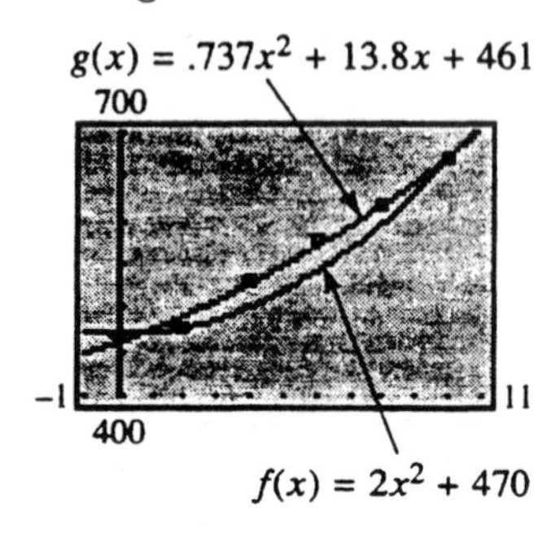

6. **(a)** Turning points are approximately
$(1.6, 3.6), (3, 1.2), (4.4, 3.6)$

(b) After 1.6 minutes the runner is 360 feet from the starting line. The runner turns and runs toward the starting line. After 3 minutes he/she is 120 feet from the starting line, turns and runs away from the starting line. After 4.4 minutes he/she is again 360 feet from the starting line and turns and runs back to the starting line.

7. **(a)** $f(x) = x^6 - 5x^5 + 3x^4 + x^3 + 40x^2 - 24x - 72$

Zeros: 3(mult. 2), -1, 2

$$\begin{array}{r|rrrrrrr} 3) & 1 & -5 & 3 & 1 & 40 & -24 & -72 \\ & & 3 & -6 & -9 & -24 & 48 & 72 \\ \hline & 1 & -2 & -3 & -8 & 16 & 24 & 0 \end{array}$$

$$\begin{array}{r|rrrrrr} 3) & 1 & -2 & -3 & -8 & 16 & 24 \\ & & 3 & 3 & 0 & -24 & -24 \\ \hline & 1 & 1 & 0 & -8 & -8 & 0 \end{array}$$

$$\begin{array}{r|rrrrr} -1) & 1 & 1 & 0 & -8 & -8 \\ & & -1 & 0 & 0 & 8 \\ \hline & 1 & 0 & 0 & -8 & 0 \end{array}$$

$$\begin{array}{r|rrrr} 2) & 1 & 0 & 0 & -8 \\ & & 2 & 4 & 8 \\ \hline & 1 & 2 & 4 & 0 \end{array}$$

Solve $x^2 + 2x + 4 = 0$:

$$x = \frac{-2 \pm \sqrt{4 - 4(4)}}{2} = \frac{-2 \pm \sqrt{-12}}{2}$$
$$= \frac{-2 \pm 2i\sqrt{3}}{2} = -1 \pm i\sqrt{3}$$

$$\boxed{\{-1 - i\sqrt{3}, \ -1 + i\sqrt{3}\}}$$

(b) End behavior ↖ ↗

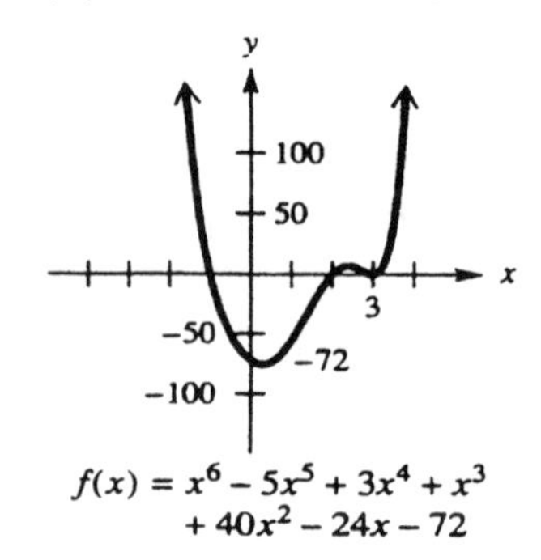

8. (a) $f(x) = 4x^4 - 21x^2 - 25$

$$0 = (4x^2 - 25)(x^2 + 1)$$

$$4x^2 = 25 \qquad x^2 = -1$$

$$x^2 = \frac{25}{4} \qquad x = \pm\sqrt{-1}$$

$$x = \pm\frac{5}{2} \qquad x = \pm i$$

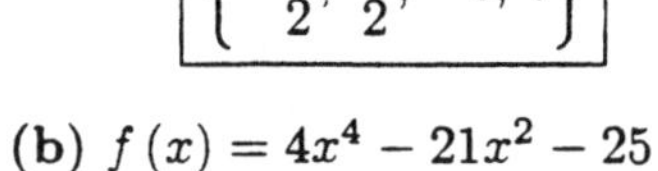

$$\left\{-\frac{5}{2}, \frac{5}{2}, -i, i\right\}$$

(b) $f(x) = 4x^4 - 21x^2 - 25$

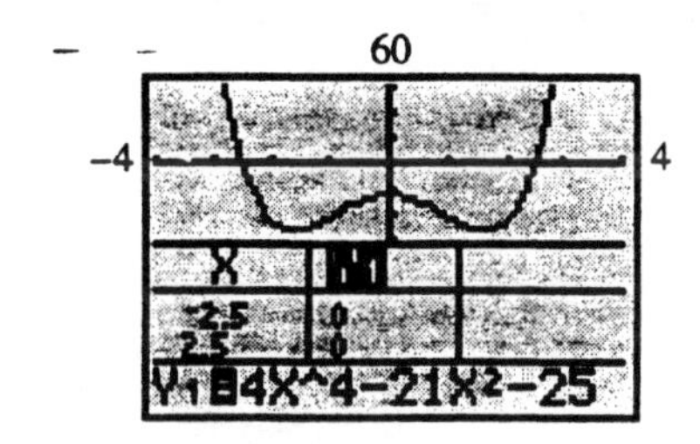

(c) It is symmetric with respect to the y-axis.

(d) $f(x) \geq 0: \left(-\infty, -\frac{5}{2}\right] \cup \left[\frac{5}{2}, \infty\right)$

$f(x) < 0: \left(-\frac{5}{2}, \frac{5}{2}\right)$

9. (a) $x^5 - 4x^4 + 2x^3 - 4x^2 + 6x - 1 = 0$

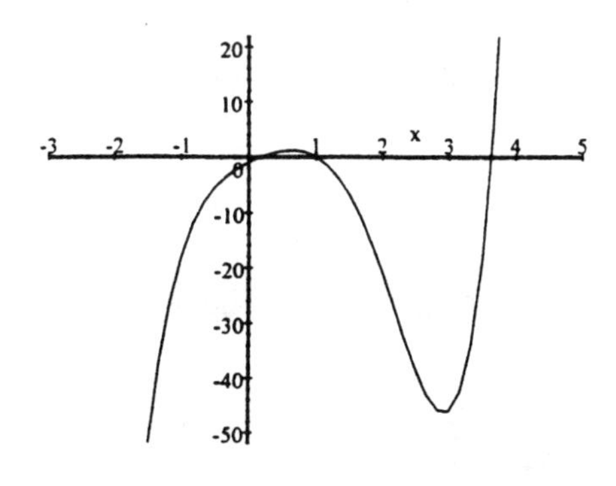

$x \approx .189,\ 1,\ \approx 3.633$

(b) A fifth degree equation has 5 zeros. Since three are real, two would be nonreal (complex).

10. (a) $\{0, 2, 4, 6, 8, 10\} \longrightarrow L_1$

$\{470, 475, 530, 580, 620, 670\} \longrightarrow L_2$

STAT / CALC / CubicReg

$f(x) = -.249x^3 + 4.47x^2 + .212x + 467$

(b) $\{0, 2, 4, 6, 8, 10\} \longrightarrow L_1$

$\{470, 475, 530, 580, 620, 670\} \longrightarrow L_2$

STAT / CALC / QuartReg

$g(x) = .0977x^4 - 2.20x^3 + 16.5x^2 - 22.1x + 470$

(c)

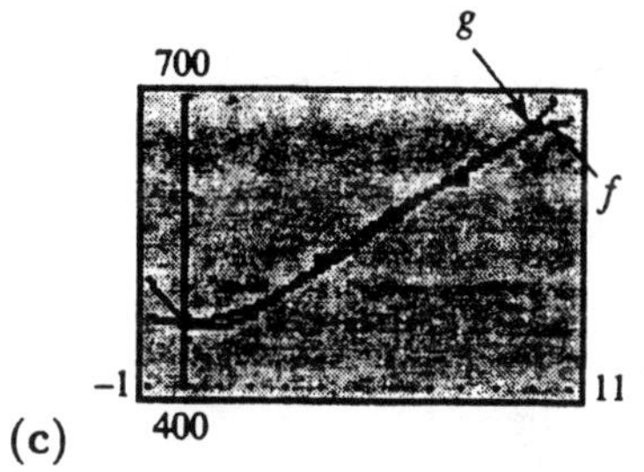

(d) In 2001, $x = 11$.

Cubic: $f(11) \approx 679$ million

Quartic: $g(11) \approx 726$ million

The quartic function is a better estimate because it continues to increase, while the cubic function turns downward. Due to September 11th, this could be debated. Airline passengers may decline in the short term.

Chapter 3 Project

Creating Your Personal Social Security Polynomial

Answers will vary depending on the social security number used. Answers are given for the illustrative social security number **539-58-0954**.

1. $SSN(x) = (x-5)(x+3)(x-9)(x+5) \times (x-8)(x+0)(x-9) \times (x+5)(x-4)$

2. There are nine terms, so the term with the highest power will be x^9. The polynomial is of degree 9, and the dominating term is x^9.

3. Polynomials of odd degree with positive coefficients have end behavior ↙↗.

4. The zeros are: $5, -3, 9, -5, 8, 0, 9, -5, 4$.

   Zeros of multiplicity one: $5, -3, 8, 0, 4$.

5. Zeros of multiplicity two: $9, -5$

6. Zeros of multiplicity three or higher: there are none.

7. $SSN(x)$:

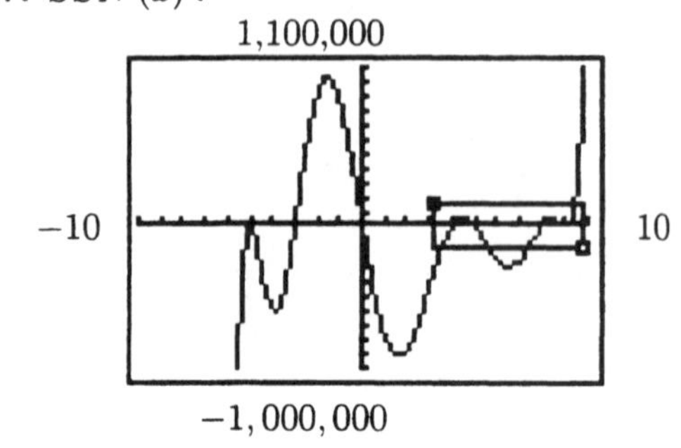

x-scale: 1; y-scale: 100,000

7. continued

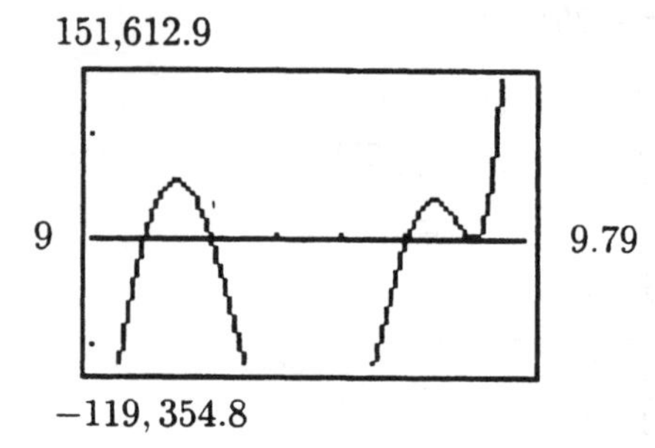

8. Coordinates of local maximums:

   $(-5,\ 0)$
   $(-1.553,\ 1{,}033{,}670.4)$
   $(4.480,\ 54{,}058.1)$,
   $(8.404,\ 37{,}044.4)$

9. Coordinates of local minimums:

   $(-3.909,\ -591{,}411.9)$
   $(1.610,\ -916{,}889.6)$
   $(6.523,\ -287{,}211.6)$
   $(9,\ 0)$

10. $SSN(x)$ domain: $(-\infty,\ \infty)$
    range: $(-\infty,\ \infty)$

11. $SSN(x)$ is increasing over the following intervals:

    $(-\infty,\ -5]$, $[-3.909,\ -1.553]$
    $[1.610,\ 4.480]$, $[6.523,\ 8.404]$
    $[9,\ \infty)$

12. $SSN(x)$ is decreasing over the following intervals:

    $[-5,\ -3.909]$, $[-1.553,\ 1.610]$
    $[4.480,\ 6.523]$, $[8.404,\ 9]$

13. In order for the polynomial graph to pass through the origin, 0 must be a zero, so one factor must be $(x \pm 0)$. Therefore, the digit 0 must appear in the social security number.

CHAPTER 4

Section 4.1

1. $f(x) = \frac{1}{x}$

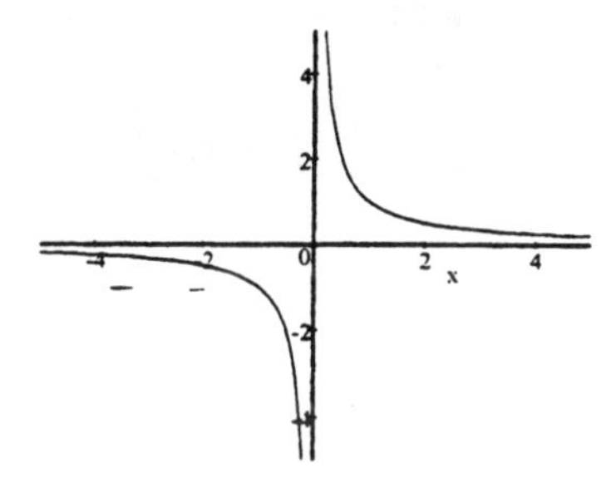

D: $(-\infty, 0) \cup (0, \infty)$
R: $(-\infty, 0) \cup (0, \infty)$

3. $f(x) = \frac{1}{x}$ never increases; decreases over $(-\infty, 0) \cup (0, \infty)$; is never constant.

5. Vertical asymptote: $x = 3$
Horizontal asymptote: $y = 2$

7. Since $f(-x) = \frac{1}{x^2} = f(x)$,
$f(x)$ is an even function; symmetric with respect to the y-axis.

9. Domain $(-\infty, 3) \cup (3, \infty)$: **A, B, C**

11. Range $(-\infty, 0) \cup (0, \infty)$: **A**

13. $f(x) = 3$: **A**

15. $f(x) = \frac{1}{x+5}$: **C**

A:

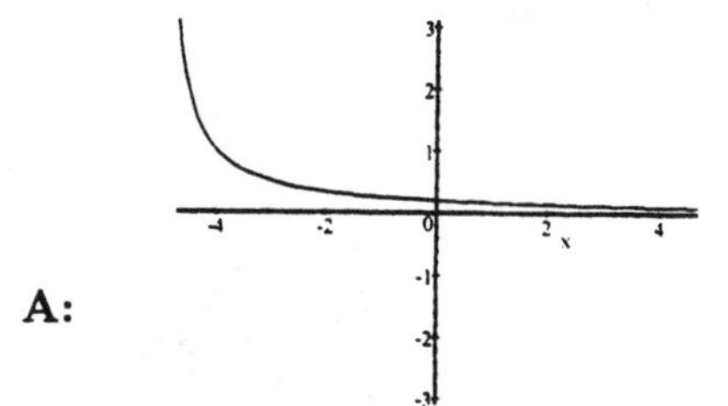

B:

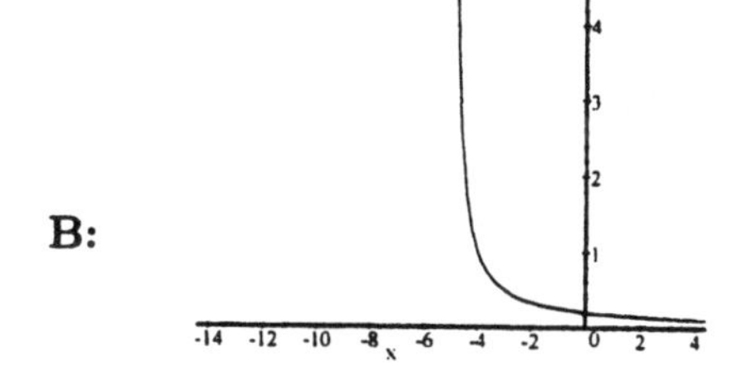

C:

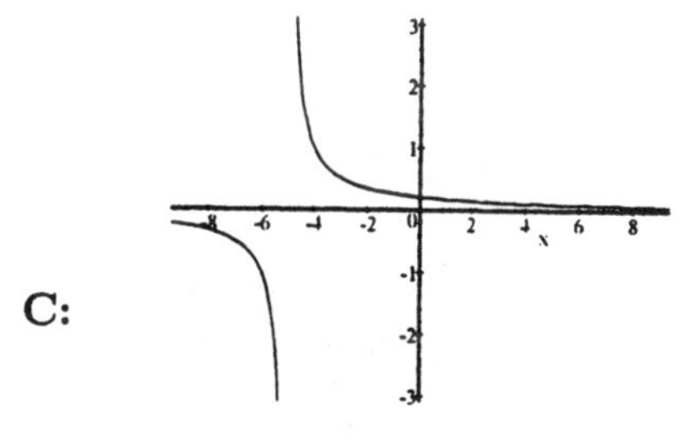

17. $y = \frac{1}{x}$

$$f(x) = 2\left(\frac{1}{x}\right) = \frac{2}{x}$$

vertical stretch by a factor of 2.

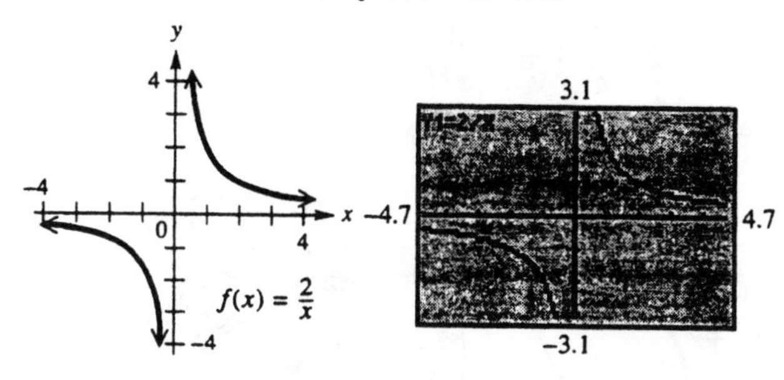

D: $(-\infty, 0) \cup (0, \infty)$;
R: $(-\infty, 0) \cup (0, \infty)$

19. $y = \dfrac{1}{x}$

$f(x) = \dfrac{1}{x+2}$; shift 2 units left.

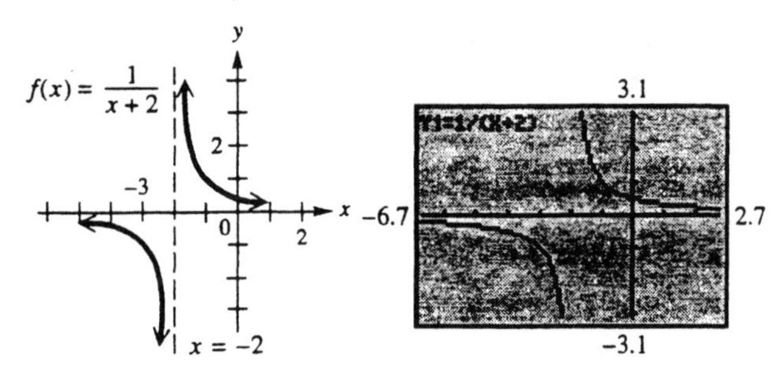

D: $(-\infty, -2) \cup (-2, \infty)$
R: $(-\infty, 0) \cup (0, \infty)$

21. $y = \dfrac{1}{x}$

$f(x) = \dfrac{1}{x} + 1$; shift 1 unit upward.

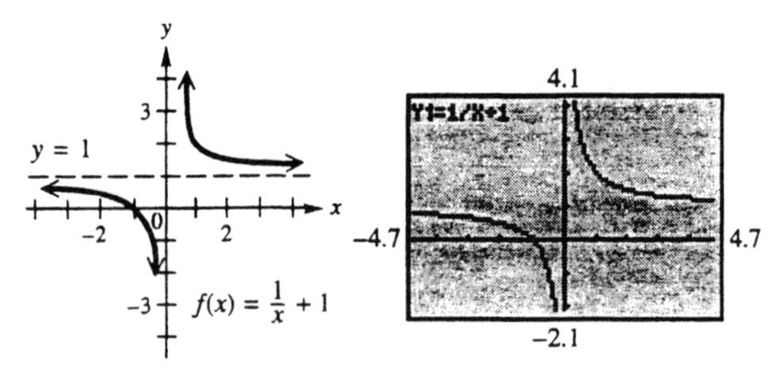

D: $(-\infty, 0) \cup (0, \infty)$
R: $(-\infty, 1) \cup (1, \infty)$

23. $y = \dfrac{1}{x^2}$

$f(x) = 2\left(\dfrac{1}{x^2}\right)$;

vertical stretch by a factor of 2.

$f(x) = -2\left(\dfrac{1}{x^2}\right) = -\dfrac{2}{x^2}$;

reflection across the x-axis.

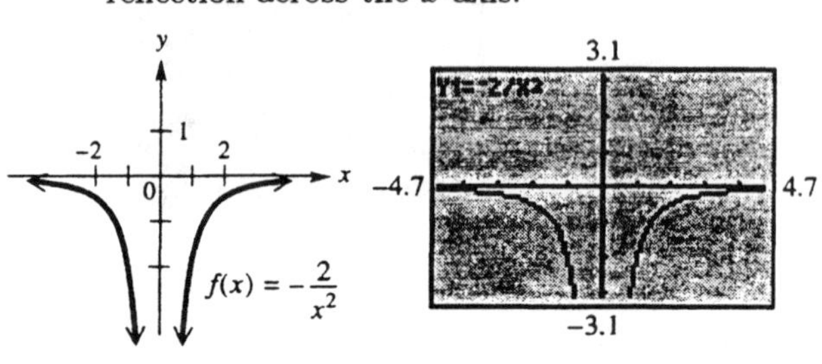

D: $(-\infty, 0) \cup (0, \infty)$; R: $(-\infty, 0)$

25. $y = \dfrac{1}{x^2}$

$f(x) = \dfrac{1}{(x-3)^2}$; shift 3 units right.

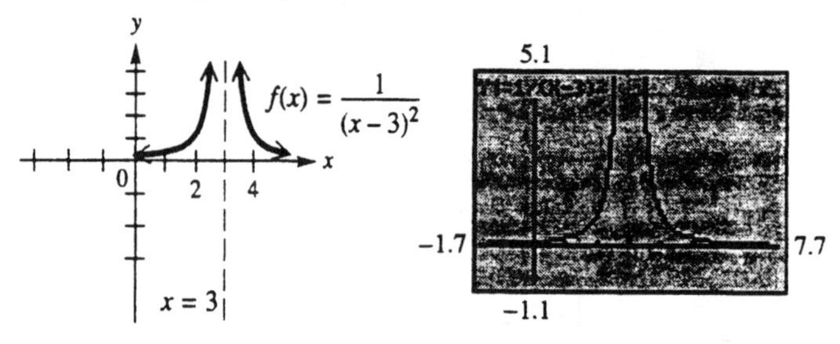

D: $(-\infty, 3) \cup (3, \infty)$; R: $(0, \infty)$

27. $y = \dfrac{1}{x^2}$

$f(x) = \dfrac{1}{(x+2)^2}$; shift 2 units left

$f(x) = \dfrac{-1}{(x+2)^2}$: reflection across the x-axis

$f(x) = \dfrac{-1}{(x+2)^2} - 3$; shift 3 units downward

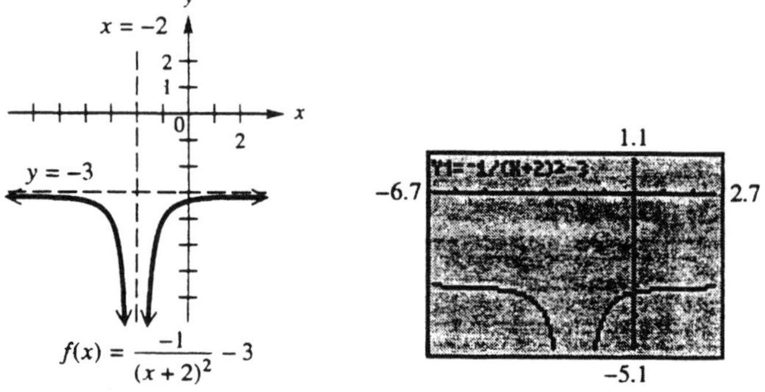

D: $(-\infty, -2) \cup (-2, \infty)$; R: $(-\infty, -3)$

29. $f(x) = \dfrac{1}{(x-2)^2}$: **C** **31.** $f(x) = \dfrac{-1}{x-2}$: **B**

33. $f(x) = \dfrac{x-1}{x-2} = 1 + \dfrac{1}{x-2}$

$$\begin{array}{r} 1 + \frac{1}{x-2} \\ x-2 \overline{)\, x - 1} \\ \underline{x - 2} \\ 1 \end{array}$$

Shift the graph of $y = \dfrac{1}{x}$ to the right 2 units and 1 unit upward.

continued

33. continued

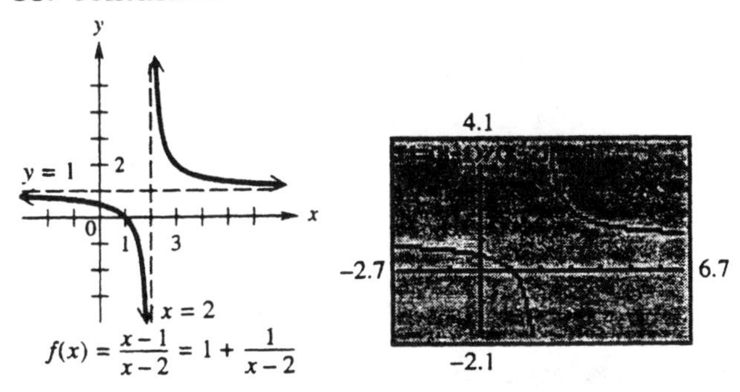

35. $f(x) = \dfrac{-2x-5}{x+3} = -2 + \dfrac{1}{x+3}$

$$\begin{array}{r} -2 + \frac{1}{x+3} \\ x+3\,\overline{)\,-2x - 5} \\ \underline{-2x - 6} \\ 1 \end{array}$$

Shift the graph of $y = \dfrac{1}{x}$ to the left 3 units and 2 units downward.

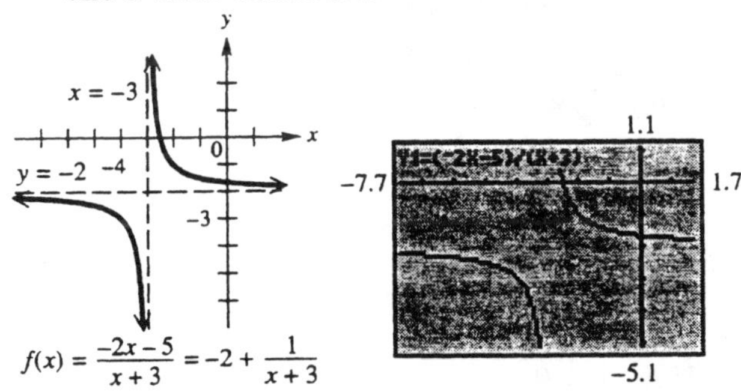

37. $f(x) = \dfrac{2x-5}{x-3} = 2 + \dfrac{1}{x-3}$

$$\begin{array}{r} 2 + \frac{1}{x-3} \\ x-3\,\overline{)\,2x - 5} \\ \underline{2x - 6} \\ 1 \end{array}$$

Shift the graph of $y = \dfrac{1}{x}$ to the right 3 units and 2 units upward.

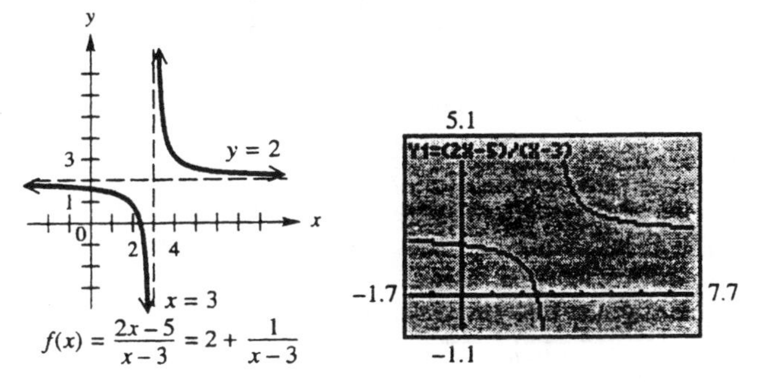

Section 4.2

Note: In the Exercises,

V.A.→ Vertical Asymptote(s);
H.A.→ Horizontal Asymptote;
O.A.→ Oblique Asymptote.

1. $f(x) = \dfrac{x+7}{x+1}$: **D**

The vertical asymptote is $x = -1$.

3. $f(x) = \dfrac{1}{x+12}$: **G**

The x-axis is its horizontal asymptote.

5. $f(x) = \dfrac{x^2-16}{x+4}$: **E**

There is a *hole* in its graph at $x = -4$.

7. $f(x) = \dfrac{x^2+3x+4}{x-5}$: **F**

The graph has an oblique asymptote.

9. $f(x) = \dfrac{3}{x-5}$

$x - 5 = 0 \rightarrow x = 5$

$f(x) : \dfrac{3/x}{x/x - 5/x}_{(\text{as } |x| \to \infty)} = \dfrac{0}{1-0} = 0$

V.A.: $x = 5$; H.A.: $y = 0$

11. $f(x) = \dfrac{4-3x}{2x+1}$

$2x + 1 = 0 \rightarrow x = -\frac{1}{2}$

$f(x) : \dfrac{\frac{4}{x} - \frac{3x}{x}}{\frac{2x}{x} + \frac{1}{x}}_{(\text{as } |x| \to \infty)} = \dfrac{0-3}{2+0} = -\dfrac{3}{2}$

V.A.: $x = -\dfrac{1}{2}$; H.A.: $y = -\dfrac{3}{2}$

13. $f(x) = \dfrac{x^2 - 1}{x + 3}$

(i) $x + 3 = 0 \to x = -3$

(ii)
$$\begin{array}{r|rrr} & x & -3 & \\ \hline x+3) & x^2 & +0x & -1 \\ & x^2 & +3x & \\ \hline & & -3x & -1 \\ & & -3x & -9 \\ \hline & & & 8 \end{array}$$

V.A.: $x = -3$; O.A: $y = x - 3$

15. $f(x) = \dfrac{x^2 - 2x - 3}{2x^2 - x - 10}$

(i) $2x^2 - x - 10 = (2x - 5)(x + 2)$

$x = \frac{5}{2}$; $x = -2$

(ii) $y = \dfrac{\frac{x^2}{x^2} - \frac{2x}{x^2} - \frac{3}{x^2}}{\frac{2x^2}{x^2} - \frac{x}{x^2} - \frac{1}{x^2}}$

$= \dfrac{1 - 0 - 0}{2 - 0 - 0} = \dfrac{1}{2}$

V.A.: $x = -2$, $x = \frac{5}{2}$

H.A.: $y = \frac{1}{2}$

17. Graph **A** has no vertical asymptotes.

(**A**) No vertical asymptotes $(x = \pm i\sqrt{2})$

(**B**) $x = \pm\sqrt{2}$

(**C**) $x = 0$

(**D**) $x = 8$

19. $f(x) = \dfrac{x + 1}{x - 4}$

(i) V.A.: $x = 4$; H.A.: $y = 1$

$\left(y = \dfrac{\frac{x}{x} + \frac{1}{x}}{\frac{x}{x} - \frac{4}{x}} = \dfrac{1 + 0}{1 - 0}\right)$

(ii) x-intercept: -1 $(x + 1 = 0 \to x = -1)$

y-intercept: $-\frac{1}{4}$ $\left(f(0) = \dfrac{0 + 1}{0 - 4} = -\dfrac{1}{4}\right)$

(iii) Crosses asymptote? No

$1 = \dfrac{x + 1}{x - 4}$

$x - 4 = x + 1$

$-4 = 1$ $\emptyset$

(iv) points: $(3, -4)$ and $(5, 6)$

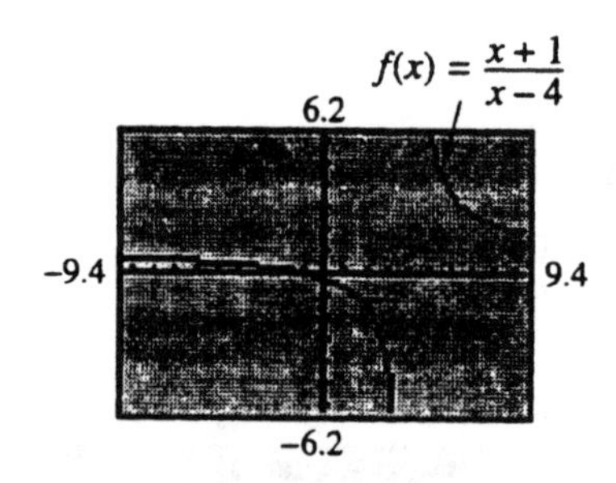

21. $f(x) = \dfrac{3x}{(x + 1)(x - 2)}$

(i) V.A.: $x = -1$, $x = 2$

H.A.: $y = 0$

(ii) x-intercept: 0 $(3x = 0)$

y-intercept: 0 $\left(f(0) = \dfrac{0}{-2} = 0\right)$

(iii) Crosses asymptote at $x = 0$:

$0 = \dfrac{3x}{(x - 1)(x - 2)}$

$0 = 3x$

$x = 0$ $(0, 0)$

(iv) points: $\left(-2, -\frac{3}{2}\right)$ and $\left(3, \frac{9}{4}\right)$

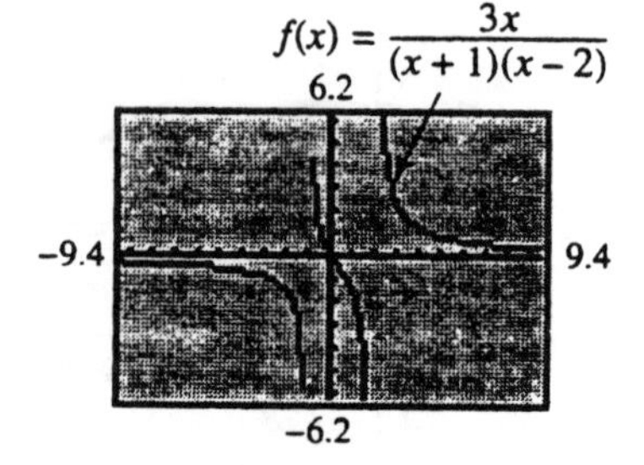

23. $f(x) = \dfrac{5x}{x^2 - 1}$

(i) V.A.: $x = -1,\ x = 1$

$\left[x^2 - 1 = (x-1)(x+1)\right]$

H.A.: $y = 0$

(ii) x-intercept: 0 $(5x = 0)$

y-intercept: 0 $(f(0) = 0)$

(iii) Crosses asymptote at $x = 0$:

$0 = \dfrac{5x}{x^2 - 1}$

$0 = 5x \qquad x = 0$

(iv) points: $\left(-2, -\dfrac{10}{3}\right), \left(2, \dfrac{10}{3}\right)$

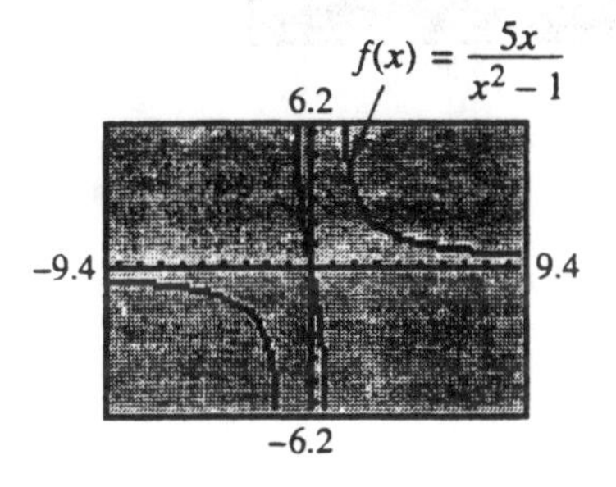

25. $f(x) = \dfrac{(x-3)(x+1)}{(x-1)^2}$

(i) V.A.: $x = 1$

H.A.: $y = 1$ $\left(y = \dfrac{\frac{x^2}{x^2} - \frac{2x}{x^2} - \frac{3}{x^2}}{\frac{x^2}{x^2} - \frac{2x}{x^2} + \frac{1}{x^2}}\right)$

(ii) x-intercept: $-1,\ 3$ (Numerator $= 0$)

y-intercept: -3 $(f(0) = -3)$

(iii) Crosses asymptote? No

(At $y = 1 \to 1 = -3 : \emptyset$)

(iv) points: $\left(-2, \dfrac{5}{9}\right), (2, -3)$

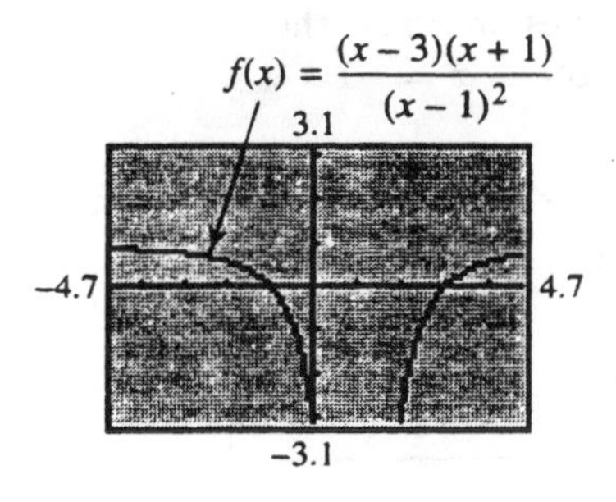

27. $f(x) = \dfrac{x}{x^2 - 9}$

(i) V.A.: $x = 3,\ x = -3$

(Denominator $= 0$)

H.A.: $y = 0$

(ii) x-intercept: 0 (Numerator $= 0$)

y-intercept: 0 $(f(0) = 0)$

(iii) Crosses asymptote at $x = 0$

$0 = \dfrac{x}{x^2 - 9} \to x = 0$

(iv) points: $\left(\pm 4, \dfrac{4}{7}\right), \left(-2, \dfrac{2}{5}\right), \left(2, -\dfrac{2}{5}\right)$

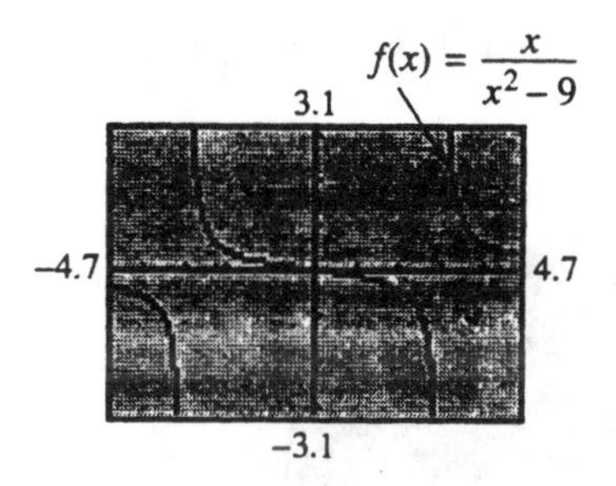

29. $f(x) = \dfrac{1}{x^2 + 1}$

(i) V.A.: none $(x^2 + 1 = 0 \to x = \pm i)$

H.A.: $y = 0$

(ii) x-intercept: none (Numerator : $1 \neq 0$)

y-intercept: 1 $(f(0) = 1)$

(iii) Crosses asymptote? No

$0 = \dfrac{1}{x^2 + 1} \to 0 = 1 : \quad \emptyset$

(iv) points: $\left(\pm 2, \dfrac{1}{5}\right), \left(\pm 1, \dfrac{1}{2}\right)$

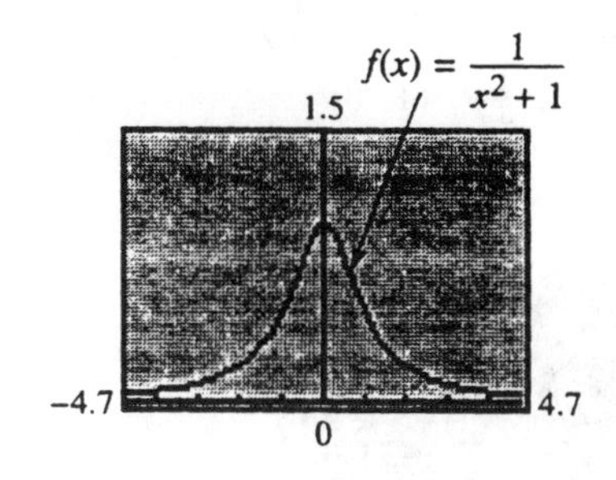

31. $f(x) = \dfrac{x^2+1}{x+3} = x - 3 + \dfrac{10}{x+3}$

$$\begin{array}{r|rrr} & x & -3 & \\ \hline x+3) & x^2 & +0x & +1 \\ & x^2 & +3x & \\ \hline & & -3x & +1 \\ & & -3x & -9 \\ \hline & & & 10 \end{array}$$

(i) V.A.: $x = -3$; O.A.: $y = x - 3$

(ii) x-intercept: none $(x^2+1=0 \to x = \pm i)$

y-intercept: $\frac{1}{3}$ $\left(f(0) = \frac{1}{3}\right)$

(iii) points: $(-4, -17)$, $(-2, 5)$, $\left(1, \frac{1}{2}\right)$

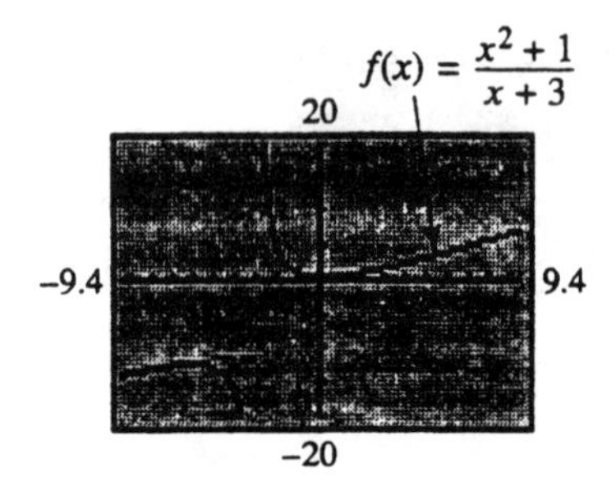

33. $f(x) = \dfrac{x^2+2x}{2x-1} = \dfrac{1}{2}x + \dfrac{5}{4} + \dfrac{5/4}{2x-1}$

$$\begin{array}{r|rrr} & \frac{1}{2}x & +\frac{5}{4} & \\ \hline 2x-1) & x^2 & +2x & +0 \\ & x^2 & -\frac{1}{2}x & \\ \hline & & \frac{5}{2}x & \\ & & \frac{5}{2}x & -\frac{5}{4} \\ \hline & & & \frac{5}{4} \end{array}$$

(i) V.A.: $x = \frac{1}{2}$; O.A.: $y = \frac{1}{2}x + \frac{5}{4}$

(ii) x-intercept: $0, -2$ $(x(x+2))$

y-intercept: 0 $\left(f(0) = \frac{0}{-1}\right)$

(iii) Points: $\left(-1, \frac{1}{3}\right)$, $(1, 3)$

f(x) = (x² + 2x)/(2x − 1)
6.2
-4.7
4.7
-3.1

35. $f(x) = \dfrac{x^2-9}{x+3} = x - 3,\ x \neq -3$

Line $y = x - 3$ with a hole at $x = -3$.

(i) No asymptotes

(ii) x-intercept: $x = 3$

y-intercept: -3

(iii) points: $(-2, -5)$, $(2, -1)$

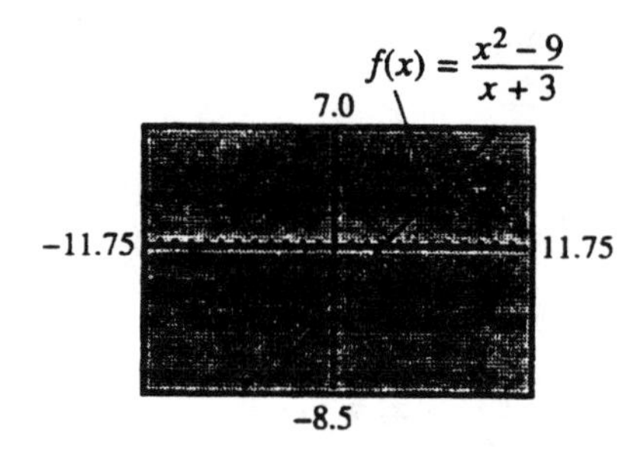

37. $Y_1 = \dfrac{x-p}{x-q}$

Since $x = 2$ gives an error, the vertical asymptote is $x = 2$:

$x - q = 0$
$2 - q = 0 \to q = 2$

When $y = 0$, $x = 4$:

$x - p = 0$
$4 - p = 0 \to 4 = p$

$\boxed{p = 4,\ q = 2}$

39. $Y_1 = \dfrac{x-p}{x-q}$

Since $x = -1$ gives an error, the vertical asymptote is $x = -1$:

$x - q = 0$
$-1 - q = 0 \Longrightarrow -1 = q$

When $y = 0$, $x = -2$:

$x - p = 0$
$-2 - p = 0$
$-2 = p$ $\boxed{p = -2,\ q = -1}$

Answers may vary in Exercises 41 - 43.

41. V.A.: $x = 2$;

one factor in denominator is $(x - 2)$.

H.A.: $y = 1$;

degree of numerator is the same as degree of denominator; ratio of leading coefficients is 1.

There is a hole in the graph at $x = -2$; one factor in both numerator and denominator is $(x + 2)$.

The x-intercept is 3; one factor in the numerator is $(x - 3)$.

$$f(x) = \frac{(x-3)(x+2)}{(x-2)(x+2)} \text{ or } f(x) = \frac{x^2 - x - 6}{x^2 - 4}$$

43. V.A.: $x = 0,\ x = 4$;

two factors in denominator; $(x)(x - 4)$.

The x-intercept is 2;

one factor in the numerator is $(x - 2)$.

H.A.: $y = 0$;

the power of the numerator is less than the power of the denominator:

$$f(x) = \frac{x-2}{x(x-4)} \text{ or } f(x) = \frac{x-2}{x^2 - 4x}$$

45. $x^2 + x + 4 = 0$

$$x = \frac{-1 \pm \sqrt{1-16}}{2} = \frac{-1 \pm i\sqrt{15}}{2}$$

Complex solutions:

$$-\frac{1}{2} + \frac{\sqrt{15}}{2}i,\quad -\frac{1}{2} - \frac{\sqrt{15}}{2}i$$

Real solution: none;
No vertical asymptotes.

47. (a) $y = -f(x)$ (b) $y = f(-x)$

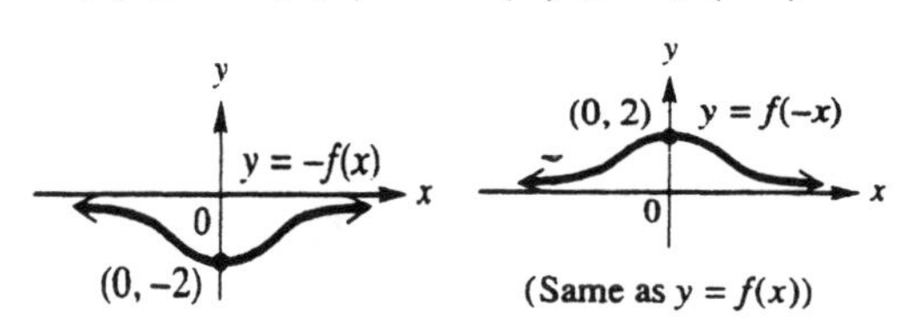

49. (a) $y = -f(x)$ (b) $y = f(-x)$

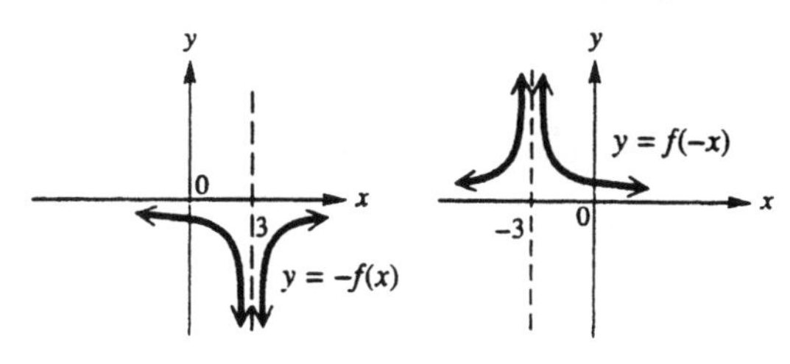

51. $f(x) = \dfrac{2x^2 + 3}{4 - x} = -2x - 8 + \dfrac{35}{4 - x}$

$$\begin{array}{r|l} & -2x - 8 \\ \hline -x + 4 & 2x^2 + 0x + 3 \\ & 2x^2 - 8x \\ \hline & \quad 8x + 3 \\ & \quad 8x - 32 \\ \hline & \qquad 35 \end{array}$$

O.A.: $y = -2x - 8$

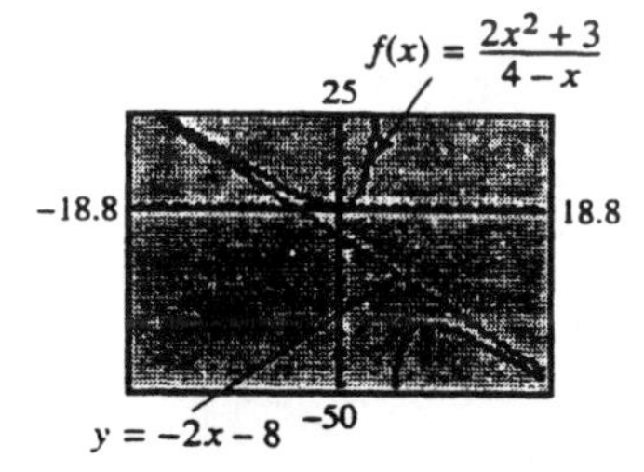

53. $f(x) = \dfrac{x - x^2}{x + 2} = -x + 3 + \dfrac{-6}{x + 2}$

$$\begin{array}{r|l} & -x + 3 \\ \hline x + 2 & -x^2 + x + 0 \\ & -x^2 - 2x \\ \hline & \quad 3x + 0 \\ & \quad 3x + 6 \\ \hline & \qquad -6 \end{array}$$

O.A.: $y = -x + 3$

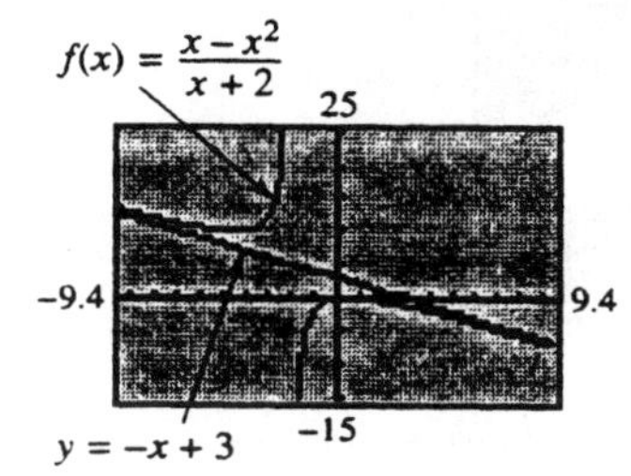

55. $f(x) = \dfrac{x^5 + x^4 + x^2 + 1}{x^4 + 1} = x + 1 + \dfrac{x^2 - x}{x^4 + 1}$

(a) oblique asymptote: $y = x + 1$

(b) intersects the asymptote at $x = 0$, $x = 1$

$$x + 1 + \frac{x^2 - x}{x^4 + 1} = x + 1$$
$$\frac{x^2 - x}{x^4 + 1} = 0$$
$$x^2 - x = 0$$
$$x(x - 1) = 0 \Longrightarrow x = 0,\ 1$$

(c) as $x \longrightarrow \infty$, $f(x) \longrightarrow x + 1$ from above

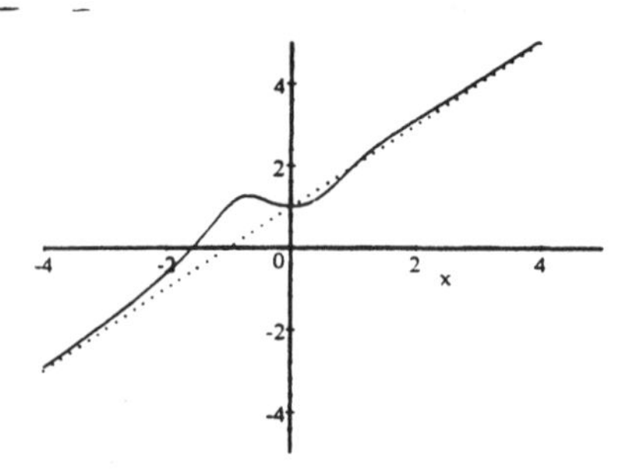

$$f(x) = x + 1 + \frac{x^2 - x}{x^4 + 1}$$

57. $f(x) = \dfrac{x^4 - 5x^2 + 4}{x^2 + x - 12}$

$$\begin{array}{r} x^2 \quad - x \quad + 8 \\ x^2 + x - 12 \overline{) \, x^4 + 0x^3 - 5x^2 + 0x + 4} \\ \underline{x^4 + 1x^3 - 12x^2} \\ -x^3 + 7x^2 + 0x \\ \underline{-x^3 - x^2 + 12x} \\ 8x^2 - 12x + 4 \\ \underline{8x^2 + 8x - 96} \\ -20x + 100 \end{array}$$

$$\boxed{Q(x) = x^2 - x + 8,\ R = -20x + 100}$$

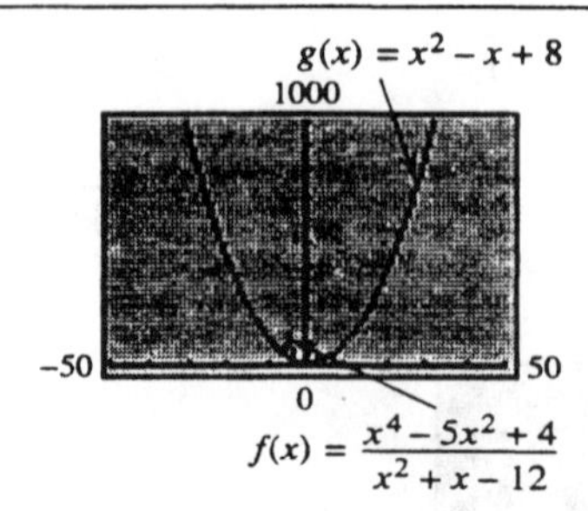

In this window, the two graphs seem to overlap, suggesting that as $|x| \to \infty$, the graph of f approaches the curve defined by Q, giving an asymptotic effect.

59. $f(x) = \dfrac{x^2 - 9}{x + 3}, x \neq -3$

(a) $f(x) = \dfrac{(x + 3)(x - 3)}{x + 3}$, $x \neq -3$

$g(x) = x - 3$

(b) f is undefined at $x = -3$, as indicated by the error message.

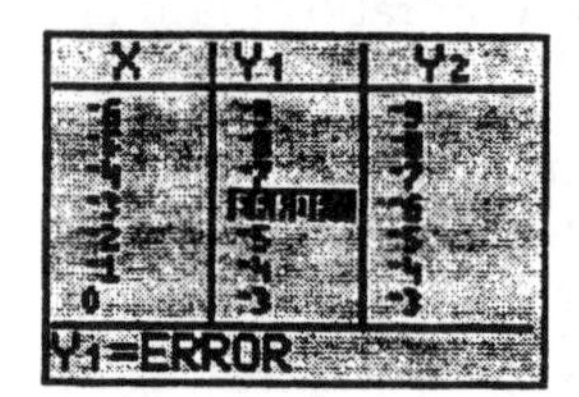

(c) The *hole* is at $(-3, -6)$.

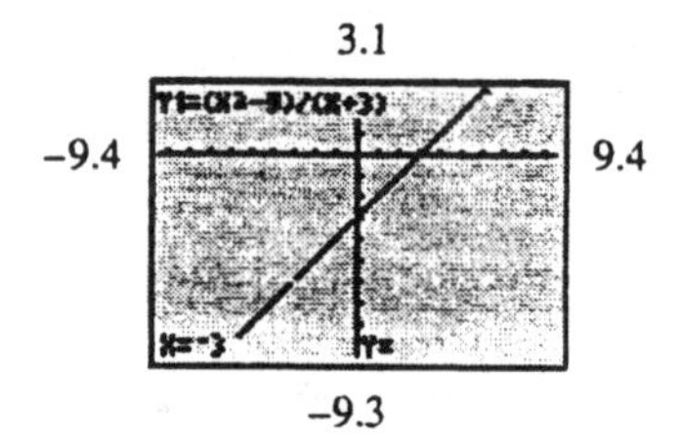

61. (a) $m < n$: The horizontal asymptote has equation $y = 0$, the x-axis.

(b) $m = n$: The horizontal asymptote has equation $y = \dfrac{a}{b}$, where a is the leading coefficient of $p(x)$ and b is the leading coefficient of $q(x)$.

(c) $m = n + 1$: The oblique asymptotehas equation $y = ax + b$, where $ax + b$ is the quotient (with remainder disregarded), found by dividing $p(x)$ by $q(x)$.

Section 4.3

1. **(a)** $f(x) = 0 : \emptyset$
(b) $f(x) < 0 : (-\infty, -2)$
(c) $f(x) > 0 : (-2, \infty)$

3. **(a)** $f(x) = 0 : \{-1\}$
(b) $f(x) < 0 : (-1, 0)$
(c) $f(x) > 0 : (-\infty, -1) \cup (0, \infty)$

5. **(a)** $f(x) = 0 : \{0\}$
(b) $f(x) < 0 : (-2, 0) \cup (2, \infty)$
(c) $f(x) > 0 : (-\infty, -2) \cup (0, 2)$

7. **(a)** $f(x) = 0 : \emptyset$
(b) $f(x) < 0 : (-\infty, 0) \cup (0, \infty)$
(c) $f(x) > 0 : \emptyset$

9. **(a)** $f(x) = 0 : \{.75\}$
(b) $f(x) < 0 : (-\infty, .75) \cup (2, \infty)$
(c) $f(x) > 0 : (.75, 2)$

11. **(a)** $\dfrac{x-3}{x+5} = 0 \quad (x \neq -5)$

$x - 3 = 0 \qquad \boxed{x = \{3\}}$

(b)–(c) Numerator:

$x - 3 = 0 \rightarrow x = 3$

Denominator:

$x + 5 = 0 \rightarrow x = -5$

		−5		3	
$(x+5)$	−		+		+
$(x-3)$	−		−		+
$(quot)$	+		−		+

continued

11. continued

$$\boxed{\begin{array}{l} \dfrac{x-3}{x+5} \leq 0 : (-5, 3] \\ \dfrac{x-3}{x+5} \geq 0 : (-\infty, -5) \cup [3, \infty) \end{array}}$$

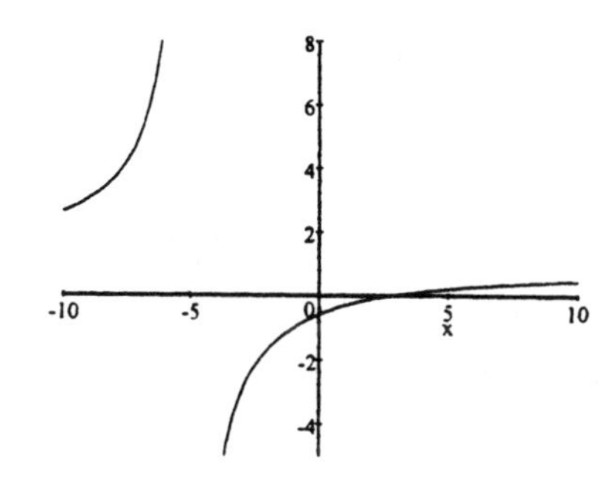

13. **(a)** $\dfrac{x-1}{x+2} = 1 \quad (x \neq -2)$

$x - 1 = x + 2$

$-1 = 2 \qquad \boxed{\emptyset}$

(b)–(c) $\dfrac{x-1}{x+2} - \dfrac{x+2}{x+2} = 0$

$\dfrac{-3}{x+2} = 0$

Denominator: $x = -2$

		−2	
$(x+2)$	−		+
$\left(\frac{-3}{x+2}\right)$	+		−

$$\boxed{\begin{array}{l} \dfrac{x-1}{x+2} > 1 : (-\infty, -2) \\ \dfrac{x-1}{x+2} < 1 : (-2, \infty) \end{array}}$$

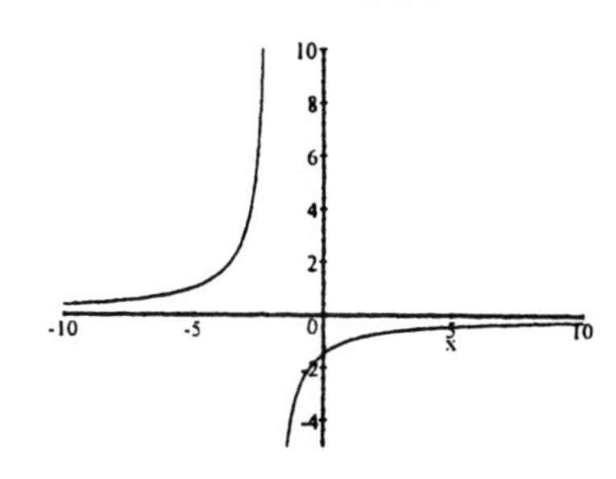

$y = \frac{x-1}{x+2} - 1$

15. (a) $\dfrac{1}{x-1} = \dfrac{5}{4} \quad (x \neq 1)$

$5x - 5 = 4$

$5x = 9, \quad \boxed{x = \left\{\dfrac{9}{5}\right\}}$

(b)–(c) $\dfrac{1}{x-1} - \dfrac{5}{4} = 0$

$\dfrac{-5x+9}{4(x-1)} = 0$

Numerator: $-5x + 9 = 0 \rightarrow x = \dfrac{9}{5}$

Denominator: $4x - 4 = 0 \rightarrow x = 1$

		1		9/5	
$(4x-4)$	$-$		$+$		$+$
$(-5x+9)$	$+$		$+$		$-$
$(quot)$	$-$		$+$		$-$

$\dfrac{1}{x-1} < \dfrac{5}{4} : (-\infty, 1) \cup \left(\dfrac{9}{5}, \infty\right)$

$\dfrac{1}{x-1} > \dfrac{5}{4} : \left(1, \dfrac{9}{5}\right)$

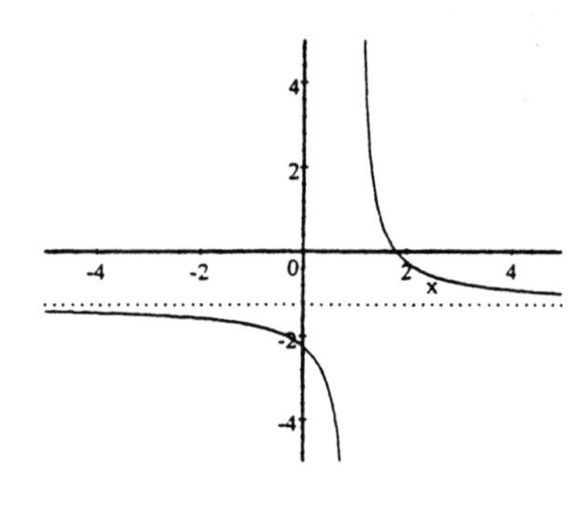

$y = \frac{1}{x-1} - \frac{5}{4}$

17. (a) $\dfrac{4}{x-2} = \dfrac{3}{x-1} \quad (x \neq 1,\ 2)$

$3x - 6 = 4x - 4$

$-x = 2 \rightarrow \boxed{x = \{-2\}}$

(b)–(c) $\dfrac{4}{x-2} - \dfrac{3}{x-1} = 0$

$\dfrac{4(x-1) - 3(x-2)}{(x-2)(x-1)} = 0$

$\dfrac{x+2}{(x-2)(x-1)} = 0$

Numerator: $x + 2 = 0 \rightarrow x = -2$

Denominator:

$x - 1 = 0 \rightarrow x = 1$

$x - 2 = 0 \rightarrow x = 2$

		-2		1		2	
$(x+2)$	$-$		$+$		$+$		$+$
$(x-2)$	$-$		$-$		$-$		$+$
$(x-1)$	$-$		$-$		$+$		$+$
$(quot)$	$-$		$+$		$-$		$+$

$\dfrac{4}{x-2} \leq \dfrac{3}{x-1} : (-\infty, -2] \cup (1, 2)$

$\dfrac{4}{x-2} \geq \dfrac{3}{x-1} : [-2, 1) \cup (2, \infty)$

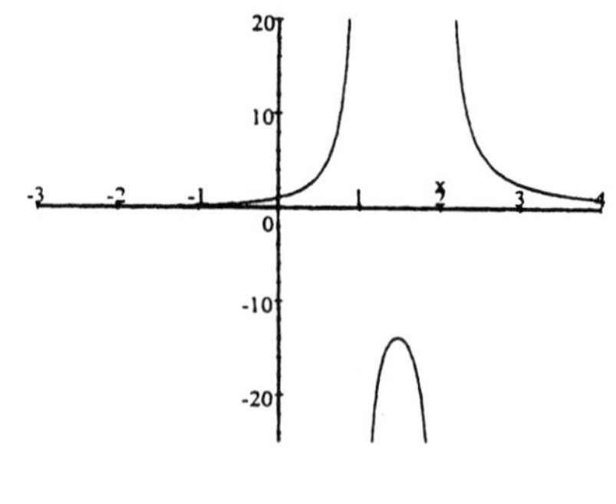

$y = \frac{4}{x-2} - \frac{3}{x-1}$

19. (a) $\frac{1}{(x-2)^2} = 0 \rightarrow 1 = 0$ $\boxed{\emptyset}$

(b)–(c) Denominator: $x - 2 = 0 \rightarrow x = 2$

$\left(\frac{1}{(x-2)^2}\right)$	$+$	$+$
	2	

$$\boxed{\begin{array}{l} \frac{1}{(x-2)^2} < 0: \emptyset \\ \frac{1}{(x-2)^2} > 0: (-\infty, 2) \cup (2, \infty) \end{array}}$$

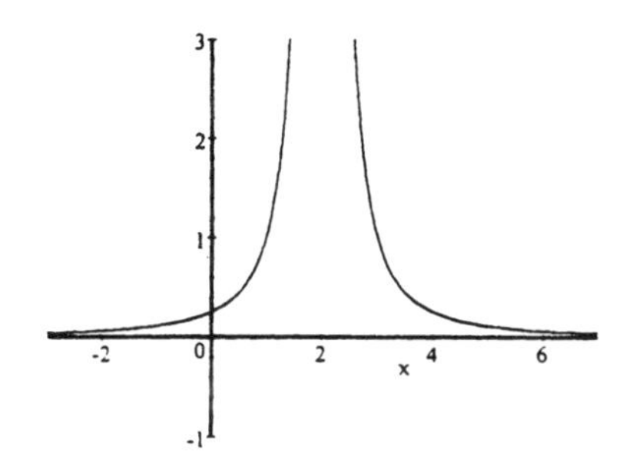

21. (a) $\frac{5}{x+1} = \frac{12}{x+1}$ $(x \neq -1)$

$5x + 5 = 12x + 12 \Longrightarrow x = -1 : \boxed{\emptyset}$

(b)–(c) $\frac{5}{x+1} - \frac{12}{x+1} \Longrightarrow \frac{-7}{x+1} = 0$

Denominator: $x + 1 = 0 \rightarrow x = -1$

$(x+1)$	$-$	$+$
$\left(\frac{-7}{x+1}\right)$	$+$	$-$
	-1	

$$\boxed{\begin{array}{l} \frac{5}{x+1} > \frac{12}{x+1} : (-\infty, -1) \\ \frac{5}{x+1} < \frac{12}{x+1} : (-1, \infty) \end{array}}$$

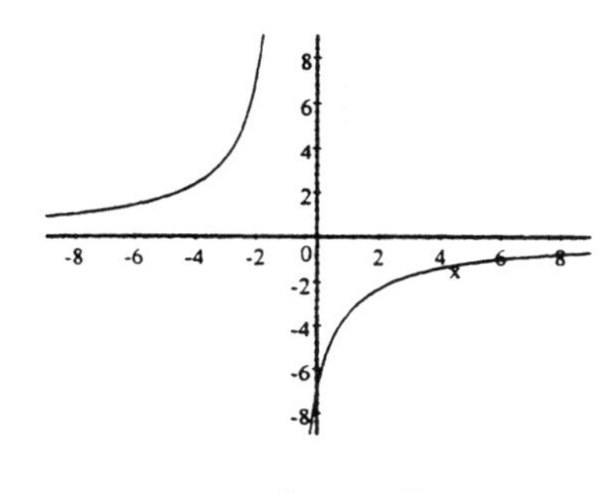

$y = \frac{5}{x+1} - \frac{12}{x+1}$

23. $\frac{-1}{x^2+2} < 0 : (-\infty, \infty)$

numerator < 0; denominator > 0,

so $\frac{-1}{x^2+2}$ will always be < 0.

25. $\frac{-5}{x^2+2} > 0 : \emptyset$

numerator < 0; denominator > 0,

so $\frac{-5}{x^2+2}$ will always be < 0.

27. $\frac{x^4+2}{-6} \geq 0 : \emptyset$

numerator > 0; denominator < 0,

so $\frac{x^4+2}{-6}$ will always be < 0, never ≥ 0.

29. $\frac{x^4+x^2+3}{x^2+2} > 0 : (-\infty, \infty)$

numerator > 0; denominator > 0,

so $\frac{x^4+x^2+3}{x^2+2}$ will always be > 0.

31. $\frac{(x-1)^2}{x^2+4} \leq 0 : \{1\}$

numerator > 0, $x \neq 1$; denominator > 0,

so $\frac{(x-1)^2}{x^2+4}$ will always be > 0, $x \neq 1$.

33. $\dfrac{2x}{x^2-1}=\dfrac{2}{x+1}-\dfrac{1}{x-1}$

$$(x^2-1)\left[\frac{2x}{x^2-1}=\frac{2}{x+1}-\frac{1}{x-1}\right]$$

$$2x=2(x-1)-1(x+1)$$

$$2x=2x-2-x-1$$

$$2x=x-3$$

$$x=-3 \qquad \{-3\}$$

$y_1=\dfrac{2x}{x^2-1}$ (dark), $y=\dfrac{2}{x+1}-\dfrac{1}{x-1}$ (light)

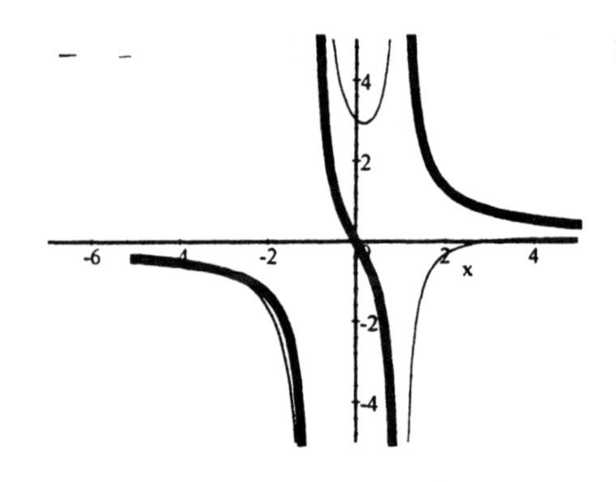

35. $\dfrac{4}{x^2-3x}-\dfrac{1}{x^2-9}=0$

$$\frac{4}{x(x-3)}-\frac{1}{(x-3)(x+3)}=0$$

$$x(x-3)(x+3)\left[\frac{4}{x(x-3)}-\frac{1}{(x-3)(x+3)}=0\right]$$

$$4(x+3)-x=0$$

$$4x+12-x=0$$

$$3x=-12 \quad\Longrightarrow x=\{-4\}$$

$$y_1=\frac{4}{x^2-3x}-\frac{1}{x^2-9}$$

37. $1-\dfrac{13}{x}+\dfrac{36}{x^2}=0$

$$x^2\left[1-\frac{13}{x}+\frac{36}{x^2}=0\right]$$

$$x^2-13x+36=0$$

$$(x-9)(x-4)=0 \qquad \boxed{x=\{4,\ 9\}}$$

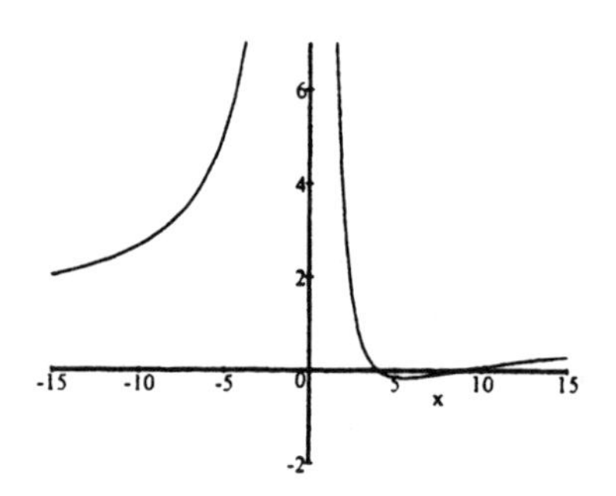

39. $1+\dfrac{3}{x}=\dfrac{5}{x^2}$

$$x^2\left[1+\frac{3}{x}-\frac{5}{x^2}=0\right]$$

$$x^2+3x-5=0$$

$$x=\frac{-3\pm\sqrt{9-4(-5)}}{2}$$

$$\boxed{x=\left\{\frac{-3\pm\sqrt{29}}{2}\right\}\approx\{-4.193,\ 1.1926\}}$$

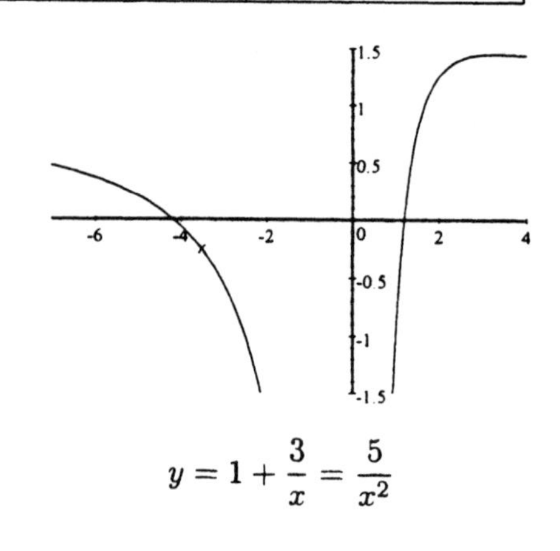

$$y=1+\frac{3}{x}=\frac{5}{x^2}$$

41. $\dfrac{x}{2-x}+\dfrac{2}{x}-5=0$

Multiply by $x(2-x)$:

$x^2+2(2-x)-5x(2-x)=0$

$x^2+4-2x-10x+5x^2=0$

$6x^2-12x+4=0$

$3x^2-6x+2=0$

$x=\dfrac{6\pm\sqrt{36-4(6)}}{6}$

$x=\dfrac{6\pm\sqrt{12}}{6}=\dfrac{6\pm2\sqrt{3}}{6}$

$\boxed{x=\left\{\dfrac{3\pm\sqrt{3}}{3}\right\}\approx\{0.423,\ 1.577\}}$

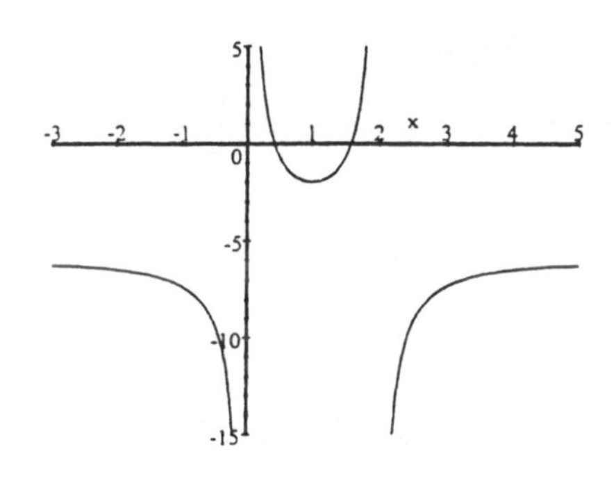

43. $x^{-4}-3x^{-2}-4=0$

$\dfrac{1}{x^4}-\dfrac{3}{x^2}-4=0$

$x^4\left[\dfrac{1}{x^4}-\dfrac{3}{x^2}-4=0\right]$

$1-3x^2-4x^4=0$

$(1-4x^2)(1+x^2)=0$

$1-4x^2=0 \quad 1+x^2=0$

$x^2=\dfrac{1}{4} \qquad x^2=-1$

$\boxed{x=\left\{\pm\dfrac{1}{2},\ \pm i\right\}}$

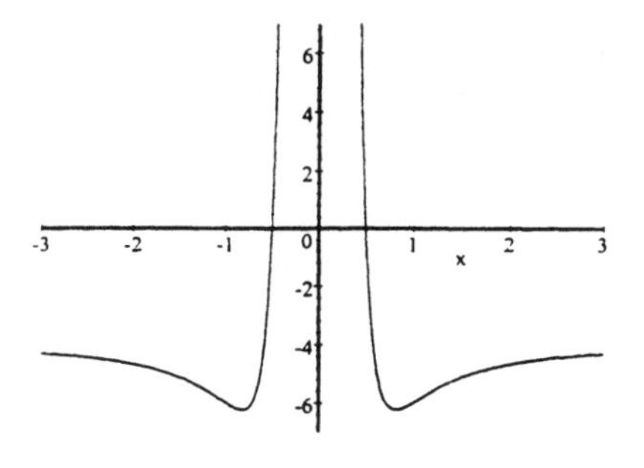

45. $\dfrac{1}{x+2}+\dfrac{3}{x+7}=\dfrac{5}{x^2+9x+14}$

Multiply by $(x+2)(x+7)$:

$x+7+3(x+2)=5$

$x+7+3x+6=5$

$4x=-8\rightarrow x=-2$

(Hole at $x=-2$; extraneous root): $\boxed{\emptyset}$

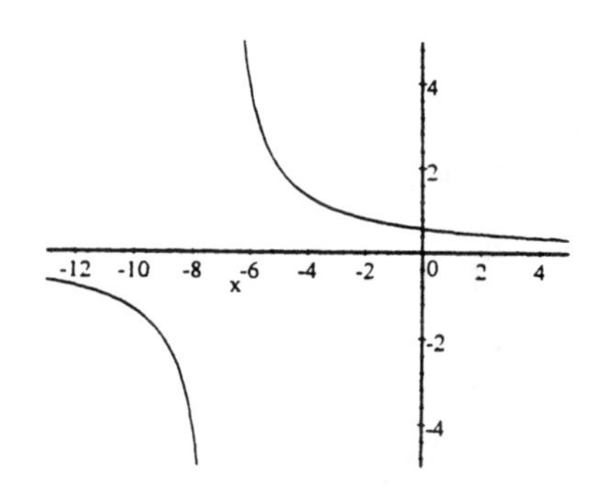

$y=\frac{1}{x+2}+\frac{3}{x+7}-\frac{5}{x^2+9x+14}$

47. $\dfrac{x}{x-3}+\dfrac{4}{x+3}=\dfrac{18}{x^2-9}$

Multiply by $(x-3)(x+3)$:

$x(x+3)+4(x-3)=18$

$x^2+3x+4x-12-18=0$

$x^2+7x-30=0$

$(x+10)(x-3)=0$

$x=-10,\ 3$

(Hole at $x=3$; extraneous root)

$\boxed{x=\{-10\}}$

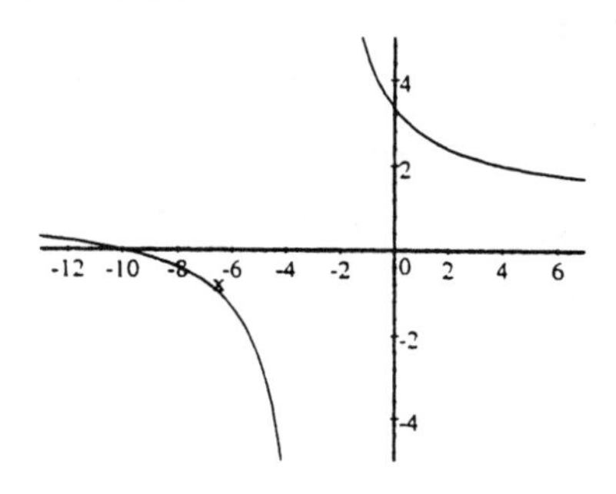

$y=\frac{x}{x-3}+\frac{4}{x+3}-\frac{18}{x^2-9}$

49. $9x^{-1} + 4(6x-3)^{-1} = 2(6x-3)^{-1}$

$\frac{9}{x} + \frac{4}{6x-3} = \frac{2}{6x-3}$

Multiply by $x(6x-3)$:

$9(6x-3) + 4x = 2x$

$54x - 27 + 4x = 2x$

$56x = 27$

$\boxed{x = \left\{\frac{27}{56}\right\}}$

This graph has 3 parts.

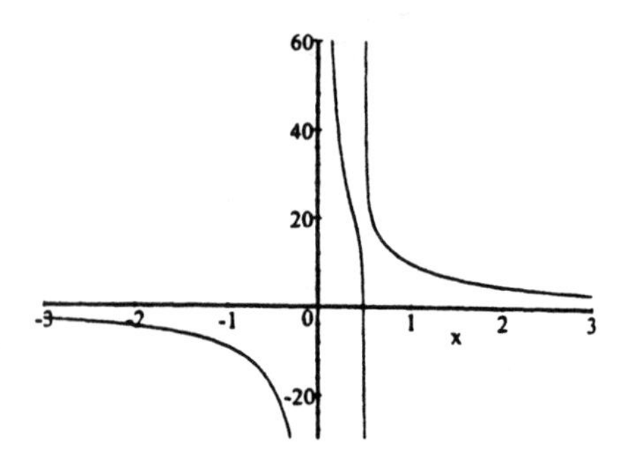

$y = \frac{9}{x} + \frac{4}{6x-3} - \frac{2}{6x-3}$

51. $\frac{x-2}{x+3} \le 2 \quad (x \ne -3)$

$(x+3)^2\left(\frac{x-2}{x+3}\right) \le 2(x+3)^2$

$(x+3)(x-2) \le 2(x^2+6x+9)$

$x^2 + x - 6 \le 2x^2 + 12x + 18$

$-x^2 - 11x - 24 \le 0$

$x^2 + 11x + 24 \ge 0$

53. $\frac{(x-3)^2}{x+1} \ge x+1 \quad (x \ne -1)$

$(x+1)^2\left[\frac{(x-3)^2}{x+1} \ge x+1\right]$

$(x+1)(x-3)^2 \ge (x+1)^3$

$(x+1)(x^2-6x+9) \ge$
$x^3 + 3x^2 + 3x + 1$

$x^3 - 5x^2 + 3x + 9 \ge$
$x^3 + 3x^2 + 3x + 1$

$-5x^2 + 3x + 9 \ge 3x^2 + 3x + 1$

$-8x^2 + 8 \ge 0$

$x^2 - 1 \le 0 \quad [\div \text{ by } -8]$

$(x-1)(x+1) \le 0$

x-values: $-1,\ 1$; from graph: $\boxed{(-1,\ 1]}$

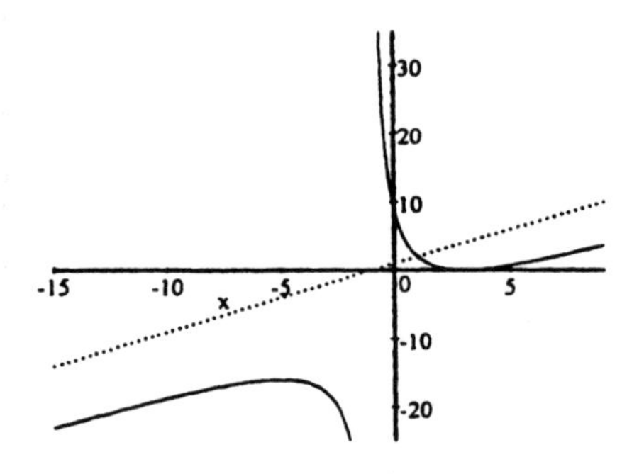

55. $\frac{x+3}{(x+1)^2} < \frac{1}{x+1} \quad (x \ne -1)$

$(x+1)^2\left[\frac{x+3}{(x+1)^2} < \frac{1}{x+1}\right]$

$x + 3 < x + 1$

$3 < 1$: $\boxed{\emptyset}$

57. $f(x) = \dfrac{\sqrt{2}x + 5}{x^3 - \sqrt{3}}$

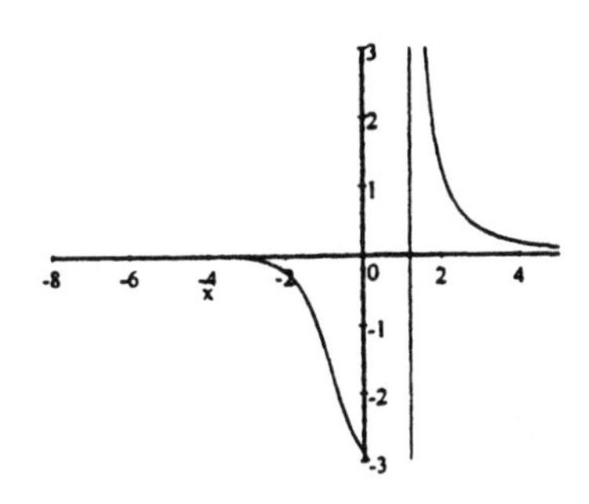

(a) $f(x) = 0$: $\boxed{\{-3.54\}}$

(b) $f(x) > 0$: $\boxed{(-\infty, -3.54) \cup (1.20, \infty)}$

(c) $f(x) < 0$: $\boxed{(-3.54, 1.20)}$

For Exercises 59 - 61:

The function and its rational approximation are graphed on the same calculator screen. From the graphs, we can see that all of the rational approximations give excellent results on the interval $[1, 15]$.

59. The rational approximation of

$f_1(x) = \sqrt{x}$ is **(C)**:

$$r_3(x) = \frac{10x^2 + 80x + 32}{x^2 + 40x + 80}$$

61. The rational approximation of

$f_3(x) = \sqrt[3]{x}$ is **(D)**:

$$r_4(x) = \frac{7x^3 + 42x^2 + 30x + 2}{2x^3 + 30x^2 + 42x + 7}$$

63. **(a)** $f(x) = \dfrac{10x + 1}{x + 1}, \quad x \geq 0$

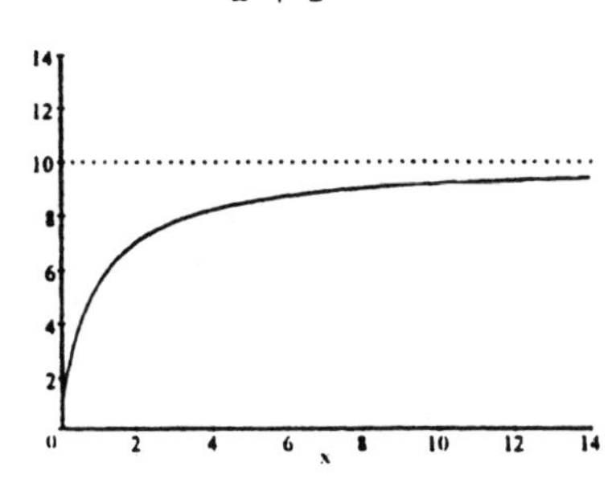

horizontal asymptote: $y = 10$

(b) $f(0) = \dfrac{10(0) + 1}{0 + 1} = \dfrac{1}{1} = 1$ million;
1,000,000 insects **continued**

63. continued

(c) After several months, the population starts to level off at 10,000,000 (it approaches the asymptote $y = 10$).

(d) The horizontal asymptote $y = 10$, represents the limiting population after a long time.

65. $f(x) = \dfrac{x - 5}{x^2 - 10x}, \quad x > 10$

(a) $.25 = \dfrac{x - 5}{x^2 - 10x}$ [15 sec = .25 min]

$.25x^2 - 2.5x = x - 5$

$.25x^2 - 3.5x + 5 = 0$

$$x = \frac{3.5 \pm \sqrt{(-3.5)^2 - 4(.25)(5)}}{2(.25)}$$

$\approx 12.38, \ 1.61$ (extraneous)

$\boxed{\text{Approximately 12.4 cars per minute.}}$

(b) 3 attendants (at 5/minute) will be needed.

67. Volume $= lwh$

$196 = 2w \cdot wh$

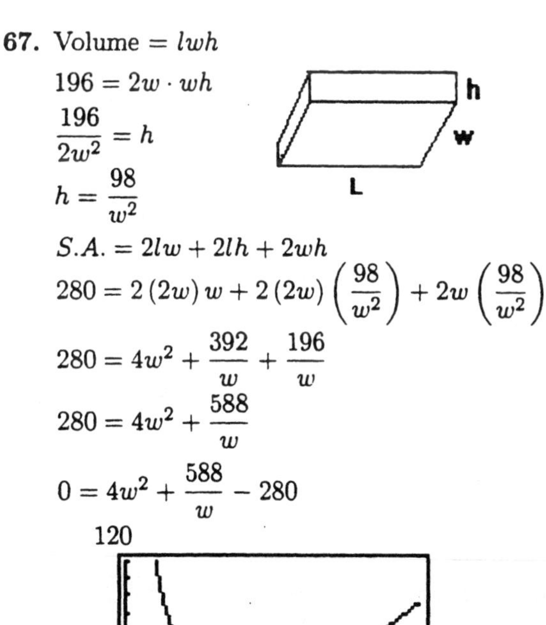

$\dfrac{196}{2w^2} = h$

$h = \dfrac{98}{w^2}$

$S.A. = 2lw + 2lh + 2wh$

$280 = 2(2w)w + 2(2w)\left(\dfrac{98}{w^2}\right) + 2w\left(\dfrac{98}{w^2}\right)$

$280 = 4w^2 + \dfrac{392}{w} + \dfrac{196}{w}$

$280 = 4w^2 + \dfrac{588}{w}$

$0 = 4w^2 + \dfrac{588}{w} - 280$

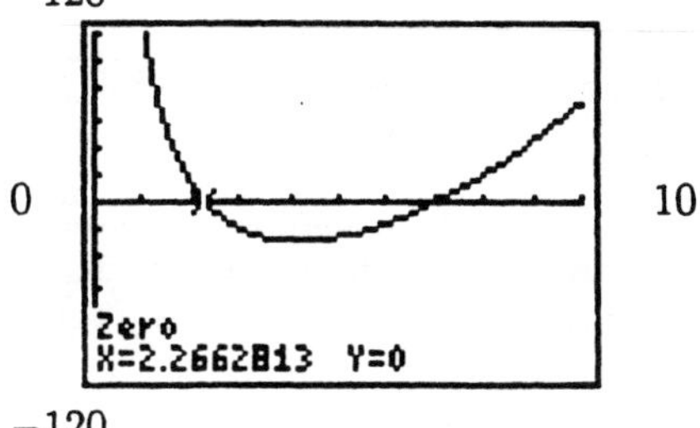

The 2 zeros are at $x \approx 2.266$ and $x = 7$

There are 2 possible solutions:

width $= 7$ inches	width ≈ 2.266 inches
length $= 14$ inches	length ≈ 4.532 inches
height $= 2$ inches	height ≈ 19.086 inches

69. $f(x) = \dfrac{2540}{x}$

(a) $f(400) = \dfrac{2540}{400} = 6.35$ inches

A curve designed for 60 mph with a radius of 400 feet should have the outer rail elevated 6.35 inches.

(b) As the radius, x, of the curve increases, the elevation of the outer rail decreases.

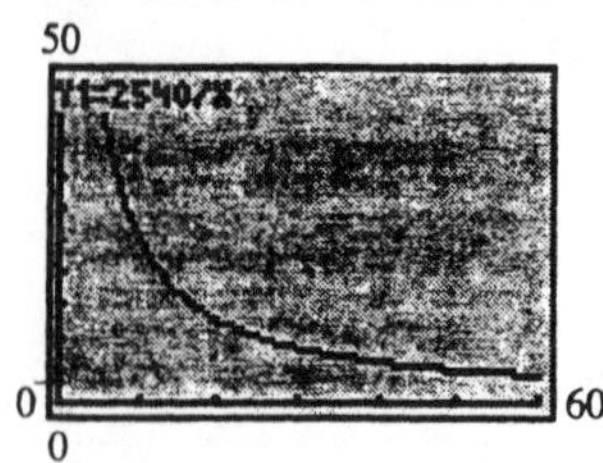

(c) The horizontal asymptote is $y = 0$. As the radius increases without limit ($x \longrightarrow \infty$), the tracks become straight and no elevation or banking is necessary ($y \longrightarrow 0$).

(d) $12.7 = \dfrac{2540}{x}$

$12.7x = 2540 \Longrightarrow x = \dfrac{2540}{12.7} = 200$ feet

71. $P(x) = \dfrac{x - 1}{x}$

(a) $P(9) = \dfrac{9 - 1}{9} = \dfrac{8}{9} \approx .89 \approx 89\%$

(b) $P(1.9) = \dfrac{1.9 - 1}{1.9} = \dfrac{.9}{1.9} \approx .47368 \approx 47\%$

73. $D(x) = \dfrac{2500}{30\,(.3 + x)}$

(a) $D(-.1) = \dfrac{2500}{30\,(.3 - .1)} = 416\frac{2}{3} \approx 417$

The braking distance for a car traveling 50 mph on a 10% downhill grade is about 417 feet.

(b) As the downhill grade, x, increases, the braking distance increases, which agrees with driving experience.

(c) As the downhill grade, x, gets close to 30%, the braking distance increases without bound, which means that stopping in time becomes impossible.

(d) $350 = \dfrac{2500}{30\,(.3 + x)}$

$3150 + 10,500x = 2500$

$10,500x = -650$

$x = \dfrac{-65}{1050} \approx -.0619 \approx -6.2\%$ (downhill)

75. $r = \dfrac{km^2}{s}$; $r = 12,\ m = 6,\ s = 4$:

$$12 = \frac{k(6)^2}{4} = \frac{36k}{4} = 9k$$

$$k = \frac{12}{9} = \frac{4}{3}$$

EQ: $r = \dfrac{4m^2}{3s}$; $m = 4,\ s = 10$

$$r = \frac{4(4)^2}{3(10)} = \frac{64}{30} = \boxed{\frac{32}{15}}$$

77. $a = \dfrac{kmn^2}{y^3}$; $a = 9,\ m = 4,\ n = 9,\ y = 3$

$$9 = \frac{k(4)(81)}{27} = 12k$$

$$k = \frac{9}{12} = \frac{3}{4}$$

EQ: $a = \dfrac{3mn^2}{4y^3}$; $m = 6,\ n = 2,\ y = 5$

$$a = \frac{3(6)(4)}{4(125)} = \boxed{\frac{18}{125}}$$

79. For $k > 0$, if y varies directly as x, when x increases, y <u>increases</u>, and when x decreases, y <u>decreases</u>.

81. $y = \dfrac{k}{x}$; $\quad \dfrac{1}{2}y = \dfrac{k}{2x}$

If x doubles, y becomes half as much.

83. $y = kx^3$; $\quad 27y = k(3x)^3 = 27kx^3$

If x triples, y becomes 27 times as much.

85. $BMI = \dfrac{kw}{h^2}$, $w =$ weight, $h =$ height

$$24 = \frac{k \cdot 177}{72^2} \qquad [6 \text{ ft} = 72 \text{ in}]$$

$$k = \frac{24 \times 72^2}{177} = \frac{41472}{59}$$

$$BMI = \frac{41472w}{59h^2} = \frac{41472(130)}{59(66)^2} \approx 20.98 \approx 21$$

87. $R = \dfrac{k}{d^2}$ $\quad .5 = \dfrac{k}{(2)^2} \Longrightarrow k = 2$

$$R = \frac{2}{d^2} = \frac{2}{3^2} = \frac{2}{9} \text{ ohm}$$

89. $w = \dfrac{k}{d^2}$, $w =$ weight, $d =$ distance from center of earth

$$160 = \frac{k}{(4000)^2}$$

$$k = 106(4000)^2 = 2.56 \times 10^9$$

$$w = \frac{2.56 \times 10^9}{d^2} = \frac{2.56 \times 10^9}{(12,000)^2} \approx 17.8 \text{ pounds}$$

91. $V = kr^2h$

$$300 = k(3)^2(10.62)$$

$$k = \frac{300}{9(10.62)} \approx 3.1387$$

$$V = 3.1387(4)^2(15.92) \approx 799.498 \approx 799.5 \text{ cubic cm}$$

93. $Y_1 = \dfrac{k}{x} \rightarrow k = xY_1$

$(3,\ 1.7) : k = 3(1.7) = \boxed{5.1}$

95. $Y_1 = \dfrac{k}{x} \rightarrow k = xY_1$

$(-2,\ .7) : k = -2(.7) = \boxed{-1.4}$

Reviewing Basic Concepts (Sections 4.1 - 4.3)

1. $y = \dfrac{1}{x+2} - 3$; Shift the graph of $f(x) = \dfrac{1}{x}$ left 2 units and downward 3 units.

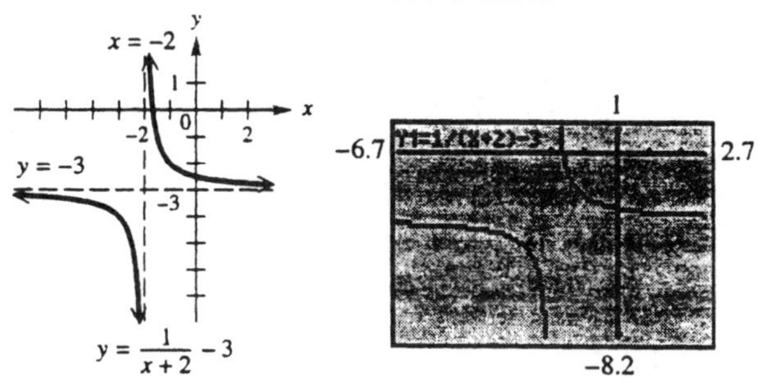

2. $f(x) = \dfrac{3}{x^2 - 1} = \dfrac{3}{(x-1)(x+1)}$

Domain: $(-\infty, -1) \cup (-1, 1) \cup (1, \infty)$

3. $f(x) = \dfrac{4x+3}{x-6}$

Vertical asymptote: $x = 6$

4. $f(x) = \dfrac{x^2+3}{x^2-4}$

Horizontal asymptote: $y = 1$

5. $f(x) = \dfrac{x^2+x+5}{x+3}$

$$\begin{array}{r} x - 2 \\ x+3 \overline{)\, x^2 + x + 5} \\ \underline{x^2 + 3x} \\ -2x + 5 \\ \underline{-2x - 6} \\ 11 \end{array}$$

Oblique Asymptote: $y = x - 2$

6. $f(x) = \dfrac{3x+6}{x-4}$

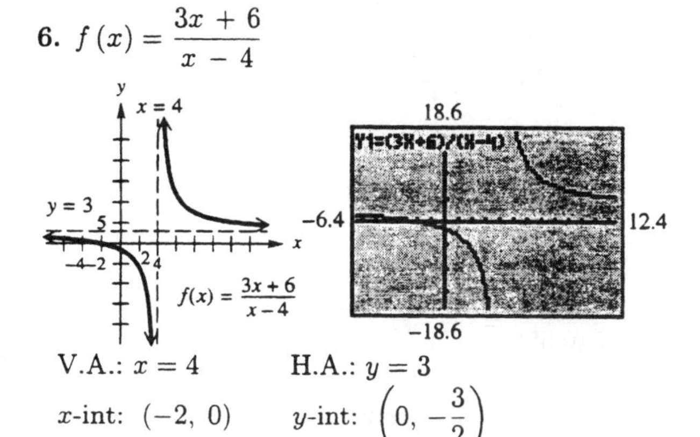

V.A.: $x = 4$ H.A.: $y = 3$

x-int: $(-2,\ 0)$ y-int: $\left(0,\ -\dfrac{3}{2}\right)$

7. **(a)** $f(x) = 0$ at $x = -2$

(b) $f(x) > 0$: $(-\infty, -2) \cup (2, \infty)$

(c) $f(x) < 0$: $(-2, 2)$

8. $\dfrac{x+4}{3x+1} > 1$ $\boxed{\left(-\frac{1}{3}, \frac{3}{2}\right)}$

$$\frac{x+4}{3x+1} - \frac{3x+1}{3x+1} > 0$$

$$\frac{x+4-3x-1}{3x+1} > 0$$

$$\frac{-2x+3}{3x+1} > 0$$

Numerator: $-2x+3=0 \rightarrow x = \dfrac{3}{2}$

Denominator: $3x+1=0 \rightarrow x = -\dfrac{1}{3}$

$(-2x+3)$	+	+	−
$(3x+1)$	−	+	+
$\left(\frac{-2x+3}{3x+1}\right)$	−	+	−
		−1/3	3/2

9. The base of this parallelogram varies inversely as its height. The constant of variation is 24.

10. $P = \dfrac{k\sqrt{t}}{L}$; $P = 5$, $t = 225\,\text{kg}$, $L = 0.60$ m:

$$5 = \frac{k\sqrt{225}}{0.60} = \frac{15k}{0.6}$$

$$3 = 15k \rightarrow k = \frac{3}{15} = \frac{1}{5}$$

EQ: $P = \dfrac{\sqrt{t}}{5L}$; $t = 196$ kg, $L = 0.65$ m:

$$P = \frac{\sqrt{196}}{5(0.65)} = \frac{14}{3.25} = \frac{1400}{325}$$

$\boxed{P = \dfrac{56}{13} \text{ or } \approx 4.3 \text{ vibrations/second}}$

Section 4.4

1. $\sqrt{169} = \sqrt{13^2} = \boxed{13}$

3. $\sqrt[5]{-32} = \sqrt[5]{(-2)^5} = \boxed{-2}$

5. $81^{3/2} = \left(81^{1/2}\right)^3 = 9^3 = \boxed{729}$

7. $125^{-2/3} = \left(125^{1/3}\right)^{-2} = 5^{-2} = \boxed{\dfrac{1}{25}}$

9. $(-1000)^{2/3} = \left(-1000^{1/3}\right)^2 = (-10)^2 = \boxed{100}$

11. $\sqrt[3]{-4} \approx \boxed{-1.587401052}$

Rational approximation

13. $\sqrt[3]{-125} = \boxed{-5}$

Exact value

15. $\sqrt[6]{9} \approx \boxed{1.44224957031}$

Rational approximation

17. $\sqrt[3]{18.609625} = \boxed{2.65}$

Exact value

19. $\sqrt[3]{-17} \approx \boxed{-2.57128159066}$

Rational approximation

21. $\sqrt[6]{\pi^2} \approx \boxed{1.46459188756}$

Rational approximation

23. $13^{-1/3} \approx \boxed{.4252903703}$

Rational approximation

25. $32^{.2} = \boxed{2}$

Exact value

27. $5^{.1} \approx \boxed{1.17461894309}$

Rational approximation

29. $\left(\frac{5}{6}\right)^{-1.3} \approx \boxed{1.26746396213}$

Rational approximation

31. $\pi^{-3} \approx \boxed{.0322515344}$

Rational approximation

33. $17^{1/17} \approx \boxed{1.18135207463}$

Rational approximation

35. (a) $\sqrt{39} \approx 6.2449979984$

(b) $\sqrt{143.8} \approx 11.9916637711$

(c) $\sqrt{9071} \approx 95.2417975471$

37. (a) $16^{-3/4} = \left(16^{1/4}\right)^{-3} = 2^{-3}$

$= \frac{1}{2^3} = \boxed{\frac{1}{8} = .125}$

(b) $\sqrt[4]{16^{-3}} = \left(\sqrt[4]{16}\right)^{-3}$

$.125 = .125$

(c) Show $.125 = \frac{1}{8}$

Relating Concepts (39 – 42):

For these problems, see the calculator graphics at the back of the textbook.

39. $6^{x}\sqrt{81} \approx 2.080083823$

40. $81^\wedge (1/6) \approx 2.080083823$

41. $y_1 = \sqrt[6]{x}$, $\Delta Tbl = 1$;

$x = 81$, $y_1 \approx 2.08008382305$

42. Graph $y_1 = \sqrt[6]{x}$; trace to $x = 81$;

$y_1 \approx 2.0800838$

43. $S(w) = 1.27w^{2/3}$

$S(4.0) = 1.27(4.0)^{2/3} \approx 3.2 \text{ feet}^2$

45. $f(x) = x^{1.5}$

$f(15) = 15^{1.5} \approx 58.1$ years

47. (a) $f(x) = ax^b$; $(1, 1960)$:

$1960 = a(1)^b = a(1) \Longrightarrow \boxed{a = 1960}$

(b) $f(x) = 1960x^b$

$(2, 850)$: $850 = 1960(2)^b$

$$\frac{850}{1960} = 2^b$$

$$\ln\frac{850}{1960} = \ln 2^b = b\ln 2$$

$$b = \frac{\ln\frac{85}{196}}{\ln 2} \approx -1.205 \approx \boxed{-1.2}$$

(c) $f(x) = 1960x^{-1.2}$

$f(4) = 1960(4)^{-1.2} \approx 371.35 \approx \boxed{371}$

If the zinc ion concentration reaches 371 mg/l, a rainbow trout will survive, on average, 4 minutes.

49. $f(x) = .445x^{1.25}$

$f(2) = .445(2)^{1.25} \approx \boxed{1.06 \text{ grams}}$

51. $\{40, 150, 400, 1000, 2000\} \longrightarrow L_1$

$\{140, 72, 44, 28, 20\} \longrightarrow L_2$

STAT / CALC / PwrReg

$y = 874.54x^{-.49789}$ $\boxed{a \approx 874.54;\ b \approx -.49789}$

53. $f(x) = \sqrt{5 + 4x}$

Domain: $5 + 4x \geq 0$

$4x \geq -5$

$x \geq -\frac{5}{4}$ $\boxed{\left[-\frac{5}{4}, \infty\right)}$

55. $f(x) = -\sqrt{6 - x}$

Domain: $6 - x \geq 0$

$-x \geq -6$

$x \leq 6$ $\boxed{(-\infty, 6]}$

57. $f(x) = \sqrt[3]{8x - 24}$; Domain: $\boxed{(-\infty, \infty)}$

59. $f(x) = \sqrt{49 - x^2}$

Domain: $49 - x^2 \geq 0$

$(7 - x)(7 + x) \geq 0$

$x = -7, 7$

$(7 + x)$	$-$	$+$	$+$
$(7 - x)$	$+$	$+$	$-$
$(prod)$	$-$	$+$	$-$
	-7	7	

Domain: $\boxed{[-7, 7]}$

61. $f(x) = \sqrt{x^3 - x}$

Domain: $x^3 - x \geq 0$

$$x\left(x^2 - 1\right) \geq 0$$
$$x(x-1)(x+1) \geq 0$$
$$x = -1,\ 0,\ 1$$

$(x+1)$	$-$	$+$	$+$	$+$
(x)	$-$	$-$	$+$	$+$
$(x-1)$	$-$	$-$	$-$	$+$
$(prod)$	$-$	$+$	$-$	$+$
	-1	0	1	

Domain: $\boxed{[-1,\ 0] \cup [1,\ \infty)}$

63. $f(x) = \sqrt{5 + 4x}$

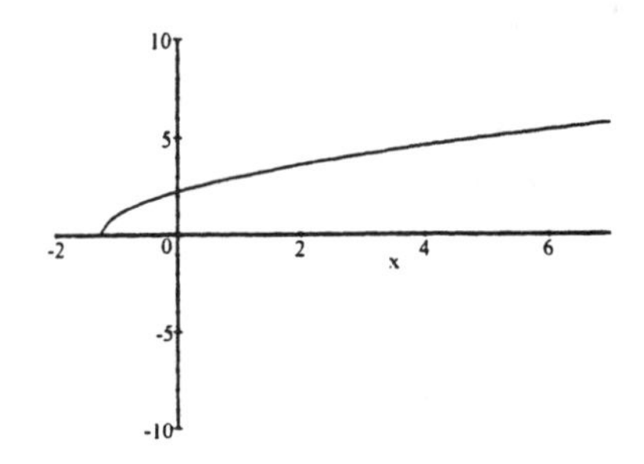

Domain: $\left[-\frac{5}{4},\ \infty\right)$ [#53]

(a) Range: $[0,\ \infty)$

(b) Increasing: $\left[-\frac{5}{4},\ \infty\right)$

(c) Decreasing: none

(d) $f(x) = 0$: $\{-1.25\}$

65. $f(x) = -\sqrt{6 - x}$

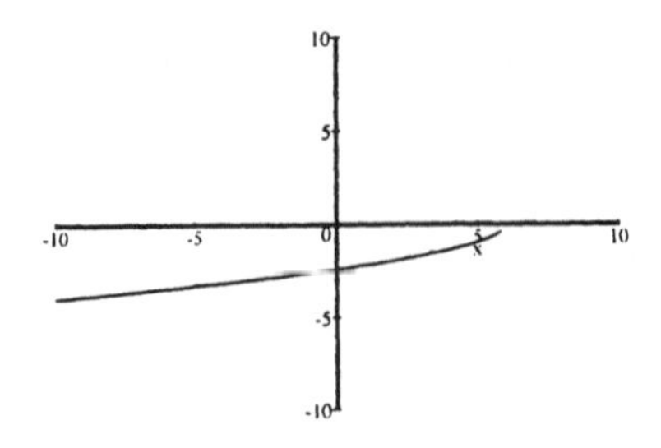

Domain: $(-\infty,\ 6]$ [#55]

(a) Range: $(-\infty,\ 0]$

(b) Increasing: $(-\infty,\ 6]$

(c) Decreasing: none

(d) $f(x) = 0$: $\{6\}$

67. $f(x) = \sqrt[3]{8x - 24}$

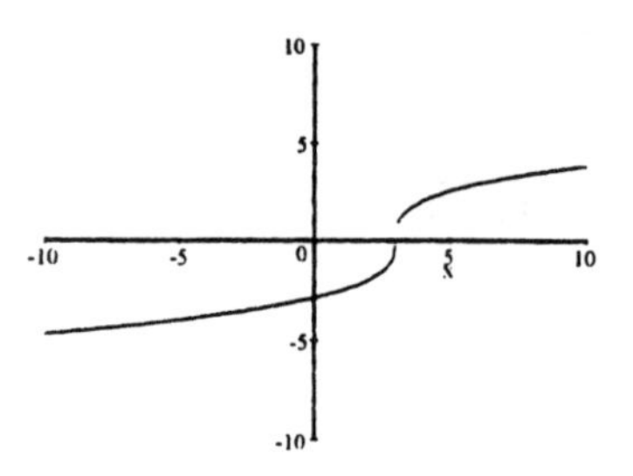

Domain: $(-\infty,\ \infty)$ [#57]

(a) Range: $(-\infty,\ \infty)$

(b) Increasing: $(-\infty,\ \infty)$

(c) Decreasing: none

(d) $f(x) = 0$: $\{3\}$

69. $f(x) = \sqrt{49 - x^2}$

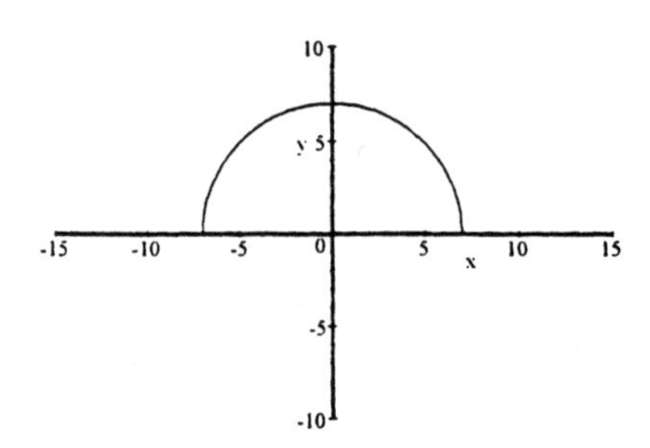

Domain: $[-7,\ 7]$ [#59]

(a) Range: $[0,\ 7]$

(b) Increasing: $[-7,\ 0]$

(c) Decreasing: $[0,\ 7]$

(d) $f(x) = 0$: $\{-7,\ 7\}$

71. $y = \sqrt{9x + 27} = \sqrt{9(x+3)}$

$y = \sqrt{x}$

Shift 3 units left: $y = \sqrt{x + 3}$

Vertically stretch by a factor of 3:

$y = 3\sqrt{x + 3} = \sqrt{9(x+3)}$

73. $y = \sqrt{4x + 16} + 4 = \sqrt{4(x+4)} + 4$

$= 2\sqrt{x + 4} + 4$

$y = \sqrt{x}$

Shift 4 units left: $y = \sqrt{x + 4}$

Vertically stretch by a factor of 2 :

$y = 2 \cdot \sqrt{x + 4} = \sqrt{4(x+4)}$

Shift 4 units up:

$y = \sqrt{4x + 16} + 4$

75. $y = \sqrt[3]{27x + 54} - 5 = \sqrt[3]{27(x+2)} - 5$

$y = \sqrt[3]{x}$

Shift 2 units left: $y = \sqrt[3]{x+2}$

Vertically stretch by a factor of 3 :

$y = 3\sqrt[3]{x+2} = \sqrt[3]{27(x+2)}$

Shift 5 units down:

$y = \sqrt[3]{27(x+2)} - 5$

77. $x^2 + y^2 = 100$: Circle

$y^2 = 100 - x^2$

$y = \pm\sqrt{100 - x^2}$

$$\boxed{\begin{aligned} y_1 &= \sqrt{100 - x^2} \\ y_2 &= -\sqrt{100 - x^2} \end{aligned}}$$

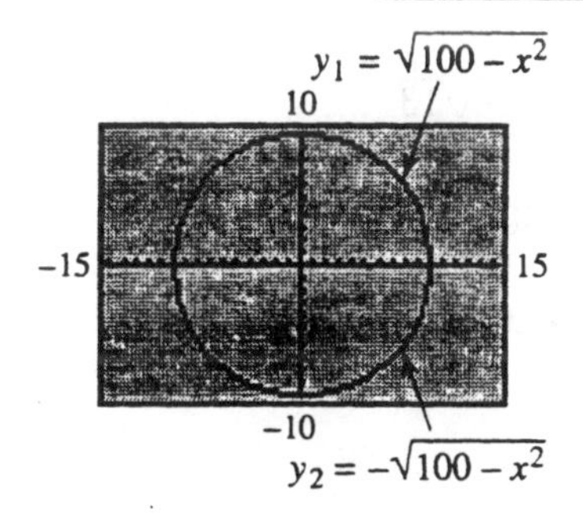

79. $(x-2)^2 + y^2 = 9$: Circle

$y^2 = 9 - (x-2)^2$

$y = \pm\sqrt{9 - (x-2)^2}$

$$\boxed{\begin{aligned} y_1 &= \sqrt{9 - (x-2)^2} \\ y_2 &= -\sqrt{9 - (x-2)^2} \end{aligned}}$$

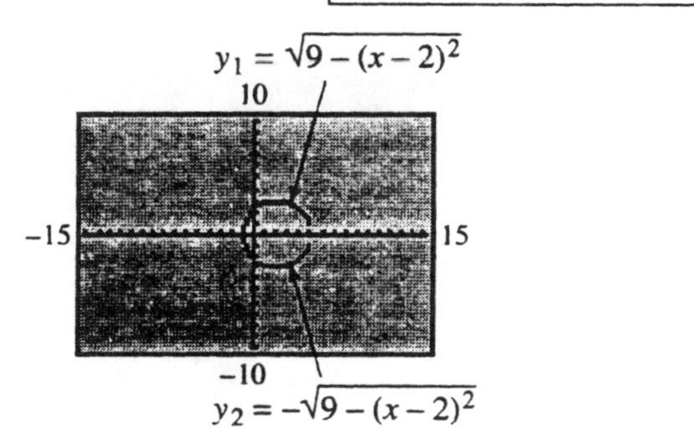

81. $x = y^2 + 6y + 9$: Horizontal parabola

$x = (y+3)^2$

$\pm\sqrt{x} = y + 3$

$-3 \pm \sqrt{x} = y$

$$\boxed{\begin{aligned} y_1 &= -3 + \sqrt{x} \\ y_2 &= -3 - \sqrt{x} \end{aligned}}$$

$y_1 = -3 + \sqrt{x}$
10
−10
10
−10
$y_2 = -3 - \sqrt{x}$

83. $x = 2y^2 + 8y + 1$: Horizontal parabola

$2y^2 + 8y + (1 - x) = 0$

$$y = \frac{-8 \pm \sqrt{64 - 8(1-x)}}{4} = \frac{-8 \pm \sqrt{56 + 8x}}{4}$$

$$y = \frac{-8 \pm \sqrt{4(14 + 2x)}}{4} = \frac{-8 \pm 2\sqrt{14 + 2x}}{4}$$

$$\boxed{\begin{aligned} y_1 &= \frac{-4 + \sqrt{14 + 2x}}{2} \\ y_2 &= \frac{-4 - \sqrt{14 + 2x}}{2} \end{aligned}}$$

$\left[y = -2 \pm \sqrt{.5x + 3.5} \text{ is equivalent}\right]$

$y_1 = -2 + \sqrt{.5x + 3.5}$
10
−10
10
−10
$y_2 = -2 - \sqrt{.5x + 3.5}$

Section 4.5

1. (a) $\sqrt{x+5} = x-1$: $\{4\}$
   (b) $\sqrt{x+5} \le x-1$: $[4, \infty)$
   (c) $\sqrt{x+5} \ge x-1$: $[-5, 4]$

3. (a) $\sqrt[3]{2-2x} = \sqrt[3]{2x+14}$: $\{-3\}$
   (b) $\sqrt[3]{2-2x} > \sqrt[3]{2x+14}$: $(-\infty, -3)$
   (c) $\sqrt[3]{2-2x} < \sqrt[3]{2x+14}$: $(-3, \infty)$

5. $y_1 = \sqrt{x}$; $y_2 = 2x - 1$

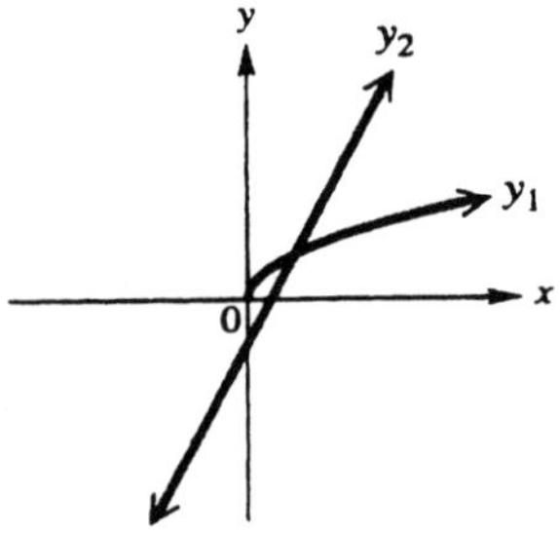

One real solution

$$\sqrt{x} = 2x-1$$
$$(\sqrt{x})^2 = (2x-1)^2$$
$$x = 4x^2 - 4x + 1$$
$$0 = 4x^2 - 5x + 1$$
$$0 = (4x-1)(x-1)$$
$$x = \frac{1}{4}, 1 \text{ possible solutions}$$

Check $\left(\frac{1}{4}\right)$: $\sqrt{\frac{1}{4}} = 2\left(\frac{1}{4}\right) - 1$

$\frac{1}{2} = \frac{1}{2} - 1 = -\frac{1}{2}$ **No**

$\frac{1}{4}$ is extraneous

Check (1): $\sqrt{1} = 2(1) - 1$

$1 = 2 - 1$ **Yes**

Solution set: $\{1\}$

7. $y_1 = \sqrt{x}$; $y_2 = -x + 3$

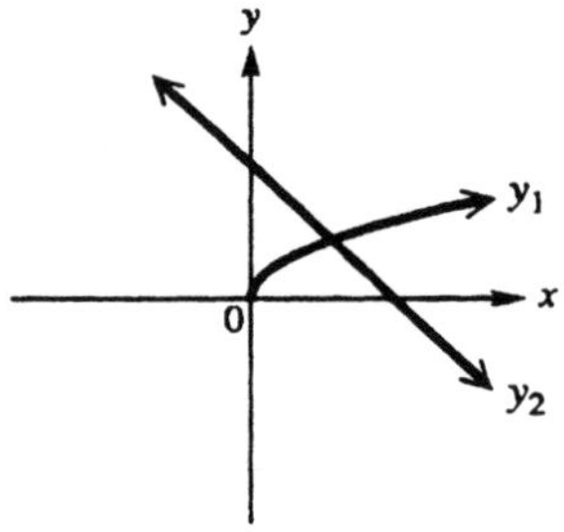

One real solution

$$\sqrt{x} = -x+3$$
$$(\sqrt{x})^2 = (-x+3)^2$$
$$x = x^2 - 6x + 9$$
$$0 = x^2 - 7x + 9$$
$$x = \frac{7 \pm \sqrt{13}}{2} \quad \text{(Quadratic formula)}$$

Check $\left(\frac{7+\sqrt{13}}{2}\right)$:

$$\sqrt{\frac{7+\sqrt{13}}{2}} = -\frac{7+\sqrt{13}}{2} + 3$$

$2.303 = -2.303$ **No**

$\frac{7+\sqrt{13}}{2}$ is extraneous

Check $\left(\frac{7-\sqrt{13}}{2}\right)$:

$$\sqrt{\frac{7-\sqrt{13}}{2}} = -\frac{7-\sqrt{13}}{2} + 3$$

$1.303 = 1.303$ **Yes**

Solution set: $\left\{\frac{7-\sqrt{13}}{2}\right\}$

9. $y_1 = \sqrt[3]{x}$; $y_2 = x^2$

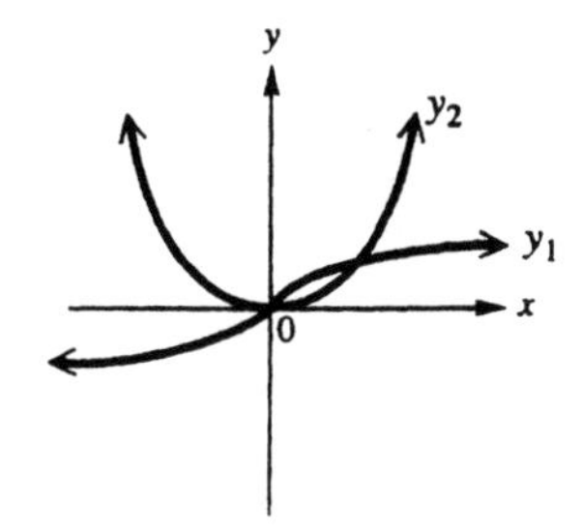

Two real solutions

$\sqrt[3]{x} = x^2$
$(\sqrt[3]{x})^3 = (x^2)^3$
$x = x^6$
$0 = x^6 - x$
$0 = x(x^5 - 1)$
$x = 0, 1$ possible solutions

Check (0) : $\sqrt[3]{0} = 0^2$
$0 = 0$ **Yes**

Check (1) : $\sqrt[3]{1} = 1^2$
$1 = 1$ **Yes**

Solution set: $\{0, 1\}$

There are no extraneous roots.

11. (a) $\sqrt{3x+7} = 2$

$(\sqrt{3x+7})^2 = (2)^2$
$3x + 7 = 4$
$3x = -3 \implies x = -1$

Check (-1) : $\sqrt{3(-1)+7} = 2$
$\sqrt{4} = 2$ **Yes**

solution set is $\{-1\}$

$y_1 = \sqrt{3x+7}$ (solid); $y_2 = 2$ (dashed)

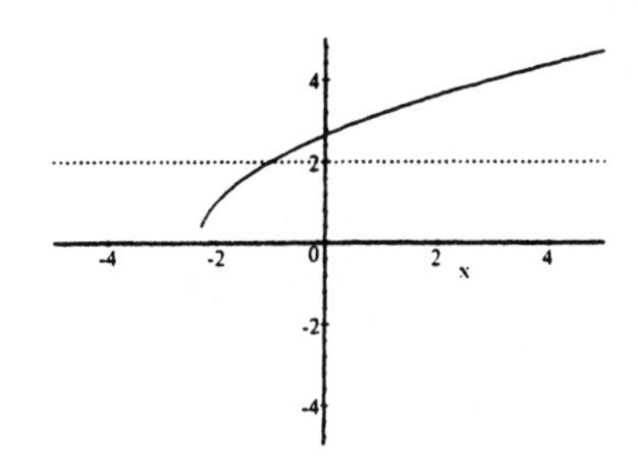

intersect at $x = -1$

domain of $y_1 : 3x + 7 \geq 0 \longrightarrow x \geq -\frac{7}{3}$

continued

11. continued

(b) $\sqrt{3x+7} > 2$: $(-1, \infty)$

(c) $\sqrt{3x+7} < 2$: $\left[-\frac{7}{3}, -1\right)$

13. (a) $\sqrt{4x+13} = 2x - 1$

$(\sqrt{4x+13})^2 = (2x-1)^2$
$4x + 13 = 4x^2 - 4x + 1$
$0 = 4x^2 - 8x - 12$
$0 = 4(x^2 - 2x - 3)$
$0 = 4(x-3)(x+1)$
$x = -1, 3$ (Possible solutions)

Check (-1) : $\sqrt{4(-1)+13} = 2(-1) - 1$
$\sqrt{9} = -3$ **No**

Check (3) : $\sqrt{4(3)+13} = 2(3) - 1$
$\sqrt{25} = 6 - 1 \Longrightarrow 5 = 5$ **Yes**

Solution set: $\{3\}$

$y_1 = \sqrt{4x+13}$ (solid); $y_2 = 2x - 1$ (dashed)

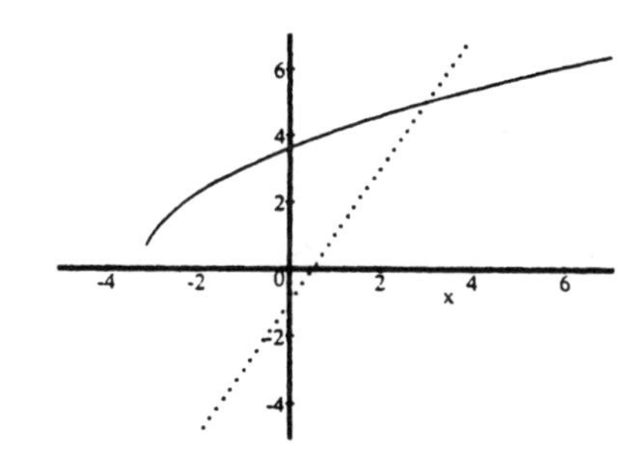

intersect at $x = 3$

domain of $y_1 : 4x + 13 \geq 0 \rightarrow x \geq -\frac{13}{4}$

(b) $\sqrt{4x+13} > 2x - 1$: $\left[-\frac{13}{4}, 3\right)$

(c) $\sqrt{4x+13} < 2x - 1$: $(3, \infty)$

15. **(a)** $\sqrt{1+5x}+2=2x$

$\sqrt{1+5x}=2x-2$

$\left(\sqrt{1+5x}\right)^2=(2x-2)^2$

$1+5x=4x^2-8x+4$

$0=4x^2-13x+3$

$0=(4x-1)(x-3)$

$x=\frac{1}{4},\ 3$ (Possible solutions)

Check $\left(\frac{1}{4}\right)$: $\sqrt{1+\frac{5}{4}}+2=2\left(\frac{1}{4}\right)$

$\frac{\sqrt{9}}{\sqrt{4}}+2=\frac{1}{2}$

$\frac{3}{2}+2=\frac{1}{2}$ **No**

Check (3): $\sqrt{1+5(3)}+2=2(3)$

$\sqrt{16}+2=6$

$6=6$ **Yes**

Solution set: $\{3\}$

$y_1=\sqrt{1+5x}+2$ (solid); $y_2=2x$ (dashed)

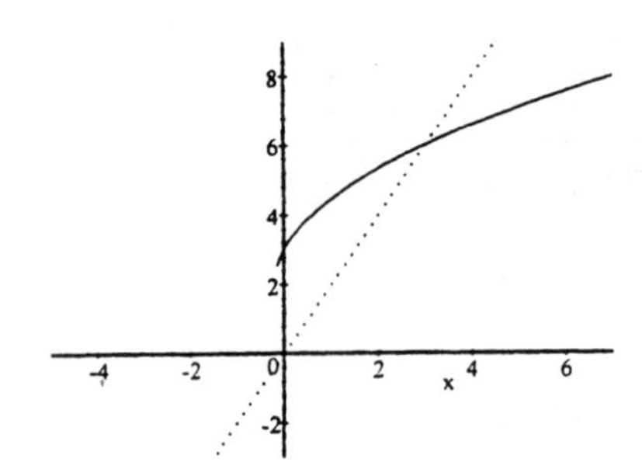

intersect at $x=3$

domain of $y_1: 1+5x\geq 0 \Longrightarrow x\geq -\frac{1}{5}$

(b) $\sqrt{5x+1}+2>2x:\left[-\frac{1}{5},\ 3\right)$

(c) $\sqrt{5x+1}+2<2x:(3,\ \infty)$

17. **(a)** $\sqrt{3x-6}+2=\sqrt{5x-6}$

$\left(2+\sqrt{3x-6}\right)^2=\left(\sqrt{5x-6}\right)^2$

$4+4\sqrt{3x-6}+3x-6=5x-6$

$4\sqrt{3x-6}=2x-4$

$\left(2\sqrt{3x-6}\right)^2=(x-2)^2$

$4(3x-6)=x^2-4x+4$

$x^2-4x+4=12x-24$

$x^2-16x+28=0$

$(x-2)(x-14)=0$

$x=2,\ 14$ (Possible solutions)

Check (2):

$\sqrt{3(2)-6}+2=\sqrt{5(2)-6}$

$\sqrt{0}+2=\sqrt{4}$

$2=2$ **Yes**

Check (14):

$\sqrt{3(14)-6}+2=\sqrt{5(14)-6}$

$\sqrt{36}+2=\sqrt{64}$

$6+2=8$ **Yes**

Solution set: $\{2,\ 14\}$

$y_1=\sqrt{3x-6}+2$ (solid)

$y_2=\sqrt{5x-6}$ (dashed)

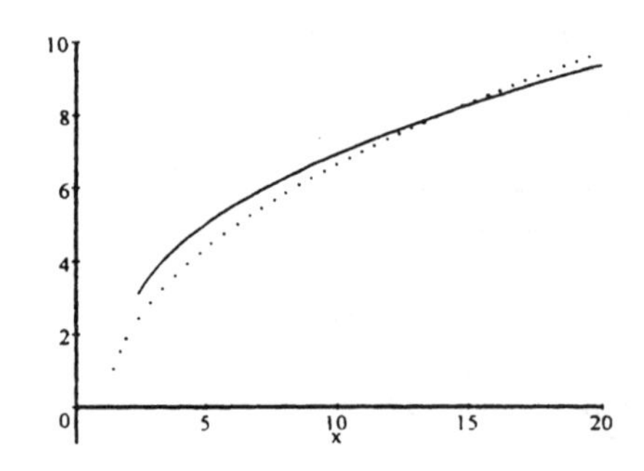

intersect at $x=2,\ 14$

domain of y_1 and y_2:

$5x-6\geq 0 \qquad 3x-6\geq 0$

$x\geq\frac{6}{5} \qquad x\geq 2$

(b) $\sqrt{3x-6}+2>\sqrt{5x-6}:(2,\ 14)$

(c) $\sqrt{3x-6}+2<\sqrt{5x-6}:(14,\ \infty)$

19. **(a)** $\sqrt[3]{x^2-2x}=\sqrt[3]{x}$

$\left(\sqrt[3]{x^2-2x}\right)^3=\left(\sqrt[3]{x}\right)^3$

$x^2-2x=x$

$x^2-3x=0$

$x(x-3)=0$

$x=0,\ 3$ (Possible solutions)

Check (0) :

$\sqrt[3]{0-2(0)}=\sqrt[3]{0}$

$0=0$ **Yes**

Check (3) :

$\sqrt[3]{(3)^2-2(3)}=\sqrt[3]{3}$

$\sqrt[3]{3}=\sqrt[3]{3}$ **Yes**

Solution set: $\{0,\ 3\}$

$y_1=\sqrt[3]{x^2-2x}$ (dark)

$y_2=\sqrt[3]{x}$ (light)

$[-5,\ 5]$ by $[-5,\ 5]$

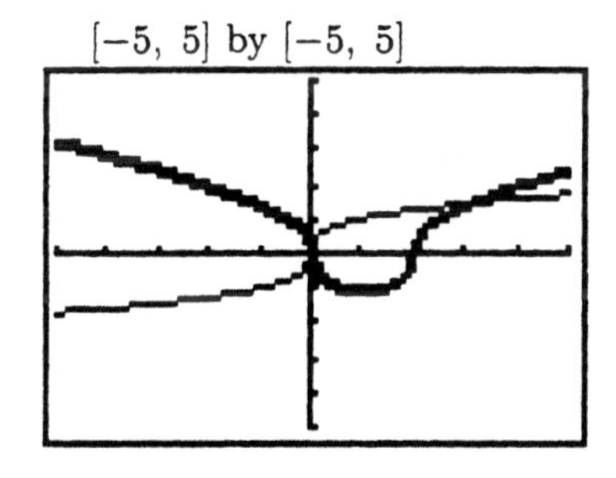

intersect at 0, 3

domain: $(-\infty,\ \infty)$

(b) $\sqrt[3]{x^2-2x}>\sqrt[3]{x}:(-\infty,\ 0)\cup(3,\ \infty)$

(c) $\sqrt[3]{x^2-2x}<\sqrt[3]{x}:(0,\ 3)$

21. **(a)** $\sqrt[4]{3x+1}=1$

$\left(\sqrt[4]{3x+1}\right)^4=1^4$

$3x+1=1$

$3x=0\Longrightarrow x=0$ (Possible solution)

Check (0) : $\sqrt[4]{3(0)+1}=1$

$\sqrt[4]{1}=1$

$1=1$ **Yes**

Solution set: $\{0\}$

(b) $y_1=\sqrt[4]{3x+1}$ (curve) ; $y_2=1$ (line)

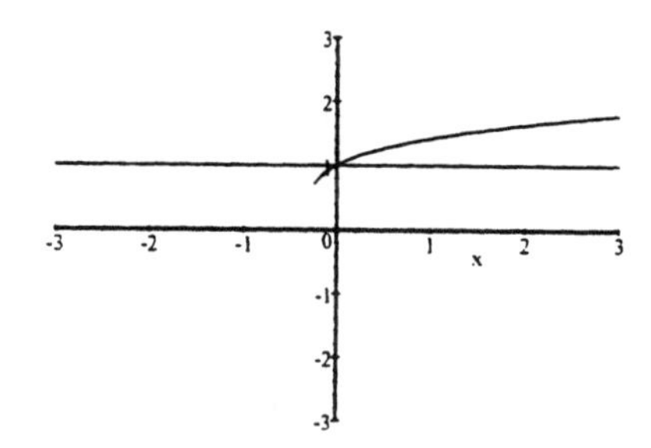

intersect at $x=0$

domain of y_1 : $3x+1\geq0\Longrightarrow x\geq-\dfrac{1}{3}$

(b) $\sqrt[4]{3x+1}>1:(0,\ \infty)$

(c) $\sqrt[4]{3x+1}<1:\left[-\dfrac{1}{3},\ 0\right)$

23. **(a)** $(2x-5)^{1/2}-2=(x-2)^{1/2}$

$\left(\sqrt{2x-5}-2\right)^2=\left(\sqrt{x-2}\right)^2$

$2x-5-4\sqrt{2x-5}+4=x-2$

$-4\sqrt{2x-5}=-x-1$

$\left(4\sqrt{2x-5}\right)^2=(x+1)^2$

$16(2x-5)=x^2+2x+1$

$32x-80=x^2+2x+1$

$0=x^2-30x+81$

$0=(x-27)(x-3)$

$x=3,\ 27$ (Possible solutions)

Check (3) :

$(2(3)-5)^{1/2}-2=(3-2)^{1/2}$

$(1)^{1/2}-2=(1)^{1/2}$

$1-2=1$ **No**

Check (27) :

$(2(27)-5)^{1/2}-2=(27-2)^{1/2}$

$(49)^{1/2}-2=(25)^{1/2}$

$7-2=5$ **Yes**

Solution set: $\{27\}$

$y_1=(2x-5)^{1/2}-2$ (solid)

$y_2=(x-2)^{1/2}$ (dashed)

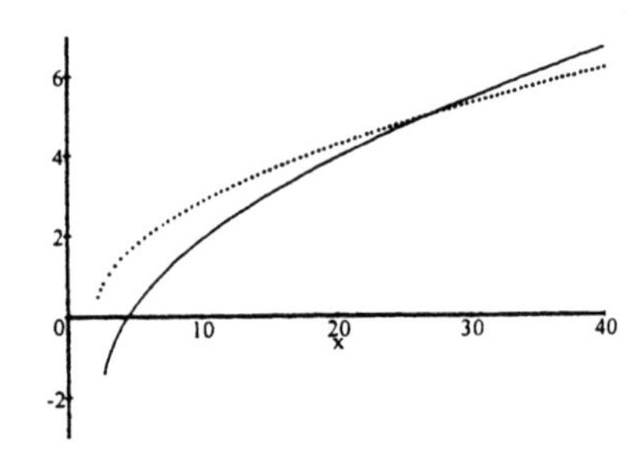

intersect at $x=27$

domain of y_1 and y_2 :

$2x-5\geq 0$ $\quad x-2\geq 0$

$x\geq \frac{5}{2}$ $\quad x\geq 2$

(b) $(2x-5)^{1/2}-2\geq (x-2)^{1/2}:[27,\infty)$

(c) $(2x-5)^{1/2}-2\leq (x-2)^{1/2}:\left[\frac{5}{2},\ 27\right]$

25. **(a)** $(x^2+6x)^{1/4}=2$

$\left[(x^2+6x)^{1/4}\right]^4=(2)^4$

$x^2+6x=16$

$x^2+6x-16=0$

$(x+8)(x-2)=0$

$x=-8,\ 2$ (Possible solutions)

Check (-8) :

$\left((-8)^2+6(-8)\right)^{1/4}=2$

$(64-48)^{1/4}=2$

$(16)^{1/4}=2$

$2=2$ **Yes**

Check (2) :

$(2^2+6(2))^{1/4}=2$

$(4+12)^{1/4}=2$

$(16)^{1/4}=2$ **Yes**

Solution set: $\{-8,\ 2\}$

$y_1=(x^2+6x)^{1/4}$ (curve); $y_2=2$ (line)

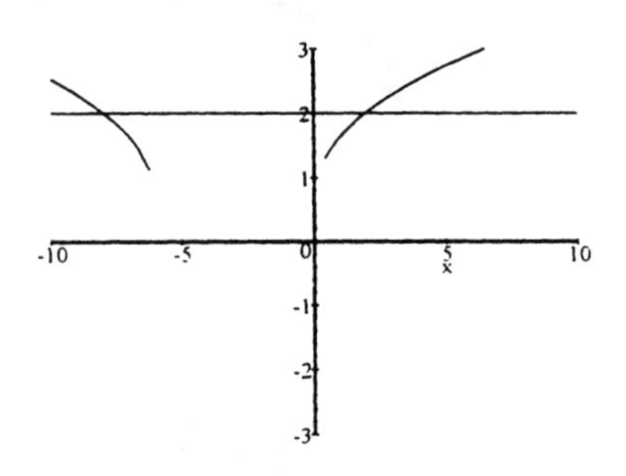

intersect at $x=-8,\ 2$

$y_1: x^2+6x\geq 0$

$x(x+6)\geq 0$

$x=0,\ -6$

$(x+6)$	−	+	+
(x)	−	−	+
$(prod)$	+	−	+
	−6	0	

domain of y_1 : $(-\infty,-6]\cup[0,\ \infty)$

(b) $(x^2+6x)^{1/4}>2:(-\infty,-8)\cup(2,\ \infty)$

(c) $(x^2+6x)^{1/4}<2:(-8,-6]\cup[0,\ 2)$

27. (a) $(2x-1)^{2/3} = x^{1/3}$

$\left[(2x-1)^{2/3}\right]^{3/2} = \left(x^{1/3}\right)^{3/2}$

$2x-1 = x^{1/2}$

$(2x-1)^2 = \left(x^{1/2}\right)^2$

$4x^2-4x+1 = x$

$4x^2-5x+1-0$

$(4x-1)(x-1) = 0$

$x = \frac{1}{4}, 1$ (possible solutions)

Check $\left(\frac{1}{4}\right)$:

$\left(2\left(\frac{1}{4}\right)-1\right)^{2/3} = \left(\frac{1}{4}\right)^{1/3}$

$\left(-\frac{1}{2}\right)^{2/3} = \left(\frac{1}{4}\right)^{1/3}$

$\left(\left(-\frac{1}{2}\right)^2\right)^{1/3} = \left(\frac{1}{4}\right)^{1/3}$ **Yes**

Check (1):

$(2(1)-1)^{2/3} = 1^{1/3}$

$1^{2/3} = 1^{1/3}$ **Yes**

Solution set: $\left\{\frac{1}{4}, 1\right\}$

$y_1 = \left((2x-1)^{1/3}\right)^2$ (solid)

$y_2 = x^{1/3}$ (dashed)

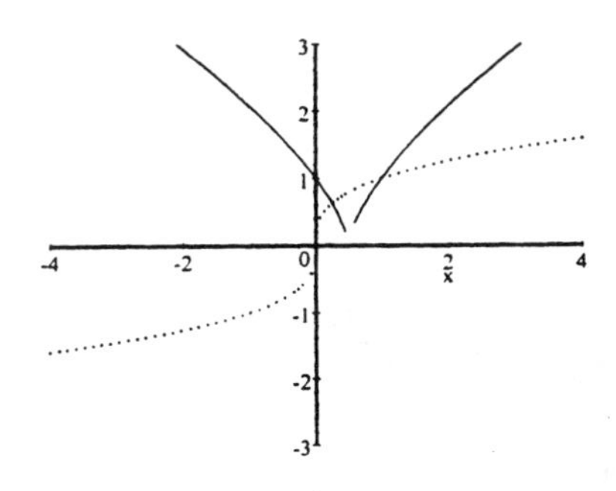

intersect at $x = \frac{1}{4}, 1$

domain of y_1 : $(-\infty, \infty)$

(b) $(2x-1)^{2/3} > x^{1/3}$: $\left(-\infty, \frac{1}{4}\right) \cup (1, \infty)$

(c) $(2x-1)^{2/3} < x^{1/3}$: $\left(\frac{1}{4}, 1\right)$

29. $T - \left(\frac{v}{4} + 7\sqrt{v}\right)\left(1 - \frac{T}{90}\right)$

(a) $T = -10,\ v = 30$:

$-10 - \left(\frac{30}{4} + 7\sqrt{30}\right)\left(1 - \frac{-10}{90}\right)$

$\approx -60.934 \approx$ $-60.9°\text{F}$

(b) $T = -40,\ v = 5$:

$-40 - \left(\frac{5}{4} + 7\sqrt{5}\right)\left(1 - \frac{-40}{90}\right)$

$\approx -64.415 \approx$ $-64.4°\text{F}$

31. (a) Temp = $-10°$, wind = 30 mph: $-63°\text{F}$

(b) Temp = $-40°$, wind = 5 mph: $-47°\text{F}$

33. $V = \frac{k}{\sqrt{d}}$; $k = 350$, $d = 6000$ km

$V = \frac{350}{\sqrt{6000}} \approx 4.518 \approx$ 4.5 km/second

35. $P = 2\pi\sqrt{\frac{L}{32}}$, $L = 5$ feet

$P = 2\pi\sqrt{\frac{5}{32}} \approx 2.484$

$\approx$ 2.5 seconds

37. $S = 30\sqrt{\frac{a}{p}}$, $a = 900$ ft, $P = 97$ ft

$S = 30\sqrt{\frac{900}{97}} \approx 91.381$

$\approx$ 91 miles per hour

39. (a) $20 - x$

(b) $0 < x < 20$

(c) $\overline{AP}$: $c^2 = a^2 + b^2$

$c^2 = 12^2 + x^2$

$c = \sqrt{12^2 + x^2}$

$\overline{PB}$: $c^2 = a^2 + b^2$

$c^2 = 16^2 + (20 - x)^2$

$c = \sqrt{16^2 + (20 - x)^2}$

(d) $f(x) = \overline{AP} + \overline{PB}, \quad 0 < x < 20$

$f(x) = \sqrt{12^2 + x^2} + \sqrt{16^2 + (20 - x)^2}$

(e) $f(4) = 35.276528 \approx 35.28$

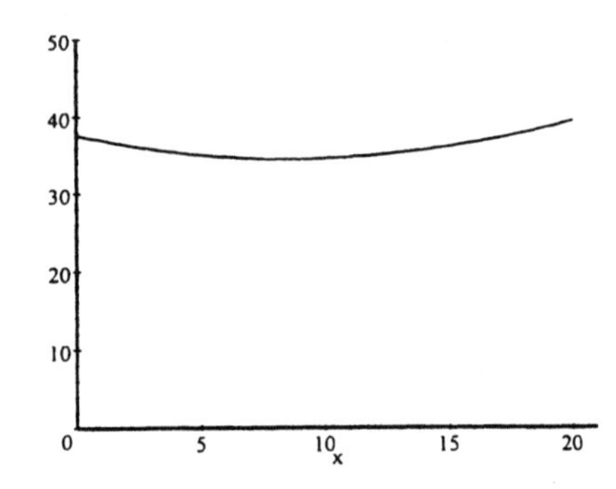

When the stake is 4 feet from the base of the 12 foot pole, ≈ 35.28 feet of wire will be used.

(f) Minimize $f(x) \approx (8.57 \text{ ft}, 34.41 \text{ ft})$

(g) We have expressed the total amount of wire to be used in terms of x, the distance from the stake at P to the base of the 12 foot pole. This amount is minimized when the stake is ≈ 8.57 feet from the 12 foot pole.

41. Total dist = river dist (d_1) + land dist (d_2)

Time $t = \dfrac{d}{r}$: $d_2 = \sqrt{3^2 + x^2}$

$$T = \frac{d_1}{r_1} + \frac{d_2}{r_2}$$

$$t(x) = \frac{8 - x}{5} + \frac{\sqrt{9 + x^2}}{2}$$

The minimum occurs at $\approx (1.31, 2.97)$;
$x = 1.31$ mi, $8 - x = 6.69$ miles

He should walk ≈ 6.69 miles up river.

43. $d = r \cdot t$; x = time in hours
$d_1 = 20x$; $d_2 = 60 - 30x$

Hypotenuse: $\sqrt{(20x)^2 + (60 - 30x)^2}$

Graph $y_1 = \sqrt{(20x)^2 + (60 - 30x)^2}$

Minimum at (1.38, 33.28)

After 1.38 hrs ($\approx$ 1:23 p.m.), the ships are 33.28 miles apart.

Relating Concepts (45 – 56)

45. $\sqrt[3]{4x - 4} = \sqrt{x + 1}$
$(4x - 4)^{1/3} = (x + 1)^{1/2}$

46. Rational exponents: $\dfrac{1}{3}$ and $\dfrac{1}{2}$; common denominator is 6.

47. $\left[(4x - 4)^{1/3}\right]^6 = \left[(x + 1)^{1/2}\right]^6$

$$(4x - 4)^2 = (x + 1)^3$$

48. $(4x - 4)^2 = (x + 1)^3$
$16x^2 - 32x + 16 = x^3 + 3x^2 + 3x + 1$
$0 = x^3 - 13x^2 + 35x - 15$

$$x^3 - 13x^2 + 35x - 15 = 0$$

49. There are 3 real roots.

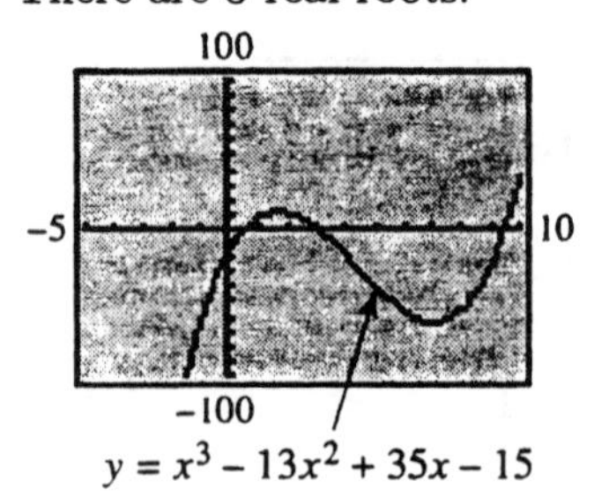

50.

3)	1	−13	35	−15
		3	−30	15
	1	−10	5	0

$P(3) = 0$

51. $P(x) = x^3 - 13x^2 + 35x - 15$

$$\boxed{P(x) = (x-3)(x^2 - 10x + 5)}$$

52. $x^2 - 10x + 5 = 0$

$$x = \frac{10 \pm \sqrt{100 - 4(5)}}{2}$$
$$= \frac{10 \pm \sqrt{80}}{2}$$
$$= \frac{10 \pm 4\sqrt{5}}{2} = \boxed{5 \pm 2\sqrt{5}}$$

53. Possible solutions:

$$\boxed{3, \quad 5 + 2\sqrt{5}, \quad 5 - 2\sqrt{5}}$$

54. $y_1 = \sqrt[3]{4x - 4}$; $y_2 = \sqrt{x+1}$

Graph $y_1 - y_2$:

$y_3 = y_1 - y_2 = \sqrt[3]{4x-4} - \sqrt{x+1}$

.5

-2 20

-.5

There are 2 real solutions.

55. $\sqrt[3]{4x - 4} = \sqrt{x+1}$

Check (3) : $\sqrt[3]{4(3) - 4} = \sqrt{(3) + 1}$

$$\sqrt[3]{8} = \sqrt{4}$$
$$2 = 2 \qquad \textbf{Yes}$$

Check $\left(5 + 2\sqrt{5}\right)$ (≈ 9.47) :

$$\sqrt[3]{4\left(5 + 2\sqrt{5}\right) - 4} = \sqrt{\left(5 + 2\sqrt{5}\right) + 1}$$
$$\sqrt[3]{20 + 8\sqrt{5} - 4} = \sqrt{6 + 2\sqrt{5}}$$
$$\sqrt[3]{16 + 8\sqrt{5}} = \sqrt{6 + 2\sqrt{5}}$$
$$\sqrt[3]{8\left(2 + \sqrt{5}\right)} = \sqrt{2\left(3 + \sqrt{5}\right)}$$
$$2\sqrt[3]{2 + \sqrt{5}} = \sqrt{2\left(3 + \sqrt{5}\right)}$$
$$3.236 = 3.236 \qquad \textbf{Yes}$$

continued

55. continued

Check $\left(5 - 2\sqrt{5}\right)$ (≈ -22.36) :

$$\sqrt[3]{4\left(5 - 2\sqrt{5}\right) - 4} = \sqrt{\left(5 - 2\sqrt{5}\right) + 1}$$

(Same steps as above)

$$2\sqrt[3]{2 - \sqrt{5}} = \sqrt{2\left(3 - \sqrt{5}\right)}$$
$$-1.236 = 1.236 \qquad \textbf{No}$$

$$\boxed{\text{Solution set: } \left\{3,\ 5 + 2\sqrt{5}\ (\approx 9.47)\right\}}$$

56. The solution set of the original equation is a subset of the solution set of the equation in **#48.** The extraneous solution, $5 - 2\sqrt{5}$, was obtained when both sides of the original equation were raised to the 6th power.

57. $\sqrt{\sqrt{x}} = x$

$$\left(\sqrt{\sqrt{x}}\right)^2 = x^2$$
$$\sqrt{x} = x^2$$
$$\left(\sqrt{x}\right)^2 = \left(x^2\right)^2$$
$$x = x^4$$
$$0 = x^4 - x = x\left(x^3 - 1\right)$$
$$0 = x(x-1)\left(x^2 + x + 1\right) : \boxed{\{0, 1\}}$$

Check (0) : $\sqrt{\sqrt{x}} = x \Longrightarrow 0 = 0$ **Yes**

Check (1) : $\sqrt{\sqrt{1}} = 1 \Longrightarrow 1 = 1$ **Yes**

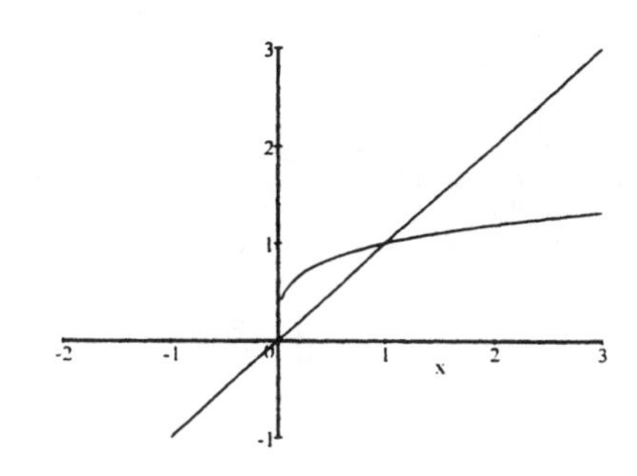

59. $\sqrt{\sqrt{28x+8}} = \sqrt{3x+2}$

$\left(\sqrt{\sqrt{28x+8}}\right)^2 = \left(\sqrt{3x+2}\right)^2$

$\sqrt{28x+8} = 3x+2$

$\left(\sqrt{28x+8}\right)^2 = (3x+2)^2$

$28x+8 = 9x^2+12x+4$

$0 = 9x^2 - 16x - 4$

$0 = (9x+2)(x-2): \boxed{\left\{-\frac{2}{9}, 2\right\}}$

Check $\left(-\frac{2}{9}\right): \sqrt{\sqrt{28\left(-\frac{2}{9}\right)+8}} = \sqrt{3\left(-\frac{2}{9}\right)+2}$

$(\approx)\ 1.15 = 1.15$ **Yes**

Check $(2): \sqrt{\sqrt{28(2)+8}} = \sqrt{3(2)+2}$

$\sqrt{\sqrt{64}} = \sqrt{8}$ **Yes**

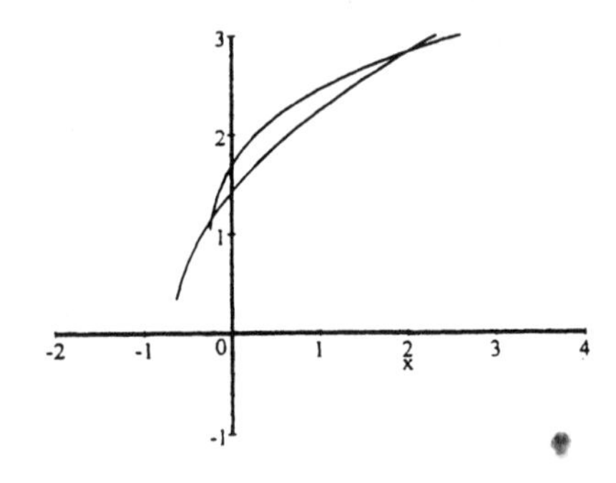

61. $\sqrt[3]{\sqrt{32x}} = \sqrt[3]{x+6}$

$\left(\sqrt[3]{\sqrt{32x}}\right)^3 = \left(\sqrt[3]{x+6}\right)^3$

$\sqrt{32x} = x+6$

$\left(\sqrt{32x}\right)^2 = (x+6)^2$

$32x = x^2+12x+36$

$0 = x^2 - 20x + 36$

$0 = (x-2)(x-18)$

$x = 2,\ 18$ $\boxed{\{2,\ 18\}}$

Check $x = 2:$

$\sqrt[3]{\sqrt{32(2)}} = \sqrt[3]{2+6}$

$\sqrt[3]{\sqrt{64}} = \sqrt[3]{8}$

$\sqrt[3]{8} = \sqrt{8}$ **Yes**

Check $x = 18:$

$\sqrt[3]{\sqrt{32(18)}} = \sqrt[3]{18+6}$

$\sqrt[3]{\sqrt{576}} = \sqrt[3]{24}$

$\sqrt[3]{24} = \sqrt{24}$ **Yes**

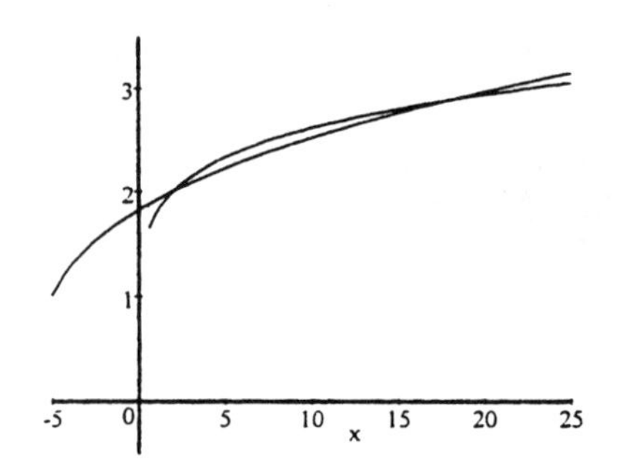

Reviewing Basic Concepts (Sections 4.4 and 4.5)

1. $y = x^{.7},\ y = x^{1.2},\ y = x^{2.4},\ x > 0$

As the exponent increases in value, the curve rises more rapidly.

2. $S(w) = .3w^{3/4}$

$S(.75) = .3(.75)^{3/4} \approx .24178 \approx .24 \text{ meters}^2$

3. $x^2 + y^2 = 16$

$y^2 = 16 - x^2$

$y = \pm\sqrt{16 - x^2}$

$y_1 = \sqrt{16 - x^2}, \quad y_2 = -\sqrt{16 - x^2}$

$y_1 = \sqrt{16 - x^2}$

6.2

−9.4 9.4

−6.2

$y_2 = -\sqrt{16 - x^2}$

4. $x = y^2 + 4y + 6$

$y^2 + 4y + 6 - x = 0$

$$y = \frac{-4 \pm \sqrt{16 - 4(6 - x)}}{2}$$

$$= \frac{-4 \pm \sqrt{16 - 24 + 4x}}{2}$$

$$= \frac{-4 \pm \sqrt{-8 + 4x}}{2}$$

$$= \frac{-4 \pm \sqrt{4(-2 + x)}}{2}$$

$$= \frac{-4 \pm 2\sqrt{x - 2}}{2} = -2 \pm \sqrt{x - 2}$$

$y_1 = -2 + \sqrt{x - 2}, \quad y_2 = -2 - \sqrt{x - 2}$

$y_1 = -2 + \sqrt{x - 2}$

6.2

−9.4 9.4

−6.2

$y_2 = -2 - \sqrt{x - 2}$

5. $\sqrt{3x + 4} = 8 - x$

$\left(\sqrt{3x + 4}\right)^2 = (8 - x)^2$

$3x + 4 = 64 - 16x + x^2$

$0 = x^2 - 19x + 60$

$0 = (x - 4)(x - 15)$

$x = 4, \ 15$ (possible solutions)

Check (4) : $\sqrt{3(4) + 4} = 8 - (4)$

$\sqrt{16} = 4$

$4 = 4$ **Yes**

Check (15) : $\sqrt{3(15) + 4} = 8 - (15)$

$\sqrt{49} = -7$

$7 = -7$ **No**

Solution set: $\{4\}$

$y_1 = \sqrt{3x + 4}$ (curve) ; $y_2 = 8 - x$ (line)

intersect at $x = 4$

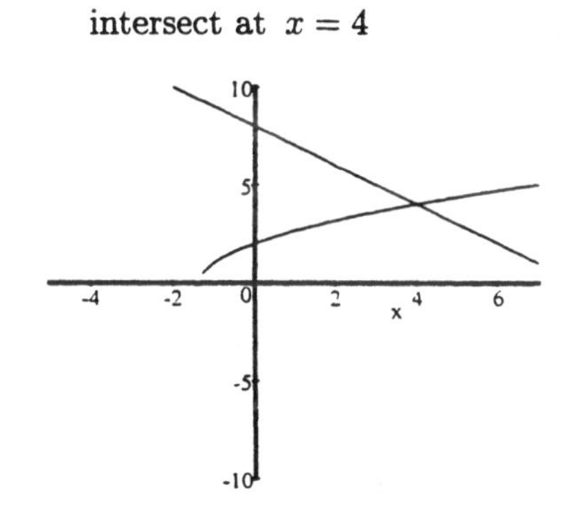

6. See #5 above:

$\sqrt{3x + 4} > 8 - x : (4, \infty)$

7. See #5 above:

domain of $y_1 : 3x + 4 \geq 0 \Longrightarrow x \geq -\frac{4}{3}$

$$\boxed{\sqrt{3x+4} < 8 - x : \left[-\frac{4}{3}, 4\right)}$$

8. $\sqrt{3x+4} + \sqrt{5x+6} = 2$

$\left(\sqrt{3x+4}\right)^2 = \left(2 - \sqrt{5x+6}\right)^2$

$3x + 4 = 4 - 4\sqrt{5x+6} + 5x + 6$

$-2x - 6 = -4\sqrt{5x+6}$

$x + 3 = 2\sqrt{5x+6}$ $\quad$ [÷ by −2]

$(x+3)^2 = \left(2\sqrt{5x+6}\right)^2$

$x^2 + 6x + 9 = 4(5x+6)$

$x^2 + 6x + 9 = 20x + 24$

$x^2 - 14x - 15 = 0$

$(x+1)(x-15) = 0$

$x = -1,\ 15$ (Possible solutions)

Check (-1) : $\sqrt{3(-1)+4} + \sqrt{5(-1)+6} = 2$

$\sqrt{1} + \sqrt{1} = 2$

$2 = 2$ $\quad$ **Yes**

Check (15) : $\sqrt{3(15)+4} + \sqrt{5(15)+6} = 2$

$\sqrt{49} + \sqrt{81} = 2$

$7 + 9 = 2$ $\quad$ **No**

$$\boxed{\text{Solution set: } \{-1\}}$$

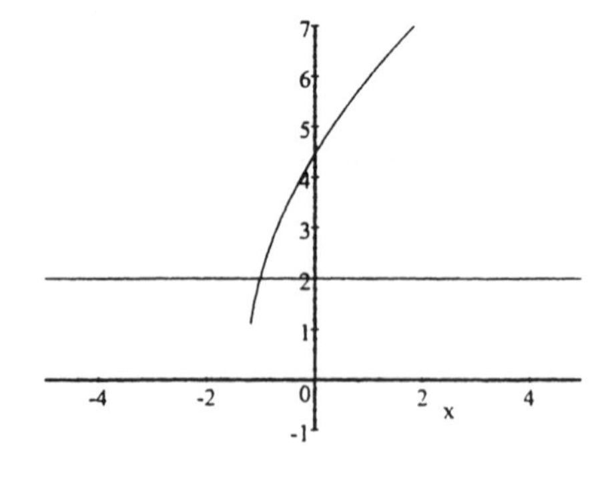

9. $f(x) = \sqrt{x - 15}$

Domain of $f(x)$: $x - 15 \geq 0$

$x \geq 15$

All values of $x < 15$ are not real numbers (not in the domain).

10. $f(x) = \dfrac{1607}{\sqrt[4]{x^3}}$

(a) 2 feet = 24 inches

$f(24) = \dfrac{1607}{\sqrt[4]{24^3}} \approx 148.2 \approx 148$

5.5 feet = 66 inches

$f(66) = \dfrac{1607}{\sqrt[4]{66^3}} \approx 69.4 \approx 69$

(b) $400 = \dfrac{1607}{\sqrt[4]{x^3}}$

$400x^{3/4} = 1607$

$x^{3/4} = \dfrac{1607}{400} = 4.0175$

$\left(x^{3/4}\right)^{4/3} = (4.0175)^{4/3}$

$x \approx 6.387 \approx 6.4$ inches

Chapter 4 Review

Note: In the Exercises,
V.A.→ Vertical Asymptote;
H.A.→ Horizontal Asymptote;
O.A. → Oblique Asymptote.

1. (a) $y = \frac{1}{x}$

$y = -\frac{1}{x}$; reflect across the x-axis

$y = -\frac{1}{x} + 6$; shift 6 units upward

(b)

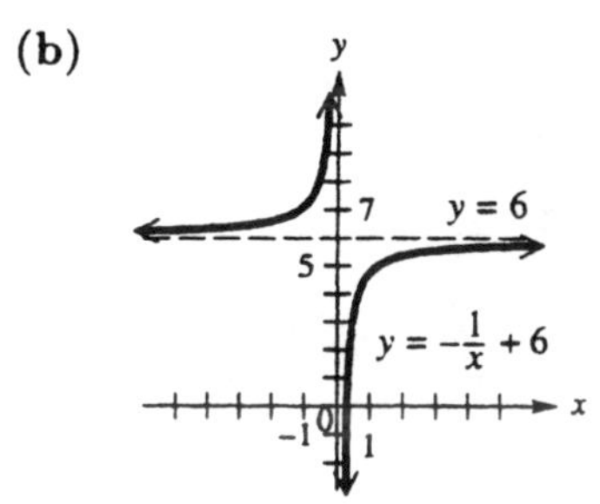

(c)

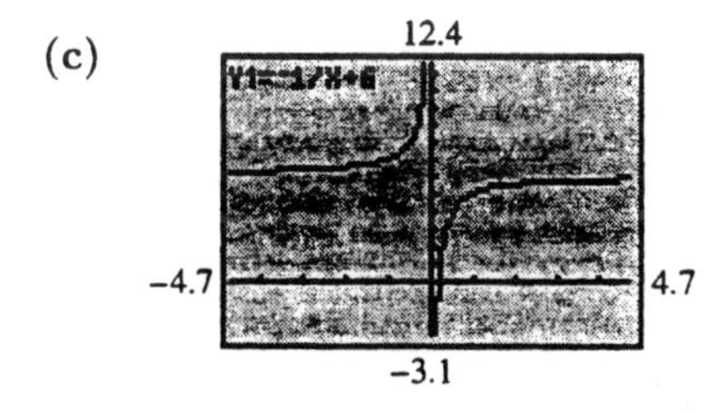

2. (a) $y = \frac{1}{x}$

$y = 4\left(\frac{1}{x}\right)$; vertical stretch by a factor of 4

$y = \frac{4}{x} - 3$; shift 3 units downward

(b)

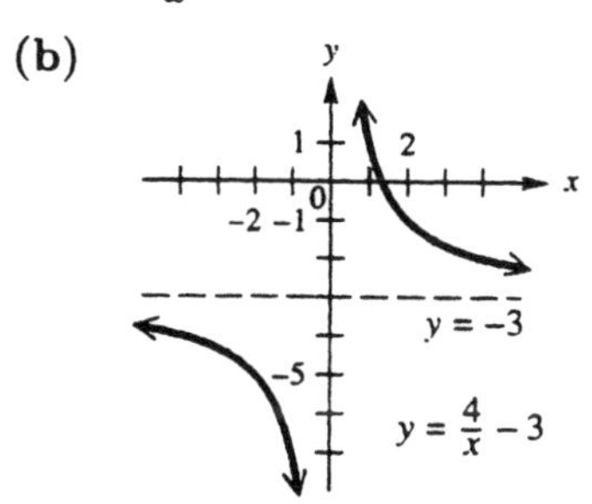

(c)

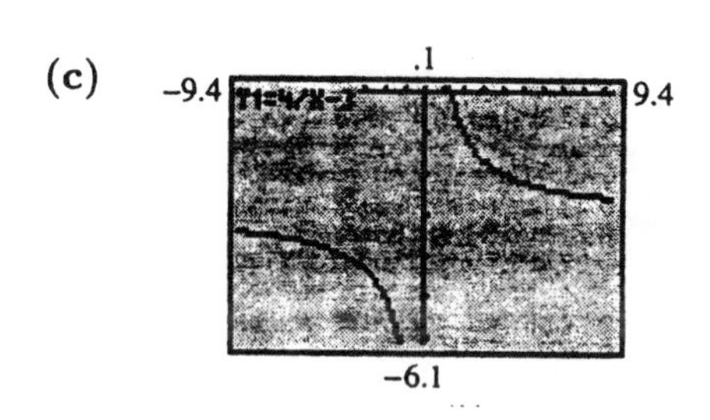

3. (a) $y = \frac{1}{x^2}$

$y = \frac{1}{(x-2)^2}$; shift 2 units right

$y = -\frac{1}{(x-2)^2}$; reflect across the x-axis

(b)

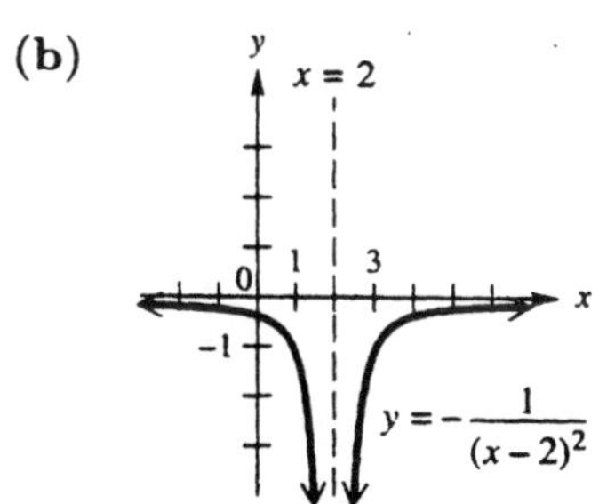

(c)

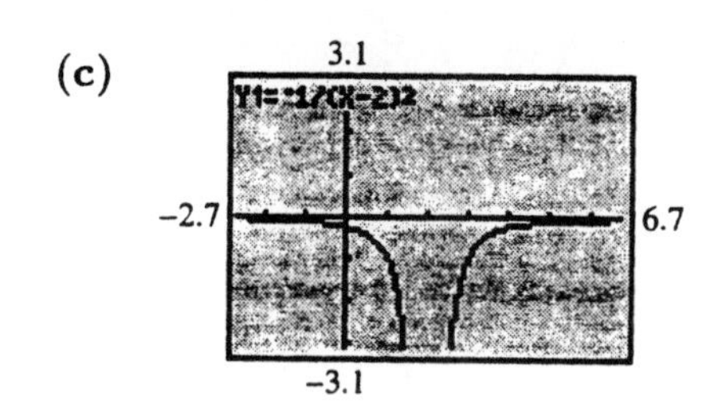

4. (a) $y = \frac{1}{x^2}$

$y = \frac{2}{x^2}$; stretch by a factor of 2

$y = \frac{2}{x^2} + 1$; shift 1 unit upward

(b)

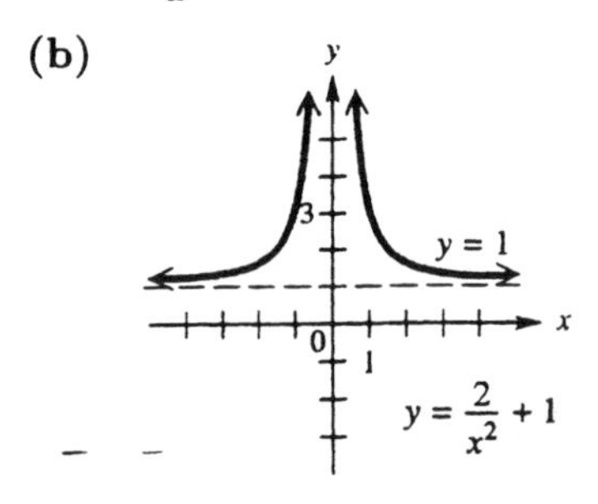

(c)

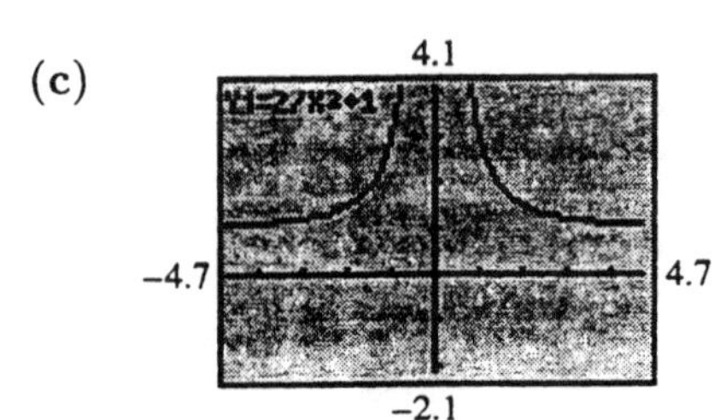

5. Oblique asymptotes occur when the numerator has a degree 1 higher than the denominator, and the denominator is not a factor of the numerator.

6. $f(x) = \frac{4x-3}{2x-1}$

V.A.: $x = \frac{1}{2}$

$[2x - 1 = 0]$

H.A.: $y = 2$

$\left[\frac{\frac{4x}{x} - \frac{3}{x}}{\frac{2x}{x} - \frac{1}{x}} = \frac{4}{2} \text{ as } x \to \infty\right]$

x-intercept: $\frac{3}{4}$ $[4x - 3 = 0]$

y-intercept: 3 [let $x = 0$]

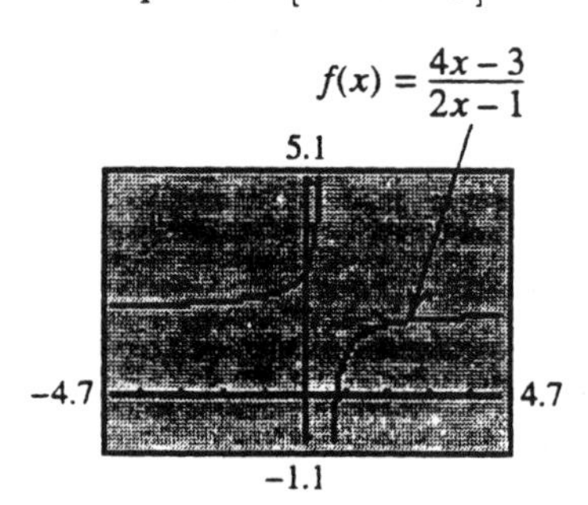

7. $f(x) = \frac{6x}{(x-1)(x+2)}$

V.A.: $x = 1$, $x = -2$

H.A.: $y = 0$

x-intercept: 0 $[6x = 0]$

y-intercept: 0 [let $x = 0$]

Points: $\left(-3, -\frac{9}{2}\right)$, $(-1, 3)$, $(2, 3)$

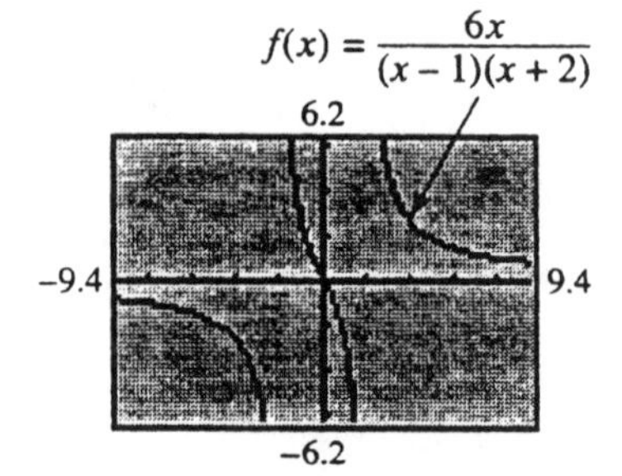

8. $f(x) = \dfrac{2x}{x^2 - 1}$

V.A.: $x = 1,\ x = -1$
$[x^2 - 1 = 0]$

H.A.: $y = 0$

x-intercept: 0 $[2x = 0]$

y-intercept: 0 [let $x = 0$]

Points: $(-2, -\frac{4}{3})$, $(2, \frac{4}{3})$

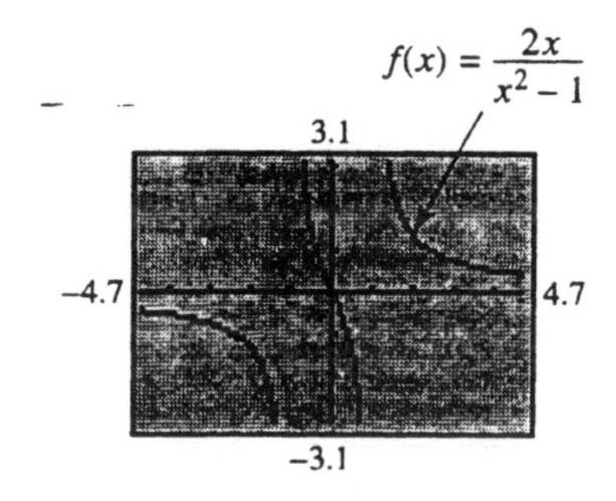

9. $f(x) = \dfrac{x^2 + 4}{x + 2} = x - 2 + \dfrac{8}{x + 2}$

$$\begin{array}{r|rrr} & x & -2 & \\ \hline x+2) & x^2 & +0x & +4 \\ & x^2 & +2x & \\ \hline & & -2x & +4 \\ & & -2x & -4 \\ \hline & & & 8 \end{array}$$

V.A.: $x = -2$ $[x + 2 = 0]$

O.A.: $y = x - 2$

x-intercept: none $[x^2 + 4 = 0]$

y-intercept: 2 [let $x = 0$]

Points: $(-3, -13)$, $(-1, 5)$

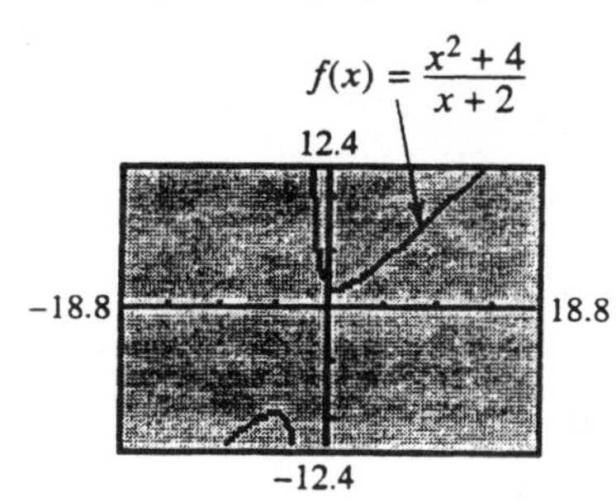

10. $f(x) = \dfrac{x^2 - 1}{x} = x - \dfrac{1}{x}$

$$\begin{array}{r|rrr} & x & & \\ \hline x) & x^2 & +0x & -1 \\ & x^2 & & \\ \hline & & & -1 \end{array}$$

V.A.: $x = 0$

O.A.: $y = x$

x-intercept: $-1,\ 1$
$[x^2 - 1 = 0]$

y-intercept: none [let $x = 0$]

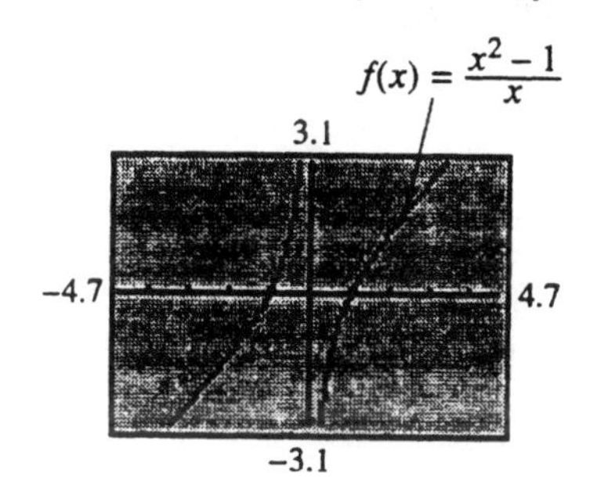

11. $f(x) = \dfrac{-2}{x^2 + 1}$

V.A.: none $[x^2 + 1 \neq 0]$

H.A.: $y = 0$

x-intercept: none

y-intercept: -2 $\left[\dfrac{-2}{0^2 + 1}\right]$

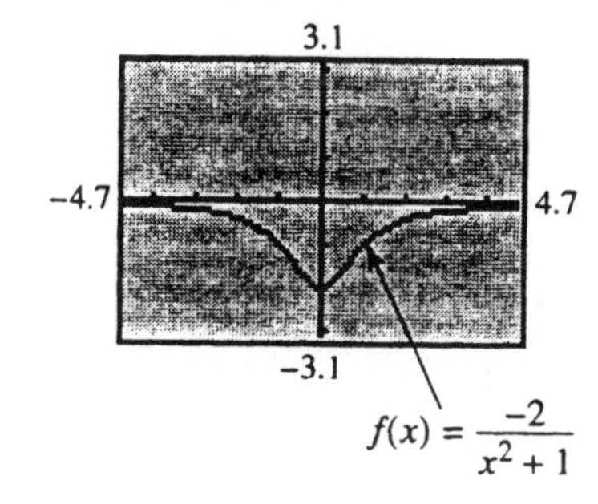

12. $f(x) = \dfrac{x^2-1}{x+1} = \dfrac{(x-1)(x+1)}{x+1}$

$= x-1,\ x \neq -1$

$f(x)$ is the graph of the line $y = x-1$ with a *hole* at $(-1, -2)$.

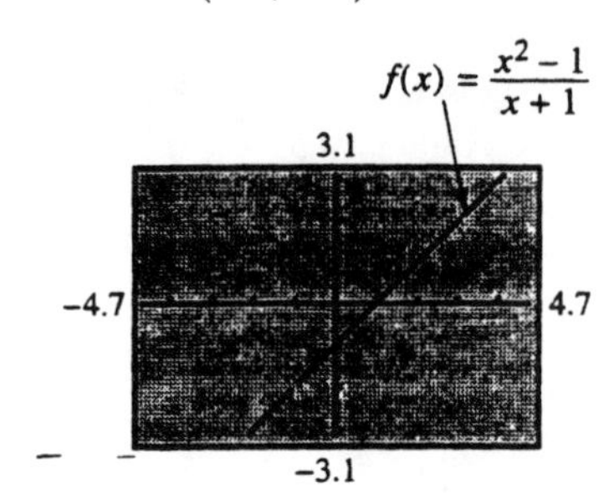

13. Since V.A. is $x = 1$, the denominator is $x-1$. Since H.A. is $y = -3$, the numerator is a first degree expression with an x-coefficient which reduces to -3. Since the x-intercept is 2, the numerator equals 0 when

$x = 2:\ -3(2) + z = 0 \rightarrow z = 6.$

So, one example: $\boxed{f(x) = \dfrac{-3x+6}{x-1}}$

Other correct functions would include multiples of 1 $\left(\dfrac{2}{2}, \dfrac{-3}{-3}, \text{etc.}\right)$.

14. The graph is the equation of the line $y = x+2$, so $x+2$ is in the numerator. There is a hole at $x = 2$, so $x-2$ is a factor in the numerator and in the denominator.

$f(x) = \dfrac{(x+2)(x-2)}{x-2} = \dfrac{x^2-4}{x-2}$

15. (a) $\dfrac{3x-2}{x+1} = 0 \quad (x \neq -1)$

$3x - 2 = 0$

$3x = 2 \quad \boxed{x = \left\{\dfrac{2}{3}\right\}}$

(b) $\dfrac{3x-2}{x+1} < 0: \boxed{\left(-1, \dfrac{2}{3}\right)}$

(c) $\dfrac{3x-2}{x+1} > 0: \boxed{(-\infty, -1) \cup \left(\dfrac{2}{3}, \infty\right)}$

16. (a) $\dfrac{5}{2x+5} = \dfrac{3}{x+2} \quad \left(x \neq \dfrac{5}{2}, -2\right)$

$5(x+2) = 3(2x+5)$

$5x + 10 = 6x + 15$

$-5 = x \quad \boxed{x = \{-5\}}$

(b) $\dfrac{5}{2x+5} < \dfrac{3}{x+2}$:

$\boxed{\left(-5, -\dfrac{5}{2}\right) \cup (-2, \infty)}$

(c) $\dfrac{5}{2x+5} > \dfrac{3}{x+2}$:

$\boxed{(-\infty, -5) \cup \left(-\dfrac{5}{2}, -2\right)}$

17. (a) $\dfrac{3}{x-2} + \dfrac{1}{x+1} = \dfrac{1}{x^2-x-2}$

Multiply by: $(x-2)(x+1)$

$3(x+1) + 1(x-2) = 1$

$3x + 3 + x - 2 = 1$

$4x = 0, \quad \boxed{x = \{0\}}$

$(x \neq -1,\ 2)$

(b) $\dfrac{3}{x-2} + \dfrac{1}{x+1} \leq \dfrac{1}{x^2-x-2}$:

$\boxed{(-\infty, -1) \cup [0, 2)}$

(c) $\dfrac{3}{x-2} + \dfrac{1}{x+1} \geq \dfrac{1}{x^2-x-2}$:

$\boxed{(-1, 0] \cup (2, \infty)}$

18. (a) $1 - \dfrac{5}{x} + \dfrac{6}{x^2} = 0$

$x^2\left[1 - \dfrac{5}{x} + \dfrac{6}{x^2} = 0\right]$

$x^2 - 5x + 6 = 0$

$(x-2)(x-3) = 0$

$\boxed{x = \{2, 3\}} \quad (x \neq 0)$

(b) $1 - \dfrac{5}{x} + \dfrac{6}{x^2} \leq 0: \boxed{[2, 3]}$

(c) $1 - \dfrac{5}{x} + \dfrac{6}{x^2} \geq 0$:

$\boxed{(-\infty, 0) \cup (0, 2] \cup [3, \infty)}$

Relating Concepts (Exercises 19 - 22)

19. $f(x) = 1 - \frac{2}{x} - \frac{2}{x^2} + \frac{3}{x^3}$

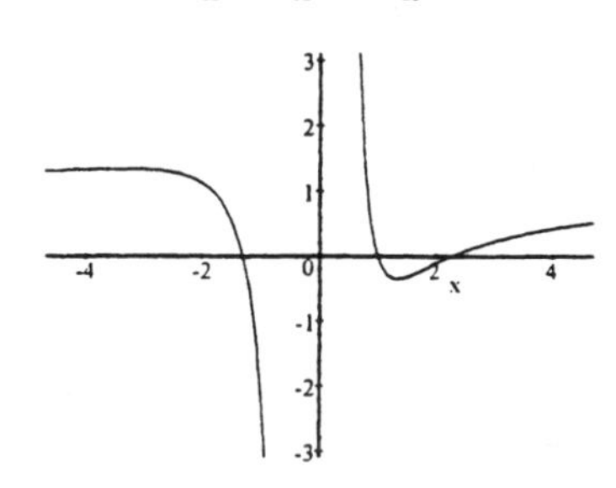

x-intercepts: $\{-1.30,\ 1,\ 2.30\}$

20. Domain of $f(x) : (-\infty, 0) \cup (0, \infty)$

21. $1 - \frac{2}{x} - \frac{2}{x^2} + \frac{3}{x^3} = 0$

$x^3 \left[1 - \frac{2}{x} - \frac{2}{x^2} + \frac{3}{x^3} = 0\right]$

$x^3 - 2x^2 - 2x + 3 = 0$

$$\begin{array}{r|rrrr} 1) & 1 & -2 & -2 & 3 \\ & & 1 & -1 & -3 \\ \hline & 1 & -1 & -3 & 0 \end{array}$$

$x^2 - x - 3 = 0$

$x = \frac{1 \pm \sqrt{1 - 4(-3)}}{2}$ $\quad$ $x = \frac{1 \pm \sqrt{13}}{2}$

22. $\frac{1 - \sqrt{13}}{2} \approx -1.30,\quad \frac{1 + \sqrt{13}}{2} \approx 2.30$

23. $f(x) = 0 : \{-2\}$

24. $f(x) > 0 : (-\infty, -2) \cup (-1, \infty)$

25. $f(x) < 0 : (-2, -1)$

26. **(a)** $C(x) = \frac{6.7x}{100 - x}$

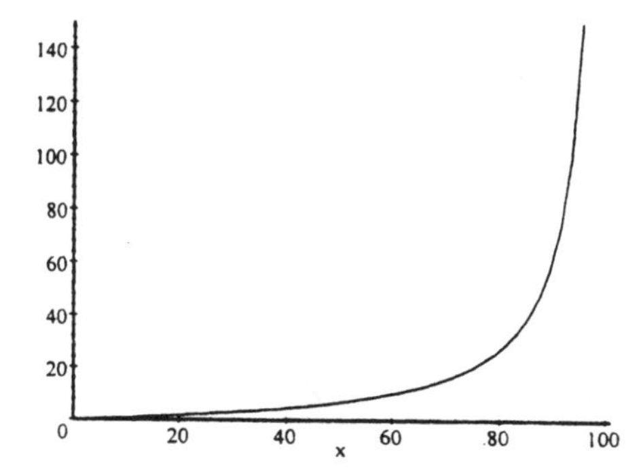

(b) $C(95) = \frac{6.7(95)}{100 - 95} = 127.3$ (thousand)

\$127,300

27. **(a)** $C(x) = \frac{10x}{49(101 - x)}$

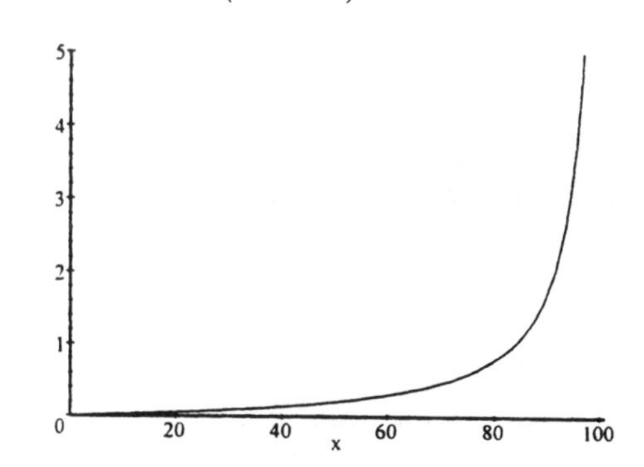

(b) $C(95) = \frac{10(95)}{49(101 - 95)} \approx 3.2312925$ (thousand)

approximately \$3231

28. $f(x) = \frac{5}{t^2 + 1},\quad t \geq 0$

(a) decreases

(b) $\frac{5}{t^2 + 1} < 1.5$

$5 < 1.5t^2 + 1.5$

$3.5 < 1.5t^2$

$2\frac{1}{3} < t^2 \longrightarrow t^2 > 2.\overline{3}$

$t > 1.528$

After 1.5 hours

29. $f(x) = \dfrac{x^2}{1600 - 40x}, \quad 0 \le x \le 40$

(a) $\dfrac{x^2}{1600 - 40x} \le 8$

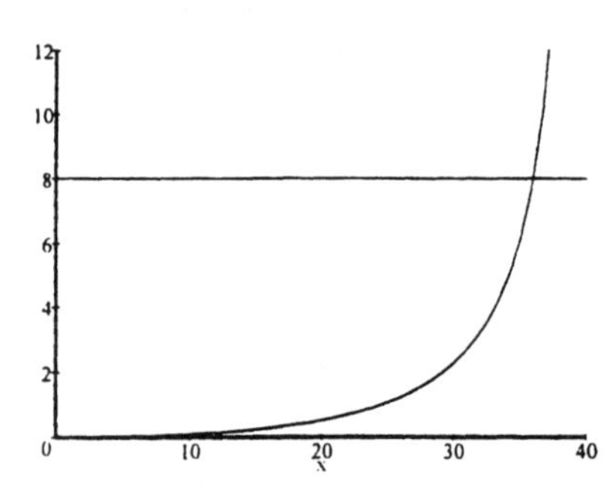

point of intersection $\approx (36,\ 8)$

$f(x) \le 8 \approx [0,\ 36]$

(b) The average line length is less than or equal to 8 cars when the average arrival rate is 36 cars per hour or less.

30. $D(x) = \dfrac{120}{x}$

(a) As the coefficient of friction becomes smaller, the braking distance increases.

(b) $\dfrac{120}{x} \ge 400 : (0,\ .3]$

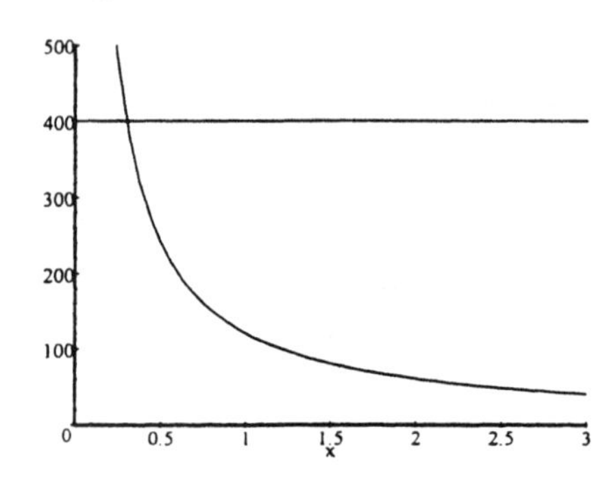

point of intersection $= (.3,\ 400)$

31. $y = \dfrac{k}{x}$

$(6,\ 5):\quad 5 = \dfrac{k}{6} \Longrightarrow k = 30$

$y = \dfrac{30}{x}$

$y = \dfrac{30}{15} \Longrightarrow y = \boxed{2}$

32. $z = \dfrac{k}{t^3}$

$(5,\ .08):\quad .08 = \dfrac{k}{5^3} \Longrightarrow k = 10$

$z = \dfrac{10}{t^3}$

$z = \dfrac{10}{2^3} = \dfrac{10}{8} \Longrightarrow z = \boxed{1.25}$

33. $m = \dfrac{knp^2}{q} \quad (m = 20,$

$n = 5,\ p = 6,\ q = 18)$

$20 = \dfrac{k(5)(36)}{18}$

$k = \dfrac{(18)(20)}{(5)(36)} = 2$

EQ: $m = \dfrac{2np^2}{q}$

$(n = 7,\ p = 11,\ q = 2)$

$m = \dfrac{2(7)(121)}{2} \qquad \boxed{m = 847}$

34. The formula for the height of a right circular cone with volume 100 is $h = \dfrac{300/\pi}{r^2}$. The height of this cone varies <u>inversely</u> as the <u>square</u> of the <u>radius</u> of its' base. The constant of variation is <u>$\dfrac{300}{\pi}$</u>.

35. $I = \dfrac{k}{d^2};\quad I = 70$ candela, $d = 5\,\text{m}$:

$70 = \dfrac{k}{(5)^2}$

$k = 25(70) = 1750$

EQ: $I = \dfrac{1750}{d^2};\quad d = 12$ m:

$I = \dfrac{1750}{144} = 12.152\overline{7}$

$\boxed{I \approx 12.15 \text{ candela}}$

36. $R = \frac{k}{d^2}$; $d = .01$ in, $R = .4$ ohm

$$.4 = \frac{k}{(.01)^2}$$

$$k = .4\,(.01)^2 = 0.00004$$

EQ: $R = \frac{0.00004}{d^2}$ $\left(\text{or } \frac{4}{10000d^2}\right)$;

$$d = .03 \text{ in},\ R = \frac{0.00004}{(.03)^2}$$

$$\boxed{R = 0.0\bar{4} \text{ or } \frac{2}{45} \text{ ohms}}$$

37. $I = kPt$; $I = \$110$, $P = \$1000$, $t = 2$ years

$$110 = k(1000)(2)$$

$$k = \frac{110}{2000} = .055$$

EQ: $I = .055Pt$; $P = \$5000$, $t = 5$ years

$$I = .055(5000)(5) = 1375 \quad \boxed{I = \$1375}$$

38. $F = \frac{kws^2}{r}$; $s = 30\,\text{mph}$, $r = 500'$

$w = 2000\,\text{lb}$, $F = 3000\,\text{lb}$:

$$3000 = \frac{k\,(2000)\,(30)^2}{500}$$

$$k = \frac{3000\,(500)}{(2000)\,(900)} = \frac{5}{6}$$

EQ: $F = \frac{5ws^2}{6r}$; $w = 2000\,\text{lb}$,

$s = 60\,\text{mph}$, $r = 800'$:

$$F = \frac{5\,(2000)\,(60)^2}{6\,(800)} = \boxed{7500 \text{ lb}}$$

39. $L = \frac{kd^4}{h^2}$; $h = 9$ m, $d = 1$ m,

$L = 8$ metric tons

$$8 = \frac{k\,(1)^4}{9^2}$$

$$k = 8 \cdot 81 = 648$$

EQ: $L = \frac{648d^4}{h^2}$; $h = 12$ m, $d = \frac{2}{3}$

$$L = \frac{648\,\left(\frac{2}{3}\right)^4}{12^2} = \frac{648\,(2^4)}{144\,(3)^4}$$

$$= \frac{(648)\,(16)}{(144)\,(81)} = \frac{8}{9} = 0.\bar{8}$$

$$\boxed{L = \frac{8}{9} \text{ metric tons}}$$

40. Let m = maximum load, w = width, h = height, l = length.

$$m = \frac{kwh^2}{l}$$

Find k if $l = 8$ m (800 cm), $w = 12$ cm, $h = 15$ cm, $m = 400$ kg:

$$400 = \frac{k\,(12)\,(15)^2}{800}$$

$$k = \frac{(400)\,(800)}{(12)\,(15)^2} = \frac{3200}{27}$$

EQ: $m = \frac{3200\,w\,h^2}{27\,l}$

Find m if $l = 16$ m (1600 cm), $w = 24$ cm, $h = 8$ cm:

$$m = \frac{3200\,(24)\,(8)^2}{27\,(1600)} = 113\tfrac{7}{9}$$

$$\boxed{\text{approximately } 113\tfrac{7}{9} \text{ kg}}$$

41. $w = \frac{k}{d^2}$; $w = 90\,\text{kg}$, $d = 6400\,\text{km}$

$$90 = k/\,(6400)^2$$

$$k = 90\,(6400)^2$$

EQ: $w = \frac{90(6400)^2}{d^2}$;

$(d = 7200 \text{ km})$

$$w = \frac{90\,(6400)^2}{(7200)^2}$$

$$\boxed{w = \frac{640}{9} \text{ kg or } 71\tfrac{1}{9} \text{ kg } (\approx 71.11 \text{ kg})}$$

42. $y = -a\sqrt{x}$

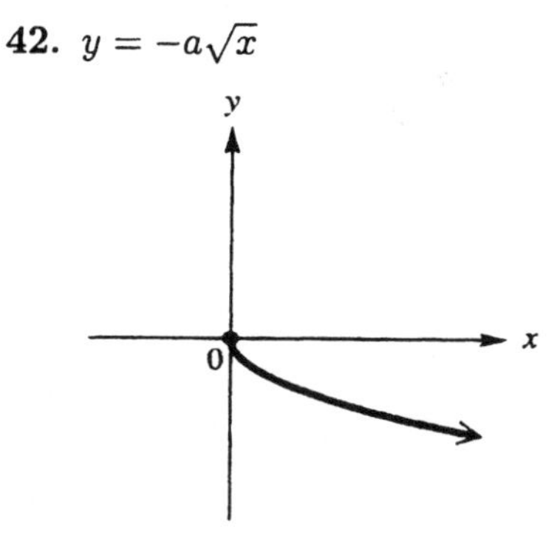

43. $y = \sqrt[3]{x+a}$

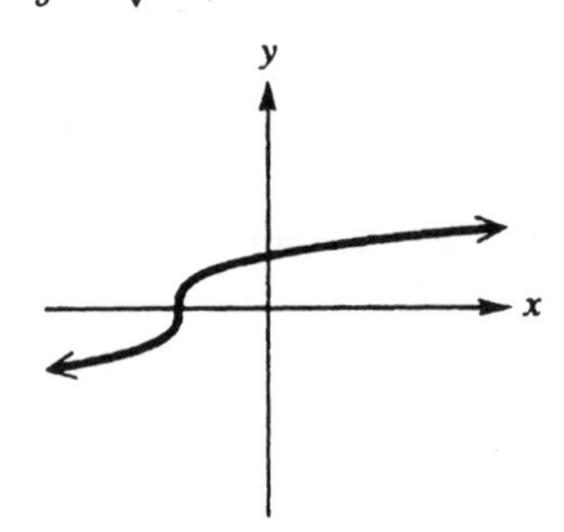

44. $y = \sqrt[3]{x-a}$

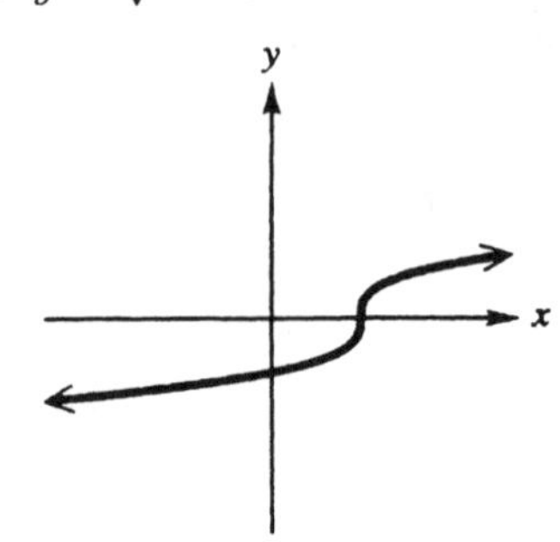

45. $y = -a\sqrt[3]{x} - b$

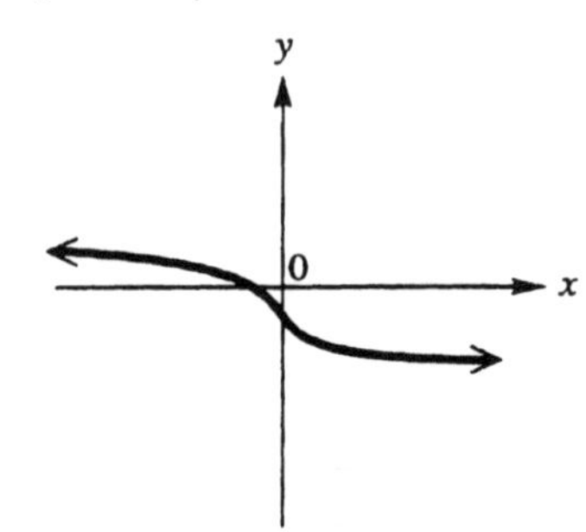

46. $y = \sqrt{x+a} + b$

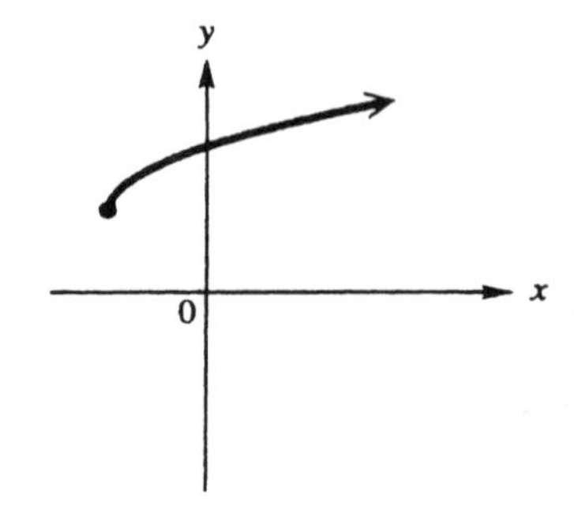

47. $-\sqrt[5]{-32} = -(-2) = \boxed{2}$

48. $36^{-(3/2)} = \left(36^{1/2}\right)^{-3} = 6^{-3} = \boxed{\dfrac{1}{216}}$

49. $-1000^{2/3} = -\left[(1000)^{1/3}\right]^2$
$= -(10)^2 = \boxed{-100}$

50. $(-27)^{-4/3} = \dfrac{1}{\left[(-27)^{1/3}\right]^4} = \dfrac{1}{[-3]^4} = \boxed{\dfrac{1}{81}}$

51. $16^{3/4} = \left[(16)^{1/4}\right]^3 = 2^3 = \boxed{8}$

52. $\sqrt[5]{84.6} \approx 2.42926041078$
Rational approximation

53. $\sqrt[4]{\dfrac{1}{16}} = 0.5;$ Exact value

54. $\left(\dfrac{1}{8}\right)^{4/3} = 0.0625$ Exact value

55. $12^{1/3} \approx 2.28942848511$
Rational approximation

Graph for problems #56–#59:

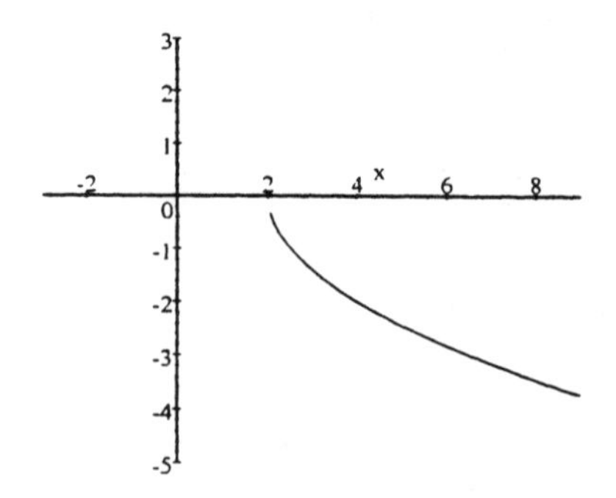

56. $f(x) = -\sqrt{2x-4}$
Domain: $2x - 4 \geq 0$
$2x \geq 4$
$x \geq 2$ $\boxed{[2, \infty)}$

57. Range: $(-\infty, 0]$

58. Increasing: none

59. Decreasing: $[2, \infty)$

60. $x^2 + y^2 = 25$

$y^2 = 25 - x^2$

$y = \pm\sqrt{25 - x^2}$

$y_1 = \sqrt{25 - x^2}$; $\quad y_2 = -\sqrt{25 - x^2}$

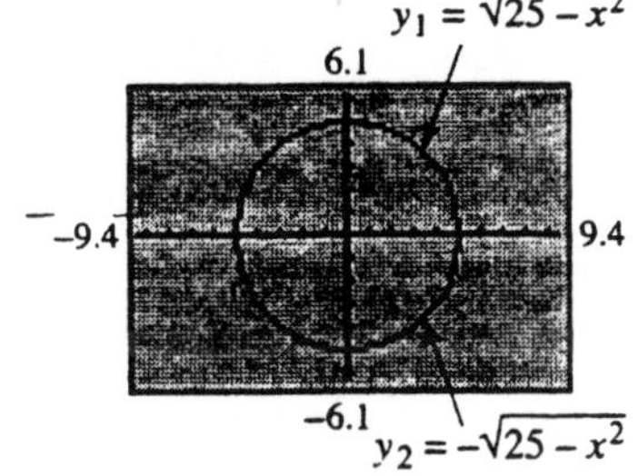

61. (a) $\sqrt{5 + 2x} = x + 1$

$\left(\sqrt{5 + 2x}\right)^2 = (x + 1)^2$

$5 + 2x = x^2 + 2x + 1$

$0 = x^2 - 4 \rightarrow \quad x = -2,\ 2$

Check (-2): $\sqrt{5 - 4} = -2 + 1$

$1 = -1$ **No**

Check (2): $\sqrt{5 + 4} = 2 + 1$

$\sqrt{9} = 3$ **Yes**

Solution set: $\{2\}$

$y_1 = \sqrt{5 + 2x}$; $y_2 = x + 1$

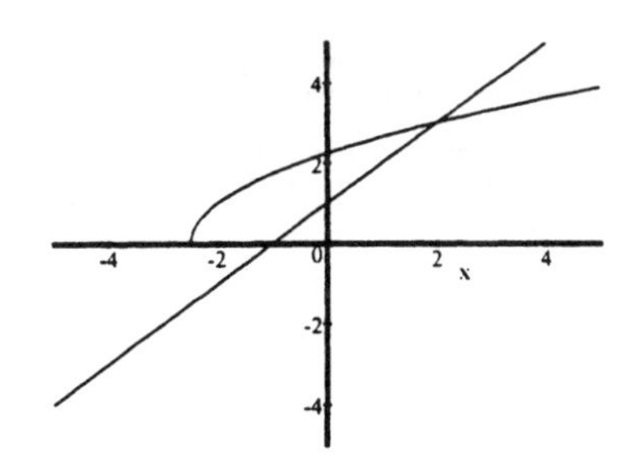

The graphs intersect at $x = 2$.

domain of y_1: $\quad 5 + 2x \geq 0$

$2x \geq -5$

$x \geq -\dfrac{5}{2}$

(b) $\sqrt{5 + 2x} > x + 1 : [-2.5,\ 2)$

(c) $\sqrt{5 + 2x} < x + 1 : (2,\ \infty)$

62. (a) $\sqrt{2x + 1} - \sqrt{x} = 1$

$\sqrt{2x + 1} = \sqrt{x} + 1$

$\left(\sqrt{2x + 1}\right)^2 = \left(\sqrt{x} + 1\right)^2$

$2x + 1 = x + 2\sqrt{x} + 1$

$x = 2\sqrt{x}$

$x^2 = \left(2\sqrt{x}\right)^2$

$x^2 = 4x$

$x^2 - 4x = 0$

$x\,(x - 4) = 0 \rightarrow \quad x = 0,\ 4$

Check (0): $\sqrt{1} - 0 = 1$ **Yes**

Check (4): $\sqrt{9} - \sqrt{4} = 1$

$3 - 2 = 1$ **Yes**

Solution set: $\{0,\ 4\}$

$y_1 = \sqrt{2x + 1} - \sqrt{x}$; $y_2 = 1$

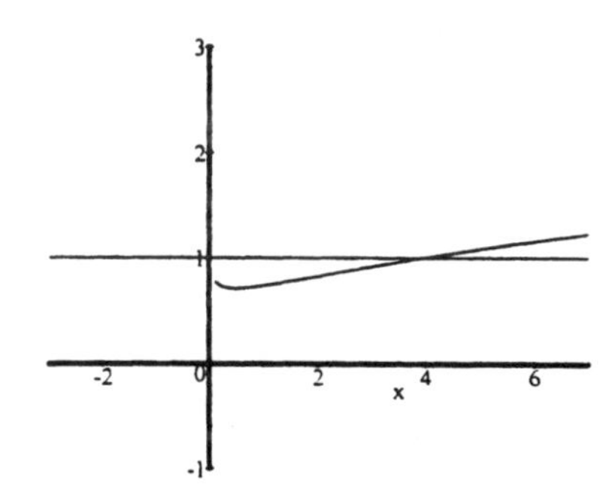

The graphs intersect at $x = 0,\ 4$.

domain of y_1:

$2x + 1 \geq 0 \quad$ and $\quad x \geq 0$

$x \geq -\dfrac{1}{2} \quad$ and $\quad x \geq 0$

$x \geq 0$

(b) $\sqrt{2x + 1} - \sqrt{x} > 1 : (4,\ \infty)$

(c) $\sqrt{2x + 1} - \sqrt{x} < 1 : (0,\ 4)$

63. (a) $\sqrt[3]{6x+2} = \sqrt[3]{4x}$

$\left(\sqrt[3]{6x+2}\right)^3 = \left(\sqrt[3]{4x}\right)^3$

$6x + 2 = 4x$

$2x = -2 \rightarrow \quad x = -1$

Check (-1):

$\sqrt[3]{-6+2} = \sqrt[3]{-4}$

$\sqrt[3]{-4} = \sqrt[3]{-4} \quad \checkmark$

Solution set: $\{-1\}$

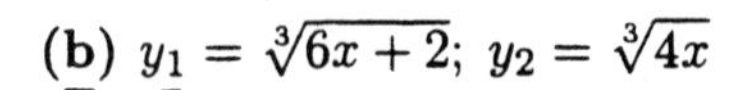

(b) $y_1 = \sqrt[3]{6x+2}$; $y_2 = \sqrt[3]{4x}$

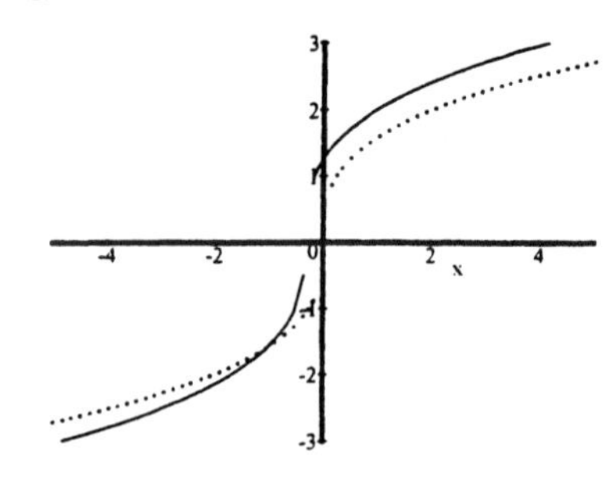

The graphs intersect at $x = -1$.
domain: $(-\infty, \infty)$

$\sqrt[3]{6x+2} \geq \sqrt[3]{4x} : [-1, \infty)$

(c) $\sqrt[3]{6x+2} \leq \sqrt[3]{4x} : (-\infty, -1]$

64. (a) $(x-2)^{2/3} - x^{1/3} = 0$

$(x-2)^{2/3} = x^{1/3}$

$\left[(x-2)^{2/3}\right]^3 = \left[x^{1/3}\right]^3$

$(x-2)^2 = x$

$x^2 - 4x + 4 = x$

$x^2 - 5x + 4 = 0$

$(x-4)(x-1) = 0$

$x = 1, 4$

Check (1):

$(1-2)^{2/3} - 1^{1/3} = 0$

$\left[(-1)^{1/3}\right]^2 - 1^{1/3} = 0$

$1 - 1 = 0 \quad \checkmark$

Check (4):

$(4-2)^{2/3} - 4^{1/3} = 0$

$2^{2/3} - 4^{1/3} = 0$

$4^{1/3} - 4^{1/3} = 0 \quad \checkmark$

Solution set: $\{1, 4\}$

(b) Graph $y_1 = (x-2)^{2/3} - x^{1/3}$

$y_2 = 0$

The graphs intersect at $x = 1, 4$.
domain: $(-\infty, \infty)$

$(x-2)^{2/3} - x^{1/3} \geq 0 :$

$(-\infty, 1] \cup [4, \infty)$

(c) $(x-2)^{2/3} - x^{1/3} \leq 0 : [1, 4]$

Relating Concepts (Exercises 65 - 68)

65. $y_1 = \sqrt{3x+12} - 4$ (solid)

$y_2 = \sqrt[3]{3x+12} - 6$ (dotted)

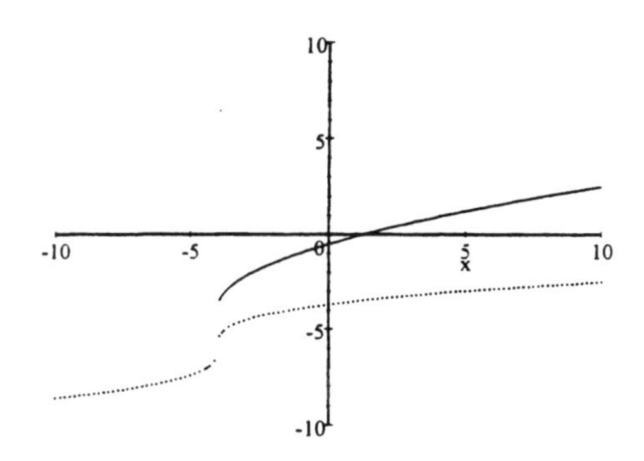

There are no solutions to $y_1 = y_2$.

66. $f(x) = y_1 - y_2$ has no x-intercepts:

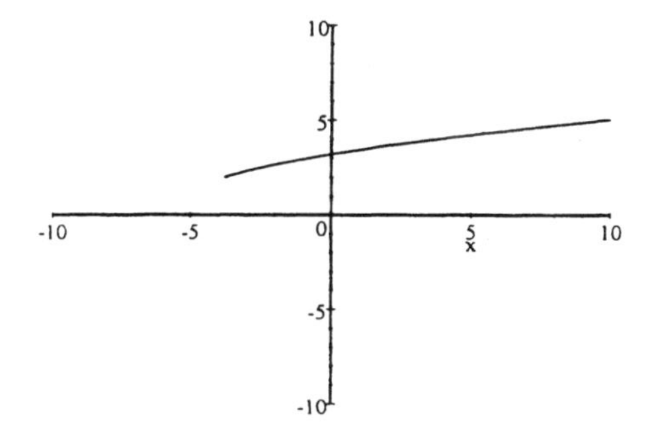

67. The graph of $-f(x) = y_2 - y_1$ is the reflection of the graph of $f(x)$ across the x-axis:

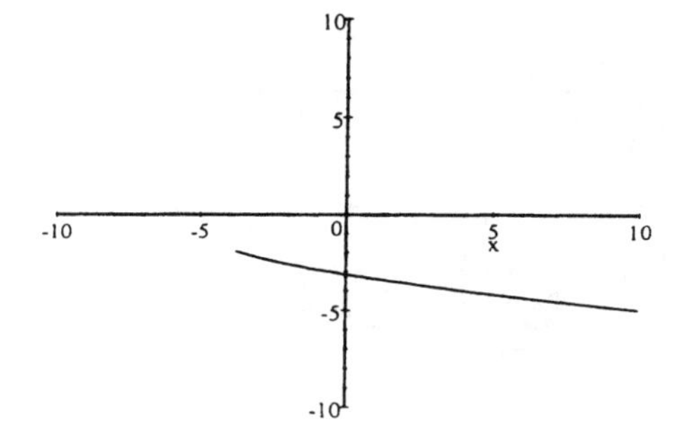

68. The graph of $f(-x)$ is the reflection of the graph of $f(x)$ across the y-axis:

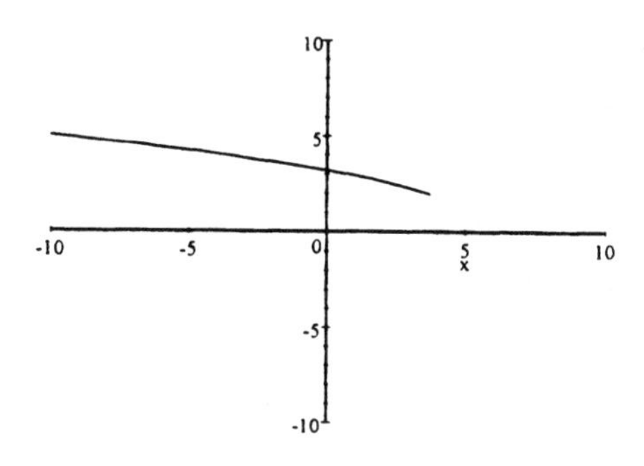

69. $L = kT^n$

(a) As the length L of the pendulum increases, so does the period of oscillation T.

(b) There are a number of ways to find n and k. One way is to realize that $k = \dfrac{L}{T^n}$ for some integer n. The ratio should be the constant k for each data point when the correct n is found. Another way is to use regression.

(c) $k = \dfrac{1.0}{1.11^n} = \dfrac{1.5}{1.36^n}$

$y_1 = \dfrac{1.0}{1.11^n}$ (solid), $y_2 = \dfrac{1.5}{1.36^n}$ (dotted)

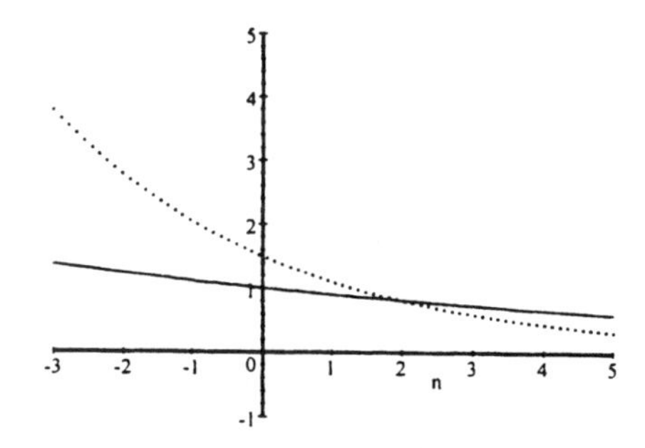

Intersection point $\approx (1.996, .812)$

$n \approx 2, \quad k = \dfrac{1.0}{1.11^{1.9961}} \approx .81$

(d) $L = .81T^2$

$5 = .81T^2$

$\dfrac{5}{.81} = T^2 \Longrightarrow T = \sqrt{\dfrac{5}{.81}} \approx 2.48$ seconds

(e) length doubles:

$L = .81T^2$

$2L = 2(.81)T^2$

$2L = .81\left(\sqrt{2}T\right)^2$

T increases by a factor of $\sqrt{2} \approx 1.414$.

70. $V = \pi r^2 h = 27\pi$

$$h = \frac{27\pi}{\pi r^2} = \frac{27}{r^2}$$

$$S.A. = 2\pi r h + \pi r^2$$

$$S(r) = 2\pi r\left(\frac{27}{r^2}\right) + \pi r^2$$

$$S(r) = \frac{54\pi}{r} + \pi r^2$$

Graph $S(r)$; minimum is at (3, 84.82)

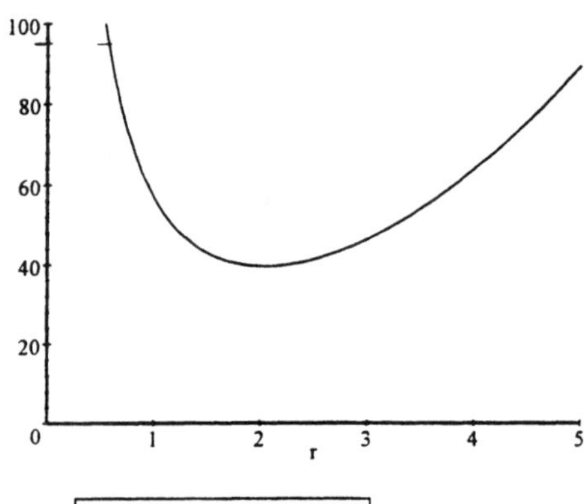

Radius = 3 inches

Chapter 4 Test

1. (a) $f(x) = -\frac{1}{x}$

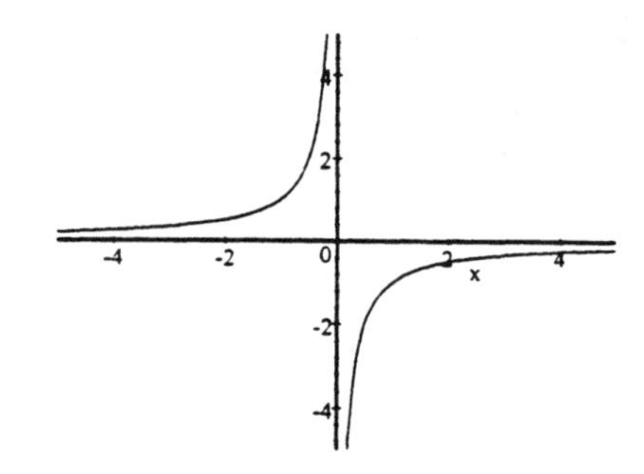

(b) reflect $y = \frac{1}{x}$ across the x-axis

(c)

3.1

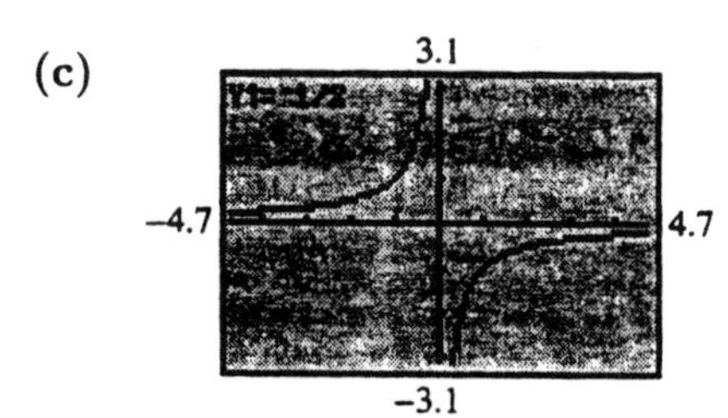

-3.1

2. (a) $f(x) = -\frac{1}{x^2} - 3$

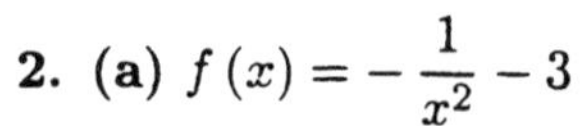

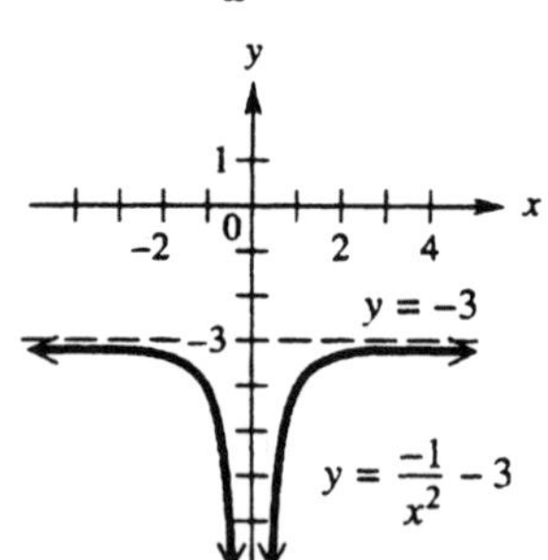

(b) reflect $y = \frac{1}{x^2}$ across the x-axis; shift 3 units downward

(c)

.1

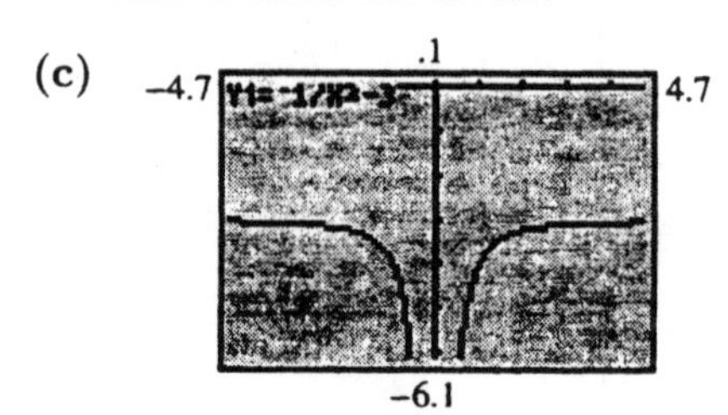

-6.1

3. $f(x) = \dfrac{x^2 + x - 6}{x^2 - 3x - 4}$

(a) V. A.: $\boxed{x = -1,\ x = 4}$

$x^2 - 3x - 4 = 0$
$(x - 4)(x + 1) = 0$

(b) H. A.: $\boxed{y = 1}$

$$\frac{\frac{x^2}{x^2} + \frac{x}{x^2} - \frac{6}{x^2}}{\frac{x^2}{x^2} - \frac{3x}{x^2} - \frac{4}{x^2}}$$

As $x \to \infty,\ y \to 1$

(c) y-intercept: $\boxed{1.5}$

$x = 0,\ y = \dfrac{-6}{-4}$

(d) x-intercepts: $\boxed{-3,\ 2}$

$x^2 + x - 6 = 0$
$(x + 3)(x - 2) = 0$

(e) Intersects asymptote at $\boxed{\left(\frac{1}{2},\ 1\right)}$

$$\frac{x^2 + x - 6}{x^2 - 3x - 4} = 1$$
$$x^2 + x - 6 = x^2 - 3x - 4$$
$$4x = 2$$
$$x = \frac{1}{2}$$

(f) $f(x)$

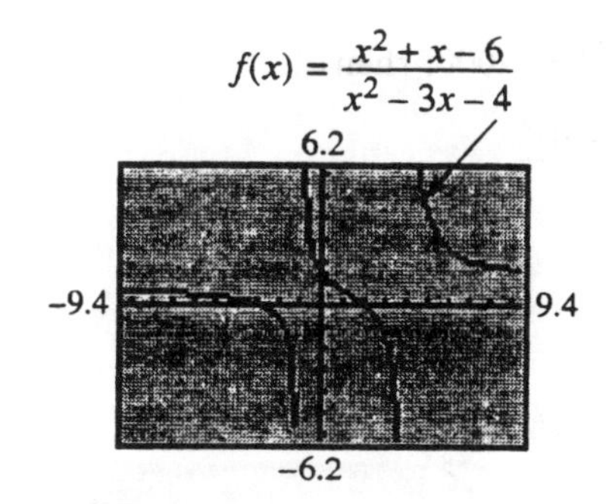

4. $f(x) = \dfrac{2x^2 + x - 3}{x - 2}$

$$\begin{array}{r} 2x + 5 \\ x - 2 \overline{)\ 2x^2 + x - 3} \\ \underline{2x^2 - 4x} \\ 5x - 3 \\ \underline{5x - 10} \\ 7 \end{array}$$

Oblique asymptote: $\boxed{y = 2x + 5}$

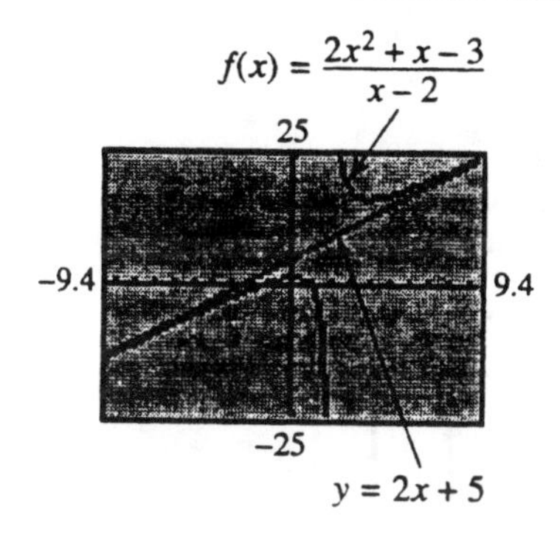

5. (a) $f(x) = \dfrac{x^2 - 16}{x + 4}$

$= \dfrac{(x - 4)(x + 4)}{x + 4}$

$= x - 4,\ x \neq -4$

The graph exhibits a *hole* when

$\boxed{x = -4}$ $(-4, -8)$

(b) $f(x) = \dfrac{x^2 - 16}{x + 4}$

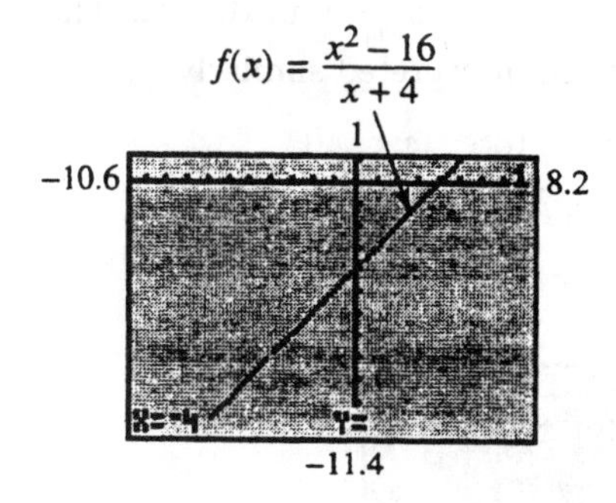

6. (a) $\dfrac{3}{x-2}+\dfrac{21}{x^2-4}=\dfrac{14}{x+2}$

Multiply both sides by (x^2-4) :

$$3(x+2)+21=14(x-2)$$
$$3x+6+21=14x-28$$
$$-11x=-55$$
$$\boxed{x=\{5\}}$$

(b) left side (dark); right side (light)
intersection point: (5, 2)
Asymptotes: $x=-2$, $x=2$

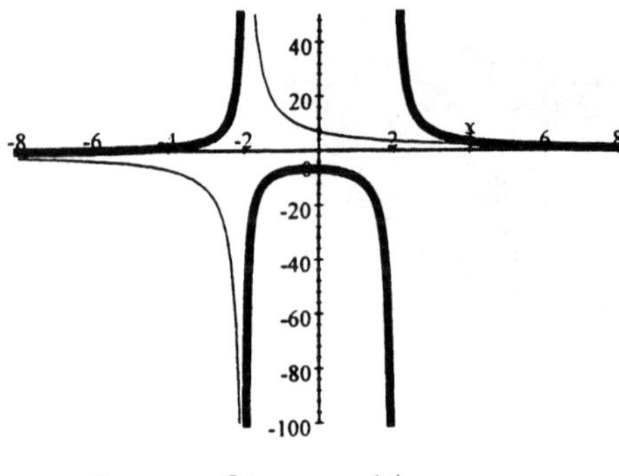

$\dfrac{3}{x-2}+\dfrac{21}{x^2-4}\geq\dfrac{14}{x+2}$:

$$\boxed{(-\infty,-2)\cup(2,\,5]}$$

7. $W(x)=\dfrac{1}{40-x}$, $0\leq x<40$

(a) $W(30)=\dfrac{1}{40-30}=.1$

$W(39)=\dfrac{1}{40-39}=1$

$W(39.9)=\dfrac{1}{40-39.9}=\dfrac{1}{.1}=10$

When the rate of vehicles is 30 vehicles/min, the average wait time is $\frac{1}{10}$ min (6 sec). The other results are interpreted similarly.

(b) Vertical asymptote: $x=40$.
As $x\longrightarrow 40$, $W(x)\longrightarrow\infty$

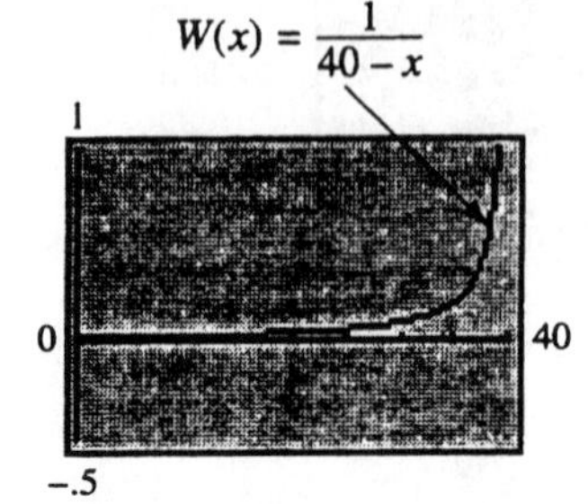

continued

7. continued

(c) $5=\dfrac{1}{40-x}$
$$200-5x=1$$
$$-5x=-199$$
$$x=\frac{199}{5}\approx 39.8 \text{ vehicles/minute}$$

8. Let p = pelidisi, w = weight, h = height. Then

$$p=\frac{k\sqrt[3]{w}}{h}$$

Find k when $p=100$, $w=48820$ gm, and $h=78.7$ cm:

$$100=\frac{k\sqrt[3]{48820}}{78.7}$$
$$k=\frac{7870}{\sqrt[3]{48820}}$$
$$\approx 215.332116$$

Find p when $w=54430$ gm and $h=88.9$ cm:

$$p\approx\frac{215.332116\sqrt[3]{54430}}{88.9}$$
$$\approx 91.79517$$

$$\boxed{\text{92; Undernourished}}$$

9. If Volume $=\pi x^2h=4000$, then

$$S(x)=\frac{8000+2\pi x^3}{x}$$

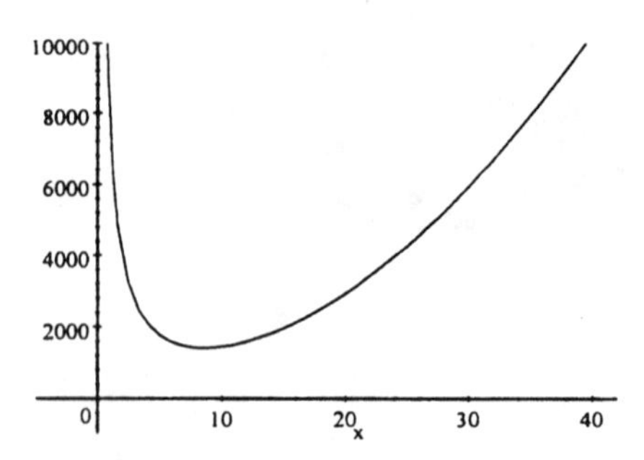

Minimum at $\approx(8.6025,\ 1394.94)$

Radius ≈ 8.6 cm Minimum amount $\approx 1394.9\ \text{cm}^2$

10. $f(x) = -\sqrt{5-x}$

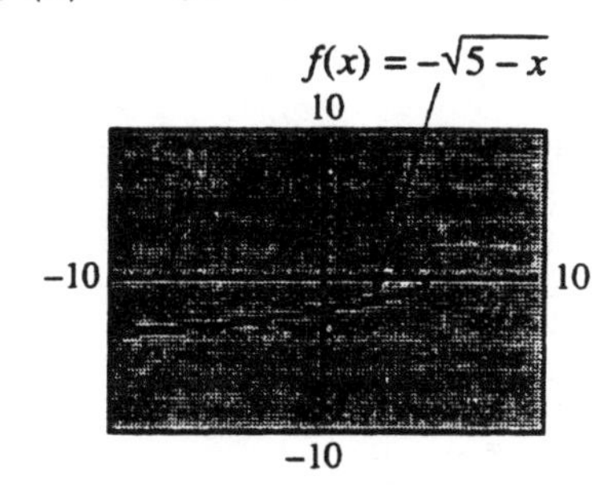

(a) Domain of $-\sqrt{5-x}$:

$$5 - x \geq 0$$
$$-x \geq -5$$
$$x \leq 5:$$

$\boxed{(-\infty, 5]}$

(b) From the graph, the Range of $-\sqrt{5-x}$:

$\boxed{(-\infty, 0]}$

(c) This function increases over its entire domain.

(d) From the graph, $f(x) = 0$: $\boxed{\{5\}}$

(e) From the graph, $f(x) < 0$: $\boxed{(-\infty, 5)}$

11. (a) $\sqrt{4-x} = x + 2$

Square both sides:

$$4 - x = x^2 + 4x + 4$$
$$0 = x^2 + 5x$$
$$0 = x(x+5)$$
$$x = 0, \ -5 \text{ (possible solutions)}$$

Check $x = 0$:

$$\sqrt{4-0} = 0 + 2$$
$$2 = 2 \qquad \textbf{Yes}$$

Check $x = -5$:

$$\sqrt{4+5} = -5 + 2$$
$$3 \neq -3 \qquad \textbf{No}$$

$\boxed{x = \{0\}}$

Graph $y_1 = \sqrt{4-x}$ (dark)
$y_2 = x + 2$ (light)

$y = \sqrt{4-x}$ $y = x + 2$
10
−10 10
Intersection X=0 Y=2
−10

(b) $\sqrt{4-x} > x + 2$: $\boxed{(-\infty, 0)}$

(c) $\sqrt{4-x} \leq x + 2$: $\boxed{[0, 4]}$

12. As the cable is moved away from Q towards R, to a point we will call S, the distance moved will be x.

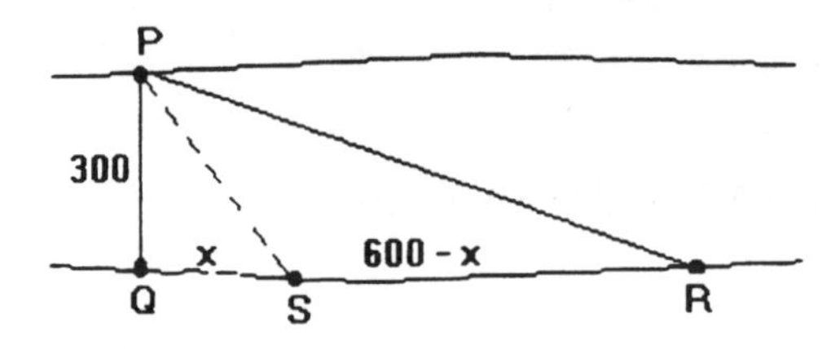

This cost will be the sum of the cost from P to S and from S to R.

To find the length of PS, use the Pythagorean theorem:

$$300^2 + x^2 = PS^2$$

$$C(x) = \left(\sqrt{300^2 + x^2}\right)125 + (600 - x)100$$

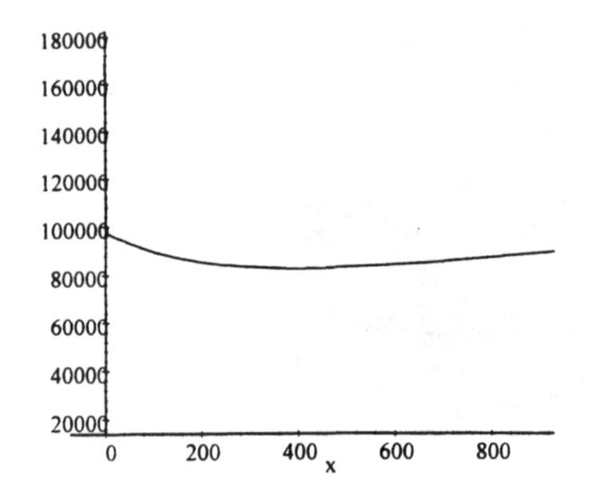

Minimum: (400, 82500)

The cable should be laid underwater from P to a point S which is on the bank 400 yds away from Q (in the direction of R).

Chapter 4 Project

How Rugged Is Your Coastline?

1. In $y = \frac{23.5}{x^{1.153}}$, the number of steps needed (y) is a function of the number of steps taken (x).

For $x = 6$, $y = \frac{23.5}{6^{1.153}} \approx 2.98$

The total distance is

$$x \cdot y = 6 \cdot 2.98 \approx 17.9 \text{ inches}$$

For $x = .1$, $y = \frac{23.5}{.1^{1.153}} \approx 334.25$

The total distance is

$$x \cdot y = .1 \cdot 334.25 \approx 33.4 \text{ inches}$$

For $x = .01$, $y = \frac{23.5}{.01^{1.153}} \approx 4754.10$

The total distance is

$$x \cdot y = .01 \cdot 4754.10 \approx 47.5 \text{ inches}$$

2. Answers will vary.

CHAPTER 5

Section 5.1

1. Since $(x_1, y_1) = (y_2, x_2)$ (the x's and y's are switched), the screens suggest they are inverse functions.

3. Since (0, 4) is on one graph, (4, 0) should be on the inverse function graph. They are not inverse functions.

5. In order for a function to have an inverse, it must be one-to-one.

7. If f and g are inverses, then $(f \circ g)(x) = \underline{x}$, and $\underline{(g \circ f)(x)} = x$.

9. If the point (a, b) lies on the graph of f, and f has an inverse, then the point $\underline{(b, a)}$ lies on the graph of f^{-1}.

11. If a function f has an inverse, then the graph of f^{-1} may be obtained by reflecting the graph of f across the line $\underline{y = x}$.

13. If $f(-4) = 16$ and $f(4) = 16$, then f does not have an inverse because it is not one-to-one.

15. one-to-one

17. not one-to-one

19. one-to-one

21. not one-to-one

23. one-to-one

25. $y = (x-2)^2$
not one-to-one

27. $y = \sqrt{36 - x^2}$
not one-to-one

29. $y = 2x^3 + 1$
one-to-one

31. $y = \dfrac{1}{x+2}$
one-to-one

33. The graph will fail the horizontal line test since the end behavior will be either $\nwarrow \nearrow$ or $\swarrow \searrow$.

35. f : tying your shoelaces
f^{-1} : untying your shoelaces

37. f : entering a room
f^{-1} : leaving a room

39. f : wrapping a package
f^{-1} : unwrapping a package

41. $f(x) = x^3$
$f(2) = 2^3 = 8$

43. $f(x) = x^3$
$f(-2) = (-2)^3 = -8$

45. $f(x) = x^3$
$f^{-1}(x) = \sqrt[3]{x}$
$f^{-1}(0) = \sqrt[3]{0} = 0$

47. Yes

49. Yes

51. Yes: $f(x) = -\dfrac{3}{11}x$

$$x = -\frac{3}{11}y$$
$$x\left(-\frac{11}{3}\right) = -\frac{3}{11}\left(-\frac{11}{3}\right)y$$
$$-\frac{11}{3}x = y$$
$$f^{-1}(x) = g(x) = -\frac{11}{3}x$$

53. Yes: $f(x) = 5x - 5$

$$x = 5y - 5$$
$$x + 5 = 5y$$
$$\frac{1}{5}(x+5) = y$$
$$f^{-1}(x) = g(x) = \frac{1}{5}x + 1$$

55. Yes: $f(x) = \dfrac{1}{x}$

$$x = \frac{1}{y}$$
$$xy = 1$$
$$y = \frac{1}{x}$$
$$f^{-1}(x) = g(x) = \frac{1}{x}$$

57. You should see the graph of the inverse function with which you started.

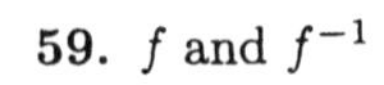

59. f and f^{-1}

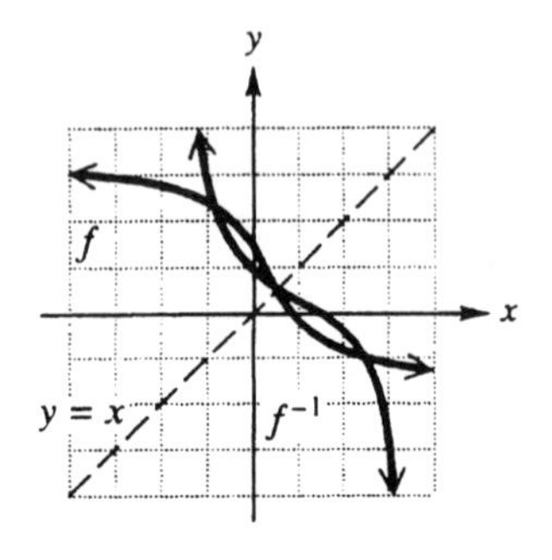

61. f and f^{-1}

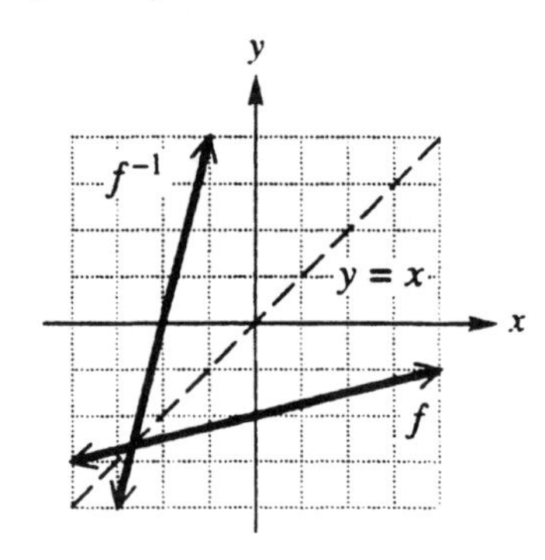

63. f and f^{-1}

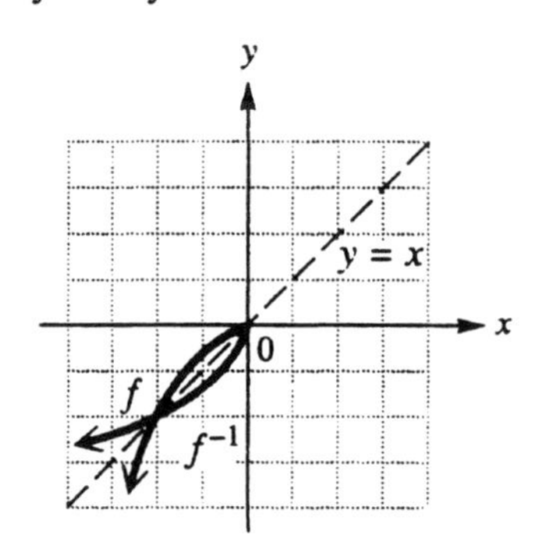

65. $f(x) = 4x - 5$

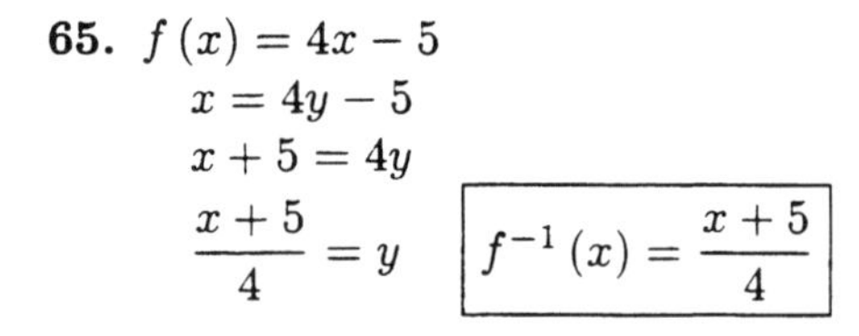

$$x = 4y - 5$$
$$x + 5 = 4y$$
$$\frac{x+5}{4} = y \qquad \boxed{f^{-1}(x) = \frac{x+5}{4}}$$

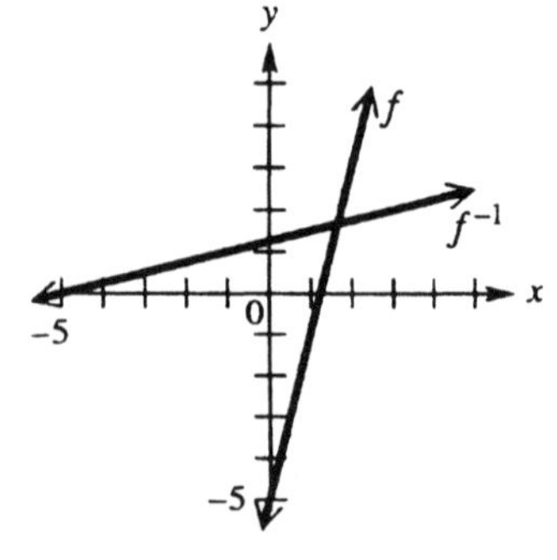

Domain and range of both f, f^{-1} : $(-\infty, \infty)$.

67. $f(x) = -x^3 - 2$

$x = -y^3 - 2$

$y^3 = -x - 2$

$y = \boxed{f^{-1}(x) = \sqrt[3]{-x-2}}$

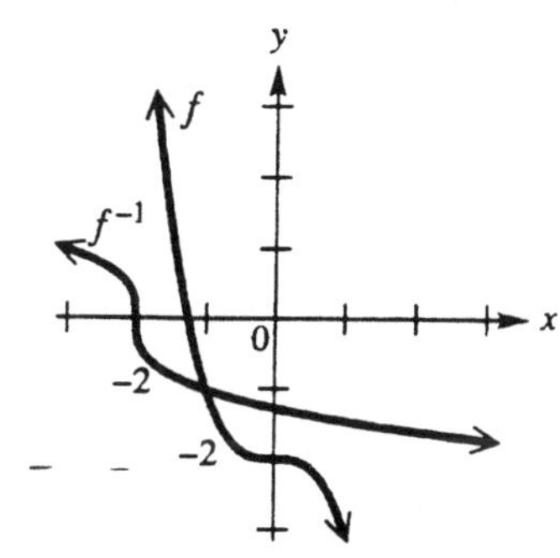

Domain and range of both f, f^{-1} : $(-\infty, \infty)$

73. $f(x) = -\sqrt{x^2 - 16}$, $x \geq 4$, $y \leq 0$

$x = -\sqrt{y^2 - 16}$

$x^2 = y^2 - 16$

$y^2 = x^2 + 16$

$y = \sqrt{x^2 + 16}$ $\boxed{f^{-1}(x) = \sqrt{x^2 + 16}, \quad x \leq 0}$

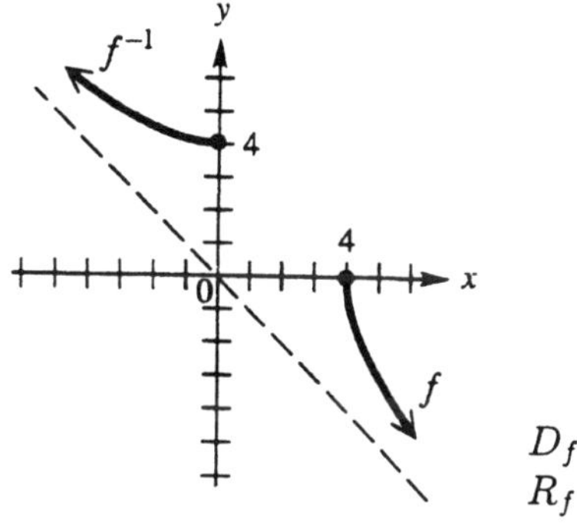

$D_f = [4, \infty) = R_{f^{-1}}$
$R_f = (-\infty, 0] = D_{f^{-1}}$

69. $y = -x^2 + 2$: not one-to-one

75. $f(5)$ is the volume of a sphere with a radius of 5 inches; so $f^{-1}(5)$ represents the radius of a sphere with volume of 5 cubic inches.

71. $f(x) = \dfrac{4}{x}$

$x = \dfrac{4}{y}$

$y = \dfrac{4}{x}$ $\boxed{f^{-1}(x) = \dfrac{4}{x}}$

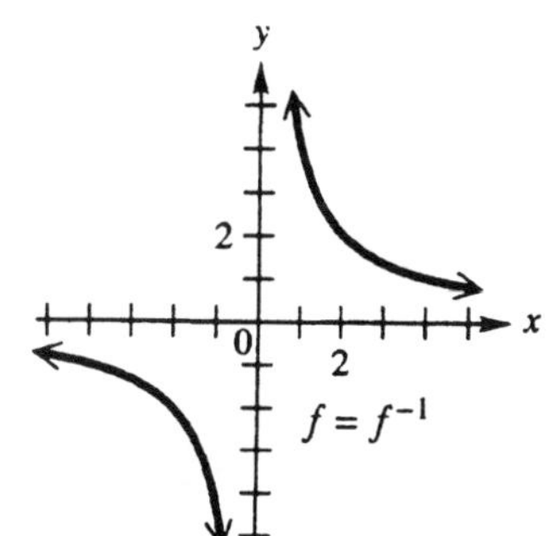

$D : x \neq 0$
$R : y \neq 0$

77. $f(2) = 3$

$f^{-1}(f(2)) = f^{-1}(3) = 2$

79. $f(x) = x^4 - 5x^2 + 6$

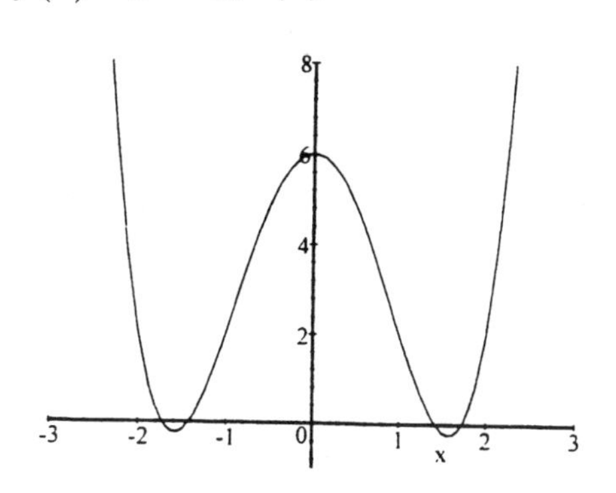

$f(x)$ is not one-to-one.

81. $f(x) = \dfrac{-x}{x-4}$

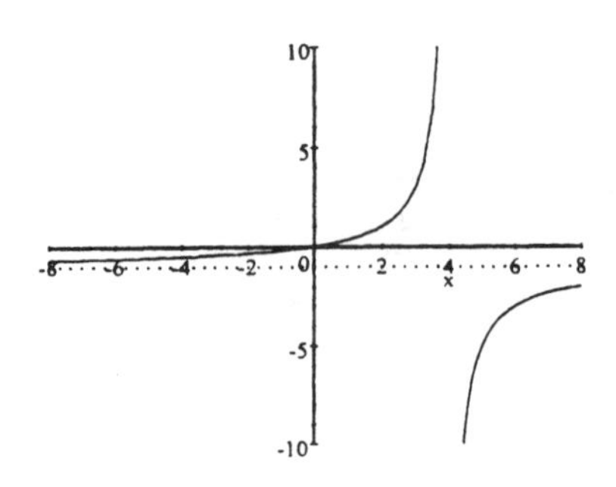

$f(x)$ is one-to-one.

$x = \dfrac{-y}{y-4}$

$xy - 4x = -y$

$xy + y = 4x$

$y(x+1) = 4x$

$y = f^{-1}(x) = \dfrac{4x}{x+1}$

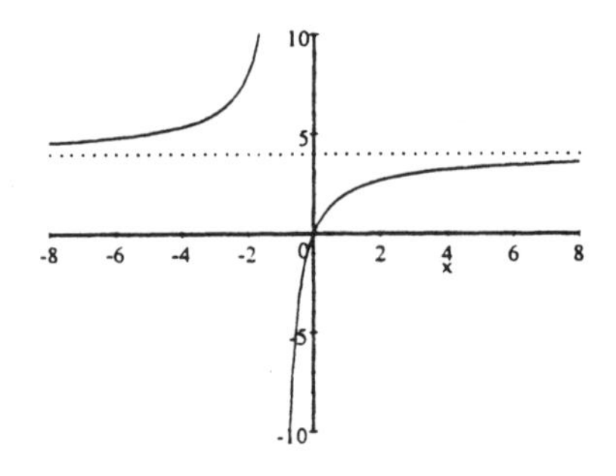

Relating Concepts (83 - 85)

83. $f(x) = y_1 = 2x - 8$

Switch x and y; solve for y :

$x = 2y - 8$
$x + 8 = 2y$
$\frac{1}{2}(x+8) = y$

$$\boxed{f^{-1}(x) = y_2 = \frac{1}{2}x + 4}$$

Enter y_1 and y_2 on your calculator.

84. $y_1 = 2x - 8$
$y_2 = \frac{1}{2}x + 4$
$y_3 = y_1(y_2(x)) = x$

This is the identity function.

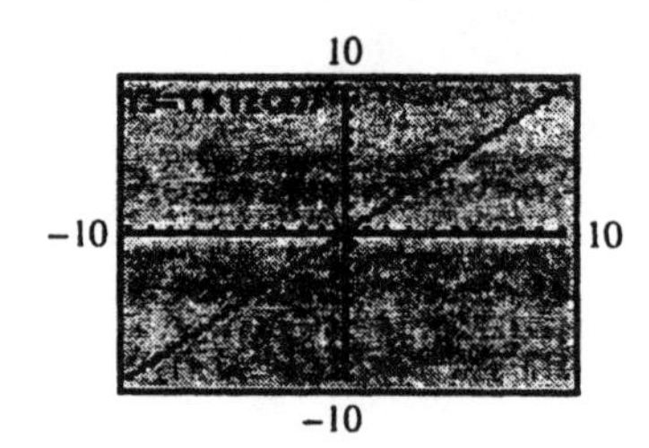

85. For each input x, the output is also x. (See the table in the answer section at the back of the textbook.)

87. $f(x) = (x-1)^2$; domain: $[1, \infty)$
(Other answers are possible.)

89. $f(x) = x^4$; domain: $[0, \infty)$
(Other answers are possible.)

91. $f(x) = -\sqrt{x^2 - 16}$; domain: $[4, \infty)$
(Other answers are possible.)

93. $f(x) = (x-1)^2, \ x \geq 1$

$x = (y-1)^2, \ y \geq 1$
$\sqrt{x} = y - 1$
$\sqrt{x} + 1 = y$ $\quad \boxed{f^{-1}(x) = \sqrt{x} + 1}$

95. $f(x) = x^4, \ x \geq 0$

$x = y^4, \ y \geq 0$
$\sqrt[4]{x} = y$ $\quad \boxed{f^{-1}(x) = \sqrt[4]{x}}$

97. $f(x) = 2x - 9$

$$x = 2y - 9$$
$$x + 9 = 2y$$
$$\frac{x+9}{2} = y \implies f^{-1}(x) = \frac{x+9}{2}$$

$f^{-1}(-5) = \frac{4}{2} = 2:$ **B**
$f^{-1}(9) = \frac{18}{2} = 9:$ **I**
$f^{-1}(5) = \frac{14}{2} = 7:$ **G**

$f^{-1}(5) = \frac{14}{2} = 7:$ **G**
$f^{-1}(9) = \frac{18}{2} = 9:$ **I**
$f^{-1}(27) = \frac{36}{2} = 18:$ **R**
$f^{-1}(15) = \frac{24}{2} = 12:$ **L**
$f^{-1}(29) = \frac{38}{2} = 19:$ **S**

$f^{-1}(-1) = \frac{8}{2} = 4:$ **D**
$f^{-1}(21) = \frac{30}{2} = 15:$ **O**
$f^{-1}(19) = \frac{28}{2} = 14:$ **N**
$f^{-1}(31) = \frac{40}{2} = 20:$ **T**

$f^{-1}(-3) = \frac{6}{2} = 3:$ **C**
$f^{-1}(27) = \frac{36}{2} = 18:$ **R**
$f^{-1}(41) = \frac{50}{2} = 25:$ **Y**

99. $f(x) = (x+1)^3$

S: $(19+1)^3 = \mathbf{8000}$
A: $(1+1)^3 = \mathbf{8}$
I: $(9+1)^3 = \mathbf{1000}$
L: $(12+1)^3 = \mathbf{2197}$
O: $(15+1)^3 = \mathbf{4096}$
R: $(18+1)^3 = \mathbf{6859}$

B: $(2+1)^3 = \mathbf{27}$
E: $(5+1)^3 = \mathbf{216}$
W: $(23+1)^3 = \mathbf{13824}$
A: $(1+1)^3 = \mathbf{8}$
R: $(18+1)^3 = \mathbf{6859}$
E: $(5+1)^3 = \mathbf{216}$

$$x = (y+1)^3$$
$$\sqrt[3]{x} = y + 1$$
$$\sqrt[3]{x} - 1 = y \implies \boxed{f^{-1}(x) = \sqrt[3]{x} - 1}$$

Section 5.2

1. $2^{\sqrt{10}} \approx 8.952419619$

3. $\left(\frac{1}{2}\right)^{\sqrt{2}} \approx .3752142272$

5. $4.1^{-\sqrt{3}} \approx .0868214883$

7. $\sqrt{7}^{\sqrt{7}} \approx 13.1207791$

9. $y = 2^x$; the point $(\sqrt{10},\ 8.9524196)$ lies on the graph of y.

11. $y = \left(\frac{1}{2}\right)^x$; the point $(\sqrt{2},\ .37521423)$ lies on the graph of y.

13. $A = 2.3 \quad (y = 2.3^x)$

15. $C = .75 \quad (y = .75^x)$

17. $E = .31 \quad (y = .31^x)$

Relating Concepts (Exercises 19 - 24)
$f(x) = a^x,\ a > 1$

19. Yes, f is a one-to-one function, so f^{-1}, the inverse function, exists.

20. $f(x) = a^x,\ a > 1$

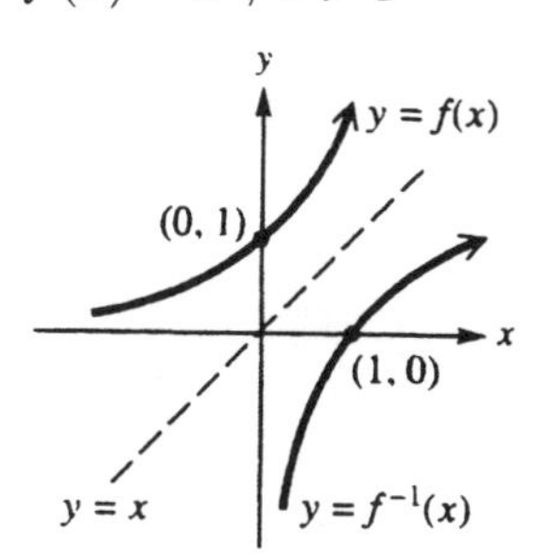

21. $y = a^x$ for f, so $f^{-1}: x = a^y$

22. $y = f^{-1}(x)$, $a = 10: \ x = 10^y$

23. $y = f^{-1}(x)$, $a = e$: $x = e^y$

24. If the point (p, q) is on the graph of f, then the point $\underline{(q, p)}$ is on the graph of f^{-1}.

25. $f(x) = 3^x$

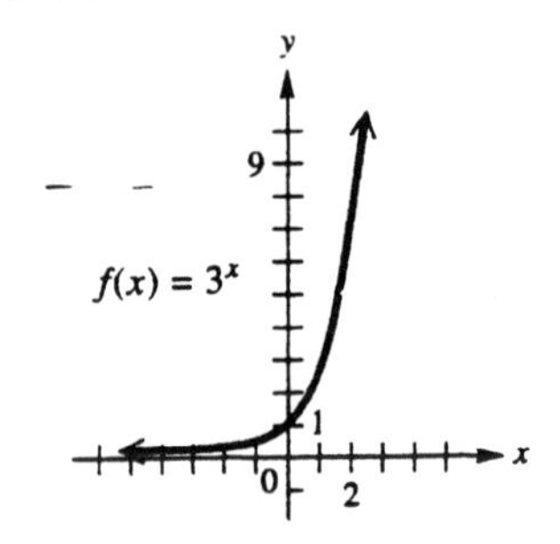

D: $(-\infty, \infty)$; R: $(0, \infty)$; $y = 0$

27. $f(x) = \left(\frac{1}{3}\right)^x$

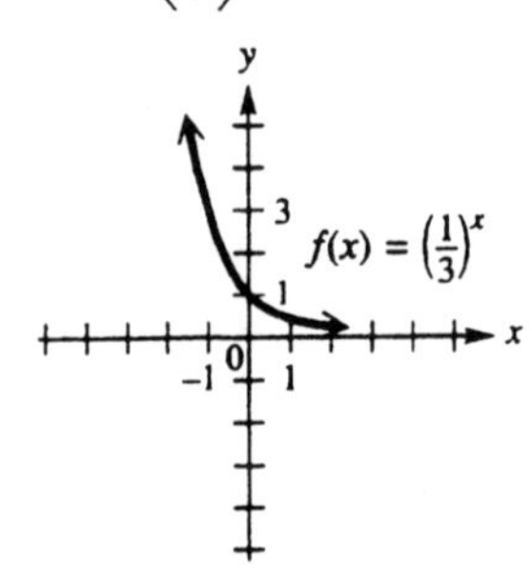

D: $(-\infty, \infty)$; R: $(0, \infty)$; $y = 0$

29. $f(x) = \left(\frac{3}{2}\right)^x$

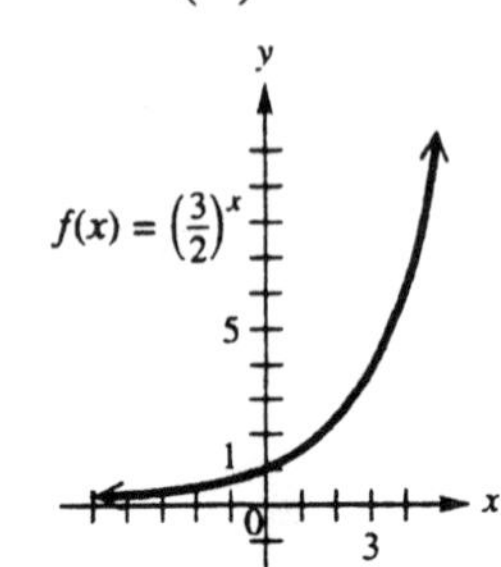

D: $(-\infty, \infty)$; R: $(0, \infty)$; $y = 0$

31. $f(x) = e^x$

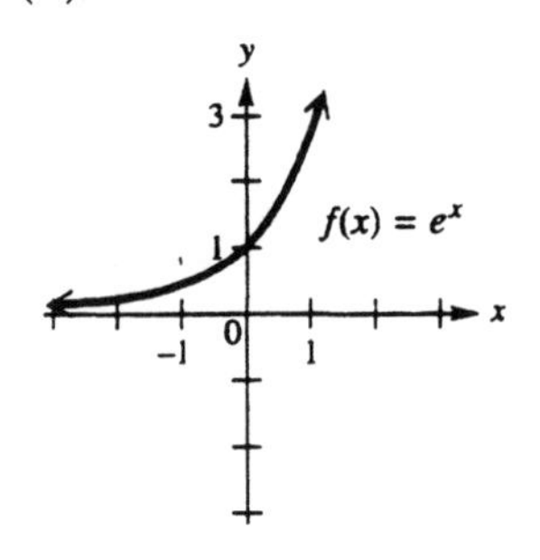

D: $(-\infty, \infty)$; R: $(0, \infty)$; $y = 0$

33. $f(x) = e^{x+1}$

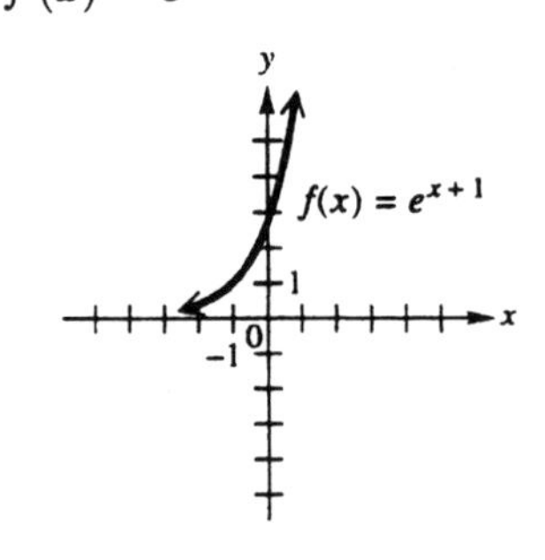

D: $(-\infty, \infty)$; R: $(0, \infty)$; $y = 0$

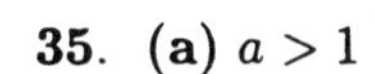

35. **(a)** $a > 1$

(b) Domain of f: $(-\infty, \infty)$
Range of f: $(0, \infty)$
Asymptote: $y = 0$

(c) $g(x) = -a^x$

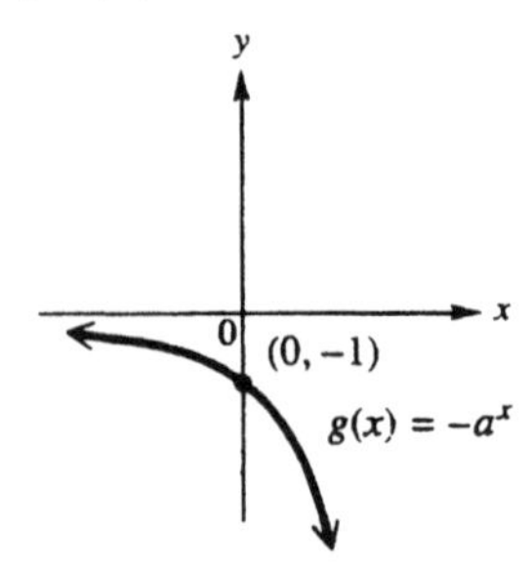

(d) Domain of g: $(-\infty, \infty)$
Range of g: $(-\infty, 0)$
Asymptote: $y = 0$

continued

35. continued

(e) $h(x) = a^{-x}$

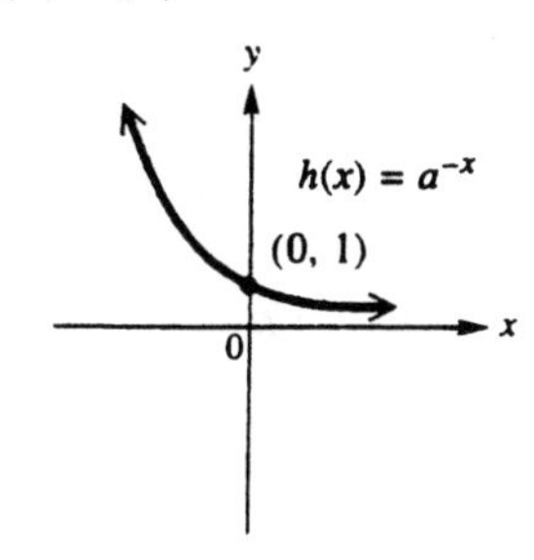

(f) Domain of h: $(-\infty, \infty)$

Range of h: $(0, \infty)$

Asymptote: $y = 0$

37. $y = 2^x$

$y = 2^{x+5}$; shift 5 units left

$y = 2^{x+5} - 3$; shift 3 units downward

39. $y = 2^x$

$y = 2^{-x}$; reflect across the y-axis

$y = 2^{-x} + 1$; shift 1 unit upward

$$y = \left(\frac{1}{2}\right)^x + 1$$

41. $y = 2^x$

$y = 3 \cdot 2^x$; vertical stretch by a factor of 3

$y = -3 \cdot 2^x$; reflect across the x-axis

43. $4^x = 2$

$(2^2)^x = 2^1$

$2^{2x} = 2^1$

$2x = 1 \longrightarrow x = \boxed{\frac{1}{2}}$

45. $\left(\frac{1}{2}\right)^x = 4$

$2^{-x} = 2^2 \rightarrow x = \boxed{\{-2\}}$

47. $2^{3-x} = 8$

$2^{3-x} = 2^3$

$3 - x = 3$

$-x = 0$

$x = \boxed{\{0\}}$

49. $\frac{1}{27} = x^{-3}$

$\frac{1}{3^3} = \frac{1}{x^3} \quad \longrightarrow \quad x = \boxed{\{3\}}$

51. (a) $27^{4x} = 9^{x+1}$

$(3^3)^{4x} = (3^2)^{x+1}$

$3^{12x} = 3^{2x+2}$

$12x = 2x + 2$

$10x = 2 \rightarrow x = \boxed{\left\{\frac{1}{5}\right\}}$

$y_1 = 27^{4x}$ (0, 1); $y_2 = 9^{x+1}$ (0, 9)

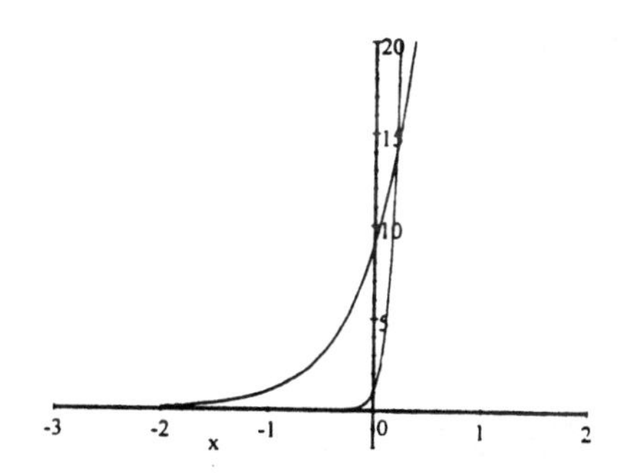

(b) $27^{4x} > 9^{x+1}$: $\boxed{\left(\frac{1}{5}, \infty\right)}$

(c) $27^{4x} < 9^{x+1}$: $\boxed{\left(-\infty, \frac{1}{5}\right)}$

53. (a) $\left(\frac{1}{2}\right)^{-x} = \left(\frac{1}{4}\right)^{x+1}$

$2^x = 4^{-(x+1)}$

$2^x = (2^2)^{-(x+1)}$

$2^x = 2^{-2x-2}$

$x = -2x - 2$

$3x = -2 \implies x = \boxed{\left\{-\frac{2}{3}\right\}}$

continued

53. continued

$y_1 = \left(\frac{1}{2}\right)^{-x}$ (increasing)

$y_2 = \left(\frac{1}{4}\right)^{x+1}$ (decreasing)

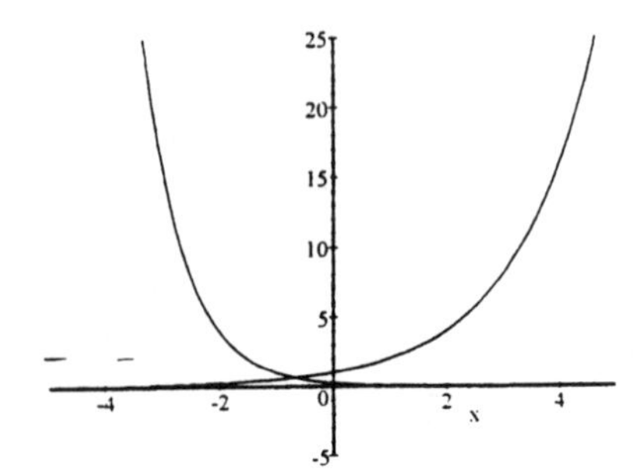

(b) $\left(\frac{1}{2}\right)^{-x} \geq \left(\frac{1}{4}\right)^{x+1} : \left[-\frac{2}{3}, \infty\right)$

(c) $\left(\frac{1}{2}\right)^{-x} \leq \left(\frac{1}{4}\right)^{x+1} : \left(-\infty, -\frac{2}{3}\right]$

55. $A = P\left(1+\frac{r}{n}\right)^{nt}$

(a) $A = 20000\left(1+\frac{.03}{1}\right)^{(1)(4)}$

$\boxed{A = \$22,510.18}$

(b) $A = 20000\left(1+\frac{.03}{2}\right)^{(2)(4)}$

$\boxed{A = \$22,529.85}$

57. (a) $A = P\left(1+\frac{r}{n}\right)^{nt}$

$A = 27500\left(1+\frac{.0395}{365}\right)^{(365)(5)}$

$\boxed{A = \$33,504.35}$

(b) $A = Pe^{rt}$

$\boxed{A = \$33,504.71}$

59. Plan A: $A = 40000\left(1+\frac{.045}{4}\right)^{(4)(3)}$

$A = \$45,746.98$

Plan B: $A = 40000e^{(.044)(3)}$

$A = \$45,644.33$

Plan A is better by \$102.65

61. $y_1 = 1000(1.05)^x$ (lower)

$y_2 = 1000\left(1+\frac{.05}{12}\right)^{12x}$ (upper)

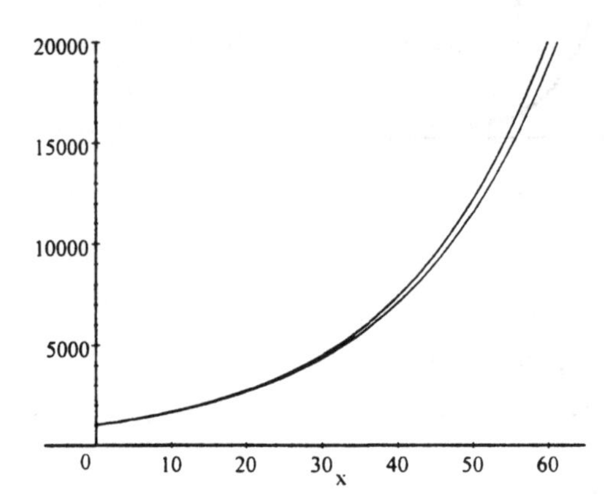

\$1000 investment:

Years	Y_1	Y_2	$Y_2 - Y_1$
1	1050.00	1051.16	\$1.16
2	1102.50	1104.94	\$2.44
5	1276.28	1283.36	\$7.08
10	1628.89	1647.01	\$18.12
20	2653.30	2712.64	\$59.34
30	4321.94	4467.74	\$145.80
40	7039.99	7358.42	\$318.43

63. $P(x) = 1013e^{-.0001341x}$

$P(1500) = 1013e^{-.0001341(1500)} \approx 828$ millibars

$P(11,000) = 1013e^{-.0001341(11,000)} \approx 232$ millibars

The calculated values are close to the actual values.

65. $f(x) = 1 - e^{-.5x}$

(a) $f(2) = 1 - e^{-.5(2)} \approx 63$

There is a 63% chance that at least one car will enter the intersection during a two minute period.

(b) As time progresses, the probability $\longrightarrow 1$.
It is almost certain that at least 1 car will enter the intersection during a 60 minute period.

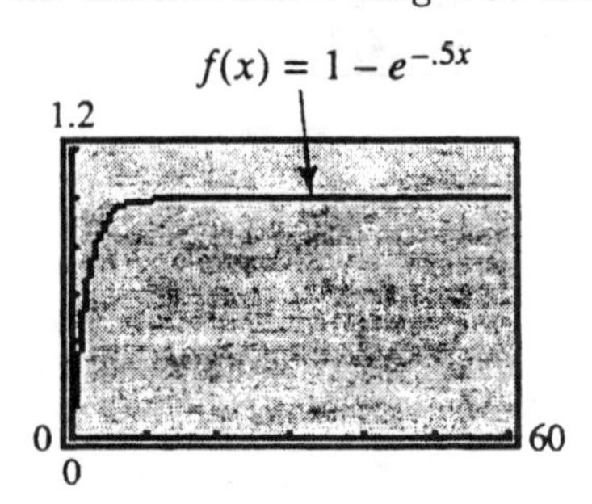

Section 5.3

1. $3^4 = 81 \Longleftrightarrow \log_3 81 = 4$

3. $\left(\frac{1}{2}\right)^{-4} = 16 \Longleftrightarrow \log_{\frac{1}{2}} 16 = -4$

5. $10^{-4} = .0001$

$\log_{10} .0001 = -4$ or

$\log .0001 = -4$

7. $e^0 = 1$

$\log_e 1 = 0$ or $\ln 1 = 0$

9. $\log_6 36 = 2 \Longleftrightarrow 6^2 = 36$

11. $\log_{\sqrt{3}} 81 = 8 \Longleftrightarrow \left(\sqrt{3}\right)^8 = 81$

13. $\log_{10} .001 = -3 \Longleftrightarrow 10^{-3} = .001$

15. $\log \sqrt{10} = .5 \Longleftrightarrow 10^{.5} = \sqrt{10}$

17. $\log_a x$ means the power to which a must be raised to obtain x.

19. $\log_5 125 = x$

$5^x = 125$

$5^x = 5^3 \qquad x = \boxed{\{3\}}$

21. $\log_x 3^{12} = 24$

$x^{24} = 3^{12}$

$\left(x^2\right)^{12} = 3^{12}$

$x^2 = 3 \qquad x = \boxed{\{\sqrt{3}\}}$

23. $\log_6 x = -3$

$x = 6^{-3} = \frac{1}{6^3} = \boxed{\left\{\frac{1}{216}\right\}}$

25. $\log_x 16 = \frac{4}{3}$

$x^{4/3} = 16$

$\left(x^{4/3}\right)^{3/4} = (16)^{3/4} \Longrightarrow x = \boxed{\{8\}}$

27. $\log_x .001 = -3$

$x^{-3} = .001 = 10^{-3} \Longrightarrow x = \boxed{\{10\}}$

29. $\log_9 \left(\frac{\sqrt[4]{27}}{3}\right) = x$

$9^x = \frac{\sqrt[4]{27}}{3}$

$\left(3^2\right)^x = \frac{3^{3/4}}{3} = 3^{-1/4}$

$2x = -\frac{1}{4} \Longrightarrow x = \boxed{\left\{-\frac{1}{8}\right\}}$

31. **(a)** $3^{\log_3 7} = 7$

(b) $4^{\log_4 9} = 9$

(c) $12^{\log_{12} 4} = 4$

(d) $a^{\log_a k} = k$

33. **(a)** $\log_3 1 = 0$

(b) $\log_4 1 = 0$

(c) $\log_{12} 1 = 0$

(d) $\log_a 1 = 0$

35. $\log 10^{4.3} = 4.3$

let $x = \log_{10} 10^{4.3}$

$10^x = 10^{4.3} \longrightarrow x = 4.3$

37. $\log 10^{\sqrt{3}} = \sqrt{3}$

let $x = \log_{10} 10^{\sqrt{3}}$

$10^x = 10^{\sqrt{3}} \longrightarrow x = \sqrt{3}$

39. $\ln e^{.5} = .5$

let $x = \log_e e^{.5}$

$e^x = e^{.5} \longrightarrow x = .5$

41. $\ln e^{\sqrt{6}} = \sqrt{6}$

let $x = \log_e e^{\sqrt{6}}$

$e^x = e^{\sqrt{6}} \longrightarrow x = \sqrt{6}$

43. $\log 43 \approx 1.633468456$

45. $\log .783 \approx -.1062382379$

47. $\log 28^3 = 3\log 28 \approx 4.341474094$

49. $\ln 43 \approx 3.761200116$

51. $\ln .783 \approx -.244622583$

53. $\ln 28^3 = 3\ln 28 \approx 9.996613531$

55. (a) $\log 2.367 \approx .3741982579$
$\log 23.67 \approx 1.374198258$
$\log 236.7 \approx 2.374198258$
$\log 2367 \approx 3.374198258$

(b) $2.367 = 2.367 \times 10^0$
$23.67 = 2.367 \times 10^1$
$236.7 = 2.367 \times 10^2$
$2367 = 2.367 \times 10^3$

(c) The decimal digits are the same; the whole number part corresponds to the exponent of 10 in scientific notation.

57. $\text{pH} = -\log[\text{H}_3\text{O}^+]$
limes: 1.6×10^{-2}
$\text{pH} = -\log(1.6 \times 10^{-2})$
$\text{pH} \approx 1.79588 \approx \boxed{1.8}$

59. $\text{pH} = -\log[\text{H}_3\text{O}^+]$
lye: 3.2×10^{-14}
$\text{pH} = -\log(3.2 \times 10^{-14})$
$\text{pH} \approx 13.49485 \approx \boxed{13.5}$

61. $\text{pH} = -\log[\text{H}_3\text{O}^+]$
wine: 3.4
$3.4 = -\log[\text{H}_3\text{O}^+]$
$-3.4 = \log[\text{H}_3\text{O}^+]$
$[\text{H}_3\text{O}^+] = 10^{-3.4}$
$= .000398 \approx \boxed{4 \times 10^{-4}}$

63. $\text{pH} = -\log[\text{H}_3\text{O}^+]$
drinking water: 6.5
$6.5 = -\log[\text{H}_3\text{O}^+]$
$-6.5 = \log[\text{H}_3\text{O}^+]$
$[\text{H}_3\text{O}^+] = 10^{-6.5}$
$= 3.162 \times 10^{-7} \approx \boxed{3.2 \times 10^{-7}}$

65. $A = Pe^{rt}$

$$3500 = 2500e^{.0425t}$$
$$\frac{3500}{2500} = e^{.0425t}$$
$$\frac{7}{5} = e^{.0425t}$$
$$\ln\frac{7}{5} = .0425t$$
$$t = \frac{\ln\frac{7}{5}}{.0425} \approx \boxed{7.9 \text{ years}}$$

67. $A = Pe^{rt}$

$$5000 = 2500\,e^{.06t}$$
$$\frac{5000}{2500} = e^{.06t}$$
$$2 = e^{.06t}$$
$$\ln 2 = .06t$$
$$t = \frac{\ln 2}{.06} \approx \boxed{11.6 \text{ years}}$$

69. $\log_4 \frac{6}{7} = \boxed{\log_4 6 - \log_4 7}$

71. $\log_3 \frac{4p}{q} = \boxed{\log_3 4 + \log_3 p - \log_3 q}$

73.
$$\log_2 \frac{2\sqrt{3}}{5p} = \log_2 2 + \log_2 3^{1/2} - \log_2 5 - \log_2 p$$
$$= \boxed{1 + \frac{1}{2}\log_2 3 - \log_2 5 - \log_2 p}$$

75. $\log_6(7m + 3q)$ cannot be rewritten.

77.
$$\log_z \frac{x^5y^3}{3} = \log_z x^5 + \log_z y^3 - \log_z 3$$
$$= \boxed{5\log_z x + 3\log_z y - \log_z 3}$$

79. $\log_p \sqrt[3]{\dfrac{m^5}{kt^2}}$

$$= \log_p \left(\frac{m^5}{kt^2}\right)^{1/3}$$

$$= \log_p \frac{m^{5/3}}{k^{1/3}t^{2/3}}$$

$$= \log_p m^{5/3} - \log_p k^{1/3} - \log_p t^{2/3}$$

$$= \boxed{\begin{array}{c}\frac{5}{3}\log_p m - \frac{1}{3}\log_p k - \frac{2}{3}\log_p t \\ \textbf{or} \\ \frac{1}{3}\left[5\log_p m - \log_p k - 2\log_p t\right]\end{array}}$$

81. $(\log_b k - \log_b m) - \log_b a = \boxed{\log_b \dfrac{k}{ma}}$

83. $\dfrac{1}{2}\log_y p^3q^4 - \dfrac{2}{3}\log_y p^4q^3$

$$= \log_y p^{3/2}q^2 - \log_y p^{8/3}q^2$$

$$= \log_y \frac{p^{3/2}q^2}{p^{8/3}q^2}$$

$$= \log_y \frac{p^{9/16}}{p^{16/6}}$$

$$= \boxed{\log_y p^{-7/6} \textbf{ or } -\log_y p^{7/6}}$$

85. $\log_b(2y+5) - \dfrac{1}{2}\log_b(y+3)$

$$= \log_b(2y+5) - \log_b(y+3)^{1/2}$$

$$= \boxed{\log_b \frac{2y+5}{\sqrt{y+3}}}$$

87. $-\dfrac{3}{4}\log_3 16p^4 - \dfrac{2}{3}\log_3 8p^3$

$$= \log_3\left(16p^4\right)^{-3/4} + \log_3\left(8p^3\right)^{-2/3}$$

$$= \log_3 2^{-3}p^{-3} + \log_3 2^{-2}p^{-2}$$

$$= \log_3\left(2^{-3}p^{-3}\right)\left(2^{-2}p^{-2}\right)$$

$$= \log_3\left(2^{-5}p^{-5}\right)$$

$$= \boxed{\log_3 \frac{1}{32p^5}}$$

89. $\log_9 12 = \dfrac{\log 12}{\log 9} \approx 1.130929754$

91. $\log_{1/2} 3 = \dfrac{\log 3}{\log \frac{1}{2}} \approx -1.584962501$

93. $\log_{200} 175 = \dfrac{\log 175}{\log 200} \approx .9747973963$

95. $\log_{5.8} 12.7 = \dfrac{\log 12.7}{\log 5.8} \approx 1.445851777$

97.
$$\begin{aligned} H &= -(P_1\log_2 P_1 + P_2\log_2 P_2 + P_3\log_2 P_3 \\ &\qquad + P_4\log_2 P_4) \\ &= -(.521\log_2 .521 + .324\log_2 .324 + .081\log_2 .081 \\ &\qquad + .074\log_2 .074) \\ &= -\left(\frac{.521\log .521}{\log 2} + \frac{.324\log .324}{\log 2} + \frac{.081\log .081}{\log 2}\right. \\ &\qquad \left. + \frac{.074\log .074}{\log 2}\right) \\ &\approx -(-.49007 - .52680 - .29370 - .27797) \\ &\approx -(-1.58854) \approx 1.59 \end{aligned}$$

Relating Concepts (99 – 104)

99. $y = 3^x$

$y = -3^x$; reflect across the x-axis

$y = -3^x + 7$; shift 7 units upward

100. $y_1 = 3^x$; $y_2 = -3^x + 7$

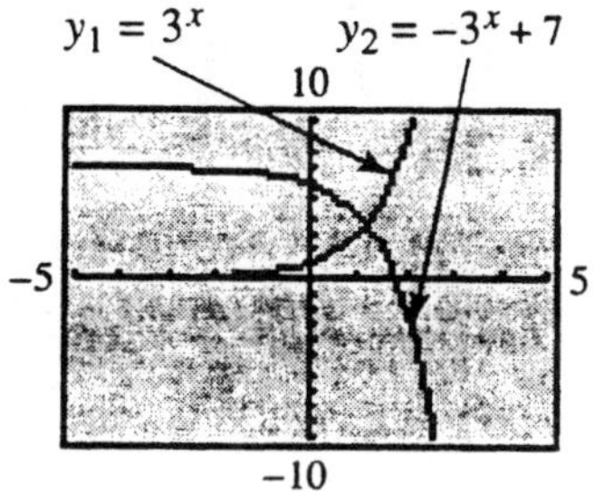

$y_1 = 3^x$ $y_2 = -3^x + 7$

10

−5 5

−10

101. x-intercept ≈ 1.7712437492

102. $0 = -3^x + 7$

$3^x = 7 \qquad \boxed{x = \{\log_3 7\}}$

103. $x = \log_3 7$

$= \dfrac{\log 7}{\log 3} = \dfrac{\ln 7}{\ln 3} \approx 1.77124374916$

104. The approximations are close enough to support the conclusion that the x-intercept is equal to $\log_3 7$.

105. Property 5:

$$\log_a \frac{x}{y} = \log_a x - \log_a y$$

Let $m = \log_a x$; $\quad n = \log_a y$,

then $a^m = x, \quad a^n = y$

$$\frac{a^m}{a^n} = \frac{x}{y}$$

$$a^{m-n} = \frac{x}{y}$$

$$\log_a \frac{x}{y} = m - n$$

Substitute for m and n :

$$\log_a \frac{x}{y} = \log_a x - \log_a y$$

Reviewing Basic Concepts (Sections 5.1 - 5.3)

1. No; each x-value must correspond to exactly one y-value, and each y-value must correspond to exactly x-value.

2. (a)

x	12	21	32	45
y	7	8	9	10

(b) $y = \dfrac{x+5}{4}$

$x = \dfrac{y+5}{4}$

$4x = y + 5$

$y = f^{-1}(x) = 4x - 5$

3. $f(x) = 2x + 3$

inverse: $x = 2y + 3$

$2y = x - 3$

$y = f^{-1}(x) = \dfrac{x-3}{2}$

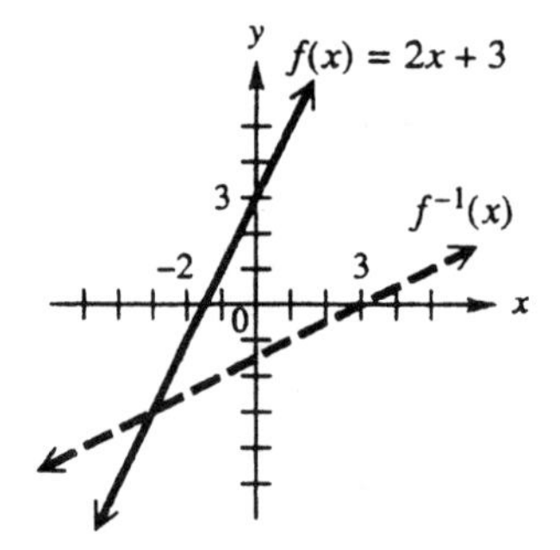

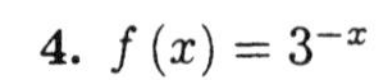
4. $f(x) = 3^{-x}$

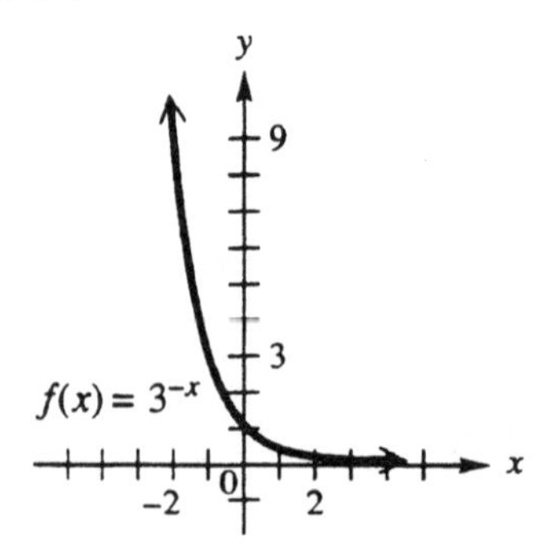

5. $4^{2x} = 8$

$(2^2)^{2x} = 2^3$

$2^{4x} = 2^3$

$4x = 3 \longrightarrow x = \left\{\frac{3}{4}\right\}$

6. $A = P\left(1 + \frac{r}{n}\right)^{nt}$

$A = 600\left(1 + \frac{.04}{4}\right)^{4(3)} = 676.10$

Interest earned $= \$676.10 - \$600 = \$76.10$

7. (a) $\log\left(\frac{1}{\sqrt{10}}\right) = \log_{10} 10^{-1/2} = -\frac{1}{2}$

(b) $2\ln e^{1.5} = \log_e e^{2(1.5)} = \log_e e^3 = 3$

(c) $\log_2 4 = 2 \quad [2^2 = 4]$

8. $\log\left(\frac{3x^2}{5y}\right) = \log 3 + \log x^2 - \log 5 - \log y$
$= \log 3 + 2\log x - \log 5 - \log y$

9. $\ln 4 + \ln x - 3\ln 2 = \ln 4 + \ln x - \ln 2^3$
$= \ln\left(\frac{4x}{2^3}\right) = \ln\left(\frac{4x}{8}\right) = \ln\left(\frac{x}{2}\right)$

10. $A = Pe^{rt}$

$1650 = 1500e^{.045t}$

$1.1 = e^{.045t}$

$.045t = \ln 1.1$

$t = \frac{\ln 1.1}{.045} \approx 2.1$ years

Section 5.4

1. f^{-1}: Domain: $(0, \infty)$
Range: $(-\infty, \infty)$
Increases on $(0, \infty)$
V.A.: $x = 0$

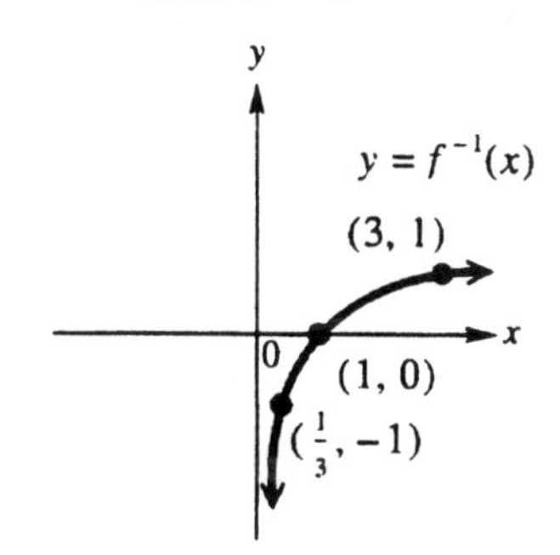

3. f^{-1}: Domain: $(0, \infty)$
Range: $(-\infty, \infty)$
Decreases on $(0, \infty)$
V.A.: $x = 0$

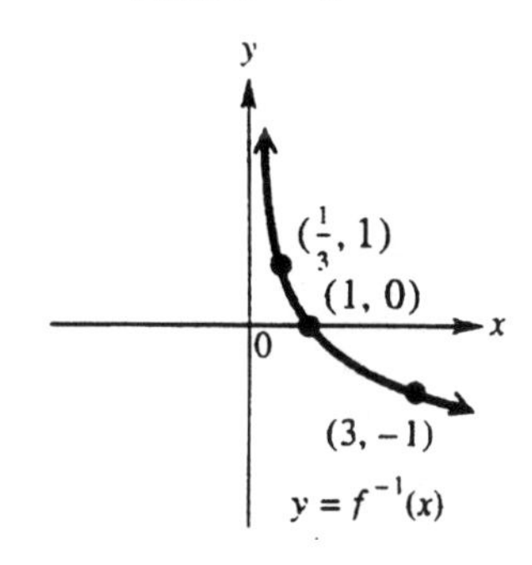

5. f^{-1}: Domain: $(1, \infty)$
Range: $(-\infty, \infty)$
Increases on $(1, \infty)$
V.A.: $x = 1$

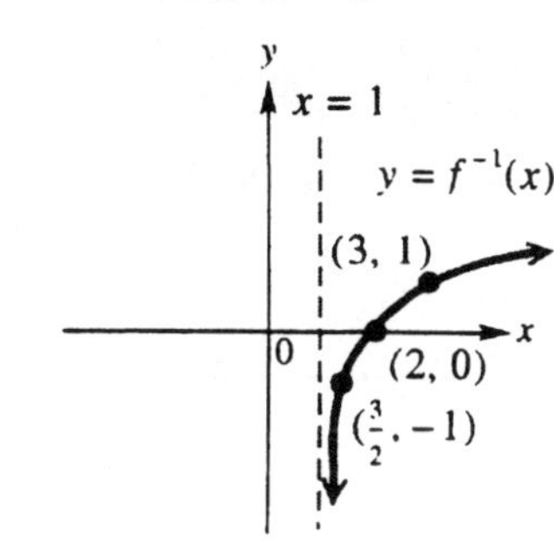

7. If f is an exponential function, f^{-1} is a <u>logarithmic</u> function.

9. $y = \log 2x$

$2x > 0$

$x > 0$ Domain: $(0, \infty)$

11. $y = \ln(x^2 + 7)$; $x^2 + 7 > 0$ (always)

Domain: $(-\infty, \infty)$

13. $y = \log_4(x^2 - 4x - 21)$

$x^2 - 4x - 21 > 0$

$(x - 7)(x + 3) > 0$

$x = -3, 7$

$(x+3)$	−	+	+
$(x-7)$	−	−	+
$(prod)$	+	−	+
		−3	7

Domain: $(-\infty, -3) \cup (7, \infty)$

15. $y = \log(x^3 - x)$

$x^3 - x > 0$

$x(x^2 - 1) > 0$

$x(x - 1)(x + 1) > 0$

$x = -1, 0, 1$

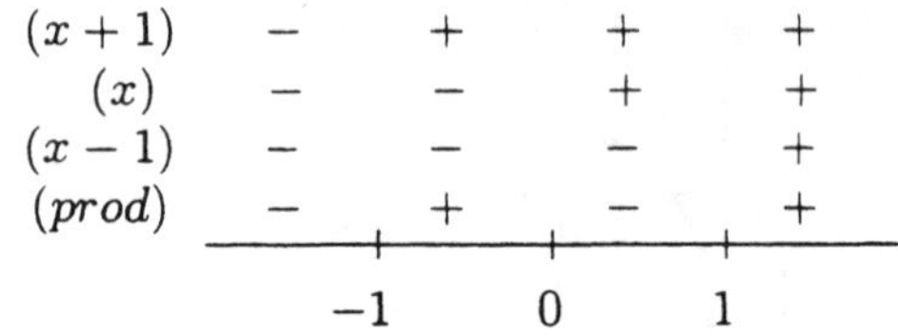

Domain: $(-1, 0) \cup (1, \infty)$

Graph for exercises 17 - 19: $f(x) = \log_2 x$

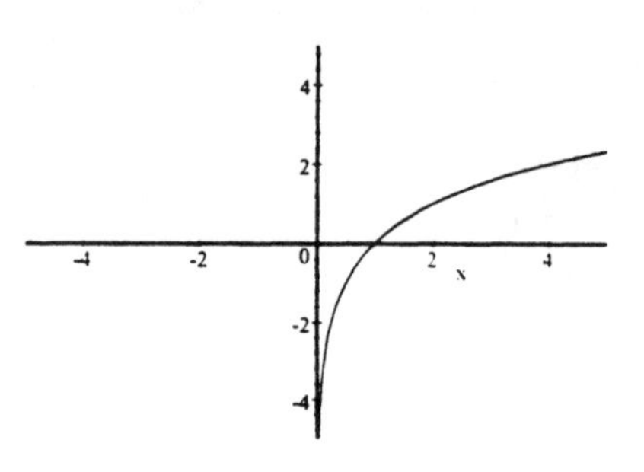

17. Shift 3 units upward: $f(x) = \log_2 x + 3$

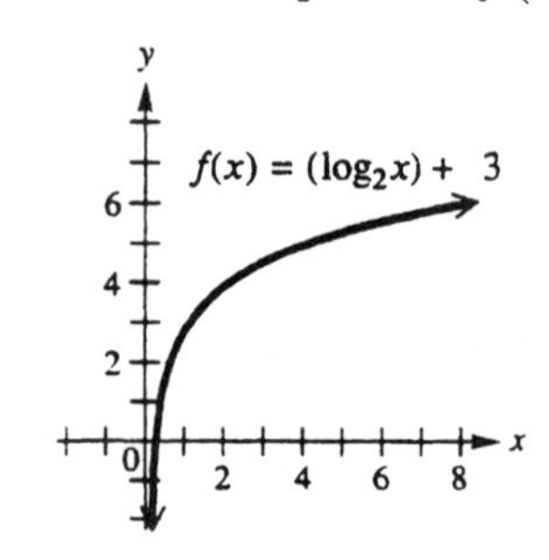

19. Shift 3 units left: $f(x) = \log_2(x + 3)$; reflect all negative y-values across the x-axis: $f(x) = |\log_2(x + 3)|$

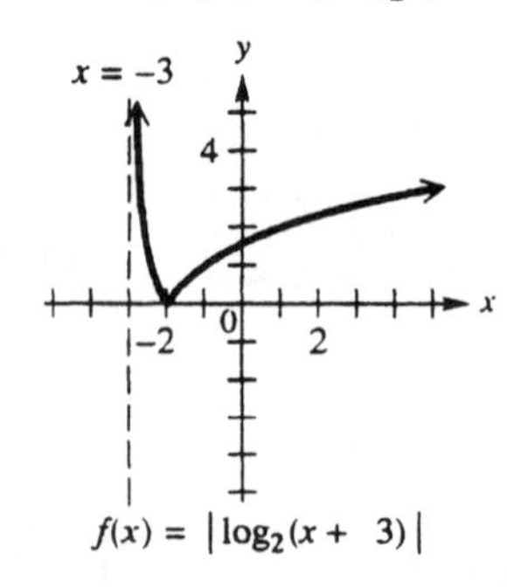

21. $f(x) = \log_{1/2} x$

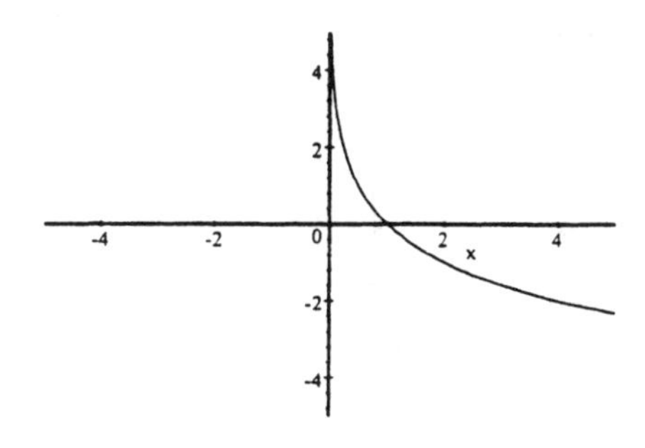

Shift 2 units right: $f(x) = \log_{1/2}(x-2)$

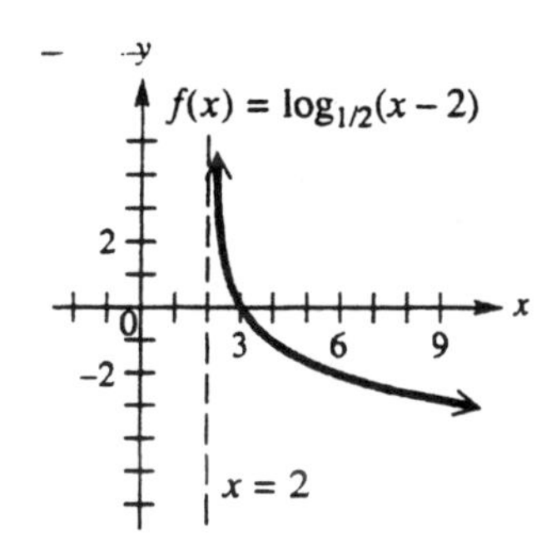

23. $y = e^x + 3$: **B**

25. $y = e^{x+3}$: **D**

27. $y = \ln x + 3$: **A**

29. $y = \ln(x-3)$: **C**

31. $y = \log_2 x$

Shift 4 units left: $y = \log_2(x+4)$

33. $y = \log_2 x$

Vertical stretch by a factor of 3:

$$y = 3\log_2 x$$

Shift 1 unit upward:

$$y = 3\log_2 x + 1$$

35. $y = \log_2 x$

Reflect across the y-axis:

$$y = \log_2(-x)$$

Shift 1 unit upward:

$$y = \log_2(-x) + 1$$

Relating Concepts (37 – 42)

37. $f(x) = \log_4(2x^2 - x)$

$$2x^2 - x > 0$$
$$x(2x-1) > 0$$
$$x = 0,\ \frac{1}{2}$$

(x)	$-$	$+$	$+$
$\left(x-\frac{1}{2}\right)$	$-$	$-$	$+$
$(prod)$	$+$	$-$	$+$
	0		.5

$$\boxed{\text{Domain: } (-\infty,\ 0) \cup \left(\frac{1}{2},\ \infty\right)}$$

38. $f(x) = \log_4(2x^2 - x) \Longrightarrow \boxed{y = \dfrac{\log(2x^2 - x)}{\log 4}}$

39. Vertical asymptotes:

$$f(x) = \frac{\log(2x^2 - x)}{\log 4}$$

($\log 4 = 0$: Never; no H.A.)

Domain: $2x^2 - x > 0$

$$x(2x-1) > 0$$
$$\left[x = 0,\ \frac{1}{2}\right]$$
$$(-\infty,\ 0) \cup \left(\frac{1}{2},\ \infty\right)$$

$$\boxed{\text{V.A.: } x = 0,\ x = \frac{1}{2}}$$

40. The domain excludes $x = 0$, therefore there can be no y-intercept, which is found by evaluating $f(0)$.

41. $f(x) = 0: \left\{-\frac{1}{2}, 1\right\}$

$f(x) < 0: \left(-\frac{1}{2}, 0\right) \cup \left(\frac{1}{2}, 1\right)$

$f(x) > 0: \left(-\infty, -\frac{1}{2}\right) \cup (1, \infty)$

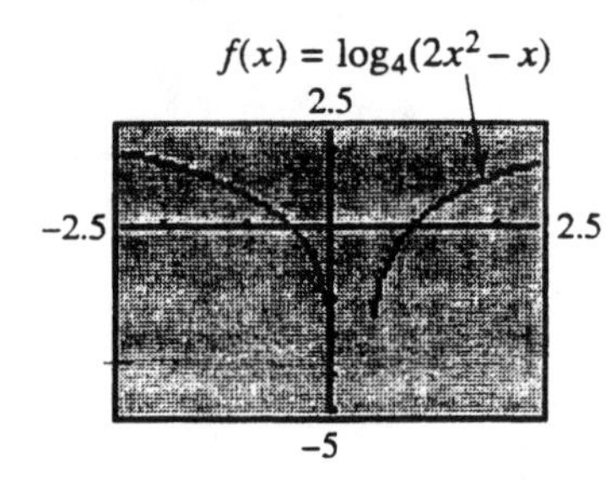

42. Range of $f: (-\infty, \infty)$

43. Domain of $y = \log x^2$ is $x^2 > 0$, or $(-\infty, 0) \cup (0, \infty)$

Domain of $y = 2\log x$ is $x > 0$, or $(0, \infty)$

$y = \log x^2$:

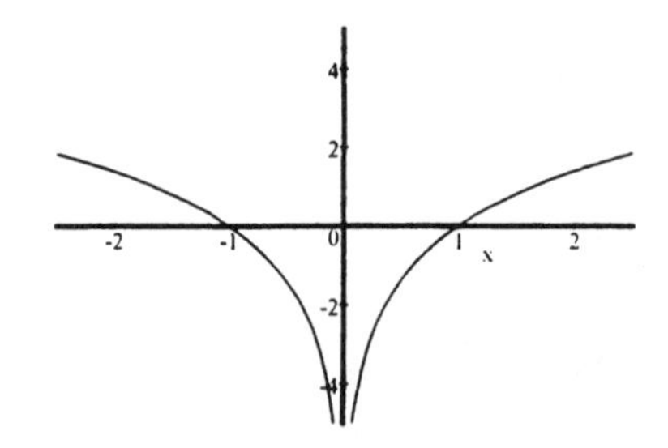

$y = 2\log x$:

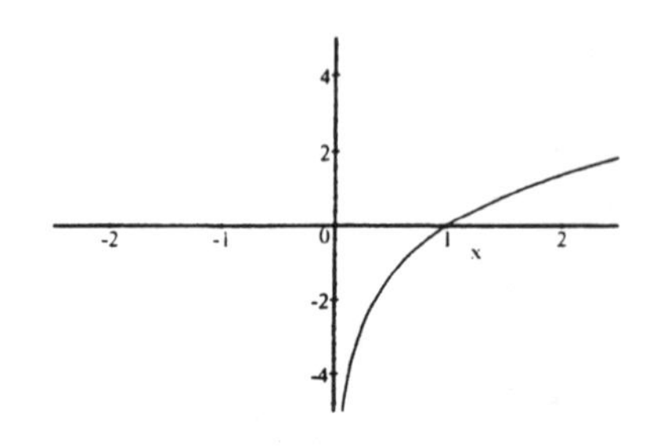

45. $\log_9 27$

(a) Let $x = \log_9 27$

$$9^x = 27$$
$$\left(3^2\right)^x = 3^3$$
$$2x = 3 \qquad \boxed{x = \left\{\frac{3}{2}\right\}}$$

(b) $\dfrac{\log 27}{\log 9} \approx \dfrac{1.43136}{.95424} = \boxed{1.5}$

(c) Graph $y = \log_9 x = \dfrac{\log x}{\log 9}$

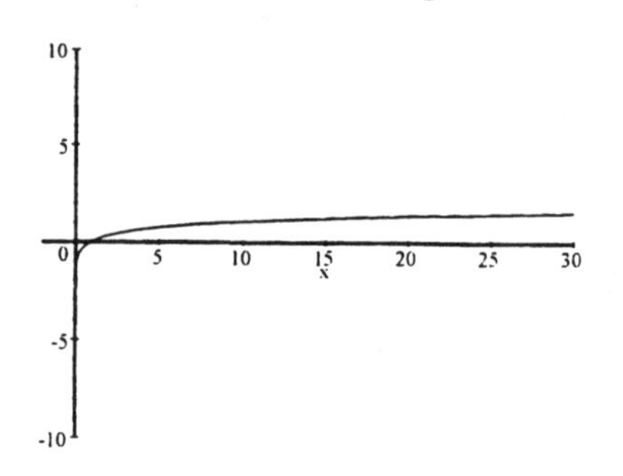

Show (27, 1.5) is on the graph.

47. $\log_{16} \frac{1}{8}$

(a) Let $x = \log_{16} \frac{1}{8}$

$$16^x = \frac{1}{8}$$
$$\left(2^4\right)^x = \frac{1}{2^3} = 2^{-3}$$
$$4x = -3 \qquad \boxed{x = \left\{-\frac{3}{4}\right\}}$$

(b) $\dfrac{\log \frac{1}{8}}{\log 16} \approx \dfrac{-.90309}{1.20412} = \boxed{-.75}$

(c) Graph $y = \log_{16} x = \dfrac{\log x}{\log 16}$

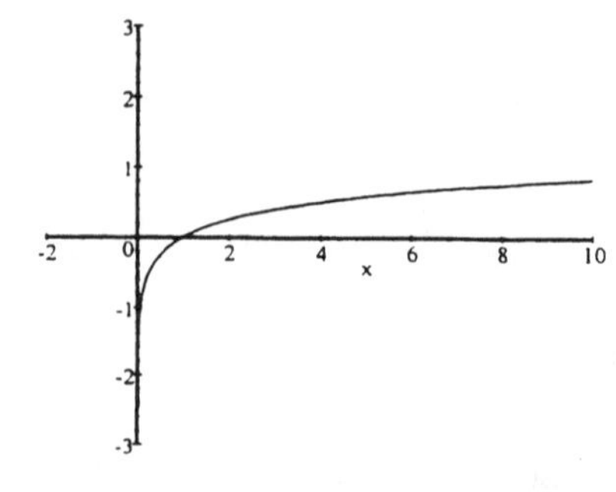

Show $\left(\frac{1}{8}, -.75\right)$ is on the graph.

49. $f(x) = 4^x - 3$

$x = 4^y - 3$

$x + 3 = 4^y$

$\log_4(x+3) = y$

$\boxed{f^{-1}(x) = \log_4(x+3)}$

Domain of f = range of f^{-1}

Range of f = domain of f^{-1}

$(0, -2) \to (-2, 0)$

f asymptote: $y = -3$

f^{-1} asymptote: $x = -3$

$f(x)$ and $f^{-1}(x)$:

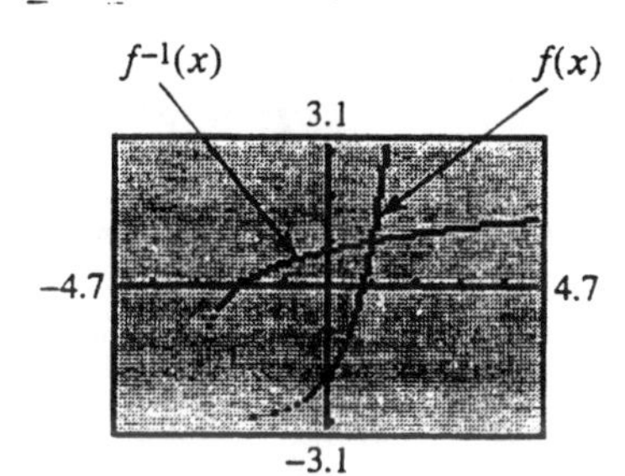

Dot Mode

51. $f(x) = -10^x + 4$

$x = -10^y + 4$

$10^y = -x + 4$

$\log(-x+4) = y$

$\boxed{f^{-1}(x) = \log(4 - x)}$

Domain of f = range of f^{-1}

Range of f = domain of f^{-1}

$(0, 3) \to (3, 0)$

f asymptote: $y = 4$

f^{-1} asymptote: $x = 4$

$f(x)$ and $f^{-1}(x)$:

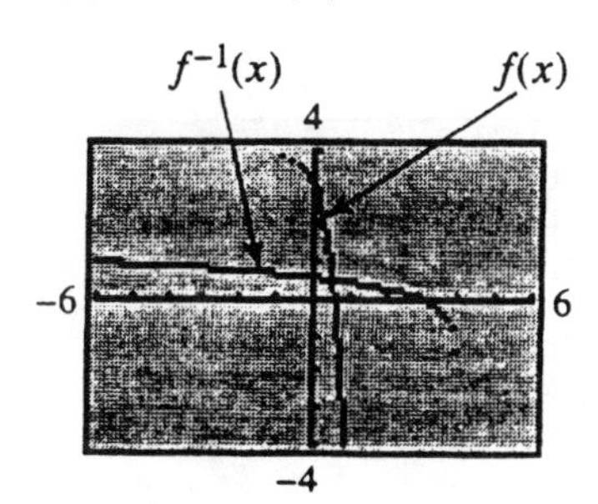

Dot Mode

53. $f(x) = -301 \ln\left(\frac{x}{207}\right)$

(a) The left side is a reflection of the right side with respect to the axis of the tower, $f(-x)$.

(b) $2x = 15.7488 \longrightarrow x = 7.8744$

$f(7.8744) = -301 \ln\left(\frac{7.8744}{207}\right) \approx 984$ feet

(c) Graph $y_1 = -301 \ln\left(\frac{x}{207}\right)$; $y_2 = 500$;

the intersection point is $\approx (39.31, 500)$,

so approximately 39 feet.

55. $f(x) = 27 + 1.105 \log(x+1)$

(a) $f(9) = 27 + 1.105 \log 10$

$= 27 + 1.105(1) = 28.105$ inches

(b) At 99 miles from the eye of a typical hurricane, the barometric pressure is 29.21 inches.

Section 5.5

1. **(a)** $f(x) = 0$: $\boxed{\{1.4036775\}}$

(b) $f(x) > 0$: $\boxed{(1.4036775, \infty)}$

3. **(a)** $f(x) = 0$: $\boxed{\{-1\}}$

(b) $f(x) > 0$: $\boxed{(-1, \infty)}$

5. $3^x = 10$

$\log 3^x = \log 10$

$x \log 3 = 1$

$$\boxed{x = \frac{1}{\log 3} \text{ or } \frac{\log 10}{\log 3}}$$

Graph $y = 3^x - 10$

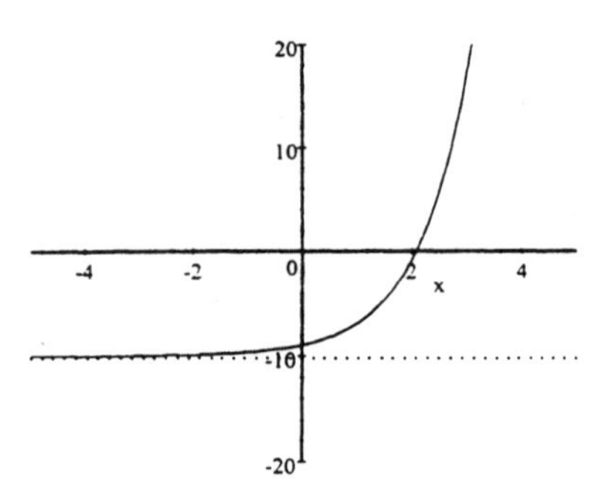

x-intercept $= \dfrac{\log 10}{\log 3} \approx 2.096.$

7. $2^{x+3} = 5^x$

$\log 2^{x+3} = \log 5^x$

$(x+3)\log 2 = x \log 5$

$x \log 2 + 3 \log 2 = x \log 5$

$3 \log 2 = x \log 5 - x \log 2$

$\log 2^3 = x(\log 5 - \log 2)$

$$\boxed{x = \frac{\log 8}{\log(5/2)} = \frac{\log(1/8)}{\log(2/5)}}$$

continued

7. continued

Graph $y = 2^{x+3} - 5^x$

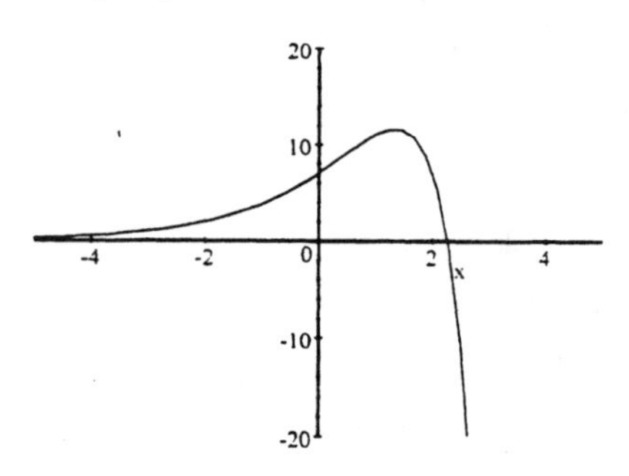

x-intercept $= \dfrac{\log 8}{\log(5/2)} \approx 2.269$

9. $e^{x-1} = 4$

$x - 1 = \ln 4$ $\boxed{x = \ln 4 + 1}$

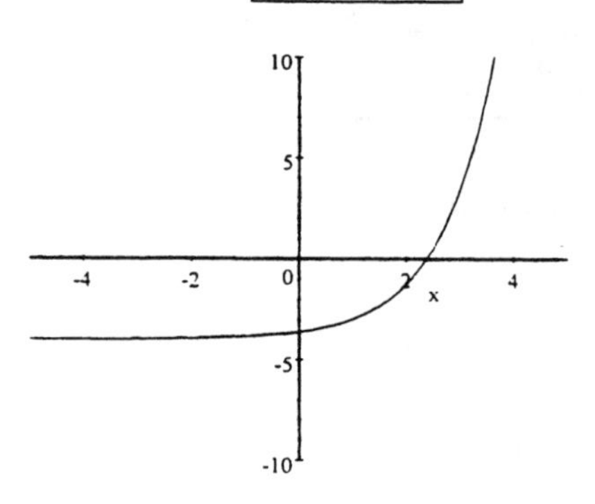

x-intercept $= \ln 4 + 1 \approx 2.3863$

11. $2e^{5x+2} = 8$

$e^{5x+2} = 4$

$5x + 2 = \ln 4$

$5x = \ln 4 - 2$ $\boxed{x = \dfrac{\ln 4 - 2}{5}}$

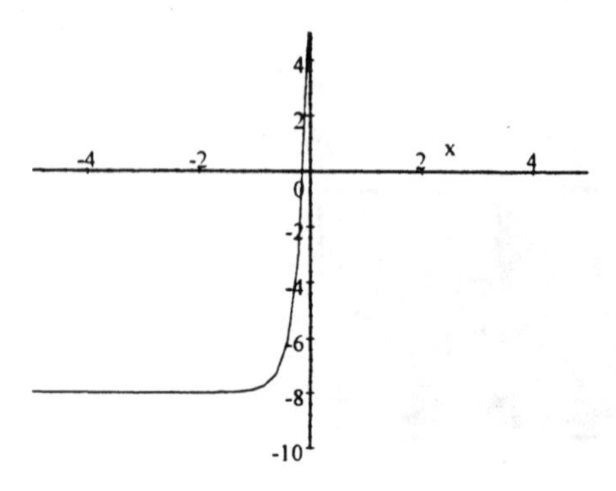

x-intercept $= \dfrac{\ln 4 - 2}{5} \approx -.1227$

13. $2^x = -3$ $\boxed{\emptyset}$

There is no power to which 2 can be raised which will give a negative answer.

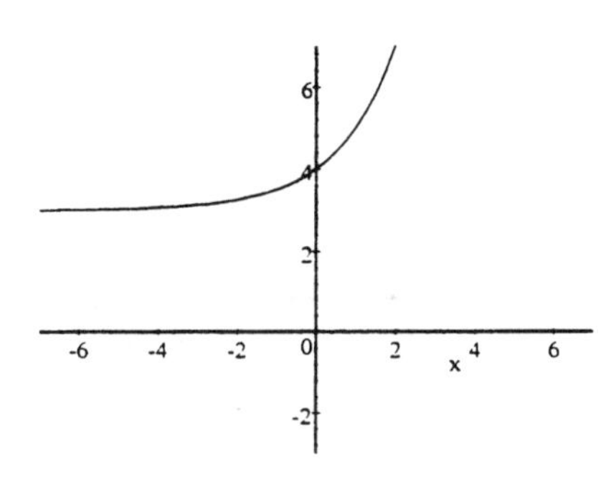

There is no x-intercept.

15.

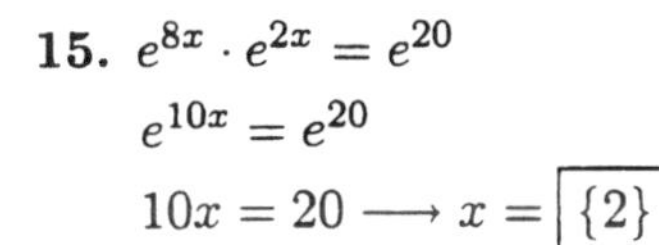

$e^{8x} \cdot e^{2x} = e^{20}$

$e^{10x} = e^{20}$

$10x = 20 \longrightarrow x = \boxed{\{2\}}$

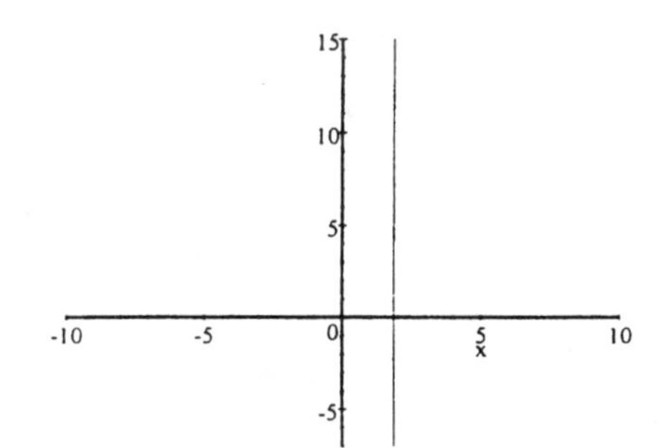

x-intercept $= 2$

17. $\log_3 (x + 50) = 2$

$x + 50 = 3^2$

$x = 9 - 50 = \boxed{\{-41\}}$

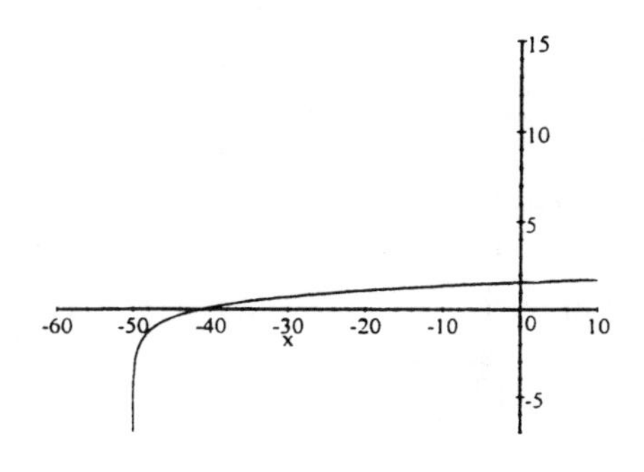

x-intercept $= -41$

19. $\log (4 - x) = -1$

$4 - x = 10^{-1}$

$-x = -4 + \frac{1}{10}$

$x = \frac{40}{10} - \frac{1}{10} = \frac{39}{10} = \boxed{\{3.9\}}$

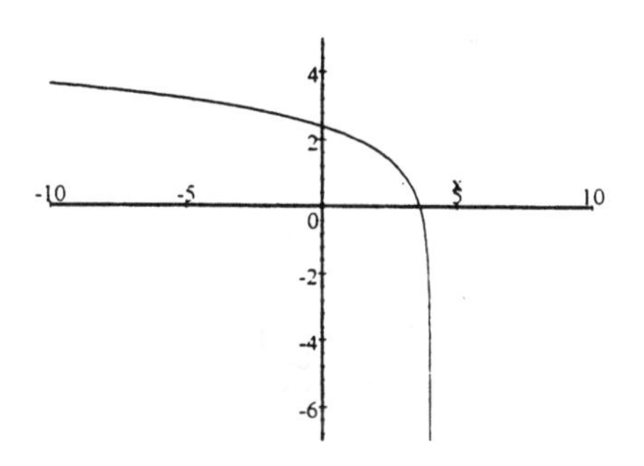

x-intercept $= 3.9$

21. $\log x + \log (x - 21) = 2$

$\log [x (x - 21)] = 2$

$x (x - 21) = 10^2$

$x^2 - 21x - 100 = 0$

$(x - 25)(x + 4) = 0$

$x = 25, \quad x = -4: \boxed{\{25\}}$

Discard $x = -4$, since

$x - 21 > 0 \Longrightarrow x > 21$

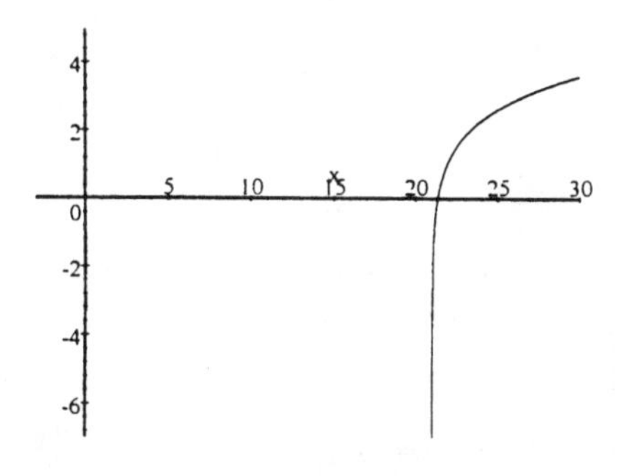

x-intercept $= 25$

23. $\ln(4x-2) = \ln 4 - \ln(x-2)$

$$\ln(4x-2) = \ln\frac{4}{x-2}$$

$$4x-2 = \frac{4}{x-2}, \quad x \neq 2, \ \frac{1}{2}$$

$$(4x-2)(x-2) = 4$$

$$4x^2 - 10x + 4 - 4 = 0$$

$$4x^2 - 10x = 0$$

$$2x(2x-5) = 0$$

$$x = 0, \ \frac{5}{2}$$

Discard $x = 0$, since

$x - 2 > 0 \to x > 2 \Longrightarrow x = \boxed{\{2.5\}}$

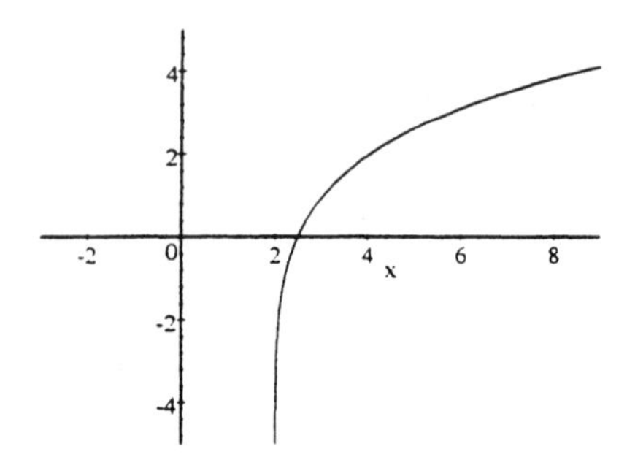

x-intercept $= 2.5$

25. $\log_5(x+2) + \log_5(x-2) = 1$

$$\log_5[(x+2)(x-2)] = 1$$

$$(x+2)(x-2) = 5^1, \quad x \neq \pm 2$$

$$x^2 - 4 = 5$$

$$x^2 = 9$$

$$x = \pm 3$$

Discard $x = -3$, since

$x - 2 > 0 \to x > 2 \qquad x = \boxed{\{3\}}$

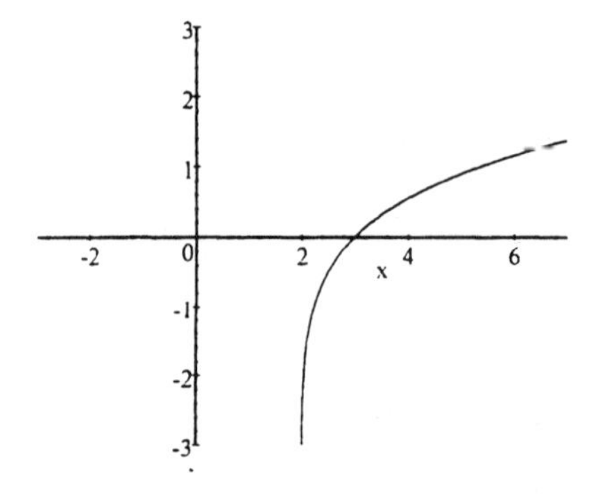

x-intercept $= 3$

27. $\ln e^x - \ln e^3 = \ln e^5$

$$x \ln e - 3\ln e = 5\ln e \quad [\ln e = 1]$$

$$x - 3 = 5 \Longrightarrow x = \boxed{\{8\}}$$

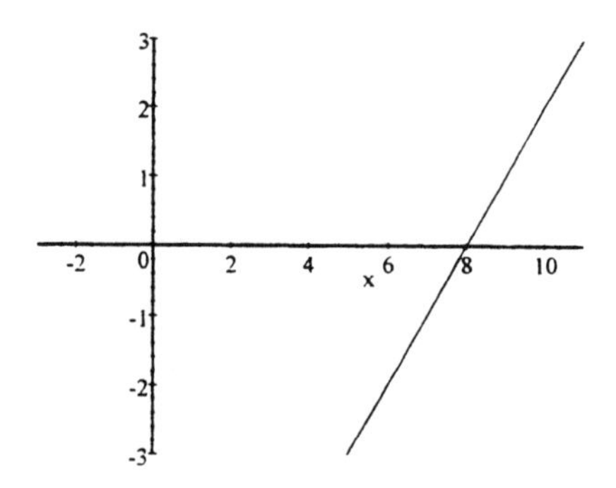

x-intercept $= 8$

29. $\log_2(\log_2 x) = 1$

$\log_2 x = 2^1$ (exponential form)

$x = 2^2 = \boxed{\{4\}}$ (exponential form)

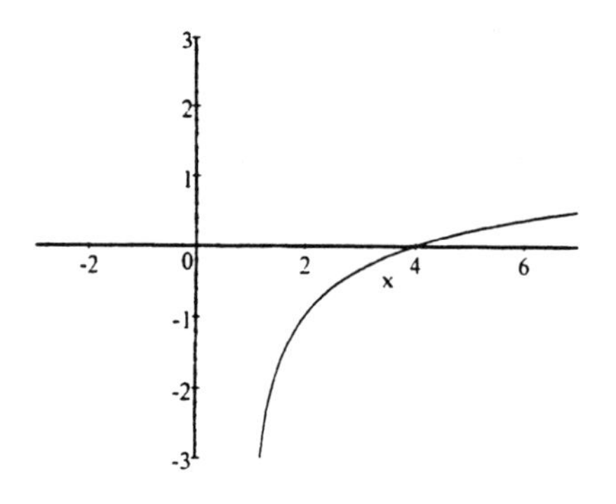

x-intercept $= 4$

31. $e^{-2\ln x} = \dfrac{1}{16}$

Solution from example 6: $x = 4$

Domain: $x > 0$

(See graph, example 6)

(a) $e^{-2\ln x} > \dfrac{1}{16}$: $\boxed{(0, 4)}$

(b) $e^{-2\ln x} < \dfrac{1}{16}$: $\boxed{(4, \infty)}$

33. The statement is incorrect. We must reject any solution that is not in the domain of any logarithmic function in the equation.

35. The graph of $y = \ln x - \ln(x+1) - \ln 5$ does not intersect the x-axis.

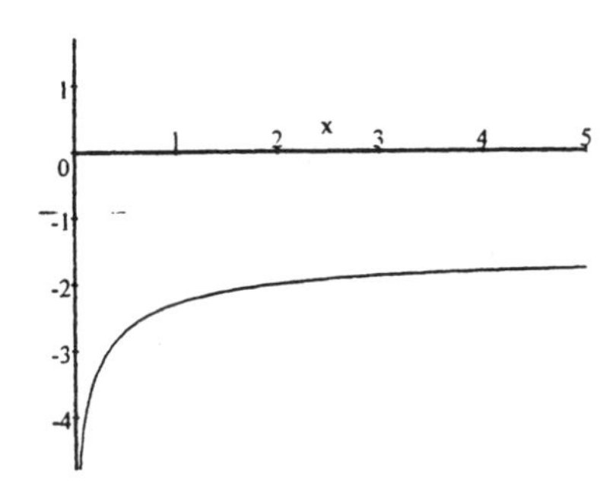

37. $y = 1.5^{\log x} - e^{.5}$

$x \approx 17.106129$ $\boxed{\approx \{17.106\}}$

39. $r = p - k \ln t$ (for t)

$$r - p = -k \ln t$$
$$\frac{r-p}{-k} = \ln t$$
$$\frac{p-r}{k} = \ln t \qquad \boxed{t = e^{\left(\frac{p-r}{k}\right)}}$$

41. $T = T_o + (T_1 - T_o)\,10^{-kt}$ (for t)

$$T - T_o = (T_1 - T_o)\,10^{-kt}$$
$$\frac{T - T_o}{T_1 - T_o} = 10^{-kt}$$
$$\log\left(\frac{T - T_o}{T_1 - T_o}\right) = -kt$$
$$\boxed{t = \left(-\frac{1}{k}\right)\log\left(\frac{T - T_o}{T_1 - T_o}\right)}$$

43. $A = T_o + Ce^{-kt}$ (for k)

$$A - T_o = Ce^{-kt}$$
$$\frac{A - T_o}{C} = e^{-kt}$$
$$\ln\left(\frac{A - T_o}{C}\right) = \ln e^{-kt}$$
$$\ln\left(\frac{A - T_o}{C}\right) = -kt \qquad \boxed{k = \frac{\ln\left(\frac{A - T_o}{C}\right)}{-t}}$$

45. $y = A + B\left(1 - e^{-Cx}\right)$ (for x)

$$y = A + B - Be^{-Cx}$$
$$y - A - B = -Be^{-Cx}$$
$$\frac{-y + A + B}{B} = e^{-Cx}$$
$$\ln\left(\frac{A + B - y}{B}\right) = \ln e^{-Cx}$$
$$\ln\left(\frac{A + B - y}{B}\right) = -Cx$$
$$\boxed{x = -\frac{\ln\left(\frac{A+B-y}{B}\right)}{C} = \frac{\ln\left(\frac{y-A-B}{-B}\right)}{-C}}$$

47. $\log A = \log B - C \log x$ (for A)

$$\log A = \log B - \log x^C$$
$$\log A = \log\left(\frac{B}{x^C}\right) \qquad \boxed{A = \frac{B}{x^C}}$$

Relating Concepts (49 - 54)

49. Because $(a^m)^n = a^{mn}$,

$(e^x)^2 = e^{x \cdot 2} = e^{2x}$

50. $(e^x)^2 - 4e^x + 3 = 0$

$(e^x - 3)(e^x - 1) = 0$

51.

$e^x - 3 = 0$	$e^x - 1 = 0$
$e^x = 3$	$e^x = 1$
$x = \ln 3$	$x = \ln 1 = 0$

$\boxed{\text{Solution set: } \{0,\ \ln 3\}}$

52. $y = (e^x)^2 - 4e^x + 3$
$\ln 1 = 0,\ \ln 3 \approx 1.099$

$\boxed{x\text{-intercept} \approx 1.099,\ 0}$

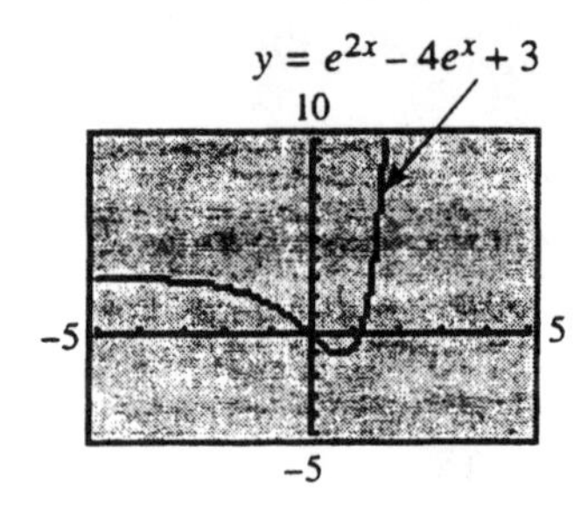

53. $e^{2x} - 4e^x + 3 > 0$: $\boxed{(-\infty, 0) \cup (\ln 3, \infty)}$

54. $e^{2x} - 4e^x + 3 < 0$: $\boxed{(0, \ln 3)}$

55. $x^2 = 2^x$;
graph $y = x^2 - 2^x$.

$\boxed{x\text{-intercepts: } \{-.767, 2, 4\}}$

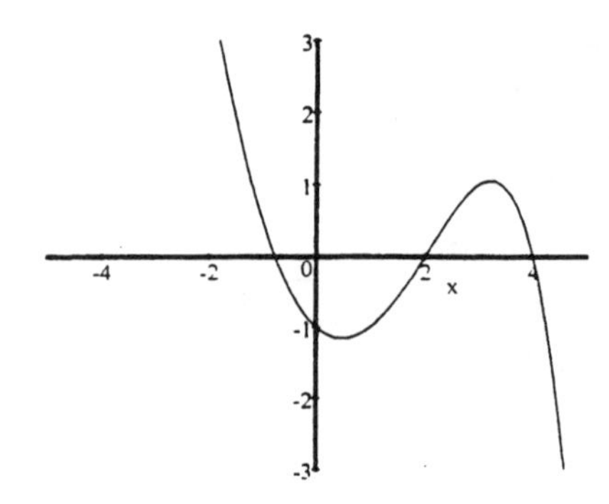

57. $\log x = x^2 - 8x + 14$;
graph $y = \log x - x^2 + 8x - 14$.

$\boxed{x\text{-intercepts: } \{2.454, 5.659\}}$

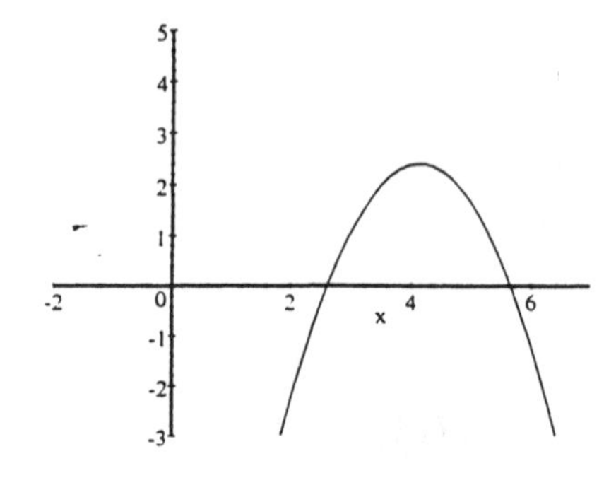

59. $e^x = \dfrac{1}{x+2}$;

graph $y = e^x - \dfrac{1}{x+2}$.

$\boxed{x\text{-intercept: } \{-.443\}}$

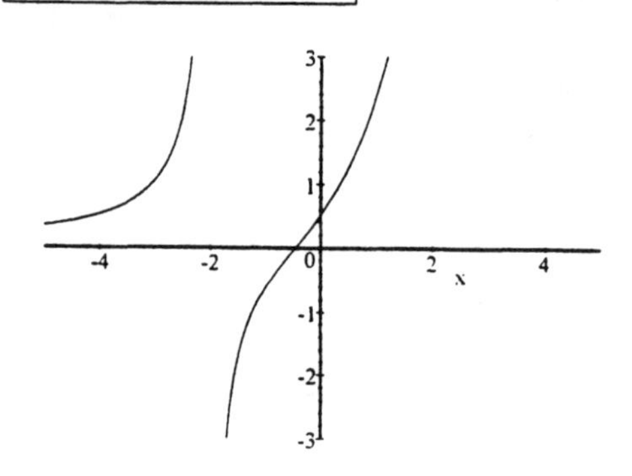

61. $\log_2 \sqrt{2x^2} - 1 = 0.5$
$\log_2 \sqrt{2x^2} = 1.5$
$2^{1.5} = \sqrt{2x^2}$
$(2^{1.5})^2 = \left(\sqrt{2x^2}\right)^2$
$2^3 = 2x^2$
$4 = x^2$
$x = \pm 2$
domain: $x^2 > 0 \to (-\infty, \infty)$

$\boxed{\text{Solution set: } \{-2, 2\}}$

63. $\ln(\ln e^{-x}) = \ln 3$
$\ln(-x) = \ln 3$
$-x = 3$

$\boxed{\text{Solution set: } \{-3\}}$

65. $y = \dfrac{2 - \log(100 - x)}{.42}$
$2 = \dfrac{2 - \log(100 - x)}{.42}$
$.84 = 2 - \log(100 - x)$
$.84 - 2 = -\log(100 - x)$
$1.16 = \log(100 - x)$
$100 - x = 10^{1.16}$
$-x = 10^{1.16} - 100$
$x = -10^{1.16} + 100 \approx \boxed{85.5\%}$

67. $f(x) = 31.5 + 1.1\log(x+1)$
$33 = 31.5 + 1.1\log(x+1)$
$1.5 = 1.1\log(x+1)$
$\frac{1.5}{1.1} = \log(x+1)$
$x + 1 = 10^{1.5/1.1}$
$x = -1 + 10^{1.5/1.1} \approx 22$ feet

Reviewing Basic Concepts (Sections 5.4 and 5.5)

1. If $f(x) = 3^x$ and $g(x) = \log_3 x$, then functions f and g are inverse functions, and their graphs are symmetric with respect to the line with equation $y = x$. The domain of f is the range of g and vice versa.

2. $f(x) = 2 - \log_2(x-1)$

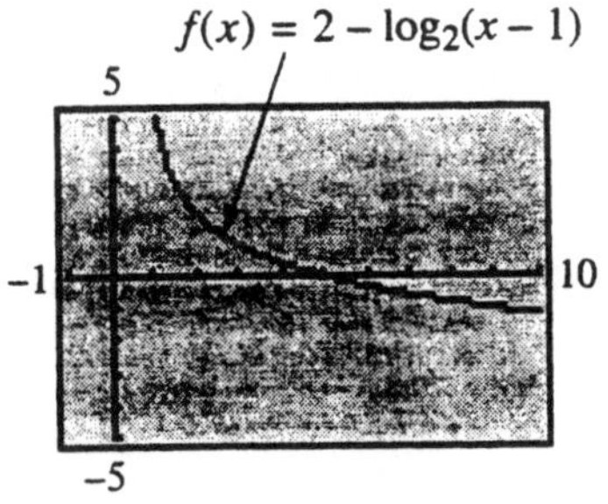

$f(x) = 2 - \log_2(x-1)$

5

−1 10

−5

3. asymptote: $x = 1$; x-intercept $(5,\ 0)$

4. $g(x) = \log_2 x$
shift 1 unit right: $f(x) = \log_2(x-1)$
reflect across x-axis: $f(x) = -\log_2(x-1)$
shift 2 units upward: $f(x) = 2 - \log_2(x-1)$

5. $x = 2 - \log_2(y-1)$
$x - 2 = -\log_2(y-1)$
$2 - x = \log_2(y-1)$
$y - 1 = 2^{2-x}$
$y = f^{-1}(x) = 1 + 2^{2-x}$

6. $3^{2x-1} = 4^x$
$\log 3^{2x-1} = \log 4^x$
$(2x-1)\log 3 = x\log 4$
$2x\log 3 - 1\log 3 = x\log 4$
$2x\log 3 - x\log 4 = \log 3$
$x(2\log 3 - \log 4) = \log 3$

$$x = \frac{\log 3}{2\log 3 - \log 4} \text{ or } \frac{\log 3}{\log\left(\frac{9}{4}\right)}$$

7. $\ln(5x) - \ln(x+2) = \ln 3$

$$\ln\left(\frac{5x}{x+2}\right) = \ln 3$$

$$\frac{5x}{x+2} = 3$$

$5x = 3x + 6 \Longrightarrow 2x = 6 \longrightarrow x = \boxed{\{3\}}$

8. $10^{5\log x} = 32$
$\log 10^{5\log x} = \log 32$
$5\log x(\log 10) = \log 32$
$5\log x(1) = \log 32$
$\log x^5 = \log 32$
$x^5 = 32 \longrightarrow x = \boxed{\{2\}}$

9. $H = 1000\left(1 - e^{-kN}\right)$

$$\frac{H}{1000} = 1 - e^{-kN}$$

$$\frac{H}{1000} - 1 = -e^{-kN}$$

$$1 - \frac{H}{1000} = e^{-kN}$$

$$\ln\left(1 - \frac{H}{1000}\right) = -kN$$

$$N = -\frac{1}{k}\ln\left(1 - \frac{H}{1000}\right)$$

10. $f(x) = 280\ln(x+1) + 1925$
$2300 = 280\ln(x+1) + 1925$
$375 = 280\ln(x+1)$

$$\frac{375}{280} = \ln(x+1)$$

$x + 1 = e^{375/280}$
$x = e^{375/280} - 1 \approx \boxed{2.8 \text{ acres}}$

Section 5.6

1. $A(t) = A_o e^{-.0001216t}$

$\frac{1}{3}A_o = A_o e^{-.0001216t}$

$\frac{1}{3} = e^{-.0001216t}$

$\ln\left(\frac{1}{3}\right) = -.0001216t$

$t = \frac{\ln\left(\frac{1}{3}\right)}{-.0001216t} \approx 9034.6$

$\boxed{\approx 9000 \text{ years ago}}$

3. $A(t) = A_o e^{-.0001216t}$

$.15A_o = A_o e^{-.0001216t}$

$.15 = e^{-.0001216t}$

$\ln .15 = -.0001216t$

$t = \frac{\ln .15}{-.0001216} \approx 15601.3$

$\boxed{\approx 16,000 \text{ years old}}$

5. $A(t) = A_o e^{-kt}$

(a) $\frac{1}{2}A_o = A_o e^{-21.7k}$

$.5 = e^{-21.7k}$

$-21.7k = \ln .5$

$k = \frac{\ln .5}{-21.7} \approx .032$

$A(t) = A_o e^{-.032t}$

(b) $400 = 500e^{-.032t}$

$.8 = e^{-.032t}$

$-.032t = \ln .8$

$t = \frac{\ln .8}{-.032} \approx 6.97$ years

(c) $A(10) = A_o e^{-.032(10)}$

$\boxed{\approx 363 \text{ grams}}$

7. (a) Since at 200 days the number of milligrams (.743) is less than 1, the half-life is less than 200 days.

(b) $A = A_o e^{-kt}$; since $A_o = 2 \longrightarrow A = 2e^{-kt}$

$(100, 1.22): \quad 1.22 = 2e^{-100k}$

$.61 = e^{-100k}$

$-100k = \ln .61$

$k = \frac{\ln .61}{-100} \approx .005$

$A = 2e^{-.005t}$

(c) $y_1 = 2e^{-.005t}; \quad y_2 = 1$

5

Intersection
X=138.62944 Y=1

0 250

0

≈ 140 days is the half-life.

9. Intensity $= \log \frac{I}{I_o}$

(a) $6.6 = \log \frac{I}{I_o}$

$10^{6.6} = \frac{I}{I_o}$

$I = 10^{6.6} I_o \approx 3,981,071.7\, I_o$

$\boxed{I \approx 4,000,000\, I_o}$

(b) $6.5 = \log \frac{I}{I_o}$

$10^{6.5} = \frac{I}{I_o}$

$I = 10^{6.5} I_o \approx 3,162,277.7\, I_o$

$\boxed{I \approx 3,200,000\, I_o}$

(c) $\frac{10^{6.6}}{10^{6.5}} \approx 1.258925$

$\left[\frac{4{,}000{,}000}{3{,}200{,}000} = 1.25\right]$

$\boxed{\approx 1.25 \text{ times as strong}}$

11. $t = T\left(\dfrac{\ln[1+8.33(A/K)]}{\ln 2}\right)$

Find t for $A/K = .103,\ \ T = 1.26\times 10^9$.

$$t = \left(1.26\times 10^9\right)\left(\frac{\ln[1+8.33(.103)]}{\ln 2}\right)$$
$$\approx 1.126\times 10^9$$

The rock sample is about 1.126 billion years old.

13. $M = 6 - \dfrac{5}{2}\log\dfrac{I}{I_0}$

$(m=1):\quad 1 = 6 - \dfrac{5}{2}\log\dfrac{I}{I_0}$

$$-5 = -\frac{5}{2}\log\frac{I}{I_0}$$
$$2 = \log\frac{I}{I_0}$$
$$10^2 = \frac{I}{I_0}$$

$I = 100\,I_o$

$(m=3):\quad 3 = 6 - \dfrac{5}{2}\log\dfrac{I}{I_0}$

$$-3 = -\frac{5}{2}\log\frac{I}{I_0}$$
$$\frac{6}{5} = \log\frac{I}{I_0}$$
$$10^{6/5} = \frac{I}{I_0}$$

$I = 15.85\,I_o$

$$\left(\text{ratio } \frac{m=1}{m=3}\right):\ \frac{100\,I_o}{15.85\,I_o} = 6.31$$

Magnitude 1 is $\approx$ 6.3 times as great as magnitude 3.

15. 100° coffee, 20° (T_o), 1 hr (t), 60° $(f(t))$

$C = 100^o - 20^o = 80^o$

(a) $f(t) = T_o + Ce^{-kt}$

$$60 = 20 + 80e^{-k(1)}$$
$$40 = 80e^{-k}$$
$$.5 = e^{-k}$$
$$\ln .5 = -k$$
$$k = -\ln .5 \approx .693$$

EQ: $f(t) = 20 + 80e^{-.693\,t}$

(b) $f(.5) = 20 + 80e^{-.693(.5)} \approx 76.6°C$

(c) $50 = 20 + 80e^{-.693\,t}$

$$30 = 80e^{-.693\,t}$$
$$.375 = e^{-.693\,t}$$
$$-.693t = \ln .375$$
$$t = \frac{\ln .375}{-.693} \approx 1.415 \text{ hours} \approx \boxed{1 \text{ hr } 25 \text{ min}}$$

17. **(a)** $\{0, 5, 10, 15, 20\} \longrightarrow L_1$

$\{.72, .88, 1.07, 1.31, 1.60\} \longrightarrow L_2$

STAT / CALC / ExpReg $\Longrightarrow y \approx .72(1.041)^x$

$a \approx .72,\ b \approx 1.041$

(b) $2013: f(13) = .72(1.041)^{13}$

$\approx$ 1.21 ppb

19. **(a)** $T(R) = 1.03R,\ \ R(x) = .2e^{.0124\,x}$

$$(T\circ R)(x) = T[R(x)] = T\left(.2e^{.0124\,x}\right) = 1.03\left(.2e^{.0124\,x}\right) = .206e^{.0124x}$$

(b) $(T\circ R)(100) = .206e^{.0124(100)} \approx .7119$

In 1900, radioactive forcing caused approximately a $.7°\,F$ increase in average global temperature.

21. (a) $A = P\left(1 + \frac{r}{n}\right)^{nt}$

$5000 = 1000\left(1 + \frac{.035}{4}\right)^{4t}$

$5 = (1.00875)^{4t}$

$\log 5 = \log 1.00875^{4t}$

$\log 5 = 4t \log 1.00875$

$t = \frac{\log 5}{4 \log 1.00875} \approx 46.18$

$\boxed{t \approx 46.2 \text{ years}}$

(b) $A = Pe^{rt}$

$5000 = 1000e^{.035t}$

$5 = e^{.035t}$

$\ln 5 = \ln e^{.035t}$

$\ln 5 = .035t$

$t = \frac{\ln 5}{.035} \approx 45.98$

$\boxed{t \approx 46.0 \text{ years}}$

23. $A = P\left(1 + \frac{r}{n}\right)^{nt}$

$30000 = 27000\left(1 + \frac{.06}{4}\right)^{4t}$

$\frac{10}{9} = (1.015)^{4t}$

$\log\left(\frac{10}{9}\right) = \log 1.015^{4t}$

$\log\left(\frac{10}{9}\right) = 4t \log 1.015$

$t = \frac{\log\left(\frac{10}{9}\right)}{4 \log 1.015} \approx 1.769$

$\boxed{t \approx 1.8 \text{ years}}$

25. 7%: $A = 60000\left(1 + \frac{.07}{4}\right)^{4(5)}$

$A = 60000\,(1.0175)^{20}$

$A = \$84886.69$

6.75%: $A = 60000e^{.0675(5)}$

$A = \$84086.38$

Better investment is 7% compounded quarterly; it will earn \$800.31 more.

27. $R = \left(1 + \frac{r}{n}\right)^{n} - 1$

$R = \left(1 + \frac{.06}{4}\right)^{4} - 1$

$R = 1.015^4 - 1$

$R \approx .06136$ $\boxed{R \approx 6.14\%}$

29. $P = A\left(1 + \frac{r}{n}\right)^{-nt}$

$P = 10,000\left(1 + \frac{.12}{2}\right)^{-2(5)}$

$P = 10,000\,(1.06)^{-10}$ $\boxed{\$5583.95}$

31. $P = A\left(1 + \frac{r}{n}\right)^{-nt}$

$25000 = 31360\left(1 + \frac{r}{1}\right)^{-1(2)}$

$\frac{25000}{31360} = (1 + r)^{-2}$

$\left(\frac{25000}{31360}\right)^{-1/2} = 1 + r$

$r \approx .1199776 \approx .12$ $\boxed{r \approx 12\%}$

33. (a) $R = \frac{P}{\frac{1 - (1 + i)^{-n}}{i}}$, $n = 12\,(4) = 48$, $i = \frac{.075}{12} = .00625$

$R = \frac{8500}{\frac{1 - (1 + .00625)^{-48}}{.00625}}$

$\boxed{\text{Payment} = \$205.52}$

(b) $I = nR - P = 48\,(205.52) - 8500$

$\boxed{I = \$1364.96}$

35. (a) $R = \dfrac{P}{\dfrac{1-(1+i)^{-n}}{i}}$ $\quad n = 12(30) = 360$ $\quad i = \dfrac{.0725}{12}$

$$R = \frac{125000}{\dfrac{1-\left(1+\frac{.0725}{12}\right)^{-360}}{\frac{.0725}{12}}}$$

$\boxed{\text{Payment} = \$852.72}$

(b) $I = nR - P = 360(852.72) - 125000$

$\boxed{I = \$107,887.89}$

37. (a) Enter $y_1 = 1500\left(1+\dfrac{.0575}{365}\right)^{365t}$

Using the table feature, the investment will triple when $y_1 = 4500$:

$t \approx 19.1078$ years

≈ 19 years $+ .1078(365)$ days

$\approx \boxed{19 \text{ years, } 39 \text{ days}}$

Analytically: $y = 4500$

$$4500 = 1500\left(1+\frac{.0575}{365}\right)^{365t}$$

$$3 = (1.000157534)^{356t}$$

$$\ln 3 = 365t \ln 1.000157534$$

$$t = \frac{\ln 3}{365 \ln 1.000157534}$$

$$\approx 19.1078$$

(b) Enter $y_1 = 2000\left(1+\dfrac{.08}{365}\right)^{365t}$

Using the table feature, the investment will be worth \$5000 when $y_1 = 5000$:

$t = 11.455$ years

$= 11$ years $+ .455(365)$ days

$= \boxed{11 \text{ years, } 166 \text{ days}}$

39. $A(t) = A_o e^{.023t}$

$A(t) = 2,400,000e^{.023t}$

(a) $A(5) = 2,400,000e^{.023(5)}$

$A(5) \approx 2,692,496$

$\boxed{A(5) \approx 2{,}700{,}000}$

(b) $A(10) = 2,400,000e^{.023(10)}$

$A(10) \approx 3,020,640$

$\boxed{A(10) \approx 3{,}000{,}000}$

(c) $A(60) = 2,400,000e^{.023(60)}$

$A(60) \approx 9,539,764$

$\boxed{A(60) \approx 9{,}500{,}000}$

41. $P(t) = 100e^{-0.1t}$

$10 = 100e^{-0.1t}$

$0.1 = e^{-0.1t}$

$\ln 0.1 = -0.1t$

$t = \dfrac{\ln 0.1}{-0.1}$ $\quad \boxed{t \approx 23 \text{ days}}$

43. $(\text{dose}_o)(\%)^{(\text{hrs}-1)} = \text{effective dose}$

$(250)(.75)^5 = \text{effective dose}$

effective dose ≈ 59.32617 $\quad \boxed{\approx 59 \text{ mg}}$

45. (a)

x	0	15	30	45	60	75	90	125
$g(x)$	7	21	57	111	136	158	164	178

(b) $g(x) = 261 - f(x)$

(c) y_1 is a better fit and models $g(x)$ better:

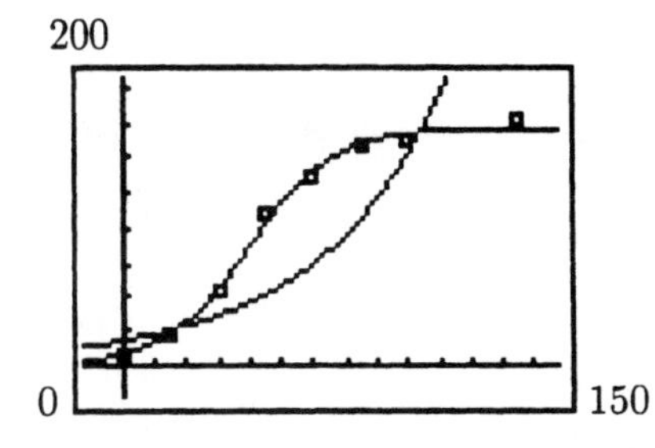

(d) $f(x) = 261 - g(x)$

$$f(x) = 261 - \frac{171}{1 + 18.6e^{-.0747x}}$$

47. (a) $\{5, 10, 15, 20\} \longrightarrow L_1$
$\{20, 41, 80, 136\} \longrightarrow L_2$

STAT / CALC / ExpReg

$f(x) = 10.98\,(1.14)^x$

(b)

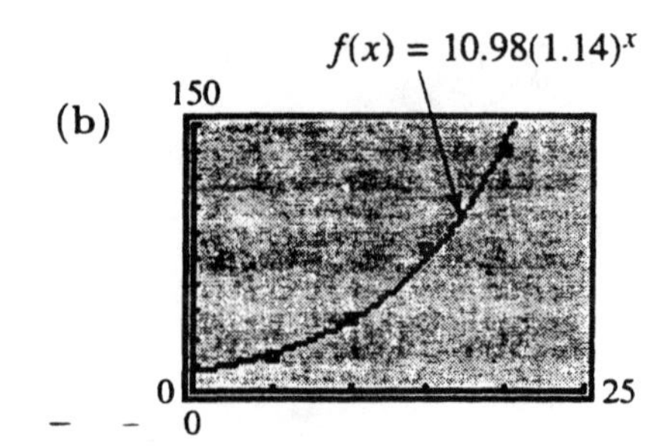

(c) $f(13) = 10.98\,(1.14)^{13} \approx \$60$ billion

49. $f(x) = 26.6e^{.131x}$

(a) $f(4) = 26.6e^{.131(4)} \approx 44.9$ billion

(b) $30 = 26.6e^{.131x}$

$$\frac{30}{26.6} = e^{.131x}$$

$$.131x = \ln\frac{30}{26.6}$$

$$x = \frac{\ln\frac{30}{26.6}}{.131} \approx .918 \approx 1$$

$1991 + 1 = 1992$

(c)

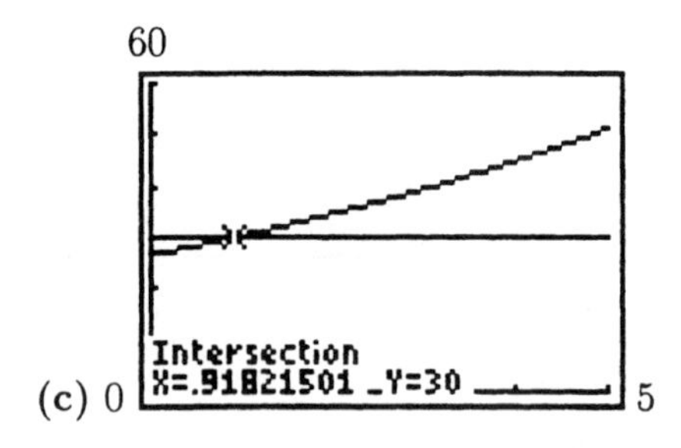

51. (a) $f(x) = Ca^{x-1989}$

$(1989,\ 131):\ \ 131 = Ca^{1989-1989}$
$\Longrightarrow C = 131$

$f(x) = 131a^{x-1989}$

$(1995,\ 34):\ \ 34 = 131a^{1995-1989}$

$$\frac{34}{131} = a^6$$

$$\left(\frac{34}{131}\right)^{1/6} = a \ \boxed{\approx .799}$$

Answers may vary.

(b) $y_1 = 131\,(.799)^{x-1989}$

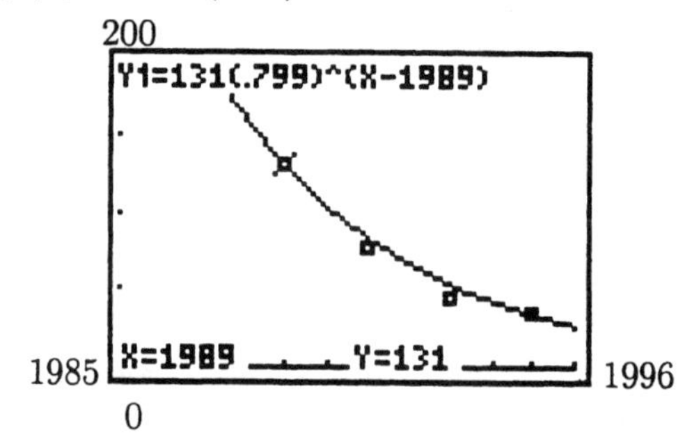

53. $f(x) = \dfrac{.9}{1 + 271e^{-.122x}}$

(a) $f(25) = \dfrac{.9}{1 + 271e^{-.122(25)}} \approx .065$

$f(65) = \dfrac{.9}{1 + 271e^{-.122(65)}} \approx .82$

Among people age 25, 6.5% have some CHD, while among people age 65, 82% have some CHD.

(b) $.50 = \dfrac{.9}{1 + 271e^{-.122x}}$

$.50 + 135.5e^{-.122x} = .9$

$135.5e^{-.122x} = .4$

$$e^{-.122x} = \frac{.4}{135.5}$$

$$-.122x = \ln\frac{.4}{135.5}$$

$$x = \frac{\ln\frac{.4}{135.5}}{-.122} \approx 47.75 \approx 48 \text{ years}$$

Chapter 5 Review

1. Not one-to-one

2. One-to-one

3. Not one-to-one

4. $f(x) = \sqrt[3]{2x-7}$
Domain: $(-\infty, \infty)$

5. Range: $(-\infty, \infty)$

6. $f(x)$ is one-to-one, and therefore has an inverse.

7. $f(x) = \sqrt[3]{2x-7}$

$$x = \sqrt[3]{2y-7}$$
$$x^3 = 2y - 7$$
$$x^3 + 7 = 2y$$
$$\frac{x^3+7}{2} = y$$
$$\boxed{f^{-1}(x) = \frac{x^3+7}{2}}$$

8. $f(x) = \sqrt[3]{2x-7}$

$$f^{-1}(x) = \frac{x^3+7}{2}$$

The graphs of f and f^{-1} are reflections of each other across the line $y = x$ and are symmetric with respect to the line $y = x$.

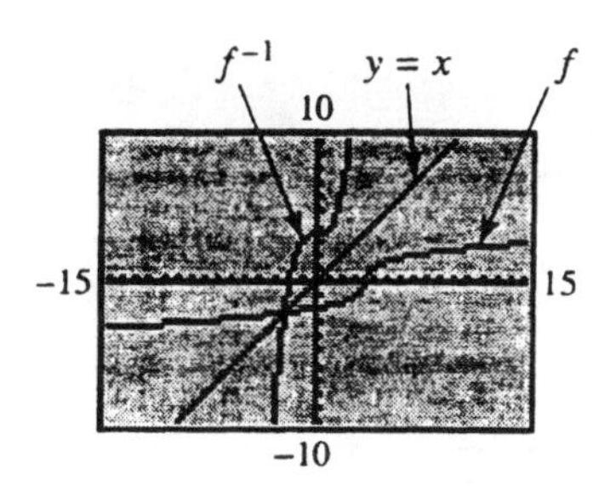

9. $f(x) = \sqrt[3]{2x-7}$; $f^{-1}(x) = \dfrac{x^3+7}{2}$

$$\begin{aligned}(f \circ f^{-1})(x) &= f\left(\frac{x^3+7}{2}\right) \\ &= \sqrt[3]{2\left(\frac{x^3+7}{2}\right) - 7} \\ &= \sqrt[3]{x^3+7-7} = x\end{aligned}$$

$$\begin{aligned}(f^{-1} \circ f)(x) &= f^{-1}\left(\sqrt[3]{2x-7}\right) \\ &= \frac{\left(\sqrt[3]{2x-7}\right)^3 + 7}{2} \\ &= \frac{2x-7+7}{2} = x\end{aligned}$$

10. $y = a^{x+2}$: **C**

11. $y = a^x + 2$: **A**

12. $y = -a^x + 2$: **D**

13. $y = a^{-x} + 2$: **B**

14. $f(x) = a^x : 0 < a < 1$

15. $f(x) = a^x$; Domain: $(-\infty, \infty)$

16. $f(x) = a^x$; Range: $(0, \infty)$

17. $f(x) = a^x$; $f(0) = 1$

18. $y = f^{-1}(x)$

19. $f^{-1}(x) = \log_a x$

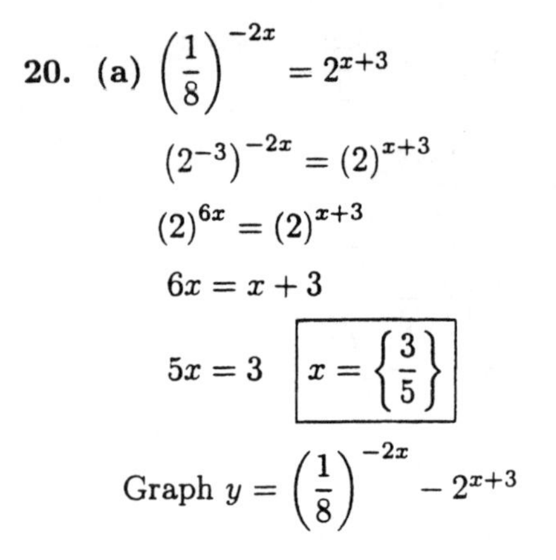

20. **(a)** $\left(\frac{1}{8}\right)^{-2x} = 2^{x+3}$

$\left(2^{-3}\right)^{-2x} = (2)^{x+3}$

$(2)^{6x} = (2)^{x+3}$

$6x = x + 3$

$5x = 3 \quad \boxed{x = \left\{\frac{3}{5}\right\}}$

Graph $y = \left(\frac{1}{8}\right)^{-2x} - 2^{x+3}$

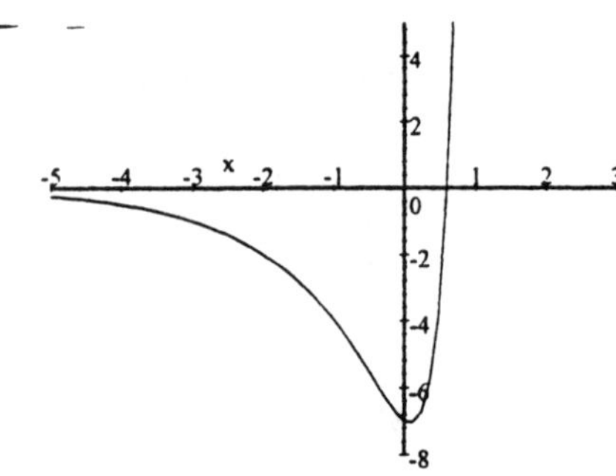

(b) $\left(\frac{1}{8}\right)^{-2x} \geq 2^{x+3}$: $\boxed{\left[\frac{3}{5}, \infty\right)}$

21. **(a)** $3^{-x} = \left(\frac{1}{27}\right)^{1-2x}$

$3^{-x} = \left(3^{-3}\right)^{1-2x}$

$3^{-x} = 3^{-3+6x}$

$-x = -3 + 6x$

$-7x = -3 \quad \boxed{x = \left\{\frac{3}{7}\right\}}$

Graph $y = 3^{-x} - \left(\frac{1}{27}\right)^{1-2x}$

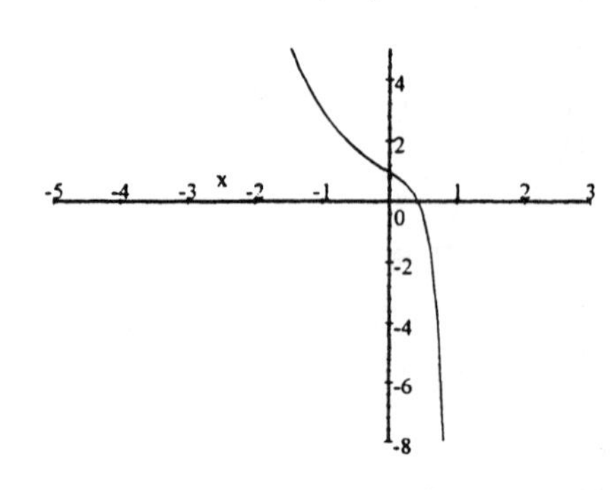

(b) $3^{-x} < \left(\frac{1}{27}\right)^{1-2x}$: $\boxed{\left(\frac{3}{7}, \infty\right)}$

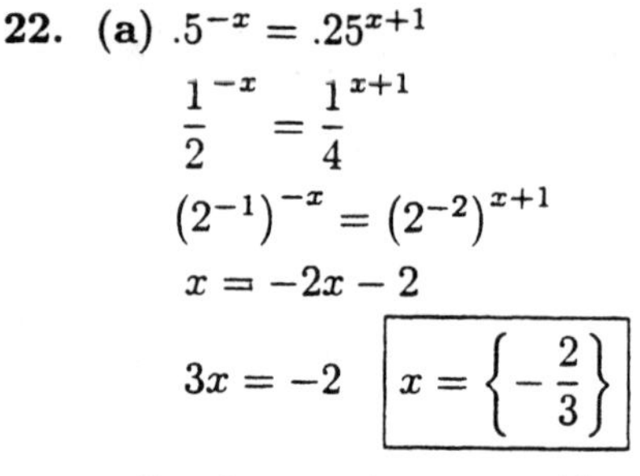

22. **(a)** $.5^{-x} = .25^{x+1}$

$\frac{1}{2}^{-x} = \frac{1}{4}^{x+1}$

$\left(2^{-1}\right)^{-x} = \left(2^{-2}\right)^{x+1}$

$x = -2x - 2$

$3x = -2 \quad \boxed{x = \left\{-\frac{2}{3}\right\}}$

Graph $y = .5^{-x} - .25^{x+1}$

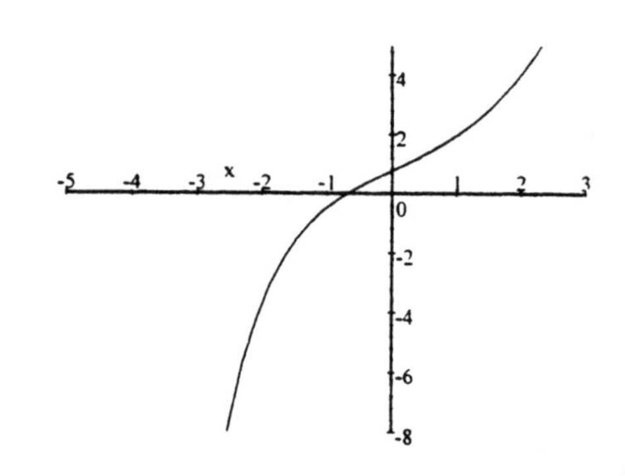

(b) $.5^{-x} > .25^{x+1}$: $\boxed{\left(-\frac{2}{3}, \infty\right)}$

23. **(a)** $.4^{x} = 2.5^{1-x}$

$\log .4^{x} = \log 2.5^{1-x}$

$x \log .4 = (1 - x) \log 2.5$

$x \log .4 = \log 2.5 - x \log 2.5$

$x \log .4 + x \log 2.5 = \log 2.5$

$x\,(\log .4 + \log 2.5) = \log 2.5$

$x = \frac{\log 2.5}{(\log .4 + \log 2.5)}$

Error (division by 0) $\boxed{\emptyset}$

Graph $y = .4^{x} - 2.5^{1-x}$

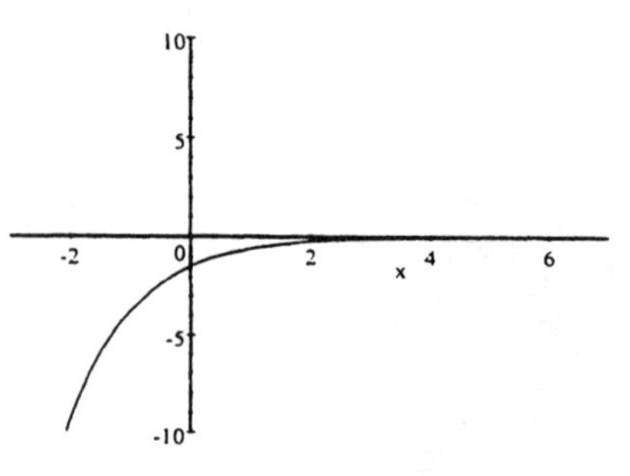

(b) $.4^{x} < 2.5^{1-x}$: $\boxed{(-\infty, \infty)}$

24. $y = x^2; \quad y = 2^x$

Graph $y = x^2 - 2^x$; check for x-intercepts.

$x = 2,\ 4,\ -.766664696$

$\boxed{(-.766664696,\ .58777475603)}$

25. $\log 58.3 \approx 1.76566855476 \approx 1.7657$

26. $\log .00233 \approx -2.63264407897 \approx -2.6326$

27. $\ln 58.3 \approx 4.06560209336 \approx 4.0656$

28. $\log_2 .00233 = \dfrac{\log .00233}{\log 2} \approx -8.7455$

29. $\log_{13} 1 = x$

$13^x = 1 \qquad \boxed{x = \{0\}}$

30. $\ln e^{\sqrt{6}} = \boxed{\sqrt{6}} \qquad [\ln e^x = x]$

31. $\log_5 5^{12} = \boxed{12} \qquad [\log_a a^b = b]$

32. $7^{\log_7 13} = \boxed{13} \qquad [a^{\log_a b} = b]$

33. $3^{\log_3 5} = \boxed{5} \qquad [a^{\log_a b} = b]$

34. $\log_4 9 = \dfrac{\log 9}{\log 4} \quad \boxed{\approx 1.58496250072}$

35. $f(x) = \log_2 x$ The x-intercept is 1, and the graph increases. The correct graph is **E**.

36. $f(x) = \log_2 (2x)$

$= \log_2 2 + \log_2 x = 1 + \log_2 x$

The graph will be similar to that of $f(x) = \log_2 x$, but will have a vertical shift up 1. The x-intercept will be $\frac{1}{2}$, since $\log_2 2\left(\frac{1}{2}\right) = \log_2 1 = 0$. The correct graph is **D**.

37. $f(x) = \log_2 \left(\dfrac{1}{x}\right) = \log_2 x^{-1} = -\log_2 x$

The graph will be similar to that of $f(x) = \log_2 x$, but will be reflected across the x-axis. The x-intercept is 1. The correct graph is **B**.

38. $f(x) = \log_2 \left(\dfrac{x}{2}\right) = \log_2 \left(\dfrac{1}{2}x\right)$

$= \log_2 \left(\dfrac{1}{2}\right) + \log_2 x$

$= -1 + \log_2 x$

The graph will be similar to that of $f(x) = \log_2 x$, but will have a vertical shift down 1. The x-intercept will be 2, since $\log_2 \left(\dfrac{2}{2}\right) = \log_2 1 = 0$. The correct graph is **C**.

39. $f(x) = \log_2 (x - 1)$

The graph will be similar to that of $f(x) = \log_2 x$, but will be shifted 1 unit to the right;. the x-intercept will be 2. The correct graph is **F**.

40. $f(x) = \log_2 (-x)$

The graph will be similar to that of $f(x) = \log_2 x$, but will be reflected across the y-axis. The x-intercept is -1 since $\log_2 (-(-1)) = \log_2 1 = 0$. The correct graph is **A**.

41. Let $f(x) = \log_a x$ be the required function. Then

$$\begin{aligned} f(81) &= 4 \\ \log_a 81 &= 4 \\ a^4 &= 81 \\ a^4 &= 3^4 \\ a &= 3 \end{aligned} \qquad \boxed{\text{The base is } 3}$$

42. Let $f(x) = a^x$ be the required function. Then

$$\begin{aligned} f(-4) &= \frac{1}{16} \\ a^{-4} &= \frac{1}{16} \\ a^{-4} &= 2^{-4} \\ a &= 2 \end{aligned} \qquad \boxed{\text{The base is } 2}$$

43. $\log_3 \left(\dfrac{mn}{5r}\right)$

$= \log_3 (mn) - \log_3 (5r)$

$= \log_3 m + \log_3 n - (\log_3 5 + \log_3 r)$

$= \log_3 m + \log_3 n - \log_3 5 - \log_3 r$

44. $\log_2\left(\dfrac{\sqrt{7}}{15}\right) = \log_2 \sqrt{7} - \log_2 15$
$= \log_2 7^{1/2} - \log_2 15$
$= \frac{1}{2}\log_2 7 - \log_2 15$

45. $\log_5\left(x^2y^4\sqrt[5]{m^3p}\right)$
$= \log_5\left[x^2y^4\left(m^3p\right)^{1/5}\right]$
$= \log_5 x^2 + \log_5 y^4 + \log_5\left(m^3p\right)^{1/5}$
$= 2\log_5 x + 4\log_5 y + \dfrac{1}{5}\log_5\left(m^3p\right)$
$= 2\log_5 x + 4\log_5 y + \dfrac{1}{5}\left(\log_5 m^3 + \log_5 p\right)$
$= 2\log_5 x + 4\log_5 y + \dfrac{3}{5}\log_5 m + \dfrac{1}{5}\log_5 p$

46. $\log_7\left(7k + 5r^2\right)$
The properties of logarithms do not apply.

47. (a) $\log(x+3) + \log x = 1,\ x > 0$
$\log[(x+3)(x)] = 1$
Rewrite in exponential form:
$x(x+3) = 10^1$
$x^2 + 3x - 10 = 0$
$(x+5)(x-2) = 0$
$x = -5,\ 2$ $\boxed{\{2\}}$
Graph $y_1 = \log(x+3) + \log x - 1$; the graph is increasing, and crosses the x-axis at 2.

(b) $\log(x+3) + \log x > 1 : (2,\ \infty)$

48. (a) $\ln e^{\ln x} - \ln(x-4) = \ln 3$
$\ln x - \ln(x-4) = \ln 3$
$\ln\left(\dfrac{x}{x-4}\right) = \ln 3$
$\dfrac{x}{x-4} = 3$
$x = 3x - 12$
$-2x = -12$ $\boxed{x = 6}$
Graph $y = \ln e^{\ln x} - \ln(x-4) - \ln 3$;
x-intercept: 6

(b) $\ln e^{\ln x} - \ln(x-4) \le \ln 3 : \boxed{[6,\ \infty)}$

49. (a) $\ln e^{\ln 2} - \ln(x-1) = \ln 5$
$\ln 2 - \ln(x-1) = \ln 5$
$\ln\left(\dfrac{2}{x-1}\right) = \ln 5$
$\dfrac{2}{x-1} = 5$
$2 = 5x - 5$
$7 = 5x$
$\boxed{x = \left\{\dfrac{7}{5}\right\} \quad (1.4)}$
Graph $y = \ln e^{\ln 2} - \ln(x-1) - \ln 5$;
x-intercept $= 1.4$

(b) $\ln e^{\ln 2} - \ln(x-1) \ge \ln 5 :$
$\boxed{(1,\ 1.4]}$ (domain: $x > 1$)

50. $8^x = 32$
$\left(2^3\right)^x = 2^5$
$2^{3x} = 2^5$
$3x = 5$ $\boxed{x = \left\{\dfrac{5}{3}\right\}}$

51. $\dfrac{8}{27} = x^{-3}$
$\left(\dfrac{2}{3}\right)^3 = x^{-3}$
$\left(\dfrac{2}{3}\right)^3 = \left(\dfrac{1}{x}\right)^3$
$\dfrac{2}{3} = \dfrac{1}{x}$ $\boxed{x = \left\{\dfrac{3}{2}\right\}}$

52. $10^{2x-3} = 17$
$\log 10^{2x-3} = \log 17$
$(2x-3)\log 10 = \log 17$
$2x - 3 = \log 17$
$2x = 3 + \log 17$
$x = \dfrac{3 + \log 17}{2}$
$x \approx 2.11522446069$ $\boxed{x \approx \{2.115\}}$

53. $e^{x+1} = 10$

$\ln e^{x+1} = \ln 10$

$(x+1)\ln e = \ln 10$

$x + 1 = \ln 10$

$x = -1 + \ln 10$

$x \approx 1.30258509299$ $\boxed{x \approx \{1.303\}}$

54. $\log_{64} x = \frac{1}{3}$

$x = 64^{1/3}$ $\boxed{x = \{4\}}$

55. $\ln 6x - \ln(x+1) = \ln 4$

$\ln \frac{6x}{x+1} = \ln 4$

$\frac{6x}{x+1} = 4$

$4x + 4 = 6x$

$4 = 2x$ $\boxed{x = \{2\}}$

56. $\log_{16} \sqrt{x+1} = \frac{1}{4}$

$16^{1/4} = \sqrt{x+1}$

$\left(16^{1/4}\right)^2 = x + 1$

$(16)^{1/2} = x + 1$

$4 = x + 1$

$x = 3$

Check $x = 3$: $\log_{16} \sqrt{3+1} = \frac{1}{4}$

$\log_{16} 2 = \frac{1}{4}$

$16^{1/4} = 2$ **Yes**

$\boxed{x = \{3\}}$

57. $\ln x + 3\ln 2 = \ln \frac{2}{x}$

$\ln x + \ln 2^3 = \ln \frac{2}{x}$

$\ln\left[(x)\left(2^3\right)\right] = \ln \frac{2}{x}$

$\ln(8x) = \ln \frac{2}{x}$

$8x = \frac{2}{x}$

$8x^2 = 2$

$x^2 = \frac{1}{4}$

$x = \pm\frac{1}{2}$ $\boxed{x = \left\{\frac{1}{2}\right\}}$

$-\frac{1}{2}$ is not in the domain of $\ln x$ or $\ln \frac{2}{x}$, so it is not a solution.

58. $\ln\left[\ln\left(e^{-x}\right)\right] = \ln 3$

$\ln[-x] = \ln 3$

$-x = 3$ $\boxed{x = \{-3\}}$

59. $N = a + b\ln\left(\frac{c}{d}\right)$

$N - a = b\ln\left(\frac{c}{d}\right)$

$\frac{N-a}{b} = \ln\left(\frac{c}{d}\right)$

$e^{(N-a)/b} = \frac{c}{d}$

$\boxed{c = de^{(N-a)/b}}$

60. $A = A_o\left(1 + \frac{r}{n}\right)^{nt}$

$\frac{A}{A_o} = \left(1 + \frac{r}{n}\right)^{nt}$

$\log \frac{A}{A_o} = \log\left(1 + \frac{r}{n}\right)^{nt}$

$\log \frac{A}{A_o} = nt \log\left(1 + \frac{r}{n}\right)$

$\frac{\log \frac{A}{A_o}}{n\log\left(1 + \frac{r}{n}\right)} = t$

61. $\log_{10} x = x - 2$; Graph $y = \log_{10} x - x + 2$:

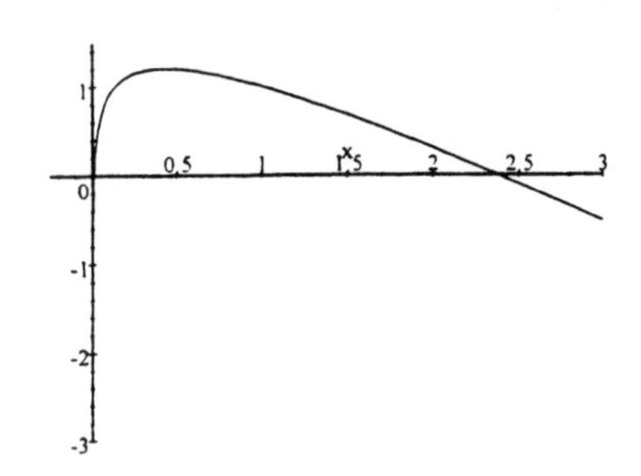

The x-intercepts are $\approx .01024,\ 2.3758$

Solution: $\boxed{\{.01,\ 2.376\}}$

62. $2^{-x} = \log_{10} x$; Graph $y = \log_{10} x - 2^{-x}$:

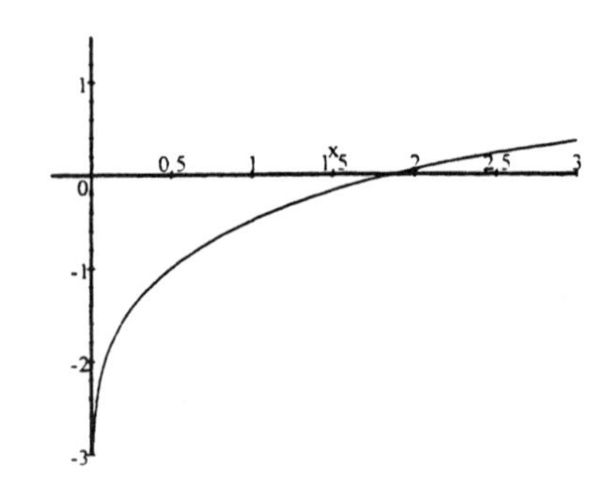

The x-intercept is ≈ 1.8741102

Solution: $\boxed{\{1.874\}}$

63. $x^2 - 3 = \log x$; Graph $y = \log x - x^2 + 3$:

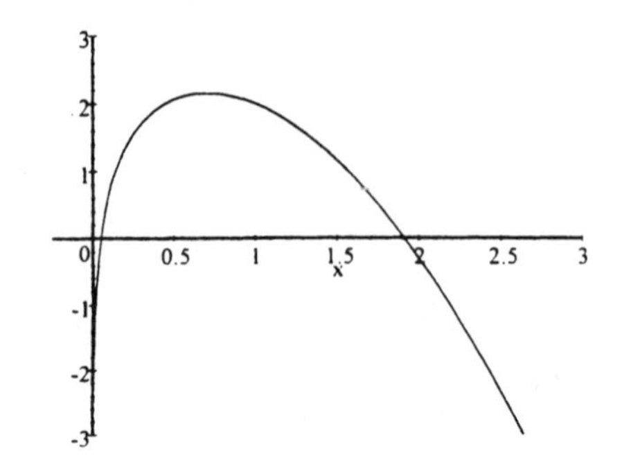

The x-intercepts are $.001, \approx 1.8045406$

Solution: $\boxed{\{.001,\ 1.805\}}$

64. $A = \$8780.\ P = \$3500,\ t = 10,\ m = 1$:

$$A = P\left(1 + \frac{r}{m}\right)^{tm}$$
$$8780 = 3500\,(1 + r)^{10}$$
$$\frac{8780}{3500} = (1 + r)^{10}$$
$$\left(\frac{8780}{3500}\right)^{.1} = 1 + r$$
$$\left(\frac{8780}{3500}\right)^{.1} - 1 = r$$
$$r = .096333405$$

The annual interest rate, to the nearest tenth, is $\boxed{9.6\%}$.

65. $A = \$58344,\ P = \$48000,$
$r = .05,\ m = 2$:

$$A = P\left(1 + \frac{r}{m}\right)^{tm}$$
$$58344 = 48000\left(1 + \frac{.05}{2}\right)^{t}$$
$$8344 = 48000\,(1.025)^{2t} \qquad 5$$
$$1.2155 = 1.025^{2t}$$
$$\ln 1.2155 = \ln 1.025^{2t}$$
$$\ln 1.2155 = 2t \ln 1.025$$
$$t = \frac{\ln 1.2155}{2 \ln 1.025} \approx 3.9516984$$
$$\boxed{\approx 4.0 \text{ years}}$$

66. **(i)** First 8 years: $P = \$12,000,$
$t = 8,\ r = .05,\ m = 1$:

$$A = P\left(1 + \frac{r}{m}\right)^{tm}$$
$$= 12000\,(1 + .05)^{8}$$
$$= \$17729.47$$

(ii) Second 6 years: $P = \$17729.47,$
$t = 6,\ r = .06,\ m = 1$:

$$A = P\left(1 + \frac{r}{m}\right)^{tm}$$
$$= 17729.47\,(1 + .06)^{6}$$
$$= \$25149.59$$

At the end of the 14 year period, about $\boxed{\$25,149.59}$ would be in the account.

67. **(a)** $A = P\left(1 + \frac{r}{n}\right)^{nt}$

$A = 2000\left(1 + \frac{.03}{4}\right)^{(4)(5)}$

$\boxed{A = \$2322.37}$

(b) $A = Pe^{rt}$

$A = 2000e^{(.03)(5)}$

$\boxed{A = \$2323.67}$

(c) $6000 = 2000e^{.03t}$

$3 = e^{.03t}$

$\ln 3 = .03t$

$t = \frac{\ln 3}{.03}$

$t \approx 36.6204$

$\boxed{t \approx 36.6 \text{ years}}$

68. **(a)** $L(x) = 3\log x$

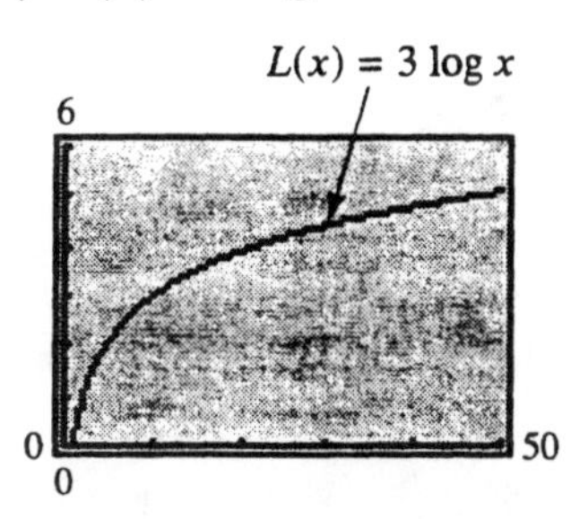

Since L is increasing, heavier planes require longer runways.

(b) $L(10) = 3\log 10 = 3 : 3000$ feet

$L(100) = 3\log 100 = 6 : 6000$ feet

No, it increases by a factor of 2 to 6000 feet.

69. $f(x) = 280\ln(x+1) + 1925$

$2200 = 280\ln(x+1) + 1925$

$275 = 280\ln(x+1)$

$\frac{275}{280} = \ln(x+1)$

$x + 1 = e^{275/280}$

$x = -1 + e^{275/280}$

$\approx \boxed{1.67 \text{ acres}}$

70. $P(x) = .04e^{-4x}$

(a) $P(.5) = .04e^{-4(.5)}$

$\boxed{P(.5) \approx .0054 \text{ grams/liter}}$

(b) $P(1) = .04e^{-4(1)}$

$\boxed{P(1) \approx .00073 \text{ grams/liter}}$

(c) $P(2) = .04e^{-4(2)}$

$\boxed{P(2) \approx .000013 \text{ grams/liter}}$

(d) $.002 = .04e^{-4x}$

$.05 = e^{-4x}$

$\ln .05 = -4x$

$x = \frac{\ln .05}{-4}$

$\boxed{x \approx .75 \text{ miles}}$

71. $p(x) = 250 - 120(2.8)^{-.5x}$

(a) $p(2) = 250 - 120(2.8)^{-.5(2)}$

$\boxed{p(2) \approx 207}$

(b) $p(10) = 250 - 120(2.8)^{-.5(10)}$

$\boxed{p(10) \approx 249}$

(c) $x = 2,\ y \approx 207$

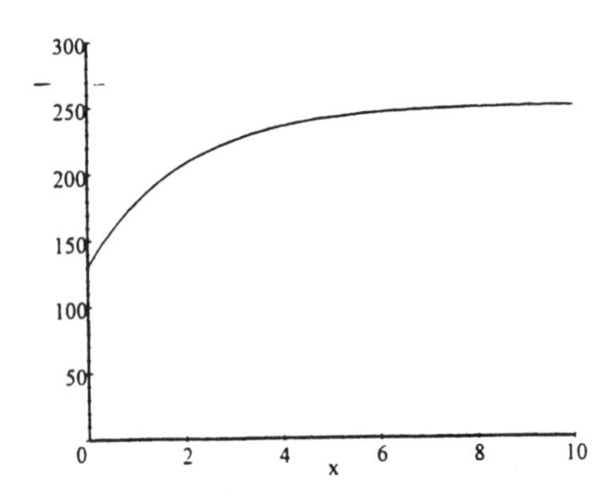

72. $v(t) = 176\left(1 - e^{-.18t}\right)$

Find the time, t, where $v(t) = 147$:

$$147 = 176\left(1 - e^{-.18t}\right)$$

$$\frac{147}{176} = 1 - e^{-.18t}$$

$$e^{-.18t} = 1 - \frac{147}{176}$$

$$e^{-.18t} = \frac{29}{176}$$

$$\ln e^{-.18t} = \ln\frac{29}{176}$$

$$-.18t = \ln\frac{29}{176}$$

$$t = \frac{\ln\frac{29}{176}}{-.18}$$

$$t \approx 10.017712$$

It will take the skydiver about 10 seconds to attain the speed of 147 ft/sec (100 mph).

Chapter 5 Test

1. (a) $y = \log_{1/2} x$: **B**

(b) $y = e^x$: **A**

(c) $y = \ln x$: **C**

(d) $y = \left(\frac{1}{2}\right)^x$: **D**

(e) inverse functions: (a) and (d), (b) and (c)

2. (a) $f(x) = -2^{x-1} + 8$

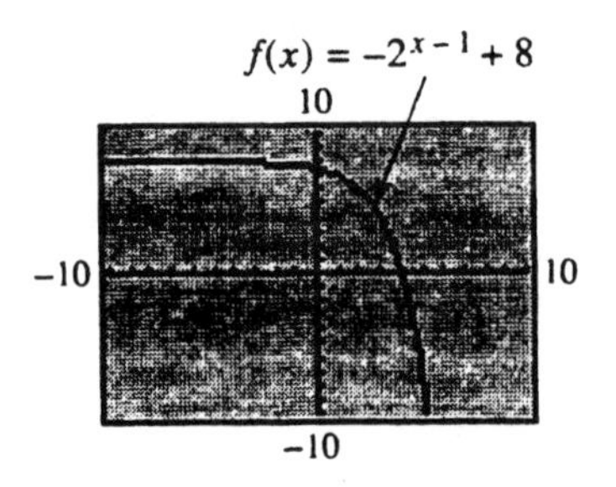

(b) Domain: $(-\infty, \infty)$; Range: $(\infty, 8)$

(c) Yes, it has a horizontal asymptote. $(y = 8)$

(d) x-intercept: let $y = 0$:

$$0 = -2^{x-1} + 8$$
$$2^{x-1} = 8$$
$$2^{x-1} = 2^3$$
$$x - 1 = 3$$
$$x = 4 \qquad \boxed{(4,\ 0)}$$

y-intercept: let $x = 0$:

$$y = -2^{-1} + 8$$
$$y = -\tfrac{1}{2} + 8$$
$$y = 7.5 \qquad \boxed{(0,\ 7.5)}$$

(e) $x = -2^{y-1} + 8$

$$x - 8 = -2^{y-1}$$
$$8 - x = 2^{y-1}$$
$$\log(8 - x) = \log 2^{y-1}$$
$$\log(8 - x) = (y - 1)\log 2$$
$$\log(8 - x) = y\log 2 - 1\log 2$$
$$\log(8 - x) + \log 2 = y\log 2$$
$$\frac{\log(8 - x) + \log 2}{\log 2} = y$$

$$\boxed{f^{-1}(x) = \frac{\log(8 - x) + \log 2}{\log 2} \quad \text{or} \quad \frac{\log(8 - x)}{\log 2} + 1}$$

3. (a) $\left(\frac{1}{8}\right)^{2x-3} = 16^{x+1}$

$\left(2^{-3}\right)^{2x-3} = \left(2^{4}\right)^{x+1}$

$2^{-6x+9} = 2^{4x+4}$

$-6x + 9 = 4x + 4$

$-10x = -5$

$x = \frac{1}{2}$ $\boxed{\{.5\}}$

4. (a) $A = P\left(1 + \frac{r}{n}\right)^{nt}$

Find A when $P = \$10,000$,
$r = 5.5\%,\ t = 4,\ n = 4$

$A = 10,000\left(1 + \frac{.055}{4}\right)^{4(4)}$

$\boxed{\$12,442.11}$

(b) $A = Pe^{rt}$

Find A when $P = \$10,000$,
$r = 5.5\%,\ t = 4$

$A = 10,000e^{.055(4)}$

$\boxed{\$12,460.77}$

5. The expression $\log_5 27$ is the exponent to which 5 must be raised in order to obtain 27. To find an approximation with the calculator, we use the change of base rule: $\log_5 27 = \frac{\log 27}{\log 5}$.

6. (a) $\log 45.6 \approx 1.659$

(b) $\ln 470 \approx 6.153$

(c) $\log_3 769 = \frac{\log 769}{\log 3} \approx 6.049$

7. $\log \frac{m^3 n}{\sqrt{y}} = \log m^3 + \log n - \log \sqrt{y}$

$= \log m^3 + \log n - \log y^{1/2}$

$= 3\log m + \log n - \frac{1}{2}\log y$

8. (a) $\log_2 x + \log_2 (x+2) = 3;\ x > 0$

$\log_2 [x(x+2)] = 3$

$x(x+2) = 2^3$

$x^2 + 2x - 8 = 0$

$(x+4)(x-2) = 0$

$x = -4,\ 2$ $\boxed{\{2\}}$

Since $x > 0$, -4 is extraneous.

(b) $y_1 = \log_2 x + \log_2 (x+2) - 3$

$= \frac{\log x}{\log 2} + \frac{\log (x+2)}{\log 2} - 3$

$y_1 = \log_2 x + \log_2 (x+2) - 3$

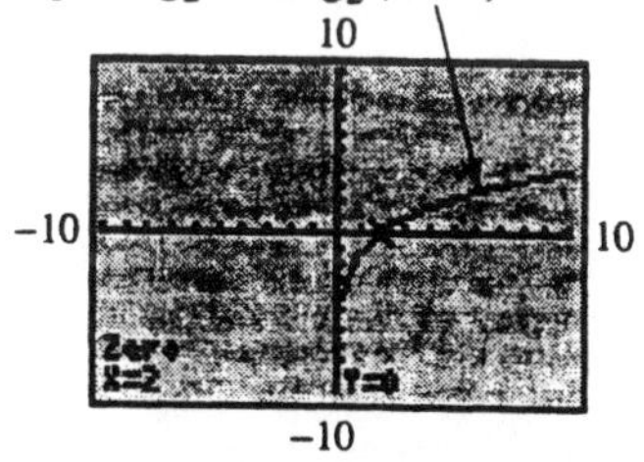

(c) $\log_2 x + \log_x (x+2) > 3 : (2,\ \infty)$

9. $6^{2-x} = 2^{3x+1}$

Take the log of both sides:

$\log 6^{2-x} = \log 2^{3x+1}$

$(2-x)\log 6 = (3x+1)\log 2$

$2\log 6 - x\log 6 = 3x\log 2 + \log 2$

$-x\log 6 - 3x\log 2 = \log 2 - 2\log 6$

$-x(\log 6 + 3\log 2) = \log 2 - 2\log 6$

$x = \frac{2\log 6 - \log 2}{\log 6 + 3\log 2} = \frac{\log 6^2 - \log 2}{\log 6 + \log 2^3}$

$= \frac{\log \frac{36}{2}}{\log (6 \cdot 8)} = \frac{\log 18}{\log 48}$

$x \approx .74663438 \approx \boxed{.747}$

10. $y = 2e^{.02t}$

(a) To find out how long it will take for the population to triple, you need to solve $2e^{.02t} = 3(2)$ for t. The correct choice is **B.**

(b) To find out when the population will reach 3 million, you need to solve $2e^{.02t} = 3$ for t. The correct choice is **D.**

(c) To find out how large the population will be in 3 years, you need to evaluate $2e^{.02(3)}$. The correct choice is **C.**

(d) To find out how large the population will be in 4 months, you need to evaluate $2e^{.02(1/3)}$. The correct choice is **A.**

11. **(a)** $A(t) = t^2 - t + 350$ (dark)

(b) $A(t) = 350 \log(t+1)$ (light)

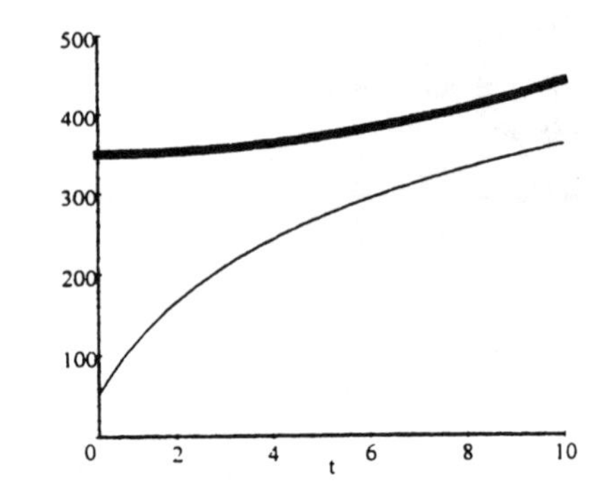

(c) $A(t) = 350(.75)^t$ (dark)

(d) $A(t) = 100(.95)^t$ (light)

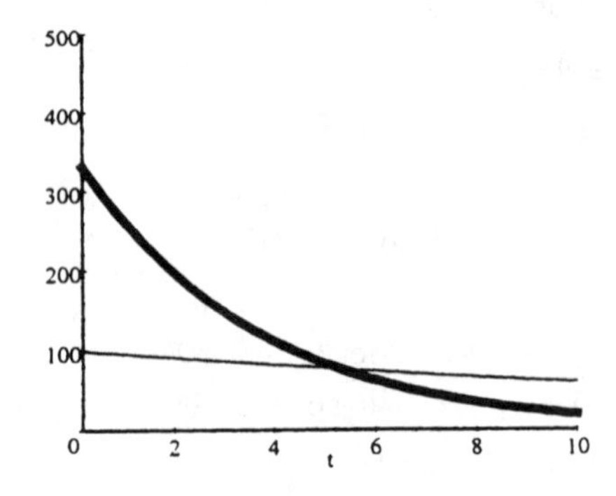

Function **(c)** best describes $A(t)$.

12. $A(t) = A_o e^{-kt}$

(a) $\frac{1}{2}A_o = A_o e^{-1600k}$

$.5 = e^{-1600k}$

$-1600k = \ln .5$

$k = \dfrac{\ln .5}{-1600} \approx .000433$

$A(t) = 2e^{-.000433\,t}$

(b) $A(9600) = 2e^{-.000433(9600)} \approx .03$ gram

(c) $.5 = 2e^{-.000433\,t}$

$.25 = e^{-.000433\,t}$

$-.000433t = \ln .25$

$t = \dfrac{\ln .25}{-.000433} \approx 3201.6$

About 3200 years

Chapter 5 Project

Modeling Motor Vehicle Sales in the United States

1. $(0,\ 13.1) : 13.1 = a\,(b)^0 \longrightarrow a = 13.1$

$(7,\ 17.4) :\ 17.4 = 13.1\,b^7$

$$\frac{17.4}{13.1} = b^7$$

$$b = \left(\frac{17.4}{13.1}\right)^{1/7} \approx 1.04138$$

$f(x) = 13.1\,(1.04138)^x$

2. $(0,\ 4.9) : 4.9 = c\,(d)^0 \longrightarrow c = 4.9$

$(7,\ 8.7) :\ 8.7 = 4.9\,d^7$

$$\frac{8.7}{4.9} = d^7$$

$$d = \left(\frac{8.7}{13.1}\right)^{1/7} \approx 1.08547$$

$g(x) = 4.9\,(1.08547)^x$

3. $f(x)$: Since $b = 1.04138 \longrightarrow 4.1\%$ increase in new motor vehicle sales.

$g(x)$: Since $b = 1.08547 \longrightarrow 8.5\%$ increase in new truck sales.

4. $1996 : x = 4$

$f(4) = 13.1\,(1.04138)^4 \approx 15.41$

≈ 15.4 million new motor vehicle sales

$g(4) = 4.9\,(1.08547)^4 \approx 6.8$

≈ 6.8 million new truck sales

These estimates are quite close to the actual values of 15.5 million and 6.9 million.

5. $2020 \longrightarrow x = 28$

$f(28) = 13.1\,(1.04138)^{28} \approx 40.769$

≈ 40.8 million new motor vehicle sales

$g(28) = 4.9\,(1.08547)^{28} \approx 48.696$

≈ 48.7 million new truck sales

New truck sales cannot exceed new motor vehicle sales since they are included in the motor vehicle sales.

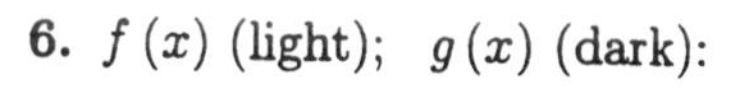

6. $f(x)$ (light); $g(x)$ (dark):

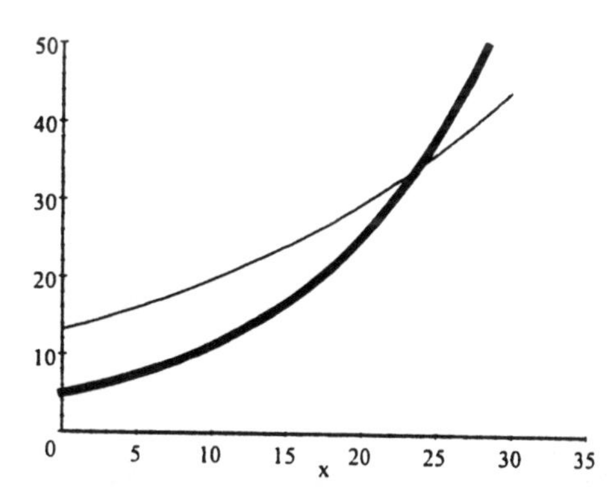

No, because over time, truck sales begin to exceed motor vehicle sales, which is impossible. Modeling functions generally provide better results when estimating values between data points rather than predicting values very far into the future.

Section 6.1

1. $(x-3)^2+(y-3)^2=0$ is the graph of a circle, center (3, 3) with radius 0, so it is the graph of the point (3, 3).

3. $x=2(y+3)^2-4$: **D**

5. $x^2=-3y$: **E**

7. $x^2+y^2=25$: **H**

9. $(x+3)^2+(y-4)^2=25$: **B**

11. Center (1, 4), $r=3$:

$$(x-h)^2+(y-k)^2=r^2$$

$$\boxed{(x-1)^2+(y-4)^2=9}$$

13. Center (0, 0), $r=1$:

$$x^2+y^2=r^2$$

$$\boxed{x^2+y^2=1}$$

15. Center $\left(\frac{2}{3}, -\frac{4}{5}\right)$, $r=\frac{3}{7}$:

$$(x-h)^2+(y-k)^2=r^2$$

$$\boxed{\left(x-\frac{2}{3}\right)^2+\left(y+\frac{4}{5}\right)^2=\frac{9}{49}}$$

17. Center (−1, 2), through (2, 6) :

$$(x-h)^2+(y-k)^2=r^2$$

$$(x+1)^2+(y-2)^2=r^2$$

$$(2,\ 6) : (2+1)^2+(6-2)^2=r^2$$

$$9+16=25=r^2$$

$$\boxed{(x+1)^2+(y-2)^2=25}$$

19. Center (−3, −2), tangent to x–axis. If the circle goes through a point on the x-axis, that point would be (−3, 0). The distance from (−3, −2) to (−3, 0) is the radius:

$$r=\sqrt{(-3-(-3))^2+(-2+0)^2}$$

$$r=\sqrt{4}=2$$

$$(x-h)^2+(y-k)^2=r^2$$

$$\boxed{(x+3)^2+(y+2)^2=4}$$

Relating Concepts (21 - 23)

21. To find the center, find the midpoint:

Midpoint: (−1, 3) and (5, −9)

$$\left(\frac{-1+5}{2}, \frac{3-9}{2}\right) \quad \boxed{(2,-3)}$$

22. Length of diameter:

$$d=\sqrt{(-1-5)^2+(3-(-9))^2}$$

$$d=\sqrt{(-6)^2+12^2}$$

$$d=\sqrt{180}$$

$$d=6\sqrt{5}$$

$$\text{Radius}=\frac{1}{2}d=\boxed{3\sqrt{5}}$$

23. $(x-h)^2+(y-k)^2=r^2$

$$(x-2)^2+(y+3)^2=\left(3\sqrt{5}\right)^2$$

$$\boxed{(x-2)^2+(y+3)^2=45}$$

25. $x^2+y^2=36$

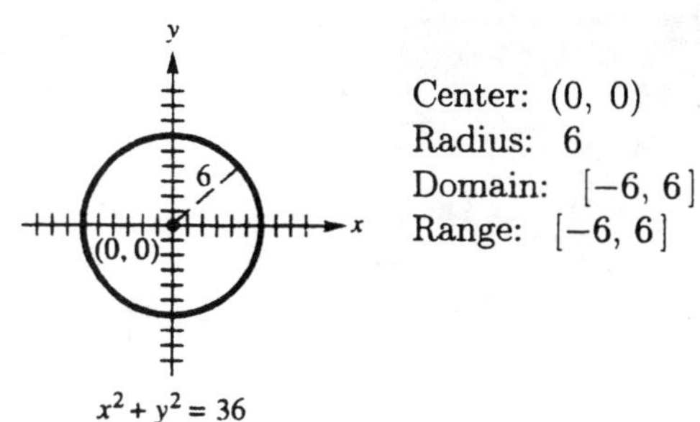

Center: (0, 0)
Radius: 6
Domain: [−6, 6]
Range: [−6, 6]

27. $(x+2)^2+(y-5)^2=16$

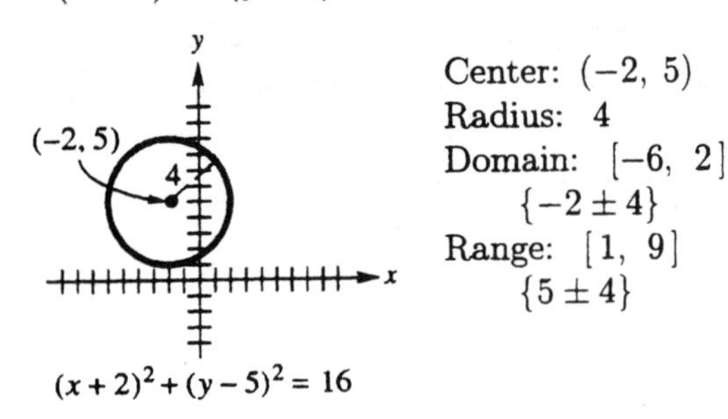

Center: $(-2,\ 5)$
Radius: 4
Domain: $[-6,\ 2]$
$\{-2 \pm 4\}$
Range: $[1,\ 9]$
$\{5 \pm 4\}$

29. $(x-4)^2+(y-3)^2=25$

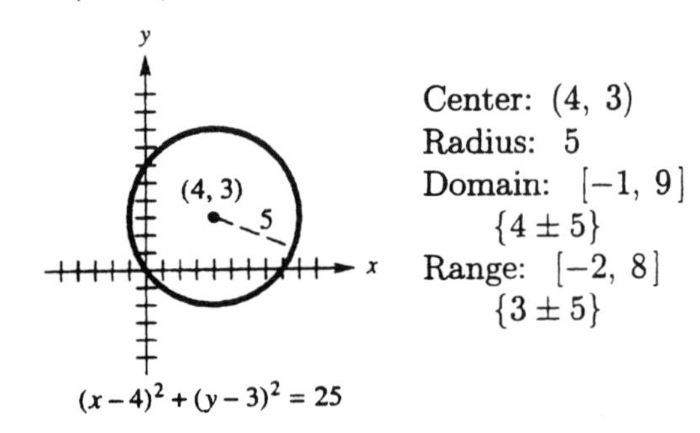

Center: $(4,\ 3)$
Radius: 5
Domain: $[-1,\ 9]$
$\{4 \pm 5\}$
Range: $[-2,\ 8]$
$\{3 \pm 5\}$

31. $x^2+y^2=81$

$$y^2 = 81 - x^2$$
$$y = \pm\sqrt{81-x^2}$$

Graph $y_1 = \sqrt{81-x^2}$
$y_2 = -\sqrt{81-x^2}$

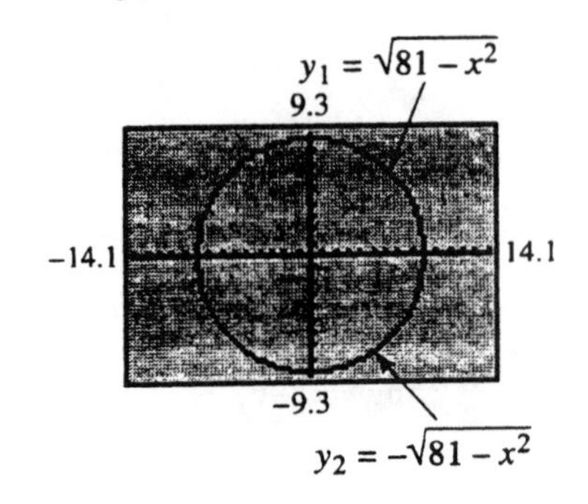

Domain: $[-9,\ 9]$; Range: $[-9,\ 9]$

33. $(x-3)^2+(y-2)^2=25$

$$(y-2)^2 = 25-(x-3)^2$$
$$y-2 = \pm\sqrt{25-(x-3)^2}$$
$$y = \pm\sqrt{25-(x-3)^2}+2$$

Graph $y_1 = 2+\sqrt{25-(x-3)^2}$
$y_2 = 2-\sqrt{25-(x-3)^2}$

$y_1 = 2+\sqrt{25-(x-3)^2}$
8.2
−9.4
9.4
−4.2
$y_2 = 2-\sqrt{25-(x-3)^2}$

Domain: $[3 \pm 5] = [-2,\ 8]$
Range: $[2 \pm 5] = [-3,\ 7]$

35. $x^2+6x+y^2+8y+9=0$

$$(x^2+6x+9)+(y^2+8y+16) = -9+9+16$$
$$(x+3)^2+(y+4)^2 = 16$$

Yes; **Center: $(-3,-4)$, $r=4$**

37. $x^2-4x+y^2+12y=-4$

$$(x^2-4x+4)+(y^2+12y+36) = -4+4+36$$
$$(x-2)^2+(y+6)^2 = 36$$

Yes; **Center: $(2,-6)$, $r=6$**

39. $4x^2+4x+4y^2-16y-19=0$

$$4(x^2+x)+4(y^2-4y) = 19$$
$$4\left(x^2+x+\frac{1}{4}\right)+4(y^2-4y+4) = 19+4\left(\frac{1}{4}\right)+4(4)$$
$$4\left(x+\frac{1}{2}\right)^2+4(y-2)^2 = 36$$
$$\left(x+\frac{1}{2}\right)^2+(y-2)^2 = 9$$

Yes; **Center: $\left(-\frac{1}{2},\ 2\right)$, Radius: 3**

41. $x^2 + 2x + y^2 - 6y + 14 = 0$

$(x^2 + 2x + 1) + (y^2 - 6y + 9) = -14 + 1 + 9$

$(x + 1)^2 + (y - 3)^2 = -4$

No, it is not a circle.

43. $x^2 - 2x + y^2 + 4y + 7 = 0$

$(x^2 - 2x + 1) + (y^2 + 4y + 4) = -7 + 1 + 4$

$(x - 1)^2 + (y + 2)^2 = -2$

No, it is not a circle.

45. $C_1: \ (x - 2)^2 + (y - 1)^2 = 25$

$C_2: \ (x + 2)^2 + (y - 2)^2 = 16$

$C_3: \ (x - 1)^2 + (y + 2)^2 = 9$

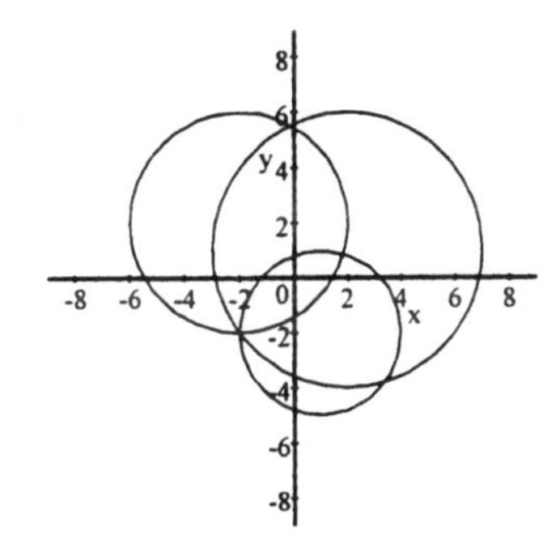

It appears that the circles intersect at $(-2, -2)$:

$C_1: \ (-2 - 2)^2 + (-2 - 1)^2 = 25$

$(-4)^2 + (-3)^2 = 25$

$16 + 9 = 25 \quad \checkmark$

$C_2: \ (-2 + 2)^2 + (-2 - 2)^2 = 16$

$0^2 + (-4)^2 = 16$

$16 = 16 \quad \checkmark$

$C_3: \ (-2 - 1)^2 + (-2 + 2)^2 = 9$

$(-3)^2 + 0^2 = 9$

$9 = 9 \quad \checkmark$

The epicenter is at $(-2, -2)$.

47. $y = (x - 2)^2 - 4$: **B**

49. $y = -(x - 2)^2 - 4$: **A**

51. $x = (y - 2)^2 - 4$: **H**

53. $x = -(y - 2)^2 - 4$: **G**

55. $x = a(y - k)^2 + h$

(a) Quad III (b) Quad II

(c) Quad IV (d) Quad I

57. $x^2 = 4y; \quad 4c = 4, \quad c = 1$

Vertex: (0, 0), opens up.

Focus: (0, 1)
Directrix: $y = -1$
Axis: y-axis

59. $x^2 = \frac{1}{9}y; \quad 4c = \frac{1}{9}, \ c = \frac{1}{36}$

Vertex: (0, 0), opens up.

Focus: $\left(0, \frac{1}{36}\right)$
Directrix: $y = -\frac{1}{36}$
Axis: y-axis

61. $y^2 = -\frac{1}{32}x; \quad 4c = -\frac{1}{32}, \quad c = -\frac{1}{128}$

Vertex: (0, 0), opens left.

Focus: $\left(-\frac{1}{128}, 0\right)$
Directrix: $x = \frac{1}{128}$
Axis: x-axis

63. $y^2 = -4x; \quad 4c = -4, \quad c = -1$

Vertex: (0, 0), opens left.

Focus: $(-1, 0)$
Directrix: $x = 1$
Axis: x-axis

65. Focus (5, 0); $c = 5$, opens right.

$y^2 = 4cx = 4(5)x$

$y^2 = 20x$

67. Focus $\left(0, \frac{1}{4}\right)$; $c = \frac{1}{4}$, opens up.

$x^2 = 4cy = 4\left(\frac{1}{4}\right)y$

$x^2 = y$

69. Through $(\sqrt{3}, 3)$, opens up.

$x^2 = 4cy$

$(\sqrt{3})^2 = 4c(3)$

$3 = 12c \Longrightarrow c = \frac{1}{4}$

$x^2 = 4\left(\frac{1}{4}\right)y \qquad \boxed{x^2 = y}$

71. Through $(-3, 3)$, opens left.

$y^2 = 4cx$

$3^2 = 4c(-3)$

$9 = -12c \Longrightarrow c = -\frac{3}{4}$

$y^2 = 4\left(-\frac{3}{4}\right)x \qquad \boxed{y^2 = -3x}$

73. Through $(3, 2)$, symmetric with respect to the x-axis.

$y^2 = 4cx \qquad (2)^2 = 4c(3)$

$4 = 12c \Longrightarrow c = \frac{1}{3}$

$y^2 = 4\left(\frac{1}{3}\right)x \qquad \boxed{y^2 = \frac{4}{3}x}$

75. $y = (x-5)^2 - 4$

Vertex: $(5, -4)$ Domain: $(-\infty, \infty)$
Axis: $x = 5$ Range: $[-4, \infty)$

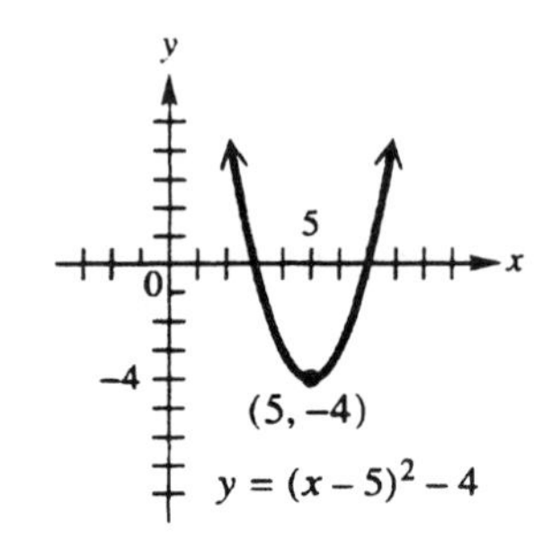

77. $y = \frac{2}{3}(x-2)^2 - 1$

Vertex: $(2, -1)$ Domain: $(-\infty, \infty)$
Axis: $x = 2$ Range: $[-1, \infty)$

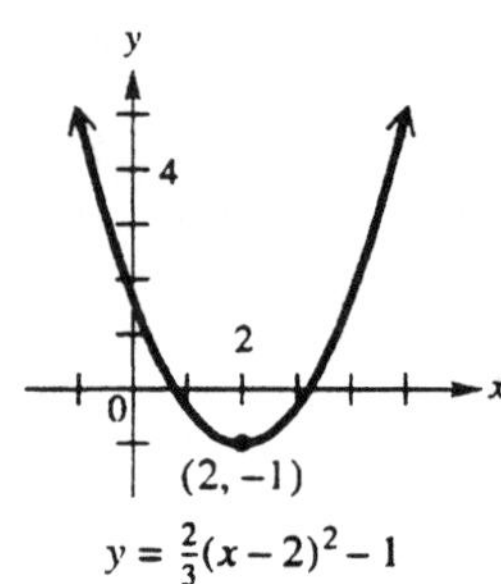

79. $y = x^2 + 6x + 5$

$y = (x^2 + 6x + 9) - 9 + 5$

$y = (x+3)^2 - 4$

Vertex: $(-3, -4)$ Domain: $(-\infty, \infty)$
Axis: $x = -3$ Range: $[-4, \infty)$

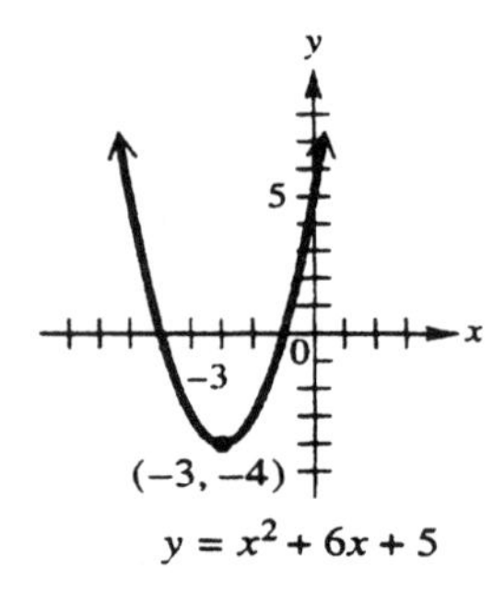

81. $y = -3x^2 + 24x - 46$

$y = -3(x^2 - 8x + 16 - 16) - 46$

$y = -3(x^2 - 8x + 16) + 48 - 46$

$y = -3(x-4)^2 + 2$

Vertex: $(4, 2)$ Domain: $(-\infty, \infty)$
Axis: $x = 4$ Range: $(-\infty, 2]$

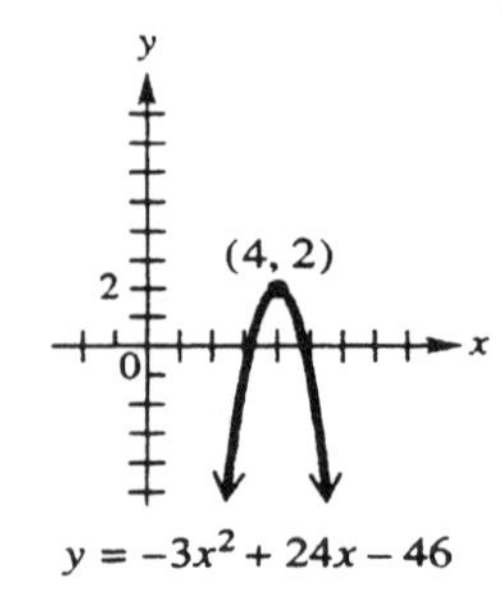

83. $x = (y+1)^2$

Vertex: $(0, -1)$ Domain: $[0, \infty)$

Axis: $y = -1$ Range: $(-\infty, \infty)$

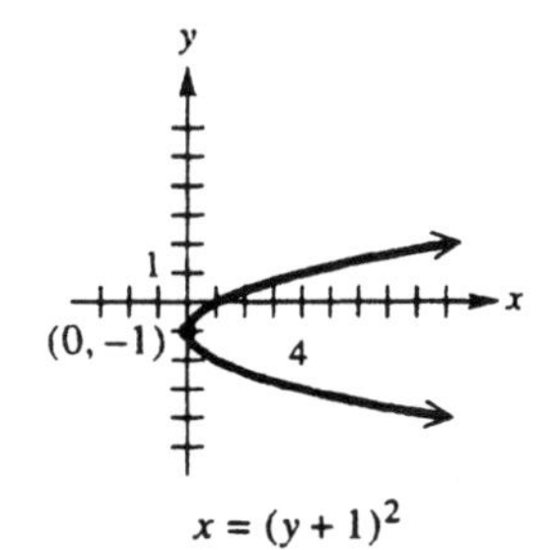

$x = (y+1)^2$

85. $x = (y+2)^2 - 1$

Vertex: $(-1, -2)$ Domain: $[-1, \infty)$

Axis: $y = -2$ Range: $(-\infty, \infty)$

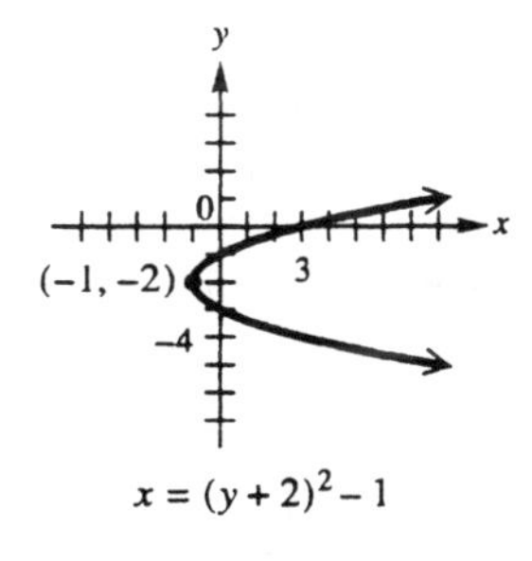

$x = (y+2)^2 - 1$

87. $x = -2(y+3)^2$

Vertex: $(0, -3)$ Domain: $(-\infty, 0]$

Axis: $y = -3$ Range: $(-\infty, \infty)$

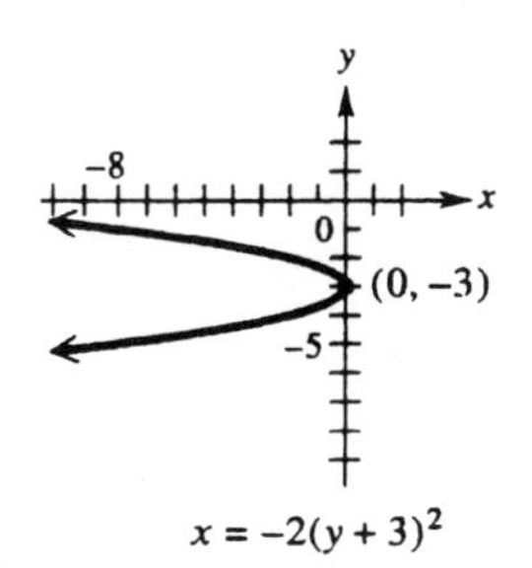

$x = -2(y+3)^2$

89. $x = y^2 + 2y - 8$

$x = (y^2 + 2y + 1) - 1 - 8$

$x = (y+1)^2 - 9$

Vertex: $(-9, -1)$ Domain: $[-9, \infty)$

Axis: $y = -1$ Range: $(-\infty, \infty)$

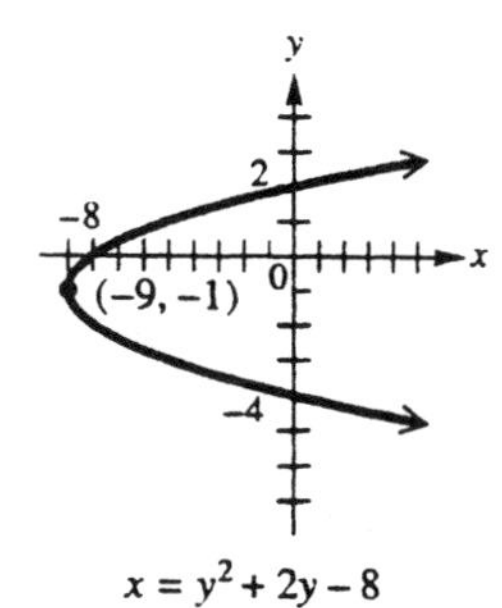

$x = y^2 + 2y - 8$

91. $x = 2y^2 - 4y + 6$

$x = 2(y^2 - 2y + 1 - 1) + 6$

$x = 2(y-1)^2 + 2(-1) + 6$

$x = 2(y-1)^2 + 4$

Vertex: $(4, 1)$ Domain: $[4, \infty)$

Axis: $y = 1$ Range: $(-\infty, \infty)$

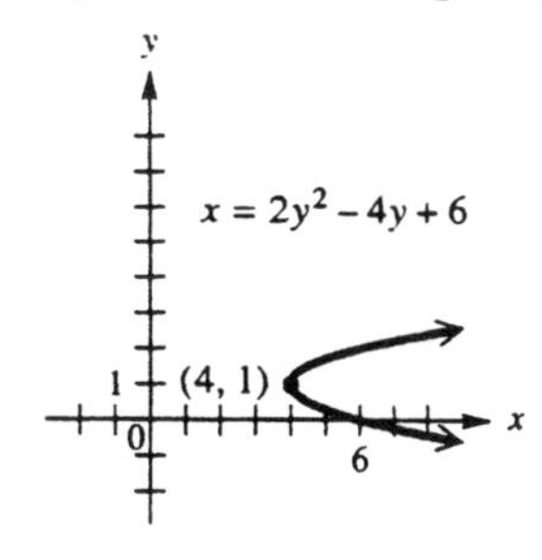

93. $2x = y^2 - 4y + 6$

$x = \frac{1}{2}y^2 - 2y + 3$

$x = \frac{1}{2}(y^2 - 4y + 4 - 4) + 3$

$x = \frac{1}{2}(y-2)^2 + \frac{1}{2}(-4) + 3$

$x = \frac{1}{2}(y-2)^2 + 1$

Vertex: $(1, 2)$ Domain: $[1, \infty)$

Axis: $y = 2$ Range: $(-\infty, \infty)$

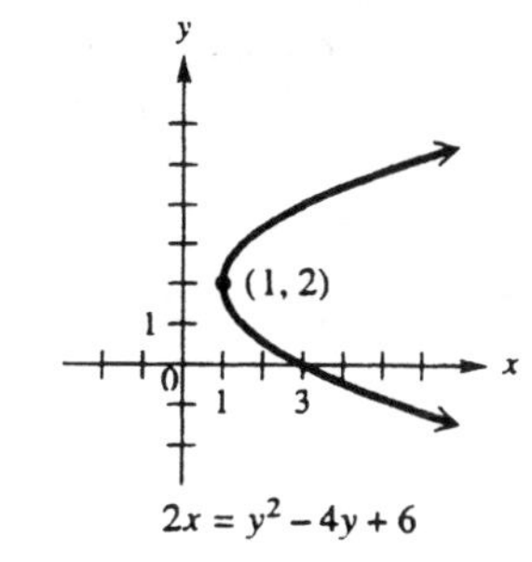

$2x = y^2 - 4y + 6$

95. $x = -3y^2 + 6y - 1$
$x = -3\left(y^2 - 2y + 1 - 1\right) - 1$
$x = -3\left(y - 1\right)^2 - 3\left(-1\right) - 1$
$x = -3\left(y - 1\right)^2 + 2$

Vertex: $(2,\ 1)$ Domain: $(-\infty,\ 2]$
Axis: $y = 1$ Range: $(-\infty,\ \infty)$

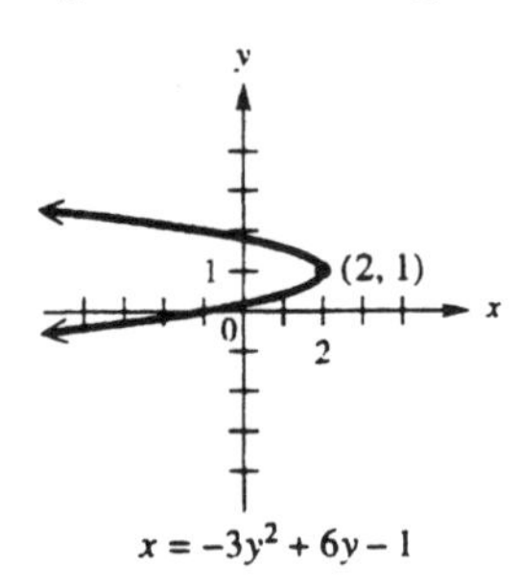

97. (a) $y = x - \dfrac{g}{1922}x^2$

For Earth, $g = 32.2$, so the equation is $y_1 = x - \dfrac{32.2}{1922}x^2 = x - \dfrac{16.1}{961}x^2$

For Mars, $g = 12.6$, so the equation is $y_2 = x - \dfrac{12.6}{1922}x^2 = x - \dfrac{6.3}{961}x^2$

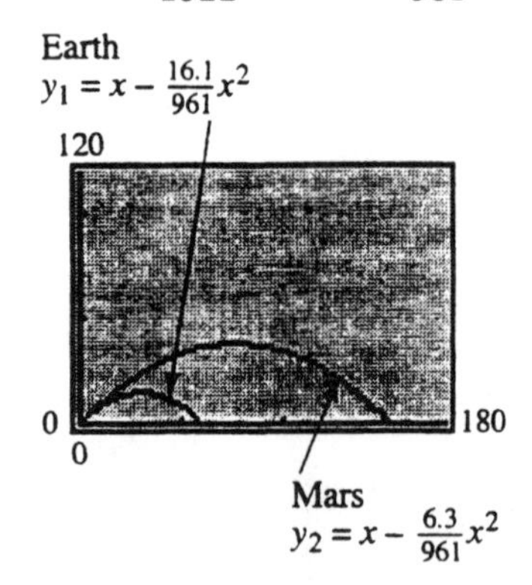

(b) From the graph, we see that the ball hits the ground (y-coordinate returns to 0) at $x \approx 153$ feet on Mars and $x \approx 60$ feet on Earth. Therefore, the difference between the horizontal distance traveled between the two balls is approximately $153 - 60 \approx \boxed{93 \text{ feet}}$.

99. Sketch a cross-section of the dish. Place this parabola on a coordinate system with the vertex at the origin. (This solution will show that the focus lies outside of the dish.)

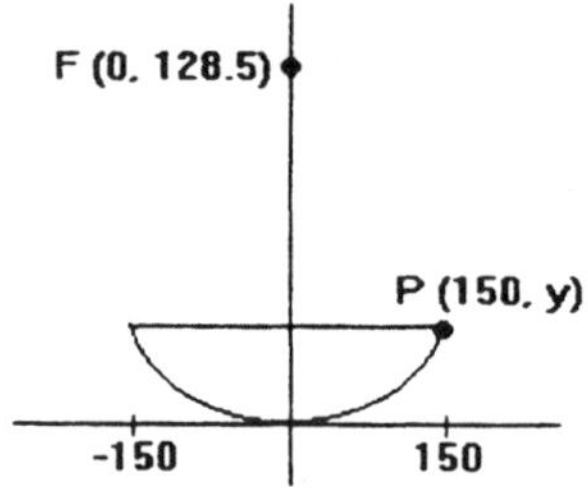

Since the parabola has vertex $(0, 0)$ and a vertical axis (the y-axis), it has an equation of the form

$$x^2 = 4cy.$$

Since the focus is $(0,\ 128.5)$, we have $c = 128.5$, and the equation is

$$x^2 = 4\left(128.5\right)y$$
$$x^2 = 514y.$$

Since the diameter (distance across the top) of the dish is 300 ft, the radius (distance halfway across the top) is 150 ft. To find the maximum depth of the dish, find the y-coordinate of point P, which lies on the parabola and has an x-coordinate of 150.

$$y = \frac{1}{514}\left(150\right)^2 \approx 43.8$$

The maximum depth of the parabolic dish is $\boxed{\text{approximately } 43.8 \text{ feet}}$.

101. Equation of the form:

$$y = a\left(x - h\right)^2 + k.$$

Let vertex be $(0,\ 10)$:

$$y = ax^2 + 10$$

through $(200,\ 210)$

$$210 = a\left(200\right)^2 + 10$$
$$200 = a\left(200\right)^2 \rightarrow a = \frac{1}{200}$$

EQ: $y = \dfrac{1}{200}x^2 + 10$;

find $(100,\ y)$:

$$y = \frac{1}{200}\left(100\right)^2 + 10$$
$$y = 50 + 10 = 60$$

$\boxed{\text{Height} = 60 \text{ feet}}$

Section 6.2

1. $\dfrac{y^2}{16} + \dfrac{x^2}{4} = 1$: **C**

3. $\dfrac{x^2}{4} - \dfrac{y^2}{16} = 1$: **F**

5. $\dfrac{(y-4)^2}{25} + \dfrac{(x+2)^2}{9} = 1$: **A**

7. $\dfrac{(x+2)^2}{9} - \dfrac{(y-4)^2}{25} = 1$: **G**

9. A circle can be interpreted as an ellipse whose 2 foci have the same coordinates; these *coinciding foci* are the center of the circle.

11. $\dfrac{x^2}{9} + \dfrac{y^2}{4} = 1$

    $c = \pm\sqrt{9-4} = \pm\sqrt{5}$

    Foci: $(-\sqrt{5},\ 0)$, $(\sqrt{5},\ 0)$

    Domain: $[-3, 3]$; Range: $[-2, 2]$

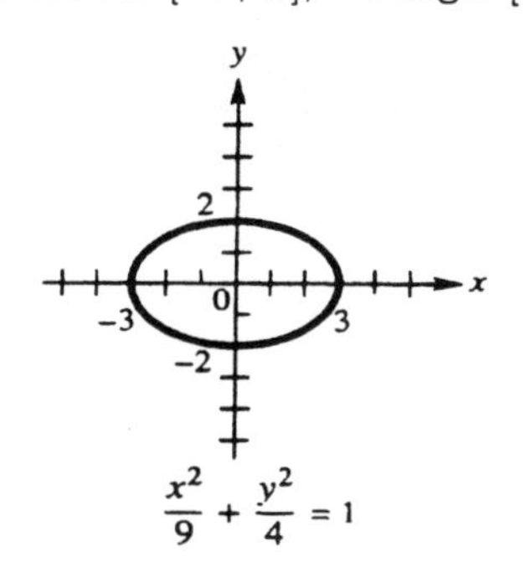

$\dfrac{x^2}{9} + \dfrac{y^2}{4} = 1$

13. $9x^2 + 6y^2 = 54 \rightarrow \dfrac{x^2}{6} + \dfrac{y^2}{9} = 1$

    $c = \pm\sqrt{9-6} = \pm\sqrt{3}$

    Foci: $(0,\ -\sqrt{3})$, $(0,\ \sqrt{3})$

    Domain: $[-\sqrt{6},\ \sqrt{6}]$; Range: $[-3, 3]$

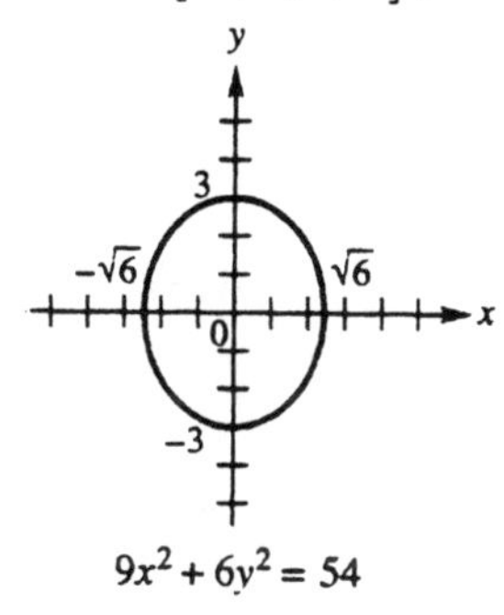

$9x^2 + 6y^2 = 54$

15. $\dfrac{25y^2}{36} + \dfrac{64x^2}{9} = 1 \rightarrow \dfrac{y^2}{\frac{36}{25}} + \dfrac{x^2}{\frac{9}{64}} = 1$

    Domain: $\left[-\dfrac{3}{8},\ \dfrac{3}{8}\right]$; Range: $\left[-\dfrac{6}{5},\ \dfrac{6}{5}\right]$

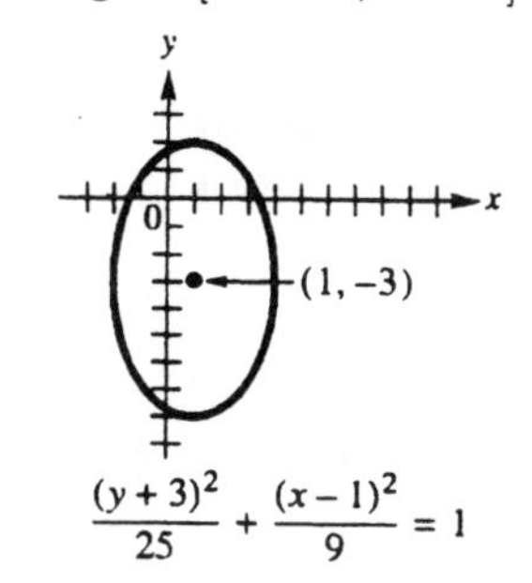

$\dfrac{25y^2}{36} + \dfrac{64x^2}{9} = 1$

17. $\dfrac{(y+3)^2}{25} + \dfrac{(x-1)^2}{9} = 1$

    Center: $(1, -3)$

    Domain: $[-3+1,\ 3+1] = [-2, 4]$

    Range: $[-5-3,\ 5-3] = [-8, 2]$

$\dfrac{(y+3)^2}{25} + \dfrac{(x-1)^2}{9} = 1$

19. $\dfrac{(x-2)^2}{16}+\dfrac{(y-1)^2}{9}=1$

Center: $(2,\ 1)$

Domain: $[-4+2,\ 4+2]=[-2,\ 6]$

Range: $[-3+1,\ 3+1]=[-2,\ 4]$

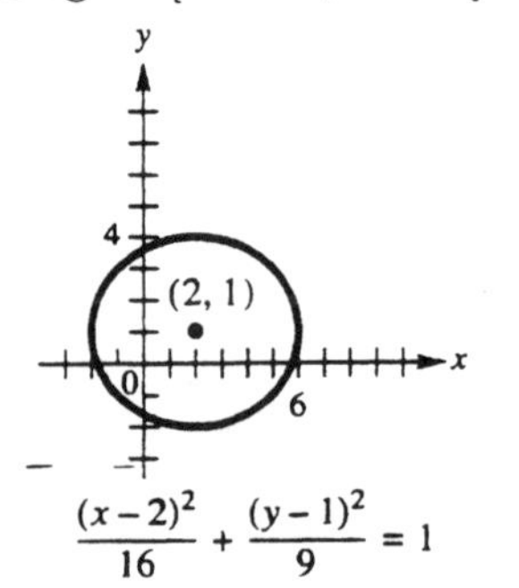

21. $\dfrac{(x+1)^2}{64}+\dfrac{(y-2)^2}{49}=1$

Center: $(-1,\ 2)$

Domain: $[-8-1,\ 8-1]=[-9,\ 7]$

Range: $[-7+2,\ 7+2]=[-5,\ 9]$

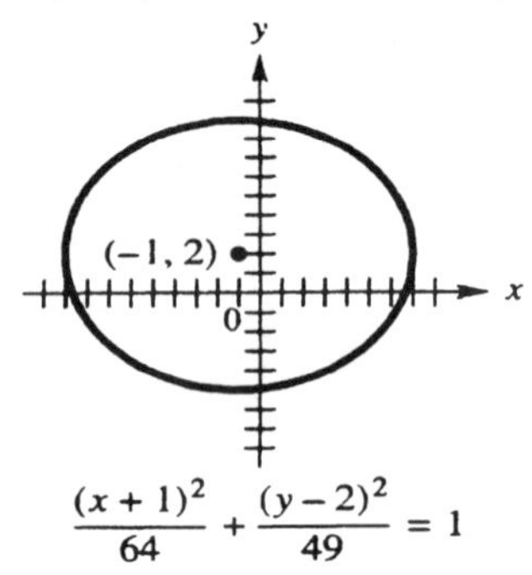

23. $\dfrac{(y+1)^2}{16}+\dfrac{(x-2)^2}{9}=1$

Since the center is at $(2,\ -1)$, the horizontal axis of symmetry is $y=-1$ and the vertical axis of symmetry is $x=2$.

Relating Concepts (24 - 29)

24. $16-\dfrac{16\,(x-2)^2}{9}\geq 0$

25. The graph is a parabola.

26. $y=16-\dfrac{16\,(x-2)^2}{9}$

$y=16-\dfrac{16(x-2)^2}{9}$

20

−10 10

−10

27. $16-\dfrac{16\,(x-2)^2}{9}\geq 0:$

$\boxed{[-1,\ 5]}$ See the graph in **26**.

28. In the figure, the domain is $[-1,\ 5]$. This corresponds to the solution set found graphically in exercises **27**.

29. $16-\dfrac{16\,(x-2)^2}{9}\geq 0$

$144-16\,(x-2)^2\geq 0$

$144-16x^2+64x-64\geq 0$

$-16x^2+64x+80\geq 0$

$-16\,(x^2-4x-5)\geq 0$

$(x+1)\,(x-5)\leq 0$

$(x+1)$	−	+	+
$(x-5)$	−	−	+
$(prod)$	+	−	+

-1 $\quad$ 5

$\boxed{[-1,\ 5]}$

31. $\dfrac{x^2}{16}-\dfrac{y^2}{9}=1$

Domain: $(-\infty,\ -4]\cup[4,\ \infty)$

Range: $(-\infty,\ \infty)$

$\dfrac{x^2}{16}-\dfrac{y^2}{9}=1$

33. $49y^2 - 36x^2 = 1764;\ \dfrac{y^2}{36} - \dfrac{x^2}{49} = 1$

Domain: $(-\infty, \infty)$

Range: $(-\infty, -6] \cup [6, \infty)$

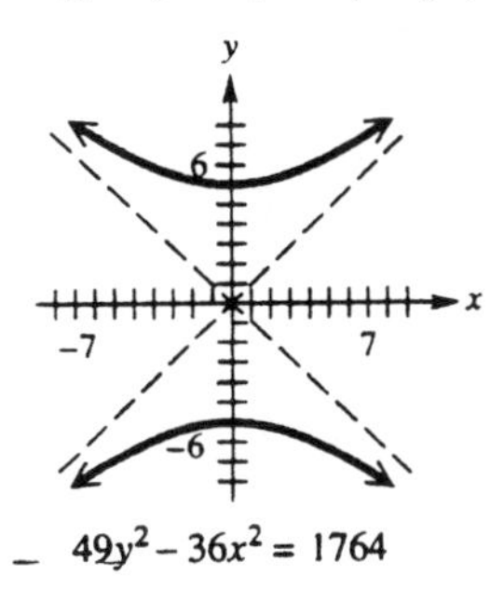

35. $\dfrac{4x^2}{9} - \dfrac{25y^2}{16} = 1;\ \dfrac{x^2}{\frac{9}{4}} - \dfrac{y^2}{\frac{16}{25}} = 1$

Domain: $\left(-\infty, -\dfrac{3}{2}\right] \cup \left[\dfrac{3}{2}, \infty\right)$

Range: $(-\infty, \infty)$

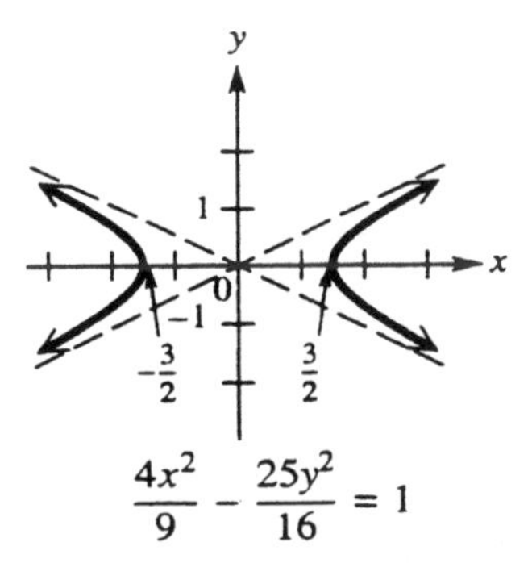

37. $\dfrac{(x-1)^2}{9} - \dfrac{(y+3)^2}{25} = 1$

Center: $(1, -3)$

Domain: $(-\infty, -2] \cup [4, \infty)$

$(-\infty, -3+1] \cup [3+1, \infty)$

Range: $(-\infty, \infty)$

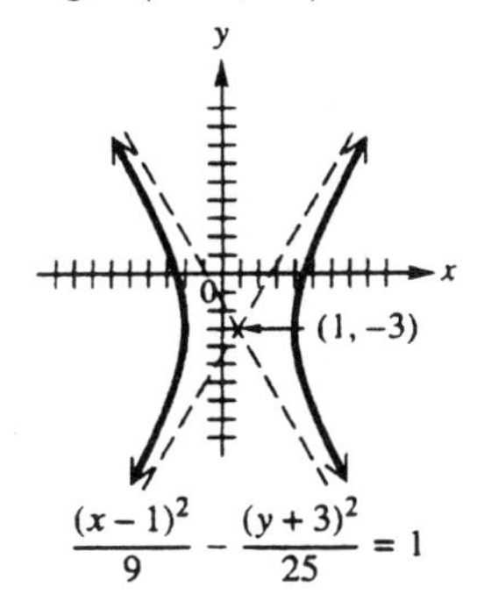

39. $\dfrac{(y-5)^2}{4} - \dfrac{(x+1)^2}{9} = 1$

Center: $(-1, 5)$

Domain: $(-\infty, \infty)$

Range: $(-\infty, 3] \cup [7, \infty)$

$(-\infty, -2+5] \cup [2+5, \infty)$

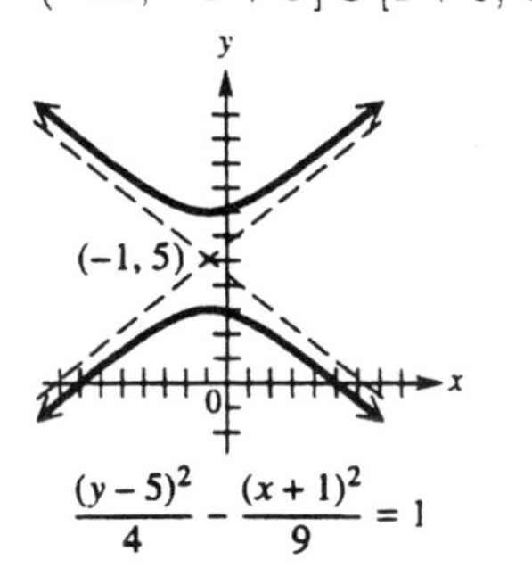

41. Ellipse: x-intercepts at ± 4; foci at $(\pm 2, 0)$.

Since the foci are on the x-axis, equidistant from the origin, the center is $(0, 0)$, the x-axis is the major axis, $c = 2$, $a = 4$ (x-int.), and b is unknown.

$$\begin{aligned} c^2 &= a^2 - b^2 \\ 4 &= 16 - b^2 \\ -12 &= -b^2 \\ b &= \pm\sqrt{12} = \pm 2\sqrt{3} \end{aligned}$$

$$\boxed{\text{EQ: } \frac{x^2}{16} + \frac{y^2}{12} = 1}$$

43. Ellipse: major axis endpoints at $(\pm 6, 0)$, $c = 4$.

Given $(\pm 6, 0)$, the origin is the center, major axis is the x-axis, $a = 6$, b is unknown.

$$\begin{aligned} c^2 &= a^2 - b^2 \\ 16 &= 36 - b^2 \\ -20 &= -b^2 \\ b &= \pm\sqrt{20} = \pm 2\sqrt{5} \end{aligned}$$

$$\boxed{\text{EQ: } \frac{x^2}{36} + \frac{y^2}{20} = 1}$$

45. Ellipse: Center $(3, -2)$, $a = 5$, $c = 3$; vertical major axis:

$$\frac{(y-k)^2}{a^2} + \frac{(x-h)^2}{b^2} = 1$$

$$\begin{aligned} c^2 &= a^2 - b^2 \\ 9 &= 25 - b^2 \\ -16 &= -b^2 \qquad b = 4 \end{aligned}$$

$$\boxed{\text{EQ: } \frac{(y+2)^2}{25} + \frac{(x-3)^2}{16} = 1}$$

47. Hyperbola: x-intercepts $(\pm 3, 0)$, foci at $(\pm 4, 0)$. Graph opens in x-direction, center $(0, 0)$; $a = 3$, $c = 4$, b is unknown:

$$\begin{aligned} c^2 &= a^2 + b^2 \\ 16 &= 9 + b^2 \\ 7 &= b^2 \qquad b = \sqrt{7} \end{aligned}$$

$$\frac{x^2}{a^2} - \frac{y^2}{b^2} = 1$$

$$\boxed{\text{EQ: } \frac{x^2}{9} - \frac{y^2}{7} = 1}$$

49. Hyperbola: asymptotes $y = \pm\frac{3}{5}x$; y-intercepts at $(0, \pm 3)$, so the center is at the origin; graph opens up/down, $a = 3$:

$$\frac{y^2}{9} - \frac{x^2}{b^2} = 1$$

Slope of asymptote:

$$\pm\frac{a}{b} = \pm\frac{3}{5} \text{ so } b = 5$$

$$\boxed{\text{EQ: } \frac{y^2}{9} - \frac{x^2}{25} = 1}$$

Relating Concepts (51 – 56)

51. $\frac{x^2}{16} + \frac{y^2}{12} = 1$

$a = 4$, $b = \sqrt{12}$

$$\begin{aligned} c^2 &= a^2 - b^2 \\ &= 16 - 12 = 4 \\ c &= \pm 2 \end{aligned}$$

Foci: $(\pm c, 0)$

$$\boxed{F_1 = (-2, 0);\ F_2 = (2, 0)}$$

52. $\frac{x^2}{16} + \frac{y^2}{12} = 1$

$$\begin{aligned} 3x^2 + 4y^2 &= 48 \\ 4y^2 &= 48 - 3x^2 \\ y^2 &= 12 - \frac{3x^2}{4} \\ y &= \pm\sqrt{12 - \frac{3x^2}{4}} \\ &= \pm\sqrt{12\left(1 - \frac{x^2}{16}\right)} \end{aligned}$$

Some coordinates: $(3,\ 2.2912878)$, $(0,\ 3.4641016)$, $(-3,\ -2.291288)$

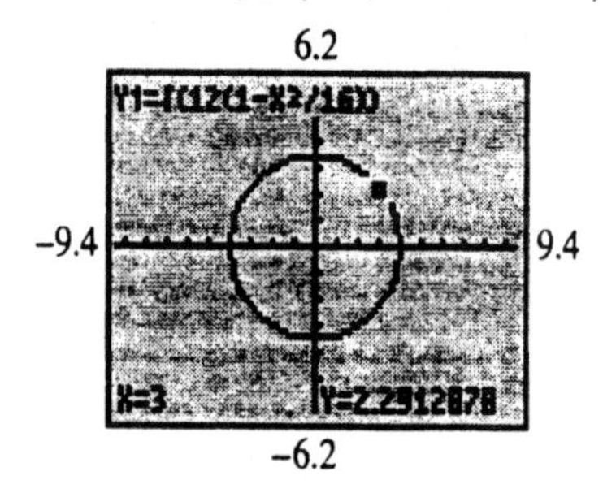

53. [Dist PF_1] + [Dist PF_2] $= 8$

$$\sqrt{(3+2)^2 + 2.2912878^2} + \sqrt{(3-2)^2 + 2.2912878^2} = 7.999999937 \approx 8$$

The points satisfy the equation.

54. **(a)** $\frac{x^2}{4} - \frac{y^2}{12} = 1$

$a = 2$, $b = \sqrt{12}$

$c^2 = a^2 + b^2 = 4 + 12 = 16$

$c = \pm 4$; Foci: $(\pm c, 0)$

$$\boxed{F_1 = (-4, 0);\ F_2 = (4, 0)}$$

continued

54. continued

(b) $-\dfrac{y^2}{12} = \dfrac{4 - x^2}{4}$

$$\frac{y^2}{12} = \frac{x^2 - 4}{4}$$

$$y^2 = 12\left(\frac{x^2 - 4}{4}\right)$$

$$y = \pm\sqrt{12\left(\frac{x^2 - 4}{4}\right)}$$

6.2

Y1=√(12(X²−4)/4)

−9.4 9.4

X=-3 Y=3.8729833

−6.2

Some coordinates: $(\pm 2, 0)$, $(4, \pm 6)$.

55. Check (4, 6) :

[Dist of P from F_1]

−[Dist of P from F_2] = 4

$$\sqrt{(4+4)^2 + 6^2} - \sqrt{(4-4)^2 + 6^2}$$
$$= \sqrt{64 + 36} - \sqrt{36}$$
$$= 10 - 6 = 4$$

The points satisfy the equation.

56. Exercise **53** demonstrates that the points on the graph satisfy the definition of the ellipse for that particular ellipse. Exercise **55** demonstrates similarly that the definition of the hyperbola is satisfied for that hyperbola.

57. Lithotropter at 12 units from source, distance from focus to focus, so $c = 6$; minor axis length is 16, so $b = 8$:

$$\begin{aligned} c^2 &= a^2 - b^2 \\ 36 &= a^2 - 64 \\ 100 &= a^2 \\ a &= 10 \end{aligned}$$

$$\frac{x^2}{a^2} + \frac{y^2}{b^2} = 1$$

EQ: $\dfrac{x^2}{100} + \dfrac{y^2}{64} = 1$

59. Major axis 620 ft $(a = 310)$;

minor axis 513 ft $\left(b = \dfrac{513}{2}\right)$.

$$\begin{aligned} c^2 &= a^2 - b^2 \\ c^2 &= 310^2 - \left(\frac{513}{2}\right)^2 \\ c^2 &= 96100 - \frac{263169}{4} \\ 4c^2 &= 384400 - 263169 \\ c^2 &= \frac{121231}{4} \\ c &= \frac{\sqrt{121231}}{2} \\ 2c &= \sqrt{121231} \end{aligned}$$

$c \approx 348.2$ feet

61. $a = 15$ (y direction); $b = 10$:

EQ: $\dfrac{y^2}{15^2} + \dfrac{x^2}{10^2} = 1$

If the center is at (0, 0), find y for $\dfrac{12}{2}$ (half of truck width):

$$\frac{y^2}{225} + \frac{36}{100} = 1$$
$$100y^2 + 8100 = 22500$$
$$100y^2 = 14400$$
$$y^2 = 144$$
$$y = 12$$

Truck must be just under 12 ft tall.

63. (a) Satellite: $\dfrac{x^2}{(4465)^2} + \dfrac{y^2}{(4462)^2} = 1$

Solve for y :

$$\frac{y^2}{(4462)^2} = 1 - \frac{x^2}{(4465)^2}$$

$$y^2 = 4462^2\left(1 - \frac{x^2}{(4465)^2}\right)$$

$$y = \pm\sqrt{4462^2\left(1 - \frac{x^2}{(4465)^2}\right)}$$

$$y_1 = 4462\sqrt{\left(1 - \frac{x^2}{(4465)^2}\right)}$$

$$y_2 = -4462\sqrt{\left(1 - \frac{x^2}{(4465)^2}\right)}$$

continued

63. continued

The graph of Earth can be represented by a circle of radius 3960 centered at one focus. To determine the foci of the orbit, we must determine c:

$$c^2 = a^2 - b^2$$
$$c^2 = 4465^2 - 4462^2$$
$$c^2 = 26781 \qquad c \approx 163.6$$

Equation of the circle will be

$(x - 163.6)^2 + y^2 = 3960^2$.

Solving for y: $y = \pm\sqrt{3960^2 - (x - 163.6)^2}$

$$y_3 = \sqrt{3960^2 - (x - 163.6)^2}$$

$$y_4 = -\sqrt{3960^2 - (x - 163.6)^2}$$

$y_3 = \sqrt{3960^2 - (x - 163.6)^2}$

$y_1 = 4462\sqrt{1 - \frac{x^2}{4465^2}}$

4500

-6750 6750

-4500

$y_4 = -y_3$ $\qquad$ $y_2 = -y_1$

(b) From the graph, we see that the distance is maximum and minimum when the orbits intersect the x-axis. The x-intercepts of the satellite's orbit are ± 4465. The x-intercepts of Earth's surface occur when

$$(x - 163.6)^2 + 0^2 = 3960^2$$
$$(x - 163.6)^2 = 3960^2$$
$$x - 163.6 = \pm 3960$$
$$x = 163.6 \pm 3960$$
$$x = 4123.6 \quad \text{or} \quad -3796.4$$

The minimum distance is

$4465 - 4123.6 = 341.4$ $\boxed{\approx 341 \text{ miles}}$

The maximum distance is

$-3796.4 - (-4465) = 668.6$ $\boxed{\approx 669 \text{ miles}}$

65. (a) Find a and b: $\frac{x^2}{a^2} - \frac{y^2}{b^2} = 1$

The asymptotes are $y = x$, $y = -x$, which have slopes of 1 and -1, so $a = b$. Look at the small triangle shown in quadrant III. The line $y = x$ intersects quadrants I and III at a 45° angle. Since the right angle vertex of the triangle lies on the line $y = x$, we know that this triangle is a 45°- 45°- 90° triangle (isosceles rt. triangle). Thus, both legs of the triangle have length d, and, by the Pythagorean theorem,

$$c^2 = d^2 + d^2$$
$$c^2 = 2d^2$$
$$c = d\sqrt{2} \qquad [\#1]$$

The coordinates of N are $(-d\sqrt{2}, 0)$. Since N is a focus of the hyperbola, c represents the distance between the center of the hyperbola, which is $(0,0)$, and either focus.

Since $a = b$, we have

$$c^2 = a^2 + b^2$$
$$c^2 = 2a^2$$
$$c = a\sqrt{2} \qquad [\#2]$$

From equations [#1] and [#2]:

$$d\sqrt{2} = a\sqrt{2}$$
$$d = a$$
$$a = b = d = 5 \times 10^{-14}$$

The equation of the trajectory of A, where $x > 0$, is given by:

$$\frac{x^2}{a^2} - \frac{y^2}{b^2} = 1$$

$$\frac{x^2}{(5 \times 10^{-14})^2} - \frac{y^2}{(5 \times 10^{-14})^2} = 1$$

$$x^2 - y^2 = (5 \times 10^{-14})^2$$
$$= 25 \times 10^{-28}$$
$$= 2.5 \times 10^{-27}$$
$$x^2 = y^2 + 2.5 \times 10^{-27}$$
$$x = \sqrt{y^2 + 2.5 \times 10^{-27}}$$

This equation represents the right half of the hyperbola, as shown in the textbook.

(b) The minimum distance between their centers is

$$c + a = d\sqrt{2} + d$$
$$= (5 \times 10^{-14})\sqrt{2} + (5 \times 10^{-14})$$
$$= 12.07107 \times 10^{-14}$$

$$\boxed{\approx 1.2 \times 10^{-13} \text{ m}}$$

67. $d\left(P, F'\right) - d(P, F) = 2a$

$b^2 = c^2 - a^2$

$$\sqrt{(x+c)^2 + y^2} - \sqrt{(x-c)^2 + y^2} = 2a$$

$$\sqrt{(x+c)^2 + y^2} = 2a + \sqrt{(x-c)^2 + y^2}$$

$$(x+c)^2 + y^2 = 4a^2 + 4a\sqrt{(x-c)^2 + y^2} + (x-c)^2 + y^2$$

$$(x+c)^2 - (x-c)^2 - 4a^2 = 4a\sqrt{(x-c)^2 + y^2}$$

$$x^2 + 2cx + c^2 - x^2 + 2cx - c^2 - 4a^2 = 4a\sqrt{(x-c)^2 + y^2}$$

$$4cx - 4a^2 = 4a\sqrt{(x-c)^2 + y^2}$$

$$cx - a^2 = a\sqrt{(x-c)^2 + y^2}$$

$$c^2x^2 - 2a^2cx + a^4 = a^2\left(x^2 - 2cx + c^2 + y^2\right)$$

$$c^2x^2 - 2a^2cx + a^4 = a^2x^2 - 2a^2cx + a^2c^2 + a^2y^2$$

$$c^2x^2 - a^2x^2 - a^2y^2 = -a^4 + a^2c^2$$

$$\frac{x^2\left(c^2 - a^2\right)}{a^2\left(c^2 - a^2\right)} - \frac{a^2y^2}{a^2\left(c^2 - a^2\right)} = \frac{a^2\left(c^2 - a^2\right)}{a^2\left(c^2 - a^2\right)}$$

$$\frac{x^2}{a^2} - \frac{y^2}{b^2} = 1$$

69. $d\left(P, F'\right) - d(P, F) = 2a$

$b^2 = c^2 - a^2; \quad c = 2,\ 2a = 2$

$$\sqrt{(x+2)^2 + y^2} - \sqrt{(x-2)^2 + y^2} = 2$$

$$\sqrt{(x+2)^2 + y^2} = 2 + \sqrt{(x-2)^2 + y^2}$$

$$(x+2)^2 + y^2 = 4 + 4\sqrt{(x-2)^2 + y^2} + (x-2)^2 + y^2$$

$$(x+2)^2 - (x-2)^2 - 4 = 4\sqrt{(x-2)^2 + y^2}$$

$$x^2 + 4x + 4 - x^2 + 4x - 4 - 4 = 4\sqrt{(x-2)^2 + y^2}$$

$$8x - 4 = 4\sqrt{(x-2)^2 + y^2}$$

$$(2x-1)^2 = \left(\sqrt{(x-2)^2 + y^2}\right)^2$$

$$4x^2 - 4x + 1 = x^2 - 4x + 4 + y^2$$

$$3x^2 - y^2 = 3$$

$$\frac{3x^2}{3} - \frac{y^2}{3} = 1$$

$$x^2 - \frac{y^2}{3} = 1$$

Reviewing Basic Concepts (Sections 6.1 and 6.2)

1. **(a)** circle: **B** **(b)** parabola: **D**
(c) ellipse: **A** **(d)** hyperbola: **C**

2. $12x^2 - 4y^2 = 48$

$$\frac{12x^2}{48} - \frac{4y^2}{48} = 1 \Longrightarrow \frac{x^2}{4} - \frac{y^2}{12} = 1$$

Center: $(0, 0)$, Vertices: $(\pm 2, 0)$

Asymptotes: $y = \pm\sqrt{3}\,x$

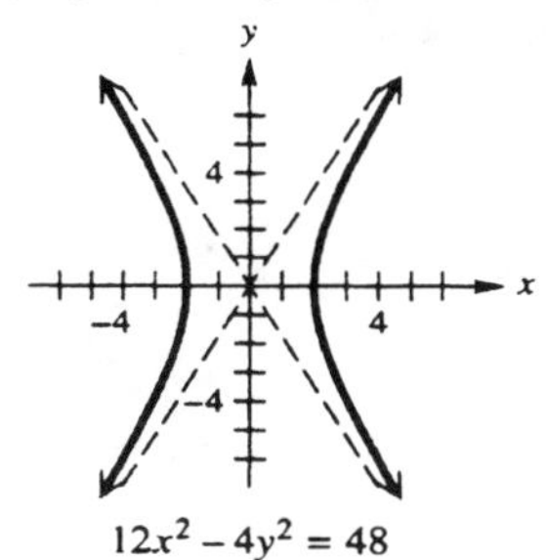

3. $y = 2x^2 + 3x - 1$

$$y = 2\left(x^2 + \frac{3}{2}x + \frac{9}{16} - \frac{9}{16}\right) - 1$$

$$y = 2\left(x + \frac{3}{4}\right)^2 + 2\left(-\frac{9}{16}\right) - 1$$

$$y = 2\left(x + \frac{3}{4}\right)^2 - \frac{17}{8}$$

Vertex: $\left(-\frac{3}{4}, -\frac{17}{8}\right)$, Axis: $x = -\frac{3}{4}$

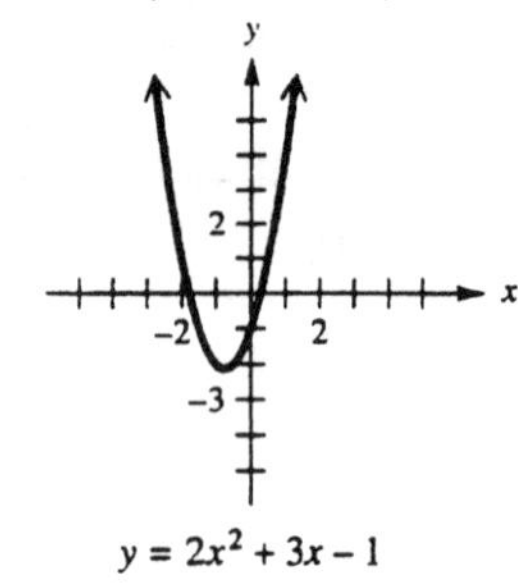

$y = 2x^2 + 3x - 1$

4. $x^2 + y^2 - 2x + 2y - 2 = 0$

$(x^2 - 2x + 1) + (y^2 + 2y + 1) = 2 + 1 + 1$

$(x - 1)^2 + (y + 1)^2 = 4$

Circle, Center $(1, -1)$, radius $= 2$

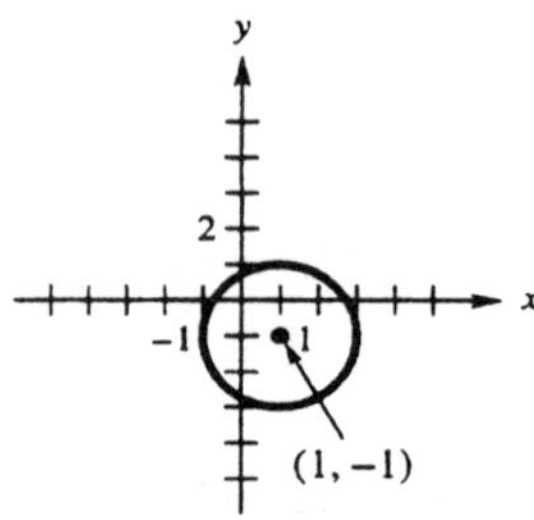

$x^2 + y^2 - 2x + 2y - 2 = 0$

5. $4x^2 + 9y^2 = 36$

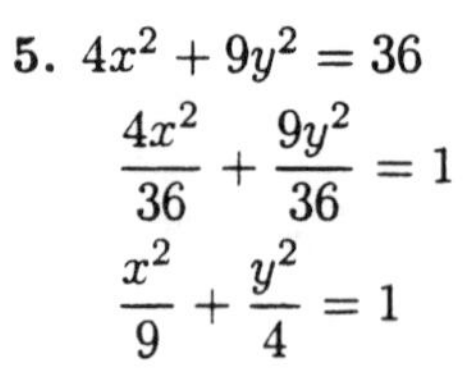

$$\frac{4x^2}{36} + \frac{9y^2}{36} = 1$$

$$\frac{x^2}{9} + \frac{y^2}{4} = 1$$

Ellipse; center: $(0, 0)$, vertices: $(\pm 3, 0)$

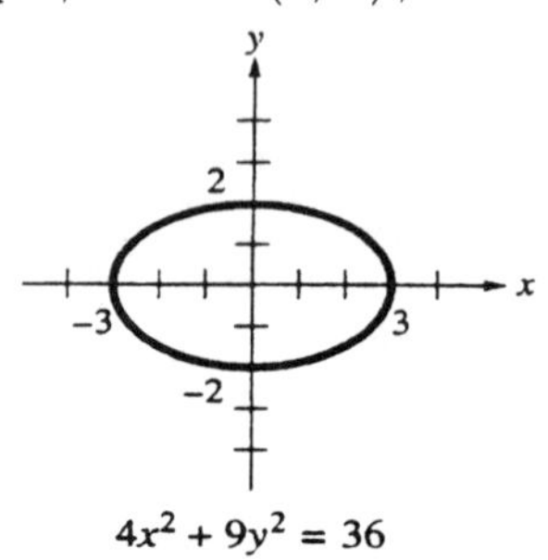

$4x^2 + 9y^2 = 36$

6. The focus lies within the curve, so if $c < a$, it is an ellipse, and if $c > a$, it is a hyperbola.

7. Circle, center at $(2, -1)$, radius $= 3$.

$(x - 2)^2 + (y + 1)^2 = 9$

8. Ellipse, major axis length 12 $(a = 6)$, foci: $(\pm 4, 0) \Longrightarrow c = 4$

$$a^2 - b^2 = c^2$$
$$36 - b^2 = 16$$
$$-b^2 = -20$$
$$b^2 = 20$$
$$\frac{x^2}{36} + \frac{y^2}{20} = 1$$

9. Hyperbola, vertices $(0, \pm 2)$, foci $(0, \pm 4)$

$a = 2, \; c = 4$ $\qquad a^2 + b^2 = c^2$

$$4 + b^2 = 16$$
$$b^2 = 12$$

$$\frac{y^2}{4} - \frac{x^2}{12} = 1$$

10. Parabola, focus $\left(0, \frac{1}{2}\right)$, vertex $(0, 0)$.

Parabola of the form:

$$x^2 = 4cy, \quad c = \frac{1}{2}$$
$$x^2 = 4\left(\frac{1}{2}\right)y$$
$$x^2 = 2y$$

Section 6.3

Conics: $Ax^2 + Dx + Cy^2 + Ey + F = 0$

1. $x^2 + y^2 = 144$

Circle $(A = C)$

3. $y = 2x^2 + 3x - 4$

Parabola $(C = 0)$

5. $x = -3(y-4)^2 + 1$

Parabola $(A = 0)$

7. $\dfrac{x^2}{49} + \dfrac{y^2}{100} = 1$

Ellipse $(A \neq C,\ AC > 0)$

9. $\dfrac{x^2}{4} - \dfrac{y^2}{16} = 1$

Hyperbola $(AC < 0)$

11. $\dfrac{x^2}{25} - \dfrac{y^2}{25} = 1$

Hyperbola $(AC < 0)$

13. $\dfrac{x^2}{4} = 1 - \dfrac{y^2}{9}$

$\dfrac{x^2}{4} + \dfrac{y^2}{9} = 1$

Ellipse $(A \neq C,\ AC > 0)$

15. $\dfrac{x^2}{4} + \dfrac{y^2}{4} = -1$

No graph

17. $x^2 = 25 - y^2$

$x^2 + y^2 = 25$

Circle $(A = C)$

19. $x^2 = 4y - 8$

Parabola $(C = 0)$

21. $\dfrac{(x-4)^2}{8} + \dfrac{(y+1)^2}{2} = 0$

1 point $(4, -1)$

23. $(x+7)^2 + (y-5)^2 + 4 = 0$

$(x+7)^2 + (y-5)^2 = -4$

No graph

25. $3x^2 + 6x + 3y^2 - 12y = 12$

$x^2 + 2x + y^2 - 4y = 4$

$(x^2 + 2x + 1) + (y^2 - 4y + 4)$
$= 4 + 1 + 4$

$(x+1)^2 + (y-2)^2 = 9$

Circle $(A = C)$

27. $x^2 - 6x + y = 0$

$y = -x^2 + 6x$

Parabola $(C = 0)$

29. $4x^2 - 8x - y^2 - 6y = 6$

$4(x^2 - 2x) - (y^2 + 6y) = 6$

$4(x^2 - 2x + 1 - 1)$
$- (y^2 + 6y + 9 - 9) = 6$

$4(x^2 - 2x + 1) - (y^2 + 6y + 9)$
$= 6 + 4 - 9$

$4(x-1)^2 - (y+3)^2 = 1$

Hyperbola $(AC < 0)$

31. $4x^2 - 8x + 9y^2 + 54y = -84$

$4(x^2 - 2x) + 9(y^2 + 6y) = -84$

$4(x^2 - 2x + 1 - 1)$
$+ 9(y^2 + 6y + 9 - 9) = -84$

$4(x^2 - 2x + 1) + 9(y^2 + 6y + 9)$
$= -84 + 4 + 81$

$4(x-1)^2 + 9(y+3)^2 = 1$

Ellipse $(A \neq C,\ AC > 0)$

33. $6x^2 - 12x + 6y^2 - 18y + 25 = 0$

$$6\left(x^2 - 2x\right) + 6\left(y^2 - 3y\right) = -25$$

$$6\left(x^2 - 2x + 1 - 1\right) + 6\left(y^2 - 3y + \frac{9}{4} - \frac{9}{4}\right) = -25$$

$$6\left(x^2 - 2x + 1\right) + 6\left(y^2 - 3y + \frac{9}{4}\right) = -25 + 6 + \frac{27}{2}$$

$$(x-1)^2 + \left(y - \frac{3}{2}\right)^2 = -\frac{11}{2}$$

No graph

35. $Ax^2 + Bx + Cy^2 + Dy + E = 0$
If $A = 0,\ C = 0$: **Line**

37. The definition of an ellipse states that "an ellipse is the set of all points in a plane the sum of whose distances from two fixed points is constant." Therefore, the set of all points in a plane for which the sum of distances from $(5,\ 0)$ to $(-5,\ 0)$ is 14 is an **Ellipse** with foci $(\pm 5,\ 0)$.

39. [Distance of P from F]
$= e$ [Distance of P] from L

This conic has $e = \frac{3}{2}$. Since $e > 1$, this is a **Hyperbola.**

41. $12x^2 + 9y^2 = 36$

$$\frac{x^2}{3} + \frac{y^2}{4} = 1$$

$$c^2 = a^2 - b^2$$
$$c^2 = 4 - 3$$
$$c^2 = 1$$
$$c = 1, \qquad a = 2$$

$$e = \frac{c}{a} \qquad \boxed{e = \frac{1}{2}}$$

43. $x^2 - y^2 = 4$

$$\frac{x^2}{4} - \frac{y^2}{4} = 1$$

$$c^2 = a^2 + b^2$$
$$c^2 = 4 + 4$$
$$c^2 = 8$$
$$c = 2\sqrt{2}, \qquad a = 2$$

$$e = \frac{c}{a} = \frac{2\sqrt{2}}{2} \qquad \boxed{e = \sqrt{2}}$$

45. $4x^2 + 7y^2 = 28$

$$\frac{x^2}{7} + \frac{y^2}{4} = 1$$

$$c^2 = a^2 - b^2$$
$$c^2 = 7 - 4$$
$$c^2 = 3$$
$$c = \sqrt{3}, \qquad a = \sqrt{7}$$

$$e = \frac{c}{a} = \frac{\sqrt{3}}{\sqrt{7}} \cdot \frac{\sqrt{7}}{\sqrt{7}} \qquad \boxed{e = \frac{\sqrt{21}}{7}}$$

47. $x^2 - 9y^2 = 18$

$$\frac{x^2}{18} - \frac{y^2}{2} = 1$$

$$c^2 = a^2 + b^2$$
$$c^2 = 18 + 2$$
$$c^2 = 20$$
$$c = 2\sqrt{5}, \qquad a = \sqrt{18}$$

$$e = \frac{c}{a} = \frac{2\sqrt{5}}{\sqrt{18}} = \frac{2\sqrt{5}}{3\sqrt{2}} \cdot \frac{\sqrt{2}}{\sqrt{2}}$$

$$= \frac{2\sqrt{10}}{6} \qquad \boxed{e = \frac{\sqrt{10}}{3}}$$

49. Focus: $(0,\ 8)$; since $e = 1$, this is a parabola. Since the vertex is at $(0,\ 0)$, and the focus at $(0,\ 8)$, the parabola opens up, $c = 8$:

$$x^2 = 4cy \rightarrow x^2 = 4\,(8)\,y$$

$$\boxed{x^2 = 32y}$$

51. Focus at $(3, 0)$, $e = \frac{1}{2}$. This is an ellipse $(0 < e < 1)$; major axis is the x-axis (focus), $c = 3$:

$$e = \frac{c}{a}$$

$$\frac{1}{2} = \frac{3}{a}, \quad a = 6$$

$$c^2 = a^2 - b^2$$

$$9 = 36 - b^2 \Longrightarrow b^2 = 27 \quad \boxed{\frac{x^2}{36} + \frac{y^2}{27} = 1}$$

53. Vertex at $(-6, 0)$, $e = 2$.
Since $e > 1$ this is a hyperbola; since the vertex is at $(-6, 0)$, the major axis is the x-axis; $a = 6$.

$$e = \frac{c}{a}$$

$$2 = \frac{c}{6}, \quad c = 12$$

$$c^2 = a^2 + b^2$$

$$144 = 36 + b^2 \Longrightarrow b^2 = 108 \quad \boxed{\frac{x^2}{36} - \frac{y^2}{108} = 1}$$

55. Since $e = 1$, this is a parabola; since the vertex is at $(0, 0)$, and the focus at $(0, -1)$, the parabola opens down, $c = -1$; equation is of the form $x^2 = 4cy$.

$$\boxed{x^2 = -4y}$$

57. Vertical major axis, (y-axis), length is 6 $(a = 3)$, $e = \dfrac{4}{5}$. Since $0 < e < 1$, this is an ellipse.

$$e = \frac{c}{a}$$

$$\frac{4}{5} = \frac{c}{3}$$

$$5c = 12, \quad c = \frac{12}{5}$$

$$c^2 = a^2 - b^2$$

$$\frac{144}{25} = 9 - b^2$$

$$144 = 225 - 25b^2$$

$$-81 = -25b^2$$

$$b^2 = \frac{81}{25} \quad \boxed{\frac{25x^2}{81} + \frac{y^2}{9} = 1}$$

59. (C): $e = 0$
(A): $0 < e < 1$
(B): $e = 1$
(D): $e > 1$

61. **(a)** *Neptune:*

$$e = \frac{c}{a}$$

$$c = ea = (.009)(30.1) = .2709$$

$$b^2 = a^2 - c^2 = (30.1)^2 - (.2709)^2 \approx 905.9366$$

$$b \approx 30.1$$

Since $c = .2709$, the graph should be translated .2709 units to the right so that the sun will be located at the origin. Its equation is essentially circular with equation:

$$\frac{(x - .2709)^2}{30.1^2} + \frac{y^2}{30.1^2} = 1$$

Pluto:

$$c = ea = (.249)(39.4) = 9.8106$$

$$b^2 = a^2 - c^2 = (39.4)^2 - (9.8106)^2 \approx 1456.1121$$

$$b \approx 38.16$$

As with Neptune, the graph should be translated 9.8106 units to the right so that the sun will be located at the origin. Its equation is:

$$\frac{(x - 9.8106)^2}{39.4^2} + \frac{y^2}{38.16^2} = 1$$

(b) In order to graph these equations on a graphing calculator, we must solve each equation for y.

Neptune:

$$\frac{(x - .2709)^2}{30.1^2} + \frac{y^2}{30.1^2} = 1$$

$$(x - .2709)^2 + y^2 = 30.1^2$$

$$y = \pm\sqrt{30.1^2 - (x - .2709)^2}$$

Graph y_1 and y_2 (dark):

continued

61. continued

Pluto:

$$\frac{(x-9.8106)^2}{39.4^2}+\frac{y^2}{38.16^2}=1$$

$$\frac{y^2}{38.16^2}=1-\frac{(x-9.8106)^2}{39.4^2}$$

$$y^2=38.16^2\left(1-\frac{(x-9.8106)^2}{39.4^2}\right)$$

$$y=\pm 38.16\sqrt{\left(1-\frac{(x-9.8106)^2}{39.4^2}\right)}$$

Graph y_3 and y_4 (light):

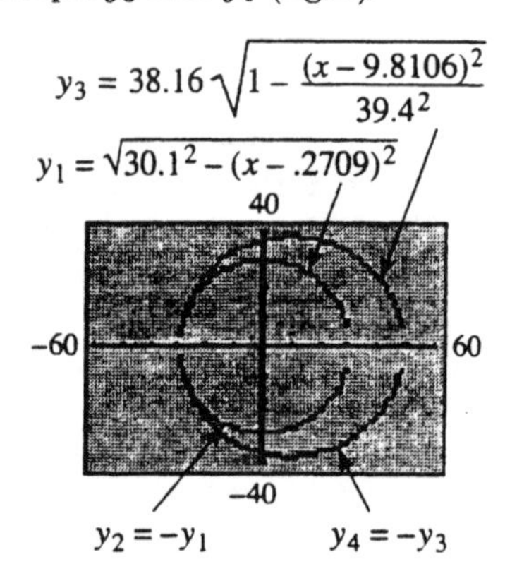

63. The shortest distance between Halley's comet and the sun is $a-c$, and the greatest distance is $a+c$. Since the eccentricity is .9673,

$$\frac{c}{a}=.9673$$
$$c=.9673a.$$

Since the greatest distance is 3281,

$$a+c=3281$$
$$a+.9673a=3281$$
$$1.9673a=3281$$
$$a\approx 1667.768$$
$$c\approx 1613.232$$
$$a-c\approx 1667.768-1613.232$$
$$\approx 54.536$$

Thus, the shortest distance between Halley's comet and the sun is approximately 55 million miles.

Section 6.4

1. $x=2t,\ y=t+1,\ t\in[-2,3]$

Solve $y=t+1$ for t: $\quad t=y-1$

$t\to x$: $\quad x=2(y-1)$

$$\frac{1}{2}x=y-1$$

$x = 2t$, $y = t + 1$, t in $[-2, 3]$

8, –8, 8, –8

$$y=\frac{1}{2}x+1,\quad x\in[-4,6]$$

3. $x=\sqrt{t},\ y=3t-4,\ t\in[0,4]$

Solve $y=3t-4$ for t:

$$3t=y+4\to t=\frac{y+4}{3}$$

$t\to x$: $\quad x=\sqrt{\frac{y+4}{3}}$

$$x^2=\frac{y+4}{3}$$
$$3x^2=y+4$$

$x=\sqrt{t}$, $y = 3t - 4$, t in $[0, 4]$

10, –6, 6, –6

$$y=3x^2-4,\quad x\in[0,2]$$

5. $x = t^3 + 1,\ y = t^3 - 1,\ t \in [-3, 3]$

Solve $y = t^3 - 1$ for t:

$$t^3 = y + 1 \rightarrow t = \sqrt[3]{y+1}$$

$t \rightarrow x$:

$$x = \left(\sqrt[3]{y+1}\right)^3 + 1$$

$$x = y + 1 + 1$$

$x = t^3 + 1, y = t^3 - 1, t$ in $[-3, 3]$

30, −30, 30, −30

$$\boxed{y = x - 2,\quad x \in [-26, 28]}$$

7. $x = 2^t,\ y = \sqrt{3t-1},\ t \in \left[\frac{1}{3}, 4\right]$

Solve $y = \sqrt{3t-1}$ for t:

$$y^2 = 3t - 1$$

$$3t = y^2 + 1$$

$$t = \frac{y^2+1}{3}$$

$t \rightarrow x$:

$$\boxed{x = 2^{(y^2+1)/3},\ y \in [0, \sqrt{11}]}$$

[Solving $x = 2^t$ for t $\left(\frac{\ln x}{\ln 2}\right)$, then substituting gives $y^2 = \frac{3 \ln x}{\ln 2} - 1$]

$x = 2^t, y = \sqrt{3t-1}, t$ in $\left[\frac{1}{3}, 4\right]$

10, −2, 30, −2

9. $x = t + 2,\ y = -\frac{1}{2}\sqrt{9-t^2},$

$t \in [-3, 3]$

Solve $x = t + 2$ for t: $t = x - 2$

$t \rightarrow y$:

$$\boxed{y = -\frac{1}{2}\sqrt{9-(x-2)^2},\ x \in [-1, 5]}$$

$x = t + 2, y = -\frac{1}{2}\sqrt{9-t^2}$, t in $[-3, 3]$

4, −6, 6, −4

11. $x = t,\ y = \frac{1}{t},$

$t \in (-\infty, 0) \cup (0, \infty)$

$t = x \rightarrow y$:

$$\boxed{y = \frac{1}{x},\ x \in (-\infty, 0) \cup (0, \infty)}$$

$x = t, y = \frac{1}{t}, t$ in $(-\infty, 0) \cup (0, \infty)$

4, −6, 6, −4

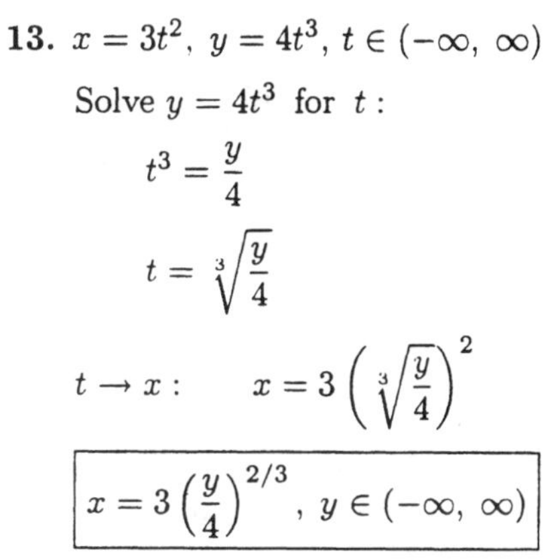

13. $x = 3t^2,\ y = 4t^3,\ t \in (-\infty, \infty)$

Solve $y = 4t^3$ for t:

$$t^3 = \frac{y}{4}$$

$$t = \sqrt[3]{\frac{y}{4}}$$

$t \rightarrow x$: $\quad x = 3\left(\sqrt[3]{\frac{y}{4}}\right)^2$

$$\boxed{x = 3\left(\frac{y}{4}\right)^{2/3},\ y \in (-\infty, \infty)}$$

15. $x = t,\ y = \sqrt{t^2+2},\ t \in (-\infty,\ \infty)$

$x \to y:$

$$y = \sqrt{x^2+2}$$

$$\boxed{y = \sqrt{x^2+2}\,,\ x \in (-\infty,\ \infty)}$$

17. $x = e^t,\ y = e^{-t},\ t \in (-\infty,\ \infty)$

Solve $x = e^t$ for t:

$$t = \ln x$$

$t \to y:$

$$y = e^{-\ln x} = e^{\ln x^{-1}} = x^{-1}$$

$$\boxed{y = \frac{1}{x},\quad x \in (0,\ \infty)}$$

19. $x = \dfrac{1}{\sqrt{t+2}},\ y = \dfrac{t}{t+2},$

$t \in (-2,\ \infty)$

Solve $x = \dfrac{1}{\sqrt{t+2}}$ for t:

$$x\sqrt{t+2} = 1$$

$$\sqrt{t+2} = \frac{1}{x}$$

$$t+2 = \frac{1}{x^2}$$

$$t = \frac{1}{x^2} - 2$$

$t \to y:$

$$y = \frac{\frac{1}{x^2} - 2}{\frac{1}{x^2} - 2 + 2} \cdot \frac{x^2}{x^2}$$

$$y = \frac{1 - 2x^2}{1}$$

$$\boxed{y = 1 - 2x^2,\ x \in (0,\ \infty)}$$

21. $x = t+2,\ y = \dfrac{1}{t+2},\ t \neq -2$

Solve $x = t+2$ for t:

$$t = x - 2$$

$t \to y:\quad \boxed{y = \dfrac{1}{x},\ x \neq 0}$

23. $x = t^2,\ y = 2\ln t,\ t \in (0,\ \infty)$

Solve $x = t^2$ for t:

$$t = \sqrt{x}$$

$t \to y:\quad y = 2\ln x^{\frac{1}{2}}$

$$y = 2\left(\frac{1}{2}\right)\ln x$$

$$\boxed{y = \ln x\,,\ x \in (0,\ \infty)}$$

25. $y = 2x + 3$

Let $x = \dfrac{1}{2}t$, then $y = t + 3$.

Let $x = \dfrac{t+3}{2}$, then $y = t + 6$.

Other answers are possible.

27. $y = \sqrt{3x+2},\ x \in \left[-\dfrac{2}{3},\ \infty\right)$

Let $x = \dfrac{1}{3}t$, then $y = \sqrt{t+2}$

$t \in [-2,\ \infty)$

Let $x = \dfrac{t-2}{3}$, then $y = \sqrt{t}$

$t \in [0,\ \infty)$

Other answers are possible.

29. $v_o = 400$ ft/sec, $\theta = 45°$

$$x = v_o\frac{\sqrt{2}}{2}t,\quad y = v_o\frac{\sqrt{2}}{2}t - 16t^2$$

(a) Find t when $y = 0$:

$$0 = \frac{400\sqrt{2}}{2}t - 16t^2$$

$$0 = 200\sqrt{2}t - 16t^2$$

$$0 = -16t^2 + 200\sqrt{2}\,t$$

$$0 = -8t\,(2t - 25\sqrt{2})$$

$$t = 0,\ 17.67767$$

$$\boxed{\text{Strikes the ground at} \approx 17.7 \text{ sec.}}$$

continued

29. continued

(b) Find x when $y = 0$:

From example 4:

$$y = x - \frac{32x^2}{(v_o)^2}$$

$$y = x - \frac{32x^2}{(400)^2}$$

$$0 = x - \frac{x^2}{5000}$$

$$0 = 5000x - x^2$$

$$-0 = -x(5000 - x)$$

$x = 0,\ 5{,}000$ $\boxed{\text{Range is 5,000 ft}}$

(c) Find the y-value of the vertex:

$$-\frac{b}{2a} = 2500 \quad \left(\frac{1}{2} \text{ of } 5000\right)$$

$$y = 2500 - \frac{32(2500)^2}{(400)^2}$$

$$y = 2500 - 1250 = 1250$$

$\boxed{\text{Maximum altitude is 1250 feet}}$

31. $x = 60t,\ y = 80t - 16t^2$

Solve $x = 60t$ for t: $t = \frac{1}{60}x$

$t \to y$: $y = 80\left(\frac{1}{60}x\right) - 16\left(\frac{1}{60}x\right)^2$

$$\boxed{y = \frac{4}{3}x - \frac{1}{225}x^2}$$

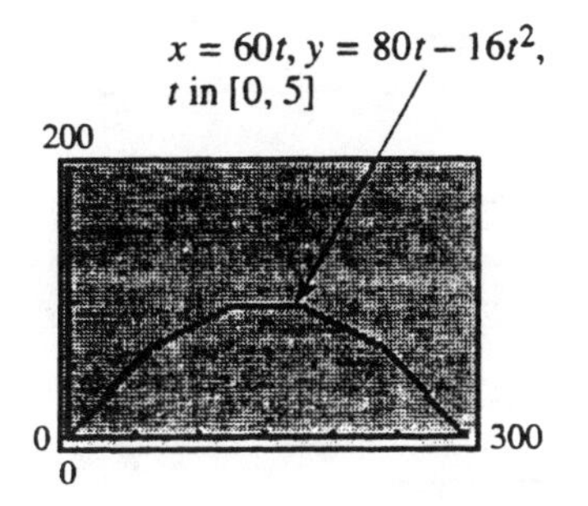

33. $x = v_o\frac{\sqrt{2}}{2}t,\quad y = v_o\frac{\sqrt{2}}{2}t - 16t^2,$

$t \in [0,\ k]$

Solve $x = v_o\frac{\sqrt{2}}{2}t$ for t:

$$t = \frac{2x}{v_o\sqrt{2}}$$

$t \to y$:

$$y = v_o\left(\frac{\sqrt{2}}{2}\right)\left(\frac{2x}{v_o\sqrt{2}}\right) - 16\left(\frac{2x}{v_o\sqrt{2}}\right)^2$$

$$\boxed{y = x - \frac{32}{(v_o)^2}x^2}$$

35. Line through $(x_1,\ y_1)$, slope m:

Cartesian equation:

$$y = m(x - x_1) + y_1$$

Many equations are possible:

(i) If $x = t$, $y = m(t - x_1) + y_1$

(ii) If $x = t + x_1$,

$$y = m(t + x_1 - x_1) + y_1$$

$$y = mt + y_1$$

Reviewing Basic Concepts (Sections 6.3 - 6.4)

1. $3x^2 + y^2 - 6x + 6y = 0$

$$3(x^2 - 2x) + y^2 + 6y = 0$$

$$3(x^2 - 2x + 1) + (y^2 + 6y + 9) = 3(1) + 9$$

$$3(x-1)^2 + (y+3)^2 = 12$$

$$\frac{(x-1)^2}{4} + \frac{(y+3)^2}{12} = 1;\ \textbf{ellipse}$$

2. $y^2 - 2x^2 + 8y - 8x - 4 = 0$

$$(y^2 + 8y) - 2(x^2 + 4x) = 4$$

$$(y^2 + 8y + 16) - 2(x^2 + 4x + 4) = 4 + 16 - 2(4)$$

$$(y+4)^2 - 2(x+2)^2 = 12$$

$$\frac{(y+4)^2}{12} - \frac{(x+2)^2}{6} = 1;\ \textbf{hyperbola}$$

3. $3y^2 + 12y + 5x = 3$

$$3\left(y^2 + 4y\right) = -5x + 3$$

$$y^2 + 4y = -\frac{5}{3}x + 1$$

$$y^2 + 4y + 4 = -\frac{5}{3}x + 1 + 4$$

$$(y+2)^2 = -\frac{5}{3}x + 5$$

$$(y+2)^2 = -\frac{5}{3}(x-3); \text{ **parabola**}$$

4. $x^2 + 25y^2 = 25$

$$\frac{x^2}{25} + \frac{y^2}{1} = 1$$

$$c^2 = a^2 - b^2$$

$$c^2 = 25 - 1 = 24$$

$$c = \sqrt{24} = 2\sqrt{6}$$

$$e = \frac{c}{a} = \boxed{\frac{2\sqrt{6}}{5}}$$

5. $8y^2 - 4x^2 = 8$

$$\frac{y^2}{1} - \frac{x^2}{2} = 1$$

$$c^2 = a^2 + b^2$$

$$c^2 = 1 + 2 = 3$$

$$c = \sqrt{3}$$

$$e = \frac{c}{a} = \frac{\sqrt{3}}{1} = \boxed{\sqrt{3}}$$

6. Parabola (eccentricity = 1), vertex $(0, 0)$, focus $(-2, 0)$, $c = -2$, opens left:

$$y^2 = 4cx$$

$$y^2 = 4(-2)x \qquad \boxed{y^2 = -8x}$$

7. Ellipse, major axis length = 10 $(a = 5)$ foci $(\pm 3, 0) \Longrightarrow c = 3$, x is major axis

$$\frac{x^2}{25} + \frac{y^2}{b^2} = 1 \qquad c^2 = a^2 - b^2$$

$$9 = 25 - b^2$$

$$b^2 = 16$$

$$\boxed{\frac{x^2}{25} + \frac{y^2}{16} = 1}$$

8. Hyperbola, foci $(0, \pm 5)$, vertices $(0, \pm 4)$ y-axis is the major axis, $a = 4$, $c = 5$

$$\frac{y^2}{16} - \frac{x^2}{b^2} = 1 \qquad c^2 = a^2 + b^2$$

$$25 = 16 + b^2$$

$$b^2 = 9$$

$$\boxed{\frac{y^2}{16} - \frac{x^2}{9} = 1}$$

9. Let the center of the base be $(0, 0)$ on the coordinate system.

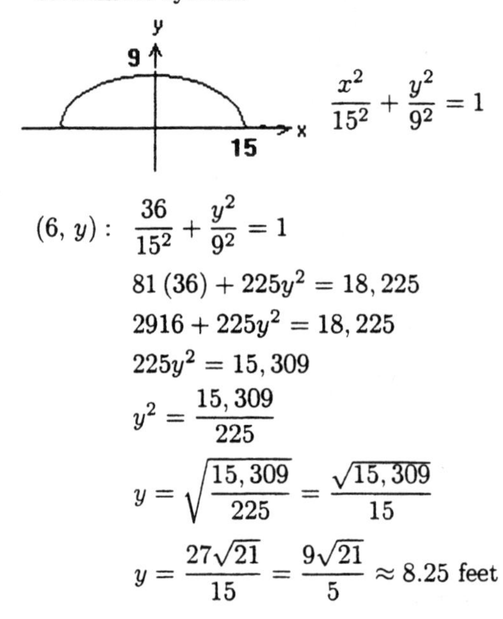

$$\frac{x^2}{15^2} + \frac{y^2}{9^2} = 1$$

$$(6, y): \quad \frac{36}{15^2} + \frac{y^2}{9^2} = 1$$

$$81(36) + 225y^2 = 18,225$$

$$2916 + 225y^2 = 18,225$$

$$225y^2 = 15,309$$

$$y^2 = \frac{15,309}{225}$$

$$y = \sqrt{\frac{15,309}{225}} = \frac{\sqrt{15,309}}{15}$$

$$y = \frac{27\sqrt{21}}{15} = \frac{9\sqrt{21}}{5} \approx 8.25 \text{ feet}$$

10. **(a)** $x = 2t, \quad y = \sqrt{t^2 + 1}, \; t \in (-\infty, \infty)$

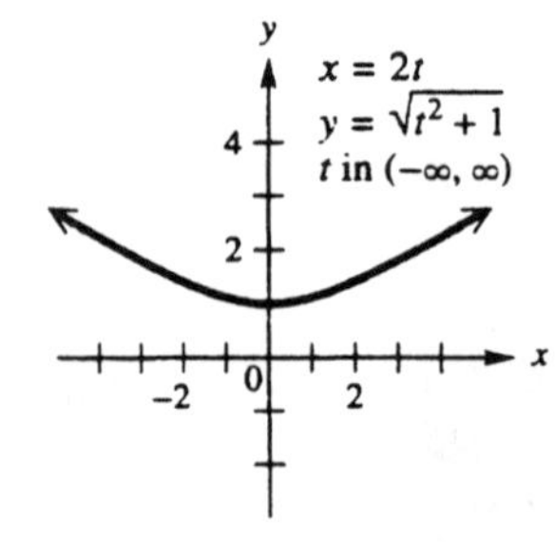

(b) Solve $x = 2t$ for $t : t = \dfrac{x}{2}$

$$t \to y: \quad y = \sqrt{\left(\frac{x}{2}\right)^2 + 1}$$

$$y = \sqrt{\frac{x^2}{4} + 1}$$

Chapter 6 Review

1. Center $(-2, 3)$, radius 5:

$(x-h)^2 + (y-k)^2 = r^2$

$\boxed{(x+2)^2 + (y-3)^2 = 25}$

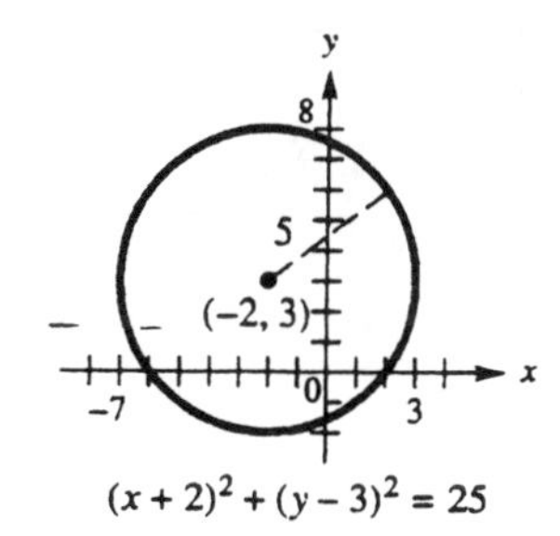

$(x+2)^2 + (y-3)^2 = 25$

Domain: $[-2 \pm 5] = [-7, 3]$
Range: $[3 \pm 5] = [-2, 8]$

2. Center $(\sqrt{5}, -\sqrt{7})$, radius $\sqrt{3}$:

$(x-h)^2 + (y-k)^2 = r^2$

$\boxed{(x-\sqrt{5})^2 + (y+\sqrt{7})^2 = 3}$

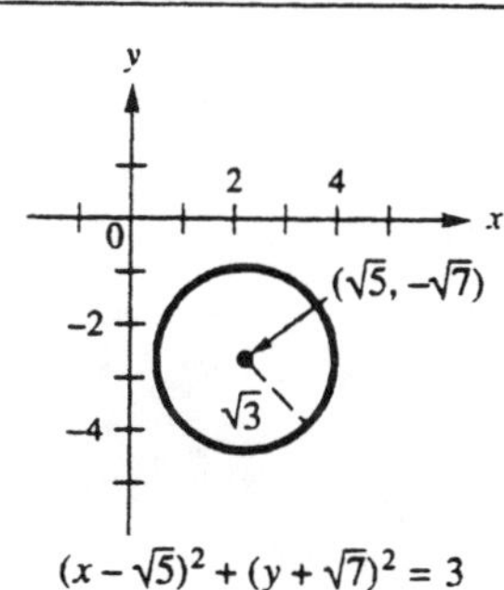

$(x-\sqrt{5})^2 + (y+\sqrt{7})^2 = 3$

Domain: $[\sqrt{5}-\sqrt{3}, \sqrt{5}+\sqrt{3}]$
Range: $[-\sqrt{7}-\sqrt{3}, -\sqrt{7}+\sqrt{3}]$

3. Center $(-8, 1)$, through $(0, 16)$:

$(x-h)^2 + (y-k)^2 = r^2$

$(x+8)^2 + (y-1)^2 = r^2$

$(0, 16): \quad (0+8)^2 + (16-1)^2 = r^2$

$64 + 225 = 289 = r^2$

$\boxed{(x+8)^2 + (y-1)^2 = 289}$

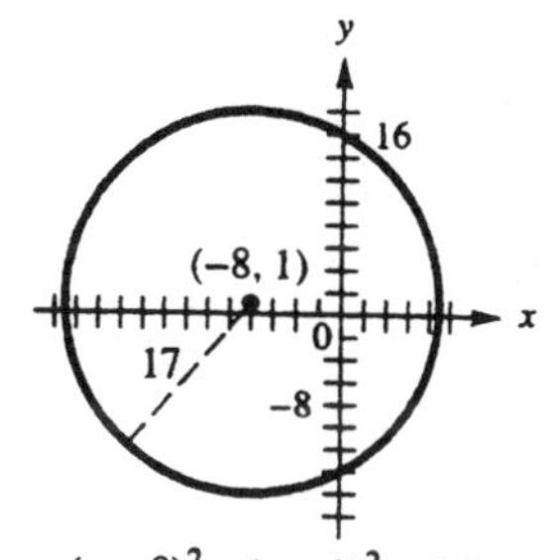

$(x+8)^2 + (y-1)^2 = 289$

Domain: $[-8 \pm 17] = [-25, 9]$
Range: $[1 \pm 17] = [-16, 18]$

4. Center $(3, -6)$, tangent to the x-axis:
If the circle is tangent to the x-axis, it must pass through $(3, 0)$; therefore the radius is the distance from $(3, 0)$ to $(3, -6) = 6$.

$(x-h)^2 + (y-k)^2 = r^2$

$\boxed{(x-3)^2 + (y+6)^2 = 36}$

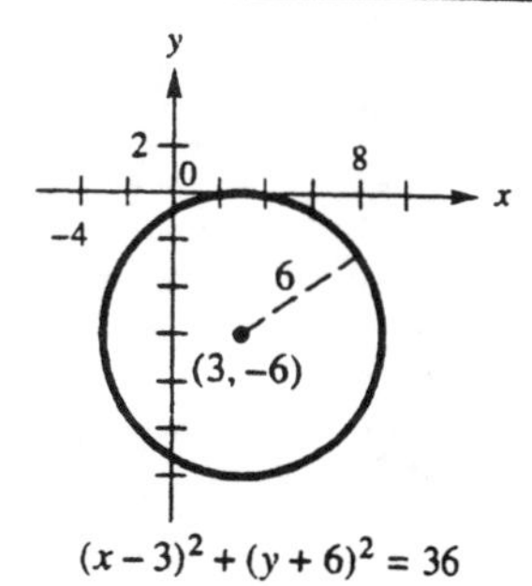

$(x-3)^2 + (y+6)^2 = 36$

Domain: $[3 \pm 6] = [-3, 9]$
Range: $[-6 \pm 6] = [-12, 0]$

5. $x^2 - 4x + y^2 + 6y + 12 = 0$

$(x^2 - 4x + 4) + (y^2 + 6y + 9)$
$= -12 + 4 + 9$

$(x-2)^2 + (y+3)^2 = 1$

$\boxed{\text{Center } (2, -3)\text{, Radius: } 1}$

6. $x^2 - 6x + y^2 - 10y + 30 = 0$

$$(x^2 - 6x + 9) + (y^2 - 10y + 25) = -30 + 9 + 25$$

$$(x-3)^2 + (y-5)^2 = 4$$

Center: $(3, 5)$, Radius: 2

7. $2x^2 + 14x + 2y^2 + 6y = -2$

$$x^2 + 7x + y^2 + 3y = -1$$

$$\left(x^2 + 7x + \frac{49}{4}\right) + \left(y^2 + 3y + \frac{9}{4}\right) = -1 + \frac{49}{4} + \frac{9}{4}$$

$$\left(x + \frac{7}{2}\right)^2 + \left(y + \frac{3}{2}\right)^2 = \frac{27}{2}$$

$$\left[\sqrt{\frac{27}{2}} = \frac{\sqrt{54}}{2} = \frac{3\sqrt{6}}{2}\right]$$

Center: $\left(-\frac{7}{2}, -\frac{3}{2}\right)$, Radius: $\frac{3\sqrt{6}}{2}$

8. $3x^2 + 3y^2 + 33x - 15y = 0$

$$x^2 + 11x + y^2 - 5y = 0$$

$$\left(x^2 + 11x + \frac{121}{4}\right) + \left(y^2 - 5y + \frac{25}{4}\right) = \frac{121}{4} + \frac{25}{4}$$

$$\left(x + \frac{11}{2}\right)^2 + \left(y - \frac{5}{2}\right)^2 = \frac{73}{2}$$

$$\left[\sqrt{\frac{73}{2}} = \frac{\sqrt{146}}{2}\right]$$

Center: $\left(-\frac{11}{2}, \frac{5}{2}\right)$, Radius: $\frac{\sqrt{146}}{2}$

9. $(x-4)^2 + (y-5)^2 = 0$; the graph consists of the single point, $(4, 5)$.

10. $y^2 = -\frac{2}{3}x$

$$4c = -\frac{2}{3}, \quad c = -\frac{1}{6}$$

Vertex at $(0, 0)$, opens left

Focus: $\left(-\frac{1}{6}, 0\right)$

Directrix: $x = \frac{1}{6}$

Axis: x-axis

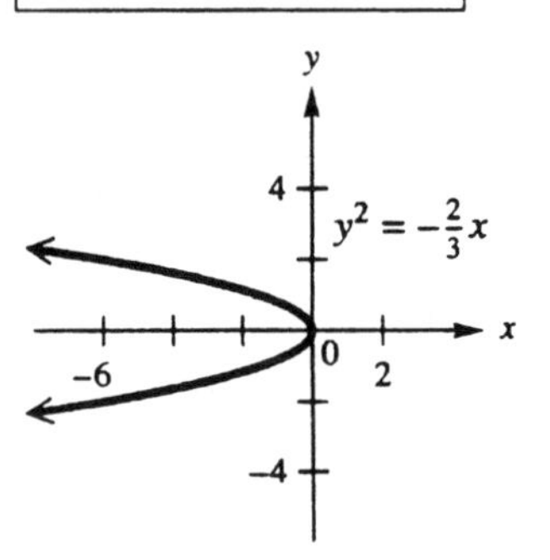

Domain: $(-\infty, 0]$; Range: $(-\infty, \infty)$

11. $y^2 = 2x$

$$4c = 2, \quad c = \frac{1}{2}$$

Vertex at $(0, 0)$ opens right:

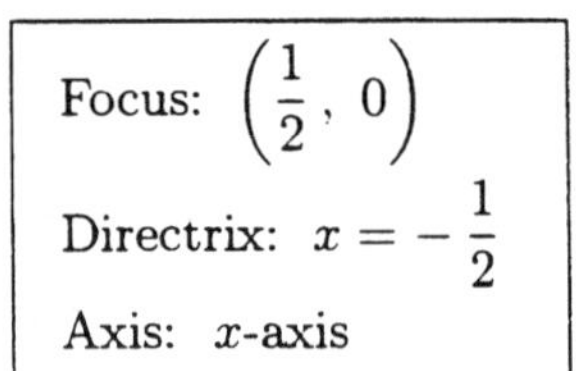

Focus: $\left(\frac{1}{2}, 0\right)$

Directrix: $x = -\frac{1}{2}$

Axis: x-axis

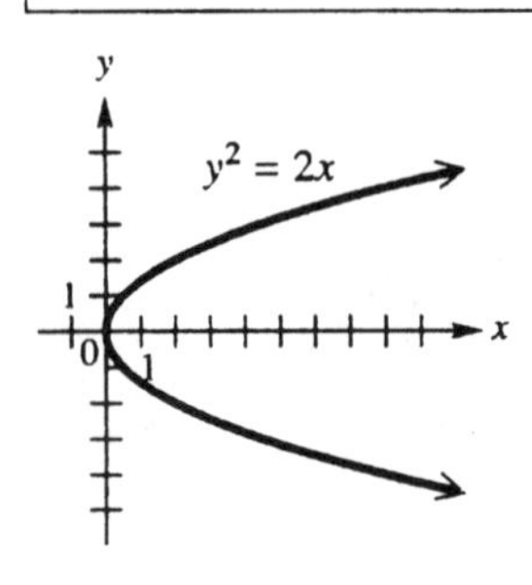

Domain: $[0, \infty)$; Range: $(-\infty, \infty)$

12. $3x^2 - y = 0$

$3x^2 = y \implies x^2 = \frac{1}{3}y$

$4c = \frac{1}{3}, \quad c = \frac{1}{12}$

Vertex at $(0, 0)$, opens upward:

Focus: $\left(0, \frac{1}{12}\right)$
Directrix: $y = -\frac{1}{12}$
Axis: y-axis

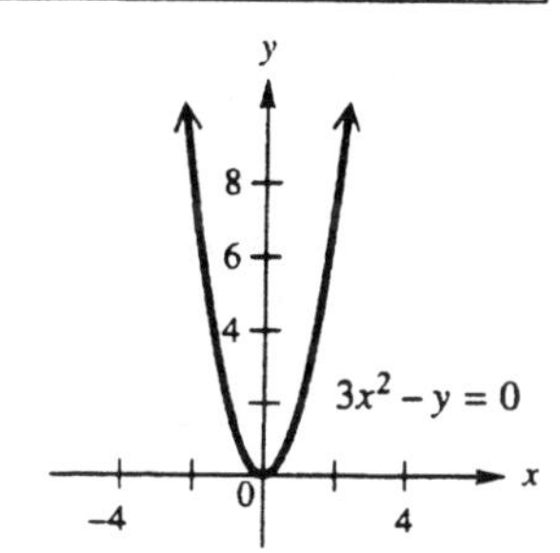

Domain: $(-\infty, \infty)$; Range: $[0, \infty)$

13. $x^2 + 2y = 0 \implies x^2 = -2y$

$4c = -2 \rightarrow c = -\frac{1}{2}$

Vertex at $(0, 0)$:

Focus: $\left(0, -\frac{1}{2}\right)$
Directrix: $y = \frac{1}{2}$
Axis: y-axis

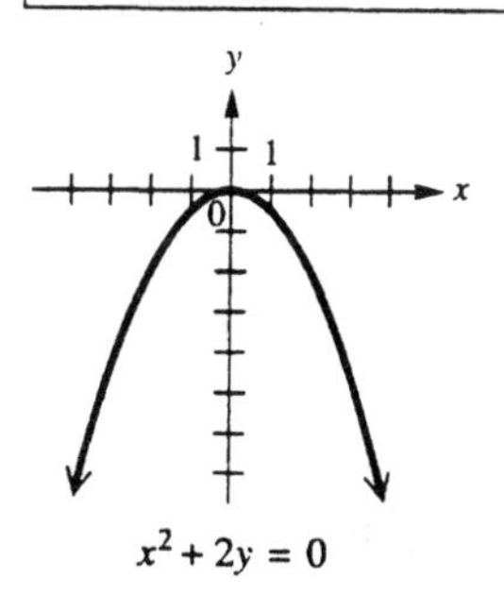

Domain: $(-\infty, \infty)$; Range: $(-\infty, 0]$

14. Vertex $(0, 0)$; focus $(4, 0)$;
parabola of the form $y^2 = 4cx$

$c =$ distance from vertex
to focus; $c = 4$ $\boxed{y^2 = 16x}$

15. Through $(2, 5)$, vertex $(0, 0)$,
opens right: parabola of the form
$y^2 = 4cx$:

$(2, 5): \quad 25 = 4c(2)$

$$c = \frac{25}{8}$$

$$y^2 = 4\left(\frac{25}{8}\right)x$$

$$\boxed{y^2 = \frac{25}{2}x}$$

16. Through $(3, -4)$, vertex $(0, 0)$,
opens downward: parabola of the form
$x^2 = 4cy$ where $c < 0$.

$(3, -4): \quad 9 = 4c(-4)$

$$c = -\frac{9}{16}$$

$$x^2 = 4\left(-\frac{9}{16}\right)y$$

$$\boxed{x^2 = -\frac{9}{4}y}$$

17. Vertex $(-5, 6)$, focus $(2, 6)$. Since the
focus is to the right of the vertex, the
parabola is:

$$(y - k)^2 = 4c(x - h);$$

c is the distance from the vertex to the
focus, so $c = 2 - (-5) = 7$.

$$(y - 6)^2 = 4(7)(x + 5)$$

$$\boxed{(y - 6)^2 = 28(x + 5)}$$

18. Vertex $(4, 3)$, focus $(4, 5)$. Since the
focus is above the vertex, the parabola is:

$$(x - h)^2 = 4c(y - k);$$

c is the distance from the vertex to the
focus, so $c = 5 - 3 = 2$.

$$(x - 4)^2 = 4(2)(y - 3)$$

$$\boxed{(x - 4)^2 = 8(y - 3)}$$

19. $\dfrac{y^2}{9} + \dfrac{x^2}{5} = 1$

Vertices: $(0, -3)$, $(0, 3)$

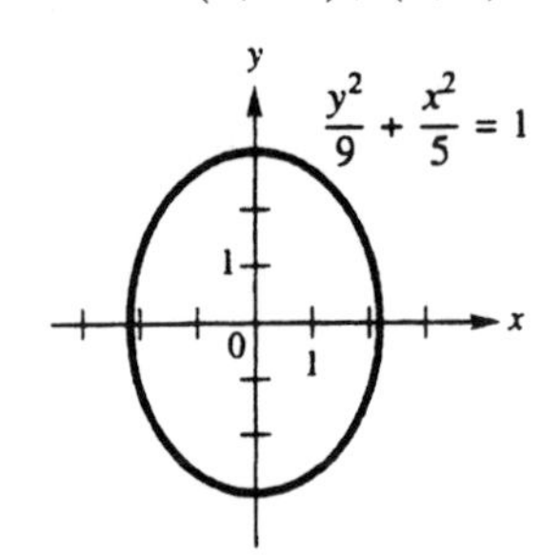

Domain: $[-\sqrt{5}, \sqrt{5}]$

Range: $[-3, 3]$

20. $\dfrac{x^2}{16} + \dfrac{y^2}{4} = 1$

Vertices: $(-4, 0)$, $(4, 0)$

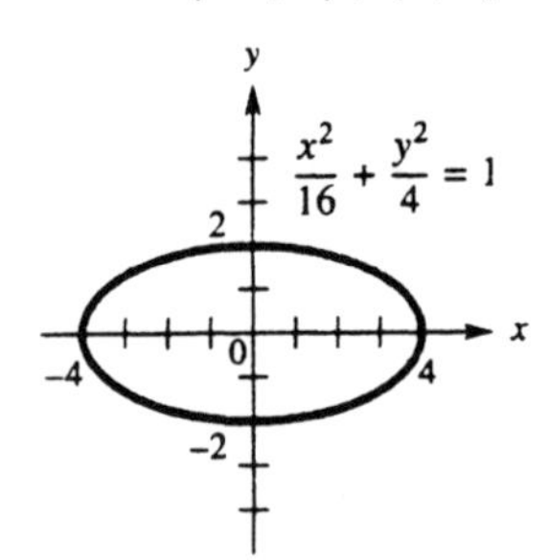

Domain: $[-4, 4]$

Range: $[-2, 2]$

21. $\dfrac{x^2}{64} - \dfrac{y^2}{36} = 1$

Vertices: $(-8, 0)$, $(8, 0)$

Asymptotes: $y = \pm\dfrac{3}{4}x$

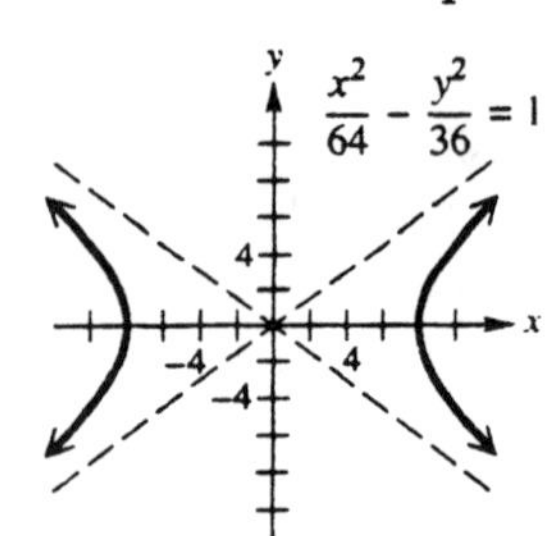

Domain: $(-\infty, -8] \cup [8, \infty)$

Range: $(-\infty, \infty)$

22. $\dfrac{y^2}{25} - \dfrac{x^2}{9} = 1$

Vertices: $(0, -5)$, $(0, 5)$

Asymptotes: $y = \pm\dfrac{5}{3}x$

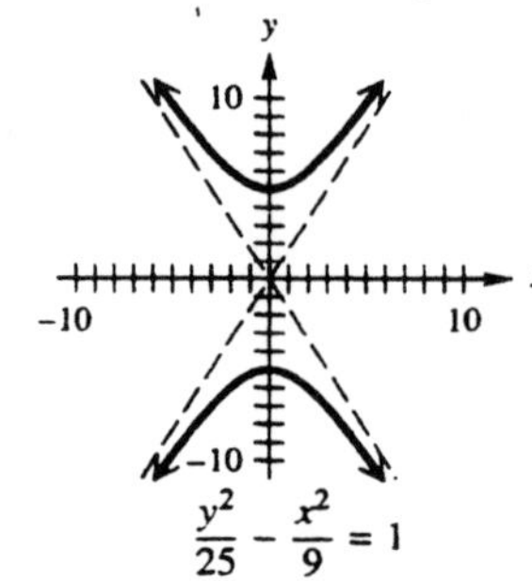

Domain: $(-\infty, \infty)$

Range: $(-\infty, -5] \cup [5, \infty)$

23. $\dfrac{(x-3)^2}{4} + (y+1)^2 = 1$

Center: $(3, -1)$

Vertices: $(1, -1)$, $(5, -1)$

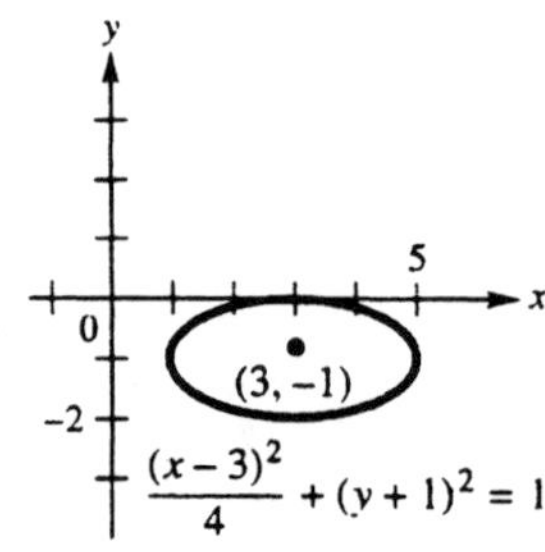

Domain: $[3 \pm 2] = [1, 5]$

Range: $[-1 \pm 1] = [-2, 0]$

24. $\dfrac{(x-2)^2}{9} + \dfrac{(y+3)^2}{4} = 1$

Center: $(2, -3)$

Vertices: $(-1, -3)$, $(5, -3)$

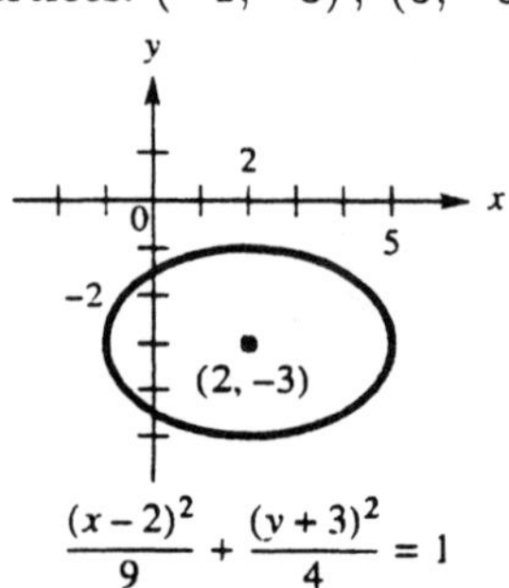

Domain: $[2 \pm 3] = [-1, 5]$

Range: $[-3 \pm 2] = [-5, -1]$

25. $\frac{(y+2)^2}{4} - \frac{(x+3)^2}{9} = 1$

Center: $(-3, -2)$
Vertices: $(-3, -4)$, $(-3, 0)$

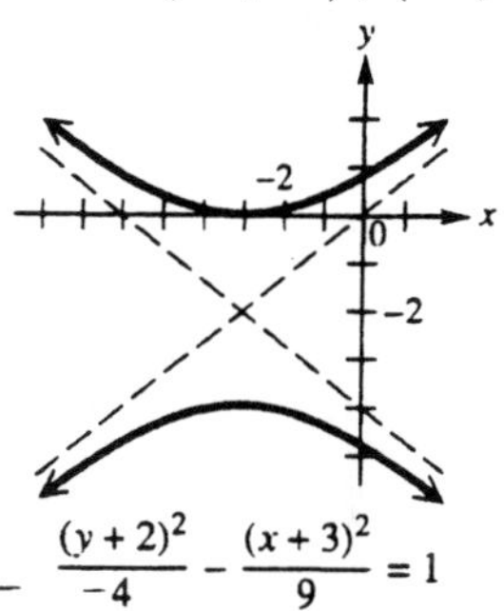

Domain: $(-\infty, \infty)$
Range: $(-\infty, -2-2] \cup [-2+2, \infty)$
$= (-\infty, -4] \cup [0, \infty)$

26. $\frac{(x+1)^2}{16} - \frac{(y-2)^2}{4} = 1$

Center: $(-1, 2)$
Vertices: $(-5, 2)$, $(3, 2)$
$(-1 \pm 4, \ 2)$;

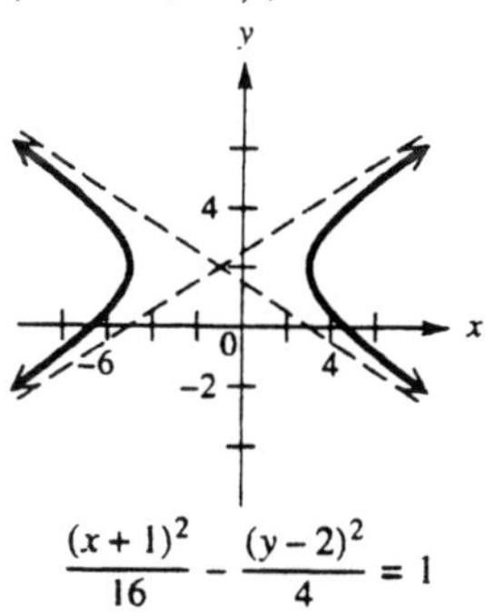

Domain: $(-\infty, -1-4] \cup [-1+4, \infty)$
$= (-\infty, -5] \cup [3, \infty)$
Range: $(-\infty, \infty)$

27. Ellipse; Vertex at $(0, 4)$, focus at $(0, 2)$, center at $(0, 0)$:

$a = 4$ (distance from center to vertex)
$c = 2$ (distance from center to focus)

$$c^2 = a^2 - b^2$$
$$4 = 16 - b^2$$
$$b^2 = 12$$

Major axis is y-axis: $\frac{y^2}{a^2} + \frac{x^2}{b^2} = 1$

$$\boxed{\frac{y^2}{16} + \frac{x^2}{12} = 1}$$

28. Ellipse; x-intercept 6, focus at $(-2, 0)$, center at $(0, 0)$:

$a = 6$ (distance from center to x-intercept)
$c = 2$ (distance from center to focus)

$$c^2 = a^2 - b^2$$
$$4 = 36 - b^2$$
$$b^2 = 32$$

Major axis is x-axis: $\frac{x^2}{a^2} + \frac{y^2}{b^2} = 1$

$$\boxed{\frac{x^2}{36} + \frac{y^2}{32} = 1}$$

29. Hyperbola; focus $(0, -5)$, center at $(0, 0)$, y-intercepts of -4, 4:

$a = 4$ ($\frac{1}{2}$ length of transverse axis)
$c = 5$ (distance from center to focus)

$$c^2 = a^2 + b^2$$
$$25 = 16 + b^2$$
$$b^2 = 9$$

Major axis is y-axis: $\frac{y^2}{a^2} - \frac{x^2}{b^2} = 1$

$$\boxed{\frac{y^2}{16} - \frac{x^2}{9} = 1}$$

30. Hyperbola; y-intercept -2, center at $(0, 0)$, passes through $(2, 3)$:

$a = 2$ (distance from center to y-intercept)

$$\frac{y^2}{4} - \frac{x^2}{b^2} = 1$$

$(2, 3)$: $\frac{3^2}{4} - \frac{2^2}{b^2} = 1$

$$9b^2 - 16 = 4b^2$$
$$5b^2 = 16 \rightarrow b^2 = \frac{16}{5}$$

$$\frac{y^2}{4} - \frac{x^2}{16/5} = 1$$

$$\boxed{\frac{y^2}{4} - \frac{5x^2}{16} = 1}$$

31. Focus at $(0, -3)$, center at $(0, 0)$,

$e = \frac{2}{3}$: since $e < 1$, this is an ellipse;

$c = 3$ (distance from center to the focus) :

$$e = \frac{c}{a}: \quad \frac{2}{3} = \frac{3}{a} \to a = \frac{9}{2}$$

$$c^2 = a^2 - b^2$$

$$9 = \tfrac{81}{4} - b^2$$

$$36 = 81 - 4b^2$$

$$4b^2 = 45 \to b^2 = \frac{45}{4}$$

y-axis is major axis:

$$\frac{y^2}{a^2} + \frac{x^2}{b^2} = 1$$

$$\frac{y^2}{81/4} + \frac{x^2}{45/4} = 1 \qquad \boxed{\frac{4y^2}{81} + \frac{4x^2}{45} = 1}$$

32. Focus at $(5, 0)$, center at $(0, 0)$, $e = \frac{5}{2}$;

Since $e > 1$, this is a hyperbola,

$c = 5$ (distance from center to focus).

$$e = \frac{c}{a}: \quad \frac{5}{2} = \frac{5}{a} \to a = 2$$

$$c^2 = a^2 + b^2$$

$$25 = 4 + b^2$$

$$b^2 = 21$$

x-axis is the major axis;

$$\frac{x^2}{a^2} - \frac{y^2}{b^2} = 1 \qquad \boxed{\frac{x^2}{4} - \frac{y^2}{21} = 1}$$

33. $x^2 + y^2 + 2x + 6y - 15 = 0$

$(x^2 + 2x + 1) + (y^2 + 6y + 9) = 15 + 1 + 9$

$(x+1)^2 + (y+3)^2 = 25$

(a) $\boxed{\text{Center at } (-1, -3)}$

(b) $\boxed{\text{Radius} = 5}$

(c) Solve for y :

$$(y+3)^2 = 25 - (x+1)^2$$

$$y + 3 = \pm\sqrt{25 - (x+1)^2}$$

$$y = \pm\sqrt{25 - (x+1)^2} - 3$$

$$\boxed{\begin{aligned} y_1 &= -3 + \sqrt{25 - (x+1)^2} \\ y_2 &= -3 - \sqrt{25 - (x+1)^2} \end{aligned}}$$

Note: $25 - (x+1)^2 = 25 - x^2 - 2x - 1$
$= 24 - 2x - x^2$

34. $\frac{x^2}{9} + \frac{y^2}{36} = 1$: **F**

35. $x = 2(y+0)^2 + 3$: **C**

36. $(x-1)^2 + (y+2)^2 = 36$: **A**

37. $\frac{x^2}{36} + \frac{y^2}{9} = 1$: **E**

38. $\frac{(y-1)^2}{36} - \frac{(x-2)^2}{36} = 1$: **B**

39. $\frac{y^2}{36} - \frac{x^2}{9} = 1$: **D**

40. $9x^2 + 25y^2 = 225$

$$\frac{9x^2}{225} + \frac{25y^2}{225} = 1$$

$$\frac{x^2}{25} + \frac{y^2}{9} = 1$$

$a^2 = 25,\ b^2 = 9$:

$$\begin{aligned} c^2 &= a^2 - b^2 \\ &= 25 - 9 \\ &= 16 \implies c = 4 \end{aligned}$$

$$e = \frac{c}{a} \qquad \boxed{e = \frac{4}{5}}$$

41. $4x^2 + 9y^2 = 36$

$$\frac{4x^2}{36} + \frac{9y^2}{36} = 1$$

$$\frac{x^2}{9} + \frac{y^2}{4} = 1$$

$a^2 = 9,\ b^2 = 4$:

$$\begin{aligned} c^2 &= a^2 - b^2 \\ &= 9 - 4 \\ &= 5 \implies c = \sqrt{5} \end{aligned}$$

$$e = \frac{c}{a} \qquad \boxed{e = \frac{\sqrt{5}}{3}}$$

42. $9x^2 - y^2 = 9$

$$\frac{x^2}{1} - \frac{y^2}{9} = 1$$

$a^2 = 1,\ b^2 = 9:$

$$\begin{aligned} c^2 &= a^2 + b^2 \\ &= 1 + 9 \\ &= 10 \implies c = \sqrt{10} \end{aligned}$$

$$e = \frac{c}{a} = \frac{\sqrt{10}}{1}$$

$$\boxed{e = \sqrt{10}}$$

43. Parabola, vertex $(-3,\ 2)$, y-intercepts at $(0,\ 5)$ and $(0,\ -1)$, of the form:

$$(y-k)^2 = 4c\,(x-h)$$
$$(y-2)^2 = 4c\,(x+3)$$

$(0,\ 5):\ (5-2)^2 = 4c\,(0+3)$

$$9 = 12c$$
$$c = \frac{9}{12} = \frac{3}{4}$$

$$(y-2)^2 = 4\left(\frac{3}{4}\right)(x+3)$$

$$\boxed{(y-2)^2 = 3\,(x+3)}$$

44. Hyperbola, foci $(0,\ \pm 12) \implies c = 12$, y is the major axis, and the center is $(0,\ 0)$;

asymptotes: $y = \pm x \implies m = \dfrac{1}{1} \implies a = b$

$$\frac{y^2}{a^2} - \frac{x^2}{b^2} = 1$$

$$c^2 = a^2 + b^2 = 2a^2$$
$$12^2 = 2a^2$$
$$a^2 = \frac{144}{2} = 72$$

$$\boxed{\frac{y^2}{72} - \frac{x^2}{72} = 1}$$

45. Ellipse, $2a = 8$

$d_1 + d_2 = 2a;\quad 2a = 8$

$$\sqrt{(x-0)^2 + (y-0)^2} + \sqrt{(x-4)^2 + (y-0)^2} = 8$$
$$\sqrt{x^2 + y^2} = 8 - \sqrt{(x-4)^2 + y^2}$$
$$x^2 + y^2 = 64 - 16\sqrt{(x-4)^2 + y^2} + (x-4)^2 + y^2$$
$$x^2 - (x-4)^2 - 64 = -16\sqrt{(x-4)^2 + y^2}$$
$$x^2 - \left(x^2 - 8x + 16\right) - 64 = -16\sqrt{(x-4)^2 + y^2}$$
$$8x - 80 = -16\sqrt{(x-4)^2 + y^2}$$
$$-x + 10 = 2\sqrt{(x-4)^2 + y^2}$$
$$x^2 - 20x + 100 = 4\left(x^2 - 8x + 16 + y^2\right)$$
$$x^2 - 20x + 100 = 4x^2 - 32x + 64 + 4y^2$$
$$3x^2 - 12x + 4y^2 = 36$$
$$3\left(x^2 - 4x\right) + 4y^2 = 36$$
$$3\left(x^2 - 4x + 4\right) + 4y^2 = 36 + 3\,(4)$$
$$3\,(x-2)^2 + 4y^2 = 48$$
$$\frac{3\,(x-2)^2}{48} + \frac{4y^2}{48} = 1$$

$$\boxed{\frac{(x-2)^2}{16} + \frac{y^2}{12} = 1}$$

46. $d\left(P,\ F'\right) - d\,(P,\ F) = 2a;\ 2a = 2$

$$\sqrt{(x-0)^2 + (y-4)^2} - \sqrt{(x-0)^2 + (y-0)^2} = 2$$
$$\sqrt{x^2 + (y-4)^2} = 2 + \sqrt{x^2 + y^2}$$
$$x^2 + (y-4)^2 = 4 + 4\sqrt{x^2 + y^2} + x^2 + y^2$$
$$x^2 + y^2 - 8y + 16 = 4 + 4\sqrt{x^2 + y^2} + x^2 + y^2$$
$$-8y + 12 = 4\sqrt{x^2 + y^2}$$
$$(-2y + 3)^2 = \left(\sqrt{x^2 + y^2}\right)^2$$
$$4y^2 - 12y + 9 = x^2 + y^2$$
$$3y^2 - 12y - x^2 = -9$$
$$3\left(y^2 - 4y\right) - x^2 = -9$$
$$3\left(y^2 - 4y + 4\right) - x^2 = -9 + 3\,(4)$$
$$3\,(y-2)^2 - x^2 = 3$$

$$\boxed{(y-2)^2 - \frac{x^2}{3} = 1}$$

47. $x = 4t - 3, \quad y = t^2$

$t \in [-3, 4] \to x \in [-15, 13]$

$x = 4t - 3, y = t^2, t$ in $[-3, 4]$

20

−20 20

−20

48. $x = t^2, \quad y = t^3$

$t \in [-2, 2] \to x \in [0, 4]$

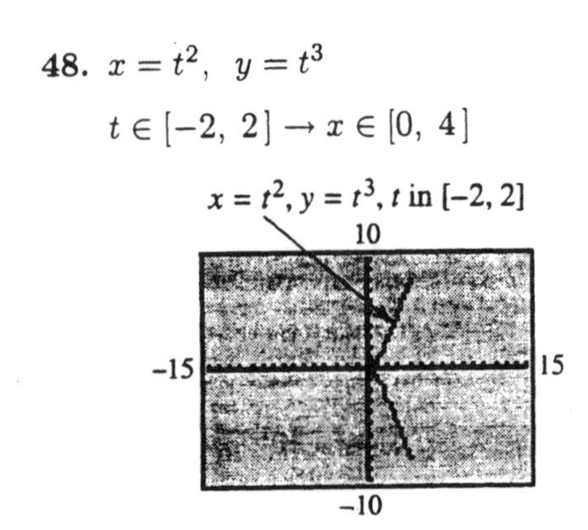

49. $x = t + \ln t, \ y = t + e^t$

$t \in (0, 2] \to x \in (-\infty, 2 + \ln 2]$

$x = t + \ln t, y = t + e^t, t$ in $(0, 2]$

10

−5 5

0

50. $x = 3t + 2, \quad y = t - 1, \quad t \in [-5, 5]$

Solve $y = t - 1$ for t: $\quad t = y + 1$

$t \Rightarrow x$:

$$x = 3(y + 1) + 2$$
$$x = 3y + 3 + 2$$

$$\boxed{x - 3y = 5, \quad x \in [-13, 17]}$$

51. $x = \sqrt{t - 1}, \quad y = \sqrt{t} \quad t \in [1, \infty)$

Solve $y = \sqrt{t}$ for t: $\quad t = y^2$

$t \Rightarrow x$: $\quad x = \sqrt{y^2 - 1}$

$$x^2 = y^2 - 1$$
$$x^2 - y^2 = -1$$

$$\boxed{y^2 - x^2 = 1, \quad x \in [0, \infty)}$$

52. $x = \dfrac{1}{t + 3}, \ y = t + 3, \ t \neq -3$

$x = \dfrac{1}{t + 3} \to t + 3 = \dfrac{1}{x}$, so

$$\boxed{y = \frac{1}{x}, \ x \neq 0}$$

53. Since the major axis has length 134.5,

$$2a = 134.5$$
$$a = 67.25$$

Use the eccentricity to find c:

$$e = \frac{c}{a} = .006775$$
$$\frac{c}{67.25} = .006775$$
$$c = .4556188 \approx .46$$

The smallest distance to the sun is
$a - c \approx 67.25 - .46 \approx 66.8$ million miles.

The greatest distance to the sun is
$a + c \approx 67.25 + .46 \approx 67.7$ million miles.

54. Since the eccentricity is .964,

$$e = \frac{c}{a} = .964 \Longrightarrow c = .964a$$

The closest distance to the sun is 89, so

$$a - c = 89$$
$$a - .964a = 89$$
$$.036a = 89$$
$$a = 2472.\overline{2}$$
$$c = .964a = 2383.\overline{2}$$

Solve for b^2:

$$c^2 = a^2 - b^2$$
$$\left(2383.\overline{2}\right)^2 = \left(2472.\overline{2}\right)^2 - b^2$$
$$5679748 = 6111883 - b^2$$
$$b^2 \approx 432,135$$
$$a^2 \approx 6,111,883$$

The equation is: $\boxed{\dfrac{x^2}{6,111,883} + \dfrac{y^2}{432,135} = 1}$

55. $\dfrac{k}{\sqrt{D}} = \dfrac{2.82 \times 10^7}{\sqrt{42.5 \times 10^6}} = \dfrac{2.82 \times 10^7}{\sqrt{42.5} \times 10^3}$

$= .432568 \times 10^4 \approx 4326$

Since $V = 2090$, we have

$$V < \frac{k}{\sqrt{D}} \quad (2090 < 4326),$$

so the shape of the satellite's trajectory was elliptic.

56. $\dfrac{k}{\sqrt{D}} = \dfrac{2.82 \times 10^7}{\sqrt{42.5 \times 10^6}} = \dfrac{2.82 \times 10^7}{\sqrt{42.5} \times 10^3}$

$= .432568 \times 10^4 \approx 4326$

The velocity must be increased from 2090 m/sec (**#55**) to at least 4326 m/sec. Thus, the minimum increase is

$$4326 - 2090 = \boxed{2236 \text{ m/sec}}$$

57. The required change in velocity is less when D is larger.

58. $Ax^2 + Cy^2 + Dx + Ey + F = 0$

Complete the square on x and y:

$$(Ax^2 + Dx) + (Cy^2 + Ey) = -F$$

$$A\left(x^2 + \frac{D}{A}x\right) + C\left(y^2 + \frac{E}{C}y\right) = -F$$

$$A\left(x^2 + \frac{D}{A}x + \frac{D^2}{4A^2} - \frac{D^2}{4A^2}\right)$$

$$+C\left(y^2 + \frac{E}{C}y + \frac{E^2}{4C^2} - \frac{E^2}{4C^2}\right) = -F$$

$$A\left(x^2 + \frac{D}{A}x + \frac{D^2}{4A^2}\right) + C\left(y^2 + \frac{E}{C}y + \frac{E^2}{4C^2}\right)$$

$$= -F + A\left(\frac{D^2}{4A^2}\right) + C\left(\frac{E^2}{4C^2}\right)$$

$$A\left(x + \frac{D}{2A}\right)^2 + C\left(y + \frac{E}{2C}\right)^2 = \frac{D^2}{4A} + \frac{E^2}{4C} - F$$

Divide both sides by AC:

$$\frac{\left(x + \frac{D}{2A}\right)^2}{C} + \frac{\left(y + \frac{E}{2C}\right)^2}{A}$$

$$= \frac{D^2}{4A^2C} + \frac{E^2}{4AC^2} - \frac{F}{AC}$$

The center of the ellipse is at:

$$\boxed{\left(-\frac{D}{2A}, -\frac{E}{2C}\right)}$$

Chapter 6 Test

1. **(a)** $4(x+3)^2 - (y+2)^2 = 16$: **B**

(b) $(x-3)^2 + (y-2)^2 = 16$: **E**

(c) $(x+3)^2 + (y-2)^2 = 16$: **F**

(d) $(x+3)^2 + y = 4$: **A**

(e) $x + (y-2)^2 = 4$: **C**

(f) $4(x+3)^2 + (y+2)^2 = 16$: **D**

2. $y^2 = \dfrac{1}{8}x$; opens right

$$4c = \frac{1}{8}$$

$$c = \frac{1}{32}$$

Focus: $\left(\dfrac{1}{32}, 0\right)$

$\left\{\dfrac{1}{32} \text{ to the right of vertex } (0,0)\right\}$

Directrix: $x = -\dfrac{1}{32}$

$\left\{\dfrac{1}{32} \text{ to the left of vertex } (0,0)\right\}$

3. $y = -\sqrt{1 - \dfrac{x^2}{36}}$

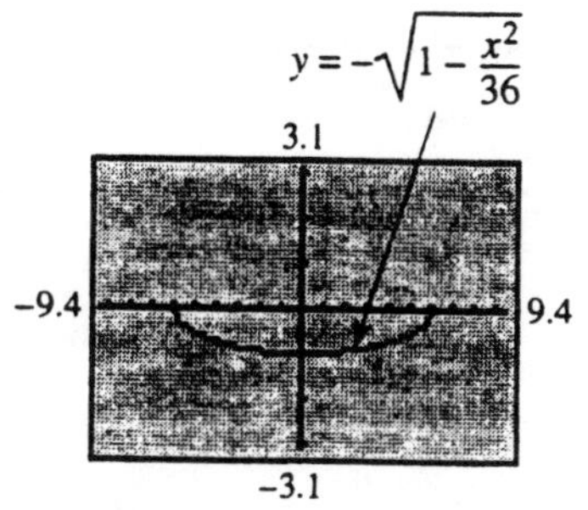

Yes, it is a function.

Domain: $[-6, 6]$; Range: $[-1, 0]$

4. $\dfrac{x^2}{25} - \dfrac{y^2}{49} = 1$

$\dfrac{y^2}{49} = \dfrac{x^2}{25} - 1$

$y^2 = 49\left(\dfrac{x^2}{25} - 1\right)$

$y = \pm\sqrt{49\left(\dfrac{x^2}{25} - 1\right)}$

$y_1 = 7\sqrt{\dfrac{x^2}{25} - 1}; \quad y_2 = -7\sqrt{\dfrac{x^2}{25} - 1}$

5. $\dfrac{y^2}{4} - \dfrac{x^2}{9} = 1$

Hyperbola (opens up/down): Center $(0,\ 0)$

Vertices: $(0,\ 2)$ and $(0,\ -2)$ $[(0,\ 0 \pm 2)]$

Foci: $\left(0,\ \sqrt{13}\right)$ and $\left(0,\ -\sqrt{13}\right)$

$a^2 + b^2 = c^2$

$4 + 9 = 13$

$c = \pm\sqrt{13}$

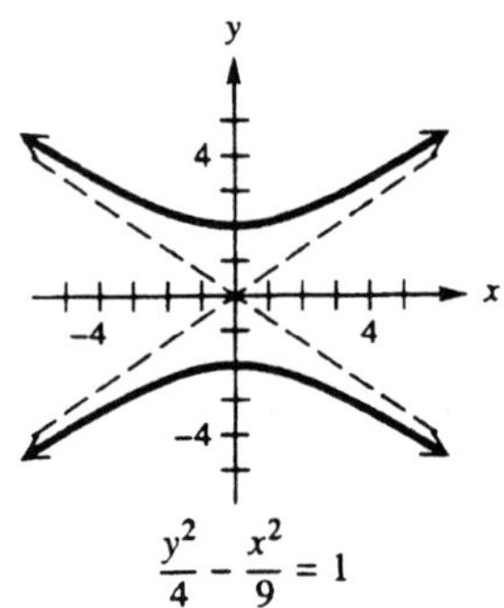

6. $x^2 + 4y^2 + 2x - 16y + 17 = 0$

$\left(x^2 + 2x + 1\right) + 4\left(y^2 - 4y + 4\right) = -17 + 1 + 16$

$(x+1)^2 + 4(y-2)^2 = 0$

This is the point $(-1,\ 2)$.

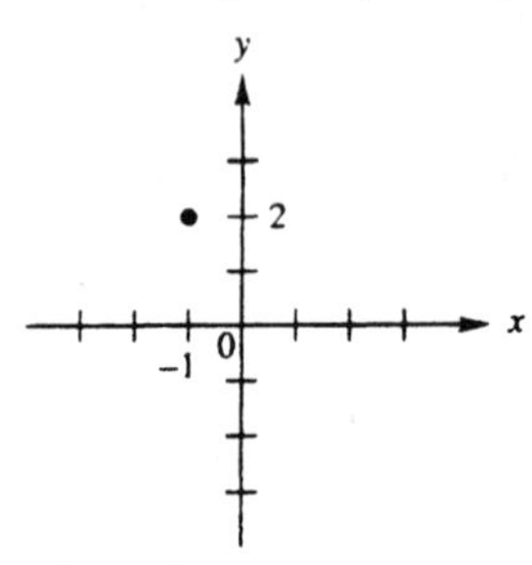

7. $y^2 - 2x - 8y + 22 = 0$

$-2x = -y^2 + 8y - 22$

$x = \dfrac{1}{2}y^2 - 4y + 11$

$x = \dfrac{1}{2}\left(y^2 - 8y + 16\right) + 11 - 8$

$x = \dfrac{1}{2}(y-4)^2 + 3$

Parabola; opens right; vertex: $(3,\ 4)$

To find the focus:

$\dfrac{1}{4c} = \dfrac{1}{2} \implies c = \dfrac{1}{2}$

Since the parabola opens right, move $\dfrac{1}{2}$ unit right from the vertex.

Parabola; Vertex: $(3,\ 4)$, Focus: $\left(\dfrac{7}{2},\ 4\right)$

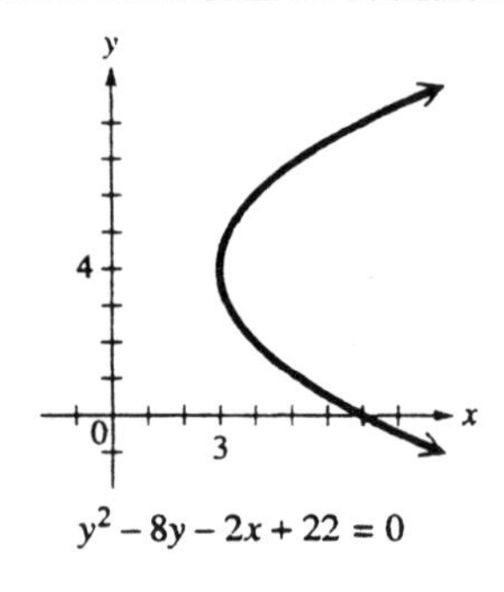

8. $x^2 + (y-4)^2 = 9$

Circle, center at $(0,\ 4)$, radius $= 3$

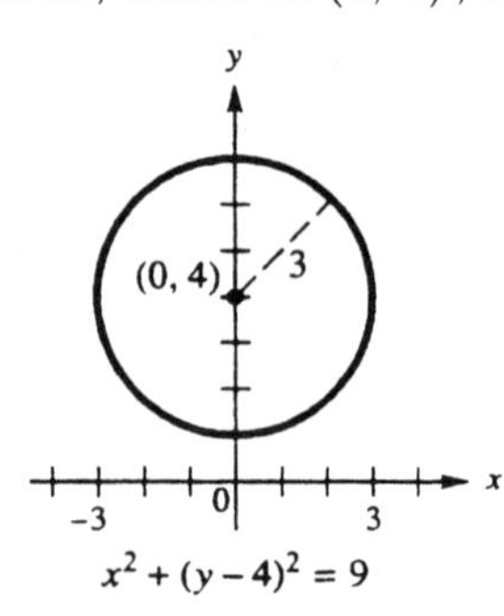

9. $\dfrac{(x-3)^2}{49}+\dfrac{(y+1)^2}{16}=1$

Ellipse, center $(3,\ -1)$

Vertices: $(3-7,\ -1)=(-4,\ -1)$

$(3+7,\ -1)=(10,\ -1)$

$c^2=a^2-b^2$

$c^2=49-16=33$

$c=\pm\sqrt{33}$

Foci: $(3-\sqrt{33},\ -1)$, $(3+\sqrt{33},\ -1)$

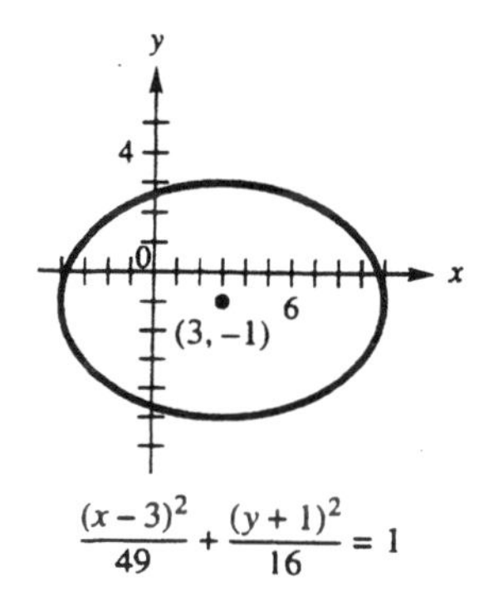

$\dfrac{(x-3)^2}{49}+\dfrac{(y+1)^2}{16}=1$

10. $x=4t^2-4,\ y=t-1,\ t\in[-1.5,\ 1.5]$

Solve $y=t-1$ for $t:\quad t=y+1$

$t\to x:\quad x=4(y+1)^2-4$

$4(y+1)^2=x+4$

$$\boxed{(y+1)^2=\frac{1}{4}(x+4)}$$

$$(y+1)^2=4\left(\frac{1}{16}\right)(x+4)$$

Parabola, vertex at $(-4,\ -1)$,

focus at $\left(-4+\dfrac{1}{16},\ -1\right)=\left(-\dfrac{63}{16},\ -1\right)$

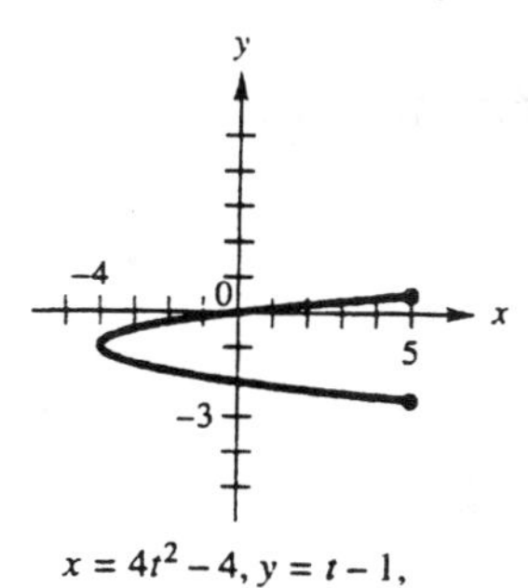

$x=4t^2-4,\ y=t-1,$
t in $[-1.5,\ 1.5]$

11. **(a)** Center $(0,\ 0)$, focus $(0,\ -2)$, $e=1$.

Since $e=1$, this is a parabola. Since the focus is *below* the center, it is of the form

$x^2=4cy$

$c=-2:\quad x^2=4(-2)y$

$$\boxed{x^2=-8y}$$

(b) Center $(0,\ 0)$, vertical major axis of length 6, $e=\dfrac{5}{6}$.

Since $e<1$, this is an ellipse of the form

$\dfrac{x^2}{b^2}+\dfrac{y^2}{a^2}=1$

Major axis length $6\to a=3$.

$e=\dfrac{5}{6}=\dfrac{c}{a}=\dfrac{c}{3}$

$6c=15\Longrightarrow c=\dfrac{5}{2}$

$c^2=a^2-b^2$

$\dfrac{25}{4}=9-b^2$

$\dfrac{25}{4}-\dfrac{36}{4}=-b^2\Longrightarrow\dfrac{11}{4}=b^2$

$$\boxed{\frac{x^2}{\frac{11}{4}}+\frac{y^2}{9}=1\quad\text{or}\quad\frac{4x^2}{11}+\frac{y^2}{9}=1}$$

12. Let the arch be placed as follows:

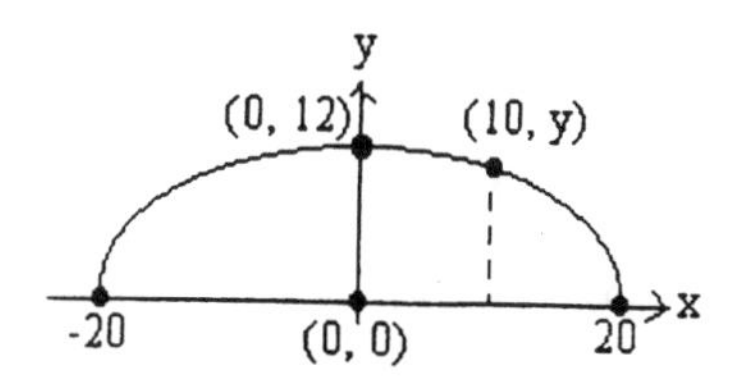

Ellipse, center $(0,\ 0)$, x is major axis, $a=20,\ b=12$.

$\dfrac{x^2}{a^2}+\dfrac{y^2}{b^2}=1$

$$\boxed{\frac{x^2}{400}+\frac{y^2}{144}=1}$$

Find y when $x=10$:

$\dfrac{100}{400}+\dfrac{y^2}{144}=1$

Multiply by 144:

$36+y^2=144$

$y^2=108$

$y=10.3923$

$\boxed{\text{Approximately 10.39 feet}}$

Chapter 6 Project

Modeling the Path of a Bouncing Ball,

1. Following the procedure in the text for the second bounce, we get the following equations and domains for the other four parabolas.

Bounce 1:

$$y = -16\,(x - .08)^2 + 2.54$$
$$(0 \le x \le .51)$$

Bounce 3:

$$y = -16\,(x - 1.53)^2 + 1.62$$
$$(1.19 < x \le 1.83)$$

Bounce 4:

$$y = -16\,(x - 2.08)^2 + 1.42$$
$$(1.83 < x \le 2.38)$$

Bounce 5:

$$y = -16\,(x - 2.63)^2 + 1.20$$
$$(2.38 < x \le 2.76)$$

Combining these facts with the equation and domain for the second bounce, the piecewise equation is

$$\begin{aligned} Y_1 = &\left[-16\,(x - .08)^2 + 2.54\right] (x \ge 0)\,(x \le .51) \\ &+ \left[-16\,(x - .85)^2 + 1.92\right] (x > .51)\,(x \le 1.19) \\ &+ \left[-16\,(x - 1.53)^2 + 1.62\right] (x > 1.19)\,(x \le 1.83) \\ &+ \left[-16\,(x - 2.08)^2 + 1.42\right] (x > 1.83)\,(x \le 2.38) \\ &+ \left[-16\,(x - 2.63)^2 + 1.20\right] (x > 2.38)\,(x \le 2.76) \end{aligned}$$

[Note: the text has open intervals for the domain, which omits the end points.]

Use the window shown with the graph of the data in Figure A to graph the piecewise function.

In the graph below, the function is graphed with the data.

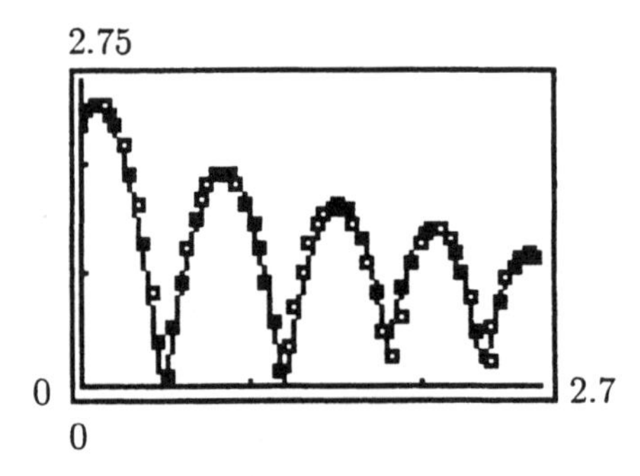

As the graph shows, the function matches the data very well.

2. In the figure, we show a partial view of the graph using the path style.

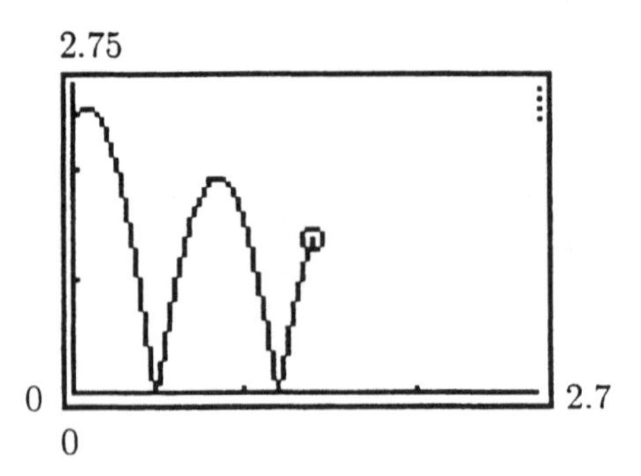

3. Answers will vary because the different groups will have different data

Section 7.1

1. The two graphs intersect at $\approx (2002,\ 3.1)$. Therefore, in approximately 2002, both projections produce the same level of migration.

3. The two graphs intersect at $\approx (2002,\ 3.1)$, which is the solution of the system.

5. t would represent time in years and y would represent the number of migrants.

7. $(1,\ 1)$: $\quad 4x - y = 3$

$$4(1) - (1) = 3$$
$$3 = 3 \ \checkmark$$

$(1,\ 1)$: $\quad -2x + 3y = 1$

$$-2(1) + 3(1) = 1$$
$$1 = 1 \ \checkmark$$

9. $(3,\ 5)$: $\quad y - x = 2$

$$5 - 3 = 2 \ \checkmark$$

$(3,\ 5)$: $\quad x^2 + y^2 = 34$

$$(3)^2 + (5)^2 = 34$$
$$9 + 25 = 34$$
$$34 = 34 \ \checkmark$$

11. $6x - y = 5 \quad [1]$

$y = x \quad [2]$

$[2] \longrightarrow [1]$: $\quad 6x - x = 5$

$$5x = 5 \qquad \mathbf{x = 1}$$

$x = 1 \longrightarrow [2]$:

$$y = 1 \qquad \mathbf{y = 1} \qquad \boxed{\{(1,\ 1)\}}$$

13. $x + 2y = -1 \quad [1]$

$2x + y = 4 \quad [2]$

Solve $[2]$ for y: $\quad y = 4 - 2x \quad [3]$

$[3] \longrightarrow [1]$: $\quad x + 2(4 - 2x) = -1$

$$x + 8 - 4x = -1$$
$$-3x = -9 \qquad \mathbf{x = 3}$$

$x = 3 \longrightarrow [3]$:

$$y = 4 - 2(3) \qquad \mathbf{y = -2} \qquad \boxed{\{(3,\ -2)\}}$$

15. $y = 2x + 3 \quad [1]$

$3x + 4y = 78 \quad [2]$

$[1] \longrightarrow [2]$: $\quad 3x + 4(2x + 3) = 78$

$$3x + 8x + 12 = 78$$
$$11x = 66 \qquad \mathbf{x = 6}$$

$x = 6 \longrightarrow [1]$:

$$y = 2(6) + 3 \qquad \mathbf{y = 15} \qquad \boxed{\{(6,\ 15)\}}$$

17. $3x - 2y = 12 \quad [1]$

$5x = 4 - 2y \quad [2]$

Solve $[1]$ for x: $\quad 3x = 12 + 2y$

$$x = 4 + \tfrac{2}{3}y \quad [3]$$

$[3] \longrightarrow [2]$:

$$5\left(4 + \tfrac{2}{3}y\right) = 4 - 2y$$
$$20 + \tfrac{10}{3}y = 4 - 2y$$
$$60 + 10y = 12 - 6y$$
$$16y = -48, \qquad \mathbf{y = -3}$$

$y = -3 \longrightarrow [3]$:

$$x = 4 + \tfrac{2}{3}(-3)$$
$$x = 4 - 2 \qquad \mathbf{x = 2} \qquad \boxed{\{(2, -3)\}}$$

19. $4x - 5y = -11 \quad [1]$

$2x + y = 5 \quad [2]$

Solve $[2]$ for y: $\quad y = 5 - 2x \quad [3]$

$[3] \longrightarrow [1]$:

$$4x - 5(5 - 2x) = -11$$
$$4x - 25 + 10x = -11$$
$$14x = 14 \qquad \mathbf{x = 1}$$

$x = 1 \longrightarrow [3]$:

$$y = 5 - 2(1) \qquad \mathbf{y = 3} \qquad \boxed{\{(1,\ 3)\}}$$

21. $4x + 5y = 7$ [1]
$9y = 31 + 2x$ [2]

Solve [2] for y:

$$y = \frac{31 + 2x}{9} \quad [3]$$

[3] ⟶ [1]:

$$4x + 5\left(\frac{31 + 2x}{9}\right) = 7$$

Multiply by 9:

$$36x + 5(31 + 2x) = 63$$
$$36x + 155 + 10x = 63$$
$$46x = -92 \quad \mathbf{x = -2}$$

$x = -2 \longrightarrow [3]$:

$$y = \frac{31 + 2(-2)}{9}$$
$$y = \frac{27}{9} \quad \mathbf{y = 3} \quad \boxed{\{(-2,\ 3)\}}$$

23. $3x - 7y = 15$ [1]
$3x + 7y = 15$ [2]

Solve [1] for x: $3x = 7y + 15$

$$x = \frac{7y + 15}{3} \quad [3]$$

[3] ⟶ [2]: $3\left(\frac{7y + 15}{3}\right) + 7y = 15$

$$7y + 15 + 7y = 15$$
$$14y = 0 \quad \mathbf{y = 0}$$

$y = 0 \longrightarrow [3]$:

$$x = \frac{7(0) + 15}{3} \quad \mathbf{x = 5} \quad \boxed{\{(5,\ 0)\}}$$

25. $4x - 5y = -11$ [1]
$2x + y = 5$ [2]

Solve [1] for y:

$$-5y = -4x - 11$$
$$y = \frac{4}{5}x + \frac{11}{5} \longrightarrow m = \frac{4}{5},\ b = \frac{11}{5}$$

Solve [2] for y:

$$y = -2x + 5 \longrightarrow m = -2, b = 5$$

Screen A appears to have lines with these characteristics.

27. An inconsistent system will end up with no variables and a false statement (ie: $-3 = 4$). A dependent system will end up with no variables and a true statement (ie: $-3 = -3$)

29. $3x - y = -4$ [1]
$x + 3y = 12$ [2]

$$\begin{array}{rl} 3x - \ \ y = \ \ -4 & [1] \\ -3x - 9y = -36 & [-3 \times [2]] \\ \hline -10y = -40 & \mathbf{y = 4} \end{array}$$

$y = 4 \longrightarrow [1]$:

$$3x - (4) = -4$$
$$3x = 0 \quad \mathbf{x = 0} \quad \boxed{\{(0,\ 4)\}}$$

31. $4x + 3y = -1$ [1]
$2x + 5y = 3$ [2]

$$\begin{array}{rl} 4x + \ \ 3y = -1 & [1] \\ -4x - 10y = -6 & [-2 \times [2]] \\ \hline -7y = -7 & \mathbf{y = 1} \end{array}$$

$y = 1 \longrightarrow [1]$:

$$4x + 3(1) = -1$$
$$4x = -4 \quad \mathbf{x = -1} \quad \boxed{\{(-1,\ 1)\}}$$

33. $12x - 5y = 9$ [1]
$3x - 8y = -18$ [2]

$$\begin{array}{rl} 12x - \ \ 5y = \ \ 9 & [1] \\ -12x + 32y = 72 & [-4 \times [2]] \\ \hline 27y = 81 & \mathbf{y = 3} \end{array}$$

$y = 3 \longrightarrow [1]$:

$$12x - 5(3) = 9$$
$$12x - 15 = 9$$
$$12x = 24 \quad \mathbf{x = 2}$$

$$\boxed{\{(2,\ 3)\}}$$

35. $4x - y = 9$ [1]
$-8x + 2y = -18$ [2]

$8x - 2y = 18$ [2 × · [1]]
$-8x + 2y = -18$ [2]
$0 = 0$ **∞ solutions**

$4x - y = 9$ [1]
$-y = -4x + 9$
$y = 4x - 9$

$\boxed{(x,\ 4x-9) \text{ or } \left(\frac{y+9}{4},\ y\right)}$

37. $9x - 5y = 1$ [1]
$-18x + 10y = 1$ [2]

$18x - 10y = 2$ [2 × [1]]
$-18x + 10y = 1$ [2]
$0 = 3$

This is a false statement. $\boxed{\emptyset}$
The system is inconsistent.

39. $3x + y = 6$ [1]
$6x + 2y = 1$ [2]

$-6x - 2y = -12$ [−2 × [1]]
$6x + 2y = 1$ [2]
$0 = -11$

This is a false statement. $\boxed{\emptyset}$
The system is inconsistent.

41. $\frac{x}{2} + \frac{y}{3} = 8$ [1]
$\frac{2x}{3} + \frac{3y}{2} = 17$ [2]

Multiply [1] × 6 and [2] × 6 :

$3x + 2y = 48$ [3]
$4x + 9y = 102$ [4]

$-12x - 8y = -192$ [−4 × [3]]
$12x + 27y = 306$ [3 × [4]]
$19y = 114$ $\mathbf{y = 6}$

$y = 6 \longrightarrow$ [3] :

$3x + 2(6) = 48$
$3x = 36$
$\mathbf{x = 12}$ $\boxed{\{(12, 6)\}}$

43. $\frac{2x-1}{3} + \frac{y+2}{4} = 4$ [1]
$\frac{x+3}{2} - \frac{x-y}{3} = 3$ [2]

Multiply [1] × 12 and [2] × 6 :

$4(2x - 1) + 3(y + 2) = 48$
$3(x + 3) - 2(x - y) = 18$

Remove parenthesis and combine like terms:

$8x + 3y = 46$ [3]
$x + 2y = 9$ [4]

$8x + 3y = 46$ [3]
$-8x - 16y = -72$ [−8 × [4]]
$-13y = -26$ $\mathbf{y = 2}$

$y = 2 \longrightarrow$ [4] :

$x + 2(2) = 9$ $\mathbf{x = 5}$ $\boxed{\{(5, 2)\}}$

45. $\sqrt{3}x - y = 5 \Longrightarrow y = \sqrt{3}x - 5$ (dashed)
$100x + y = 9 \Longrightarrow y = 9 - 100x$ (solid)

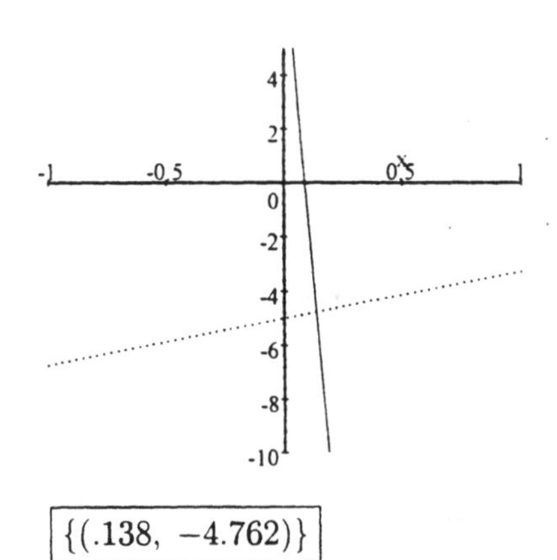

$\boxed{\{(.138,\ -4.762)\}}$

47. $\sqrt{5}x + \sqrt[3]{6}y = 9 \Longrightarrow y = \frac{-\sqrt{5}x + 9}{\sqrt[3]{6}}$ (solid)
$\sqrt{2}x + \sqrt[5]{9}y = 12 \Longrightarrow y = \frac{-\sqrt{2}x + 12}{\sqrt[5]{9}}$ (dashed)

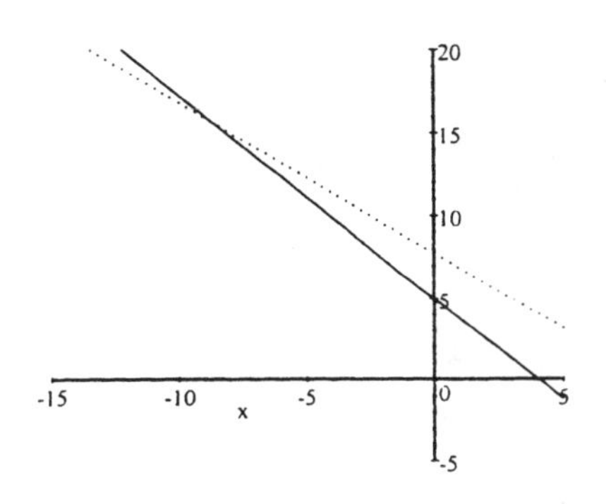

$\boxed{\{(-8.708,\ 15.668)\}}$

49. If (3, 5) then (−5, −3) :

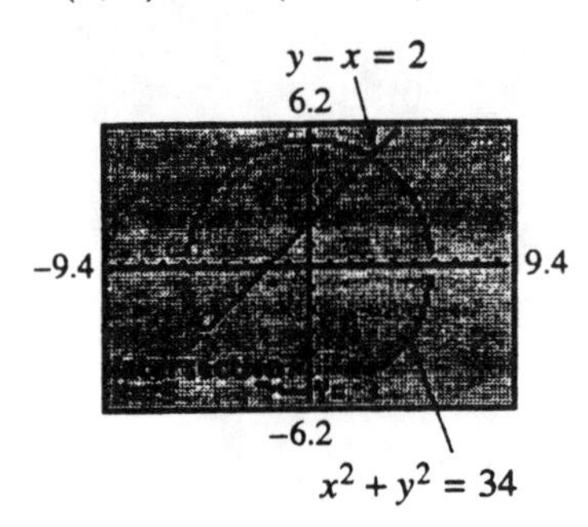

51. A line and a circle; no points:

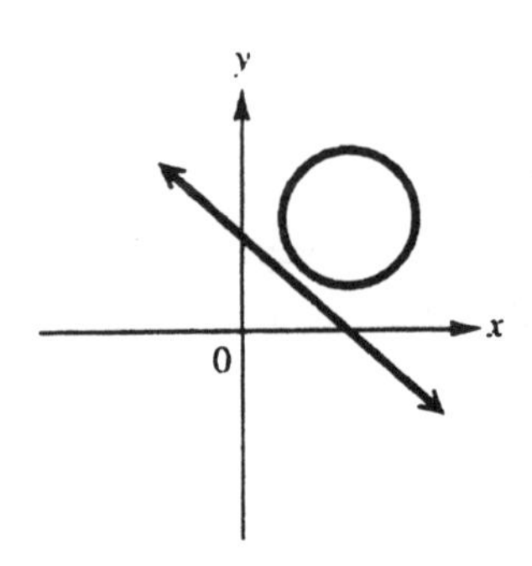

53. A line and a circle; two points:

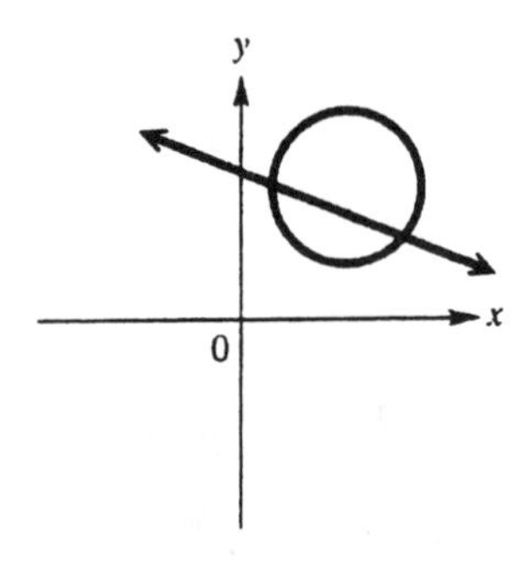

55. A line and a hyperbola; one point:

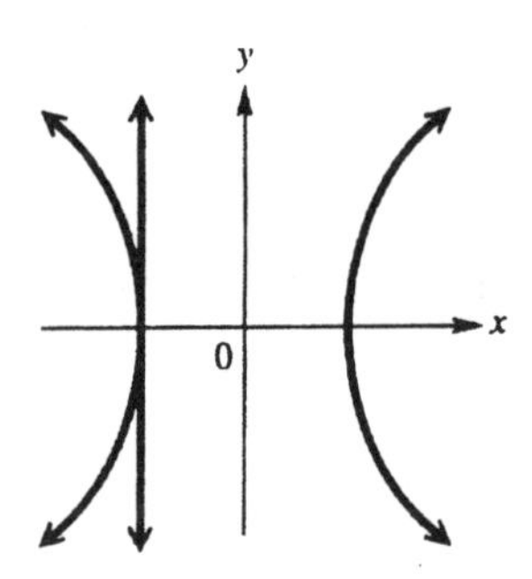

57. A circle and an ellipse; 4 points:

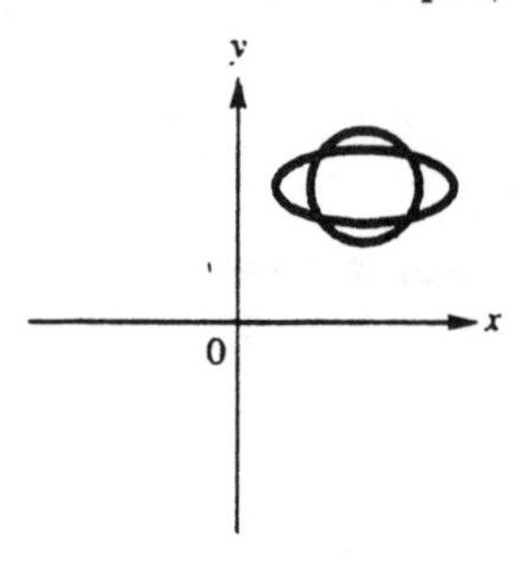

59. A parabola and a hyperbola; 4 points:

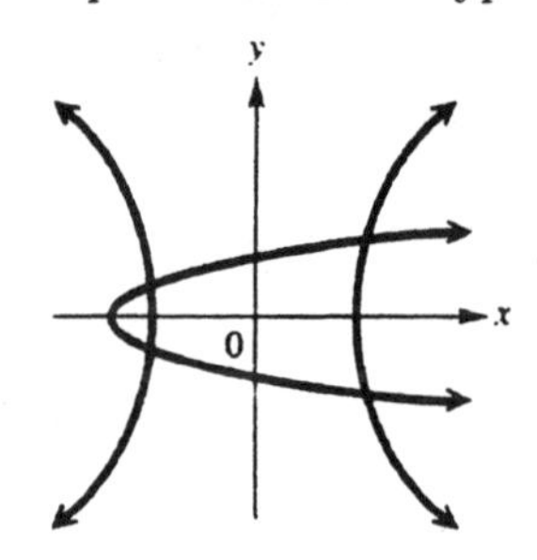

61. $y = -x^2 + 2 \quad [1]$

$x - y = 0 \quad [2]$

$[1] \longrightarrow [2]:$

$x - (-x^2 + 2) = 0$

$x^2 + x - 2 = 0$

$(x + 2)(x - 1) = 0$

$\mathbf{x = -2,\ 1}$

$x = -2 \longrightarrow [1]:$

$y = -(-2)^2 + 2$

$y = -4 + 2 = \mathbf{-2}$ $\boxed{(-2, -2)}$

$x = 1 \longrightarrow [1]:$

$y = -(1^2) + 2$

$y = -1 + 2 = \mathbf{1}$ $\boxed{(1, 1)}$

63. $3x^2 + 2y^2 = 5 \quad [1]$

$x - y = -2 \Longrightarrow x = y - 2 \quad [2]$

$[2] \longrightarrow [1]:$

$3(y - 2)^2 + 2y^2 = 5$

$3(y^2 - 4y + 4) + 2y^2 = 5$

$3y^2 - 12y + 12 + 2y^2 - 5 = 0$

$5y^2 - 12y + 7 = 0$

$(5y - 7)(y - 1) = 0$

$\mathbf{y = \frac{7}{5},\ 1}$

continued

63. continued

$y = \frac{7}{5} \longrightarrow [2]:$

$x = \frac{7}{5} - 2 = -\mathbf{\frac{3}{5}}$ $\boxed{\left(-\frac{3}{5}, \frac{7}{5}\right)}$

$y = 1 \longrightarrow [2]:$

$x = 1 - 2 = \mathbf{-1}$ $\boxed{(-1, 1)}$

65. $x^2 + y^2 = 10$ [1]

$2x^2 - y^2 = 17$ [2]

$-\ 3x^2 = 27$

$x^2 = 9 \rightarrow \mathbf{x = \pm 3}$

$x = 3 \longrightarrow [1]: \quad 3^2 + y^2 = 10$

$y^2 = 1 \rightarrow \mathbf{y = \pm 1}$

$\boxed{(3, 1), (3, -1)}$

$x = -3 \longrightarrow [1]: \quad (-3)^2 + y^2 = 10$

$y^2 = 1 \rightarrow \mathbf{y = \pm 1}$

$\boxed{(-3, -1), (-3, 1)}$

67. $x^2 + 2y^2 = 9$ [1]

$3x^2 - 4y^2 = 27$ [2]

$2x^2 + 4y^2 = 18$ [2 × [1]]

$3x^2 - 4y^2 = 27$ [2]

$5x^2 = 45$

$x^2 = 9 \rightarrow \mathbf{x = \pm 3}$

$x = 3 \longrightarrow [1]: \quad 3^2 + 2y^2 = 9$

$2y^2 = 0 \rightarrow \mathbf{y = 0}$

$x = -3 \longrightarrow [1]: \quad (-3)^2 + 2y^2 = 9$

$2y^2 = 0 \rightarrow \mathbf{y = 0}$

$\boxed{(3, 0), (-3, 0)}$

69. $2x^2 + 2y^2 = 20$ [1]

$3x^2 + 3y^2 = 30$ [2]

$x^2 + y^2 = 10$ $[\frac{1}{2} \times [1]]$

$-x^2 - y^2 = -10$ $[-\frac{1}{3} \times [2]]$

$0 = 0$

Same line, ∞ **solutions**:

$y^2 = 10 - x^2$

$y = \pm\sqrt{10 - x^2}$ $\boxed{(x, \pm\sqrt{10 - x^2})}$

71. $y = |x - 1|$ [1]

$y = x^2 - 4$ [2]

$[1] \longrightarrow [2]:$

$|x - 1| = x^2 - 4$

$x - 1 = x^2 - 4$

$x^2 - x - 3 = 0$

$x = \frac{1 \pm \sqrt{1 - 4(-3)}}{2}$ $\mathbf{x = \frac{1 \pm \sqrt{13}}{2}}$

or $x - 1 = -x^2 + 4$

$x^2 + x - 5 = 0$

$x = \frac{-1 \pm \sqrt{1 - 4(-5)}}{2}$ $\mathbf{x = \frac{-1 \pm \sqrt{21}}{2}}$

(i) $x = \frac{1+\sqrt{13}}{2} \longrightarrow [2]: \quad y = \left(\frac{1 + \sqrt{13}}{2}\right)^2 - 4$

$= \frac{1 + 2\sqrt{13} + 13}{4} - \frac{16}{4}$

$= \frac{2\sqrt{13} - 2}{4} = \frac{\sqrt{13} - 1}{2}$

$\boxed{\left(\frac{1 + \sqrt{13}}{2}, \frac{-1 + \sqrt{13}}{2}\right)}$

(ii) $x = \frac{1-\sqrt{13}}{2} \longrightarrow [2]: \quad y = \left(\frac{1 - \sqrt{13}}{2}\right)^2 - 4$

$= \frac{1 - 2\sqrt{13} + 13}{4} - \frac{16}{4}$

$= \frac{-2\sqrt{13} - 2}{4} < 0$

From [1], $y > 0$: $\emptyset$

(iii) $x = \frac{-1+\sqrt{21}}{2} \longrightarrow [2]: y = \left(\frac{-1 + \sqrt{21}}{2}\right)^2 - 4$

$= \frac{1 - 2\sqrt{21} + 21}{4} - \frac{16}{4}$

$= \frac{-2\sqrt{21} + 6}{4} < 0$

From [1], $y > 0$: $\emptyset$

(iv) $x = \frac{-1-\sqrt{21}}{2} \longrightarrow [2]: y = \left(\frac{-1 - \sqrt{21}}{2}\right)^2 - 4$

$= \frac{1 + 2\sqrt{21} + 21}{4} - \frac{16}{4}$

$= \frac{2\sqrt{21} + 6}{4} = \frac{\sqrt{21} + 3}{2}$

$\boxed{\left(\frac{-1 - \sqrt{21}}{2}, \frac{3 + \sqrt{21}}{2}\right)}$

73. $y_1 = \log(x+5)$

$y_2 = x^2$

Graph y_1 and y_2; intersection at

$\boxed{(-.79,\ .62),\ (.88,\ .77)}$

75. $y_1 = e^{x+1}$

$2x + y = 3$

$y_2 = 3 - 2x$

Graph y_1 and y_2; intersection at

$\boxed{(.06,\ 2.88)}$

Relating Concepts (77 – 82):

77. $x^3 - 85x + 300$

5)	1	0	−85	300
		5	25	−300
	1	5	−60	$\boxed{0}$

5 is a solution.

78. $x^3 - 85x + 300$

$(x-5)(x^2 + 5x - 60)$

79. $x^2 + 5x - 60 = 0$

$$x = \frac{-5 \pm \sqrt{25 - 4(-60)}}{2} = \frac{-5 \pm \sqrt{25 + 240}}{2} = \frac{-5 \pm \sqrt{265}}{2}$$

Positive solution: $\dfrac{-5 + \sqrt{265}}{2}$

80. $y = x^3 - 85x + 300$

negative x-intercept ≈ -10.64.

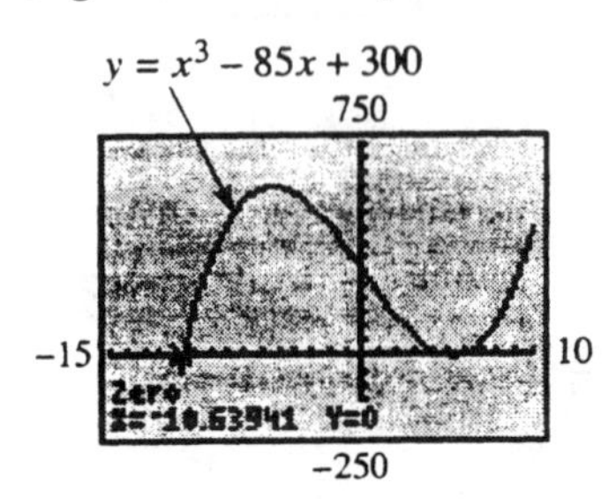

81. Negative solution: $\dfrac{-5 - \sqrt{265}}{2}$

82. Because x represents a length, it must be a positive number.

83. **(a)** Let x represent the number of bank robberies in 1994 and y the number in 1995. Then,

$x + y = 13,787 \Longrightarrow y = 13,787 - x$

$x - y = 271 \Longrightarrow y = x - 271$

(b) $x + y = 13,787$ [1]

$x - y = \ \ 271$ [2]

$2x = 14,058$ $\quad$ $\mathbf{x = 7029}$

$y = (7029) - 271$ $\quad$ $\mathbf{y = 6758}$

(c) The intersection point is $(7029,\ 6758)$. There were 7029 bank robberies in 1994 and 6758 bank robberies in 1995.

85. Let x represent the number of foreign tourists and y the number of American tourists. Then,

$x + y = 6.2$ [1]

$y = 4x$ [2]

[2] $\longrightarrow$ [1] :

$x + 4x = 6.2$

$5x = 6.2$ $\quad$ $\mathbf{x = 1.24}$

$x = 1.24 \longrightarrow$ [2] :

$y = 4(1.24)$ $\quad$ $\mathbf{y = 4.96}$

There were ≈ 1.2 million foreign tourists and ≈ 5.0 million American tourists.

87. Let $H = 180$ in each equation and solve for y :

$$180 = .491x + .468y + 11.2$$
$$.468y = 180 - 11.2 - .491x$$
$$y = \frac{168.8 - .491x}{.468}$$
$$180 = -.981x + 1.872y + 26.4$$
$$1.872y = 180 - 26.4 + .981x$$
$$y = \frac{153.6 + .981x}{1.872}$$

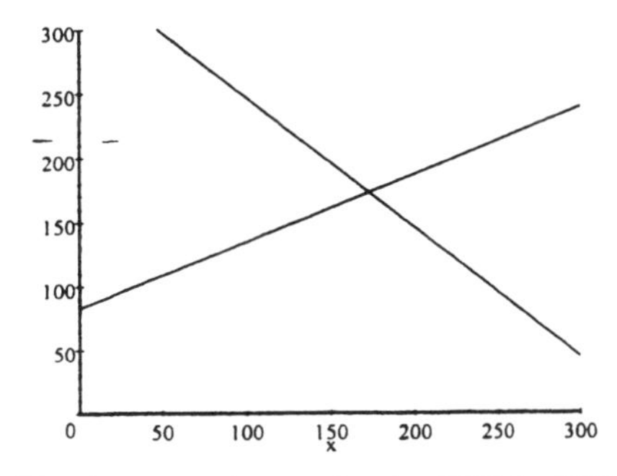

The intersection point is $(177.1,\ 174.9)$.
This means that if an athlete achieves a maximum heart rate of 180 beats per minute (bpm), then 5 seconds after stopping, his heart rate would be ≈ 177.1 bpm, and after 10 seconds it would be ≈ 174.9bpm.

89. **(a)** $f(x) = -6.393x + 894.9$
$g(x) = 19.14x + 746.9$

Set the two equations = to each other;solve for x :

$$-6.393x + 894.9 = 19.14x + 746.9$$
$$-6.393x - 19.14x = 746.9 - 894.9$$
$$-25.533x = -148$$
$$x = \frac{-148}{-25.533} \approx 5.8$$

Substitute $x = 5.8$ into the first equation:

$$f(5.8) = -6.393(5.8) + 894.9 \approx 857.8$$

Solution set: $\{(5.8,\ 857.8)\}$

(b) $1990 + 5.8 = 1995.8$

During 1995, 857.8 million pounds of both canned tuna and fresh shrimp were available.

(c)

y = −6.393x + 894.9
y = 19.14x + 746.9
1000
Intersection
X=5.7964203 Y=857.84348
0 20
500

91. $V = \pi r^2 h = 50;\quad S.A. = 2\pi r h = 65$

Solve each equation for h :

$$\pi r^2 h = 50 \Longrightarrow h = \frac{50}{\pi r^2}$$
$$2\pi r h = 65 \Longrightarrow h = \frac{65}{2\pi r}$$

Set the equations equal to each other:

$$\frac{50}{\pi r^2} = \frac{65}{2\pi r}$$
$$100\pi r = 65\pi r^2$$
$$0 = 65\pi r^2 - 100\pi r$$
$$0 = 5\pi r(13r - 20)$$
$$r = 0,\quad r = \tfrac{20}{13} \approx 1.538 \text{ inches}$$
$$h = \frac{65}{2\pi(20/13)} \approx 6.724 \text{ inches}$$

93.

$$\begin{aligned} w_1 + \sqrt{2}\,w_2 &= 300 \\ \sqrt{3}\,w_1 - \sqrt{2}\,w_2 &= 0 \\ \hline (1 + \sqrt{3})\,w_1 &= 300 \end{aligned}$$

$$w_1 = \frac{300}{(1 + \sqrt{3})} \approx 109.8 \text{ pounds}$$

Substitute w_1 into the first equation:

$$\frac{300}{(1 + \sqrt{3})} + \sqrt{2}\,w_2 = 300$$
$$\sqrt{2}\,w_2 = 300 - \frac{300}{(1 + \sqrt{3})}$$
$$\sqrt{2}\,w_2 = \frac{300(1 + \sqrt{3}) - 300}{1 + \sqrt{3}} = \frac{300\sqrt{3}}{1 + \sqrt{3}}$$
$$w_2 = \frac{300\sqrt{3}}{1 + \sqrt{3}} \times \frac{1}{\sqrt{2}}$$
$$= \frac{300\sqrt{3}}{\sqrt{2} + \sqrt{6}} \approx 134.5 \text{ pounds}$$

95. The total number of vehicles entering intersection **A** is 500 + 150 = 650 vehicles per hour. The expression $x+y$ represents the number of vehicles leaving intersection **A** each hour. Therefore, we have $x+y=650$.

The total number of vehicles leaving intersection **B** is 50 + 400 = 450 vehicles per hour. There are 100 vehicles entering intersection **B** from the south and y vehicles entering intersection **B** from the west. Thus, $y+100=450$.

Solve the system:

$x+y=650$

$y+100=450 \Longrightarrow \mathbf{y=350}$

$x+350=650 \Longrightarrow \mathbf{x=300}$

At intersection **A**, a stoplight should allow for 300 vehicles per hour to travel south and 350 vehicles per hour to continue traveling east.

97. $C=4x+125$

$R=9x-200$

Find where $C=R$:

$4x+125=9x-200$

$325=5x$

$x=65$

$C=4(65)+125$

$=260+125=385$

$\boxed{x=65; \quad R=C=\$385}$

99. $p=630-\frac{3}{4}q$

$p=\frac{3}{4}q$

Find where $p=p$:

$630-\frac{3}{4}q=\frac{3}{4}q$

$630=\frac{6}{4}q$

$420=q$

$p=\frac{3}{4}q=\frac{3}{4}(420)=315$

$\boxed{p=\$315, \quad q=420}$

Relating Concepts (101 – 106):

101. Let $t=\frac{1}{x}$, $u=\frac{1}{y}$

$5\left(\frac{1}{x}\right)+15\left(\frac{1}{y}\right)=16$

$5\left(\frac{1}{x}\right)+4\left(\frac{1}{y}\right)=5$

$$\boxed{\begin{array}{l}5t+15u=16\\5t+4u=5\end{array}}$$

102. $5t+15u=16$ [1]

$5t+4u=5$ [2]

$$\begin{array}{rl}5t+15u=16 & [1]\\ -5t-4u=-5 & [-1\times[2]]\\ \hline 11u=11 & \boxed{u=1}\end{array}$$

$5t+15(1)=16$ [1]

$5t=1 \qquad \boxed{t=\frac{1}{5}}$

103. $t=\frac{1}{x}, \quad t=\frac{1}{5}$

$\frac{1}{x}=\frac{1}{5} \qquad \boxed{x=5}$

$u=\frac{1}{y}, \quad u=1$

$\frac{1}{y}=1$

$\boxed{y=1}$

104. $\frac{5}{x}+\frac{15}{y}=16$

$5y+15x=16xy$

$5y-16xy=-15x$

$y(5-16x)=-15x$

$\boxed{y=\frac{-15x}{5-16x}}$

105. $\dfrac{5}{x} + \dfrac{4}{y} = 5$

$$5y + 4x = 5xy$$
$$5y - 5xy = -4x$$
$$y(5 - 5x) = -4x$$
$$\boxed{y = \frac{-4x}{5 - 5x}}$$

106. $y_1 = \dfrac{-15x}{5 - 16x}$; $y_2 = \dfrac{-4x}{5 - 5x}$

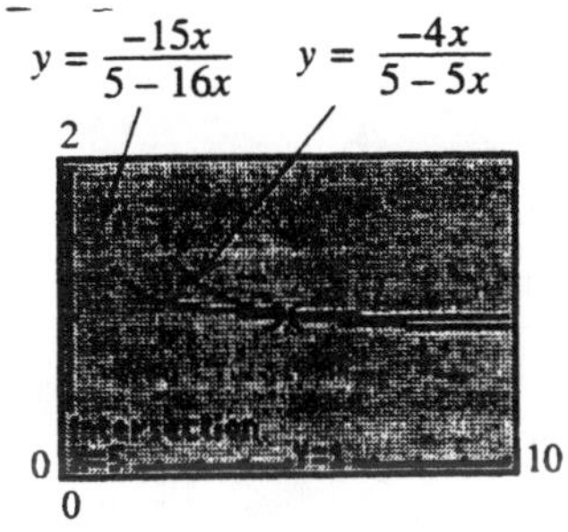

$y = \dfrac{-15x}{5-16x}$ $y = \dfrac{-4x}{5-5x}$

2

0 10

0

Dot Mode

Point of intersection: (5, 1)

107. $\dfrac{2}{x} + \dfrac{1}{y} = \dfrac{3}{2}$ [1]

$\dfrac{3}{x} - \dfrac{1}{y} = 1$ [2]

Let $t = \dfrac{1}{x}$, $u = \dfrac{1}{y}$:

$2t + u = \dfrac{3}{2}$ [3]

$3t - u = 1$ [4]

$$5t = \frac{5}{2}$$
$$t = \frac{1}{2}$$
$$\frac{1}{x} = \frac{1}{2} \qquad \mathbf{x = 2}$$

$t = \frac{1}{2} \longrightarrow [3]$:

$$2\left(\frac{1}{2}\right) + u = \frac{3}{2}$$
$$u = \frac{1}{2}$$
$$\frac{1}{y} = \frac{1}{2} \qquad \mathbf{y = 2} \qquad \boxed{(2, 2)}$$

Section 7.2

1. **(a)** Multiply equation (2) by 2 and add to equation (1) to get $7x - z = 0$.

(b) Multiply equation (1) by -2 and add to equation (3) to get $-3y + 2z = 2$.

(c) Multiply equation (2) by 4 and add to equation (3) to get $14x - 3y = 2$.

3. $\left(\dfrac{1}{2}, -\dfrac{3}{4}, \dfrac{1}{6}\right)$:

(i) $2x + 8y - 6z = -6$

$$2\left(\frac{1}{2}\right) + 8\left(-\frac{3}{4}\right) - 6\left(\frac{1}{6}\right) = -6$$
$$1 - 6 - 1 = -6 \longrightarrow -6 = -6$$

(ii) $x + y + z = -\dfrac{1}{12}$

$$\frac{1}{2} - \frac{3}{4} + \frac{1}{6} = -\frac{1}{12}$$
$$\frac{6 - 9 + 2}{12} = -\frac{1}{12} \longrightarrow -\frac{1}{12} = -\frac{1}{12}$$

(iii) $x + 3z = 1$

$$\frac{1}{2} + 3\left(\frac{1}{6}\right) = 1$$
$$\frac{1}{2} + \frac{1}{2} = 1 \longrightarrow 1 = 1$$

5.
$x + y + z = 2$ [1]
$2x + y - z = 5$ [2]
$x - y + z = -2$ [3]

$-x - y - z = -2$ [$-1 \times$ [1]]
$2x + y - z = 5$ [2]

$x - 2z = 3$ [4]

$x + y + z = 2$ [1]
$x - y + z = -2$ [3]

$2x + 2z = 0$ [5]

$x - 2z = 3$ [4]
$2x + 2z = 0$ [5]

$3x = 3 \rightarrow$ $\mathbf{x = 1}$

$x = 1 \longrightarrow [4]$: $1 - 2z = 3$

$-2z = 2 \rightarrow$ $\mathbf{z = -1}$

$x = 1,\ z = -1 \longrightarrow [1]$:

$1 + y - 1 = 2 \rightarrow$ $\mathbf{y = 2}$ $\boxed{\{(1, 2, -1)\}}$

7.
$$\begin{aligned} x+3y+4z&=14 &&[1]\\ 2x-3y+2z&=10 &&[2]\\ 3x-y+z&=9 &&[3] \end{aligned}$$

$$\begin{aligned} x+3y+4z&=14 &&[1]\\ 2x-3y+2z&=10 &&[2]\\ \hline 3x+6z&=24\\ x+2z&=8 &&[4] \end{aligned}$$

$$\begin{aligned} x+3y+4z&=14 &&[1]\\ 9x-3y+3z&=27 &&[3\times[3]]\\ \hline 10x+7z&=41 &&[5] \end{aligned}$$

$$\begin{aligned} -10x-20z&=-80 &&[-10\times[4]]\\ 10x+7z&=41 &&[5]\\ \hline -13z&=-39\rightarrow \quad \mathbf{z=3} \end{aligned}$$

$z=3\longrightarrow[4]:\quad x+2(3)=8\rightarrow\quad \mathbf{x=2}$

$x=2,\ z=3\longrightarrow[1]:$

$2+3y+4(3)=14$

$3y=0\rightarrow\quad \mathbf{y=0}\quad \boxed{\{(2,\ 0,\ 3)\}}$

9.
$$\begin{aligned} x+2y+3z&=8 &&[1]\\ 3x-y+2z&=5 &&[2]\\ -2x-4y-6z&=5 &&[3] \end{aligned}$$

$$\begin{aligned} 2x+4y+6z&=16 &&[2\times[1]]\\ -2x-4y-6z&=5 &&[3]\\ \hline 0&=22 \end{aligned}$$

No solution: $\emptyset$

11.
$$\begin{aligned} x+4y-z&=6 &&[1]\\ 2x-y+z&=3 &&[2]\\ 3x+2y+3z&=16 &&[3] \end{aligned}$$

$$\begin{aligned} x+4y-z&=6 &&[1]\\ 2x-y+z&=3 &&[2]\\ \hline 3x+3y&=9\\ x+y&=3 &&[4] \end{aligned}$$

$$\begin{aligned} 3x+12y-3z&=18 &&[3\times[1]]\\ 3x+2y+3z&=16 &&[3]\\ \hline 6x+14y&=34\\ 3x+7y&=17 &&[5] \end{aligned}$$

$$\begin{aligned} -3x-3y&=-9 &&[-3\times[4]]\\ 3x+7y&=17 &&[5]\\ \hline 4y&=8\rightarrow\quad \mathbf{y=2} \end{aligned}$$

$y=2\longrightarrow[4]:\quad x+2=3\rightarrow\quad \mathbf{x=1}$

$x=1,\ y=2\longrightarrow[1]:\quad 1+4(2)-z=6$

$-z=-3\rightarrow\quad \mathbf{z=3}\quad \boxed{\{(1,\ 2,\ 3)\}}$

13.
$$\begin{aligned} 5x+y-3z&=-6 &&[1]\\ 2x+3y+z&=5 &&[2]\\ -3x-2y+4z&=3 &&[3] \end{aligned}$$

$$\begin{aligned} -15x-3y+9z&=18 &&[-3\times[1]]\\ 2x+3y+z&=5 &&[2]\\ \hline -13x+10z&=23 &&[4] \end{aligned}$$

$$\begin{aligned} 10x+2y-6z&=-12 &&[2\times[1]]\\ -3x-2y+4z&=3 &&[3]\\ \hline 7x-2z&=-9 &&[5] \end{aligned}$$

$$\begin{aligned} -13x+10z&=23 &&[4]\\ 35x-10z&=-45 &&[5\times[5]]\\ \hline 22x&=-22\rightarrow\quad \mathbf{x=-1} \end{aligned}$$

$x=-1\longrightarrow[5]:$

$7(-1)-2z=-9$

$-2z=-2\rightarrow\quad \mathbf{z=1}$

$x=-1,\ z=1\longrightarrow[2]:$

$2(-1)+3y+1=5$

$3y-1=5$

$3y=6\rightarrow\quad \mathbf{y=2}\quad \boxed{\{(-1,\ 2,\ 1)\}}$

15.
$$\begin{aligned} x-3y-2z&=-3 &&[1]\\ 3x+2y-z&=12 &&[2]\\ -x-y+4z&=3 &&[3] \end{aligned}$$

$$\begin{aligned} x-3y-2z&=-3 &&[1]\\ -x-y+4z&=3 &&[3]\\ \hline -4y+2z&=0\\ -2y+z&=0 &&[4] \end{aligned}$$

$$\begin{aligned} 3x+2y-z&=12 &&[2]\\ -3x-3y+12z&=9 &&[3\times[3]]\\ \hline -y+11z&=21 &&[5] \end{aligned}$$

$$\begin{aligned} -2y+z&=0 &&[4]\\ 2y-22z&=-42 &&[-2\times[5]]\\ \hline -21z&=-42\rightarrow\quad \mathbf{z=2} \end{aligned}$$

$z=2\longrightarrow[4]:\quad -2y+2=0$

$-2y=-2\rightarrow\quad \mathbf{y=1}$

$y=1,\ z=2\longrightarrow[3]:$

$-x-1+4(2)=3$

$-x+7=3$

$-x=-4\rightarrow\quad \mathbf{x=4}\quad \boxed{\{(4,\ 1,\ 2)\}}$

17.
$$\begin{aligned} 2x + 6y - z &= 6 & [1] \\ 4x - 3y + 5z &= -5 & [2] \\ 6x + 9y - 2z &= 11 & [3] \end{aligned}$$

$$\begin{aligned} -4x - 12y + 2z &= -12 & [-2 \times [1]] \\ 4x - 3y + 5z &= -5 & [2] \\ \hline -15y + 7z &= -17 & [4] \end{aligned}$$

$$\begin{aligned} -6x - 18y + 3z &= -18 & [-3 \times [1]] \\ 6x + 9y - 2z &= 11 & [3] \\ \hline -9y + z &= -7 & [5] \end{aligned}$$

$$\begin{aligned} -15y + 7z &= -17 & [4] \\ 63y - 7z &= 49 & [-7 \times [5]] \\ \hline \end{aligned}$$

$$48y = 32 \rightarrow \quad \mathbf{y} = \frac{32}{48} = \mathbf{\frac{2}{3}}$$

$$y = \frac{2}{3} \longrightarrow [5]: \quad -9\left(\frac{2}{3}\right) + z = -7$$

$$-6 + z = -7 \rightarrow \quad \mathbf{z = -1}$$

$$y = \frac{2}{3},\ z = -1 \longrightarrow [1]:$$

$$2x + 6\left(\frac{2}{3}\right) - (-1) = 6$$

$$2x + 4 + 1 = 6$$

$$2x = 1 \rightarrow \quad \mathbf{x} = \mathbf{\frac{1}{2}} \qquad \boxed{\left\{\left(\frac{1}{2}, \frac{2}{3}, -1\right)\right\}}$$

19.
$$\begin{aligned} x + z &= 4 & [1] \\ x + y &= 4 & [2] \\ y + z &= 4 & [3] \end{aligned}$$

$$\begin{aligned} x + z &= 4 & [1] \\ -x - y &= -4 & [-1 \times [2]] \\ \hline z - y &= 0 & [4] \end{aligned}$$

$$\begin{aligned} y + z &= 4 & [3] \\ -y + z &= 0 & [4] \\ \hline 2z &= 4 \rightarrow & \mathbf{z = 2} \end{aligned}$$

$$z = 2 \longrightarrow [1]: \quad x + 2 = 4 \rightarrow \quad \mathbf{x = 2}$$

$$x = 2 \longrightarrow [2]: \quad 2 + y = 4 \rightarrow \quad \mathbf{y = 2}$$

$$\boxed{\{(2,\ 2,\ 2)\}}$$

21.
$$\begin{aligned} 2x + y - z &= -4 & [1] \\ y + 2z &= 12 & [2] \\ 2x - z &= -4 & [3] \end{aligned}$$

$$\begin{aligned} 2x + y - z &= -4 & [1] \\ -2x + 0y + z &= 4 & [-1 \times [3]] \\ \hline \mathbf{y} &= \mathbf{0} \end{aligned}$$

$$y = 0 \longrightarrow [2]: \quad 0 + 2z = 12 \qquad \mathbf{z = 6}$$

$$z = 6 \longrightarrow [3]: \quad 2x - (6) = -4 \qquad \mathbf{x = 1}$$

$$\boxed{\{(1,\ 0,\ 6)\}}$$

23.
$$\begin{aligned} 2x + 3y + 4z &= 3 & [1] \\ 6x + 3y + 8z &= 6 & [2] \\ 6y - 4z &= 1 & [3] \end{aligned}$$

$$\begin{aligned} -6x - 9y - 12z &= -9 & [-3 \times [1]] \\ 6x + 3y + 8z &= 6 & [2] \\ \hline -6y - 4z &= -3 & [4] \end{aligned}$$

$$\begin{aligned} -6y - 4z &= -3 & [4] \\ 6y - 4z &= 1 & [3] \\ \hline -8z &= -2 & \mathbf{z} = \mathbf{\frac{1}{4}} \end{aligned}$$

$$z = \frac{1}{4} \longrightarrow [3]: \quad 6y - 4\left(\tfrac{1}{4}\right) = 1$$

$$6y - 1 = 1$$

$$6y = 2 \qquad \mathbf{y} = \mathbf{\frac{1}{3}}$$

$$y = \frac{1}{3},\ z = \frac{1}{4} \longrightarrow [1]:$$

$$2x + 3\left(\frac{1}{3}\right) + 4\left(\frac{1}{4}\right) = 3$$

$$2x + 1 + 1 = 3$$

$$2x = 1 \qquad \mathbf{x} = \mathbf{\frac{1}{2}} \qquad \boxed{\left\{\left(\frac{1}{2}, \frac{1}{3}, \frac{1}{4}\right)\right\}}$$

25. $\frac{1}{x}+\frac{1}{y}-\frac{1}{z}=\frac{1}{4}$

$\frac{2}{x}-\frac{1}{y}+\frac{3}{z}=\frac{9}{4}$

$-\frac{1}{x}-\frac{2}{y}+\frac{4}{z}=1$

Let $t=\frac{1}{x},\ u=\frac{1}{y},\ v=\frac{1}{z}$

$t+u-v=\frac{1}{4}$ [1]

$2t-u+3v=\frac{9}{4}$ [2]

$-t-2u+4v=1$ [3]

$t+u-v=\frac{1}{4}$ [1]

$-t-2u+4v=1$ [3]

$-u+3v=\frac{5}{4}$ [4]

$2t-u+3v=\frac{9}{4}$ [2]

$-2t-4u+8v=2$ [2 × [3]]

$-5u+11v=\frac{17}{4}$ [5]

$5u-15v=-\frac{25}{4}$ [−5 × [4]]

$-5u+11v=\frac{17}{4}$ [5]

$-4v=-\frac{8}{4}$

$v=\frac{1}{2}\rightarrow \mathbf{z=2}$

$v=\frac{1}{2}\longrightarrow$ [4] :

$-u+\frac{3}{2}=\frac{5}{4}$

$-u=-\frac{1}{4}$

$u=\frac{1}{4}\rightarrow \mathbf{y=4}$

$v=\frac{1}{2},\ u=\frac{1}{4}\longrightarrow$ [3] :

$-t-2\left(\frac{1}{4}\right)+4\left(\frac{1}{2}\right)=1$

$-2t-1+4=2$

$-2t=-1$

$t=\frac{1}{2}\rightarrow \mathbf{x=2}$

$\boxed{\{(2,\ 4,\ 2)\}}$

27. $\frac{2}{x}-\frac{2}{y}+\frac{1}{z}=-1$

$\frac{4}{x}+\frac{1}{y}-\frac{2}{z}=-9$

$\frac{1}{x}+\frac{1}{y}-\frac{3}{z}=-9$

Let $t=\frac{1}{x},\ u=\frac{1}{y},\ v=\frac{1}{z}$

$2t-2u+v=-1$ [1]

$4t+u-2v=-9$ [2]

$t+u-3v=-9$ [3]

$2t-2u+v=-1$ [1]

$8t+2u-4v=-18$ [2 × [2]]

$10t-3v=-19$ [4]

$2t-2u+v=-1$ [1]

$2t+2u-6v=-18$ [2 × [3]]

$4t-5v=-19$ [5]

$20t-6v=-38$ [2 × [4]]

$-20t+25v=95$ [−5 × [5]]

$19v=57$

$v=3\rightarrow \mathbf{z=\frac{1}{3}}$

$v=3\longrightarrow$ [5] :

$4t-15=-19$

$4t=-4$

$t=-1\rightarrow \mathbf{x=-1}$

$t=-1,\ v=3\longrightarrow$ [1] :

$2(-1)-2u+3=-1$

$-2u=-2$

$u=1\rightarrow \mathbf{y=1}$

$\boxed{\left\{\left(-1,\ 1,\ \frac{1}{3}\right)\right\}}$

29. $x+y+z=4$

(a) $x-y+z=0$
$x+y-z=0$

(b) $x+y+z=0$
$x-y-z=1$

(c) $2x+2y+2z=8$
$x-y-z=1$

There are many other examples.

31. For example, two perpendicular walls meeting in a corner with the floor intersect in a point.

33.

$$\begin{aligned} x - 2y + 3z &= 6 \quad [1] \\ 2x - y + 2z &= 5 \quad [2] \end{aligned}$$

$$\begin{aligned} -2x + 4y - 6z &= -12 \quad [-2 \times [1]] \\ 2x - y + 2z &= 5 \quad [2] \\ \hline 3y - 4z &= -7 \\ 3y &= 4z - 7 \\ \mathbf{y} &= \mathbf{\frac{4z-7}{3}} \quad [3] \end{aligned}$$

$[3] \longrightarrow [2]:$

$$\begin{aligned} 2x - \frac{4z-7}{3} + 2z &= 5 \\ 6x - 4z + 7 + 6z &= 15 \\ 6x + 2z &= 8 \\ 6x &= 8 - 2z \\ \mathbf{x} = \frac{8-2z}{6} &= \mathbf{\frac{4-z}{3}} \end{aligned}$$

$$\boxed{\left\{\left(\frac{4-z}{3}, \frac{4z-7}{3}, z\right)\right\}}$$

35.

$$\begin{aligned} 5x - 4y + z &= 9 \quad [1] \\ x + y &= 15 \quad [2] \end{aligned}$$

$$\begin{aligned} 5x - 4y + z &= 9 \quad [1] \\ 4x + 4y &= 60 \quad [4 \times [2]] \\ \hline 9x + z &= 69 \\ 9x = 69 - z \rightarrow \quad \mathbf{x} &= \mathbf{\frac{69-z}{9}} \quad [3] \end{aligned}$$

$[3] \longrightarrow [2]:$

$$\begin{aligned} \frac{69-z}{9} + y &= 15 \\ 69 - z + 9y &= 135 \\ -z + 9y &= 66 \\ 9y = z + 66 \rightarrow \quad \mathbf{y} &= \mathbf{\frac{z+66}{9}} \end{aligned}$$

$$\boxed{\left\{\left(\frac{69-z}{9}, \frac{z+66}{9}, z\right)\right\}}$$

37.

$$\begin{aligned} x - y + z &= -6 \quad [1] \\ 4x + y + z &= 7 \quad [2] \\ \hline 5x + 2z &= 1 \\ 5x = -2z + 1 \rightarrow \quad \mathbf{x} &= \mathbf{\frac{1-2z}{5}} \end{aligned}$$

$x \longrightarrow [1]:$

$$\begin{aligned} \frac{1-2z}{5} - y + z &= -6 \\ 1 - 2z - 5y + 5z &= -30 \\ -5y + 3z &= -31 \\ -5y = -3z - 31 \rightarrow \quad \mathbf{y} &= \mathbf{\frac{3z+31}{5}} \end{aligned}$$

$$\boxed{\left\{\left(\frac{1-2z}{5}, \frac{3z+31}{5}, z\right)\right\}}$$

39. Let $x = \#$ of gallons of \$9 grade,
$y = \#$ of gallons of \$13 grade,
$z = \#$ of gallons of \$4.50 grade.

$$\begin{aligned} x + y + z &= 300 \quad [1] \\ 9x + 3y + 4.5z &= 6(300) \quad [2] \\ z &= 2y \quad [3] \end{aligned}$$

$[3] \longrightarrow [1]:$

$$\begin{aligned} x + y + 2y &= 300 \\ x + 3y &= 300 \quad [4] \end{aligned}$$

$[3] \longrightarrow [2]:$

$$\begin{aligned} 9x + 3y + 4.5(2y) &= 1800 \\ 9x + 12y &= 1800 \quad [5] \end{aligned}$$

$$\begin{aligned} 9x + 12y &= 1800 \quad [5] \\ -9x - 27y &= -2700 \quad [-9 \times [4]] \\ \hline -15y &= -900 \quad \mathbf{y = 60} \end{aligned}$$

$y = 60 \longrightarrow [3]:$

$z = 2(60) \quad \mathbf{z = 120}$

$y = 60,\ z = 120 \longrightarrow [1]:$

$x + 60 + 120 = 300 \quad \mathbf{x = 120}$

\$9 water: 120 gallons
\$3 water: 60 gallons
\$4.50 water: 120 gallons

41. Let x = price of Up close, y = price of Middle, z = price of Farther back.

$$\begin{aligned} x &= 6+y & [1] \\ y &= 3+z & [2] \\ 2x &= 3+3z & [3] \end{aligned}$$

[1] & [2] ⟶ [3] :

$2(6+y) = 3+3z$
$2(6+(3+z)) = 3+3z$
$2(9+z) = 3+3z$
$18+2z = 3+3z$
$\mathbf{z = 15}$

$z = 15 \longrightarrow [2]$:

$y = 3+15 \rightarrow \mathbf{y = 18}$

$y = 18 \longrightarrow [1]$:

$x = 6+18 \rightarrow \mathbf{x = 24}$

Up close: \$24, *Middle:* \$18, *Farther back:* \$15

43. Let a = measure of largest angle, b = measure of medium angle, c = measure of smallest angle.

$$\begin{aligned} a+b+c &= 180 & [1] \\ a &= 2b-55 & [2] \\ c &= b-25 & [3] \end{aligned}$$

[2] and [3] ⟶ [1] :

$(2b-55)+b+(b-25) = 180$
$4b-80 = 180$
$4b = 260$
$\mathbf{b = 65}$

$b = 65 \longrightarrow [2]$:

$a = 2(65)-55$
$\mathbf{a = 75}$

$b = 65 \longrightarrow [3]$:

$c = 65-25$
$\mathbf{c = 40}$

75°, 65°, 40°

45. Let x be the amount at 4%, y be the amount at 4.5%, and z be the amount at 2.5%.

You then have the following system:

$$\begin{aligned} x+y+z &= 10,000 & [1] \\ .04x+.045y+.025z &= 415 & [2] \\ y &= 2x & [3] \end{aligned}$$

[3] ⟶ [1] :

$x+2x+z = 10,000$
$3x+z = 10,000 \quad [4]$

[3] ⟶ [1000 × [2]] :

$40x+45(2x)+25z = 415,000$
$130x+25z = 415,000$
$26x+5z = 83,000 \quad [5]$

$$\begin{aligned} -15x-5z &= -50,000 & [-5 \times [4]] \\ 26x+5z &= 83,000 & [5] \\ \hline 11x &= 33,000 \\ \mathbf{x} &= \mathbf{3,000} \end{aligned}$$

$x = 3,000 \longrightarrow [3]$:

$y = 2(3,000)$
$\mathbf{y = 6,000}$

$x,\ y \longrightarrow [1]$:

$3,000+6,000+z = 10,000$
$\mathbf{z = 1,000}$

\$3,000 at 4% \$6,000 at 4.5% \$1,000 at 2.5%

47. Let $x =$ # of EZ models
$y =$ # of Compact models
$z =$ # of Commercial models.

You then have the following system:

$$10x + 20y + 60z = 440 \quad [1]$$
$$10x + 8y + 28z = 248 \quad [2]$$

$[1] - [2]$:

$$12y + 32z = 192$$
$$12y = -32z + 192$$
$$y = -\frac{8}{3}z + 16$$

Substitute y into $[1]$:

$$10x + 20\left(-\frac{8}{3}z + 16\right) + 60z = 440$$
$$10x - \frac{160}{3}z + 320 + 60z = 440$$
$$30x - 160z + 960 + 180z = 1320$$
$$30x = -20z + 360$$
$$x = -\frac{2}{3}z + 12$$

z must be a multiple of 3 for x and y to be whole numbers:

If

$z = 0$:	12 EZ, 16 Comp., 0 Commer.
$z = 3$:	10 EZ, 8 Comp., 3 Commer.
$z = 6$:	8 EZ, 0 Comp., 6 Commer.
$z \geq 9$:	Not possible; $y < 0$

49. $\mathbf{y = ax^2 + bx + c}$

$(2, 9)$:

$$9 = a(2)^2 + b(2) + c$$
$$4a + 2b + c = 9 \quad [1]$$

$(-2, 1)$:

$$1 = a(-2)^2 + b(-2) + c$$
$$4a - 2b + c = 1 \quad [2]$$

$(-3, 4)$:

$$4 = a(-3)^2 + b(-3) + c$$
$$9a - 3b + c = 4 \quad [3]$$

$$4a + 2b + c = 9 \quad [1]$$
$$-4a + 2b - 4b = 8 - 1 \quad [-1\ 5 \times [2]]$$

$$\mathbf{b = 2}$$

$$4a - 2b + c = 1 \quad [2]$$
$$-9a + 3b - c = -4 \quad [-1 \times [3]]$$
$$-5a + b = -3 \quad [4]$$

$b = 2 \longrightarrow [4]$:

$$-5a + 2 = -3$$
$$-5a = -5$$
$$\mathbf{a = 1}$$

$a = 1,\ b = 2 \longrightarrow [1]$:

$$4(1) + 2(2) + c = 9$$
$$8 + c = 9$$
$$\mathbf{c = 1}$$

$$\boxed{y = x^2 + 2x + 1}$$

10.36
Y1=X²+2X+1
−3.5 2.5
−.36

51. $\mathbf{y = ax^2 + bx + c}$

$(2, 14)$:

$$14 = a(2)^2 + b(2) + c$$
$$4a + 2b + c = 14 \quad [1]$$

$(0, 0)$:

$$0 = a(0)^2 + b(0) + c \quad \mathbf{c = 0}$$

$(-1, 1)$:

$$1 = a(-1)^2 + b(-1) + c$$
$$a - b + c = 1 \quad [2]$$

$$\begin{array}{rl} 4a + 2b + 0 = 14 & [1] \\ 2a - 2b + 0 = -2 & [2 \times [2]] \\ \hline 6a = 12 & \\ \mathbf{a = 2} & \end{array}$$

$a = 2 \longrightarrow [1]$:

$$4(2) + 2b + 0 = 14$$
$$2b = 6$$
$$\mathbf{b = 3}$$

$a = 2,\ b = 3,\ c = 0$

$$\boxed{y = 2x^2 + 3x}$$

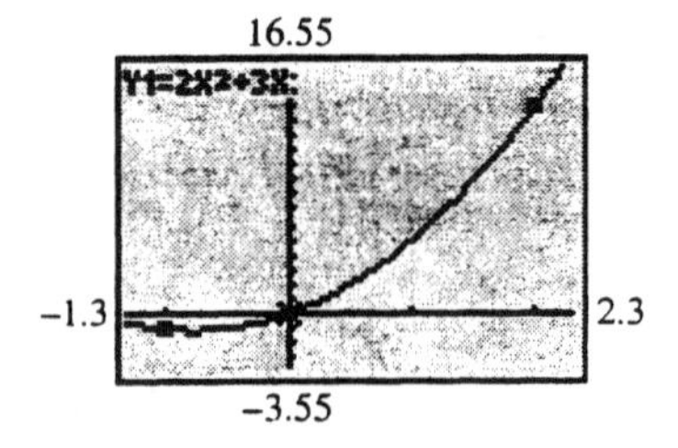

53. $\mathbf{x^2 + y^2 + ax + by + c = 0}$

$(2, 1)$: $2^2 + 1^2 + a(2) + b(1) + c = 0$
$$4 + 1 + 2a + b + c = 0$$
$$2a + b + c = -5 \quad [1]$$

$(-1, 0)$: $(-1)^2 + 0^2 - 1a + b(0) + c = 0$
$$-a + c = -1 \quad [2]$$

$(3, 3)$: $3^2 + 3^2 + a(3) + b(3) + c = 0$
$$9 + 9 + 3a + 3b + c = 0$$
$$3a + 3b + c = -18 \quad [3]$$

$$\begin{array}{rl} -6a - 3b - 3c = \ \ 15 & [-3 \times [1]] \\ 3a + 3b + \ c = -18 & [3] \\ \hline -3a - 2c = \ -3 & [4] \end{array}$$

$$\begin{array}{rl} -2a + 2c = -2 & [2 \times [2]] \\ -3a - 2c = -3 & [4] \\ \hline -5a = -5 \rightarrow \quad \mathbf{a = 1} & \end{array}$$

$a = 1 \longrightarrow [2]: -1 + c = -1 \rightarrow \quad \mathbf{c = 0}$

$a = 1,\ c = 0 \longrightarrow [1]$:

$$2 + b + 0 = -5 \rightarrow \quad \mathbf{b = -7}$$

$$\boxed{x^2 + y^2 + x - 7y = 0}$$

55. $s(t) = at^2 + bt + c$

$(0, 5)$: $\mathbf{5 = c}$
$(1, 23)$: $23 = a + b + c \quad [1]$
$(2, 37)$: $37 = 4a + 2b + c \quad [2]$

$c = 5 \longrightarrow [1]$:

$$a + b + 5 = 23$$
$$a + b = 18 \quad [3]$$

$c = 5 \longrightarrow [2]$:

$$4a + 2b + 5 = 37$$
$$4a + 2b = 32$$
$$2a + \ b = 16 \quad [4]$$

$$\begin{array}{rl} -a - b = -18 & [-1 \times [3]] \\ 2a + b = \ \ 16 & [4] \\ \hline \mathbf{a = -2} & \end{array}$$

$a = -2 \longrightarrow [3]$:

$$-2 + b = 18 \quad \mathbf{b = 20}$$

$a = -2,\ b = 20,\ c = 5$

$$\boxed{s(t) = -2t^2 + 20t + 5}$$

$$s(8) = -2(8)^2 + 20(8) + 5$$
$$s(8) = -128 + 160 + 5 \quad \boxed{s(8) = 37}$$

Section 7.3

1. $\begin{bmatrix} 2 & 4 \\ 4 & 7 \end{bmatrix}$

$$-2R_1 \to \begin{bmatrix} -4 & -8 \\ 4 & 7 \end{bmatrix}$$

3. $\begin{bmatrix} 1 & 5 & 6 \\ -2 & 3 & -1 \\ 4 & 7 & 0 \end{bmatrix}$

$$R_1 + R_2 \to \begin{bmatrix} 1 & 5 & 6 \\ -1 & 8 & 5 \\ 4 & 7 & 0 \end{bmatrix}$$

5. $\begin{bmatrix} -3 & 1 & -4 \\ 2 & 1 & 3 \\ -7 & 5 & 2 \end{bmatrix}$

$$-5R_2 + R_3 \to \begin{bmatrix} -3 & 1 & -4 \\ 2 & 1 & 3 \\ -17 & 0 & -13 \end{bmatrix}$$

7. $\begin{aligned} 2x + 3y &= 11 \\ x + 2y &= 8 \end{aligned}$

$$\left[\begin{array}{cc|c} 2 & 3 & 11 \\ 1 & 2 & 8 \end{array}\right]$$

9. $\begin{aligned} x + 5y &= 6 \\ x &= 3 \end{aligned}$

$$\left[\begin{array}{cc|c} 1 & 5 & 6 \\ 1 & 0 & 3 \end{array}\right]$$

11. $\begin{aligned} 2x + y + z &= 3 \\ 3x - 4y + 2z &= -7 \\ x + y + z &= 2 \end{aligned}$

$$\left[\begin{array}{ccc|c} 2 & 1 & 1 & 3 \\ 3 & -4 & 2 & -7 \\ 1 & 1 & 1 & 2 \end{array}\right]$$

13. $\begin{aligned} x + y &= 2 \\ 2y + z &= -4 \\ z &= 2 \end{aligned}$

$$\left[\begin{array}{ccc|c} 1 & 1 & 0 & 2 \\ 0 & 2 & 1 & -4 \\ 0 & 0 & 1 & 2 \end{array}\right]$$

15. $\left[\begin{array}{cc|c} 2 & 1 & 1 \\ 3 & -2 & -9 \end{array}\right]$

$$\begin{aligned} 2x + y &= 1 \\ 3x - 2y &= -9 \end{aligned}$$

17. $\left[\begin{array}{ccc|c} 1 & 0 & 0 & 2 \\ 0 & 1 & 0 & 3 \\ 0 & 0 & 1 & -2 \end{array}\right]$

$$\begin{aligned} x &= 2 \\ y &= 3 \\ z &= -2 \end{aligned}$$

19. $\left[\begin{array}{ccc|c} 3 & 2 & 1 & 1 \\ 0 & 2 & 4 & 22 \\ -1 & -2 & 3 & 15 \end{array}\right]$

$$\begin{aligned} 3x + 2y + z &= 1 \\ 2y + 4z &= 22 \\ -x - 2y + 3z &= 15 \end{aligned}$$

21. $\begin{aligned} x + y &= 5 \\ x - y &= -1 \end{aligned}$ $\qquad \left[\begin{array}{cc|c} 1 & 1 & 5 \\ 1 & -1 & -1 \end{array}\right]$

$$-R_1 + R_2 \to \left[\begin{array}{cc|c} 1 & 1 & 5 \\ 0 & -2 & -6 \end{array}\right]$$

$-2y = -6 \qquad \mathbf{y = 3}$

$x + 3 = 5 \qquad \mathbf{x = 2} \qquad \boxed{\{(2, 3)\}}$

23. $\begin{aligned} x + y &= -3 \\ 2x - 5y &= -6 \end{aligned}$ $\qquad \left[\begin{array}{cc|c} 1 & 1 & -3 \\ 2 & -5 & -6 \end{array}\right]$

$$-2R_1 + R_2 \to \left[\begin{array}{cc|c} 1 & 1 & -3 \\ 0 & -7 & 0 \end{array}\right]$$

$$-\tfrac{1}{7}R_2 \to \left[\begin{array}{cc|c} 1 & 1 & -3 \\ 0 & 1 & 0 \end{array}\right]$$

$\mathbf{y = 0}: \quad x + 0 = -3$

$\mathbf{x = -3} \qquad \boxed{\{(-3, 0)\}}$

25. $\begin{aligned} 2x - 3y &= 10 \\ 2x + 2y &= 5 \end{aligned}$ $\quad \left[\begin{array}{cc|c} 2 & -3 & 10 \\ 2 & 2 & 5 \end{array}\right]$

$-R_1 + R_2 \to \left[\begin{array}{cc|c} 2 & -3 & 10 \\ 0 & 5 & -5 \end{array}\right]$

$\left.\begin{array}{l} \frac{1}{2}R_1 \\ \frac{1}{5}R_2 \end{array}\right\} \to \left[\begin{array}{cc|c} 1 & -\frac{3}{2} & 5 \\ 0 & 1 & -1 \end{array}\right]$

$\mathbf{y = -1}: \quad x - \frac{3}{2}(-1) = 5$

$2x + 3 = 10$

$2x = 7 \quad \mathbf{x = \frac{7}{2}}$ $\quad \boxed{\left\{\left(\frac{7}{2}, -1\right)\right\}}$

27. $\begin{aligned} 2x - 3y &= 2 \\ 4x - 6y &= 1 \end{aligned}$ $\quad \left[\begin{array}{cc|c} 2 & -3 & 2 \\ 4 & -6 & 1 \end{array}\right]$

$-2R_1 + R_2 \to \left[\begin{array}{cc|c} 2 & -3 & 2 \\ 0 & 0 & -3 \end{array}\right]$

$0 \neq -3 \quad \boxed{\emptyset}$

29. $\begin{aligned} 6x - 3y &= 1 \\ -12x + 6y &= -2 \end{aligned}$

$\left[\begin{array}{cc|c} 6 & -3 & 1 \\ -12 & 6 & -2 \end{array}\right]$

$2R_1 + R_2 \to \left[\begin{array}{cc|c} 6 & -3 & 1 \\ 0 & 0 & 0 \end{array}\right]$

$0 = 0 \quad \infty$ **solutions**

$-3y = 1 - 6x$

$y = \dfrac{6x - 1}{3} \quad \left[\text{or} \quad x = \dfrac{3y + 1}{6}\right]$

$\boxed{\left(x, \dfrac{6x - 1}{3}\right) \text{ or } \left(\dfrac{3y + 1}{6}, y\right)}$

31. $\begin{aligned} x + y &= -1 \\ y + z &= 4 \\ x + z &= 1 \end{aligned}$ $\quad \left[\begin{array}{ccc|c} 1 & 1 & 0 & -1 \\ 0 & 1 & 1 & 4 \\ 1 & 0 & 1 & 1 \end{array}\right]$

$-R_1 + R_3 \to \left[\begin{array}{ccc|c} 1 & 1 & 0 & -1 \\ 0 & 1 & 1 & 4 \\ 0 & -1 & 1 & 2 \end{array}\right]$

$R_2 + R_3 \to \left[\begin{array}{ccc|c} 1 & 1 & 0 & -1 \\ 0 & 1 & 1 & 4 \\ 0 & 0 & 2 & 6 \end{array}\right]$

$\frac{1}{2}R_3 \to \left[\begin{array}{ccc|c} 1 & 1 & 0 & -1 \\ 0 & 1 & 1 & 4 \\ 0 & 0 & 1 & 3 \end{array}\right]$

(i) $\mathbf{z = 3}$

(ii) $y + z = 4$

$y + 3 = 4 \quad \mathbf{y = 1}$

(iii) $x + y = -1$

$x + 1 = -1 \quad \mathbf{x = -2}$ $\quad \boxed{\{(-2, 1, 3)\}}$

33. $\begin{aligned} x + y - z &= 6 \\ 2x - y + z &= -9 \\ x - 2y + 3z &= 1 \end{aligned}$

$\left[\begin{array}{ccc|c} 1 & 1 & -1 & 6 \\ 2 & -1 & 1 & -9 \\ 1 & -2 & 3 & 1 \end{array}\right]$

$\left.\begin{array}{r} -2R_1 + R_2 \\ -R_1 + R_3 \end{array}\right\} \to \left[\begin{array}{ccc|c} 1 & 1 & -1 & 6 \\ 0 & -3 & 3 & -21 \\ 0 & -3 & 4 & -5 \end{array}\right]$

$-R_2 + R_3 \to \left[\begin{array}{ccc|c} 1 & 1 & -1 & 6 \\ 0 & -3 & 3 & -21 \\ 0 & 0 & 1 & 16 \end{array}\right]$

$-\frac{1}{3}R_2 \to \left[\begin{array}{ccc|c} 1 & 1 & -1 & 6 \\ 0 & 1 & -1 & 7 \\ 0 & 0 & 1 & 16 \end{array}\right]$

(i) $\mathbf{z = 16}$

(ii) $y - z = 7$

$y - 16 = 7 \quad \mathbf{y = 23}$

(iii) $x + y - z = 6$

$x + 23 - 16 = 6$

$x + 7 = 6 \quad \mathbf{x = -1}$ $\quad \boxed{\{(-1, 23, 16)\}}$

35.
$$\begin{aligned} -x+y &= -1 \\ y-z &= 6 \\ x+z &= -1 \end{aligned} \quad \left[\begin{array}{ccc|c} -1 & 1 & 0 & -1 \\ 0 & 1 & -1 & 6 \\ 1 & 0 & 1 & -1 \end{array}\right]$$

$$R_1+R_3 \rightarrow \left[\begin{array}{ccc|c} -1 & 1 & 0 & -1 \\ 0 & 1 & -1 & 6 \\ 0 & 1 & 1 & -2 \end{array}\right]$$

$$-R_2+R_3 \rightarrow \left[\begin{array}{ccc|c} -1 & 1 & 0 & -1 \\ 0 & 1 & -1 & 6 \\ 0 & 0 & 2 & -8 \end{array}\right]$$

$$\left.\begin{array}{l} -R_1 \\ \frac{1}{2}R_3 \end{array}\right\} \rightarrow \left[\begin{array}{ccc|c} 1 & -1 & 0 & 1 \\ 0 & 1 & -1 & 6 \\ 0 & 0 & 1 & -4 \end{array}\right]$$

(i) $\mathbf{z = -4}$

(ii) $y - z = 6$
$y + 4 = 6 \qquad \mathbf{y = 2}$

(iii) $x - y = 1$
$x - 2 = 1 \qquad \mathbf{x = 3}$

$\boxed{\{(3,\ 2,\ -4)\}}$

37.
$$\begin{aligned} 2x - y + 3z &= 0 \\ x + 2y - z &= 5 \\ 2y + z &= 1 \end{aligned}$$

$$\left[\begin{array}{ccc|c} 2 & -1 & 3 & 0 \\ 1 & 2 & -1 & 5 \\ 0 & 2 & 1 & 1 \end{array}\right]$$

$$R_1 \leftrightarrow R_2 \rightarrow \left[\begin{array}{ccc|c} 1 & 2 & -1 & 5 \\ 2 & -1 & 3 & 0 \\ 0 & 2 & 1 & 1 \end{array}\right]$$

$$-2R_1+R_2 \rightarrow \left[\begin{array}{ccc|c} 1 & 2 & -1 & 5 \\ 0 & -5 & 5 & -10 \\ 0 & 2 & 1 & 1 \end{array}\right]$$

$$2R_2+5R_3 \rightarrow \left[\begin{array}{ccc|c} 1 & 2 & -1 & 5 \\ 0 & -5 & 5 & -10 \\ 0 & 0 & 15 & -15 \end{array}\right]$$

$$\left.\begin{array}{l} -\frac{1}{5}R_2 \\ \frac{1}{15}R_3 \end{array}\right\} \rightarrow \left[\begin{array}{ccc|c} 1 & 2 & -1 & 5 \\ 0 & 1 & -1 & 2 \\ 0 & 0 & 1 & -1 \end{array}\right]$$

(i) $\mathbf{z = -1}$

(ii) $y - z = 2$
$y + 1 = 2 \qquad \mathbf{y = 1}$

(iii) $x + 2y - z = 5$
$x + 2 + 1 = 5 \qquad \mathbf{x = 2}$

$\boxed{\{(2,\ 1,\ -1)\}}$

39.
$$\begin{aligned} 2.1x + .5y + 1.7z &= 4.9 \\ -2x + 1.5y - 1.7z &= 3.1 \\ 5.8x - 4.6y + .8z &= 9.3 \end{aligned}$$

$$\text{RREF}\left[\begin{array}{ccc|c} 2.1 & .5 & 1.7 & 4.9 \\ -2 & 1.5 & -1.7 & 3.1 \\ 5.8 & -4.6 & .8 & 9.3 \end{array}\right]$$

$$= \left[\begin{array}{ccc|c} 1 & 0 & 0 & 5.21127 \\ 0 & 1 & 0 & 3.73944 \\ 0 & -0 & 1 & -4.65493 \end{array}\right]$$

$\boxed{\text{Solution set: } \approx \{(5.211,\ 3.739,\ -4.655)\}}$

41.
$$\begin{aligned} 53x + 95y + 12z &= 108 \\ 81x - 57y - 24z &= -92 \\ -9x + 11y - 78z &= 21 \end{aligned}$$

$$\text{RREF}\left[\begin{array}{ccc|c} 53 & 95 & 12 & 108 \\ 81 & -57 & -24 & -92 \\ -9 & 11 & -78 & 21 \end{array}\right]$$

$$= \left[\begin{array}{ccc|c} 1 & 0 & 0 & -.24997 \\ 0 & 1 & 0 & 1.2838 \\ 0 & -0 & 1 & -.05934 \end{array}\right]$$

$\boxed{\text{Solution set: } \approx \{(-.250,\ 1.284,\ -.059)\}}$

43. Both shorthand methods utilize writing only the coefficients and no variables, which is possible because the order of the variables is determined beforehand.

45. $x - 3y + 2z = 10$
$2x - y - z = 8$

$$\left[\begin{array}{ccc|c} 1 & -3 & 2 & 10 \\ 2 & -1 & -1 & 8 \end{array}\right]$$

$$-2R_1 + R_2 \rightarrow \left[\begin{array}{ccc|c} 1 & -3 & 2 & 10 \\ 0 & 5 & -5 & -12 \end{array}\right]$$

$$\tfrac{1}{5}R_2 \rightarrow \left[\begin{array}{ccc|c} 1 & -3 & 2 & 10 \\ 0 & 1 & -1 & -\frac{12}{5} \end{array}\right]$$

(i) $y - z = -\dfrac{12}{5}$

$5y - 5z = -12$

$5y = 5z - 12 \qquad \mathbf{y = \dfrac{5z - 12}{5}}$

(ii) $x - 3\left(\dfrac{5z - 12}{5}\right) + 2z = 10$

$5x - 15z + 36 + 10z = 50$

$5x - 5z = 14$

$5x = 5z + 14 \qquad \mathbf{x = \dfrac{5z + 14}{5}}$

$$\boxed{\left(\frac{5z + 14}{5}, \frac{5z - 12}{5}, z\right)}$$

47. $x + 2y - z = 0$
$3x - y + z = 6$
$-2x - 4y + 2z = 0$

$$\left[\begin{array}{ccc|c} 1 & 2 & -1 & 0 \\ 3 & -1 & 1 & 6 \\ -2 & -4 & 2 & 0 \end{array}\right]$$

$$\left.\begin{array}{r} -3R_1 + R_2 \\ 2R_1 + R_3 \end{array}\right\} \rightarrow \left[\begin{array}{ccc|c} 1 & 2 & -1 & 0 \\ 0 & -7 & 4 & 6 \\ 0 & 0 & 0 & 0 \end{array}\right]$$

$0 = 0:$ ∞ **solutions**

(i) $-7y + 4z = 6$

$-7y = 6 - 4z \qquad \mathbf{y = \dfrac{4z - 6}{7}}$

(ii) $x + 2\left(\dfrac{4z - 6}{7}\right) - z = 0$

$7x + 8z - 12 - 7z = 0$

$7x + z = 12$

$7x = 12 - z \qquad \mathbf{x = \dfrac{12 - z}{7}}$

$$\boxed{\left(\frac{12 - z}{7}, \frac{4z - 6}{7}, z\right)}$$

49. $x - 2y + z = 5$
$-2x + 4y - 2z = 2$
$2x + y - z = 2$

$$\left[\begin{array}{ccc|c} 1 & -2 & 1 & 5 \\ -2 & 4 & -2 & 2 \\ 2 & 1 & -1 & 2 \end{array}\right]$$

$$\left.\begin{array}{r} 2R_1 + R_2 \\ -2R_1 + R_3 \end{array}\right\} \rightarrow \left[\begin{array}{ccc|c} 1 & -2 & 1 & 5 \\ 0 & 0 & 0 & 12 \\ 0 & 5 & -3 & -8 \end{array}\right]$$

2nd row: $0 = 12$; **no solutions** $\boxed{\emptyset}$

51. $x + 3y - 2z - w = 9$
$4x + y + z + 2w = 2$
$-3x - y + z - w = -5$
$x - y - 3z - 2w = 2$

$$\left[\begin{array}{cccc|c} 1 & 3 & -2 & -1 & 9 \\ 4 & 1 & 1 & 2 & 2 \\ -3 & -1 & 1 & -1 & -5 \\ 1 & -1 & -3 & -2 & 2 \end{array}\right]$$

$$\left.\begin{array}{r} -4R_1 + R_2 \\ 3R_1 + R_3 \\ -R_1 + R_4 \end{array}\right\} \rightarrow \left[\begin{array}{cccc|c} 1 & 3 & -2 & -1 & 9 \\ 0 & -11 & 9 & 6 & -34 \\ 0 & 8 & -5 & -4 & 22 \\ 0 & -4 & -1 & -1 & -7 \end{array}\right]$$

$$\left.\begin{array}{r} R_3 + 2R_4 \\ 8R_2 + 11R_3 \end{array}\right\} \rightarrow \left[\begin{array}{cccc|c} 1 & 3 & -2 & -1 & 9 \\ 0 & -11 & 9 & 6 & -34 \\ 0 & 0 & 17 & 4 & -30 \\ 0 & 0 & -7 & -6 & 8 \end{array}\right]$$

$$7R_3 + 17R_4 \rightarrow \left[\begin{array}{cccc|c} 1 & 3 & -2 & -1 & 9 \\ 0 & -11 & 9 & 6 & -34 \\ 0 & 0 & 17 & 4 & -30 \\ 0 & 0 & 0 & -74 & -74 \end{array}\right]$$

$$-\tfrac{1}{11}R_2,\ \tfrac{1}{17}R_3,\ -\tfrac{1}{74}R_4 \rightarrow$$

$$\left[\begin{array}{cccc|c} 1 & 3 & -2 & -1 & 9 \\ 0 & 1 & \frac{-9}{11} & \frac{-6}{11} & \frac{34}{11} \\ 0 & 0 & 1 & \frac{4}{17} & -\frac{30}{17} \\ 0 & 0 & 0 & 1 & 1 \end{array}\right]$$

continued

51. continued

(i) $\mathbf{w = 1}$

(ii) $z + \frac{4}{17} = -\frac{30}{17}$

$z = -\frac{34}{17}$

$\mathbf{z = -2}$

(iii) $y - \frac{9}{11}(-2) - \frac{6}{11} = \frac{34}{11}$

$y + \frac{18-6}{11} = \frac{34}{11}$

$y = \frac{34-12}{11} = \frac{22}{11}$ $\quad \mathbf{y = 2}$

(iv) $x + 3(2) - 2(-2) - 1 = 9$

$x + 9 = 9$ $\quad \mathbf{x = 0}$

$\boxed{\{(0,\ 2,\ -2,\ 1)\}}$

53. (a) $F = aN + bR + c$

$$\begin{aligned} 1300 &= 1800a + 5000b + c \\ 5300 &= 3200a + 12000b + c \\ 6500 &= 4500a + 13000b + c \end{aligned}$$

$$\text{RREF}\left[\begin{array}{ccc|c} 1800 & 5000 & 1 & 1300 \\ 3200 & 12000 & 1 & 5300 \\ 4500 & 13000 & 1 & 6500 \end{array}\right]$$

$$= \left[\begin{array}{ccc|c} 1 & 0 & 0 & .5714286 \\ 0 & 1 & 0 & .4571429 \\ 0 & 0 & 1 & -2014.2857 \end{array}\right]$$

$\boxed{F = .5714N + .4571R - 2014}$

(b) Let $N = 3500$ and $R = 12,500$:

$F = .5714\,(3500) + .4571\,(12500) - 2014$

$F = 5699.65$

This model predicts monthly food costs of approximately \$5700.

55. Let x represent the number of model 201, y, the number of model 301:

$$\begin{aligned} 2x + 3y &= 34 \\ 25x + 30y &= 365 \end{aligned} \quad \left[\begin{array}{cc|c} 2 & 3 & 34 \\ 25 & 30 & 365 \end{array}\right]$$

$$\left.\begin{array}{l} \frac{1}{2}R_1 \\ \frac{1}{5}R_2 \end{array}\right\} \rightarrow \left[\begin{array}{cc|c} 1 & \frac{3}{2} & 17 \\ 5 & 6 & 73 \end{array}\right]$$

$$-5R_1 + R_2 \rightarrow \left[\begin{array}{cc|c} 1 & \frac{3}{2} & 17 \\ 0 & -\frac{3}{2} & -12 \end{array}\right]$$

$$-\tfrac{2}{3}R_2 \rightarrow \left[\begin{array}{cc|c} 1 & \frac{3}{2} & 17 \\ 0 & 1 & 8 \end{array}\right]$$

(i) $\mathbf{y = 8}$

(ii) $x + \frac{3}{2}y = 17$

$x + \frac{3}{2}(8) = 17$

$x + 12 = 17$ $\quad \mathbf{x = 5}$

Model 201: 5 bicycles
Model 301: 8 bicycles

57. Let x = the amount of money at 8%, y = the amount of money at 10%, and z = the amount of money at 9%.

$x + y + z = 25,000$

$y = 2000 + \frac{1}{2}x$

$.08x + .10y + .09z = 2220$

$$\left[\begin{array}{ccc|c} 1 & 1 & 1 & 25,000 \\ -\frac{1}{2} & 1 & 0 & 2,000 \\ .08 & .10 & .09 & 2,220 \end{array}\right]$$

$$\left.\begin{array}{l} \frac{1}{2}R_1 + R_2 \\ 100R_3 \end{array}\right\} \rightarrow$$

$$\left[\begin{array}{ccc|c} 1 & 1 & 1 & 25,000 \\ 0 & \frac{3}{2} & \frac{1}{2} & 14,500 \\ 8 & 10 & 9 & 222,000 \end{array}\right]$$

continued

57. continued

$$\left.\begin{array}{l}\frac{2}{3}R_2 \\ -8R_1 + R_3\end{array}\right\} \rightarrow$$

$$\left[\begin{array}{ccc|c} 1 & 1 & 1 & 25,000 \\ 0 & 1 & \frac{1}{3} & 9666\frac{2}{3} \\ 0 & 2 & 1 & 22,000 \end{array}\right]$$

$R_2 - \frac{1}{2}R_3 \rightarrow$

$$\left[\begin{array}{ccc|c} 1 & 1 & 1 & 25,000 \\ 0 & 1 & \frac{1}{3} & 9666\frac{2}{3} \\ 0 & 0 & -\frac{1}{6} & -1333\frac{1}{3} \end{array}\right]$$

$$-6R_3 \rightarrow \left[\begin{array}{ccc|c} 1 & 1 & 1 & 25,000 \\ 0 & 1 & \frac{1}{3} & 9666\frac{2}{3} \\ 0 & 0 & 1 & 8000 \end{array}\right]$$

(i) $\mathbf{z = 8,000}$

(ii) $y + \frac{1}{3}(8000) = 9666\frac{2}{3}$

$3y + 8000 = 29,000$

$3y = 21,000 \quad \mathbf{y = 7000}$

(iii) $x + 7000 + 8000 = 25,000$

$\mathbf{x = 10,000}$

$10,000 at 8%, $7,000 at 10% $8,000 at 9%

59. (a) Let $f(x) = ax^2 + bx + c$:

$$1990^2a + 1990b + c = 11$$
$$2010^2a + 2010b + c = 10$$
$$2030^2a + 2030b + c = 6$$

(b) RREF $\left[\begin{array}{ccc|c} 1990^2 & 1990 & 1 & 11 \\ 2010^2 & 2010 & 1 & 10 \\ 2030^2 & 2030 & 1 & 6 \end{array}\right]$

$$= \left[\begin{array}{ccc|c} 1 & 0 & 0 & -.00375 \\ 0 & 1 & 0 & 14.95 \\ 0 & 0 & 1 & -14889.125 \end{array}\right]$$

$$\boxed{f(x) = -.00375x^2 + 14.95x - 14889.125}$$

(c) $f(x) = -.00375x^2 + 14.95x - 14,889.125$

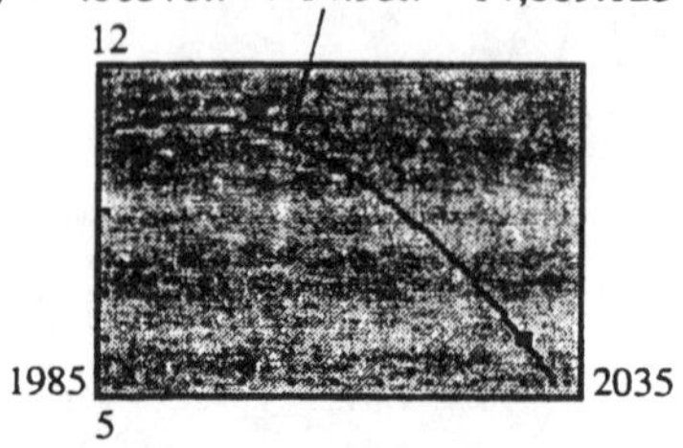

Window: 1985 to 2035, 5 to 12

(d) Answers may vary; for example in 2015 the ratio could be $f(2015) \approx 9.3$.

61. (a) Let $f(x) = ax^2 + bx + c$:

$$3^2a + 3b + c = 2.2$$
$$18^2a + 18b + c = 10.4$$
$$26^2a + 26b + c = 12.8$$

(b) RREF $\left[\begin{array}{ccc|c} 9 & 3 & 1 & 2.2 \\ 324 & 18 & 1 & 10.4 \\ 676 & 26 & 1 & 12.8 \end{array}\right]$

$$= \left[\begin{array}{ccc|c} 1 & 0 & 0 & -.010725 \\ 0 & 1 & 0 & .771884 \\ 0 & 0 & 1 & -.019130 \end{array}\right]$$

$$\boxed{f(x) = -.010725x^2 + .77188x - .019130}$$

(c) $f(x) = -.010725x^2 + .77188x - .019130$

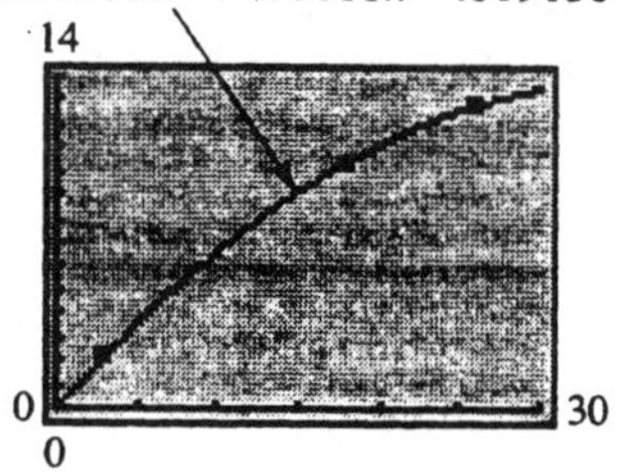

Window: 0 to 30, 0 to 14

(d) Answers may vary; for example in 1990, this percentage was $f(20) \approx 11.1$.

63. (a) At intersection **A** incoming traffic is equal to $x + 5$. The outgoing traffic is given by $y + 7$. Therefore, $x + 5 = y + 7$. The incoming traffic at intersection **B** is $z + 6$ and the outgoing traffic is $x + 3$, so $z + 6 = x + 3$. Finally at intersection **C**, the incoming flow is $y + 3$ and the outgoing flow is $z + 4$, so $y + 3 = z + 4$.

(b) These 3 equations can be written:

$x - y = 2$

$x - z = 3$

$y - z = 1$ and can be represented by:

$$\left[\begin{array}{ccc|c} 1 & -1 & 0 & 2 \\ 1 & 0 & -1 & 3 \\ 0 & 1 & -1 & 1 \end{array}\right]$$

$$-R_1 + R_2 \rightarrow \left[\begin{array}{ccc|c} 1 & -1 & 0 & 2 \\ 0 & 1 & -1 & 1 \\ 0 & 1 & -1 & 1 \end{array}\right]$$

$$-R_2 + R_3 \rightarrow \left[\begin{array}{ccc|c} 1 & -1 & 0 & 2 \\ 0 & 1 & -1 & 1 \\ 0 & 0 & 0 & 0 \end{array}\right]$$

continued

63. continued

There are infinite solutions:

$y = z + 1$ and $x = z + 3,\ z \geq 0$:

$\{(z+3,\ z+1,\ z) \mid z \geq 0\}$

(c) There are infinitely many solutions since some cars could be driving around the block continually.

65. (a) $a + bA + cP + dW = F$

$$a + 871b + 11.5c + 3d = 239$$
$$a + 847b + 12.2c + 2d = 234$$
$$a + 685b + 10.6c + 5d = 192$$
$$a + 969b + 14.2c + 1d = 343$$

(b) $\left[\begin{array}{cccc|c} 1 & 871 & 11.5 & 3 & 239 \\ 1 & 847 & 12.2 & 2 & 234 \\ 1 & 685 & 10.6 & 5 & 192 \\ 1 & 969 & 14.2 & 1 & 343 \end{array}\right]$

$a \approx -715.457,\ b \approx .34756,$
$c \approx 48.6585,\ d \approx 30.71951$

(c) $F = -715.457 + .34756A + 48.6585P + 30.71951W$

(d) $F = -715.457 + .34756(960) + 48.6585(12.6) + 30.71951(3) = 323.45623$

Approximately 323, which is close to the actual value of 320.

Reviewing Basic Concepts (Sections 7.1 - 7.3)

1. $2x - 3y = 18$ [1]
$5x + 2y = 7$ [2]

$4x - 6y = 36$ [2 × [1]]
$15x + 6y = 21$ [3 × [2]]
$19x = 57$ **x = 3**

$x = 3 \longrightarrow$ [1]: $2(3) - 3y = 18$
$-3y = 12$ **y = −4**

$\boxed{\{(3, -4)\}}$

2. $2x + y = -4 \longrightarrow y_1 = -2x - 4$
$-x + 2y = 2 \longrightarrow y_2 = \frac{1}{2}x + 1$

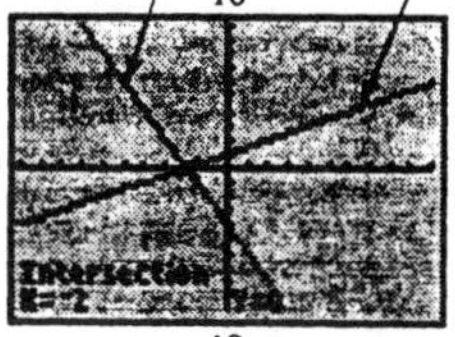

$2x + y = -4$ $-x + 2y = 2$

$\boxed{(-2, 0)}$

3. $5x + 10y = 10$ [1]
$x + 2y = 2$ [2]

Solve [2] for x: $x = 2 - 2y$ [3]

[3] $\longrightarrow$ [1]: $5(2 - 2y) + 10y = 10$
$10 - 10y + 10y = 10$
$0 = 0$ **∞ solutions**

$(-2y + 2,\ y)$ or $\left(x,\ \dfrac{2-x}{2}\right)$

4. $x - y = 6$ [1]
$x - y = 4$ [2]

$x - y = 6$ [1]
$-x + y = -4$ [−1 × [2]]
$0 = 2$ No solution $\boxed{\emptyset}$

5. $6x + 2y = 10$ [1]
$2x^2 - 3y = 11$ [2]

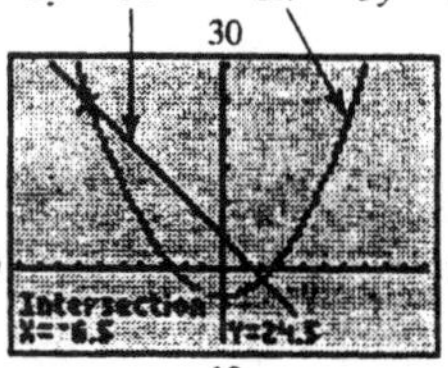

$6x + 2y = 10$ $2x^2 - 3y = 11$

The other point of intersection is (2, −1).

Solve [1] for y: $2y = -6x + 10$
$y = -3x + 5$ [3]

[3] $\longrightarrow$ [2]: $2x^2 - 3(-3x + 5) = 11$
$2x^2 + 9x - 15 = 11$
$2x^2 + 9x - 26 = 0$
$(2x + 13)(x - 2) = 0$
x = −6.5, x = 2

$x = -6.5 \longrightarrow$ [3]: $y = -3(-6.5) + 5$
$y = 19.5 + 5$ **y = 24.5**

$x = 2 \longrightarrow$ [3]: $y = -3(2) + 5$
$y = -6 + 5$ **y = −1**

$\boxed{(2, -1) \text{ and } (-6.5, 24.5)}$

6. $\begin{aligned} x+y+z &= 1 \\ -x+y+z &= 5 \\ y+2z &= 5 \end{aligned}$

$$\left[\begin{array}{ccc|c} 1 & 1 & 1 & 1 \\ -1 & 1 & 1 & 5 \\ 0 & 1 & 2 & 5 \end{array}\right]$$

$$R_1+R_2: \left[\begin{array}{ccc|c} 1 & 1 & 1 & 1 \\ 0 & 2 & 2 & 6 \\ 0 & 1 & 2 & 5 \end{array}\right]$$

$$R_2-2R_3: \left[\begin{array}{ccc|c} 1 & 1 & 1 & 1 \\ 0 & 2 & 2 & 6 \\ 0 & 0 & -2 & -4 \end{array}\right]$$

(i) $-2z=-4 \longrightarrow \mathbf{z=2}$

(ii) $2y+2z=6$
$2y+4=6$
$2y=2 \longrightarrow \mathbf{y=1}$

(iii) $x+y+z=1$
$x+1+2=1 \longrightarrow \mathbf{x=-2}$

$\boxed{\{(-2,\ 1,\ 2)\}}$

7. $\begin{aligned} 2x+4y+4z &= 4 \\ x+3y+\ z &= 4 \\ -x+3y+2z &= -1 \end{aligned}$

$$\left[\begin{array}{ccc|c} 2 & 4 & 4 & 4 \\ 1 & 3 & 1 & 4 \\ -1 & 3 & 2 & -1 \end{array}\right]$$

$$\left.\begin{array}{r} \frac{1}{2}R_1 \\ R_2+R_3 \end{array}\right\} \rightarrow \left[\begin{array}{ccc|c} 1 & 2 & 2 & 2 \\ 1 & 3 & 1 & 4 \\ 0 & 6 & 3 & 3 \end{array}\right]$$

$$\left.\begin{array}{r} -R_1+R_2 \\ \frac{1}{3}R_3 \end{array}\right\} \rightarrow \left[\begin{array}{ccc|c} 1 & 2 & 2 & 2 \\ 0 & 1 & -1 & 2 \\ 0 & 2 & 1 & 1 \end{array}\right]$$

$$-2R_2+R_3 \rightarrow \left[\begin{array}{ccc|c} 1 & 2 & 2 & 2 \\ 0 & 1 & -1 & 2 \\ 0 & 0 & 3 & -3 \end{array}\right]$$

$$-2R_2+R_1 \rightarrow \left[\begin{array}{ccc|c} 1 & 0 & 4 & -2 \\ 0 & 1 & -1 & 2 \\ 0 & 0 & 3 & -3 \end{array}\right]$$

$$\tfrac{1}{3}R_3 \rightarrow \left[\begin{array}{ccc|c} 1 & 0 & 4 & -2 \\ 0 & 1 & -1 & 2 \\ 0 & 0 & 1 & -1 \end{array}\right]$$

$$\left.\begin{array}{r} -4R_3+R_1 \\ R_3+R_2 \end{array}\right\} \rightarrow \left[\begin{array}{ccc|c} 1 & 0 & 0 & 2 \\ 0 & 1 & 0 & 1 \\ 0 & 0 & 1 & -1 \end{array}\right]$$

$x=2,\ y=1,\ z=-1$ $\boxed{\{(2,\ 1,\ -1)\}}$

8. $$\left[\begin{array}{ccc|c} 2 & 1 & 2 & 10 \\ 1 & 0 & 2 & 5 \\ 1 & -2 & 2 & 1 \end{array}\right]$$

$$\left.\begin{array}{r} R_1-2R_2 \\ R_1-2R_3 \end{array}\right\} \rightarrow \left[\begin{array}{ccc|c} 2 & 1 & 2 & 10 \\ 0 & 1 & -2 & 0 \\ 0 & 5 & -2 & 8 \end{array}\right]$$

$$-5R_2+R_3 \rightarrow \left[\begin{array}{ccc|c} 2 & 1 & 2 & 10 \\ 0 & 1 & -2 & 0 \\ 0 & 0 & 8 & 8 \end{array}\right]$$

(i) $8z=8 \longrightarrow \mathbf{z=1}$

(ii) $y-2z=0 \Rightarrow y-2=0 \longrightarrow \mathbf{y=2}$

(iii) $2x+y+2z=10$
$2x+2+2=10 \longrightarrow \mathbf{x=3}$

$\boxed{\{(3,\ 2,\ 1)\}}$

9. Let x represent # of sets with stereo sound.
Let y represent # of sets without stereo sound.

Then, $x+y=32{,}000{,}000$ [1]
$10x=19y$ [2]

Solve [1] for x : $x=32{,}000{,}000-y$ [3]

[3] $\longrightarrow$ [2] : $10\,(32{,}000{,}000-y)=19y$
$320{,}000{,}000-10y=19y$
$320{,}000{,}000=29y$
$y=11.034{,}482.76$

$y \longrightarrow$ [3] : $x=32{,}000{,}000-11{,}034{,}483$
$=20{,}965{,}517$

$\boxed{\begin{array}{c}\approx 11.03 \text{ million with stereo sound,} \\ \approx 21 \text{ million without stereo sound.}\end{array}}$

10. Let $x=$ amount at 8%, $y=$ amount at 11%, and $z=$ amount at 14%. Then,

$x+y+z=5000$ [1]
$.08x+.11y+.14z=595$ [2]
$z=x+y$ [3]

Rewrite [3] as $x+y-z=0$ [4]

$$\text{RREF} \left[\begin{array}{ccc|c} 1 & 1 & 1 & 5000 \\ .08 & .11 & .14 & 595 \\ 1 & 1 & -1 & 0 \end{array}\right] = \left[\begin{array}{ccc|c} 1 & 0 & 0 & 1000 \\ 0 & 1 & 0 & 1500 \\ 0 & 0 & 1 & 2500 \end{array}\right]$$

$\boxed{\begin{array}{c}\text{At 8\%: \$1000, at 11\%: \$1500,} \\ \text{at 14\%: \$2500}\end{array}}$

Section 7.4

1. $\begin{bmatrix} -3 & 6 \\ 7 & -4 \end{bmatrix}$ 2 × 2 square matrix

3. $\begin{bmatrix} -6 & 8 & 0 & 0 \\ 4 & 1 & 9 & 2 \\ 3 & -5 & 7 & 1 \end{bmatrix}$ 3 × 4 matrix

5. $\begin{bmatrix} 2 \\ 4 \end{bmatrix}$ 2 × 1 column matrix

7. $[-9]$ 1 × 1 matrix
square, row, column

9. $\begin{bmatrix} w & x \\ y & z \end{bmatrix} = \begin{bmatrix} 3 & 2 \\ -1 & 4 \end{bmatrix}$
$w = 3,\ x = 2,\ y = -1,\ z = 4$

11. $\begin{bmatrix} 0 & 5 & x \\ -1 & 3 & y+2 \\ 4 & 1 & z \end{bmatrix} = \begin{bmatrix} 0 & w+3 & 6 \\ -1 & 3 & 0 \\ 4 & 1 & 8 \end{bmatrix}$
$w + 3 = 5 \Rightarrow \mathbf{w = 2},\quad \mathbf{x = 6}$
$y + 2 = 0 \Rightarrow \mathbf{y = -2},\quad \mathbf{z = 8}$

13. $\begin{bmatrix} -7+z & 4r & 8s \\ 6p & 2 & 5 \end{bmatrix} + \begin{bmatrix} -9 & 8r & 3 \\ 2 & 5 & 4 \end{bmatrix}$

$= \begin{bmatrix} 2 & 36 & 27 \\ 20 & 7 & 12a \end{bmatrix}$

$-7 + z - 9 = 2 \Rightarrow \mathbf{z = 18}$
$4r + 8r = 36 \Rightarrow \mathbf{r = 3}$
$8s + 3 = 27 \Rightarrow 8s = 24 \Rightarrow \mathbf{s = 3}$
$6p + 2 = 20 \Rightarrow 6p = 18 \Rightarrow \mathbf{p = 3}$
$5 + 4 = 12a \Rightarrow 9 = 12a \Rightarrow \mathbf{a = \frac{3}{4}}$

15. Only matrices of the same dimensions may be added; add corresponding elements. The sum will be a matrix of the same dimensions as the original matrix.

17. $\begin{bmatrix} 6 & -9 & 2 \\ 4 & 1 & 3 \end{bmatrix} + \begin{bmatrix} -8 & 2 & 5 \\ 6 & -3 & 4 \end{bmatrix}$

$= \begin{bmatrix} 6-8 & -9+2 & 2+5 \\ 4+6 & 1-3 & 3+4 \end{bmatrix}$

$= \begin{bmatrix} -2 & -7 & 7 \\ 10 & -2 & 7 \end{bmatrix}$

19. $\begin{bmatrix} -6 & 8 \\ 0 & 0 \end{bmatrix} - \begin{bmatrix} 0 & 0 \\ -4 & -2 \end{bmatrix}$

$= \begin{bmatrix} -6-0 & 8-0 \\ 0-(-4) & 0-(-2) \end{bmatrix} = \begin{bmatrix} -6 & 8 \\ 4 & 2 \end{bmatrix}$

21. $\begin{bmatrix} 6 & -2 \\ 5 & 4 \end{bmatrix} + \begin{bmatrix} -1 & 7 \\ 7 & -4 \end{bmatrix}$

$= \begin{bmatrix} 6-1 & -2+7 \\ 5+7 & 4-4 \end{bmatrix} = \begin{bmatrix} 5 & 5 \\ 12 & 0 \end{bmatrix}$

23. $\begin{bmatrix} -8 & 4 & 0 \\ 2 & 5 & 0 \end{bmatrix} + \begin{bmatrix} 6 & 3 \\ 8 & 9 \end{bmatrix}$
$(2 \times 3) + (2 \times 2)$; Cannot add.

25. $\begin{bmatrix} 9 & 4 & 1 & -2 \\ 5 & -6 & 3 & 4 \\ 2 & -5 & 1 & 2 \end{bmatrix} - \begin{bmatrix} -2 & 5 & 1 & 3 \\ 0 & 1 & 0 & 2 \\ -8 & 3 & 2 & 1 \end{bmatrix}$

$+ \begin{bmatrix} 2 & 4 & 0 & 3 \\ 4 & -5 & 1 & 6 \\ 2 & -3 & 0 & 8 \end{bmatrix}$

$= \begin{bmatrix} 9+2+2 & 4-5+4 & 1-1+0 & -2-3+3 \\ 5-0+4 & -6-1-5 & 3-0+1 & 4-2+6 \\ 2+8+2 & -5-3-3 & 1-2+0 & 2-1+8 \end{bmatrix}$

$= \begin{bmatrix} 13 & 3 & 0 & -2 \\ 9 & -12 & 4 & 8 \\ 12 & -11 & -1 & 9 \end{bmatrix}$

27. $3\begin{bmatrix} 6 & -1 & 4 \\ 2 & 8 & -3 \\ -4 & 5 & 6 \end{bmatrix} + 5\begin{bmatrix} -2 & -8 & -6 \\ 4 & 1 & 3 \\ 2 & -1 & 5 \end{bmatrix}$

$= \begin{bmatrix} 18 & -3 & 12 \\ 6 & 24 & -9 \\ -12 & 15 & 18 \end{bmatrix} + \begin{bmatrix} -10 & -40 & -30 \\ 20 & 5 & 15 \\ 10 & -5 & 25 \end{bmatrix}$

$= \begin{bmatrix} 18-10 & -3-40 & 12-30 \\ 6+20 & 24+5 & -9+15 \\ -12+10 & 15-5 & 18+25 \end{bmatrix}$

$= \begin{bmatrix} 8 & -43 & -18 \\ 26 & 29 & 6 \\ -2 & 10 & 43 \end{bmatrix}$

29. $A = \begin{bmatrix} -2 & 4 \\ 0 & 3 \end{bmatrix}, \quad 2A = \begin{bmatrix} -4 & 8 \\ 0 & 6 \end{bmatrix}$

31. $A = \begin{bmatrix} -2 & 4 \\ 0 & 3 \end{bmatrix}, \quad B = \begin{bmatrix} -6 & 2 \\ 4 & 0 \end{bmatrix}$

$$2A - B = \begin{bmatrix} -4+6 & 8-2 \\ 0-4 & 6-0 \end{bmatrix}$$

$$= \begin{bmatrix} 2 & 6 \\ -4 & 6 \end{bmatrix}$$

33. $A = \begin{bmatrix} -2 & 4 \\ 0 & 3 \end{bmatrix}, \quad B = \begin{bmatrix} -6 & 2 \\ 4 & 0 \end{bmatrix}$

$$5A + \tfrac{1}{2}B = \begin{bmatrix} -10-3 & 20+1 \\ 0+2 & 15+0 \end{bmatrix}$$

$$= \begin{bmatrix} -13 & 21 \\ 2 & 15 \end{bmatrix}$$

35. $[A] = ([A] + [B]) - [B]$

$$= \begin{bmatrix} 6 & 12 & 0 \\ -10 & -4 & 11 \end{bmatrix} - \begin{bmatrix} 4 & 6 & -5 \\ -6 & 3 & 2 \end{bmatrix}$$

$$= \begin{bmatrix} 2 & 6 & 5 \\ -4 & -7 & 9 \end{bmatrix}$$

37. $A: 4 \times 2; \quad B: 2 \times 4$

$AB: 4 \times 4; \quad BA: 2 \times 2$

39. $A: 3 \times 5; \quad B: 5 \times 2$

$AB: 3 \times 2; \quad BA: \emptyset$

41. $A: 4 \times 3; \quad B: 2 \times 5$

$AB: \emptyset; \quad BA: \emptyset$

43. The product MN of two matrices can be found only if the number of columns of M equal the number of rows of N.

45. $\begin{bmatrix} p & q \\ r & s \end{bmatrix} \begin{bmatrix} a & c \\ b & d \end{bmatrix} = \begin{bmatrix} pa+qb & pc+qd \\ ra+sb & rc+sd \end{bmatrix}$

47. $\begin{bmatrix} 3 & -4 & 1 \\ 5 & 0 & 2 \end{bmatrix} \begin{bmatrix} -1 \\ 4 \\ 2 \end{bmatrix}$

$$= \begin{bmatrix} 3(-1) - 4(4) + 1(2) \\ 5(-1) + 0(4) + 2(2) \end{bmatrix}$$

$$= \begin{bmatrix} -3 - 16 + 2 \\ -5 + 0 + 4 \end{bmatrix} = \begin{bmatrix} -17 \\ -1 \end{bmatrix}$$

49. $\begin{bmatrix} 5 & 2 \\ -1 & 4 \end{bmatrix} \begin{bmatrix} 3 & -2 \\ 1 & 0 \end{bmatrix}$

$$= \begin{bmatrix} 5 \cdot 3 + 2 \cdot 1 & 5(-2) + 2 \cdot 0 \\ -1 \cdot 3 + 4 \cdot 1 & -1(-2) + 4 \cdot 0 \end{bmatrix}$$

$$= \begin{bmatrix} 15+2 & -10+0 \\ -3+4 & 2+0 \end{bmatrix}$$

$$= \begin{bmatrix} 17 & -10 \\ 1 & 2 \end{bmatrix}$$

51. $\begin{bmatrix} 2 & 2 & -1 \\ 3 & 0 & 1 \end{bmatrix} \begin{bmatrix} 0 & 2 \\ -1 & 4 \\ 0 & 2 \end{bmatrix}$

$$= \begin{bmatrix} 0-2+0 & 4+8-2 \\ 0+0+0 & 6+0+2 \end{bmatrix}$$

$$= \begin{bmatrix} -2 & 10 \\ 0 & 8 \end{bmatrix}$$

53. $\begin{bmatrix} -2 & -3 & -4 \\ 2 & -1 & 0 \\ 4 & -2 & 3 \end{bmatrix} \begin{bmatrix} 0 & 1 & 4 \\ 1 & 2 & -1 \\ 3 & 2 & -2 \end{bmatrix}$

$$= \begin{bmatrix} 0-3-12 & -2-6-8 & -8+3+8 \\ 0-1+0 & 2-2+0 & 8+1+0 \\ 0-2+9 & 4-4+6 & 16+2-6 \end{bmatrix}$$

$$= \begin{bmatrix} -15 & -16 & 3 \\ -1 & 0 & 9 \\ 7 & 6 & 12 \end{bmatrix}$$

55. $\begin{bmatrix} -2 & 4 & 1 \end{bmatrix} \begin{bmatrix} 3 & -2 & 4 \\ 2 & 1 & 0 \\ 0 & -1 & 4 \end{bmatrix}$

$$= \begin{bmatrix} -6+8+0 & 4+4-1 & -8+0+4 \end{bmatrix}$$

$$= \begin{bmatrix} 2 & 7 & -4 \end{bmatrix}$$

57. $BA = \begin{bmatrix} 5 & 1 \\ 0 & -2 \\ 3 & 7 \end{bmatrix} \begin{bmatrix} 4 & -2 \\ 3 & 1 \end{bmatrix}$

$= \begin{bmatrix} 5(4)+1(3) & 5(-2)+1(1) \\ 0(4)+(-2)(3) & 0(-2)+(-2)(1) \\ 3(4)+7(3) & 3(-2)+7(1) \end{bmatrix}$

$= \begin{bmatrix} 23 & -9 \\ -6 & -2 \\ 33 & 1 \end{bmatrix}$

59. $BC = \begin{bmatrix} 5 & 1 \\ 0 & -2 \\ 3 & 7 \end{bmatrix} \begin{bmatrix} -5 & 4 & 1 \\ 0 & 3 & 6 \end{bmatrix}$

$= \begin{bmatrix} 5(-5)+1(0) & 5(4)+1(3) & 5(1)+1(6) \\ 0(-5)-2(0) & 0(4)-2(3) & 0(1)-2(6) \\ 3(-5)+7(0) & 3(4)+7(3) & 3(1)+7(6) \end{bmatrix}$

$= \begin{bmatrix} -25 & 23 & 11 \\ 0 & -6 & -12 \\ -15 & 33 & 45 \end{bmatrix}$

61. $AB = \begin{bmatrix} 4 & -2 \\ 3 & 1 \end{bmatrix} \begin{bmatrix} 5 & 1 \\ 0 & -2 \\ 3 & 7 \end{bmatrix} = \emptyset$

Cannot multiply $(2 \times 2) \cdot (3 \times 2)$

63. $A^2 = \begin{bmatrix} 4 & -2 \\ 3 & 1 \end{bmatrix} \begin{bmatrix} 4 & -2 \\ 3 & 1 \end{bmatrix}$

$= \begin{bmatrix} 4(4)-2(3) & 4(-2)-2(1) \\ 3(4)+1(3) & 3(-2)+1(1) \end{bmatrix}$

$= \begin{bmatrix} 10 & -10 \\ 15 & -5 \end{bmatrix}$

65. $AB \neq BA$; $BC \neq CB$; $AC \neq CA$. Matrix multiplication is **not** commutative.

67. **(a)** $R = \begin{bmatrix} 11,375 & 316 & 83,000 \\ 6970 & 115 & 73,000 \\ 5446 & 159 & 35,700 \\ 4534 & 141 & 36,700 \\ 4059 & 9 & 27,364 \end{bmatrix}$

$W = \begin{bmatrix} 15,307 & 511 & 90,000 \\ 13,363 & 436 & 85,000 \\ 11,778 & 372 & 77,000 \\ 10,395 & 321 & 68,000 \\ 9235 & 282 & 61,900 \end{bmatrix}$

(b) $R + W = \begin{bmatrix} 26,682 & 827 & 173,000 \\ 20,333 & 551 & 158,000 \\ 17,224 & 531 & 112,700 \\ 14,929 & 462 & 104,700 \\ 13,294 & 291 & 89,264 \end{bmatrix}$

158,000 represents the total number of employees for the two companies in 1997.

(c) $W - R = \begin{bmatrix} 3932 & 195 & 7,000 \\ 6393 & 321 & 12,000 \\ 6332 & 213 & 41,300 \\ 5861 & 180 & 31,300 \\ 5176 & 273 & 34,536 \end{bmatrix}$

195 represents the amount in 1998 that Walgreen's net income exceeded Rite Aid's net income in millions of dollars.

69. (a) $d_{n+1} = -.05m_n + 1.05d_n$ (hundreds)

The deer population will grow at 105/year.

(b) $$\begin{bmatrix} m_{n+1} \\ d_{n+1} \end{bmatrix} = \begin{bmatrix} .51 & .4 \\ -.05 & 1.05 \end{bmatrix} \begin{bmatrix} 2000 \\ 5000 \end{bmatrix} = \begin{bmatrix} 3020 \\ 5150 \end{bmatrix}$$

After 1 year there will be 3020 mountain lions and 515,000 deer.

$$\begin{bmatrix} .51 & .4 \\ -.05 & 1.05 \end{bmatrix} \begin{bmatrix} 3020 \\ 5150 \end{bmatrix} = \begin{bmatrix} 3600.2 \\ 5256.5 \end{bmatrix}$$

After 2 years there will be 3600 mountain lions and $\approx$ 525,700 deer.

(c) $$\begin{bmatrix} m_{n+1} \\ d_{n+1} \end{bmatrix} = \begin{bmatrix} .51 & .4 \\ -.05 & 1.05 \end{bmatrix} \begin{bmatrix} 4000 \\ 5000 \end{bmatrix} = \begin{bmatrix} 4040 \\ 5050 \end{bmatrix}$$

$$\begin{bmatrix} m_{n+2} \\ d_{n+2} \end{bmatrix} = \begin{bmatrix} .51 & .4 \\ -.05 & 1.05 \end{bmatrix} \begin{bmatrix} 4040 \\ 5050 \end{bmatrix} = \begin{bmatrix} 4080.4 \\ 5100.5 \end{bmatrix}$$

lions: after 1 year $= \dfrac{4040}{4000} = 1.01$

after 2 years $= \dfrac{4080.4}{4040} = 1.01$

deer: after 1 year $= \dfrac{505,000}{500,000} = 1.01$

after 2 years $= \dfrac{510,050}{500,000} = 1.01$

For Exercises 71 - 77:

$$A = \begin{bmatrix} a_{11} & a_{12} \\ a_{21} & a_{22} \end{bmatrix} \quad B = \begin{bmatrix} b_{11} & b_{12} \\ b_{21} & b_{22} \end{bmatrix}$$

$$C = \begin{bmatrix} c_{11} & c_{12} \\ c_{21} & c_{22} \end{bmatrix}$$

71. $$A + B = \begin{bmatrix} a_{11} + b_{11} & a_{12} + b_{12} \\ a_{21} + b_{21} & a_{22} + b_{22} \end{bmatrix} = B + A$$

73. $(AB)C$

$$= \begin{bmatrix} a_{11}b_{11} + a_{12}b_{21} & a_{11}b_{12} + a_{12}b_{22} \\ a_{21}b_{11} + a_{22}b_{21} & a_{21}b_{12} + a_{22}b_{22} \end{bmatrix} \begin{bmatrix} c_{11} & c_{12} \\ c_{21} & c_{22} \end{bmatrix}$$

$$= \begin{bmatrix} (a_{11}b_{11}c_{11} + a_{12}b_{21}c_{11}) + (a_{11}b_{12}c_{21} + a_{12}b_{22}c_{21}) & (a_{11}b_{11}c_{12} + a_{12}b_{21}c_{12}) + (a_{11}b_{12}c_{22} + a_{12}b_{22}c_{22}) \\ (a_{21}b_{11}c_{11} + a_{22}b_{21}c_{11}) + (a_{21}b_{12}c_{21} + a_{22}b_{22}c_{21}) & (a_{21}b_{11}c_{12} + a_{22}b_{21}c_{12}) + (a_{21}b_{12}c_{22} + a_{22}b_{22}c_{22}) \end{bmatrix}$$

$A(BC)$

$$= \begin{bmatrix} a_{11} & a_{12} \\ a_{21} & a_{22} \end{bmatrix} \begin{bmatrix} b_{11}c_{11} + b_{12}c_{21} & b_{11}c_{12} + b_{12}c_{22} \\ b_{21}c_{11} + b_{22}c_{21} & b_{21}c_{12} + b_{22}c_{22} \end{bmatrix}$$

$$= \begin{bmatrix} (a_{11}b_{11}c_{11} + a_{11}b_{12}c_{21}) + (a_{12}b_{21}c_{11} + a_{12}b_{22}c_{21}) & (a_{11}b_{11}c_{12} + a_{11}b_{12}c_{22}) + (a_{12}b_{21}c_{12} + a_{12}b_{22}c_{22}) \\ (a_{21}b_{11}c_{11} + a_{21}b_{12}c_{21}) + (a_{22}b_{21}c_{11} + a_{22}b_{22}c_{21}) & (a_{21}b_{11}c_{12} + a_{21}b_{12}c_{22}) + (a_{22}b_{21}c_{12} + a_{22}b_{22}c_{22}) \end{bmatrix}$$

Therefore, $(AB)C = A(BC)$

75. $$c(A + B) = \begin{bmatrix} c(a_{11} + b_{11}) & c(a_{12} + b_{12}) \\ c(a_{21} + b_{21}) & c(a_{22} + b_{22}) \end{bmatrix}$$

$$= \begin{bmatrix} ca_{11} + cb_{11} & ca_{12} + cb_{12} \\ ca_{21} + cb_{21} & ca_{22} + cb_{22} \end{bmatrix} = cA + cB$$

77. $$(cA)d = \begin{bmatrix} ca_{11} & ca_{12} \\ ca_{21} & ca_{22} \end{bmatrix} \cdot d$$

$$= \begin{bmatrix} cda_{11} & cda_{12} \\ cda_{21} & cda_{22} \end{bmatrix} = (cd)A$$

Section 7.5

1. $\det\begin{bmatrix} -5 & 9 \\ 4 & -1 \end{bmatrix} = 5 - 36 = \boxed{-31}$

3. $\det\begin{bmatrix} -1 & -2 \\ 5 & 3 \end{bmatrix} = -3 - (-10) = \boxed{7}$

5. $\det\begin{bmatrix} 9 & 3 \\ -3 & -1 \end{bmatrix} = -9 - (-9) = \boxed{0}$

7. $\det\begin{bmatrix} 3 & 4 \\ 5 & -2 \end{bmatrix} = -6 - 20 = \boxed{-26}$

9. $a_{21} : (-1)^3 \begin{bmatrix} 0 & 1 \\ 2 & 1 \end{bmatrix} = -1\,(0-2) = \boxed{-2}$

$a_{22} : (-1)^4 \begin{bmatrix} -2 & 1 \\ 4 & 1 \end{bmatrix} = 1\,(-2-4) = \boxed{-6}$

$a_{23} : (-1)^5 \begin{bmatrix} -2 & 0 \\ 4 & 2 \end{bmatrix} = -1\,(-4-0) = \boxed{4}$

11. $a_{21} : (-1)^3 \begin{bmatrix} 2 & -1 \\ 4 & 1 \end{bmatrix} = -1\,(2+4) = \boxed{-6}$

$a_{22} : (-1)^4 \begin{bmatrix} 1 & -1 \\ -1 & 1 \end{bmatrix} = 1\,(1-1) = \boxed{0}$

$a_{23} : (-1)^5 \begin{bmatrix} 1 & 2 \\ -1 & 4 \end{bmatrix} = -1\,(4+2) = \boxed{-6}$

13.

```
[A]
      [[4  -7 8]
       [2  1  3]
       [-6 3  0]]
det([A])
             186
█
```

15. $\det\begin{bmatrix} 1 & 2 & 0 \\ -1 & 2 & -1 \\ 0 & 1 & 4 \end{bmatrix}$

Expand on 1st column:

$$= 1\begin{vmatrix} 2 & -1 \\ 1 & 4 \end{vmatrix} - (-1)\begin{vmatrix} 2 & 0 \\ 1 & 4 \end{vmatrix} + 0\begin{vmatrix} 2 & 0 \\ 2 & -1 \end{vmatrix}$$

$$= 1\,(8+1) + 1\,(8-0) + 0$$

$$= 9 + 8 = \boxed{17}$$

17. $\det\begin{bmatrix} 10 & 2 & 1 \\ -1 & 4 & 3 \\ -3 & 8 & 10 \end{bmatrix}$

Expand on 1st row:

$$= 10\begin{vmatrix} 4 & 3 \\ 8 & 10 \end{vmatrix} - 2\begin{vmatrix} -1 & 3 \\ -3 & 10 \end{vmatrix} + 1\begin{vmatrix} -1 & 4 \\ -3 & 8 \end{vmatrix}$$

$$= 10\,(40-24) - 2\,(-10+9) + 1\,(-8+12)$$

$$= 10\,(16) - 2\,(-1) + 1\,(4)$$

$$= 160 + 2 + 4 = \boxed{166}$$

19. $\det\begin{bmatrix} 1 & -2 & 3 \\ 0 & 0 & 0 \\ 1 & 10 & -12 \end{bmatrix}$

Expanding on 2nd row gives $\boxed{0}$

21. $\det\begin{bmatrix} 3 & 3 & -1 \\ 2 & 6 & 0 \\ -6 & -6 & 2 \end{bmatrix}$

Expand on 3rd column:

$$= -1\begin{vmatrix} 2 & 6 \\ -6 & -6 \end{vmatrix} - 0\begin{vmatrix} 3 & 3 \\ -6 & -6 \end{vmatrix} + 2\begin{vmatrix} 3 & 3 \\ 2 & 6 \end{vmatrix}$$

$$= -1\,(-12+36) - 0 + 2\,(18-6)$$

$$= -1\,(24) + 2\,(12) = \boxed{0}$$

23.

```
[A]
[[.4  -.8 .6  ]
 [.3  .9  .7  ]
 [3.1 4.1 -2.8]]
det([A])
            -5.5
```

25. $\det\begin{bmatrix} 5 & x \\ -3 & 2 \end{bmatrix} = 6$

$10 - (-3x) = 6$

$3x = -4 \rightarrow x = -\frac{4}{3}$ $\boxed{\left\{-\frac{4}{3}\right\}}$

27. $\det\begin{bmatrix} x & 3 \\ x & x \end{bmatrix} = 4$

$x^2 - 3x = 4$

$x^2 - 3x - 4 = 0$

$(x-4)(x+1) = 0$

$x = 4, -1$ $\boxed{\{4, -1\}}$

29. $\det\begin{bmatrix} -2 & 0 & 1 \\ -1 & 3 & x \\ 5 & -2 & 0 \end{bmatrix} = 3$

Expand on 1st row:

$$-2\begin{vmatrix} 3 & x \\ -2 & 0 \end{vmatrix} + 0 + 1\begin{vmatrix} -1 & 3 \\ 5 & -2 \end{vmatrix} = 3$$

$-2(0+2x) + 1(2-15) = 3$

$-4x + 2 - 15 = 3$

$-4x = 16 \rightarrow x = -4$ $\boxed{\{-4\}}$

31. $\det\begin{bmatrix} 5 & 3x & -3 \\ 0 & 2 & -1 \\ 4 & -1 & x \end{bmatrix} = -7$

Expand on 1st column:

$$5\begin{vmatrix} 2 & -1 \\ -1 & x \end{vmatrix} + 0 + 4\begin{vmatrix} 3x & -3 \\ 2 & -1 \end{vmatrix} = -7$$

$5(2x-1) + 4(-3x+6) = -7$

$10x - 5 - 12x + 24 = -7$

$-2x = -26 \rightarrow x = 13$ $\boxed{\{13\}}$

33. $P(0, 0), Q(0, 2), R(1, 4)$

$$A = \frac{1}{2}\det\begin{bmatrix} 0 & 0 & 1 \\ 0 & 2 & 1 \\ 1 & 4 & 1 \end{bmatrix}$$

$= \frac{1}{2}|-2| =$ $\boxed{1 \text{ square unit}}$

35. $P(2, 5), Q(-1, 3), R(4, 0)$

$$A = \frac{1}{2}\det\begin{bmatrix} 2 & 5 & 1 \\ -1 & 3 & 1 \\ 4 & 0 & 1 \end{bmatrix}$$

$= \frac{1}{2}|19| =$ $\boxed{\frac{19}{2} \text{ square units}}$

37. $P(1, 2), Q(4, 3), R(3, 5)$

$$A = \frac{1}{2}\det\begin{bmatrix} 1 & 2 & 1 \\ 4 & 3 & 1 \\ 3 & 5 & 1 \end{bmatrix}$$

$= \frac{1}{2}|7| =$ $\boxed{\frac{7}{2} \text{ square units}}$

39.

```
[A]
 [[3   -6 5   -1]
  [0   2  -1  3 ]
  [-6  4  2   0 ]
  [-7  3  1   1 ]]
det([A])
                298
```

41. $\det\begin{bmatrix} 4 & 0 & 0 & 2 \\ -1 & 0 & 3 & 0 \\ 2 & 4 & 0 & 1 \\ 0 & 0 & 1 & 2 \end{bmatrix}$

Expand on 2nd column:

$$= -0\begin{vmatrix} -1 & 3 & 0 \\ 2 & 0 & 1 \\ 0 & 1 & 2 \end{vmatrix} + 0\begin{vmatrix} 4 & 0 & 2 \\ 2 & 0 & 1 \\ 0 & 1 & 2 \end{vmatrix}$$

$$-4\begin{vmatrix} 4 & 0 & 2 \\ -1 & 3 & 0 \\ 0 & 1 & 2 \end{vmatrix} + 0\begin{vmatrix} 4 & 0 & 2 \\ -1 & 3 & 0 \\ 2 & 0 & 1 \end{vmatrix}$$

$$= -4\begin{vmatrix} 4 & 0 & 2 \\ -1 & 3 & 0 \\ 0 & 1 & 2 \end{vmatrix}$$

Expand on 3rd row:

$$= -4\left[0\begin{vmatrix} 0 & 2 \\ 3 & 0 \end{vmatrix} - 1\begin{vmatrix} 4 & 2 \\ -1 & 0 \end{vmatrix} + 2\begin{vmatrix} 4 & 0 \\ -1 & 3 \end{vmatrix}\right]$$

$= -4[0 - 1(0+2) + 2(12+0)]$

$= -4[-2+24] =$ $\boxed{-88}$

43. $\det\begin{bmatrix} 1 & 0 & 0 \\ 1 & 0 & 1 \\ 3 & 0 & 0 \end{bmatrix} = \boxed{0}$

(Column 2 is all 0's)
Determinant Theorem 1.

45. $\det\begin{bmatrix} 6 & 8 & -12 \\ -1 & 0 & 2 \\ 4 & 0 & -8 \end{bmatrix}$

Multiply column 1 by 2 and add the result to column 3:

$$\det\begin{bmatrix} 6 & 8 & -12 \\ -1 & 0 & 2 \\ 4 & 0 & -8 \end{bmatrix} = \det\begin{bmatrix} 6 & 8 & 0 \\ -1 & 0 & 0 \\ 4 & 0 & 0 \end{bmatrix}$$

Column 3 is all zeros:

$$\det\begin{bmatrix} 6 & 8 & -12 \\ -1 & 0 & 2 \\ 4 & 0 & -8 \end{bmatrix} = \boxed{0}$$

Determinant Theorem 1.

47. $\det\begin{bmatrix} -4 & 1 & 4 \\ 2 & 0 & 1 \\ 0 & 2 & 4 \end{bmatrix}$

Multiply column 2 by 4 and add the result to column 1.
Multiply column 2 by -4 and add the result to column 3.

$$\det\begin{bmatrix} -4 & 1 & 4 \\ 2 & 0 & 1 \\ 0 & 2 & 4 \end{bmatrix} = \det\begin{bmatrix} 0 & 1 & 0 \\ 2 & 0 & 1 \\ 8 & 2 & -4 \end{bmatrix}$$

Expand about row 1:

$$\det\begin{bmatrix} 0 & 1 & 0 \\ 2 & 0 & 1 \\ 8 & 2 & -4 \end{bmatrix}$$

$$= 0\,(0-2) - 1\,(-8-8) + 0\,(4-0) = \boxed{16}$$

Relating Concepts (49 – 52):

49. $\det\begin{bmatrix} x & y & 1 \\ 2 & 3 & 1 \\ -1 & 4 & 1 \end{bmatrix} = 0$

Expand on 3rd column:

$$= 1\begin{vmatrix} 2 & 3 \\ -1 & 4 \end{vmatrix} - 1\begin{vmatrix} x & y \\ -1 & 4 \end{vmatrix} + 1\begin{vmatrix} x & y \\ 2 & 3 \end{vmatrix}$$

$$= 1\,(8+3) - (4x+y) + (3x-2y)$$

$$= 11 - 4x - y + 3x - 2y = -x - 3y + 11$$

Set $= 0$: $\boxed{\begin{array}{c} -x - 3y + 11 = 0 \\ \text{or} \\ x + 3y - 11 = 0 \end{array}}$

50. Line through $(2,\ 3)$ and $(-1,\ 4)$:

$$m = \frac{4-3}{-1-2} = -\frac{1}{3}$$

$$y - 3 = -\frac{1}{3}(x-2)$$

$$3y - 9 = -x + 2$$

$$x + 3y - 11 = 0$$

This is equivalent to the equation found in **49.**

51. $(x_1,\ y_1)$ and $(x_2,\ y_2)$

$$y - y_1 = \frac{y_2 - y_1}{x_2 - x_1}(x - x_1)$$

52. $\det \begin{bmatrix} x & y & 1 \\ x_1 & y_1 & 1 \\ x_2 & y_2 & 1 \end{bmatrix} = 0$

Expand on 3rd column:

$$0 = 1 \begin{vmatrix} x_1 & y_1 \\ x_2 & y_2 \end{vmatrix} - 1 \begin{vmatrix} x & y \\ x_2 & y_2 \end{vmatrix} + 1 \begin{vmatrix} x & y \\ x_1 & y_1 \end{vmatrix}$$

$$0 = 1\,(x_1 y_2 - x_2 y_1) - (x y_2 - x_2 y) + (x y_1 - x_1 y)$$

$$0 = x_1 y_2 - x_2 y_1 - x y_2 + x_2 y + x y_1 - x_1 y$$

$$y x_2 - y x_1 - y_1 x_2 = y_2 x - y_2 x_1 - y_1 x$$

Add $y_1 x_1$ to both sides:

$$y x_2 - y x_1 - y_1 x_2 + y_1 x_1 = y_2 x - y_2 x_1 - y_1 x + y_1 x$$

$$(y - y_1)(x_2 - x_1) = (y_2 - y_1)(x - x_1)$$

$$y - y_1 = \frac{y_2 - y_1}{x_2 - x_1}(x - x_1)$$

When the determinant is expanded, the equation is the same as the answer in **51.**

53. $4x + 3y - 2z = 1$
$7x - 4y + 3z = 2$
$-2x + y - 8z = 0$

$$D = \begin{bmatrix} 4 & 3 & -2 \\ 7 & -4 & 3 \\ -2 & -1 & -8 \end{bmatrix}$$

$$D_x = \begin{bmatrix} 1 & 3 & -2 \\ 2 & -4 & 3 \\ 0 & -1 & -8 \end{bmatrix}$$

$$D_y = \begin{bmatrix} 4 & 1 & -2 \\ 7 & 2 & 3 \\ -2 & 0 & -8 \end{bmatrix}$$

$$D_z = \begin{bmatrix} 4 & 3 & 1 \\ 7 & -4 & 2 \\ -2 & -1 & 0 \end{bmatrix}$$

(a) **IV** (b) **I** (c) **III** (d) **II**

55. $x + y = 4$
$2x - y = 2$

$$D = \det \begin{bmatrix} 1 & 1 \\ 2 & -1 \end{bmatrix} = -3$$

$$D_x = \det \begin{bmatrix} 4 & 1 \\ 2 & -1 \end{bmatrix} = -6$$

$$D_y = \det \begin{bmatrix} 1 & 4 \\ 2 & 2 \end{bmatrix} = -6$$

$$x = \frac{D_x}{D} = \frac{-6}{-3} = 2$$

$$y = \frac{D_y}{D} = \frac{-6}{-3} = 2 \qquad \boxed{\{(2,\ 2)\}}$$

57. $4x + 3y = -7$
$2x + 3y = -11$

$$D = \det \begin{bmatrix} 4 & 3 \\ 2 & 3 \end{bmatrix} = 6$$

$$D_x = \det \begin{bmatrix} -7 & 3 \\ -11 & 3 \end{bmatrix} = 12$$

$$D_y = \det \begin{bmatrix} 4 & -7 \\ 2 & -11 \end{bmatrix} = -30$$

$$x = \frac{D_x}{D} = \frac{12}{6} = 2$$

$$y = \frac{D_y}{D} = \frac{-30}{6} = -5 \qquad \boxed{\{(2, -5)\}}$$

59. $3x + 2y = 4$
$6x + 4y = 8$

$$D = \det \begin{bmatrix} 3 & 2 \\ 6 & 4 \end{bmatrix} = 0$$

$$\left[\begin{array}{cc|c} 3 & 2 & 4 \\ 6 & 4 & 8 \end{array}\right]$$

$$-2R_1 + R_2 \rightarrow \left[\begin{array}{cc|c} 3 & 2 & 4 \\ 0 & 0 & 0 \end{array}\right]$$

$0 = 0$, ∞ solutions:

$$3x = 4 - 2y \qquad x = \frac{4 - 2y}{3}$$

$$\boxed{\left(\frac{4 - 2y}{3},\ y\right)}$$

61. $2x - 3y = -5$
$x + 5y = 17$

$$D = \det\begin{bmatrix} 2 & -3 \\ 1 & 5 \end{bmatrix} = 13$$

$$D_x = \det\begin{bmatrix} -5 & -3 \\ 17 & 5 \end{bmatrix} = 26$$

$$D_y = \det\begin{bmatrix} 2 & -5 \\ 1 & 17 \end{bmatrix} = 39$$

$$x = \frac{D_x}{D} = \frac{26}{13} = 2$$
$$y = \frac{D_y}{D} = \frac{39}{13} = 3$$

$\boxed{\{(2,\ 3)\}}$

63. $4x - y + 3z = -3$
$3x + y + z = 0$
$2x - y + 4z = 0$

$$D = \det\begin{bmatrix} 4 & -1 & 3 \\ 3 & 1 & 1 \\ 2 & -1 & 4 \end{bmatrix} = 15$$

$$D_x = \det\begin{bmatrix} -3 & -1 & 3 \\ 0 & 1 & 1 \\ 0 & -1 & 4 \end{bmatrix} = -15$$

$$D_y = \det\begin{bmatrix} 4 & -3 & 3 \\ 3 & 0 & 1 \\ 2 & 0 & 4 \end{bmatrix} = 30$$

$$D_z = \det\begin{bmatrix} 4 & -1 & -3 \\ 3 & 1 & 0 \\ 2 & -1 & 0 \end{bmatrix} = 15$$

$$x = \frac{D_x}{D} = \frac{-15}{15} = -1$$
$$y = \frac{D_y}{D} = \frac{30}{15} = 2$$
$$z = \frac{D_z}{D} = \frac{15}{15} = 1$$

$\boxed{\{(-1,\ 2,\ 1)\}}$

65. $2x - y + 4z = -2$
$3x + 2y - z = -3$
$x + 4y + 2z = 17$

$$D = \det\begin{bmatrix} 2 & -1 & 4 \\ 3 & 2 & -1 \\ 1 & 4 & 2 \end{bmatrix} = 63$$

$$D_x = \det\begin{bmatrix} -2 & -1 & 4 \\ -3 & 2 & -1 \\ 17 & 4 & 2 \end{bmatrix} = -189$$

$$D_y = \det\begin{bmatrix} 2 & -2 & 4 \\ 3 & -3 & -1 \\ 1 & 17 & 2 \end{bmatrix} = 252$$

$$D_z = \det\begin{bmatrix} 2 & -1 & -2 \\ 3 & 2 & -3 \\ 1 & 4 & 17 \end{bmatrix} = 126$$

$$x = \frac{D_x}{D} = \frac{-189}{63} = -3$$
$$y = \frac{D_y}{D} = \frac{252}{63} = 4$$
$$z = \frac{D_z}{D} = \frac{126}{63} = 2$$

$\boxed{\{(-3,\ 4,\ 2)\}}$

67. $5x - y = -4$
$3x + 2z = 4$
$4y + 3z = 22$

$$D = \det\begin{bmatrix} 5 & -1 & 0 \\ 3 & 0 & 2 \\ 0 & 4 & 3 \end{bmatrix} = -31$$

$$D_x = \det\begin{bmatrix} -4 & -1 & 0 \\ 4 & 0 & 2 \\ 22 & 4 & 3 \end{bmatrix} = 0$$

$$D_y = \det\begin{bmatrix} 5 & -4 & 0 \\ 3 & 4 & 2 \\ 0 & 22 & 3 \end{bmatrix} = -124$$

$$D_z = \det\begin{bmatrix} 5 & -1 & -4 \\ 3 & 0 & 4 \\ 0 & 4 & 22 \end{bmatrix} = -62$$

$$x = \frac{D_x}{D} = \frac{0}{-31} = 0$$
$$y = \frac{D_y}{D} = \frac{-121}{-31} = 4$$
$$z = \frac{D_z}{D} = \frac{-62}{-31} = 2$$

$\boxed{\{(0,\ 4,\ 2)\}}$

69. $2x - y + 3z = 1$ [1]
$-2x + y - 3z = 2$ [2]
$5x - y + z = 2$ [3]

$$D = \det \begin{bmatrix} 2 & -1 & 3 \\ -2 & 1 & -3 \\ 5 & -1 & 1 \end{bmatrix} = 0$$

[1] + [2] $\Rightarrow 0 = 3$

No solution $\boxed{\emptyset}$

71. $3x + 2y - w = 0$
$2x + z + 2w = 5$
$x + 2y - z = -2$
$2x - y + z + w = 2$

$$D = \det \begin{bmatrix} 3 & 2 & 0 & -1 \\ 2 & 0 & 1 & 2 \\ 1 & 2 & -1 & 0 \\ 2 & -1 & 1 & 1 \end{bmatrix} = -9$$

$$D_x = \det \begin{bmatrix} 0 & 2 & 0 & -1 \\ 5 & 0 & 1 & 2 \\ -2 & 2 & -1 & 0 \\ 2 & -1 & 1 & 1 \end{bmatrix} = 9$$

$$D_y = \det \begin{bmatrix} 3 & 0 & 0 & -1 \\ 2 & 5 & 1 & 2 \\ 1 & -2 & -1 & 0 \\ 2 & 2 & 1 & 1 \end{bmatrix} = -18$$

$$D_z = \det \begin{bmatrix} 3 & 2 & 0 & -1 \\ 2 & 0 & 5 & 2 \\ 1 & 2 & -2 & 0 \\ 2 & -1 & 2 & 1 \end{bmatrix} = -45$$

$$D_w = \det \begin{bmatrix} 3 & 2 & 0 & 0 \\ 2 & 0 & 1 & 5 \\ 1 & 2 & -1 & -2 \\ 2 & -1 & 1 & 2 \end{bmatrix} = -9$$

$$x = \frac{D_x}{D} = \frac{9}{-9} = -1$$

$$y = \frac{D_y}{D} = \frac{-18}{-9} = 2$$

$$z = \frac{D_z}{D} = \frac{-45}{-9} = 5$$

$$w = \frac{D_w}{D} = \frac{-9}{-9} = 1$$

$\boxed{\{(-1, 2, 5, 1)\}}$

73. If $D = 0$, Cramer's rule cannot be applied because there is no unique solution (no solution or ∞ solutions).

Section 7.6

1. Yes (i) $\begin{bmatrix} 5 & 7 \\ 2 & 3 \end{bmatrix} \begin{bmatrix} 3 & -7 \\ -2 & 5 \end{bmatrix}$

$$= \begin{bmatrix} 15-14 & -35+35 \\ 6-6 & -14+15 \end{bmatrix} = \begin{bmatrix} 1 & 0 \\ 0 & 1 \end{bmatrix}$$

(ii) $\begin{bmatrix} 3 & -7 \\ -2 & 5 \end{bmatrix} \begin{bmatrix} 5 & 7 \\ 2 & 3 \end{bmatrix}$

$$= \begin{bmatrix} 15-14 & 21-21 \\ -10+10 & -14+15 \end{bmatrix} = \begin{bmatrix} 1 & 0 \\ 0 & 1 \end{bmatrix}$$

3. No $\begin{bmatrix} -1 & 2 \\ 3 & -5 \end{bmatrix} \begin{bmatrix} -5 & -2 \\ -3 & -1 \end{bmatrix}$

$$= \begin{bmatrix} 5-6 & 2-2 \\ -15+15 & -6+5 \end{bmatrix} = \begin{bmatrix} -1 & 0 \\ 0 & -1 \end{bmatrix}$$

5. No $\begin{bmatrix} 0 & 1 & 0 \\ 0 & 0 & -2 \\ 1 & -1 & 0 \end{bmatrix} \begin{bmatrix} 1 & 0 & 1 \\ 1 & 0 & 0 \\ 0 & -1 & 0 \end{bmatrix}$

$$= \begin{bmatrix} 1 & 0 & 0 \\ 0 & 2 & 0 \\ 0 & 0 & 1 \end{bmatrix}$$

7. $A = \begin{bmatrix} -1 & -1 & -1 \\ 4 & 5 & 0 \\ 0 & 1 & -3 \end{bmatrix}$; $B = \begin{bmatrix} 15 & 4 & -5 \\ -12 & -3 & 4 \\ -4 & -1 & 1 \end{bmatrix}$

Yes: $AB = \begin{bmatrix} 1 & 0 & 0 \\ 0 & 1 & 0 \\ 0 & 0 & 1 \end{bmatrix} = BA$

9. The inverse will not exist if the determinant is equal to 0.

11. $A = \begin{bmatrix} 3 & 7 \\ 2 & 5 \end{bmatrix}$

$$A^{-1} = \frac{1}{15-14} \begin{bmatrix} 5 & -7 \\ -2 & 3 \end{bmatrix} = \begin{bmatrix} 5 & -7 \\ -2 & 3 \end{bmatrix}$$

13. $A=\begin{bmatrix}-1 & -2\\ 3 & 4\end{bmatrix}$

$$A^{-1}=\frac{1}{-4+6}\begin{bmatrix}4 & 2\\ -3 & -1\end{bmatrix}$$

$$=\frac{1}{2}\begin{bmatrix}4 & 2\\ -3 & -1\end{bmatrix}=\begin{bmatrix}2 & 1\\ -\frac{3}{2} & -\frac{1}{2}\end{bmatrix}$$

15. $A=\begin{bmatrix}-6 & 4\\ -3 & 2\end{bmatrix}$; determinant $A=0$

No inverse

17. $A=\begin{bmatrix}.6 & .2\\ .5 & .1\end{bmatrix}$

$$A^{-1}=\frac{1}{.06-.1}\begin{bmatrix}.1 & -.2\\ -.5 & .6\end{bmatrix}$$

$$=-25\begin{bmatrix}.1 & -.2\\ -.5 & .6\end{bmatrix}=\begin{bmatrix}-2.5 & 5\\ 12.5 & -15\end{bmatrix}$$

Note: Problems 19 – 29 are worked on a calculator.

19. $A=\begin{bmatrix}1 & 0 & 0\\ 0 & -1 & 0\\ 1 & 0 & 1\end{bmatrix}$ $(\det A=-1)$

$$A^{-1}=\begin{bmatrix}1 & 0 & 0\\ 0 & -1 & 0\\ -1 & 0 & 1\end{bmatrix}$$

21. $A=\begin{bmatrix}-1 & -1 & -1\\ 4 & 5 & 0\\ 0 & 1 & -3\end{bmatrix}$ $(\det A=-1)$

$$A^{-1}=\begin{bmatrix}15 & 4 & -5\\ -12 & -3 & 4\\ -4 & -1 & 1\end{bmatrix}$$

23. $A=\begin{bmatrix}-.4 & .1 & .2\\ 0 & .6 & .8\\ .3 & 0 & -.2\end{bmatrix}$ $(\det A=.036)$

$$A^{-1}=\begin{bmatrix}-\frac{10}{3} & \frac{5}{9} & -\frac{10}{9}\\ \frac{20}{3} & \frac{5}{9} & \frac{80}{9}\\ -5 & \frac{5}{6} & -\frac{20}{3}\end{bmatrix}$$

25. $A=\begin{bmatrix}2 & 1 & 2\\ 5 & 10 & 5\\ 3 & 6 & 3\end{bmatrix}$ $(\det A=0)$

No inverse

27. $A=\begin{bmatrix}\sqrt{2} & .5\\ -17 & \frac{1}{2}\end{bmatrix}$ $(\det A\approx 9.207)$

$$A^{-1}\approx\begin{bmatrix}.0543058761 & -.0543058761\\ 1.846399787 & .153600213\end{bmatrix}$$

29. $A=\begin{bmatrix}1.4 & .5 & .59\\ .84 & 1.36 & .62\\ .56 & .47 & 1.3\end{bmatrix}$ $(\det A\approx 1.478)$

$$A^{-1}\approx\begin{bmatrix}.9987635516 & -.252092087 & -.3330564627\\ -.5037783375 & 1.007556675 & -.2518891688\\ -.2481013617 & -.255676976 & 1.003768868\end{bmatrix}$$

Relating Concepts (31 - 40)

$$A=\begin{bmatrix}2 & 4\\ 1 & -1\end{bmatrix},\quad A^{-1}=\begin{bmatrix}x & y\\ z & w\end{bmatrix}$$

31. $A\,A^{-1}=\begin{bmatrix}2x+4z & 2y+4w\\ x-z & y-w\end{bmatrix}$

32. $\begin{bmatrix}2x+4z & 2y+4w\\ x-z & y-w\end{bmatrix}=\begin{bmatrix}1 & 0\\ 0 & 1\end{bmatrix}$

$$\begin{aligned}2x+4z&=1 & [1]\\ 2y+4w&=0 & [2]\\ x-z&=0 & [3]\\ y-w&=1 & [4]\end{aligned}$$

33. $[1] - 2 \times [3]$:

$$\begin{array}{r} 2x + 4z = 1 \\ -2x + 2z = 0 \\ \hline 6z = 1 \end{array} \quad \rightarrow \quad z = \frac{1}{6}$$

$$x - \frac{1}{6} = 0 \rightarrow x = \frac{1}{6}$$

$\frac{1}{2}[2] - [4]$:

$$\begin{array}{r} y + 2w = 0 \\ -y + w = -1 \\ \hline 3w = -1 \end{array} \quad \rightarrow \quad w = -\frac{1}{3}$$

$$y + \frac{1}{3} = 1 \rightarrow y = \frac{2}{3}$$

$$\boxed{x = \frac{1}{6},\ y = \frac{2}{3},\ z = \frac{1}{6},\ w = -\frac{1}{3}}$$

34. $A^{-1} = \begin{bmatrix} \frac{1}{6} & \frac{2}{3} \\ \frac{1}{6} & -\frac{1}{3} \end{bmatrix}$

35. $ad - bc$ is the determinant of A.

36. $\begin{bmatrix} \dfrac{d}{\det A} & \dfrac{-b}{\det A} \\ \dfrac{-c}{\det A} & \dfrac{a}{\det A} \end{bmatrix}$

37. $A^{-1} = \dfrac{1}{\det A}\begin{bmatrix} d & -b \\ -c & a \end{bmatrix}$

38. Interchange the entries in row 1, column 1 and row 2, column 2. Then change the entries in row 1, column 2 and row 2, column 1 to their negatives. Then multiply the resulting matrix by the scalar $\dfrac{1}{\det A}$.

39. $A = \begin{bmatrix} 4 & 2 \\ 7 & 3 \end{bmatrix}$ $\quad \det A = 12 - 14 = -2$

$$A^{-1} = \frac{1}{-2}\begin{bmatrix} 3 & -2 \\ -7 & 4 \end{bmatrix} = \begin{bmatrix} -\frac{3}{2} & 1 \\ \frac{7}{2} & -2 \end{bmatrix}$$

40. The inverse of a 2×2 matrix does not exist if the determinant of A has the value <u>0</u>.

41. $2x - y = -8$
$3x + y = -2$

$$\begin{bmatrix} 2 & -1 \\ 3 & 1 \end{bmatrix}\begin{bmatrix} x \\ y \end{bmatrix} = \begin{bmatrix} -8 \\ -2 \end{bmatrix}$$

$$\begin{bmatrix} x \\ y \end{bmatrix} = \begin{bmatrix} 2 & -1 \\ 3 & 1 \end{bmatrix}^{-1}\begin{bmatrix} -8 \\ -2 \end{bmatrix}$$

$$= \frac{1}{5}\begin{bmatrix} 1 & 1 \\ -3 & 2 \end{bmatrix}\begin{bmatrix} -8 \\ -2 \end{bmatrix}$$

$$= \begin{bmatrix} \frac{1}{5} & \frac{1}{5} \\ -\frac{3}{5} & \frac{2}{5} \end{bmatrix}\begin{bmatrix} -8 \\ -2 \end{bmatrix}$$

$$= \begin{bmatrix} -\frac{8}{5} - \frac{2}{5} \\ \frac{24}{5} - \frac{4}{5} \end{bmatrix} = \begin{bmatrix} -2 \\ 4 \end{bmatrix} \quad \boxed{\{(-2, 4)\}}$$

43. $2x + 3y = -10$
$3x + 4y = -12$

$$\begin{bmatrix} 2 & 3 \\ 3 & 4 \end{bmatrix}\begin{bmatrix} x \\ y \end{bmatrix} = \begin{bmatrix} -10 \\ -12 \end{bmatrix}$$

$$\begin{bmatrix} x \\ y \end{bmatrix} = \begin{bmatrix} 2 & 3 \\ 3 & 4 \end{bmatrix}^{-1}\begin{bmatrix} -10 \\ -12 \end{bmatrix}$$

$$= -1\begin{bmatrix} 4 & -3 \\ -3 & 2 \end{bmatrix}\begin{bmatrix} -10 \\ -12 \end{bmatrix}$$

$$= \begin{bmatrix} -4 & 3 \\ 3 & -2 \end{bmatrix}\begin{bmatrix} -10 \\ -12 \end{bmatrix}$$

$$= \begin{bmatrix} 40 - 36 \\ -30 + 24 \end{bmatrix} = \begin{bmatrix} 4 \\ -6 \end{bmatrix} \quad \boxed{\{(4, -6)\}}$$

45. $2x - 5y = 10$
$4x - 5y = 15$

$$\begin{bmatrix} 2 & -5 \\ 4 & -5 \end{bmatrix}\begin{bmatrix} x \\ y \end{bmatrix} = \begin{bmatrix} 10 \\ 15 \end{bmatrix}$$

$$\begin{bmatrix} x \\ y \end{bmatrix} = \begin{bmatrix} 2 & -5 \\ 4 & -5 \end{bmatrix}^{-1}\begin{bmatrix} 10 \\ 15 \end{bmatrix}$$

$$= \frac{1}{10}\begin{bmatrix} -5 & 5 \\ -4 & 2 \end{bmatrix}\begin{bmatrix} 10 \\ 15 \end{bmatrix}$$

$$= \begin{bmatrix} -.5 & .5 \\ -.4 & .2 \end{bmatrix}\begin{bmatrix} 10 \\ 15 \end{bmatrix}$$

$$= \begin{bmatrix} -5 + 7.5 \\ -4 + 3 \end{bmatrix} = \begin{bmatrix} 2.5 \\ -1 \end{bmatrix}$$

$$\boxed{(2.5, -1) \text{ or } \left(\frac{5}{2}, -1\right)}$$

Note: Problems 47 – 65 are worked on a calculator.

47.
$$\begin{aligned} 2x + 4z &= 14 \\ 3x + y + 5z &= 19 \\ -x + y - 2z &= -7 \end{aligned}$$

$$\begin{bmatrix} x \\ y \\ z \end{bmatrix} = \begin{bmatrix} 2 & 0 & 4 \\ 3 & 1 & 5 \\ -1 & 1 & -2 \end{bmatrix}^{-1} \cdot \begin{bmatrix} 14 \\ 19 \\ -7 \end{bmatrix}$$

$\boxed{\{(3,\ 0,\ 2)\}}$

49.
$$\begin{aligned} x + 3y + z &= 2 \\ x - 2y + 3z &= -3 \\ 2x - 3y - z &= 34 \end{aligned}$$

$$\begin{bmatrix} x \\ y \\ z \end{bmatrix} = \begin{bmatrix} 1 & 3 & 1 \\ 1 & -2 & 3 \\ 2 & -3 & -1 \end{bmatrix}^{-1} \cdot \begin{bmatrix} 2 \\ -3 \\ 34 \end{bmatrix}$$

$\boxed{\left\{\left(12,\ -\frac{15}{11},\ -\frac{65}{11}\right)\right\}}$

51.
$$\begin{aligned} x + 3y - 2z - w &= 9 \\ 4x + y + z + 2w &= 2 \\ -3x - y + z - w &= -5 \\ x - y - 3z - 2w &= 2 \end{aligned}$$

$$A = \begin{bmatrix} 1 & 3 & -2 & -1 \\ 4 & 1 & 1 & 2 \\ -3 & -1 & 1 & -1 \\ 1 & -1 & -3 & -2 \end{bmatrix}$$

$$\begin{bmatrix} x \\ y \\ z \\ w \end{bmatrix} = A^{-1} \cdot \begin{bmatrix} 9 \\ 2 \\ -5 \\ 2 \end{bmatrix}$$

$\boxed{\{(0,\ 2,\ -2,\ 1)\}}$

53.
$$\begin{aligned} x - \sqrt{2}\,y &= 2.6 \\ .75x + y &= -7 \end{aligned}$$

$$\begin{bmatrix} x \\ y \end{bmatrix} = \begin{bmatrix} 1 & -\sqrt{2} \\ .75 & 1 \end{bmatrix}^{-1} \cdot \begin{bmatrix} 2.6 \\ -7 \end{bmatrix}$$

$\boxed{\{(-3.542308934,\ -4.343268299)\}}$

55.
$$\begin{aligned} \pi x + e\,y + \sqrt{2}\,z &= 1 \\ ex + \pi y + \sqrt{2}\,z &= 2 \\ \sqrt{2}\,x + ey + \pi z &= 3 \end{aligned}$$

$$\begin{bmatrix} x \\ y \\ z \end{bmatrix} = \begin{bmatrix} \pi & e & \sqrt{2} \\ e & \pi & \sqrt{2} \\ \sqrt{2} & e & \pi \end{bmatrix}^{-1} \cdot \begin{bmatrix} 1 \\ 2 \\ 3 \end{bmatrix}$$

$\boxed{\{(-.9704156959,\ 1.391914631,\ .1874077432)\}}$

57. $P(x) = ax^3 + bx^2 + cx + d$

$(\mathbf{-1,\ 14}) : -a + b - c + d = 14$

$(\mathbf{1.5,\ 1.5}) : 3.375a + 2.25b + 1.5c + d = 1.5$

$(\mathbf{2, -1}) : 8a + 4b + 2c + d = -1$

$(\mathbf{3, -18}) : 27a + 9b + 3c + d = -18$

$$A = \begin{bmatrix} -1 & 1 & -1 & 1 \\ 3.375 & 2.25 & 1.5 & 1 \\ 8 & 4 & 2 & 1 \\ 27 & 9 & 3 & 1 \end{bmatrix}$$

$$\begin{bmatrix} a \\ b \\ c \\ d \end{bmatrix} = A^{-1} \cdot \begin{bmatrix} 14 \\ 1.5 \\ -1 \\ -18 \end{bmatrix}$$

$a = -2,\ b = 5,\ c = -4,\ d = 3$

$\boxed{P(x) = -2x^3 + 5x^2 - 4x + 3}$

59. $P(x) = ax^4 + bx^3 + cx^2 + dx + e$

$(\mathbf{-2,\ 13}) : 16a - 8b + 4c - 2d + e = 13$

$(\mathbf{-1,\ 2}) : a - b + c - d + e = 2$

$(\mathbf{0, -1}) : e = -1$

$(\mathbf{1,\ 4}) : a + b + c + d + e = 4$

$(\mathbf{2,\ 41}) : 16a + 8b + 4c + 2d + e = 41$

$$A = \begin{bmatrix} 16 & -8 & 4 & -2 & 1 \\ 1 & -1 & 1 & -1 & 1 \\ 0 & 0 & 0 & 0 & 1 \\ 1 & 1 & 1 & 1 & 1 \\ 16 & 8 & 4 & 2 & 1 \end{bmatrix}$$

$$\begin{bmatrix} a \\ b \\ c \\ d \\ e \end{bmatrix} = A^{-1} \cdot \begin{bmatrix} 13 \\ 2 \\ -1 \\ 4 \\ 41 \end{bmatrix}$$

$a = 1,\ b = 2,\ c = 3,\ d = -1,\ e = -1$

$\boxed{P(x) = x^4 + 2x^3 + 3x^2 - x - 1}$

61. (a) Let x be the cost of a soft drink and let y be the cost of a box of popcorn. Then:

$$3x + 2y = 8.5$$
$$4x + 3y = 12$$

$$Ax = B: \begin{bmatrix} 3 & 2 \\ 4 & 3 \end{bmatrix} \cdot \begin{bmatrix} x \\ y \end{bmatrix} = \begin{bmatrix} 8.5 \\ 12 \end{bmatrix}$$

$$\begin{bmatrix} x \\ y \end{bmatrix} = \begin{bmatrix} 3 & 2 \\ 4 & 3 \end{bmatrix}^{-1} \cdot \begin{bmatrix} 8.5 \\ 12 \end{bmatrix} = \begin{bmatrix} 1.5 \\ 2 \end{bmatrix}$$

The cost of a soft drink is \$1.50 and the cost of popcorn is \$2.00

(b) No. There are two variables. This typically requires two equations. The linear system would be dependent because the two equations would be the same. A^{-1} would not exist. If we attempt to calculate A^{-1} with a calculator, an error message tells us that the matrix is singular.

63. (a) The following three equations must be solved:

$$113a + 308b + c = 10,170$$
$$133a + 622b + c = 15,305$$
$$155a + 1937b + c = 21,289$$

$$\begin{bmatrix} a \\ b \\ c \end{bmatrix} = \begin{bmatrix} 113 & 308 & 1 \\ 133 & 622 & 1 \\ 155 & 1937 & 1 \end{bmatrix}^{-1} \cdot \begin{bmatrix} 10170 \\ 15305 \\ 21289 \end{bmatrix}$$

$$= \begin{bmatrix} 251.3175021 \\ .3460189769 \\ -18335.45158 \end{bmatrix}$$

$$T \approx 251A + .346I - 18,300$$

(b) $T \approx 251\,(118) + .346\,(311) - 18,300$
$\approx 11,426$

11,426 is quite close to the actual value of 11,314.

65. (a) The following three equations must be solved:

$$a + 1500b + 8c = 122$$
$$a + 2000b + 5c = 130$$
$$a + 2200b + 10c = 158$$

$$\begin{bmatrix} a \\ b \\ c \end{bmatrix} = \begin{bmatrix} 1 & 1500 & 8 \\ 1 & 2000 & 5 \\ 1 & 2200 & 10 \end{bmatrix}^{-1} \cdot \begin{bmatrix} 122 \\ 130 \\ 158 \end{bmatrix}$$

$$= \begin{bmatrix} 30 \\ .04 \\ 4 \end{bmatrix}$$

$$P = 30 + .04S + 4C$$

(b) $P = 30 + .04\,(1800) + 4\,(7) = 130$

\$130,000

67. $A^{-1} = \begin{bmatrix} 5 & -9 \\ -1 & 2 \end{bmatrix}$

$$A = \left(A^{-1}\right)^{-1} = \begin{bmatrix} 2 & 9 \\ 1 & 5 \end{bmatrix}$$

69. $A^{-1} = \begin{bmatrix} \frac{2}{3} & -\frac{1}{3} & 0 \\ \frac{1}{3} & -\frac{5}{3} & 1 \\ \frac{1}{3} & \frac{1}{3} & 0 \end{bmatrix}$

$$A = \left(A^{-1}\right)^{-1} = \begin{bmatrix} 1 & 0 & 1 \\ -1 & 0 & 2 \\ -2 & 1 & 3 \end{bmatrix}$$

71. $A = \begin{bmatrix} a & 0 & 0 \\ 0 & b & 0 \\ 0 & 0 & c \end{bmatrix}$

This is a shortened method:

$$[A\,/\,I] = \left[\begin{array}{ccc|ccc} a & 0 & 0 & 1 & 0 & 0 \\ 0 & b & 0 & 0 & 1 & 0 \\ 0 & 0 & c & 0 & 0 & 1 \end{array}\right]$$

Since a, b, c are all non-zero, $\frac{1}{a}, \frac{1}{b}, \frac{1}{c}$ all exist.

$$\left.\begin{matrix} \frac{1}{a}R_1 \\ \frac{1}{b}R_2 \\ \frac{1}{c}R_3 \end{matrix}\right\} \rightarrow \left[\begin{array}{ccc|ccc} 1 & 0 & 0 & \frac{1}{a} & 0 & 0 \\ 0 & 1 & 0 & 0 & \frac{1}{b} & 0 \\ 0 & 0 & 1 & 0 & 0 & \frac{1}{c} \end{array}\right]$$

$$A^{-1} = \begin{bmatrix} \frac{1}{a} & 0 & 0 \\ 0 & \frac{1}{b} & 0 \\ 0 & 0 & \frac{1}{c} \end{bmatrix}$$

Reviewing Basic Concepts (Sections 7.4 - 7.6)

$$A=\begin{bmatrix}-5 & 4\\ 2 & -1\end{bmatrix},\quad B=\begin{bmatrix}0 & -2\\ 3 & -4\end{bmatrix},$$

$$C=\begin{bmatrix}2 & -3 & 1\\ -2 & 1 & 0\\ 0 & -1 & 4\end{bmatrix},\quad D=\begin{bmatrix}0 & 4 & 1\\ 0 & 2 & 2\\ 1 & 1 & 1\end{bmatrix}$$

1. $A-B=\begin{bmatrix}-5-0 & 4+2\\ 2-3 & -1+4\end{bmatrix}=\begin{bmatrix}-5 & 6\\ -1 & 3\end{bmatrix}$

2. $-3B=-3\begin{bmatrix}0 & -2\\ 3 & -4\end{bmatrix}=\begin{bmatrix}0 & 6\\ -9 & 12\end{bmatrix}$

3. $A^2=\begin{bmatrix}-5 & 4\\ 2 & -1\end{bmatrix}\begin{bmatrix}-5 & 4\\ 2 & -1\end{bmatrix}$

$=\begin{bmatrix}-5(-5)+4(2) & -5(4)+4(-1)\\ 2(-5)-1(2) & 2(4)-1(-1)\end{bmatrix}$

$=\begin{bmatrix}33 & -24\\ -12 & 9\end{bmatrix}$

4. $CD=\begin{bmatrix}2 & -3 & 1\\ -2 & 1 & 0\\ 0 & -1 & 4\end{bmatrix}\begin{bmatrix}0 & 4 & 1\\ 0 & 2 & 2\\ 1 & 1 & 1\end{bmatrix}$

on calculator: $\begin{bmatrix}1 & 3 & -3\\ 0 & -6 & 0\\ 4 & 2 & 2\end{bmatrix}$

5. $\det A=-5(-1)-2(4)=5-8=\mathbf{-3}$

6. Expand on column 1:

$\det C=2(4+0)+2(-12+1)+0(0+1)$
$=2(4)+2(-11)=8-22=\mathbf{-14}$

7. $A^{-1}=\dfrac{1}{5-8}\begin{bmatrix}-1 & -4\\ -2 & -5\end{bmatrix}=\begin{bmatrix}\frac{1}{3} & \frac{4}{3}\\ \frac{2}{3} & \frac{5}{3}\end{bmatrix}$

8. $C^{-1}=\begin{bmatrix}-\frac{2}{7} & -\frac{11}{14} & \frac{1}{14}\\ -\frac{4}{7} & -\frac{4}{7} & \frac{1}{7}\\ -\frac{1}{7} & -\frac{1}{7} & \frac{2}{7}\end{bmatrix}$ (calculator)

9. $\dfrac{\sqrt{3}}{2}(w_1+w_2)=100$

$\dfrac{\sqrt{3}}{2}w_1+\dfrac{\sqrt{3}}{2}w_2=100$

$w_1-w_2=0$

Using Cramer's Rule:

$D=\begin{bmatrix}\frac{\sqrt{3}}{2} & \frac{\sqrt{3}}{2}\\ 1 & -1\end{bmatrix}=-\sqrt{3}$

$D_{w_1}=\begin{bmatrix}100 & \frac{\sqrt{3}}{2}\\ 0 & -1\end{bmatrix}=-100$

$D_{w_2}=\begin{bmatrix}\frac{\sqrt{3}}{2} & 100\\ 1 & 0\end{bmatrix}=-100$

$w_1=\dfrac{-100}{-\sqrt{3}}=\dfrac{100}{\sqrt{3}}\cdot\dfrac{\sqrt{3}}{\sqrt{3}}=\dfrac{100\sqrt{3}}{3}\approx 57.7$

$w_2=\dfrac{-100}{-\sqrt{3}}=\dfrac{100}{\sqrt{3}}\cdot\dfrac{\sqrt{3}}{\sqrt{3}}=\dfrac{100\sqrt{3}}{3}\approx 57.7$

Both w_1 and w_2 are approximately 57.7 pounds.

10. $2x+y+2z=10$
$y+2z=4$
$x-2y+2z=1$

$$\begin{bmatrix}x\\ y\\ z\end{bmatrix}=\begin{bmatrix}2 & 1 & 2\\ 0 & 1 & 2\\ 1 & -2 & 2\end{bmatrix}^{-1}\cdot\begin{bmatrix}10\\ 4\\ 1\end{bmatrix}=\begin{bmatrix}3\\ 2\\ 1\end{bmatrix}$$

$\{(3,\ 2,\ 1)\}$

Section 7.7

1.

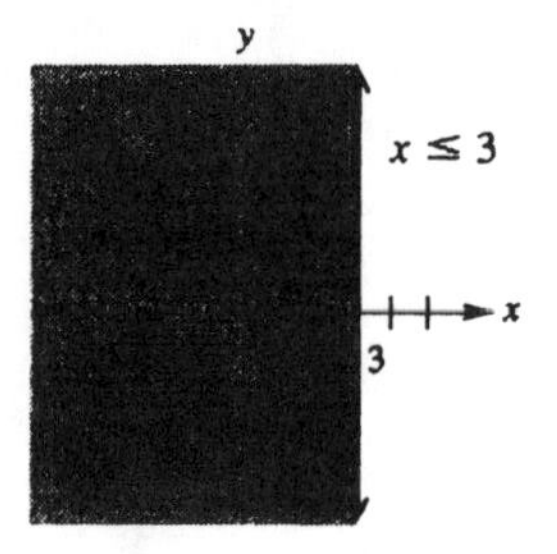

3.

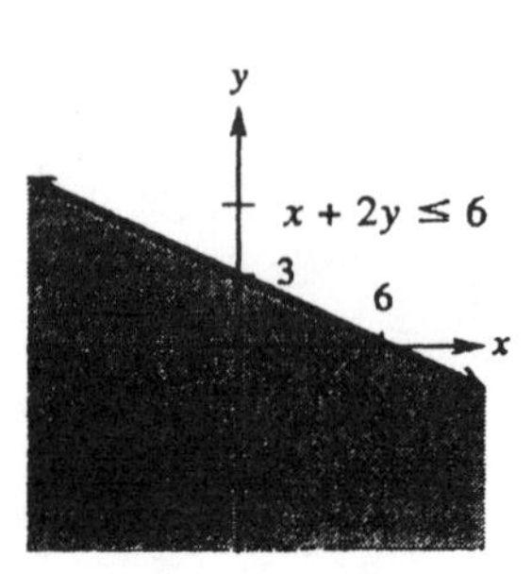

5.

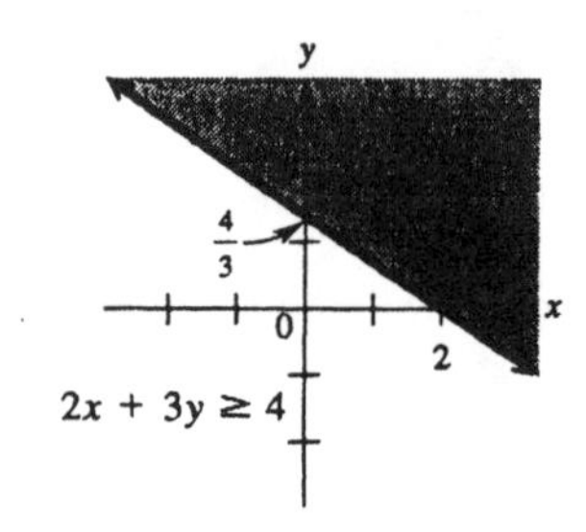

7.

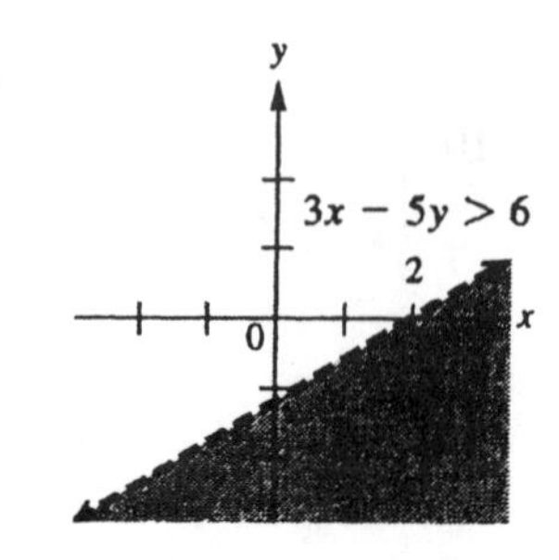

9.

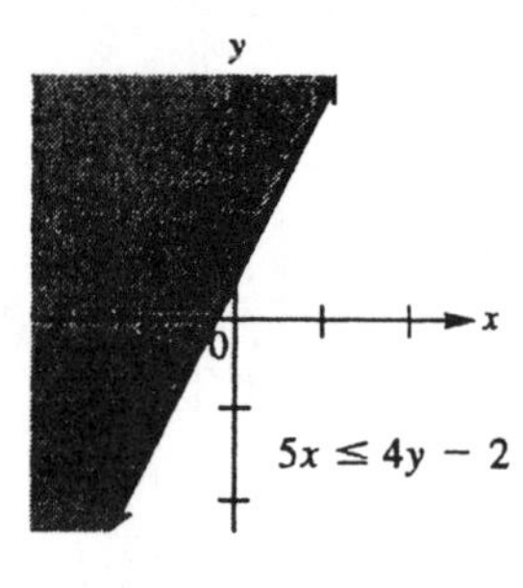

11.

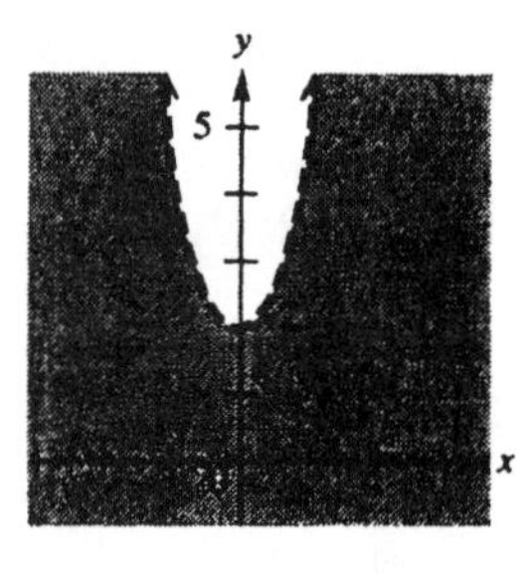

13.

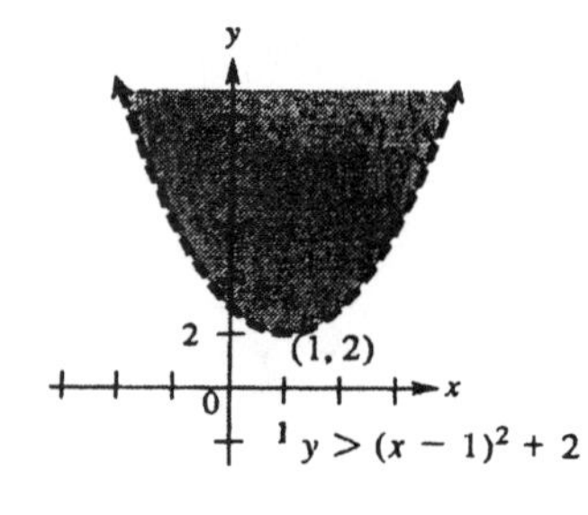

15.

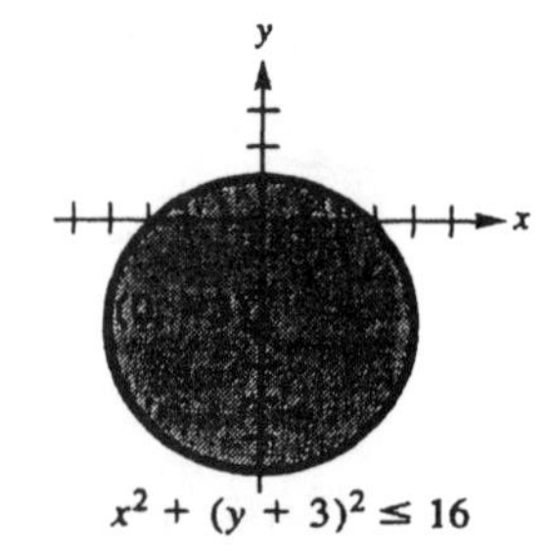

17. The boundary of an inequality is solid if the inequality is $\geq$ or $\leq$; the boundary is dashed if the inequality is $>$ or $<$.

19. $Ax + By \geq C$
$By \geq -Ax + C$
If $B > 0$, $y \geq -\frac{A}{B}x + \frac{C}{B}$
Shade above the line.

21. $(x-5)^2 + (y-2)^2 < 4$: **B**
The region inside a circle with center $(5, 2)$ and radius 2.

23. $y \leq 3x - 6$: **C**

25. $y \leq -3x - 6$: **A**

27.

x + y ≥ 0
2x − y ≥ 3

29.

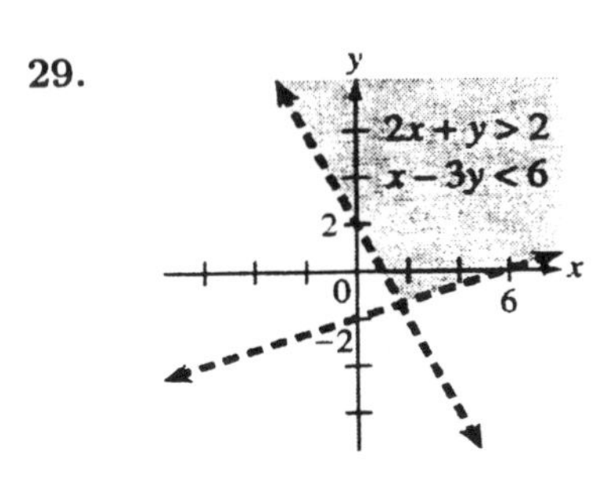

31.

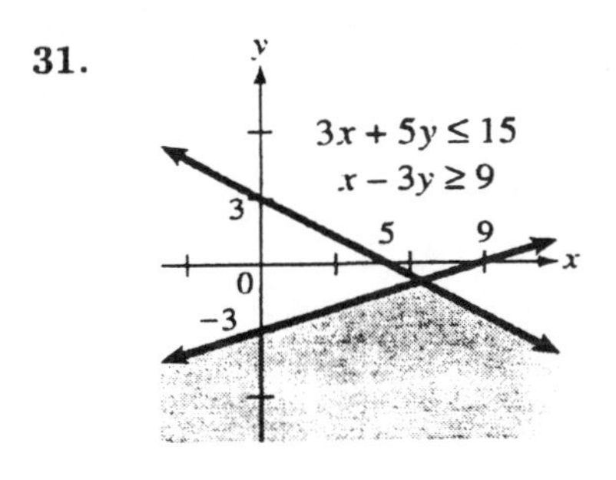

33.

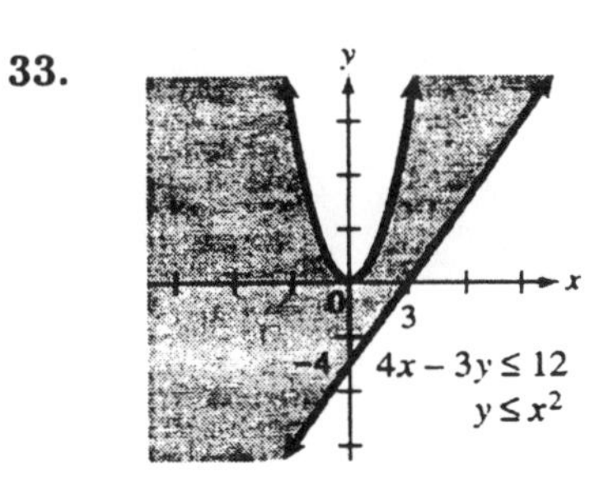

35.

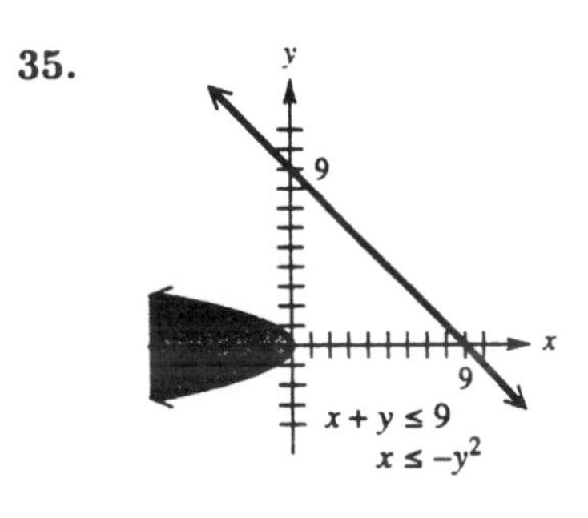

37.

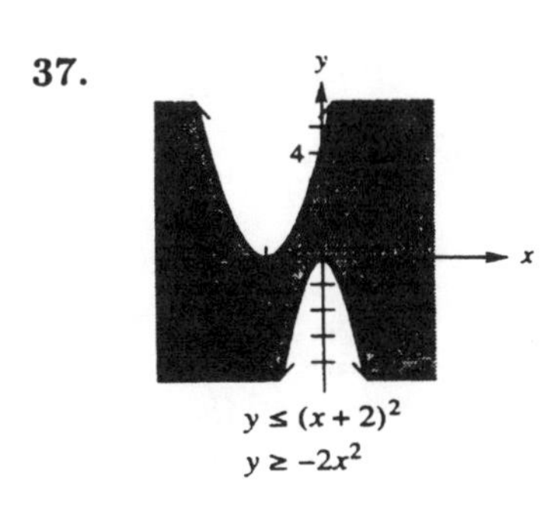

39.

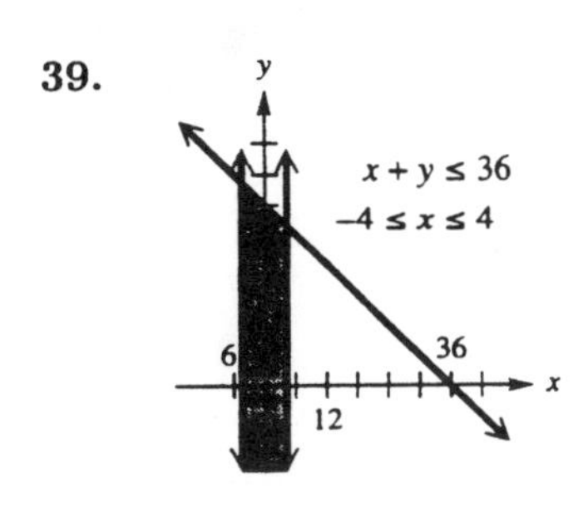

41.

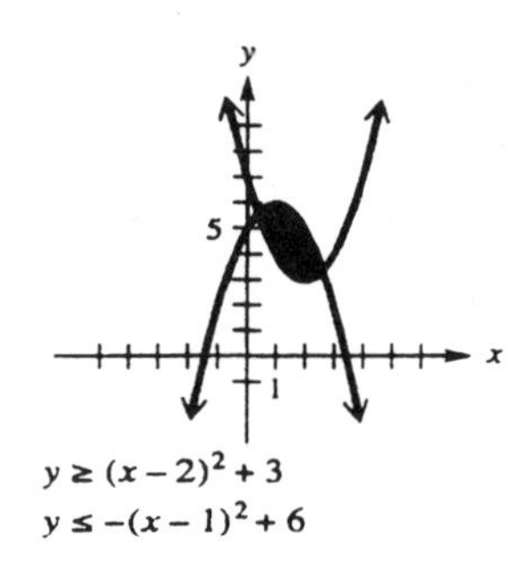

43.

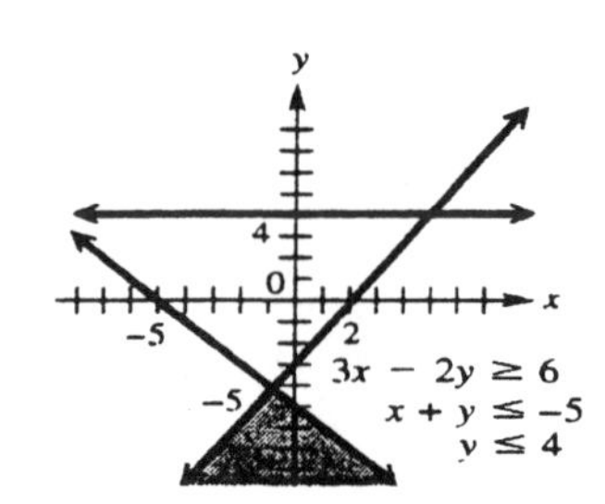

45.

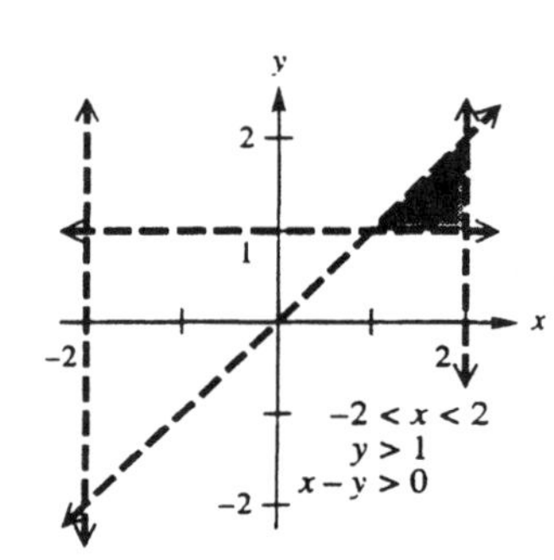

47.

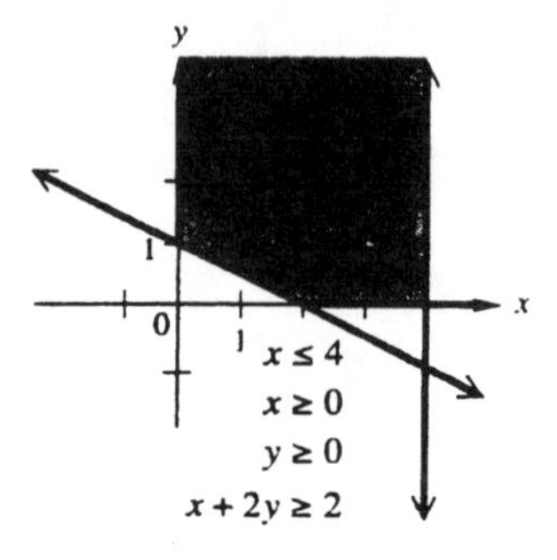

49.

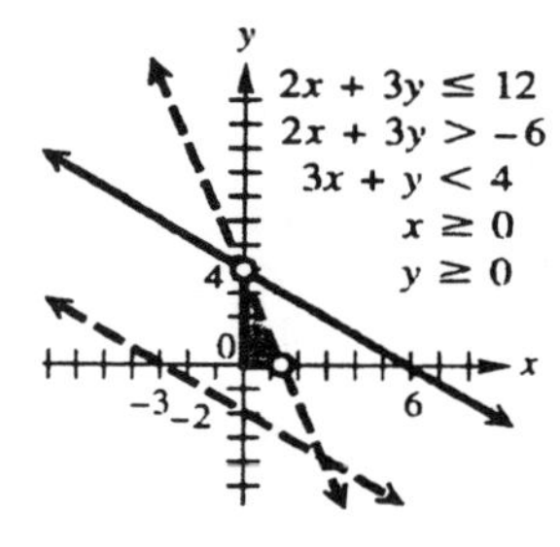

51.

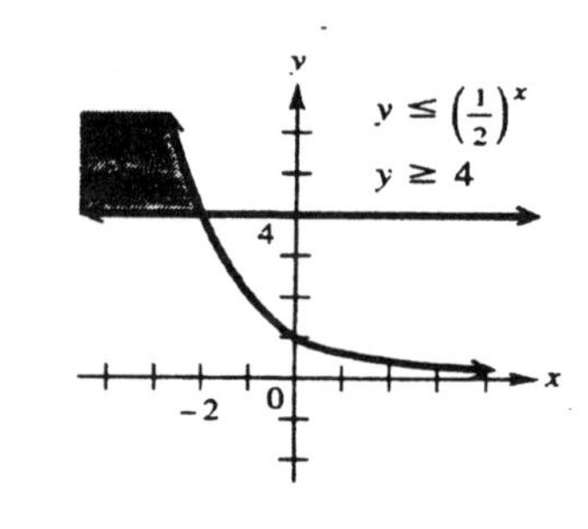

53.

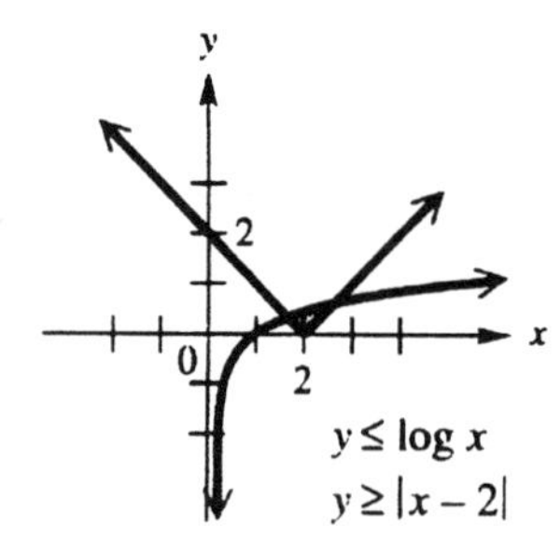

55. $x^2 + 4y^2 < 36$
$y < x$

D: All points inside the ellipse, below the line.

57. A: $y \geq x$
$y \leq 2x - 3$

59. B: $x^2 + y^2 \leq 16$
$y \geq 0$

61. $3x + 2y \geq 6 \rightarrow y \geq -\frac{3}{2}x + 3$

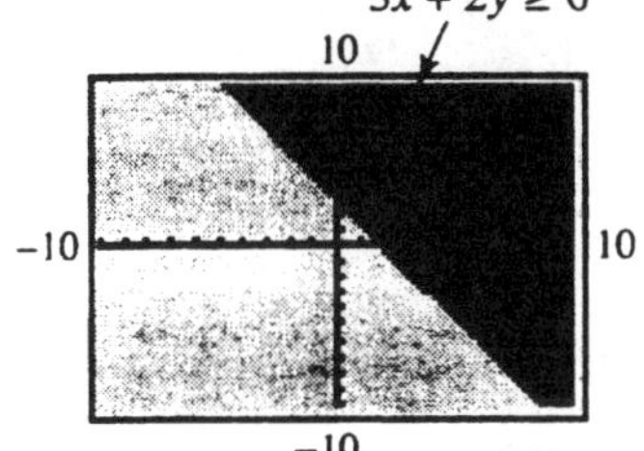

63. $x + y \geq 2 \rightarrow y \geq -x + 2$

$x + y \leq 6 \rightarrow y \leq -x + 6$

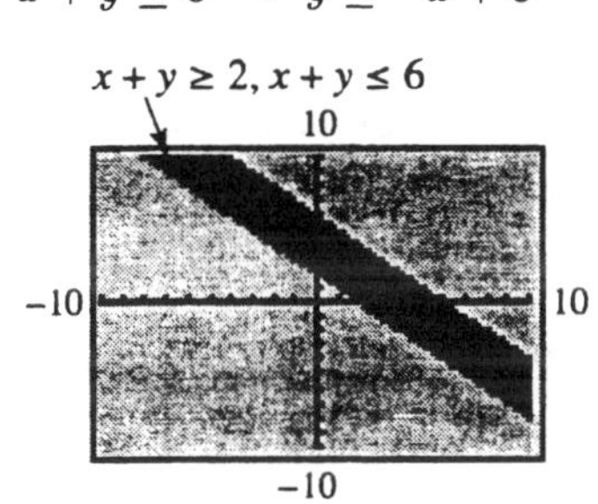

65.

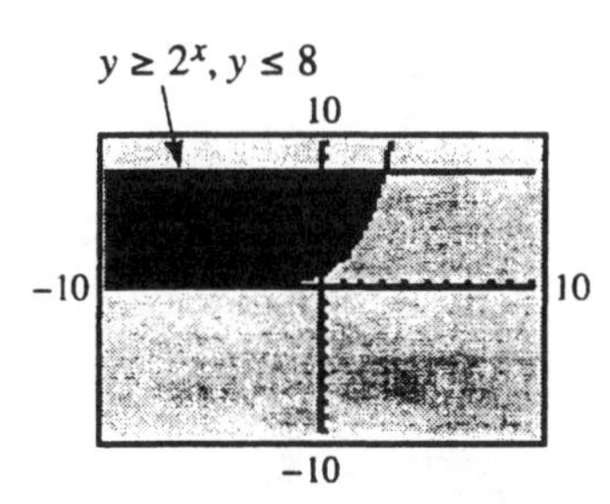

67. Since we are in the first quadrant, $x \geq 0$, $y \geq 0$. The lines $x + 2y - 8 = 0$ and $x + 2y = 12$ are parallel, with $x + 2y = 12$ having the greater y-intercept. Therefore, we must shade below $x + 2y = 12$ and above $x + 2y - 8 = 0$. The system is:

$$x + 2y - 8 \geq 0$$
$$x + 2y \leq 12$$
$$x \geq 0, \quad y \leq 0$$

69.

	$3x + 5y$
(1, 1) :	$3(1) + 5(1) = 8$
(2, 7) :	$3(2) + 5(7) = 41$
(5, 10) :	$3(5) + 5(10) = 65$
(6, 3) :	$3(6) + 5(3) = 33$

Maximum value 65 at (5, 10)
Minimum value 8 at (1, 1).

71.

	$3x + 5y$
(1, 10) :	$3(1) + 5(10) = 53$
(7, 9) :	$3(7) + 5(9) = 66$
(7, 6) :	$3(7) + 5(6) = 51$
(1, 0) :	$3(1) + 5(0) = 3$

Maximum value 66 at (7, 9).
Minimum value 3 at (1, 0).

73.

	$10y$
(1, 10) :	$10(10) = 100$
(7, 9) :	$10(9) = 90$
(7, 6) :	$10(6) = 60$
(1, 0) :	$10(0) = 0$

Maximum value 100 at (1, 10).
Minimum value 0 at (1, 0).

75. Let $x = \#$ of hat units,
$y = \#$ of whistle units:

Objective function: $3x + 2y$

Restraints: $2x + 4y \leq 12$
$x + y \leq 5$
$x \geq 0,\ y \geq 0$

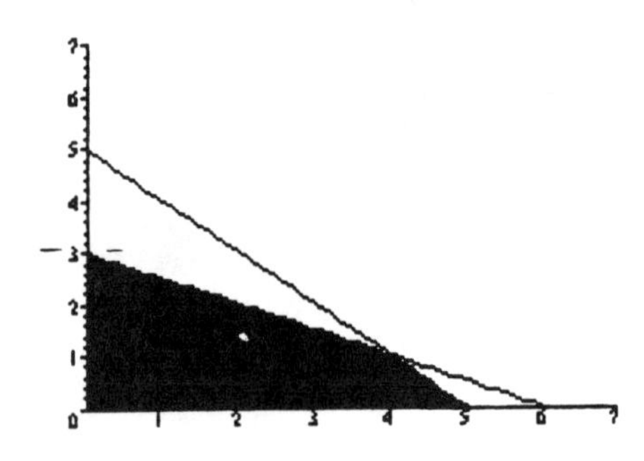

Point of intersection:

$$\begin{aligned} 2x + 4y &= 12 \\ x + y &= 5 \end{aligned}$$

$$\begin{aligned} -x - 2y &= -6 \\ x + y &= 5 \\ \hline -y &= -1 \end{aligned}$$

$$y = 1,\ x = 4$$

	$3x + 2y$
(0, 0) :	0
(0, 3) :	6
(4, 1) :	14
(5, 0) :	15

Maximum of 15 inquiries: display 5 hat units, 0 whistle units.

77. Let $x = \#$ of refrigerators shipped to A,
$y = \#$ of refrigerators shipped to B:

Objective function: $12x + 10y$

Restraints: $x + y \geq 100$
$0 \leq x \leq 75$
$0 \leq y \leq 80$

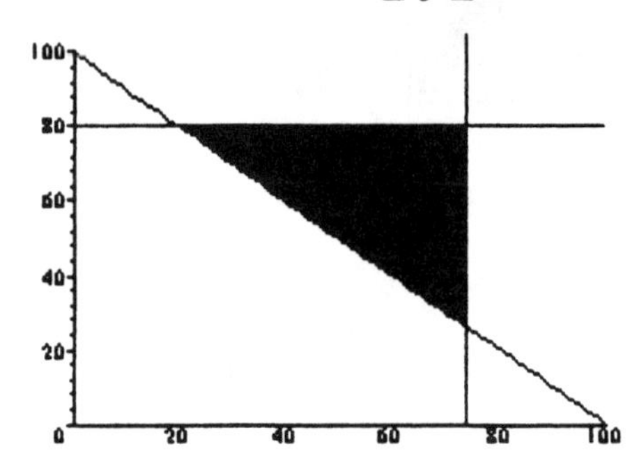

Points of intersection:

$x + y = 100,\ y = 80 \rightarrow (20, 80)$

$x = 75,\ y = 80 \rightarrow (75, 80)$

$x + y = 100,\ x = 75 \rightarrow (75, 25)$

	$12x + 10y$
(20, 80) :	$240 + 800 = 1040$
(75, 80) :	$900 + 800 = 1700$
(75, 25) :	$900 + 250 = 1150$

Minimum cost of \$1,040; ship 20 to A, 80 to B.

79. Let $x = \#$ of gallons (millions) of gasoline, $y = \#$ of gallons (millions) of fuel oil;

Objective function: $1.9x + 1.5y$

Restraints: $y \leq \frac{1}{2}x$

$y \geq 3$ million

$x \leq 6.4$ million

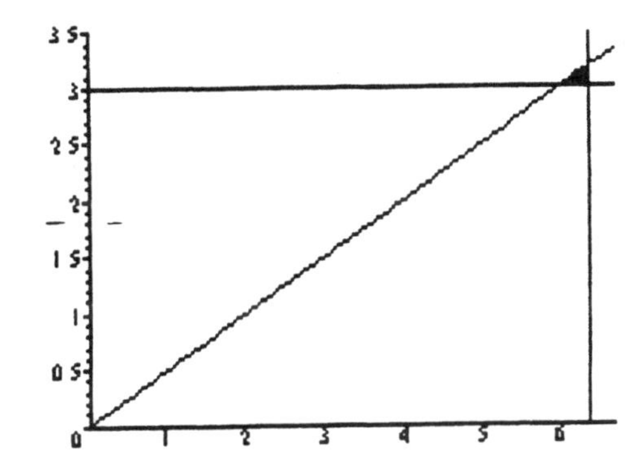

Points of intersection:

(i) $y = \frac{1}{2}x,\ y = 3 \rightarrow (6,\ 3)$

(ii) $y = \frac{1}{2}x,\ x = 6.4 \rightarrow (6.4,\ 3.2)$

(iii) $y = 3,\quad x = 6.4 \rightarrow (6.4,\ 3)$

[Points in millions]	$1.9x + 1.5y$
(6, 3) :	15.9
(6.4, 3.2) :	16.96
(6.4, 3) :	16.66

For a maximum revenue of \$16,960,000 :
6,400,000 gallons of gasoline,
3,200,000 gallons of fuel oil.

81. Let $x = \#$ of medical kits; $y = \#$ of water.

Objective function: $4x + 10y$

Restraints:

$$x + y \leq 6000$$
$$10x + 20y \leq 80,000$$
$$x \geq 0,\quad y \geq 0$$

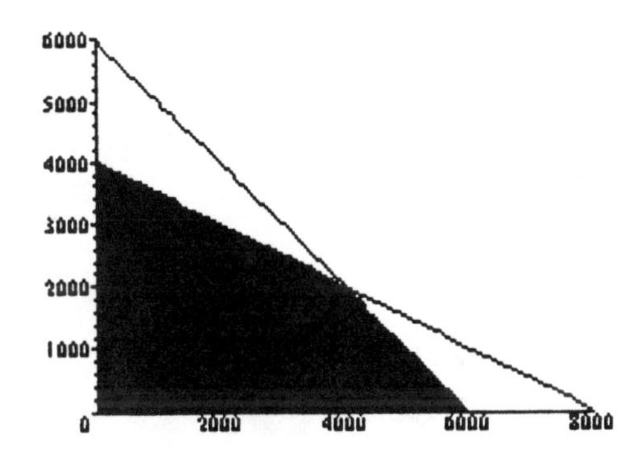

Point of intersection:

$$x + y = 6000$$
$$10x + 20y = 80000$$

$$-10x - 10y = -60000$$
$$10x + 20y = 80000$$
$$10y = 20000$$

$$y = 2000, \quad x = 4000$$

	$4x + 10y$
(0, 0) :	0
(0, 4000) :	40,000
(4000, 2000) :	36,000
(6000, 0) :	24,000

To maximize # of people aided:
0 medical kits and
4,000 containers of water
to aid 400 people.

Section 7.8

1. $\dfrac{5}{3x(2x+1)} = \dfrac{A}{3x} + \dfrac{B}{2x+1}$

Multiply by $3x(2x+1)$:

$$5 = A(2x+1) + B(3x)$$

Let $x = 0$: $5 = A(1) \rightarrow A = 5$

Let $x = -\dfrac{1}{2}$: $5 = B\left(-\dfrac{3}{2}\right) \rightarrow B = -\dfrac{10}{3}$

$$\boxed{\dfrac{5}{3x} + \dfrac{-10}{3(2x+1)}}$$

3. $\dfrac{4x+2}{(x+2)(2x-1)} = \dfrac{A}{x+2} + \dfrac{B}{2x-1}$

Multiply by $(x+2)(2x-1)$:

$$4x+2 = A(2x-1) + B(x+2)$$

Let $x = -2$: $-6 = A(-5) \rightarrow A = \dfrac{6}{5}$

Let $x = \dfrac{1}{2}$: $4 = B\left(\dfrac{5}{2}\right) \rightarrow B = 4 \times \dfrac{2}{5} = \dfrac{8}{5}$

$$\boxed{\dfrac{6}{5(x+2)} + \dfrac{8}{5(2x-1)}}$$

5. $\dfrac{x}{x^2+4x-5}$

$\dfrac{x}{(x+5)(x-1)} = \dfrac{A}{x+5} + \dfrac{B}{x-1}$

Multiply by $(x+5)(x-1)$:

$$x = A(x-1) + B(x+5)$$

Let $x = -5$: $-5 = A(-6) \rightarrow A = \dfrac{5}{6}$

Let $x = 1$: $1 = B(6) \rightarrow B = \dfrac{1}{6}$

$$\boxed{\dfrac{5}{6(x+5)} + \dfrac{1}{6(x-1)}}$$

7. $\dfrac{2x}{(x+1)(x+2)^2} = \dfrac{A}{x+1} + \dfrac{B}{x+2} + \dfrac{C}{(x+2)^2}$

Multiply by $(x+1)(x+2)^2$:

$$2x = A(x+2)^2 + B(x+1)(x+2) + C(x+1)$$

Let $x = -1$: $-2 = A(1) \rightarrow A = -2$

Let $x = -2$: $-4 = C(-1) \rightarrow C = 4$

Let $x = 0$, $A = -2$, $C = 4$:

$$0 = -2(4) + B(2) + 4(1)$$
$$4 = 2B \rightarrow B = 2$$

$$\boxed{\dfrac{-2}{x+1} + \dfrac{2}{x+2} + \dfrac{4}{(x+2)^2}}$$

9. $\dfrac{4}{x(1-x)} = \dfrac{A}{x} + \dfrac{B}{1-x}$

Multiply by $x(1-x)$:

$$4 = A(1-x) + Bx$$

Let $x = 0$: $4 = A(1) \rightarrow A = 4$

Let $x = 1$: $4 = B(1) \rightarrow B = 4$

$$\boxed{\dfrac{4}{x} + \dfrac{4}{1-x}}$$

11. $\dfrac{4x^2-x-15}{x(x+1)(x-1)} = \dfrac{A}{x} + \dfrac{B}{x+1} + \dfrac{C}{x-1}$

Multiply by $x(x-1)(x+1)$:

$$4x^2 - x - 15 = A(x-1)(x+1) + Bx(x-1) + Cx(x+1)$$

Let $x = 0$: $-15 = A(-1) \rightarrow A = 15$

Let $x = 1$: $-12 = C(1)(2) \rightarrow C = -6$

Let $x = -1$: $-10 = B(-1)(-2) \rightarrow B = -5$

$$\boxed{\dfrac{15}{x} + \dfrac{-5}{x+1} + \dfrac{-6}{x-1}}$$

13. $\dfrac{x^2}{x^2+2x+1} = 1 + \dfrac{-2x-1}{(x+1)^2}$

$$\begin{array}{r} 1 \\ x^2+2x+1 \enclose{longdiv}{x^2+0x+0} \\ \underline{x^2+2x+1} \\ -2x-1 \end{array}$$

$$\frac{-2x-1}{(x+1)^2} = \frac{A}{x+1} + \frac{B}{(x+1)^2}$$

Multiply by $(x+1)^2$:

$-2x-1 = A(x+1) + B$

Let $x=-1$: $1 = B$

Let $x=0$, $B=1$:

$-1 = A+1 \to A = -2$

$$\boxed{1 + \frac{-2}{x+1} + \frac{1}{(x+1)^2}}$$

15. $\dfrac{2x^5+3x^4-3x^3-2x^2+x}{2x^2+5x+2}$

$$\begin{array}{r} x^3 - x^2 \\ 2x^2+5x+2 \enclose{longdiv}{2x^5+3x^4-3x^3-2x^2+x} \\ \underline{2x^5+5x^4+2x^3} \\ -2x^4-5x^3-2x^2 \\ \underline{-2x^4-5x^3-2x^2} \\ 0+x \end{array}$$

$$x^3 - x^2 + \frac{x}{2x^2+5x+2}$$

Partial fraction decomposition:

$$\frac{x}{2x^2+5x+2} = \frac{A}{2x+1} + \frac{B}{x+2}$$

Multiply by $(2x+1)(x+2)$:

$x = A(x+2) + B(2x+1)$

Let $x = -\frac{1}{2}$: $-\frac{1}{2} = A\left(\frac{3}{2}\right) \to A = -\frac{1}{3}$

Let $x=-2$: $-2 = B(-3) \to B = \frac{2}{3}$

$$\boxed{x^3 - x^2 + \frac{-1}{3(2x+1)} + \frac{2}{3(x+2)}}$$

17. $\dfrac{x^3+4}{9x^3-4x} = \dfrac{1}{9} + \dfrac{\frac{4}{9}x+4}{9x^3-4x}$

$$\begin{array}{r} \frac{1}{9} \\ 9x^3-4x \enclose{longdiv}{x^3 \qquad +4} \\ \underline{x^3 - \frac{4}{9}x} \\ \frac{4}{9}x+4 \end{array}$$

Partial fraction decomposition:

$$\frac{\frac{4}{9}x+4}{9x^3-4x} = \frac{A}{x} + \frac{B}{3x+2} + \frac{C}{3x-2}$$

Multiply by $9x^3-4x$:

$\frac{4}{9}x + 4 = A(9x^2-4) + Bx(3x-2) + Cx(3x+2)$

Let $x=0$: $4 = A(-4) \to A = -1$

Let $x = -\frac{2}{3}$: $-\frac{8}{27} + 4 = B\left(-\frac{2}{3}\right)(-4)$

$\frac{100}{27} = \frac{8}{3}B \to B = \frac{25}{18}$

Let $x = \frac{2}{3}$: $\frac{8}{27} + 4 = C\left(\frac{2}{3}\right)(4)$

$\frac{116}{27} = \frac{8}{3}C \to C = \frac{29}{18}$

$$\boxed{\frac{1}{9} + \frac{-1}{x} + \frac{25}{18(3x+2)} + \frac{29}{18(3x-2)}}$$

19. $\dfrac{-3}{x^2(x^2+5)} = \dfrac{A}{x} + \dfrac{B}{x^2} + \dfrac{Cx+D}{x^2+5}$

Multiply by $x^2(x^2+5)$:

$$\begin{aligned} -3 &= Ax(x^2+5) + B(x^2+5) + (Cx+D)x^2 \\ &= Ax^3 + 5Ax + Bx^2 + 5B + Cx^3 + Dx^2 \end{aligned}$$

Equate x^3 coefficients: $0 = A + C$ [*i*]

Equate x^2 coefficients: $0 = B + D$ [*ii*]

Equate x coefficients: $0 = 5A \to A = 0$

Equate constants: $-3 = 5B \to B = -\frac{3}{5}$

Substitute A into [*i*]: $C = 0$

Substitute B into [*ii*]:

$0 = -\frac{3}{5} + D \to D = \frac{3}{5}$

$$\boxed{\frac{-3}{5x^2} + \frac{3}{5(x^2+5)}}$$

21. $\dfrac{3x-2}{(x+4)(3x^2+1)} = \dfrac{A}{x+4} + \dfrac{Bx+C}{3x^2+1}$

Multiply by $(x+4)(3x^2+1)$:

$$3x-2 = A(3x^2+1) + (Bx+C)(x+4)$$
$$= 3Ax^2 + A + Bx^2 + 4Bx + Cx + 4C$$

Let $x=-4$: $-14 = 49A \to A = -\dfrac{2}{7}$

Equate x^2 coefficients:

$$0 = 3A + B = -\frac{6}{7} + B \to B = \frac{6}{7}$$

Equate x coefficients:

$$3 = 4B + C = \frac{24}{7} + C \to C = -\frac{3}{7}$$

$$\frac{-2}{7(x+4)} + \frac{\frac{6}{7}x - \frac{3}{7}}{3x^2+1}$$

$$\boxed{\frac{-2}{7(x+4)} + \frac{6x-3}{7(3x^2+1)}}$$

23. $\dfrac{1}{x(2x+1)(3x^2+4)} = \dfrac{A}{x} + \dfrac{B}{2x+1} + \dfrac{Cx+D}{3x^2+4}$

Multiply by $x(2x+1)(3x^2+4)$:

$$1 = A(2x+1)(3x^2+4) + Bx(3x^2+4) + (Cx+D)(x)(2x+1) \quad [i]$$

Let $x=0$: $1 = A(1)(4) \to A = \dfrac{1}{4}$

Let $x = -\dfrac{1}{2}$: $1 = B\left(-\dfrac{1}{2}\right)\left(\dfrac{19}{4}\right) \to B = -\dfrac{8}{19}$

Multiply $[i]$ out:

$$1 = A(6x^3+3x^2+8x+4) + 3Bx^3 + 4Bx + 2Cx^3 + Cx^2 + 2Dx^2 + Dx$$

$$1 = 6Ax^3 + 3Ax^2 + 8Ax + 4A + 3Bx^3 + 4Bx + 2Cx^3 + Cx^2 + 2Dx^2 + Dx$$

Equate x^3 coefficients:

$$0 = 6A + 3B + 2C$$
$$0 = 6\left(\frac{1}{4}\right) + 3\left(-\frac{8}{19}\right) + 2C$$
$$0 = \frac{57-48}{38} + 2C \to C = -\frac{9}{76}$$

continued

23. continued

Equate x^2 coefficients:

$$0 = 3A + C + 2D$$
$$0 = \frac{3}{4} - \frac{9}{76} + 2D$$
$$0 = \frac{57-9}{76} + 2D$$
$$2D = -\frac{48}{76} \to D = -\frac{24}{76}$$

$$\frac{\frac{1}{4}}{x} + \frac{-\frac{8}{19}}{2x+1} + \frac{-\frac{9}{76}x - \frac{24}{76}}{3x^2+4}$$

$$\boxed{\frac{1}{4x} + \frac{-8}{19(2x+1)} + \frac{-9x-24}{76(3x^2+4)}}$$

25. $\dfrac{3x-1}{x(2x^2+1)^2} = \dfrac{A}{x} + \dfrac{Bx+C}{2x^2+1} + \dfrac{Dx+E}{(2x^2+1)^2}$

Multiply by $x(2x^2+1)^2$:

$$3x-1 = A(2x^2+1)^2 + (Bx+C)(x)(2x^2+1) + (Dx+E)(x) \quad [i]$$

Let $x=0$: $-1 = A(1) \to A = -1$

Multiply $[i]$ out:

$$3x-1 = A(4x^4+4x^2+1) + 2Bx^4 + Bx^2 + Cx + 2Cx^3 + Dx^2 + Ex$$

$$3x-1 = 4Ax^4 + 4Ax^2 + A + 2Bx^4 + Bx^2 + Cx + 2Cx^3 + Dx^2 + Ex$$

Equate x^4 coefficients $(A=-1)$:

$$0 = 4A + 2B$$
$$0 = -4 + 2B \to B = 2$$

Equate x^3 coefficients: $0 = 2C \to C = 0$

Equate x^2 coefficients $(A=-1,\ B=2)$:

$$0 = 4A + B + D$$
$$0 = -4 + 2 + D \to D = 2$$

Equate x coefficients $(C=0)$:

$$3 = C + E$$
$$3 = 0 + E \to E = 3$$

$$\boxed{\frac{-1}{x} + \frac{2x}{2x^2+1} + \frac{2x+3}{(2x^2+1)^2}}$$

27. $\dfrac{-x^4 - 8x^2 + 3x - 10}{(x+2)(x^2+4)^2}$

$= \dfrac{A}{x+2} + \dfrac{Bx+C}{x^2+4} + \dfrac{Dx+E}{(x^2+4)^2}$

Multiply by $(x+2)(x^2+4)^2$:

$$-x^4 - 8x^2 + 3x - 10 = A(x^2+4)^2 + (Bx+C)(x+2)(x^2+4) + (Dx+E)(x+2) \quad [i]$$

Let $x = -2$:

$$--16 - 32 - 6 - 10 = A(64)$$
$$-64 = 64A \to A = -1$$

Multiply [i] out:

$$-x^4 - 8x^2 + 3x - 10 = Ax^4 + 8Ax^2 + 16A + Bx^4 + 2Bx^3 + 4Bx^2 + 8Bx + Cx^3 + 2Cx^2 + 4Cx + 8C + Dx^2 + 2Dx + Ex + 2E$$

Equate x^4 coefficients $(A = -1)$:

$$-1 = A + B$$
$$-1 = -1 + B \to B = 0$$

Equate x^3 coefficients $(B = 0)$:

$$0 = 2B + C$$
$$0 = 0 + C \quad \to C = 0$$

Equate x^2 coefficients

$(A = -1,\ B = 0,\ C = 0)$:

$$-8 = 8A + 4B + 2C + D$$
$$-8 = -8 + D \quad \to D = 0$$

Equate x coefficients

$(B = 0,\ C = 0,\ D = 0)$:

$$3 = 8B + 4C + 2D + E$$
$$3 = 0 + 0 + 0 + E \quad \to E = 3$$

$$\boxed{\dfrac{-1}{x+2} + \dfrac{3}{(x^2+4)^2}}$$

29. $\dfrac{5x^5 + 10x^4 - 15x^3 + 4x^2 + 13x - 9}{x^3 + 2x^2 - 3x}$

$= 5x^2 + \dfrac{4x^2 + 13x - 9}{x^3 + 2x^2 - 3x}$

$$\begin{array}{r} 5x^2 \qquad\qquad\qquad\qquad\qquad\qquad\qquad \\ x^3 + 2x^2 - 3x \overline{)5x^5 + 10x^4 - 15x^3 + 4x^2 + 13x - 9} \\ \underline{5x^5 + 10x^4 - 15x^3} \qquad\qquad\qquad\quad \\ 0 + 4x^2 + 13x - 9 \end{array}$$

Partial fraction decomposition:

$$\dfrac{4x^2 + 13x - 9}{x^3 + 2x^2 - 3x} = \dfrac{A}{x} + \dfrac{B}{x+3} + \dfrac{C}{x-1}$$

Multiply by $x(x+3)(x-1)$:

$$4x^2 + 13x - 9 = A(x+3)(x-1) + Bx(x-1) + Cx(x+3)$$

Let $x = 0$: $-9 = A(-3) \to A = 3$

Let $x = 1$: $8 = C(4) \to C = 2$

Let $x = -3$:

$$36 - 39 - 9 = B(-3)(-4)$$
$$-12 = B(12) \to B = -1$$

$$\boxed{5x^2 + \dfrac{3}{x} + \dfrac{-1}{x+3} + \dfrac{2}{x-1}}$$

31. $\dfrac{4x^2 - 3x - 4}{x^3 + x^2 - 2x} = \dfrac{2}{x} + \dfrac{-1}{x-1} + \dfrac{3}{x+2}$

Graph $y_1 = \dfrac{4x^2 - 3x - 4}{x^3 + x^2 - 2x}$

and $y_2 = \dfrac{2}{x} + \dfrac{-1}{x-1} + \dfrac{3}{x+2}$

The graphs coincide;
the partial decomposition is correct.

33. $\dfrac{x^3 - 2x}{(x^2 + 2x + 2)^2} = \dfrac{x - 2}{x^2 + 2x + 2} + \dfrac{2}{(x^2 + 2x + 2)^2}$

Graph $y_1 = \dfrac{x^3 - 2x}{(x^2 + 2x + 2)^2}$

and $y_2 = \dfrac{x - 2}{x^2 + 2x + 2} + \dfrac{2}{(x^2 + 2x + 2)^2}$

The graphs do not coincide;
the partial decomposition is not correct.

$$\left[\frac{x^3 - 2x}{(x^2 + 2x + 2)^2} = \frac{x - 2}{x^2 + 2x + 2} + \frac{4}{(x^2 + 2x + 2)^2}\right]$$

Reviewing Basic Concepts (Sections 7.7 and 7.8)

1. $-2x - 3y \leq 6$

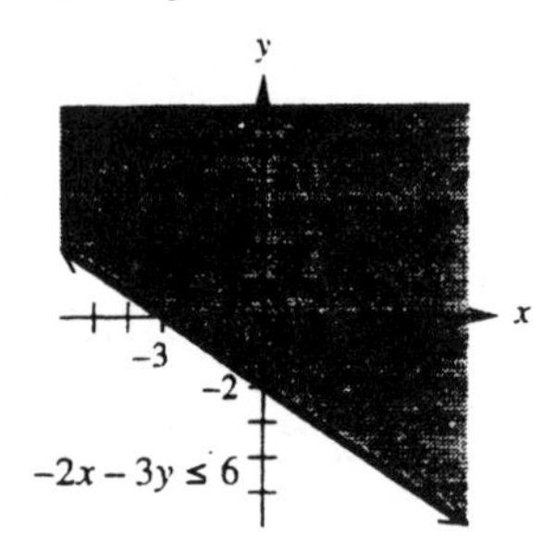

2. $x - y < 5$
$x + y \geq 3$

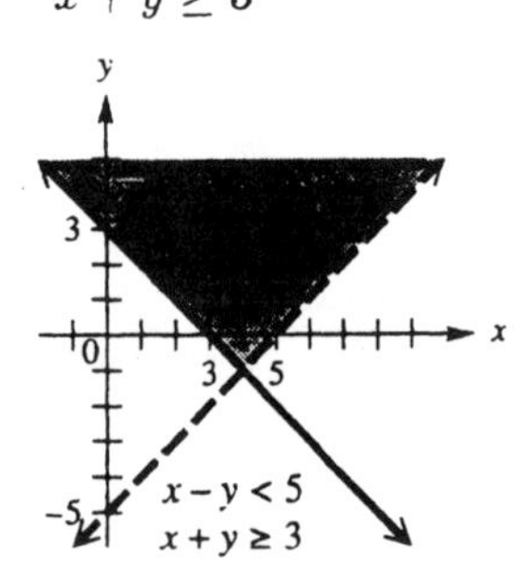

3. $y \geq x^2 - 2$
$x + 2y \geq 4$

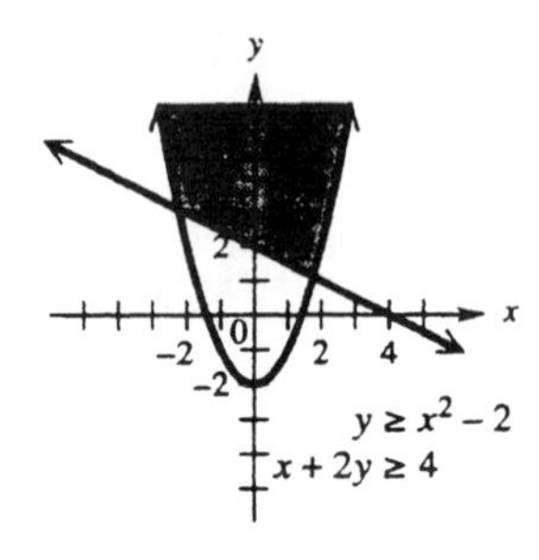

4. $x^2 + y^2 \leq 25$
$x^2 + y^2 \geq 9$

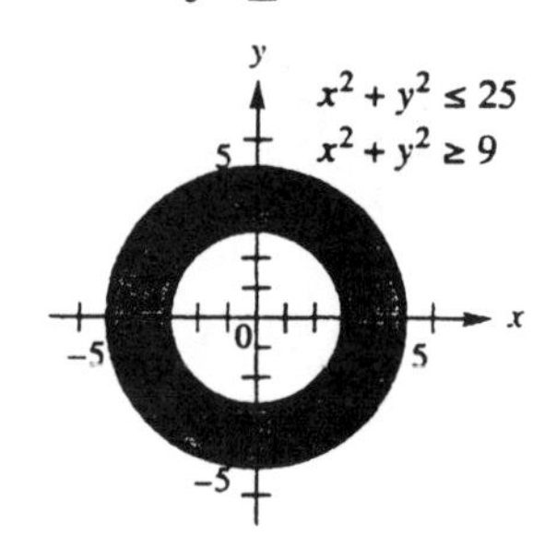

5. A: above line: $y > x - 3$
below parabola: $y < -x^2 + 4$

6. $x + y \geq 4$
$2x + y \leq 8$
$x \geq 0, \quad y \geq 0$

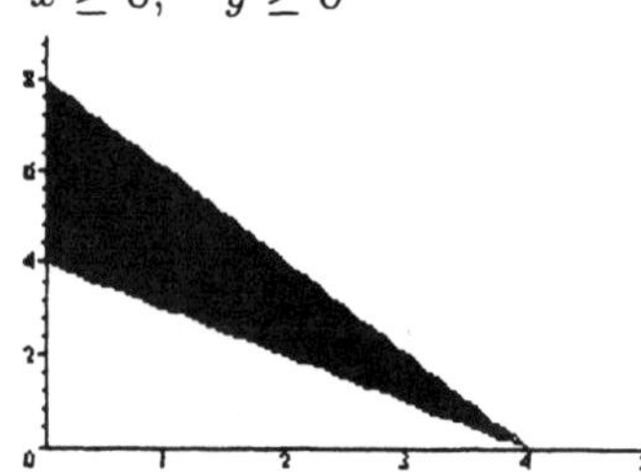

intersection points:

$x = 0, \quad x + y = 4 \rightarrow (0,\ 4)$
$x = 0, \quad 2x + y = 8 \rightarrow (0,\ 8)$

$$\begin{array}{r} x + y = 4 \\ -\,(2x + y = 8) \\ \hline -x = -4 \rightarrow x = 4 \end{array}$$

$x = 4 \Rightarrow x + y = 4 \rightarrow y = 0 \rightarrow (4,\ 0)$

	$2x + 3y$
(0, 4)	12
(0, 8)	24
(4, 0)	8

Minimum value is 8 at $(4,\ 0)$.

7.

	$3x + 5y$
$(1,\ 1)$	$3(1) + 5(1) = 8$
$(2,\ 7)$	$3(2) + 5(7) = 41$
$(5,\ 10)$	$3(5) + 5(10) = 65$
$(6,\ 3)$	$3(6) + 5(3) = 33$

Maximum value is 65 at $(5,\ 10)$.

Minimum value is 8 at $(1,\ 1)$.

8. Let x = number of pounds of substance X, y = number of pounds of substance Y.

Objective function: $2x + 3y$

Restraints:
$$.2x + .5y \geq 251$$
$$.5x + .3y \geq 200$$
$$x \geq 0, \quad y \geq 0$$

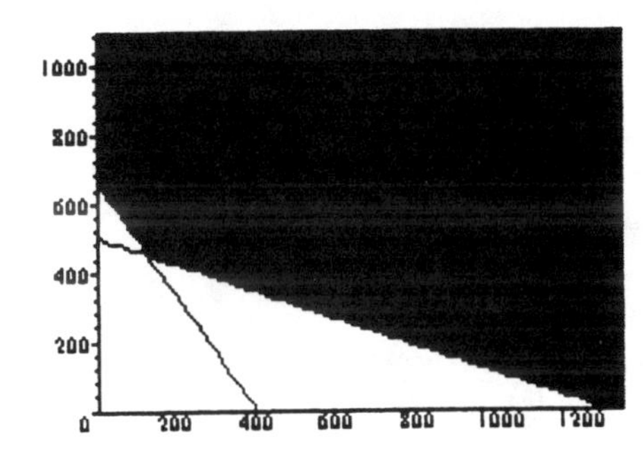

Points of intersection:

$$x = 0 : .5(0) + .3y = 200 \rightarrow \left(0,\ \frac{2000}{3}\right)$$

$$y = 0 : .2x + .5(0) = 251 \rightarrow (1255,\ 0)$$

$$10(.2x + .5y = 251)$$
$$2x + 5y = 2510 \qquad [1]$$

$$10(.5x + .3y = 200)$$
$$5x + 3y = 2000 \qquad [2]$$

$$-10x - 25y = -12550 \qquad [-5 \times [1]]$$
$$10x + 6y = 4000 \qquad [2 \times [2]]$$
$$-19y = -8550 \rightarrow y = 450$$

$y = 450 \rightarrow [1]$:
$$2x + 5(450) = 2510$$
$$2x = 260 \rightarrow x = 130 \rightarrow (130,\ 450)$$

	$2x + 3y$
$(0,\ \frac{2000}{3})$:	2000
$(1255,\ 0)$:	2510
$(130,\ 450)$:	1610

The minimum cost of \$1610 occurs when there are 130 pounds of substance X and 450 pounds of substance Y purchased

9. $$\frac{10x + 13}{x^2 - x - 20} = \frac{A}{x - 5} + \frac{B}{x + 4}$$

Multiply by $(x - 5)(x + 4)$:
$$10x + 13 = A(x + 4) + B(x - 5)$$

Let $x = -4$: $\quad -27 = B(-9) \rightarrow B = 3$

Let $x = 5$: $\quad 63 = A(9) \rightarrow A = 7$

$$\boxed{\frac{7}{x - 5} + \frac{3}{x + 4}}$$

10. $$\frac{2x^2 - 15x - 32}{(x - 1)(x^2 + 6x + 8)} = \frac{A}{x - 1} + \frac{B}{x + 2} + \frac{C}{x + 4}$$

Multiply by $(x - 1)(x + 2)(x + 4)$:

$$2x^2 - 15x - 32 = A(x + 2)(x + 4) + B(x - 1)(x + 4) + C(x - 1)(x + 2)$$

Let $x = -2$: $\quad 6 = B(-6) \rightarrow B = -1$

Let $x = -4$: $\quad 60 = C(10) \rightarrow C = 6$

Let $x = 1$: $\quad -45 = A(15) \rightarrow A = -3$

$$\boxed{\frac{-3}{x - 1} + \frac{-1}{x + 2} + \frac{6}{x + 4}}$$

Chapter 7 Review

1. $4x - 3y = -1$ [1]
$3x + 5y = 50$ [2]

Solve [1] for x:

$4x = 3y - 1 \rightarrow x = \dfrac{3y-1}{4}$ [3]

$[3] \rightarrow [2]: \quad 3\left(\dfrac{3y-1}{4}\right) + 5y = 50$

$3(3y-1) + 20y = 200$

$9y - 3 + 20y = 200$

$29y = 203 \rightarrow y = 7$

$y = 7 \rightarrow [3]:$

$x = \dfrac{3(7)-1}{4} = \dfrac{20}{4} = 5$ $\boxed{\{(5, 7)\}}$

2. $\dfrac{x}{2} - \dfrac{y}{5} = \dfrac{11}{10}$ [1]
$2x - \dfrac{4y}{5} = \dfrac{22}{5}$ [2]

Solve [1] for x:

$5x - 2y = 11 \rightarrow x = \dfrac{2y+11}{5}$ [3]

$[3] \rightarrow 5 \times [2]: \quad 10x - 4y = 22$

$10\left(\dfrac{2y+11}{5}\right) - 4y = 22$

$2(2y+11) - 4y = 22$

$4y + 22 - 4y = 22$

$0 = 0$ $\quad\infty$ solutions

Dependent system:

$\boxed{\left(x, \dfrac{5x-11}{2}\right) \text{ or } \left(\dfrac{2y+11}{5}, y\right)}$

3. $4x + 5y = 5$ [1]
$3x + 7y = -6$ [2]

Solve [1] for y:

$5y = 5 - 4x \rightarrow y = \dfrac{5-4x}{5}$ [3]

$[3] \rightarrow [2]: \quad 3x + 7\left(\dfrac{5-4x}{5}\right) = -6$

$15x + 7(5-4x) = -30$

$15x + 35 - 28x = -30$

$-13x = -65 \rightarrow x = 5$

$x = 5 \rightarrow [3]:$

$y = \dfrac{5-4(5)}{5} = \dfrac{5-20}{5} = -3$

$\boxed{\{(5, -3)\}}$

4. $y = x^2 - 1$ [1]
$x + y = 1$ [2]

$[1] \rightarrow [2]:$

$x + (x^2 - 1) = 1$

$x^2 + x - 2 = 0$

$(x+2)(x-1) = 0 \rightarrow x = -2, 1$

$x = -2 \rightarrow [1]: y = (-2)^2 - 1 = 3$ $\boxed{\{(-2, 3)\}}$

$x = 1 \rightarrow [1]: y = 1^2 - 1 = 0$ $\boxed{\{(1, 0)\}}$

5. $x^2 + y^2 = 2$ [1]
$3x + y = 4$ [2]

Solve [2] for y: $y = 4 - 3x$

$y = 4 - 3x \rightarrow [1]:$

$x^2 + (4-3x)^2 = 2$

$x^2 + 16 - 24x + 9x^2 = 2$

$10x^2 - 24x + 14 = 0$

$2(5x-7)(x-1) = 0 \rightarrow x = \dfrac{7}{5}, 1$

$x = \dfrac{7}{5} \rightarrow [2]:$

$y = 4 - 3\left(\dfrac{7}{5}\right) = \dfrac{20}{5} - \dfrac{21}{5}$

$y = -\dfrac{1}{5}$ $\boxed{\left\{\left(\dfrac{7}{5}, -\dfrac{1}{5}\right)\right\}}$

$x = 1 \rightarrow [2]:$

$y = 4 - 3(1) = 1$ $\boxed{\{(1, 1)\}}$

6. $x^2 + 2y^2 = 22$ [1]
$2x^2 - y^2 = -1$ [2]

$x^2 + 2y^2 = 22$ [1]
$4x^2 - 2y^2 = -2$ [2 × [2]]

$5x^2 = 20$

$x^2 = 4 \rightarrow x = -2, 2$

$x = 2 \rightarrow [2]: \quad 2(2)^2 - y^2 = -1$

$8 - y^2 = -1$

$9 = y^2$

$y = 3, -3$

$\boxed{\{(2, 3)\}, \{(2, -3)\}}$

$x = -2 \rightarrow [2]: \quad 2(-2)^2 - y^2 = -1$

$8 - y^2 = -1$

$9 = y^2$

$y = 3, -3$

$\boxed{\{(-2, 3)\}, \{(-2, -3)\}}$

7. $x^2 - 4y^2 = 19$ [1]
$x^2 + y^2 = 29$ [2]

$$\begin{array}{rl} x^2 - 4y^2 = 19 & [1] \\ -x^2 - y^2 = -29 & [-1 \times [2]] \\ \hline -5y^2 = -10 & \\ y^2 = 2 \rightarrow y = \pm\sqrt{2} & \end{array}$$

$y = \sqrt{2} \rightarrow [2]$:

$$x^2 + \left(\sqrt{2}\right)^2 = 29$$
$$x^2 = 27$$
$$x = \pm\sqrt{27} = \pm 3\sqrt{3}$$

$$\boxed{\{(3\sqrt{3},\ \sqrt{2})\},\ \{(-3\sqrt{3},\ \sqrt{2})\}}$$

$y = -\sqrt{2} \rightarrow [2]$:

$$x^2 + \left(-\sqrt{2}\right)^2 = 29$$
$$x^2 = 27$$
$$x = \pm\sqrt{27} = \pm 3\sqrt{3}$$

$$\boxed{\{(3\sqrt{3},\ -\sqrt{2})\},\ \{(-3\sqrt{3},\ -\sqrt{2})\}}$$

8. $xy = 4$ [1]
$x - 6y = 2$ [2]

Solve [2] for x :

$$x = 6y + 2$$

$x = 6y + 2 \rightarrow [1]$:

$$(6y + 2)\,y = 4$$
$$6y^2 + 2y - 4 = 0$$
$$2\,(3y - 2)\,(y + 1) = 0$$
$$y = \frac{2}{3},\ -1$$

$y = \frac{2}{3} \rightarrow [2]$:

$$x = 6\left(\frac{2}{3}\right) + 2$$
$$x = 4 + 2 = 6$$

$$\boxed{\left\{\left(6,\ \frac{2}{3}\right)\right\}}$$

$y = -1 \rightarrow [2]$:

$$x = 6\,(-1) + 2$$
$$x = -4$$

$$\boxed{\{(-4, -1)\}}$$

9. $x^2 + 2xy + y^2 = 4$ [1]
$x = 3y - 2$ [2]

$[2] \rightarrow [1]$:

$$(3y - 2)^2 + 2\,(3y - 2)\,y + y^2 = 4$$
$$9y^2 - 12y + 4 + 6y^2 - 4y + y^2 - 4 = 0$$
$$16y^2 - 16y = 0$$
$$16y\,(y - 1) = 0 \rightarrow y = 0,\ 1$$

$y = 0 \rightarrow [2] : x = 3\,(0) - 2 = -2$ $\boxed{\{(-2,\ 0)\}}$

$y = 1 \rightarrow [2] : x = 3\,(1) - 2 = 1$ $\boxed{\{(1,\ 1)\}}$

10. $x^2 + y^2 = 144$ [1]
$x + 2y = 8$ [2]

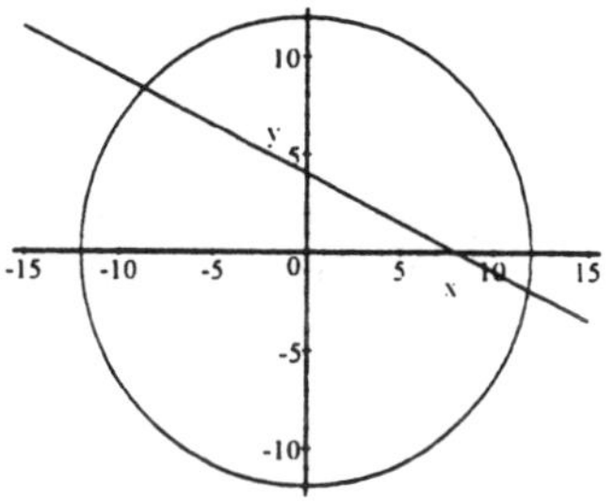

(a) Yes, they have points in common.

(b) $(11.8, -1.9)$, $(-8.6, 8.3)$

(c) $x = 8 - 2y$ [2]

$[2] \rightarrow [1]$:

$$(8 - 2y)^2 + y^2 = 144$$
$$64 - 32y + 4y^2 + y^2 - 144 = 0$$
$$5y^2 - 32y - 80 = 0$$

$$y = \frac{32 \pm \sqrt{1024 - 4(-400)}}{10} = \frac{32 \pm \sqrt{2624}}{10}$$
$$= \frac{32 \pm 8\sqrt{41}}{10} = \frac{16 \pm 4\sqrt{41}}{5}$$

$$x = 8 - 2\left(\frac{16 + 4\sqrt{41}}{5}\right)$$
$$= \frac{40 - 32 - 8\sqrt{41}}{5} = \frac{8 - 8\sqrt{41}}{5}$$

$$\boxed{\left(\frac{8 - 8\sqrt{41}}{5},\ \frac{16 + 4\sqrt{41}}{5}\right)}$$

If $y = \dfrac{16 - 4\sqrt{41}}{5}$:

$$\boxed{\left(\frac{8 + 8\sqrt{41}}{5},\ \frac{16 - 4\sqrt{41}}{5}\right)}$$

11. (a) $x^2 + y^2 = 2$

$y^2 = 2 - x^2$

$y = \pm\sqrt{2 - x^2}$

$y_1 = \sqrt{2 - x^2}, \quad y_2 = -\sqrt{2 - x^2}$

(b) $3x + y = 4$

$y_3 = -3x + 4$

(c) The standard viewing window ($[-10, 10]$ by $[-10, 10]$) should show the intersection; other settings are possible.

12. No, two linear equations in two variables cannot intersect in two points; they will intersect in at most one point. If the lines coincide, there are infinitely many solutions.

13. No, a system consisting of two equations in three variables cannot have a unique solution, because two distinct planes intersect in a line, so there are infinitely many solutions. If the planes coincide, there are infinitely many solutions. If the planes are parallel, there are no solutions.

14.
$$\begin{aligned} 2x - 3y + z &= -5 \quad [1] \\ x + 4y + 2z &= 13 \quad [2] \\ 5x + 5y + 3z &= 14 \quad [3] \end{aligned}$$

$$\left[\begin{array}{ccc|c} 2 & -3 & 1 & -5 \\ 1 & 4 & 2 & 13 \\ 5 & 5 & 3 & 14 \end{array}\right]$$

$$R_1 \leftrightarrow R_2 \rightarrow \left[\begin{array}{ccc|c} 1 & 4 & 2 & 13 \\ 2 & -3 & 1 & -5 \\ 5 & 5 & 3 & 14 \end{array}\right]$$

$$\left.\begin{array}{l} -2R_1 + R_2 \\ -5R_1 + R_3 \end{array}\right\} \rightarrow \left[\begin{array}{ccc|c} 1 & 4 & 2 & 13 \\ 0 & -11 & -3 & -31 \\ 0 & -15 & -7 & -51 \end{array}\right]$$

$$-\frac{1}{11}R_2 \rightarrow \left[\begin{array}{ccc|c} 1 & 4 & 2 & 13 \\ 0 & 1 & \frac{3}{11} & \frac{3}{11} \\ 0 & -15 & -7 & -51 \end{array}\right]$$

$$15R_2 + R_3 \rightarrow \left[\begin{array}{ccc|c} 1 & 4 & 2 & 13 \\ 0 & 1 & \frac{3}{11} & \frac{31}{11} \\ 0 & 0 & -\frac{32}{11} & -\frac{96}{11} \end{array}\right]$$

(i) $-\dfrac{32}{11}z = -\dfrac{96}{11}$ $\quad$ **z = 3**

(ii) $y + \dfrac{3}{11}z = \dfrac{31}{11}$

$y + \dfrac{3}{11}(3) = \dfrac{31}{11}$

$y = \dfrac{22}{11}$ $\quad$ **y = 2**

(iii) $x + 4y + 2z = 13$

$x + 4(2) + 2(3) = 13$ $\quad$ **x = −1**

$\boxed{\{(-1,\ 1,\ 3)\}}$

15.
$$\begin{aligned} x - 3y &= 12 \quad [1] \\ 2y + 5z &= 1 \quad [2] \\ 4x + z &= 25 \quad [3] \end{aligned}$$

$$\left[\begin{array}{ccc|c} 1 & -3 & 0 & 12 \\ 0 & 2 & 5 & 1 \\ 4 & 0 & 1 & 25 \end{array}\right]$$

$$-4R_1 + R_3 \rightarrow \left[\begin{array}{ccc|c} 1 & -3 & 0 & 12 \\ 0 & 2 & 5 & 1 \\ 1 & 12 & 1 & -23 \end{array}\right]$$

$$\frac{1}{2}R_2 \rightarrow \left[\begin{array}{ccc|c} 1 & -3 & 0 & 12 \\ 0 & 1 & 2.5 & .5 \\ 1 & 12 & 1 & -23 \end{array}\right]$$

$$-12R_2 + R_3 \rightarrow \left[\begin{array}{ccc|c} 1 & -3 & 0 & 12 \\ 0 & 1 & 2.5 & .5 \\ 0 & 0 & -29 & -29 \end{array}\right]$$

$$-\frac{1}{29}R_3 \rightarrow \left[\begin{array}{ccc|c} 1 & -3 & 0 & 12 \\ 0 & 1 & 2.5 & .5 \\ 0 & 0 & 1 & 1 \end{array}\right]$$

(i) $\mathbf{z = 1}$

(ii) $y + 2.5z = .5$
$y + 2.5(1) = .5 \qquad \mathbf{y = -2}$

(iii) $x - 3y = 12$
$x - 3(-2) = 12 \qquad \mathbf{x = 6}$

$\boxed{\{(6,\ -2,\ 1)\}}$

16.
$$\begin{aligned} x + y - z &= 5 \quad [1] \\ 2x + y + 3z &= 2 \quad [2] \\ 4x - y + 2z &= -1 \quad [3] \end{aligned}$$

$$\left[\begin{array}{ccc|c} 1 & 1 & -1 & 5 \\ 2 & 1 & 3 & 2 \\ 4 & -1 & 2 & -1 \end{array}\right]$$

$$\left.\begin{array}{r} -2R_1 + R_2 \\ -4R_1 + R_3 \end{array}\right\} \rightarrow \left[\begin{array}{ccc|c} 1 & 1 & -1 & 5 \\ 0 & -1 & 5 & -8 \\ 0 & -5 & 6 & -21 \end{array}\right]$$

$$-R_2 \rightarrow \left[\begin{array}{ccc|c} 1 & 1 & -1 & 5 \\ 0 & 1 & -5 & 8 \\ 0 & -5 & 6 & -21 \end{array}\right]$$

$$5R_2 + R_3 \rightarrow \left[\begin{array}{ccc|c} 1 & 1 & -1 & 5 \\ 0 & 1 & -5 & 8 \\ 0 & 0 & -19 & 19 \end{array}\right]$$

$$-\frac{1}{19}R_3 \rightarrow \left[\begin{array}{ccc|c} 1 & 1 & -1 & 5 \\ 0 & 1 & -5 & 8 \\ 0 & 0 & 1 & -1 \end{array}\right]$$

(i) $\mathbf{z = -1}$

(ii) $y - 5z = 8$
$y - 5(-1) = 8 \qquad \mathbf{y = 3}$

(iii) $x + y - z = 5$
$x + (3) - (-1) = 5 \qquad \mathbf{x = 1}$

$\boxed{\{(1,\ 3,\ -1)\}}$

17.
$$\begin{aligned} 5x - 3y + 2z &= -5 \quad [1] \\ 2x + y - z &= 4 \quad [2] \\ -4x - 2y + 2z &= -1 \quad [3] \end{aligned}$$

$$\left[\begin{array}{ccc|c} 5 & -3 & 2 & -5 \\ 2 & 1 & -1 & 4 \\ -4 & -2 & 2 & -1 \end{array}\right]$$

$$\frac{1}{5}R_1 \rightarrow \left[\begin{array}{ccc|c} 1 & -.6 & .4 & -1 \\ 2 & 1 & -1 & 4 \\ -4 & -2 & 2 & -1 \end{array}\right]$$

$$\left.\begin{array}{r} -2R_1 + R_2 \\ 4R_1 + R_3 \end{array}\right\} \rightarrow \left[\begin{array}{ccc|c} 1 & -.6 & .4 & -1 \\ 0 & 2.2 & -1.8 & 6 \\ 0 & -4.4 & 3.6 & -5 \end{array}\right]$$

$$\frac{5}{11}R_2 \rightarrow \left[\begin{array}{ccc|c} 1 & -.6 & .4 & -1 \\ 0 & 1 & -\frac{9}{11} & \frac{30}{11} \\ 0 & -4.4 & 3.6 & -5 \end{array}\right]$$

$$4.4R_2 + R_3 \rightarrow \left[\begin{array}{ccc|c} 1 & -.6 & .4 & -1 \\ 0 & 1 & -\frac{9}{11} & \frac{30}{11} \\ 0 & 0 & 0 & 7 \end{array}\right]$$

Inconsistent; $\emptyset$

18.
$$\begin{aligned} 2x + 3y &= 10 \\ -3x + y &= 18 \end{aligned} \qquad \left[\begin{array}{cc|c} 2 & 3 & 10 \\ -3 & 1 & 18 \end{array}\right]$$

$$\frac{1}{2}R_1 \rightarrow \left[\begin{array}{cc|c} 1 & \frac{3}{2} & 5 \\ -3 & 1 & 18 \end{array}\right]$$

$$3R_1 + R_2 \rightarrow \left[\begin{array}{cc|c} 1 & \frac{3}{2} & 5 \\ 0 & \frac{11}{2} & 33 \end{array}\right]$$

$$\frac{2}{11}R_2 \rightarrow \left[\begin{array}{cc|c} 1 & \frac{3}{2} & 5 \\ 0 & 1 & 6 \end{array}\right]$$

$$-\frac{3}{2}R_2 + R_1 \rightarrow \left[\begin{array}{cc|c} 1 & 0 & -4 \\ 0 & 1 & 6 \end{array}\right]$$

$\{(-4, 6)\}$

19.
$$\begin{aligned} 3x + y &= -7 \\ x - y &= -5 \end{aligned} \qquad \left[\begin{array}{cc|c} 3 & 1 & -7 \\ 1 & -1 & -5 \end{array}\right]$$

$$R_2 + R_1 \rightarrow \left[\begin{array}{cc|c} 4 & 0 & -12 \\ 1 & -1 & -5 \end{array}\right]$$

$$\frac{1}{4}R_1 \rightarrow \left[\begin{array}{cc|c} 1 & 0 & -3 \\ 1 & -1 & -5 \end{array}\right]$$

$$R_1 - R_2 \rightarrow \left[\begin{array}{cc|c} 1 & 0 & -3 \\ 0 & 1 & 2 \end{array}\right]$$

$\{(-3, 2)\}$

20.
$$\begin{aligned} x - z &= -3 \\ y + z &= 6 \\ 2x - 3z &= -9 \end{aligned} \qquad \left[\begin{array}{ccc|c} 1 & 0 & -1 & -3 \\ 0 & 1 & 1 & 6 \\ 2 & 0 & -3 & -9 \end{array}\right]$$

$$-2R_1 + R_3 \rightarrow \left[\begin{array}{ccc|c} 1 & 0 & -1 & -3 \\ 0 & 1 & 1 & 6 \\ 0 & 0 & -1 & -3 \end{array}\right]$$

$$-R_3 \rightarrow \left[\begin{array}{ccc|c} 1 & 0 & -1 & -3 \\ 0 & 1 & 1 & 6 \\ 0 & 0 & 1 & 3 \end{array}\right]$$

$$\left.\begin{array}{r} R_3 + R_1 \\ -R_3 + R_2 \end{array}\right\} \rightarrow \left[\begin{array}{ccc|c} 1 & 0 & 0 & 0 \\ 0 & 1 & 0 & 3 \\ 0 & 0 & 1 & 3 \end{array}\right]$$

$\{(0, 3, 3)\}$

21.
$$\begin{aligned} 2x - y + 4z &= -1 \\ -3x + 5y - z &= 5 \\ 2x + 3y + 2z &= 3 \end{aligned}$$

$$\left[\begin{array}{ccc|c} 2 & -1 & 4 & -1 \\ -3 & 5 & -1 & 5 \\ 2 & 3 & 2 & 3 \end{array}\right]$$

$$\left.\begin{array}{r} \frac{3}{2}R_1 + R_2 \\ -R_1 + R_3 \end{array}\right\} \rightarrow \left[\begin{array}{ccc|c} 2 & -1 & 4 & -1 \\ 0 & \frac{7}{2} & 5 & \frac{7}{2} \\ 0 & 4 & -2 & 4 \end{array}\right]$$

$$\frac{2}{7}R_2 \rightarrow \left[\begin{array}{ccc|c} 2 & -1 & 4 & -1 \\ 0 & 1 & \frac{10}{7} & 1 \\ 0 & 4 & -2 & 4 \end{array}\right]$$

$$-4R_2 + R_3 \rightarrow \left[\begin{array}{ccc|c} 2 & -1 & 4 & -1 \\ 0 & 1 & \frac{10}{7} & 1 \\ 0 & 0 & -\frac{54}{7} & 0 \end{array}\right]$$

$$\left.\begin{array}{r} R_2 + R_1 \\ -\frac{7}{54}R_3 \end{array}\right\} \rightarrow \left[\begin{array}{ccc|c} 2 & 0 & \frac{38}{7} & 0 \\ 0 & 1 & \frac{10}{7} & 1 \\ 0 & 0 & 1 & 0 \end{array}\right]$$

$$\left.\begin{array}{r} -\frac{38}{7}R_3 + R_1 \\ -\frac{10}{7}R_3 + R_2 \end{array}\right\} \rightarrow \left[\begin{array}{ccc|c} 2 & 0 & 0 & 0 \\ 0 & 1 & 0 & 1 \\ 0 & 0 & 1 & 0 \end{array}\right]$$

$$\frac{1}{2}R_1 \rightarrow \left[\begin{array}{ccc|c} 1 & 0 & 0 & 0 \\ 0 & 1 & 0 & 1 \\ 0 & 0 & 1 & 0 \end{array}\right]$$

$\{(0, 1, 0)\}$

22. $A = \begin{bmatrix} -5 & 4 & 9 \\ 2 & -1 & -2 \end{bmatrix}, \quad B = \begin{bmatrix} 1 & -2 & 7 \\ 4 & -5 & -5 \end{bmatrix}$

$$A + B = \begin{bmatrix} -5+1 & 4-2 & 9+7 \\ 2+4 & -1-5 & -2-5 \end{bmatrix} = \begin{bmatrix} -4 & 2 & 16 \\ 6 & -6 & -7 \end{bmatrix}$$

23. $\begin{bmatrix} 3 \\ 2 \\ 5 \end{bmatrix} - \begin{bmatrix} 8 \\ -4 \\ 6 \end{bmatrix} + \begin{bmatrix} 1 \\ 0 \\ 2 \end{bmatrix}$

$$= \begin{bmatrix} 3-8+1 \\ 2+4+0 \\ 5-6+2 \end{bmatrix} = \begin{bmatrix} -4 \\ 6 \\ 1 \end{bmatrix}$$

24. $\begin{bmatrix} 2 & 5 & 8 \\ 1 & 9 & 2 \end{bmatrix} - \begin{bmatrix} 3 & 4 \\ 7 & 1 \end{bmatrix}$ $\boxed{\emptyset}$

$(2 \times 3) - (2 \times 2)$; Cannot subtract

25. $3\begin{bmatrix} 2 & 4 \\ -1 & 4 \end{bmatrix} - 2\begin{bmatrix} 5 & 8 \\ 2 & -2 \end{bmatrix}$

$$= \begin{bmatrix} 6 & 12 \\ -3 & 12 \end{bmatrix} - \begin{bmatrix} 10 & 16 \\ 4 & -4 \end{bmatrix} = \begin{bmatrix} 6-10 & 12-16 \\ -3-4 & 12+4 \end{bmatrix} = \begin{bmatrix} -4 & -4 \\ -7 & 16 \end{bmatrix}$$

26. $-1\begin{bmatrix} 3 & -5 & 2 \\ 1 & 7 & -4 \end{bmatrix} + 5\begin{bmatrix} 0 & 2 \\ -1 & 3 \end{bmatrix}$ $\boxed{\emptyset}$

$(2 \times 3) + (2 \times 2)$; Cannot add

27. $10\begin{bmatrix} 2x+3y & 4x+y \\ x-5y & 6x+2y \end{bmatrix} + 2\begin{bmatrix} -3x-y & x+6y \\ 4x+2y & 5x-y \end{bmatrix}$

$$= \begin{bmatrix} 20x+30y-6x-2y & 40x+10y+2x+12y \\ 10x-50y+8x+4y & 60x+20y+10x-2y \end{bmatrix}$$

$$= \begin{bmatrix} 14x+28y & 42x+22y \\ 18x-46y & 70x+18y \end{bmatrix}$$

28. The sum of two $m \times n$ matrices is found by adding corresponding elements.

29. $A = \begin{bmatrix} 3 & -1 \\ 7 & 2 \end{bmatrix}, \quad B = \begin{bmatrix} -8 & 6 \\ 5 & 2 \end{bmatrix}$

$$BA = \begin{bmatrix} -8 & 6 \\ 5 & 2 \end{bmatrix} \cdot \begin{bmatrix} 3 & -1 \\ 7 & 2 \end{bmatrix} = \begin{bmatrix} -24+42 & 8+12 \\ 15+14 & -5+4 \end{bmatrix} = \begin{bmatrix} 18 & 20 \\ 29 & -1 \end{bmatrix}$$

30. $\begin{bmatrix} 3 & 2 & -1 \\ 4 & 0 & 6 \end{bmatrix}\begin{bmatrix} -2 & 0 \\ 0 & 2 \\ 3 & 1 \end{bmatrix}$

$$= \begin{bmatrix} -6+0-3 & 0+4-1 \\ -8+0+18 & 0+0+6 \end{bmatrix} = \begin{bmatrix} -9 & 3 \\ 10 & 6 \end{bmatrix}$$

31. $\begin{bmatrix} 1 & -2 & 4 & 2 \\ 0 & 1 & -1 & 8 \end{bmatrix}\begin{bmatrix} -1 \\ 2 \\ 0 \\ 1 \end{bmatrix}$

$$= \begin{bmatrix} -1-4+0+2 \\ 0+2+0+8 \end{bmatrix} = \begin{bmatrix} -3 \\ 10 \end{bmatrix}$$

32. $\begin{bmatrix} 1 & 2 & 5 \\ -3 & 4 & 7 \\ 0 & 2 & -1 \end{bmatrix}\begin{bmatrix} 4 & 2 & 3 \\ 10 & -5 & 6 \end{bmatrix}$

Cannot multiply: $\boxed{\emptyset}$

$[(3 \times 3) \times (2 \times 3)]$

33. $\begin{bmatrix} 4 & 2 & 3 \\ 10 & -5 & 6 \end{bmatrix}\begin{bmatrix} 1 & 2 & 5 \\ -3 & 4 & 7 \\ 0 & 2 & -1 \end{bmatrix}$

$$= \begin{bmatrix} 4-6+0 & 8+8+6 & 20+14-3 \\ 10+15+0 & 20-20+12 & 50-35-6 \end{bmatrix}$$

$$= \begin{bmatrix} -2 & 22 & 31 \\ 25 & 12 & 9 \end{bmatrix}$$

34. $\begin{bmatrix} 3 & -1 & 0 \end{bmatrix}\begin{bmatrix} 1 & 3 & 2 \\ 2 & -4 & 0 \\ 5 & 7 & 3 \end{bmatrix}$

$= \begin{bmatrix} 3-2+0 & 9+4+0 & 6+0+0 \end{bmatrix}$

$= \begin{bmatrix} 1 & 13 & 6 \end{bmatrix}$

35. Yes, A and B are inverses.

```
[A][B]
          [[1 0]
           [0 1]]
[B][A]
          [[1 0]
           [0 1]]
```

36. Yes

(i) $AB = \begin{bmatrix} 1 & 0 \\ 2 & -3 \end{bmatrix}\begin{bmatrix} 1 & 0 \\ \frac{2}{3} & -\frac{1}{3} \end{bmatrix}$

$= \begin{bmatrix} 1+0 & 0+0 \\ 2-2 & 0+1 \end{bmatrix}$

$= \begin{bmatrix} 1 & 0 \\ 0 & 1 \end{bmatrix}$

(ii) $BA = \begin{bmatrix} 1 & 0 \\ \frac{2}{3} & -\frac{1}{3} \end{bmatrix}\begin{bmatrix} 1 & 0 \\ 2 & -3 \end{bmatrix}$

$= \begin{bmatrix} 1+0 & 0+0 \\ \frac{2}{3}-\frac{2}{3} & 0+1 \end{bmatrix}$

$= \begin{bmatrix} 1 & 0 \\ 0 & 1 \end{bmatrix}$

37. No

$AB = \begin{bmatrix} 2 & 0 & 6 \\ 0 & 1 & 0 \\ 1 & 0 & 1 \end{bmatrix}\begin{bmatrix} -1 & 0 & \frac{3}{2} \\ 0 & 1 & 0 \\ \frac{1}{4} & 0 & -1 \end{bmatrix}$

$= \begin{bmatrix} -2+0+\frac{3}{2} & 0+0+0 & 3+0-6 \\ 0+0+0 & 0+1+0 & 0+0+0 \\ -1+0+\frac{1}{4} & 0+0+0 & \frac{3}{2}+0-1 \end{bmatrix}$

$= \begin{bmatrix} -\frac{1}{2} & 0 & -3 \\ 0 & 1 & 0 \\ -\frac{3}{4} & 0 & \frac{1}{2} \end{bmatrix}$

38. Yes

(i) $AB = \begin{bmatrix} 1 & 0 & 2 \\ 0 & 2 & 4 \\ 0 & 0 & 1 \end{bmatrix}\begin{bmatrix} 1 & 0 & -2 \\ 0 & \frac{1}{2} & -2 \\ 0 & 0 & 1 \end{bmatrix}$

$= \begin{bmatrix} 1+0+0 & 0+0+0 & -2+0+2 \\ 0+0+0 & 0+1+0 & 0-4+4 \\ 0+0+0 & 0+0+0 & 0+0+1 \end{bmatrix}$

$= \begin{bmatrix} 1 & 0 & 0 \\ 0 & 1 & 0 \\ 0 & 0 & 1 \end{bmatrix}$

(ii) $BA = \begin{bmatrix} 1 & 0 & -2 \\ 0 & \frac{1}{2} & -2 \\ 0 & 0 & 1 \end{bmatrix}\begin{bmatrix} 1 & 0 & 2 \\ 0 & 2 & 4 \\ 0 & 0 & 1 \end{bmatrix}$

$= \begin{bmatrix} 1+0+0 & 0+0+0 & 2+0-2 \\ 0+0+0 & 0+1+0 & 0+2-2 \\ 0+0+0 & 0+0+0 & 0+0+1 \end{bmatrix}$

$= \begin{bmatrix} 1 & 0 & 0 \\ 0 & 1 & 0 \\ 0 & 0 & 1 \end{bmatrix}$

39. A does not have an inverse; the determinant is equal to 0: A^{-1} does not exist.

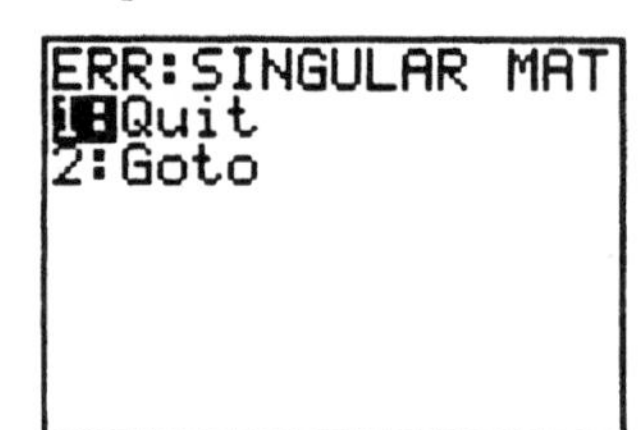

40. $\begin{bmatrix} -4 & 2 \\ 0 & 3 \end{bmatrix}^{-1} = \frac{1}{-12}\begin{bmatrix} 3 & -2 \\ 0 & -4 \end{bmatrix}$

$= \begin{bmatrix} -\frac{1}{4} & \frac{1}{6} \\ 0 & \frac{1}{3} \end{bmatrix}$

41. $\begin{bmatrix} 2 & 0 \\ -1 & 5 \end{bmatrix}^{-1} = \frac{1}{10}\begin{bmatrix} 5 & 0 \\ 1 & 2 \end{bmatrix}$

$= \begin{bmatrix} \frac{1}{2} & 0 \\ \frac{1}{10} & \frac{1}{5} \end{bmatrix}$

42. $\begin{bmatrix} 2 & 0 & 4 \\ 1 & -1 & 0 \\ 0 & 1 & -2 \end{bmatrix}^{-1}$ Calculator:

$$= \begin{bmatrix} .25 & .5 & .5 \\ .25 & -.5 & .5 \\ .125 & -.25 & -.25 \end{bmatrix}$$

or $\begin{bmatrix} \frac{1}{4} & \frac{1}{2} & \frac{1}{2} \\ \frac{1}{4} & -\frac{1}{2} & \frac{1}{2} \\ \frac{1}{8} & -\frac{1}{4} & -\frac{1}{4} \end{bmatrix}$

43. $\begin{bmatrix} 2 & -1 & 0 \\ 1 & 0 & 1 \\ 1 & -2 & 0 \end{bmatrix}^{-1}$ Calculator:

$$= \begin{bmatrix} \frac{2}{3} & 0 & -\frac{1}{3} \\ \frac{1}{3} & 0 & -\frac{2}{3} \\ -\frac{2}{3} & 1 & \frac{1}{3} \end{bmatrix}$$

44. $\begin{bmatrix} 2 & 3 & 5 \\ -2 & -3 & -5 \\ 1 & 4 & 2 \end{bmatrix}^{-1}$ (det = 0)

No inverse

45. $\begin{aligned} x + y &= 4 \\ 2x + 3y &= 10 \end{aligned}$ (det = 1)

$$\begin{bmatrix} x \\ y \end{bmatrix} = \begin{bmatrix} 1 & 1 \\ 2 & 3 \end{bmatrix}^{-1} \begin{bmatrix} 4 \\ 10 \end{bmatrix}$$

$$= \begin{bmatrix} 3 & -1 \\ -2 & 1 \end{bmatrix} \begin{bmatrix} 4 \\ 10 \end{bmatrix}$$

$$= \begin{bmatrix} 2 \\ 2 \end{bmatrix} \quad \boxed{\{(2,\ 2)\}}$$

46. $\begin{aligned} 5x - 3y &= -2 \\ 2x + 7y &= -9 \end{aligned}$ (det = 41)

$$\begin{bmatrix} x \\ y \end{bmatrix} = \begin{bmatrix} 5 & -3 \\ 2 & 7 \end{bmatrix}^{-1} \begin{bmatrix} -2 \\ -9 \end{bmatrix}$$

$$= \begin{bmatrix} \frac{7}{41} & \frac{3}{41} \\ -\frac{2}{41} & \frac{5}{41} \end{bmatrix} \begin{bmatrix} -2 \\ -9 \end{bmatrix}$$

$$= \begin{bmatrix} -1 \\ -1 \end{bmatrix} \quad \boxed{\{(-1, -1)\}}$$

47. $\begin{aligned} 2x + y &= 5 \\ 3x - 2y &= 4 \end{aligned}$ (det = −7)

$$\begin{bmatrix} x \\ y \end{bmatrix} = \begin{bmatrix} 2 & 1 \\ 3 & -2 \end{bmatrix}^{-1} \begin{bmatrix} 5 \\ 4 \end{bmatrix}$$

$$= \begin{bmatrix} \frac{2}{7} & \frac{1}{7} \\ \frac{3}{7} & -\frac{2}{7} \end{bmatrix} \begin{bmatrix} 5 \\ 4 \end{bmatrix}$$

$$= \begin{bmatrix} 2 \\ 1 \end{bmatrix} \quad \boxed{\{(2,\ 1)\}}$$

48. $\begin{aligned} x - 2y &= 7 \\ 3x + y &= 7 \end{aligned}$ (det = 7)

$$\begin{bmatrix} x \\ y \end{bmatrix} = \begin{bmatrix} 1 & -2 \\ 3 & 1 \end{bmatrix}^{-1} \begin{bmatrix} 7 \\ 7 \end{bmatrix}$$

$$= \begin{bmatrix} \frac{1}{7} & \frac{2}{7} \\ -\frac{3}{7} & \frac{1}{7} \end{bmatrix} \begin{bmatrix} 7 \\ 7 \end{bmatrix}$$

$$= \begin{bmatrix} 3 \\ -2 \end{bmatrix} \quad \boxed{\{(3, -2)\}}$$

49. $\begin{aligned} x + 2y &= -1 \\ 3y - z &= -5 \\ x + 2y - z &= -3 \end{aligned}$

$$\begin{bmatrix} x \\ y \\ z \end{bmatrix} = \begin{bmatrix} 1 & 2 & 0 \\ 0 & 3 & -1 \\ 1 & 2 & -1 \end{bmatrix}^{-1} \begin{bmatrix} -1 \\ -5 \\ -3 \end{bmatrix}$$

$$= \begin{bmatrix} 1 \\ -1 \\ 2 \end{bmatrix} \quad \boxed{\{(1, -1,\ 2)\}}$$

50.
$$\begin{aligned} 3x-2y+4z &= 1 \\ 4x+y-5z &= 2 \\ -6x+4y-8z &= -2 \end{aligned}$$

$$\det \begin{bmatrix} 3 & -2 & 4 \\ 4 & 1 & -5 \\ -6 & 4 & -8 \end{bmatrix} = 0$$

$$\left[\begin{array}{ccc|c} 3 & -2 & 4 & 1 \\ 4 & 1 & -5 & 2 \\ -6 & 4 & -8 & -2 \end{array}\right]$$

$$2R_1+R_3 \rightarrow \left[\begin{array}{ccc|c} 3 & -2 & 4 & 1 \\ 4 & 1 & -5 & 2 \\ 0 & 0 & 0 & 0 \end{array}\right]$$

∞ solutions, dependent equations.

$$R_1+2R_2 \rightarrow \left[\begin{array}{ccc|c} 3 & -2 & 4 & 1 \\ 11 & 0 & -6 & 5 \\ 0 & 0 & 0 & 0 \end{array}\right]$$

(i) $11x-6z=5$

$11x=6z+5$

$$x=\frac{6z+5}{11}$$

(ii) $3\left(\frac{6z+5}{11}\right)-2y+4z=1$

$18z+15-22y+44z=11$

$-22y=-62z-4$

$$y=\frac{62z+4}{22}=\frac{31z+2}{11}$$

$$\boxed{\left(\frac{6z+5}{11},\ \frac{31z+2}{11},\ z\right)}$$

51.
$$\begin{aligned} x+y+z &= 1 \\ 2x-y &= -2 \\ 3y+z &= 2 \end{aligned}$$

$$\begin{bmatrix} x \\ y \\ z \end{bmatrix} = \begin{bmatrix} 1 & 1 & 1 \\ 2 & -1 & 0 \\ 0 & 3 & 1 \end{bmatrix}^{-1} \begin{bmatrix} 1 \\ -2 \\ 2 \end{bmatrix}$$

$$= \begin{bmatrix} -1 \\ 0 \\ 2 \end{bmatrix} \qquad \boxed{\{(-1,\ 0,\ 2)\}}$$

52.
$$\begin{aligned} x &= -3 \\ y+z &= 6 \\ 2x-3z &= -9 \end{aligned}$$

$$\begin{bmatrix} x \\ y \\ z \end{bmatrix} = \begin{bmatrix} 1 & 0 & 0 \\ 0 & 1 & 1 \\ 2 & 0 & -3 \end{bmatrix}^{-1} \begin{bmatrix} -3 \\ 6 \\ -9 \end{bmatrix}$$

$$= \begin{bmatrix} -3 \\ 5 \\ 1 \end{bmatrix} \qquad \boxed{\{(-3,\ 5,\ 1)\}}$$

53. $\det \begin{bmatrix} -1 & 8 \\ 2 & 9 \end{bmatrix} = -9-16 = \boxed{-25}$

54. $\det \begin{bmatrix} -2 & 4 \\ 0 & 3 \end{bmatrix} = -6-0 = \boxed{-6}$

55. $\det \begin{bmatrix} -2 & 4 & 1 \\ 3 & 0 & 2 \\ -1 & 0 & 3 \end{bmatrix}$

Expand on 2nd column:

$$= -4 \begin{vmatrix} 3 & 2 \\ -1 & 3 \end{vmatrix} + 0 - 0$$

$$= -4(9+2) = -4(11) = \boxed{-44}$$

56. $\det \begin{bmatrix} -1 & 2 & 3 \\ 4 & 0 & 3 \\ 5 & -1 & 2 \end{bmatrix}$

Expand on 2nd column:

$$= -2 \begin{vmatrix} 4 & 3 \\ 5 & 2 \end{vmatrix} - (-1) \begin{vmatrix} -1 & 3 \\ 4 & 3 \end{vmatrix}$$

$$= -2(8-15)+1(-3-12)$$

$$= -2(-7)+1(-15) = 14-15 = \boxed{-1}$$

57. $\det \begin{bmatrix} -3 & 2 \\ 1 & x \end{bmatrix} = 5$

$-3x-2=5$

$-3x=7 \qquad \boxed{\left\{-\frac{7}{3}\right\}}$

58. $\det\begin{bmatrix} 3x & 7 \\ -x & 4 \end{bmatrix} = 8$

$12x + 7x = 8$

$19x = 8$ $\boxed{\left\{\frac{8}{19}\right\}}$

59. $\det\begin{bmatrix} 2 & 5 & 0 \\ 1 & 3x & -1 \\ 0 & 2 & 0 \end{bmatrix} = 4$

Expand on 3rd column:

$0 - (-1)\begin{vmatrix} 2 & 5 \\ 0 & 2 \end{vmatrix} + 0 = 4$

$4 = 4$

Any real number will work.

$\boxed{\{\text{All real numbers}\}}$

60. $\det\begin{bmatrix} 6x & 2 & 0 \\ 1 & 5 & 3 \\ x & 2 & -1 \end{bmatrix} = 2x$

Expand on 1st row:

$6x\begin{vmatrix} 5 & 3 \\ 2 & -1 \end{vmatrix} - 2\begin{vmatrix} 1 & 3 \\ x & -1 \end{vmatrix} + 0 = 2x$

$6x(-5-6) - 2(-1-3x) = 2x$
$-66x + 2 + 6x = 2x$
$-60x + 2 = 2x$
$-62x = -2$
$x = \frac{-2}{-62}$ $\boxed{\left\{\frac{1}{31}\right\}}$

61. $3x - y = 28$
$2x + y = 2$

(a) $D = \det\begin{bmatrix} 3 & -1 \\ 2 & 1 \end{bmatrix} = 5$

(b) $D_x = \det\begin{bmatrix} 28 & -1 \\ 2 & 1 \end{bmatrix} = 30$

(c) $D_y = \det\begin{bmatrix} 3 & 28 \\ 2 & 2 \end{bmatrix} = -50$

(d) $x = \frac{D_x}{D} = \frac{30}{5} = 6$

$y = \frac{D_y}{D} = \frac{-50}{5} = -10$ $\boxed{\{(6, -10)\}}$

62. $3x - y = 28$
$2x + y = 2$

(a) $A = \begin{bmatrix} 3 & -1 \\ 2 & 1 \end{bmatrix}$, **(b)** $B = \begin{bmatrix} 28 \\ 2 \end{bmatrix}$

(c) To solve for x and y, multiply A^{-1} by B, resulting in $x = 6$, $y = -10$.

$\begin{bmatrix} x \\ y \end{bmatrix} = A^{-1}B$

63. If $D = 0$, then there would be division by zero, which is undefined. When $D = 0$, there are either no solutions or infinitely many solutions.

64. $3x + y = -1$
$5x + 4y = 10$

$D = \det\begin{bmatrix} 3 & 1 \\ 5 & 4 \end{bmatrix} = 7$

$D_x = \det\begin{bmatrix} -1 & 1 \\ 10 & 4 \end{bmatrix} = -14$

$D_y = \det\begin{bmatrix} 3 & -1 \\ 5 & 10 \end{bmatrix} = 35$

$x = \frac{D_x}{D} = \frac{-14}{7} = -2$

$y = \frac{D_y}{D} = \frac{35}{7} = 5$ $\boxed{\{(-2, 5)\}}$

65. $3x + 7y = 2$
$5x - y = -22$

$D = \det\begin{bmatrix} 3 & 7 \\ 5 & -1 \end{bmatrix} = -38$

$D_x = \det\begin{bmatrix} 2 & 7 \\ -22 & -1 \end{bmatrix} = 152$

$D_y = \det\begin{bmatrix} 3 & 2 \\ 5 & -22 \end{bmatrix} = -76$

$x = \frac{D_x}{D} = \frac{152}{-38} = -4$

$y = \frac{D_y}{D} = \frac{-76}{-38} = 2$ $\boxed{\{(-4, 2)\}}$

66. $2x - 5y = 8$
$3x + 4y = 10$

$$D = \det\begin{bmatrix} 2 & -5 \\ 3 & 4 \end{bmatrix} = 23$$

$$D_x = \det\begin{bmatrix} 8 & -5 \\ 10 & 4 \end{bmatrix} = 82$$

$$D_y = \det\begin{bmatrix} 2 & 8 \\ 3 & 10 \end{bmatrix} = -4$$

$$x = \frac{D_x}{D} = \frac{82}{23}$$

$$y = \frac{D_y}{D} = \frac{-4}{23}$$

$$\boxed{\left\{\left(\frac{82}{23}, -\frac{4}{23}\right)\right\}}$$

67. $3x + 2y + z = 2$
$4x - y + 3z = -16$
$x + 3y - z = 12$

$$D = \det\begin{bmatrix} 3 & 2 & 1 \\ 4 & -1 & 3 \\ 1 & 3 & -1 \end{bmatrix} = 3$$

$$D_x = \det\begin{bmatrix} 2 & 2 & 1 \\ -16 & -1 & 3 \\ 12 & 3 & -1 \end{bmatrix} = -12$$

$$D_y = \det\begin{bmatrix} 3 & 2 & 1 \\ 4 & -16 & 3 \\ 1 & 12 & -1 \end{bmatrix} = 18$$

$$D_z = \det\begin{bmatrix} 3 & 2 & 2 \\ 4 & -1 & -16 \\ 1 & 3 & 12 \end{bmatrix} = 6$$

$$x = \frac{D_x}{D} = \frac{-12}{3} = -4$$

$$y = \frac{D_y}{D} = \frac{18}{3} = 6$$

$$z = \frac{D_z}{D} = \frac{6}{3} = 2$$

$$\boxed{\{(-4, 6, 2)\}}$$

68.
$$\begin{aligned} 5x - 2y - z &= 8 & [1] \\ -5x + 2y + z &= -8 & [2] \\ x - 4y - 2z &= 0 & [3] \end{aligned}$$

$D = \det(\text{coeff}) = 0$

Interchange equations:

$$\left[\begin{array}{ccc|c} 1 & -4 & -2 & 0 \\ 5 & -2 & -1 & 8 \\ -5 & 2 & 1 & -8 \end{array}\right] \begin{matrix} [3] \\ [1] \\ [2] \end{matrix}$$

$R_2 + R_3 \rightarrow 0\ 0\ 0 = 0$

Dependent; ∞ solutions

$$R_1 + 2R_3 \rightarrow \left[\begin{array}{ccc|c} 1 & -4 & -2 & 0 \\ 5 & -2 & -1 & 8 \\ -9 & 0 & 0 & -16 \end{array}\right]$$

(i) $-9x = -16 \rightarrow x = \dfrac{16}{9}$

(ii) $\frac{16}{9} - 4y - 2z = 0$

$16 - 36y - 18z = 0$

$-36y = 18z - 16$

$$y = \frac{18z - 16}{-36} = \frac{8 - 9z}{18}$$

$$\boxed{\left\{\left(\frac{16}{9}, \frac{8 - 9z}{18}, z\right)\right\}}$$

69. $-x + 3y - 4z = 2$
$2x + 4y + z = 3$
$3x - z = 9$

$$D = \det\begin{bmatrix} -1 & 3 & -4 \\ 2 & 4 & 1 \\ 3 & 0 & -1 \end{bmatrix} = 67$$

$$D_x = \det\begin{bmatrix} 2 & 3 & -4 \\ 3 & 4 & 1 \\ 9 & 0 & -1 \end{bmatrix} = 172$$

$$D_y = \det\begin{bmatrix} -1 & 2 & -4 \\ 2 & 3 & 1 \\ 3 & 9 & -1 \end{bmatrix} = -14$$

$$D_z = \det\begin{bmatrix} -1 & 3 & 2 \\ 2 & 4 & 3 \\ 3 & 0 & 9 \end{bmatrix} = -87$$

$$x = \frac{D_x}{D} = \frac{172}{67}$$

$$y = \frac{D_y}{D} = \frac{-14}{67}$$

$$z = \frac{D_z}{D} = \frac{-87}{67}$$

$$\boxed{\left\{\left(\frac{172}{67}, -\frac{14}{67}, -\frac{87}{67}\right)\right\}}$$

70. Let $x =$ the amount of rice in cups and $y =$ the amount of soybeans in cups.

$$15x + 22.5y = 9.5 \quad [1]$$
$$810x + 270y = 324 \quad [2]$$

$-12 \times [1] + [2]:$

$$-180x - 270y = -114$$
$$810x + 270y = 324$$
$$630x = 210 \quad \rightarrow \quad x = \frac{1}{3}$$

$$x \rightarrow [1]: \ 15\left(\frac{1}{3}\right) + 22.5y = 9.5$$
$$5 + 22.5y = 9.5$$
$$22.5y = 4.5$$
$$y = .20 \quad \left(\frac{1}{5}\right)$$

$\boxed{\frac{1}{3} \text{ cup of rice, } \frac{1}{5} \text{ cup of soybeans}}$

71. Let $x =$ # of CDs and $y =$ #of 3.5 in diskettes.

$$x + y = 100 \quad [1]$$
$$.40x + .30y = 38.00 \quad [2]$$

$-30 \times [1] + 100 \times [2]:$

$$-30x - 30y = -3000$$
$$40x + 30y = 3800$$
$$10x = 800 \quad \rightarrow \quad x = 80$$

$$x \rightarrow [1]: 80 + y = 100 \quad \rightarrow \quad y = 20$$

$\boxed{\text{80 CDs, 20 3.5 inch diskettes}}$

72. Let $x =$ # lbs of \$4.60 tea
$y =$ # lbs of \$5.75 tea
$z =$ # lbs of \$6.50 tea

$$x + y + z = 20 \quad [1]$$
$$4.6x + 5.75y + 6.5z = 20(5.25) = 105 \quad [2]$$
$$x = y + z \quad [3]$$

Rewrite $[3]: x - y - z = 0$

$$A = \begin{bmatrix} 1 & 1 & 1 \\ 4.6 & 5.75 & 6.5 \\ 1 & -1 & -1 \end{bmatrix}$$

$$\begin{bmatrix} x \\ y \\ z \end{bmatrix} = A^{-1} \times \begin{bmatrix} 20 \\ 105 \\ 0 \end{bmatrix}$$

$$x = 10, \ y = 8, \ z = 2$$

$\boxed{\text{\$4.60: 10 pounds; \$5.75: 8 pounds; \$6.50: 2 pounds}}$

73. Let $x =$ amount of 5% solution (in ml)
$y =$ amount of 15% solution (in ml)
$z =$ amount of 10% solution (in ml)

$$x + y + z = 20 \quad [1]$$
$$.05x + .15y + .10z = .08(20) = 1.6 \quad [2]$$
$$x = y + z + 2 \quad [3]$$

Rewrite [2] and [3] in standard form:

$$x - y - z = 2$$
$$5x + 15y + 10z = 160$$

$$A = \begin{bmatrix} 1 & 1 & 1 \\ 5 & 15 & 10 \\ 1 & -1 & -1 \end{bmatrix}$$

$$\begin{bmatrix} x \\ y \\ z \end{bmatrix} = A^{-1} \times \begin{bmatrix} 20 \\ 160 \\ 2 \end{bmatrix}$$

$$x = 11, \ y = 3, \ z = 6$$

$\boxed{\text{5\%: 11 ml; 15\%: 3 ml; 10\%: 6 ml}}$

74. (a) $P = a + bA + cW$

From the table:

$$113 = a + 39b + 142c$$
$$138 = a + 53b + 181c$$
$$152 = a + 65b + 191c$$

$$\text{RREF}\left[\begin{array}{ccc|c} 1 & 39 & 142 & 113 \\ 1 & 53 & 181 & 138 \\ 1 & 65 & 191 & 152 \end{array}\right]$$

$$\approx \left[\begin{array}{ccc|c} 1 & 0 & 0 & 32.780488 \\ 0 & 1 & 0 & .9024390 \\ 0 & 0 & 1 & .3170732 \end{array}\right]$$

$$a \approx 32.78, \ b \approx .9024, \ c \approx .3171$$

$\boxed{P \approx 32.78 + .9024A + .3171W}$

(b) $P \approx 32.78 + .9024(55) + .3171(175)$

$\boxed{P \approx 138}$

75. $P(x) = ax^2 + bx + c$

$$\begin{aligned}(-6, 4) &: 36a - 6b + c = 4\\(-4, -2) &: 16a - 4b + c = -2\\(2, 4) &: 4a + 2b + c = 4\end{aligned}$$

$$D = \det\begin{bmatrix} 36 & -6 & 1 \\ 16 & -4 & 1 \\ 4 & 2 & 1 \end{bmatrix} = -96$$

$$D_a = \det\begin{bmatrix} 4 & -6 & 1 \\ -2 & -4 & 1 \\ 4 & 2 & 1 \end{bmatrix} = -48$$

$$D_b = \det\begin{bmatrix} 36 & 4 & 1 \\ 16 & -2 & 1 \\ 4 & 4 & 1 \end{bmatrix} = -192$$

$$D_c = \det\begin{bmatrix} 36 & -6 & 4 \\ 16 & -4 & -2 \\ 4 & 2 & 4 \end{bmatrix} = 192$$

$$a = \frac{D_a}{D} = \frac{-48}{-96} = \frac{1}{2}$$

$$b = \frac{D_b}{D} = \frac{-192}{-96} = 2$$

$$c = \frac{D_c}{D} = \frac{192}{-96} = -2$$

$$\boxed{P(x) = \frac{1}{2}x^2 + 2x - 2}$$

76. $P(x) = ax^3 + bx^2 + cx + d$

$$\begin{aligned}(-2, 1) &: -8a + 4b - 2c + d = 1\\(-1, 6) &: -a + b - c + d = 6\\(2, 9) &: 8a + 4b + 2c + d = 9\\(3, 26) &: 27a + 9b + 3c + d = 26\end{aligned}$$

$$\text{RREF}\left[\begin{array}{cccc|c} -8 & 4 & -2 & 1 & 1 \\ -1 & 1 & -1 & 1 & 6 \\ 8 & 4 & 2 & 1 & 9 \\ 27 & 9 & 3 & 1 & 26 \end{array}\right]$$

$$= \left[\begin{array}{cccc|c} 1 & 0 & 0 & 0 & 1 \\ 0 & 1 & 0 & 0 & 0 \\ 0 & 0 & 1 & 0 & -2 \\ 0 & 0 & 0 & 1 & 5 \end{array}\right]$$

$a = 1,\ b = 0,\ c = -2,\ d = 5$

$$\boxed{P(x) = x^3 - 2x + 5}$$

77. $x + y \le 6$

$2x - y \ge 3$

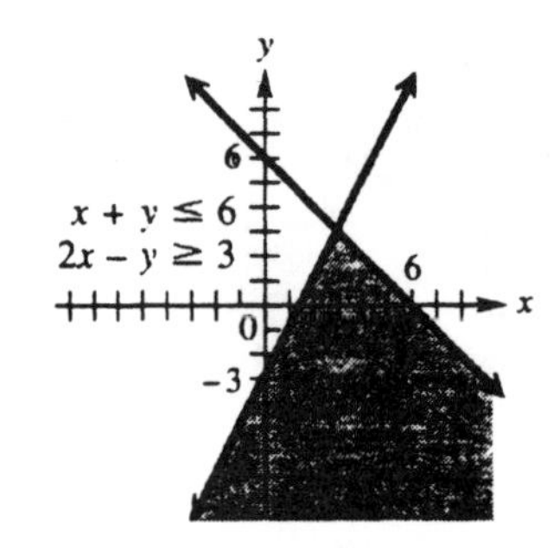

78. $y \le \frac{1}{3}x - 2$

$y^2 \le 16 - x^2$

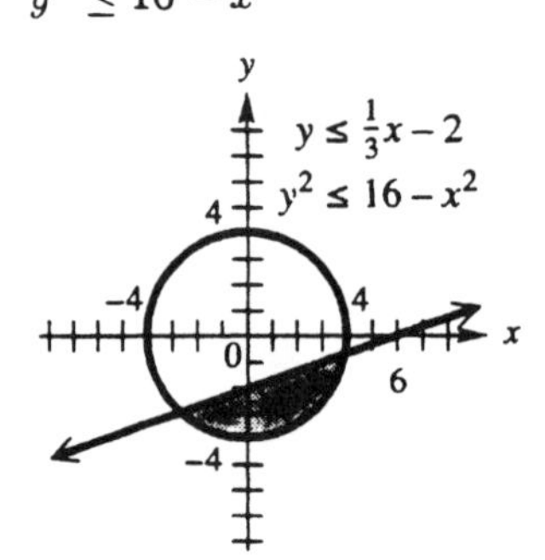

79. $3x + 2y \le 12$

$5x + y \ge 5$

$x \ge 0, \quad y \ge 0$

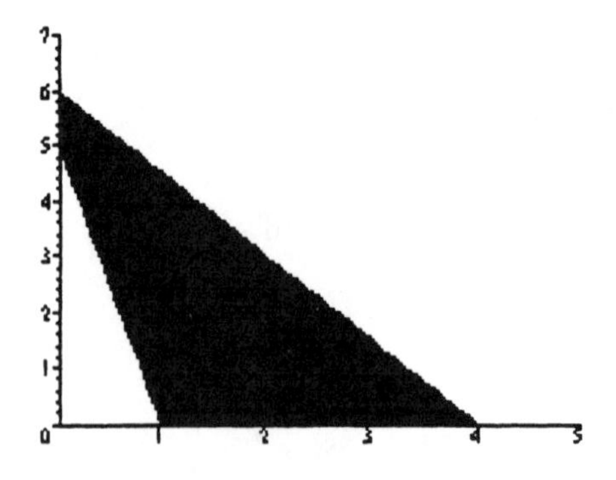

	$2x + 4y$
$(0, 5)$:	20
$(0, 6)$:	24
$(1, 0)$:	2
$(4, 0)$:	8

Maximum value is 24 at $(0, 6)$.

80. $x + y \le 50$
$2x + y \ge 20$
$x + 2y \ge 30$
$x \ge 0, \quad y \ge 0$

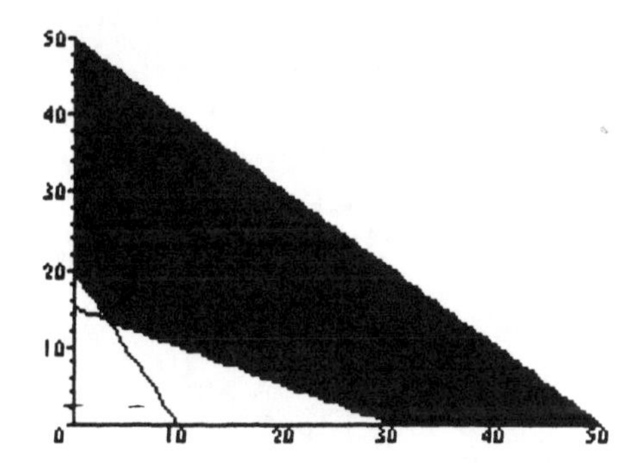

Point of intersection:

$$2x + y = 20$$
$$x + 2y = 30$$

$$\begin{array}{r} 2x + y = 20 \\ -2x - 4y = -60 \\ \hline \end{array}$$

$$-3y = -40 \rightarrow y = \frac{40}{3}$$

$$2x + \frac{40}{3} = 20$$
$$6x + 40 = 60$$
$$6x = 20 \rightarrow x = \frac{10}{3}$$

	$4x + 2y$
$(0, 50)$:	100
$(0, 20)$:	40
$\left(\frac{10}{3}, \frac{40}{3}\right)$:	40
$(30, 0)$:	120
$(50, 0)$:	200

Minimum value of 40 at $(0, 20)$ or $\left(\frac{10}{3}, \frac{40}{3}\right)$.

[Note: values on the line $2x + y = 20$ also give minimums of 40 between the two points.]

81. Let x represent the number of radios produced daily. Let y represent the number of CD players produced daily.

$5 \le x \le 25$
$0 \le y \le 30$
$x \le y$

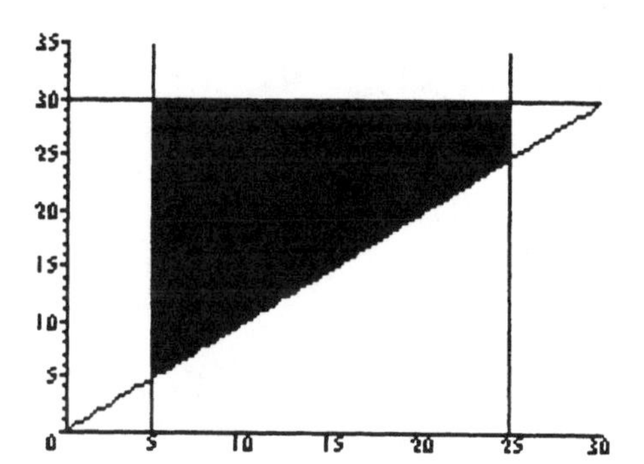

Point of intersection:

$x = 5,\ y = 30 \rightarrow (5,\ 30)$
$x = 5,\ y = x \rightarrow (5,\ 5)$
$x = 25,\ y = 30 \rightarrow (25,\ 30)$
$x = 25,\ y = x \rightarrow (25,\ 25)$

	$15x + 35y$
$(5,\ 30)$:	1125
$(5,\ 5)$:	250
$(25,\ 30)$:	1425
$(25,\ 25)$:	1250

Maximum value of 1425 at (25, 30)

Maximum profit is \$1425, when 25 radios and 30 CD players are manufactured.

82. $\dfrac{5x - 2}{x^2 - 4} = \dfrac{A}{x - 2} + \dfrac{B}{x + 2}$

Multiply by $(x - 2)(x + 2)$:

$$5x - 2 = A(x + 2) + B(x - 2)$$

Let $x = -2$: $\quad -12 = B(-4) \rightarrow B = 3$

Let $x = 2$: $\quad 8 = A(4) \rightarrow A = 2$

$$\boxed{\frac{2}{x - 2} + \frac{3}{x + 2}}$$

83. $\dfrac{x+2}{x^3+2x^2+x}=\dfrac{A}{x}+\dfrac{B}{x+1}+\dfrac{C}{(x+1)^2}$

Multiply by $x(x+1)^2$:

$$x+2=A(x+1)^2+Bx(x+1)+Cx$$

Let $x=0$: $2=A(1)\to A=2$

Let $x=-1$: $1=C(-1)\to C=-1$

Let $x=1$, $A=2$, $C=-1$:

$$3=2(4)+B(1)(2)-1$$
$$3=8+2B-1$$
$$B=-2$$

$$\boxed{\frac{2}{x}+\frac{-2}{x+1}+\frac{-1}{(x+1)^2}}$$

84. $\dfrac{x+2}{x^3-x^2+4x}=\dfrac{A}{x}+\dfrac{Bx+C}{x^2-x+4}$

Multiply by $x(x^2-x+4)$:

$$x+2=A(x^2-x+4)+(Bx+C)(x)$$

$$x+2=Ax^2-Ax+4A+Bx^2+Cx$$

Let $x=0$: $2=4A$

$$A=\frac{1}{2}$$

Set x^2 coefficients equal:

$$0=A+B=\frac{1}{2}+B$$

$$B=-\frac{1}{2}$$

Set x coefficients equal:

$$1=-A+C=-\frac{1}{2}+C$$

$$C=\frac{3}{2}$$

$$\frac{\frac{1}{2}}{x}+\frac{-\frac{1}{2}x+\frac{3}{2}}{x^2-x+4}=\frac{\frac{1}{2}}{x}+\frac{\frac{1}{2}(-x+3)}{x^2-x+4}$$

$$\boxed{\frac{1}{2x}+\frac{-x+3}{2(x^2-x+4)}}$$

Chapter 7 Test

1. $x^2-4y^2=-15$
   $3x+y=1$

   (a) The first is the equation of a hyperbola; the second is the equation of a line.

   (b) The two graphs could intersect in 0, 1, or 2 points.

   (c) $y=1-3x$

$$x^2-4(1-3x)^2=-15$$
$$x^2-4(1-6x+9x^2)=-15$$
$$x^2-4+24x-36x^2=-15$$
$$-35x^2+24x+11=0$$
$$(35x+11)(-x+1)=0$$
$$x=-\tfrac{11}{35},\quad 1$$

$$x=1\to y: y=1-3=-2$$

$$x=-\tfrac{11}{35}\to y: y=1-3\left(-\tfrac{11}{35}\right)=\tfrac{68}{35}$$

$$\boxed{\left(-\frac{11}{35},\ \frac{68}{35}\right),\quad (1,-2)}$$

   (d) There are two points of intersection.

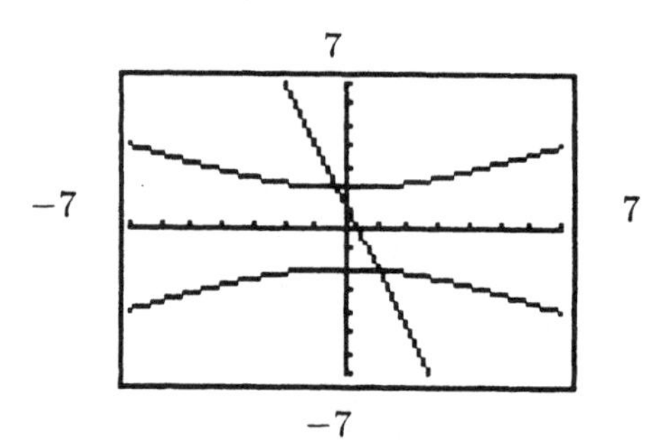

2. $\begin{aligned}2x+\ y+z&=3\\ x+2y-z&=3\\ 3x-\ y+z&=5\end{aligned}\longrightarrow\left[\begin{array}{rrr|r}2&1&1&3\\1&2&-1&3\\3&-1&1&5\end{array}\right]$

$$R_1\longleftrightarrow R_2:\left[\begin{array}{rrr|r}1&2&-1&3\\2&1&1&3\\3&-1&1&5\end{array}\right]$$

$$\left.\begin{array}{r}-2R_1+R_2\\-3R_1+R_3\end{array}\right\}\to\left[\begin{array}{rrr|r}1&2&-1&3\\0&-3&3&-3\\0&-7&4&-4\end{array}\right]$$

$$7R_2-3R_3\to\left[\begin{array}{rrr|r}1&2&-1&3\\0&-3&3&-3\\0&0&9&-9\end{array}\right]$$

continued

2. continued

(i) $9z = -9 \rightarrow z = -1$

(ii) $-3y + 3(-1) = -3$

$-3y = 0 \rightarrow y = 0$

(iii) $x + 2y - z = 3$

$x + 2(0) - (-1) = 3 \rightarrow x = 2$

Solution set: $\boxed{(2,\ 0,\ -1)}$

3. (a) $3\begin{bmatrix} 2 & 3 \\ 1 & -4 \\ 5 & 9 \end{bmatrix} - \begin{bmatrix} -2 & 6 \\ 3 & -1 \\ 0 & 8 \end{bmatrix}$

$$= \begin{bmatrix} 6 & 9 \\ 3 & -12 \\ 15 & 27 \end{bmatrix} - \begin{bmatrix} -2 & 6 \\ 3 & -1 \\ 0 & 8 \end{bmatrix}$$

$$= \begin{bmatrix} 6+2 & 9-6 \\ 3-3 & -12+1 \\ 15-0 & 27-8 \end{bmatrix}$$

$$= \begin{bmatrix} 8 & 3 \\ 0 & -11 \\ 15 & 19 \end{bmatrix}$$

(b) $\begin{bmatrix} 1 \\ 2 \end{bmatrix} + \begin{bmatrix} 4 \\ -6 \end{bmatrix} + \begin{bmatrix} 2 & 8 \\ -7 & 5 \end{bmatrix}$

$(1 \times 2) + (1 \times 2) + (2 \times 2)$

Cannot add $\boxed{\emptyset}$

(c) $\begin{bmatrix} 2 & 1 & -3 \\ 4 & 0 & 5 \end{bmatrix} \begin{bmatrix} 1 & 3 \\ 2 & 4 \\ 3 & -2 \end{bmatrix}$

$$= \begin{bmatrix} 2+2-9 & 6+4+6 \\ 4+0+15 & 12+0-10 \end{bmatrix}$$

$$= \begin{bmatrix} -5 & 16 \\ 19 & 2 \end{bmatrix}$$

4. $A:\ n \times n; \quad B:\ n \times n$

(a) AB can be found; it will be $n \times n$.

(b) BA can be found; it will be $n \times n$.

(c) Does $AB = BA$?
Not necessarily, since matrix multiplication is not commutative.

(d) $A:\ n \times n; \quad C:\ m \times n$

AC cannot be found.
CA can be found; it will be $m \times n$.

5. (a) $\det \begin{bmatrix} 4 & 9 \\ -5 & -11 \end{bmatrix} = -44 + 45 = \boxed{1}$

(b) $\det \begin{bmatrix} 2 & 0 & 8 \\ -1 & 7 & 9 \\ 12 & 5 & -3 \end{bmatrix}$

Expand on 2nd column:

$$= 0\begin{vmatrix} -1 & 9 \\ 12 & -3 \end{vmatrix} + 7\begin{vmatrix} 2 & 8 \\ 12 & -3 \end{vmatrix} - 5\begin{vmatrix} 2 & 8 \\ -1 & 9 \end{vmatrix}$$

$$= 0 + 7(-6 - 96) - 5(18 + 8)$$

$$= 7(-102) - 5(26)$$

$$= \boxed{-844}$$

6. $2x - 3y = -33$
$4x + 5y = 11$

$$D = \det\begin{bmatrix} 2 & -3 \\ 4 & 5 \end{bmatrix} = 22$$

$$D_x = \det\begin{bmatrix} -33 & -3 \\ 11 & 5 \end{bmatrix} = -132$$

$$D_y = \det\begin{bmatrix} 2 & -33 \\ 4 & 11 \end{bmatrix} = 154$$

$$x = \frac{D_x}{D} = \frac{-132}{22} = -6$$

$$y = \frac{D_y}{D} = \frac{154}{22} = 7$$

$\boxed{(-6,\ 7)}$

7. $x + y - z = -4$ [1]
$2x - 3y - z = 5$ [2]
$x + 2y + 2z = 3$ [3]

$$A = \begin{bmatrix} 1 & 1 & -1 \\ 2 & -3 & -1 \\ 1 & 2 & 2 \end{bmatrix}$$

$$X = \begin{bmatrix} x \\ y \\ z \end{bmatrix} \quad B = \begin{bmatrix} -4 \\ 5 \\ 3 \end{bmatrix}$$

(b) $$A^{-1} = \begin{bmatrix} \frac{1}{4} & \frac{1}{4} & \frac{1}{4} \\ \frac{5}{16} & -\frac{3}{16} & \frac{1}{16} \\ -\frac{7}{16} & \frac{1}{16} & \frac{5}{16} \end{bmatrix}$$

(c) $$\begin{bmatrix} x \\ y \\ z \end{bmatrix} = A^{-1} \times B \quad \boxed{\{(1, -2, 3)\}}$$

(d) $.5x + y + z = 1.5$ [1]
$2x - 3y - z = 5$ [2]
$x + 2y + 2z = 3$ [3]

$$A = \begin{bmatrix} .5 & 1 & 1 \\ 2 & -3 & -1 \\ 1 & 2 & 2 \end{bmatrix}$$

$\det A = 0$; A^{-1} does not exist.

8. $f(x) = ax^2 + bx + c$

$(24, 48.9) : 24^2a + 24b + c = 48.9$
$(60, 62.8) : 60^2a + 60b + c = 62.8$
$(96, 48.8) : 96^2a + 96b + c = 48.8$

$$\text{RREF} \left[\begin{array}{ccc|c} 24^2 & 24 & 1 & 48.9 \\ 60^2 & 60 & 1 & 62.8 \\ 96^2 & 96 & 1 & 48.8 \end{array}\right]$$

$$\approx \left[\begin{array}{ccc|c} 1 & 0 & 0 & -.010764 \\ 0 & 1 & 0 & 1.2903 \\ 0 & 0 & 1 & 24.133 \end{array}\right]$$

$a \approx -.010764,\ b \approx 1.2903,\ c \approx 24.133$

$$\boxed{f(x) = -.010764x^2 + 1.2903x + 24.133}$$

[163.8, 103] by [46.4, 64.2]

$f(x) \approx -.010764x^2 + 1.2903x + 24.133$

75

20 100

45

9. B: $y > 2 - x$
$y < x^2 - 5$

A: Above line, inside parabola
B: Above line, outside parabola
C: Below line, outside parabola
D: Below line, inside parabola

10. Let $x = \#$ of type X cabinets,
$y = \#$ of type Y cabinets:

$100x + 200y \leq 1400$
$6x + 8y \leq 72$
$x \geq 0, \quad y \geq 0$

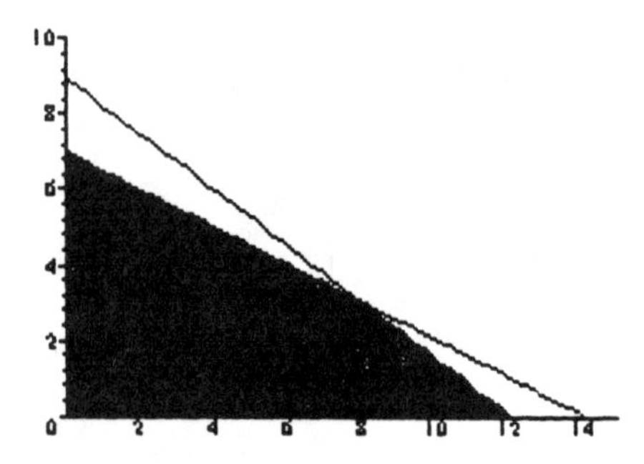

Point of intersection:

$100x + 200y = 1400 \rightarrow$
$x + 2y = 14$ [1]
$6x + 8y = 72$ [2]

$-6x - 12y = -84$ [$-6 \times$ [1]]
$6x + 8y = 72$ [2]
$-4y = -12 \rightarrow y = 3$

$y = 3 \rightarrow$ [1] :
$x + 2(3) = 14 \rightarrow x = 8$
(8, 3)

	$8x + 12y$
(0, 0) :	0
(12, 0) :	96
(8, 3) :	100
(0, 7) :	84

Maximum storage space is 100 cubic feet when there are 8 cabinets of type X and 3 cabinets of type Y.

11. $\dfrac{7x-1}{x^2-x-6}=\dfrac{A}{x-3}+\dfrac{B}{x+2}$

Multiply by $(x-3)(x+2)$:

$$7x-1=A(x+2)+B(x-3)$$

Let $x=-2$: $-15=B(-5)\rightarrow B=3$

Let $x=3$: $20=A(5)\rightarrow A=4$

$$\boxed{\frac{4}{x-3}+\frac{3}{x+2}}$$

12. $\dfrac{x^2-11x+6}{(x+2)(x-2)^2}=\dfrac{A}{x+2}+\dfrac{B}{x-2}+\dfrac{C}{(x-2)^2}$

Multiply by $(x+2)(x-2)^2$:

$$x^2-11x+6$$
$$=A(x-2)^2+B(x-2)(x+2)+C(x+2)$$

Let $x=-2$: $32=A(16)\rightarrow A=2$

Let $x=2$: $-12=C(4)\rightarrow C=-3$

Let $x=0$, $A=2$, $C=-3$:

$$6=4A-4B+2C$$
$$6=8-4B-6$$
$$6=2-4B$$
$$4=-4B\rightarrow B=-1$$

$$\boxed{\frac{2}{x+2}+\frac{-1}{x-2}+\frac{-3}{(x-2)^2}}$$

Chapter 7 Project

Finding a polynomial that goes through any number of given points.

1. Using, for example, 585:

$(1,5)$, $(2,8)$, $(3,5)$

$ax^2+bx+c=y$

System of equations:

$$a(1)^2+b(1)+c=5$$
$$a(2)^2+b(2)+c=8$$
$$a(3)^2+b(3)+c=5$$

$$[A]=\begin{bmatrix}1&1&1\\4&2&1\\9&3&1\end{bmatrix};\ [B]=\begin{bmatrix}5\\8\\5\end{bmatrix}$$

$$[A]^{-1}[B]=[X]$$

$$[X]=\begin{bmatrix}-3\\12\\-4\end{bmatrix}$$

Equation: $-3x^2+12x-4=y$

Follow the procedures outlined to obtain the scatterplot and the graph of the polynomial.

2. – 3. Answers will vary. Make sure students follow the steps outlined in the explanation for these activities.

Section 8.1

1. (a) Complement of $30°$:

$$90° - 30° = \boxed{60°}$$

(b) Supplement of $30°$:

$$180° - 30° = \boxed{150°}$$

3. (a) Complement of $45°$:

$$90° - 45° = \boxed{45°}$$

(b) Supplement of $45°$:

$$180° - 45° = \boxed{135°}$$

5. (a) Complement of $\frac{\pi}{4}$:

$$\frac{\pi}{2} - \frac{\pi}{4} = \frac{2\pi - \pi}{4} = \boxed{\frac{\pi}{4}}$$

(b) Supplement of $\frac{\pi}{4}$:

$$\pi - \frac{\pi}{4} = \frac{4\pi - \pi}{4} = \boxed{\frac{3\pi}{4}}$$

7. (a) Complement of x :

$$90° - x° = \boxed{(90 - x)°}$$

(b) Supplement of x :

$$180° - x° = \boxed{(180 - x)°}$$

9. The hour hand is at 25 minutes; the minute hand is at 60 minutes.

$$\frac{25 \text{ min}}{60 \text{ min}} = \frac{x \text{ degrees}}{360 \text{ degrees}} \Longrightarrow x = \boxed{150°}$$

11. The two angles are supplementary:

$$\begin{aligned} 7x + 11x &= 180 \\ 18x &= 180 \\ x &= 10 \end{aligned}$$

The measures of the two angles are:

$$\begin{aligned} 7x &= 7(10) = 70° \\ 11x &= 11(10) = 110° \end{aligned}$$

13. The two angles are supplementary:

$$\begin{aligned} (6x - 4) + (8x - 12) &= 180 \\ 14x - 16 &= 180 \\ 14x &= 196 \\ x &= 14 \end{aligned}$$

The measures of the two angles are:

$$\begin{aligned} 6x - 4 &= 6(14) - 4 = 80° \\ 8x - 12 &= 8(14) - 12 = 100° \end{aligned}$$

15. $62° \, 18' + 21° \, 41' = 83° \, 59'$

17. $71° \, 18' - 47° \, 29' = 70° \, 78' - 47° \, 29' = 23° \, 49'$

19. $90° - 72° \, 58' \, 11'' = 89° \, 59' \, 60'' - 72° \, 58' \, 11'' = 17° \, 01' \, 49''$

21. $20° \, 54' = \left(20 + \frac{54}{60}\right)^{\circ} = 20.9°$

23. $91° \, 35' \, 54'' = \left(91 + \frac{35}{60} + \frac{54}{3600}\right)^{\circ} \approx 91.598°$

25.
$$\begin{aligned} 31.4296° &= 31° + (.4296)\,60' \\ &= 31° \, 25.776' \\ &= 31° \, 25' + (.776)\,60'' \\ &\approx 31° \, 25' \, 47'' \end{aligned}$$

27.
$$\begin{aligned} 89.9004° &= 89° + (.9004)\,60' \\ &= 89° \, 54.024' \\ &= 89° \, 54' + (.024)\,60'' \\ &\approx 89° \, 54' \, 1'' \end{aligned}$$

Note: Exercises 29 - 33; other answers are possible.

29. $75°$; $435°$; $-285°$; Quadrant I

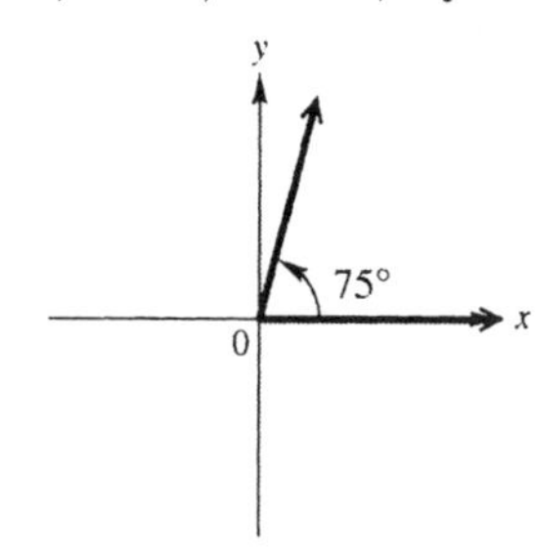

31. 174°; 534°; −186°; Quadrant II

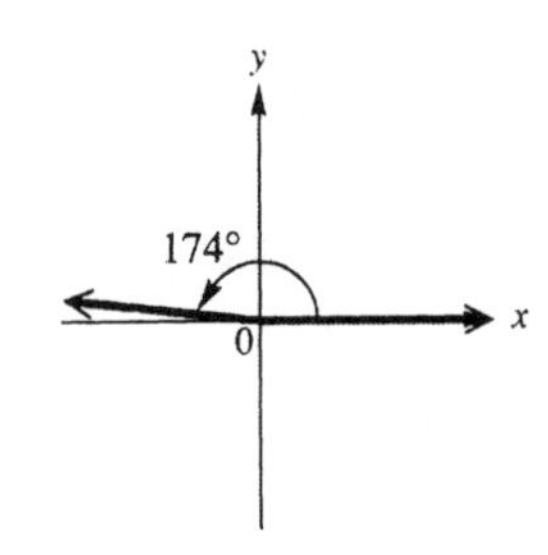

33. −61°; 299°; −421°; Quadrant IV

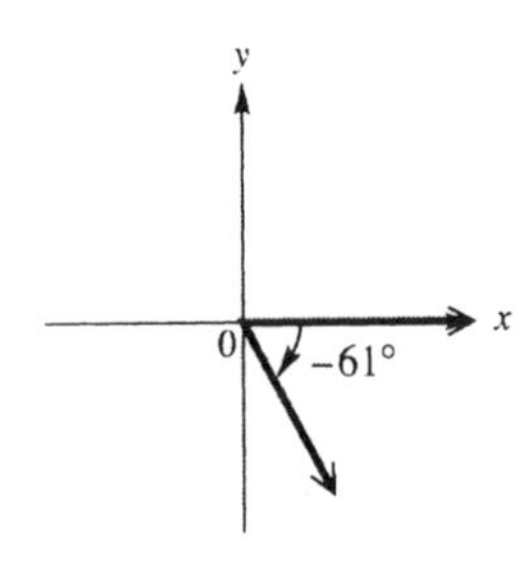

35. −40° coterminal to **320°**
(−40° + 360°)

37. 450° coterminal to **90°**
(450° − 360°)

39. $-\frac{\pi}{4}$ coterminal to $\boldsymbol{\frac{7\pi}{4}}$
$\left(-\frac{\pi}{4}+2\pi\right)$

41. $-\frac{3\pi}{2}$ coterminal to $\boldsymbol{\frac{\pi}{2}}$
$\left(-\frac{3\pi}{2}+2\pi\right)$

43. 30° ; $30° + n \cdot 360°,\ n \in \{\text{Integers}\}$

45. −90° ; $-90° + n \cdot 360°,\ n \in \{\text{Integers}\}$

47. $\frac{\pi}{4}$; $\frac{\pi}{4} + 2n\pi,\ n \in \{\text{Integers}\}$

49. $-\frac{3\pi}{4}$; $-\frac{3\pi}{4} + 2n\pi,\ n \in \{\text{Integers}\}$

51. $\frac{60°}{1} \cdot \frac{\pi}{180°} = \frac{60\pi}{180} = \frac{\pi}{3}$

$\boxed{60° = \frac{\pi}{3}}$

53. $\frac{150°}{1} \cdot \frac{\pi}{180°} = \frac{150\pi}{180} = \frac{5\pi}{6}$

$\boxed{150° = \frac{5\pi}{6}}$

55. $\frac{\pi}{3} \cdot \frac{180°}{\pi} = \boxed{60°}$

57. $\frac{7\pi}{4} \cdot \frac{180°}{\pi} = \frac{7\,(90°)}{2\pi} = \boxed{315°}$

59. $\frac{11\pi}{6} \cdot \frac{180°}{\pi} = 11\,(30°) = \boxed{330°}$

61. $\frac{39°}{1} \cdot \frac{\pi}{180°} = \frac{39\pi}{180} = \frac{13\pi}{60} \approx \boxed{.68}$

63. $\frac{139°\ 10'}{1} \cdot \frac{\pi}{180°} = \frac{139°\ 10'\pi}{180°} \approx \boxed{2.43}$

65. $\frac{64.29°}{1} \cdot \frac{\pi}{180°} = \frac{64.29\pi}{180} \approx \boxed{1.12}$

67. $2 \cdot \frac{180°}{\pi} \approx 114.591559° \approx \boxed{144°\ 35'}$

69. $1.74 \cdot \frac{180°}{\pi} \approx 99.69465635° \approx \boxed{99°\ 42'}$

71.

$30° = \frac{\pi}{6}$ radians	$210° = \frac{7\pi}{6}$ radians
$45° = \frac{\pi}{4}$ radians	$225° = \frac{5\pi}{4}$ radians
$60° = \frac{\pi}{3}$ radians	$240° = \frac{4\pi}{3}$ radians
$120° = \frac{2\pi}{3}$ radians	$300° = \frac{5\pi}{3}$ radians
$135° = \frac{3\pi}{4}$ radians	$315° = \frac{7\pi}{4}$ radians
$150° = \frac{5\pi}{6}$ radians	$330° = \frac{11\pi}{6}$ radians
$180° = \pi$ radians	

73. $s = r\theta = 4 \cdot \dfrac{\pi}{2} = \boxed{2\pi}$

75. $r = \dfrac{s}{\theta} = \dfrac{6\pi}{3\pi/4} = \dfrac{6\pi}{1} \cdot \dfrac{4}{3\pi} = \boxed{8}$

77. $\theta = \dfrac{s}{r} = \dfrac{3}{3} = \boxed{1}$

79. $r = 12.3$ cm, $\theta = \dfrac{2\pi}{3}$

$$\begin{aligned} s &= r\theta \\ &= 12.3\left(\frac{2\pi}{3}\right) \\ &\approx 25.7610598 \\ &\approx \boxed{25.8 \text{ cm}} \end{aligned}$$

81. $r = 4.82$ m, $\theta = 60°$

$$\begin{aligned} s &= r\theta \\ &= 4.82\left(\frac{60}{1} \cdot \frac{\pi}{180}\right) \\ &\approx 5.04749 \\ &\approx \boxed{5.05 \text{ m}} \end{aligned}$$

83. Madison: 44° N,
Dallas: 33° N

$$\theta = 44° - 33° = 11°$$

$$\begin{aligned} s &= r\theta \\ &= 6400\left(\frac{11}{1} \cdot \frac{\pi}{180}\right) \\ &\approx 1228.71179 \\ &\approx \boxed{1200 \text{ km}} \end{aligned}$$

85. New York: 41° N,
Lima: 12° S

$$\theta = 41° + 12° = 53°$$

$$\begin{aligned} s &= r\theta \\ &= 6400\left(\frac{53}{1} \cdot \frac{\pi}{180}\right) \\ &\approx 5920.1568 \\ &\approx \boxed{5900 \text{ km}} \end{aligned}$$

87. Pulley: 9.27 inches, angle: $71°50'$.

$$\begin{aligned} 71\frac{50}{60} \cdot \frac{\pi}{180} &= \frac{4310}{60} \cdot \frac{\pi}{180} \\ &\approx 1.253728 \end{aligned}$$

(a) $$\begin{aligned} s &= r\theta \\ &= (9.27)(1.253728) \\ &\approx \boxed{11.6 \text{ inches}} \end{aligned}$$

(b) If $s = 6$ in, $r = 9.27$ in,

$$\begin{aligned} \theta &= \frac{s}{r} = \frac{6}{9.27} \\ &= 0.647249 \text{ radians} \\ &= (0.647249)\left(\frac{180}{\pi}\right) \\ &\approx 37.084636° \approx \boxed{37°05'} \end{aligned}$$

89. smaller wheel: $60.0° = \dfrac{\pi}{3}$

$$s = r\theta = 5.23 \cdot \frac{\pi}{3} = \frac{5.23\pi}{3} \approx 5.477$$

larger wheel:

$$\theta = \frac{s}{r} \approx \frac{5.477}{8.16} \cdot \frac{180°}{\pi} \approx \boxed{38.5°}$$

91. The chain moves the same distance as the arc length of the larger gear ($180° = \pi$):

$$s = r\theta = 4.72\pi$$

The small gear rotates through an angle of

$$\theta = \frac{s}{r} = \frac{4.72\pi}{1.38} \approx 3.42\pi$$

θ is the same for the wheel and the small gear;

wheel: $$\begin{aligned} s &= r\theta \approx (13.6)(3.42\pi) \\ &\approx \boxed{146 \text{ inches}} \end{aligned}$$

93. $r = 29.2$ m, $\theta = \dfrac{5\pi}{6}$

$$\begin{aligned} &= 5\frac{1}{2}r^2\theta \\ &= \frac{1}{2}(29.2)^2\left(\frac{5\pi}{6}\right) \\ &\approx 1116.103 \approx \boxed{1120 \text{ m}^2} \end{aligned}$$

95. $r = 12.7$ cm, $\theta = 81°$

$$A = \frac{1}{2}r^2\theta$$
$$= \frac{1}{2}(12.7)^2\left(\frac{81}{1}\cdot\frac{\pi}{180}\right)$$
$$\approx 114.009 \approx \boxed{114 \text{ cm}^2}$$

97. $r = 6$, $s = 2\pi$

$$\theta = \frac{s}{r} = \frac{2\pi}{6} = \frac{\pi}{3}$$
$$A = \frac{1}{2}r^2\theta$$
$$= \frac{1}{2}(6)^2\left(\frac{\pi}{3}\right) = \frac{36\pi}{6} = \boxed{6\pi}$$

99. radius $= 240$ miles

$$A = \frac{1}{2}r^2\theta$$
$$= \frac{1}{2}(240)^2\left(\frac{1}{48}\cdot 2\pi\right)$$
$$\approx 3769.9 \approx \boxed{3700 \text{ mi}^2}$$

101. $A = A_L - A_S$

$$= \frac{1}{2}(10)^2\left(\frac{95\pi}{180}\right) - \frac{1}{2}(3)^2\left(\frac{95\pi}{180}\right)$$
$$= \frac{475\pi}{18} - \frac{95\pi}{40}$$
$$\approx 75.442 \approx \boxed{75.4 \text{ in}^2}$$

103. Diameter $= 50$ m, 26 sectors:

(a) $\theta = \left(\frac{360}{26}\right)^\circ = \left(\frac{180}{13}\right)^\circ \approx \boxed{13.85°}$

(b) $A = \frac{1}{2}r^2\theta$

$$= \frac{1}{2}(25)^2\left(\frac{180}{13}\cdot\frac{\pi}{180}\right)$$
$$\approx 75.519 \approx \boxed{76 \text{ m}^2}$$

105. The area of a circle of radius r is given by $A = \pi r^2$.

107. $\omega = 15$ radians/second, $t = 1$ second, $r = 13$ inches $= \frac{13}{12}$ feet.

$$\theta = \omega t = 15\,(1) = 15$$
$$v = r\omega = \frac{13}{12}\cdot 15 = \boxed{16.25 \text{ ft/sec}}$$
$$v = \frac{16.25 \text{ ft}}{1 \text{ sec}}\cdot\frac{1 \text{ mi}}{5280 \text{ ft}}\cdot\frac{3600 \text{ sec}}{1 \text{ hr}}$$
$$= 11.0795 \approx \boxed{11.1 \text{ mi/hr}}$$

109. If $r = 12.96$ cm, $s = 56$ cm, $t = 18$ sec:

$$\theta = \frac{s}{r} = \frac{56}{12.96} \text{ radians}$$
$$\omega = \frac{\theta}{t} = \left(\frac{56}{12.96}\right)\left(\frac{1}{18}\right)$$
$$= \boxed{.24 \text{ radian/second}}$$

111. 5000 rev/min, $r = 5$ in

$$\omega = \frac{\theta}{t} = \frac{2\pi}{1}\cdot\frac{5000}{60}$$
$$= \boxed{\frac{500\pi}{3} \text{ radians/second}}$$
$$v = r\omega = 5\left(\frac{500\pi}{3}\right)$$
$$= \frac{2500\pi}{3} \text{ inches/second}$$

113. (a) $\theta = \frac{1}{365}(2\pi) = \boxed{\frac{2\pi}{365} \text{ radian}}$

(b) $\omega = \frac{2\pi}{365}$ radian/day

$$= \frac{2\pi}{365}\cdot\frac{1}{24} \text{ radian/hour}$$
$$= \boxed{\frac{\pi}{4380} \text{ radian/hour}}$$

(c) $v = r\omega$

$$= 93{,}000{,}000\left(\frac{\pi}{4380}\right)$$
$$\approx 66{,}705$$
$$\approx \boxed{66{,}700 \text{ miles/hour}}$$

115. $r = \frac{s}{\theta} = \frac{496 \text{ miles}}{(7°\,12')\left(\frac{\pi}{180°}\right)} \approx \boxed{3947 \text{ miles}}$

$$C = 2\pi r \approx 2\pi\,(3947) \approx \boxed{24{,}800 \text{ miles}}$$

Section 8.2

1. $-\pi$ corresponds to $(-1,\ 0)$.

$\cos s = -1, \quad \sin s = 0$

undefined: $\csc s,\ \cot s$

3. 2π corresponds to $(1,\ 0)$.

$\cos s = 1, \quad \sin s = 0$

undefined: $\csc s,\ \cot s$

5. $\frac{3\pi}{2}$ corresponds to $(0,\ -1)$.

$\cos s = 0, \quad \sin s = -1$

undefined: $\sec s,\ \tan s$

7. $\frac{\pi}{2}$ corresponds to $(0,\ 1)$.

$\cos s = 0, \quad \sin s = 1$

undefined: $\sec s,\ \tan s$

9. $s = -5\pi$

$\cos s = -1$	$\sin s = 0$
$\tan s = 0$	$\cot s$ is undefined
$\sec s = -1$	$\csc s$ is undefined

11. $s = 6\pi$

$\cos s = 1$	$\sin s = 0$
$\tan s = 0$	$\cot s$ is undefined
$\sec s = 1$	$\csc s$ is undefined

13. $s = -\frac{5\pi}{2}$

$\cos s = 0$	$\sin s = -1$
$\tan s$ is undefined	$\cot s = 0$
$\sec s$ is undefined	$\csc s = -1$

15. $s = \frac{5\pi}{2}$

$\cos s = 0$	$\sin s = 1$
$\tan s$ is undefined	$\cot s = 0$
$\sec s$ is undefined	$\csc s = 1$

17. Quadrant I

$\cos .75 \approx .7316888689$

$\sin .75 \approx .681638760$

$\tan .75 \approx .9315964599$

$\cot .75 = (\tan .75)^{-1} \approx 1.073426149$

$\sec .75 = (\cos .75)^{-1} \approx 1.366701125$

$\csc .75 = (\sin .75)^{-1} \approx 1.467052724$

19. Quadrant II

$\cos -4.25 \approx -.4460874899$

$\sin -4.25 \approx .8949893582$

$\tan -4.25 \approx -2.006309028$

$\cot -4.25 = (\tan -4.25)^{-1} \approx -.4984277029$

$\sec -4.25 = (\cos -4.25)^{-1} \approx -2.241712719$

$\csc -4.25 = (\sin -4.25)^{-1} \approx 1.117331721$

21. Quadrant III

$\cos -2.25 \approx -.6281736227$

$\sin -2.25 \approx -.77807319689$

$\tan -2.25 \approx 1.2386276162$

$\cot -2.25 = (\tan -2.25)^{-1} \approx .80734515112$

$\sec -2.25 = (\cos -2.25)^{-1} \approx -1.591916572$

$\csc -2.25 = (\sin -2.25)^{-1} \approx -1.285226125$

23. Quadrant IV

$\cos 5.5 \approx .7086697743$

$\sin 5.5 \approx -.70554032557$

$\tan 5.5 \approx -.99558405221$

$\cot 5.5 = (\tan 5.5)^{-1} \approx -1.0044355349$

$\sec 5.5 = (\cos 5.5)^{-1} \approx 1.4110944706$

$\csc 5.5 = (\sin 5.5)^{-1} \approx -1.4173534294$

Relating Concepts (Exercises 25 - 30)

Answers will vary depending on the name used.

25. If your first name has 5 letters, $s = 5$.

$\cos 5 \approx .28366218546$

26. If your last name has 7 letters, $n = 7$.

$$\cos(5 + 2(7)\pi) = \cos(5 + 14\pi) \approx .28366218546$$

27. The results of **25** and **26** are the same. The real numbers s and $s + 2n\pi$ correspond to the same point on the unit circle, because its circumference is 2π.

28. If your last name has 6 letters, $s = 6$.
$\sin 6 \approx -.279415498199$

29. If your first name has 4 letters, $n = 4$.

$$\sin(6 + 2(4)\pi) = \sin(6 + 8\pi) \approx -.279415498199$$

30. The results of **28** and **29** are the same. The real numbers s and $s + 2n\pi$ correspond to the same point on the unit circle, because its circumference is 2π.

Note: Exercises 31 - 41 refer to Figure 36.

31. $\boxed{\sin \frac{7\pi}{6} = -\frac{1}{2}}$ Since $s = \frac{7\pi}{6}$ has coordinates $\left(-\frac{\sqrt{3}}{2}, -\frac{1}{2}\right)$, $\sin s = y$.

33. $\boxed{\tan \frac{3\pi}{4} = -1}$ Since $s = \frac{3\pi}{4}$ has coordinates $\left(-\frac{\sqrt{2}}{2}, \frac{\sqrt{2}}{2}\right)$, $\tan s = \frac{y}{x}$

35. $\boxed{\sec \frac{2\pi}{3} = -2}$ Since $s = \frac{2\pi}{3}$ has coordinates $\left(-\frac{1}{2}, \frac{\sqrt{3}}{2}\right)$, $\sec s = \frac{1}{x}$.

37. $\boxed{\cot \frac{5\pi}{6} = -\sqrt{3}}$ Since $s = \frac{5\pi}{6}$ has coordinates $\left(-\frac{\sqrt{3}}{2}, \frac{1}{2}\right)$, $\cot s = \frac{x}{y}$.
$(-\sqrt{3} \approx -1.732050808)$

39. $\boxed{\sin\left(-\frac{5\pi}{6}\right) = -\frac{1}{2}}$

Since $-\frac{5\pi}{6} = \frac{7\pi}{6}$, and $s = \frac{7\pi}{6}$ has coordinates $\left(-\frac{\sqrt{3}}{2}, -\frac{1}{2}\right)$, $\sin s = y$.

41. $\boxed{\sec \frac{23\pi}{6} = \frac{2\sqrt{3}}{3}}$

Since $\frac{23\pi}{6} = \frac{11\pi}{6}$, and $s = \frac{11\pi}{6}$ has coordinates $\left(\frac{\sqrt{3}}{2}, -\frac{1}{2}\right)$, $\sec s = \frac{1}{x}$.
$\left(\frac{2\sqrt{3}}{3} \approx 1.154700538\right)$

43. $\cos\left(-\frac{13\pi}{6}\right) = \cos\left(-\frac{\pi}{6}\right) = \cos \frac{11\pi}{6} = \frac{\sqrt{3}}{2}$

45. $\tan\left(-\frac{13\pi}{4}\right) = \tan\left(-\frac{5\pi}{4}\right) = \tan \frac{3\pi}{4} = -1$

47. $\left(\frac{3}{5}, y\right) \quad y > 0$

$$\cos^2 s + \sin^2 s = 1$$

$$\left(\frac{3}{5}\right)^2 + y^2 = 1$$

$$y^2 = 1 - \frac{9}{25} = \frac{16}{25} \qquad \boxed{y = \frac{4}{5}}$$

$\cos s = x = \frac{3}{5}$ $\qquad \cot s = \frac{x}{y} = \frac{3}{4}$

$\sin s = y = \frac{4}{5}$ $\qquad \sec s = \frac{1}{x} = \frac{5}{3}$

$\tan s = \frac{y}{x} = \frac{4}{3}$ $\qquad \csc s = \frac{1}{y} = \frac{5}{4}$

49. $\left(x, \dfrac{24}{25}\right) \quad x < 0$

$$\cos^2 s + \sin^2 s = 1$$
$$x^2 + \left(\frac{24}{25}\right)^2 = 1$$
$$x^2 = 1 - \frac{576}{625} = \frac{49}{625} \qquad \boxed{x = -\frac{7}{25}}$$

$$\cos s = x = -\frac{7}{25} \qquad \cot s = \frac{x}{y} = -\frac{7}{24}$$
$$\sin s = y = \frac{24}{25} \qquad \sec s = \frac{1}{x} = -\frac{25}{7}$$
$$\tan s = \frac{y}{x} = -\frac{24}{7} \qquad \csc s = \frac{1}{y} = \frac{25}{24}$$

51. $\left(-\dfrac{1}{3}, y\right) \quad y < 0$

$$\cos^2 s + \sin^2 s = 1$$
$$\left(-\frac{1}{3}\right)^2 + y^2 = 1$$
$$y^2 = 1 - \frac{1}{9} = \frac{8}{9} \qquad \boxed{y = -\frac{2\sqrt{2}}{3}}$$

$$\cos s = x = -\frac{1}{3} \qquad \cot s = \frac{x}{y} = \frac{\sqrt{2}}{4}$$
$$\sin s = y = -\frac{2\sqrt{2}}{3} \qquad \sec s = \frac{1}{x} = -3$$
$$\tan s = \frac{y}{x} = 2\sqrt{2} \qquad \csc s = \frac{1}{y} = -\frac{3\sqrt{2}}{4}$$

53. $\sin s = \dfrac{1}{2}, \quad \cos s = \dfrac{\sqrt{3}}{2} \Rightarrow x = \dfrac{\sqrt{3}}{2}, \ y = \dfrac{1}{2}$

$$\tan s = \frac{\sqrt{3}}{3} \qquad \sec s = \frac{2\sqrt{3}}{3}$$
$$\cot s = \sqrt{3} \qquad \csc s = 2$$

55. $\sin s = \dfrac{4}{5}, \quad \cos s = -\dfrac{3}{5} \Rightarrow x = -\dfrac{3}{5}, \ y = \dfrac{4}{5}$

$$\tan s = -\frac{4}{3} \qquad \sec s = -\frac{5}{3}$$
$$\cot s = -\frac{3}{4} \qquad \csc s = \frac{5}{4}$$

57. $\sin s = -\dfrac{\sqrt{3}}{2}, \quad \cos s = \dfrac{1}{2} \Rightarrow x = \dfrac{1}{2}, \ y = -\dfrac{\sqrt{3}}{2}$

$$\tan s = -\sqrt{3} \qquad \sec s = 2$$
$$\cot s = -\frac{\sqrt{3}}{3} \qquad \csc s = -\frac{2\sqrt{3}}{3}$$

59. $\sin s > 0:$ I, II
$\cos s < 0:$ II, III

Quadrant II

61. $\sec s < 0:$ II, III
$\csc s < 0:$ III, IV

Quadrant III

63. $\cos s > 0:$ I, IV
$\sin s < 0:$ III, IV

Quadrant IV

65. $\cos s \neq 0,$

$$\cos^2 s + \sin^2 s = 1$$
$$\frac{\cos^2 s}{\cos^2 s} + \frac{\sin^2 s}{\cos^2 s} = \frac{1}{\cos^2 s}$$
$$1 + \tan^2 s = \sec^2 s$$

Reviewing Basic Concepts (Sections 8.1 and 8.2)

1. **(a)** $35°$; complement $= 90° - 35° = 55°$
supplement $= 180° - 35° = 145°$

(b) $\dfrac{\pi}{4}$; complement $= \dfrac{\pi}{2} - \dfrac{\pi}{4} = \dfrac{2\pi - \pi}{4} = \dfrac{\pi}{4}$

supplement $= \pi - \dfrac{\pi}{4} = \dfrac{4\pi - \pi}{4} = \dfrac{3\pi}{4}$

2. $35.25° = 35° + .25\,(60)\,' = 35°\,15'0''$

3. $59°\,35'\,30'' = (59 + \dfrac{35}{60} + \dfrac{30}{3600})°$

$= 59.591\overline{6}\,°$

4. (a) $560° = 560° - 360° = 200°$

(b) $-\dfrac{2\pi}{3} = 2\pi + \left(-\dfrac{2\pi}{3}\right) = \dfrac{6\pi - 2\pi}{3} = \dfrac{4\pi}{3}$

5. (a) $240° \cdot \dfrac{\pi}{180°} = \dfrac{4\pi}{3}$

(b) $\dfrac{3\pi}{4} \cdot \dfrac{180°}{\pi} = 135°$

6. $r = 3$ cm, $\theta = 120° \cdot \dfrac{\pi}{180°} = \dfrac{2\pi}{3}$

(a) $s = r\theta = 3 \cdot \dfrac{2\pi}{3} = 2\pi$ cm

(b)
$$\begin{aligned} A &= \frac{1}{2}r^2\theta \\ &= \frac{1}{2}(3)^2\left(\frac{2\pi}{3}\right) \\ &= 3\pi \text{ cm}^2 \end{aligned}$$

7. (a) $-2\pi : (1, 0)$

(b) $\dfrac{5\pi}{4} : \left(-\dfrac{\sqrt{2}}{2}, -\dfrac{\sqrt{2}}{2}\right)$

for $\dfrac{\pi}{4}$, $x = \dfrac{\sqrt{2}}{2}$, $y = \dfrac{\sqrt{2}}{2}$

$\dfrac{5\pi}{4}$ is in Quadrant III

(c) $\dfrac{5\pi}{2} = 2\pi + \dfrac{\pi}{2} : (0, 1)$

8. $-\dfrac{5\pi}{2} + 4\pi = -\dfrac{5\pi}{2} + \dfrac{8\pi}{2} = \dfrac{3\pi}{2}$

$\cos\left(-\dfrac{5\pi}{2}\right) = \cos\dfrac{3\pi}{2} = 0$

$\sin\left(-\dfrac{5\pi}{2}\right) = \sin\dfrac{3\pi}{2} = -1$

$\cot\left(-\dfrac{5\pi}{2}\right) = \cot\dfrac{3\pi}{2} = 0$

$\csc\left(-\dfrac{5\pi}{2}\right) = \csc\dfrac{3\pi}{2} = -1$

$\tan\left(-\dfrac{5\pi}{2}\right) = \tan\dfrac{3\pi}{2}$ is undefined

$\sec\left(-\dfrac{5\pi}{2}\right) = \sec\dfrac{3\pi}{2}$ is undefined

9. (a) $\cos\left(\dfrac{7\pi}{6}\right) = -\cos\dfrac{\pi}{6} = -\dfrac{\sqrt{3}}{2}$

$\sin\left(\dfrac{7\pi}{6}\right) = -\sin\dfrac{\pi}{6} = -\dfrac{1}{2}$

$\tan\left(\dfrac{7\pi}{6}\right) = \tan\dfrac{\pi}{6} = \dfrac{\sqrt{3}}{3}$

$\cot\left(\dfrac{7\pi}{6}\right) = \cot\dfrac{\pi}{6} = \sqrt{3}$

$\sec\left(\dfrac{7\pi}{6}\right) = -\sec\dfrac{\pi}{6} = -\dfrac{2\sqrt{3}}{3}$

$\csc\left(\dfrac{7\pi}{6}\right) = -\csc\dfrac{\pi}{6} = -2$

(b) $\cos\left(-\dfrac{2\pi}{3}\right) = \cos\dfrac{4\pi}{3} = -\dfrac{1}{2}$

$\sin\left(-\dfrac{2\pi}{3}\right) = \sin\dfrac{4\pi}{3} = -\dfrac{\sqrt{3}}{2}$

$\tan\left(-\dfrac{2\pi}{3}\right) = \tan\dfrac{4\pi}{3} = \sqrt{3}$

$\cot\left(-\dfrac{2\pi}{3}\right) = \cot\dfrac{4\pi}{3} = \dfrac{\sqrt{3}}{3}$

$\sec\left(-\dfrac{2\pi}{3}\right) = \sec\dfrac{4\pi}{3} = -2$

$\csc\left(-\dfrac{2\pi}{3}\right) = \csc\dfrac{4\pi}{3} = -\dfrac{2\sqrt{3}}{3}$

10. $\cos 2.25 \approx -.6281736227$

$\sin 2.25 \approx .7780731969$

$\tan 2.25 \approx -1.238627616$

$\cot 2.25 \approx -.8073451511$

$\sec 2.25 \approx -1.591916572$

$\csc 2.25 \approx 1.285226125$

Section 8.3

1. $y = \sin x$: **G**

3. $y = -\sin x$: **E**

5. $y = \sin 2x$: **B**

7. $y = 2\sin x$: **F**

9. $y = \sin\left(x - \frac{\pi}{4}\right)$: **D**

11. $y = \cos\left(x - \frac{\pi}{4}\right)$: **H**

13. $y = 1 + \sin x$: **B**

15. $y = 1 + \cos x$: **F**

17. $y = 3\sin(2x - 4)$: **B**

Amplitude: 3
Period: π
Phase shift: 2

19. $y = 4\sin(3x - 2)$: **C**

Amplitude: 4

Period: $\frac{2\pi}{3}$

Phase shift: $\frac{2}{3}$

21. Amplitude: 4;

Period: $4\pi \Longrightarrow \frac{4\pi}{1} = \frac{2\pi}{b} \Longrightarrow b = \frac{1}{2}$

$y = 4\sin\left(\frac{1}{2}x\right)$

Other answers are possible.

Relating Concepts (23 – 30):

$$\mathbf{f(x) = -5 + 3\sin\left[2\left(x - \frac{\pi}{2}\right)\right]}$$

23. **(a)** Because the maximum value of the sine function is <u>1</u>, the maximum value of $\sin\left[2\left(x - \frac{\pi}{2}\right)\right]$ is <u>1</u>, the maximum value of $3\sin\left[2\left(x - \frac{\pi}{2}\right)\right]$ is <u>3</u>, and thus the maximum value of $-5 + 3\sin\left[2\left(x - \frac{\pi}{2}\right)\right]$ is <u>-2</u>.

(b) Because the minimum value of the sine function is <u>-1</u>, the minimum value of $\sin\left[2\left(x - \frac{\pi}{2}\right)\right]$ is <u>-1</u>, the minimum value of $3\sin\left[2\left(x - \frac{\pi}{2}\right)\right]$ is <u>-3</u>, and thus the minimum value of $-5 + 3\sin\left[2\left(x - \frac{\pi}{2}\right)\right]$ is <u>-8</u>.

24. Range of $f(x)$: $[-8, -2]$

25. The standard trig window has a range of $[-4, 4]$. The range of $f(x)$ is $[-8, -2]$, so many values would not appear on the screen.

26. $Y_{\max} \geq -2$, $Y_{\min} \leq -8$

27. The period of $f(x)$ is $\pi \left(\frac{2\pi}{2}\right)$, and $[-2\pi, 2\pi]$ has a distance of 4π. Since $\frac{4\pi}{\pi} = 4$, there will be 4 periods.

28. Graph of $f(x)$: $(2\pi \approx 6.3)$

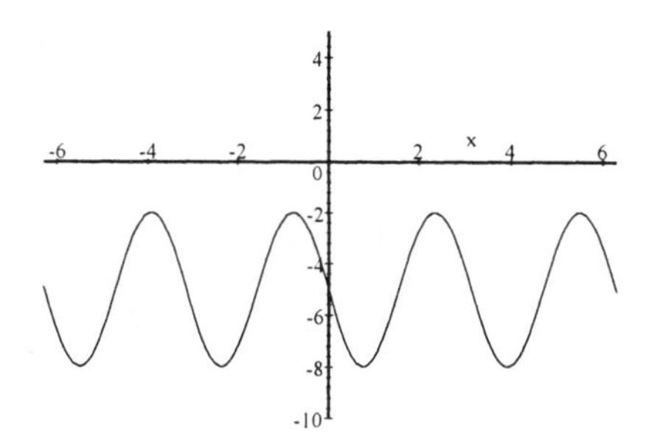

29. $Y_{\min} = -8;\quad Y_{\max} = -2$

30. $f(-2) = -5 + 3\sin 2\left(-2 - \frac{\pi}{2}\right)$

≈ -7.270407486

$f(-2+\pi) \approx -7.270407486$

Since the period of f is π,
$f(p) = f(p+\pi)$ for all values of p.

31. $y = 2\sin(x - \pi)$

(a) amplitude: 2
(b) period: 2π
(c) phase shift: π
(d) vertical translation: none
(e) range: $[-2,\ 2]$

33. $y = 4\cos\left(\frac{1}{2}x + \frac{\pi}{2}\right)$

(a) amplitude: 4

(b) period: $\dfrac{2\pi}{1/2} = 4\pi$

(c) phase shift: $\dfrac{-\pi/2}{1/2} = -\pi$

(d) vertical translation: none

(e) range: $[-4,\ 4]$

35. $y = 2 - \sin\left(3x - \frac{\pi}{5}\right)$

(a) amplitude: 1

(b) period: $\dfrac{2\pi}{3}$

(c) phase shift: $\dfrac{\pi/5}{3} = \dfrac{\pi}{15}$

(d) vertical translation: up 2 units

(e) range: $[1,\ 3]$

37. $y = \sin\left(x - \frac{\pi}{4}\right)$

Amplitude: 1

Period: 2π

Vertical translation: none

Phase shift: $\frac{\pi}{4}$ units to the right

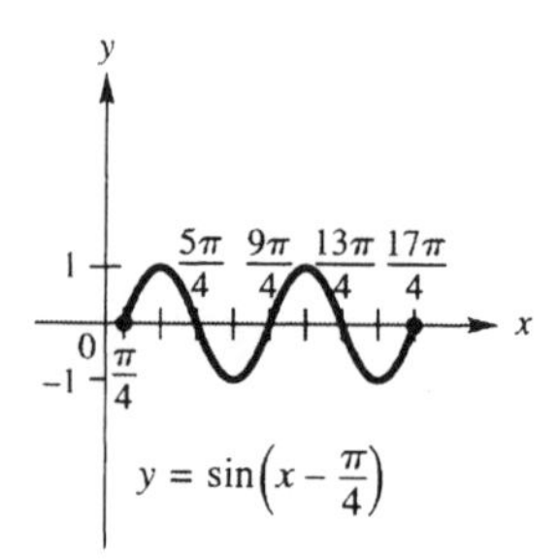

39. $y = 2\cos\left(x - \frac{\pi}{3}\right)$

Amplitude: 2

Period: 2π

Vertical translation: none

Phase shift: $\frac{\pi}{3}$ units to the right

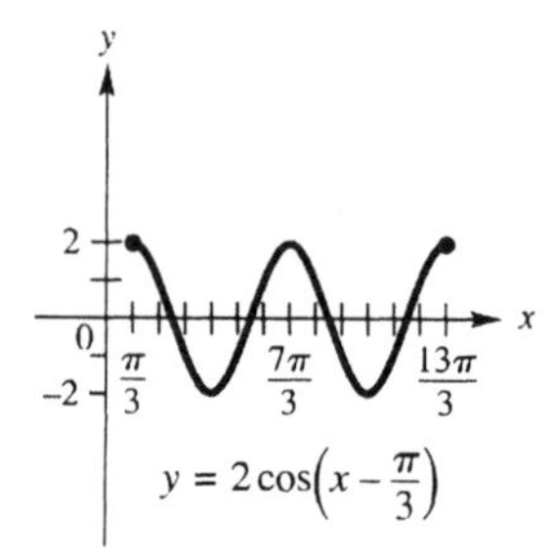

41. $y = -4\sin(2x - \pi)$

$= -4\sin\left[2\left(x - \frac{\pi}{2}\right)\right]$

Amplitude: $|-4| = 4$

Period: $\dfrac{2\pi}{2} = \pi$

Vertical translation: none

Phase shift: $\frac{\pi}{2}$ units to the right

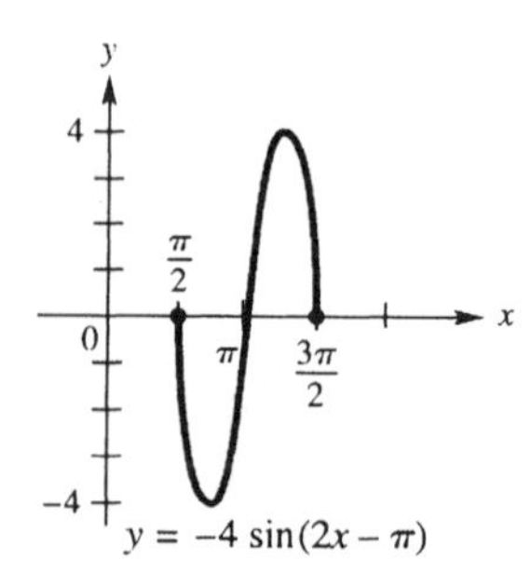

43. $y = \frac{1}{2}\cos\left(\frac{1}{2}x - \frac{\pi}{4}\right)$

$= \frac{1}{2}\cos\left[\frac{1}{2}\left(x - \frac{\pi}{2}\right)\right]$

Amplitude: $\frac{1}{2}$

Period: $\frac{2\pi}{1/2} = 4\pi$

Vertical translation: none

Phase shift: $\frac{\pi}{2}$ units to the right

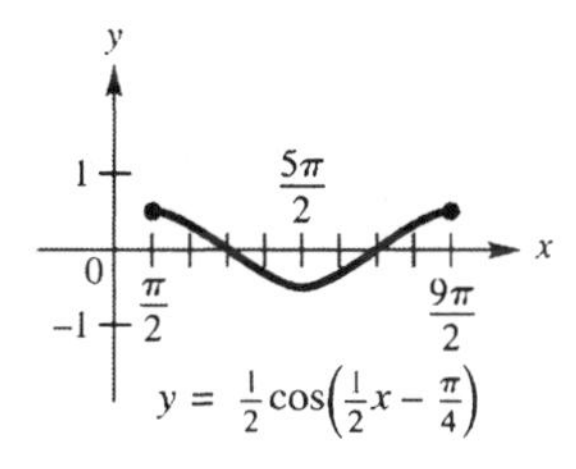

45. $y = 1 - \frac{2}{3}\sin\frac{3}{4}x$

Amplitude: $\left|-\frac{2}{3}\right| = \frac{2}{3}$

Period: $\frac{2\pi}{3/4} = \frac{8\pi}{3}$

Vertical translation: up 1 unit

Phase shift: none

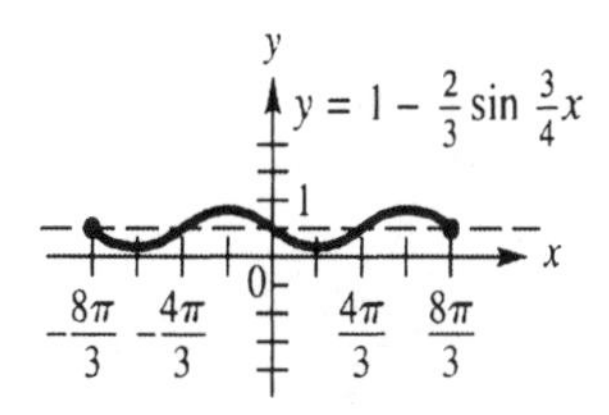

47. $y = 1 - 2\cos\frac{1}{2}x$

Amplitude: $|-2| = 2$

Period: $\frac{2\pi}{1/2} = 4\pi$

Vertical translation: up 1 unit

Phase shift: none

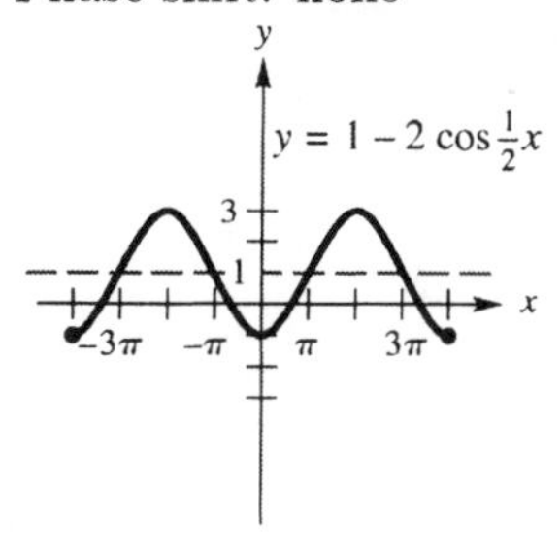

49. $y = -3 + 2\sin\left(x + \frac{\pi}{2}\right)$

Amplitude: 2

Period: 2π

Vertical translation: down 3 units

Phase shift: $\frac{\pi}{2}$ units to the left

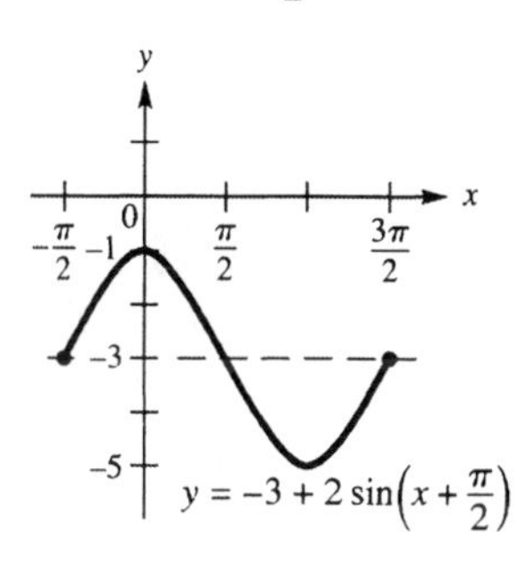

51. $y = \frac{1}{2} + \sin\left[2\left(x + \frac{\pi}{4}\right)\right]$

Amplitude: 1

Period: $\frac{2\pi}{2} = \pi$

Vertical translation: up $\frac{1}{2}$ unit

Phase shift: $\frac{\pi}{4}$ units to the left

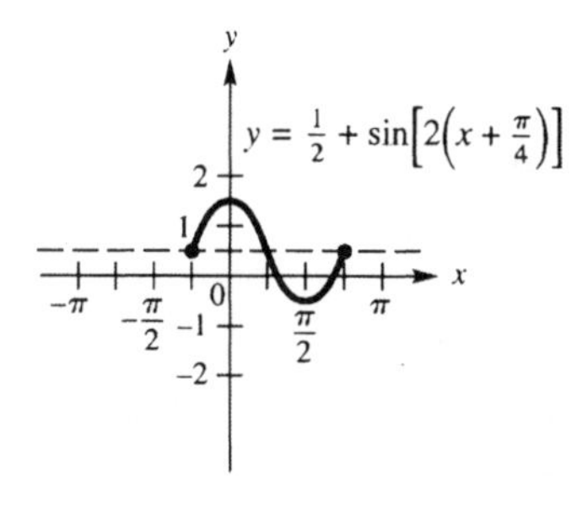

53. The graph repeats each day, so the period is 24 hours.

55. Approximately 6 P.M.; approximately .2 feet.

57. Approximately 2 A.M.; approximately 2.6 feet.

59. **(a)** The highest temperature is 80°; The lowest temperature is 50°.

(b) The amplitude is $\frac{1}{2}(80° - 50°) = 15°$.

(c) The period is about 35,000 years.

(d) The trend of the temperature now is downward.

61. **(a)** The latest time that the animals begin their evening activity is $\approx 8:00$ P.M., the earliest time is $\approx 4:00$ P.M.

$$4:00 \le y \le 8:00$$

Since there is a difference of 4 hours in these times, the amplitude is

$$\frac{1}{2}(4) = 2 \text{ hours.}$$

(b) The length of this period is 1 year.

63. $E = 3.8\cos 40\pi t$

(a) Amplitude: 3.8; Period: $\frac{2\pi}{40\pi} = \frac{1}{20}$

(b) Frequency:

$$\frac{1}{\text{Period}} = \frac{1}{1/20} = 20 \text{ cycles/second}$$

(c) $t = .02 : E = 3.8\cos 40\pi(.02) \approx -3.074$

$t = .04 : E = 3.8\cos 40\pi(.04) \approx 1.174$

$t = .08 : E = 3.8\cos 40\pi(.08) \approx -3.074$

$t = .12 : E = 3.8\cos 40\pi(.12) \approx -3.074$

$t = .14 : E = 3.8\cos 40\pi(.14) \approx 1.174$

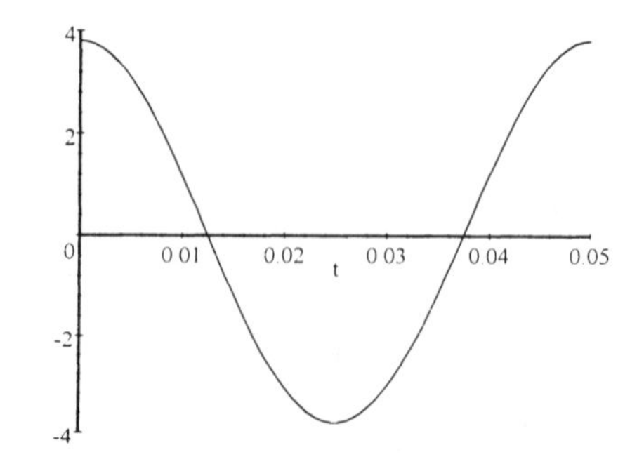

65. **(a)** $C(x) = .04x^2 + .6x + 330 + 7.5\sin(2\pi x)$

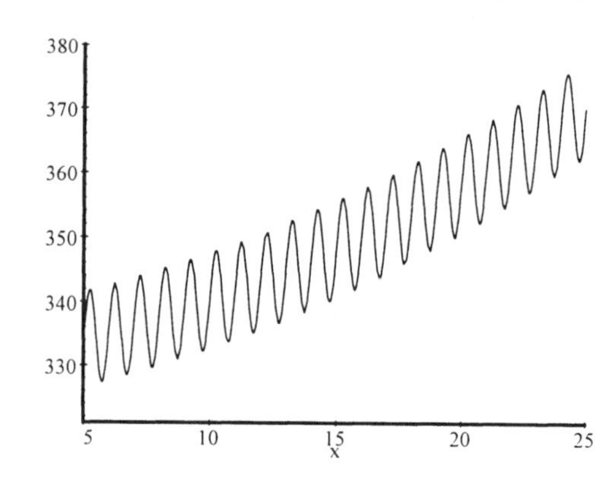

The amplitude of the oscillations of this graph are larger than those in **#64**. Graph **#65** is also above **#64**.

(b) Since the air is much thinner (volume is less), the carbon dioxide will have more effect percentage wise.

(c) If $x = 1970$ instead of 0:

$$C(x) = .04(x - 1970)^2 + .6(x - 1970) + 330 + 7.5\sin(2\pi(x - 1970))$$

67. **(a)** We can predict the average yearly temperature by finding the mean of the average monthly temperatures:

$$\frac{51 + 55 + 63 + .. + 59 + 52}{12} = \frac{845}{12} \approx 70.4°\text{F}$$

which is very close to the actual value of 70°F.

(b)

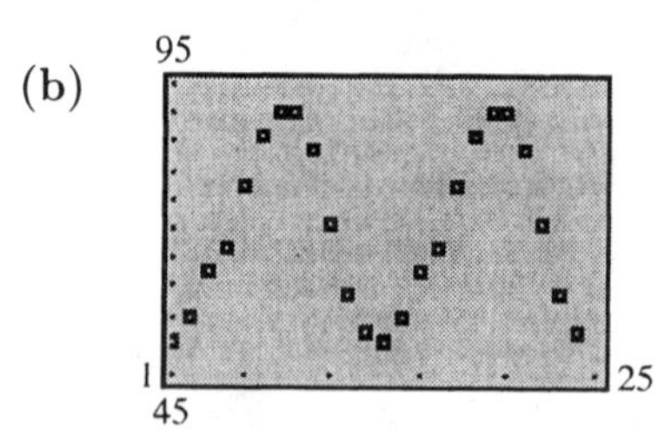

continued

67. continued

(c) $a \approx 19.5,\ b = \dfrac{\pi}{6},\ c = 70.5,\ d = 7.2$

$$f(x) = 19.5 \cos\left[\frac{\pi}{6}(x - 7.2)\right] + 70.5$$

(d) The function gives an excellent model.

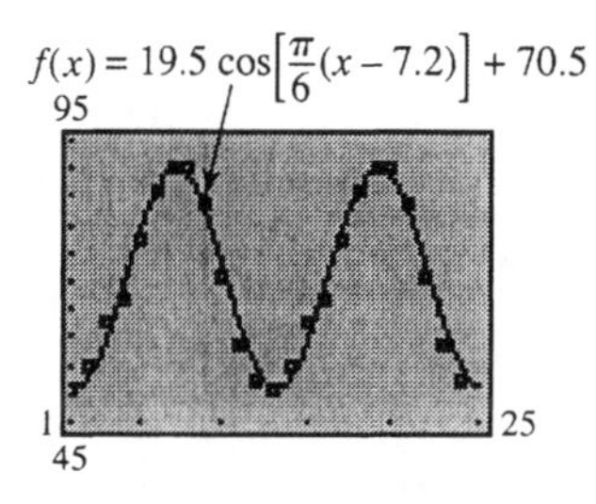

(e) Regression equation:

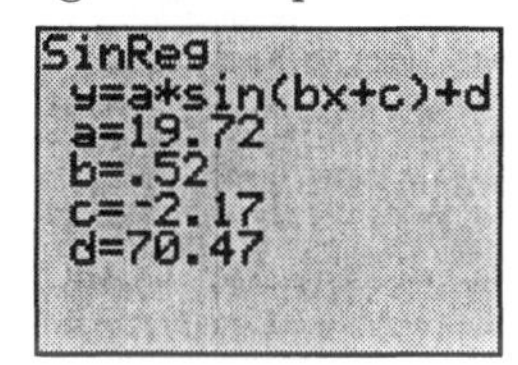

TI-83 fixed to the nearest hundredth

69. Regression equation:

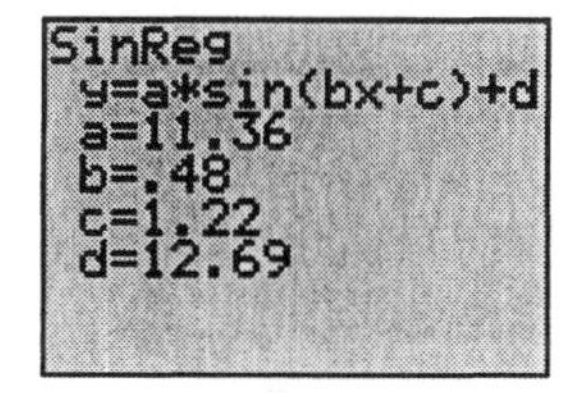

TI-83 fixed to the nearest hundredth

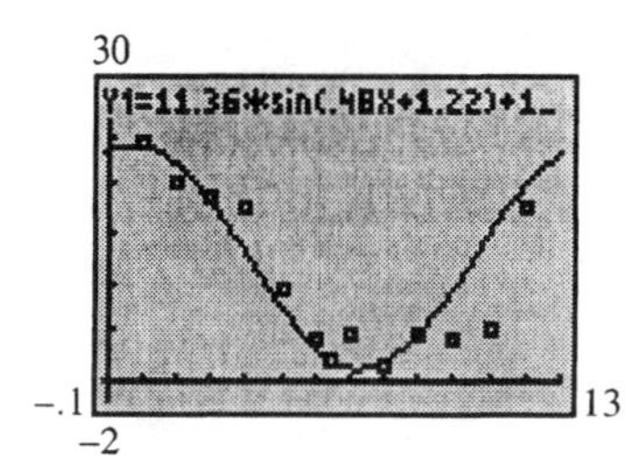

Section 8.4

1. $y = -\csc x$: **B**

3. $y = -\tan x$: **E**

5. $y = \tan\left(x - \dfrac{\pi}{4}\right)$: **D**

7. True

9. True

11. False; $\tan(-x) = -\tan x$ for all s in the domain.

13. $y = 2 \csc \dfrac{1}{2}x$

(a) period: $4\pi \quad \left(2\pi \div \dfrac{1}{2} = 2\pi \cdot 2\right)$

(b) phase shift: none

(c) range: $(-\infty, -2] \cup [2, \infty)$

15. $y = -2 \sec\left(x + \dfrac{\pi}{2}\right)$

(a) period: 2π

(b) phase shift: $-\dfrac{\pi}{2} \quad \left(\dfrac{\pi}{2} \text{ units to the left}\right)$

(c) range: $(-\infty, -2] \cup [2, \infty)$

17. $y = \dfrac{5}{2} \cot\left[\dfrac{1}{3}\left(x - \dfrac{\pi}{2}\right)\right]$

(a) period: $3\pi \quad \left(\pi \div \dfrac{1}{3} = \pi \cdot 3\right)$

(b) phase shift: $\dfrac{\pi}{2} \quad \left(\dfrac{\pi}{2} \text{ units to the right}\right)$

(c) range: $(-\infty, \infty)$

19. $y = \dfrac{1}{2} \sec(2x + \pi)$

(a) period: $\pi \quad (2\pi \div 2)$

(b) phase shift: $-\dfrac{\pi}{2} \quad \left(\dfrac{\pi}{2} \text{ units to the left}\right)$

(c) range: $\left(-\infty, -\dfrac{1}{2}\right] \cup \left[\dfrac{1}{2}, \infty\right)$

21. $y = -1 - \tan\left(x + \dfrac{\pi}{4}\right)$

(a) period: π

(b) phase shift: $-\dfrac{\pi}{4}$ $\left(\dfrac{\pi}{4} \text{ units to the left}\right)$

(c) range: $(-\infty,\ \infty)$

23. $y = \csc\left(x - \dfrac{\pi}{4}\right)$

Period: 2π

Phase shift: $\dfrac{\pi}{4}$ units to the right

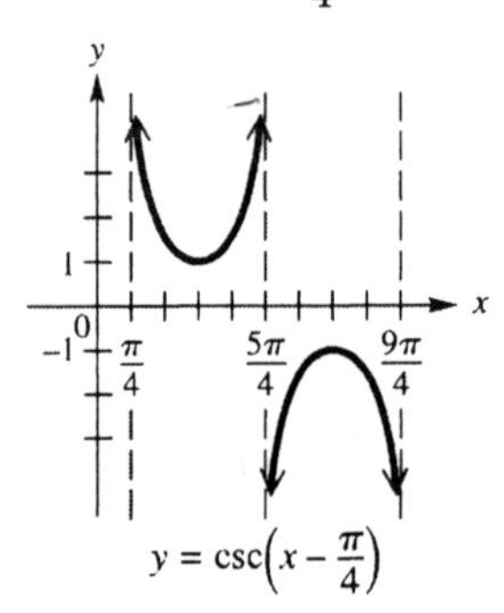

25. $y = \sec\left(x + \dfrac{\pi}{4}\right)$

Period: 2π

Phase shift: $\dfrac{\pi}{4}$ units to the left

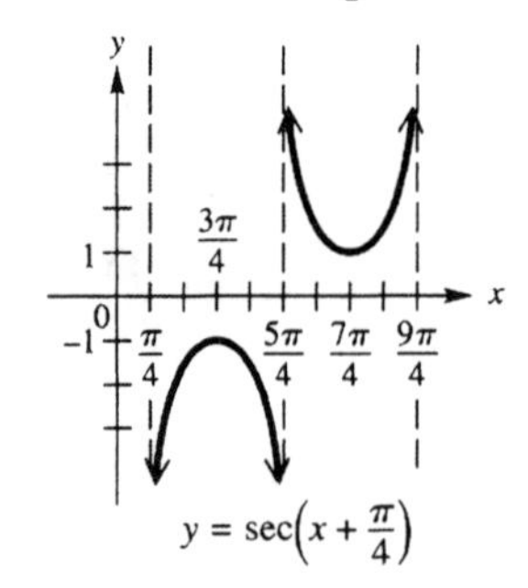

27. $y = \sec\left(\dfrac{1}{2}x + \dfrac{\pi}{3}\right)$

Period: $\dfrac{2\pi}{1/2} = 4\pi$

Phase shift: $\dfrac{\pi}{3} \div \dfrac{1}{2} = \dfrac{2\pi}{3}$ units to the left

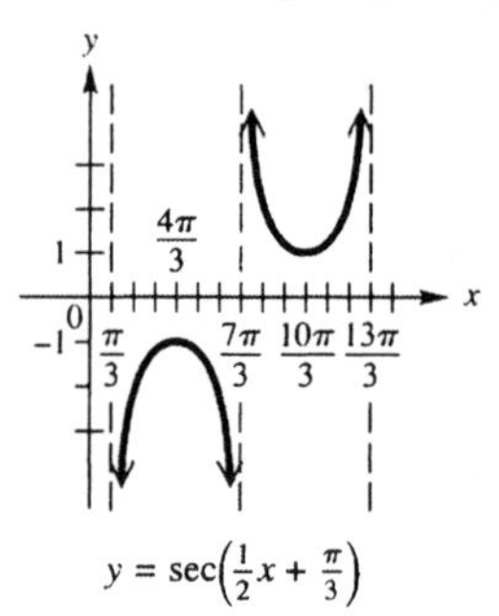

29. $y = 2 + 3\sec(2x - \pi)$

Period: $\dfrac{2\pi}{2} = \pi$

Phase shift: $\dfrac{\pi}{2}$ units to the right

Move up 2 units

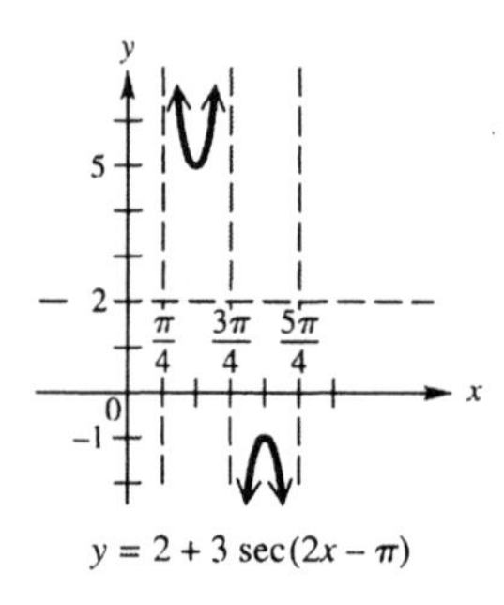

31. $y = 1 - \dfrac{1}{2}\csc\left(x - \dfrac{3\pi}{4}\right)$

Period: 2π; Move up 1 unit

Phase shift: $\dfrac{3\pi}{4}$ units to the right

Reflected about x-axis; y's halved

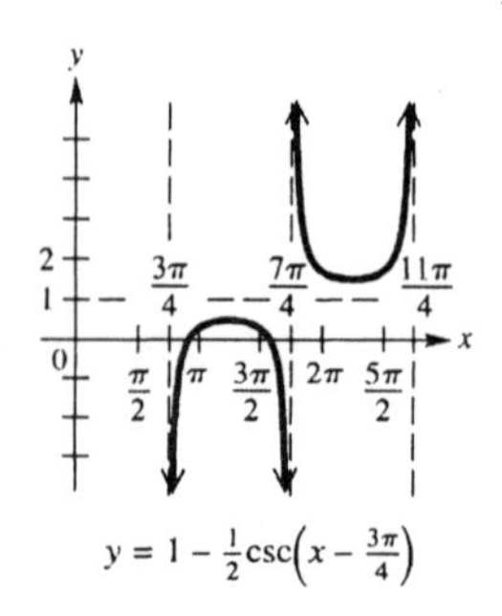

33. $y = \tan(2x - \pi)$

Period: $\frac{\pi}{2}$

Phase shift: $\frac{\pi}{2}$ units to the right

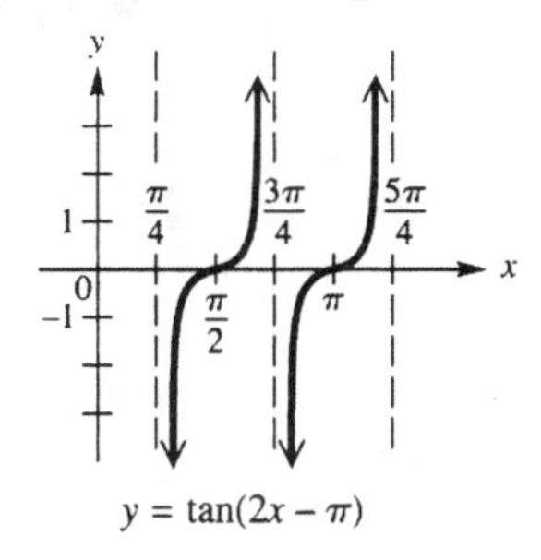

$y = \tan(2x - \pi)$

35. $y = \cot\left(3x + \frac{\pi}{4}\right)$

Period: $\frac{\pi}{3}$

Phase shift: $\frac{\pi}{4} \div 3 = \frac{\pi}{12}$ units to the left

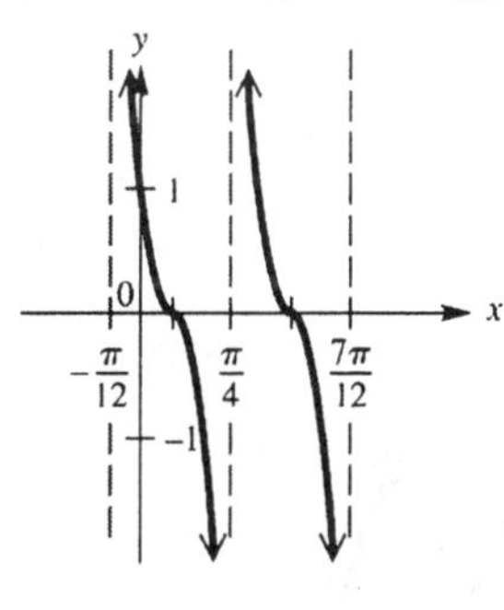

$y = \cot\left(3x + \frac{\pi}{4}\right)$

37. $y = 1 + \tan x$

Period: π
Move up 1 unit

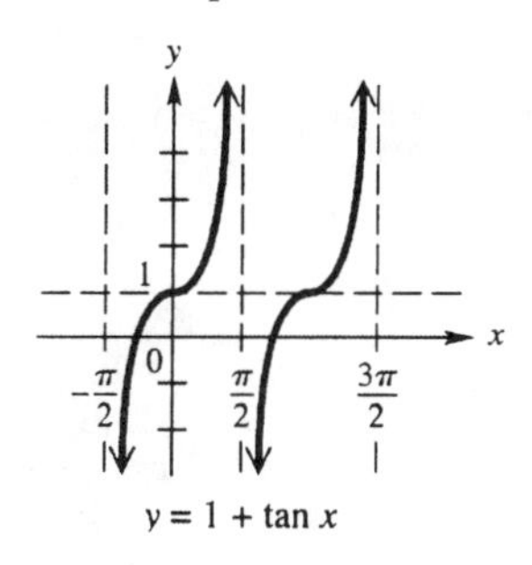

$y = 1 + \tan x$

39. $y = 1 - \cot x$

Period: π
Reflected about the x-axis
Move up 1 unit

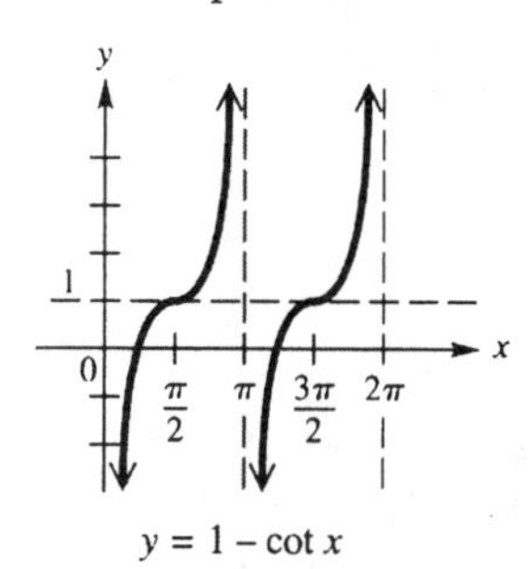

$y = 1 - \cot x$

41. $y = -1 + 2\tan x$

Period: π
y's double
Move down 1 unit

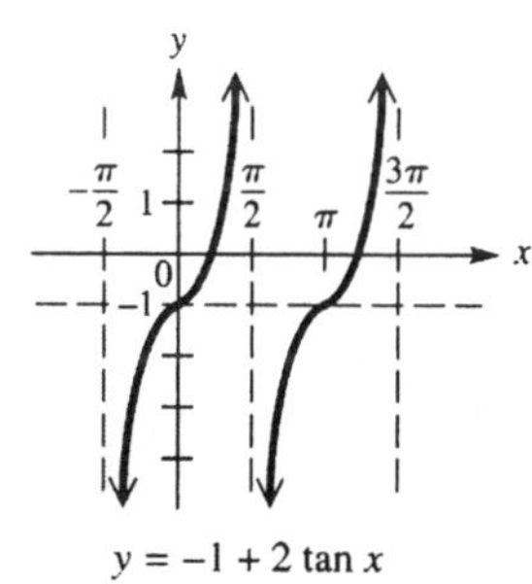

$y = -1 + 2\tan x$

43. $y = -1 + \frac{1}{2}\cot(2x - 3\pi)$

Period: $\frac{\pi}{2}$

Phase shift: $\frac{3\pi}{2}$ units to the right

y's halved; Move down 1 unit

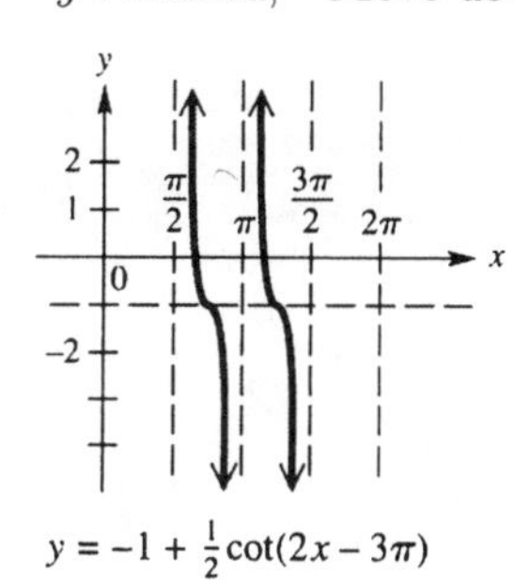

$y = -1 + \frac{1}{2}\cot(2x - 3\pi)$

45. $y = \tan x$ (solid); $y = x$ (dotted)

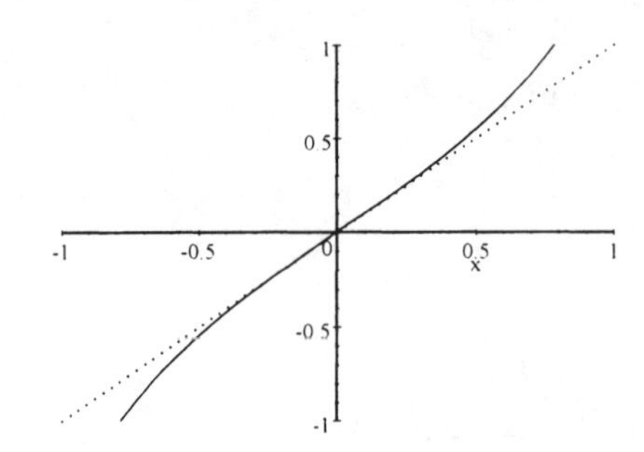

For values in the interval $-1 \leq x \leq 1$, x and $\tan x$ are approximately equal. As x gets closer to zero, $\tan x$ gets closer to the value of x.

47. $d = 4 \tan 2\pi t$

(a) $t = 0 : d = 4 \tan [2\pi (0)]$
$= 4 \tan 0$
$= \boxed{0 \text{ m}}$

(b) $t = .4 : d = 4 \tan [2\pi (.4)]$
$= 4 \tan (.8\pi)$
≈ -2.9062
$\approx \boxed{-2.9 \text{ m}}$

(c) $t = .8 : d = 4 \tan [2\pi (.8)]$
$= 4 \tan (1.6\pi)$
≈ -12.311
$\approx \boxed{-12.3 \text{ m}}$

(d) $t = 1.2 : d = 4 \tan [2\pi (1.2)]$
$= 4 \tan (2.4\pi)$
≈ 12.311
$\approx \boxed{12.3 \text{ m}}$

(e) $t = .25 : d = 4 \tan [2\pi (.25)]$
$= 4 \tan \frac{\pi}{2}$
undefined

49. $y_1 = \sin x$ (solid); $y_2 = \sin 2x$ (dashed); $y_1 + y_2$ (dark)

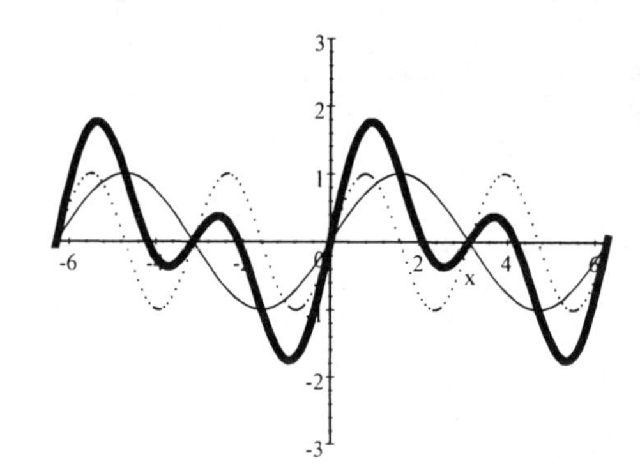

$$y_1\left(\frac{\pi}{6}\right) + y_1\left(\frac{\pi}{6}\right) = .5 + \frac{\sqrt{3}}{2} \approx 1.366025404$$

$$(y_1 + y_2)\left(\frac{\pi}{6}\right) \approx 1.366025404$$

51. $y_1 = \sin 2x$ (solid); $y_2 = \cos \frac{1}{2}x$ (dashed); $y_1 + y_2$ (dark)

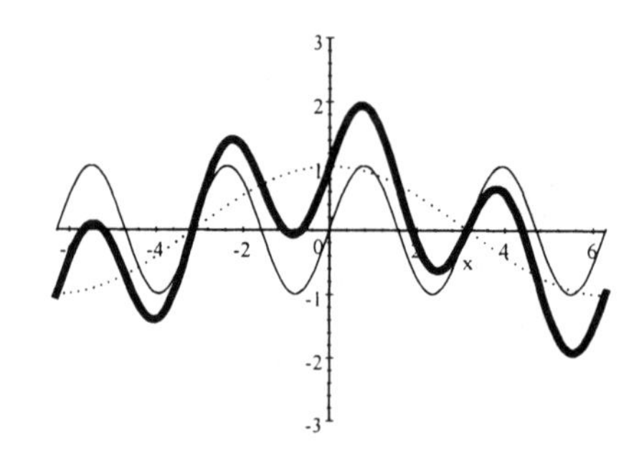

$$y_1\left(\frac{\pi}{6}\right) + y_1\left(\frac{\pi}{6}\right) \approx \frac{\sqrt{3}}{2} + .9659258263$$
$$\approx 1.83195123$$

$$(y_1 + y_2)\left(\frac{\pi}{6}\right) \approx 1.83195123$$

Reviewing Basic Concepts (Sections 8.3 and 8.4)

1. $y = -\cos x$ Period: 2π
Amplitude $= |-1| = 1$

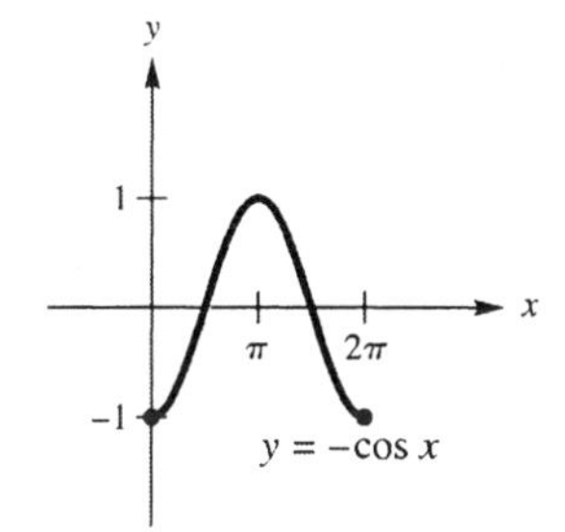

2. $f(x) = -2\sin(4x + \pi)$

(a) amplitude: $|-2| = 2$

(b) period: $\frac{2\pi}{4} = \frac{\pi}{2}$

(c) phase shift: $-\frac{\pi}{4}$ ($\frac{\pi}{4}$ units to the left)

(d) range: $[-2, 2]$

(e) y-intercept: 0

(f) least positive x-intercept: $\frac{\pi}{4}$

3. $y = \csc\left(x - \frac{\pi}{4}\right)$

Period: 2π

Phase shift: $\frac{\pi}{4}$ units to the right

Domain: $\left\{x | x \neq \frac{\pi}{4} + n\pi,\ n \in \text{integer}\right\}$

Range: $(-\infty, -1] \cup [1, \infty)$

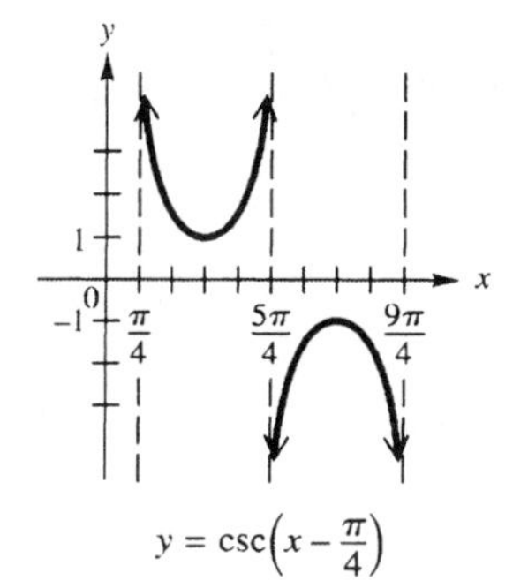

4. $y = \sec\left(x + \frac{\pi}{4}\right)$

Period: 2π

Phase shift: $\frac{\pi}{4}$ units to the left

Domain: $\left\{x | x \neq \frac{\pi}{4} + n\pi,\ n \in \text{integer}\right\}$

Range: $(-\infty, -1] \cup [1, \infty)$

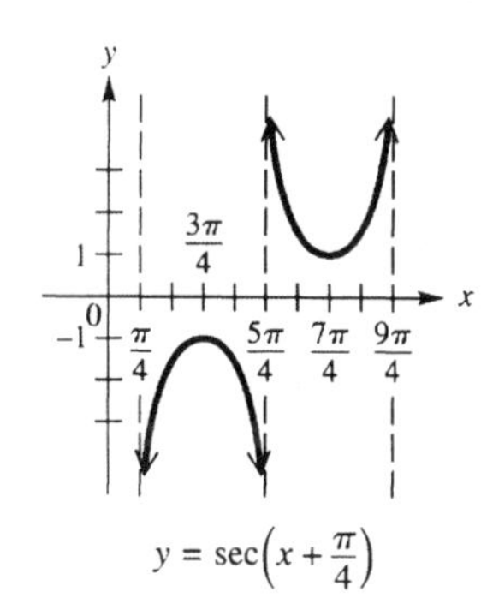

5. $y = 2\tan x$

Period: π

Domain: $\left\{x | x \neq \frac{\pi}{2} + n\pi,\ n \in \text{integer}\right\}$

Range: $(-\infty, \infty)$

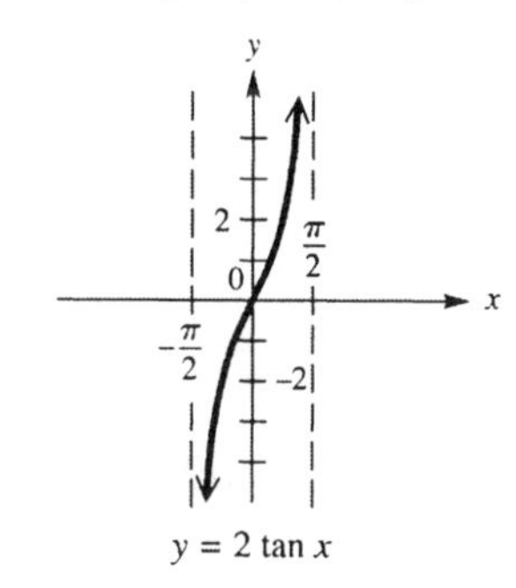

6. $y = 2\cot x$

Period: π

Domain: $\{x | x \neq n\pi,\ n \in \text{integer}\}$

Range: $(-\infty, \infty)$

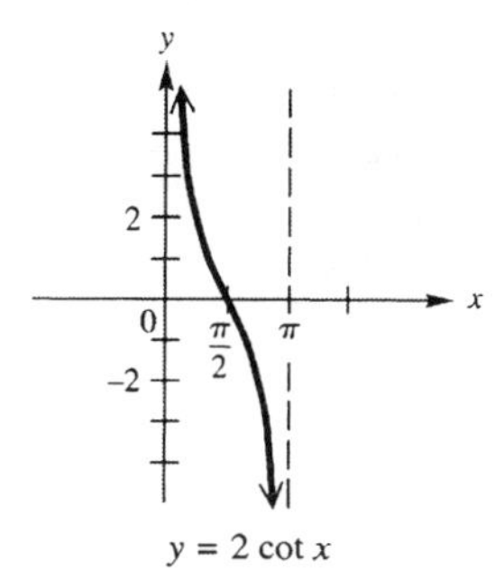

7. $y = 3\sin(\pi x + \pi)$ Period: $\dfrac{2\pi}{\pi} = 2$

Amplitude $= 3$

Phase shift: $-\dfrac{\pi}{\pi} = -1$

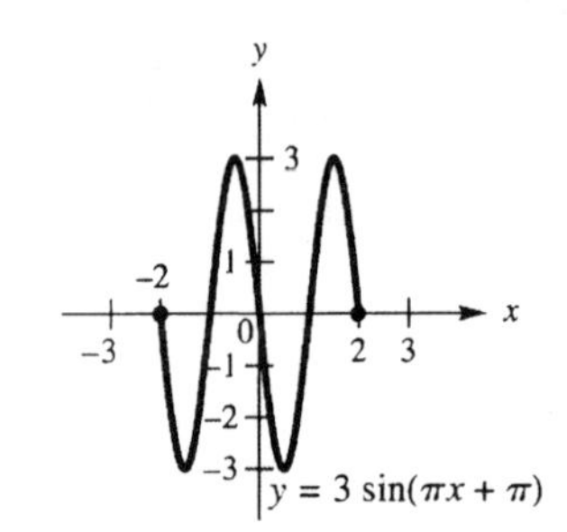

8. $f(t) = 6.5\sin\left[\dfrac{\pi}{6}(x - 3.65)\right] + 12.4$

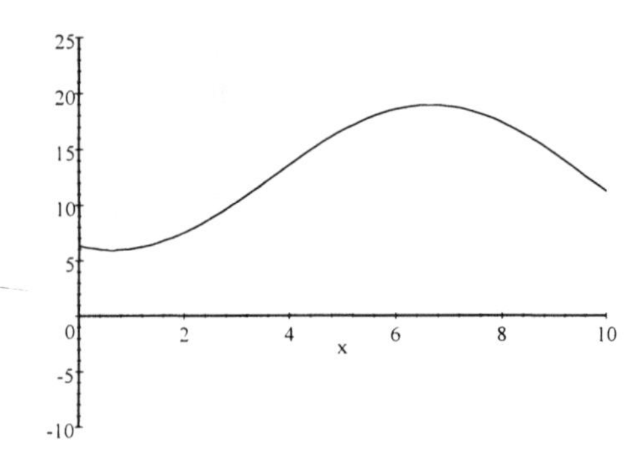

(a) Maximum $= 6.5 + 12.4 = 18.9$ hours

Minimum $= -6.5 + 12.4 = 5.9$ hours

(b) The amplitude, 6.5, represents half the difference in daylight hours between the longest and shortest day.

The period, $\dfrac{2\pi}{\pi/6} = \dfrac{2\pi}{1} \times \dfrac{6}{\pi} = 12$, represents 12 months or 1 year.

Section 8.5

1. $(5, -12)$

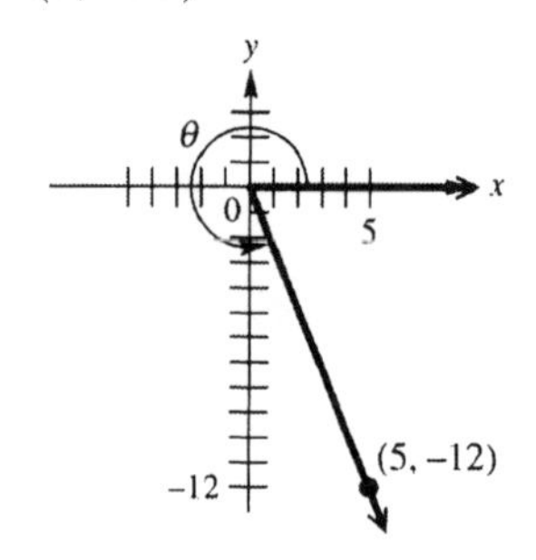

3. $(-3,\ 4)$; $r = \sqrt{(-3)^2 + 4^2}$

$= \sqrt{9 + 16} = \sqrt{25} = 5$

$$\sin\theta = \frac{y}{r} = \mathbf{\frac{4}{5}}$$

$$\cos\theta = \frac{x}{r} = \frac{-3}{5} = -\mathbf{\frac{3}{5}}$$

$$\tan\theta = \frac{y}{x} = \frac{4}{-3} = -\mathbf{\frac{4}{3}}$$

$$\cot\theta = \frac{x}{y} = \frac{-3}{4} = -\mathbf{\frac{3}{4}}$$

$$\sec\theta = \frac{r}{x} = \frac{5}{-3} = -\mathbf{\frac{5}{3}}$$

$$\csc\theta = \frac{r}{y} = \mathbf{\frac{5}{4}}$$

5. $(0,\ 2)$; $r = \sqrt{(0)^2 + 2^2} = \sqrt{4} = 2$

$$\sin\theta = \frac{y}{r} = \frac{2}{2} = \mathbf{1}$$

$$\cos\theta = \frac{x}{r} = \frac{0}{2} = \mathbf{0}$$

$$\tan\theta = \frac{y}{x} = \frac{2}{0} \quad \text{undefined}$$

$$\cot\theta = \frac{x}{y} = \frac{0}{2} = \mathbf{0}$$

$$\sec\theta = \frac{r}{x} = \frac{2}{0} \quad \text{undefined}$$

$$\csc\theta = \frac{r}{y} = \frac{2}{2} = \mathbf{1}$$

7. $(1, \sqrt{3})$; $r = \sqrt{1^2 + (\sqrt{3})^2} = \sqrt{4} = 2$

$$\sin\theta = \frac{y}{r} = \frac{\sqrt{3}}{2} = \mathbf{\frac{\sqrt{3}}{2}}$$
$$\cos\theta = \frac{x}{r} = \frac{1}{2} = \mathbf{\frac{1}{2}}$$
$$\tan\theta = \frac{y}{x} = \frac{\sqrt{3}}{1} = \mathbf{\sqrt{3}}$$
$$\cot\theta = \frac{x}{y} = \frac{1}{\sqrt{3}} = \mathbf{\frac{\sqrt{3}}{3}}$$
$$\sec\theta = \frac{r}{x} = \frac{2}{1} = \mathbf{2}$$
$$\csc\theta = \frac{r}{y} = \frac{2}{\sqrt{3}} = \mathbf{\frac{2\sqrt{3}}{3}}$$

9. $\sin\theta = \frac{y}{r}$ and $\csc\theta = \frac{r}{y}$ are reciprocals of each other. Reciprocals always have the same sign.

11. (x, y) in QII: $(-, +)$

$\frac{x}{r}$ $\left(\frac{-}{+}\right)$: **negative**

13. (x, y) in QIV: $(+, -)$

$\frac{y}{x}$ $\left(\frac{-}{+}\right)$: **negative**

15. $2x + y = 0 \Longrightarrow y = -2x,\ x \geq 0$

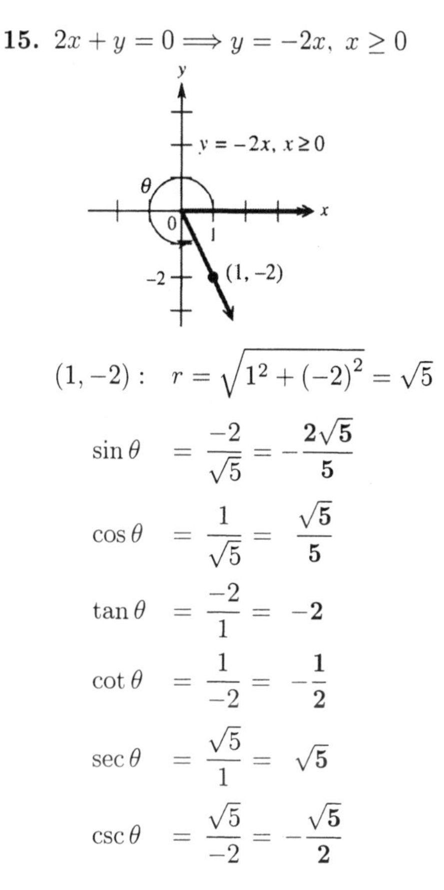

$(1, -2)$: $r = \sqrt{1^2 + (-2)^2} = \sqrt{5}$

$$\sin\theta = \frac{-2}{\sqrt{5}} = \mathbf{-\frac{2\sqrt{5}}{5}}$$
$$\cos\theta = \frac{1}{\sqrt{5}} = \mathbf{\frac{\sqrt{5}}{5}}$$
$$\tan\theta = \frac{-2}{1} = \mathbf{-2}$$
$$\cot\theta = \frac{1}{-2} = \mathbf{-\frac{1}{2}}$$
$$\sec\theta = \frac{\sqrt{5}}{1} = \mathbf{\sqrt{5}}$$
$$\csc\theta = \frac{\sqrt{5}}{-2} = \mathbf{-\frac{\sqrt{5}}{2}}$$

17. $-6x - y = 0 \Longrightarrow y = -6x,\ x \leq 0$

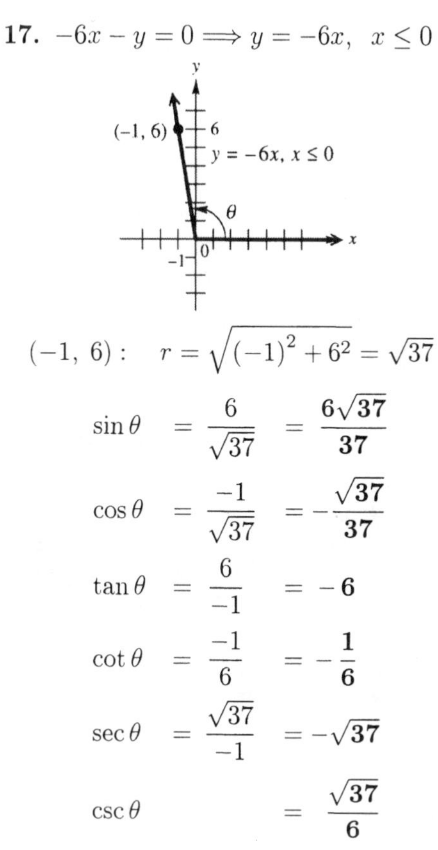

$(-1, 6)$: $r = \sqrt{(-1)^2 + 6^2} = \sqrt{37}$

$$\sin\theta = \frac{6}{\sqrt{37}} = \mathbf{\frac{6\sqrt{37}}{37}}$$
$$\cos\theta = \frac{-1}{\sqrt{37}} = \mathbf{-\frac{\sqrt{37}}{37}}$$
$$\tan\theta = \frac{6}{-1} = \mathbf{-6}$$
$$\cot\theta = \frac{-1}{6} = \mathbf{-\frac{1}{6}}$$
$$\sec\theta = \frac{\sqrt{37}}{-1} = \mathbf{-\sqrt{37}}$$
$$\csc\theta = \mathbf{\frac{\sqrt{37}}{6}}$$

19. $\theta = 540° = 540° - 360° = 180°$

$\sin 540° = \sin 180° = 0$

$\cos 540° = \cos 180° = -1$

$\tan 540° = \tan 180° = 0$

$\csc 540° = \dfrac{1}{\sin 540°}$ is undefined

$\sec 540° = \dfrac{1}{\cos 540°} = -1$

$\cot 540° = \dfrac{1}{\tan 540°}$ is undefined

21. $4 \csc 270° + 3 \cos 180°$
$= 4(-1) + 3(-1) = \boxed{-7}$

23. $2 \sec 0° + 4 \cot^2 90° + \cos 360°$
$= 2(1) + 4(0)^2 + 1 = \boxed{3}$

25. $\sin^2 360° + \cos^2 360°$
$= (0)^2 + (1)^2 = \boxed{1}$

27. $\tan(n \cdot 180°) = 0$ since

$n = 0 : \tan 0° = 0$
$n = 1 : \tan 180° = 0$
$n = 2 : \tan 360° = 0$, etc.

29. $\sin(n \cdot 180°) = 0$ since

$n = 0 : \sin 0° = 0$
$n = 1 : \sin 180° = 0$
$n = 2 : \sin 360° = 0$, etc.

31. $\tan 25° \approx .4663077 \approx \cot 65°$

Tangents and cotangents of complementary angles are equal.

33. $\cos 20° \approx .9396926 \approx \cos(-20)°$

Cosines of an angle and its opposite are equal.

35. $\cos 40° \approx .766$ so $T = 40°$

37. $\cos T = \sin T$ when $T = 45°$

39. $\cos T$ decreases as T increases from 90° to 180°.
$\sin T$ decreases as T increases from 90° to 180°.

41. If $a = -1$, then $\dfrac{1}{-1} = -1$.
$\cos 180° = \sec 180° = -1$, so $\theta = 180°$.

43. $\tan \beta = -\dfrac{1}{5}$

$\cot \beta = \dfrac{1}{\tan \beta} = \boxed{-5}$

45. $\cot \theta = -\dfrac{\sqrt{5}}{3}$

$\tan \theta = \dfrac{1}{\cot \theta} = -\dfrac{3}{\sqrt{5}} = \boxed{-\dfrac{3\sqrt{5}}{5}}$

47. $\sec \alpha = 9.80425133$

$$\begin{aligned}\cos \alpha &= \frac{1}{\sec \alpha} \\ &= 9.80425133^{-1} \\ &= \boxed{.10199657}\end{aligned}$$

49. $\cos \theta = \dfrac{3}{2}$ is not possible.
$-1 \leq \cos \theta \leq 1$ for all values of θ.

51. $\cot \theta = \dfrac{\sqrt{3}}{3}$

$$\begin{aligned}\tan \theta = \frac{1}{\cot \theta} &= \frac{1}{\sqrt{3}/3} \\ &= \frac{3}{\sqrt{3}} \cdot \frac{\sqrt{3}}{\sqrt{3}} = \boxed{\sqrt{3}}\end{aligned}$$

53. $\tan(3\theta - 4°) = \dfrac{1}{\cot(5\theta - 8°)}$

$\cot(3\theta - 4°) = \cot(5\theta - 8°)$

$3\theta - 4° = 5\theta - 8°$

$2\theta = 4°$

$\boxed{\theta = 2°}$

55. $\sin \theta > 0 :$ QI, QII
$\cos \theta < 0 :$ QII, QIII

Quadrant II

57. $\tan \theta > 0 :$ QI, QIII
$\cot \theta > 0 :$ QI, QIII

Quadrants I or III

59. 129°; Quadrant II

sin 129° : +
cos 129° : −
tan 129° : −

61. 298°; Quadrant IV

sin 298° : −
cos 298° : +
tan 298° : −

63. $-82° = 360° - 82° = 278°$; Quadrant IV

sin 52° : −
cos 52° : +
tan 52° : −

65. $\sin 30° = \frac{1}{2}$; $\cos 30° = \frac{\sqrt{3}}{2}$; $\tan 30° = \frac{\sqrt{3}}{3}$

tan 30° is greater.

67. $\sec 33° > \sin 33°$ since $\sec 33° > 1$ and $\sin 33° < 1$.

69. $\cos\theta = -1.001$ is impossible since $-1 \le \cos\theta \le 1$.

71. $\cot\theta = -12.1$ is possible since the range of $\cot\theta$ is $(-\infty, \infty)$.

73. $\tan\theta = 1$ is possible since the range of $\tan\theta$ is $(-\infty, \infty)$.

75. $\tan\theta = 2$ and $\cot\theta = -2$ is impossible since $\cot\theta = \frac{1}{\tan\theta} = \frac{1}{2}$.

77. If $\cos\theta = -\frac{1}{4}$, $\theta \in$ Quadrant II:

$$\sin^2\theta + \cos^2\theta = 1$$
$$\sin^2\theta + \left(-\frac{1}{4}\right)^2 = 1$$
$$\sin^2\theta = 1 - \frac{1}{16} = \frac{15}{16}$$
$$\sin\theta = \pm\frac{\sqrt{15}}{4}$$

Quadrant II: $\boxed{\sin\theta = \frac{\sqrt{15}}{4}}$

79. If $\tan\theta = \frac{\sqrt{7}}{3}$, $\theta \in$ Quadrant III:

$$\tan^2\theta + 1 = \sec^2\theta$$
$$\left(\frac{\sqrt{7}}{3}\right)^2 + 1 = \sec^2\theta$$
$$\frac{7}{9} + 1 = \sec^2\theta$$
$$\sec^2\theta = \frac{16}{9}$$
$$\sec\theta = \pm\frac{4}{3}$$

Quadrant III: $\boxed{\sec\theta = -\frac{4}{3}}$

81. If $\sec\theta = 2$, $\theta \in$ Quadrant IV:

$$\cos\theta = \frac{1}{\sec\theta} = \frac{1}{2}$$
$$\sin^2\theta + \cos^2\theta = 1$$
$$\sin^2\theta + \left(\frac{1}{2}\right)^2 = 1$$
$$\sin^2\theta = 1 - \frac{1}{4} = \frac{3}{4}$$
$$\sin\theta = \pm\frac{\sqrt{3}}{2}$$

Quadrant IV: $\boxed{\sin\theta = -\frac{\sqrt{3}}{2}}$

83. If $\sin\theta = .49268329$, $\theta \in$ Quadrant II:

$$\sin^2\theta + \cos^2\theta = 1$$
$$(.49268329)^2 + \cos^2\theta = 1$$
$$\cos^2\theta = 1 - .24273682$$
$$\cos^2\theta = .7572631758$$
$$\cos\theta = \pm.8702086967$$

Quadrant II: $\cos\theta = -.8702086967$

$$\tan\theta = \frac{\sin\theta}{\cos\theta} = \frac{.49268329}{-.8702086967}$$

$\boxed{\tan\theta \approx -.56616682}$

85. $\tan^2 x + 1 = \sec^2 x$

$$2^2 + 1 = \sec^2 x$$
$$\boxed{5} = \sec^2 x$$

87. If $\tan\theta = -\dfrac{15}{8}$, $\theta \in$ Quadrant II:

(i) $\tan^2\theta + 1 = \sec^2\theta$

$$\left(-\frac{15}{8}\right)^2 + 1 = \sec^2\theta$$

$$\sec^2\theta = \frac{289}{64}$$

$$\sec\theta = \pm\frac{17}{8}$$

Quadrant II: $\boxed{\sec\theta = -\dfrac{17}{8}}$

(ii) $\cos\theta = \dfrac{1}{\sec\theta}$ $\quad\boxed{\cos\theta = -\dfrac{8}{17}}$

(iii) $\cot\theta = \dfrac{1}{\tan\theta}$ $\quad\boxed{\cot\theta = -\dfrac{8}{15}}$

(iv) $\dfrac{\sin\theta}{\cos\theta} = \tan\theta$

$$\sin\theta = \left(-\frac{8}{17}\right)\left(-\frac{15}{8}\right)$$

$$\boxed{\sin\theta = \frac{15}{17}}$$

(v) $\csc\theta = \dfrac{1}{\sin\theta}$ $\quad\boxed{\csc\theta = \dfrac{17}{15}}$

89. If $\tan\theta = \sqrt{3}$, $\theta \in$ Quadrant III:

(i) $\cot\theta = \dfrac{1}{\tan\theta} = \dfrac{1}{\sqrt{3}}$

$$\boxed{\cot\theta = \frac{\sqrt{3}}{3}}$$

(ii) $\tan^2\theta + 1 = \sec^2\theta$

$$(\sqrt{3})^2 + 1 = \sec^2\theta$$

$$\sec^2\theta = 4$$

$$\sec\theta = \pm 2$$

Quadrant III: $\boxed{\sec\theta = -2}$

(iii) $\cos\theta = \dfrac{1}{\sec\theta}$ $\quad\boxed{\cos\theta = -\dfrac{1}{2}}$

continued

89. continued

(iv) $\dfrac{\sin\theta}{\cos\theta} = \tan\theta$

$$\sin\theta = \cos\theta\,\tan\theta = \left(-\frac{1}{2}\right)(\sqrt{3})$$

$$\boxed{\sin\theta = -\frac{\sqrt{3}}{2}}$$

(v) $\csc\theta = \dfrac{1}{\sin\theta} = -\dfrac{2}{\sqrt{3}}$

$$\boxed{\csc\theta = -\frac{2\sqrt{3}}{3}}$$

91. If $\cot\theta = -1.49586$ (Quadrant IV):

(i) $\tan\theta = (\cot\theta)^{-1}$

$$\boxed{\tan\theta = -.668512}$$

(ii) $\tan^2\theta + 1 = \sec^2\theta$
$(-.668512)^2 + 1 = \sec^2\theta$
$\sec^2\theta \approx 1.446908$
$\sec\theta \approx \pm 1.20287$

Quadrant IV: $\boxed{\sec\theta \approx 1.20287}$

(iii) $\cos\theta = (\sec\theta)^{-1}$

$$\boxed{\cos\theta \approx .831342}$$

(iv) $\sin\theta = \cos\theta \cdot \tan\theta$
$= (.831342)(-.668512)$

$$\boxed{\sin\theta \approx -.555762}$$

(v) $\csc\theta = (\sin\theta)^{-1}$

$$\boxed{\csc\theta \approx -1.79933}$$

93. $x^2 + y^2 = r^2$

$$\frac{x^2}{y^2} + \frac{y^2}{y^2} = \frac{r^2}{y^2}$$

$$\cot^2\theta + 1 = \csc^2\theta$$

$$1 + \cot^2\theta = \csc^2\theta$$

95. False; $\sin\theta + \cos\theta \neq 1$. For example,
$\theta = 30° : \sin 30° + \cos 30°$

$$= \frac{1}{2} + \frac{\sqrt{3}}{2} = \frac{1+\sqrt{3}}{2} \neq 1.$$

Section 8.6

1. $$\sin A = \frac{\text{opp}}{\text{hyp}} = \frac{21}{29}$$
$$\cos A = \frac{\text{adj}}{\text{hyp}} = \frac{20}{29}$$
$$\tan A = \frac{\text{opp}}{\text{adj}} = \frac{21}{20}$$
$$\cot A = \frac{\text{adj}}{\text{opp}} = \frac{20}{21}$$
$$\sec A = \frac{\text{hyp}}{\text{adj}} = \frac{29}{20}$$
$$\csc A = \frac{\text{hyp}}{\text{opp}} = \frac{29}{21}$$

3. $$\sin A = \frac{\text{opp}}{\text{hyp}} = \frac{n}{p}$$
$$\cos A = \frac{\text{adj}}{\text{hyp}} = \frac{m}{p}$$
$$\tan A = \frac{\text{opp}}{\text{adj}} = \frac{n}{m}$$
$$\cot A = \frac{\text{adj}}{\text{opp}} = \frac{m}{n}$$
$$\sec A = \frac{\text{hyp}}{\text{adj}} = \frac{p}{m}$$
$$\csc A = \frac{\text{hyp}}{\text{opp}} = \frac{p}{n}$$

5. **(a)** $\tan 30° = \frac{\sqrt{3}}{3}$

(b) .5773502692

$$\left(\tan 30° \text{ or } \frac{\sqrt{3}}{3}\right)$$

7. **(a)** $\sin 30° = \frac{1}{2}$

(b) rational

9. **(a)** $\sec 30° = \frac{2}{\sqrt{3}} = \frac{2\sqrt{3}}{3}$

(b) 1.154700538

$$\left((\cos 30°)^{-1} \text{ or } \frac{2\sqrt{3}}{3}\right)$$

11. **(a)** $\csc 45° = \sqrt{2}$

(b) 1.414213562

$$\left((\sin 45°)^{-1} \text{ or } \sqrt{2}\right)$$

13. **(a)** $\cos 45° = \frac{\sqrt{2}}{2}$

(b) .7071067812

$$\left(\cos 45° \text{ or } \frac{\sqrt{2}}{2}\right)$$

15. **(a)** $\sin \frac{\pi}{3} = \frac{\sqrt{3}}{2}$

(b) .8660254038

$$\left(\sin \frac{\pi}{3} \text{ or } \frac{\sqrt{3}}{2}\right)$$

17. **(a)** $\tan \frac{\pi}{3} = \sqrt{3}$

(b) 1.732050808

$$\left(\tan \frac{\pi}{3} \text{ or } \sqrt{3}\right)$$

19. **(a)** $\sec \frac{\pi}{3} = 2$

(b) rational

21. The exact value of $\sin 45°$ is $\frac{\sqrt{2}}{2}$. The decimal value given is an approximation.

23. $\cot 73° = \tan(90° - 73°) = \boxed{\tan 17°}$

25. $\sin 38° = \cos(90° - 38°) = \boxed{\cos 52°}$

27. $$\begin{aligned}\tan 25°43' &= \cot\left(90° - 25°43'\right)\\ &= \cot\left(89°60' - 25°43'\right)\\ &= \boxed{\cot 64°17'}\end{aligned}$$

29. $\cos\frac{\pi}{5} = \sin\left(\frac{\pi}{2} - \frac{\pi}{5}\right)$

$= \sin\left(\frac{5\pi}{10} - \frac{2\pi}{10}\right) = \boxed{\sin\frac{3\pi}{10}}$

31. $\tan .5 = \boxed{\cot\left(\frac{\pi}{2} - .5\right)}$

33. $\theta = 98°;\quad \theta' = 180° - 98° = 82°$

35. $\theta = 230°;\quad \theta' = 230° - 180° = 50°$

37. $\theta = -135°;\quad \theta' = 180° + (-135°) = 45°$

39. $\theta = 750°;\quad \theta' = 750° - 720° = 30°$

41. $\theta = \frac{4\pi}{3};\quad \theta' = \frac{4\pi}{3} - \pi = \frac{4\pi - 3\pi}{3} = \frac{\pi}{3}$

43. $\theta = -\frac{4\pi}{3} = 2\pi - \frac{4\pi}{3} = \frac{2\pi}{3}$

$\theta' = \pi - \frac{2\pi}{3} = \frac{3\pi - 2\pi}{3} = \frac{\pi}{3}$

45. It is easy to find one-half of 2, which is 1. This is, then, the measure of the side opposite the 30° angle, and the ratios are easily found. Yes, any positive number could have been used.

47. $30°:\quad \frac{1}{2}\quad \frac{\sqrt{3}}{2}\quad \frac{\sqrt{3}}{3}\quad \sqrt{3}\quad \frac{2\sqrt{3}}{3}\quad 2$

49. $60°:\quad \frac{\sqrt{3}}{2}\quad \frac{1}{2}\quad \sqrt{3}\quad \frac{\sqrt{3}}{3}\quad 2\quad \frac{2\sqrt{3}}{3}$

51. $135°:\quad \frac{\sqrt{2}}{2}\quad -\frac{\sqrt{2}}{2}\quad -1\quad -1\quad -\sqrt{2}\quad \sqrt{2}$

53. $210°:\quad -\frac{1}{2}\quad -\frac{\sqrt{3}}{2}\quad \frac{\sqrt{3}}{3}\quad \sqrt{3}\quad -\frac{2\sqrt{3}}{3}\quad -2$

55.
$\sin 300° = -\sin 60° = -\frac{\sqrt{3}}{2}$

$\cos 300° = \cos 60° = \frac{1}{2}$

$\tan 300° = -\tan 60° = -\sqrt{3}$

$\cot 300° = -\cot 60° = -\frac{\sqrt{3}}{3}$

$\sec 300° = \sec 60° = 2$

$\csc 300° = -\csc 60° = -\frac{2\sqrt{3}}{3}$

57.
$\sin 405° = \sin 45° = \frac{\sqrt{2}}{2}$

$\cos 405° = \cos 45° = \frac{\sqrt{2}}{2}$

$\tan 405° = \tan 45° = 1$

$\cot 405° = \cot 45° = 1$

$\sec 405° = \sec 45° = \sqrt{2}$

$\csc 405° = \csc 45° = \sqrt{2}$

59.
$\sin\frac{11\pi}{6} = -\sin\frac{\pi}{6} = -\frac{1}{2}$

$\cos\frac{11\pi}{6} = \cos\frac{\pi}{6} = \frac{\sqrt{3}}{2}$

$\tan\frac{11\pi}{6} = -\tan\frac{\pi}{6} = -\frac{\sqrt{3}}{3}$

$\cot\frac{11\pi}{6} = -\cot\frac{\pi}{6} = -\sqrt{3}$

$\sec\frac{11\pi}{6} = \sec\frac{\pi}{6} = \frac{2\sqrt{3}}{3}$

$\csc\frac{11\pi}{6} = -\csc\frac{\pi}{6} = -2$

61.
$\sin\left(-\frac{7\pi}{4}\right) = \sin\frac{\pi}{4} = \frac{\sqrt{2}}{2}$

$\cos\left(-\frac{7\pi}{4}\right) = \cos\frac{\pi}{4} = \frac{\sqrt{2}}{2}$

$\tan\left(-\frac{7\pi}{4}\right) = \tan\frac{\pi}{4} = 1$

$\cot\left(-\frac{7\pi}{4}\right) = \cot\frac{\pi}{4} = 1$

$\sec\left(-\frac{7\pi}{4}\right) = \sec\frac{\pi}{4} = \sqrt{2}$

$\csc\left(-\frac{7\pi}{4}\right) = \csc\frac{\pi}{4} = \sqrt{2}$

63. $\tan 29° \approx \boxed{.5543090515}$

65. $\cot 41°24' = \cot 41.4° = (\tan 41.4°)^{-1}$
$\approx \boxed{1.134277349}$

67. $\sec 183°48' = \sec 183.8° = [\cos(183.8°)]^{-1}$
$\approx \boxed{-1.002203376}$

69. $\tan\left(-80°6'\right) = \tan(-80.1°)$
$\approx \boxed{-5.729741647}$

71. $\sin 2.5 \approx \boxed{.5984721441}$

73. $\tan 5 \approx \boxed{-3.380515006}$

75. $\frac{7\pi}{6}$ in Quadrant III

(a) $\sin\frac{7\pi}{6} = -\sin\frac{\pi}{6}$

(b) $-\sin\frac{\pi}{6} = -\frac{1}{2}$

(c) $\sin\frac{7\pi}{6} = -\sin\frac{\pi}{6} = -.5$

77. $\frac{3\pi}{4}$ in Quadrant II

(a) $\tan\frac{3\pi}{4} = -\tan\frac{\pi}{4}$

(b) $-\tan\frac{\pi}{4} = -1$

(c) $\tan\frac{3\pi}{4} = -\tan\frac{\pi}{4} = -1$

79. $\frac{7\pi}{6}$ in Quadrant III

(a) $\cos\frac{7\pi}{6} = -\cos\frac{\pi}{6}$

(b) $-\cos\frac{\pi}{6} = -\frac{\sqrt{3}}{2}$

(c) $\cos\frac{7\pi}{6} = -\cos\frac{\pi}{6} \approx -.8660254038$

81. $\sin\theta = \frac{1}{2}$

$\sin\theta$ is positive in Quads I and II;
since $\sin 30° = \frac{1}{2}$, the ref $\measuredangle = 30°$:

Quadrant I: $\boxed{30°}$

Quadrant II: $180° - 30° = \boxed{150°}$

83. $\tan\theta = -\sqrt{3}$

$\tan\theta$ is negative in Quadrants II and IV;
since $\tan 60° = \sqrt{3}$, the ref $\measuredangle = 60°$:

Quadrant II: $180° - 60° = \boxed{120°}$

Quadrant IV: $360° - 60° = \boxed{300°}$

85. $\cot\theta = -\frac{\sqrt{3}}{3}$

$\cot\theta$ is negative in Quadrants II and IV;
since $\cot 60° = \frac{\sqrt{3}}{3}$, the ref $\measuredangle = 60°$:

Quadrant II: $180° - 60° = \boxed{120°}$

Quadrant IV: $360° - 60° = \boxed{300°}$

87. $\cos\theta \approx .68716510$

Calculator: $\cos^{-1} .68716510$

Quadrant I: $\boxed{\theta \approx 46.59388121°}$

Quadrant IV: $360° - \theta'$

$\boxed{\theta \approx 313.4061188°}$

89. $\sin\theta \approx .41298643$

Calculator: $\sin^{-1} .41298643$

Quadrant I: $\boxed{\theta \approx 24.39257624°}$

Quadrant II: $180° - \theta'$

$\boxed{\theta \approx 155.6074238°}$

91. $\tan\theta \approx .87692035$

Calculator: $\tan^{-1} .87692035$

Quadrant I: $\boxed{\theta \approx 41.24818261°}$

Quadrant III: $180° + \theta'$

$\boxed{\theta \approx 221.2481826°}$

93. $\tan\theta \approx .21264138$

calculator: $\tan^{-1} .21264138$

Quadrant I: $\boxed{\theta \approx .2095206607}$

Quadrant III: $\pi + \theta$: $\boxed{\theta \approx 3.351113314}$

95. $\cot\theta \approx .29949853$

calculator: $\tan^{-1}(.29949853)^{-1}$

[Since $\cot\theta = (\tan\theta)^{-1}$
$(\tan\theta)^{-1} = .29949853$
$\tan\theta = (.29949853)^{-1}$
$\theta = \tan^{-1}(.29949853)^{-1}$]

Quadrant I: $\boxed{\theta \approx 1.27979966}$

Quadrant III: $\pi + \theta = \boxed{\theta \approx 4.421392314}$

Relating Concepts (97 - 108):

Graph for #97 – #108:

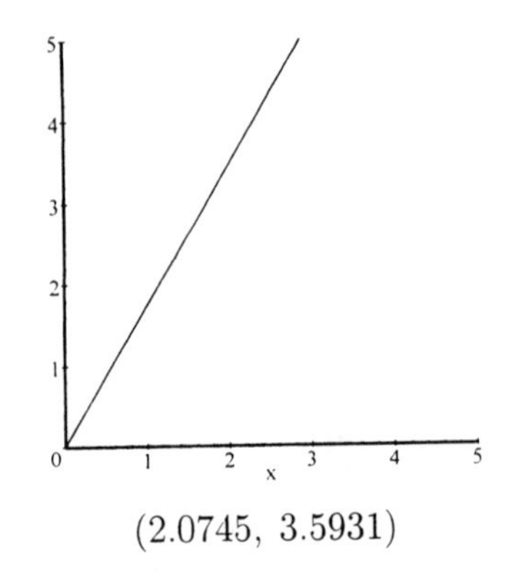

(2.0745, 3.5931)

97. $\sqrt{(x_1)^2 + (y_1)^2} = r$. Using the above point, $r \approx 4.14897$, represents the distance from the origin to the point (x_1, y_1).

98. $\tan^{-1}\left(\frac{y_1}{x_1}\right) = \tan^{-1}\left(\frac{3.5931}{2.0745}\right)$

$\approx 59.9997° \approx \boxed{60°}$

99. $\sin^{-1}\left(\frac{y_1}{r}\right) = \sin^{-1}\left(\frac{3.5931}{4.14897}\right)$

$\approx 59.9996° \approx \boxed{60°}$

100. $\cos^{-1}\left(\frac{x_1}{r}\right) = \cos^{-1}\left(\frac{2.0745}{4.14897}\right)$

$\approx 59.9996° \approx \boxed{60°}$

101. All angles appear to be 60°; the angle on the screen appears to be $\frac{2}{3}$ of the way from the x-axis to the y-axis, or 60°. It is the measure of the angle formed by the positive x-axis and the ray $y = \sqrt{3}\,x$, $x \geq 0$.

102. $\frac{y_1}{x_1} \approx 1.732032$

$(1.732032)^2 \approx 2.999934 \approx \boxed{3}$

If $x = 1$, $y = \sqrt{3}$, so $\frac{y_1}{x_1} = \sqrt{3}$ (exact).

103. The slope of a line passing through the origin is equal to the tangent of the angle it forms with the positive x-axis.

104. $\left(\frac{x_1}{r}\right)^2 + \left(\frac{y_1}{r}\right)^2 \approx 1$,

which illustrates the identity $\cos^2\theta + \sin^2\theta = 1$.

105. $\csc 60° = \frac{1}{\sin 60°} = \frac{2\sqrt{3}}{3} \approx 1.154700538$

$\frac{r}{y_1} = \frac{4.14897}{3.5931} \approx 1.154704851$

Both are approximately the same.

106. $y_1 = (\sqrt{3})\,x$ (dotted)

$y_2 = \sqrt{1 - x^2}$ (solid)

Point of intersection: (.5, .86602540)

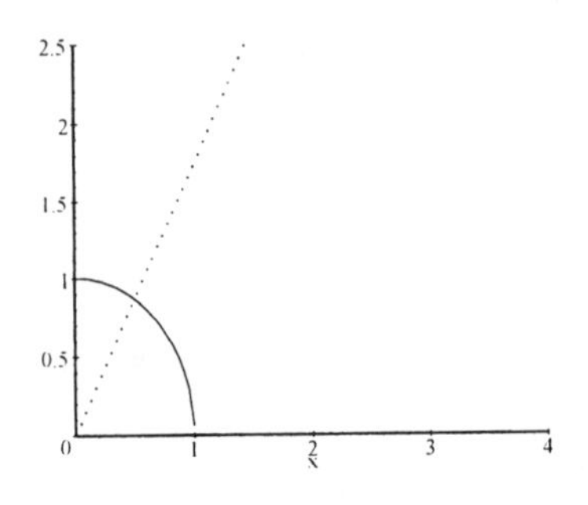

107. $\cos 60° = .5$; this is the x-coordinate of the point found in **106**. Because $r = 1$, $\cos 60° = \frac{x}{1} = x = .5$.

108. $\sin 60° \approx .8660254038$; this is the y-coordinate of the point found in **106**. Because $r = 1$,
$\sin 60° = \frac{y}{1} = y \approx .86602540$.

109. $\frac{c_1}{c_2} = \frac{\sin\theta_1}{\sin\theta_2}$ $\quad \theta_1 = 46°,\ \theta_2 = 31°,\ c_1 = 3 \times 10^8$ m/s

$$\frac{3 \times 10^8}{c_2} = \frac{\sin 46°}{\sin 31°}$$

$$c_2 = \frac{(3 \times 10^8)\sin 31°}{\sin 46°} \approx 214796154 \approx 2 \times 10^8 \text{ m/s}$$

111. $\frac{c_1}{c_2} = \frac{\sin\theta_1}{\sin\theta_2}$ $\quad \theta_1 = 40°,\ c_1 = 3 \times 10^8$ m/s, $c_2 = 1.5 \times 10^8$ m/s

$$\frac{3 \times 10^8}{1.5 \times 10^8} = \frac{\sin 40°}{\sin\theta_2}$$

$$(3 \times 10^8)\sin\theta_2 = (1.5 \times 10^8)\sin 40°$$

$$\sin\theta_2 = \frac{(1.5 \times 10^8)\sin 40°}{3 \times 10^8}$$

$$\theta_2 = \sin^{-1}\left(\frac{(1.5 \times 10^8)\sin 40°}{3 \times 10^8}\right)$$

$$\theta_2 \approx 18.747° \approx \boxed{19°}$$

113. $\frac{c_1}{c_2} = \frac{\sin\theta_1}{\sin\theta_2}$ $\quad \theta_1 = 90°,\ c_1 = 3 \times 10^8$ m/s, $c_2 = 2.254 \times 10^8$ m/s

$$\frac{3 \times 10^8}{2.254 \times 10^8} = \frac{\sin 90°}{\sin\theta_2}$$

$$(3 \times 10^8)\sin\theta_2 = (2.254 \times 10^8)\sin 90°$$

$$\sin\theta_2 = \frac{(2.254 \times 10^8)\sin 90°}{3 \times 10^8}$$

$$\theta_2 = \sin^{-1}\left(\frac{(2.254 \times 10^8)(1)}{3 \times 10^8}\right)$$

$$\theta_2 \approx 48.706° \approx \boxed{48.7°}$$

115. **(a)** $\frac{55 \text{ mi}}{1 \text{ hr}} \cdot \frac{5280 \text{ ft}}{1 \text{ mi}} \cdot \frac{1 \text{ hr}}{3600 \text{ sec}} = 80\frac{2}{3}$ ft/sec

$\frac{30 \text{ mi}}{1 \text{ hr}} \cdot \frac{5280 \text{ ft}}{1 \text{ mi}} \cdot \frac{1 \text{ hr}}{3600 \text{ sec}} = 44$ ft/sec

Let $\theta = 3.5°,\ K_1 = .4,\ K_2 = .02$

$$D = \frac{1.05\left(V_1{}^2 - V_2{}^2\right)}{64.4\left(K_1 + K_2 + \sin\theta\right)}$$

$$= \frac{1.05\left(\left(80\frac{2}{3}\right)^2 - 44^2\right)}{64.4\left(.4 + .02 + \sin 3.5°\right)}$$

$$\approx 154.9303 \approx \boxed{155 \text{ feet}}$$

(b) Let $\theta = -2°,\ K_1 = .4,\ K_2 = .02$

$$D = \frac{1.05\left(V_1{}^2 - V_2{}^2\right)}{64.4\left(K_1 + K_2 + \sin\theta\right)}$$

$$= \frac{1.05\left(\left(80\frac{2}{3}\right)^2 - 44^2\right)}{64.4\left(.4 + .02 + \sin(-2°)\right)}$$

$$\approx 193.5313 \approx \boxed{194 \text{ feet}}$$

(c) At a given speed, less distance is required to stop when traveling uphill compared to downhill. Yes, this agrees with experience.

Section 8.7

1. $A = 36°20'$, $c = 964$ m

(i) to find B:

$B = 90° - 36°20'$

$B = 89°60' - 36°20'$ $\boxed{B = 53°40'}$

(ii) to find a:

$\sin A = \frac{a}{c}$

$\sin 36°20' = \frac{a}{964}$

$a = 964 \sin 36°20'$

$a \approx 571.1526$ $\boxed{a \approx 571 \text{ m}}$

(iii) to find b:

$\cos A = \frac{b}{c}$

$\cos 36°20' = \frac{b}{964}$

$b = 964 \cos 36°20'$

$b \approx 776.5827$ $\boxed{b \approx 777 \text{ m}}$

3. $N = 51.2°$, $m = 124$ m

(i) to find M:

$M = 90.0° - 51.2°$ $\boxed{M = 38.8°}$

(ii) to find n:

$\tan N = \frac{n}{m}$

$\tan 51.2° = \frac{n}{124}$

$n = 124 \tan 51.2°$

$n \approx 154.2249$ $\boxed{n \approx 154 \text{ m}}$

(iii) to find p:

$\cos N = \frac{m}{p}$

$\cos 51.2° = \frac{124}{p}$

$p = \frac{124}{\cos 51.2°}$

$p \approx 197.8922$ $\boxed{p \approx 198 \text{ m}}$

5. $B = 42.0892°$, $b = 56.851$ cm

(i) to find A:

$A = 90.0° - 42.0892°$ $\boxed{A = 47.9108°}$

(ii) to find a:

$\tan B = \frac{b}{a}$

$a = \frac{b}{\tan B} = \frac{56.851}{\tan 42.0892°}$

$a \approx 62.942095$ $\boxed{a \approx 62.942 \text{ cm}}$

(iii) to find c:

$\sin B = \frac{b}{c}$

$c = \frac{b}{\sin B} = \frac{56.851}{\sin 42.0892°}$

$c \approx 84.81594$ $\boxed{c \approx 84.816 \text{ cm}}$

7. $A = 28.00°$, $c = 17.4$ ft

(i) to find B:

$B = 90.00° - 28.00°$ $\boxed{B = 62.00°}$

(ii) to find a:

$\sin A = \frac{a}{c}$

$\sin 28.00° = \frac{a}{17.4}$

$a = 17.4 \sin 28.00°$

$a \approx 8.1688$ $\boxed{a \approx 8.17 \text{ ft}}$

(iii) to find b:

$\cos A = \frac{b}{c}$

$\cos 28.00° = \frac{b}{17.4}$

$b = 17.4 \cos 28.00°$

$b \approx 15.3633$ $\boxed{b \approx 15.4 \text{ ft}}$

9. $B = 73.00^\circ$, $b = 128$ in

(i) to find A :

$A = 90.00^\circ - 73.00^\circ$ $\boxed{A = 17.00^\circ}$

(ii) to find a :

$\tan B = \dfrac{b}{a}$

$a = \dfrac{b}{\tan B} = \dfrac{128}{\tan 73.00^\circ}$

$a \approx 39.1335$ $\boxed{a \approx 39.1 \text{ in}}$

(iii) to find c :

$\sin B = \dfrac{b}{c}$

$c = \dfrac{b}{\sin B} = \dfrac{128}{\sin 73.00^\circ}$

$c \approx 133.849$ $\boxed{c \approx 134 \text{ in}}$

11. $a = 76.4$ yd, $b = 39.3$ yd

(i) to find c :

$$c^2 = a^2 + b^2 = (76.4)^2 + (39.3)^2 = 7381.45$$

$c \approx 85.9154$ $\boxed{c \approx 85.9 \text{ yd}}$

(ii) to find A :

$\tan A = \dfrac{a}{b}$

$A = \tan^{-1}\left(\dfrac{76.4}{39.3}\right)$

$A \approx 62.7788^\circ$ or $62^\circ 46' 44''$

$\boxed{A \approx 62.8^\circ \text{ or } 62^\circ 50'}$

(iii) to find B :

$\tan B = \dfrac{b}{a}$

$B = \tan^{-1}\left(\dfrac{39.3}{76.4}\right)$

$B \approx 27.2212^\circ$ or $27^\circ 13' 16''$

$\boxed{B \approx 27.2^\circ \text{ or } 27^\circ 10'}$

13. Finding the other acute angle would require the least amount of work.

15. $\tan \theta = \dfrac{y}{x}$

$\theta = \tan^{-1}\left(\dfrac{3.68}{4.6}\right)$

$\theta \approx 38.6598^\circ$

17. Because AD and BC are parallel, angle DAB is congruent to angle ABC, as they are alternate interior angles of the transversal AB. A theorem of elementary geometry assures us of this.

19. It is measured clockwise from the north.

21. $\sin 43^\circ 50' = \dfrac{h}{13.5}$

$h = 13.5 \sin 43^\circ 50'$

$h \approx 9.34959996$

$\boxed{\text{Distance} \approx 9.35 \text{ m}}$

23. $\tan 23.4^\circ = \dfrac{5.75}{x}$

$x = \dfrac{5.75}{\tan 23.4^\circ} \approx 13.2875$

$\boxed{\text{Shadow length} \approx 13.3 \text{ feet}}$

25. $h_1 = 30$ since the window is 30.0 feet above the ground.

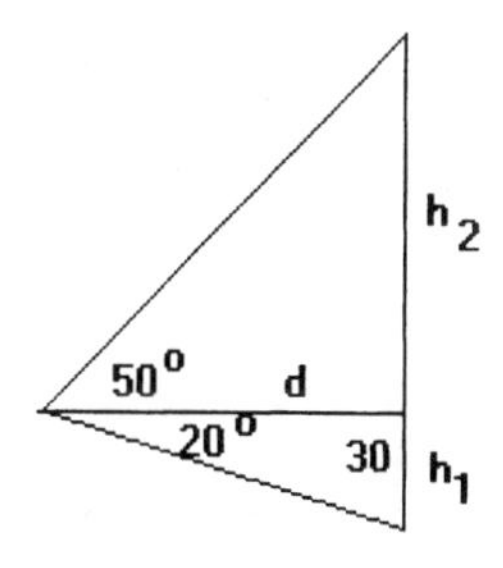

First, find d: $\tan 20.0^\circ = \dfrac{30.0}{d}$

$d = \dfrac{30.0}{\tan 20.0^\circ} \approx 82.424$

Then, h_2 : $\tan 50^\circ = \dfrac{h_2}{d} \approx \dfrac{h_2}{82.424}$

$h_2 \approx 82.424 \tan 50.0^\circ \approx 98.229$

height: $h_1 + h_2 \approx 30.0 + 98.229$

$\approx 128.229 \approx \boxed{128 \text{ ft}}$

27. First, find the complementary angle to the angle of depression:

$$\tan C = \frac{12.02}{5.93}$$

$$C = \tan^{-1}\left(\frac{12.02}{5.93}\right)$$

$$C \approx 63.74^\circ$$

Angle of depression:

$90^\circ - 63.74^\circ = 26.26^\circ$ or $26^\circ 15' 36''$

$\boxed{\approx 26.3^\circ \text{ or } 26^\circ 20'}$

29. (i) $\tan A = \dfrac{1.0837}{1.4923}$

$$A = \tan^{-1}\left(\frac{1.0837}{1.4923}\right)$$

$$A \approx 35.9869^\circ$$

$\boxed{A \approx 35.987^\circ \text{ or } 35^\circ 59' 10''}$

(ii) $B = 90^\circ - A$

$$B \approx 90^\circ - 35.987^\circ$$

$\boxed{B \approx 54.013^\circ \text{ or } 54^\circ 00' 50''}$

31. Let x be the base of the smaller right triangle.

(i) $\tan 21^\circ 10' = \dfrac{h}{135 + x}$

$$h = (135 + x)\tan 21^\circ 10'$$

(ii) $\tan 35^\circ 30' = \dfrac{h}{x}$

$$h = x \tan 35^\circ 30'$$

(iii) Set (i) = (ii) :

$$(135 + x)\tan 21^\circ 10' = x \tan 35^\circ 30'$$

$$135 \tan 21^\circ 10' = x \tan 35^\circ 30' - x \tan 21^\circ 10'$$

$$135 \tan 21^\circ 10' = x\left(\tan 35^\circ 30' - \tan 21^\circ 10'\right)$$

$$x = \frac{135 \tan 21^\circ 10'}{\tan 35^\circ 30' - \tan 21^\circ 10'}$$

$$x \approx 160.30258$$

(iv) $h = x \tan 35^\circ 30'$

$$h = 160.30258 \tan 35^\circ 30'$$

$$h \approx 114.343$$

$\boxed{\text{Height of pryamid} \approx 114 \text{ feet}}$

33. Let h = height of the house.

(i) $\tan 18^\circ 10' = \dfrac{h}{28.0}$

$$h = 28.0 \tan 18^\circ 10'$$

(ii) $\tan 27^\circ 10' = \dfrac{h + x}{28.0}$

$$h + x = 28.0 \tan 27^\circ 10'$$

(iii) Substitute (i) $\Rightarrow$ (ii) for h :

$$28.0 \tan 18^\circ 10' + x = 28.0 \tan 27^\circ 10'$$

$$x = 28.0 \tan 27^\circ 10' - 28.0 \tan 18^\circ 10'$$

$$x \approx 5.18157$$

$\boxed{\text{Height of antenna} \approx 5.18 \text{ m}}$

35. First, solve for the common hypotenuse by using the top triangle:

$$\cos 30^\circ 50' = \frac{198.4}{c}$$

$$c = \frac{198.4}{\cos 30^\circ 50'} \approx 231.05719$$

The smaller angle in the bottom triangle is equal to $21^\circ 30'$:

$$\sin 21^\circ 30' = \frac{x}{c}$$

$$x = c \sin 21^\circ 30' \approx 84.6827$$

$\boxed{x \approx 84.7 \text{ m}}$

37. The $\angle$ between the two ships is 90° $[180^\circ - \left(28^\circ 10' + 61^\circ 50'\right)]$; the first ship sails 96 miles (4×24), and the second ship sails 112 miles (4×28).

$$c^2 = a^2 + b^2$$

$$c^2 = 96^2 + 112^2$$

$$c^2 = 21760$$

$$c \approx 147.51$$

$\boxed{\text{Ships are} \approx 148 \text{ miles apart.}}$

39. From **38,** using angle $A = 53°40'$:

$$\cos 53°40' = \frac{d}{2.50}$$

$$d = 2.50 \cos 53°40'$$

$$d \approx 1.4812$$

$\boxed{\text{Distance} \approx 1.48\text{miles}}$

41. The angle at the top of the triangle is $90°\ (63° + 27°)$:

$$c^2 = a^2 + b^2$$

$$c^2 = 50^2 + 140^2$$

$$c = \sqrt{22100} \approx 148.66$$

$\boxed{\text{Distance traveled} \approx 150 \text{ km}}$

43. Let h = height of searchlight.

$$\tan 30° = \frac{h}{1000}$$

$$h = 1000 \tan 30° = \frac{1000\sqrt{3}}{3} \approx 577.35$$

Cloud ceiling $= 6 + 577.35 \approx \boxed{583 \text{ feet}}$

45. To find $\overline{RS}$: $\tan 32°\,10' = \dfrac{\overline{RS}}{53.1}$

$$\overline{RS} = 53.1 \tan 32°\,10'$$

$$\overline{RS} \approx 33.3957 \approx \boxed{33.4 \text{ m}}$$

47. $d = R\left(1 - \cos\dfrac{\beta}{2}\right)$

(a) $S = 336$ feet, $R = 600$ feet

$$\beta \approx \frac{57.3\,S}{R} \approx \frac{57.3\,(336)}{600} \approx 32.088°$$

$$d = 600\,(1 - \cos 16.044°) \approx 23.3702 \quad \boxed{\approx 23.4 \text{ feet}}$$

(b) $S = 485$ feet, $R = 600$ feet

$$\beta \approx \frac{57.3\,S}{R} \approx \frac{57.3\,(485)}{600} \approx 46.3175°$$

$$d = 600\,(1 - \cos 23.15875°) \approx 48.34877 \quad \boxed{\approx 48.3 \text{ feet}}$$

(c) The faster the speed, the more land needs to be cleared on the inside of the curve.

49. **(a)** $\tan\theta = \dfrac{y}{x}$

(b) $x\tan\theta = y$

$$x = \frac{y}{\tan\theta}$$

51. **a:** $\cos 60° = \dfrac{a}{24}$

$$a = 24\cos 60° = 24\left(\frac{1}{2}\right) = \mathbf{12}$$

b: $\sin 60° = \dfrac{b}{24}$

$$b = 24\sin 60° = 24\left(\frac{\sqrt{3}}{2}\right) = \mathbf{12\sqrt{3}}$$

c: $\cos 45° = \dfrac{b}{c}$; $\quad b = 12\sqrt{3}$

$$c = \frac{12\sqrt{3}}{\cos 45°} = \frac{12\sqrt{3}}{\sqrt{2}\,/\,2} = 12\sqrt{3}\cdot\frac{2}{\sqrt{2}}\cdot\frac{\sqrt{2}}{\sqrt{2}} = \mathbf{12\sqrt{6}}$$

d: $\sin 45° = \dfrac{d}{c}$; $\quad c = 12\sqrt{6}$

$$d = c\sin 45° = 12\sqrt{6}\cdot\frac{\sqrt{2}}{2} = 6\sqrt{12} = 6\,(2\sqrt{3}) = \mathbf{12\sqrt{3}}$$

53. **a:** $\sin 60° = \dfrac{7}{a}$

$$a = \frac{7}{\sin 60°} = \frac{7}{\sqrt{3}\,/\,2}$$

$$= 7 \cdot \frac{2}{\sqrt{3}} \cdot \frac{\sqrt{3}}{\sqrt{3}} = \mathbf{\frac{14\sqrt{3}}{3}}$$

n: $n = a$ (45° triangle), $n = \mathbf{\dfrac{14\sqrt{3}}{3}}$

m: $\cos 60° = \dfrac{m}{a}; \quad a = \dfrac{14\sqrt{3}}{3}$

$$m = a\cos 60° = \frac{14\sqrt{3}}{3} \cdot \frac{1}{2} = \mathbf{\frac{7\sqrt{3}}{3}}$$

q: $\cos 45° = \dfrac{n}{q}; \quad n = \dfrac{14\sqrt{3}}{3}$

$$q = \frac{n}{\cos 45°} = \frac{14\sqrt{3}}{3} \div \frac{\sqrt{2}}{2}$$

$$\frac{14\sqrt{3}}{3} \cdot \frac{2}{\sqrt{2}} \cdot \frac{\sqrt{2}}{\sqrt{2}} = \frac{28\sqrt{6}}{6} = \mathbf{\frac{14\sqrt{6}}{3}}$$

55. Bisect the upper angle, which will divide the triangle in half.

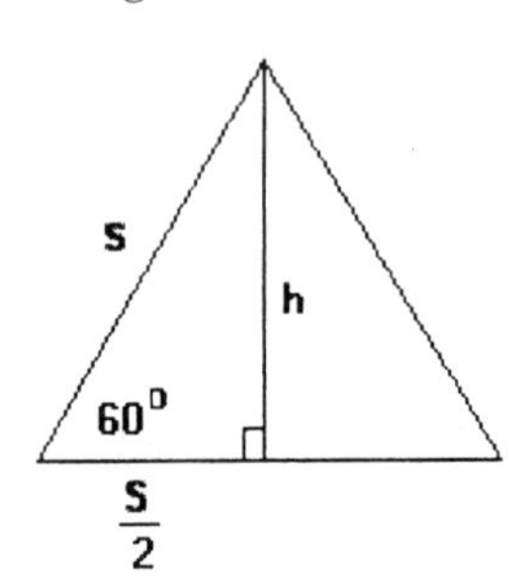

$$h: \sin 60° = \frac{h}{s}$$

$$h = s\sin 60° = \frac{s\sqrt{3}}{2}$$

So, area of small triangle:

$$A = \frac{1}{2}bh = \frac{1}{2} \cdot \frac{s}{2} \cdot \frac{s\sqrt{3}}{2} = \frac{s^2\sqrt{3}}{8}$$

Area of larger triangle:

$$A = 2 \cdot \frac{s^2\sqrt{3}}{8} = \frac{s^2\sqrt{3}}{4}$$

Section 8.8

1. **(a)** $s(t) = a\cos(\omega t)$

$(0,\ 2):\ \ 2 = a\cos 0 = a(1)$
$a = 2$

Since the period is .5 seconds,

$$\frac{2\pi}{\omega} = \frac{1}{2} \longrightarrow \omega = 4\pi$$

Thus $s(t) = a\cos(\omega t) = 2\cos(4\pi t)$.

(b) $s(1) = 2\cos(4\pi(1)) = 2$

The spring is compressed 2 inches one second after the weight is released. The weight is moving neither upward nor downward.

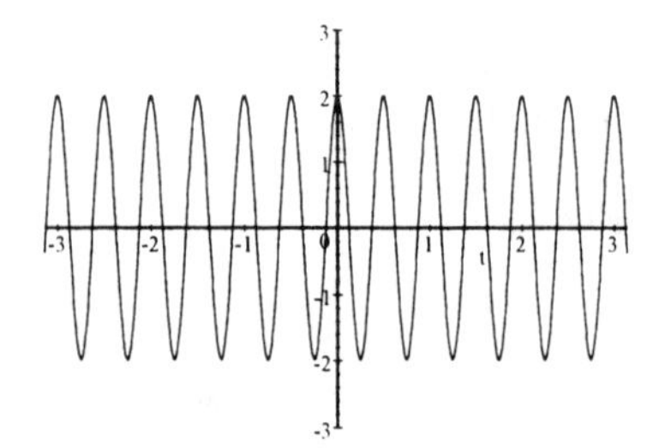

3. **(a)** $s(t) = a\cos(\omega t)$

$(0, -3):\ \ -3 = a\cos 0 = a(1)$
$a = -3$

Since the period is .8 seconds,

$$\frac{2\pi}{\omega} = \frac{8}{10} \longrightarrow \omega = \frac{20\pi}{8} = 2.5\pi$$

Thus $s(t) = a\cos(\omega t) = -3\cos(2.5\pi\, t)$.

(b) $s(1) = -3\cos(2.5\pi\,(1)) = -3(0) = 0$

The spring is at its natural length one second after the weight is released. The weight is moving upward.

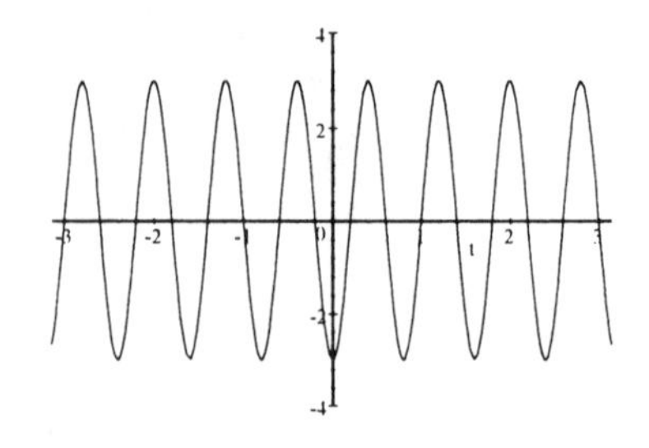

5. First note that

$b = \omega = 2\pi(27.5) = 55\pi$

Thus, $s(t) = a\cos(55\pi t)$.

$$s(t) = a\cos(\omega t)$$
$$.21 = a\cos 0 = a(1)$$
$$a = .21$$

The equation is $s(t) = .21\cos(55\pi t)$.

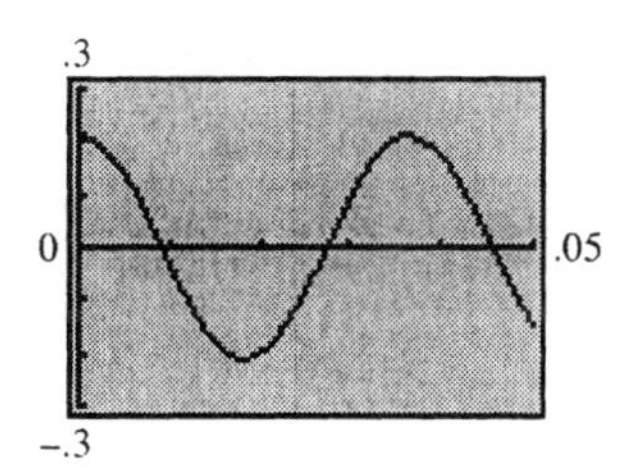

7. First note that

$b = \omega = 2\pi(55) = 110\pi$

Thus, $s(t) = a\cos(110\pi t)$.

$$s(t) = a\cos(110\pi t)$$
$$.14 = a\cos 0 = a(1)$$
$$a = .14$$

The equation is $s(t) = .14\cos(110\pi t)$.

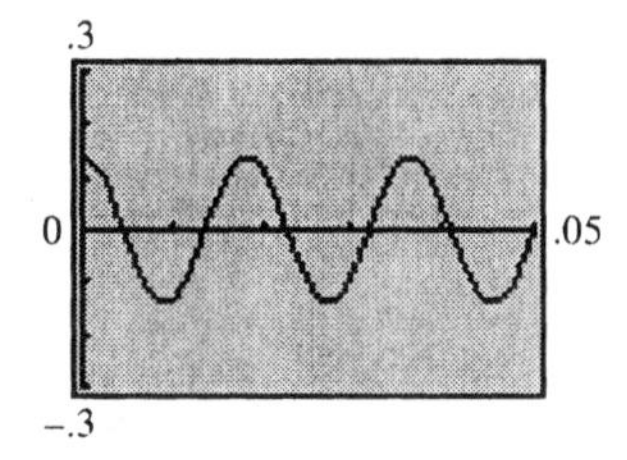

9. **(a)** $s(t) = a\sin\omega t;\ a = 2,\ \omega = |2| = 2$

$s(t) = 2\sin 2t$

$$\text{Amplitude} = |a| = |2| = 2$$
$$\text{period} = \frac{2\pi}{2} = \pi$$
$$\text{frequency} = \frac{\omega}{2\pi} = \frac{2}{2\pi} = \frac{1}{\pi}$$

(b) $s(t) = a\sin\omega t;\ a = 2,\ \omega = 4$

$s(t) = 2\sin 4t$

$$\text{Amplitude} = |a| = |2| = 2$$
$$\text{period} = \frac{2\pi}{4} = \frac{\pi}{2}$$
$$\text{frequency} = \frac{\omega}{2\pi} = \frac{4}{2\pi} = \frac{2}{\pi}$$

11. $g = 32$, Period $= 1$; $\quad \frac{2\pi}{\omega} = 1$

$$\omega = 2\pi$$

$$\omega = \sqrt{\frac{g}{l}}$$
$$2\pi = \sqrt{\frac{32}{l}}$$
$$(2\pi)^2 = \left(\sqrt{\frac{32}{l}}\right)^2$$
$$4\pi^2 = \frac{32}{l}$$
$$l = \frac{32}{4\pi^2} = \frac{8}{\pi^2}$$

13. $k = 2,\ m = 1$

(a) $\omega = \sqrt{\frac{k}{m}} = \sqrt{\frac{2}{1}} = \sqrt{2}$

$$\text{Amplitude} = a = \frac{1}{2}$$
$$\text{Period} = \frac{2\pi}{\omega} = \frac{2\pi}{\sqrt{2}} = \sqrt{2}\pi$$
$$\text{Frequency} = \frac{\omega}{2\pi} = \frac{\sqrt{2}}{2\pi}$$

(b) $s(t) = a\sin\omega t$

$$s(t) = \frac{1}{2}\sin\sqrt{2}\,t$$

15. $a = -4,\ \omega = 10$

(a) Maximum height $=$ amplitude $= 4$ inches

(b) Frequency $= \frac{\omega}{2\pi} = \frac{10}{2\pi} = \frac{5}{\pi}$ cycles/second

Period $= \frac{2\pi}{\omega} = \frac{\pi}{5}$ seconds

(c)
$$s(t) = a\sin\omega t$$
$$4 = -4\cos 10t$$
$$\cos 10t = -1$$
$$10t = \pi$$
$$t = \frac{\pi}{10}$$

The weight first reaches its maximum height after $\frac{\pi}{10}$ second.

(d) $s(1.466) = -4\cos(10(1.466)) \approx 2$

After 1.466 seconds, the weight is about 2 inches above the equilibrium position.

17. $a = -2$

(a) $\text{Period} = \dfrac{2\pi}{\omega} = \dfrac{1}{3}$

$\omega = 6\pi$

$$s(t) = a\cos\omega t$$
$$s(t) = -2\cos 6\pi t$$

(b) $\text{Frequency} = \dfrac{\omega}{2\pi} = \dfrac{6\pi}{2\pi} = 3 \text{ cycles/sec}$

Reviewing Basic Concepts (Section 8.5 and 8.8)

1. $x = -2,\ y = 5,\ r = \sqrt{(-2)^2 + 5^2} = \sqrt{29}$

$$\begin{aligned}
\sin\theta &= \frac{y}{r} = \frac{5}{\sqrt{29}} = \frac{5\sqrt{29}}{29} \\
\cos\theta &= \frac{x}{r} = \frac{-2}{\sqrt{29}} = -\frac{2\sqrt{29}}{29} \\
\tan\theta &= \frac{y}{x} = \frac{5}{-2} = -\frac{5}{2} \\
\cot\theta &= \frac{x}{y} = \frac{-2}{5} = -\frac{2}{5} \\
\sec\theta &= \frac{r}{x} = \frac{\sqrt{29}}{-2} = -\frac{\sqrt{29}}{2} \\
\csc\theta &= \frac{r}{y} = \frac{\sqrt{29}}{5}
\end{aligned}$$

2. $\sin 270° = -1$ $\quad \cos 270° = 0$

$\tan 270°$ is undefined $\quad \cot 270° = 0$

$\sec 270°$ is undefined $\quad \csc 270° = -1$

3. **(a)** $\cos\theta = \dfrac{3}{2}$; impossible since $-1 \le \cos\theta \le 1$.

(b) $\tan\theta = 300$; possible since the range of $\tan\theta$ is $(-\infty, \infty)$.

(c) $\csc\theta = 5$; possible since $\csc\theta \le -1$ or $\csc\theta \ge 1$.

4. $\sin\theta = -\dfrac{2}{3}$, quadrant III.

$$y = -2,\ r = 3,\ x = -\sqrt{3^2 - (-2)^2} = -\sqrt{5}$$

$$\begin{aligned}
\cos\theta &= \frac{x}{r} = \frac{-\sqrt{5}}{3} = -\frac{\sqrt{5}}{3} \\
\tan\theta &= \frac{y}{x} = \frac{-2}{-\sqrt{5}} = \frac{2\sqrt{5}}{5} \\
\cot\theta &= \frac{x}{y} = \frac{-\sqrt{5}}{-2} = \frac{\sqrt{5}}{2} \\
\sec\theta &= \frac{r}{x} = \frac{3}{-\sqrt{5}} = -\frac{3\sqrt{5}}{5} \\
\csc\theta &= \frac{r}{y} = \frac{3}{-2} = -\frac{3}{2}
\end{aligned}$$

5. $\sin A = \dfrac{8}{17} \quad \cos A = \dfrac{15}{17} \quad \tan A = \dfrac{8}{15}$

$\cot A = \dfrac{15}{8} \quad \sec A = \dfrac{17}{15} \quad \csc A = \dfrac{17}{8}$

6.

θ	$30° = \frac{\pi}{6}$	$45° = \frac{\pi}{4}$	$60° = \frac{\pi}{3}$
$\sin\theta$	$\frac{1}{2}$	$\frac{\sqrt{2}}{2}$	$\frac{\sqrt{3}}{2}$
$\cos\theta$	$\frac{\sqrt{3}}{2}$	$\frac{\sqrt{2}}{2}$	$\frac{1}{2}$
$\tan\theta$	$\frac{\sqrt{3}}{3}$	1	$\sqrt{3}$
$\cot\theta$	$\sqrt{3}$	1	$\frac{\sqrt{3}}{3}$
$\sec\theta$	$\frac{2\sqrt{3}}{3}$	$\sqrt{2}$	2
$\csc\theta$	2	$\sqrt{2}$	$\frac{2\sqrt{3}}{3}$

7. **(a)** $\sin 27° = \cos(90° - 27°) = \cos 63°$

(b)
$$\begin{aligned}
\tan\frac{\pi}{5} &= \cot\left(\frac{\pi}{2} - \frac{\pi}{5}\right) \\
&= \cot\frac{5\pi - 2\pi}{10} = \cot\frac{3\pi}{10}
\end{aligned}$$

8. **(a)** 100° : Reference angle $= 180^\circ - 100^\circ = 80^\circ$

(b) $-365^\circ : 720^\circ - 365^\circ = 355^\circ$

Reference angle $= 360^\circ - 355^\circ = 5^\circ$

(c) $\frac{8\pi}{3} : \frac{8\pi}{3} - 2\pi = \frac{8\pi}{3} - \frac{6\pi}{3} = \frac{2\pi}{3}$

Reference angle $= \pi - \frac{2\pi}{3} = \frac{3\pi - 2\pi}{3} = \frac{\pi}{3}$

9. 315°; quadrant IV,

Reference angle $= 360^\circ - 315^\circ = 45^\circ$

$$\sin 315^\circ = -\sin 45^\circ = -\frac{\sqrt{2}}{2}$$

$$\cos 315^\circ = \cos 45^\circ = \frac{\sqrt{2}}{2}$$

$$\tan 315^\circ = -\tan 45^\circ = -1$$

$$\cot 315^\circ = -\cot 45^\circ = -1$$

$$\sec 315^\circ = \sec 45^\circ = \sqrt{2}$$

$$\csc 315^\circ = -\csc 45^\circ = -\sqrt{2}$$

10. **(a)** $\sin 46^\circ 30' \approx .725374371$

(b) $\tan(-100^\circ) \approx 5.67128182$

(c) $\csc 4 = (\sin 4)^{-1} \approx -1.321348709$

11. $\tan\theta = -\frac{\sqrt{3}}{3}$; Reference angle $= 30^\circ$

$\tan\theta < 0$ in quadrants II and IV:

quadrant II: $\theta = 180^\circ - 30^\circ = 150^\circ$

quadrant IV: $\theta = 360^\circ - 30^\circ = 330^\circ$

12. $\sin\theta = .68163876 \Longrightarrow \sin^{-1} .68163876 = .75$

$\sin\theta > 0$ in quadrants I and II:

quadrant II: $\pi - .75 \approx 2.391592654$

13. $\tan 13.7^\circ = \frac{h}{x}$

$$x = \frac{h}{\tan 13.7^\circ} \quad [1] \quad \tan 10.4^\circ = \frac{h}{x+5} \quad [2]$$

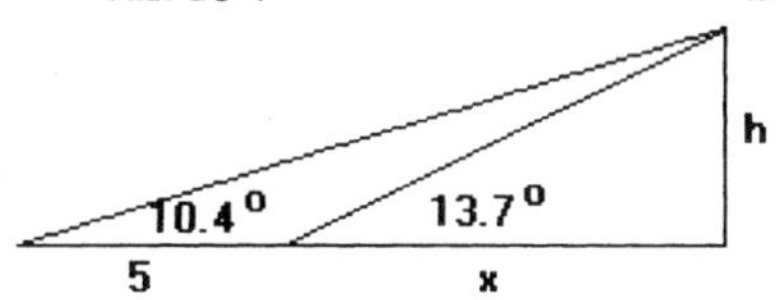

$$[1] \to [2] : \tan 10.4^\circ = \frac{h}{\frac{h}{\tan 13.7^\circ} + 5}$$

Multiply by $\tan 13.7^\circ$:

$$\tan 10.4^\circ = \frac{h}{\frac{h}{\tan 13.7^\circ} + 5} \cdot \frac{\tan 13.7^\circ}{\tan 13.7^\circ}$$

$$\tan 10.4^\circ = \frac{h \tan 13.7^\circ}{h + 5\tan 13.7^\circ}$$

$$\tan 10.4^\circ (h + 5\tan 13.7^\circ) = h \tan 13.7^\circ$$

$$h \tan 10.4^\circ + 5\tan 13.7^\circ \cdot \tan 10.4^\circ = h \tan 13.7^\circ$$

$$h \tan 10.4^\circ - h\tan 13.7^\circ = -5\tan 13.7^\circ \cdot \tan 10.4^\circ$$

$$h(\tan 10.4^\circ - \tan 13.7^\circ) = -5\tan 13.7^\circ \cdot \tan 10.4^\circ$$

$$h = \frac{-5\tan 13.7^\circ \cdot \tan 10.4^\circ}{\tan 10.4^\circ - \tan 13.7^\circ}$$

$$h \approx \frac{3.713588 \text{ miles}}{1} \cdot \frac{5280 \text{ ft}}{1 \text{ mile}}$$

$$h \approx 19{,}607.7 \approx 19{,}600 \text{ feet}$$

14. $s(t) = -4\cos 8\pi t$

(a) The amplitude $= |-4| = 4$, so the maximum height is 4 inches.

(b)

$$s(t) = -4\cos 8\pi t$$
$$4 = -4\cos 8\pi t$$
$$\cos 8\pi t = -1$$
$$8\pi t = \pi$$
$$t = \frac{\pi}{8\pi} = \frac{1}{8}$$

The maximum height is reached after $\frac{1}{8}$ second.

(c) Period $= \frac{2\pi}{8\pi} = \frac{1}{4}$

Frequency $= \frac{1}{\text{period}} = 4$ cycles/sec

Chapter 8 Review

1. $-174° + 360° = 186°$

2. Let n represent any integer. Any angle coterminal with $270°$ would be $270° + n \cdot 360°$.

3. Rotates 320 times/minute:

Each rotation is $360°$:

$$\begin{aligned}\omega &= 320\,(360°)\\ &= 115,200°/\text{min}\\ &= \frac{115,200°}{1 \text{ min}} \cdot \frac{1 \text{ min}}{60 \text{ sec}}\\ &= 1920°/\text{sec}\\ &= 1920°\left(\frac{2}{3}\right) \text{ per } \frac{2}{3} \text{ sec}\\ &= \boxed{1280°} \text{ per } \frac{2}{3} \text{ second}\end{aligned}$$

4. Rotates 650 times/minute:

Each rotation is $360°$:

$$\begin{aligned}\omega &= 650\,(360°)\\ &= 234,000°/\text{min}\\ &= \frac{234,000°}{1 \text{ min}} \cdot \frac{1 \text{ min}}{60 \text{ sec}}\\ &= 3900°/\text{sec}\\ &= 3900°\,(2.4)\,/2.4 \text{ sec}\\ &= \boxed{9360°} \text{ per 2.4 second}\end{aligned}$$

5. Since $1° = \frac{1°}{1} \cdot \frac{\pi}{180°} \approx .0175$ radians

1 radian is larger than 1 degree.

6. **(a)** $\frac{\pi}{2} < 3 < \pi$, so the terminal side is in quadrant II.

(b) $\pi < 4 < \frac{3\pi}{2}$, so the terminal side is in quadrant III.

(c) $-\frac{\pi}{2} > -2 > -\pi$, so the terminal side is in quadrant III.

7. $120° = \frac{120}{1} \cdot \frac{\pi}{180°} = \frac{2\pi}{3}$

8. $800° = \frac{800}{1} \cdot \frac{\pi}{180°} = \frac{40\pi}{9}$

9. $\frac{5\pi}{4} = \frac{5\pi}{4} \cdot \frac{180°}{\pi} = 225°$

10. $-\frac{6\pi}{5} = -\frac{6\pi}{5} \cdot \frac{180°}{\pi} = -216°$

11. $\frac{20}{60} = \frac{1}{3}$ rotation

$$\theta = \frac{1}{3}\,(2\pi) = \frac{2\pi}{3}$$

$$s = r\theta = 2\left(\frac{2\pi}{3}\right) = \frac{4\pi}{3} \text{ inches}$$

12. $\theta = 3\,(2\pi) = 6\pi$

$$s = r\theta = 2\,(6\pi) = 12\pi \text{ inches}$$

13. $r = 15.2$ cm, $\theta = \frac{3\pi}{4}$

$$\begin{aligned}s &= r\theta\\ &= 15.2\left(\frac{3\pi}{4}\right)\\ &= 11.4\pi \approx \boxed{35.8 \text{ cm}}\end{aligned}$$

14. $r = 28.69$ in, $\theta = \frac{7\pi}{4}$

$$\begin{aligned}A &= \frac{1}{2}r^2\theta\\ &= \frac{1}{2}\,(28.69)^2\left(\frac{7\pi}{4}\right)\\ &\approx \boxed{2263 \text{ square inches}}\end{aligned}$$

15. Because the central angle is very small, the arc length is approximately equal to the length of the inscribed chord.

Let h = height of tree, $r = 2000$, $\theta = 1°10'$

$$\begin{aligned}h &= r\theta\\ &= 2000\left(1 + \frac{10}{60}\right)^{\circ}\left(\frac{\pi}{180°}\right)\\ &\approx \boxed{41 \text{ yards}}\end{aligned}$$

or $\tan 1°10' = \frac{h}{2000} \Rightarrow h \approx \boxed{41 \text{ yards}}$

16. $s = 4,\ r = 8$

$$\theta = \frac{s}{r} = \frac{4}{8} = \frac{1}{2} \text{ radians}$$

$$\begin{aligned} A &= \frac{1}{2}r^2\theta \\ &= \frac{1}{2}(8)^2\left(\frac{1}{2}\right) \\ &= \boxed{16 \text{ square units}} \end{aligned}$$

17. $\cos\frac{2\pi}{3} = -\cos\frac{\pi}{3} = -\frac{1}{2}$

18. $$\begin{aligned} \tan\left(-\frac{7\pi}{3}\right) &= \tan\left(\frac{6\pi}{3} - \frac{7\pi}{3}\right) \\ &= \tan\left(-\frac{\pi}{3}\right) \\ &= -\tan\frac{\pi}{3} = -\sqrt{3} \end{aligned}$$

19. $$\begin{aligned} \csc\left(-\frac{11\pi}{6}\right) &= \csc\left(\frac{12\pi}{6} - \frac{11\pi}{6}\right) \\ &= \csc\frac{\pi}{6} = 2 \end{aligned}$$

20. $\cos(-.2443) \approx -.97030688 \approx -.9703$

21. $\cot 3.0543 \approx -11.426605 \approx -11.4266$

22. $\sin s = .49244294$

$s = \sin^{-1} .49244294 \approx .51489440 \approx .5149$

23. Let $|s|$ = the length of the arc.

$$\tan s = \frac{Y}{X} = \frac{-.5250622}{.85106383} \approx -.6170$$

$$s = \tan^1 -.6170 \approx -.5528$$

The length of the shortest arc of the circle from (1, 0) to (.85106383, −.5250622) is .5528

24. $y = 4\sin 2x$: **B**

The amplitude is 4 and the period is $\frac{2\pi}{2} = \pi$.

25. $y = 2\sin x$

amplitude: 2
period: 2π
vertical translation: none
phase shift: none

26. $y = \tan 3x$

amplitude: not applicable
period: $\frac{\pi}{3}$
vertical translation: none
phase shift: none

27. $y = -\frac{1}{2}\cos 3x$

amplitude: $\frac{1}{2}$
period: $\frac{2\pi}{3}$
vertical translation: none
phase shift: none

28. $y = 2\sin 5x$

amplitude: 2
period: $\frac{2\pi}{5}$
vertical translation: none
phase shift: none

29. $y = 1 + 2\sin\frac{1}{4}x$

amplitude: 2
period: $\frac{2\pi}{1/4} = 8\pi$
vertical translation: up 1 unit
phase shift: none

30. $y = 3 - \frac{1}{4}\cos\frac{2}{3}x$

amplitude: $\left|-\frac{1}{4}\right| = \frac{1}{4}$
period: $\frac{2\pi}{2/3} = 3\pi$
vertical translation: up 3 units
phase shift: none

31. $y = 3\cos\left(x + \frac{\pi}{2}\right)$

amplitude: 3
period: 2π
vertical translation: none
phase shift: $\frac{\pi}{2}$ units to the left

32. $y = -\sin\left(x - \frac{3\pi}{4}\right)$

amplitude: $|-1| = 1$
period: 2π
vertical translation: none
phase shift: $\frac{3\pi}{4}$ units to the right

33. $y = \frac{1}{2}\csc\left(2x - \frac{\pi}{4}\right)$

amplitude: not applicable
period: $\frac{2\pi}{2} = \pi$
vertical translation: none
phase shift: $\frac{\pi/4}{2} = \frac{\pi}{8}$ units to the right

34. $y = a \sin bx$:

If $\left(\frac{6\pi}{5}, 1\right)$ is on the graph, then $\left(-\frac{4\pi}{5}, 1\right)$ is also on the graph since

$$1 = f\left(\frac{6\pi}{5}\right) = f\left(\frac{6\pi}{5} - 2\pi\right) = f\left(-\frac{4\pi}{5}\right).$$

35. $y = 3\cos 2x$

Amplitude: 3

Period $= \frac{2\pi}{2} = \pi$

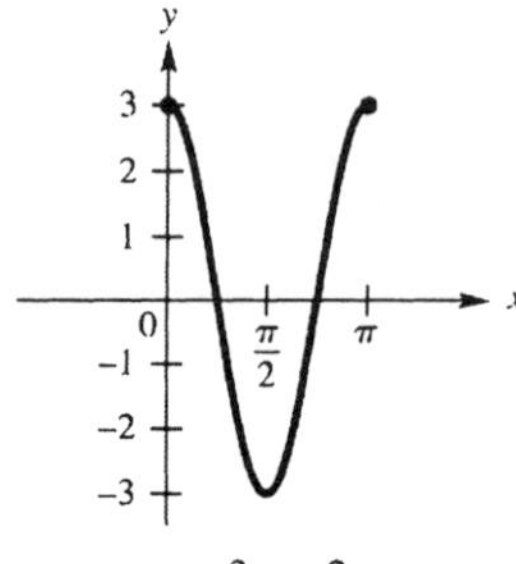

$y = 3\cos 2x$

36. $y = \frac{1}{2}\cot 3x$

Period $= \frac{\pi}{3}$

Asymptotes: $x = 0$, $x = \frac{\pi}{3}$

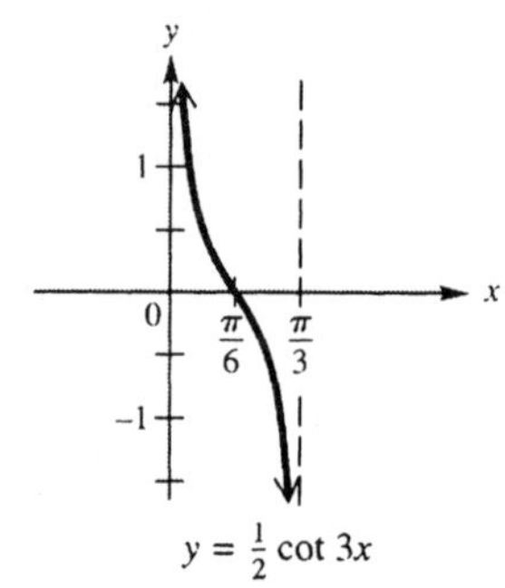

$y = \frac{1}{2}\cot 3x$

37. $y = \cos\left(x - \frac{\pi}{4}\right)$

Amplitude: 1

Period $= 2\pi$

Phase shift: $\frac{\pi}{4}$ units to the right

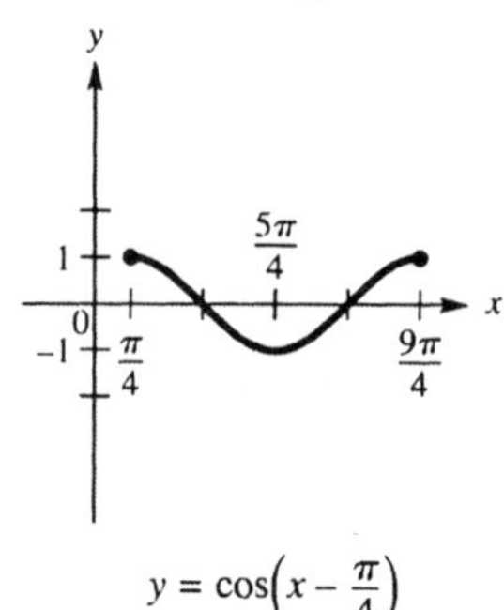

$y = \cos\left(x - \frac{\pi}{4}\right)$

38. $y = \tan\left(x - \frac{\pi}{2}\right)$

Period $= \pi$

Asymptotes: $x = 0$, $x = \pi$

Phase shift: $\frac{\pi}{2}$ units to the right

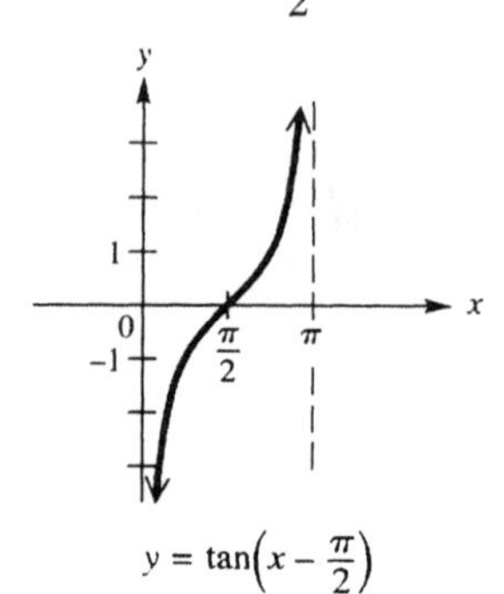

$y = \tan\left(x - \frac{\pi}{2}\right)$

39. $y = 1 + 2\cos 3x$

Amplitude: 2

Period $= \frac{2\pi}{3}$

Vertical translation: 1 unit up

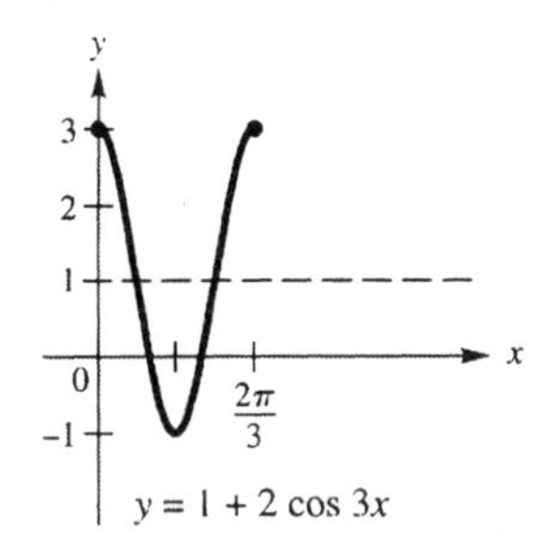

40. $y = -1 - 3\sin 2x$

Amplitude: 3

Period $= \frac{2\pi}{2} = \pi$

Vertical translation: down 1 unit

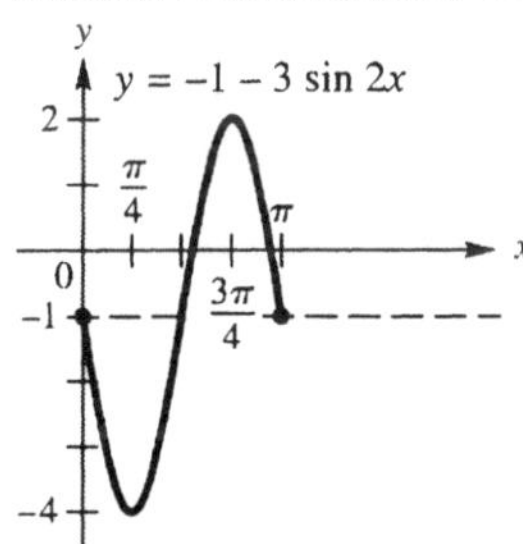

41. Period is π ($\tan x$ or $\cot x$)

x-intercepts: $n\pi$, $n \in$ I. $\boxed{\text{tangent}}$

42. Period is 2π, so it is not tangent or cotangent; passes through $(0, 0)$: $\boxed{\text{sine}}$

43. Period is 2π, so it is not tangent or cotangent passes through $\left(\frac{\pi}{2}, 0\right)$: $\boxed{\text{cosine}}$

44. Period is 2π, so it is not tangent or cotangent domain: $\{x \mid x \neq n\pi,\ n \in \text{I}\}$

$\boxed{\text{cosecant}}$

45. Period is π, so it is tangent or cotangent; decreasing $(0, \pi)$: $\boxed{\text{cotangent}}$

46. Period is 2π, so it is not tangent or cotangent; vertical asymptote is $x = (2n+1)\frac{\pi}{2}$, $n \in$ I: $\boxed{\text{secant}}$

47. $y = a\sin b(x - c)$

Amplitude $= 3 \Rightarrow a = 3$

Period $= \pi$

$\frac{2\pi}{b} = \pi \Rightarrow b = 2$

$(0, 0) \rightarrow \left(\frac{\pi}{4}, 0\right)$

Phase shift $= \frac{\pi}{4}$

$\boxed{y = 3\sin\left[2\left(x - \frac{\pi}{4}\right)\right]}$

Other answers are possible.

48. $y = a\sin b(x - c)$

Amplitude $= 4 \Rightarrow a = 4$

Period $= 4\pi$

$\frac{2\pi}{b} = 4\pi \Rightarrow b = \frac{1}{2}$

No phase shift

$\boxed{y = 4\sin\frac{1}{2}x}$

Other answers are possible.

49. $y = a\sin b(x - c)$

Amplitude $= \frac{1}{3} \Rightarrow a = \frac{1}{3}$

Period $= 4$

$\frac{2\pi}{b} = 4 \Rightarrow b = \frac{\pi}{2}$

No phase shift

$\boxed{y = \frac{1}{3}\sin\frac{\pi}{2}x}$

Other answers are possible.

50. $y = a \sin b(x - c)$

Amplitude $= \pi \Rightarrow a = \pi$

Period $= 2$

$$\frac{2\pi}{b} = 2 \Rightarrow b = \pi$$

Phase shift $= \frac{1}{2}$

$$\boxed{y = \pi \sin\left[\pi\left(x - \frac{1}{2}\right)\right]}$$

Other answers are possible.

51. **(a)** Let January correspond to $x = 1$, February to $x = 2$, ... , and December of the 2nd year to $x = 24$. The data appear to follow the pattern of a translated sine graph.

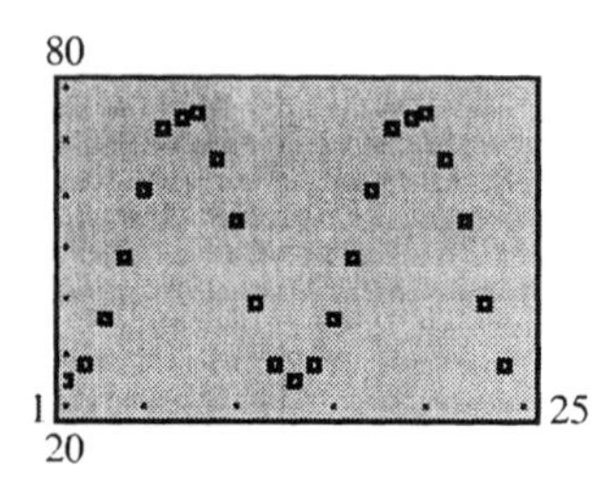

(b) $f(x) = a \sin b(x - d) + c$

The maximum average monthly temperature is 75° F, and the minimum is 25° F. Let the amplitude a be $\frac{75 - 25}{2} = 25$.

Since the period is $12 = \frac{2\pi}{b}$, $b = \frac{\pi}{6}$.

The data are centered vertically around the line $y = \frac{75 + 25}{2} = 50$, so $c = 50$.

The minimum temperature occurs in January. Thus, when $x = 1$, $b(x - d) = -\frac{\pi}{2}$ since the sine function is minimum at $-\frac{\pi}{2}$.

Solving for d: $\frac{\pi}{6}(1 - d) = -\frac{\pi}{2} \Longrightarrow d = 4$.

d can be adjusted slightly to give a better visual fit. Try $d = 4.2$. Thus,

$$\begin{aligned} f(x) &= a \sin b(x - d) + c \\ &= 25 \sin\left[\frac{\pi}{6}(x - 4.2)\right] + 50 \end{aligned}$$

(c) a controls the amplitude, b the period, c the vertical shift, and d the phase shift.

continued

51. continued

(d) The function gives an excellent model.

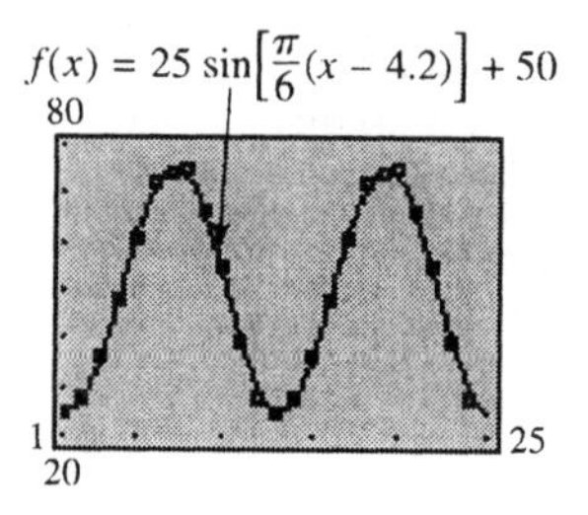

(e) Regression equation:

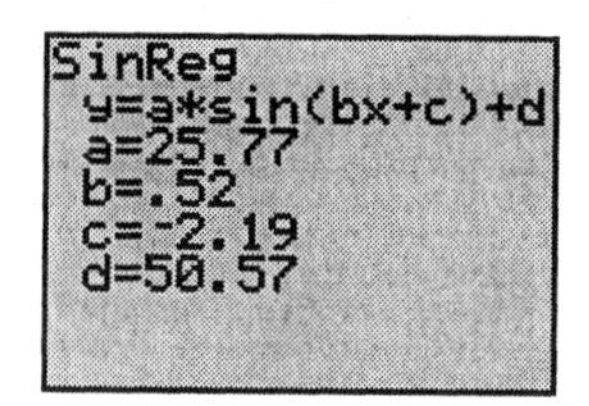

52. $P(t) = 7(1 - \cos 2\pi t)(t + 10) + 100e^{.2t}$

(a) January 1, base year: $t = 0$

$$\begin{aligned} P(0) &= 7(1 - \cos 0)(10) + 100e^{0} \\ &= 7(1 - 1)(10) + 100(1) \\ &= 100 \end{aligned}$$

(b) July 1, base year: $t = .5$

$$\begin{aligned} P(.5) &= 7(1 - \cos \pi)(10.5) + 100e^{.2(.5)} \\ &= 7(1 - (-1))(10.5) + 100e^{.1} \\ &\approx 258 \end{aligned}$$

(c) July 1, following year: $t = 1$

$$\begin{aligned} P(1) &= 7(1 - \cos 2\pi)(11) + 100e^{.2} \\ &= 7(1 - 1)(11) + 100e^{.2} \\ &\approx 122 \end{aligned}$$

(d) July 1, following year: $t = 1.5$

$$\begin{aligned} P(1.5) &= 7(1 - \cos 3\pi)(11.5) + 100e^{.2(1.5)} \\ &= 7(1 - (-1))(11.5) + 100e^{.3} \\ &\approx 296 \end{aligned}$$

53. $(-3,-3)$; $\quad r=\sqrt{(-3)^2+(-3)^2}$
$=\sqrt{9+9}=\sqrt{18}=3\sqrt{2}$

$$\sin\theta = \frac{y}{r} = \frac{-3}{3\sqrt{2}} = -\frac{\sqrt{2}}{2}$$
$$\cos\theta = \frac{x}{r} = \frac{-3}{3\sqrt{2}} = -\frac{\sqrt{2}}{2}$$
$$\tan\theta = \frac{y}{x} = \frac{-3}{-3} = 1$$
$$\cot\theta = \frac{x}{y} = \frac{-3}{-3} = 1$$
$$\sec\theta = \frac{r}{x} = \frac{3\sqrt{2}}{-3} = -\sqrt{2}$$
$$\csc\theta = \frac{r}{y} = \frac{3\sqrt{2}}{-3} = -\sqrt{2}$$

54. $(1,-\sqrt{3})$; $\quad r=\sqrt{(1)^2+(-\sqrt{3})^2}$
$=\sqrt{1+3}=\sqrt{4}=2$

$$\sin\theta = \frac{y}{r} = \frac{-\sqrt{3}}{2} = -\frac{\sqrt{3}}{2}$$
$$\cos\theta = \frac{x}{r} = \frac{1}{2} = \frac{1}{2}$$
$$\tan\theta = \frac{y}{x} = \frac{-\sqrt{3}}{1} = -\sqrt{3}$$
$$\cot\theta = \frac{x}{y} = \frac{1}{-\sqrt{3}} = -\frac{\sqrt{3}}{3}$$
$$\sec\theta = \frac{r}{x} = \frac{2}{1} = 2$$
$$\csc\theta = \frac{r}{y} = \frac{2}{-\sqrt{3}} = -\frac{2\sqrt{3}}{3}$$

55. $\sin 180^\circ = 0$
$\cos 180^\circ = -1$
$\tan 180^\circ = 0$
$\cot 180^\circ$ is undefined
$\sec 180^\circ = -1$
$\csc 180^\circ$ is undefined

56. $(3,\ -4)$; $\quad r=\sqrt{(3)^2+(-4^2)}$
$=\sqrt{9+16}=\sqrt{25}=5$

$$\sin\theta = \frac{y}{r} = \frac{-4}{5} = -\frac{4}{5}$$
$$\cos\theta = \frac{x}{r} = \frac{3}{5} = \frac{3}{5}$$
$$\tan\theta = \frac{y}{x} = \frac{-4}{3} = -\frac{4}{3}$$
$$\cot\theta = \frac{x}{y} = \frac{3}{-4} = -\frac{3}{4}$$
$$\sec\theta = \frac{r}{x} = \frac{5}{3} = \frac{5}{3}$$
$$\csc\theta = \frac{r}{y} = \frac{5}{-4} = -\frac{5}{4}$$

57. $(9,\ -2)$; $\quad r=\sqrt{(9)^2+(-2^2)}$
$=\sqrt{81+4}=\sqrt{85}$

$$\sin\theta = \frac{y}{r} = \frac{-2}{\sqrt{85}} = -\frac{2\sqrt{85}}{85}$$
$$\cos\theta = \frac{x}{r} = \frac{9}{\sqrt{85}} = \frac{9\sqrt{85}}{85}$$
$$\tan\theta = \frac{y}{x} = \frac{-2}{9} = -\frac{2}{9}$$
$$\cot\theta = \frac{x}{y} = \frac{9}{-2} = -\frac{9}{2}$$
$$\sec\theta = \frac{r}{x} = \frac{\sqrt{85}}{9} = \frac{\sqrt{85}}{9}$$
$$\csc\theta = \frac{r}{y} = \frac{\sqrt{85}}{-2} = -\frac{\sqrt{85}}{2}$$

58. $(-2\sqrt{2},\ 2\sqrt{2})$; $\quad r=\sqrt{(-2\sqrt{2})^2+\left(2\sqrt{2}^2\right)}$
$=\sqrt{8+8}=\sqrt{16}=4$

$$\sin\theta = \frac{y}{r} = \frac{2\sqrt{2}}{4} = \frac{\sqrt{2}}{2}$$
$$\cos\theta = \frac{x}{r} = \frac{-2\sqrt{2}}{4} = -\frac{\sqrt{2}}{2}$$
$$\tan\theta = \frac{y}{x} = \frac{2\sqrt{2}}{-2\sqrt{2}} = -1$$
$$\cot\theta = \frac{x}{y} = \frac{-2\sqrt{2}}{2\sqrt{2}} = -1$$
$$\sec\theta = \frac{r}{x} = \frac{4}{-2\sqrt{2}} = -\sqrt{2}$$
$$\csc\theta = \frac{r}{y} = \frac{4}{2\sqrt{2}} = \sqrt{2}$$

59. If the terminal side of a quadrantal angle lies along the y-axis, a point on the terminal side would be of the form $(0, k)$, where k is a real number, $k \neq 0$.

$$\begin{aligned}
\sin\theta &= \frac{y}{r} = \frac{k}{r} \\
\cos\theta &= \frac{x}{r} = \frac{0}{r} = 0 \\
\tan\theta &= \frac{y}{x} = \frac{k}{0}, \text{ undefined} \\
\cot\theta &= \frac{x}{y} = \frac{0}{k} = 0 \\
\sec\theta &= \frac{r}{x} = \frac{r}{0}, \text{ undefined} \\
\csc\theta &= \frac{r}{y} = \frac{r}{k}
\end{aligned}$$

The tangent and secant are undefined.

60. For any angle θ, $\sec\theta \leq -1$ or $\sec\theta \geq 1$. Therefore $\sec\theta = -\frac{2}{3}$ is impossible.

61. $\tan\theta$ can take on all values, so $\tan\theta = 1.4$ is possible.

62. $\sin\theta = \frac{\sqrt{3}}{5}$, $\cos\theta < 0$

$$\begin{aligned}
r^2 &= x^2 + y^2 \\
25 &= x^2 + 3 \\
x^2 &= 22 \\
x &= \pm\sqrt{22}
\end{aligned}$$

if $\cos\theta < 0$, $x = -\sqrt{22}$

$$\begin{aligned}
\cos\theta &= \frac{x}{r} = \frac{-\sqrt{22}}{5} = -\frac{\sqrt{22}}{5} \\
\tan\theta &= \frac{y}{x} = \frac{\sqrt{3}}{-\sqrt{22}} \cdot \frac{\sqrt{22}}{\sqrt{22}} = -\frac{\sqrt{66}}{22} \\
\cot\theta &= \frac{x}{y} = \frac{-\sqrt{22}}{\sqrt{3}} \cdot \frac{\sqrt{3}}{\sqrt{3}} = -\frac{\sqrt{66}}{3} \\
\sec\theta &= \frac{r}{x} = \frac{5}{-\sqrt{22}} \cdot \frac{\sqrt{22}}{\sqrt{22}} = -\frac{5\sqrt{22}}{22} \\
\csc\theta &= \frac{r}{y} = \frac{5}{\sqrt{3}} \cdot \frac{\sqrt{3}}{\sqrt{3}} = \frac{5\sqrt{3}}{3}
\end{aligned}$$

63. $\cos\theta = \frac{x}{r} = \frac{-5}{8}$, θ in Quadrant III

$$\begin{aligned}
r^2 &= x^2 + y^2 \\
64 &= 25 + y^2 \\
y^2 &= 39 \\
y &= \pm\sqrt{39}
\end{aligned}$$

Since θ is in Quadrant III, $y = -\sqrt{39}$

$$\begin{aligned}
\sin\theta &= \frac{y}{r} = \frac{-\sqrt{39}}{8} = -\frac{\sqrt{39}}{8} \\
\tan\theta &= \frac{y}{x} = \frac{-\sqrt{39}}{-5} = \frac{\sqrt{39}}{5} \\
\cot\theta &= \frac{x}{y} = \frac{-5}{-\sqrt{39}} \cdot \frac{\sqrt{39}}{\sqrt{39}} = \frac{5\sqrt{39}}{39} \\
\sec\theta &= \frac{r}{x} = \frac{8}{-5} = -\frac{8}{5} \\
\csc\theta &= \frac{r}{y} = \frac{8}{-\sqrt{39}} \cdot \frac{\sqrt{39}}{\sqrt{39}} = -\frac{8\sqrt{39}}{39}
\end{aligned}$$

64. The sine is negative in quadrants III and IV. The cosine is positive in quadrants I and IV. Since $\sin\theta < 0$ and $\cos\theta > 0$, θ lies in quadrant IV.

Since $\tan\theta = \frac{\sin\theta}{\cos\theta}$, the sign of $\tan\theta$ is negative.

65. $\tan\theta = 1.6778490$

$$\frac{1}{\tan\theta} = \frac{1}{1.6778490}$$

$$\cot\theta = \frac{1}{1.6778490} \approx .5960011896$$

66.

$$\begin{aligned}
\sin A &= \frac{\text{side opposite}}{\text{hypotenuse}} = \frac{60}{61} \\
\cos A &= \frac{\text{side adjacent}}{\text{hypotenuse}} = \frac{11}{61} \\
\tan A &= \frac{\text{side opposite}}{\text{side adjacent}} = \frac{60}{11} \\
\cot A &= \frac{\text{side adjacent}}{\text{side opposite}} = \frac{11}{60} \\
\sec A &= \frac{\text{hypotenuse}}{\text{side adjacent}} = \frac{61}{11} \\
\csc A &= \frac{\text{hypotenuse}}{\text{side opposite}} = \frac{61}{60}
\end{aligned}$$

67. $\sin A = \dfrac{\text{side opposite}}{\text{hypotenuse}} = \dfrac{40}{58} = \dfrac{20}{29}$

$\cos A = \dfrac{\text{side adjacent}}{\text{hypotenuse}} = \dfrac{42}{58} = \dfrac{21}{29}$

$\tan A = \dfrac{\text{side opposite}}{\text{side adjacent}} = \dfrac{40}{42} = \dfrac{20}{21}$

$\cot A = \dfrac{\text{side adjacent}}{\text{side opposite}} = \dfrac{42}{40} = \dfrac{21}{20}$

$\sec A = \dfrac{\text{hypotenuse}}{\text{side adjacent}} = \dfrac{58}{42} = \dfrac{29}{21}$

$\csc A = \dfrac{\text{hypotenuse}}{\text{side opposite}} = \dfrac{58}{40} = \dfrac{29}{20}$

68. $\cos A = \dfrac{b}{c}$; $\sin B = \dfrac{b}{c}$

69. 300°: reference angle = 360° − 300° = 60°

Since 300° is in quadrant IV, the sine, tangent, cotangent and cosecant are negative.

$\sin 300° = -\sin 60° = -\dfrac{\sqrt{3}}{2}$

$\cos 300° = \cos 60° = \dfrac{1}{2}$

$\tan 300° = -\tan 60° = -\sqrt{3}$

$\cot 300° = -\cot 60° = -\dfrac{\sqrt{3}}{3}$

$\sec 300° = \sec 60° = 2$

$\csc 300° = -\csc 60° = -\dfrac{2\sqrt{3}}{3}$

70. −225° is coterminal with −225° + 360° = 135°
reference angle = 180° − 135° = 45°

Since −225° is in quadrant II, the cosine, tangent, cotangent and secant are negative.

$\sin(-225°) = \sin 45° = \dfrac{\sqrt{2}}{2}$

$\cos(-225°) = -\cos 45° = -\dfrac{\sqrt{2}}{2}$

$\tan(-225°) = -\tan 45° = -1$

$\cot(-225°) = -\cot 45° = -1$

$\sec(-225°) = -\sec 45° = -\sqrt{2}$

$\csc(-225°) = \csc 45° = \sqrt{2}$

71. −390° is coterminal with −390° + 720° = 330°
reference angle = 360° − 330° = 30°

Since −390° is in quadrant IV, the sine, tangent, cotangent and cosecant are negative.

$\sin(-390°) = -\sin 30° = -\dfrac{1}{2}$

$\cos(-390°) = \cos 30° = \dfrac{\sqrt{3}}{2}$

$\tan(-390°) = -\tan 30° = -\dfrac{\sqrt{3}}{3}$

$\cot(-390°) = -\cot 30° = -\sqrt{3}$

$\sec(-390°) = \sec 30° = \dfrac{2\sqrt{3}}{3}$

$\csc(-390°) = -\csc 30° = -2$

72. $\sin 72°30' = \sin\left(72 + \dfrac{30}{60}\right)^{\circ} = \sin 72.5°$

$\approx \boxed{.95371695}$

73. $\sec 222°30' = \sec\left(222 + \dfrac{30}{60}\right)^{\circ} = \sec 222.5°$

$= \dfrac{1}{\cos 222.5°}$ or $(\cos 222.5°)^{-1}$

$\approx \boxed{-1.3563417}$

74. $\cot 305.6° = \dfrac{1}{\tan 305.6°}$ or $(\tan 305.6°)^{-1}$

$\approx \boxed{-.71592968}$

75. $\tan 11.7689° \approx \boxed{.20834446}$

76. If $\theta = 135°$, $\theta' = 45°$
If $\theta = 45°$, $\theta' = 45°$
If $\theta = 300°$, $\theta' = 60°$
If $\theta = 140°$, $\theta' = 40°$

Of these reference angles, 40° is the only one which is not a special angle, so **D,** tan 140°, is the only one which cannot be determined exactly.

77. $\sin\theta = .8254121$
$\theta = \sin^{-1} .8254121 \approx 55.673870° \approx 55.7$

78. $\cot\theta = 1.1249386$

$$\frac{1}{\tan\theta} = 1.1249386$$

$$\tan\theta = (1.1249386)^{-1}$$

$$\theta = \tan^{-1}(1.1249386)^{-1} \approx 41.635092° \approx 41.6$$

79. $\cot 25° = \dfrac{1}{\tan 25°}$

$\tan^{-1} 25$ give the angle whose tangent is 25.

They are not the same.

80. $\theta = 1997°$

$$\cos 1997° = -.956304756$$

$$\sin 1997° = -.292371705$$

Since sine and cosine are both negative, the angle θ is in quadrant **III**.

81. $A = 58°30'$, $c = 748$

(i) to find B:

$$B = 90° - 58°30'$$

$$B = 89°60' - 58°30' \quad \boxed{B = 31°30'}$$

(ii) to find a:

$$\sin A = \frac{a}{c}$$

$$\sin 58°30' = \frac{a}{748}$$

$$a = 748\sin 58°30' \approx 637.775 \quad \boxed{a \approx 638}$$

(iii) to find b:

$$\cos A = \frac{b}{c}$$

$$\cos 58°30' = \frac{b}{748}$$

$$b = 748\cos 58°30' \approx 390.829 \quad \boxed{b \approx 391}$$

82. $A = 39.72°$, $b = 38.97$

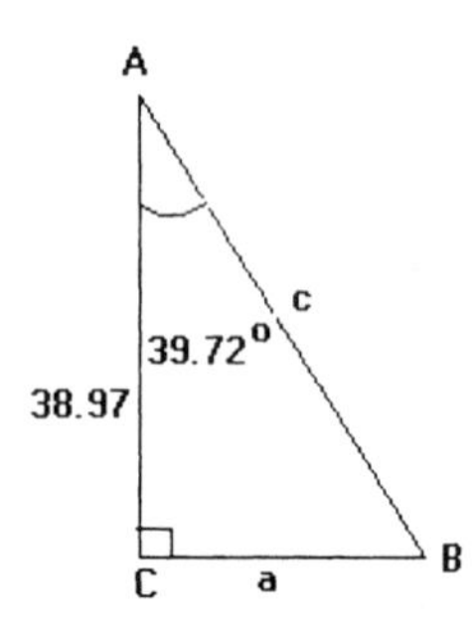

continued

81. continued

(i) to find B:

$$B = 90° - 39.72° \quad \boxed{B = 50.28°}$$

(ii) to find a:

$$\tan A = \frac{a}{b}$$

$$\tan 39.72° = \frac{a}{38.97}$$

$$a = 38.97\tan 39.72°$$

$$a \approx 32.3765 \quad \boxed{a \approx 32.38 \text{ m}}$$

(iii) to find c:

$$\cos A = \frac{b}{c}$$

$$\cos 39.72° = \frac{38.97}{c}$$

$$c = \frac{38.97}{\cos 39.72°} \approx 50.665 \quad \boxed{c \approx 50.66 \text{ m}}$$

83. $\tan 38°20' = \dfrac{h}{93.2}$

$$h = 93.2\tan 38°20' \approx 73.69300534 \approx \boxed{73.7 \text{ feet}}$$

84. The angle at the point on the ground is 29.5° (alternate interior angles).

$$\tan 29.5° = \frac{h}{36.0}$$

$$h = 36.0\tan 29.5° \approx 20.36782 \approx \boxed{20.4 \text{ m}}$$

85. The angle at C is 90° $(344° - 254°)$; the angle at B is 42°

$$([254° - 180°] - 32°):$$

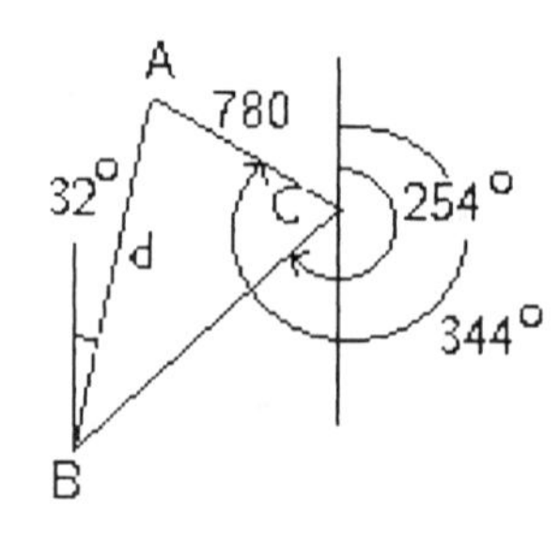

$$\sin 42° = \frac{780}{d}$$

$$d = \frac{780}{\sin 42°} \approx 1165.692$$

$\boxed{\text{The distance from } A \text{ to } B \text{ is} \approx 1200 \text{ m.}}$

86. A right triangle is formed since the angle where the ship turns is $90°$ $(55° + 35°)$:

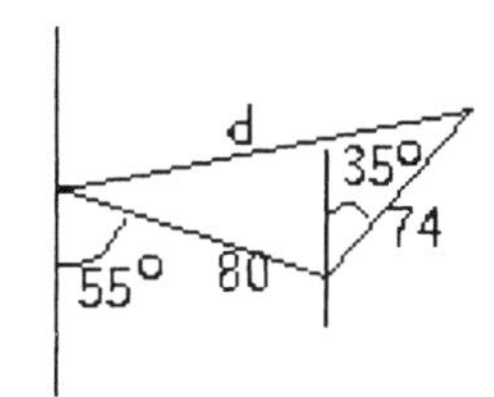

$$d^2 = 80^2 + 74^2 = 11876$$
$$d \approx 108.97706$$

The ship is $\approx$ 110 km from the pier.

87. A right triangle is formed with the bottom angle $= 36°$ $(360° - 324°)$, and the adjacent side $= 110$ $(2 \cdot 55)$:

$$\cos 36° = \frac{110}{x}$$
$$x = \frac{110}{\cos 36°} \approx 135.96748$$

The cars are $\approx$ 140 miles apart.

88. Two right triangles are formed. Let x be the distance the boat travels, and y be the distance from shore at the 2nd observation.

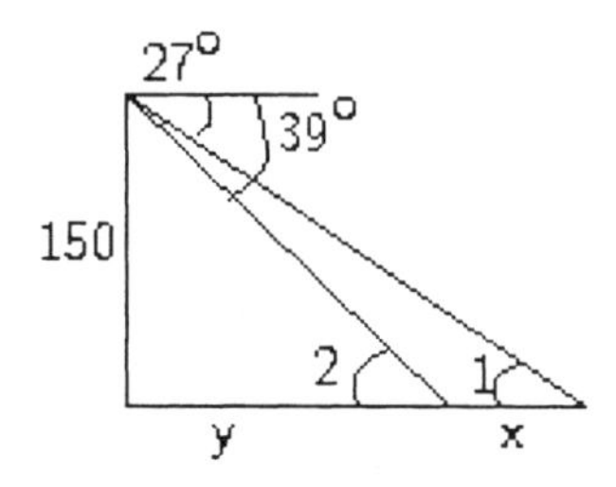

(i) Larger right $\triangle$:

$$\angle 1 = 27° \text{ (alt. int. } \angle's)$$
$$\tan 27° = \frac{150}{x+y}$$
$$x + y = \frac{150}{\tan 27°}$$
$$y = \frac{150}{\tan 27°} - x$$

continued

88. continued

(ii) Smaller right $\triangle$:

$$\angle 2 = 39° \text{ (alt. int. } \angle's)$$
$$\tan 39° = \frac{150}{y}$$
$$y = \frac{150}{\tan 39°}$$

(iii) Set **(i)** = **(ii)**:

$$\frac{150}{\tan 27°} - x = \frac{150}{\tan 39°}$$
$$x = \frac{150}{\tan 27°} - \frac{150}{\tan 39°} \approx 109.157$$

The boat travels $\approx$ 109 feet

89. Let x be the length of the base of the smaller right $\triangle$:

(i) $\tan 29° = \dfrac{h}{392 + x}$

$$392 + x = \frac{h}{\tan 29°}$$
$$x = \frac{h}{\tan 29°} - 392$$

(ii) $\tan 49° = \dfrac{h}{x}$

$$x = \frac{h}{\tan 49°}$$

(iii) Set **(i)** = **(ii)**:

$$\frac{h}{\tan 29°} - 392 = \frac{h}{\tan 49°}$$
$$\frac{h}{\tan 29°} - \frac{h}{\tan 49°} = 392$$
$$h \tan 49° - h \tan 29° = 392 \tan 29° \tan 49°$$
$$h(\tan 49° - \tan 29°) = 392 \tan 29° \tan 49°$$
$$h = \frac{392 \tan 29° \tan 49°}{\tan 49° - \tan 29°} \approx 419.3585$$

$h \approx 419$

90. (a) From the figure in the text,

$$\sin\theta = \frac{x_Q - x_P}{d}$$

$$x_Q = x_P + d\sin\theta$$

Similarly,

$$\cos\theta = \frac{y_Q - y_P}{d}$$

$$y_Q = y_P + d\cos\theta$$

(b) Let $(x_P, y_P) = (123.62, 337.95)$,

$\theta = 17°19'22''$, and $d = 193.86$.

$$\begin{aligned} x_Q &= x_P + d\sin\theta \\ &= 123.62 + 193.86\sin 17°19'22'' \\ &= 123.62 + 193.86\sin 17.3228° \\ &\approx 181.34 \end{aligned}$$

$$\begin{aligned} y_Q &= y_P + d\cos\theta \\ &= 337.95 + 193.86\cos 17°19'22'' \\ &= 337.95 + 193.86\cos 17.3228° \\ &\approx 523.02 \end{aligned}$$

The coordinates of Q are $(181.34, 523.02)$

91. (a) Length of shorter leg of right triangle: $h_2 - h_1$.

(b) When $h_2 = 55$ and $h_1 = 5$,

$d = (55 - 5)\cot\theta$

$d = 50\cot\theta$.

The period is π, but the graph wanted is d for $0 < \theta < \frac{\pi}{2}$.

The asymptote is the line $\theta = 0$.

When $\theta = \frac{\pi}{4}$, $d = 50(1) = 50$.

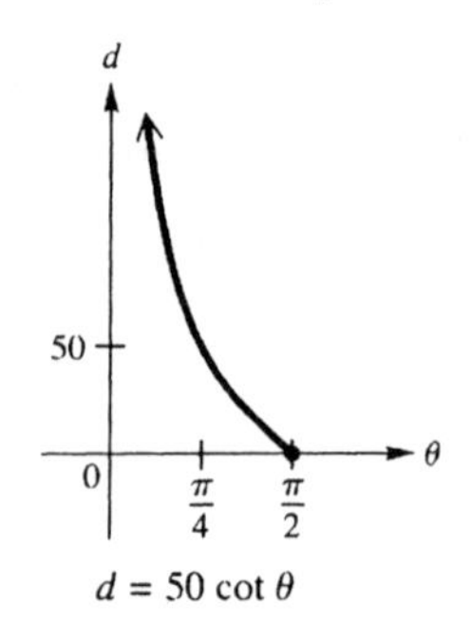

$d = 50\cot\theta$

92. (a) $\csc\theta = \frac{d}{h} \Longrightarrow d = h\csc\theta$

(b) $\csc\theta = \frac{2h}{h} = 2$

$\sin\theta = \frac{1}{2}$

$\theta = \sin^{-1}\left(\frac{1}{2}\right) = \frac{\pi}{6}$

d is double h when the sun is 30° above the horizon.

(c) $\csc\frac{\pi}{2} = 1$ and $\csc\frac{\pi}{3} \approx 1.15$.
When the sun is lower in the sky, $\theta = \frac{\pi}{3}$, sunlight is filtered by more atmosphere. There is less ultraviolet light reaching the earth's surface, and therefore, there is less likelihood of becoming sunburned. In the case of $\frac{\pi}{3}$, sunlight passes through 15% more atmosphere.

93. $\tan\theta = \frac{Y}{X} = \frac{5.1961524}{3} \approx 1.73205$

$$\begin{aligned} \theta &= \tan^{-1} 1.73205 \quad \text{(in degree mode)} \\ &= 60° \\ &= \frac{60°}{1} \cdot \frac{\pi}{180°} = \frac{\pi}{3} \text{ radians} \end{aligned}$$

94. $s(t) = 3\cos 2t$

Amplitude $= 3$

Period $= \frac{2\pi}{2} = \pi$

Frequency $= \frac{1}{P} = \frac{1}{\pi}$

95. $s(t) = 4\sin\pi t$

Amplitude $= 4$

Period $= \frac{2\pi}{\pi} = 2$

Frequency $= \frac{1}{P} = \frac{1}{2}$

96. The period is the time to complete one cycle. The amplitude is the maximum distance (on either side) from the initial point.

97. The frequency is the number of cycles in one unit of time.

$s(1.5) = 4\sin 1.5\pi = 4(-1) = -4$

$s(2) = 4\sin 2\pi = 4(0) = 0$

$s(3.25) = 4\sin 3.25\pi \approx -2.82843 = -2\sqrt{2}$

Chapter 8 Test

1. $360° + (-157°) = 203°$

2. $\dfrac{450 \text{ rev}}{\text{min}} \cdot \dfrac{1 \text{ min}}{60 \text{ sec}} = 7.5 \text{ rev/second}$

$7.5 \text{ rev} = 7.5\,(360°) = 2700°$

The point on the edge of the tire moves 2700° in one second.

3. **(a)** $120° = \dfrac{120°}{1} \cdot \dfrac{\pi}{180°} = \dfrac{2\pi}{3}$ radians

(b) $\dfrac{9\pi}{10} = \dfrac{9\pi}{10} \cdot \dfrac{180°}{\pi} = 162°$

4. **(a)** $s = 200$ cm, $r = 150$ cm,

$$\theta = \frac{s}{r} = \frac{200}{150} = \frac{4}{3}$$

(b) $A = \dfrac{1}{2}r^2\theta$

$= \dfrac{1}{2}(150)^2\left(\dfrac{4}{3}\right)$

$= 15,000 \text{ cm}^2$

5. $\theta = 36° = \dfrac{36°}{1} \cdot \dfrac{\pi}{180°} = .2\pi, \quad r = 12$ in

$s = r\theta = 12\,(.2\pi) = 2.4\pi$ in $(\approx 7.54$ in$)$

6. $y = -1 + 2\sin(x + \pi)$

Amplitude: 2

Period $= 2\pi$

Vertical translation: 1 unit down

Phase shift: π units to the left

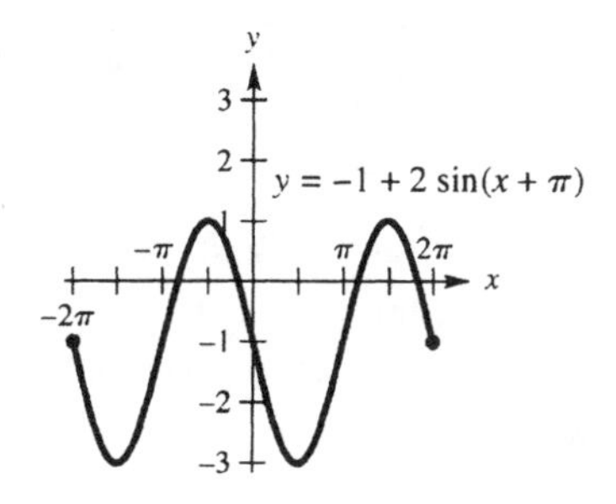

7. $y = -\cos 2x$

Amplitude: $|-1| = 1$; Period $= \dfrac{2\pi}{2} = \pi$

Vertical translation: none

Phase shift: none; Reflection about x-axis

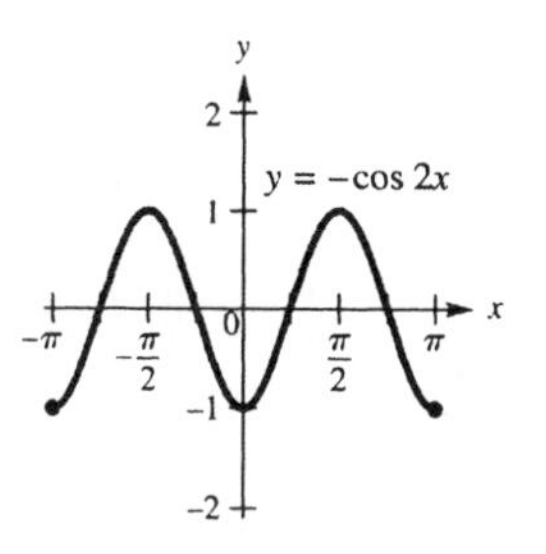

8. $y = \tan\left(x - \dfrac{\pi}{2}\right) \quad (2\pi \approx 6.3)$

Period: π; Phase shift: $\dfrac{\pi}{2}$ units to the right

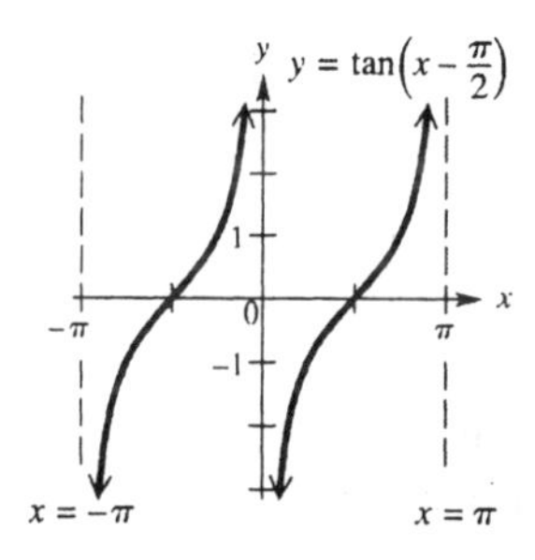

9. **(a)** Average monthly temperature:

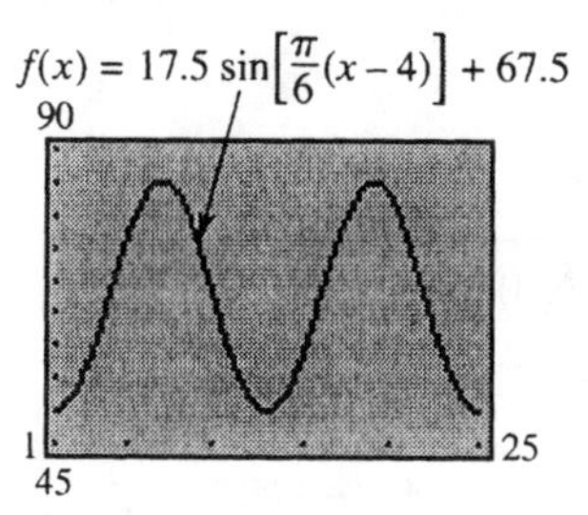

(b) $f(x) = 17.5\sin\left[\dfrac{\pi}{6}(x - 4)\right] + 67.5$

Amplitude $= 17.5$

Period: $\dfrac{2\pi}{\pi/6} = \dfrac{2\pi}{1} \cdot \dfrac{6}{\pi} = 12$

Phase shift: 4 units to the right

Vertical translation: 67.5 units up

(c) Average monthly temperature is $\approx 52°$F

(d) Minimum temperature is 50°F in January; the maximum temperature is 85°F in July.

(e) Average yearly temperature in Austin is $\approx 67.5°$; this is the vertical translation.

10. **(a)** $\theta = -150°$; coterminal:

$360° + (-150°) = 210°$

(b) $\theta' = 30°$; $x = -\sqrt{3}$, $y = -1$

(c) $\sin(-150°) = -\sin 30° = -\frac{1}{2}$

$\cos(-150°) = -\cos 30° = -\frac{\sqrt{3}}{2}$

$\tan 150° = \tan 30° = \frac{1}{\sqrt{3}} \cdot \frac{\sqrt{3}}{\sqrt{3}} = \frac{\sqrt{3}}{3}$

$\cot 150° = \cot 30° = \sqrt{3}$

$\sec 150° = -\sec 30° = -\frac{2}{\sqrt{3}} \cdot \frac{\sqrt{3}}{\sqrt{3}} = -\frac{2\sqrt{3}}{3}$

$\csc 150° = -\csc 30° = -2$

(d) $-150° = \frac{150°}{1} \cdot \frac{\pi}{180°} = -\frac{5\pi}{6}$

11. If $\cos\theta < 0$, then θ is in quadrant II or III.
If $\cot\theta > 0$, then θ is in quadrant I or III.
Therefore, θ terminates in quadrant III.

12. $(2, -5)$; $r = \sqrt{(2)^2 + (-5)^2} = \sqrt{4 + 25} = \sqrt{29}$

$$\sin\theta = \frac{y}{r} = \frac{-5}{\sqrt{29}} \cdot \frac{\sqrt{29}}{\sqrt{29}} = -\frac{5\sqrt{29}}{29}$$

$$\cos\theta = \frac{x}{r} = \frac{2}{\sqrt{29}} \cdot \frac{\sqrt{29}}{\sqrt{29}} = \frac{2\sqrt{29}}{29}$$

$$\tan\theta = \frac{y}{x} = \frac{-5}{2} = -\frac{5}{2}$$

13. $\cos\theta = \frac{4}{5}$, $\theta \in$ Quadrant IV

$$r^2 = x^2 + y^2$$
$$5^2 = 4^2 + y^2$$
$$y^2 = 9$$
$$y = \pm 3$$

Since $\theta \in$ Quad IV, $y = -3$

$$\sin\theta = \frac{y}{r} = -\frac{3}{5}$$

$$\tan\theta = \frac{y}{x} = -\frac{3}{4}$$

$$\cot\theta = \frac{x}{y} = -\frac{4}{3}$$

$$\sec\theta = \frac{r}{x} = \frac{5}{4}$$

$$\csc\theta = \frac{r}{y} = -\frac{5}{3}$$

14. To find w: $\sin 30° = \frac{4}{w}$

$w = \frac{4}{\sin 30°} = \frac{4}{1/2} = 8$

To find x: $\tan 45° = \frac{4}{x}$

$x = \frac{4}{\tan 45°} = \frac{4}{1} = 4$

To find y: $\cos 30° = \frac{y}{w} = \frac{y}{8}$

$y = 8\cos 30° = 8 \cdot \frac{\sqrt{3}}{2} = 4\sqrt{3}$

To find z: $\sin 45° = \frac{4}{z}$

$z = \frac{4}{\sin 45°} = \frac{4}{1/\sqrt{2}} = 4\sqrt{2}$

15. $\cot(-750°) = \cot(3 \cdot 360° - 750°)$

$= \cot 330°$ (coterminal)

$= \cot(360° - 330°)$

$= -\cot 30°$ (quadrant IV)

$= -\sqrt{3}$

16. **(a)** $\sin 78^\circ 21' = \sin\left(78 + \frac{21}{60}\right)^\circ$

$= \sin 78.35^\circ \approx .97939940$

(b) $\tan 11.7689^\circ \approx .20834446$

(c) $\sec 58.9041^\circ = \frac{1}{\cos 58.9041^\circ}$

$(\cos 58.9041^\circ)^{-1} \approx 1.9362132$

17. $\sin s = .82584121$

$s = \sin^{-1} .82584121 \approx .97169234$

18. $A = 58^\circ 30'$, $c = 748$

(i) to find B :

$B = 90^\circ - 58^\circ 30'$

$B = 89^\circ 60' - 58^\circ 30'$ $\boxed{B = 31^\circ 30'}$

(ii) to find a : $\sin A = \frac{a}{c}$

$\sin 58^\circ 30' = \frac{a}{748}$

$a = 748 \sin 58^\circ 30'$

$a \approx 637.7748$ $\boxed{a \approx 638}$

(iii) to find b : $\cos A = \frac{b}{c}$

$\cos 58^\circ 30' = \frac{b}{748}$

$b = 748 \cos 58^\circ 30'$

$b \approx 390.8289$ $\boxed{b \approx 391}$

19. Let x be the height of the flag pole.

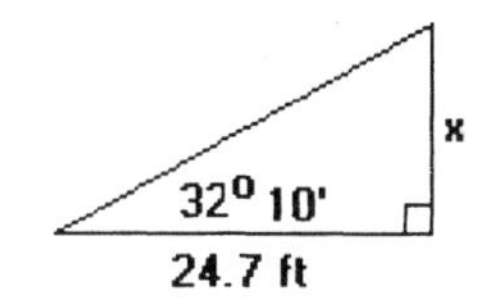

$\tan 32^\circ 10' = \frac{x}{24.7}$

$x = 24.7 \tan 32^\circ 10' \approx 15.5$ feet

20. The angle from the South line to $\overline{AB} = 55^\circ$;
the angle from the North line to $\overline{BC} = 35^\circ$.

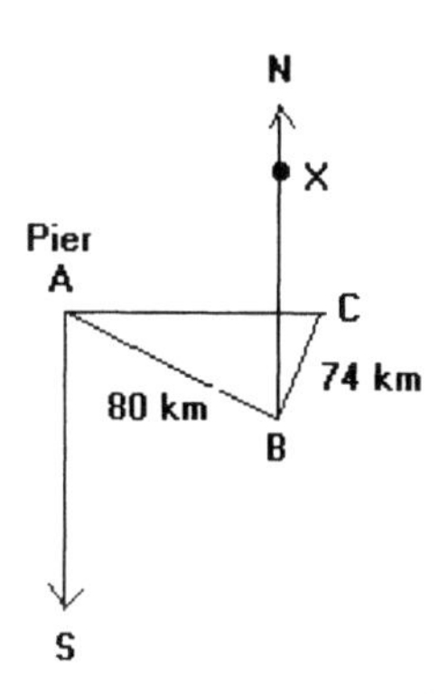

Angle ABX is 55° (alternate interior angles of parallel lines cut by a transversal are congruent). Angle ABC is $55^\circ + 35^\circ = 90^\circ$, therefore triangle ABC is a right triangle.

$AC^2 = 80^2 + 74^2$

$AC = \sqrt{80^2 + 74^2} \approx 108.977$

$AC \approx 110$ km

21. $s(t) = -3 \cos 2\pi t$

The amplitude is $|-3| = 3$, so the maximum height $= 3$ inches.

22. $s(t) = -3 \cos 2\pi t$

$3 = -3 \cos 2\pi t$

$-1 = \cos 2\pi t$

$2\pi t = \cos^{-1}(-1) = \pi$

$t = \frac{\pi}{2\pi} = \frac{1}{2}$

After $\frac{1}{2}$ second.

Chapter 8 Project

Modeling Sunset Times

1. Enter the data into two lists in your graphing calculator.

L1	L2	L3 1
1	4.77	------
2	5.32	
3	5.87	
4	6.4	
5	6.92	
6	7.38	
7	7.55	

L1(7)=7

L1	L2	L3 1
8	7.23	
9	6.53	
10	5.7	
11	4.97	
12	4.58	
13	4.77	
14	5.32	

L1(14) =14

L1	L2	L3 1
15	5.87	
16	6.4	
17	6.92	
18	7.38	
19	7.55	
20	7.23	
21	6.53	

L1(21) =21

L1	L2	L3 1
19	7.55	
20	7.23	
21	6.53	
22	5.7	
23	4.97	
24	4.58	

L1(25) =

Now, get a scatterplot of the data.

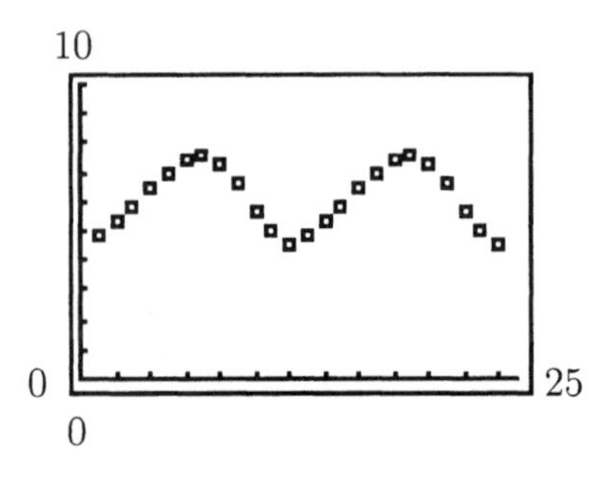

2. The minimum sunset time is 4:35 or 4.58; the maximum is 7:33 or 7.55. Thus, the amplitude is
$$\frac{7.55 - 4.58}{2} = 1.48 \quad \text{or} \quad 1{:}29.$$
In the equation $y = a \sin b(x - d) + c$, the amplitude is a, so $a \approx 1.48$, or about 1 minute, 29 seconds.

3. The times repeat every 12 months, so the period is
$$b = \frac{2\pi}{12} \approx .52\,.$$

4. We want our cycle to start with March 21, which corresponds to month $3\frac{21}{31}$, so the phase shift will be
$$d = 3 + \frac{21}{31} \approx 3.68$$

5. The vertical shift is the average of the maximum and minimum sunset times, so
$$c = \frac{4.58 + 7.55}{2} = 6.065$$
$$c \approx 6.07$$

6. Using the results of items **1** – **5** (rounded to nearest hundredth), we have the function defined by
$$y = a \sin b(x - d) + c$$
$$y = 1.48 \sin[.52(x - 3.68)] + 6.07$$
$$y = 1.48 \sin(.52x - 1.91) + 6.07$$

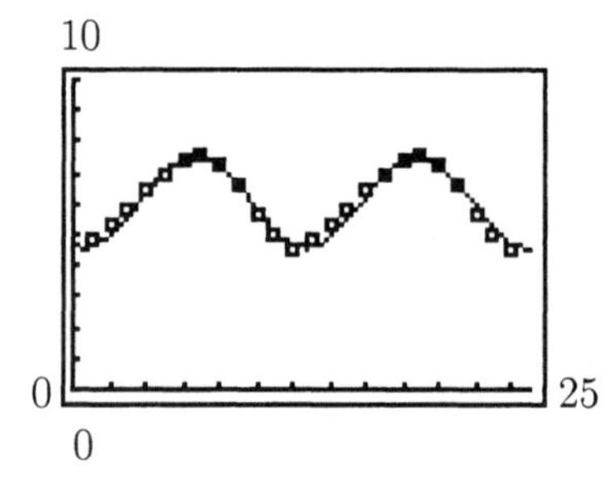

Be sure to put your calculator in radian mode before graphing the function in the same window as the scatterplot.

7. The TI-83 regression equation gives
$$y = 1.42 \sin(.52x - 1.81) + 6.10,$$
which is very close to the equation we found above. The two forms are nearly equivalent, but not exactly.

Section 9.1

1. $\sin x = -\sin(-x)$: Odd

3. $\tan x = -\tan(-x)$: Odd

5. $\sec x = \sec(-x)$: Even

7. $\dfrac{\cos x}{\sin x} = \cot x$ **(B)**

9. $\cos(-x) = \cos x$ **(E)**

11. $1 = \sin^2 x + \cos^2 x$ **(A)**

13. $\sec^2 x - 1 = \tan^2 x = \dfrac{\sin^2 x}{\cos^2 x}$ **(A)**

15. $1 + \sin^2 x = \left(\csc^2 x - \cot^2 x\right) + \sin^2 x$ **(D)**

17. $\cos(-4.38) = \cos 4.38$
$[\cos(-\theta) = \cos\theta]$

19. $\sin(-.5) = -\sin .5$
$[\sin(-\theta) = -\sin\theta]$

21. $\tan\left(-\frac{\pi}{7}\right) = -\tan\frac{\pi}{7}$
$[\tan(-\theta) = -\tan\theta]$

23. The correct identity is $1 + \cot^2 x = \csc^2 x$; the function must have the argument 'x'. (or θ, t, etc)

25. $\sin\theta$ in terms of $\cot\theta$ and $\sec\theta$:

(i) $$\sin\theta = \frac{1}{\csc\theta} = \frac{1}{\pm\sqrt{1+\cot^2\theta}} = \boxed{\pm\frac{\sqrt{1+\cot^2\theta}}{1+\cot^2\theta}}$$

(ii) $$\sin\theta = \cos\theta \cdot \frac{\sin\theta}{\cos\theta} = \cos\theta \cdot \tan\theta = \frac{1}{\sec\theta}\left(\pm\sqrt{\sec^2\theta - 1}\right) = \boxed{\pm\frac{\sqrt{\sec^2\theta - 1}}{\sec\theta}}$$

27. $\tan\theta$ in terms of $\sin\theta$, $\cos\theta$, $\sec\theta$ and $\csc\theta$:

(i) $$\tan\theta = \frac{\sin\theta}{\cos\theta} = \frac{\sin\theta}{\pm\sqrt{1-\sin^2\theta}} = \boxed{\pm\frac{\sin\theta\sqrt{1-\sin^2\theta}}{1-\sin^2\theta}}$$

(ii) $$\tan\theta = \frac{\sin\theta}{\cos\theta} = \boxed{\pm\frac{\sqrt{1-\cos^2\theta}}{\cos\theta}}$$

(iii) $$\tan\theta = \boxed{\pm\sqrt{\sec^2\theta - 1}}$$

(iv) $$\tan\theta = \frac{1}{\cot\theta} = \frac{1}{\pm\sqrt{\csc^2\theta - 1}} = \boxed{\pm\frac{\sqrt{\csc^2\theta - 1}}{\csc^2\theta - 1}}$$

29. $\sec\theta$ in terms of $\sin\theta$, $\tan\theta$, $\cot\theta$ and $\csc\theta$:

(i) $\sec\theta = \dfrac{1}{\cos\theta} = \dfrac{1}{\pm\sqrt{1-\sin^2\theta}}$

$= \boxed{\pm\dfrac{\sqrt{1-\sin^2\theta}}{1-\sin^2\theta}}$

(ii) $\sec\theta = \boxed{\pm\sqrt{\tan^2\theta+1}}$

(iii) $\sec\theta = \pm\sqrt{\tan^2\theta+1} = \pm\sqrt{\dfrac{1}{\cot^2\theta}+1}$

$= \pm\sqrt{\dfrac{1+\cot^2\theta}{\cot^2\theta}}$

$= \boxed{\pm\dfrac{\sqrt{1+\cot^2\theta}}{\cot\theta}}$

(iv) $\sec\theta = \dfrac{1}{\cos\theta} = \dfrac{1}{\pm\sqrt{1-\sin^2\theta}}$

$= \dfrac{1}{\pm\sqrt{1-\dfrac{1}{\csc^2\theta}}} = \dfrac{1}{\pm\sqrt{\dfrac{\csc^2\theta-1}{\csc^2\theta}}}$

$= \dfrac{\pm\sqrt{\csc^2\theta}}{\sqrt{\csc^2\theta-1}} = \dfrac{\pm\csc\theta}{\sqrt{\csc^2\theta-1}}$

$= \boxed{\pm\dfrac{\csc\theta\sqrt{\csc^2\theta-1}}{\csc^2\theta-1}}$

31. $\tan\theta\cos\theta = \dfrac{\sin\theta}{\cos\theta}\cdot\dfrac{\cos\theta}{1}$

$= \boxed{\sin\theta}$

33. $\dfrac{\sin\beta\tan\beta}{\cos\beta} = \dfrac{\sin\beta}{\cos\beta}\cdot\tan\beta$

$= \tan\beta\cdot\tan\beta$

$= \boxed{\tan^2\beta}$

35. $\sec^2 x - 1 = (\tan^2 x + 1) - 1$

$= \boxed{\tan^2 x}$

37. $\dfrac{\sin^2 x}{\cos^2 x} + \sin x\csc x$

$= \tan^2 x + \dfrac{\sin x}{1}\cdot\dfrac{1}{\sin x}$

$= \tan^2 x + 1$

$= \boxed{\sec^2 x}$

Relating Concepts (39 – 44):

39. $y = (\sec x + \tan x)(1-\sin x)$ $\quad (2\pi \approx 6.3)$

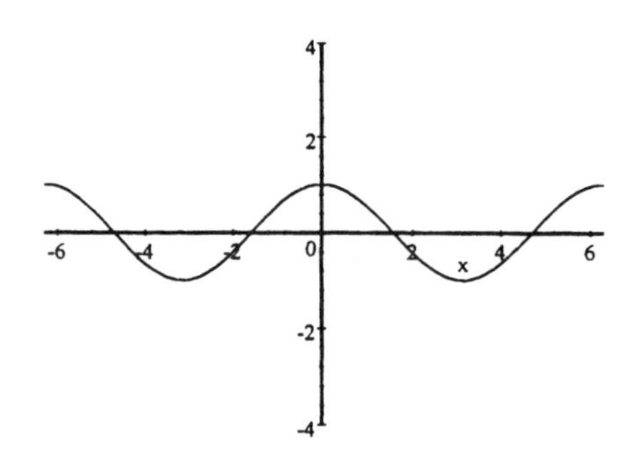

40. The graph looks like the graph of $y = \cos x$.

41. $y = \cos x$ $\quad (2\pi \approx 6.3)$

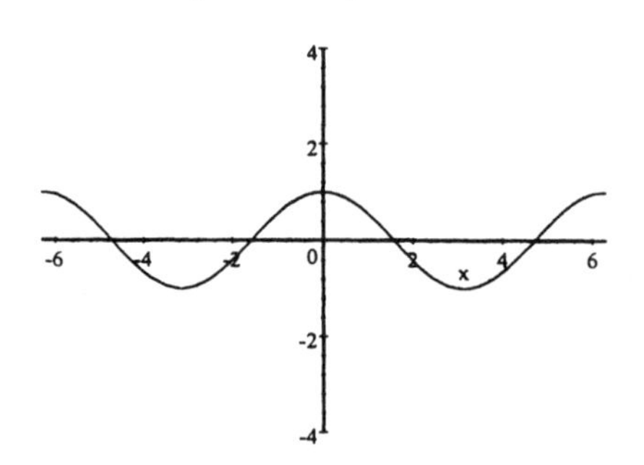

42. Yes, it suggests the identity $(\sec x + \tan x)(1-\sin x) = \cos x$.

43. Verify $(\sec x + \tan x)(1-\sin x) = \cos x$

$(\sec x + \tan x)(1-\sin x)$

$= \left(\dfrac{1}{\cos x} + \dfrac{\sin x}{\cos x}\right)(1-\sin x)$

$= \dfrac{1+\sin x}{\cos x}\cdot\dfrac{1-\sin x}{1} = \dfrac{1-\sin^2 x}{\cos x}$

$= \dfrac{\cos^2 x}{\cos x} = \cos x$

44. $y = \dfrac{\cos x + 1}{\sin x + \tan x}$

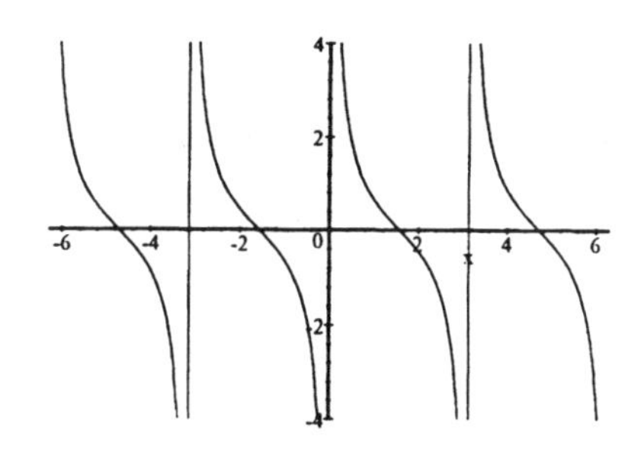

The graph looks like $y = \cot x$, so the identity is $\dfrac{\cos x + 1}{\sin x + \tan x} = \cot x$

Verify the above identity:

$$\begin{aligned}
&\frac{\cos x + 1}{\sin x + \tan x} = \\
&= \frac{\cos x + 1}{\sin x + \frac{\sin x}{\cos x}} \\
&= \frac{\cos x + 1}{1} \div \frac{\sin x \cos x + \sin x}{\cos x} \\
&= \frac{\cos x + 1}{1} \cdot \frac{\cos x}{\sin x (\cos x + 1)} \\
&= \frac{\cos x}{\sin x} = \cot x
\end{aligned}$$

45. Verify $\dfrac{\cot \theta}{\csc \theta} = \cos \theta$

$$\begin{aligned}
\frac{\cot \theta}{\csc \theta} &= \frac{\frac{\cos \theta}{\sin \theta}}{\frac{1}{\sin \theta}} \\
&= \frac{\cos \theta}{1} \\
&= \cos \theta
\end{aligned}$$

47. Verify $\cos^2 \theta \left(\tan^2 \theta + 1\right) = 1$

$$\begin{aligned}
\cos^2 \theta \left(\tan^2 \theta + 1\right) &= \cos^2 \theta \left(\sec^2 \theta\right) \\
&= \cos^2 \theta \left(\frac{1}{\cos^2 \theta}\right) \\
&= 1
\end{aligned}$$

49. Verify $\dfrac{\tan^2 \gamma + 1}{\sec \gamma} = \sec \gamma$

$$\begin{aligned}
\frac{\tan^2 \gamma + 1}{\sec \gamma} &= \frac{\sec^2 \gamma}{\sec \gamma} \\
&= \sec \gamma
\end{aligned}$$

51. Verify $\sin^2 \alpha + \tan^2 \alpha + \cos^2 \alpha = \sec^2 \alpha$

$$\begin{aligned}
&\sin^2 \alpha + \tan^2 \alpha + \cos^2 \alpha \\
&\quad = \sin^2 \alpha + \cos^2 \alpha + \tan^2 \alpha \\
&\quad = 1 + + \tan^2 \alpha \\
&\quad = \sec^2 \alpha
\end{aligned}$$

53. Verify $\dfrac{\sin^2 \gamma}{\cos \gamma} = \sec \gamma - \cos \gamma$

$$\begin{aligned}
\frac{\sin^2 \gamma}{\cos \gamma} &= \frac{1 - \cos^2 \gamma}{\cos \gamma} \\
&= \frac{1}{\cos \gamma} - \frac{\cos^2 \gamma}{\cos \gamma} \\
&= \sec \gamma - \cos \gamma
\end{aligned}$$

55. Verify $\dfrac{\cos \theta}{\sin \theta \cot \theta} = 1$

$$\begin{aligned}
\frac{\cos \theta}{\sin \theta \cot \theta} &= \frac{\cos \theta}{\sin \theta \cdot \frac{\cos \theta}{\sin \theta}} \\
&= \frac{\cos \theta}{\cos \theta} \\
&= 1
\end{aligned}$$

57. Verify $\tan^2 \gamma \sin^2 \gamma = \tan^2 \gamma + \cos^2 \gamma - 1$

$$\begin{aligned}
&\tan^2 \gamma \sin^2 \gamma \\
&\quad = \left(\sec^2 \gamma - 1\right)\left(1 - \cos^2 \gamma\right) \\
&\quad = \sec^2 \gamma - \sec^2 \gamma \cos^2 \gamma - 1 + \cos^2 \gamma \\
&\quad = \sec^2 \gamma - 1 + \cos^2 \gamma - \sec^2 \gamma \cos^2 \gamma \\
&\quad = \tan^2 \gamma + \cos^2 \gamma - 1
\end{aligned}$$

59. Verify $\dfrac{(\sec\theta-\tan\theta)^2+1}{\sec\theta\csc\theta-\tan\theta\csc\theta}=2\tan\theta$

$$\frac{(\sec\theta-\tan\theta)^2+1}{\sec\theta\csc\theta-\tan\theta\csc\theta}$$
$$=\frac{\sec^2\theta-2\sec\theta\tan\theta+\tan^2\theta+1}{\csc\theta(\sec\theta-\tan\theta)}$$
$$=\frac{\sec^2\theta-2\sec\theta\tan\theta+\sec^2\theta}{\csc\theta(\sec\theta-\tan\theta)}$$
$$=\frac{2\sec^2\theta-2\sec\theta\tan\theta}{\csc\theta(\sec\theta-\tan\theta)}$$
$$=\frac{2\sec\theta(\sec\theta-\tan\theta)}{\csc\theta(\sec\theta-\tan\theta)}$$
$$=\frac{2\sec\theta}{\csc\theta}=2\left(\frac{1}{\cos\theta}\cdot\frac{\sin\theta}{1}\right)$$
$$=2\tan\theta$$

61. Verify $\dfrac{1}{\tan\alpha-\sec\alpha}+\dfrac{1}{\tan\alpha+\sec\alpha}=-2\tan\alpha$

$$\frac{1}{\tan\alpha-\sec\alpha}+\frac{1}{\tan\alpha+\sec\alpha}$$
$$=\frac{(\tan\alpha+\sec\alpha)+(\tan\alpha-\sec\alpha)}{\tan^2\alpha-\sec^2\alpha}$$
$$=\frac{2\tan\alpha}{\tan^2\alpha-(\tan^2\alpha+1)}$$
$$=\frac{2\tan\alpha}{-1}=-2\tan\alpha$$

63. An equation that is true for one value, may or may not be an identity; in this case, $\frac{\pi}{2}$ is merely a solution of this equation. To be an identity, an equation must be true for every value of θ for which the functions involved are defined.

65. $\sin^2 x=1-\cos^2 x$

$\sin x=\pm\sqrt{1-\cos^2 x}$

When $\sin x\geq 0$, $\sin x=\sqrt{1-\cos^2 x}$

67. $I=k\cos^2\theta$

(a) $I=k\left(1-\sin^2\theta\right)$

(b) For $\theta=2\pi n$, $n\in$ integer, $\cos^2\theta=1$, its maximum value, and I attains a maximum value of k.

69. $P(t)=k\cos^2(4\pi t)$; $K(t)=k\sin^2(4\pi t)$; $E(t)=P(t)+K(t)$

(a) $k=2$

The total mechanical energy E is always 2. The spring has maximum potential energy when it is fully stretched but not moving. The spring has maximum kinetic energy when it is not stretched but is moving fastest.

(b)

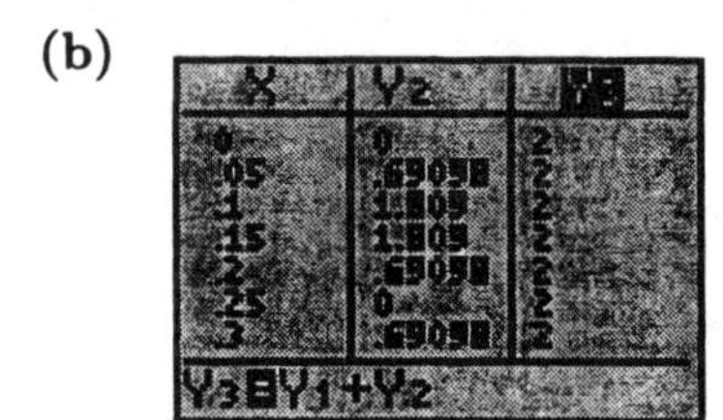

Let $Y_1=P(t)$, $Y_2=k(t)$, $Y_3=E(t)=2$ for all inputs. The spring is stretched the most (has greatest potential energy) when $t=.25$, $.5$, $.75$, etc. At these times kinetic energy is 0.

(c)
$$\begin{aligned}E(t)&=P(t)+k(t)\\&=2\cos^2(4\pi t)+2\sin^2(4\pi t)\\&=2\left(\cos^2(4\pi t)+\sin^2(4\pi t)\right)\\&=2(1)\\&=2\end{aligned}$$

Section 9.2

1. $\cos(x+y) = \cos x \cos y - \sin x \sin y$: **F**

3. $\sin(x+y) = \sin x \cos y + \cos x \sin y$: **C**

5. $$\begin{aligned}\sin\frac{\pi}{12} &= \sin\left(\frac{\pi}{3}-\frac{\pi}{4}\right)\\ &= \sin\frac{\pi}{3}\cos\frac{\pi}{4}-\cos\frac{\pi}{3}\sin\frac{\pi}{4}\\ &= \frac{\sqrt{3}}{2}\cdot\frac{\sqrt{2}}{2}-\frac{1}{2}\cdot\frac{\sqrt{2}}{2}\\ &= \boxed{\frac{\sqrt{6}-\sqrt{2}}{4}}\end{aligned}$$

7. $$\begin{aligned}\sin\left(-\frac{5\pi}{12}\right) &= -\sin\frac{5\pi}{12}\\ &= -\sin\left(\frac{\pi}{6}+\frac{\pi}{4}\right)\\ &= -\left(\sin\frac{\pi}{6}\cos\frac{\pi}{4}+\cos\frac{\pi}{6}\sin\frac{\pi}{4}\right)\\ &= -\left(\frac{1}{2}\cdot\frac{\sqrt{2}}{2}+\frac{\sqrt{3}}{2}\cdot\frac{\sqrt{2}}{2}\right)\\ &= \boxed{\frac{-\sqrt{2}-\sqrt{6}}{4}}\end{aligned}$$

9. $$\begin{aligned}\sin\frac{13\pi}{12} &= \sin\left(\frac{5\pi}{4}-\frac{\pi}{6}\right)\\ &= \sin\frac{5\pi}{4}\cos\frac{\pi}{6}-\cos\frac{5\pi}{4}\sin\frac{\pi}{6}\\ &= \frac{-\sqrt{2}}{2}\cdot\frac{\sqrt{3}}{2}-\frac{-\sqrt{2}}{2}\cdot\frac{1}{2}\\ &= \boxed{\frac{-\sqrt{6}+\sqrt{2}}{4}}\end{aligned}$$

11. $$\begin{aligned}\cos 75^\circ &= \cos(45^\circ+30^\circ)\\ &= \cos 45^\circ\cos 30^\circ-\sin 45^\circ\sin 30^\circ\\ &= \frac{\sqrt{2}}{2}\cdot\frac{\sqrt{3}}{2}-\frac{\sqrt{2}}{2}\cdot\frac{1}{2}\\ &= \boxed{\frac{\sqrt{6}-\sqrt{2}}{4}}\end{aligned}$$

13. $$\begin{aligned}\tan 105^\circ &= \tan(60^\circ+45^\circ)\\ &= \frac{\tan 60^\circ+\tan 45^\circ}{1-\tan 60^\circ\tan 45^\circ}\\ &= \frac{\sqrt{3}+1}{1-\sqrt{3}}\cdot\frac{1+\sqrt{3}}{1+\sqrt{3}}\\ &= \frac{2\sqrt{3}+4}{-2}=\boxed{-\sqrt{3}-2}\end{aligned}$$

15. $$\begin{aligned}\cos(-15^\circ) &= \cos 15^\circ\\ &= \cos(60^\circ-45^\circ)\\ &= \cos 60^\circ\cos 45^\circ+\sin 60^\circ\sin 45^\circ\\ &= \frac{1}{2}\cdot\frac{\sqrt{2}}{2}+\frac{\sqrt{3}}{2}\cdot\frac{\sqrt{2}}{2}\\ &= \boxed{\frac{\sqrt{2}+\sqrt{6}}{4}}\end{aligned}$$

17. $$\begin{aligned}\cos\frac{\pi}{3}\cos\frac{2\pi}{3}-\sin\frac{\pi}{3}\sin\frac{2\pi}{3} &= \cos\left(\frac{\pi}{3}+\frac{2\pi}{3}\right)\\ &= \cos\pi=-1\end{aligned}$$

19. $$\begin{aligned}\sin 76^\circ\cos 31^\circ-\cos 76^\circ\sin 31^\circ &= \sin(76^\circ-31^\circ)\\ &= \sin 45^\circ=\frac{\sqrt{2}}{2}\end{aligned}$$

21. $$\begin{aligned}\frac{\tan 80^\circ+\tan 55^\circ}{1-\tan 80^\circ\tan 55^\circ} &= \tan(80^\circ+55^\circ)\\ &= \tan 135^\circ=-1\end{aligned}$$

23. $$\begin{aligned}\cos\left(\frac{\pi}{2}-x\right) &= \cos\frac{\pi}{2}\cos x+\sin\frac{\pi}{2}\sin x\\ &= (0)(\cos x)+(1)(\sin x)\\ &= \boxed{\sin x}\end{aligned}$$

25. $$\begin{aligned}\cos\left(\frac{3\pi}{2}+x\right) &= \cos\frac{3\pi}{2}\cos x-\sin\frac{3\pi}{2}\sin x\\ &= (0)(\cos x)-(-1)(\sin x)\\ &= \boxed{\sin x}\end{aligned}$$

27. $\sin(\pi + x) = \sin\pi\cos x + \cos\pi\sin x$
$= (0)(\cos x) + (-1)(\sin x)$
$= \boxed{-\sin x}$

29. $\cos(135° - x) = \cos 135° \cos x + \sin 135° \sin x$
$= -\frac{\sqrt{2}}{2} \cdot \cos x + \frac{\sqrt{2}}{2} \cdot \sin x$
$= \boxed{\frac{\sqrt{2}(\sin x - \cos x)}{2}}$

31. $\sin(45° + x) = \sin 45° \cos x + \cos 45° \sin x$
$= \frac{\sqrt{2}}{2} \cdot \cos x + \frac{\sqrt{2}}{2} \cdot \sin x$
$= \boxed{\frac{\sqrt{2}(\cos x + \sin x)}{2}}$

33. $\tan(\pi - x) = \frac{\tan\pi - \tan x}{1 + \tan\pi \tan x}$
$= \frac{0 - \tan x}{1 + (0)\tan x} = \boxed{-\tan x}$

35. You can find $\cot(A+B)$ by finding $\tan(A+B)$ and then using the reciprocal identity, $\cot\theta = \frac{1}{\tan\theta}$; other cofunctions may be found in a similar manner.

Relating Concepts (36 – 41):

36. $y_1 = \cos\left(x + \frac{\pi}{2}\right)$

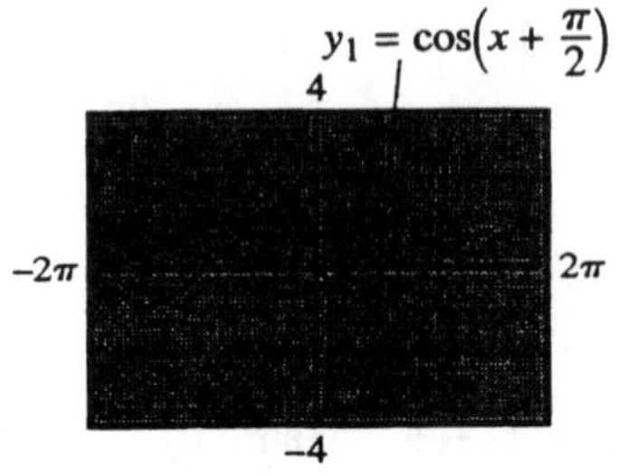

$y_1 = \cos\left(x + \frac{\pi}{2}\right)$

4

-2π 2π

-4

37. $y = \cos x$

Shift y left $\frac{\pi}{2}$ units to obtain

$$y = \cos\left(x + \frac{\pi}{2}\right).$$

38. $y_1 = \cos\left(x + \frac{\pi}{2}\right)$

Let $x = 1$:

$y_1 = \cos\left(1 + \frac{\pi}{2}\right) \approx \boxed{-.841471}$

39. $\cos\left(x + \frac{\pi}{2}\right) = \cos x \cos\frac{\pi}{2} - \sin x \sin\frac{\pi}{2}$
$= \cos x \cdot 0 - \sin x \cdot 1$
$= \boxed{-\sin x}$

40. $y_2 = -\sin x$

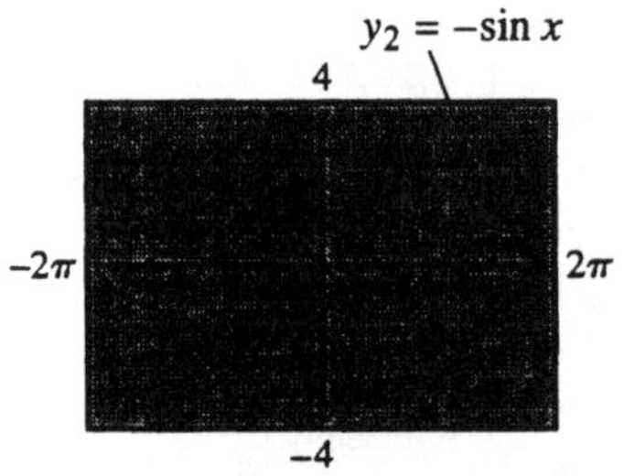

$y_2 = -\sin x$

4

-2π 2π

-4

It is the same graph as $y = \cos\left(x + \frac{\pi}{2}\right)$.

41. $y_2 = -\sin x$
$= -\sin(1)$
$\approx -.841471$

The result is the same as #38.

43. $\sin A = \frac{3}{5}$, $\sin B = -\frac{12}{13}$,
$A \in$ Quad I, $B \in$ Quad III

(i) $\cos^2 A = 1 - \sin^2 A = 1 - \frac{9}{25} = \frac{16}{25}$

$\cos A = \frac{4}{5}$ (Quadrant I)

(ii) $\cos^2 B = 1 - \sin^2 B = 1 - \frac{144}{169} = \frac{25}{169}$

$\cos B = -\frac{5}{13}$ (Quadrant III)

(a) $\sin(A+B) = \sin A \cos B + \cos A \sin B = \frac{3}{5} \cdot \left(-\frac{5}{13}\right) + \frac{4}{5} \cdot \left(-\frac{12}{13}\right) = \frac{-15 - 48}{65}$ $\boxed{\sin(A+B) = -\frac{63}{65}}$

(b) $\sin(A-B) = \sin A \cos B - \cos A \sin B = \frac{3}{5} \cdot \left(-\frac{5}{13}\right) - \frac{4}{5} \cdot \left(-\frac{12}{13}\right) = \frac{-15 + 48}{65}$ $\boxed{\sin(A-B) = \frac{33}{65}}$

(c) $\tan A = \frac{\sin A}{\cos A} = \frac{3}{4}$; $\tan B = \frac{\sin B}{\cos B} = \frac{12}{5}$

$$\tan(A+B) = \frac{\tan A + \tan B}{1 - \tan A \tan B} = \frac{\frac{3}{4} + \frac{12}{5}}{1 - \frac{3}{4} \cdot \frac{12}{5}} \cdot \frac{20}{20} = \frac{15 + 48}{20 - 36}$$

$\boxed{\tan(A+B) = -\frac{63}{16}}$

(d) $\tan(A-B) = \frac{\tan A - \tan B}{1 + \tan A \tan B} = \frac{15 - 48}{20 + 36}$ [See **(c)**]

$\boxed{\tan(A-B) = -\frac{33}{56}}$

(e) $A + B$: Since sine < 0 and tangent < 0 : $\boxed{\text{Quadrant IV}}$

(f) $A - B$: Since sine > 0 and tangent < 0 : $\boxed{\text{Quadrant II}}$

45. $\cos A = -\frac{15}{17}$, $\sin B = \frac{4}{5}$,
$A \in$ Quad II, $B \in$ Quad I

(i) $\sin^2 A = 1 - \cos^2 A = 1 - \frac{225}{289} = \frac{64}{289}$

$\sin A = \frac{8}{17}$ (Quadrant II)

(ii) $\cos^2 B = 1 - \sin^2 B = 1 - \frac{16}{25} = \frac{9}{25}$

$\cos B = \frac{3}{5}$ (Quadrant I)

(a) $\sin(A+B) = \sin A \cos B + \cos A \sin B = \frac{8}{17} \cdot \frac{3}{5} + \left(-\frac{15}{17}\right) \cdot \frac{4}{5} = \frac{24 - 60}{85}$ $\boxed{\sin(A+B) = -\frac{36}{85}}$

(b) $\sin(A-B) = \sin A \cos B - \cos A \sin B = \frac{8}{17} \cdot \frac{3}{5} - \left(-\frac{15}{17}\right) \cdot \frac{4}{5} = \frac{24 + 60}{85}$ $\boxed{\sin(A-B) = \frac{84}{85}}$

(c) $\tan A = \frac{\sin A}{\cos A} = -\frac{8}{15}$; $\tan B = \frac{\sin B}{\cos B} = \frac{4}{3}$

$$\tan(A+B) = \frac{\tan A + \tan B}{1 - \tan A \tan B} = \frac{-\frac{8}{15} + \frac{4}{3}}{1 - \left(-\frac{8}{15}\right) \cdot \frac{4}{3}} \cdot \frac{45}{45} = \frac{-24 + 60}{45 + 32}$$

$\boxed{\tan(A+B) = \frac{36}{77}}$

(d) $\tan(A-B) = \frac{\tan A - \tan B}{1 + \tan A \tan B} = \frac{-24 - 60}{45 - 32}$ [See **(c)**]

$\boxed{\tan(A-B) = -\frac{84}{13}}$

(e) $A + B$: Since sine < 0 and tangent > 0 : $\boxed{\text{Quadrant III}}$

(f) $A - B$: Since sine > 0 and tangent < 0 : $\boxed{\text{Quadrant II}}$

47. Verify $\tan(x-y)-\tan(y-x)=\dfrac{2(\tan x-\tan y)}{1+\tan x\tan y}$

$$\begin{aligned}
&\tan(x-y)-\tan(y-x)\\
&=\frac{\tan x-\tan y}{1+\tan x\tan y}-\frac{\tan y-\tan x}{1+\tan y\tan x}\\
&=\frac{\tan x-\tan y-(\tan y-\tan x)}{1+\tan x\tan y}\\
&=\frac{\tan x-\tan y-\tan y+\tan x}{1+\tan x\tan y}\\
&=\frac{2\tan x-2\tan y}{1+\tan x\tan y}\\
&=\frac{2(\tan x-\tan y)}{1+\tan x\tan y}
\end{aligned}$$

49. Verify $\dfrac{\sin(A+B)}{\cos A\cos B}=\tan A+\tan B$

$$\begin{aligned}
\frac{\sin(A+B)}{\cos A\cos B}&=\frac{\sin A\cos B+\cos A\sin B}{\cos A\cos B}\\
&=\frac{\sin A\cos B}{\cos A\cos B}+\frac{\cos A\sin B}{\cos A\cos B}\\
&=\frac{\sin A}{\cos A}+\frac{\sin B}{\cos B}\\
&=\tan A+\tan B
\end{aligned}$$

51. Verify $\dfrac{\tan(A+B)-\tan B}{1+\tan(A+B)\tan B}=\tan A$

$$\begin{aligned}
&\frac{\tan(A+B)-\tan B}{1+\tan(A+B)\tan B}\\
&=\frac{\frac{\tan A+\tan B}{1-\tan A\tan B}-\tan B}{1+\left(\frac{\tan A+\tan B}{1-\tan A\tan B}\right)(\tan B)}
\end{aligned}$$

Multiply by $\dfrac{1-\tan A\tan B}{1-\tan A\tan B}$:

$$\begin{aligned}
&=\frac{\tan A+\tan B-\tan B(1-\tan A\tan B)}{1-\tan A\tan B+\tan B(\tan A+\tan B)}\\
&=\frac{\tan A+\tan B-\tan B+\tan A\tan^2 B}{1-\tan A\tan B+\tan B\tan A+\tan^2 B}\\
&=\frac{\tan A+\tan A\tan^2 B}{1+\tan^2 B}\\
&=\frac{\tan A(1+\tan^2 B)}{1+\tan^2 B}=\tan A
\end{aligned}$$

53. Since there are 60 cycles per second, the number of cycles in .05 seconds is given by
(.05 sec) (60 cycles/sec) = $\boxed{3 \text{ cycles}}$

55. $F=\dfrac{.6W\sin(\theta+90^\circ)}{\sin 12^\circ}$

(a) $F=\dfrac{.6(170)\sin(30^\circ+90^\circ)}{\sin 12^\circ}$
≈ 424.8659171
≈ 425 pounds

(b) $F=\dfrac{.6W}{\sin 12^\circ}(\sin\theta\cos 90^\circ+\cos\theta\sin 90^\circ)$
$\approx 2.8858W\,(\sin\theta\,(0)+\cos\theta\,(1))$
$\approx 2.9W\cos\theta$

(c) F is a maximum when $\sin(\theta+90^\circ)=1$
$\Longrightarrow \sin^{-1}1=\theta+90^\circ$
$\Longrightarrow 90^\circ=\theta+90^\circ$
$\Longrightarrow \theta=0^\circ$

57. (a) $a=.4$ lb per ft^2, $\lambda=4.9$ ft, $c=1026$ ft per sec, $r=10$ ft

$$\begin{aligned}
P&=\frac{a}{r}\cos\left[\frac{2\pi r}{\lambda}-ct\right]\\
&=\frac{.4}{10}\cos\left[\frac{20\pi}{4.9}-1026t\right]
\end{aligned}$$

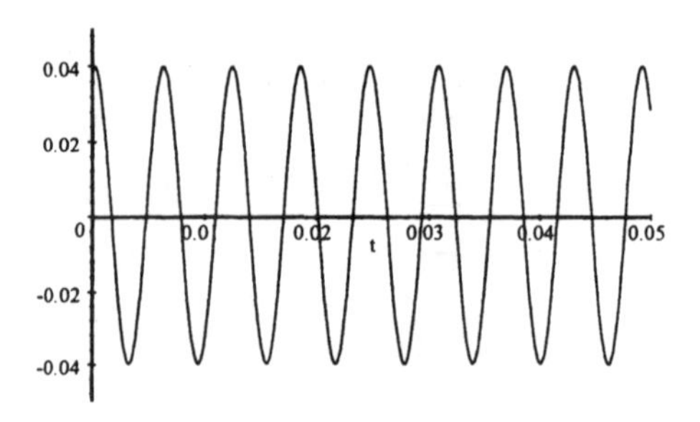

The pressure P is oscillating.

continued

57. continued

(b) $a = 3$ lb per ft^2, $\lambda = 4.9$ ft, $c = 1026$ ft per sec, $t = 10$ ft

$$P = \frac{a}{r}\cos\left[\frac{2\pi r}{\lambda} - ct\right] = \frac{3}{r}\cos\left[\frac{2\pi r}{4.9} - 10,260\right]$$

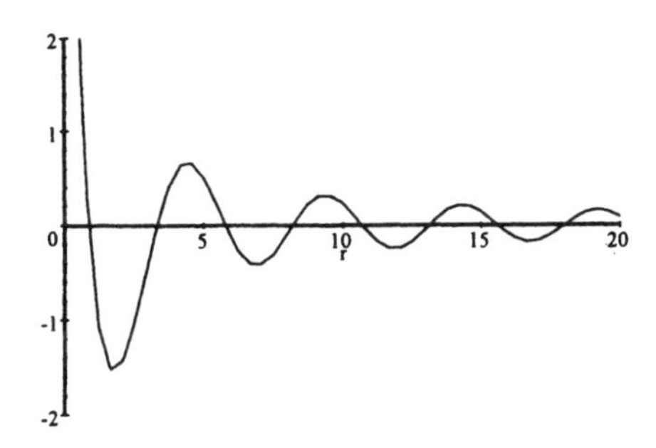

The pressure P oscillates, and amplitude decreases as r increases.

(c) $P = \frac{a}{r}\cos\left[\frac{2\pi r}{\lambda} - ct\right]$

Let $r = n\lambda$:

$$\begin{aligned} P &= \frac{a}{n\lambda}\cos\left[\frac{2\pi n\lambda}{\lambda} - ct\right] \\ &= \frac{a}{n\lambda}\cos[2\pi n - ct] \\ &= \frac{a}{n\lambda}[\cos(2\pi n)\cos(ct) + \sin(2\pi n)\sin(ct)] \\ &= \frac{a}{n\lambda}[(1)\cos(ct) + (0)\sin(ct)] \\ &= \frac{a}{n\lambda}\cos(ct) \end{aligned}$$

Reviewing Basic Concepts (Sections 9.1 and 9.2)

1. $$\begin{aligned} \frac{\csc x}{\cot x} - \frac{\cot x}{\csc x} &= \frac{1}{\sin x}\cdot\frac{\sin x}{\cos x} - \frac{\cos x}{\sin x}\cdot\frac{\sin x}{1} \\ &= \frac{1}{\cos x} - \cos x = \frac{1-\cos^2 x}{\cos x} = \boxed{\frac{\sin^2 x}{\cos x}} \end{aligned}$$

2. $\left[\frac{\pi}{4} - \frac{\pi}{3} = \frac{3\pi}{12} - \frac{4\pi}{12} = -\frac{\pi}{12}\right]$

$$\begin{aligned} \tan\left(-\frac{\pi}{12}\right) &= \tan\left(\frac{\pi}{4} - \frac{\pi}{3}\right) \\ &= \frac{\tan(\pi/4) - \tan(\pi/3)}{1 + \tan(\pi/4)\tan(\pi/3)} \\ &= \frac{1-\sqrt{3}}{1+1(\sqrt{3})} = \frac{1-\sqrt{3}}{1+\sqrt{3}} \quad \left(= \frac{\sqrt{3}-3}{3+\sqrt{3}}\right) \end{aligned}$$

3. $$\begin{aligned} &\cos 18^\circ \cos 108^\circ + \sin 18^\circ \sin 108^\circ \\ &= \cos(18^\circ - 108^\circ) \\ &= \cos(-90^\circ) = \cos 90^\circ = \boxed{0} \end{aligned}$$

4. $$\begin{aligned} \sin\left(x - \frac{\pi}{4}\right) &= \sin x \cos\frac{\pi}{4} - \cos x \sin\frac{\pi}{4} \\ &= \sin x\left(\frac{\sqrt{2}}{2}\right) - \cos x\left(\frac{\sqrt{2}}{2}\right) \\ &= \frac{\sqrt{2}}{2}(\sin x - \cos x) \end{aligned}$$

5. $\sin A = \frac{2}{3}$, $A \in$ quadrant II

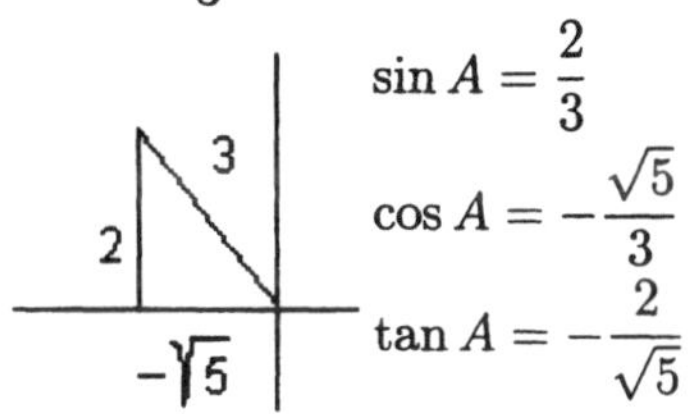

$\sin A = \frac{2}{3}$

$\cos A = -\frac{\sqrt{5}}{3}$

$\tan A = -\frac{2}{\sqrt{5}}$

$\cos B = -\frac{1}{2}$, $B \in$ quadrant III

-1

-√3

2

$\sin B = -\frac{\sqrt{3}}{2}$

$\cos B = -\frac{1}{2}$

$\tan B = \sqrt{3}$

$$\begin{aligned} \sin(A+B) &= \sin A\cos B + \cos A \sin B \\ &= \frac{2}{3}\left(-\frac{1}{2}\right) + \left(-\frac{\sqrt{5}}{3}\right)\left(-\frac{\sqrt{3}}{2}\right) \\ &= -\frac{2}{6} + \frac{\sqrt{15}}{6} = \frac{-2+\sqrt{15}}{6} \end{aligned}$$

$$\begin{aligned} \cos(A-B) &= \cos A\cos B + \sin A \sin B \\ &= \left(-\frac{\sqrt{5}}{3}\right)\left(-\frac{1}{2}\right) + \frac{2}{3}\left(-\frac{\sqrt{3}}{2}\right) \\ &= \frac{\sqrt{5}}{6} - \frac{2\sqrt{3}}{6} = \frac{\sqrt{5}-2\sqrt{3}}{6} \end{aligned}$$

$$\begin{aligned} \tan(A-B) &= \frac{\tan A - \tan B}{1 + \tan A \tan B} \\ &= \frac{-2/\sqrt{5} - \sqrt{3}}{1 + (-2/\sqrt{5})(\sqrt{3})}\cdot\frac{\sqrt{5}}{\sqrt{5}} \\ &= \frac{-2-\sqrt{15}}{\sqrt{5}-2\sqrt{3}} \end{aligned}$$

6. Verify $\csc^2\theta - \cot^2\theta = 1$

$$\csc^2\theta - \cot^2\theta = 1 + \cot^2\theta - \cot^2\theta = 1$$

7. Verify $\dfrac{\sin t}{1-\cos t} = \dfrac{1+\cos t}{\sin t}$

$$\begin{aligned}\frac{\sin t}{1-\cos t} &= \frac{\sin t}{1-\cos t}\cdot\frac{1+\cos t}{1+\cos t}\\ &= \frac{\sin t\,(1+\cos t)}{1-\cos^2 t}\\ &= \frac{\sin t\,(1+\cos t)}{\sin^2 t} = \frac{1+\cos t}{\sin t}\end{aligned}$$

8. Verify $\dfrac{\cot A - \tan A}{\csc A \sec A} = \cos^2 A - \sin^2 A$

$$\begin{aligned}\frac{\cot A - \tan A}{\csc A \sec A} &= \frac{\frac{\cos A}{\sin A} - \frac{\sin A}{\cos A}}{\frac{1}{\sin A}\cdot\frac{1}{\cos A}}\cdot\frac{\sin A\cos A}{\sin A\cos A}\\ &= \frac{\cos^2 A - \sin^2 A}{1} = \cos^2 A - \sin^2 A\end{aligned}$$

9. Verify $\dfrac{\sin(x-y)}{\sin x\,\sin y} = \cot y - \cot x$

$$\begin{aligned}\frac{\sin(x-y)}{\sin x\,\sin y} &= \frac{\sin x\cos y - \cos x\sin y}{\sin x\,\sin y}\\ &= \frac{\sin x\cos y}{\sin x\,\sin y} - \frac{\cos x\sin y}{\sin x\,\sin y}\\ &= \frac{\cos y}{\sin y} - \frac{\cos x}{\sin x} = \cot y - \cot x\end{aligned}$$

10. $e = 20\sin\left(\dfrac{\pi t}{4} - \dfrac{\pi}{2}\right)$

$$\begin{aligned}&= 20\left[\sin\frac{\pi t}{4}\cos\frac{\pi}{2} - \cos\frac{\pi t}{4}\sin\frac{\pi}{2}\right]\\ &= 20\left[\sin\frac{\pi t}{4}(0) - \cos\frac{\pi t}{4}(1)\right]\\ &= -20\cos\frac{\pi t}{4}\end{aligned}$$

Section 9.3

1. $\sin\theta = \dfrac{2}{5}$, $\cos\theta < 0$ (Quadrant II)

(i) $\cos^2\theta = 1 - \sin^2\theta$

$$= 1 - \frac{4}{25} = \frac{21}{25}$$

$$\cos\theta = -\frac{\sqrt{21}}{5}$$

$$\begin{aligned}\sin 2\theta &= 2\sin\theta\cos\theta\\ &= 2\left(\frac{2}{5}\right)\left(-\frac{\sqrt{21}}{5}\right) = -\frac{4\sqrt{21}}{25}\end{aligned}$$

$$\boxed{\sin 2\theta = -\frac{4\sqrt{21}}{25}}$$

$$\begin{aligned}\cos 2\theta &= \cos^2\theta - \sin^2\theta\\ &= \frac{21}{25} - \frac{4}{25} \qquad \boxed{\cos 2\theta = \frac{17}{25}}\end{aligned}$$

3. $\tan x = 2$, $\cos x > 0$ (Quadrant I)

(i) $\sec^2 x = 1 + \tan^2 x$

$$= 1 + 4 = 5 \qquad \sec x = \sqrt{5}$$

(ii) $\cos x = \dfrac{1}{\sec x} = \dfrac{1}{\sqrt{5}} = \dfrac{\sqrt{5}}{5}$

(iii) $\sin^2 x = 1 - \cos^2 x$

$$= 1 - \frac{5}{25} = \frac{20}{25} \qquad \sin x = \frac{2\sqrt{5}}{5}$$

$$\begin{aligned}\sin 2x &= 2\sin x\cos x\\ &= 2\left(\frac{2\sqrt{5}}{5}\right)\left(\frac{\sqrt{5}}{5}\right) = \frac{20}{25}\end{aligned}$$

$$\boxed{\sin 2x = \frac{4}{5}}$$

$$\begin{aligned}\cos 2x &= \cos^2 x - \sin^2 x\\ &= \frac{5}{25} - \frac{20}{25} = -\frac{15}{25}\end{aligned}$$

$$\boxed{\cos 2x = -\frac{3}{5}}$$

5. $\sin\alpha = -\frac{\sqrt{5}}{7}$, $\cos\alpha > 0$ (Quadrant IV)

(i) $\cos^2\alpha = 1 - \sin^2\alpha$

$$= 1 - \left(-\frac{\sqrt{5}}{7}\right)^2$$

$$= 1 - \frac{5}{49} = \frac{44}{49}$$

$$\cos\alpha = \sqrt{\frac{44}{49}} = \frac{2\sqrt{11}}{7}$$

$$\sin 2\alpha = 2\sin\alpha\cos\alpha$$

$$= 2\left(-\frac{\sqrt{5}}{7}\right)\left(\frac{2\sqrt{11}}{7}\right)$$

$$\boxed{\sin 2\alpha = -\frac{4\sqrt{55}}{49}}$$

$$\cos 2\alpha = 1 - 2\sin^2\alpha$$

$$= 1 - 2\left(-\frac{\sqrt{5}}{7}\right)^2$$

$$= 1 - 2\left(\frac{5}{49}\right) \qquad \boxed{\cos 2\alpha = \frac{39}{49}}$$

7. $\cos^2 15° - \sin^2 15° = \cos[2(15°)] = \cos 30° = \frac{\sqrt{3}}{2}$

9. $1 - 2\sin^2 15° = \cos[2(15°)] = \cos 30° = \frac{\sqrt{3}}{2}$

11. $2\cos^2 67.5° - 1 = \cos[2(67.5°)] = \cos 135° = -\frac{\sqrt{2}}{2}$

13. $\frac{\tan 51°}{1 - \tan^2 51°} = \frac{1}{2} \cdot \frac{2\tan 51°}{1 - \tan^2 51°}$

$$= \frac{1}{2}\tan[2 \cdot 51°]$$

$$= \frac{1}{2}\tan 102°$$

15. $\frac{1}{4} - \frac{1}{2}\sin^2 47.1° = \frac{1}{4}\left[4\left(\frac{1}{4} - \frac{1}{2}\sin^2 47.1°\right)\right]$

$$= \frac{1}{4}\left[1 - 2\sin^2 47.1°\right]$$

$$= \frac{1}{4}\cos[2(47.1°)] = \frac{1}{4}\cos 94.2°$$

17. The graph of $\cos^4 x - \sin^4 x$ looks like $\cos 2x$.

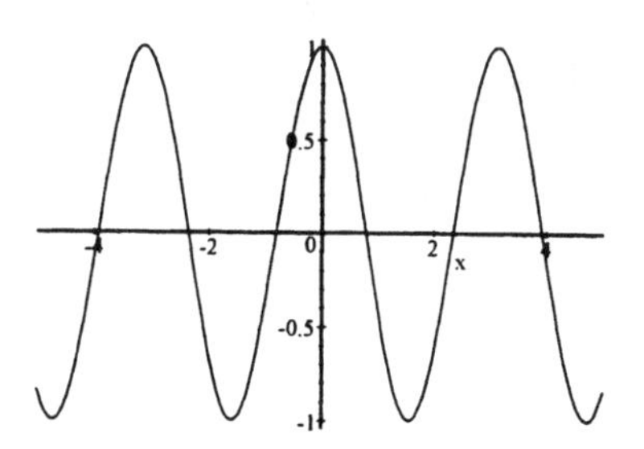

Verify $\cos^4 x - \sin^4 x = \cos 2x$

$$\cos^4 x - \sin^4 x$$

$$= \left(\cos^2 x - \sin^2 x\right)\left(\cos^2 x + \sin^2 x\right)$$

$$= \left(\cos^2 x - \sin^2 x\right)(1)$$

$$= \cos 2x$$

19. $\cos 3x = \cos(2x + x)$

$$= \cos 2x\cos x - \sin 2x\sin x$$

$$= \left(2\cos^2 x - 1\right)\cos x - (2\sin x\cos x)\sin x$$

$$= 2\cos^3 x - \cos x - 2\sin^2 x\cos x$$

$$= 2\cos^3 x - \cos x - 2\left(1 - \cos^2 x\right)\cos x$$

$$= 2\cos^3 x - \cos x - 2\cos x + 2\cos^3 x$$

$$\boxed{\cos 3x = 4\cos^3 x - 3\cos x}$$

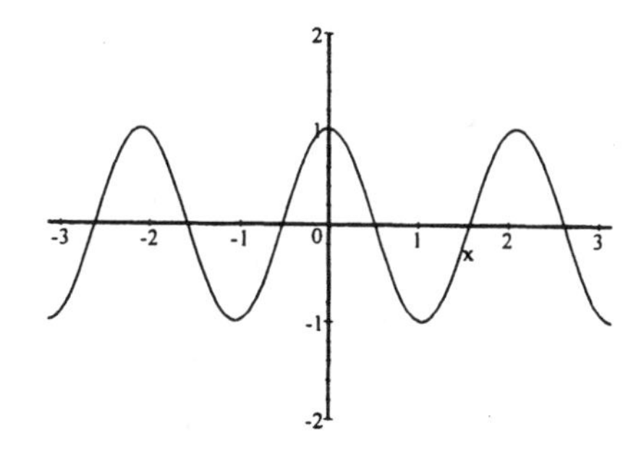

21. $\tan 4x = \tan 2(2x) = \dfrac{2\tan 2x}{1-\tan^2 2x}$

$$= \frac{2\left(\frac{2\tan x}{1-\tan^2 x}\right)}{1-\left(\frac{2\tan x}{1-\tan^2 x}\right)^2}$$

Multiply by $\dfrac{(1-\tan^2 x)^2}{(1-\tan^2 x)^2}$:

$$= \frac{4\tan x\,(1-\tan^2 x)}{(1-\tan^2 x)^2 - 4\tan^2 x}$$

$$= \frac{4\tan x - 4\tan^3 x}{1-2\tan^2 x+\tan^4 x-4\tan^2 x}$$

$$\boxed{\tan 4x = \frac{4\tan x - 4\tan^3 x}{1-6\tan^2 x+\tan^4 x}}$$

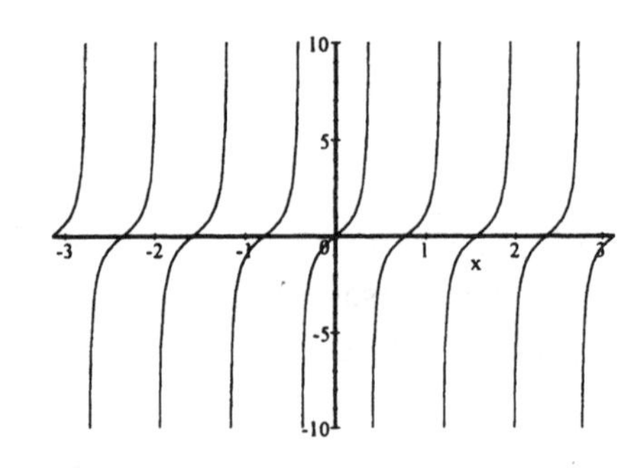

23. $\sin\dfrac{\pi}{12} = \sin\left(\dfrac{\pi/6}{2}\right)$

$\left[\sin\dfrac{\pi}{12} \in \text{Quadrant I}\right]$

$$= \sqrt{\frac{1-\cos\frac{\pi}{6}}{2}} = \sqrt{\frac{1-\frac{\sqrt{3}}{2}}{2}\cdot\frac{2}{2}}$$

$$= \sqrt{\frac{2-\sqrt{3}}{4}} = \boxed{\frac{\sqrt{2-\sqrt{3}}}{2}}$$

25. $\tan\left(-\dfrac{\pi}{8}\right) = -\tan\left(\dfrac{\pi}{8}\right)$

$$= -\tan\left(\frac{\pi/4}{2}\right) = -\left(\frac{1-\cos\frac{\pi}{4}}{\sin\frac{\pi}{4}}\right)$$

$$= -\frac{1-\frac{\sqrt{2}}{2}}{\frac{\sqrt{2}}{2}}\cdot\frac{2}{2} = -\frac{2-\sqrt{2}}{\sqrt{2}}\cdot\frac{\sqrt{2}}{\sqrt{2}}$$

$$= -\frac{2\sqrt{2}-2}{2} = \boxed{1-\sqrt{2}}$$

Alternate method:

$$-\sqrt{\frac{1-\cos\left(-\frac{\pi}{4}\right)}{1+\cos\left(-\frac{\pi}{4}\right)}} = -\sqrt{3-2\sqrt{2}}$$

27. $\sin 67.5° = \sin\left(\dfrac{135°}{2}\right)$

$[\sin 67.5° \in \text{Quadrant I}]$

$$= \sqrt{\frac{1-\cos 135°}{2}} = \sqrt{\frac{1-\left(-\frac{\sqrt{2}}{2}\right)}{2}\cdot\frac{2}{2}}$$

$$= \frac{2+\sqrt{2}}{4} = \boxed{\frac{\sqrt{2+\sqrt{2}}}{2}}$$

29. $\cos x = \dfrac{1}{4}$, $x \in \text{Quadrant I}$

$$\cos\frac{x}{2} = \sqrt{\frac{1+\cos x}{2}}$$

$$= \sqrt{\frac{1+\frac{1}{4}}{2}\cdot\frac{4}{4}}$$

$$= \sqrt{\frac{4+1}{8}}$$

$$= \frac{\sqrt{5}}{\sqrt{8}}\cdot\frac{\sqrt{2}}{\sqrt{2}}$$

$$= \boxed{\frac{\sqrt{10}}{4}} \quad \left[\frac{x}{2} \in \text{Quadrant I}\right]$$

31. $\sin x = \frac{3}{5}, \ \frac{\pi}{2} < x < \pi$ (Quadrant II)

$$\cos x = -\sqrt{1-\sin^2 x} = -\sqrt{1-\frac{9}{25}} = -\sqrt{\frac{16}{25}} = -\frac{4}{5}$$

$$\tan\frac{x}{2} = \frac{1-\cos\theta}{\sin\theta} = \frac{1-\left(-\frac{4}{5}\right)}{\frac{3}{5}}\cdot\frac{5}{5} = \frac{5+4}{3} = \boxed{3} \quad \left[\frac{\pi}{4} < \frac{x}{2} < \frac{\pi}{2}\right]$$

33. $\tan x = \frac{\sqrt{7}}{3}, \ \pi < x < \frac{3\pi}{2}$ (Quadrant III)

$$\sec x = -\sqrt{1+\tan^2 x} = -\sqrt{1+\frac{7}{9}} = -\sqrt{\frac{16}{9}} = -\frac{4}{3}$$

$$\cos x = -\frac{3}{4}$$

$$\sin x = -\sqrt{1-\cos^2 x} = -\sqrt{1-\frac{9}{16}} = -\sqrt{\frac{7}{16}} = -\frac{\sqrt{7}}{4}$$

$$\tan\frac{x}{2} = \frac{1-\cos x}{\sin x} = \frac{1-\left(-\frac{3}{4}\right)}{\frac{-\sqrt{7}}{4}}\cdot\frac{4}{4} = \frac{4+3}{-\sqrt{7}}\cdot\frac{\sqrt{7}}{\sqrt{7}} = \frac{7\sqrt{7}}{-7} = \boxed{-\sqrt{7}} \quad \left[\frac{\pi}{2} < \frac{x}{2} < \frac{3\pi}{4}\right]$$

35.
$$\tan\frac{A}{2} = \frac{\sin A}{1+\cos A} = \frac{\sin A}{1+\cos A}\cdot\frac{1-\cos A}{1-\cos A} = \frac{\sin A\,(1-\cos A)}{1-\cos^2 A} = \frac{\sin A\,(1-\cos A)}{\sin^2 A} = \frac{1-\cos A}{\sin A}$$

37. $\tan 22.5^\circ = \sqrt{3-2\sqrt{2}}$

$\tan 22.5^\circ = \sqrt{2}-1$

$$\left(\sqrt{3-2\sqrt{2}}\right)^2 = 3-2\sqrt{2}$$

$$\left(\sqrt{2}-1\right)^2 = 2-2\sqrt{2}+1 = 3-2\sqrt{2}$$

39. $\sqrt{\frac{1-\cos 40^\circ}{2}} = \sin\frac{40^\circ}{2} = \sin 20^\circ$

41. $\sqrt{\frac{1-\cos 147^\circ}{1+\cos 147^\circ}} = \tan\frac{147^\circ}{2} = \tan 73.5^\circ$

43. $\frac{1-\cos 59.74^\circ}{\sin 59.74^\circ} = \tan\frac{59.74^\circ}{2} = \tan 29.87^\circ$

45. Verify $\sin 4\alpha = 4\sin\alpha\cos\alpha\cos 2\alpha$

$$\sin 4\alpha = 2\sin 2\alpha\cos 2\alpha = 2\,(2\sin\alpha\cos\alpha)\cos 2\alpha = 4\sin\alpha\cos\alpha\cos 2\alpha$$

47. Verify $\frac{2\cos 2\alpha}{\sin 2\alpha} = \cot\alpha - \tan\alpha$

$$\frac{2\cos 2\alpha}{\sin 2\alpha} = \frac{2\left(\cos^2\alpha - \sin^2\alpha\right)}{2\sin\alpha\cos\alpha} = \frac{\cos^2\alpha}{\sin\alpha\cos\alpha} - \frac{\sin^2\alpha}{\sin\alpha\cos\alpha} = \frac{\cos\alpha}{\sin\alpha} - \frac{\sin\alpha}{\cos\alpha} = \cot\alpha - \tan\alpha$$

49. Verify $\sin 2\alpha\cos 2\alpha = \sin 2\alpha - 4\sin^3\alpha\cos\alpha$

$$\sin 2\alpha\cos 2\alpha = \sin 2\alpha\left(1-2\sin^2\alpha\right) = \sin 2\alpha - 2\sin 2\alpha\sin^2\alpha = \sin 2\alpha - 2\,(2\sin\alpha\cos\alpha)\sin^2\alpha = \sin 2\alpha - 4\sin^3\alpha\cos\alpha$$

51. Verify $\tan s + \cot s = 2\csc 2s$

$$\text{Left: } \tan s + \cot s = \frac{\sin s}{\cos s} + \frac{\cos s}{\sin s} = \frac{\sin^2 s + \cos^2 s}{\sin s \cos s} = \frac{1}{\sin s \cos s}$$

$$\text{Right: } 2\csc 2s = 2\left(\frac{1}{2\sin 2s}\right) = 2\left(\frac{1}{2\sin s \cos s}\right) = \frac{1}{\sin s \cos s}$$

53. Verify $\sec^2 \frac{x}{2} = \frac{2}{1+\cos x}$

$$\sec^2 \frac{x}{2} = \left(\frac{1}{\cos \frac{x}{2}}\right)^2 = \left(\pm\frac{1}{\sqrt{\frac{1+\cos x}{2}}}\right)^2 = \frac{1}{\frac{1+\cos x}{2}} = \frac{2}{1+\cos x}$$

55. $2\sin 58^\circ \cos 102^\circ$

$$= 2\left[\frac{1}{2}\left[\sin(58^\circ + 102^\circ) + \sin(58^\circ - 102^\circ)\right]\right]$$
$$= \sin 160^\circ + \sin(-44^\circ)$$
$$= \sin 160^\circ - \sin 44^\circ$$

57. $2\cos 85^\circ \sin 140^\circ$

$$= \frac{1}{2}\left[2\left[\sin(85^\circ + 140^\circ) - \sin(85^\circ - 140^\circ)\right]\right]$$
$$= \sin 225^\circ - \sin(-55^\circ)$$
$$= \sin 225^\circ + \sin 55^\circ$$

59. $\cos 4x - \cos 2x$

$$= -2\sin\left(\frac{4x+2x}{2}\right)\sin\left(\frac{4x-2x}{2}\right)$$
$$= -2\sin 3x \sin x$$

61. $\sin 25^\circ + \sin(-48^\circ)$

$$= 2\sin\left(\frac{25^\circ + (-48^\circ)}{2}\right)\cos\left(\frac{25^\circ - (-48^\circ)}{2}\right)$$
$$= 2\sin(-11.5^\circ)\cos 36.5^\circ$$
$$= -2\sin 11.5^\circ \cos 36.5^\circ$$

63. $\cos 4x + \cos 8x$

$$= 2\cos\left(\frac{4x+8x}{2}\right)\cos\left(\frac{4x-8x}{2}\right)$$
$$= 2\cos 6x \cos(-2x)$$
$$= 2\cos 6x \cos 2x$$

65. **(a)** Since R is the radius of the circle, the dashed line has length $R-b$, so $\cos\frac{\theta}{2} = \frac{R-b}{R}$.

(b) $$\tan\frac{\theta}{4} = \frac{1-\cos\frac{\theta}{2}}{\sin\frac{\theta}{2}} = \frac{1-\frac{R-b}{R}}{\frac{50}{R}} = \frac{R-(R-b)}{50} = \frac{b}{50}$$

(c) $$\tan\frac{\theta}{4} = \frac{b}{50} = \frac{12}{50}$$
$$\frac{\theta}{4} = \tan^{-1}\left(\frac{12}{50}\right)$$
$$\theta = 4\tan^{-1}\left(\frac{12}{50}\right) \approx 53.98^\circ \approx 54^\circ$$

67. (a) $W = VI$

$= [163\sin(120\pi t)][1.23\sin(120\pi t)]$

$= 200.49\sin^2(120\pi t)$

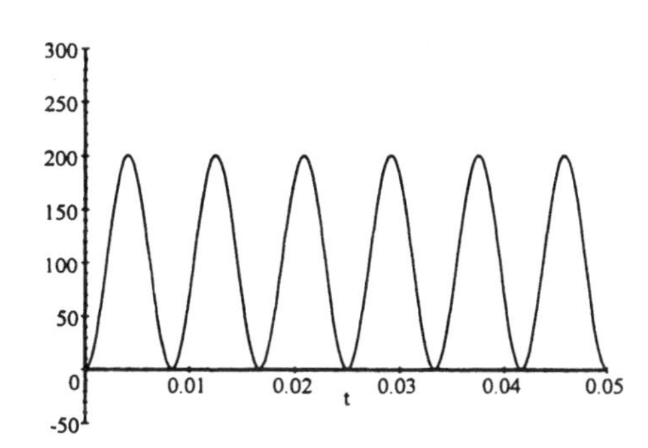

(b) The minimum wattage is 0 watts and occurs whenever $\sin(120\pi t) = 0$. The maximum wattage will occur when $\sin(120\pi t) = 1$. This would be $200.49(1) = 200.49$ watts.

(c) Using the identity,

$$\cos 2A = 1 - 2\sin^2 A$$
$$\sin^2 A = \frac{1}{2}[1 - \cos 2A]$$

$$W = 200.49\sin^2(120\pi t)$$
$$= 200.49\left[\frac{1}{2}(1 - \cos(240\pi t))\right]$$
$$= -100.245\cos(240\pi t) + 100.245$$

Thus, $a = -100.245$, $\omega = 240\pi$, $c = 100.245$.

(d) $W = 200.49\sin^2(120\pi t)$ and $W = -100.245\cos(240\pi t) + 100.245$ have the same graphs.

(e) Graph W and $y = 100.245$ together. The cosine (or sine) graph of W appears to be vertically centered about this line. An estimate for the average wattage consumed is 100.245 watts. The light bulb would be rated about 100 watts. (For sinusoidal current, the average wattage consumed by an electrical device will be equal to half of the peak wattage.)

Section 9.4

1. $y = \sin^{-1} x$

(a) domain: $[-1, 1]$

(b) range: $\left[-\frac{\pi}{2}, \frac{\pi}{2}\right]$

(c) For this function, as x increases, y increases. Therefore, it is an increasing function.

(d) $\arcsin(-2)$ is not defined since -2 is not in the domain.

3. $y = \tan^{-1} x = \arctan x$

(a) domain: $(-\infty, \infty)$

(b) range: $\left(-\frac{\pi}{2}, \frac{\pi}{2}\right)$

(c) For this function, as x increases, y increases. Therefore, it is an increasing function.

(d) No, since the domain is $\left(-\frac{\pi}{2}, \frac{\pi}{2}\right)$, $\arctan x$ is defined for all numbers.

5. $\sec^{-1} a = y$

$$a = \sec y = \frac{1}{\cos y}$$
$$\frac{1}{a} = \cos y$$
$$\boxed{y = \cos^{-1}\left(\frac{1}{a}\right)}$$

7. $y = \tan^{-1} 1$ $\boxed{y = \frac{\pi}{4}}$

$\left(\text{range: } \left(-\frac{\pi}{2}, \frac{\pi}{2}\right)\right)$

9. $y = \cos^{-1}(-1)$ $\boxed{y = \pi}$

(range: $[0, \pi]$)

11. $y = \sin^{-1}(-1)$ $\boxed{y = -\frac{\pi}{2}}$

$\left(\text{range: } \left[-\frac{\pi}{2}, \frac{\pi}{2}\right]\right)$

13. $y = \arctan 0$ $\boxed{y = 0}$

$\left(\text{range: } \left(-\frac{\pi}{2}, \frac{\pi}{2}\right)\right)$

15. $y = \arccos 0$ $\boxed{y = \frac{\pi}{2}}$

(range: $[0, \pi]$)

17. $y = \sin^{-1}\left(\frac{\sqrt{2}}{2}\right)$ $\boxed{y = \frac{\pi}{4}}$

$\left(\text{range: } \left[-\frac{\pi}{2}, \frac{\pi}{2}\right]\right)$

19. $y = \arccos\left(-\frac{\sqrt{3}}{2}\right)$ $\boxed{y = \frac{5\pi}{6}}$

(range: $[0, \pi]$)

21. $y = \cot^{-1}(-1)$ $\boxed{y = \frac{3\pi}{4}}$

(range: $(0, \pi)$)

23. $y = \csc^{-1}(-2)$ $\boxed{y = -\frac{\pi}{6}}$

$\left(\text{range: } \left[-\frac{\pi}{2}, 0\right) \cup \left(0, \frac{\pi}{2}\right]\right)$

25. $y = \text{arcsec}\left(\frac{2\sqrt{3}}{3}\right)$ $\boxed{y = \frac{\pi}{6}}$

$\left(\text{range: } \left[0, \frac{\pi}{2}\right) \cup \left(\frac{\pi}{2}, \pi\right]\right)$

27. $\theta = \arctan(-1)$ $\boxed{\theta = -45°}$

(range: $(-90°, 90°)$)

29. $\theta = \arcsin\left(-\frac{\sqrt{3}}{2}\right)$ $\boxed{\theta = -60°}$

(range: $[-90°, 90°]$)

31. $\theta = \cot^{-1}\left(-\frac{\sqrt{3}}{3}\right)$ $\boxed{\theta = 120°}$

(range: $(0°, 180°)$)

33. $\theta = \csc^{-1}(-2)$ $\boxed{\theta = -30°}$

(range: $[-90°, 0°) \cup (0°, 90°]$)

35. $\theta = \sin^{-1}(-.13349122)$
$\theta \approx -7.6713835°$

37. $\theta = \arccos(-.39876459)$
$\theta \approx 113.500970°$

39. $\theta = \csc^{-1} 1.9422833$
$\csc\theta = 1.9422833$
$\sin\theta = (1.9422833)^{-1}$
$\theta = \sin(1.9422833)^{-1}$
$\theta \approx 30.987961°$

41. $y = \arctan 1.1111111$
$y \approx .83798122$

43. $y = \cot^{-1}(-.92170128)$
$\cot y = -.92170128$
$\tan y = (-.92170128)^{-1}$
$y = \tan(-.92170128)^{-1} \approx -.8261201193 + \pi$
≈ 2.315472534

45. $y = \arcsin .92837781 \approx 1.1900238$

47. $y = \cot^{-1} x$

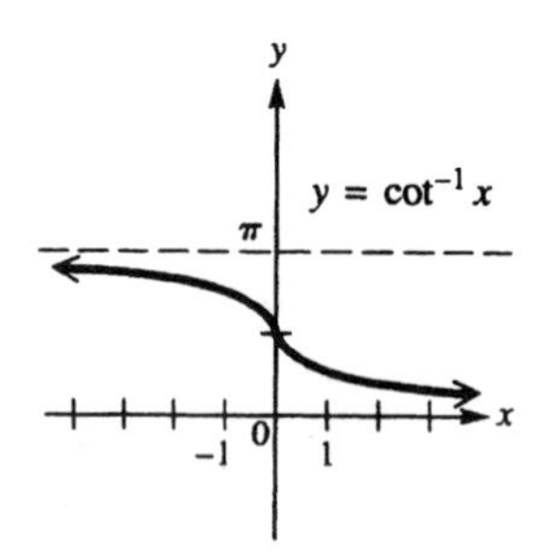

49. $y = \sec^{-1} x$

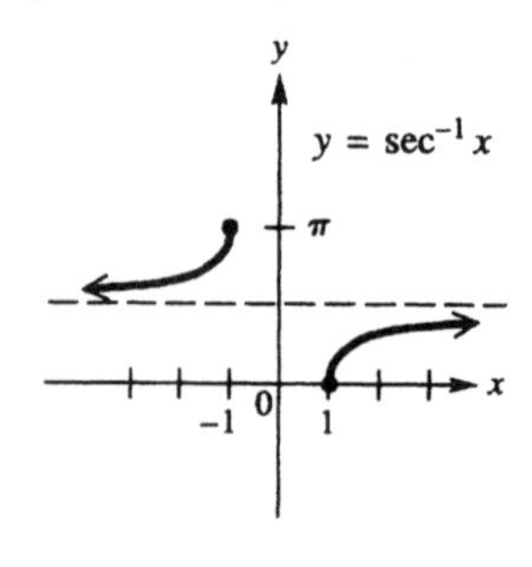

51. $y = \operatorname{arcsec} \frac{1}{2}x$

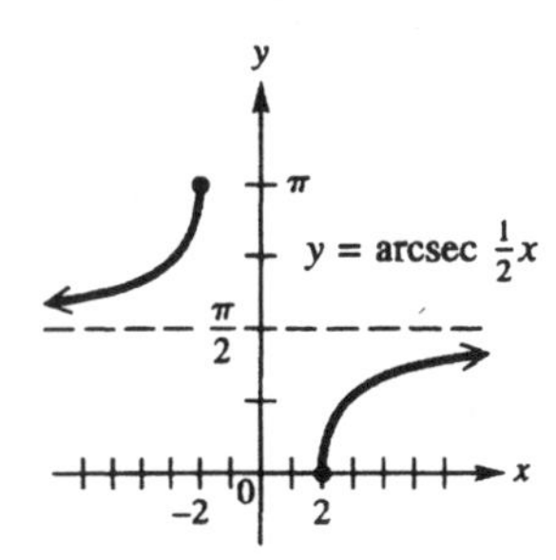

53. $\tan^{-1} 1.003$ exists because the domain of $\tan^{-1} x$ is $(-\infty, \infty)$. However, we cannot find the inverse sine unless the number is in the interval $[-1, 1]$.

Relating Concepts (Exercises 54 - 56)

54. $f(x) = 3x - 2, \quad f^{-1}(x) = \dfrac{x+2}{3}$

$$\begin{aligned} f\left[f^{-1}(x)\right] &= f\left(\frac{x+2}{3}\right) \\ &= 3\left(\frac{x+2}{3}\right) - 2 \\ &= x + 2 - 2 = x \end{aligned}$$

$$\begin{aligned} f^{-1}[f(x)] &= f^{-1}(3x-2) \\ &= \frac{(3x-2)+2}{3} \\ &= \frac{3x}{3} = x \end{aligned}$$

In both cases, the result is x. The graph is a straight line, $y = x$, bisecting quadrants I and III.

55. $y = \tan\left(\tan^{-1} x\right)$ is the graph of $y = x$:

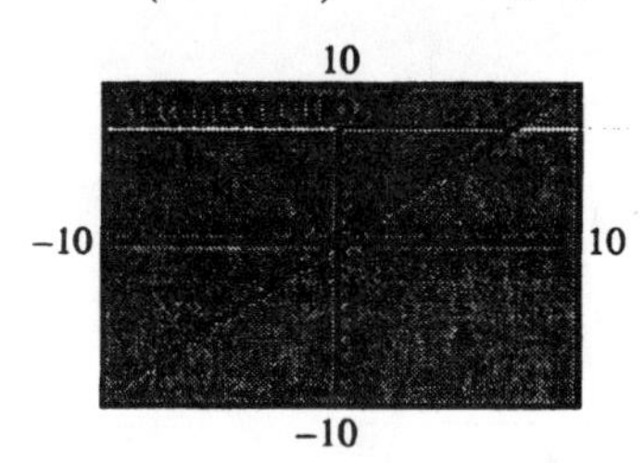

It is the same graph as in Exercise 54.

56. $y = \tan^{-1}(\tan x)$

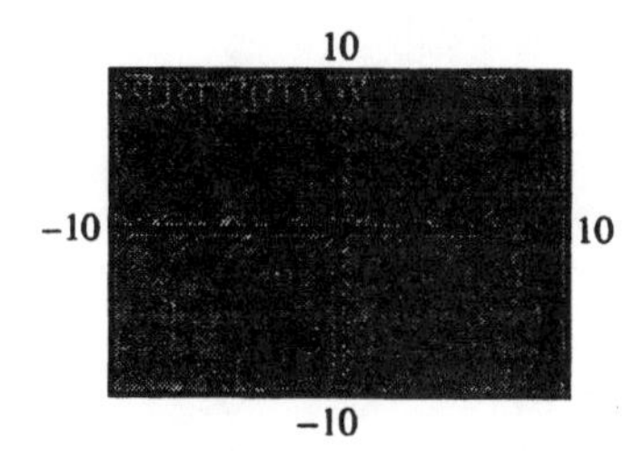

The range of $\tan^{-1} x$ is $\left(-\frac{\pi}{2}, \frac{\pi}{2}\right)$, not $(-\infty, \infty)$, so the graphs do not agree.

57. $\tan\left(\arccos \dfrac{3}{4}\right)$

$\cos\theta = \dfrac{3}{4}$, $\theta \in$ Quadrant I

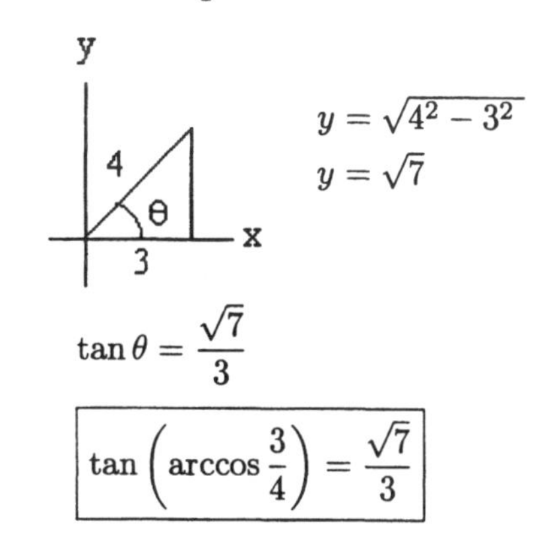

$y = \sqrt{4^2 - 3^2}$

$y = \sqrt{7}$

$\tan\theta = \dfrac{\sqrt{7}}{3}$

$$\boxed{\tan\left(\arccos \frac{3}{4}\right) = \frac{\sqrt{7}}{3}}$$

59. $\cos\left(\tan^{-1}(-2)\right)$

$\tan\theta = -2$, $\theta \in$ Quadrant IV

y, 1, x, θ, -2

$r = \sqrt{1^2 + (-2)^2}$

$r = \sqrt{5}$

$\cos\theta = \dfrac{1}{\sqrt{5}}$

$$\boxed{\cos\left(\tan^{-1}(-2)\right) = \frac{\sqrt{5}}{5}}$$

61. $\sin\left(2\tan^{-1}\frac{12}{5}\right)$

$\tan\theta = \frac{12}{5}$, $\theta \in$ Quadrant I

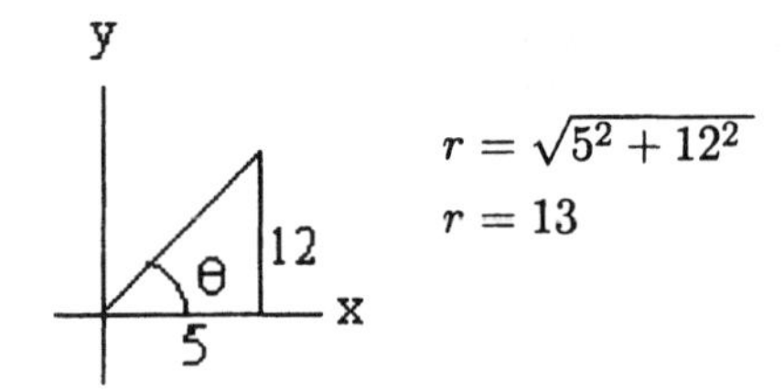

$r = \sqrt{5^2 + 12^2}$
$r = 13$

$$\begin{aligned}\sin 2\theta &= 2\sin\theta\cos\theta \\ &= 2\left(\frac{12}{13}\right)\left(\frac{5}{13}\right) = \frac{120}{169}\end{aligned}$$

$$\boxed{\sin\left(2\tan^{-1}\frac{12}{5}\right) = \frac{120}{169}}$$

63. $\cos\left(2\arctan\frac{4}{3}\right)$

$\tan\theta = \frac{4}{3}$, $\theta \in$ Quadrant I

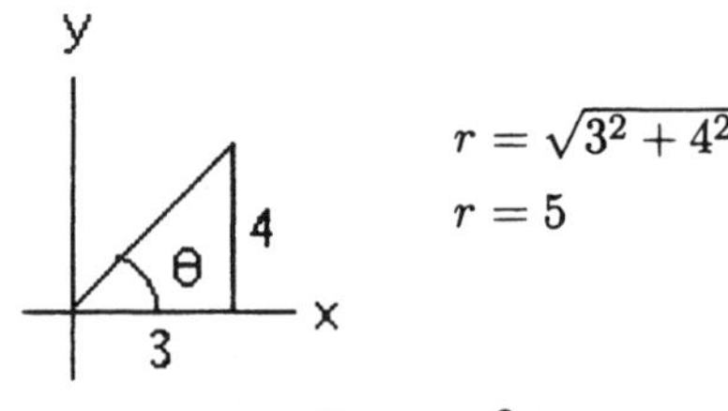

$r = \sqrt{3^2 + 4^2}$
$r = 5$

$$\begin{aligned}\cos 2\theta &= \cos^2\theta - \sin^2\theta \\ &= \left(\frac{3}{5}\right)^2 - \left(\frac{4}{5}\right)^2 \\ &= \frac{9 - 16}{25} = -\frac{7}{25}\end{aligned}$$

$$\boxed{\cos\left(2\arctan\frac{4}{3}\right) = -\frac{7}{25}}$$

65. $\sin\left(2\cos^{-1}\frac{1}{5}\right)$

$\cos\theta = \frac{1}{5}$, $\theta \in$ Quadrant I

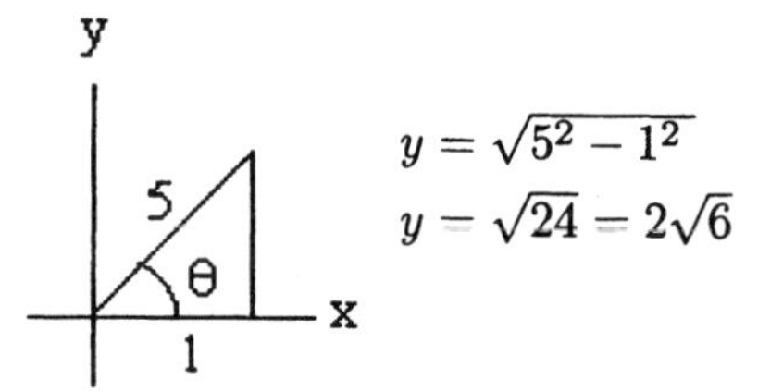

$y = \sqrt{5^2 - 1^2}$
$y = \sqrt{24} = 2\sqrt{6}$

$$\begin{aligned}\sin 2\theta &= 2\sin\theta\cos\theta \\ &= 2\left(\frac{2\sqrt{6}}{5}\right)\left(\frac{1}{5}\right) = \frac{4\sqrt{6}}{25}\end{aligned}$$

$$\boxed{\sin\left(2\cos^{-1}\frac{1}{5}\right) = \frac{4\sqrt{6}}{25}}$$

67. $\sec\left(\sec^{-1} 2\right)$

$\sec\theta = 2$, $\theta \in$ Quadrant I

$$\boxed{\sec\left(\sec^{-1} 2\right) = 2}$$

69. $\cos\left(\tan^{-1}\frac{5}{12} - \tan^{-1}\frac{3}{4}\right)$

$\tan\theta = \frac{5}{12}$, $\theta \in$ Quadrant I

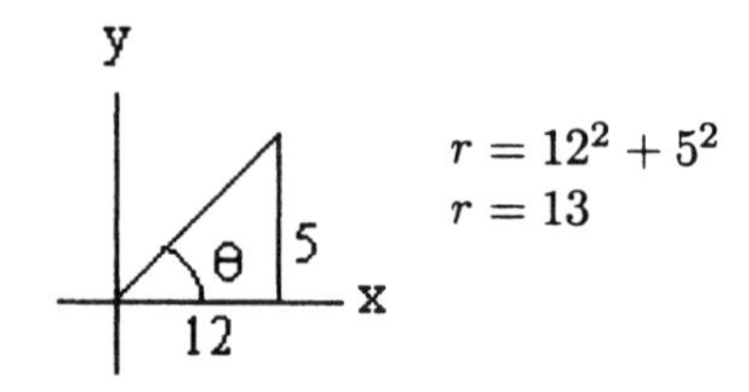

$r = 12^2 + 5^2$
$r = 13$

$\tan\beta = \frac{3}{4}$, $\beta \in$ Quadrant I

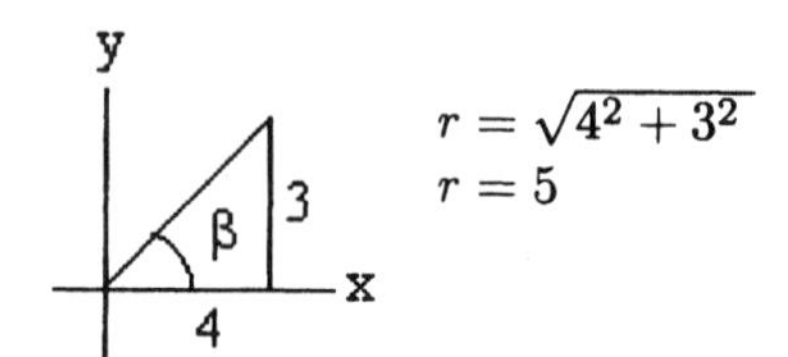

$r = \sqrt{4^2 + 3^2}$
$r = 5$

$$\begin{aligned}\cos(\theta - \beta) &= \cos\theta\cos\beta + \sin\theta\sin\beta \\ &= \frac{12}{13}\cdot\frac{4}{5} + \frac{5}{13}\cdot\frac{3}{5} \\ &= \frac{48 + 15}{65}\end{aligned}$$

$$\boxed{\cos\left(\tan^{-1}\frac{5}{12} - \tan^{-1}\frac{3}{4}\right) = \frac{63}{65}}$$

71. $\sin\left(\sin^{-1}\frac{1}{2} + \tan^{-1}(-3)\right)$

$\sin\theta = \frac{1}{2}$, $\theta \in$ Quadrant I

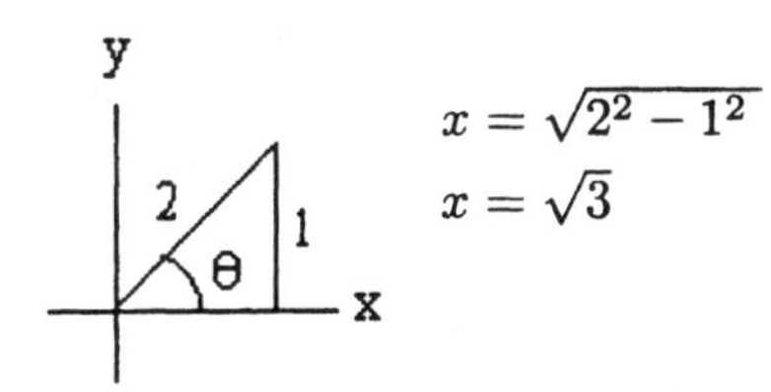

$x = \sqrt{2^2 - 1^2}$

$x = \sqrt{3}$

$\tan\beta = -3$, $\beta \in$ Quadrant IV

$r = \sqrt{(-3)^2 + 1^2}$

$r = \sqrt{10}$

$$\begin{aligned}\sin(\theta + \beta) &= \sin\theta\cos\beta + \cos\theta\sin\beta \\ &= \frac{1}{2}\cdot\frac{1}{\sqrt{10}} + \frac{\sqrt{3}}{2}\cdot\frac{-3}{\sqrt{10}} \\ &= \frac{1 - 3\sqrt{3}}{2\sqrt{10}}\end{aligned}$$

$$\boxed{\sin\left(\sin^{-1}\frac{1}{2} + \tan^{-1}(-3)\right) = \frac{\sqrt{10} - 3\sqrt{30}}{20}}$$

73. $\cos\left(\tan^{-1}.5\right) \approx .894427191$

75. $\tan(\arcsin .12251014) \approx .1234399811$

77. $\sin(\arccos u)$

$\cos\theta = u$, $\theta \in$ Quadrant I

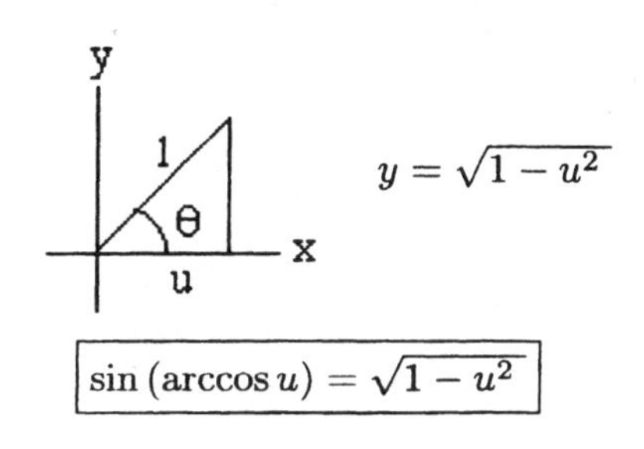

$y = \sqrt{1 - u^2}$

$$\boxed{\sin(\arccos u) = \sqrt{1 - u^2}}$$

79. $\cot(\arcsin u)$

$\sin\theta = u$, $\theta \in$ Quadrant I

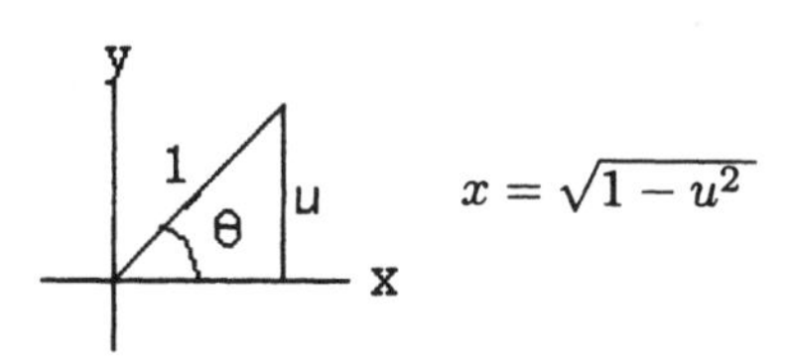

$x = \sqrt{1 - u^2}$

$$\boxed{\cot(\arcsin u) = \frac{\sqrt{1 - u^2}}{u}}$$

81. $\sin\left(\sec^{-1}\frac{u}{2}\right)$

$\sec\theta = \frac{u}{2}$, $\theta \in$ Quadrant I

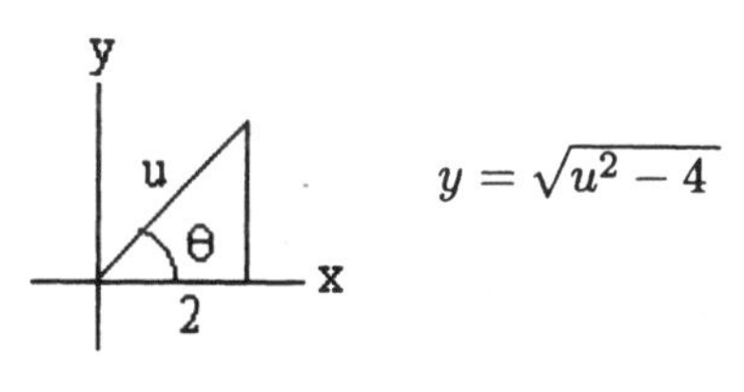

$y = \sqrt{u^2 - 4}$

$$\boxed{\sin\left(\sec^{-1}\frac{u}{2}\right) = \frac{\sqrt{u^2 - 4}}{u}}$$

83. $\tan\left(\sin^{-1}\frac{u}{\sqrt{u^2 + 2}}\right)$

$\sin\theta = \frac{u}{\sqrt{u^2 + 2}}$, $\theta \in$ Quadrant I

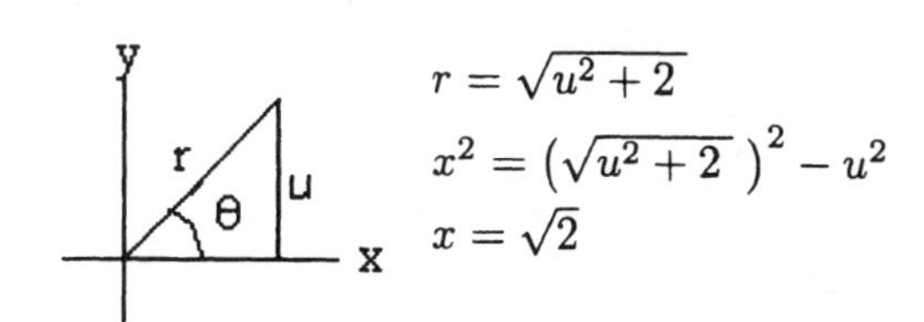

$r = \sqrt{u^2 + 2}$

$x^2 = \left(\sqrt{u^2 + 2}\right)^2 - u^2$

$x = \sqrt{2}$

$\tan\theta = \frac{u}{\sqrt{2}}$

$$\boxed{\tan\left(\sin^{-1}\frac{u}{\sqrt{u^2 + 2}}\right) = \frac{u\sqrt{2}}{2}}$$

85. $\sec\left(\text{arccot}\,\dfrac{\sqrt{4-u^2}}{u}\right)$

$\cot\theta = \dfrac{\sqrt{4-u^2}}{u}$, $\theta \in$ Quadrant I

y r u θ x x

$x = \sqrt{4-u^2}$

$r^2 = \left(\sqrt{4-u^2}\right)^2 + u^2$

$r = 2$

$\sec\theta = \dfrac{2}{\sqrt{4-u^2}}$

$$\boxed{\sec\left(\text{arccot}\,\frac{\sqrt{4-u^2}}{u}\right) = \frac{2\sqrt{4-u^2}}{4-u^2}}$$

87. $\theta = \arcsin\left(\sqrt{\dfrac{v^2}{2v^2+64h}}\right)$

(a) $\theta = \arcsin\left(\sqrt{\dfrac{v^2}{2v^2+0}}\right)$

$= \arcsin\left(\sqrt{\dfrac{v^2}{2v^2}}\right)$

$= \arcsin\sqrt{\dfrac{1}{2}} = \arcsin\dfrac{1}{\sqrt{2}} = 45°$

(b) $\theta = \arcsin\left(\sqrt{\dfrac{v^2}{2v^2+64(6)}}\right)$

$= \arcsin\left(\sqrt{\dfrac{v^2}{2v^2+384}}\right)$

As $v \longrightarrow \infty$, $\sqrt{\dfrac{v^2}{2v^2+384}} \longrightarrow \sqrt{\dfrac{1}{2}}$

$\theta = \arcsin\sqrt{\dfrac{1}{2}} = \arcsin\dfrac{1}{\sqrt{2}} = 45°$

The equation of the asymptote is $\theta = 45°$ or $y = x$.
($\sin 45° = \cos 45°$, so $y = x$)

89. $\theta = \tan^{-1}\left(\dfrac{3x}{x^2+4}\right)$

(a) $x = 3: \theta = \tan^{-1}\left(\dfrac{9}{9+4}\right) = \tan^{-1}\dfrac{9}{13}$

$\approx 34.70 \approx \boxed{35°}$

(b) $x = 6: \theta = \tan^{-1}\left(\dfrac{18}{36+4}\right) = \tan^{-1}\dfrac{9}{20}$

$\approx 24.23 \approx \boxed{24°}$

(c) $x = 9: \theta = \tan^{-1}\left(\dfrac{27}{81+4}\right) = \tan^{-1}\dfrac{27}{85}$

$\approx 17.62 \approx \boxed{18°}$

(d) $\tan(\theta+\alpha) = \dfrac{3+1}{x} = \dfrac{4}{x}$; $\tan\alpha = \dfrac{1}{x}$

3 1 θ α x

$\tan(\theta+\alpha) = \dfrac{\tan\theta + \tan\alpha}{1 - \tan\theta\tan\alpha}$

$\dfrac{4}{x} = \dfrac{\tan\theta + \frac{1}{x}}{1 - (\tan\theta)\left(\frac{1}{x}\right)}$

$\dfrac{4}{x} = \dfrac{x\tan\theta + 1}{x - \tan\theta}$

$4(x - \tan\theta) = x(x\tan\theta + 1)$

$4x - 4\tan\theta = x^2\tan\theta + x$

$4x - x = x^2\tan\theta + 4\tan\theta$

$3x = \tan\theta\,(x^2+4)$

$\tan\theta = \dfrac{3x}{x^2+4}$

$\theta = \tan^{-1}\left(\dfrac{3x}{x^2+4}\right)$

(e) $y_1 = \tan^{-1}\left(\dfrac{3x}{x^2+4}\right)$

1 0.5 -6 -4 -2 0 2 x 4 6 -0.5 -1

The maximum value occurs when $x \approx 2$ ft.

91. Since the diameter of the earth is 7927 miles at the equator, the radius of the earth is 3963.5 miles. Then

$$\cos\theta = \frac{3963.5}{20,000 + 3963.5} = \frac{3963.5}{23963.5}$$

$$\theta = \cos^{-1}\frac{3963.5}{23963.5} \approx 80.48°$$

The percent of the equator that can be seen by the satellite is

$$\frac{2\theta}{360}\cdot 100 = \frac{2\,(80.48)}{360}\cdot 100 \approx 44.7\%.$$

Reviewing Basic Concepts (Sections 9.3 and 9.4)

1. $\cos 2x = -\frac{5}{12},\quad \frac{\pi}{2} < x < \pi$

$$\pi < 2x < 2\pi$$

$$\cos 2x = 1 - 2\sin^2 x$$

$$-\frac{5}{12} = 1 - 2\sin^2 x$$

$$2\sin^2 x = \frac{17}{12}$$

$$\sin^2 x = \frac{17}{24}$$

$$\sin x = \frac{\sqrt{17}}{\sqrt{24}} \quad \text{(Quadrant II)}$$

$$\sin x = \frac{\sqrt{17}}{2\sqrt{6}}\cdot\frac{\sqrt{6}}{\sqrt{6}} = \frac{\sqrt{102}}{12}$$

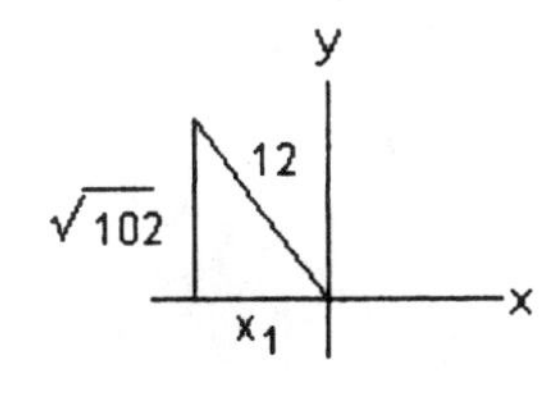

$$x_1 = \sqrt{12^2 - (\sqrt{102})^2}$$

$$x_1 = \sqrt{144 - 102} = \sqrt{42}$$

$$\text{So } \tan x = -\frac{\sqrt{102}}{\sqrt{42}}\cdot\frac{\sqrt{42}}{\sqrt{42}}$$

$$= -\frac{\sqrt{4284}}{42} = -\frac{6\sqrt{119}}{42}$$

$$= -\frac{\sqrt{119}}{7}$$

2. $\sin\theta = \frac{1}{3}$, θ in Quadrant III

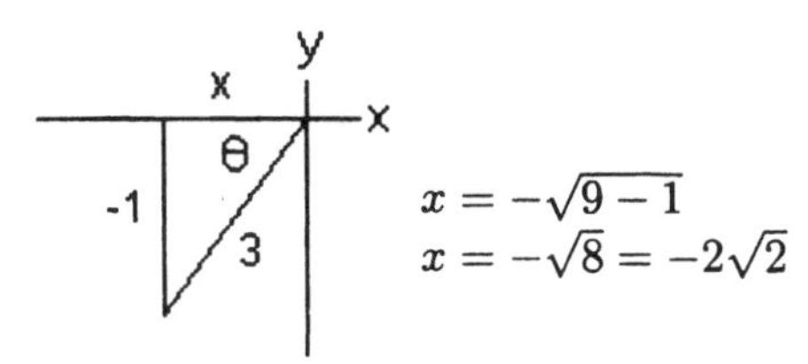

$$x = -\sqrt{9-1}$$

$$x = -\sqrt{8} = -2\sqrt{2}$$

$$\sin 2\theta = 2\sin\theta\cos\theta$$

$$= 2\left(-\frac{1}{3}\right)\left(-\frac{2\sqrt{2}}{3}\right) = \frac{4\sqrt{2}}{9}$$

$$\cos 2\theta = 2\cos^2\theta - 1$$

$$= 2\left(-\frac{2\sqrt{2}}{3}\right)^2 - 1$$

$$= 2\left(\frac{8}{9}\right) - 1$$

$$= \frac{16}{9} - \frac{9}{9} = \frac{7}{9}$$

$$\tan 2\theta = \frac{\sin 2\theta}{\cos 2\theta} = \frac{4\sqrt{2}/9}{7/9} = \frac{4\sqrt{2}}{7}$$

3. $\sin 75° = \sin\frac{150°}{2} = \sqrt{\frac{1-\cos 150°}{2}}$

$$= \sqrt{\frac{1-(-\sqrt{3}/2)}{2}\cdot\frac{2}{2}}$$

$$= \frac{\sqrt{2+\sqrt{3}}}{2}$$

or: $\sin 75° = \sin(30° + 45°)$

$$= \sin 30°\cos 45° + \cos 30°\sin 45°$$

$$= \frac{1}{2}\cdot\frac{\sqrt{2}}{2} + \frac{\sqrt{3}}{2}\cdot\frac{\sqrt{2}}{2}$$

$$= \frac{\sqrt{2}}{4} + \frac{\sqrt{6}}{4} = \frac{\sqrt{2}+\sqrt{6}}{4}$$

4. $2\sin 25°\cos 150°$

$$= 2\left[\frac{1}{2}\left[\sin(25° + 150°) + \sin(25° - 150°)\right]\right]$$

$$= \sin 175° + \sin(-125°)$$

$$= \sin 175° - \sin 125°$$

5. **(a)** Verify $\sin^2 \frac{x}{2} = \frac{\tan x - \sin x}{2 \tan x}$

Left: $\sin^2 \frac{x}{2} = \left(\pm\sqrt{\frac{1-\cos x}{2}}\right)^2 = \frac{1-\cos x}{2}$

Right: $\frac{\tan x - \sin x}{2\tan x} = \frac{\frac{\sin x}{\cos x} - \sin x}{2\,\frac{\sin x}{\cos x}} \cdot \frac{\cos x}{\cos x}$

$$= \frac{\sin x - \sin x \cos x}{2 \sin x}$$

$$= \frac{\sin x\,(1-\cos x)}{2\sin x}$$

$$= \frac{1-\cos x}{2}$$

(b) Verify $\frac{\sin 2x}{2\sin x} = \cos^2\frac{x}{2} - \sin^2\frac{x}{2}$

Left: $\frac{\sin 2x}{2\sin x} = \frac{2\sin x\cos x}{2\sin x} = \cos x$

Right: $\cos^2\frac{x}{2} - \sin^2\frac{x}{2}$

$$= \left(\pm\sqrt{\frac{1+\cos x}{2}}\right)^2 - \left(\pm\sqrt{\frac{1-\cos x}{2}}\right)^2$$

$$= \frac{1+\cos x}{2} - \frac{1-\cos x}{2}$$

$$= \frac{1+\cos x - (1-\cos x)}{2}$$

$$= \frac{2\cos x}{2} = \cos x$$

6. **(a)** $y = \arccos\frac{\sqrt{3}}{2} = \frac{\pi}{6}$

(b) $y = \sin^{-1}\left(-\frac{\sqrt{2}}{2}\right) = -\frac{\pi}{4}$

7. $\theta = \sec^{-1} 2$

(a) $\theta = \arccos .5 = 60°$

(b) $\theta = \cot^{-1}(-1) = 135°$

8. $y = 2\csc^{-1} x$

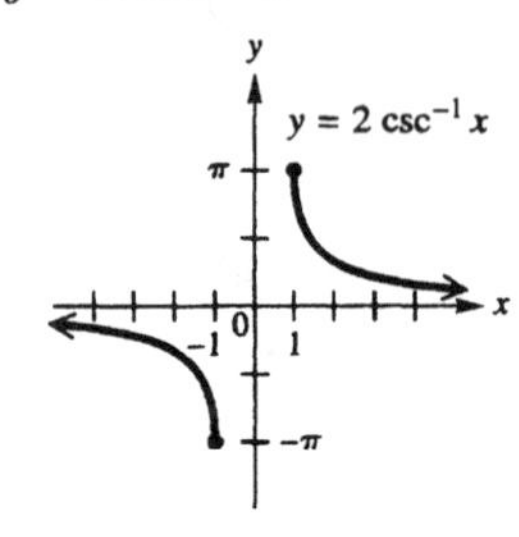

9. **(a)** $\cot\left(\arcsin\left(-\frac{2}{3}\right)\right)$:

$\arcsin\left(-\frac{2}{3}\right)$:

$x = \sqrt{9-4} = \sqrt{5}$

$\cot\left(\arcsin\left(-\frac{2}{3}\right)\right) = \frac{\sqrt{5}}{-2} = -\frac{\sqrt{5}}{2}$

(b) $\cos\left(\tan^{-1}\frac{5}{12} - \sin^{-1}\frac{3}{5}\right)$:

Let $\theta = \tan^{-1}\frac{5}{12}$:

$$r = \sqrt{12^2+5^2} = \sqrt{144+25} = \sqrt{169} = 13$$

Let $\beta = \sin^{-1}\frac{3}{5}$:

$$x = \sqrt{5^2-3^2} = \sqrt{25-9} = \sqrt{16} = 4$$

$$\cos(\theta-\beta) = \cos\theta\cos\beta + \sin\alpha\sin\beta$$

$$= \frac{12}{13}\cdot\frac{4}{5} + \frac{5}{13}\cdot\frac{3}{5}$$

$$= \frac{48}{65} + \frac{15}{65} = \frac{63}{65}$$

10. $\sin(\operatorname{arccot} u)$:

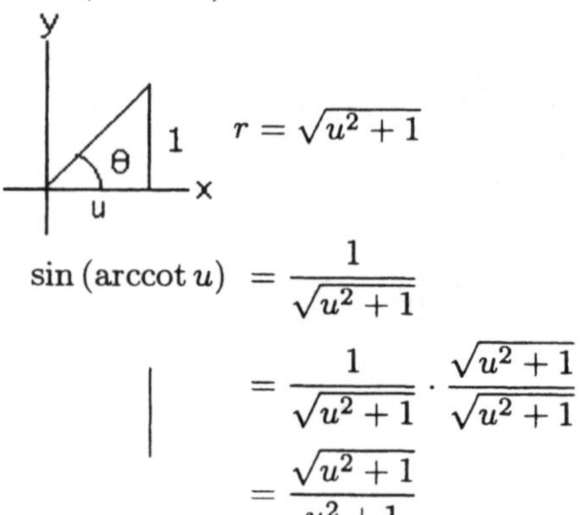

$r = \sqrt{u^2+1}$

$$\sin(\operatorname{arccot} u) = \frac{1}{\sqrt{u^2+1}}$$

$$= \frac{1}{\sqrt{u^2+1}} \cdot \frac{\sqrt{u^2+1}}{\sqrt{u^2+1}}$$

$$= \frac{\sqrt{u^2+1}}{u^2+1}$$

Section 9.5

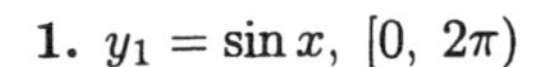

1. $y_1 = \sin x,\ [0,\ 2\pi)$

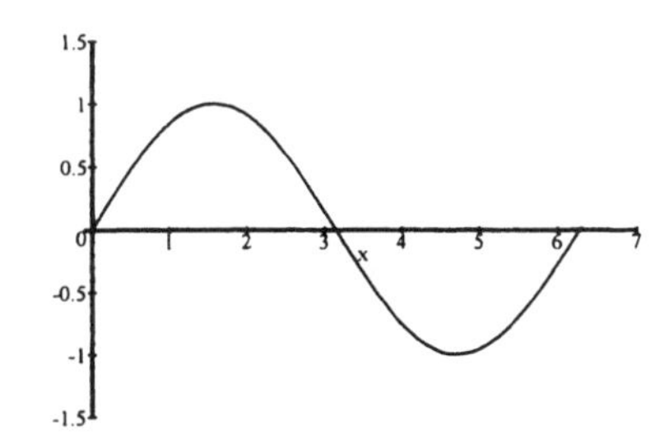

The equation $\sin x - b = 0$ is equivalent to $\sin x = b$. Consider the intersection of y_1 with $y_2 = b$.

two solutions: $-1 < b < 1$

one solution: $b = -1,\ 1$

no solutions: $b < -1$ or $b > 1$

3. $2\cot x + 1 = -1$

$2\cot x = -2$

$\cot x = -1$

$x = \dfrac{3\pi}{4},\ \dfrac{7\pi}{4}$

Solution set: $\boxed{\left\{\dfrac{3\pi}{4},\ \dfrac{7\pi}{4}\right\}}$

5. $2\sin x + 3 = 4$

$2\sin x = 1$

$\sin x = \dfrac{1}{2}$

$x = \dfrac{\pi}{6},\ \dfrac{5\pi}{6}$

Solution set: $\boxed{\left\{\dfrac{\pi}{6},\ \dfrac{5\pi}{6}\right\}}$

7. $(\cot x - 1)\left(\sqrt{3}\cot x + 1\right) = 0$

(i) $\cot x - 1 = 0$

$\cot x = 1 \longrightarrow x = \dfrac{\pi}{4},\ \dfrac{5\pi}{4}$

(ii) $\sqrt{3}\cot x + 1 = 0$

$\sqrt{3}\cot x = -1$

$\cot x = -\dfrac{1}{\sqrt{3}} \longrightarrow x = \dfrac{2\pi}{3},\ \dfrac{5\pi}{3}$

Solution set: $\boxed{\left\{\dfrac{\pi}{4},\ \dfrac{2\pi}{3},\ \dfrac{5\pi}{4},\ \dfrac{5\pi}{3}\right\}}$

9. $\cos^2 x + 2\cos x + 1 = 0$

$(\cos x + 1)^2 = 0$

$\cos x + 1 = 0$

$\cos x = -1$

$x = \pi$ Solution set: $\boxed{\{\pi\}}$

11. $\left(\cot x - \sqrt{3}\right)\left(2\sin x + \sqrt{3}\right) = 0$

$\cot x = \sqrt{3}$ $\sin x = -\dfrac{\sqrt{3}}{2}$

$x = \dfrac{\pi}{6},\ \dfrac{7\pi}{6}$ $x = \dfrac{4\pi}{3},\ \dfrac{5\pi}{3}$

Solution set: $\boxed{\left\{\dfrac{\pi}{6},\ \dfrac{7\pi}{6},\ \dfrac{4\pi}{3},\ \dfrac{5\pi}{3}\right\}}$

13. $\tan\theta - \cot\theta = 0$

$\tan\theta - \dfrac{1}{\tan\theta} = 0$

$\tan^2\theta - 1 = 0$

$\tan^2\theta = 1$

$\tan\theta = \pm 1$

Solution set: $\boxed{\left\{\dfrac{\pi}{4},\ \dfrac{3\pi}{4},\ \dfrac{5\pi}{4},\ \dfrac{7\pi}{4}\right\}}$

15. $\cos^2 x = \sin^2 x$

$\dfrac{\cos^2 x}{\sin^2 x} = 1$

$\tan^2 x = 1$

$\tan x = \pm 1$

$x = \dfrac{\pi}{4},\ \dfrac{3\pi}{4},\ \dfrac{5\pi}{4},\ \dfrac{7\pi}{4}$

Solution set: $\boxed{\left\{\dfrac{\pi}{4},\ \dfrac{3\pi}{4},\ \dfrac{5\pi}{4},\ \dfrac{7\pi}{4}\right\}}$

17. $\csc^2 x = 2\cot x$

$\csc^2 x - 2\cot x = 0$

$1 + \cot^2 x - 2\cot x = 0$

$\cot^2 x - 2\cot x + 1 = 0$

$(\cot x - 1)^2 = 0$

$\cot x = 1$

$x = \dfrac{\pi}{4},\ \dfrac{5\pi}{4}$

Solution set: $\boxed{\left\{\dfrac{\pi}{4},\ \dfrac{5\pi}{4}\right\}}$

19. $f(x) = 2\sin x + 1$

(a) $2\sin x + 1 = 0$

$2\sin x = -1 \Longrightarrow \sin x = -\dfrac{1}{2}$

Ref $\angle : \dfrac{\pi}{6}$; Quadrants III, IV

$f(x) = 0 : \boxed{\left\{\dfrac{7\pi}{6}, \dfrac{11\pi}{6}\right\}}$

(b) $f(x) = 2\sin x + 1$

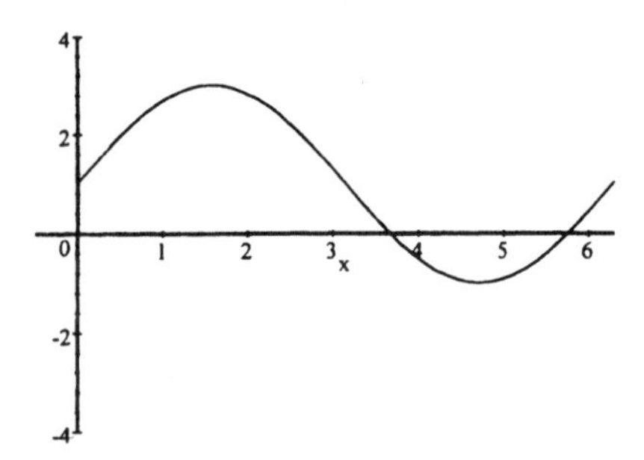

$f(x) > 0 : \boxed{\left[0, \dfrac{7\pi}{6}\right) \cup \left(\dfrac{11\pi}{6}, 2\pi\right)}$

(c) $f(x) < 0 : \boxed{\left(\dfrac{7\pi}{6}, \dfrac{11\pi}{6}\right)}$

21. $f(x) = \sec^2 x - 1$

(a) $\sec^2 x - 1 = 0$

$\sec^2 x = 1$

$\sec x = \pm 1$

Ref $\angle : 0, \pi$

$f(x) = 0 : \boxed{\{0, \pi\}}$

(b) $f(x) = \sec^2 x - 1$

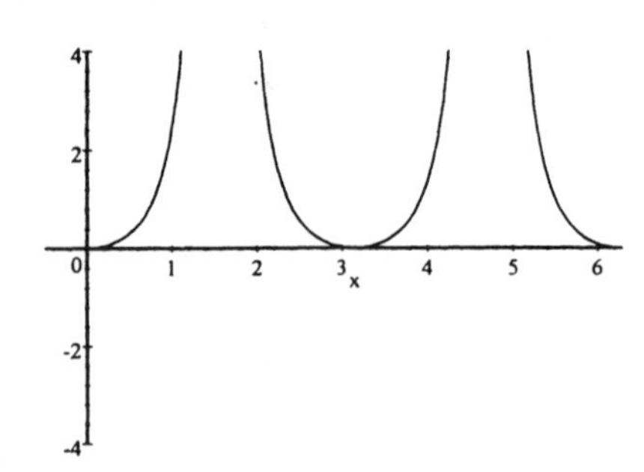

$f(x) > 0 :$

$\boxed{\left(0, \dfrac{\pi}{2}\right) \cup \left(\dfrac{\pi}{2}, \pi\right) \cup \left(\pi, \dfrac{3\pi}{2}\right) \cup \left(\dfrac{3\pi}{2}, 2\pi\right)}$

(c) $f(x) < 0 : \boxed{\emptyset}$

23. $f(x) = 2\sin^2 x + 3\sin x + 1$

(a) $2\sin^2 x + 3\sin x + 1 = 0$

$(2\sin x + 1)(\sin x + 1) = 0$

(i) $2\sin x + 1 = 0$

$\sin x = -\dfrac{1}{2}$

Ref $\angle : \dfrac{\pi}{6}$; Quadrants III, IV

$x = \dfrac{7\pi}{6}, \dfrac{11\pi}{6}$

(ii) $\sin x + 1 = 0$

$\sin x = -1$

$x = \dfrac{3\pi}{2}$

$f(x) = 0 : \boxed{\left\{\dfrac{7\pi}{6}, \dfrac{3\pi}{2}, \dfrac{11\pi}{6}\right\}}$

(b) $f(x) = 2\sin^2 x + 3\sin x + 1$

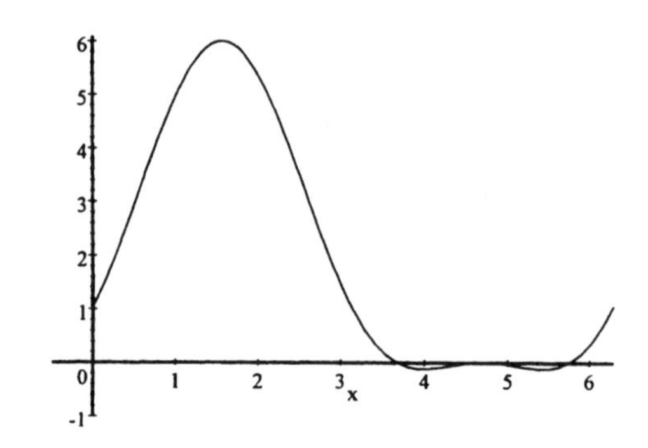

(c) $f(x) > 0 : \boxed{\left[0, \dfrac{7\pi}{6}\right) \cup \left(\dfrac{11\pi}{6}, 2\pi\right)}$

(d) $f(x) < 0 : \boxed{\left(\dfrac{7\pi}{6}, \dfrac{3\pi}{2}\right) \cup \left(\dfrac{3\pi}{2}, \dfrac{11\pi}{6}\right)}$

25. $f(x) = \sin^2 x \cos^2 x$

(a) $\sin^2 x \cos^2 x = 0$

$(1 - \cos^2 x)(\cos^2 x) = 0$

(i) $\cos^2 x = 1$

$\cos x = \pm 1$

$x = 0,\ \pi$

(ii) $\cos^2 x = 0$

$\cos x = 0$

$x = \frac{\pi}{2},\ \frac{3\pi}{2}$

$f(x) = 0:$ $\boxed{\left\{0,\ \frac{\pi}{2},\ \pi,\ \frac{3\pi}{2}\right\}}$

(b) $f(x) = 2\cos^2 x - \sqrt{3}\cos x$

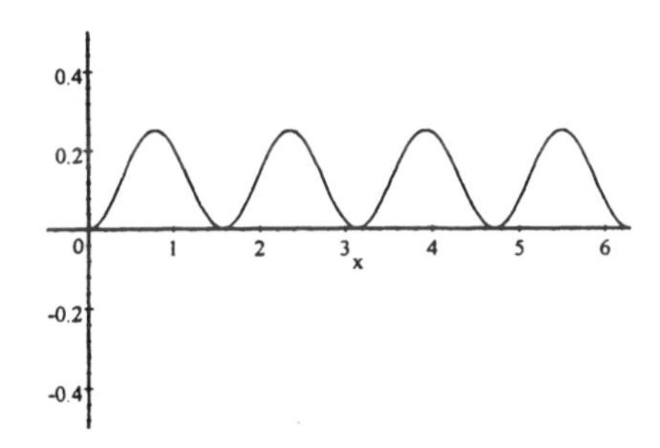

$f(x) > 0:$

$\boxed{\left(0,\ \frac{\pi}{2}\right) \cup \left(\frac{\pi}{2},\ \pi\right) \cup \left(\pi,\ \frac{3\pi}{2}\right) \cup \left(\frac{3\pi}{2},\ 2\pi\right)}$

(c) $f(x) < 0:$ $\boxed{\emptyset}$

Relating Basic Concepts (Exercises 27-31)

27. $\tan^3 x = 3\tan x$

$\tan^3 x - 3\tan x = 0$

28. $\tan^3 x - 3\tan x = 0$

$\tan x\left(\tan^2 x - 3\right) = 0$

$\tan x = 0 \qquad \tan^2 x = 3$

$x = 0,\ \pi \qquad \tan x = \pm\sqrt{3}$

$x = \frac{\pi}{3},\ \frac{2\pi}{3},\ \frac{4\pi}{3},\ \frac{5\pi}{3}$

$x = \left\{0,\ \frac{\pi}{3},\ \frac{2\pi}{3},\ \pi,\ \frac{4\pi}{3},\ \frac{5\pi}{3}\right\}$

29. $x = 0:$ $(\tan 0)^3 - 3\tan 0 = 0$

$0 = 0$ ✓

$x = \frac{\pi}{3}:$ $\left(\tan\frac{\pi}{3}\right)^3 - 3\tan\frac{\pi}{3} = 0$

$\left(\sqrt{3}\right)^3 - 3\sqrt{3} = 3\sqrt{3} - 3\sqrt{3} = 0$

$0 = 0$ ✓

$x = \frac{2\pi}{3}:$ $\left(\tan\frac{2\pi}{3}\right)^3 - 3\tan\frac{2\pi}{3} = 0$

$\left(-\sqrt{3}\right)^3 - 3\left(-\sqrt{3}\right) = -3\sqrt{3} + 3\sqrt{3} = 0$

$0 = 0$ ✓

$x = \pi:$ $(\tan\pi)^3 - 3\tan\pi = 0$

$0 = 0$ ✓

$x = \frac{4\pi}{3}:$ $\left(\tan\frac{4\pi}{3}\right)^3 - 3\tan\frac{4\pi}{3} = 0$

$\left(\sqrt{3}\right)^3 - 3\sqrt{3} = 3\sqrt{3} - 3\sqrt{3} = 0$

$0 = 0$ ✓

$x = \frac{5\pi}{3}:$ $\left(\tan\frac{5\pi}{3}\right)^3 - 3\tan\frac{5\pi}{3} = 0$

$\left(-\sqrt{3}\right)^3 - 3\left(-\sqrt{3}\right) = -3\sqrt{3} + 3\sqrt{3} = 0$

$0 = 0$ ✓

30. $\frac{\tan^3 x}{\tan x} = \frac{3\tan x}{\tan x}$

$\tan^2 x = 3$

$\tan x = \pm\sqrt{3}$

$x = \frac{\pi}{3},\ \frac{2\pi}{3},\ \frac{4\pi}{3},\ \frac{5\pi}{3}$

31. The answers do not agree. The solutions 0 and π were lost when dividing by $\tan x$.

33. $9\sin^2 x = 6\sin x + 1$

$9\sin^2 x - 6\sin x - 1 = 0$

$$\begin{aligned}\sin x &= \frac{6 \pm \sqrt{36 - 4(9)(-1)}}{2(9)} \\ &= \frac{6 \pm \sqrt{72}}{18} = \frac{6 \pm 6\sqrt{2}}{18} \\ &= \frac{1 \pm \sqrt{2}}{3}\end{aligned}$$

(i) $x = \sin^{-1}\left(\frac{1+\sqrt{2}}{3}\right)$

Ref $\angle$: .9352; Quadrants I, II

$x \approx .935,\ 2.206$

(ii) $x = \sin^{-1}\left(\frac{1-\sqrt{2}}{3}\right)$

Ref $\angle$: .1385; Quadrants II, IV

$x \approx 3.280,\ 6.1447$

$f(x) = 0$: $\boxed{\{.94,\ 2.21,\ 3.28,\ 6.14\}}$

35. $3\cot^2 x - 3\cot x = 1$

$3\cot^2 x - 3\cot x - 1 = 0$

$$\begin{aligned}\cot x &= \frac{3 \pm \sqrt{9 - 4(-3)}}{6} \\ &= \frac{3 \pm \sqrt{21}}{6} \\ \tan x &= \frac{6}{3 \pm \sqrt{21}}\end{aligned}$$

(i) $x = \tan^{-1}\left(\frac{6}{3+\sqrt{21}}\right)$

Ref $\angle$: .6694; Quadrants I, III

$x \approx .67,\ 3.81$

(i) $x = \tan^{-1}\left(\frac{6}{3-\sqrt{21}}\right)$

Ref $\angle$: 1.313; Quadrants II, IV

$x \approx 1.83,\ 4.97$

$f(x) = 0$: $\boxed{\{.67,\ 1.83,\ 3.81,\ 4.97\}}$

37. $\sin^2 x - 2\sin x + 3 = 0$

$$\begin{aligned}\sin x &= \frac{2 \pm \sqrt{4 - 4(3)}}{2} = \frac{2 \pm \sqrt{-8}}{2} \\ &= \frac{2 \pm 2i\sqrt{2}}{2} = 1 \pm i\sqrt{2}\end{aligned}$$

No such angle: $\boxed{\emptyset}$

39. $\cot\theta + 2\csc\theta = 3$

$\frac{\cos\theta}{\sin\theta} + \frac{2}{\sin\theta} = 3$

$\cos\theta + 2 = 3\sin\theta$

$(\cos\theta + 2)^2 = (3\sin\theta)^2$

$\cos^2\theta + 4\cos\theta + 4 = 9\sin^2\theta$

$\cos^2\theta + 4\cos\theta + 4 = 9\left(1 - \cos^2\theta\right)$

$\cos^2\theta + 4\cos\theta + 4 = 9 - 9\cos^2\theta$

$10\cos^2\theta + 4\cos\theta - 5 = 0$

$$\begin{aligned}\cos\theta &= \frac{-4 \pm \sqrt{16 + 200}}{20} = \frac{-4 \pm \sqrt{216}}{20} \\ &= \frac{-4 \pm 6\sqrt{6}}{20} = \frac{-2 \pm 3\sqrt{6}}{10}\end{aligned}$$

If $\cos\theta = \frac{-2 + 3\sqrt{6}}{10} \approx .53484692$

$\theta = \cos^{-1} .53484692$

Ref $\angle$: 57.67°; Quadrants I, IV

$\theta \approx 57.67°,\ 302.33°$

If $\cos\theta = \frac{-2 - 3\sqrt{6}}{10} \approx -.9348692$

$\theta = \cos^{-1}(-.9348692)$

Ref $\angle$: 20.8°; Quadrants II, III

$\theta \approx 159.20°,\ 200.80°$

Check each equation since the equation was squared:

$\cos 57.67° + 2 = 3\sin 57.67°$ ✓

$\cos 302.33° + 2 = 3\sin 302.33°$ X

$\cos 159.20° + 2 = 3\sin 159.20°$ ✓

$\cos 200.80° + 2 = 3\sin 200.80°$ X

$\theta = \{57.7°,\ 159.2°\}$

41. $\cot x + 2\csc x = 3$

Graph $y = (\tan x)^{-1} + 2(\sin x)^{-1} - 3$

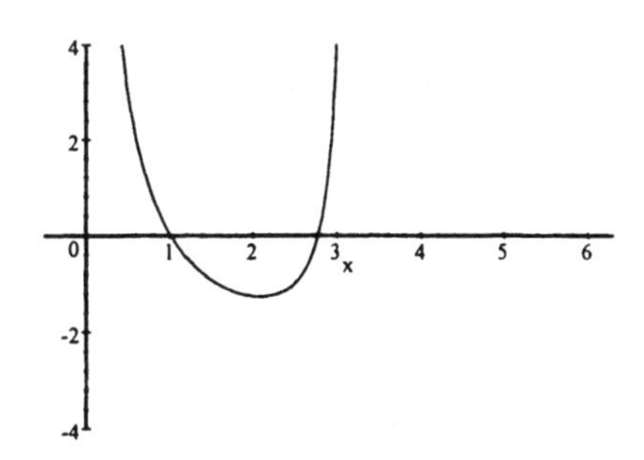

x-intercepts: $\boxed{\{1.01,\ 2.78\}}$

43. $\sin^3 x + \sin x = 1$

Graph $y = \sin^3 x + \sin x - 1$

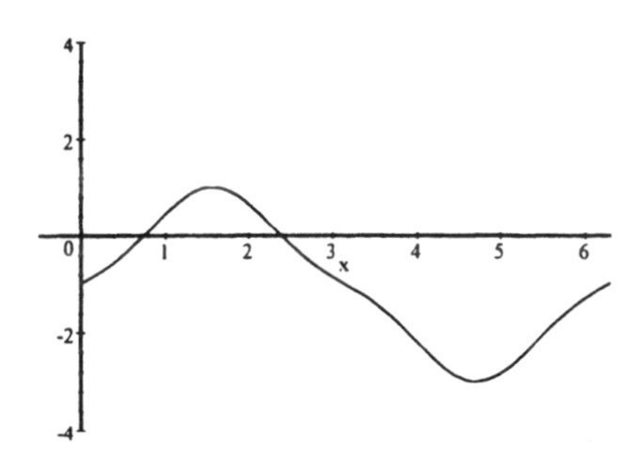

x-intercepts: $\boxed{\{0.75,\ 2.39\}}$

45. $e^x = \sin x + 3$

Graph $y = e^x - \sin x - 3$

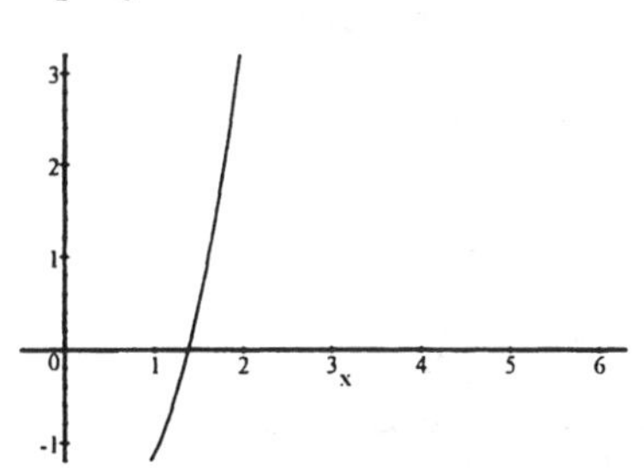

x-intercept: $\boxed{\{1.38\}}$

47. Dividing by $\sin x$ in the second step causes one solution to be lost.

$$\sin^2 x - \sin x = 0$$
$$\sin x(\sin x - 1) = 0$$
$$\sin x = 0 \qquad \sin x = 1$$
$$x = 0,\ \pi \qquad x = \frac{\pi}{2}$$

49. Substituting 1.5 for m in the equation and solving yields the following:

$$\sin\frac{\alpha}{2} = \frac{1}{1.5}$$
$$\sin\frac{\alpha}{2} = \frac{2}{3}$$
$$\frac{\alpha}{2} = \sin^{-1}\left(\frac{2}{3}\right)$$
$$\frac{\alpha}{2} \approx 41.8^\circ \qquad \boxed{\alpha \approx 83.6^\circ}$$

51. $.342D\cos\theta + h\cos^2\theta = \dfrac{16D^2}{V^2}$

$V = 60,\ D = 80,\ h = 2$

$$.342(80)\cos\theta + 2\cos^2\theta = \frac{16(80)^2}{(60)^2}$$
$$27.36\cos\theta + 2\cos^2\theta = \frac{256}{9}$$
$$2\cos^2\theta + 27.36\cos\theta - \frac{256}{9} = 0$$

$$\cos\theta = \frac{-27.36 \pm \sqrt{748.5696 - 4(2)\left(-\frac{256}{9}\right)}}{4}$$
$$= \frac{-27.36 \pm 31.243}{4}$$
$$= .97075,\ -14.65\ (\emptyset)$$

$$\theta = \cos^{-1} .97075 \approx 13.892^\circ \qquad \boxed{\theta \approx 14^\circ}$$

Section 9.6

1. $2x = \frac{2\pi}{3}, 2\pi, \frac{8\pi}{3}$

Divide by 2:

$x = \frac{\pi}{3}, \pi, \frac{4\pi}{3}$

3. (a) $\cos 2x = \frac{\sqrt{3}}{2}$

$2x = \frac{\pi}{6}, \frac{11\pi}{6}, \frac{13\pi}{6}, \frac{23\pi}{6}$

$[0 \le x < 2\pi \Rightarrow 0 \le 2x < 4\pi]$

$$x = \left\{\frac{\pi}{12}, \frac{11\pi}{12}, \frac{13\pi}{12}, \frac{23\pi}{12}\right\}$$

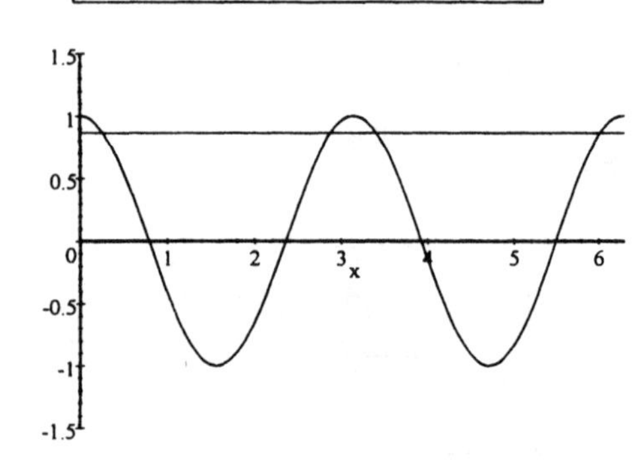

(b) $\cos 2x > \frac{\sqrt{3}}{2} : \left[0, \frac{\pi}{12}\right) \cup \left(\frac{11\pi}{12}, \frac{13\pi}{12}\right) \cup \left(\frac{23\pi}{12}, 2\pi\right)$

5. (a) $\sin 3x = -1$

$3x = \frac{3\pi}{2}, \frac{7\pi}{2}, \frac{11\pi}{2}$

$[0 \le x < 2\pi \Rightarrow 0 \le 3x < 6\pi]$

$$x = \left\{\frac{\pi}{2}, \frac{7\pi}{6}, \frac{11\pi}{6}\right\}$$

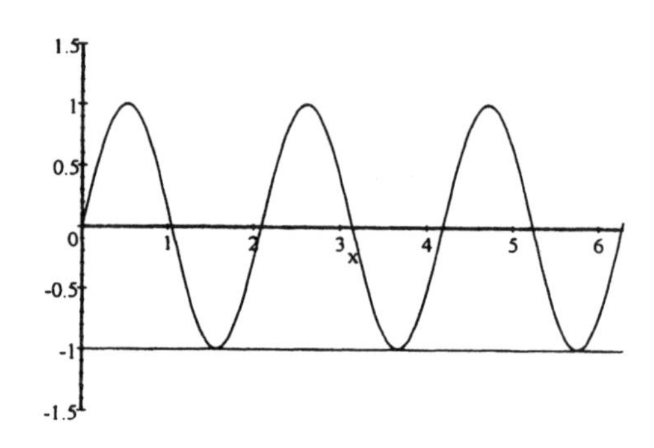

(b) $\sin 3x < -1 : \emptyset$

7. (a) $\sqrt{2}\cos 2x = -1$

$\cos 2x = -\frac{1}{\sqrt{2}}$

$2x = \frac{3\pi}{4}, \frac{5\pi}{4}, \frac{11\pi}{4}, \frac{13\pi}{4}$

$[0 \le x < 2\pi \Rightarrow 0 \le 2x < 4\pi]$

$$x = \left\{\frac{3\pi}{8}, \frac{5\pi}{8}, \frac{11\pi}{8}, \frac{13\pi}{8}\right\}$$

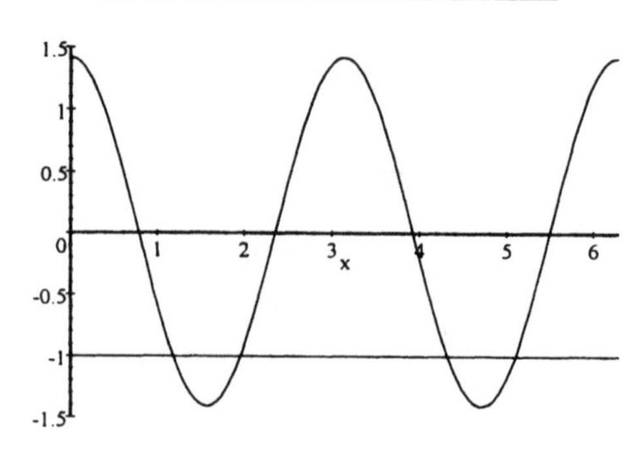

(b) $\sqrt{2}\cos 2x \le -1 :$ $\left[\frac{3\pi}{8}, \frac{5\pi}{8}\right] \cup \left[\frac{11\pi}{8}, \frac{13\pi}{8}\right]$

9. (a) $\sin\frac{x}{2} = \sqrt{2} - \sin\frac{x}{2}$

$$2\sin\frac{x}{2} = \sqrt{2}$$

$$\sin\frac{x}{2} = \frac{\sqrt{2}}{2}$$

$$\frac{x}{2} = \frac{\pi}{4}, \frac{3\pi}{4} \quad \left[0 \le x < 2\pi \Rightarrow 0 \le \tfrac{x}{2} < \pi\right]$$

$$\boxed{x = \left\{\frac{\pi}{2}, \frac{3\pi}{2}\right\}}$$

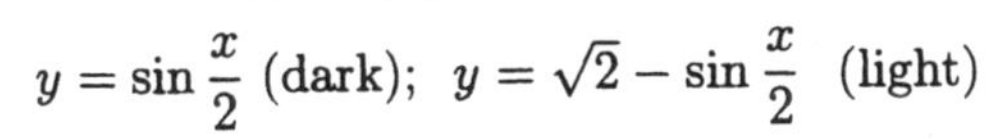

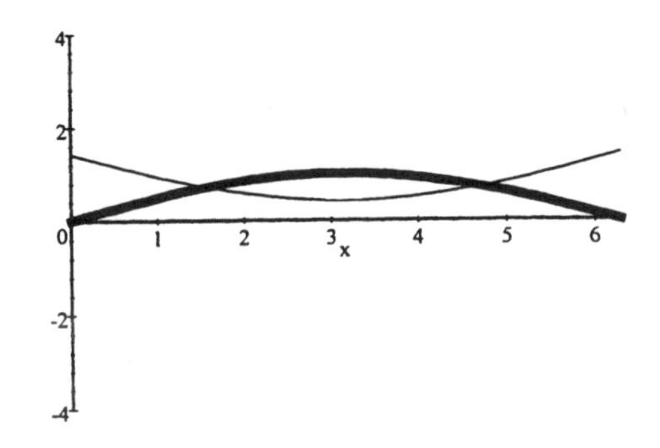

(b) $\boxed{\sin\frac{x}{2} > \sqrt{2} - \sin\frac{x}{2} : \left(\frac{\pi}{2}, \frac{3\pi}{2}\right)}$

11. $\sin\frac{x}{2} = \cos\frac{x}{2}$

$$\left(\sin\frac{x}{2}\right)^2 = \left(\cos\frac{x}{2}\right)^2$$

$$\frac{1-\cos x}{2} = \frac{1+\cos x}{2}$$

$$1 - \cos x = 1 + \cos x$$

$$-2\cos x = 0$$

$$\cos x = 0 \qquad x = \frac{\pi}{2}, \frac{3\pi}{2}$$

Since both sides of equation are squared, check solutions:

$$x = \frac{\pi}{2} : \sin\frac{\pi}{4} = \cos\frac{\pi}{4} \ \checkmark$$

$$x = \frac{3\pi}{2} : \sin\frac{3\pi}{4} \ne \cos\frac{3\pi}{4}$$

$$\boxed{x = \left\{\frac{\pi}{2}\right\}}$$

13. $\sin^2\left(\frac{x}{2}\right) - 1 = 0$

$$\left(\sqrt{\frac{1-\cos x}{2}}\right)^2 - 1 = 0$$

$$\frac{1-\cos x}{2} - 1 = 0$$

$$1 - \cos x - 2 = 0$$

$$-\cos x = 1$$

$$\cos x = -1 \qquad \boxed{x = \{\pi\}}$$

15. $\sin 2x = 2\cos^2 x$

$$2\sin x\cos x = 2\cos^2 x$$

$$2\sin x\cos x - 2\cos^2 x = 0$$

$$2\cos x\,(\sin x - \cos x) = 0$$

(i) $2\cos x = 0 \rightarrow x = \frac{\pi}{2}, \frac{3\pi}{2}$

(ii) $\sin x - \cos x = 0$

$$\sin x = \cos x$$

$$x = \frac{\pi}{4}, \frac{5\pi}{4}$$

$$\boxed{x = \left\{\frac{\pi}{4}, \frac{\pi}{2}, \frac{5\pi}{4}, \frac{3\pi}{2}\right\}}$$

17. $\cos x - 1 = \cos 2x$

$$\cos x - 1 = 2\cos^2 x - 1$$

$$\cos x - 2\cos^2 x = 0$$

$$(\cos x)(1 - 2\cos x) = 0$$

(i) $\cos x = 0 \rightarrow x = \frac{\pi}{2}, \frac{3\pi}{2}$

(ii) $\cos x = \frac{1}{2} \rightarrow x = \frac{\pi}{3}, \frac{5\pi}{3}$

$$\boxed{x = \left\{\frac{\pi}{3}, \frac{\pi}{2}, \frac{3\pi}{2}, \frac{5\pi}{3}\right\}}$$

19. $\sqrt{2}\sin 3\theta - 1 = 0, \quad 0° \le 3\theta < 1080°$

$$\sin 3\theta = \frac{1}{\sqrt{2}}$$

$$3\theta = 45°,\ 135°,\ 405°,\ 495°\ 765°,\ 855°$$

$$\boxed{\theta = \{15°,\ 45°,\ 135°,\ 165°\ 255°,\ 285°\}}$$

21. $\cos\frac{\theta}{2} = 1, \quad 0° \le \frac{\theta}{2} < 180°$

$\frac{\theta}{2} = 0°$ $\boxed{\theta = \{0°\}}$

23. $2\sqrt{3}\sin\frac{\theta}{2} = 3, \quad 0° \le \frac{\theta}{2} < 180°$

$\sin\frac{\theta}{2} = \frac{3}{2\sqrt{3}} \cdot \frac{\sqrt{3}}{\sqrt{3}} = \frac{\sqrt{3}}{2}$

$\frac{\theta}{2} = 60°,\ 120°$

$\boxed{\theta = \{120°,\ 240°\}}$

25. $2\sin\theta = 2\cos 2\theta$

$2\sin\theta = 2\left(1 - 2\sin^2\theta\right)$

$2\sin\theta = 2 - 4\sin^2\theta$

$4\sin^2\theta + 2\sin\theta - 2 = 0$

$2\left(2\sin^2\theta + \sin\theta - 1\right) = 0$

$2(2\sin\theta - 1)(\sin\theta + 1)$

$\sin\theta = \frac{1}{2}$ $\qquad \sin\theta = -1$

$\theta = 30°,\ 150°$ $\qquad \theta = 270°$

$\boxed{\theta = \{30°,\ 150°,\ 270°\}}$

27. $2 - \sin 2\theta = 4\sin 2\theta, \quad 0° \le 2\theta < 720°$

$2 = 5\sin 2\theta$

$\sin 2\theta = \frac{2}{5}$

Ref $\angle : 2\theta = \sin^{-1}\frac{2}{5} \approx 23.6°$

$2\theta = 23.6°,\ 156.4°,\ 383.6°,\ 516.4°$

$\boxed{\theta = \{11.8°,\ 78.2°,\ 191.8°,\ 258.2°\}}$

29. $2\cos^2 2\theta = 1 - \cos 2\theta, \quad 0° \le 2\theta < 720°$

$2\cos^2 2\theta + \cos 2\theta - 1 = 0$

$(2\cos 2\theta - 1)(\cos 2\theta + 1) = 0$

$\cos 2\theta = \frac{1}{2}, \quad \cos 2\theta = -1$

If $\cos 2\theta = \frac{1}{2}$

$2\theta = 60°,\ 300°,\ 420°,\ 660°$

$\theta = 30°,\ 150°,\ 210°,\ 330°$

If $\cos 2\theta = -1$

$2\theta = 180°,\ 540°$

$\theta = 90°,\ 270°$

$\boxed{\theta = \{30°,\ 90°,\ 150°,\ 210°,\ 270°,\ 330°\}}$

31. $\sin x + \sin 3x = \cos x$

Graph $y = \sin x + \sin 3x - \cos x$:

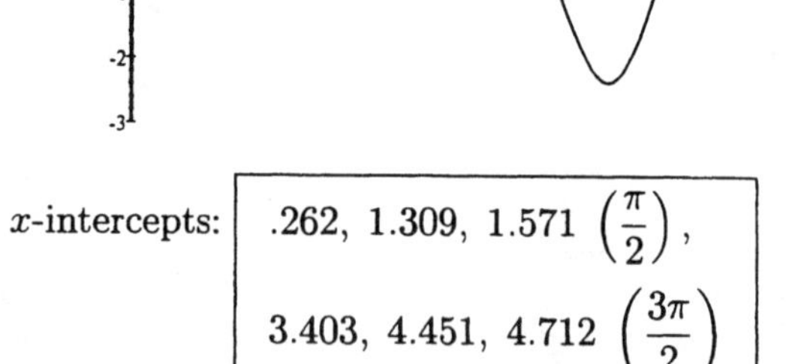

x-intercepts: $\boxed{.262,\ 1.309,\ 1.571\ \left(\frac{\pi}{2}\right),\quad 3.403,\ 4.451,\ 4.712\ \left(\frac{3\pi}{2}\right)}$

33. $\cos 2x + \cos x = 0$

Graph $y = \cos 2x + \cos x$:

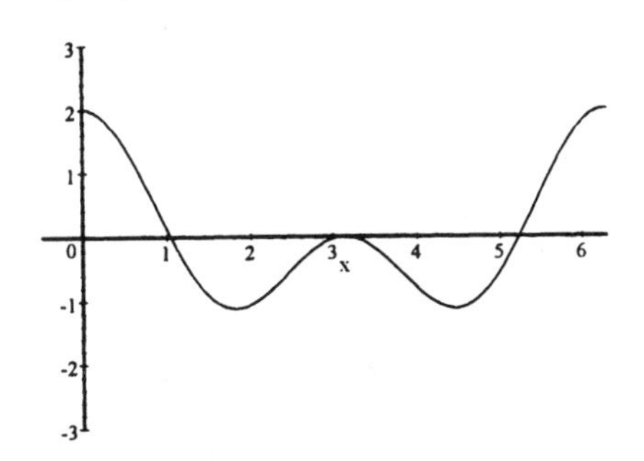

x-intercepts:

$$\boxed{1.047\ \left(\frac{\pi}{3}\right),\ 3.142\ (\pi),\ 5.236\ \left(\frac{5\pi}{3}\right)}$$

35. $\cos\frac{x}{2} = 2\sin 2x$

Graph $y = \cos\frac{x}{2} - 2\sin 2x$:

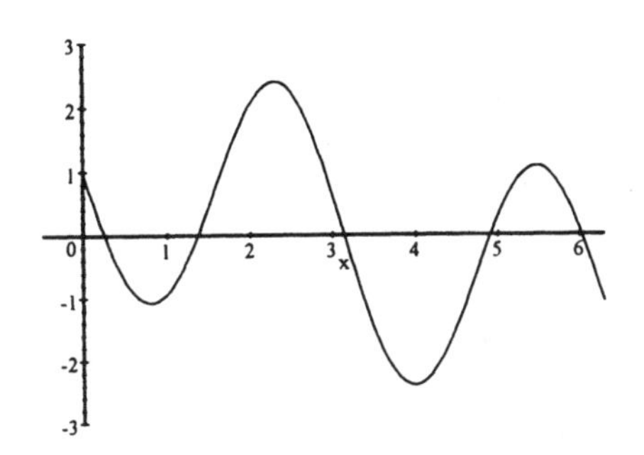

x-intercepts:

$$\boxed{.259,\ 1.372,\ 3.142\ (\pi),\ 4.911,\ 6.024}$$

37. $\frac{\tan 2\theta}{2} \neq \frac{2\tan\theta}{2} \neq \tan\theta$

Graph $y = \tan 2\theta - 2$:

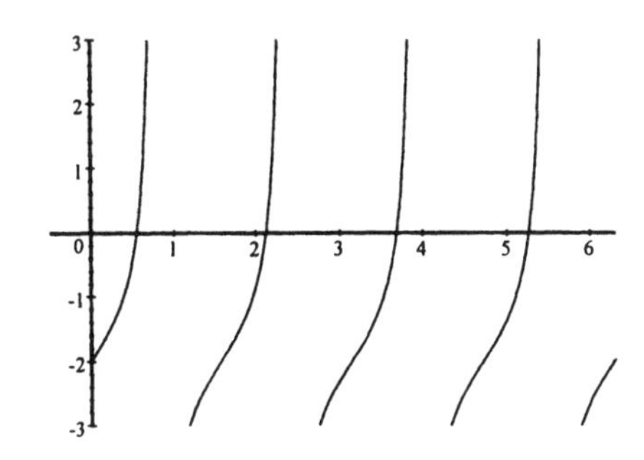

$\theta \approx .5536,\ 2.1244,\ 3.695,\ 5.266$

$[2\theta \approx 1.107,\ 4.249,\ 7.390,\ 10.532]$

39. $e = 20\sin\left(\frac{\pi t}{4} - \frac{\pi}{2}\right)$

(a) $e = 0$:

$$0 = 20\sin\left(\frac{\pi t}{4} - \frac{\pi}{2}\right)$$

Since $\arcsin 0 = 0$, solve the following equation.

$$\frac{\pi t}{4} - \frac{\pi}{2} = 0$$

$$\frac{\pi t}{4} = \frac{\pi}{2} \longrightarrow t = \boxed{2 \text{ seconds}}$$

(b) $e = 10\sqrt{3}$:

$$10\sqrt{3} = 20\sin\left(\frac{\pi t}{4} - \frac{\pi}{2}\right)$$

$$\frac{\sqrt{3}}{2} = \sin\left(\frac{\pi t}{4} - \frac{\pi}{2}\right)$$

Since $\arcsin\frac{\sqrt{3}}{2} = \frac{\pi}{3}$, solve the following equation.

$$\frac{\pi t}{4} - \frac{\pi}{2} = \frac{\pi}{3}$$

$$\frac{\pi t}{4} = \frac{10\pi}{12} \longrightarrow t = \boxed{\frac{10}{3} \text{ seconds}}$$

41. $P = A \sin(2\pi f t + \phi)$

(a) Middle C: $f = 261.63$ cycles/sec

$$A = .004,\ \phi = \frac{\pi}{7}$$

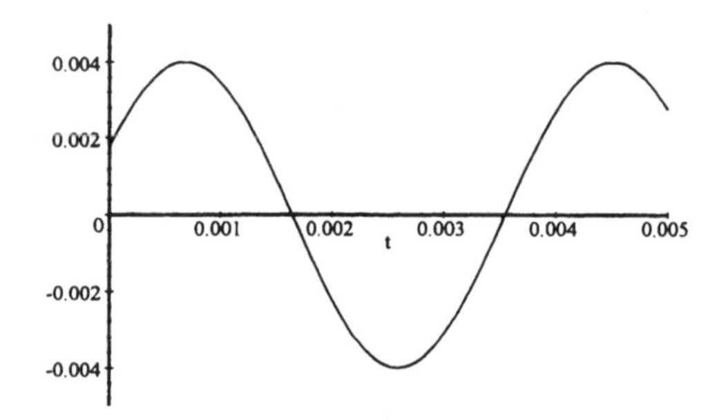

(b) $0 = .004 \sin\left[2\pi(261.63)t + \dfrac{\pi}{7}\right]$

$0 = \sin(1643.87t + .45)$

$1643.87t + .45 = n\pi,\ n \in$ integer

$$t = \frac{n\pi - .45}{1643.87}$$

If $n = 0$, then $t = -.000274$

If $n = 1$, then $t = .00164$

If $n = 2$, then $t = .00355$

If $n = 3$, then $t = .00546$

The only solutions for $t \in [0,\ .005]$ are $\boxed{.00164,\ .00355}$

(c) x-intercepts $\approx .00164,\ .00355$

$\boxed{P < 0: [.00164,\ .00355]}$

(d) $P < 0$ implies that there is a decrease in pressure so an eardrum would be vibrating outward.

43. (a) $P = .005 \sin 440\pi t + .005 \sin 446\pi t$

3 beats per second.

For $x = t$,
$P(t) = .005 \sin 440\pi t + .005 \sin 446\pi t$

.01
.15
1.15
−.01

(b) $P = .005 \sin 440\pi t + .005 \sin 432\pi t$

4 beats per second.

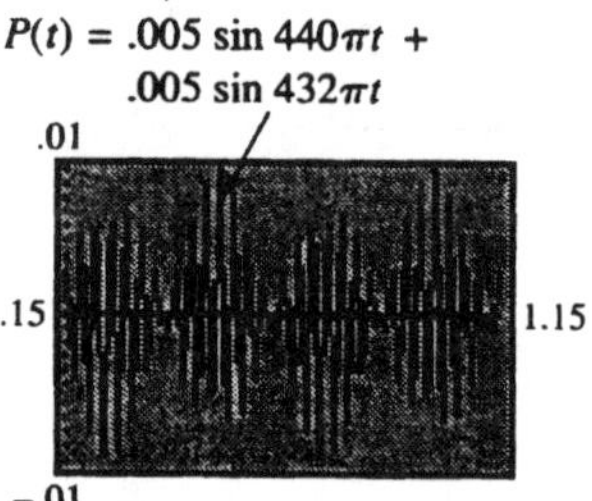

(c) The number of beats per second is equal to the absolute value of the difference in the frequencies of the two tones.

Reviewing Basic Concepts (Sections 9.5 and 9.6)

1. $\cos 2x = \dfrac{\sqrt{3}}{2},\quad 0 \le x < 4\pi$

$$2x = \frac{\pi}{6},\ \frac{11\pi}{6},\ \frac{13\pi}{6},\ \frac{23\pi}{6}$$

$$x = \left\{\frac{\pi}{12},\ \frac{11\pi}{12},\ \frac{13\pi}{12},\ \frac{23\pi}{12}\right\}$$

2. $2 \sin x + 1 = 0$

$$\sin x = -\frac{1}{2} \qquad x = \left\{\frac{7\pi}{6},\ \frac{11\pi}{6}\right\}$$

3. $(\tan x - 1)(\cos x - 1) = 0$

$\tan x = 1 \qquad \cos x = 1$

$x = \dfrac{\pi}{4},\ \dfrac{5\pi}{4} \qquad x = 0$

$$x = \left\{0,\ \frac{\pi}{4},\ \frac{5\pi}{4}\right\}$$

4. $2\cos^2 x = \sqrt{3}\cos x$

$2\cos^2 x - \sqrt{3}\cos x = 0$

$\cos x\left(2\cos x - \sqrt{3}\right) = 0$

$\cos x = 0 \qquad \cos x = \dfrac{\sqrt{3}}{2}$

$x = \dfrac{\pi}{2},\ \dfrac{3\pi}{2} \qquad x = \dfrac{\pi}{6},\ \dfrac{11\pi}{6}$

$x = \left\{\dfrac{\pi}{6},\ \dfrac{\pi}{2},\ \dfrac{3\pi}{2},\ \dfrac{11\pi}{6}\right\}$

5. $3\cot^2\theta - 3\cot\theta = 1$

$3\cot^2\theta - 3\cot\theta - 1 = 0$

$\cot\theta = \dfrac{3 \pm \sqrt{9 - 4(-3)}}{6} = \dfrac{3 \pm \sqrt{21}}{6}$

$\tan\theta = \dfrac{6}{3 \pm \sqrt{21}}$

If $\tan\theta = \dfrac{6}{3+\sqrt{21}} \approx .79128785$

Ref $\angle \approx 38.4°$, Quadrants I, III

$\theta \approx 38.4°,\ 218.4°$

If $\tan\theta = \dfrac{6}{3-\sqrt{21}} \approx -3.79128785$

Ref $\angle \approx 75.2°$, Quadrants II, IV

$\theta \approx 104.8°,\ 284.8°$

$\theta = \{38.4°,\ 104.8°,\ 218.4°,\ 284.8°\}$

6. $4\cos^2\theta + 4\cos\theta - 1 = 0$

$\cos\theta = \dfrac{-4 \pm \sqrt{16 - 4(-4)}}{8} = \dfrac{-4 \pm \sqrt{32}}{8}$

$= \dfrac{-4 \pm 4\sqrt{2}}{8} = \dfrac{-1 \pm \sqrt{2}}{2}$

If $\cos\theta = \dfrac{-1+\sqrt{2}}{2} \approx .20710678$

Ref $\angle \approx 78.0°$, Quadrants I, IV

$\theta \approx 78.0°,\ 282.0°$

If $\cos\theta = \dfrac{-1-\sqrt{2}}{2} \approx -1.207 : \quad \emptyset$

$\theta = \{78.0°,\ 282.0°\}$

7. $2\sin\theta - 1 = \csc\theta$

$2\sin\theta - 1 = \dfrac{1}{\sin\theta}$

$2\sin^2\theta - \sin\theta = 1$

$2\sin^2\theta - \sin\theta - 1 = 0$

$(2\sin\theta + 1)(\sin\theta - 1) = 0$

$\sin\theta = -\dfrac{1}{2} \qquad \sin\theta = 1$

$\theta = 210°,\ 330° \qquad \theta = 90°$

$\theta = \{90°,\ 210°,\ 330°\}$

8. $\sec^2\dfrac{\theta}{2} = 2, \quad 0° \le \dfrac{\theta}{2} < 180°$

$\cos^2\dfrac{\theta}{2} = \dfrac{1}{2}$

$\cos\dfrac{\theta}{2} = \pm\dfrac{1}{\sqrt{2}} = \pm\dfrac{\sqrt{2}}{2}$

$\dfrac{\theta}{2} = 45°,\ 135° \qquad \theta = \{90°,\ 270°\}$

9. $x^2 + \sin x - x^3 - \cos x = 0$

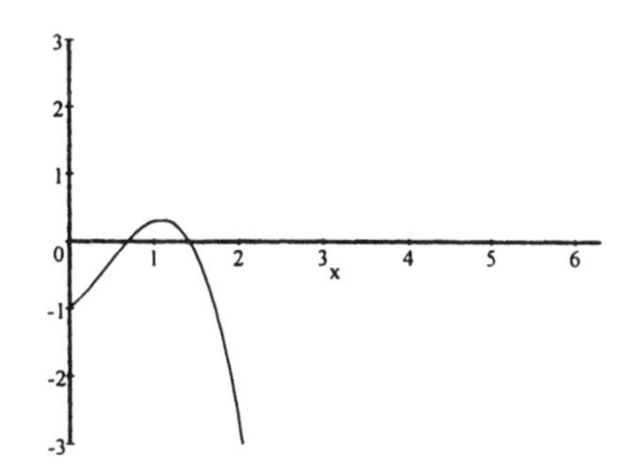

Zeros at $\approx \{.68058878,\ 1.4158828\}$

10. $x^3 - \cos^2 x = \dfrac{1}{2}x - 1$

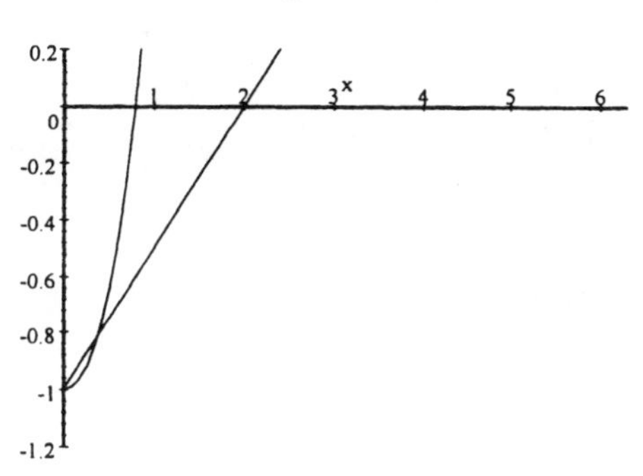

Intersection at $x = \{0,\ \approx .37600772\}$

Chapter 9 Review

1. $f(-x) = -f(x)$; this is an odd function, so $(x, y) \to (-x, -y)$. The following trigonometric functions are odd:

   sine, tangent, cotangent, and cosecant.

2. $f(-x) = f(x)$; this is an even function, so $(x, y) \to (-x, y)$. The following trigonometric functions are even:

   cosine and secant.

3. $\cos(-3) = \cos 3$
   $[\cos(-x) = \cos x]$

4. $\sin(-3) = -\sin 3$
   $[\sin(-x) = -\sin x]$

5. $\tan(-3) = -\tan 3$
   $[\tan(-x) = -\tan x]$

6. $\sec(-3) = \sec 3$
   $[\sec(-x) = \sec x]$

7. $\csc(-3) = -\csc 3$
   $[\csc(-x) = -\csc x]$

8. $\cot(-3) = -\cot 3$
   $[\cot(-x) = -\cot x]$

9. $\sec x = \dfrac{1}{\cos x}$ (B)

10. $\tan x = \dfrac{\sin x}{\cos x}$ (C)

11. $\cot x = \dfrac{\cos x}{\sin x}$ (F)

12. $\tan^2 x + 1 = \dfrac{1}{\cos^2 x}$ (E)

13. $\tan^2 x = \dfrac{1}{\cot^2 x}$ (D)

14. $$\begin{aligned}\sec^2\theta - \tan^2\theta &= \frac{1}{\cos^2\theta} - \frac{\sin^2\theta}{\cos^2\theta}\\ &= \frac{1-\sin^2\theta}{\cos^2\theta} = \frac{\cos^2\theta}{\cos^2\theta} = 1\end{aligned}$$

15. $$\begin{aligned}\frac{\cot\theta}{\sec\theta} &= \frac{\cos\theta \,/\, \sin\theta}{1 \,/\, \cos\theta}\\ &= \frac{\cos\theta}{\sin\theta}\cdot\frac{\cos\theta}{1}\\ &= \frac{\cos^2\theta}{\sin\theta}\end{aligned}$$

16. $$\begin{aligned}\tan^2\theta\left(1+\cot^2\theta\right) &= \frac{\sin^2\theta}{\cos^2\theta}\left(\csc^2\theta\right)\\ &= \frac{\sin^2\theta}{\cos^2\theta}\left(\frac{1}{\sin^2\theta}\right)\\ &= \frac{1}{\cos^2\theta}\end{aligned}$$

17. $$\begin{aligned}\csc\theta + \cot\theta &= \frac{1}{\sin\theta} + \frac{\cos\theta}{\sin\theta}\\ &= \frac{1+\cos\theta}{\sin\theta}\end{aligned}$$

18. $\cos x = \dfrac{3}{5}$, $x \in$ Quadrant IV

$$\begin{aligned}\sin x &= -\sqrt{1-\cos^2 x}\\ &= -\sqrt{1-\frac{9}{25}}\\ &= -\sqrt{\frac{16}{25}} = -\frac{4}{5}\end{aligned}$$

$$\tan x = \frac{\sin x}{\cos x} = \frac{-4/5}{3/5} = -\frac{4}{3}$$

$$\sec x = \frac{1}{\cos x} = \frac{5}{3}$$

$$\csc x = \frac{1}{\sin x} = -\frac{5}{4}$$

$$\cot x = \frac{1}{\tan x} = -\frac{3}{4}$$

19. $\tan x = -\frac{5}{4},\ \frac{\pi}{2} < x < \pi$

$$\begin{aligned}\sec x &= -\sqrt{1+\tan^2 x}\\ &= -\sqrt{1+\frac{25}{16}}\\ &= -\sqrt{\frac{41}{16}} = -\frac{\sqrt{41}}{4}\end{aligned}$$

$$\cos x = \frac{1}{\sec x} = \frac{-4}{\sqrt{41}}\cdot\frac{\sqrt{41}}{\sqrt{41}} = -\frac{4\sqrt{41}}{41}$$

$$\cot x = \frac{1}{\tan x} = -\frac{4}{5}$$

$$\begin{aligned}\sin x = &= \sqrt{1-\cos^2 x}\\ &= \sqrt{1-\left(-\frac{4\sqrt{41}}{41}\right)^2}\\ &= \sqrt{1-\frac{656}{1681}} = \sqrt{\frac{1025}{1681}}\\ &= \frac{5\sqrt{41}}{41}\end{aligned}$$

$$\begin{aligned}\csc x = \frac{1}{\sin x} &= \frac{41}{5\sqrt{41}}\cdot\frac{\sqrt{41}}{\sqrt{41}}\\ &= \frac{41\sqrt{41}}{5\cdot 41} = \frac{\sqrt{41}}{5}\end{aligned}$$

20. $\cos 210° = \cos 150° \cos 60° - \sin 150° \sin 60°$: **E**

21. $\sin 35° = \cos 55°$: **B**

22. $\tan(-35°) = \cot 125°$: **J**

23. $-\sin 35° = \sin(-35°)$: **A**

24. $\cos 35° = \cos(-35°)$: **I**

25. $\cos 75° = \cos\frac{150°}{2} = \sqrt{\frac{1+\cos 150°}{2}}$: **C**

26. $\sin 75° = \sin 15° \cos 60° + \cos 15° \sin 60°$: **H**

27. $\sin 300° = 2\sin 150° \cos 150°$: **D**

28. $\cos 300° = \cos(2\cdot 150°)$
$= \cos^2 150° - \sin^2 150°$: **G**

29. Verify $\sin^2 x - \sin^2 y = \cos^2 y - \cos^2 x$

$$\begin{aligned}&\sin^2 x - \sin^2 y\\ &\quad= \left(1-\cos^2 x\right) - \left(1-\cos^2 y\right)\\ &\quad= -\cos^2 x + \cos^2 y\\ &\quad= \cos^2 y - \cos^2 x\end{aligned}$$

30. Verify $2\cos^3 x - \cos x = \frac{\cos^2 x - \sin^2 x}{\sec x}$

$$\begin{aligned}&2\cos^3 x - \cos x\\ &\quad= \cos x\left(2\cos^2 x - 1\right)\\ &\quad= \frac{1}{\sec x}(\cos 2x)\\ &\quad= \frac{\cos^2 x - \sin^2 x}{\sec x}\end{aligned}$$

31. Verify $-\cot\frac{x}{2} = \frac{\sin 2x + \sin x}{\cos 2x - \cos x}$

Right: $\frac{\sin 2x + \sin x}{\cos 2x - \cos x}$

$$\begin{aligned}&= \frac{2\sin x\cos x + \sin x}{2\cos^2 x - 1 - \cos x}\\ &= \frac{\sin x\,(2\cos x + 1)}{(2\cos x + 1)(\cos x - 1)}\\ &= \frac{\sin x}{\cos x - 1}\end{aligned}$$

Left: $-\cot\frac{x}{2} = -\frac{1}{\tan\frac{x}{2}} = -\frac{1}{\frac{1-\cos x}{\sin x}}$

$$= \frac{-\sin x}{1-\cos x} = \frac{\sin x}{\cos x - 1}$$

32. Verify $\dfrac{\sin^2 x}{2 - 2\cos x} = \cos^2 \dfrac{x}{2}$

Left: $$\frac{\sin^2 x}{2 - 2\cos x} = \frac{1 - \cos^2 x}{2(1 - \cos x)} = \frac{(1 - \cos x)(1 + \cos x)}{2(1 - \cos x)} = \frac{1 + \cos x}{2}$$

Right: $$\cos^2 \frac{x}{2} = \left(\pm\sqrt{\frac{1 + \cos x}{2}}\right)^2 = \frac{1 + \cos x}{2}$$

33. Verify $\dfrac{\sin 2x}{\sin x} = \dfrac{2}{\sec x}$

$$\frac{\sin 2x}{\sin x} = \frac{2\sin x \cos x}{\sin x} = 2\cos x = \frac{2}{\sec x}$$

34. Verify: $2\cos A - \sec A = \cos A - \dfrac{\tan A}{\csc A}$

Right: $$\cos A - \frac{\tan A}{\csc A} = \cos A - \frac{\sin A / \cos A}{1 / \sin A} = \cos A - \frac{\sin^2 A}{\cos A} = \frac{\cos^2 A - \sin^2 A}{\cos A} = \frac{\cos 2A}{\cos A}$$

Left: $$2\cos A - \sec A = 2\cos A - \frac{1}{\cos A} = \frac{2\cos^2 A - 1}{\cos A} = \frac{\cos 2A}{\cos A}$$

35. Verify $\dfrac{2\tan B}{\sin 2B} = \sec^2 B$

$$\frac{2\tan B}{\sin 2B} = \frac{\frac{2\sin B}{\cos B}}{2\sin B\cos B} = \frac{2\sin B}{\cos B}\cdot\frac{1}{2\sin B\cos B} = \frac{1}{\cos^2 B} = \sec^2 B$$

36. Verify $1 + \tan^2\alpha = 2\tan\alpha\csc 2\alpha$

$$2\tan\alpha\csc 2\alpha = \frac{2\sin\alpha}{\cos\alpha}\cdot\frac{1}{\sin 2\alpha} \quad \frac{2\sin\alpha}{\cos\alpha(2\sin\alpha\cos\alpha)} = \frac{1}{\cos^2\alpha} = \sec^2\alpha = 1 + \tan^2\alpha$$

37. Verify $\dfrac{\sin t}{1 - \cos t} = \cot\dfrac{t}{2}$

$$\cot\frac{t}{2} = \frac{1}{\tan\frac{t}{2}} = \frac{1}{\frac{1 - \cos t}{\sin t}} = \frac{\sin t}{1 - \cos t}$$

38. Verify $\dfrac{2\cot x}{\tan 2x} = \csc^2 x - 2$

$$\frac{2\cot x}{\tan 2x} = \frac{2\cos x}{\sin x} \div \frac{\sin 2x}{\cos 2x} = \frac{2\cos x}{\sin x}\cdot\frac{1 - 2\sin^2 x}{2\sin x\cos x} = \frac{1 - 2\sin^2 x}{\sin^2 x} = \frac{1}{\sin^2 x} - \frac{2\sin^2 x}{\sin^2 x} = \csc^2 x - 2$$

39. Verify $\tan\theta\sin 2\theta = 2 - 2\cos^2\theta$

$$\begin{aligned}\tan\theta\sin 2\theta &= \frac{\sin\theta}{\cos\theta}(2\sin\theta\cos\theta)\\ &= 2\sin^2\theta\\ &= 2\left(1-\cos^2\theta\right)\\ &= 2-2\cos^2\theta\end{aligned}$$

40. Verify $2\tan x\csc 2x - \tan^2 x = 1$

$$\begin{aligned}&2\tan x\csc 2x - \tan^2 x\\ &\quad= \tan x\,(2\csc 2x - \tan x)\\ &\quad= \tan x\left(\frac{2}{\sin 2x} - \frac{\sin x}{\cos x}\right)\\ &\quad= \tan x\left(\frac{2}{2\sin x\cos x} - \frac{\sin x}{\cos x}\right)\\ &\quad= \tan x\left(\frac{2-2\sin^2 x}{2\sin x\cos x}\right)\\ &\quad= \tan x\left(\frac{2\left(1-\sin^2 x\right)}{2\sin x\cos x}\right)\\ &\quad= \frac{\sin x}{\cos x}\cdot\left(\frac{\cos^2 x}{\sin x\cos x}\right)\\ &\quad= \frac{\sin x}{\cos x}\cdot\frac{\cos x}{\sin x} = 1\end{aligned}$$

41. $\sin^{-1}\dfrac{\sqrt{2}}{2} = \dfrac{\pi}{4}$

$\left(\text{range: } \left[-\frac{\pi}{2}, \frac{\pi}{2}\right]\right)$

42. $\arccos\left(-\dfrac{1}{2}\right) = \dfrac{2\pi}{3}$

(range: $[0, \pi]$)

43. $\arctan\dfrac{\sqrt{3}}{3} = \dfrac{\pi}{6}$

$\left(\text{range: } \left(-\frac{\pi}{2}, \frac{\pi}{2}\right)\right)$

44. $\sec^{-1}(-2) = \dfrac{2\pi}{3}$

$\left(\text{range: } \left[0, \frac{\pi}{2}\right) \cup \left(\frac{\pi}{2}, \pi\right]\right)$

45. $\operatorname{arccsc}\dfrac{2\sqrt{3}}{3} = \dfrac{\pi}{3}$

$\left(\text{range: } \left[-\frac{\pi}{2}, 0\right) \cup \left(0, \frac{\pi}{2}\right]\right)$

46. $\cot^{-1}(-1) = \dfrac{3\pi}{4}$

(range: $(0, \pi)$)

47. $\theta = \arccos\dfrac{1}{2}$

$\cos\theta = \dfrac{1}{2}, \quad \theta = 60°$

48. $\theta = \arcsin\left(-\dfrac{\sqrt{3}}{2}\right)$

$\sin\theta = -\dfrac{\sqrt{3}}{2}, \quad \theta = -60°$

49. $\theta = \tan^{-1} 0$

$\tan\theta = 0, \quad \theta = 0°$

50. $\theta = \arcsin(-.66045320)$

$\theta \approx -41.33444556°$

51. $\theta = \cot^{-1} 4.5046388$

$\cot\theta = 4.5046388$

$\tan\theta = 4.5046388^{-1}$

$\theta = \tan^{-1} 4.5046388^{-1}$

$\theta \approx 12.51631252°$

52. There is no real number whose sine value is -3; the domain of sin is $[-1,\ 1]$.

53. $\cos x$ is defined for every real number, but $\arccos x$ is defined only on the interval $[-1,\ 1]$; $\arccos(\cos x) \neq x$ for values of x less than -1, for example.

54. $\sin\left(\sin^{-1}\frac{1}{2}\right)$

$\sin\theta = \frac{1}{2},\ \theta \in$ Quadrant I

$$\boxed{\sin\left(\sin^{-1}\frac{1}{2}\right) = \frac{1}{2}}$$

55. $\sin\left(\cos^{-1}\frac{3}{4}\right)$

$\cos\theta = \frac{3}{4},\ \theta \in$ Quadrant I:

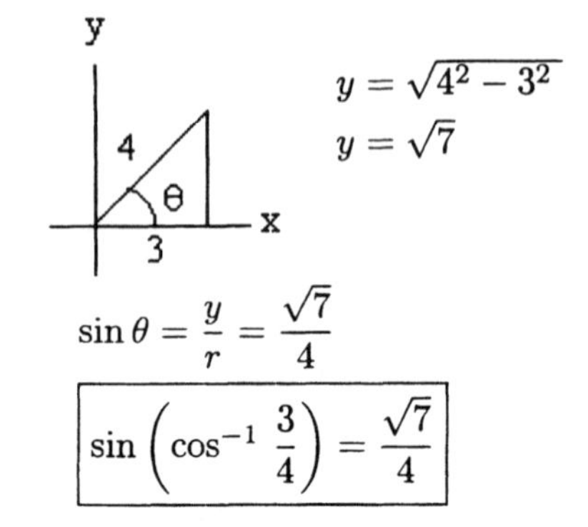

$y = \sqrt{4^2 - 3^2}$

$y = \sqrt{7}$

$\sin\theta = \frac{y}{r} = \frac{\sqrt{7}}{4}$

$$\boxed{\sin\left(\cos^{-1}\frac{3}{4}\right) = \frac{\sqrt{7}}{4}}$$

56. $\cos(\arctan 3)$

$\tan\theta = 3,\ \theta \in$ Quadrant I:

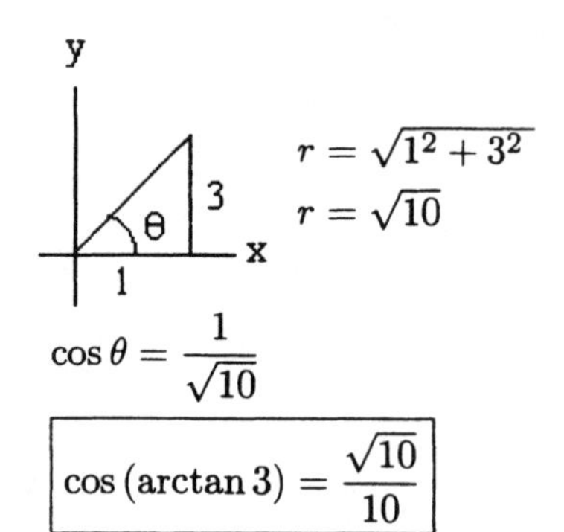

$r = \sqrt{1^2 + 3^2}$

$r = \sqrt{10}$

$\cos\theta = \frac{1}{\sqrt{10}}$

$$\boxed{\cos(\arctan 3) = \frac{\sqrt{10}}{10}}$$

57. $\sec\left(2\sin^{-1}\left(-\frac{1}{3}\right)\right)$

$\sin\theta = -\frac{1}{3},\ \theta \in$ Quadrant IV:

$x = \sqrt{3^2 - (-1)^2}$

$x = \sqrt{8} = 2\sqrt{2}$

$$\begin{aligned}\sec 2\theta &= \frac{1}{\cos 2\theta} \\ &= \frac{1}{1 - 2\sin^2\theta} \\ &= \frac{1}{1 - 2\left(\frac{-1}{3}\right)^2}\cdot\frac{9}{9} \\ &= \frac{9}{9-2} = \frac{9}{7}\end{aligned}$$

$$\boxed{\sec\left(2\sin^{-1}\left(-\frac{1}{3}\right)\right) = \frac{9}{7}}$$

58. $\cos^{-1}\left(\cos\frac{3\pi}{2}\right) = \cos^{-1}(0) = \boxed{\frac{\pi}{2}}$

(range: $[0,\ \pi]$)

59. $\tan\left(\sin^{-1}\frac{3}{5}+\cos^{-1}\frac{5}{7}\right)$

$\sin\theta=\frac{3}{5}$, $\theta\in$ Quadrant I

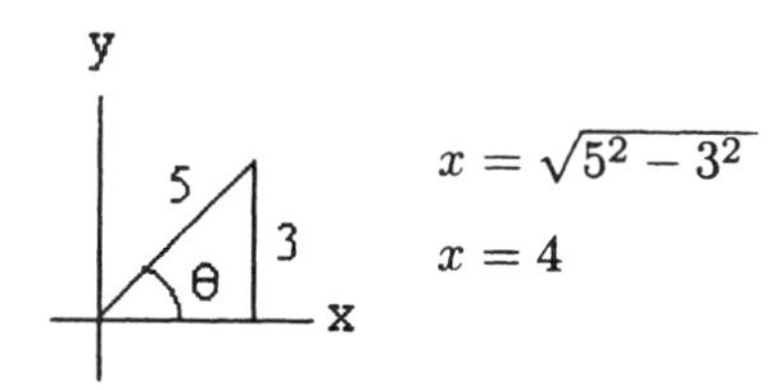

$x=\sqrt{5^2-3^2}$

$x=4$

$\cos\beta=\frac{5}{7}$, $\beta\in$ Quadrant I

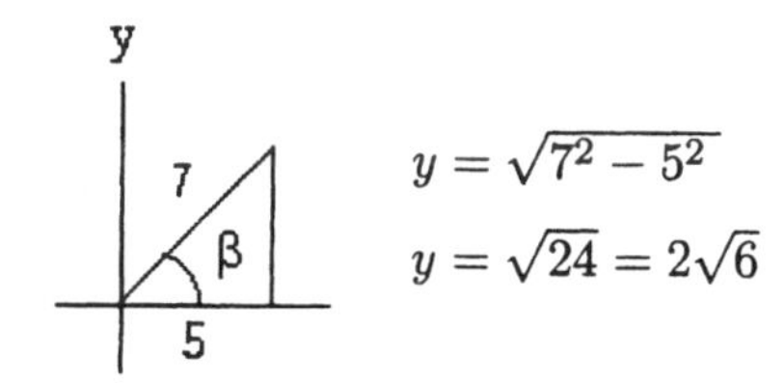

$y=\sqrt{7^2-5^2}$

$y=\sqrt{24}=2\sqrt{6}$

$$\begin{aligned}\tan(\theta+\beta)&=\frac{\tan\theta+\tan\beta}{1-\tan\theta\tan\beta}\\&=\frac{\frac{3}{4}+\frac{2\sqrt{6}}{5}}{1-\left(\frac{3}{4}\right)\left(\frac{2\sqrt{6}}{5}\right)}\cdot\frac{20}{20}\\&=\frac{15+8\sqrt{6}}{20-6\sqrt{6}}\cdot\frac{20+6\sqrt{6}}{20+6\sqrt{6}}\\&=\frac{588+250\sqrt{6}}{184}=\boxed{\frac{294+125\sqrt{6}}{92}}\end{aligned}$$

60. $\sin\left(\tan^{-1}u\right)$

$\tan\theta=u$, $\theta\in$ Quadrant I

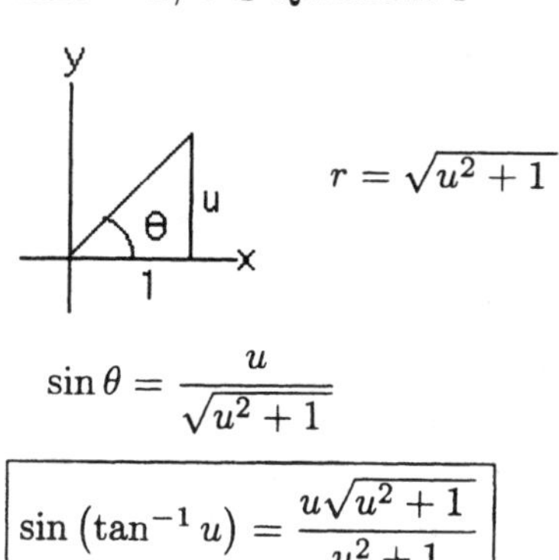

$r=\sqrt{u^2+1}$

$\sin\theta=\frac{u}{\sqrt{u^2+1}}$

$$\boxed{\sin\left(\tan^{-1}u\right)=\frac{u\sqrt{u^2+1}}{u^2+1}}$$

61. $\cos\left(\arctan\frac{u}{\sqrt{1-u^2}}\right)$

$\tan\theta=\frac{u}{\sqrt{1-u^2}}$, $\theta\in$ Quadrant I

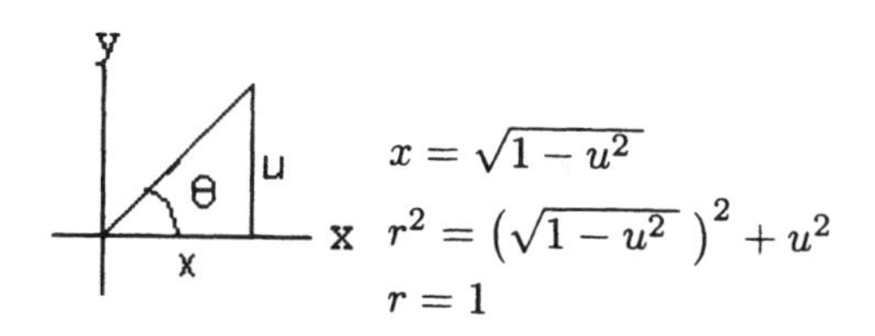

$x=\sqrt{1-u^2}$

$r^2=\left(\sqrt{1-u^2}\right)^2+u^2$

$r=1$

$\cos\theta=\sqrt{1-u^2}$

$$\boxed{\cos\left(\arctan\frac{u}{\sqrt{1-u^2}}\right)=\sqrt{1-u^2}}$$

62. $\tan\left(\arccos\frac{u}{\sqrt{u^2+1}}\right)$

$\cos\theta=\frac{u}{\sqrt{u^2+1}}$, $\theta\in$ Quadrant I

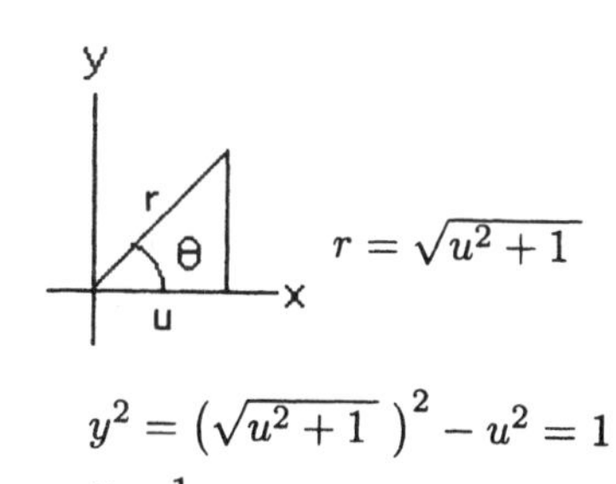

$r=\sqrt{u^2+1}$

$y^2=\left(\sqrt{u^2+1}\right)^2-u^2=1$

$y=1$

$$\boxed{\tan\left(\arccos\frac{u}{\sqrt{u^2+1}}\right)=\frac{1}{u}}$$

63. $\sin^2 x=1$

$\sin x=\pm1$

$x=\left\{\frac{\pi}{2},\ \frac{3\pi}{2}\right\}$

64. $2\tan x-1=0$

$2\tan x=1$

$\tan x=\frac{1}{2}$

Ref $\angle: x\approx .463647609$, Quadrants I, III

$x\approx\{.463647609,\ 3.605240263\}$

65. $3\sin^2 x - 5\sin x + 2 = 0$

$(3\sin x - 2)(\sin x - 1) = 0$

$\sin x = \frac{2}{3} \qquad \sin x = 1$

If $\sin x = \frac{2}{3}$

$x = \sin^{-1}\frac{2}{3} \qquad$ Quadrants I, II

Ref $\angle$: $x = \sin^{-1}\frac{2}{3}$

$x \approx .7297276562$

$x \approx .729726562,\ 2.411864997$

If $\sin x = 1, \quad x = \frac{\pi}{2}$

$x \approx \left\{.729726562, \frac{\pi}{2}, 2.411864997\right\}$

66. $\tan x = \cot x$

$\tan x = \frac{1}{\tan x}$

$\tan^2 x = 1$

$\tan x = \pm 1$

$x = \left\{\frac{\pi}{4}, \frac{3\pi}{4}, \frac{5\pi}{4}, \frac{7\pi}{4}\right\}$

67. $5\cot^2 x + 3\cot x = 2$

$5\cot^2 x + 3\cot x - 2 = 0$

$(5\cot x - 2)(\cot x + 1) = 0$

$\cot x = \frac{2}{5} \qquad \cot x = -1$

If $\cot x = \frac{2}{5}$, $\tan x = \frac{5}{2}$

Ref $\angle$: $x \approx 1.19028995$, Quadrants I, III

$x \approx 1.19028994,\ 4.331882603$

If $\cot x = -1, \quad x = \frac{3\pi}{4}, \frac{7\pi}{4}$

$x \approx \left\{1.19028995, \frac{3\pi}{4}, 4.331882603, \frac{7\pi}{4}\right\}$

68. $\sec\frac{x}{2} = \cos\frac{x}{2}$

$\frac{1}{\cos\frac{x}{2}} = \cos\frac{x}{2}$

$1 = \cos^2\frac{x}{2}$

$1 = \left(\pm\sqrt{\frac{1+\cos x}{2}}\right)^2$

$1 = \frac{1+\cos x}{2}$

$2 = 1 + \cos x$

$\cos x = 1 \qquad x = \{0\}$

69. $\sin 2x = \cos 2x + 1$

$\sin 2x - \cos 2x - 1 = 0$

$2\sin x\cos x - (2\cos^2 x - 1) - 1 = 0$

$2\sin x\cos x - 2\cos^2 x = 0$

$2\cos x(\sin x - \cos x) = 0$

(i) $\cos x = 0 \rightarrow x = \frac{\pi}{2}, \frac{3\pi}{2}$

(ii) $\cos x = \sin x$

$\cos^2 x = \sin^2 x$

$\cos^2 x = 1 - \cos^2 x$

$2\cos^2 x = 1$

$\cos^2 x = \frac{1}{2}$

$\cos x = \pm\frac{1}{\sqrt{2}}$

$x = \frac{\pi}{4}, \frac{3\pi}{4}, \frac{5\pi}{4}, \frac{7\pi}{4}$

$\cos x$ and $\sin x$ have different signs in Quadrants II and IV, so only $\frac{\pi}{4}$ and $\frac{5\pi}{4}$ will check.

$$\boxed{\left\{\frac{\pi}{4}, \frac{\pi}{2}, \frac{5\pi}{4}, \frac{3\pi}{2}\right\}}$$

70. $2\sin 2x = 1, \quad 0 \le 2x < 4\pi$

$\sin 2x = \frac{1}{2}$

$2x = \frac{\pi}{6}, \frac{5\pi}{6}, \frac{13\pi}{6}, \frac{17\pi}{6}$

$$\boxed{\left\{\frac{\pi}{12}, \frac{5\pi}{12}, \frac{13\pi}{12}, \frac{17\pi}{12}\right\}}$$

71. $\sin 2x + \sin 4x = 0,\ 0 \le 2x < 4\pi$

$\sin 2x + 2\sin 2x\cos 2x = 0$

$\sin 2x\,(1 + 2\cos 2x) = 0$

(i) $\sin 2x = 0$

$2x = 0,\ \pi,\ 2\pi,\ 3\pi$

$x = 0,\ \frac{\pi}{2},\ \pi,\ \frac{3\pi}{2}$

(ii) $2\cos 2x = -1$

$\cos 2x = -\frac{1}{2}$

$2x = \frac{2\pi}{3},\ \frac{4\pi}{3},\ \frac{8\pi}{3},\ \frac{10\pi}{3}$

$x = \frac{\pi}{3},\ \frac{2\pi}{3},\ \frac{4\pi}{3},\ \frac{5\pi}{3}$

$\boxed{\left\{0,\ \frac{\pi}{3},\ \frac{\pi}{2},\ \frac{2\pi}{3},\ \pi,\ \frac{4\pi}{3},\ \frac{3\pi}{2},\ \frac{5\pi}{3}\right\}}$

72. $\cos x - \cos 2x = 2\cos x$

$-\cos x - \left(2\cos^2 x - 1\right) = 0$

$-\cos x - 2\cos^2 x + 1 = 0$

$2\cos^2 x + \cos x - 1 = 0$

$(2\cos x - 1)(\cos x + 1) = 0$

(i) $\cos x = \frac{1}{2}$

$x = \frac{\pi}{3},\ x = \frac{5\pi}{3}$

(ii) $\cos x = -1$

$x = \pi$

$\boxed{\left\{\frac{\pi}{3},\ \pi,\ \frac{5\pi}{3}\right\}}$

73. $\tan 2x = \sqrt{3}$

$[\,0 \le x < 2\pi \Rightarrow 0 \le 2x < 4\pi\,]$

$2x = \frac{\pi}{3},\ \frac{4\pi}{3},\ \frac{7\pi}{3},\ \frac{10\pi}{3}$

$\boxed{\left\{\frac{\pi}{6},\ \frac{2\pi}{3},\ \frac{7\pi}{6},\ \frac{5\pi}{3}\right\}}$

74. $\cos^2\frac{x}{2} - 2\cos\frac{x}{2} + 1 = 0$

$\left(\cos\frac{x}{2} - 1\right)\left(\cos\frac{x}{2} - 1\right) = 0$

$\cos\frac{x}{2} = 1$

$\left[0 \le x < 2\pi \Rightarrow 0 \le \frac{x}{2} < \pi\right]$

$\frac{x}{2} = 0 \quad \boxed{\{0\}}$

75. **From #69:** $\sin 2x - \cos 2x - 1 = 0$

$x = \left\{\frac{\pi}{4},\ \frac{\pi}{2},\ \frac{5\pi}{4},\ \frac{3\pi}{2}\right\}$

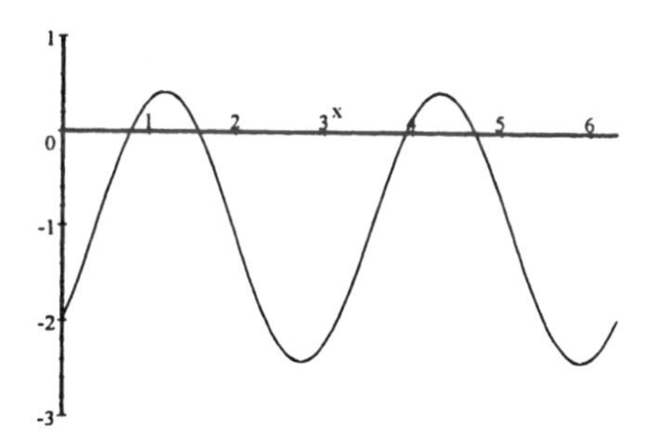

(a) $\sin 2x > \cos 2x + 1$

$\left(\frac{\pi}{4},\ \frac{\pi}{2}\right) \cup \left(\frac{5\pi}{4},\ \frac{3\pi}{2}\right)$

(b) $\sin 2x < \cos 2x + 1$

$\left[0,\ \frac{\pi}{4}\right) \cup \left(\frac{\pi}{2},\ \frac{5\pi}{4}\right) \cup \left(\frac{3\pi}{2},\ 2\pi\right)$

76. **(a)** Let α be the angle to the left of θ; $\tan\alpha = \frac{5}{x}$

Then $\tan(\alpha + \theta) = \frac{5 + 10}{x}$

$\alpha + \theta = \arctan\frac{15}{x}$

$\theta = \arctan\frac{15}{x} - \alpha$

$\theta = \arctan\frac{15}{x} - \arctan\frac{5}{x}$

continued

76. continued

(b) The maximum occurs at $x \approx 8.6602567$.

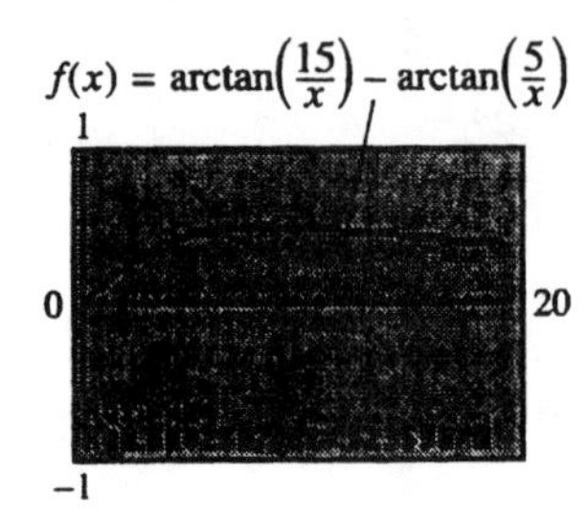

There may be a discrepancy in the final digits.

77. $y = \frac{1}{3} \sin \frac{4\pi t}{3}$

(a) $3y = \sin \frac{4\pi t}{3}$

$\frac{4\pi t}{3} = \sin^{-1} 3y$

$\boxed{t = \frac{3 \sin^{-1} 3y}{4\pi}}$

(b) $t = \frac{3 \sin^{-1} 3\,(.3)}{4\pi}$

$= .267325 \approx \boxed{.27 \text{ second}}$

78. $i = I_{\text{max}} \sin 2\pi \text{ ft}$

$40 = 100 \sin (2\pi \cdot 60)\, t$

$.4 = \sin 120\pi\, t$

$120\pi\, t = \sin^{-1} .4$

$t = \frac{\sin^{-1} .4}{120\pi}$

$= .00109 \approx \boxed{.001 \text{ second}}$

79. $i = I_{\text{max}} \sin 2\pi \text{ ft}$

$50 = 100 \sin (2\pi \cdot 120)\, t$

$.5 = \sin 240\pi\, t$

$240\pi\, t = \sin^{-1} .5$

$t = \frac{\sin^{-1} .5}{240\pi}$

$= .000694 \approx \boxed{.0007 \text{ second}}$

80. $\frac{c_1}{c_2} = \frac{\sin \theta_1}{\sin \theta_2}$

$.752 = \frac{\sin \theta_1}{\sin \theta_2}$

If $\theta_2 = 90°$:

$.752 = \frac{\sin \theta_1}{\sin 90°}$

$= \frac{\sin \theta_1}{1} = \sin \theta_1$

$\theta_1 = \sin^{-1} .752 \approx \boxed{48.8°}$

81. If $\theta_1 > 48.8°$, then $\theta_2 > 90°$ and the light beam will stay completely under the water.

Relating Concepts (82 – 86):

82. $\tan \theta = \frac{b}{a}$

Slope of $l : m = \frac{\Delta y}{\Delta x} = \frac{b}{a}$

Conclusion: $m = \tan \theta$

83. In the triangle formed,

$\alpha + \theta + (180° - \beta) = 180°$

$\alpha + \theta = \beta$

84. $\alpha + \theta = \beta$

$\theta = \beta - \alpha$

85. $\tan \theta = \tan (\beta - \alpha)$

$\tan \theta = \frac{\tan \beta - \tan \alpha}{1 + \tan \beta \tan \alpha}$

86. $\tan \alpha = m_1; \quad \tan \beta = m_2$

$\tan \theta = \frac{m_2 - m_1}{1 + m_1 m_2}$

Chapter 9 Test

1. $\sin y = -\frac{2}{3}$, $\pi < y < \frac{3\pi}{2}$

$$r^2 = x^2 + y^2$$
$$3^2 = x^2 + (-2)^2$$
$$x^2 = 9 - 4 = 5$$
$$x = -\sqrt{5} \quad \text{(Quadrant III)}$$
$$\cos y = -\frac{\sqrt{5}}{3}$$

$\cos x = -\frac{1}{5}$, $\frac{\pi}{2} < x < \pi$

$$r^2 = x^2 + y^2$$
$$5^2 = (-1)^2 + y^2$$
$$y^2 = 25 - 1 = 24$$
$$y = \sqrt{24} = 2\sqrt{6} \quad \text{(Quadrant II)}$$
$$\sin x = \frac{2\sqrt{6}}{5}$$

(a) $\sin(x+y) = \sin x \cos y + \cos x \sin y$

$$= \left(\frac{2\sqrt{6}}{5}\right)\left(-\frac{\sqrt{5}}{3}\right) + \left(-\frac{1}{5}\right)\left(-\frac{2}{3}\right)$$
$$= \frac{-2\sqrt{30}}{15} + \frac{2}{15} = \boxed{\frac{2 - 2\sqrt{30}}{15}}$$

(b) $\cos(x-y) = \cos x \cos y + \sin x \sin y$

$$= \left(-\frac{1}{5}\right)\left(-\frac{\sqrt{5}}{3}\right) + \left(\frac{2\sqrt{6}}{5}\right)\left(-\frac{2}{3}\right)$$
$$= \frac{\sqrt{5}}{15} - \frac{4\sqrt{6}}{15} = \boxed{\frac{\sqrt{5} - 4\sqrt{6}}{15}}$$

(c) $\tan\frac{y}{2} = \frac{1 - \cos y}{\sin y}$

$$= \frac{1 - \left(-\frac{\sqrt{5}}{3}\right)}{-\frac{2}{3}} \cdot \frac{3}{3}$$
$$= \frac{3 + \sqrt{5}}{-2} = \boxed{\frac{-3 - \sqrt{5}}{2}}$$

(d) $\cos 2x = 2\cos^2 x - 1$

$$= 2\left(-\frac{1}{5}\right)^2 - 1$$
$$= \frac{2}{25} - \frac{25}{25} = \boxed{-\frac{23}{25}}$$

2. $\tan^2 x - \sec^2 x$

$$= \frac{\sin^2 x}{\cos^2 x} - \frac{1}{\cos^2 x}$$
$$= \frac{\sin^2 x - 1}{\cos^2 x} = -\frac{\cos^2 x}{\cos^2 x} = -1$$

3. $y = \sec x - \sin x \tan x$

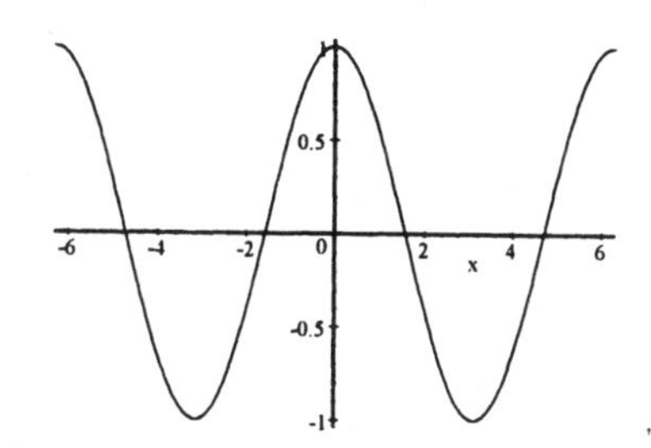

The graph looks like $\cos x$, so the identity is $\sec x - \sin x \tan x = \cos x$.

Verify $\sec x - \sin x \tan x = \cos x$

$$\sec x - \sin x \tan x = \frac{1}{\cos x} - \frac{\sin x}{1} \cdot \frac{\sin x}{\cos x}$$
$$= \frac{1 - \sin^2 x}{\cos x}$$
$$= \frac{\cos^2 x}{\cos x} = \cos x$$

4. (a) Verify $\sec^2 B = \frac{1}{1 - \sin^2 B}$

$$\sec^2 B = \frac{1}{\cos^2 B} = \frac{1}{1 - \sin^2 B}$$

(b) Verify $\cos 2A = \frac{\cot A - \tan A}{\csc A \sec A}$

$$\frac{\cot A - \tan A}{\csc A \sec A} = \frac{\frac{\cos A}{\sin A} - \frac{\sin A}{\cos A}}{\frac{1}{\sin A} \cdot \frac{1}{\cos A}}$$
$$= \frac{\cos^2 A - \sin^2 A}{\sin A \cos A} \cdot \frac{\sin A \cos A}{1}$$
$$= \cos^2 A - \sin^2 A = \cos 2A$$

5. (a) $\cos(270° - \theta)$

$= \cos 270° \cos\theta + \sin 270° \sin\theta$

$= 0 \cdot \cos\theta + (-1)\sin\theta$

$= -\sin\theta$

(b) $\sin(\pi + \theta) = \sin\pi\cos\theta + \cos\pi\sin\theta$

$= 0 \cdot \cos\theta + (-1)\sin\theta$

$= -\sin\theta$

6. $f(x) = \sin^{-1}(x - 1)$

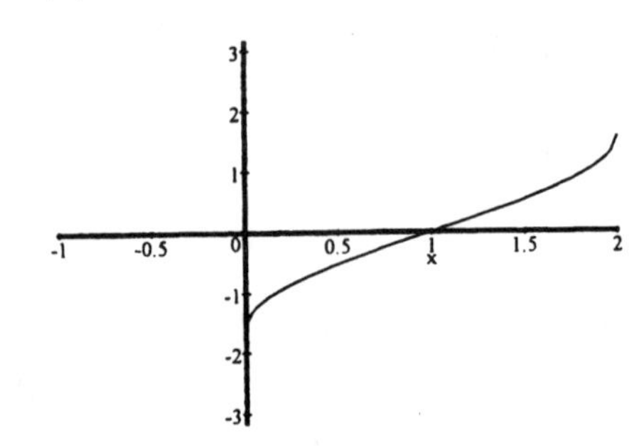

(a) domain: $[0,\ 2]$; range: $\left[-\frac{\pi}{2},\ \frac{\pi}{2}\right]$

(b) x-intercept: 1

y-intercept: $-\frac{\pi}{2}$

(c) $y = \sin^{-1} 2$ is not possible, since the range of $y = \sin x$ is $[-1,\ 1]$.
(Or the domain of $\sin^{-1} x$ is $[-1,\ 1]$.)

7. (a) $\arccos\left(-\frac{1}{2}\right) = \frac{2\pi}{3}$

(b) $\tan^{-1} 0 = 0$

(c) $\csc^{-1}\frac{2\sqrt{3}}{3} = y$

$\csc y = \frac{2\sqrt{3}}{3}$

$\sin y = \frac{3}{2\sqrt{3}} \cdot \frac{\sqrt{3}}{\sqrt{3}} = \frac{\sqrt{3}}{2}$

$y = \frac{\pi}{3}$

continued

7. continued

(d) $\cos\left(\arcsin\frac{2}{3}\right)$

Let $\theta = \arcsin\frac{2}{3}$, $\theta \in$ Quadrant I

$x = \sqrt{3^2 - 2^2}$

$x = \sqrt{5}$

$\cos\left(\arcsin\frac{2}{3}\right) = \frac{\sqrt{5}}{3}$

(e) $\sin\left(2\cos^{-1}\frac{1}{3}\right)$

Let $\beta = \cos^{-1}\frac{1}{3}$, $\beta \in$ Quadrant I

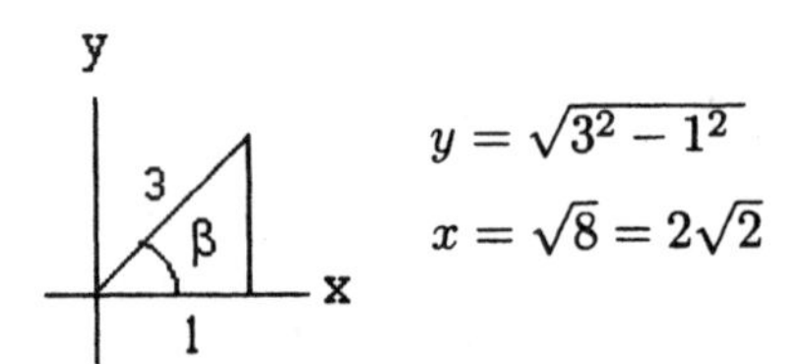

$y = \sqrt{3^2 - 1^2}$

$x = \sqrt{8} = 2\sqrt{2}$

$\sin 2\beta = 2\sin\beta\cos\beta$

$= 2 \cdot \frac{2\sqrt{2}}{3} \cdot \frac{1}{3} = \frac{4\sqrt{2}}{9}$

8. $\tan(\arcsin u)$

$\sin\theta = u,\ \theta \in$ Quadrant I

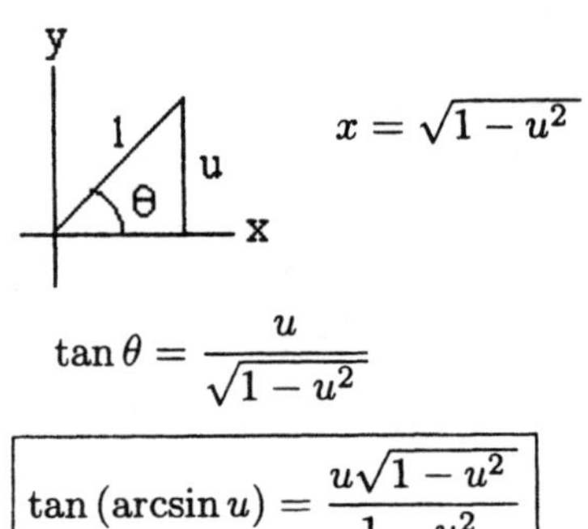

$x = \sqrt{1 - u^2}$

$\tan\theta = \frac{u}{\sqrt{1 - u^2}}$

$$\boxed{\tan(\arcsin u) = \frac{u\sqrt{1 - u^2}}{1 - u^2}}$$

9. **(a)** $\sin^2\theta = \cos^2\theta + 1 \qquad [0,\ 2\pi)$

$1 - \cos^2\theta = \cos^2\theta + 1$

$0 = 2\cos^2\theta$

$\cos\theta = 0 \qquad \theta = \left\{\frac{\pi}{2},\ \frac{3\pi}{2}\right\}$

(b) $\csc^2\theta - 2\cot\theta = 4 \qquad [0°,\ 360°)$

$1 + \cot^2\theta - 2\cot\theta = 4$

$\cot^2\theta - 2\cot\theta - 3 = 0$

$(\cot\theta - 3)(\cot\theta + 1) = 0$

$\cot\theta = 3 \qquad \cot\theta = -1$

If $\cot\theta = 3, \quad \theta = \tan^{-1}\frac{1}{3}$

$\theta \approx 18.4°,\ 198.4°$

If $\cot\theta = -1, \quad \theta = 135°,\ 315°$

$\theta = \{18.4°,\ 135°,\ 198.4°,\ 315°\}$

(c) $\cos x = \cos 2x \qquad [0,\ 2\pi)$

$\cos x = 2\cos^2 x - 1$

$0 = 2\cos^2 x - \cos x - 1$

$0 = (2\cos x + 1)(\cos x - 1)$

$\cos x = -\frac{1}{2} \qquad \cos x = 1$

If $\cos x = -\frac{1}{2},\ x = \frac{2\pi}{3},\ \frac{4\pi}{3}$

If $\cos x = 1,\ x = 0$

$x = \left\{0,\ \frac{2\pi}{3},\ \frac{4\pi}{3}\right\}$

(d) $2\sqrt{3}\sin\frac{\theta}{2} = 3 \qquad 0° \le \theta < 360°$

$0° \le \frac{\theta}{2} < 180°$

$\sin\frac{\theta}{2} = \frac{3}{2\sqrt{3}} \cdot \frac{\sqrt{3}}{\sqrt{3}} = \frac{\sqrt{3}}{2}$

$\frac{\theta}{2} = 60°,\ 120° \qquad \theta = \{120°,\ 240°\}$

10. $T(t) = 50 + 50\cos\left(\frac{\pi}{6}t\right)$, t is the time in months, $t = 0 \rightarrow$ July.

(a) Since t will repeat every 12 months, the domain for 1 year will be $[0,\ 11]$.

$50 + 50\cos\left(\frac{\pi}{6}t\right)$ is limited by the range of cosine: at the minimum (-1),

$50 + 50\cos(-1) = 0,$

continued

10. continued

and at the maximum, $50 + 50\cos(1) = 100$.

The range will be $[0,\ 100]$ in hundreds.

(b) $T(t) = 50 + 50\cos\left(\frac{\pi}{6}t\right)$

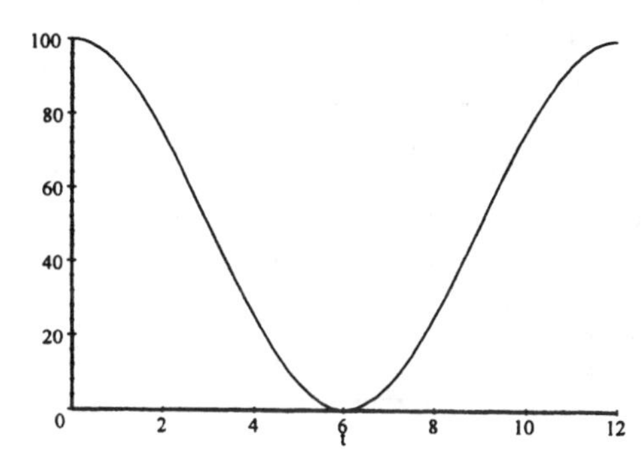

(c) The maximum of 10,000 animals occurs at 0 or 12 months (July). The minimum of 0 animals occurs at 6 months (January).

(d) $T(t) = 50 + 50\cos\left(\frac{\pi}{6}t\right)$

$T(3) = 50 + 50\cos\left(\frac{3\pi}{6}\right)$

$= 50 + 50(0)$

$= 50 \text{ hundreds } = 5000$

There are 5000 animals at 3 months (October) and at 9 months (April).

(e) 7500 animals when $T(t) = 75$:

$75 = 50 + 50\cos\left(\frac{\pi}{6}t\right)$

$25 = 50\cos\left(\frac{\pi}{6}t\right)$

$\frac{1}{2} = \cos\left(\frac{\pi}{6}t\right)$

$\frac{\pi}{6}t = \frac{\pi}{3},\ \frac{5\pi}{3} \qquad t = 2,\ 10$

At 2 months (September) and at 10 months (May), there will be 7500 animals.

(f) $T = 50 + 50\cos\left(\frac{\pi}{6}t\right)$

$T - 50 = 50\cos\left(\frac{\pi}{6}t\right)$

$\frac{T-50}{50} = \cos\left(\frac{\pi}{6}t\right)$

$\frac{\pi}{6}t = \arccos\left(\frac{T-50}{50}\right)$

$t = \frac{6}{\pi}\arccos\left(\frac{T-50}{50}\right)$

Chapter 9 Project

Modeling a Damped Pendulum

1. The list is shown in the first two screens and the scatterplot in the third.

L1	L2	----- 1
.8	3.81	------
1.7	2.17	
2.6	3.61	
3.5	2.33	
4.4	3.52	
5.3	2.42	
	3.43	

L1(7)=6.2

L1	L2	----- 1
5.3	2.42	
6.2	3.43	
7.1	2.5	
8	3.37	
8.9	2.56	
	3.32	
------	------	

L1(11) =9.8

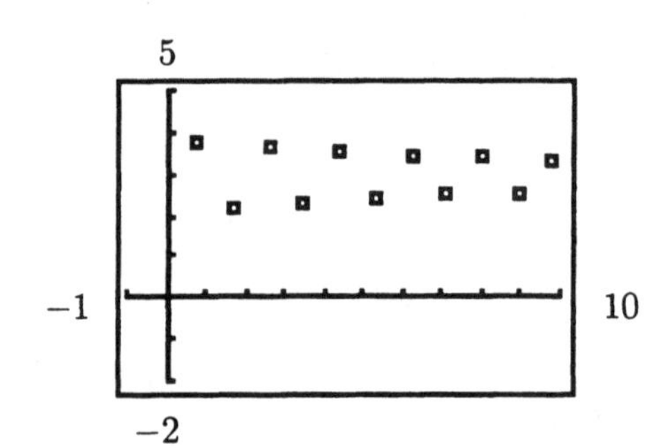

2. (a) From the list, 3.81 is the greatest height and 2.17 is the least height, so the amplitude is

$$a = \frac{3.81 - 2.17}{2} = .82$$

(b) The time difference between successive maximums (or minimums) is 1.8 seconds, so the period is

$$b = \frac{2\pi}{1.8} \approx 3.49066 \approx 3.49$$

(c) The smallest time corresponding to the mean height can be found by taking half the difference of the time for the first maximum and the first minimum and subtracting that from the time at the 1st maximum.

$$\frac{1.7 - .8}{2} = .45, \text{ and so}$$

$$d = .8 - .45 = .35$$

(d) The amplitude diminishing rate is

$$r = \frac{3.61}{3.81} \approx .9475066 \approx .95$$

To find c, the vertical shift, find the mean height by averaging the maximum and minimum heights:

$$c = \frac{3.81 + 2.17}{2} = 2.99$$

With these values, the function is defined by

$$y = r^x a \sin[b(x - d)] + c$$

$$y = .95^x \cdot .82 \sin[3.49(x - .35)] + 2.99$$

3. The function is a reasonably good fit for the data.

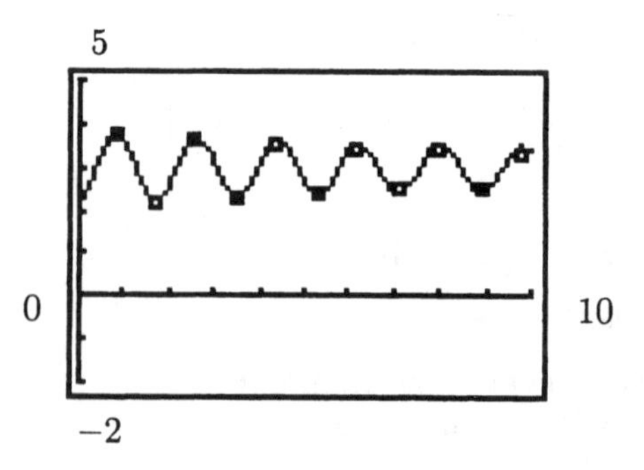

Section 10.1

1. **C** is not valid since **B** and **D** are valid.

For #3 - #9: the ambiguous case occurs when we are given SSA.

3. No

5. No

7. Yes

9. Yes

11. $A = 37°,\ B = 48°,\ c = 18$ m

(i) $C = 180° - (37° + 48°)$ $\boxed{C = 95°}$

(ii) $\dfrac{a}{\sin A} = \dfrac{c}{\sin C}$

$\dfrac{a}{\sin 37°} = \dfrac{18}{\sin 95°}$

$a = \dfrac{18 \sin 37°}{\sin 95°}$

$a \approx 10.874$ $\boxed{a \approx 11 \text{ m}}$

(iii) $\dfrac{b}{\sin B} = \dfrac{c}{\sin C}$

$\dfrac{b}{\sin 48°} = \dfrac{18}{\sin 95°}$

$b = \dfrac{18 \sin 48°}{\sin 95°}$

$b \approx 13.4277$ $\boxed{b \approx 13 \text{ m}}$

13. $A = 27.2°,\ C = 115.5°,\ c = 76.0$ feet

(i) $B = 180° - (27.2° + 115.5°)$

$\boxed{B = 37.3°}$

(ii) $\dfrac{a}{\sin A} = \dfrac{c}{\sin C}$

$\dfrac{a}{\sin 27.2°} = \dfrac{76.0}{\sin 115.5°}$

$a = \dfrac{76.0 \sin 27.2°}{\sin 115.5°}$

$a \approx 38.489$ $\boxed{a \approx 38.5 \text{ ft}}$

(iii) $\dfrac{b}{\sin B} = \dfrac{c}{\sin C}$

$\dfrac{b}{\sin 37.3°} = \dfrac{76.0}{\sin 115.5°}$

$b = \dfrac{76.0 \sin 37.3°}{\sin 115.5°}$

$b \approx 51.026$ $\boxed{b \approx 51.0 \text{ ft}}$

15. $A = 37°,\ B = 48°,\ c = 18$ m

(i) $C = 180° - (37° + 48°)$

$\boxed{C = 95°}$

(ii) $\dfrac{a}{\sin A} = \dfrac{c}{\sin C}$

$\dfrac{a}{\sin 37°} = \dfrac{18}{\sin 95°}$

$a = \dfrac{18 \sin 37°}{\sin 95°}$

$a \approx 10.874$ $\boxed{a \approx 11 \text{ m}}$

(iii) $\dfrac{b}{\sin B} = \dfrac{c}{\sin C}$

$\dfrac{b}{\sin 48°} = \dfrac{18}{\sin 95°}$

$b = \dfrac{18 \sin 48°}{\sin 95°}$

$b \approx 13.4277$ $\boxed{b \approx 13 \text{ m}}$

17. $C = 74.08°,\ B = 69.38°,\ c = 45.38$ m

(i) $A = 180° - (74.08° + 69.38°)$

$$\boxed{A = 36.54°}$$

(ii) $\dfrac{a}{\sin A} = \dfrac{c}{\sin C}$

$$\frac{a}{\sin 36.54°} = \frac{45.38}{\sin 74.08°}$$

$$a = \frac{45.38 \sin 36.54°}{\sin 74.08°}$$

$$a \approx 28.0961 \qquad \boxed{a \approx 28.10 \text{ m}}$$

(iii) $\dfrac{b}{\sin B} = \dfrac{c}{\sin C}$

$$\frac{b}{\sin 69.38°} = \frac{45.38}{\sin 74.08°}$$

$$b = \frac{45.38 \sin 69.38°}{\sin 74.08°}$$

$$b \approx 44.1668 \qquad \boxed{b \approx 44.17 \text{ m}}$$

19. $B = 38°40',\ C = 91°40',\ a = 19.7$ cm

(i) $A = 180° - \left(38°40' + 91°40'\right)$

$$\boxed{A = 49°40'}$$

(ii) $\dfrac{a}{\sin A} = \dfrac{b}{\sin B}$

$$\frac{19.7}{\sin 49°40'} = \frac{b}{\sin 38°40'}$$

$$b = \frac{19.7 \sin 38°40'}{\sin 49°40'}$$

$$b \approx 16.1465 \qquad \boxed{b \approx 16.1 \text{ cm}}$$

(iii) $\dfrac{a}{\sin A} = \dfrac{c}{\sin C}$

$$\frac{19.7}{\sin 49°40'} = \frac{c}{\sin 91°40'}$$

$$c = \frac{19.7 \sin 91°40'}{\sin 49°40'}$$

$$c \approx 25.8322 \qquad \boxed{c \approx 25.8 \text{ cm}}$$

21. $A = 35.3°,\ B = 52.8°,\ b = 675$ ft

(i) $C = 180° - (35.3° + 52.8°)$

$$\boxed{C = 91.9°}$$

(ii) $\dfrac{a}{\sin A} = \dfrac{b}{\sin B}$

$$\frac{a}{\sin 35.3°} = \frac{675}{\sin 52.8°}$$

$$a = \frac{675 \sin 35.3°}{\sin 52.8°}$$

$$a \approx 489.691$$

$$\boxed{a \approx 490 \text{ feet}}$$

(iii) $\dfrac{b}{\sin B} = \dfrac{c}{\sin C}$

$$\frac{675}{\sin 52.8°} = \frac{c}{\sin 91.9°}$$

$$c = \frac{675 \sin 91.9°}{\sin 52.8°}$$

$$c \approx 846.9599$$

$$\boxed{c \approx 847 \text{ feet}}$$

23. $A = 29.7°,\ a = 27.2$ ft, $b = 41.5$ ft

$$\frac{a}{\sin A} = \frac{b}{\sin B}$$

$$\frac{27.2}{\sin 29.7°} = \frac{41.5}{\sin B}$$

$$\sin B = \frac{41.5 \sin 29.7°}{27.2}$$

$$\sin B \approx .7559387772$$

$$B \approx 49.1°$$

$$\boxed{B_1 \approx 49.1°}$$

$$\boxed{B_2 \approx 130.9°} \quad (180° - B_1)$$

[ck: $A + B_2 = 160.6° < 180°$]

$$C_1 = 180° - (A + B_1)$$

$$\boxed{C_1 \approx 101.2°}$$

$$C_2 = 180° - (A + B_2)$$

$$\boxed{C_2 \approx 19.4°}$$

25. $B = 74.3°$, $a = 859$ m, $b = 783$ m

$$\frac{a}{\sin A} = \frac{b}{\sin B}$$
$$\frac{859}{\sin A} = \frac{783}{\sin 74.3°}$$
$$\sin A = \frac{859 \sin 74.3°}{783}$$
$$\sin A \approx 10.561$$

A is undefined. **No such triangle.**

27. $A = 142.13°$, $a = 7.297$ ft, $b = 5.432$ ft

$$\frac{a}{\sin A} = \frac{b}{\sin B}$$
$$\frac{7.297}{\sin 142.13°} = \frac{5.432}{\sin B}$$
$$\sin B = \frac{5.432 \sin 142.13°}{7.297}$$
$$\sin B \approx .456975804539 \qquad \boxed{B \approx 27.19°}$$
$$B_2 \approx 152.81° \quad (180° - B_1)$$

[does not ck: $A + B_2 = 294.94° \not< 180°$]

$$C = 180° - (A + B) \qquad \boxed{C \approx 10.68°}$$

29. $A = 42.5°$, $a = 15.6$ ft, $b = 8.14$ ft

$$\frac{a}{\sin A} = \frac{b}{\sin B}$$
$$\frac{15.6}{\sin 42.5°} = \frac{8.14}{\sin B}$$
$$\sin B = \frac{8.14 \sin 42.5°}{15.6}$$
$$\sin B \approx 0.3525195$$
$$B \approx 20.6415° \qquad \boxed{B_1 \approx 20.6°}$$
$$B_2 \approx 159.4° \quad (180° - B_1)$$

[Is B_2 possible? **No**
$A + B_2 = 201.9° > 180°$]

$$\boxed{C \approx 116.9°} \quad (180° - (A + B))$$
$$\frac{c}{\sin C} = \frac{a}{\sin A}$$
$$\frac{c}{\sin 116.9°} = \frac{15.6}{\sin 42.5°}$$
$$c = \frac{15.6 \sin 116.9°}{\sin 42.5°}$$
$$c \approx 20.5924 \qquad \boxed{c \approx 20.6 \text{ ft}}$$

31. $B = 72.2°$, $b = 78.3$ m, $c = 145$ m:

$$\frac{b}{\sin B} = \frac{c}{\sin C}$$
$$\frac{78.3}{\sin 72.2°} = \frac{145}{\sin C}$$
$$\sin C = \frac{145 \sin 72.2°}{78.3}$$
$$\sin C \approx 1.763202579$$
$$\emptyset \quad (-1 \le \sin\theta \le 1)$$
$$\boxed{\text{No such triangle}}$$

33. $A = 38°40'$, $a = 9.72$ km, $b = 11.8$ km

$$\frac{a}{\sin A} = \frac{b}{\sin B}$$
$$\frac{9.72}{\sin 38°40'} = \frac{11.8}{\sin B}$$
$$\sin B = \frac{11.8 \sin 38°40'}{9.72}$$
$$\sin B \approx 0.758488114$$
$$B \approx 49.3311° \quad \left(49°19'52''\right)$$
$$\boxed{B_1 \approx 49°20'}$$
$$\boxed{B_2 \approx 130°40'} \quad (180° - B_1)$$

[ck: $A + B_2 = 169°20' < 180°$]

$$C_1 = 180° - (A + B_1) \qquad \boxed{C_1 \approx 92°00'}$$
$$C_2 = 180° - (A + B_2) \qquad \boxed{C_2 \approx 10°40'}$$
$$\frac{a}{\sin A} = \frac{c_1}{\sin C_1}$$
$$\frac{9.72}{\sin 38°40'} \approx \frac{c_1}{\sin 92°}$$
$$c_1 \approx \frac{9.72 \sin 92°}{\sin 38°40'}$$
$$c_1 \approx 15.5478 \qquad \boxed{c_1 \approx 15.5 \text{ km}}$$
$$\frac{a}{\sin A} = \frac{c_2}{\sin C_2}$$
$$\frac{9.72}{\sin 38°40'} \approx \frac{c_2}{\sin 10°40'}$$
$$c_2 \approx \frac{9.72 \sin 10°40'}{\sin 38°40'}$$
$$c_2 \approx 2.8796 \qquad \boxed{c_2 \approx 2.88 \text{ km}}$$

35. $B = 32°50'$, $a = 7540$ cm, $b = 5180$ cm

$$\frac{a}{\sin A} = \frac{b}{\sin B}$$

$$\frac{7540}{\sin A} = \frac{5180}{\sin 32°50'}$$

$$\sin A = \frac{7540 \sin 32°50'}{5180}$$

$$\sin A \approx 0.7892213202$$

$$A \approx 52.1128° \quad \left(52°06'46''\right)$$

$$\boxed{A_1 \approx 52°10'}$$

$$\boxed{A_2 \approx 127°50'} \quad (180° - A_1)$$

$$[\text{ck: } A_2 + B = 160°40' < 180°]$$

$$C_1 = 180° - (A_1 + B)$$

$$\boxed{C_1 \approx 95°00'}$$

$$C_2 = 180° - (A_2 + B)$$

$$\boxed{C_2 \approx 19°20'}$$

$$\frac{b}{\sin B} = \frac{c_1}{\sin C_1}$$

$$\frac{5180}{\sin 32°50'} \approx \frac{c_1}{\sin 95°}$$

$$c_1 \approx \frac{5180 \sin 95°}{\sin 32°50'}$$

$$c_1 \approx 9517.37$$

$$\boxed{c_1 \approx 9520 \text{ cm}}$$

$$\frac{b}{\sin B} = \frac{c_2}{\sin C_2}$$

$$\frac{5180}{\sin 32°50'} \approx \frac{c_2}{\sin 19°20'}$$

$$c_2 \approx \frac{5180 \sin 19°20'}{\sin 32°50'}$$

$$c_2 \approx 3162.89$$

$$\boxed{c_2 \approx 3160 \text{ cm}}$$

37. The Pythagorean theorem can only be used for right triangles.

39. The law of sines compares in triangles, the ratio of a side to the sine of the angle opposite it; therefore, at least one angle and the side opposite must be known to use it.

41. $B = 112°10'$, $C = 15°20'$, $BC = 354$ m

First, solve for A :

$$A = 180° - \left(112°10' + 15°20'\right)$$

$$A = 52°\,30'$$

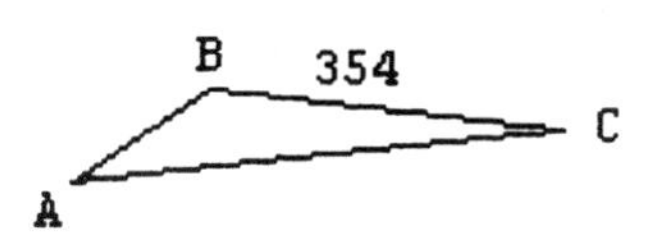

$$\frac{BC}{\sin A} = \frac{AB}{\sin C}$$

$$\frac{354}{\sin 52°\,30'} = \frac{AB}{\sin 15°20'}$$

$$AB = \frac{354 \sin 15°20'}{\sin 52°\,30'}$$

$$AB \approx 117.992$$

$$\boxed{AB \approx 118 \text{ m}}$$

43. A triangle is formed with
$A = 42.3°$ $(90° - 47.7°)$, and
$B = 32.5°$ $(302.5° - 270°)$.
Call the $3^{rd} \angle$ T, the location of the transmitter; $T = 105.2°$
$(180° - (42.3° + 32.5°))$

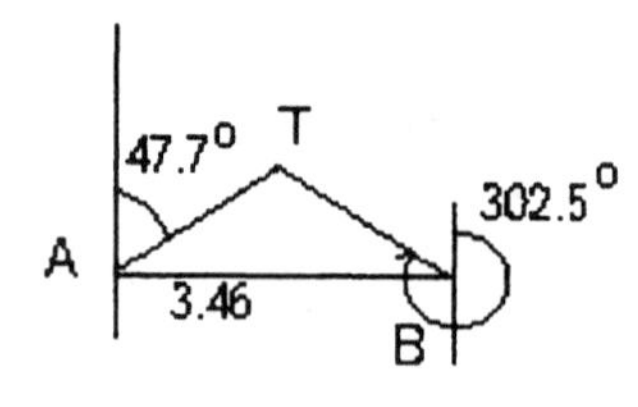

$$\frac{AB}{\sin T} = \frac{AT}{\sin B}$$

$$\frac{3.46}{\sin 105.2°} = \frac{AT}{\sin 32.5°}$$

$$AT = \frac{3.46 \sin 32.5°}{\sin 105.2°}$$

$$AT \approx 1.92645$$

$$\boxed{\text{Distance} \approx 1.93 \text{ miles}}$$

45. $\dfrac{12.0}{\sin 70.4°} = \dfrac{x}{\sin 54.8°}$

$x = \dfrac{12.0 \sin 54.8°}{\sin 70.4°}$

$x = 10.4089$

$\boxed{x \approx 10.4 \text{ inches}}$

47. A triangle is formed with sides 4.3 (1.6 + 2.7) and 5.2 (1.6 + 3.6) :

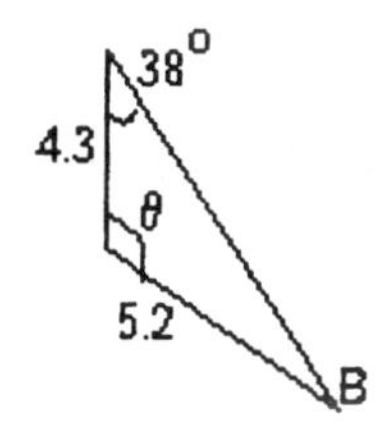

(i) Find the bottom angle B :

$\dfrac{4.3}{\sin B} = \dfrac{5.2}{\sin 38°}$

$\sin B = \dfrac{4.3 \sin 38°}{5.2}$

$\sin B = 0.5091046815$

$B \approx 30.604°$

$B \approx 31°$

(ii) $\theta = 180° - (38° + 31°)$

$\boxed{\theta \approx 111°}$

49. A triangle is formed with

A (1st sighting) $= 143°$ $(180° - 37°)$

B (2nd sighting) $= 25°$

$C = 12°$ $(180° - (143° + 25°))$

side $AB = 2.5$; find AC, BC :

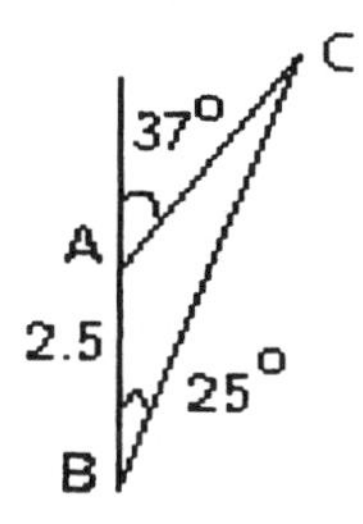

AC : $\dfrac{AB}{\sin C} = \dfrac{AC}{\sin B}$

$\dfrac{2.5}{\sin 12°} = \dfrac{AC}{\sin 25°}$

$AC = \dfrac{2.5 \sin 25°}{\sin 12°}$

$AC \approx 5.0817$

$\boxed{AC \approx 5.1 \text{ miles}}$

BC : $\dfrac{AB}{\sin C} = \dfrac{BC}{\sin A}$

$\dfrac{2.5}{\sin 12°} = \dfrac{BC}{\sin 143°}$

$BC = \dfrac{2.5 \sin 143°}{\sin 12°}$

$BC \approx 7.2364$

$\boxed{BC \approx 7.2 \text{ miles}}$

51. The triangle formed has

$A = 112.6°\quad (180° - (45° + 22.4°))$

$B = 34.4°\quad (180° - (135° + 10.6°))$

$AB = 25.5$

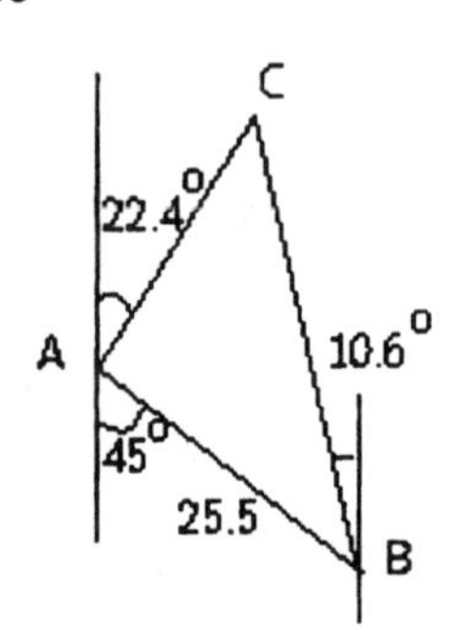

(i) Find C:

$C = 180° - (112.6° + 34.4°) = 33°$

(ii) $\dfrac{AC}{\sin B} = \dfrac{AB}{\sin C}$

$\dfrac{AC}{\sin 34.4°} = \dfrac{25.5}{\sin 33°}$

ria $\quad AC = \dfrac{25.5 \sin 34.4°}{\sin 33°} \approx 26.4518$

Distance of ship from Station A $\approx$ 26.5 km

53. The triangle formed has $T_1 = 28.1°$, $T_2 = 100.5°$ $(180° - 79.5°)$, $T_3 = 51.4°$ $(180° - (28.1° + 100.5°))$, and the side opposite, $T_3 = 1.73$.

$\dfrac{1.73}{\sin 51.4°} = \dfrac{x}{\sin 100.5°}$

$x = \dfrac{1.73 \sin 100.5°}{\sin 51.4°} \approx 2.17657$

Distance $\approx$ 2.18 km

55. (i) Find W:

$\dfrac{11.2}{\sin W} = \dfrac{28.6}{\sin 25.5°}$

$\sin W = \dfrac{11.2 \sin 25.5°}{28.6}$

$\sin W \approx 0.1685917582$

$W \approx 9.70595° \approx 9.71°$

(ii) Find P:

$P \approx 180° - (25.5° + 9.71°) \approx 144.79°$

(iii) Find CW:

$\dfrac{CW}{\sin P} = \dfrac{11.2}{\sin W}$

$CW \approx \dfrac{11.2 \sin 144.79°}{\sin 9.71°} \approx 38.28759$

$CW \approx 38.3$ cm

57. (a) Two triangles: $4 < a < 5$

(b) Exactly one t ngle: $a = 4$ or $a \geq 5$

(c) No triangle: $a < 4$

59. $a = \sqrt{5},\ c = 2\sqrt{5},\ A = 30°$

$\dfrac{\sqrt{5}}{\sin 30°} = \dfrac{2\sqrt{5}}{\sin C}$

$\sin C = \dfrac{2\sqrt{5} \sin 30°}{\sqrt{5}} = 2\left(\dfrac{1}{2}\right) = 1$

$C = 90°$, A right triangle.

61. The longest side is always opposite the largest angle. Since $A = 103°20'$, B and C must be acute angles, so b cannot be the longest side.

63. $\dfrac{21.9}{\sin 38°50'} = \dfrac{78.3}{\sin \theta}$

$\sin \theta \approx 2.2419$

Since $-1 \leq \sin \theta \leq 1$, there is no such triangle. Such a piece of property could not exist.

Relating Concepts (Exercises 65-69)

65. $y = \sin x$ is increasing from $\left[0, \frac{\pi}{2}\right]$.

66. If $B < A$, then $\sin B < \sin A$ because $y = \sin x$ is increasing on $\left[0, \frac{\pi}{2}\right]$.

67. $\frac{a}{\sin A} = \frac{b}{\sin B}$

$b = \frac{a \sin B}{\sin A}$

68. $b = \frac{a \sin B}{\sin A} = a \frac{\sin B}{\sin A}$

Since $\frac{\sin B}{\sin A} < 1$

$b = a \frac{\sin B}{\sin A} < a \cdot 1 = a,$

so $b < a$

69. If $B < A$ then $b < a$, but $b > a$ in triangle ABC, where $A = 83°$, $a = 14,\ b = 20$.

71. We must find the length of CD.

$\phi = 35° - 30° = 5°;\ \alpha = 30°;$

$\beta = 30°$, side $AB = 5000$ ft.

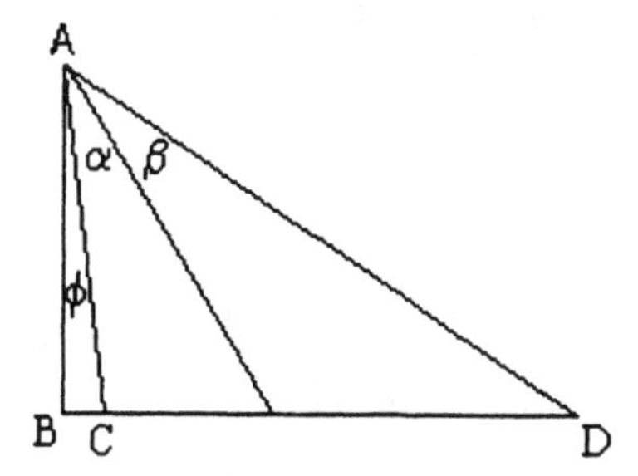

Since $\triangle\ ABC$ is a right triangle:

$$\cos 5° = \frac{5000°}{AC}$$

$$AC = \frac{5000}{\cos 5°} \approx 5019 \text{ ft.}$$

The angular coverage of the lens is $60°$, so angle $CAD = 60°$. From geometry,

$\measuredangle\ ACB = 85°\ (180° - (90° + 5°))$

$\measuredangle\ ACD = 95°\ (180° - 85°)$

$\measuredangle\ ADB = 25°\ (180° - (60° + 95°))$

We use the law of sines to solve for the length of CD:

$$\frac{CD}{\sin 60°} = \frac{AC}{\sin 25°}$$

$$CD = \frac{5019 \sin 60°}{\sin 25°} \approx 10,285 \text{ ft}$$

The photograph would cover a horizontal distance of

$\boxed{\approx 10,285 \text{ ft}}$ (or ≈ 1.95 miles)

Section 10.2

1. a, b, C : **(a)** SAS, **(b)** law of cosines

3. a, b, A : **(a)** SSA, **(b)** law of sines

5. A, B, c : **(a)** ASA, **(b)** law of sines

7. a, B, C : **(a)** ASA, **(b)** law of sines

9. $A = 61°$, $b = 4$, $c = 6$

$$a^2 = b^2 + c^2 - 2bc\cos A$$
$$= 4^2 + 6^2 - 2(4)(6)\cos 61°$$
$$\approx 28.7291382282$$

$a \approx 5.3599569241$ $\boxed{a \approx 5.4}$

$$\frac{5.3599569241}{\sin 61°} = \frac{4}{\sin B}$$

$\sin B \approx .652706519493$ $\boxed{B \approx 40.7°}$

$C \approx 180° - (61° + 40.7°)$ $\boxed{C \approx 78.3°}$

11. $a = 4$, $b = 10$, $c = 8$

$$\cos B = \frac{a^2 + c^2 - b^2}{2ac}$$
$$= \frac{4^2 + 8^2 - 10^2}{2(4)(8)} \approx -.3125$$

$B \approx 108.209957°$ $\boxed{B \approx 108.2°}$

$$\frac{10}{\sin 108.209957°} = \frac{4}{\sin A}$$

$\sin A \approx .379967104$ $\boxed{A \approx 22.3°}$

$C \approx 180° - (108.2° + 22.3°)$ $\boxed{C \approx 49.5°}$

13. $a = 5$, $b = 7$, $c = 9$

$$\cos C = \frac{a^2 + b^2 - c^2}{2ab}$$
$$= \frac{5^2 + 7^2 - 9^2}{2(5)(7)} = -.1$$

$C \approx 95.73917048°$ $\boxed{C \approx 95.7°}$

$$\frac{9}{\sin 95.73917048°} = \frac{5}{\sin A}$$

$\sin A \approx .552770798$ $\boxed{A \approx 33.6°}$

$B \approx 180° - (95.7° + 33.6°)$ $\boxed{B \approx 50.7°}$

15. $C = 28.3°$, $b = 5.71$ in, $a = 4.21$ in

$$c^2 = a^2 + b^2 - 2ab\cos C$$
$$= (4.21)^2 + (5.71)^2 - 2(4.21)(5.71)\cos 28.3°$$
$$\approx 7.99643$$

$c \approx 2.8278$ $\boxed{c \approx 2.83 \text{ inches}}$

$$\frac{c}{\sin C} = \frac{a}{\sin A}$$
$$\frac{2.8278}{\sin 28.3°} = \frac{4.21}{\sin A}$$

$\sin A \approx 0.70582$ $\boxed{A \approx 44.9°}$

$B = 180° - (A + C)$ $\boxed{B \approx 106.8°}$

17. $C = 45.6°$, $b = 8.94$ m, $a = 7.23$ m

$$c^2 = a^2 + b^2 - 2ab\cos C$$
$$c^2 = 7.23^2 + 8.94^2 - 2(7.23)(8.94)\cos 45.6°$$
$$\approx 41.74934$$

$c \approx 6.461373$ $\boxed{c \approx 6.46 \text{ m}}$

$$\frac{c}{\sin C} = \frac{a}{\sin A}$$
$$\frac{6.461373}{\sin 45.6°} = \frac{7.23}{\sin A}$$

$\sin A \approx 0.79946$ $\boxed{A \approx 53.1°}$

$B = 180° - (A + C)$ $\boxed{B \approx 81.3°}$

19. $a = 9.3$ cm, $b = 5.7$ cm, $c = 8.2$ cm; largest $\angle : A$

$$\cos A = \frac{b^2 + c^2 - a^2}{2bc}$$
$$= \frac{5.7^2 + 8.2^2 - 9.3^2}{2(5.7)(8.2)} \approx 0.141635$$

$A \approx 81.857°$ or $81°51'$ $\boxed{A \approx 82°}$

$$\frac{a}{\sin A} = \frac{b}{\sin B}$$
$$\frac{9.3}{\sin 82°} = \frac{5.7}{\sin B}$$

$\sin B \approx 0.60694$

$B \approx 37.37°$ or $37°22'$ $\boxed{B \approx 37°}$

$C = 180° - (A + B)$ $\boxed{C \approx 61°}$

21. $a = 42.9$ m, $b = 37.6$ m, $c = 62.7$ m, largest $\angle : C$

$$\cos C = \frac{a^2 + b^2 - c^2}{2ab}$$
$$= \frac{42.9^2 + 37.6^2 - 62.7^2}{2\,(42.9)\,(37.6)} \approx -0.209889$$

$C \approx 102.116°$ or $102°07'$ $\boxed{C \approx 102°10'}$

$$\frac{c}{\sin C} = \frac{b}{\sin B}$$
$$\frac{62.7}{\sin 102°10'} = \frac{37.6}{\sin B}$$
$\sin B \approx 0.58621$

$B \approx 35.889°$ or $35°53'$ $\boxed{B \approx 35°50'}$

$A = 180° - (B + C)$ $\boxed{A \approx 42°00'}$

23. $AB = 1240$ ft, $AC = 876$ ft, $BC = 965$ ft, largest $\angle : C$

$$\cos C = \frac{AC^2 + BC^2 - AB^2}{2\,(AC)\,(BC)}$$
$$= \frac{876^2 + 965^2 - 1240^2}{2\,(876)\,(965)} \approx 0.09523$$

$C \approx 84.536°$ or $84°32'$ $\boxed{C \approx 84°30'}$

$$\frac{AB}{\sin C} = \frac{AC}{\sin B}$$
$$\frac{1240}{\sin 84°30'} = \frac{876}{\sin B}$$
$\sin B \approx 0.703199$

$B \approx 44.684°$ or $44°41'$ $\boxed{B \approx 44°40'}$

$A = 180° - (B + C)$ $\boxed{A \approx 50°50'}$

25. $A = 80°\,40'$, $b = 143$ cm, $c = 89.6$ cm

$$a^2 = b^2 + c^2 - 2bc\cos A$$
$$= 143^2 + 89.6^2 - 2\,(143)\,(89.6)\cos 80°\,40'$$
$$\approx 24321.2534137$$

$a \approx 155.95273$ $\boxed{a \approx 156 \text{ cm}}$

$$\frac{a}{\sin A} = \frac{b}{\sin B}$$
$$\frac{155.95273}{\sin 80°\,40'} = \frac{143}{\sin B}$$
$\sin B \approx .9048056$

$B \approx 64.797°$ or $64°48'$ $\boxed{B \approx 64°\,50'}$

$C = 180° - (A + B)$ $\boxed{C \approx 34°\,30'}$

27. $B = 74.80°$, $a = 8.919$ in, $c = 6.427$ in

$$b^2 = a^2 + c^2 - 2ac\cos B$$
$$= 8.919^2 + 6.427^2 - 2\,(8.919)\,(6.427)\cos 74.80°$$
$$\approx 90.79625724$$

$b \approx 9.5287$ $\boxed{b \approx 9.529 \text{ in}}$

$$\frac{b}{\sin B} = \frac{a}{\sin A}$$
$$\frac{9.529}{\sin 74.80°} = \frac{8.919}{\sin A}$$
$\sin A \approx .9032408557$

$A \approx 64.587°$ $\boxed{A \approx 64.59°}$

$C = 180° - (A + B)$ $\boxed{C \approx 40.61°}$

29. $A = 112.8°$, $b = 6.28$ m, $c = 12.2$ m

$$a^2 = b^2 + c^2 - 2bc\cos A$$
$$= 6.28^2 + 12.2^2 - 2\,(6.28)\,(12.2)\cos 112.8°$$
$$\approx 247.6581883$$

$a \approx 15.73715$ $\boxed{a \approx 15.7 \text{ m}}$

$$\frac{a}{\sin A} = \frac{b}{\sin B}$$
$$\frac{15.73715}{\sin 112.8°} = \frac{6.28}{\sin B}$$
$\sin B \approx .3678747799$

$B \approx 21.584°$ $\boxed{B \approx 21.6°}$

$C = 180° - (A + B)$ $\boxed{C \approx 45.6°}$

31. $a = 3.0$ ft, $b = 5.0$ ft, $c = 6.0$ ft

$$\cos C = \frac{a^2 + b^2 - c^2}{2ab}$$
$$= \frac{3.0^2 + 5.0^2 - 6.0^2}{2\,(3.0)\,(5.0)} = -.06666\ldots$$

$C \approx 93.82°$ $\boxed{C \approx 94°}$

$$\frac{6.0}{\sin 94°} = \frac{3.0}{\sin A}$$
$\sin A \approx .498782025$

$A \approx 29.92°$ $\boxed{A \approx 30°}$

$B \approx 180° - (94° + 30°)$

$\boxed{B \approx 56°}$

33. The value of cosine θ will be greater than one; your calculator will give you an error message (or a complex number) when using the inverse cosine function.

35. $AB^2 = AC^2 + BC^2 - 2(AC)(BC)\cos C$

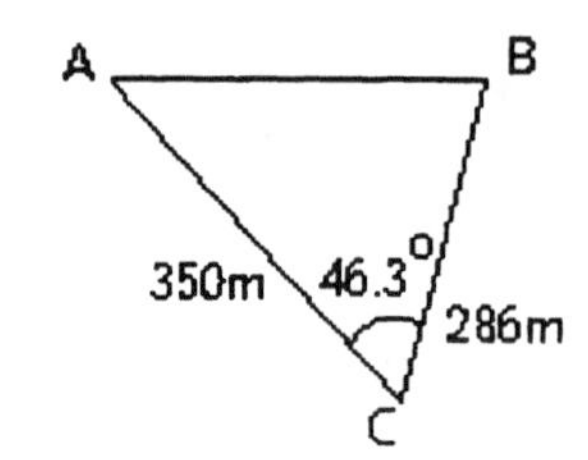

$$AB^2 = 350^2 + 286^2 - 2(350)(286)\cos 46.3°$$
$$\approx 65981.3413$$

$AB \approx 256.8683$ $\boxed{AB \approx 257 \text{ m}}$

37. Find AC (b) in the following triangle:

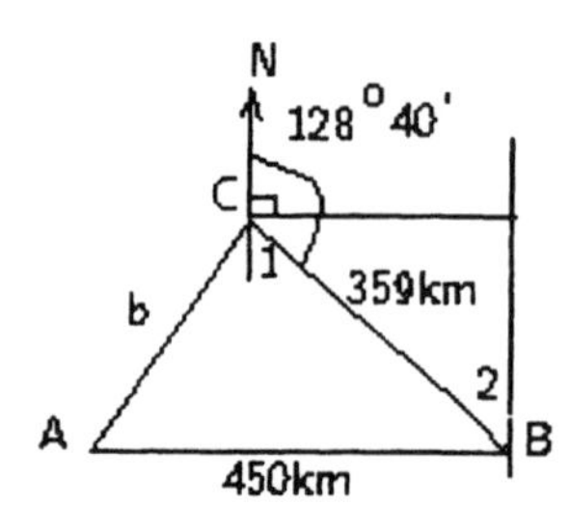

Angle 1 = 51°20′ (180° − 128°40′)
Angle 2 = Angle 1
Angle ABC = 38°40′ (90° − Angle 2)

$$b^2 = a^2 + c^2 - 2ac\cos B$$
$$= 359^2 + 450^2 - 2(359)(450)\cos 38°40'$$
$$\approx 79106.449981$$

$b \approx 281.2587$ $\boxed{b \approx 281 \text{ km}}$

39. A triangle is formed with angles 44°40′ (90° − 45°20′) and 38°40′ (308°40′ − 270°); the side between them is 15.2 miles.

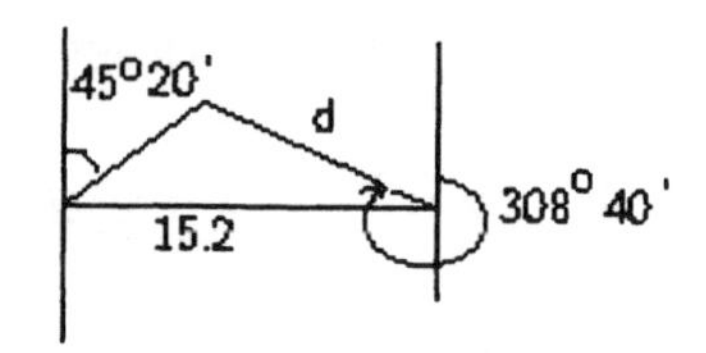

Find the 3rd angle:

$$180° - 44°40' - 38°40' = 96°40'$$

Find the distance d:

$$\frac{15.2}{\sin 96°40'} = \frac{d}{\sin 44°40'}$$

$d \approx 10.758$ $\boxed{d \approx 10.8 \text{ miles}}$

41. After 3 hours, the distances traveled are 108.6 km and 136.8 km.

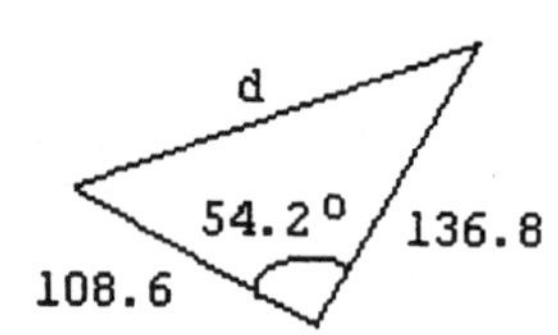

$$d^2 = 108.6^2 + 136.8^2 - 2(108.6)(136.8)\cos 54°10'$$
$$\approx 13113.35865$$

$d \approx 114.5136$ $\boxed{d \approx 115 \text{ km}}$

43. $d^2 = 10^2 + 10^2 - 2(10)(10)\cos 128°$
$$\approx 323.1322951$$

$d \approx 17.97588$ $\boxed{d \approx 18 \text{ feet}}$

45. $L^2 = 3800^2 + 2900^2 - 2(3800)(2900)\cos 110°$
$$\approx 30388123.96$$

$L \approx 5512.5424$ $\boxed{L \approx 5500 \text{ m}}$

47. Find d in the following triangle:

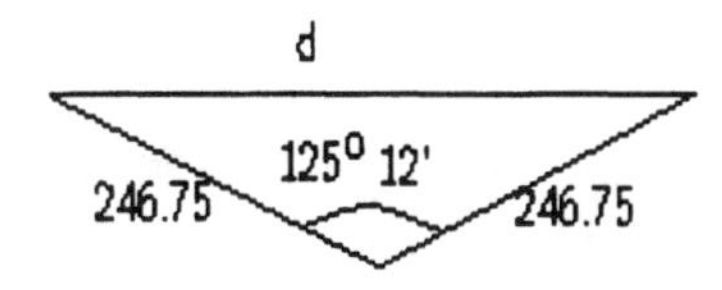

$$d^2 = 246.75^2 + 246.75^2 - 2(246.75)(246.75)\cos 125°12'$$
$$\approx 191963.9366$$
$$d \approx 438.13689 \qquad \boxed{d \approx 438.14 \text{ feet}}$$

49. $$\cos\alpha = \frac{17^2 + 21^2 - 9^2}{2(17)(21)}$$
$$\approx 0.90896359$$
$$\alpha \approx 24.6375° \quad (24°38')$$
α to nearest degree: $25°$

Bearing from A: $325° + 25° = \boxed{350°}$

51. Let x equal the distance from the tracking station to the satellite at 12:03 pm. The distance from the center of the earth to the satellite is 8000 km $(6400 + 1600)$. Let θ be the angle made by the satellite from noon to 12:03 pm; find the angle in radians:

$$\frac{2\pi}{2 \text{ hrs}} = \frac{\theta}{3 \text{ min}}$$
$$\frac{2\pi}{120} = \frac{\theta}{3}$$
$$\theta = \frac{3(2\pi)}{120} = \frac{\pi}{20}$$
$$x^2 = 6400^2 + 8000^2 - 2(6400)(8000)\cos\frac{\pi}{20}$$
$$\approx 3820713.923$$
$$x \approx 1954.6647 \qquad \boxed{x \approx 2000 \text{ km}}$$

53. Find $\angle 1$ in the parallelogram:

$$\cos\theta = \frac{32.5^2 + 25.9^2 - 57.8^2}{2(32.5)(25.9)}$$
$$\approx -0.9585863$$
$$\theta \approx 163.453° \qquad \boxed{\theta \approx 163.5°}$$

55. $$x^2 = 25^2 + 25^2 - 2(25)(25)\cos 52°$$
$$\approx 480.423155843$$
$$x \approx 21.9185573 \qquad \boxed{x \approx 22 \text{ feet}}$$

Relating Concepts (Exercises 57 - 60)

57. Since A is obtuse, $90° < A < 180°$. The cosine of a quadrant II angle is negative.

58. $a^2 = b^2 + c^2 - 2bc\cos A$. Since $\cos A < 0$,
$a^2 = b^2 + c^2 +$ some positive quantity.
Thus, $a^2 > b^2 + c^2$.

59. $a^2 > b^2 + c^2$, a, b, c are positive.
Since $b^2 + c^2 > b^2 \Longrightarrow a^2 > b^2 \Longrightarrow a > b$.
Since $b^2 + c^2 > c^2 \Longrightarrow a^2 > c^2 \Longrightarrow a > c$.

60. $A = 103°$, $a = 25$, $c = 30$.
Since A is obtuse, a must be the longest side.

61. $A = 42.5°$, $b = 13.6\,\text{m}$, $c = 10.1\,\text{m}$

$$Area = \frac{1}{2}bc\sin A$$
$$= \frac{1}{2}(13.6)(10.1)\sin 42.5° \approx 46.3995$$
$$\boxed{\text{Area} \approx 46.4 \text{ m}^2}$$

63. $a = 12$ m, $b = 16$ m, $c = 25$ m

(i) $$s = \frac{1}{2}(a + b + c)$$
$$= \frac{1}{2}(12 + 16 + 25) = 26.5$$

(ii) $$s - a = 26.5 - 12 = 14.5$$
$$s - b = 26.5 - 16 = 10.5$$
$$s - c = 26.5 - 25 = 1.5$$

(iii) $$Area = \sqrt{s(s-a)(s-b)(s-c)}$$
$$= \sqrt{26.5(14.5)(10.5)(1.5)}$$
$$= \sqrt{6051.9375} \approx 77.7942$$
$$\boxed{\text{Area} \approx 78 \text{ m}^2}$$

65. $a = 76.3$ ft, $b = 109$ ft, $c = 98.8$ ft

(i) $s = \frac{1}{2}(a + b + c)$

$= \frac{1}{2}(76.3 + 109 + 98.8) = 142.05$

(ii) $s - a = 142.05 - 76.3 = 65.75$

$s - b = 142.05 - 109 = 33.05$

$s - c = 142.05 - 98.8 = 43.25$

(iii) $Area = \sqrt{s(s-a)(s-b)(s-c)}$

$= \sqrt{142.05(65.75)(33.05)(43.25)}$

$= \sqrt{13350409} \approx 3653.82115$

$\boxed{\text{Area} \approx 3650 \text{ ft}^2}$

67. $a = 25.4$ yd, $b = 38.2$ yd, $c = 19.8$ yd

(i) $s = \frac{1}{2}(a + b + c)$

$= \frac{1}{2}(25.4 + 38.2 + 19.8) = 41.7$

(ii) $s - a = 41.7 - 25.4 = 16.3$

$s - b = 41.7 - 38.2 = 3.5$

$s - c = 41.7 - 19.8 = 21.9$

(iii) $Area = \sqrt{s(s-a)(s-b)(s-c)}$

$= \sqrt{41.7(16.3)(3.5)(21.9)}$

$= \sqrt{52099.7715} \approx 228.2537$

$\boxed{\text{Area} \approx 228 \text{ yd}^2}$

69. $Area = \frac{1}{2}bc\sin A$

$= \frac{1}{2}(16.1)(15.2)\sin 125° \approx 100.23144$

$\boxed{\text{Area} \approx 100 \text{ m}^2}$

71. (i) $s = \frac{1}{2}(a + b + c)$

$= \frac{1}{2}(75 + 68 + 85) = 114$

(ii) $s - a = 114 - 75 = 39$

$s - b = 114 - 68 = 46$

$s - c = 114 - 85 = 29$

(iii) $Area = \sqrt{s(s-a)(s-b)(s-c)}$

$= \sqrt{114(39)(46)(29)}$

$= \sqrt{5930964} \approx 2435.357$

(iv) If each can covers 75 m^2 :

$\frac{2435.357}{75} = 32.47$

$\boxed{\text{33 cans of paint will be needed.}}$

73. Sides of length 9, 10, 17.

$P = 9 + 10 + 17 = 36$

$A = \sqrt{18(18-9)(18-10)(18-17)}$

$= \sqrt{18(9)(8)(1)} = \sqrt{1296} = 36$

Relating Concepts (75 – 82):

75. $P(2, 5)$, $Q(-1, 3)$, $R(4, 0)$:

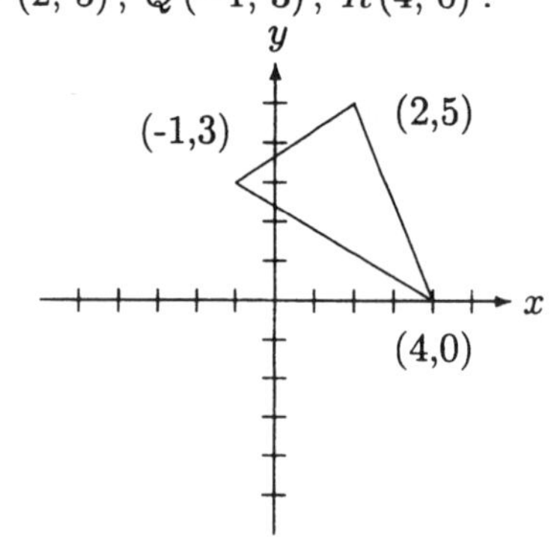

76. (i) $P(2, 5),\ Q(-1, 3)$:

$$PQ = \sqrt{(2+1)^2 + (5-3)^2} = \sqrt{9+4} = \boxed{\sqrt{13}}$$

(ii) $Q(-1, 3),\ R(4, 0)$:

$$QR = \sqrt{(-1-4)^2 + (3-0)^2} = \sqrt{25+9} = \boxed{\sqrt{34}}$$

(iii) $P(2, 5),\ R(4, 0)$:

$$PR = \sqrt{(2-4)^2 + (5-0)^2} = \sqrt{4+25} = \boxed{\sqrt{29}}$$

77. $s = \dfrac{1}{2}(a+b+c) = \dfrac{\sqrt{13}+\sqrt{34}+\sqrt{29}}{2}$

78. $s = \dfrac{\sqrt{13} + \sqrt{34} + \sqrt{29}}{2}$

(i)
$$s - a = \frac{\sqrt{13} + \sqrt{34} + \sqrt{29}}{2} - \frac{2\sqrt{13}}{2} = \frac{-\sqrt{13} + \sqrt{34} + \sqrt{29}}{2}$$

$$s - b = \frac{\sqrt{13} + \sqrt{34} + \sqrt{29}}{2} - \frac{2\sqrt{34}}{2} = \frac{\sqrt{13} - \sqrt{34} + \sqrt{29}}{2}$$

$$s - c = \frac{\sqrt{13} + \sqrt{34} + \sqrt{29}}{2} - \frac{2\sqrt{29}}{2} = \frac{\sqrt{13} + \sqrt{34} - \sqrt{29}}{2}$$

(ii) $s(s-a)$

$$= \frac{\sqrt{13} + \sqrt{34} + \sqrt{29}}{2} \cdot \frac{-\sqrt{13} + \sqrt{34} + \sqrt{29}}{2}$$
$$\approx 28.20031847$$

$(s-b)(s-c)$

$$= \frac{\sqrt{13} - \sqrt{34} + \sqrt{29}}{2} \cdot \frac{\sqrt{13} + \sqrt{34} - \sqrt{29}}{2}$$
$$\approx 3.200318468$$

(iii)
$$Area = \sqrt{s(s-a)(s-b)(s-c)} = \sqrt{28.20032 \times 3.20032} = \sqrt{90.25} = \boxed{9.5\ (\text{units}^2)}$$

79.
$$Area = \frac{1}{2}\det\begin{bmatrix} 2 & 5 & 1 \\ -1 & 3 & 1 \\ 4 & 0 & 1 \end{bmatrix} = \frac{1}{2}(19) = \boxed{9.5\ (\text{units}^2)}$$
Yes

80.
$$\cos PQR = \frac{PQ^2 + QR^2 - PR^2}{2(PQ)(QR)} = \frac{(\sqrt{13})^2 + (\sqrt{34})^2 - (\sqrt{29})^2}{2(\sqrt{13})(\sqrt{34})} \approx 0.4280863447$$

$PQR \approx 64.65382406$ $\quad \boxed{PQR \approx 64.65^\circ}$

81.
$$Area = \frac{1}{2}ab\sin C = \frac{1}{2}\left(\sqrt{13}\cdot\sqrt{34}\cdot\sin 64.65^\circ\right) \approx 9.4996996 \approx \boxed{9.5\ \text{units}^2}$$
Yes

82. Given three distinct points, there are at least three different ways to calculate the area of the triangle formed; they must lead to the same answer, because the area is unique.

Section 10.3

1. Equal vectors: m and p, n and r

3. Positive scalar multiples:

$m = 2t,\ p = 2t,\ t = \frac{1}{2}m,$

$t = \frac{1}{2}p,\ m = 1p,\ n = 1r$

5. $-b$:

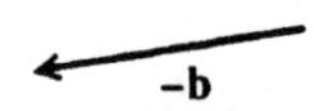

7. $3a$:

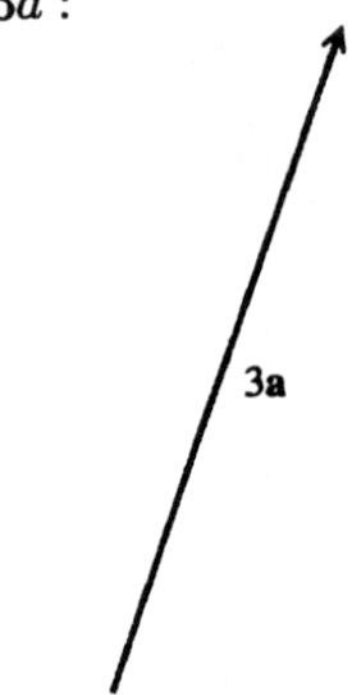

9. $a + b$

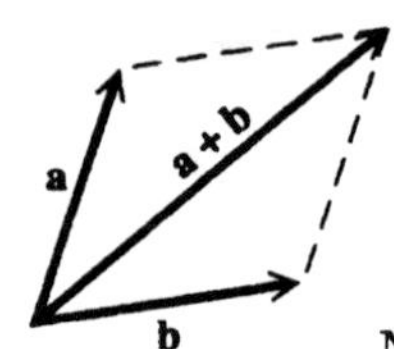

New picture

11. $a - c$:

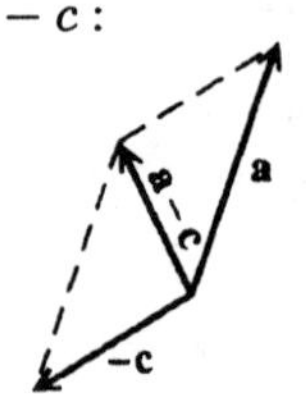

13. $a + (b + c)$:

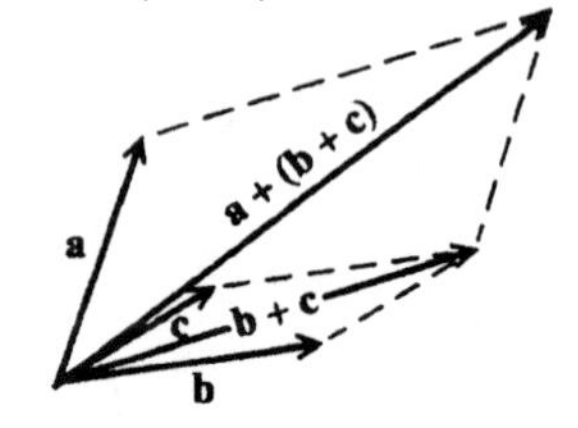

15. $c + d$:

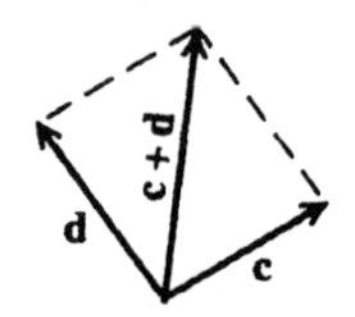

17. (a) $\mathbf{a}+\mathbf{b} = \langle -8,\ 8\rangle + \langle 4,\ 8\rangle = \langle -4,\ 16\rangle$

(b) $\mathbf{a}-\mathbf{b} = \langle -8,\ 8\rangle - \langle 4,\ 8\rangle = \langle -12,\ 0\rangle$

(c) $-\mathbf{a} = \langle 8,\ -8\rangle$

19. (a) $\mathbf{a}+\mathbf{b} = \langle 4,\ 8\rangle + \langle 4,\ -8\rangle = \langle 8,\ 0\rangle$

(b) $\mathbf{a}-\mathbf{b} = \langle 4,\ 8\rangle - \langle 4,\ -8\rangle = \langle 0,\ 16\rangle$

(c) $-\mathbf{a} = \langle -4,\ -8\rangle$

21. (a) $\mathbf{a}+\mathbf{b} = \langle -8,\ 4\rangle + \langle 8,\ 8\rangle = \langle 0,\ 12\rangle$

(b) $\mathbf{a}-\mathbf{b} = \langle -8,\ 4\rangle - \langle 8,\ 8\rangle = \langle -16,\ 4\rangle$

(c) $-\mathbf{a} = \langle 8,\ -4\rangle$

23. $a = 2i,\ b = i + j$

(a) $2\mathbf{a} = 2(2i) = 4i$

(b) $2\mathbf{a}+3\mathbf{b} = 4i + 3(i+j)$
$= 4i + 3i + 3j = 7i + 3j$

(c) $\mathbf{b}-3\mathbf{a} = i + j - 3(2i)$
$= i + j - 6i = -5i + j$

25. $a = \langle -1,\ 2\rangle,\ b = \langle 3,\ 0\rangle$

(a) $2\mathbf{a} = 2\langle -1,\ 2\rangle = \langle -2,\ 4\rangle$

(b) $2\mathbf{a}+3\mathbf{b} = \langle -2,\ 4\rangle + 3\langle 3,\ 0\rangle$
$= \langle -2,\ 4\rangle + \langle 9,\ 0\rangle = \langle 7,\ 4\rangle$

(c) $\mathbf{b}-3\mathbf{a} = \langle 3,\ 0\rangle - 3\langle -1,\ 2\rangle$
$= \langle 3,\ 0\rangle + \langle 3,\ -6\rangle = \langle 6,\ -6\rangle$

27. $\mathbf{u} = \langle -2,\ 5\rangle,\ \mathbf{v} = \langle 4,\ 3\rangle$

$\mathbf{u}+\mathbf{v} = \langle -2+4,\quad 5+3\rangle = \boxed{\langle 2,\ 8\rangle}$

29. $\mathbf{u} = \langle -2,\ 5\rangle,\ \mathbf{v} = \langle 4,\ 3\rangle$

$\mathbf{v}-\mathbf{u} = \langle 4-(-2),\quad 3-5\rangle$
$= \boxed{\langle 6,\ -2\rangle}$

31. $\mathbf{u} = \langle -2,\ 5\rangle,\ \mathbf{v} = \langle 4,\ 3\rangle$

$$-5\mathbf{v} = \langle -5\cdot 4,\quad -5\cdot 3\rangle = \boxed{\langle -20,\ -15\rangle}$$

33. $|v| = 4,\ \theta = 40°$

horizontal: $x = |v|\cos\theta$

$x = 4\cos 40° \approx \boxed{3.06}$

vertical: $y = |v|\sin\theta$

$y = 4\sin 40° \approx \boxed{2.57}$

35. $|v| = 5,\ \theta = -35°$

horizontal: $x = |v|\cos\theta$

$x = 5\cos(-35°) \approx \boxed{4.10}$

vertical: $y = |v|\sin\theta$

$y = 5\sin(-35°) \approx \boxed{-2.87}$

37. Adjacent angle $= 180° - 40° = 140°$

$|v^2| = 60^2 + 40^2 - 2(60)(40)\cos 140°$

$|v^2| \approx 8877.013$

$|v| \approx \boxed{94.2 \text{ pounds}}$

39. Adjacent angle $= 180° - 110° = 70°$

$|v^2| = 15^2 + 25^2 - 2(15)(25)\cos 70°$

$|v^2| \approx 593.485$

$|v| \approx \boxed{24.4 \text{ pounds}}$

Use the following definition for 41– 47:

If $\mathbf{v} = \langle a,\ b\rangle$, then $\mathbf{v} = ai + bj$

41. $\mathbf{v} = \langle -5,\ 8\rangle \rightarrow \mathbf{v} = \boxed{-5i + 8j}$

43. $\mathbf{v} = \langle 2,\ 0\rangle \rightarrow \mathbf{v} = 2i + 0j = \boxed{2i}$

45. $|v| = 8,\ \theta = 45°$

$$x = |v|\cos\theta = 8\cos 45° = 8\left(\tfrac{\sqrt{2}}{2}\right) = 4\sqrt{2}$$

$$y = |v|\sin\theta = 8\sin 45° = 8\left(\tfrac{\sqrt{2}}{2}\right) = 4\sqrt{2}$$

$$\mathbf{v} = \langle 4\sqrt{2},\ 4\sqrt{2}\rangle = \boxed{4\sqrt{2}i + 4\sqrt{2}j}$$

47. $|v| = .6,\ \theta = 115°$

$$x = |v|\cos\theta = .6\cos 115° \approx -.25357 \approx -.254$$

$$y = |v|\sin\theta = .6\sin 115° \approx .54378 \approx .544$$

$$\mathbf{v} = \langle -.254,\ .544\rangle = \boxed{-.254i + .544j}$$

49. Let $\mathbf{u} = \langle 1,\ 1\rangle$ (Quadrant I)

Magnitude:

$$|u| = \sqrt{1^2 + 1^2} = \boxed{\sqrt{2}}$$

Direction angle:

$$\cos\theta = \frac{a}{|u|} = \frac{1}{\sqrt{2}}$$

$$\left[\text{or}\quad \tan\theta = \frac{1}{1}\right]\quad \theta = \boxed{45°}$$

51. Let $\mathbf{u} = \langle 8\sqrt{2},\ -8\sqrt{2}\rangle$ (Quadrant IV)

Magnitude:

$$|u| = \sqrt{\left(8\sqrt{2}\right)^2 + \left(-8\sqrt{2}\right)^2} = \sqrt{128 + 128} = \sqrt{256} = \boxed{16}$$

Direction angle:

$$\cos\theta = \frac{a}{|u|} = \frac{8\sqrt{2}}{16} = \frac{\sqrt{2}}{2}$$

$\widehat{\theta} = 45°$, Quadrant IV

$\theta = \boxed{315°}$

53. Let $\mathbf{u} = \langle 15,\ -8\rangle$ (Quadrant IV)

Magnitude:

$$|u| = \sqrt{15^2 + (-8)^2} = \sqrt{225 + 64} = \sqrt{289} = \boxed{17}$$

Direction angle:

$$\cos\theta = \frac{a}{|u|} = \frac{15}{17}$$

$\widehat{\theta} \approx 28.072°$, Quadrant IV

$\theta \approx \boxed{331.9°}$

55. Let $\mathbf{u} = \langle -6,\ 0\rangle$ (Negative x-axis)

Magnitude:

$$|u| = \sqrt{(-6)^2 + 0^2} = \sqrt{36+0} = \boxed{6}$$

Direction angle:

$$\cos\theta = \frac{a}{|u|} = \frac{-6}{6} = -1$$

$$\theta = \boxed{180^\circ}$$

57. $\langle 6,\ -1\rangle \cdot \langle 2,\ 5\rangle$

$$= (6)(2) + (-1)(5) = 12 - 5 = \boxed{7}$$

59. $\langle 2,\ -3\rangle \cdot \langle 6,\ 5\rangle$

$$= (2)(6) + (-3)(5) = 12 - 15 = \boxed{-3}$$

61. $\langle 4,\ 0\rangle \cdot \langle 5,\ -9\rangle$

$$= (4)(5) + (0)(-9) = 20 + 0 = \boxed{20}$$

63. $\mathbf{u} = \langle 2,\ 1\rangle,\ \mathbf{v} = \langle -3,\ 1\rangle$

$$\cos\theta = \frac{\mathbf{u}\cdot\mathbf{v}}{|u|\,|v|} = \frac{\langle 2,\ 1\rangle\cdot\langle -3,\ 1\rangle}{|\langle 2,\ 1\rangle|\,|\langle -3,\ 1\rangle|} = \frac{2(-3)+1(1)}{\sqrt{4+1}\sqrt{9+1}} = \frac{-6+1}{\sqrt{5}\sqrt{10}} = \frac{-5}{5\sqrt{2}} \times \frac{\sqrt{2}}{\sqrt{2}} = \frac{-\sqrt{2}}{2}$$

$$\theta = \cos^{-1}\left(\frac{-\sqrt{2}}{2}\right) \qquad \boxed{\theta = 135^\circ}$$

65. $\mathbf{u} = \langle 1,\ 2\rangle,\ \mathbf{v} = \langle -6,\ 3\rangle$

$$\cos\theta = \frac{\mathbf{u}\cdot\mathbf{v}}{|u|\,|v|} = \frac{\langle 1,\ 2\rangle\cdot\langle -6,\ 3\rangle}{|\langle 1,\ 2\rangle|\,|\langle -6,\ 3\rangle|} = \frac{1(-6)+2(3)}{\sqrt{1+4}\sqrt{36+9}} = \frac{-6+6}{\sqrt{5}\sqrt{45}} = \frac{0}{\sqrt{225}} = 0$$

$$\theta = \cos^{-1} 0 \qquad \boxed{\theta = 90^\circ}$$

67. $\mathbf{u} = \langle 3,\ 4\rangle,\ \mathbf{v} = \langle 0,\ 1\rangle$

$$\cos\theta = \frac{\mathbf{u}\cdot\mathbf{v}}{|u|\,|v|} = \frac{\langle 3,\ 4\rangle\cdot\langle 0,\ 1\rangle}{|\langle 3,\ 4\rangle|\,|\langle 0,\ 1\rangle|} = \frac{3(0)+4(1)}{\sqrt{9+16}\sqrt{0+1}} = \frac{0+4}{\sqrt{25}\sqrt{1}} = \frac{4}{5(1)} = \frac{4}{5}$$

$$\theta = \cos^{-1}\left(\frac{4}{5}\right) \approx 36.869898^\circ \qquad \boxed{\theta \approx 36.87^\circ}$$

69. $\mathbf{u} = \langle -2,\ 1\rangle,\ \mathbf{v} = \langle 3,\ 4\rangle$

$$(3\mathbf{u})\cdot\mathbf{v} = \langle -6,\ 3\rangle\cdot\langle 3,\ 4\rangle = (-6)(3) + (3)(4) = -18 + 12 = \boxed{-6}$$

71. $\mathbf{u} = \langle -2,\ 1\rangle,\ \mathbf{v} = \langle 3,\ 4\rangle,$ $\mathbf{w} = \langle -5,\ 12\rangle$

$\mathbf{u}\cdot\mathbf{v} - \mathbf{u}\cdot\mathbf{w}$

$$= \langle -2,\ 1\rangle\cdot\langle 3,\ 4\rangle - \langle -2,\ 1\rangle\cdot\langle -5,\ 12\rangle = [(-2)(3) + 1(4)] - [(-2)(-5) + 1(12)] = [-6+4] - [10+12] = -2 - 22 = \boxed{-24}$$

73. $\langle 1,\ 2\rangle \cdot \langle -6,\ 3\rangle$
$= (1)(-6) + (2)(3)$
$= -6 + 6 = 0$

Orthogonal

75. $\langle 1,\ 0\rangle \cdot \langle \sqrt{2},\ 0\rangle$
$= (1)(\sqrt{2}) + (0)(0) = \sqrt{2}$

Not Orthogonal

77. $\langle \sqrt{5},\ -2\rangle \cdot \langle -5,\ 2\sqrt{5}\rangle$
$= (\sqrt{5})(-5) + (-2)(2\sqrt{5})$
$= -5\sqrt{5} - 4\sqrt{5} = -9\sqrt{5}$

Not Orthogonal

79. The angle between the wind and the plane bearing is
88.7° (175.3° − (266.6° − 180°)),
the other angle of the parallelogram formed is 91.3°. To find the resulting bearing, add 86.6° (266.6° − 180°) to θ :

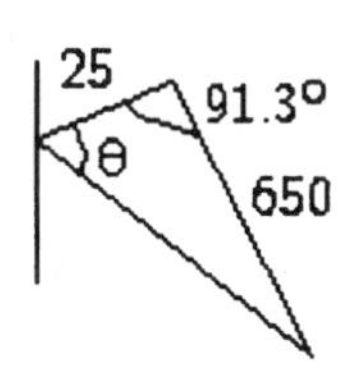

$$|v|^2 = 25^2 + 650^2 - 2(25)(650)\cos 91.3°$$
$$\approx 423862.3383$$
$$|v| \approx 651.047$$
$$\frac{650}{\sin\theta} = \frac{651.047}{\sin 91.3°}$$
$$\sin\theta \approx .99813484$$
$$\theta \approx 86.5000$$

Bearing: $\theta + 86.6° = 173.1000$

The resulting bearing is 173.1°.

81. The triangle formed has two sides of length 18.5 and 47.8; the angle where the ship turns is 52° :
$[(360° - 317°) + (189° - 180°)]$.
Let $|d|$ represent the distance vector:

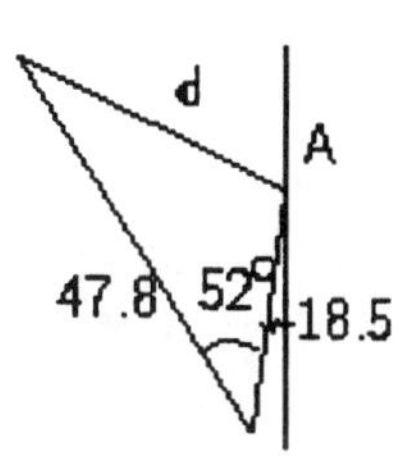

$$|d|^2 = 47.8^2 + 18.5^2 - 2(47.8)(18.5)\cos 52°$$
$$\approx 1538.2311$$
$$|d| \approx 39.22$$

The distance from point A is ≈ 39.2 km.

83. First, solve for the bearing:

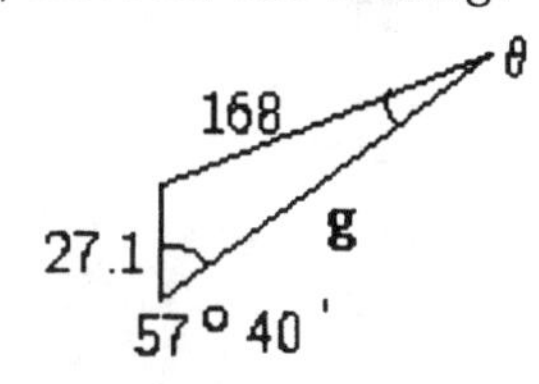

$$\frac{168}{\sin 57°40'} = \frac{27.1}{\sin\theta}$$
$$\sin\theta \approx .13629861$$
$$\theta \approx 7.8337° \approx 7°50'$$

$\angle G = 180° - (57°40' + 7°50') = 114°30'$

Bearing $= 180° - 114°30' = 65°30'$

Solve for the groundspeed, $|g|$:

$$\frac{168}{\sin 57°40'} = \frac{|g|}{\sin 114°30'}$$
$$|g| \approx 180.92591$$

Bearing: 65°30'
Groundspeed: 181 mph

85. The parallelogram formed has sides of length 23 and 192; the angles are 78° and 102° (180° − 78°).

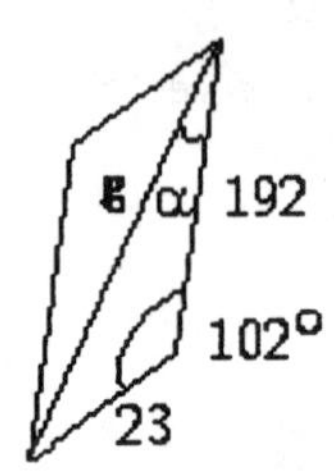

Find the groundspeed, $|g|$:

$$|g|^2 = 23^2 + 192^2 - 2(23)(192)\cos 102^\circ$$
$$\approx 39229.276$$
$$|g| \approx 198.0638 \approx 198$$

Find angle α :

$$\frac{\sin\alpha}{23} = \frac{\sin 102^\circ}{198}$$
$$\sin\alpha \approx .113623$$
$$\alpha \approx 6.524^\circ \approx 7^\circ$$

Bearing: $180^\circ + \alpha = 187^\circ$

Bearing = 187°
Groundspeed = 198 mph

87. Let **r** be the vertical component of the person exerting a 114-lb force and **s** be the vertical component of the person exerting a 150 lb force.

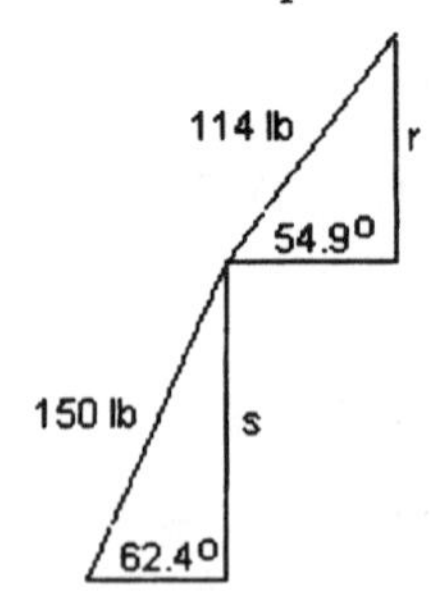

The weight of the box is the sum of the magnitudes of the vertical components of the two vectors representing the forces exerted by the two people.

$$|r| = 114\sin 54.9^\circ \approx 93.27$$
$$|s| = 150\sin 62.4^\circ \approx 132.93$$

Weight of the box $= |r| + |s| \approx 226$ lb.

89. Let $|x|$ = the weight of the boat.

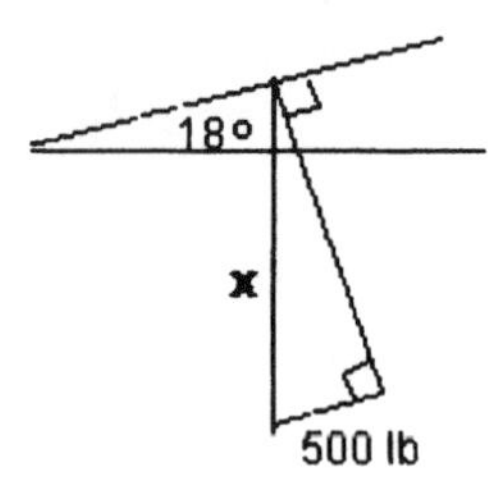

$$\frac{500}{|x|} = \sin 18^\circ$$
$$|x| = \frac{500}{\sin 18^\circ} \approx 1618$$
$$|x| \approx 1600 \text{ pounds}$$

91. Find the magnitude of the second force and of the resultant. Use the parallelogram rule: **v** is the resultant and **x** is the second force.

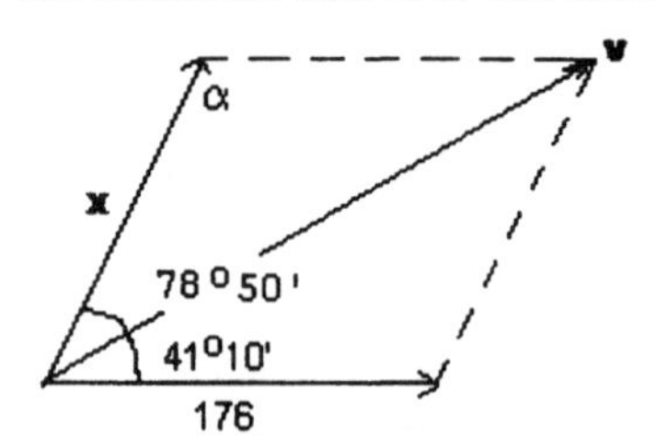

$\alpha = 180^\circ - 78^\circ\,50' = 101^\circ\,10'$.

The angle between the second force and the resultant is

$\beta = 78^\circ\,50' - 41^\circ\,10' = 37^\circ\,40'$.

Use the law of sines to find $|v|$.

$$\frac{|v|}{\sin\alpha} = \frac{176}{\sin\beta}$$
$$|v| = \frac{176\sin 101^\circ\,10'}{\sin 37^\circ\,40'} \approx 283 \text{ lb}$$

Use the law of sines to find $|x|$.

$$\frac{|x|}{\sin\alpha} = \frac{176}{\sin\beta}$$
$$|x| = \frac{176\sin 41^\circ\,10'}{\sin 37^\circ\,40'} \approx 190 \text{ lb}$$

Relating Concepts (93 – 98):

93. The parallelogram formed has sides of length 3 and 12, and angles of 150° (260° – 110°) and 30° (180° – 150°).

Find $\mathbf{u}+\mathbf{v}$:

$$(\mathbf{u}+\mathbf{v})^2 = 3^2+12^2-2\,(3)\,(12)\cos 30^\circ \approx 90.6461709275$$
$$\mathbf{u}+\mathbf{v} \approx 9.52082827$$

Find the $\angle$ cut by the diagonal:

$$\frac{\mathbf{u}+\mathbf{v}}{\sin 30^\circ} = \frac{3}{\sin\alpha}$$
$$\sin\alpha \approx .15754978$$
$$\alpha \approx 9.0646784^\circ$$

Direction $\angle$ of $\mathbf{u}+\mathbf{v} = 110^\circ + \alpha$

$$\boxed{\begin{array}{l}\text{Magnitude: } 9.52082827\\ \text{Direction } \angle: 119.0646784^\circ\end{array}}$$

94.
$$\begin{aligned}\mathbf{u} &= \langle r\cos\theta,\ r\sin\theta\rangle\\ &= \langle 12\cos 110^\circ,\ 12\sin 110^\circ\rangle\\ &= \boxed{\langle -4.10424172,\ 11.27631145\rangle}\end{aligned}$$

95.
$$\begin{aligned}\mathbf{v} &= \langle r\cos\theta,\ r\sin\theta\rangle\\ &= \langle 3\cos 260^\circ,\ 3\sin 260^\circ\rangle\\ &= \boxed{\langle -.520944533,\ -2.954423259\rangle}\end{aligned}$$

96. $\mathbf{u}+\mathbf{v}$
$$\begin{aligned}&= \langle -4.10424172+(-.520944533),\\ &\qquad 11.27631145+(-2.954423259)\rangle\\ &= \boxed{\langle -4.625186253,\ 8.321888191\rangle}\end{aligned}$$

97. Magnitude: $\sqrt{\mathbf{u}+\mathbf{v}}$

$$\sqrt{-4.625186253^2+8.32188191^2}$$
$$\sqrt{90.6461709384}$$
$$\boxed{9.52082827}$$

Direction angle:

$$\tan\theta = \frac{8.321888191}{-4.625186253}$$
$$\theta = \tan^{-1}\left(\frac{8.321888191}{-4.625186253}\right)$$
$$\theta = -60.9353216127^\circ$$

Quadrant II:

$$180^\circ + \theta = \boxed{119.0646784^\circ}$$

98. The answers in exercises **93** and **97** are the same; preference of method is an individual choice.

99. (a) $\mathbf{v} = 6\mathbf{i}+8\mathbf{j}$

$$\sqrt{6^2+8^2} = \sqrt{36+64} = \sqrt{100} = 10 \text{ mph}$$

(b) $3\mathbf{v} = 3\,(6\mathbf{i}+8\mathbf{j}) = 18\mathbf{i}+24\mathbf{j}$

$$\begin{aligned}\sqrt{18^2+24^2} &= \sqrt{324+576} = \sqrt{900}\\ &= 30 \text{ mph in the direction of } \mathbf{v}.\end{aligned}$$

(c) $\mathbf{u} = -8\mathbf{i}+8\mathbf{j}$

$$\begin{aligned}\sqrt{(-8)^2+8^2} &= \sqrt{64+64} = \sqrt{128}\\ &\approx 11.3 \text{ mphSE}\end{aligned}$$

101. $|\mathbf{a}-\mathbf{b}|^2 = |\mathbf{a}|^2+|\mathbf{b}|^2-2\,|\mathbf{a}|\,|\mathbf{b}|\cos\theta$

$$\left(\sqrt{(a_1-b_1)^2+(a_2-b_2)^2}\right)^2 = \left(\sqrt{a_1{}^2+a_2{}^2}\right)^2 + \left(\sqrt{b_1{}^2+b_2{}^2}\right)^2 - 2\,|\mathbf{a}|\,|\mathbf{b}|\cos\theta$$
$$a_1{}^2-2a_1b_1+b_1{}^2+a_2{}^2-2a_2b_2+b_2{}^2 = a_1{}^2+b_1{}^2+a_2{}^2+b_2{}^2 - 2\,|\mathbf{a}|\,|\mathbf{b}|\cos\theta$$
$$-2a_1b_1-2a_2b_2 = -2\,|\mathbf{a}|\,|\mathbf{b}|\cos\theta$$
$$a_1b_1+a_2b_2 = |\mathbf{a}|\,|\mathbf{b}|\cos\theta$$
$$\mathbf{a}\cdot\mathbf{b} = |\mathbf{a}|\,|\mathbf{b}|\cos\theta$$

Reviewing Basic Concepts (Sections 10.1 - 10.3)

1. $A = 44°,\ C = 62°,\ a = 12$

$$\frac{12}{\sin 44°} = \frac{c}{\sin 62°}$$

$$c = \frac{12 \sin 62°}{\sin 44°} \qquad \boxed{c \approx 15.3}$$

$$B = 180° - (A + C) \qquad \boxed{B \approx 74°}$$

$$\frac{12}{\sin 44°} = \frac{b}{\sin 74°}$$

$$b = \frac{12 \sin 74°}{\sin 44°} \qquad \boxed{b \approx 16.6}$$

2. $A = 32°,\ a = 6,\ b = 8;$ two solutions

$$\frac{6}{\sin 32°} = \frac{8}{\sin B°}$$

$$\sin B = \frac{8 \sin 32°}{6} \approx .706559019$$

$$B \approx 44.956° \qquad \boxed{B_1 \approx 45.0°}$$

$$C_1 = 180° - (A + B_1) \qquad \boxed{C_1 \approx 103°}$$

$$\frac{6}{\sin 32°} = \frac{c_1}{\sin 103°}$$

$$c_1 = \frac{6 \sin 103°}{\sin 32°} \qquad \boxed{c_1 \approx 11.0}$$

$$B_2 = 180° - B_1 \qquad \boxed{B_2 \approx 135.0°}$$

$$C_2 = 180° - (A + B) \qquad \boxed{C_2 \approx 13.0°}$$

$$\frac{6}{\sin 32°} = \frac{c_2}{\sin 13°}$$

$$c_2 = \frac{6 \sin 13°}{\sin 32°} \qquad \boxed{c_2 \approx 2.5}$$

3. $C = 41°,\ a = 7,\ c = 12;$ one solution

$$\frac{12}{\sin 41°} = \frac{7}{\sin A°}$$

$$\sin A = \frac{7 \sin 41°}{12} \approx .3827011002$$

$$\boxed{A \approx 22.5°}$$

$$B = 180° - (A + C) \qquad \boxed{B \approx 116.5°}$$

$$\frac{12}{\sin 41°} = \frac{b}{\sin 116.5°}$$

$$b = \frac{12 \sin 116.5°}{\sin 41°} \qquad \boxed{b \approx 16.4}$$

$[A_2 = 180° - 22.5° \qquad A_2 \approx 157.5°$

$B_2 = 180° - (A_2 + C) \qquad B_2 \approx -18.5°$

2nd triangle does not exist.]

4. (a)

$$\begin{aligned} b^2 &= a^2 + c^2 - 2ac \cos B \\ &= 8.1^2 + 8.3^2 - 2(8.1)(8.3)\cos 51° \\ &\approx 49.88158022 \end{aligned}$$

$$b \approx 7.0627 \qquad \boxed{b \approx 7.1}$$

$$\frac{7.0627}{\sin 51°} = \frac{8.1}{\sin A}$$

$$\sin A = \frac{8.1 \sin 51°}{7.0627} \approx .891285526$$

$$\boxed{A \approx 63.0°}$$

$$\frac{7.0627}{\sin 51°} = \frac{8.3}{\sin C}$$

$$\sin C = \frac{8.3 \sin 51°}{7.0627} \approx .91329258$$

$$\boxed{C \approx 66.0°}$$

(b) $\cos A = \dfrac{a^2 - b^2 - c^2}{-2bc}$

$$\cos A \approx -.354166667 \qquad \boxed{A \approx 110.7°}$$

$$\frac{14}{\sin 110.7°} = \frac{9}{\sin B}$$

$$\sin B = \frac{9 \sin 110.7°}{14} \approx .601356877$$

$$\boxed{B \approx 37.0°}$$

$$C = 180° - (A + B) \qquad \boxed{C \approx 32.3°}$$

5.

$$\begin{aligned} A &= \frac{1}{2} ab \sin C \\ &= \frac{1}{2}(4.5)(5.2)\sin 55° \\ &\approx 9.584 \quad \boxed{\approx 9.6} \end{aligned}$$

6. $a = 6,\ b = 7,\ c = 9$

$$\begin{aligned} s &= \frac{1}{2}(a + b + c) \\ &= \frac{1}{2}(6 + 7 + 9) = 11 \end{aligned}$$

$$\begin{aligned} A &= \sqrt{s(s-a)(s-b)(s-c)} \\ &= \sqrt{11(11-6)(11-7)(11-9)} \\ &= \sqrt{11(5)(4)(2)} \\ &= \sqrt{440} \approx 20.976 \approx \boxed{21} \end{aligned}$$

7. $v = 2i - j, \quad u = -3i + 2j$

(a) $\begin{aligned} 2\mathbf{v} + \mathbf{u} &= 2(2i - j) + (-3i + 2j) \\ &= 4i - 2j - 3i + 2j = i \end{aligned}$

(b) $2\mathbf{v} = 2(2i - j) = 4i - 2j$

(c) $\begin{aligned} \mathbf{v} - 3\mathbf{u} &= 2i - j - 3(-3i + 2j) \\ &= 2i - j + 9i - 6j \\ &= 11i - 7j \end{aligned}$

8. $\mathbf{a} = \langle 3, -2 \rangle, \quad \mathbf{b} = \langle -1, 3 \rangle$

$\mathbf{a} \cdot \mathbf{b} = 3(-1) + (-2)(3) = -3 - 6 = -9$

$$\cos\theta = \frac{\mathbf{a} \cdot \mathbf{b}}{|\mathbf{a}|\ |\mathbf{b}|} = \frac{-9}{\sqrt{9+4} \cdot \sqrt{1+9}} = \frac{-9}{\sqrt{13} \cdot \sqrt{10}} = \frac{-9}{\sqrt{130}} \approx -.7893522174$$

$\theta \approx 142.125° \approx \boxed{142.1°}$

9. Adjacent angle $= 180° - 52° = 128°$

$|v^2| = 100^2 + 130^2 - 2(100)(130)\cos 128°$

$|v^2| \approx 42907.19836$

$|v| \approx \boxed{207.1 \text{ pounds}}$

10. Let h be the perpendicular height of the airplane. Then

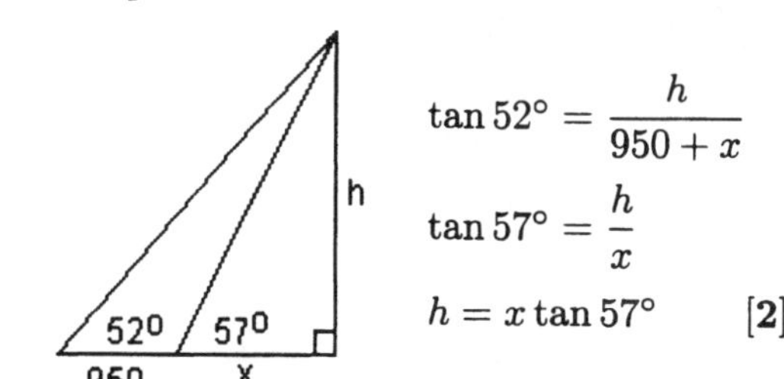

$$\tan 52° = \frac{h}{950 + x} \quad [1]$$

$$\tan 57° = \frac{h}{x}$$

$$h = x \tan 57° \quad [2]$$

$[2] \longrightarrow [1]: \quad \tan 52° = \dfrac{x \tan 57°}{950 + x}$

$$\begin{aligned} 950 \tan 52° + x \tan 52° &= x \tan 57° \\ x \tan 52° - x \tan 57° &= -950 \tan 52° \\ x(\tan 52° - \tan 57°) &= -950 \tan 52° \\ x = \frac{-950 \tan 52°}{\tan 52° - \tan 57°} &\approx 4678.089277 \end{aligned}$$

$[2]: h = x \tan 57°$

$\approx 4678.089277 \tan 57° \approx 7203.626$

$\boxed{\text{The height is} \approx 7200 \text{ feet.}}$

Section 10.4

1. The modulus of a complex number represents the <u>magnitude</u> (length)of the vector representing it in the complex plane.

3. $6 - 5i$

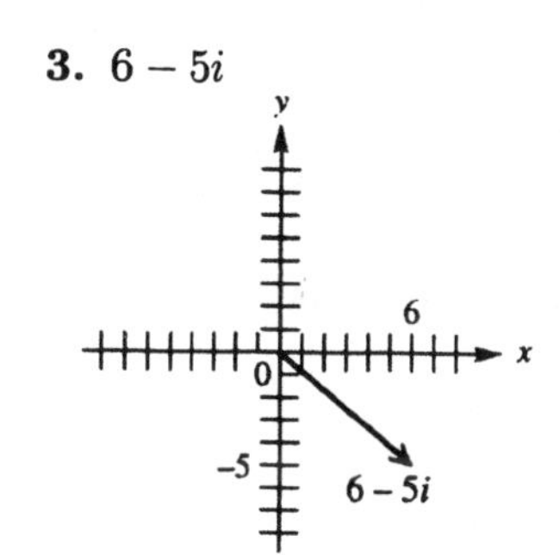

5. $-4i$

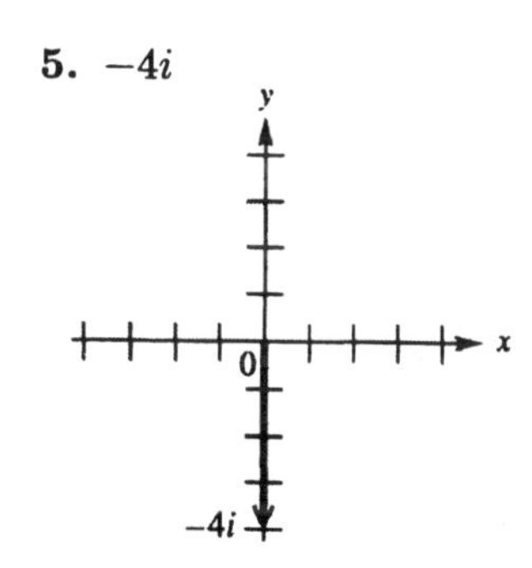

7. -8

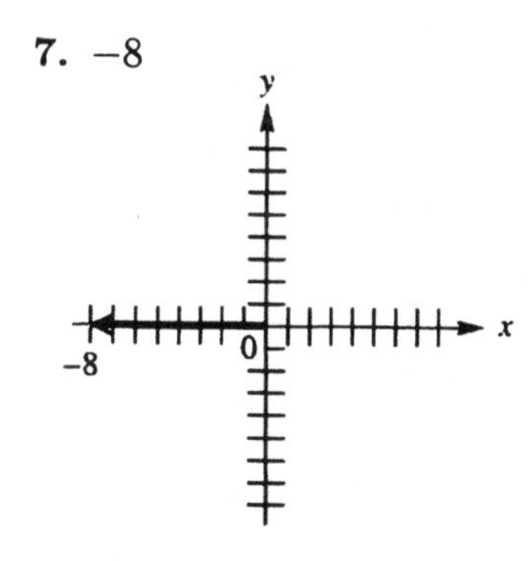

9. $x = 1, \; y = -4 \Longrightarrow 1 - 4i$

11. If $b = 0$ $(bi = 0)$, then a complex number is also a real number.

13. A complex number of the form $a + bi$ will have its corresponding vector lying on the y-axis provided $a = \underline{0}$.

15. $(4 - 3i) + (-1 + 2i) = \boxed{3 - i}$

17. $-3+3i = \boxed{-3+3i}$

19. $(2+6i)+(-2i) = \boxed{2+4i}$

21. $1+i \Longrightarrow x=1,\ y=1$

$$r=\sqrt{x^2+y^2}=\sqrt{1^2+1^2}=\boxed{\sqrt{2}}$$

23. $12-5i \Longrightarrow x=12,\ y=-5$

$$r=\sqrt{x^2+y^2}=\sqrt{12^2+(-5)^2}=\sqrt{169}=\boxed{13}$$

25. $-6 \Longrightarrow x=-6,\ y=0$

$$r=\sqrt{x^2+y^2}=\sqrt{(-6)^2}=\boxed{6}$$

27. $2-3i \Longrightarrow x=2,\ y=-3$

$$r=\sqrt{x^2+y^2}=\sqrt{2^2+(-3)^2}=\boxed{\sqrt{13}}$$

29. $2(\cos 45^\circ + i\sin 45^\circ) = 2\left(\dfrac{\sqrt{2}}{2}+\dfrac{i\sqrt{2}}{2}\right) = \boxed{\sqrt{2}+i\sqrt{2}}$

31. $10\,\text{cis}\,90^\circ = 10(\cos 90^\circ + i\sin 90^\circ) = 10(0+1i) = \boxed{10i}$

33. $4(\cos 240^\circ + i\sin 240^\circ) = 4\left(-\dfrac{1}{2}-\dfrac{i\sqrt{3}}{2}\right) = \boxed{-2-2i\sqrt{3}}$

35. $\cos\dfrac{\pi}{6}+i\sin\dfrac{\pi}{6} = \boxed{\dfrac{\sqrt{3}}{2}+\dfrac{1}{2}i}$

37. $5\,\text{cis}\left(-\dfrac{\pi}{6}\right) = 5\left(\cos\left(-\dfrac{\pi}{6}\right)+i\sin\left(-\dfrac{\pi}{6}\right)\right) = 5\left(\dfrac{\sqrt{3}}{2}-\dfrac{1}{2}i\right) = \boxed{\dfrac{5\sqrt{3}}{2}-\dfrac{5}{2}i}$

39. $\sqrt{2}\,\text{cis}\,\pi = \sqrt{2}(\cos\pi + i\sin\pi) = \sqrt{2}(-1+i(0)) = \boxed{-\sqrt{2}}$

41. $3-3i:\quad r=\sqrt{x^2+y^2}=\sqrt{3^2+(-3)^2}=\sqrt{18}=3\sqrt{2}$

$$\tan\theta=\frac{y}{x}=\frac{-3}{3}=-1$$
$$\theta=135^\circ,\ -45^\circ$$

Since $3-3i$ $(+,\,-)\in$ Quadrant IV:

$3-3i = 3\sqrt{2}\,\text{cis}(-45^\circ) = \boxed{3\sqrt{2}(\cos(-45^\circ)+i\sin(-45^\circ))}$

43. $1+i\sqrt{3}:\quad r=\sqrt{x^2+y^2}=\sqrt{1^2+(\sqrt{3})^2}=\sqrt{4}=2$

$$\tan\theta=\frac{y}{x}=\frac{\sqrt{3}}{1}=\sqrt{3}$$
$$\theta=60^\circ,\ -120^\circ$$

Since $1+i\sqrt{3}$ $(+,\,+)\in$ Quadrant I:

$1+i\sqrt{3} = 2\,\text{cis}^\circ\,60^\circ = \boxed{2(\cos 60^\circ + i\sin 60^\circ)}$

45. $-2i\quad (0-2i):\quad r=\sqrt{x^2+y^2}=\sqrt{(0)^2+(-2)^2}=\sqrt{4}=2$

$$\tan\theta=\frac{y}{x}=\frac{-2}{0}:\ \emptyset$$
$$\theta=90^\circ,\ -90^\circ$$

Since $(0-2i)$ is on the negative y-axis,

$-2i = 2\,\text{cis}(-90^\circ) = \boxed{2(\cos(-90^\circ)+i\sin(-90^\circ))}$

47. $4\sqrt{3}+4i:\ \ r=\sqrt{x^2+y^2}$

$$=\sqrt{\left(4\sqrt{3}\right)^2+4^2}$$
$$=\sqrt{48+16}=\sqrt{64}$$
$$=8$$

$$\tan\theta=\frac{y}{x}=\frac{4}{4\sqrt{3}}=\frac{\sqrt{3}}{3}$$

$$\theta=\frac{\pi}{6},\ \frac{7\pi}{6}\ \text{or}\ -\frac{5\pi}{6}$$

Since $4\sqrt{3}+4\ (+,+)\in$ Quadrant I:

$$4\sqrt{3}+4=8\,\text{cis}\,\frac{\pi}{6}$$
$$=\boxed{8\left(\cos\frac{\pi}{6}+i\sin\frac{\pi}{6}\right)}$$

49. $-\sqrt{2}+i\sqrt{2}:\ \ r=\sqrt{x^2+y^2}$

$$=\sqrt{\left(-\sqrt{2}\right)^2+\left(\sqrt{2}\right)^2}$$
$$=\sqrt{2+2}=\sqrt{4}=2$$

$$\tan\theta=\frac{y}{x}=\frac{\sqrt{2}}{-\sqrt{2}}=-1$$

$$\theta=\frac{3\pi}{4},\ -\frac{\pi}{4}$$

Since $-\sqrt{2}+i\sqrt{2}\ (-,+)\in$ Quadrant II:

$$-\sqrt{2}+i\sqrt{2}=2\,\text{cis}\,\frac{3\pi}{4}$$
$$=\boxed{2\left(\cos\frac{3\pi}{4}+i\sin\frac{3\pi}{4}\right)}$$

51. $-4\quad(-4+0i):\ \ r=\sqrt{x^2+y^2}$

$$=\sqrt{(-4)^2+0^2}$$
$$=\sqrt{16}=4$$

$$\tan\theta=\frac{y}{x}=\frac{0}{-4}=0$$
$$\theta=0,\ \pi$$

Since $(-4+0i)$ is on the negative x-axis,

$$-4=4\,\text{cis}\,\pi=\boxed{4\left(\cos\pi+i\sin\pi\right)}$$

53. $z=-.2i$; Yes

$$z^2-1=(-.2i)^2-1=-1.04$$
$$\left(z^2-1\right)^2-1=(-1.04)^2-1=-.0816$$
$$\left[\left(z^2-1\right)^2-1\right]^2=(-.0816)^2-1$$
$$=-.99334144$$

55. $3+5i;\ \ r=\sqrt{3^2+5^2}$

$$=\sqrt{34}\approx 5.830951895$$

$$\tan\theta=\frac{5}{3},\quad \theta\approx 1.030376827$$

$$3+5i\approx 5.830951895\ \text{cis}\ 1.030376827$$

Note: Exercises 57 - 67, the compact form, $r\,cis\,\theta$**, is used, rather than** $r\left(\cos\theta+i\sin\theta\right)$.

57. $(r_1\,\text{cis}\,\theta_1)\,(r_2\,\text{cis}\,\theta_2)=r_1r_2\,\text{cis}\,(\theta_1+\theta_2)$

$$(3\,\text{cis}\,60^\circ)\,(2\,\text{cis}\,90^\circ)$$
$$=(3)\,(2)\,\text{cis}\,(60^\circ+90^\circ)$$
$$=\mathbf{6}\,\text{cis}\,\mathbf{150^\circ}$$
$$=6\left(\cos150^\circ+i\sin150^\circ\right)$$
$$=6\left(-\frac{\sqrt{3}}{2}+\frac{1}{2}i\right)$$
$$=\boxed{-3\sqrt{3}+3i}$$

59. $(r_1\,\text{cis}\,\theta_1)\,(r_2\,\text{cis}\,\theta_2)=r_1r_2\,\text{cis}\,(\theta_1+\theta_2)$

$$(2\,\text{cis}\,45^\circ)\,(2\,\text{cis}\,225^\circ)$$
$$=(2)\,(2)\,\text{cis}\,(45^\circ+225^\circ)$$
$$=\mathbf{4}\,\text{cis}\,\mathbf{270^\circ}$$
$$=4\left(\cos270^\circ+i\sin270^\circ\right)$$
$$=4\,(0-i)=\boxed{-4i}$$

61. $(r_1 \operatorname{cis}\theta_1)(r_2 \operatorname{cis}\theta_2) = r_1 r_2 \operatorname{cis}(\theta_1+\theta_2)$

$$\left(5 \operatorname{cis}\frac{\pi}{2}\right)\left(3 \operatorname{cis}\frac{\pi}{4}\right)$$
$$= (5)(3)\operatorname{cis}\left(\frac{\pi}{2}+\frac{\pi}{4}\right)$$
$$= \mathbf{15} \operatorname{cis}\frac{\mathbf{3\pi}}{\mathbf{4}}$$
$$= 15\left(\cos\frac{3\pi}{4} + i\sin\frac{3\pi}{4}\right)$$
$$= 15\left(-\frac{\sqrt{2}}{2}+\frac{\sqrt{2}}{2}i\right)$$
$$= \boxed{-\frac{15\sqrt{2}}{2}+\frac{15\sqrt{2}}{2}i}$$

63. $(r_1 \operatorname{cis}\theta_1)(r_2 \operatorname{cis}\theta_2) = r_1 r_2 \operatorname{cis}(\theta_1+\theta_2)$

$$\left(\sqrt{3} \operatorname{cis}\frac{\pi}{4}\right)\left(\sqrt{3} \operatorname{cis}\frac{5\pi}{4}\right)$$
$$= (\sqrt{3})(\sqrt{3})\operatorname{cis}\left(\frac{\pi}{4}+\frac{5\pi}{4}\right)$$
$$= \mathbf{3} \operatorname{cis}\frac{\mathbf{3\pi}}{\mathbf{2}}$$
$$= 3\left(\cos\frac{3\pi}{2} + i\sin\frac{3\pi}{2}\right)$$
$$= 3(0-1i) = \boxed{-3i}$$

65. $\dfrac{r_1 \operatorname{cis}\theta_1}{r_2 \operatorname{cis}\theta_2} = \dfrac{r_1}{r_2}\operatorname{cis}(\theta_1-\theta_2)$

$$\frac{4\operatorname{cis}120^\circ}{2\operatorname{cis}150^\circ} = \frac{4}{2}\operatorname{cis}(120^\circ-150^\circ)$$
$$= \mathbf{2}\operatorname{cis}(\mathbf{-30^\circ})$$
$$= 2(\cos(-30^\circ)+i\sin(-30^\circ))$$
$$= 2(\cos 30^\circ - i\sin 30^\circ)$$
$$= 2\left(\frac{\sqrt{3}}{2}-\frac{1}{2}i\right) = \boxed{\sqrt{3}-i}$$

67. $\dfrac{r_1 \operatorname{cis}\theta_1}{r_2 \operatorname{cis}\theta_2} = \dfrac{r_1}{r_2}\operatorname{cis}(\theta_1-\theta_2)$

$$\frac{16\operatorname{cis}300^\circ}{8\operatorname{cis}60^\circ} = \frac{16}{8}\operatorname{cis}(300^\circ-60^\circ)$$
$$= \mathbf{2}\operatorname{cis}\mathbf{240^\circ}$$
$$= 2(\cos 240^\circ + i\sin 240^\circ)$$
$$= 2\left(-\frac{1}{2}-\frac{\sqrt{3}}{2}i\right)$$
$$= \boxed{-1-i\sqrt{3}}$$

69. $\dfrac{r_1 \operatorname{cis}\theta_1}{r_2 \operatorname{cis}\theta_2} = \dfrac{r_1}{r_2}\operatorname{cis}(\theta_1-\theta_2)$

$$\frac{3\operatorname{cis}\frac{61\pi}{36}}{9\operatorname{cis}\frac{13\pi}{36}} = \frac{3}{9}\operatorname{cis}\left(\frac{61\pi}{36}-\frac{13\pi}{36}\right)$$
$$= \frac{\mathbf{1}}{\mathbf{3}}\operatorname{cis}\frac{\mathbf{4\pi}}{\mathbf{3}}$$
$$= \frac{1}{3}\left(\cos\frac{4\pi}{3}+i\sin\frac{4\pi}{3}\right)$$
$$= \frac{1}{3}\left(-\frac{1}{2}-\frac{\sqrt{3}}{2}i\right)$$
$$= \boxed{-\frac{1}{6}-\frac{\sqrt{3}}{6}i}$$

Relating Concepts (71 – 77):

71. $w = -1+i,\ z = -1-i$

$$wz = (-1+i)(-1-i)$$
$$= 1+i-i-i^2$$
$$= 1+1 = \boxed{2}$$

72. $w = -1+i;\quad r = \sqrt{x^2+y^2}$

$$= \sqrt{(-1)^2+1^2}$$
$$= \sqrt{2}$$

$$\tan\theta = \frac{y}{x} = \frac{1}{-1} = -1$$
$$\theta = 135^\circ,\ 315^\circ \quad (-45^\circ)$$
$$(-1,\ 1) \in \text{Quadrant II},\quad \theta = 135^\circ$$
$$\boxed{w = \sqrt{2}\operatorname{cis}135^\circ}$$

$z = -1-i;\quad r = \sqrt{x^2+y^2}$

$$= \sqrt{(-1)^2+(-1)^2}$$
$$= \sqrt{2}$$

$$\tan\theta = \frac{y}{x} = \frac{-1}{-1} = 1$$
$$\theta = 45^\circ,\ 225^\circ$$
$$(-1,-1) \in \text{Quadrant III},\quad \theta = 225^\circ$$
$$\boxed{z = \sqrt{2}\operatorname{cis}225^\circ \text{ or } \sqrt{2}\operatorname{cis}(-135^\circ)}$$

73. $\left(\sqrt{2}\text{ cis }135°\right)\left(\sqrt{2}\text{ cis }225°\right)$
$$= \left(\sqrt{2}\right)\left(\sqrt{2}\right)\text{cis}\,(135° + 225°)$$
$$= 2\,\text{cis}\,360° = \boxed{2\,\text{cis}\,0°}$$

74. $2\text{ cis }360° = 2\left(\cos 0° + i\sin 0°\right)$
$$= 2\,(1 + 0i) = \boxed{2}$$

The results are the same.

75. $w = -1 + i,\ z = -1 - i$
$$\frac{w}{z} = \frac{-1+i}{-1-i}\cdot\frac{-1+i}{-1+i}$$
$$= \frac{1-2i+i^2}{1-i^2}$$
$$= \frac{-2i}{2} = \boxed{-i}$$

76. $w = -1 + i,\ z = -1 - i$
$$\frac{w}{z} = \frac{\sqrt{2}\,\text{cis}\,135°}{\sqrt{2}\,\text{cis}\,225°}$$
$$= \frac{\sqrt{2}}{\sqrt{2}}\,\text{cis}\,(135° - 225°)$$
$$= \boxed{\text{cis}\,(-90°)\ \text{ or }\ \text{cis}\,270°}$$

77. $\text{cis}\,(-90°) = \cos(-90°) + i\sin(-90°)$
$$= 0 + i\,(-1) = \boxed{-i}$$

The results are the same.

79. To square $r\,\text{cis}\,\theta$, square $|r|$ and double θ.

81. $I = \dfrac{E}{R + (X_L - X_C)\,i}$
$$= \frac{12\,\text{cis}\,25°}{3 + (4-6)\,i} = \frac{12\,\text{cis}\,25°}{3 - 2\,i}$$

$3 - 2i:\ \ r = \sqrt{9+4} = \sqrt{13}$
$$\tan\theta = \frac{-2}{3}\ \text{(Quadrant IV)}$$
$$\theta = \tan^{-1}\left(-\frac{2}{3}\right) \approx 326.31°$$

$$I = \frac{12\,\text{cis}\,25°}{\sqrt{13}\,\text{cis}\,326.31°}$$
$$= \frac{12}{\sqrt{13}}\,\text{cis}\,(25° - 326.31°)$$
$$= \frac{12}{\sqrt{13}}\,\text{cis}\,(-301.31°) = \frac{12}{\sqrt{13}}\,\text{cis}\,58.69°$$
$$\approx 3.3282\,(.519668 + .854368\,i)$$
$$\approx 1.7296 + 2.8435\,i \approx \boxed{1.7 + 2.8\,i}$$

83. $Z_1 = 50 + 25i,\ Z_2 = 60 + 20i$
$$Z = \frac{1}{\frac{1}{Z_1} + \frac{1}{Z_2}} = \frac{Z_1\,Z_2}{Z_2 + Z_1}$$
$$= \frac{(50+25i)\,(60+20i)}{(60+20i) + (50+25i)}$$
$$= \frac{3000 + 1000i + 1500i + 500i^2}{110 + 45i}$$
$$= \frac{2500 + 2500i}{110 + 45i} = \frac{5\,(500 + 500i)}{5\,(22 + 9i)}$$
$$= \frac{500 + 500i}{22 + 9i} \times \frac{22 - 9i}{22 - 9i}$$
$$= \frac{11000 - 4500i + 11000i - 4500i^2}{484 - 81i^2}$$
$$= \frac{15500 + 6500i}{565}$$
$$= 27.43363 + 11.50442i$$
$$\boxed{\approx 27.43 + 11.5i}$$

Section 10.5

1. $[r(\cos\theta + i\sin\theta)]^n = r^n(\cos n\theta + i\sin n\theta)$

$$\begin{aligned}[3(\cos 30^\circ + i\sin 30^\circ)]^3 &= 3^3 \operatorname{cis}(3\cdot 30^\circ)\\ &= 27(\cos 90^\circ + i\sin 90^\circ)\\ &= 27(0+1i) = \boxed{27i}\end{aligned}$$

3. $[r(\cos\theta + i\sin\theta)]^n = r^n(\cos n\theta + i\sin n\theta)$

$$\begin{aligned}\left(\cos\frac{\pi}{4} + i\sin\frac{\pi}{4}\right)^8 &= 1^8 \operatorname{cis}\left(8\cdot\frac{\pi}{4}\right)\\ &= \operatorname{cis} 2\pi\\ &= \cos 0 + i\sin 0\\ &= 1 + 0i = \boxed{1}\end{aligned}$$

5. $[r(\cos\theta + i\sin\theta)]^n = r^n(\cos n\theta + i\sin n\theta)$

$$\begin{aligned}[3\operatorname{cis} 100^\circ]^3 &= 3^3\operatorname{cis}(3\cdot 100^\circ)\\ &= 27(\cos 300^\circ + i\sin 300^\circ)\\ &= 27\left(\frac{1}{2} - \frac{\sqrt{3}}{2}i\right)\\ &= \boxed{\frac{27}{2} - \frac{27\sqrt{3}}{2}i}\end{aligned}$$

7. $(\sqrt{3}+i)^5: \quad x = \sqrt{3},\ y = 1$

$$r = \sqrt{(\sqrt{3})^2 + 1} = 2$$

$$\tan\theta = \frac{1}{\sqrt{3}}, \quad \text{Quadrant I}$$

$$\theta = 30^\circ$$

$$\begin{aligned}[2\operatorname{cis} 30^\circ]^5 &= 2^5\operatorname{cis}(5\cdot 30^\circ)\\ &= 32(\cos 150^\circ + i\sin 150^\circ)\\ &= 32\left(-\frac{\sqrt{3}}{2} + \frac{1}{2}i\right)\\ &= \boxed{-16\sqrt{3} + 16i}\end{aligned}$$

9. $(2 - 2i\sqrt{3})^4: \quad x = 2,\ y = -2\sqrt{3}$

$$r = \sqrt{2^2 + (-2\sqrt{3})^2} = \sqrt{4+12} = 4$$

$$\tan\theta = \frac{-2\sqrt{3}}{2} = -\sqrt{3}, \quad \text{Quad IV}$$

$$\theta = -60^\circ$$

$$\begin{aligned}[4\operatorname{cis}(-60^\circ)]^4 &= 4^4\operatorname{cis}(4\cdot -60^\circ)\\ &= 256\operatorname{cis}(-240^\circ)\\ &= 256(\cos 120^\circ + i\sin 120^\circ)\\ &= 256\left(-\frac{1}{2} + \frac{\sqrt{3}}{2}i\right)\\ &= \boxed{-128 + 128i\sqrt{3}}\end{aligned}$$

11. $(-2-2i)^5: \quad x = -2,\ y = -2$

$$r = \sqrt{(-2)^2 + (-2)^2} = \sqrt{8} = 2\sqrt{2}$$

$$\tan\theta = \frac{-2}{-2} = 1, \text{ Quad III, } \theta = 225^\circ$$

$$\begin{aligned}[2\sqrt{2}\operatorname{cis}(225^\circ)]^5 &= (2\sqrt{2})^5\operatorname{cis}(5\cdot 225^\circ)\\ &= 128\sqrt{2}\operatorname{cis} 1125^\circ\\ &= 128\sqrt{2}(\cos 45^\circ + i\sin 45^\circ)\\ &= 128\sqrt{2}\left(\frac{\sqrt{2}}{2} + \frac{\sqrt{2}}{2}i\right)\\ &= \boxed{128 + 128i}\end{aligned}$$

13. $[r(\cos\theta + i\sin\theta)]^{1/n} = r^{1/n}(\cos\alpha + i\sin\alpha)$.

where $\alpha = \dfrac{\theta}{n} + \dfrac{360^\circ\cdot k}{n}$

$1^{1/3} = (1+0i)^{1/3} \quad x = 1,\ y = 0,\ r = 1$

$\tan\theta = 0, \quad \theta = 0^\circ$

$[1(\cos 0^\circ + i\sin 0^\circ)]^{1/3}$

$$r^{1/3} = 1^{1/3} = 1$$

$$\alpha = \frac{0^\circ}{3} + \frac{360^\circ k}{3} = 0^\circ + 120^\circ k$$

if $k = 0, 1, 2 \rightarrow \alpha = 0^\circ, 120^\circ, 240^\circ$

$\boxed{\text{cube roots: } \operatorname{cis} 0^\circ,\ \operatorname{cis} 120^\circ,\ \operatorname{cis} 240^\circ}$

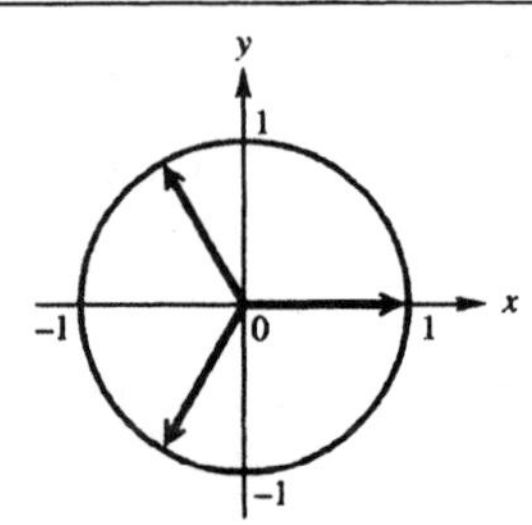

15. $[r(\cos\theta + i\sin\theta)]^{1/n} = r^{1/n}(\cos\alpha + i\sin\alpha)$

where $\alpha = \dfrac{\theta}{n} + \dfrac{360° \cdot k}{n}$

$[8(\cos 60° + i\sin 60°)]^{1/3}$

$r^{1/3} = 8^{1/3} = 2$

$\alpha = \dfrac{60°}{3} + \dfrac{360°k}{3} = 20° + 120°k$

if $k = 0, 1, 2 \rightarrow \alpha = 20°, 140°, 260°$

The cube roots are:
$2\,\text{cis}\,20°,\ 2\,\text{cis}\,140°,\ 2\,\text{cis}\,260°$

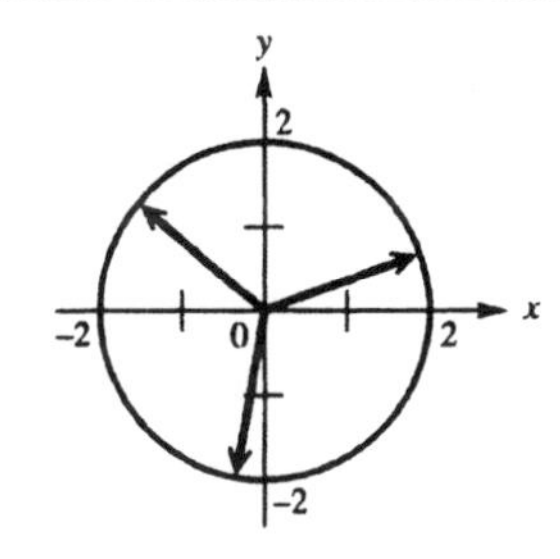

17. $[r(\cos\theta + i\sin\theta)]^{1/n} = r^{1/n}(\cos\alpha + i\sin\alpha)$

where $\alpha = \dfrac{\theta}{n} + \dfrac{360° \cdot k}{n}$

$(-8i)^{1/3} = (0 - 8i)^{1/3}$

$x = 0,\ y = -8,\ r = 8$

$\tan\theta$ is $\emptyset$, $\theta = 270°$

$[8(\cos 270° + i\sin 270°)]^{1/3}$

$r^{1/3} = 8^{1/3} = 2$

$\alpha = \dfrac{270°}{3} + \dfrac{360°k}{3} = 90° + 120°k$

if $k = 0, 1, 2 \rightarrow \alpha = 90°, 210°, 330°$

The cube roots are:
$2\,\text{cis}\,90°,\ 2\,\text{cis}\,210°,\ 2\,\text{cis}\,330°$

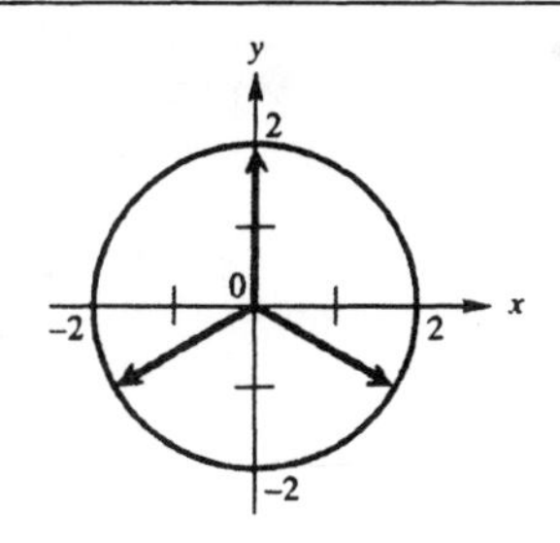

19. $[r(\cos\theta + i\sin\theta)]^{1/n} = r^{1/n}(\cos\alpha + i\sin\alpha)$

where $\alpha = \dfrac{\theta}{n} + \dfrac{360° \cdot k}{n}$

$(-64)^{1/3} = (-64 + 0i)^{1/3}$

$x = -64,\ y = 0,\ r = 64$

$\tan\theta = 0,\ \theta = 180°$

$[64(\cos 180° + i\sin 180°)]^{1/3}$

$r^{1/3} = (64)^{1/3} = 4$

$\alpha = \dfrac{180°}{3} + \dfrac{360°k}{3} = 60° + 120°k$

if $k = 0, 1, 2 \rightarrow \alpha = 60°, 180°, 300°$

The cube roots are:
$4\,\text{cis}\,60°,\ 4\,\text{cis}\,180°,\ 4\,\text{cis}\,300°$

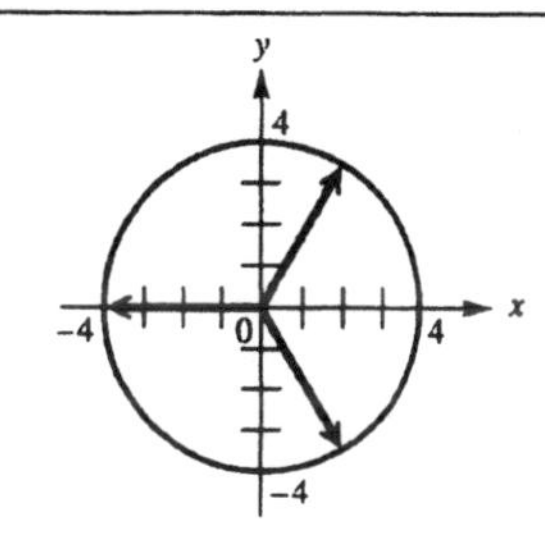

21. $[r(\cos\theta + i\sin\theta)]^{1/n} = r^{1/n}(\cos\alpha + i\sin\alpha)$

where $\alpha = \dfrac{\theta}{n} + \dfrac{360° \cdot k}{n}$

$(1 + i\sqrt{3})^{1/3}$ $x = 1,\ y = \sqrt{3},\ r = 2$

$\tan\theta = \dfrac{\sqrt{3}}{1}$, Quad I, $\theta = 60°$

$[2(\cos 60° + i\sin 60°)]^{1/3}$

$r^{1/3} = 2^{1/3} = \sqrt[3]{2}$

$\alpha = \dfrac{60°}{3} + \dfrac{360°k}{3} = 20° + 120°k$

if $k = 0, 1, 2 \rightarrow \alpha = 20°, 140°, 260°$

The cube roots are:
$\sqrt[3]{2}\,\text{cis}\,20°,\ \sqrt[3]{2}\,\text{cis}\,140°,\ \sqrt[3]{2}\,\text{cis}\,260°$

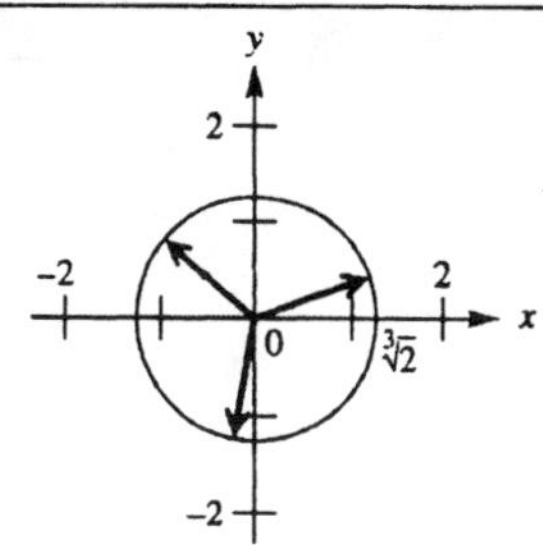

23. $[r(\cos\theta + i\sin\theta)]^{1/n} = r^{1/n}(\cos\alpha + i\sin\alpha)$

where $\alpha = \dfrac{\theta}{n} + \dfrac{360°\cdot k}{n}$

$(-2\sqrt{3}+2i)^{1/3}$ $\quad x = -2\sqrt{3},\ y = 2,$

$r = \sqrt{(-2\sqrt{3})^2 + 4} = 4$

$\tan\theta = \dfrac{2}{-2\sqrt{3}} = -\dfrac{1}{\sqrt{3}}$, Quad II, $\theta = 150°$

$[4(\cos 150° + i\sin 150°)]^{1/3}$

$r^{1/3} = 4^{1/3} = \sqrt[3]{4}$

$\alpha = \dfrac{150°}{3} + \dfrac{360°k}{3} = 50° + 120°k$

if $k = 0, 1, 2 \to \alpha = 50°, 170°, 290°$

The cube roots are:
$\sqrt[3]{4}\text{ cis }50°,\ \sqrt[3]{4}\text{ cis }170°,\ \sqrt[3]{4}\text{ cis }290°$

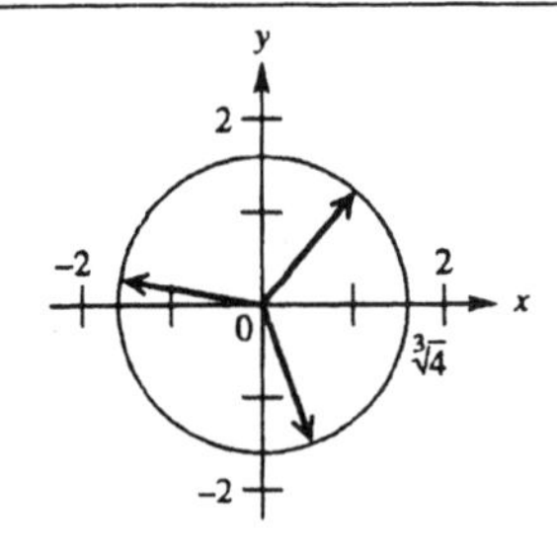

25. $[r(\cos\theta + i\sin\theta)]^{1/n} = r^{1/n}(\cos\alpha + i\sin\alpha)$

$[4(\cos 120° + i\sin 120°)]^{1/2}$

$r^{1/2} = 4^{1/2} = 2$

$\alpha = \dfrac{120°}{2} + \dfrac{360°k}{2} = 60° + 180°k$

if $k = 0, 1 \to \alpha = 60°, 240°$

The square roots are:

$2\text{ cis }60° = 2\left(\dfrac{1}{2} + i\dfrac{\sqrt{3}}{2}\right) = \boxed{1 + i\sqrt{3}}$

$2\text{ cis }240° = 2\left(-\dfrac{1}{2} - i\dfrac{\sqrt{3}}{2}\right) = \boxed{-1 - i\sqrt{3}}$

27. $[r(\cos\theta + i\sin\theta)]^{1/n} = r^{1/n}(\cos\alpha + i\sin\alpha)$

$(\cos 180° + i\sin 180°)^{1/3}$

$r^{1/3} = 1^{1/3} = 1$

$\alpha = \dfrac{180°}{3} + \dfrac{360°k}{3} = 60° + 120°k$

if $k = 0, 1, 2 \to \alpha = 60°, 180°, 300°$

The cube roots are:

$1\text{ cis }60° = \boxed{\dfrac{1}{2} + \dfrac{\sqrt{3}}{2}i}$

$1\text{ cis }180° = -1 - i(0) = \boxed{-1}$

$1\text{ cis }300° = \boxed{\dfrac{1}{2} - \dfrac{\sqrt{3}}{2}i}$

29. $i^{1/2} = (0 + i)^{1/2}$

$x = 0,\ y = 1,\ r = 1$

$\tan\theta$ is undefined; $\theta = 90°$

$[1(\cos 90° + i\sin 90°)]^{1/2}$

$r^{1/2} = 1^{1/2} = 1$

$\alpha = \dfrac{90°}{2} + \dfrac{360°k}{2} = 45° + 180°k$

if $k = 0, 1 \longrightarrow \alpha = 45°, 225°$

$\text{cis }45° = \dfrac{\sqrt{2}}{2} + \dfrac{\sqrt{2}}{2}i$

$\text{cis }225° = -\dfrac{\sqrt{2}}{2} - \dfrac{\sqrt{2}}{2}i$

31. $(64i)^{1/3} = (0 + 64i)^{1/3}$

$x = 0,\ y = 64,\ r = 64$

$\tan\theta$ is undefined; $\theta = 90°$

$[64(\cos 90° + i\sin 90°)]^{1/3}$

$r^{1/3} = 64^{1/3} = 4$

$\alpha = \dfrac{90°}{3} + \dfrac{360°k}{3} = 30° + 120°k$

if $k = 0,\ 1,\ 2 \longrightarrow \alpha = 30°,\ 150°,\ 270°$

$4\,\text{cis}\,30° = 4\left(\dfrac{\sqrt{3}}{2} + \dfrac{1}{2}i\right) = 2\sqrt{3} + 2i$

$4\,\text{cis}\,150° = 4\left(-\dfrac{\sqrt{3}}{2} + \dfrac{1}{2}i\right) = -2\sqrt{3} + 2i$

$4\,\text{cis}\,270° = 4(0 - i) = -4i$

33. $81^{1/4} = (81 + 0i)^{1/4}$

$x = 81,\ y = 0,\ r = 81$

$\tan\theta = 0,\ \theta = 0°$

$81[\cos 0° + i\sin 0°]^{1/4}$

$r^{1/4} = 81^{1/4} = 3$

$\alpha = \dfrac{0°}{4} + \dfrac{360°k}{4} = 0° + 90°k$

if $k = 0,\ 1,\ 2,\ 3$

$\alpha = 0°,\ 90°,\ 180°,\ 270°$

$3\,\text{cis}\ 0° = 3(1 + 0i) = 3$

$3\,\text{cis}\ 90° = 3(0 + i) = 3i$

$3\,\text{cis}\ 180° = 3(-1 + 0i) = -3$

$3\,\text{cis}\ 270° = 3(0 - i) = -3i$

35. $(1 + 0i)$: $x = 1,\ y = 0,\ r = 1$

$\tan\theta = 0,\ \theta = 0°$

(a) $(1 + 0i)^{1/4} = [\cos 0° + i\sin 0°]^{1/4}$

$r^{1/4} = 1^{1/4} = 1$

$\alpha = \dfrac{0°}{4} + \dfrac{360°k}{4} = 0° + 90°k$

if $k = 0,\ 1,\ 2,\ 3$

$\alpha = 0°,\ 90°,\ 180°,\ 270°$

$1^{1/4} = \text{cis}\,0°,\ \text{cis}\,90°,\ \text{cis}\,180°,\ \text{cis}\,270°$

$= \boxed{1,\ i,\ -1,\ -i}$

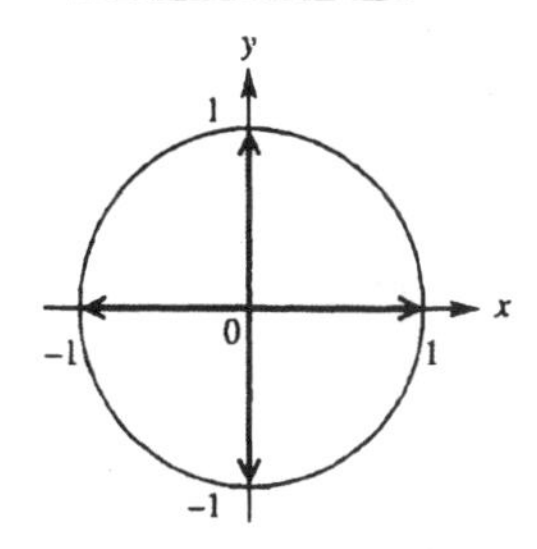

(b) $(1 + 0i)^{1/6} = [\cos 0° + i\sin 0°]^{1/6}$

$r^{1/6} = 1^{1/6} = 1$

$\alpha = \dfrac{0°}{6} + \dfrac{360°k}{6} = 0° + 60°k$

if $k = 0,\ 1,\ 2,\ 3,\ 4,\ 5$

$\alpha = 0°,\ 60°,\ 120°,\ 180°,\ 240°,\ 300°$

$1^{1/6} =$ $\text{cis}\,0°,\ \text{cis}\,60°,\ \text{cis}\,120°$
$\text{cis}\,180°,\ \text{cis}\,240°,\ \text{cis}\,300°$

$$\boxed{\begin{array}{l} 1,\ \dfrac{1}{2} + \dfrac{\sqrt{3}}{2}i,\ -\dfrac{1}{2} + \dfrac{\sqrt{3}}{2}i, \\ -1,\ -\dfrac{1}{2} - \dfrac{\sqrt{3}}{2}i,\ \dfrac{1}{2} - \dfrac{\sqrt{3}}{2}i \end{array}}$$

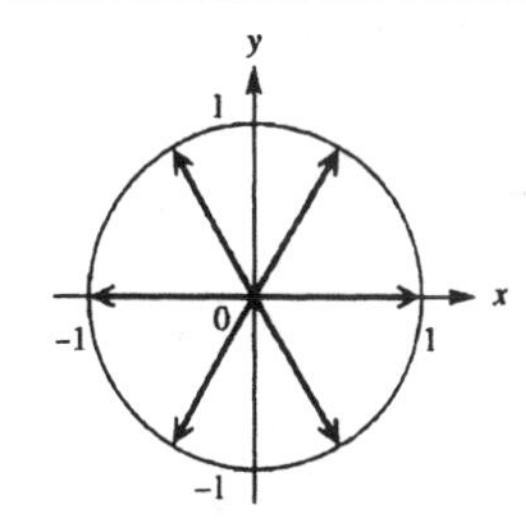

37. A positive real number must have a positive real n^{th} root since $\theta = 0°$. In computing $\dfrac{0° + 360°k}{n}$ for $k = 0$, we obtain $0°$; thus one of the roots will always be a positive number $(n\,\text{cis}\,0° = n(1 + 0i) = n)$.

Relating Concepts (39 – 44):

39. $x^8+8=\boxed{(x+2)\left(x^2-2x+4\right)}$

40. $x+2=0 \quad \boxed{x=-2}$

41. $x^2-2x+4=0$

$$x=\frac{2\pm\sqrt{4-4(4)}}{2}=\frac{2\pm\sqrt{-12}}{2}$$

$$=\frac{2\pm 2i\sqrt{3}}{2}=1\pm i\sqrt{3}$$

$$\boxed{\left(1+i\sqrt{3}\right),\ \left(1-i\sqrt{3}\right)}$$

42. $(-8)^{1/3}=(-8+0i)^{1/3}$

$x=-8,\ y=0,\ r=8$

$\tan\theta=0,\ \theta=180°$

$[8(\cos 180°+i\sin 180°)]^{1/3}$

$r^{1/3}=8^{1/3}=2$

$$\alpha=\frac{180°}{3}+\frac{360°k}{3}=60°+120°k$$

if $k=0,\ 1,\ 2\rightarrow\alpha=60°,\ 180°,\ 300°$

$$\boxed{\begin{array}{l}\text{The cube roots of } (-8)^{1/3} \text{ are:}\\ 2\operatorname{cis}60°,\ 2\operatorname{cis}180°,\ 2\operatorname{cis}300°\end{array}}$$

43. $2(\cos 60°+i\sin 60°)$

$$=2\left(\frac{1}{2}+\frac{\sqrt{3}}{2}i\right)=\boxed{1+i\sqrt{3}}$$

$2(\cos 180°+i\sin 180°)$

$$=2(-1+0i)=\boxed{-2}$$

$2(\cos 300°+i\sin 300°)$

$$=2\left(\frac{1}{2}-\frac{\sqrt{3}}{2}i\right)=\boxed{1-i\sqrt{3}}$$

44. The results of exercises **#40 - #41** and **#43** are the same.

45. $x^4+1=0$

$x^4=-1\rightarrow x=(-1)^{1/4}$

(i) Change -1 to polar form:

$-1=(-1+0i)$

$x=-1,\ y=0,$

$r=1,\ \theta=180°$

$-1=1(\cos 180°+i\sin 180°)$

(ii) $[\cos 180°+i\sin 180°]^{1/4}$

$r^{1/4}=1^{1/4}=1$

$$\alpha=\frac{180°}{4}+\frac{360°k}{4}=45°+90°k$$

if $k=0,\ 1,\ 2,\ 3$

$\alpha=45°,\ 135°,\ 225°,\ 315°$

(iii) $x=\boxed{\begin{array}{l}\cos 45°+i\sin 45°\\ \cos 135°+i\sin 135°\\ \cos 225°+i\sin 225°\\ \cos 315°+i\sin 315°\end{array}}$

47. $x^5-i=0$

$x^5=i\rightarrow x=(i)^{1/5}$

(i) Change i to polar form:

$i=(0+1i)$

$x=0,\ y=1,$

$r=1,\ \theta=90°$

$i=1(\cos 90°+i\sin 90°)$

(ii) $[\cos 90°+i\sin 90°]^{1/5}$

$r^{1/5}=1^{1/5}=1$

$$\alpha=\frac{90°}{5}+\frac{360°k}{5}=18°+72°k$$

if $k=0,\ 1,\ 2,\ 3,\ 4$

$\alpha=18°,\ 90°,\ 162°,\ 234°,\ 306°$

(iii) $x=\boxed{\begin{array}{l}\cos 18°+i\sin 18°\\ \cos 90°+i\sin 90°\\ \cos 162°+i\sin 162°\\ \cos 234°+i\sin 234°\\ \cos 306°+i\sin 306°\end{array}}$

49. $x^3 + 1 = 0$
$x^3 = -1 \longrightarrow x = (-1)^{1/3}$

(i) Change -1 to polar form:

$-1 = (-1 + 0i)$
$x = -1,\ y = 0,\ r = 1,$
$\tan\theta = 0,\ \theta = 180°$
$-1 = 1(\cos 180° + i \sin 180°)$

(ii) $[\cos 180° + i \sin 180°]^{1/3}$

$r^{1/3} = 1^{1/3} = 1$

$$\alpha = \frac{180°}{3} + \frac{360° k}{3} = 60° + 120° k$$

if $k = 0,\ 1,\ 2$

$\alpha = 60°,\ 180°,\ 300°$

(iii) $x =$ $\cos 60° + i \sin 60°$, $\cos 180° + i \sin 180°$, $\cos 300° + i \sin 300°$

51. $x^3 - 8 = 0$
$x^3 = 8 \longrightarrow x = (8)^{1/3}$

(i) Change 8 to polar form:

$8 = (8 + 0i)$
$x = 8,\ y = 0,\ r = 8,$
$\tan\theta = 0,\ \theta = 0°$
$8 = 8(\cos 0° + i \sin 0°)$

(ii) $[8(\cos 0° + i \sin 0°)]^{1/3}$

$r^{1/3} = 8^{1/3} = 2$

$$\alpha = \frac{0°}{3} + \frac{360° k}{3} = 0° + 120° k$$

if $k = 0,\ 1,\ 2$

$\alpha = 0°,\ 120°,\ 240°$

(iii) $x =$ $2(\cos 0° + i \sin 0°)$, $2(\cos 120° + i \sin 120°)$, $2(\cos 240° + i \sin 240°)$

53. (a) Let $f(z) = \dfrac{2z^3 + 1}{3z^2}$ and $z_1 = i$. Then,

$$z_2 = f(z_1) = \frac{2i^3 + 1}{3i^2} = -\frac{1}{3} + \frac{2}{3} i.$$

Similarly,

$$z_3 = f(z_2) = \frac{2{z_2}^3 + 1}{3{z_2}^2}$$
$$\approx -.58222 + .92444\, i$$

$$z_4 = f(z_3) = \frac{2{z_3}^3 + 1}{3{z_3}^2}$$
$$\approx -.50879 + .868165\, i$$

The values of z seem to approach

$$w_2 = -\frac{1}{2} + \frac{\sqrt{3}}{2} i.$$

Color the pixel at (0, 1) blue.

(b) Let $f(z) = \dfrac{2z^3 + 1}{3z^2}$ and $z_1 = 2 + i$.

Then,

$$z_2 = f(z_1) = \frac{2(2+i)^3 + 1}{3(2+i)^2}$$
$$\approx 1.373333 + .61333\, i.$$

$$z_3 = f(z_2) = \frac{2{z_2}^3 + 1}{3{z_2}^2}$$
$$\approx 1.01389 + .299161\, i$$

$$z_4 = f(z_3) = \frac{2{z_3}^3 + 1}{3{z_3}^2}$$
$$\approx .926439 + .0375086\, i$$

$$z_5 = f(z_4) = \frac{2{z_4}^3 + 1}{3{z_4}^2}$$
$$\approx 1.00409 + .00633912\, i$$

The values of z seem to approach

$w_1 = 1.$

Color the pixel at (2, 1) red.

continued

53. continued

(c) Let $f(z) = \dfrac{2z^3+1}{3z^2}$ and $z_1 = -1 - i$.
Then,

$$z_2 = f(z_1) = \frac{2(-1-i)^3+1}{3(-1-i)^2}$$
$$= -\frac{2}{3} - \frac{5}{6}i$$
$$\approx 1.33333 + .833333\,i.$$

$$z_3 = f(z_2) = \frac{2{z_2}^3+1}{3{z_2}^2}$$
$$\approx -.508691 - .841099\,i$$

$$z_4 = f(z_3) = \frac{2{z_3}^3+1}{3{z_3}^2}$$
$$\approx -.499330 - .866269\,i$$

The values of z seem to approach

$$w_3 = -\frac{1}{2} - \frac{\sqrt{3}}{2}i.$$

Color the pixel at $(-1, -1)$ yellow.

55. Using the trace function, we find that three of the tenth roots of 1 are:

1

$.80901699 + .58778525\,i$

$.30901699 + .95105622\,i$

Reviewing Basic Concepts (Sections 10.4 and 10.5)

1. $2(\cos 60^\circ + i\sin 60^\circ) = 2\left(\frac{1}{2} + i\frac{\sqrt{3}}{2}\right)$
$$= \boxed{1 + i\sqrt{3}}$$

2. $3 - 4i$

$$r = \sqrt{3^2 + (-4)^2} = \sqrt{9+16} = \boxed{5}$$

3. $-\sqrt{2} + i\sqrt{2}, \quad 0^\circ \le \theta < 360^\circ$

$$r = \sqrt{\left(-\sqrt{2}\right)^2 + \left(\sqrt{2}\right)^2} = \sqrt{2+2} = 2$$
$$\tan\theta = \frac{\sqrt{2}}{-\sqrt{2}} = -1$$

Ref angle: 45°, quadrant II; $\theta = 135^\circ$

$$-\sqrt{2} + i\sqrt{2} = \boxed{2(\cos 135^\circ + i\sin 135^\circ)}$$

4. $z_1 = 4(\cos 135^\circ + i\sin 135^\circ)$
$z_2 = 4(\cos 45^\circ + i\sin 45^\circ)$

$$z_1 z_2 = 4\cdot 2\left[\cos(135^\circ + 45^\circ) + i\sin(135^\circ + 45^\circ)\right]$$
$$= \boxed{8(\cos 180^\circ + i\sin 180^\circ)}$$
$$= 8(-1 + 0i) = \boxed{-8}$$

$$\frac{z_1}{z_2} = \frac{4}{2}\left[\cos(135^\circ - 45^\circ) + i\sin(135^\circ - 45^\circ)\right]$$
$$= \boxed{2(\cos 90^\circ + i\sin 90^\circ)}$$
$$= 2(0 + i) = \boxed{2i}$$

5. $[4\,\text{cis}17^\circ]^3 = 4^3\left[\cos(3\cdot 17^\circ) + i\sin(3\cdot 17^\circ)\right]$
$$= \boxed{64(\cos 51^\circ + i\sin 51^\circ)}$$

6. cube roots of -64: $\quad -64 + 0i$

$$r = \sqrt{(-64)^2} = 64$$
$$\tan\theta = 0 \Longrightarrow \theta = 180^\circ$$
$$-64 + 0i = 64\ \text{cis}\ 180^\circ$$
$$\alpha = \frac{180^\circ}{3} + \frac{360^\circ k}{3} = 60^\circ + 120^\circ k$$
$$k = 1,\ 2,\ 3 \to \alpha = 60^\circ,\ 180^\circ,\ 300^\circ$$

$$\left[64(\cos\theta + i\sin\theta)\right]^{1/3} = \sqrt[3]{64}\left[\cos\alpha + i\sin\alpha\right]$$
$$= 4(\cos\alpha + i\sin\alpha)$$

$$4(\cos 60^\circ + i\sin 60^\circ) = 4\left(\frac{1}{2} + i\frac{\sqrt{3}}{2}\right) = \boxed{2 + 2i\sqrt{3}}$$

$$4(\cos 180^\circ + i\sin 180^\circ) = 4(-1 + 0i) = \boxed{-4}$$

$$4(\cos 300^\circ + i\sin 300^\circ) = 4\left(\frac{1}{2} - i\frac{\sqrt{3}}{2}\right) = \boxed{2 - 2i\sqrt{3}}$$

7. $(2i)^{1/2} = (0+2i)^{1/2}$

$r = \sqrt{2^2} = 2$

$\tan\theta$ is undefined $\Longrightarrow \theta = 90^\circ$

$0 + 2i = 2 \text{ cis} 0^\circ$

$\alpha = \frac{90^\circ}{2} + \frac{360^\circ k}{2} = 45^\circ + 180^\circ k$

$k = 1,\ 2 \rightarrow \alpha = 45^\circ,\ 225^\circ$

$$(0+2i)^{1/2} = 2(\cos 45^\circ + i\sin 45^\circ) = 2\left(\frac{\sqrt{2}}{2} + i\frac{\sqrt{2}}{2}\right) = \boxed{\sqrt{2} + i\sqrt{2}}$$

$$(0+2i)^{1/2} = 2(\cos 225^\circ + i\sin 225^\circ) = 2\left(-\frac{\sqrt{2}}{2} - i\frac{\sqrt{2}}{2}\right) = \boxed{-\sqrt{2} - i\sqrt{2}}$$

8. $x^3 = -1 = (-1+0i)$

$r = \sqrt{(-1)^2} = 1$

$\tan\theta = 0 \Longrightarrow \theta = 180^\circ$

$(-1+0i)^{1/3}$:

$\alpha = \frac{180^\circ}{3} + \frac{360^\circ k}{3} = 60^\circ + 120^\circ k$

$k = 1,\ 2,\ 3 \rightarrow \alpha = 60^\circ,\ 180^\circ,\ 300^\circ$

$[1(\cos\alpha + i\sin\alpha)]^{1/3} = \cos\frac{\alpha}{3} + i\sin\frac{\alpha}{3}$

$\cos 60^\circ + i\sin 60^\circ = \boxed{\text{cis } 60^\circ}$

$\cos 180^\circ + i\sin 180^\circ = -1 + 0i = \boxed{\text{cis } 180^\circ}$

$\cos 300^\circ + i\sin 300^\circ = \boxed{\text{cis } 300^\circ}$

Section 10.6

1. **(a)** $(5,\ 135^\circ)$ Quadrant II
(b) $(2,\ 60^\circ)$ Quadrant I
(c) $(6,\ -30^\circ)$ Quadrant IV
(d) $(4.6,\ 213^\circ)$ Quadrant III

Graph for #3 – #11 odd:

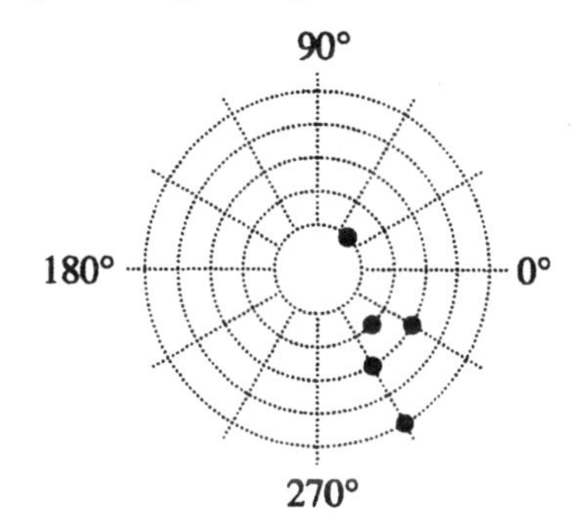

3. $(1,\ 45^\circ) = \boxed{(1,\ 405^\circ),\ (-1, -135^\circ)\ (-1,\ 225^\circ),\ (1, -315^\circ)}$

5. $(-2,\ 135^\circ) = \boxed{(-2, -225^\circ),\ (2, -45^\circ),\ (2,\ 315^\circ),\ (-2,\ 495^\circ)}$

7. $(5, -60^\circ) = \boxed{(5,\ 300^\circ),\ (5, -420^\circ),\ (-5,\ 120^\circ),\ (-5, -240^\circ)}$

9. $(-3, -210^\circ) = \boxed{(-3,\ 150^\circ),\ (3,\ 330^\circ),\ (3, -30^\circ),\ (-3, -570^\circ)}$

11. $(3,\ 300^\circ) = \boxed{(3,\ 660^\circ),\ (-3, -240^\circ),\ (-3,\ 120^\circ),\ (3, -60^\circ)}$

Graph for #13 – #21 odd:

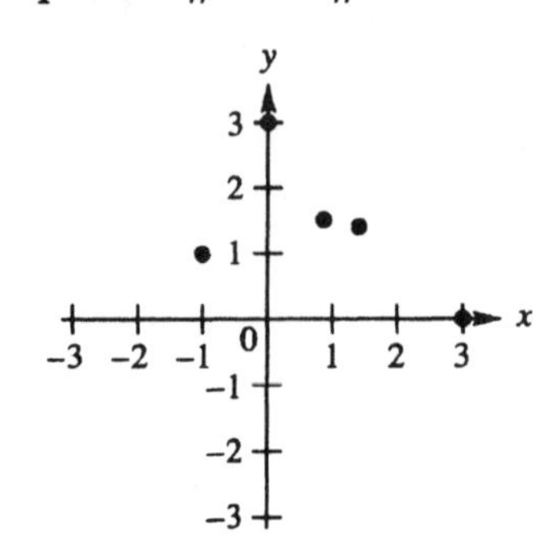

13. $(-1,\ 1)$ $\quad r = \sqrt{(-1)^2 + 1^2} = \pm\sqrt{2}$

$\tan\theta = -1$, Quadrant II

$\theta = 135°$

$(\sqrt{2},\ 135°)$ or $(-\sqrt{2},\ 315°)$

15. $(0,\ 3)$ $\quad r = \sqrt{0^2 + 3^2} = \pm 3$

$\tan\theta : \emptyset, \quad \theta = 90°$

$(3,\ 90°)$ or $(-3,\ 270°)$

17. $(\sqrt{2},\ \sqrt{2})$ $\quad r = \sqrt{(\sqrt{2})^2 + (\sqrt{2})^2} = \pm 2$

$\tan\theta = \dfrac{\sqrt{2}}{\sqrt{2}} = 1$, Quadrant I $\longrightarrow \theta = 45°$

$(2,\ 45°)$ or $(-2,\ 225°)$

19. $\left(\dfrac{\sqrt{3}}{2}, \dfrac{3}{2}\right)$ $\quad r = \sqrt{\left(\dfrac{\sqrt{3}}{2}\right)^2 + \left(\dfrac{3}{2}\right)^2}$

$= \sqrt{\dfrac{3}{4} + \dfrac{9}{4}} = \pm\dfrac{\sqrt{12}}{2} \pm \sqrt{3}$

$\tan\theta = \dfrac{3/2}{\sqrt{3}/2} = \sqrt{3}$, Quadrant I $\longrightarrow \theta = 60°$

$(\sqrt{3},\ 60°)$ or $(-\sqrt{3},\ 240°)$

21. $(3.\ 0)$, $\ r = \sqrt{3^2 + 0^2} = \pm 3$, $\tan\theta = 0$, $\theta = 0°$

$(3,\ 0°)$ or $(-3,\ 180°)$

23. $r = 2 + 2\cos\theta$ $\quad$ cardioid

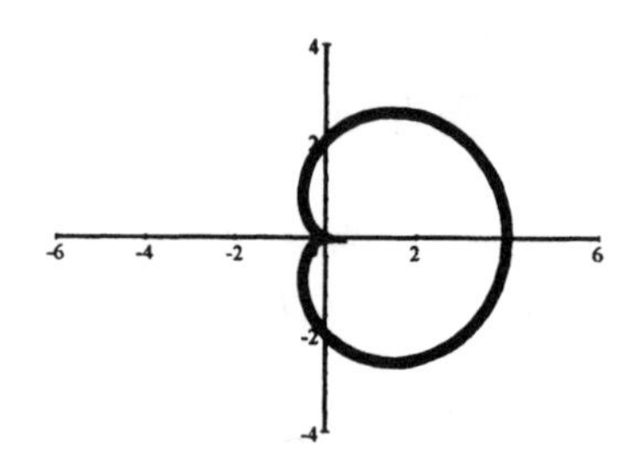

25. $r = 1 + 2\sin\theta$ $\quad$ limacon with a loop

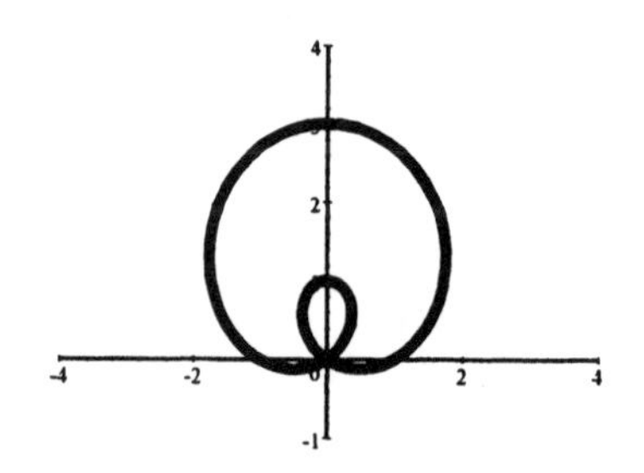

27. $r = 4\cos 2\theta$ $\quad$ 4-leaved rose

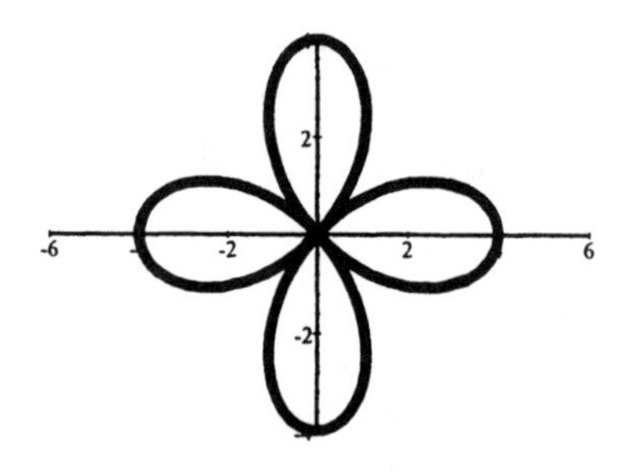

29. $r^2 = 4\cos 2\theta$ $\quad$ lemniscate

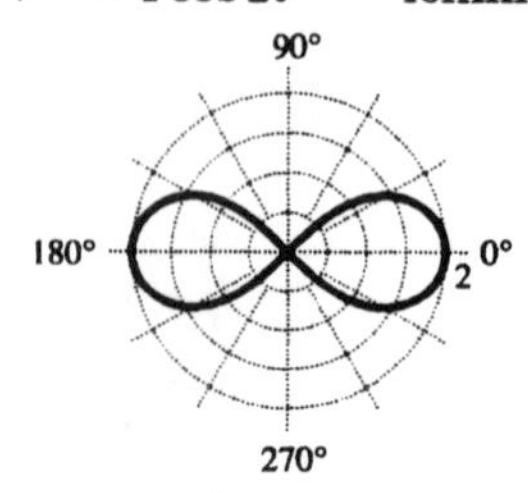

31. $r = 4(1 - \cos\theta)$ cardioid

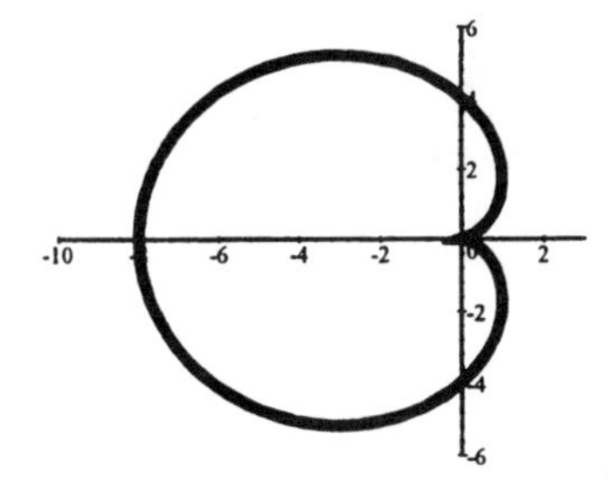

33. $r = 2\sin\theta\tan\theta$ cissoid

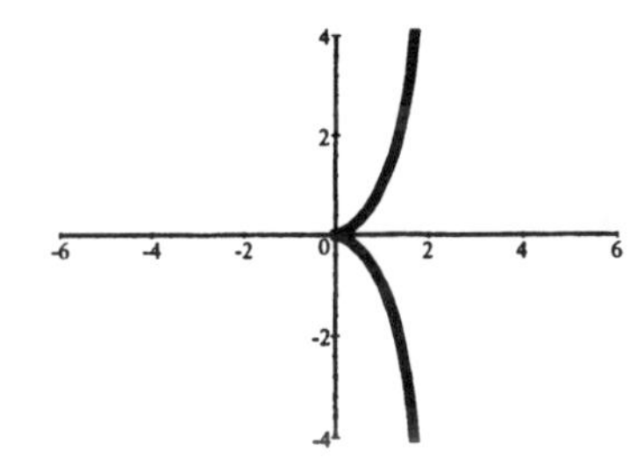

Relating Concepts (35 – 42):

35. (a) Symmetry with respect to the x-axis:
If $(r,\ \theta)$ then $(r,\ -\theta)$.

(b) Symmetry with respect to the y-axis:
If $(r,\ \theta)$ then $(r,\ \pi - \theta)$ or $(-r,\ -\theta)$.

(c) Symmetry with respect to the origin:
If $(r,\ \theta)$ then $(r,\ \pi + \theta)$ or $(-r,\ \theta)$.

36. The graph of $r = f(\theta)$ is symmetric with respect to the polar axis if substitution of $\underline{-\theta}$ for θ leads to an equivalent equation.

37. The graph of $r = f(\theta)$ is symmetric with respect to the vertical line $\theta = \frac{\pi}{2}$ if substitution of $\underline{\pi - \theta}$ for θ leads to an equivalent equation.

38. Alternatively, the graph of $r = f(\theta)$ is symmetric with respect to the vertical line $\theta = \frac{\pi}{2}$ if substitution of $\underline{-r}$ for r and $\underline{-\theta}$ for θ leads to an equivalent equation.

39. The graph of $r = f(\theta)$ is symmetric with respect to the pole if substitution of $\underline{-r}$ for r leads to an equivalent equation.

40. Alternatively, the graph of $r = f(\theta)$ is symmetric with respect to the pole if substitution of $\underline{\pi + \theta}$ for θ leads to an equivalent equation.

41. In general, the completed statements in Exercises **35 – 40** mean that the graphs of polar equations of the form $r = a \pm b\cos\theta$ (where a may be 0) are symmetric with respect to the polar axis.

42. In general, the completed statements in Exercises **35 – 40** mean that the graphs of polar equations of the form $r = a \pm b\sin\theta$ (where a may be 0) are symmetric with respect to the line $\theta = \frac{\pi}{2}$.

43. To graph $(r,\ \theta)$, $r < 0$, you could locate θ, add 180° to it, and move $|r|$ units along the terminal ray of $\theta + 180°$ in standard position.

45. θ is a quadrantal angle and is coterminal with 0°, 90°, 180°, 270°.

47. The graph in example 3, figure 62 is of $r = 1 + \cos\theta$. If you graph $r = 1 - \cos\theta$, the largest part of the graph would be on the left of the vertical axis ($\theta = 90°$) (or reflect $r = 1 + \cos\theta$ across the line $\theta = \frac{\pi}{2}$.)

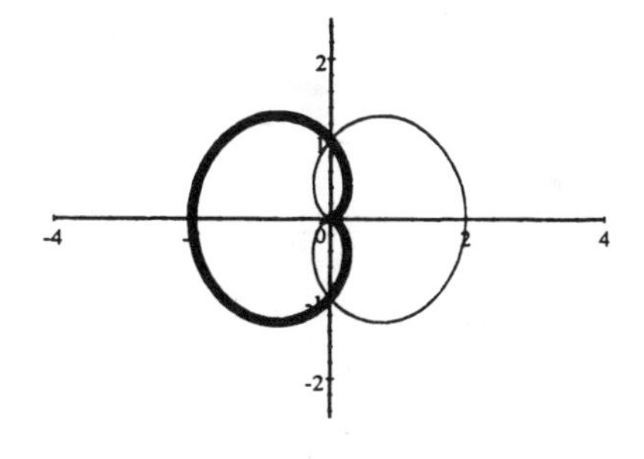

49. In rose curve graphs, $r = a\cos n\theta$ and $r = a\sin n\theta$, the value of a determines the length of the petals and the value of n determines the number of petals (n petals if n is odd; $2n$ petals if n is even).

51. $r = 2\sin\theta$

$r^2 = 2r\sin\theta$

$x^2 + y^2 = 2y$

$x^2 + \left(y^2 - 2y + 1\right) = 1$

$\boxed{x^2 + (y-1)^2 = 1}$

Circle, center $(0,\ 1)$, $r = 1$.

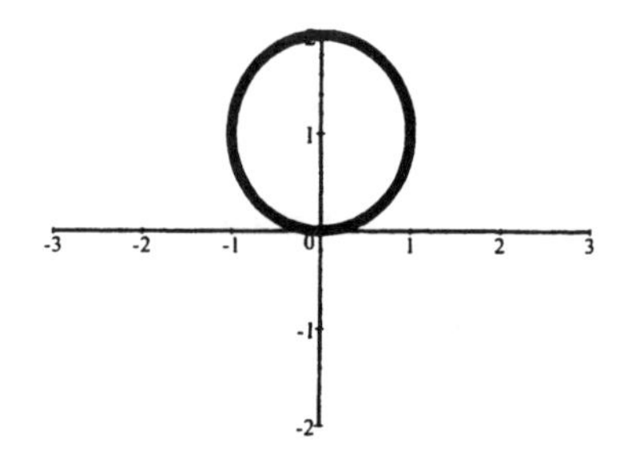

53. $r = \dfrac{2}{1-\cos\theta}$

$r\,(1-\cos\theta) = 2$

$r - r\cos\theta = 2$

$\sqrt{x^2+y^2} - x = 2$

$\sqrt{x^2+y^2} = x + 2$

$x^2 + y^2 = x^2 + 4x + 4$

$y^2 = 4x + 4$

$\boxed{y^2 = 4\,(x+1)}$

Parabola, vertex $(-1,\ 0)$

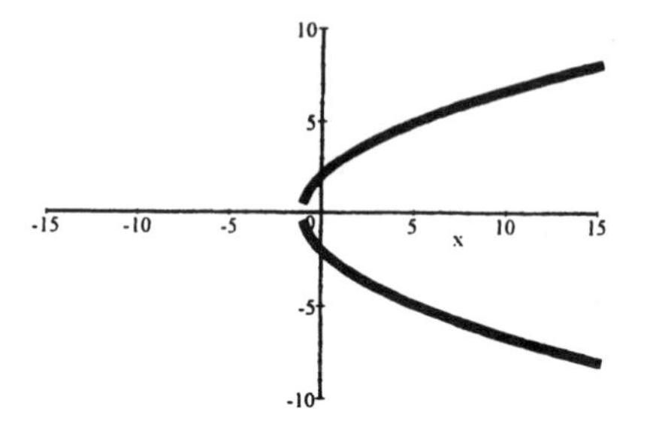

55. $r + 2\cos\theta = -2\sin\theta$

$r^2 + 2r\cos\theta = -2r\sin\theta$

$x^2 + y^2 + 2x + 2y = 0$

$\left(x^2 + 2x + 1\right) + \left(y^2 + 2y + 1\right) = 1 + 1$

$\boxed{(x+1)^2 + (y+1)^2 = 2}$

Circle, center $(-1, -1)$, $r = \sqrt{2}$.

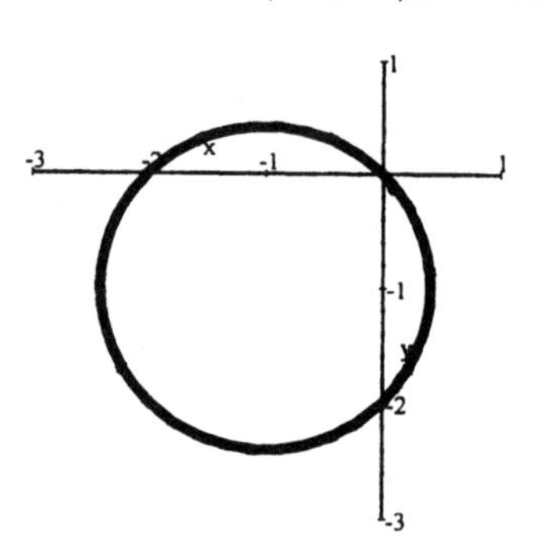

57. $r = 2\sec\theta$

$r = \dfrac{2}{\cos\theta}$

$r\cos\theta = 2$ $\quad\boxed{x = 2}$

Vertical line

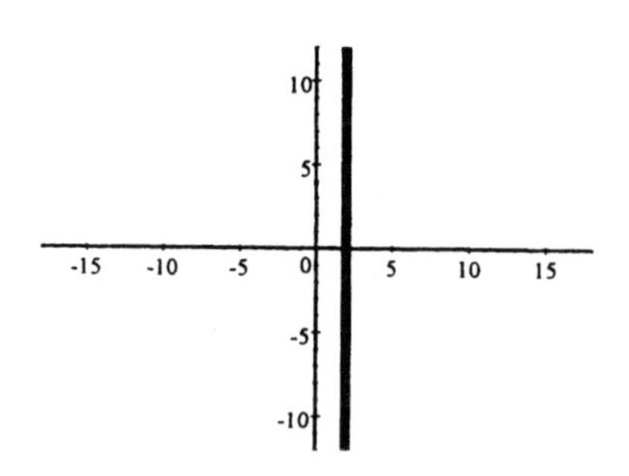

59. $r(\cos\theta + \sin\theta) = 2$

$r\cos\theta + r\sin\theta = 2$

$\boxed{x + y = 2}$

Line through $(0,\ 2)$, $(2,\ 0)$.

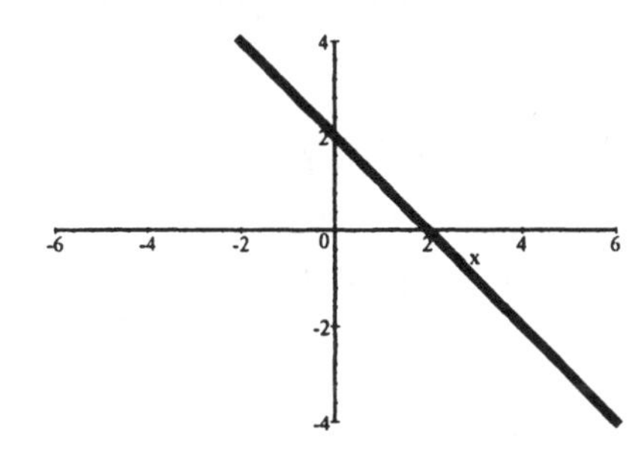

61. $x + y = 4$

$r\cos\theta + r\sin\theta = 4$

$r(\cos\theta + \sin\theta) = 4$

$\boxed{r = \dfrac{4}{\cos\theta + \sin\theta}}$

63. $x^2 + y^2 = 16$

$r^2 = 16$ $\quad\boxed{r = 4}$

$[r$ is always positive$]$

65. $y = 2$

$r\sin\theta = 2$

$r = \dfrac{2}{\sin\theta}$ $\quad\boxed{r = 2\csc\theta}$

67. $r = a\theta,\ \ a = 1$

$\mathbf{r} = \boldsymbol{\theta},\ \ 0 \le \theta \le 4\pi$

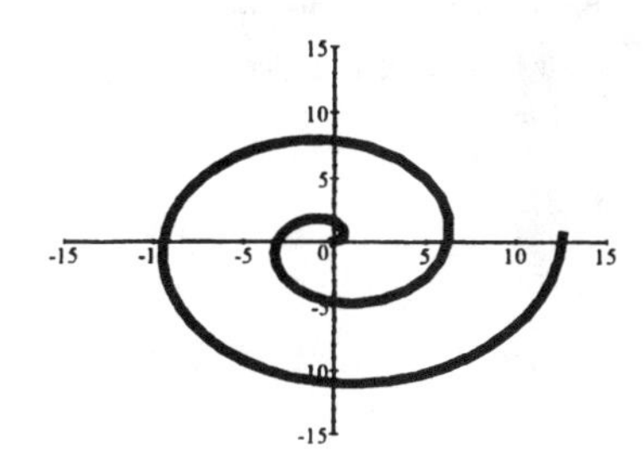

69. $r = a\theta,\ \ a = 1.5$

$\mathbf{r} = \mathbf{1.5}\,\boldsymbol{\theta},\ \ -4\pi \le \theta \le 4\pi$

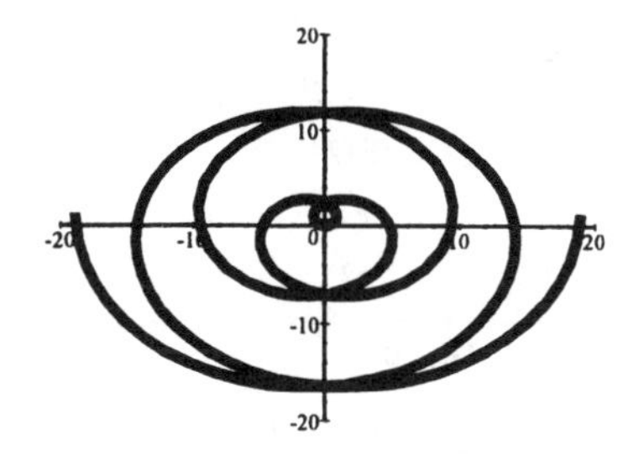

71. This is a matter of preference, but most people would prefer the rectangular form.

73. $r = 4\sin\theta,\ r = 1 + 2\sin\theta,\ 0 \le \theta < \pi$

$4\sin\theta = 1 + 2\sin\theta$

$2\sin\theta = 1$

$\sin\theta = \dfrac{1}{2} \to \theta = \dfrac{\pi}{6}$ or $\dfrac{5\pi}{6}$

The points of intersection are

$$\left(4\sin\frac{\pi}{6},\ \frac{\pi}{6}\right) = \left(2,\ \frac{\pi}{6}\right)$$

$$\left(4\sin\frac{5\pi}{6},\ \frac{5\pi}{6}\right) = \left(2,\ \frac{5\pi}{6}\right)$$

75. $r = 2 + \sin\theta,\ r = 2 + \cos\theta,\ 0 \le \theta < \pi$

$2 + \sin\theta = 2 + \cos\theta$

$\sin\theta = \cos\theta \to \theta = \dfrac{\pi}{4}$ or $\dfrac{5\pi}{4}$

$r = 2 + \sin\dfrac{\pi}{4} = 2 + \dfrac{\sqrt{2}}{2} = \dfrac{4 + \sqrt{2}}{2}$

$r = 2 + \sin\dfrac{5\pi}{4} = 2 - \dfrac{\sqrt{2}}{2} = \dfrac{4 - \sqrt{2}}{2}$

The points of intersection are

$$\left(2 + \sin\frac{\pi}{4},\ \frac{\pi}{4}\right) = \left(\frac{4 + \sqrt{2}}{2},\ \frac{\pi}{4}\right)$$

$$\left(2 + \sin\frac{5\pi}{4},\ \frac{5\pi}{4}\right) = \left(\frac{4 - \sqrt{2}}{2},\ \frac{5\pi}{4}\right)$$

77. (a) Plot the following polar equations on the same polar axis:

Mercury: $r = \dfrac{.39\,(1 - .206^2)}{1 + .206\cos\theta}$

Venus: $r = \dfrac{.78\,(1 - .007^2)}{1 + .007\cos\theta}$

Earth: $r = \dfrac{1\,(1 - .017^2)}{1 + .017\cos\theta}$

Mars: $r = \dfrac{1.52\,(1 - .093^2)}{1 + .093\cos\theta}$

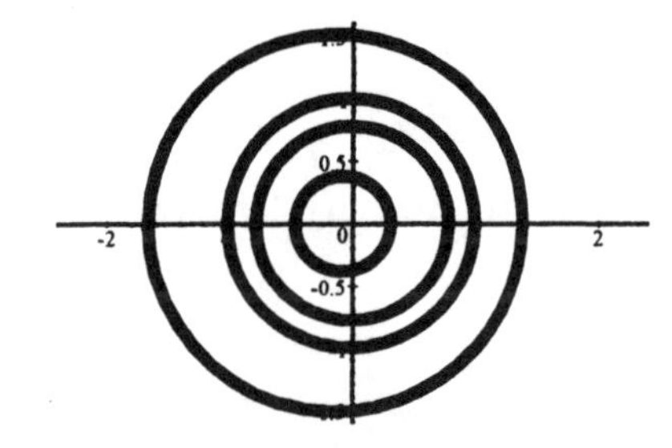

(b) Plot the following polar equations on the same polar axis:

Earth: $r = \dfrac{1\,(1 - .017^2)}{1 + .017\cos\theta}$

Jupiter: $r = \dfrac{5.2\,(1 - .048^2)}{1 + .048\cos\theta}$

Uranus: $r = \dfrac{19.2\,(1 - .047^2)}{1 + .047\cos\theta}$

Pluto: $r = \dfrac{39.4\,(1 - .249^2)}{1 + .249\cos\theta}$

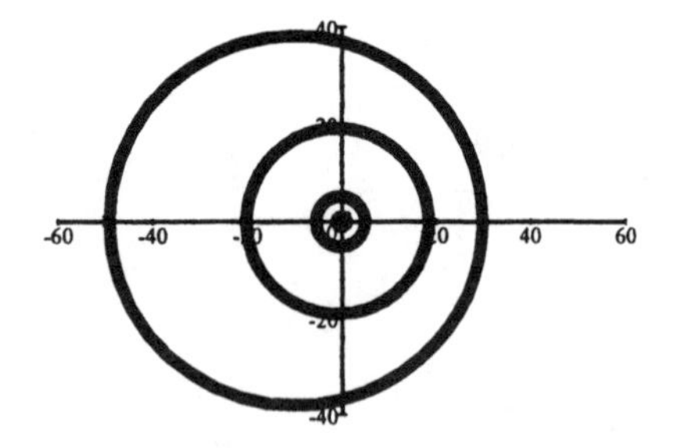

From the graph, Earth is closest to the sun of these four planets.

continued

77. continued

(c) We must determine if the orbit of Pluto is always outside the orbits of the other planets. Since Neptune is closest to Pluto, plot the orbits of Neptune and Pluto on the same polar axes.

Neptune: $r = \dfrac{30.1\,(1 - .009^2)}{1 + .009\cos\theta}$

Pluto: $r = \dfrac{39.4\,(1 - .249^2)}{1 + .249\cos\theta}$

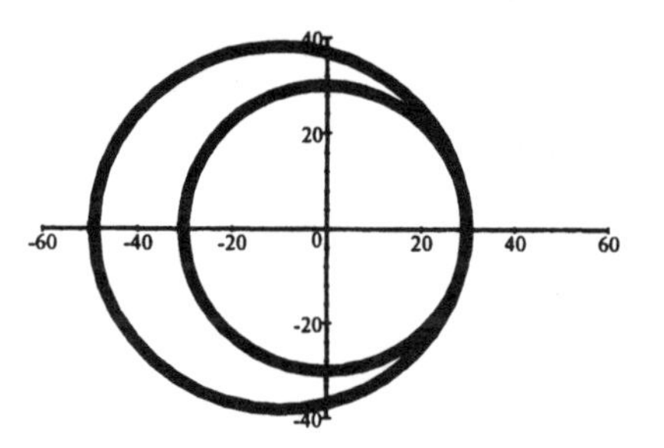

The graph shows that their orbits are very close near the polar axis. use ZOOM to determine that the orbit of Pluto does indeed pass inside the orbit of Neptune. Therefore, there are times when Neptune not Pluto, is the farthest planet from the sun. (However, Pluto's average distance from the sun is considerably greater than Neptune's average distance.)

79. The radio signal is received inside the "figure eight". This region is generally in a southwest-northeast direction from the two towers with maximum distance 150 miles.

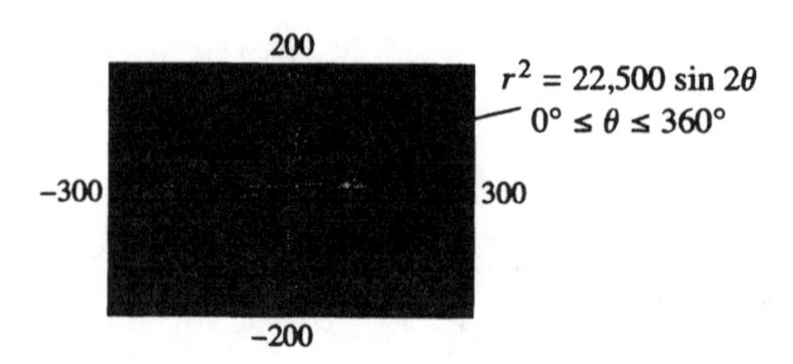

Section 10.7

1. $x = 3\sin t,\ y = 3\cos t;\quad -\pi \le t \le \pi$

$x^2 = 9\sin^2 t \qquad y^2 = 9\cos^2 t$

$x^2 + y^2 = 9\sin^2 t + 9\cos^2 t$

$x^2 + y^2 = 9\left(\sin^2 t + \cos^2 t\right)$

$\boxed{x^2 + y^2 = 9}$ Circle, radius 3

3. $x = 2\cos^2 t,\ y = 2\sin^2 t;\quad 0 \le t \le \dfrac{\pi}{2}$

$x + y = 2\cos^2 t + 2\sin^2 t$

$x + y = 2\left(\sin^2 t + \cos^2 t\right)$

$x + y = 2 \quad \boxed{y = -x + 2}$ Line segment

5. $x = 3\tan t,\ y = 2\sec t;\quad -\dfrac{\pi}{2} < t < \dfrac{\pi}{2}$

$x^2 = 9\tan^2 t \qquad y^2 = 4\sec^2 t$

$\dfrac{x^2}{9} = \tan^2 t \qquad \dfrac{y^2}{4} = \sec^2 t$

$\tan^2 t + 1 = \sec^2 t$

$\dfrac{x^2}{9} + 1 = \dfrac{y^2}{4} \quad \boxed{\dfrac{y^2}{4} - \dfrac{x^2}{9} = 1}$

Hyperbola; upper branch

7. **(a)** $x = 3\cos t,\ y = 3\sin t$ traces a circle of radius 3 once.

(b) $x = 3\cos 2t,\ y = 3\sin 2t$ traces a a circle of radius 3 twice.

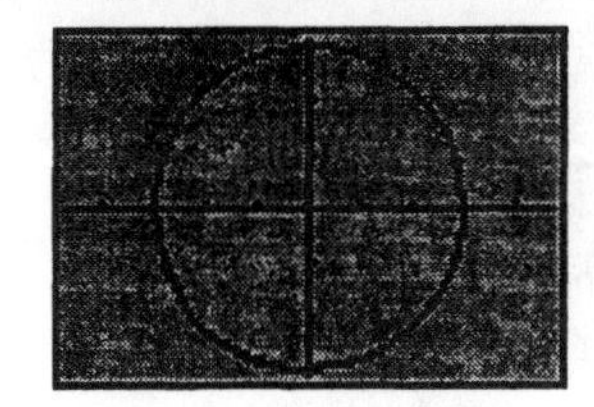

9. **(a)** $x = 3\cos t,\ y = 3\sin t$ traces a circle of radius 3 once counterclockwise starting at (3, 0).

(b) $x = 3\sin t,\ y = 3\cos s$ traces a a circle of radius 3 once clockwise starting at (0, 3).

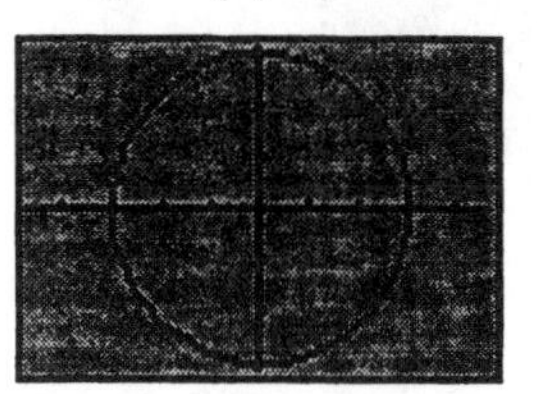

11. $x = \sin t,\ y = \csc t = \dfrac{1}{\sin t},\quad t \in (0,\ \pi)$

$x = \dfrac{1}{y}$ or $\boxed{y = \dfrac{1}{x},\quad x \in (0,\ 1]}$

$y = \frac{1}{x}$ for x in (0, 1)

(1, 1)

13. $x = 2 + \sin t, \qquad y = 1 + \cos t, \quad t \in [0,\ 2\pi]$

$x - 2 = \sin t \qquad y - 1 = \cos t$

$(x-2)^2 = \sin^2 t \qquad (y-1)^2 = \cos^2 t$

$(x-2)^2 + (y-1)^2 = \sin^2 t + \cos^2 t$

$\boxed{(x-2)^2 + (y-1)^2 = 1,\quad x \in [1,\ 3]}$

$(x-2)^2 + (y-1)^2 = 1$ for x in [1, 3]

15. $x = 2 + \cos t,\ y = \sin t - 1; \quad 0 \le t \le 2\pi$

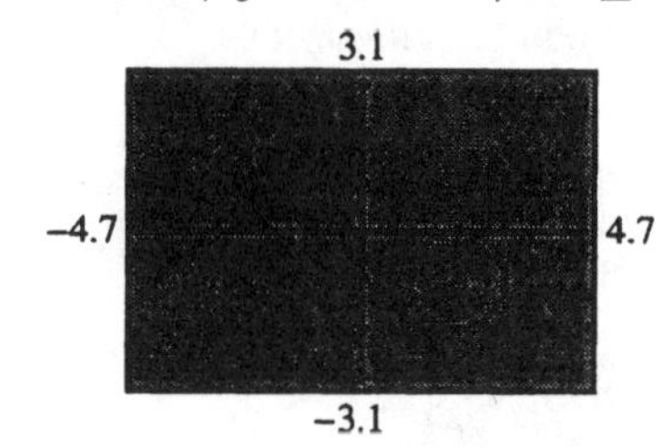

17. $x = \cos^3 t,\ y = \sin^3 t; \quad 0 \le t \le 2\pi$

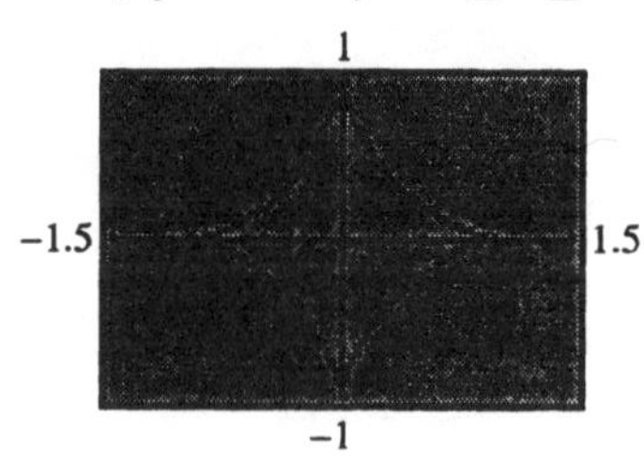

19. $x = |3 \sin t|\,,\ y = |3 \cos t|\,; \quad 0 \le t \le 2\pi$

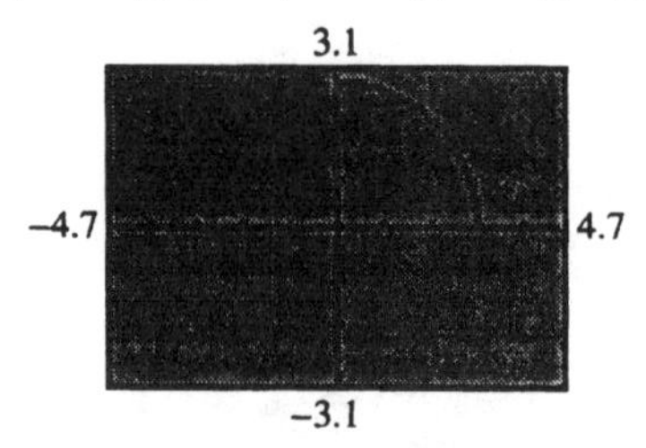

21. $x = t - \sin t,\ y = 1 - \cos; \quad 0 \le t \le 4\pi$

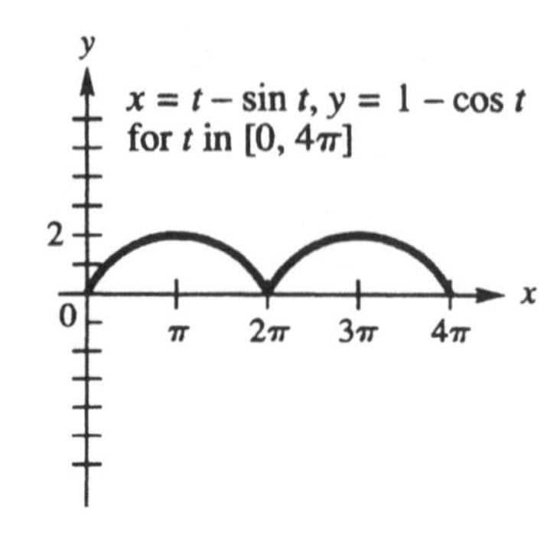

23. $x_1 = 1 \qquad y_1 = 1 + \dfrac{t}{\pi}$ **F**

$x_2 = 1 + \dfrac{t}{3\pi} \qquad y_2 = 2$

$x_3 = 1 + \dfrac{t}{2\pi} \qquad y_3 = 3$

25. $x_1 = 1 \qquad y_1 = 1 + \dfrac{t}{\pi}$ **D**

$x_2 = 1 + 1.3 \sin(.5t) \qquad y_2 = 2 + \cos(.5t)$

27. Answers may vary. For $0 \le t \le 1$:

$x_1 = 0 \qquad y_1 = 2t$

$x_2 = t \qquad y_2 = 0$

29. Answers may vary. For $0 \le t \le \pi$:

$x_1 = \sin t \qquad y_1 = \cos t$

$x_2 = 0 \qquad y_2 = t - 2$

31. Answers may vary.

33. $x = 2 \cos t, \quad y = 3 \sin 2t$

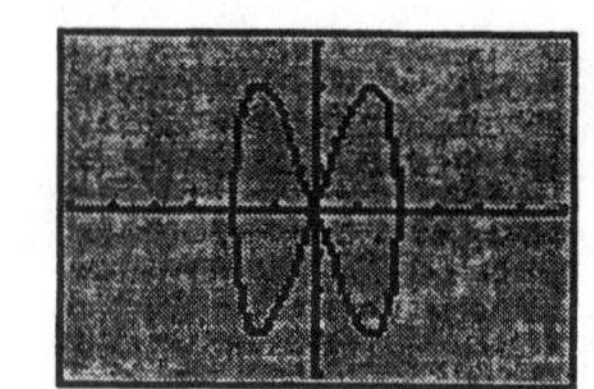

35. $x = 3\sin 4t, \quad y = 3\cos 3t$

37. **(a)** $x = (48\cos 60°)\,t$

$$= 48\left(\frac{1}{2}\right)t = 24t$$

$$y = (48\sin 60°)\,t - 16t^2$$

$$= 48\left(\frac{\sqrt{3}}{2}\right)t - 16t^2$$

$$= 24\sqrt{3}\,t - 16t^2$$

(b) $x = 24t\ , \quad y = -16t^2 + 24\sqrt{3}\,t$

$$t = \frac{x}{24}$$

$$t \longrightarrow y: \quad y = -16\left(\frac{x}{24}\right)^2 + 24\sqrt{3}\left(\frac{x}{24}\right)$$

$$y = -\frac{16x^2}{576} + \sqrt{3}\,x$$

$$y = -\frac{x^2}{36} + \sqrt{3}\,x$$

(c) Graph y and locate the positive zero:

$x \approx 62.3538 \approx$ 62 feet.

$t = \dfrac{x}{24} \approx \dfrac{62.3538}{24} \approx$ 2.6 seconds.

39. **(a)** $x = (88\cos 20°)\,t$

$$y = (88\sin 20°)\,t - 16t^2 + 2$$

$$= -16t^2 + (88\sin 20°)\,t + 2$$

(b) $x = (88\cos 20°)\,t$

$$t = \frac{x}{88\cos 20°}$$

$t \longrightarrow y:$

$$y = -16\left(\frac{x}{88\cos 20°}\right)^2 + (88\sin 20°)\left(\frac{x}{88\cos 20°}\right) + 2$$

$$y = \frac{-16x^2}{7744\cos^2 20°} + (\tan 20°)\,x + 2$$

$$y = \frac{-x^2}{484\cos^2 20°} + (\tan 20°)\,x + 2$$

(c) Graph y and locate the positive zero:

$x \approx 160.868 \approx$ 161 feet.

$t \approx \dfrac{160.868}{88\cos 20°} \approx$ 1.9 seconds.

41. $x = (88\cos 45°)\,t = 88\left(\dfrac{\sqrt{2}}{2}\right)t$

$$= 44\sqrt{2}\,t \longrightarrow t = \frac{x}{44\sqrt{2}}$$

$$y = (88\sin 45°)\,t - 2.66t^2 + 0$$

$$= \frac{88\left(\frac{\sqrt{2}}{2}\right)x}{44\sqrt{2}} - 2.66\left(\frac{x}{44\sqrt{2}}\right)^2$$

$$= x - \frac{2.66x^2}{3872}$$

Graph y: the positive zero is at

$x \approx 1455.639 \approx$ 1456 feet.

43. **(a)**

$x = 82.69265063t$
$y = -16t^2 + 30.09777261t$

30
0
0
200

(b) $x = 82.69265063t = (88\cos\theta)\,t$

$$\cos\theta = \frac{82.69265063t}{88} \longrightarrow \theta \approx 20.0°$$

(c) $x = (88\cos 20.0°)\,t$

$$y = -16t^2 + (88\sin 20.0°)\,t$$

45. $r = a\theta,\ r^2 = x^2 + y^2,\ \theta \in (-\infty,\ \infty)$

$$r = a\theta = \frac{x}{\cos\theta} \Longrightarrow x = a\theta\cos\theta$$

$$r = a\theta = \frac{y}{\sin\theta} \Longrightarrow y = a\,\theta\sin\theta$$

Reviewing Basic Concepts (Sections 10.6 and 10.7)

1. $(-2,\ 130°)$ lies in quadrant IV since it equals $(2,\ -50°)$.

2. $(-2,\ 2)\quad r = \sqrt{(-2)^2 + 2^2} = \sqrt{8} = 2\sqrt{2}$

$$\tan\theta = \frac{2}{-2} = -1,\ \text{quadrant II}$$

$$\theta = 135°$$

$\left(2\sqrt{2},\ 135°\right)$ or $\left(-2\sqrt{2},\ -45°\right)$

Answers may vary.

3. $r = 2 - 2\cos\theta$ cardioid

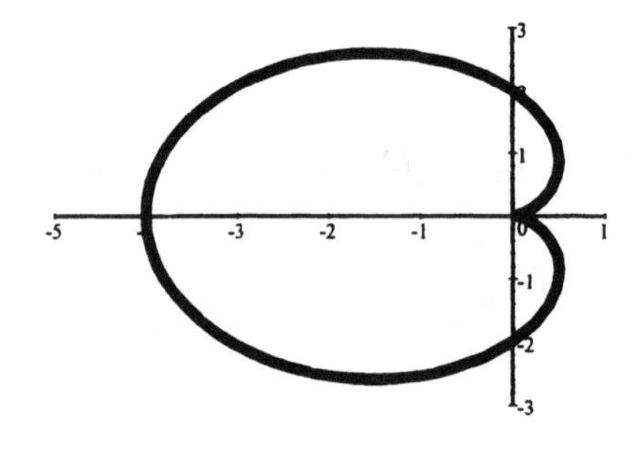

4. $r = 2\cos\theta$

$$r^2 = 2r\cos\theta$$
$$x^2 + y^2 = 2x$$
$$x^2 - 2x + y^2 = 0$$
$$x^2 - 2x + 1 + y^2 = 1$$
$$(x-1)^2 + y^2 = 1$$

5. $x + y = 6$

$$r\cos\theta + r\sin\theta = 6$$
$$r(\cos\theta + \sin\theta) = 6$$
$$r = \frac{6}{\cos\theta + \sin\theta}$$

6. $x = 2\cos t \qquad y = 4\sin t,\qquad 0 \le t \le 2\pi$

$$x^2 = 4\cos^2 t \qquad y^2 = 16\sin^2 t$$

$$\frac{x^2}{4} = \cos^2 t \qquad \frac{y^2}{16} = \sin^2 t$$

$$\frac{x^2}{4} + \frac{y^2}{16} = \cos^2 t + \sin^2 t$$

$$\frac{x^2}{4} + \frac{y^2}{16} = 1$$

7. $x = 2 - \sin t,\quad y = \cos t - 1,\quad 0 \le t \le 2\pi$

2

−3 3

$x = 2 - \sin t$
$y = \cos t - 1$
$0 \le t < 2\pi$

−2

8. $x = (88\cos 45°)\,t = 88\left(\frac{\sqrt{2}}{2}\right)t$

$$= 44\sqrt{2}\,t \longrightarrow t = \frac{x}{44\sqrt{2}}$$

$$y = (88\sin 45°)\,t - 16t^2 + 50$$

$$= \frac{88\left(\frac{\sqrt{2}}{2}\right)x}{44\sqrt{2}} - 16\left(\frac{x}{44\sqrt{2}}\right)^2 + 50$$

$$= \frac{-16x^2}{3872} + x + 50$$

$$= -\frac{x^2}{242} + x + 50$$

Graph y: the positive zero is at

$x \approx 284.53 \approx \boxed{285 \text{ feet}}$.

Chapter 10 Review

1. $C = 74°10'$, $c = 96.3$ m, $B = 39°30'$

$$\frac{c}{\sin C} = \frac{b}{\sin B}$$

$$\frac{96.3}{\sin 74°10'} = \frac{b}{\sin 39°30'}$$

$b \approx 63.67$ $\boxed{b \approx 63.7\,\text{m}}$

2. $A = 100.2°$, $a = 165$ m, $B = 25.0°$

$$\frac{a}{\sin A} = \frac{b}{\sin B}$$

$$\frac{165}{\sin 100.2°} = \frac{b}{\sin 25.0°}$$

$b \approx 70.8518$ $\boxed{b \approx 70.9\,\text{m}}$

3. $a = 86.14$ in, $b = 253.2$ in, $c = 241.9$ in

$$\cos A = \frac{b^2 + c^2 - a^2}{2bc}$$

$$= \frac{253.2^2 + 241.9^2 - 86.14^2}{2\,(253.2)\,(241.9)}$$

≈ 0.940469

$A \approx 19.8695°$ $(19°52'10'')$

$\boxed{A \approx 19.87° \quad \text{or} \quad 19°52'}$

4. $a = 14.8$ ft, $b = 19.7$ ft, $c = 31.8$ ft

$$\cos B = \frac{a^2 + c^2 - b^2}{2ac}$$

$$= \frac{14.8^2 + 31.8^2 - 19.7^2}{2\,(14.8)\,(31.8)}$$

≈ 0.8947284

$B \approx 26.5264°$ $(26°32')$

$\boxed{B \approx 26.5° \quad \text{or} \quad 26°30'}$

5. $A = 129°40'$, $a = 127$ ft, $b = 69.8$ ft

$$\frac{a}{\sin A} = \frac{b}{\sin B}$$

$$\frac{127}{\sin 129°40'} = \frac{69.8}{\sin B}$$

$\sin B \approx 0.423071$

$B \approx 25.0286°$ $(25°01'43'')$

$\boxed{B \approx 25.0° \quad \text{or} \quad 25°00'}$

6. $B = 39°50'$, $a = 340$ cm, $b = 268$ cm

$$\frac{a}{\sin A} = \frac{b}{\sin B}$$

$$\frac{340}{\sin A} = \frac{268}{\sin 39°50'}$$

$\sin A \approx 0.812646$

$A_1 \approx 54.355°$ $(54°21')$

$A_2 \approx 125°39'$ $(180° - A_1)$

$[\text{ck: } A_2 + B < 180°]$

$\boxed{A \approx 54°20' \quad \text{or} \quad 125°40'}$

7. $B = 120.7°$, $a = 127$ ft, $c = 69.8$ ft

$$b^2 = a^2 + c^2 - 2ac\cos B$$

$$= 127^2 + 69.8^2 - 2\,(127)\,(69.8)\cos 120.7°$$

≈ 30052.5575

$b \approx 173.3567$ $\boxed{b \approx 173\text{ ft}}$

8. $A = 46.2°$, $b = 184$ cm, $c = 192$ cm

$$a^2 = b^2 + c^2 - 2bc\cos A$$

$$= 184^2 + 192^2 - 2\,(184)\,(192)\cos 46.2°$$

≈ 21815.93191

$a \approx 147.702$ $\boxed{a \approx 148\text{ cm}}$

9. $b = 840.6$ m, $c = 715.9$ m, $A = 149.3°$

$$Area = \frac{1}{2}bc\sin A$$

$$= \frac{1}{2}\,(840.6)\,(715.9)\sin 149.3°$$

≈ 153618.67

$\boxed{\text{Area} \approx 153{,}600\text{ m}^2}$

10. $a = 6.90$ ft, $b = 10.2$ ft, $C = 35°10'$

$$Area = \frac{1}{2}ab\sin C$$

$$= \frac{1}{2}\,(6.90)\,(10.2)\sin 35°10'$$

≈ 20.2679

$\boxed{\text{Area} \approx 20.3\text{ ft}^2}$

11. $a = .913\,\text{km},\ b = .816\,\text{km},\ c = .582\,\text{km}$

(i) $s = \frac{1}{2}(a+b+c)$

$= \frac{1}{2}(.913 + .816 + .582)$

$= 1.1555$

(ii) $s - a = 1.1555 - .913 = .2425$

$s - b = 1.1555 - .816 = .3395$

$s - c = 1.1555 - .582 = .5735$

(iii) $A = \sqrt{s(s-a)(s-b)(s-c)}$

$= \sqrt{1.1555 \cdot .2425 \cdot .3395 \cdot .5735}$

$= \sqrt{0.0545576} \approx .233576$

Area $\approx .234\ \text{km}^2$

12. $a = 43\,\text{m},\ b = 32\,\text{m},\ c = 51\,\text{m}$

(i) $s = \frac{1}{2}(a+b+c)$

$= \frac{1}{2}(43 + 32 + 51) = 63$

(ii) $s - a = 63 - 43 = 20$

$s - b = 63 - 32 = 31$

$s - c = 63 - 51 = 12$

(iii) $A = \sqrt{s(s-a)(s-b)(s-c)}$

$= \sqrt{63(20)(31)(12)}$

$= \sqrt{468720} \approx 684.6313$

Area $\approx 680\ \text{m}^2$

13. Let h = height of balloon; two right triangles are formed.

$C = 180° - (47°20' + 24°50')$

$= 180° - 72°10' = 107°50'$

Use the law of sines to solve for BC:

$$\frac{\sin 107°50'}{8.4} = \frac{\sin 24°50'}{BC}$$

$BC \approx 3.70589632$

Use the left right triangle to solve for h:

$$\sin 47°20' \approx \frac{h}{3.70589632}$$

$h \approx 3.70589632\ \sin 47°20'$

≈ 2.72497895

The balloon height is ≈ 2.7 miles.

14. Find the angle at home plate; extend the line from home to 2nd base, forming a right $\triangle$ with sides 60′ and 60′.

$$\tan\theta = \frac{60}{60} = 1,\ \theta = 45°$$

$c^2 = a^2 + b^2 - 2ab\cos C$

$= 60^2 + 46^2 - 2(60)(46)\cos 45°$

≈ 1812.7706

$c \approx 42.5766$

The distance from the pitcher's mound to 3rd base is ≈ 43 feet.

15. Find x (YZ):

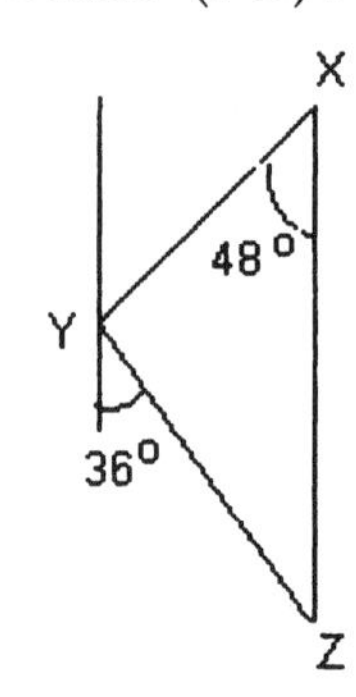

The angle at $Y = 96°$

$(180° - 48° - 36°)$

$$\frac{x}{\sin 48°} = \frac{10}{\sin 96°}$$

$x \approx 7.47238$

$x \approx 7$ km

16. Find b in the following $\triangle$:

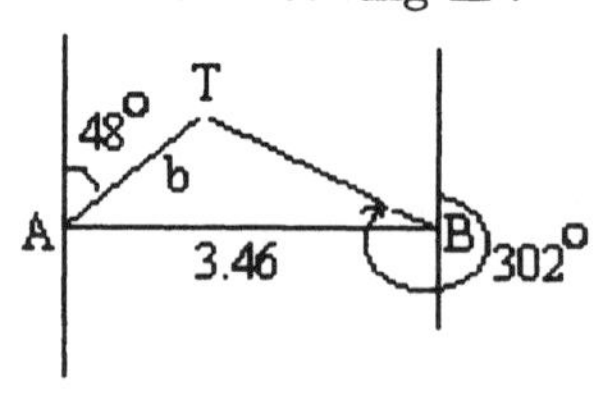

The $\triangle$ formed has $A = 42°$ $(90° - 48°)$, $B = 32°$ $(302° - 270°)$, and $T = 106°$ $(180° - A - B)$:

$$\frac{b}{\sin 32°} = \frac{3.46}{\sin 106°}$$

$b \approx 1.9074$

The distance between the transmitter and $A \approx 1.91$ miles.

17. $B = 58.4°,\ C = 27.9°,\ BC = 125$ ft

$A = 180° - (B + C) = 93.7°$

$$\frac{BC}{\sin A} = \frac{AB}{\sin C}$$

$$\frac{125}{\sin 93.7°} = \frac{AB}{\sin 27.9°}$$

$AB \approx 58.6134$ $\boxed{AB \approx 58.6 \text{ feet}}$

18. Let l = length of brace; 3rd angle $= 43°$

$[180° - (22° + 115°)]$

$$\frac{8.0}{\sin 43°} = \frac{l}{\sin 115°}$$

$l \approx 10.631$ $\boxed{\text{Length} \approx 11 \text{ feet}}$

19. Let a = distance between wire;

$$a^2 = 15.0^2 + 12.2^2 - 2\,(15.0)\,(12.2)\cos 70.3°$$

$$\approx 250.4631$$

$a \approx 15.826$

$\boxed{\text{The ends of the wire should be} \approx 15.8 \text{ feet apart.}}$

20. $$\cos C = \frac{AC^2 + BC^2 - AB^2}{2\,(AC)\,(BC)}$$

$$= \frac{102^2 + 135^2 - 150^2}{2\,(102)\,(135)}$$

$$\approx .222549$$

$C \approx 77.1412°$ $\boxed{C \approx 77.1°}$

21. Given a, A, and C in triangle ABC does **not** lead to an ambiguous case.

$B = 180° - (A + C)$

The ambiguous case can exist only if you are given two sides and the angle opposite the smaller side.

22. **No**, a triangle ABC cannot exist if $a = 4.7$, $b = 2.3$, $c = 7.0$. The sum of the lengths of each pair of sides must be greater than the length of the third side. Here, $4.7 + 2.3 = 7.0$, which is not greater than the length of the third side.

23. $\dfrac{\sin A}{a} = \dfrac{\sin B}{b}$ $a = 10,\ B = 30°$

$$\sin A = \frac{a \sin B}{b}$$

$$\sin A = \frac{10 \sin 30°}{b}$$

$$\sin A = \frac{10\left(\frac{1}{2}\right)}{b}$$

$$\sin A = \frac{5}{b}$$

if $b = 5 : \sin A = 1,\ A = 90°$

if $b < 5 : \sin A > 1$, does not exist

if $5 < b < 10 : A$ can be 2 angles

if $b = 10 : A = 30°$ or $150°$ (not possible)

if $b > 10 : A < 30°$ or $> 150°$ (not possible)

(a) Exactly 1 value: $b = 5$ or $b \geq 10$

(b) Two values: $5 < b < 10$

(c) No value: $b < 5$

24. $c^2 = a^2 + b^2 - 2ab \cos 90°$

$c^2 = a^2 + b^2 - 2ab\,(0)$

$c^2 = a^2 + b^2$

The Pythagorean Theorem

25. $\mathbf{a} + 3\mathbf{c}$:

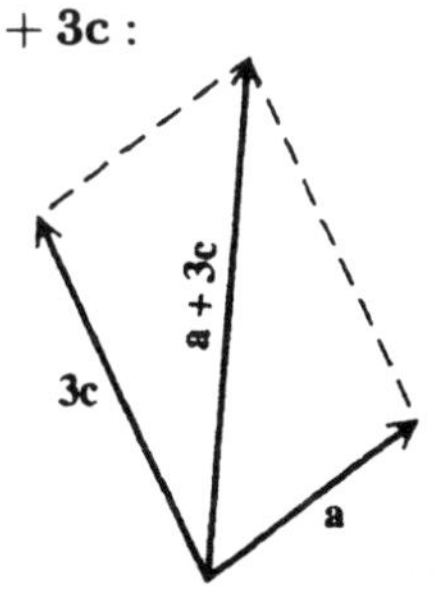

26. $\mathbf{u} = \langle 21,\ -20 \rangle$ Quadrant IV

Magnitude:

$$|u| = \sqrt{21^2 + (-20)^2} = \sqrt{841} = \boxed{29}$$

Direction angle:

$$\cos\theta = \frac{a}{|u|} = \frac{21}{29} \to \theta \approx 43.6°$$

Quadrant IV: $360° = \theta \approx \boxed{316.4°}$

27. $\mathbf{u} = \langle -9,\ 12\rangle$ Quadrant II

Magnitude:

$$|u| = \sqrt{(-9)^2 + 12^2} = \sqrt{225} = \boxed{15}$$

Direction angle:

$$\cos\theta = \frac{a}{|u|} = \frac{-9}{15} \longrightarrow \theta \approx \boxed{126.9^\circ}$$

28. $|v| = 50,\ \theta = 45^\circ$

$$x = r\cos\theta = 50\cos 45^\circ = 50\left(\frac{\sqrt{2}}{2}\right) = 25\sqrt{2}$$

$$y = r\sin\theta = 50\sin 45^\circ = 50\left(\frac{\sqrt{2}}{2}\right) = 25\sqrt{2}$$

$\boxed{\text{horizontal: } 25\sqrt{2};\quad \text{vertical: } 25\sqrt{2}}$

29. $|v| = 69.2,\ \theta = 75^\circ$

$$x = r\cos\theta = 69.2\cos 75^\circ \approx 17.9103$$

$$y = r\sin\theta = 69.2\sin 75^\circ \approx 66.8421$$

$\boxed{\text{horizontal} \approx 17.9;\quad \text{vertical} \approx 66.8}$

30. $|v| = 964,\ \theta = 154^\circ 20'$

$$x = r\cos\theta = 964\cos 154^\circ 20' \approx -868.88$$

$$y = r\sin\theta = 964\sin 154^\circ 20' \approx 417.542$$

$\boxed{\text{horizontal} \approx -869;\quad \text{vertical} \approx 418}$

31. **a)** $u\cdot v = \langle 6,\ 2\rangle\cdot\langle 3, -2\rangle = 6(3) + 2(-2) = 18 - 4 = \boxed{14}$

b)
$$\cos\theta = \frac{u\cdot v}{|u|\ |v|} = \frac{\langle 6,\ 2\rangle\cdot\langle 3, -2\rangle}{|\langle 6,\ 2\rangle|\ |\langle 3, -2\rangle|} = \frac{14}{\sqrt{36+4}\ \sqrt{9+4}} = \frac{14}{\sqrt{40}\ \sqrt{13}} = \frac{14}{2\sqrt{130}} \times \frac{\sqrt{130}}{\sqrt{130}} = \frac{7\sqrt{130}}{130}$$

$$\theta = \cos^{-1}\left(\frac{7\sqrt{130}}{130}\right) \approx 52.12502 \approx \boxed{52.13^\circ}$$

32. **a)** $u\cdot v = \langle 2\sqrt{3},\ 2\rangle\cdot\langle 5,\ 5\sqrt{3}\rangle = (2\sqrt{3})(5) + 2(5\sqrt{3}) = 10\sqrt{3} + 10\sqrt{3} = \boxed{20\sqrt{3}}$

b)
$$\cos\theta = \frac{u\ \cdot\ v}{|u|\ |v|} = \frac{20\sqrt{3}}{|\langle 2\sqrt{3},\ 2\rangle|\ |\langle 5,\ 5\sqrt{3}\rangle|} = \frac{20\sqrt{3}}{(\sqrt{12+4})\ (\sqrt{25+75})} = \frac{20\sqrt{3}}{4\times 10} = \frac{\sqrt{3}}{2}$$

$$\theta = \cos^{-1}\left(\frac{\sqrt{3}}{2}\right) = \boxed{30^\circ}$$

33. Yes; nonzero vectors **u** and **v** are orthogonal vectors if and only if $\mathbf{u}\cdot\mathbf{v} = 0$.

$$\langle 5, -1\rangle\cdot\langle -2, -10\rangle = \langle (5)(-2) + (-1)(-10)\rangle = -10 + 10 = 0$$

34. Let $\mathbf{x}$ = the resultant vector, θ is the angle opposite $\mathbf{x}$.

$$\theta = 180^\circ - 15^\circ - 10^\circ = 155^\circ$$

$$|x|^2 = 12^2 + 18^2 - 2(12)(18)\cos 155^\circ$$

$$|x|^2 \approx 859.52496$$

$$\mathbf{x} \approx 29.318 \approx \boxed{29\text{ lb}}$$

35. Let $\mathbf{x}$ = the resultant vector.

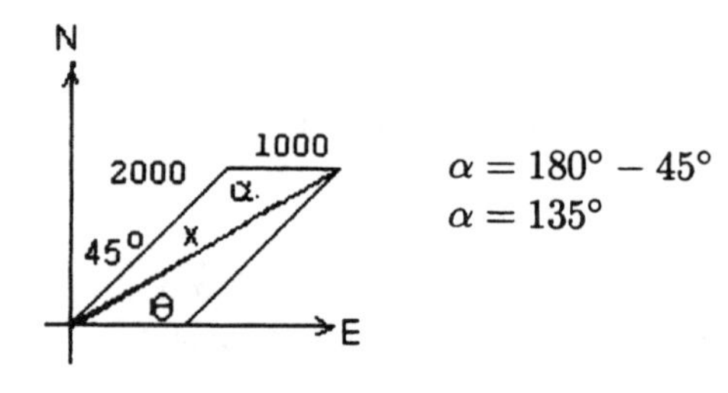

$$\alpha = 180^\circ - 45^\circ$$
$$\alpha = 135^\circ$$

$$|\mathbf{x}|^2 = 2000^2 + 1000^2 - 2(2000)(1000)\cos 135^\circ$$

$$|\mathbf{x}| \approx 2797.9327 \approx \boxed{2800\text{ newtons}}$$

Using the law of sines, we get

$$\frac{\sin 135^\circ}{2797.93} = \frac{\sin\alpha}{2000}$$

$$\alpha \approx 30.36123^\circ \approx \boxed{30.4^\circ}$$

36. Let $|g|$ = groundspeed; the parallelogram has an $\angle$ of 82° $[(212° - 180°) + (360° - 310°)]$.

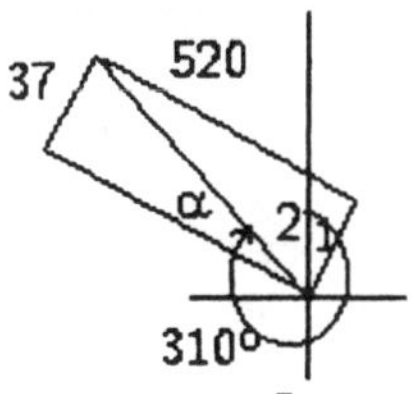

1 = 212° - 180°
2 = 360° - 310°

$$\frac{520}{\sin 82°} = \frac{37}{\sin\alpha}$$

$\sin\alpha = .0704614$

$\alpha \approx 4.04° \approx 4°$

Bearing $= 310° - \alpha$
$= 306°$

The 3rd angle of the triangle will be 94° $(180° - (4° + 82°))$.

$$\frac{520}{\sin 82°} = \frac{|g|}{\sin 94°}$$

$|g| \approx 523.83 \approx 524$

Heading of 306° Actual speed is 524 mph

37. A parallelogram is formed with sides of length 3.2 and 5.1, and angles of 12° and 168° $(180° - 12°)$. Use the law of cosines to solve for diagonal $|v|$:

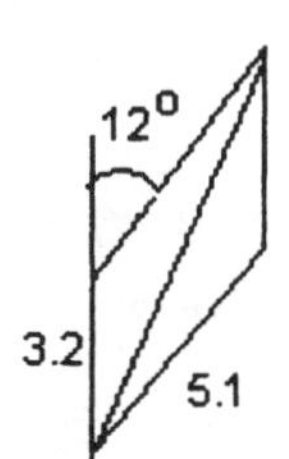

$|v|^2 = 3.2^2 + 5.1^2 - 2(3.2)(5.1)\cos 168°$
≈ 68.176738

$|v| \approx 8.257 \approx 8.3$

$$\frac{\sin 168°}{8.3} = \frac{\sin\alpha}{3.2}$$

$\sin\alpha \approx .0801587 \rightarrow \alpha \approx 4.598°$

Bearing $= 12° - 4.598°$
$= 7.402° \quad (7°24'7'')$

Resulting Bearing = 7.4° or 7°20′ Resulting speed = 8.3 mph

38. $5i$

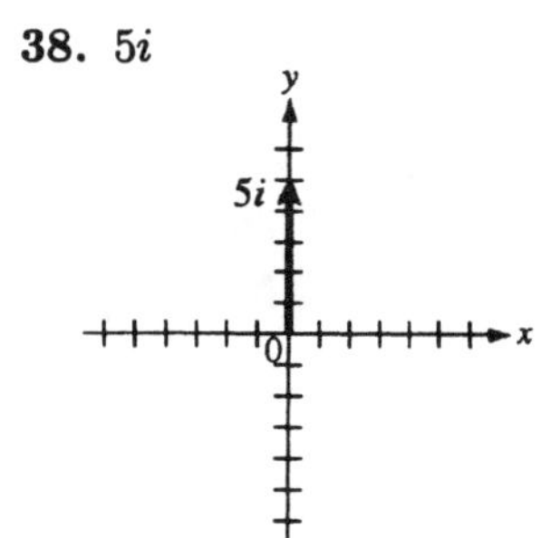

39. $-4 + 2i$

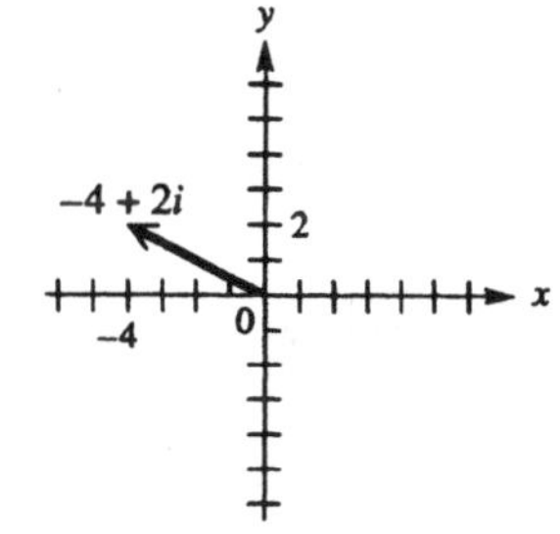

40. $(7 + 3i) + (-2 + i) = (7 - 2) + (3 + 1)i$
$= 5 + 4i$

41. $(2 - 4i) + (5 + i) = (2 + 5) + (-4 + 1)i$
$= 7 - 3i$

42. $-2 + 2i: \quad x = -2,\ y = 2$

$r = \sqrt{x^2 + y^2}$
$= \sqrt{(-2)^2 + (2)^2}$
$= \sqrt{8} = 2\sqrt{2}$

$$\tan\theta = \frac{y}{x} = \frac{2}{-2} = -1$$

$\theta = 135°,\ -45°$

Since $-2 + 2i$ $(-, +) \in$ Quadrant II:

$\boxed{2\sqrt{2}\,\text{cis}\,135°}$

43. $3(\cos 90° + i\sin 90°)$
$= 3(0 + 1i) = \boxed{3i}$

44. $2\,\text{cis}\,225° = 2(\cos 225° + i\sin 225°)$

$$= 2\left(-\frac{\sqrt{2}}{2} - \frac{\sqrt{2}}{2}i\right)$$

$= \boxed{-\sqrt{2} - i\sqrt{2}}$

45. $-4+4i\sqrt{3}$

$x=-4,\ y=4\sqrt{3},\ r=8$

$\tan\theta=-\sqrt{3},\ \theta=120^\circ \quad \boxed{8\operatorname{cis}120^\circ}$

46. $(r_1\operatorname{cis}\theta_1)(r_2\operatorname{cis}\theta_2)=r_1r_2\operatorname{cis}(\theta_1+\theta_2)$

$$\begin{aligned}(5\operatorname{cis}90^\circ)(6\operatorname{cis}180^\circ) &= (5)(6)[\operatorname{cis}(90^\circ+180^\circ)]\\ &= \mathbf{30}\operatorname{cis}\mathbf{270^\circ}\\ &= 30(\cos 270^\circ+i\sin 270^\circ)\\ &= 30(0-i)=\boxed{-30i}\end{aligned}$$

47. $(r_1\operatorname{cis}\theta_1)(r_2\operatorname{cis}\theta_2)=r_1r_2\operatorname{cis}(\theta_1+\theta_2)$

$$\begin{aligned}(3\operatorname{cis}135^\circ)(2\operatorname{cis}105^\circ) &= (3)(2)[\operatorname{cis}(135^\circ+105^\circ)]\\ &= \mathbf{6}\operatorname{cis}\mathbf{240^\circ}\\ &= 6(\cos 240^\circ+i\sin 240^\circ)\\ &= 6\left(-\frac{1}{2}-\frac{\sqrt{3}}{2}i\right)=\boxed{-3-3i\sqrt{3}}\end{aligned}$$

48. $\dfrac{r_1\operatorname{cis}\theta_1}{r_2\operatorname{cis}\theta_2}=\dfrac{r_1}{r_2}\operatorname{cis}(\theta_1-\theta_2)$

$$\begin{aligned}\frac{2\operatorname{cis}60^\circ}{8\operatorname{cis}300^\circ} &= \frac{2}{8}[\operatorname{cis}(60^\circ-300^\circ)]\\ &= \frac{\mathbf{1}}{\mathbf{4}}\operatorname{cis}(\mathbf{-240^\circ})\\ &= \frac{1}{4}[\cos(-240^\circ)+i\sin(-240^\circ)]\\ &= \frac{1}{4}\left(-\frac{1}{2}+\frac{\sqrt{3}}{2}i\right)\\ &= \boxed{-\frac{1}{8}+\frac{\sqrt{3}}{8}i}\end{aligned}$$

49. $\dfrac{r_1\operatorname{cis}\theta_1}{r_2\operatorname{cis}\theta_2}=\dfrac{r_1}{r_2}\operatorname{cis}(\theta_1-\theta_2)$

$$\begin{aligned}\frac{4\operatorname{cis}270^\circ}{2\operatorname{cis}90^\circ} &= \frac{4}{2}[\operatorname{cis}(270^\circ-90^\circ)]\\ &= \mathbf{2}\operatorname{cis}\mathbf{180^\circ}\\ &= 2[\cos 180^\circ+i\sin 180^\circ]\\ &= 2(-1+0i)=\boxed{-2}\end{aligned}$$

50. $(\sqrt{3}+i)^3:\quad x=\sqrt{3},\ y=1$

$$r=\sqrt{(\sqrt{3})^2+1}=2$$

$$\begin{aligned}\tan\theta &= \frac{1}{\sqrt{3}},\ \text{Quadrant I}\\ \theta &= 30^\circ\end{aligned}$$

$$\begin{aligned}[2\operatorname{cis}30^\circ]^3 &= 2^3\operatorname{cis}(3\cdot 30^\circ)\\ &= 8(\cos 90^\circ+i\sin 90^\circ)\\ &= 8(0+1i)=\boxed{8i}\end{aligned}$$

51. $(2-2i)^5:\quad x=2,\ y=-2$

$$r=\sqrt{2^2+(-2)^2}=\sqrt{8}=2\sqrt{2}$$

$$\begin{aligned}\tan\theta &= \frac{-2}{2}=-1,\ \text{Quadrant IV}\\ \theta &= -45^\circ\end{aligned}$$

$$\begin{aligned}[2\sqrt{2}\operatorname{cis}(-45^\circ)]^5 &= (2\sqrt{2})^5\operatorname{cis}(5\cdot -45^\circ)\\ &= 128\sqrt{2}\operatorname{cis}(-225^\circ)\\ &= 128\sqrt{2}(\cos 135^\circ+i\sin 135^\circ)\\ &= 128\sqrt{2}\left(-\frac{\sqrt{2}}{2}+\frac{\sqrt{2}}{2}i\right)\\ &= \boxed{-128+128i}\end{aligned}$$

52. $[r(\cos\theta+i\sin\theta)]^n=r^n(\cos n\theta+i\sin n\theta)$

$$\begin{aligned}(\cos 100^\circ+i\sin 100^\circ)^6 &= 1^6\operatorname{cis}(6\cdot 100^\circ)\\ &= \operatorname{cis}600^\circ\\ &= \cos 240^\circ+i\sin 240^\circ\\ &= \boxed{-\frac{1}{2}-\frac{\sqrt{3}}{2}i}\end{aligned}$$

53. $[r(\cos\theta+i\sin\theta)]^n=r^n(\cos n\theta+i\sin n\theta)$

$$\begin{aligned}[\operatorname{cis}20^\circ]^3 &= 1^3\operatorname{cis}(3\cdot 20^\circ)\\ &= \cos 60^\circ+i\sin 60^\circ\\ &= \boxed{\frac{1}{2}+\frac{\sqrt{3}}{2}i}\end{aligned}$$

54. $r^{1/n}(\cos\theta + i\sin\theta) = r^{1/n}(\cos\alpha + i\sin\alpha)$

where $\alpha = \dfrac{\theta}{n} + \dfrac{360° \cdot k}{n}$

$(-27)^{1/3} = (0 - 27i)^{1/3}$

$x = 0,\ y = -27,\ r = 27$

$\tan\theta : \emptyset, \quad \theta = 270°$

$[27(\cos 270° + i\sin 270°)]^{1/3}$

$r^{1/3} = (27)^{1/3} = 3$

$\alpha = \dfrac{270°}{3} + \dfrac{360°k}{3} = 90° + 120°k$

if $k = 0,\ 1,\ 2$

$\alpha = 90°,\ 210°,\ 330°$

$\boxed{3\,\text{cis}\,90°,\ 3\,\text{cis}\,210°,\ 3\,\text{cis}\,330°}$

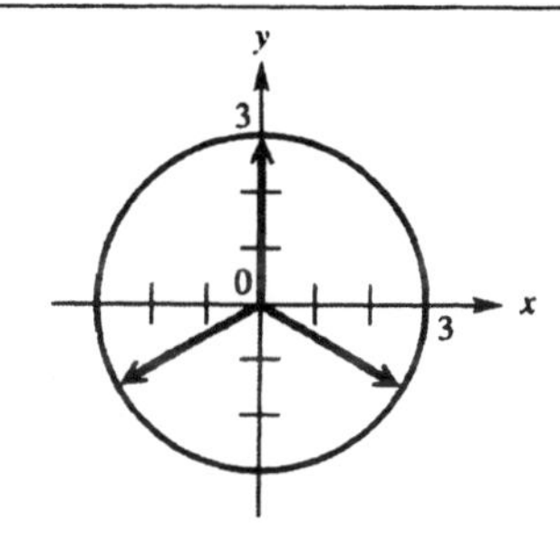

55. $(16i)^{1/4} = (0 + 16i)^{1/4}$

$x = 0,\ y = 16,\ r = 16$

$\tan\theta : \emptyset, \quad \theta = 90°$

$(16i)^{1/4} = (16\,\text{cis}\,90°)^{1/4}$

$r^{1/4} = 16^{1/4} = 2$

$\alpha = \dfrac{90°}{4} + \dfrac{360°k}{4} = 22.5° + 90°k$

if $k = 0,\ 1,\ 2,\ 3$

$\alpha = 22.5°,\ 112.5°,\ 202.5°,\ 292.5°$

$\boxed{\begin{array}{l} 2\,\text{cis}\,22.5°,\ 2\,\text{cis}\,112.5°, \\ 2\,\text{cis}\,202.5°,\ 2\,\text{cis}\,292.5° \end{array}}$

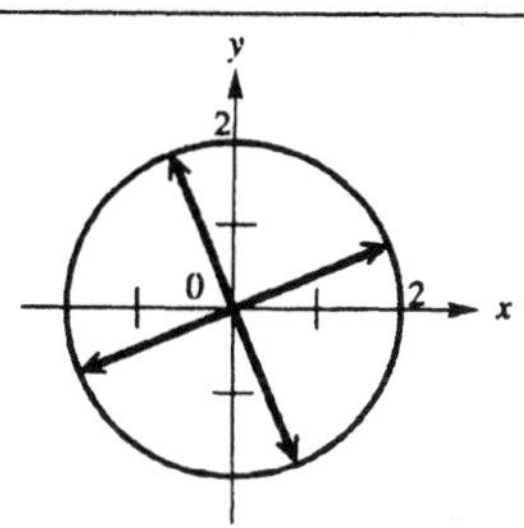

56. $32^{1/5} = (32 + 0i)^{1/5}$

$x = 32,\ y = 0,\ r = 32$

$\tan\theta = 0, \quad \theta = 0°$

$32^{1/5} = (32\,\text{cis}\,0°)^{1/5}$

$r^{1/5} = 32^{1/5} = 2$

$\alpha = \dfrac{0°}{5} + \dfrac{360°k}{5} = 0° + 72°k$

if $k = 0,\ 1,\ 2,\ 3,\ 4$

$\alpha = 0°,\ 72°,\ 144°,\ 216°,\ 288°$

$\boxed{\begin{array}{l} 2\,\text{cis}\,0°,\ 2\,\text{cis}\,72°,\ 2\,\text{cis}\,144°, \\ 2\,\text{cis}\,216°,\ 2\,\text{cis}\,288° \end{array}}$

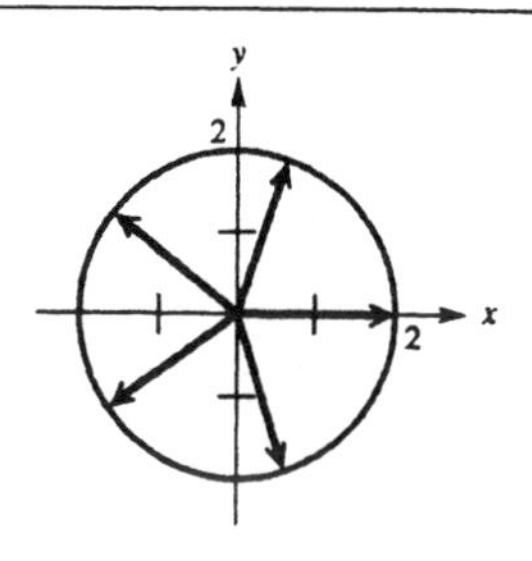

57. $x^4 + i = 0$

$x^4 = -i$

$x = (-i)^{1/4} = (0 - 1i)^{1/4}$

$x = 0,\ y = -1,\ r = 1,\ \theta = 270°$

$(-i)^{1/4} = (1\,\text{cis}\,270°)^{1/4}$

$r^{1/4} = 1 \quad \alpha = \dfrac{270°}{4} + \dfrac{360°k}{4}$

$= 67.5° + 90°k$

if $k = 0,\ 1,\ 2,\ 3$

$\alpha = 67.5°,\ 157.5°,\ 247.5°,\ 337.5°$

$\boxed{\begin{array}{l} \text{cis}\,67.5°,\ \text{cis}\,157.5°, \\ \text{cis}\,247.5°,\ \text{cis}\,337.5°, \end{array}}$

58. $(12,\ 225°)$

$x = r\cos\theta$

$= 12\cos 225°$

$= 12\left(-\dfrac{\sqrt{2}}{2}\right)$

$= -6\sqrt{2}$

$y = r\sin\theta$

$= 12\sin 225°$

$= 12\left(-\dfrac{\sqrt{2}}{2}\right)$

$= -6\sqrt{2}$

$\boxed{(12,\ 225°) = \left(-6\sqrt{2},\ -6\sqrt{2}\right)}$

59. $\left(-8, -\frac{\pi}{3}\right)$

$$x = r\cos\theta = -8\cos\left(-\frac{\pi}{3}\right) = -8\left(\frac{1}{2}\right) = -4$$

$$y = r\sin\theta = -8\sin\left(-\frac{\pi}{3}\right) = -8\left(-\frac{\sqrt{3}}{2}\right) = 4\sqrt{3}$$

$$\boxed{\left(-8, -\frac{\pi}{3}\right) = \left(-4,\ 4\sqrt{3}\right)}$$

60. $(-6,\ 6)$ $\quad r = \sqrt{(-6)^2 + 6^2} = \sqrt{72} = 6\sqrt{2}$

$\tan\theta = \frac{6}{-6} = -1$, Quadrant II

$\theta = 135°$ $\quad \boxed{(6\sqrt{2},\ 135°)}$

61. $(0, -5)$ $\quad r = \sqrt{0^2 + (-5)^2} = \sqrt{25} = 5$

$\tan\theta : \emptyset$, $\theta = -90°$ $\quad \boxed{(5, -90°)}$

62. $r = 4\cos\theta$; circle

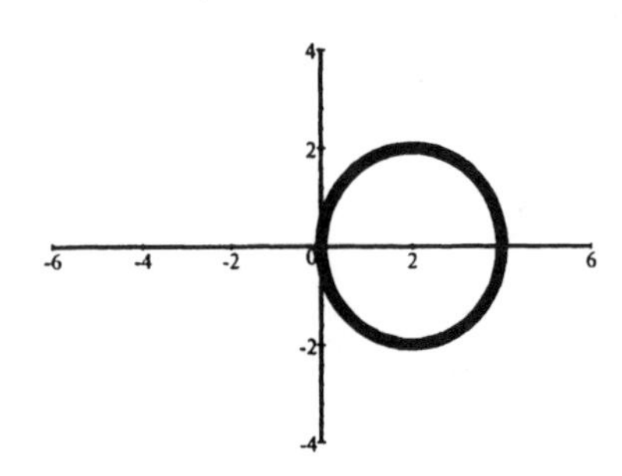

63. $r = 1 - 2\sin\theta$; limacon with loop

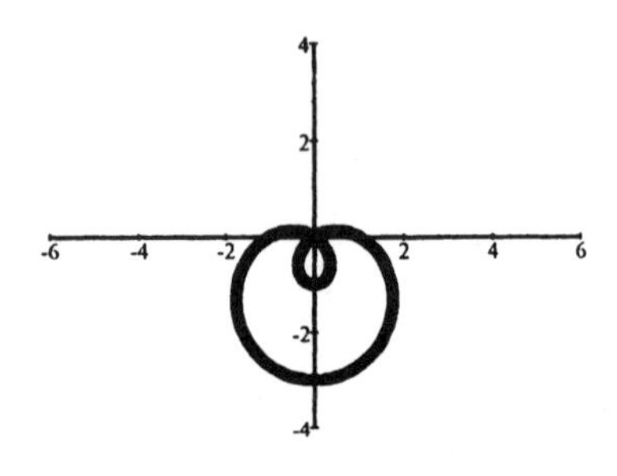

64. $r = 2\sin 4\theta$; 8-leaved rose

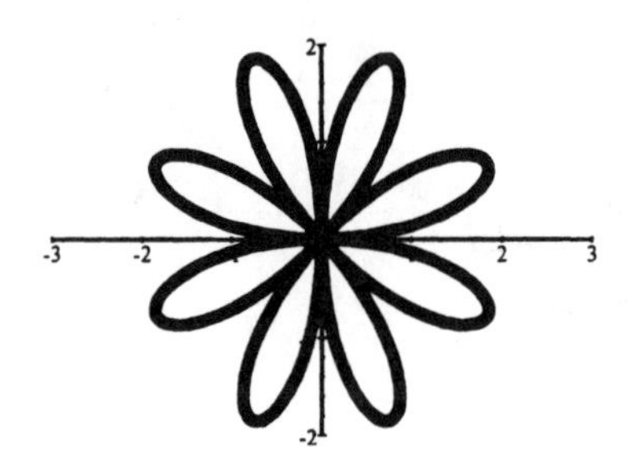

65. $r = \frac{3}{1 + \cos\theta}$

$$r(1 + \cos\theta) = 3$$
$$r + r\cos\theta = 3$$
$$\sqrt{x^2 + y^2} + x = 3$$
$$\sqrt{x^2 + y^2} = -x + 3$$
$$x^2 + y^2 = x^2 - 6x + 9$$
$$\boxed{y^2 + 6x - 9 = 0}$$

66. $r = \frac{4}{2\sin\theta - \cos\theta}$

$$2r\sin\theta - r\cos\theta = 4$$
$$\boxed{2y - x = 4}$$

67. $r = \sin\theta + \cos\theta$

$$r^2 = r\sin\theta + r\cos\theta$$
$$x^2 + y^2 = y + x$$
$$\boxed{x^2 - x + y^2 - y = 0}$$

68. $r = 2$

$$\sqrt{x^2 + y^2} = 2 \quad \boxed{x^2 + y^2 = 4}$$

69. $x = -3$

$$r\cos\theta = -3 \quad \boxed{r = \frac{-3}{\cos\theta}}$$

70. $y = x$

$$r\sin\theta = r\cos\theta$$
$$\frac{r\sin\theta}{r\cos\theta} = 1 \quad \boxed{\tan\theta = 1}$$

71. $y = x^2$

$r\sin\theta = r^2\cos^2\theta$

$\sin\theta = r\cos^2\theta$

$$\boxed{r = \frac{\sin\theta}{\cos^2\theta} \quad \text{or} \quad \frac{\tan\theta}{\cos\theta}}$$

72. $x = y^2$

$r\cos\theta = r^2\sin^2\theta$

$\cos\theta = r\sin^2\theta$

$$\boxed{r = \frac{\cos\theta}{\sin^2\theta} \quad \text{or} \quad \frac{\cot\theta}{\sin\theta}}$$

73. Since θ is coterminal with $\theta + k\cdot 360°$, $k \in I$,

$45°$ is coterminal with $45° + (-1)\,360° = -315°$

$90°$ is coterminal with $90° + (-1)\,360° = -270°$

Therefore, the quotients are the same.

$$\frac{2\,\text{cis}\,45°}{5\,\text{cis}\,90°} = \frac{2\,\text{cis}\,(-315°)}{5\,\text{cis}\,(-270°)}$$

74. $(a+bi) - (c+di) = a - c$ only if $bi = di$, which means $b = d$. To be a non-real number, $b \neq d$.

75. $x = \cos 2t \qquad t \in (-\pi,\ \pi)$

$x = \cos^2 t - \sin^2 t$

$y = \sin t$

$y^2 = \sin^2 t$

identity: $\cos^2 t + \sin^2 t = 1$

$\cos^2 t - \sin^2 t + 2\sin^2 t = 1$

$x + 2y^2 = 1$

$2y^2 = 1 - x$

$y^2 = \frac{1}{2}(1 - x)$

$$\boxed{y^2 = -\frac{1}{2}(x-1),\quad x \in [-1,\ 1]}$$

76. $x = 5\tan x, \qquad t \in \left(-\frac{\pi}{2},\ \frac{\pi}{2}\right)$

$x^2 = 25\tan^2 x$

$\frac{x^2}{25} = \tan^2 x$

$y = 3\sec x$

$y^2 = 9\sec^2 x$

$\frac{y^2}{9} = \sec^2 x$

identity: $\tan^2 x + 1 = \sec^2 x$

$\frac{x^2}{25} + 1 = \frac{y^2}{9}$

$\frac{x^2}{25} - \frac{y^2}{9} = -1$

$\frac{y^2}{9} - \frac{x^2}{25} = 1$

hyperbola; top half:

$$y = 3\sqrt{1 + \frac{x^2}{25}}, \qquad x \in (-\infty,\ \infty)$$

77. $x = t + \cos t, \quad y = \sin t, \quad t \in [0,\ 2\pi]$

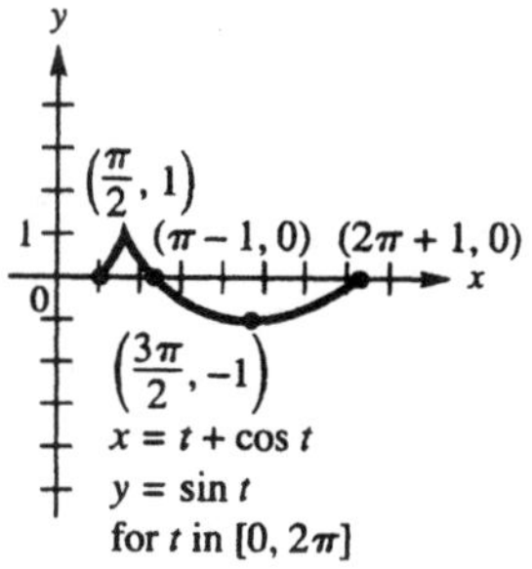

78. $x = (118\cos 27°)\,t \longrightarrow t = \frac{x}{118\cos 27°}$

$y = (118\sin 27°)\,t - 16t^2 + 3.2$

$\;\; = -16t^2 + (118\sin 27°)\,t + 3.2$

$t \longrightarrow y:$

$$y = -16\left(\frac{x}{118\cos 27°}\right)^2 + (118\sin 27°)\left(\frac{x}{118\cos 27°}\right) + 3.2$$

$$y = \frac{-x^2}{870.25\cos^2 27°} + x\tan 27° + 3.2$$

Graph y and locate the positive zero:

$x \approx 358.196 \approx \boxed{360 \text{ feet}}$.

Chapter 10 Test

1. **(a)** $A = 25.2°$, $a = 6.92$ yd, $b = 4.82$ yd

$$\frac{a}{\sin A} = \frac{b}{\sin B}$$
$$\frac{6.92}{\sin 25.2°} = \frac{4.82}{\sin B}$$
$$\sin B = \frac{4.82 \sin 25.2°}{6.92}$$
$$\sin B \approx .2965688$$
$$B \approx 17.25163$$
$$B_1 = 17.3°$$
$$B_2 = 162.7° \quad (180° - B_1)$$

Is B_2 possible? **No**

$$A + B_2 > 180°$$
$$C = 180° - (17.3° + 25.2°)$$
$$\boxed{C = 137.5°}$$

(b) $C = 118°$, $b = 132$ km, $a = 75.1$ km

$$c^2 = a^2 + b^2 - 2ab\cos C$$
$$= 75.1^2 + 132^2 - 2(75.1)(132)\cos 118°$$
$$\approx 32371.94099$$
$$c \approx 179.92204 \qquad \boxed{c \approx 180 \text{ km}}$$

(c) $a = 17.3$ ft, $b = 22.6$ ft, $c = 29.8$ ft

$$\cos B = \frac{a^2 + c^2 - b^2}{2ac}$$
$$= \frac{17.3^2 + 29.8^2 - 22.6^2}{2(17.3)(29.8)}$$
$$\approx .656176$$
$$B \approx 48.991116° \qquad \boxed{B \approx 49.0°}$$

2. $\frac{\sin A}{a} = \frac{\sin B}{b}$

$a = 10, \ B = 150°$

$$\sin A = \frac{a \sin B}{b}$$
$$\sin A = \frac{10 \sin 150°}{b}$$
$$\sin A = \frac{10\left(\frac{1}{2}\right)}{b}$$
$$\sin A = \frac{5}{b}$$

if $b = 5 : \sin A = 1, \ A = 90°$

not possible; $(A + B > 180°)$

if $b < 5 : \sin A > 1$, does not exist

if $5 < b < 10$: not possible;

$(A + B > 180°)$

if $b = 10 : \ A = 30°$ not possible;

$(A + B = 180°)$

if $b > 10 : A < 30°$

or $> 150°$ (not possible)

(a) Exactly 1 value: $b > 10$

(b) Two values: not possible

(c) No value: $b \leq 10$

3. In any triangle, the sum of the length of any two sides must be greater than the length of the remaining side.

4. (a) After 3 hours, the distances traveled are 108.6 km and 136.8 km.

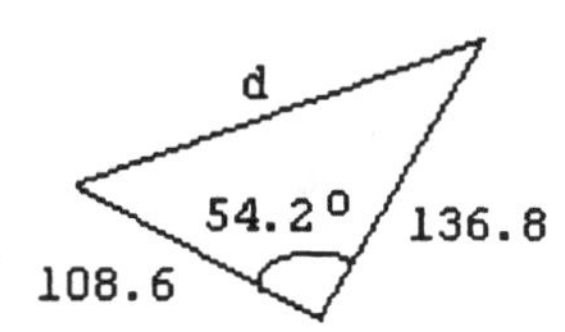

$$d^2 = 108.6^2 + 136.8^2 - 2\,(108.6)\,(136.8)\cos 54.2°$$
$$\approx 13127.376$$
$$d \approx 114.57476$$
$$\boxed{d \approx 115 \text{ km}}$$

(b) Find the length of the diagonal of the square:

$$d^2 = 90^2 + 90^2$$
$$d = \sqrt{16200}$$
$$d \approx 127.2792$$
$$(127.2792 - 60.5 \approx 66.7792)$$

$$\boxed{\text{Distance from pitchers mound to 2nd base} \approx 66.8 \text{ feet.}}$$

(c)
$$\mathbf{u} = \langle r\cos\theta,\ r\sin\theta\rangle$$
$$= \langle 569\cos 127.5°,\ 569\sin 127.5°\rangle$$
$$\approx \langle -346.385,\ 451.418\rangle$$
$$\approx \boxed{-346,\ 451}$$

(d) A parallelogram is formed with sides 475 and 586, and angles of 78.2° and 101.8° $(180° - 82°50')$. Use the law of cosines to solve for $|v|$:

$$|v|^2 = 475^2 + 586^2 - 2\,(475)\,(586)\cos 101.8°$$
$$= 682863.95206$$
$$|v| = 826.35583$$

$$\boxed{\text{Magnitude} \approx 826 \text{ pounds}}$$

5. $\mathbf{u} = \langle -1,\ 3\rangle$ and $\mathbf{v} = \langle 2,\ -6\rangle$

(a) $\mathbf{u} + \mathbf{v} = \langle -1+2,\ 3-6\rangle = \boxed{\langle 1,\ -3\rangle}$

(b)
$$-3\mathbf{v} = -3\,\langle 2,\ -6\rangle$$
$$= \langle -3\,(2),\ (-3)\,(-6)\rangle = \boxed{\langle -6,\ 18\rangle}$$

(c)
$$\mathbf{u}\cdot\mathbf{v} = (-1)\,(2) + (3)\,(-6)$$
$$= -2 - 18 = \boxed{-20}$$

(d) Angle between $\mathbf{u}$ and $\mathbf{v}$:

$$\cos\theta = \frac{\mathbf{u}\cdot\mathbf{v}}{|u|\ |v|} = \frac{\langle -1,\ 3\rangle\cdot\langle 2,\ -6\rangle}{|\langle -1,\ 3\rangle|\ |\langle 2,\ -6\rangle|}$$
$$= \frac{-20}{\sqrt{1+9}\times\sqrt{4+36}}$$
$$= \frac{-20}{\sqrt{10}\times\sqrt{40}} = \frac{-20}{\sqrt{400}}$$
$$= \frac{-20}{20} = -1$$
$$\cos\theta = -1 \rightarrow \theta = \boxed{180°}$$

6. (a)
$$4\operatorname{cis} 240° = 4\,(\cos 240° + i\sin 240°)$$
$$= 4\,(-\cos 60° + i\,(-\sin 60°))$$
$$= 4\left(-\frac{1}{2} + i\left(-\frac{\sqrt{3}}{2}\right)\right)$$
$$= \boxed{-2 - 2i\sqrt{3}}$$

(b) $-4 + 4i\sqrt{3}$: $x = -4,\ y = 4\sqrt{3}$

$$r = \sqrt{x^2 + y^2}$$
$$= \sqrt{(-4)^2 + \left(4\sqrt{3}\right)^2}$$
$$= \sqrt{16+48} = \sqrt{64} = 8$$

$$\tan\theta = \frac{y}{x} = \frac{4\sqrt{3}}{-4} = -\sqrt{3}$$
$$\theta = 120°,\ -60°$$

Since $-4 + 4i\sqrt{3}$ $(-,\ +) \in$ Quadrant II:

$$\boxed{8\operatorname{cis} 120°}$$

(c) Resultant of $-2 - 2i\sqrt{3}$ and $-4 + 4i\sqrt{3}$:

$$\left(-2 - 2i\sqrt{3}\right) + \left(-4 + 4i\sqrt{3}\right)$$
$$= \boxed{-6 + 2i\sqrt{3}}$$

7. **(a)** $(r_1 \operatorname{cis}\theta_1)(r_2 \operatorname{cis}\theta_2) = r_1 r_2 \operatorname{cis}(\theta_1+\theta_2)$

$$\begin{aligned}(3\operatorname{cis}30^\circ)(5\operatorname{cis}90^\circ) &= (3)(5)[\operatorname{cis}(30^\circ+90^\circ)] \\ &= \mathbf{15}\operatorname{cis}\mathbf{120^\circ} \\ &= 15(\cos 120^\circ + i\sin 120^\circ) \\ &= 15\left(-\frac{1}{2}+\frac{\sqrt{3}}{2}i\right) \\ &= \boxed{-\frac{15}{2}+\frac{15\sqrt{3}}{2}i}\end{aligned}$$

(b) $\dfrac{r_1 \operatorname{cis}\theta_1}{r_2 \operatorname{cis}\theta_2} = \dfrac{r_1}{r_2}\operatorname{cis}(\theta_1-\theta_2)$

$$\begin{aligned}\frac{2\operatorname{cis}315^\circ}{4\operatorname{cis}45^\circ} &= \frac{2}{4}[\operatorname{cis}(315^\circ-45^\circ)] \\ &= \tfrac{1}{2}\operatorname{cis}(\mathbf{270^\circ}) \\ &= \tfrac{1}{2}[\cos(270^\circ)+i\sin(270^\circ)] \\ &= \tfrac{1}{2}(0-1i) = \boxed{-\tfrac{1}{2}i}\end{aligned}$$

(c) $(1-i\sqrt{3})^5: \quad x=1,\ y=-\sqrt{3}$

$$r=\sqrt{1^2+(-\sqrt{3})^2}=\sqrt{4}=2$$

$$\begin{aligned}\tan\theta &= \frac{-\sqrt{3}}{1} \\ &= -\sqrt{3},\ \text{Quadrant IV} \\ \theta &= -60^\circ\end{aligned}$$

$$\begin{aligned}[2\operatorname{cis}(-60^\circ)]^5 &= (2)^5\operatorname{cis}(5\cdot -60^\circ) \\ &= 32\operatorname{cis}(-300^\circ) \\ &= 32(\cos 60^\circ + i\sin 60^\circ) \\ &= 32\left(\frac{1}{2}+\frac{\sqrt{3}}{2}i\right) \\ &= \boxed{16+16i\sqrt{3}}\end{aligned}$$

(d) $(\sqrt{3}+i)^{1/4}; \quad x=\sqrt{3},\ y=1,$

$$r=\sqrt{1^2+(-\sqrt{3})^2}=2$$

$$\tan\theta=\frac{\sqrt{3}}{3}\rightarrow\theta=30^\circ$$

$$(\sqrt{3}+i)^{1/4}=(2\operatorname{cis}30^\circ)^{1/4}$$

$$r^{1/4}=2^{1/4}=\sqrt[4]{2}$$

$$\alpha=\frac{30^\circ}{4}+\frac{360^\circ k}{4}=7.5^\circ+90^\circ k$$

if $k=0,\ 1,\ 2,\ 3$

$\alpha = 7.5^\circ,\ 97.5^\circ,\ 187.5^\circ,\ 277.5^\circ$

$$\boxed{\begin{aligned}&\sqrt[4]{2}\operatorname{cis}7.5^\circ,\ \sqrt[4]{2}\operatorname{cis}97.5^\circ,\\ &\sqrt[4]{2}\operatorname{cis}187.5^\circ,\ \sqrt[4]{2}\operatorname{cis}277.5^\circ\end{aligned}}$$

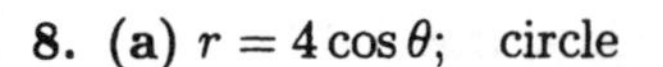

8. **(a)** $r=4\cos\theta$; circle

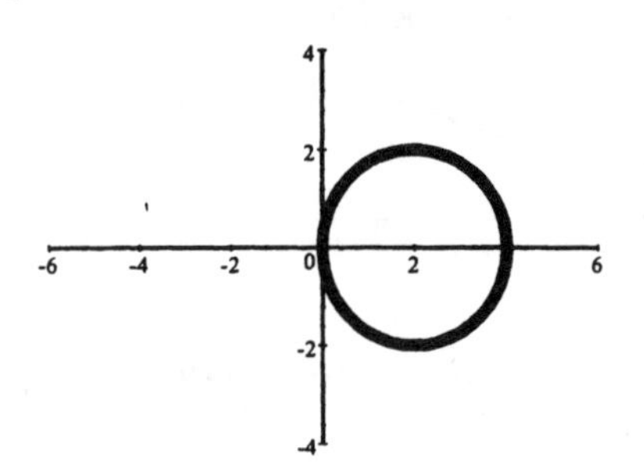

(b)
$$\begin{aligned}r &= 4\cos\theta \\ r^2 &= 4r\cos\theta \\ x^2+y^2 &= 4x \\ &\boxed{x^2-4x+y^2=0}\end{aligned}$$

(c) Yes, the graph of $x^2-4x+y^2=0$ is a circle with center at (2, 0) and radius 2.

9. $-x+2y=4$

$$\begin{aligned}-(r\cos\theta)+2(r\sin\theta) &= 4 \\ r(-\cos\theta+2\sin\theta) &= 4\end{aligned}$$

$$\boxed{r=\frac{4}{2\sin\theta-\cos\theta}}$$

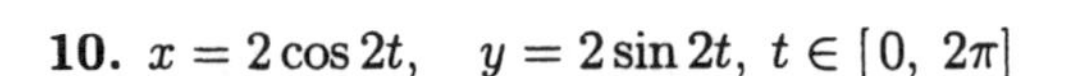

10. $x=2\cos 2t,\quad y=2\sin 2t,\ t\in[0,\ 2\pi]$

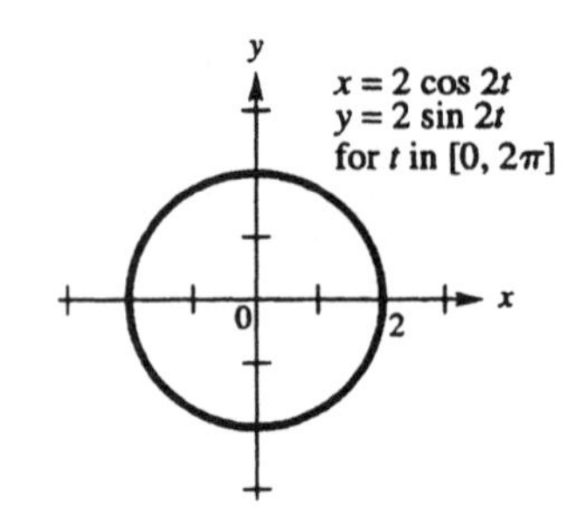

Chapter 10 Project

When Is a Circle Really a Polygon? Investigating Polar Circles

1. Use a θstep of 1 to get a 360-gon.

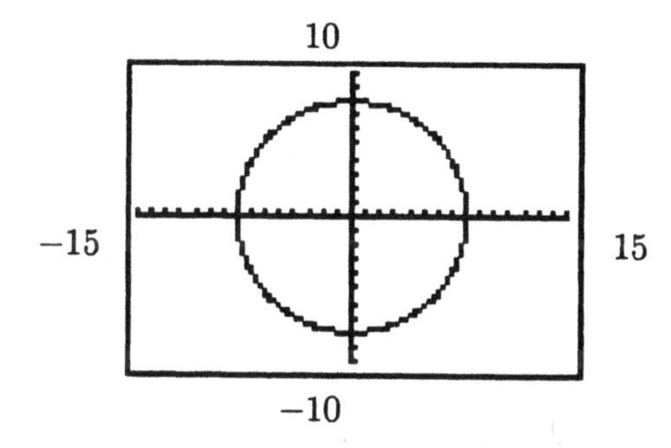

We show only the calculations needed to graph each figure.

2. **Polygons**

Square $\theta\text{step} = \dfrac{360°}{4} = 90°$

Pentagon $\theta\text{step} = \dfrac{360°}{5} = 72°$

Hexagon $\theta\text{step} = \dfrac{360°}{6} = 60°$

Septagon $\theta\text{step} = \dfrac{360°}{7} = 51.43°$

Octagon $\theta\text{step} = \dfrac{360°}{8} = 45°$

Nonagon $\theta\text{step} = \dfrac{360°}{9} = 40°$

Decagon $\theta\text{step} = \dfrac{360°}{10} = 36°$

3. **Star Polygon**

One possibility is:

5-pointed star $\theta\,\text{step} = \dfrac{360°}{5} \cdot 2 = 144°$

$\theta\,\text{max} = 360° \times 2 = 720°$

9-pointed star $\theta\,\text{step} = \dfrac{360°}{9} \cdot 4 = 160°$

$\theta\,\text{max} = 360° \times 4 = 1440°$

11-pointed star $\theta\,\text{step} = \dfrac{360°}{11} \cdot 5 = 163.64°$

$\theta\,\text{max} = 360° \times 5\frac{1}{11} = 1833°$

Section 11.1

1. $a_n = 4n + 10$

$a_1 = 4(1) + 10 = \mathbf{14}$
$a_2 = 4(2) + 10 = \mathbf{18}$
$a_3 = 4(3) + 10 = \mathbf{22}$
$a_4 = 4(4) + 10 = \mathbf{26}$
$a_5 = 4(5) + 10 = \mathbf{30}$

3. $a_n = 2^{n-1}$

$a_1 = 2^{(1)-1} = 2^0 = \mathbf{1}$
$a_2 = 2^{(2)-1} = 2^1 = \mathbf{2}$
$a_3 = 2^{(3)-1} = 2^2 = \mathbf{4}$
$a_4 = 2^{(4)-1} = 2^3 = \mathbf{8}$
$a_5 = 2^{(5)-1} = 2^4 = \mathbf{16}$

5. $a_n = \left(\frac{1}{3}\right)^n (n-1)$

$$a_1 = \left(\frac{1}{3}\right)^1 (1-1) = \left(\frac{1}{3}\right)(0) = \mathbf{0}$$
$$a_2 = \left(\frac{1}{3}\right)^2 (2-1) = \left(\frac{1}{9}\right)(1) = \mathbf{\frac{1}{9}}$$
$$a_3 = \left(\frac{1}{3}\right)^3 (3-1) = \left(\frac{1}{27}\right)(2) = \mathbf{\frac{2}{27}}$$
$$a_4 = \left(\frac{1}{3}\right)^4 (4-1) = \left(\frac{1}{81}\right)(3) = \mathbf{\frac{1}{27}}$$
$$a_5 = \left(\frac{1}{3}\right)^5 (5-1) = \left(\frac{1}{243}\right)(4) = \mathbf{\frac{4}{243}}$$

7. $a_n = (-1)^n (2n)$

$a_1 = (-1)^1 (2) = \mathbf{-2}$
$a_2 = (-1)^2 (4) = \mathbf{4}$
$a_3 = (-1)^3 (6) = \mathbf{-6}$
$a_4 = (-1)^4 (8) = \mathbf{8}$
$a_5 = (-1)^5 (10) = \mathbf{-10}$

9. $a_n = \frac{4n-1}{n^2+2}$

$$a_1 = \frac{4(1)-1}{1^2+2} = \frac{3}{3} = \mathbf{1}$$
$$a_2 = \frac{4(2)-1}{2^2+2} = \frac{7}{6} = \mathbf{\frac{7}{6}}$$
$$a_3 = \frac{4(3)-1}{3^2+2} = \frac{11}{11} = \mathbf{1}$$
$$a_4 = \frac{4(4)-1}{4^2+2} = \frac{15}{18} = \mathbf{\frac{5}{6}}$$
$$a_5 = \frac{4(5)-1}{5^2+2} = \frac{19}{27} = \mathbf{\frac{19}{27}}$$

11. The nth term is the term that appears in the nth position. For example, when $n = 1$, the first term is obtained. When $n = 2$, the second term is obtained, etc.

13. The sequence of days of the week: finite

15. 1, 2, 3, 4: finite

17. 1, 2, 3, 4, ... : infinite

19. $a_n = 3 \cdot a_{n-1}$; $a_1 = 3$
$2 \le n \le 10$
finite

21. $a_1 = -2$, $a_n = a_{n-1} + 3$ for $n > 1$

$a_1 = \mathbf{-2}$
$a_2 = -2 + 3 = \mathbf{1}$
$a_3 = 1 + 3 = \mathbf{4}$
$a_4 = 4 + 3 = \mathbf{7}$

23. $a_1 = 1$, $a_2 = 1$,
$a_n = a_{n-1} + a_{n-2}$ for $n \ge 3$

$a_1 = \mathbf{1}$
$a_2 = \mathbf{1}$
$a_3 = 1 + 1 = \mathbf{2}$
$a_4 = 2 + 1 = \mathbf{3}$

25. $\sum_{i=1}^{5}(2i+1)$

$= [2(1)+1]+[2(2)+1]+[2(3)+1] + [2(4)+1]+[2(5)+1]$

$= 3+5+7+9+11 = \boxed{35}$

27. $\sum_{j=1}^{4}\frac{1}{j} = \frac{1}{1}+\frac{1}{2}+\frac{1}{3}+\frac{1}{4}$

$= \frac{12+6+4+3}{12} = \boxed{\frac{25}{12}}$

29. $\sum_{i=1}^{4} i^i = (1)^1+(2)^2+(3)^3+(4)^4$

$= 1+4+27+256 = \boxed{288}$

31. $\sum_{k=1}^{6}(-1)^k\cdot k = (-1)^1(1)+(-1)^2(2) + (-1)^3(3)+(-1)^4(4) + (-1)^5(5)+(-1)^6(6)$

$= -1+2-3+4-5+6 = \boxed{3}$

33. $\sum_{i=2}^{5}(6-3i)$

$= (6-3(2))+(6-3(3))+(6-3(4)) + (6-3(5))$

$= 0-3-6-9 = \boxed{-18}$

35. $\sum_{i=-2}^{3} 2(3)^i = 2(3)^{-2}+2(3)^{-1}+2(3)^0 + 2(3)^1+2(3)^2+2(3)^3$

$= \frac{2}{9}+\frac{2}{3}+2+6+18+54$

$= \boxed{80\tfrac{8}{9} \text{ or } \frac{728}{9}}$

37. $\sum_{i=-1}^{5}(i^2-2i) = \left[(-1)^2-2(-1)\right]+\left[(0)^2-2(0)\right] + \left[(1)^2-2(1)\right]+\left[(2)^2-2(2)\right] + \left[(3)^2-2(3)\right]+\left[(4)^2-2(4)\right] + \left[(5)^2-2(5)\right]$

$= (1+2)+0+(1-2)+(4-4)+(9-6) + (16-8)+(25-10) = \boxed{28}$

39. $\sum_{i=1}^{5}(2x_i+3)$; $x_1=-2,\ x_2=-1,\ x_3=0,\ x_4=1,\ x_5=2$

$\sum_{i=1}^{5}(2x_i+3) = (2(-2)+3)+(2(-1)+3) + (2(0)+3)+(2(1)+3)+(2(2)+3)$

$= \boxed{-1+1+3+5+7}$

41. $\sum_{i=1}^{3}(3x_i-x_i^2)$; $x_1=-2,\ x_2=-1,\ x_3=0$

$\sum_{i=1}^{3}(3x_i-x_i^2) = \left(3(-2)-(-2)^2\right)+\left(3(-1)-(-1)^2\right) + \left(3(0)-(0)^2\right)$

$= (-6-4)+(-3-1)+0$

$= \boxed{-10-4+0}$

43. $\sum_{i=2}^{5}\frac{x_i+1}{x_i+2}$; $x_1=-2,\ x_2=-1,\ x_3=0,\ x_4=1,\ x_5=2$

$\sum_{i=2}^{5}\frac{x_i+1}{x_i+2} = \frac{-1+1}{-1+2}+\frac{0+1}{0+2}+\frac{1+1}{1+2}+\frac{2+1}{2+2}$

$= \frac{0}{1}+\frac{1}{2}+\frac{2}{3}+\frac{3}{4}$

$= \boxed{0+\frac{1}{2}+\frac{2}{3}+\frac{3}{4}}$

45. $f(x)=4x-7,\ \Delta x=.5$

$x_1=0,\ x_2=2,\ x_3=4,\ x_4=6$

$\sum_{i=1}^{4} f(x_i)\,\Delta x$

$= f(x_1)\,\Delta x+f(x_2)\,\Delta x+f(x_3)\,\Delta x + f(x_4)\,\Delta x$

$= (4x_1-7)(.5)+(4x_2-7)(.5) + (4x_3-7)(.5)+(4x_4-7)(.5)$

$= (4(0)-7)(.5)+(4(2)-7)(.5) + (4(4)-7)(.5)+(4(6)-7)(.5)$

$= (-7)(.5)+(1)(.5)+(9)(.5)+(17)(.5)$

$= \boxed{-3.5+.5+4.5+8.5}$

47. $f(x) = 2x^2,\ \Delta x = .5$
$x_1 = 0,\ x_2 = 2,\ x_3 = 4,\ x_4 = 6$

$$\sum_{i=1}^{4} f(x_i)\,\Delta x$$
$$= f(x_1)\,\Delta x + f(x_2)\,\Delta x + f(x_3)\,\Delta x + f(x_4)\,\Delta x$$
$$= (2x_1)^2(.5) + (2x_2)^2(.5) + (2x_3)^2(.5) + (2x_4)^2(.5)$$
$$= (2(0))^2(.5) + (2(2))^2(.5) + (2(4))^2(.5) + (2(6))^2(.5)$$
$$= 0 + (8)(.5) + (32)(.5) + (72)(.5)$$
$$= \boxed{0 + 4 + 16 + 36}$$

49. $f(x) = \dfrac{-2}{x+1},\quad \Delta x = .5$
$x_1 = 0,\ x_2 = 2,\ x_3 = 4,\ x_4 = 6$

$$\sum_{i=1}^{4} f(x_i)\,\Delta x$$
$$= f(x_1)\,\Delta x + f(x_2)\,\Delta x + f(x_3)\,\Delta x + f(x_4)\,\Delta x$$
$$= \left(\frac{-2}{x_1+1}\right)(.5) + \left(\frac{-2}{x_2+1}\right)(.5) + \left(\frac{-2}{x_3+1}\right)(.5) + \left(\frac{-2}{x_4+1}\right)(.5)$$
$$= \frac{-2}{1}\cdot\frac{1}{2} + \frac{-2}{3}\cdot\frac{1}{2} + \frac{-2}{5}\cdot\frac{1}{2} + \frac{-2}{7}\cdot\frac{1}{2}$$
$$= \boxed{-1 - \frac{1}{3} - \frac{1}{5} - \frac{1}{7}}$$

51. $\displaystyle\sum_{i=1}^{100} 6 = 100(6) = \boxed{600}$

53. $\displaystyle\sum_{i=1}^{15} i^2 = \frac{15(15+1)(30+1)}{6} = \boxed{1240}$

55. $$\sum_{i=1}^{5}(5i+3) = \sum_{i=1}^{5}5i + \sum_{i=1}^{5}3$$
$$= 5\left[\frac{(5)(6)}{2}\right] + 5(3)$$
$$= 5(15) + 15 = \boxed{90}$$

57. $$\sum_{i=1}^{5}(4i^2 - 2i + 6) = \sum_{i=1}^{5}4i^2 - \sum_{i=1}^{5}2i + \sum_{i=1}^{5}6$$
$$= 4\left[\frac{(5)(6)(11)}{6}\right] - 2\left[\frac{(5)(6)}{2}\right] + 5(6)$$
$$= 4(55) - 2(15) + 30 = \boxed{220}$$

59. $$\sum_{i=1}^{4}(3i^3 + 2i - 4)$$
$$= \sum_{i=1}^{4}3i^3 + \sum_{i=1}^{4}2i - \sum_{i=1}^{4}4$$
$$= \frac{3(4^2)(5)^2}{4} + \frac{2(4)(5)}{2} - 4(4)$$
$$= 300 + 20 - 16 = \boxed{304}$$

61. $a_n = \dfrac{n+4}{2n}$

Using the sequence graphing capability of a graphing calculator, the given sequence appears to converge to $\frac{1}{2}$.

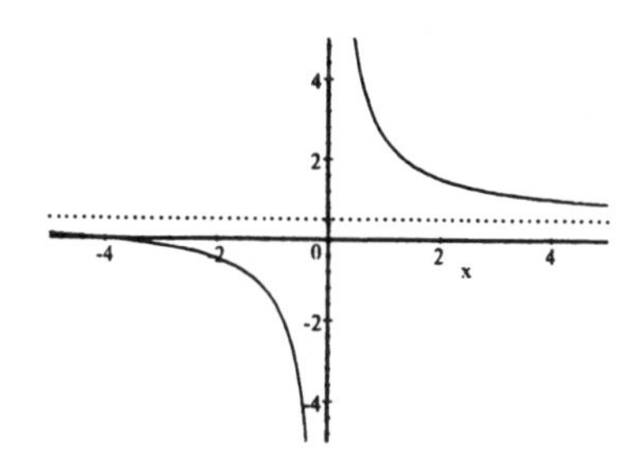

63. $a_n = 2e^n$

Using the sequence graphing capability of a graphing calculator, the given sequence appears to diverge.

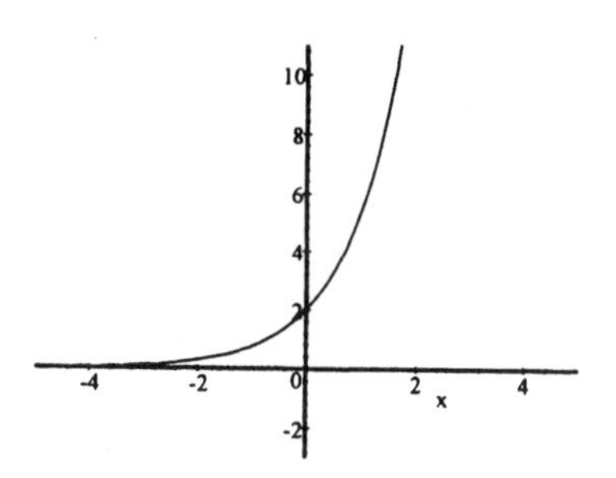

65. $a_n = \left(1 + \frac{1}{n}\right)^n$

Using the sequence graphing capability of a graphing calculator, the given sequence appears to converge to $e \approx 2.71828$.

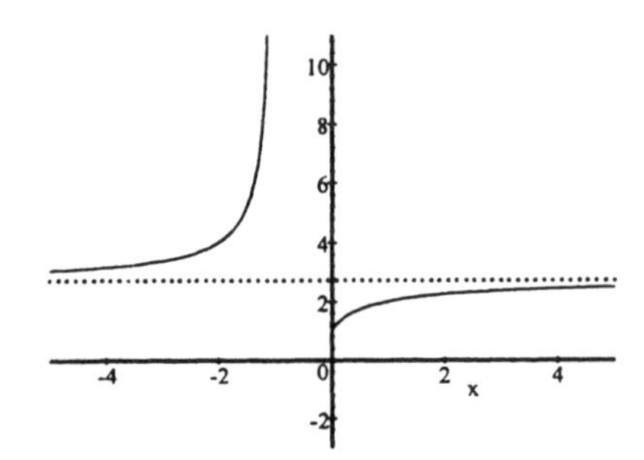

67. $\frac{\pi^4}{90} = \frac{1}{1^4} + \frac{1}{2^4} + \frac{1}{3^4} + \frac{1}{4^4} + \frac{1}{5^4} + \ldots + \frac{1}{n^4} + \cdots$

Sum the first 6 terms:

$$\frac{1}{1^4} + \frac{1}{2^4} + \frac{1}{3^4} + \frac{1}{4^4} + \frac{1}{5^4} + \frac{1}{6^4}$$

$$= \frac{1}{1} + \frac{1}{16} + \frac{1}{81} + \frac{1}{256} + \frac{1}{625} + \frac{1}{1296}$$

$$\approx 1.081123534$$

$$\frac{\pi^4}{90} \approx 1.081123534$$

$$\pi^4 \approx 97.30111806$$

$$\pi \approx 3.140721718$$

Calculator: $\pi = 3.141592654$
Accurate to 3 decimal places rounded.

69. (a) Since the number of bacteria doubles every 40 minutes, it follows that:

$$N_{j+1} = 2N_j \text{ for } j \geq 1.$$

(b) Two hours is 120 minutes.

$$N_j = 40(j-1)$$
$$120 = 40(j-1)$$
$$3 = j - 1 \Rightarrow j = 4$$

$N_1 = 230$, $N_2 = 460$,
$N_3 = 920$, $N_4 = 1840$

If there are initially 230 bacteria, then there will be 1840 bacteria after 2 hours.

continued

69. continued

(c) We must graph the sequence $N_{j+1} = 2N_j$ for $j = 1, 2, 3, \ldots, 7$ if $N_1 = 230$.

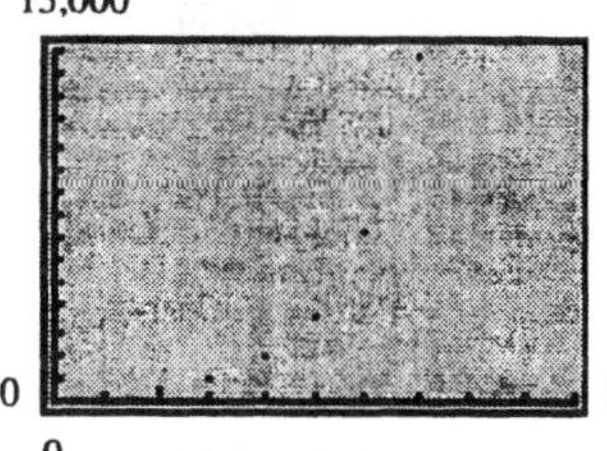

(d) The growth is very rapid. Since there is a doubling of the bacteria at equal intervals, their growth is exponential.

71. Let $a = 1$

$$e = 1 + 1 + \frac{1^2}{2!} + \frac{1^3}{3!} + \frac{1^4}{4!} + \frac{1^5}{5!} + \frac{1^6}{6!} + \frac{1^7}{7!}$$

$$= 2 + \frac{1}{2} + \frac{1}{6} + \frac{1}{24} + \frac{1}{120} + \frac{1}{720} + \frac{1}{5040}$$

$$\approx 2.718253968$$

calculator: $e = 2.718281828$

Eight terms of the series provide an approximation accurate to 4 decimal places.

73. $a_1 = k, \quad a_n = \frac{1}{2}\left(a_{n-1} + \frac{k}{a_{n-1}}\right)$

Let $a_1 = 2$. Then,

$$a_2 = \frac{1}{2}\left(2 + \frac{2}{2}\right) = 1.5$$

$$a_3 = \frac{1}{2}\left(1.5 + \frac{2}{1.5}\right) \approx 1.416666667$$

$$a_4 = \frac{1}{2}\left(1.416666667 + \frac{2}{1.416666667}\right)$$
$$\approx 1.414215686$$

$$a_5 = \frac{1}{2}\left(1.414215686 + \frac{2}{1.414215686}\right)$$
$$\approx 1.414213562$$

$$a_6 = \frac{1}{2}\left(1.414213727 + \frac{2}{1.414213727}\right)$$
$$\approx 1.414213562$$

Since $\sqrt{2} \approx 1.414213562$, this is a very accurate approximation.

Section 11.2

1. 2, 5, 8, 11, ...

$2 + d = 5$ $\boxed{d = 3}$

3. 3, −2, −7, −12, ...

$3 + d = -2$ $\boxed{d = -5}$

5. $x + 3y,\ 2x + 5y,\ 3x + 7y,\ \ldots$

$$x + 3y + d = 2x + 5y$$
$$d = 2x + 5y - x - 3y$$
$$\boxed{d = x + 2y}$$

7. $a_1 = 8,\ d = 6$

$$a_n = a_1 + (n-1)\,d$$
$$a_n = 8 + 6\,(n-1)$$
$$a_n = 6n + 2$$

$a_1 = \mathbf{8}$
$a_2 = 6\,(2) + 2 = \mathbf{14}$
$a_3 = 6\,(3) + 2 = \mathbf{20}$
$a_4 = 6\,(4) + 2 = \mathbf{26}$
$a_5 = 6\,(5) + 2 = \mathbf{32}$

9. $a_1 = 5,\ d = -2$

$$a_n = a_1 + (n-1)\,d$$
$$a_n = 5 - 2\,(n-1)$$
$$a_n = -2n + 7$$

$a_1 = \mathbf{5}$
$a_2 = -2\,(2) + 7 = \mathbf{3}$
$a_3 = -2\,(3) + 7 = \mathbf{1}$
$a_4 = -2\,(4) + 7 = \mathbf{-1}$
$a_5 = -2\,(5) + 7 = \mathbf{-3}$

11. $a_3 = 10,\ d = -2$

$$a_n = a_1 + (n-1)\,d$$
$$a_3 = a_1 + (3-1)\,(-2)$$
$$10 = a_1 - 4$$

$a_1 = \mathbf{14}$
$a_2 = 14 - 2 = \mathbf{12}$
$a_3 = 12 - 2 = \mathbf{10}$
$a_4 = 10 - 2 = \mathbf{8}$
$a_5 = 8 - 2 = \mathbf{6}$

13. $a_1 = 5,\ d = 2$

$$a_8 = a_1 + (8-1)\,d$$
$$a_8 = 5 + 7\,(2)$$
$$\boxed{a_8 = 19}$$

$$a_n = a_1 + (n-1)\,d$$
$$a_n = 5 + 2\,(n-1)$$
$$\boxed{a_n = 2n + 3}$$

15. $a_3 = 2,\ d = 1$

$$a_3 = a_1 + (3-1)\,(1)$$
$$2 = a_1 + (2)$$
$$a_1 = 0$$

$$a_8 = a_1 + (8-1)\,d$$
$$a_8 = 0 + 7\,(1)$$
$$\boxed{a_8 = 7}$$

$$a_n = a_1 + (n-1)\,d$$
$$a_n = 0 + (n-1)$$
$$\boxed{a_n = n - 1}$$

17. $a_1 = 8,\ a_2 = 6$

$$d = a_2 - a_1$$
$$d = 6 - 8 = -2$$

$$a_8 = a_1 + (8-1)\,d$$
$$a_8 = 8 + 7\,(-2)$$
$$\boxed{a_8 = -6}$$

$$a_n = a_1 + (n-1)\,d$$
$$a_n = 8 + (n-1)\,(-2)$$
$$\boxed{a_n = -2n + 10}$$

19. $a_{10} = 6,\ a_{12} = 15$

$$2d = a_{12} - a_{10}$$
$$2d = 15 - 6 = 9$$
$$d = \tfrac{9}{2} = 4.5$$

$$a_{10} = a_1 + (10-1)\,(4.5)$$
$$6 = a_1 + 40.5$$
$$a_1 = -34.5$$

$$a_8 = a_1 + (8-1)\,d$$
$$a_8 = -34.5 + 7\,(4.5)$$
$$\boxed{a_8 = -3}$$

$$a_n = a_1 + (n-1)\,d$$
$$a_n = -34.5 + 4.5\,(n-1)$$
$$\boxed{a_n = 4.5n - 39}$$

21. $a_1 = x,\ a_2 = x + 3$

$$d = a_2 - a_1$$
$$d = x + 3 - x = 3$$

$$a_8 = a_1 + (8-1)\,d$$
$$a_8 = x + 7\,(3)$$

$$\boxed{a_8 = x + 21}$$

$$a_n = a_1 + (n-1)\,d$$
$$a_n = x + (n-1)\,(3)$$

$$\boxed{a_n = x + 3n - 3}$$

23. $a_5 = 27,\ a_{15} = 87$

$$a_n = a_1 + (n-1)\,d$$
$$n = 5 : 27 = a_1 + 4d \quad [1]$$
$$n = 15 : 87 = a_1 + 14d \quad [2]$$

$[2] + (-1 \times [1]):$

$$87 = a_1 + 14d$$
$$-27 = -a_1 - 4d$$
$$60 = 10d \rightarrow d = 6$$

$d = 6 \rightarrow [1]:$

$$27 = a_1 + 4\,(6)$$
$$27 = a_1 + 24 \qquad \boxed{a_1 = 3}$$

25. $S_{16} = -160,\ a_{16} = -25$

$$S_n = \frac{n}{2}\,(a_1 + a_n)$$

$$S_{16} = \frac{16}{2}\,(a_1 + a_{16})$$

$$-160 = 8\,(a_1 - 25)$$
$$-160 = 8a_1 - 200$$
$$40 = 8a_1 \qquad \boxed{a_1 = 5}$$

27. $a_1 = 8,\ d = 3$

$$S_n = \frac{n}{2}\,[2a_1 + (n-1)\,d]$$

$$S_{10} = \frac{10}{2}\,[2 \cdot 8 + 9 \cdot 3]$$
$$= 5\,[16 + 27] = \boxed{215}$$

29. $a_3 = 5,\ a_4 = 8$

$$d = a_4 - a_3 = 8 - 5 = 3$$

$$a_n = a_1 + (n-1)\,d,\ n = 3$$
$$5 = a_1 + 2\,(3)$$
$$a_1 = -1$$

$$S_n = \frac{n}{2}\,[2a_1 + (n-1)\,d]$$

$$S_{10} = \frac{10}{2}\,[2\,(-1) + 9 \cdot 3]$$
$$= 5\,[-2 + 27] = \boxed{125}$$

31. $5,\ 9,\ 13, \ldots \quad d = 4$

$$S_n = \frac{n}{2}\,[2a_1 + (n-1)\,d]$$

$$S_{10} = \frac{10}{2}\,[2\,(5) + 9 \cdot 4]$$
$$= 5\,[10 + 36] = \boxed{230}$$

33. $a_1 = 10,\ a_{10} = 5.5$

$$9d = a_{10} - a_1 = 5.5 - 10 = -4.5$$
$$d = -.5$$

$$S_n = \frac{n}{2}\,[2a_1 + (n-1)\,d]$$

$$S_{10} = \frac{10}{2}\,[2\,(10) + 9\,(-.5)]$$
$$= 5\,[20 - 4.5] = \boxed{77.5}$$

$$\left[\text{or } S_{10} = \frac{10}{2}\,(10 + 5.5)\right]$$

35. $S_{20} = 1090,\ a_{20} = 102$

$$a_n = a_1 + (n-1)\,d$$
$$n = 20 : 102 = a_1 + 19d \quad [1]$$

$$S_n = \frac{n}{2}\,[2a_1 + (n-1)\,d]$$

$$n = 20 : \quad 1090 = 10\,[2a_1 + 19d]$$
$$109 = 2a_1 + 19d \quad [2]$$

$[2] - [1]:$

$$109 = 2a_1 + 19d$$
$$-102 = -a_1 - 19d$$
$$7 = a_1$$

$a_1 \rightarrow [1]:$

$$102 = 7 + 19d$$
$$95 = 19d$$
$$d = 5 \qquad \boxed{a_1 = 7,\ d = 5}$$

37. $S_{12} = -108,\ a_{12} = -19$

$$S_n = \frac{n}{2}(a_1 + a_n)$$
$$-108 = \frac{12}{2}(a_1 + (-19))$$
$$-18 = a_1 - 19$$
$$a_1 = 1$$
$$a_n = a_1 + (n-1)d$$
$$-19 = 1 + 11d$$
$$-20 = 11d$$
$$d = -\frac{20}{11}$$
$$\boxed{a_1 = 1,\ d = -\frac{20}{11}}$$

39. $(1,-2), (2,-1), (3,0), (4,1), (5,2), (6,3)$

Note that $y = x - 3$, so

$$a_n = n - 3$$

Domain: $\{1, 2, 3, 4, 5, 6\}$

Range: $\{-2, -1, 0, 1, 2, 3\}$

41. $(1,2.5), (2,2), (3,1.5), (4,1), (5,.5), (6,0)$

Note that $y = -.5x + 3$, so

$$a_n = 3 - .5n$$

Domain: $\{1, 2, 3, 4, 5, 6\}$

Range: $\{0, .5, 1, 1.5, 2, 2.5\}$

43. $(1,10), (2,-10), (3,-30), (4,-50), (5,-70)$

Note that $y = -20x + 30$, so

$$a_n = 30 - 20n$$

Domain: $\{1, 2, 3, 4, 5\}$

Range: $\{-70, -50, -30, -10, 10\}$

45. $a_1 = 3,\ a_8 = 17$

$$S_n = \frac{n}{2}(a_1 + a_n)$$
$$S_8 = \frac{8}{2}(3 + 17) = \boxed{80}$$

47. $a_1 = 1,\ a_{50} = 50$

$$S_n = \frac{n}{2}(a_1 + a_n)$$
$$S_{50} = \frac{50}{2}(1 + 50) = \boxed{1275}$$

49. $a_1 = -7,\ d = 3,\ a_n = 101$

$$a_n = a_1 + (n-1)d$$
$$101 = -7 + (n-1)(3)$$
$$108 = 3n - 3$$
$$111 = 3n \Longrightarrow n = 37$$
$$S_n = \frac{n}{2}(a_1 + a_n)$$
$$S_{37} = \frac{37}{2}(-7 + 101) = \boxed{1739}$$

51. $n = 40,\ a_1 = 5(1) = 5,$
$a_{40} = 5(40) = 200$

$$S_n = \frac{n}{2}(a_1 + a_n)$$
$$S_{40} = \frac{40}{2}(5 + 200) = \boxed{4100}$$

[Note: #53 – #59 may be worked in two ways; the first is demonstrated in the first four problems; the second in the last four.]

53. $\sum_{i=1}^{3}(i+4)$ represents the sum of the first 3 terms of the arithmetic sequence having

$a_1 = 1 + 4 = 5$
$n = 3$
$a_n = a_3 = 3 + 4 = 7$

$$S_n = \frac{n}{2}(a_1 + a_n)$$
$$S_3 = \frac{3}{2}(5 + 7) = \boxed{18}$$

55. $\sum_{j=1}^{10}(2j+3)$ represents the sum of the 1st 10 terms of the arithmetic sequence having

$a_1 = 2 \cdot 1 + 3 = 5$
$n = 10$
$a_n = a_{10} = 2 \cdot 10 + 3 = 23$

$$S_n = \frac{n}{2}(a_1 + a_n)$$
$$S_3 = \frac{10}{2}(5 + 23) = \boxed{140}$$

57. $\sum_{i=1}^{12}(-5-8i)$

$$a_1 = -5 - 8 = -13$$
$$d = -8$$
$$S_n = \frac{n}{2}\left[2a_1 + (n-1)d\right]$$
$$S_{12} = \frac{12}{2}\left[2(-13) + 11(-8)\right]$$
$$= 6(-26 - 88)$$
$$= 6(-114) = \boxed{-684}$$

59. $\sum_{i=1}^{1000} i \qquad a_1 = 1,\ d = 1$

$$S_n = \frac{n}{2}\left[2a_1 + (n-1)d\right]$$
$$S_{1000} = \frac{1000}{2}\left[2(1) + 999(1)\right]$$
$$= 500\left[2 + 999\right]$$
$$= 500(1001) = \boxed{500,500}$$

Relating Concepts (61 – 64):

61. $f(x) = mx + b$

$$f(1) = m + b$$
$$f(2) = 2m + b$$
$$f(3) = 3m + b$$

62. Yes, $f(1),\ f(2),\ f(3),\ \ldots$ is an arithmetic sequence.

63. $d = a_{n+1} - a_n$

$$= (2m + b) - (m + b)$$
$$= m$$

64. $a_n = a_1 + (n-1)d$

$$a_n = (m + b) + (n-1)(m)$$
$$a_n = m + b + mn - m$$
$$a_n = mn + b$$

65. $a_n = 4.2n + 9.73$

Using the sequence feature of a graphing calculator, we obtain

$$\boxed{S_{10} = 328.3}$$
$$\boxed{S_{10} = 824.9}$$

67. $a_n = \sqrt{8}n + \sqrt{3}$

Using the sequence feature of a graphing calculator, we obtain

$$\boxed{S_{10} = 172.884}$$

69. Sum of all integers from 51 to 71:

$$\sum_{i=51}^{71} i = \sum_{i=1}^{71} i - \sum_{i=1}^{50} i$$
$$= S_{71} - S_{50}$$

$a_1 = 1,\ d = 1,\ a_{50} = 50,\ a_{71} = 71$

$$S_n = \frac{n}{2}(a_1 + a_n)$$
$$S_{71} = \frac{71}{2}(1 + 71) = 2556$$
$$S_{51} = \frac{50}{2}(1 + 50) = 1275$$
$$\sum_{i=51}^{71} i = 2556 - 1275 = \boxed{1281}$$

71. In every 12 hour cycle, the clock will chime $1 + 2 + 3 + \ldots + 12$

$a_1 = 1,\ a_{12} = 12,\ n = 12$

$$S_n = \frac{n}{2}(a_1 + a_n)$$
$$S_{12} = \frac{12}{2}(1 + 12) = 78$$

Since there are two 12-hour cycles in 1 day, every day the clock will chime 2(78) = 156 times.

Since this month has 30 days, the clock will chime 156 × 30 times, or

$$\boxed{\text{4680 times}}$$

73. $a_1 = 49,000, \quad d = 580, \quad n = 11$

$$a_n = a_1 + (n-1)d$$
$$a_{11} = 49,000 + (10)580 = 54,800$$

Five years from now the maximum population will be 54,800.

75. $a_1 = 18, \quad a_{31} = 28, \quad n = 31$

$$S_n = \frac{n}{2}(a_1 + a_n)$$

$$S_{31} = \frac{31}{2}(18 + 28) = 713$$

713 inches of material would be needed.

77. Assume that $a_1, a_2, a_3, \ldots$ is an arithmetic sequence. Consider the sequence ${a_1}^2, {a_2}^2, {a_3}^2, \ldots$; find differences of successive terms:

$$\begin{aligned} {a_2}^2 - {a_1}^2 &= (a_1 + d)^2 - {a_1}^2 \\ &= {a_1}^2 + 2a_1 d + d^2 - {a_1}^2 \\ &= 2a_1 d + d^2 \end{aligned}$$

$$\begin{aligned} {a_3}^2 - {a_2}^2 &= (a_1 + 2d)^2 - (a_1 + d)^2 \\ &= {a_1}^2 + 4a_1 d + 4d^2 \\ &\quad - \left({a_2}^2 + 2a_2 d + d^2\right) \\ &= 2a_1 d + 3d^2 \end{aligned}$$

For the sequence of squared terms to be arithmetic, these two differences must have the same value. That is,

$$\begin{aligned} 2a_1 d + d^2 &= 2a_1 d + 3d^2 \\ d^2 &= 3d^2 \\ 0 &= 2d^2 \\ 0 &= d \end{aligned}$$

For $d = 0$ to be true in the sequence $a_1, a_2, a_3, \ldots$, all terms of the sequence must be the same constant.

79. Consider the arithmetic sequence $a_1, a_2, a_3, a_4, a_5, \ldots$. then consider the sequence $a_1, a_3, a_5, \ldots$

$$a_3 - a_1 = (a_1 + 2d) - a_1 = \mathbf{2d}$$

$$a_5 - a_3 = (a_1 + 4d) - (a_1 + 2d) = \mathbf{2d}$$

$$\begin{aligned} a_n - a_{n-2} &= [a_1 + (n-1)d] - [a_1 + (n-2-1)d] \\ &= a_1 + nd - d - (a_1 + nd - 3d) \\ &= a_1 + nd - d - a_1 - nd + 3d \\ &= \mathbf{2d} \end{aligned}$$

Since there is a common difference, $2d$, between these terms, $a_1, a_3, a_5, \ldots$ is an arithmetic sequence.

Section 11.3

Note: methods may vary.

1. $a_1 = \frac{5}{3}, \ r = 3, \ n = 4$

$$a_n = a_1 r^{n-1}$$

$$\begin{aligned} a_1 &= & &= \mathbf{\tfrac{5}{3}} \\ a_2 &= \tfrac{5}{3}(3)^1 & &= \mathbf{5} \\ a_3 &= \tfrac{5}{3}(3)^2 & &= \mathbf{15} \\ a_4 &= \tfrac{5}{3}(3)^3 & &= \mathbf{45} \end{aligned}$$

3. $a_4 = 5, \quad a_5 = 10, \quad n = 5$

First find r: $\quad r = \frac{a_5}{a_4} = \frac{10}{5} = 2$

$$a_4 = a_1 r^3$$

$$5 = a_1\left(2^3\right)$$

$$a_1 = \mathbf{\frac{5}{8}}$$

$$\begin{aligned} a_2 &= \tfrac{5}{8}\left(2^1\right) = \mathbf{\tfrac{5}{4}} \\ a_3 &= \tfrac{5}{8}\left(2^2\right) = \mathbf{\tfrac{5}{2}} \\ a_4 &= \tfrac{5}{8}\left(2^3\right) = \mathbf{5} \\ a_5 &= \tfrac{5}{8}\left(2^4\right) = \mathbf{10} \end{aligned}$$

5. $a_1 = 5, \ r = -2$

$$a_n = a_1 r^{n-1}$$

$a_5 = 5(-2)^4$ $\quad \boxed{a_5 = 80}$

$$\boxed{a_n = 5(-2)^{n-1}}$$

7. $a_2 = -4,\ r = 3$

First find a_1:

$$a_1 = \frac{a_2}{r} = -\frac{4}{3}$$

$$a_n = a_1 r^{n-1}$$

$$a_5 = -\frac{4}{3}(3)^4 \quad \boxed{a_5 = -108}$$

$$\boxed{a_n = -\frac{4}{3}(3)^{n-1}}$$

Or: $a_n = -4(3)^{-1}(3)^{n-1}$

$$\boxed{a_n = -4(3)^{n-2}}$$

9. $a_4 = 243,\ r = -3$

First, find a_1 :

$$a_n = a_1 r^{n-1}$$

$n = 4$: $243 = a_1(-3)^3$

$$a_1 = -\frac{243}{27} = -9$$

$$a_5 = -9(-3)^4 \quad \boxed{a_5 = -729}$$

$$\boxed{a_n = -9(-3)^{n-1}}$$

Or: $a_n = -(-3)^2(-3)^{n-1}$

$$\boxed{a_n = -(-3)^{n+1}}$$

11. $-4, -12, -36, -108, \ldots$

Since each term is multiplied by 3 ($r = 3$), the sequence is geometric with $a_1 = -4$.

$$a_5 = a_1 r^4 = -4(3)^4$$

or $a_5 = a_4 r = -108(3)$

$$\boxed{a_5 = -324}$$

$$a_n = a_1 r^{n-1} \quad \boxed{a_n = -4(3)^{n-1}}$$

13. $\frac{4}{5}, 2, 5, \frac{25}{2}, \ldots$

Find r : $r = \frac{a_{n+1}}{a_n} = \frac{5}{2}$

$$a_5 = a_1 r^4 = \frac{4}{5}\left(\frac{5}{2}\right)^4$$

or $a_5 = a_4 r = \frac{25}{2}\left(\frac{5}{2}\right)$

$$\boxed{a_5 = \frac{125}{4}}$$

$$a_n = a_1 r^{n-1} \quad \boxed{a_n = \frac{4}{5}\left(\frac{5}{2}\right)^{n-1}}$$

15. $10, -5, \frac{5}{2}, -\frac{5}{4}, \ldots$

Find r : $r = \frac{a_{n+1}}{a_n} = \frac{-5}{10} = -\frac{1}{2}$

$$a_5 = a_1 r^4 = 10\left(-\frac{1}{2}\right)^4$$

or $a_5 = a_4 r = -\frac{5}{4}\left(-\frac{1}{2}\right)$

$$\boxed{a_5 = \frac{5}{8}}$$

$$a_n = a_1 r^{n-1} \quad \boxed{a_n = 10\left(-\frac{1}{2}\right)^{n-1}}$$

Relating Concepts (17 – 20):

17. $5,\ x,\ .6$ If the sequence is arithmetic, $x - 5 = .6 - x$.

18. $x - 5 = .6 - x$
$2x = 5.6$
$x = 2.8$

$a_1 = 5$
$a_2 = x = 2.8$
$a_3 = .6$

19. 5, x, .6 If the sequence is geometric,
$\frac{x}{5} = \frac{.6}{x}$.

20. $\frac{x}{5} = \frac{.6}{x}$
$x^2 = 3 \rightarrow x = \pm\sqrt{3}$
The positive solution is $\sqrt{3}$.
$a_1 = 5$
$a_2 = x = \sqrt{3}$
$a_3 = .6$

21. $a_3 = 5,\ a_8 = \frac{1}{625}$

$$a_n = a_1\, r^{n-1}$$

$n = 3 : 5 = a_1 r^2$

$a_1 = \frac{5}{r^2}$ [1]

$n = 8 : \frac{1}{625} = a_1 r^7$ [2]

[1] → [2] :

$\frac{1}{625} = \frac{5}{r^2}\, r^7 = 5r^5$

$\frac{1}{625} \times \frac{1}{5} = r^5$

$\frac{1}{3125} = r^5 \longrightarrow r = \frac{1}{5}$

$r \rightarrow [1] :\quad a_1 = \frac{5}{1/25} = 125$

$$\boxed{a_1 = 125,\ r = \frac{1}{5}}$$

23. $a_4 = -\frac{1}{4},\ a_9 = -\frac{1}{128}$

$$a_n = a_1\, r^{n-1}$$

$n = 4 : -\frac{1}{4} = a_1 r^3$

$a_1 = -\frac{1}{4r^3}$ [1]

$n = 9 : -\frac{1}{128} = a_1 r^8$ [2]

[1] → [2] :

$-\frac{1}{128} = -\frac{1}{4r^3}\, r^8 = -\frac{1}{4} r^5$

$-\frac{1}{128} \times -4 = r^5$

$\frac{1}{32} = r^5$

$r = \frac{1}{2}$

$r \rightarrow [1] :\quad a_1 = -\frac{1}{4\left(\frac{1}{2}\right)^3} = -2$

$$\boxed{a_1 = -2,\ r = \frac{1}{2}}$$

25. 2, 8, 32, 128, ...

$r = \frac{a_2}{a_1} = \frac{8}{2} = 4$

$S_n = \frac{a_1\,(1 - r^n)}{1 - r}$

$S_5 = \frac{2\,(1 - 4^5)}{1 - 4}$

$= \frac{2\,(-1023)}{-3} = \boxed{682}$

27. 18, -9, $\frac{9}{2}$, $-\frac{9}{4}$, ...

$r = \frac{a_2}{a_1} = \frac{-9}{18} = -\frac{1}{2}$

$S_n = \frac{a_1\,(1 - r^n)}{1 - r}$

$S_5 = \frac{18\left(1 - \left(-\frac{1}{2}\right)^5\right)}{1 - \left(-\frac{1}{2}\right)} = \frac{18\left(1 + \frac{1}{32}\right)}{\frac{3}{2}}$

$= 18\left(\frac{33}{32}\right)\left(\frac{2}{3}\right) = \boxed{\frac{99}{8}}$

29. $a_1 = 8.423, \quad r = 2.859$

$$S_n = \frac{a_1(1-r^n)}{1-r}$$

$$S_5 = \frac{8.423\left(1-(2.859)^5\right)}{1-2.859}$$

$$\approx 860.95136376 \approx \boxed{860.95}$$

31. $\sum_{i=1}^{5} 3^i$

For this geometric series, $a_1 = 3$, $r = 3,\ n = 5$.

$$S_n = \frac{a_1(1-r^n)}{1-r}$$

$$S_5 = \frac{3(1-3^5)}{1-3} = \frac{3(1-243)}{-2} = \boxed{363}$$

33. $\sum_{j=1}^{6} 48\left(\frac{1}{2}\right)^j$

For this geometric series,

$$a_1 = 48\left(\frac{1}{2}\right) = 24,\ r = \frac{1}{2},\ n = 6$$

$$S_n = \frac{a_1(1-r^n)}{1-r}$$

$$S_6 = \frac{24\left(1-\left(\frac{1}{2}\right)^6\right)}{1-\frac{1}{2}} = \frac{24\left(1-\frac{1}{64}\right)}{\frac{1}{2}}$$

$$= \frac{24}{1}\cdot\frac{63}{64}\cdot\frac{2}{1} = \boxed{\frac{189}{4}}$$

35. $\sum_{k=4}^{10} 2^k$

For this geometric series, $a_1 = 2^4 = 16$, $r = 2,\ \ n = 10-4+1 = 7$.

$$S_n = \frac{a_1(1-r^n)}{1-r}$$

$$S_7 = \frac{16(1-2^7)}{1-2}$$

$$= \frac{16(1-128)}{-1} = \boxed{2032}$$

37. An infinite geometric sequence will have a sum when $-1 < r < 1$, or $|r| < 1$.

39. $.8 + .08 + .008 + .0008 + \ldots$

$$a_1 = .8,\ r = \frac{a_2}{a_1} = .1$$

$$S_\infty = \frac{a_1}{1-r} = \frac{.8}{1-.1} = \boxed{\frac{8}{9}}$$

41. $.45 + .0045 + .000045 + \ldots$

$$a_1 = .45,\ r = \frac{a_2}{a_1} = .01$$

$$S_\infty = \frac{a_1}{1-r} = \frac{.45}{1-.01} = \frac{45}{99} = \boxed{\frac{5}{11}}$$

43. 12, 24, 48, 96, ...

$$r = \frac{a_2}{a_1} = \frac{24}{12} = 2$$

The sum would not converge ($|r| > 1$).

45. $-48, -24, -12, -6, \ldots$

$$r = \frac{a_2}{a_1} = \frac{-24}{-48} = \frac{1}{2}$$

The sum would converge ($|r| < 1$).

47. $16 + 2 + \frac{1}{4} + \frac{1}{32} + \ldots$

$$r = \frac{a_2}{a_1} = \frac{2}{16} = \frac{1}{8}$$

$$S_\infty = \frac{a_1}{1-r} = \frac{16}{1-\frac{1}{8}} = \frac{16}{\frac{7}{8}}$$

$$= \frac{16}{1}\cdot\frac{8}{7} = \boxed{\frac{128}{7}}$$

49. $100 + 10 + 1 + \ldots$

$$r = \frac{a_2}{a_1} = \frac{10}{100} = \frac{1}{10}$$

$$S_\infty = \frac{a_1}{1-r} = \frac{100}{1-\frac{1}{10}} = \frac{100}{\frac{9}{10}}$$

$$= \frac{100}{1}\cdot\frac{10}{9} = \boxed{\frac{1000}{9}}$$

51. $\frac{4}{3}+\frac{2}{3}+\frac{1}{3}+\ldots$

$r=\frac{a_2}{a_1}=\frac{2}{3}\times\frac{3}{4}=\frac{1}{2}$

$$S_\infty=\frac{a_1}{1-r}=\frac{\frac{4}{3}}{1-\frac{1}{2}}=\frac{\frac{4}{3}}{\frac{1}{2}}$$

$$=\frac{4}{3}\cdot\frac{2}{1}=\boxed{\frac{8}{3}}$$

53. $\sum_{i=1}^{\infty}3\left(\frac{1}{4}\right)^{i-1}$

$a_1=3\left(\frac{1}{4}\right)^0=3,\quad r=\frac{1}{4}$

$$S_\infty=\frac{a_1}{1-r}=\frac{3}{1-\frac{1}{4}}=\frac{3}{\frac{3}{4}}$$

$$=\frac{3}{1}\cdot\frac{4}{3}=\boxed{4}$$

55. $\sum_{k=1}^{\infty}(.3)^k$

$a_1=(.3)^1=.3,\quad r=.3$

$$S_\infty=\frac{a_1}{1-r}=\frac{.3}{1-.3}=\frac{.3}{.7}=\boxed{\frac{3}{7}}$$

Relating Concepts (57 - 60):

57. $g(x)=ab^x$

$g(1)=ab^1=ab$
$g(2)=ab^2$
$g(3)=ab^3$

58. The sequence $g(1),\ g(2),\ g(3),$ is a geometric sequence because each term after the first is a constant multiple of the preceding term. The common ratio is $\frac{ab^2}{ab}=b.$

59. From **#57**, $a_1=ab$. From **#58**, $r=b$. Therefore,

$a_n=a_1r^{n-1}$
$a_n=ab(b)^{n-1}$
$a_n=ab^n$

60. In both geometric sequences and exponential functions, the independent variable is in the exponent.

61. $\sum_{i=1}^{10}(1.4)^i$

Using the sequence feature of a graphing calculator, we obtain

$$S_{10}\approx 97.739.$$

63. $\sum_{j=3}^{8}2(.4)^j$

Using the sequence feature of a graphing calculator, we obtain

$$S_6\approx .212.$$

65. The payments plus interest form a geometric sequence with $r=1.08$:

$1000,\ 1000(1.08),\ldots$

Find S_9 to find the total interest:

$$S_n=\frac{a_1(1-r^n)}{1-r}$$

$$S_9=\frac{1000\left(1-(1.08)^9\right)}{1-1.08}$$

$$\approx\boxed{\$12,487.56}$$

67. The payments plus interest form a geometric sequence with $r=1.06$:

$2430,\ 2430(1.06),\ldots$

Find S_{10} to find the total interest:

$$S_n=\frac{a_1(1-r^n)}{1-r}$$

$$S_{10}=\frac{2430\left(1-(1.06)^{10}\right)}{1-1.06}$$

$$\approx\boxed{\$32,029.33}$$

69. **(a)** $a_n = 1276\,(.916)^n$

$a_1 = 1276\,(.916)^1$ $\boxed{a_1 \approx 1169,\ r = .916}$

(b) $a_{10} = 1276\,(.916)^{10} \approx 531$

$a_{20} = 1276\,(.916)^{20} \approx 221$

A person 10 years from retirement should have savings of 531% of her annual salary. A person 20 years from retirement should have savings of 221% of her annual salary.

71. **(a)** $a_n = a_1 \cdot 2^{n-1}$

(b) If $a_1 = 100$, we have

$a_n = 100 \cdot 2^{n-1}$

Since $100 = 10^2$ and $1,000,000 = 10^6$ we need to solve the equation

$10^6 = 10^2 \cdot 2^{n-1}$

Divide both sides by 10^2 :

$$10^4 = 2^{n-1}$$
$$\log 10^4 = \log 2^{n-1}$$
$$4 = (n-1)\log 2$$
$$4 = n\log 2 - \log 2$$
$$4 + \log 2 = n \log 2$$
$$\frac{4 + \log 2}{\log 2} = n$$
$$n \approx 14.2877 \approx 15$$

Since the number of bacteria is increasing, the first value of n where $a_n > 1,000,000$ is 15.

(c) Since a_n represents the number of bacteria after $40\,(n-1)$ minutes, a_{15} represents the number after

$40\,(15-1) = 40\,(14) = 560$ minutes

$= \boxed{\text{9 hours, 20 minutes}}$

73. A geometric sequence is formed which describes the strength of the mixture:

100, 80, 74, ... where $r = .8$

$$\left(80\% = \frac{80L}{100L}\right)$$

$a_n = a_1 r^{n-1}, \quad n = 10$

$a_{10} = 100\,(.8)^9 = 13.4217728$

$\boxed{\text{There will be} \approx 13.4\% \text{ of chemical mixed with the water.}}$

75. A machine that loses 20% of its value yearly retains 80% of its value. At the end of 6 years, the value of the machine will be

$\$100{,}000(.80)^6 = \boxed{\$26,214.40}$

77. $40,\ 40\,(.8),\ 40\,(.8)^2, \ldots$

$$S_\infty = \frac{a_1}{1-r} = \frac{40}{1-.8} = 200$$

$\boxed{\text{It will swing 200 cm.}}$

79. Use the formula for the sum of the first n terms of a geometric sequence with $a_1 = 2$, $r = 2$, and $n = 5$.

$$S_n = \frac{a_1\,(1-r^n)}{1-r}$$

$$S_5 = \frac{2\,(1-2^5)}{1-2} = 62$$

$\boxed{\text{Going back 5 generations, the total number of ancestors is 62.}}$

Next, use the same formula with $a_1 = 2$, $r = 2$, and $n = 10$.

$$S_n = \frac{a_1\,(1-r^n)}{1-r}$$

$$S_{10} = \frac{2\,(1-2^{10})}{1-2} = 2046$$

$\boxed{\text{Going back 10 generations, the total number of ancestors is 2046.}}$

81. When the midpoints of the sides of an equilateral triangle are connected, the length of a side of the new triangle is one-half the length of a side of the original triangle. Use the formula for the nth term of a geometric sequence with $a_1 = 2$, $r = \frac{1}{2}$, and $n = 8$.

$$a_n = a_1 r^{n-1}$$

$$a_8 = 2\left(\frac{1}{2}\right)^7 = 2\left(\frac{1}{128}\right) = \frac{1}{64}$$

$\boxed{\text{The eighth triangle has sides of length } \frac{1}{64} \text{ meter.}}$

83. The first option is modeled by the sequence

$$a_n = 5000 + 10,000\,(n-1)$$
$$a_n = 5000 + 10,000(29)$$
$$a_n = 295,000$$

$$S_{30} = \frac{30}{2}(5,000 + 295,000) = 4,500,000$$

The second option is modeled by the sequence

$a_n = .01\,(2)^{n-1}$, with sum

$$S_n = \sum_{i=1}^{n} a_i$$

$$S_{30} = \sum_{i=1}^{30} .01\,(2)^{i-1} = 10,737,418.23$$

The first option pays \$4,500,000, while the second option pays \$10,737,418.23. The second option pays better.

85. Since $a_1, a_2, a_3, \ldots$ forms a geometric sequence, it has a common ratio

$$\frac{a_{n+1}}{a_n} = r_1.$$

Since $b_1, b_2, b_3, \ldots$ forms a geometric sequence, it has a common ratio

$$\frac{b_{n+1}}{b_n} = r_2.$$

Show $d_1, d_2, d_3, \ldots$ forms a geometric sequence, where

$$d_n = ca_nb_n.$$

$d_1, d_2, d_3, \ldots$ has a common ratio $r = r_1r_2$, since

$$\frac{d_{n+1}}{d_n} = \frac{ca_{n+1}b_{n+1}}{ca_nb_n} = \frac{c}{c} \times \frac{a_{n+1}}{a_n} \times \frac{b_{n+1}}{b_n} = r_1r_2$$

Thus, it forms a geometric sequence.

Reviewing Basic Concepts (Sections 8.1 - 8.3)

1. $a_n = (-1)^{n-1}(4n)$

$$a_1 = (-1)^0(4(1)) = \mathbf{4}$$
$$a_2 = (-1)^1(4(2)) = \mathbf{-8}$$
$$a_3 = (-1)^2(4(3)) = \mathbf{12}$$
$$a_4 = (-1)^3(4(4)) = \mathbf{-16}$$
$$a_5 = (-1)^4(4(5)) = \mathbf{20}$$

2. $\sum_{i=1}^{5}(3i+1) =$

$$(3(1)+1) + (3(2)+1) + (3(3)+1) + (3(4)+1) + (3(5)+1)$$
$$= 4 + 7 + 10 + 13 + 16 = \boxed{50}$$

3. $a_n = 1 - \dfrac{2}{n}$

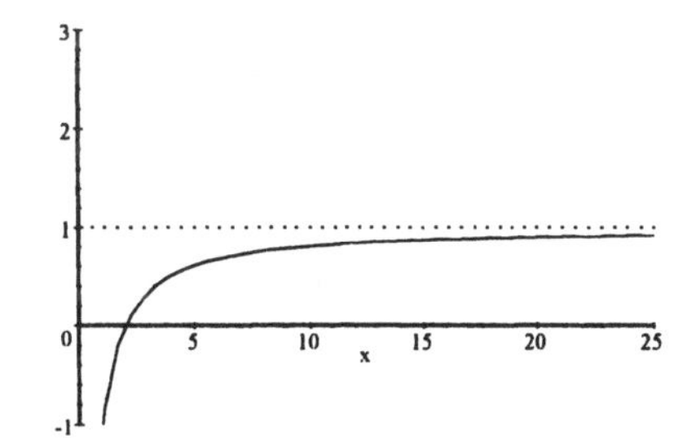

The series converges to 1.

4. $a_1 = 8,\ d = -2$

$$a_n = a_1 + (n-1)\,d$$
$$a_n = 8 - 2\,(n-1)$$
$$a_n = -2n + 10$$

$$a_1 = \mathbf{8}$$
$$a_2 = -2\,(2) + 10 = \mathbf{6}$$
$$a_3 = -2\,(3) + 10 = \mathbf{4}$$
$$a_4 = -2\,(4) + 10 = \mathbf{2}$$
$$a_5 = -2\,(5) + 10 = \mathbf{0}$$

5. $a_5 = 5,\ a_8 = 17$

$$3d = a_8 - a_5$$
$$3d = 17 - 5 = 12$$
$$d = 4$$

$$a_8 = a_1 + (8-1)\,(d)$$
$$17 = a_1 + 7\,(4)$$
$$\boxed{a_1 = -11}$$

6. $a_1 = 2,\ d = 5$

$$a_n = a_1 + (n-1)\,d$$
$$a_{10} = 2 + 9\,(5) = 47$$
$$S_n = \frac{n}{2}\,[a_1 + a_n]$$
$$S_{10} = \frac{10}{2}\,[2 + 47] = \boxed{245}$$

7. $a_1 = -2,\ r = -3$

$$a_n = a_1 r^{n-1}$$
$$a_3 = -2\,(-3)^2 \quad \boxed{a_3 = -18}$$
$$\boxed{a_n = -2\,(-3)^{n-1}}$$

8. $5 + 3 + \frac{9}{5} + \ldots + \frac{243}{625}$; geometric

$$r = \frac{a_2}{a_1} = \frac{3}{5}$$

Find n: $\frac{243}{625} = 5\left(\frac{3}{5}\right)^{n-1}$

$$\frac{243}{3125} = \left(\frac{3}{5}\right)^{n-1}$$
$$\left(\frac{3}{5}\right)^5 = \left(\frac{3}{5}\right)^{n-1} \Longrightarrow n = 6$$
$$S_n = \frac{a_1\,(1 - r^n)}{1 - r}$$
$$S_6 = \frac{5\left(1 - \left(\frac{3}{5}\right)^6\right)}{1 - \left(\frac{3}{5}\right)} = \frac{5\left(1 - \frac{729}{15625}\right)}{\frac{2}{5}} = \boxed{\frac{7448}{625}}$$

9. $\sum_{k=1}^{\infty} 3\left(\frac{2}{3}\right)^k$

$$a_1 = 3\left(\frac{2}{3}\right) = 2, \quad r = \frac{2}{3}$$
$$S_\infty = \frac{a_1}{1 - r} = \frac{2}{1 - \left(\frac{2}{3}\right)} = \frac{2}{\frac{1}{3}} = \boxed{6}$$

10. $S_n = \frac{a_1\,(1 - r^n)}{1 - r}$, $r = 1+$ interest rate

Payments = \$500 for 13 years at 9% interest compounded annually, so

$$a_1 = 500,\ r = 1.09,\ n = 13$$
$$S_{13} = \frac{500\,(1 - 1.09^{13})}{1 - 1.09} \approx \$11,476.69$$

Section 11.4

1. $\frac{6!}{3!\,3!} = \frac{6 \cdot 5 \cdot 4 \cdot 3!}{3 \cdot 2 \cdot 1 \cdot 3!} = \boxed{20}$

3. $\frac{7!}{3!\,4!} = \frac{7 \cdot 6 \cdot 5 \cdot 4!}{3 \cdot 2 \cdot 1 \cdot 4!} = \boxed{35}$

5. $\binom{8}{3} = \frac{8!}{3!\,5!} = \frac{8 \cdot 7 \cdot 6 \cdot 5!}{3 \cdot 2 \cdot 1 \cdot 5!} = \boxed{56}$

7. $\binom{10}{8} = \frac{10!}{8!\,2!} = \frac{10 \cdot 9 \cdot 8!}{8! \cdot 2 \cdot 1} = \boxed{45}$

9. $\binom{13}{13} = \frac{13!}{13!\,1!} = \frac{13!}{13! \cdot 1} = \boxed{1}$

11.
$$\binom{n}{n-1} = \frac{n!}{(n-1)!\,1!} = \frac{n \cdot (n-1)!}{(n-1)!} = \boxed{n}$$

13.
$${}_8C_3 = \frac{8!}{3!\,(8-3)!} = \frac{8!}{3!\,5!} = \frac{8 \cdot 7 \cdot 6 \cdot 5!}{3 \cdot 2 \cdot 1 \cdot 5!} = \boxed{56}$$

15.
$${}_{100}C_2 = \frac{100!}{2!\,(100-2)!} = \frac{100!}{2!\,98!} = \frac{100 \cdot 99 \cdot 98!}{2 \cdot 1 \cdot 98!} = \boxed{4950}$$

17. $(x + y)^8$ has 9 terms.

19.
$$(2x + 3y)^4 = (2x)^4 + \binom{4}{1}(2x)^3\,(3y) + \binom{4}{2}(2x)^2\,(3y)^2 + \binom{4}{3}(2x)\,(3y)^3 + (3y)^4$$

first term: $16x^4$; last term: $81y^4$

21. $(x+y)^6 = x^6 + \binom{6}{1}x^5y + \binom{6}{2}x^4y^2 + \binom{6}{3}x^3y^3 + \binom{6}{4}x^2y^4 + \binom{6}{5}xy^5 + y^6$

$$= \boxed{x^6 + 6x^5y + 15x^4y^2 + 20x^3y^3 + 15x^2y^4 + 6xy^5 + y^6}$$

23. $(p-q)^5 = p^5 + \binom{5}{1}p^4(-q) + \binom{5}{2}p^3(-q)^2 + \binom{5}{3}p^2(-q)^3 + \binom{5}{4}p(-q)^4 + (-q)^5$

$$= \boxed{p^5 - 5p^4q + 10p^3q^2 - 10p^2q^3 + 5pq^4 - q^5}$$

25. $(r^2+s)^5 = (r^2)^5 + \binom{5}{1}(r^2)^4 s + \binom{5}{2}(r^2)^3 s^2 + \binom{5}{3}(r^2)^2 s^3 + \binom{5}{4}(r^2)s^4 + s^5$

$$= \boxed{r^{10} + 5r^8s + 10r^6s^2 + 10r^4s^3 + 5r^2s^4 + s^5}$$

27. $(p+2q)^4 = p^4 + \binom{4}{1}p^3(2q) + \binom{4}{2}p^2(2q)^2 + \binom{4}{3}p(2q)^3 + (2q)^4$

$$= p^4 + 4p^3(2q) + 6p^2(4q^2) + 4p(8q^3) + 16q^4$$

$$= \boxed{p^4 + 8p^3q + 24p^2q^2 + 32pq^3 + 16q^4}$$

29. $(7p+2q)^4 = (7p)^4 + \binom{4}{1}(7p)^3(2q) + \binom{4}{2}(7p)^2(2q)^2 + \binom{4}{3}(7p)(2q)^3 + (2q)^4$

$$= 2401p^4 + 4(343p^3)(2q) + 6(49p^2)(4q^2) + 4(7p)(8q^3) + 16q^4$$

$$= \boxed{2401p^4 + 2744p^3q + 1176p^2q^2 + 224pq^3 + 16q^4}$$

31. $(3x-2y)^6 = (3x)^6 + \binom{6}{1}(3x)^5(-2y) + \binom{6}{2}(3x)^4(-2y)^2 + \binom{6}{3}(3x)^3(-2y)^3 + \binom{6}{4}(3x)^2(-2y)^4 + \binom{6}{5}(3x)(-2y)^5 + (-2y)^6$

$$= 729x^6 + 6(243x^5)(-2y) + 15(81x^4)(4y^2) + 20(27x^3)(-8y^3) + 15(9x^2)(16y^4) + 6(3x)(-32y^5) + 64y^6$$

$$= \boxed{729x^6 - 2916x^5y + 4860x^4y^2 - 4320x^3y^3 + 2160x^2y^4 - 576xy^5 + 64y^6}$$

33. $\left(\frac{m}{2}-1\right)^6$

$$= \left(\frac{m}{2}\right)^6 + \binom{6}{1}\left(\frac{m}{2}\right)^5(-1) + \binom{6}{2}\left(\frac{m}{2}\right)^4(-1)^2 + \binom{6}{3}\left(\frac{m}{2}\right)^3(-1)^3 + \binom{6}{4}\left(\frac{m}{2}\right)^2(-1)^4 + \binom{6}{5}\left(\frac{m}{2}\right)(-1)^5 + (-1)^6$$

$$= \frac{m^6}{64} + 6\left(\frac{m^5}{32}\right)(-1) + 15\left(\frac{m^4}{16}\right)(1) + 20\left(\frac{m^3}{8}\right)(-1) + 15\left(\frac{m^2}{4}\right)(1) + 6\left(\frac{m}{2}\right)(-1) + 1$$

$$= \boxed{\frac{1}{64}m^6 - \frac{3}{16}m^5 + \frac{15}{16}m^4 - \frac{5}{2}m^3 + \frac{15}{4}m^2 - 3m + 1}$$

35. $\left(\sqrt{2}r + \frac{1}{m}\right)^4 = (\sqrt{2}r)^4 + \binom{4}{1}(\sqrt{2}r)^3\left(\frac{1}{m}\right) + \binom{4}{2}(\sqrt{2}r)^2\left(\frac{1}{m}\right)^2 + \binom{4}{3}(\sqrt{2}r)\left(\frac{1}{m}\right)^3 + \left(\frac{1}{m}\right)^4$

$$= 4r^4 + 4(2\sqrt{2}r^3)\left(\frac{1}{m}\right) + 6(2r^2)\left(\frac{1}{m^2}\right) + 4(\sqrt{2}r)\left(\frac{1}{m^3}\right) + \frac{1}{m^4}$$

$$= \boxed{4r^4 + \frac{8\sqrt{2}r^3}{m} + \frac{12r^2}{m^2} + \frac{4\sqrt{2}r}{m^3} + \frac{1}{m^4}}$$

37. $\binom{n}{r-1}\, \mathbf{x}^{n-(r-1)}\, \mathbf{y}^{r-1}$

6th term of $(4h-j)^8$:

$$n=8,\ r=6,$$
$$\mathbf{x}=4h,\ \mathbf{y}=-j$$

$$\binom{8}{5}(4h)^3(-j)^5 = 56\,(64h^3)\,(-j^5)$$

$$= \boxed{-3584\,h^3\,j^5}$$

39. $\binom{n}{r-1}\, \mathbf{x}^{n-(r-1)}\mathbf{y}^{r-1}$

15th term of $(a^2+b)^{22}$:

$$n=22,\ r=15,$$
$$\mathbf{x}=a^2,\ \mathbf{y}=b$$

$$\binom{22}{14}(a^2)^8(b)^{14} = 319,770\,(a^2)^8\,(b)^{14}$$

$$= \boxed{319,770\,a^{16}\,b^{14}}$$

41. $\binom{n}{r-1}\, \mathbf{x}^{n-(r-1)}\mathbf{y}^{r-1}$

15th term of $(x-y^3)^{20}$:

$$n=20,\ r=15,\ \mathbf{x}=x,\ \mathbf{y}=-y^3$$

$$\binom{20}{14}(x)^6(-y^3)^{14} = 38,760\,(x^6)\,(y^{42})$$

$$= \boxed{38,760\,x^6\,y^{42}}$$

43. $(3x^7+2y^3)^8$ Since this expansion has 9 terms, the middle term is the 5th :

$$\binom{n}{r-1}\, \mathbf{x}^{n-(r-1)}\mathbf{y}^{r-1}$$

$$n=8,\ r=5,\ \mathbf{x}=3x^7,\ \mathbf{y}=2y^3$$

$$\binom{8}{4}(3x^7)^4(2y^3)^4 = 70\,(81x^{28})\,(16y^{12})$$

$$= \boxed{90,720\,x^{28}\,y^{12}}$$

45. If the coefficients of the 5th and 8th terms in the expansion of $(x+y)^n$ are the same, then the symmetry of the expansion can be used to determine n. There are 4 terms before the 5th term, so there must be 4 terms after the 8th term. This means that there are 12 terms. This in turn means that $n=11$, since $(x+y)^{11}$ is the expansion that has 12 terms.

Relating Concepts (47 – 50):

47. Using a calculator, we obtain the exact value,

$$10! = 3,628,800.$$

Using Stirling's formula, we obtain the approximate value

$$10! \approx \sqrt{2\pi\,(10)}\cdot 10^{10}\cdot e^{-10}$$
$$\approx 3,598,695.619$$

48. $$\frac{3,628,800-3,598,695.619}{10!}$$

$$\approx .00830 \approx \boxed{.830\%}$$

49. Using a calculator, we obtain the exact value,

$$12! = 479,001,600.$$

Using Stirling's formula, we obtain the approximate value

$$12! \approx \sqrt{2\pi\,(12)}\cdot 12^{12}\cdot e^{-12}$$
$$\approx 475,687,486.5$$

Find the percent error:

$$\frac{479,001,600-475,687,486.5}{12!}$$

$$\approx .00692 \approx \boxed{.692\%}$$

50. Using a calculator, we obtain the exact value,

$13! = 6,277,020,800.$

Using Stirling's formula, we obtain the approximate value

$$13! \approx \sqrt{2\pi(13)} \cdot 13^{13} \cdot e^{-13}$$
$$\approx 6,187,239,475$$

Find the percent error:

$$\frac{6,277,020,800 - 6,187,239,475}{13!}$$
$$\approx .00639 \approx \boxed{.639\%}$$

As n gets larger, the percent error decreases.

51. $(1.02)^{-3} = (1 + .02)^{-3}$

$$= 1 + (-3)(.02) + \frac{-3(-4)}{2!}(.02)^2 + ...$$
$$= 1 - .06 + .0024 - ...$$
$$\approx 0.9424 \approx \boxed{0.942}$$

53. $(1.01)^{3/2} = (1 + .01)^{1.5}$

$$= 1 + (1.5)(.01) + \frac{1.5(.5)}{2!}(.01)^2 + ...$$
$$= 1 + .015 + .0000375 + ...$$
$$\approx 1.0150375 \approx \boxed{1.015}$$

Section 11.5

1. A proof by mathematical induction allows us to prove that a statement is true for all <u>positive integers</u>.

3. $2^n > 2n$:

Try $n = 1 : 2^1 > 2(1)$ False
Try $n = 2 : 2^2 > 2(2)$ False
Try $n = 3 : 2^3 > 2(3)$ True

$2^n > 2n$ will be true for all values of n which are greater than 2, so the statement is not true for $n = 1,\ 2$.

5. $3 + 6 + 9 + ... + 3n = \dfrac{3n(n+1)}{2}$

(i) Show true for $n = 1$:

$$3(1) = \frac{3(1)(2)}{2}$$
$$3 = 3$$

(ii) Assume S_k is true:

$$3 + 6 + ... + 3k = \frac{3k(k+1)}{2}$$

Show S_{k+1} is true:

$$3 + 6 + ... + 3(k+1) = \frac{3(k+1)(k+2)}{2}$$

Add $3(k+1)$ to both sides of S_k :

$$3 + 6 + ... + 3k + 3(k+1)$$
$$= \frac{3k(k+1)}{2} + 3(k+1)$$
$$= \frac{3k(k+1) + 6(k+1)}{2}$$
$$= \frac{(k+1)(3k+6)}{2}$$
$$= \frac{3(k+1)(k+2)}{2}$$

Since S_k implies S_{k+1}, the statement is true for every positive value of n.

7. $5 + 10 + 15 + ... + 5n = \dfrac{5n(n+1)}{2}$

(i) Show true for $n = 1$:

$$5(1) = \frac{5(1)(2)}{2}$$
$$5 = 5$$

(ii) Assume S_k is true:

$$5 + 10 + ... + 5k = \frac{5k(k+1)}{2}$$

Show S_{k+1} is true:

$$5 + 10 + ... + 5(k+1) = \frac{5(k+1)(k+2)}{2}$$

Add $5(k+1)$ to both sides of S_k:

$$5 + 10 + ... + 5k + 5(k+1)$$
$$= \frac{5k(k+1)}{2} + 5(k+1)$$
$$= \frac{5k(k+1) + 10(k+1)}{2}$$
$$= \frac{5(k+1)(k+2)}{2}$$

Since S_k implies S_{k+1}, the statement is true for every positive value of n.

9. $3 + 3^2 + 3^3 + ... + 3^n = \dfrac{3(3^n - 1)}{2}$

(i) Show true for $n = 1$:

$$3^1 = \frac{3(3^1 - 1)}{2}$$
$$3 = \frac{3(2)}{2} = 3$$

(ii) Assume S_k is true:

$$3 + 3^2 + ... + 3^k = \frac{3(3^k - 1)}{2}$$

continued

9. continued

Show S_{k+1} is true:

$$3 + 3^2 + 3^3 + ... + 3^{k+1} = \frac{3(3^{k+1} - 1)}{2}$$

Add 3^{k+1} to both sides of S_k:

$$3 + 3^2 + ... + 3^k + 3^{k+1}$$
$$= \frac{3(3^k - 1)}{2} + 3^{k+1}$$
$$= \frac{3(3^k - 1) + 2(3^{k+1})}{2}$$
$$= \frac{3^{k+1} - 3 + 2(3^{k+1})}{2}$$
$$= \frac{3(3^{k+1}) - 3}{2}$$
$$= \frac{3(3^{k+1} - 1)}{2}$$

Since S_k implies S_{k+1}, the statement is true for every positive value of n.

11. $1^3 + 2^3 + 3^3 + ... + n^3 = \dfrac{n^2(n+1)^2}{4}$

(i) Show true for $n = 1$:

$$1^3 = \frac{1^2 \cdot 2^2}{4}$$
$$1 = 1$$

(ii) Assume S_k is true:

$$1^3 + 2^3 + ... + k^3 = \frac{k^2(k+1)^2}{4}$$

Show S_{k+1} is true:

$$1^3 + 2^3 + ... + (k+1)^3 = \frac{(k+1)^2(k+2)^2}{4}$$

Add $(k+1)^3$ to both sides of S_k:

$$1^3 + 2^3 + ... + k^3 + (k+1)^3$$
$$= \frac{k^2(k+1)^2}{4} + (k+1)^3$$
$$= \frac{k^2(k+1)^2 + 4(k+1)^3}{4}$$
$$= \frac{(k+1)^2(k^2 + 4k + 4)}{4}$$
$$= \frac{(k+1)^2(k+2)^2}{4}$$

Since S_k implies S_{k+1}, the statement is true for every positive value of n.

13. $\dfrac{1}{1\cdot 2}+\dfrac{1}{2\cdot 3}+...+\dfrac{1}{n(n+1)}=\dfrac{n}{n+1}$

(i) Show true for $n=1$:

$$\frac{1}{1\cdot 2}=\frac{1}{1+1}$$
$$\frac{1}{2}=\frac{1}{2}$$

(ii) Assume S_k is true:

$$\frac{1}{1\cdot 2}+\frac{1}{2\cdot 3}+...+\frac{1}{k(k+1)}=\frac{k}{k+1}$$

Show S_{k+1} is true:

$$\frac{1}{1\cdot 2}+\frac{1}{2\cdot 3}+\frac{1}{(k+1)(k+2)}=\frac{k+1}{k+2}$$

Add $\dfrac{1}{(k+1)(k+2)}$ to both sides of S_k:

$$\frac{1}{1\cdot 2}+\frac{1}{2\cdot 3}+...+\frac{1}{k(k+1)}+\frac{1}{(k+1)(k+2)}$$
$$=\frac{k}{k+1}+\frac{1}{(k+1)(k+2)}$$
$$=\frac{k(k+2)+1}{(k+1)(k+2)}$$
$$=\frac{k^2+2k+1}{(k+1)(k+2)}$$
$$=\frac{(k+1)(k+1)}{(k+1)(k+2)}$$
$$=\frac{k+1}{k+2}$$

Since S_k implies S_{k+1}, the statement is true for every positive value of n.

15. $\dfrac{1}{1\cdot 4}+\dfrac{1}{4\cdot 7}+...+\dfrac{1}{(3n-2)(3n+1)}=\dfrac{n}{3n+1}$

(i) Show true for $n=1$:

$$\frac{1}{1\cdot 4}=\frac{1}{3+1}$$
$$\frac{1}{4}=\frac{1}{4}$$

(ii) Assume S_k is true:

$$\frac{1}{1\cdot 4}+\frac{1}{4\cdot 7}+...+\frac{1}{(3k-2)(3k+1)}=\frac{k}{3k+1}$$

Show S_{k+1} is true:

$$\frac{1}{1\cdot 4}+\frac{1}{4\cdot 7}+...+\frac{1}{[3(k+1)-2][3(k+1)+1]}=\frac{k+1}{3(k+1)+1}$$

Add $\dfrac{1}{[3(k+1)-2][3(k+1)+1]}$

to both sides of S_k:

$$\frac{1}{1\cdot 4}+\frac{1}{4\cdot 7}+...+\frac{1}{[3(k+1)-2][3(k+1)+1]}$$
$$=\frac{k}{3k+1}+\frac{1}{[3(k+1)-2][3(k+1)+1]}$$
$$=\frac{k}{3k+1}+\frac{1}{(3k+1)(3k+4)}$$
$$=\frac{k(3k+4)+1}{(3k+1)(3k+4)}$$
$$=\frac{3k^2+4k+1}{(3k+1)(3k+4)}$$
$$=\frac{(3k+1)(k+1)}{(3k+1)(3k+4)}$$
$$=\frac{k+1}{3k+4}=\frac{k+1}{3(k+1)+1}$$

Since S_k implies S_{k+1}, the statement is true for every positive value of n.

17. $\frac{4}{5}+\frac{4}{5^2}+...+\frac{4}{5^n}=1-\frac{1}{5^n}$

(i) Show true for $n=1$:

$$\frac{4}{5}=1-\frac{1}{5}$$
$$\frac{4}{5}=\frac{4}{5}$$

(ii) Assume S_k is true:

$$\frac{4}{5}+\frac{4}{5^2}+...+\frac{4}{5^k}=1-\frac{1}{5^k}$$

Show S_{k+1} is true:

$$\frac{4}{5}+\frac{4}{5^2}+...+\frac{4}{5^{k+1}}=1-\frac{1}{5^{k+1}}$$

Add $\frac{4}{5^{k+1}}$ to both sides of S_k:

$$\begin{aligned}\frac{4}{5}+\frac{4}{5^2}+...+\frac{4}{5^k}+\frac{4}{5^{k+1}} &= 1-\frac{1}{5^k}+\frac{4}{5^{k+1}}\\ &= 1+\frac{-1\left(5^{k+1}\right)+4\left(5^k\right)}{\left(5^k\right)\left(5^{k+1}\right)}\\ &= 1+\frac{5^k\left(-5+4\right)}{\left(5^k\right)\left(5^{k+1}\right)}\\ &= 1+\frac{-1}{5^{k+1}}\\ &= 1-\frac{1}{5^{k+1}}\end{aligned}$$

Since S_k implies S_{k+1}, the statement is true for every positive value of n.

19. $3^n > 6n$

If $n=1: 3^1 > 6(1)$ or $3>6$, which is false.

If $n=2: 3^2 > 6(2)$ or $9>12$, which is false.

If $n=3: 3^3 > 6(3)$ or $27>18$, which is true.

For $n \geq 3$, the statement is true. The statement is false for $n=1$ or 2.

21. $2^n > n^2$

If $n=1: 2^1 > 1^2$ or $2>1$, which is true.

If $n=2: 2^2 > 2^2$ or $4>4$, which is false.

If $n=3; 2^3 > 3^2$ or $8>9$, which is false.

If $n=4: 2^4 > 4^2$ or $16>16$, which is false.

If $n=5: 2^5 > 5^2$ or $32>25$, which is true.

For $n \geq 5$, the statement is true. The statement is false for $n=2$, 3, or 4.

23. $(a^m)^n = a^{mn}$

(Assume that a and m are constant.)

(i) Show true for $n=1$:

$$(a^m)^1 = a^{m\cdot 1}$$
$$a^m = a^m$$

(ii) Assume true for S_k:

$$(a^m)^k = a^{mk}$$

Show S_{k+1} is true:

$$(a^m)^{k+1} = a^{m(k+1)}$$

L: $(a^m)^{k+1} = (a^m)^k (a^m)^1$ $\quad [a^m \cdot a^n = a^{m+n}]$

$= a^{mk}\cdot a^m$ $\quad [S_k$ is true$]$

$= a^{mk+m}$ $\quad [a^m \cdot a^n = a^{m+n}]$

$= a^{m(k+1)}$ **R**

Since S_k implies S_{k+1}, the statement is true for every positive value of n.

25. If $n \geq 3,\ 2^n > 2n$

(i) Show true for $n = 3$:

$$2^3 > 2\,(3)$$
$$8 > 6$$

(ii) Assume true for S_k:

$$2^k > 2k$$

Show S_{k+1} is true:

$$2^{k+1} > 2\,(k+1)$$

Multiply both sides of S_k by 2:

$$2\left(2^k\right) > 2\,(2k)$$

$$2^{k+1} > 2k + 2k$$

Since $k \geq 3, 2k > 2$

$$2^{k+1} > 2k + 2$$

$$2^{k+1} > 2\,(k+1)$$

Since S_k implies S_{k+1}, the statement is true for every positive value of n.

27. If $a > 1$, then $a^n > 1$

(i) Show true for $k = 1$:

If $a > 1$, then $a^1 > 1$
If $a > 1$, then $a > 1$

(ii) Assume true for S_k:

If $a > 1$, then $a^k > 1$

Show S_{k+1} is true

If $a > 1$, then $a^{k+1} > 1$

Multiply both sides of S_k by a:

If $a > 1$, then

$$a \cdot a^k > a \cdot 1$$
$$a^1 \cdot a^k > a \cdot 1$$
$$a^{k+1} > a \qquad [a^m \cdot a^n = a^{mn}]$$

Since $a > 1$, $a^{k+1} > 1$ and S_{k+1} is true. Since S_k implies S_{k+1}, the statement is true for every positive value of n.

29. If $0 < a < 1$, then $a^n < a^{n-1}$

(i) Show true for $n = 1$:

If $0 < a < 1$, then $a^1 < a^{1-1}$
If $0 < a < 1$, then $a < 1$

(ii) Assume true for S_k:

If $0 < a < 1$, then $a^k < a^{k-1}$

Show S_{k+1} is true:

If $0 < a < 1$, then $a^{k+1} < a^k$

Multiply both sides of S_k by a:

Since $a > 0$, the inequality symbol will not change.

If $0 < a < 1$, then

$$a \cdot a^k < a \cdot a^{k-1}$$
$$a^{k+1} < a^{1+(k-1)} \qquad [a^m \cdot a^n = a^{m+n}]$$
$$a^{k+1} < a^{k-1+1}$$
$$a^{k+1} < a^k$$

Since S_k implies S_{k+1}, the statement is true for every positive value of n.

31. If $n \geq 4,\ n! > 2^n$

(i) Show true for $n = 4$:

$$4! > 2^4 \Rightarrow 24 > 16$$

(ii) Assume true for S_k: $k! > 2^k$

Show S_{k+1} is true: $(k+1)! > 2^{k+1}$

Multiply both sides of S_k by $k+1$:

$$(k+1)\,k! > (k+1)\,2^k$$
$$(k+1)! > (k+1)\,2^k$$

Since $k+1 > 2,\ (k+1)\,2^k > 2 \cdot 2^k$

$$(k+1)\,2^k > 2^{k+1}$$
$$(k+1)! > (k+1)\,2^k > 2^{k+1}$$
$$(k+1)! > 2^{k+1}$$

Since S_k implies S_{k+1}, the statement is true for every positive value of n.

33. Let $S_n = \dfrac{n^2 - n}{2}$ be the number of handshakes of n people. Since 2 is the smallest number of people who can shake hands, we need to prove this statement for every positive integer $n \geq 2$.

(i) Show true for $n = 2$:

S_2 is the statement that 2 people can shake hands in 1 way, since

$$\frac{2^2 - 2}{2} = 1.$$

(ii) Assume true for S_k:

S_k is the statement that 2 people can shake hands in $\dfrac{k^2 - k}{2}$ ways.

Show S_{k+1} is true:

$$S_{k+1} = \frac{(k+1)^2 - (k+1)}{2}$$

If 1 more person joins the k people, this $(k+1)$ *st* person will shake hands with the previous k people one time each. Thus, there will be k additional handshakes. Thus S_{k+1} is

$$\frac{k^2 - k}{2} + k = \frac{k^2 - k + 2k}{2} = \frac{k^2 + k}{2}$$
$$= \frac{k^2 + 2k + 1 - k - 1}{2}$$
$$= \frac{(k+1)^2 - (k+1)}{2}$$

Since S_k implies S_{k+1}, the statement is true for every positive value of n.

35. Let $a_n =$ the number of sides of the nth figure.

$a_1 = 3$
$a_2 = 3 \cdot 4$ since each side will develop into 4 sides.
$a_3 = 3 \cdot 4^2$, and so on.

This gives a geometric sequence with $a_1 = 3$, $r = 4$:

$$a_n = 3 \cdot 4^{n-1}$$

To find the perimeter of each figure, multiply the number of sides by the length of each side.

In each figure, the lengths of the sides are $\frac{1}{3}$ the lengths of the sides in the preceding figure.

continued

35. continued

Thus, if $P_n =$ perimeter of nth figure,

$$P_1 = 3\left(4^0\right) = 3$$
$$P_2 = 3\left(4^1\right)\left(\frac{1}{3}\right) = 4$$
$$P_3 = 3\left(4^2\right)\left(\frac{1}{3}\right)^2 = \frac{16}{3}, \text{ and so on.}$$

This gives a geometric sequence with

$$P_1 = 3 \text{ and } r = \frac{4}{3}.$$

Thus, $P_n = 3\left(\dfrac{4}{3}\right)^{n-1}$.

This result may also be written as

$$P_n = \frac{3^1 \cdot 4^{n-1}}{3^{n-1}} = \frac{4^{n-1}}{3^{-1} \cdot 3^{n-1}} = \frac{4^{n-1}}{3^{n-2}}$$

37. With 1 ring, 1 move is required. With 2 rings, 3 moves are required. Note that $3 = 2 + 1$. With 3 rings, 7 moves are required. Note that $7 = 2^2 + 2 + 1$. With n rings

$$2^{n-1} + 2^{n-2} + \ldots + 2^1 + 1 = 2^n - 1$$

moves are required.

Let S_n be the statement:

For n rings, the number of required moves is $2^n - 1$.

We will prove that S_n is true for every positive integer n.

(i) S_1 is the statement:

For 1 ring, the number of required moves is $2^1 - 1 = 2 - 1 = 1$. The statement is true.

(ii) Show that if S_k is true then S_{k+1} is true.

Assume $k+1$ rings are on the first peg; since S_k is true, the top k rings can be moved to the second peg in $2^k - 1$ moves. Now move the bottom ring to the third peg. Since S_k is true, move the k rings on the second peg on top of the largest ring on the third peg in $2^k - 1$ moves. The total number of moves is

$$(2^k - 1) + 1 + (2^k - 1) = 2 \cdot 2^k - 1$$
$$= 2^{k+1} - 1$$

Therefore, S_{k+1} is true. By steps (i) and (ii), S_n is true for every positive integer n.

Reviewing Basic Concepts (Sections 8.4 and 8.5)

1. $_5C_3 = \dfrac{5!}{3!\,(5-3)!} = \dfrac{5!}{3!\,2!}$

$= \dfrac{5\cdot 4\cdot 3!}{2\cdot 1\cdot 3!} = \boxed{10}$

2. In the expansion of $(x+y)^n$, there are $n+1$ terms.

3. $(a+2b)^4$

$= a^4 + \binom{4}{1}a^3(2b) + \binom{4}{2}a^2(2b)^2 + \binom{4}{3}a(2b)^3 + (2b)^4$

$= a^4 + 4a^3(2b) + 6a^2(4b^2) + 4a(8b^3) + 16b^4$

$= \boxed{a^4 + 8a^3b + 24a^2b^2 + 32ab^3 + 16b^4}$

4. $\left(\dfrac{1}{m} - n^2\right)^3$

$= \left(\dfrac{1}{m}\right)^3 + \binom{3}{1}\left(\dfrac{1}{m}\right)^2(-n^2) + \binom{3}{2}\left(\dfrac{1}{m}\right)(-n^2)^2 + (-n^2)^3$

$= \left(\dfrac{1}{m}\right)^3 + 3\left(\dfrac{1}{m}\right)^2(-n^2) + 3\left(\dfrac{1}{m}\right)(n^4) + (-n^6)$

$= \boxed{\dfrac{1}{m^3} - \dfrac{3n^2}{m^2} + \dfrac{3n^4}{m} - n^6}$

5. $\binom{n}{r-1}\mathbf{x}^{n-(r-1)}\mathbf{y}^{r-1}$

3rd term of $(x-2y)^6$:

$n=6,\ r=3,\ \mathbf{x}=x,\ \mathbf{y}=-2y$

$\binom{6}{2}(x)^4(-2y)^2 = 15x^4(4y^2)$

$= \boxed{60x^4y^2}$

6. $4+8+12+\ \dots\ +4n = 2n(n+1)$

(i) Show true for $n=1$:

$4(1) = 2(1)(1+1)$
$4 = 4$

(ii) Assume S_k is true:

$4+8+12+\ \dots\ +4k = 2k(k+1)$

Show S_{k+1} is true:

$4+8+12+\ \dots\ +4k+4(k+1)$
$= 2(k+1)((k+1)+1)$

Add $4(k+1)$ to both sides of S_k:

$4+8+12+\ \dots\ +4k+4(k+1)$
$= 2k(k+1)+4(k+1)$
$= 2k^2+2k+4k+4$
$= 2k^2+6k+4$
$= 2(k^2+3k+2)$
$= 2(k+1)(k+2)$
$= 2(k+1)(k+1+1)$

Since S_k implies S_{k+1}, the statement is true for every positive value of n.

7. $n^2 \le 2^n$ for $n \ge 4$

Let S_n be the following statement:

If $n\ge 4,\ n^2 \le 2^n$

(i) $S_4 : 4^2 \le 2^4$
$16 \le 16$, which is true.

(ii) If S_k is true, then S_{k+1} is true.

$S_k : k^2 \le 2^k \Rightarrow S_{k+1} : (k+1)^2 \le 2^{k+1}$

Multiply both sides of S_k by 2:

$2k^2 \le 2\cdot 2^k$
$2k^2 \le 2^{k+1}$

Since $k\ge 4,\quad 2k \ge 8 \rightarrow k > 1$

Since $k \ge 4,\quad k^2 \ge 4k$

$2^{k+1} \ge 2k^2$
$\ge k^2+k^2$
$\ge k^2+4k$
$\ge k^2+2k+2k$
$\ge k^2+2k+1$
$\ge (k+1)^2$

Section 11.6

1. $P(12, 8)$

$$= \frac{12!}{(12-8)!} = \frac{12!}{4!}$$
$$= \frac{12 \cdot 11 \cdot 10 \cdot 9 \cdot 8 \cdot 7 \cdot 6 \cdot 5 \cdot 4!}{4!}$$
$$= \boxed{19,958,400}$$

[Calculator: $12 \,{}_nP_r\, 8$]

3. $$P(9, 2) = \frac{9!}{(9-2)!} = \frac{9!}{7!} = \frac{9 \cdot 8 \cdot 7!}{7!} = \boxed{72}$$

[Calculator: $9 \,{}_nP_r\, 2$]

5. $$P(5, 1) = \frac{5!}{(5-1)!} = \frac{5!}{4!} = \frac{5 \cdot 4!}{4!} = \boxed{5}$$

[Calculator: $5 \,{}_nP_r\, 1$]

7. $$C(4, 2) = \frac{4!}{(4-2)!\,2!} = \frac{4 \cdot 3 \cdot 2 \cdot 1}{(2 \cdot 1)(2 \cdot 1)} = \boxed{6}$$

[Calculator: $4 \,{}_nC_r\, 2$]

9. $$C(6, 0) = \frac{6!}{(6-0)!\,0!} = \frac{6!}{6!} = \boxed{1}$$

[Calculator: $6 \,{}_nC_r\, 0$]

11. $$\binom{12}{4} = \frac{12!}{(12-4)!\,4!} = \frac{12 \cdot 11 \cdot 10 \cdot 9 \cdot 8!}{8!\,(4 \cdot 3 \cdot 2 \cdot 1)} = \boxed{495}$$

[Calculator: $12 \,{}_nC_r\, 4$]

13. $${}_{20}P_5 = \frac{20!}{(20-5)!} = \frac{20!}{15!} = \frac{20 \cdot 19 \cdot 18 \cdot 17 \cdot 16 \cdot 15!}{15!} = \boxed{1,860,480}$$

15. $${}_{15}P_8 = \frac{15!}{(15-8)!} = \frac{15!}{7!} = \frac{15 \cdot 14 \cdot 13 \cdot 12 \cdot 11 \cdot 10 \cdot 9 \cdot 8 \cdot 7!}{7!} = \boxed{259,459,200}$$

17. $${}_{20}C_5 = \frac{20!}{(20-5)!\,5!} = \frac{20 \cdot 19 \cdot 18 \cdot 17 \cdot 16 \cdot 15!}{15!\,(5 \cdot 4 \cdot 3 \cdot 2 \cdot 1)} = \boxed{15,504}$$

19. $$\binom{15}{8} = \frac{15!}{(15-8)!\,8!} = \frac{15 \cdot 14 \cdot 13 \cdot 12 \cdot 11 \cdot 10 \cdot 9 \cdot 8!}{(7 \cdot 6 \cdot 5 \cdot 4 \cdot 3 \cdot 2 \cdot 1)\,8!} = \boxed{6435}$$

21. **(a)** a telephone number (order matters): permutation

(b) a social security number (order matters): permutation

(c) a hand of cards in poker (order does not matter): combination

(d) a committee of politicians (order does not matter): combination

(e) the *combination* on a combination lock (order matters): permutation

(f) a lottery choice of six numbers where the order does not matter: combination

(g) an automobile license plate (order matters): permutation

23. $5 \cdot 3 \cdot 2 = \boxed{30}$

There are 30 different homes available if a builder offers a choice of 5 basic plans, 3 roof styles, and 2 exterior finishes.

25. **(a)** 1st letter: one of 2
2nd letter: one of 25
3rd letter: one of 24
4th letter: one of 23

Since $2 \cdot 25 \cdot 24 \cdot 23 = \boxed{27,600}$,
there are 27,600 different call letters without repeats.

(b) With repeats, the count is

$2 \cdot 26 \cdot 26 \cdot 26 = \boxed{35,152}$

(c) 1st letter: one of 2
2nd letter: one of 24
(Cannot repeat the 1st letter or be R.)
3rd letter: one of 23
(Cannot repeat the 1st or 2nd letters or be R.)
4th letter: one of 1
(Must be R.)

Since $2 \cdot 24 \cdot 23 \cdot 1 = \boxed{1104}$,
there are 1104 different such call letters.

27. Use the multiplication principle of counting:

$3 \cdot 5 = \boxed{15}$

There are 15 different first– and middle–name arrangements.

29. **(a)** The 1st three positions could each be any one of 26 letters, and the 2nd three positions could each be any one of the 10 numbers.

$$26 \cdot 26 \cdot 26 \cdot 10 \cdot 10 \cdot 10 = \boxed{17,576,000}$$

17,576,000 license plates were possible.

(b) $10 \cdot 10 \cdot 10 \cdot 26 \cdot 26 \cdot 26 = \boxed{17,576,000}$

17,576,000 additional license plates were made possible.

(c) $26 \cdot 10 \cdot 10 \cdot 10 \cdot 26 \cdot 26 \cdot 26 = \boxed{456,976,000}$

456,976,000 plates were provided by prefixing the previous pattern with an additional letter.

31. They can be seated in

$6 \cdot 5 \cdot 4 \cdot 3 \cdot 2 \cdot 1$ or $6!$ ways:

$P(6,6) = (6\,{}_nP_r\,6) = \boxed{720}$

33. The student has 6 choices for the 1st course, 5 choices for the 2nd, and 4 choices for the 3rd.

$6 \cdot 5 \cdot 4 = \boxed{120}$

35. 3 out of 15:

$P(15,3) = 15\,{}_nP_r\,3 = 15 \cdot 14 \cdot 13 = \boxed{2730}$

37. 5 players → 5 positions:

$P(5,5) = 5\,{}_nP_r\,5 = 5! = \boxed{120}$

10 players → 5 positions:

$$\begin{aligned} P(10,5) &= 10\,{}_nP_r\,5 \\ &= \frac{10!}{5!} = 10 \cdot 9 \cdot 8 \cdot 7 \cdot 6 \\ &= \boxed{30,240} \end{aligned}$$

39. Committee of 4 from 30: $C(30,4)$

$$\begin{aligned} 30\,{}_nC_r\,4 &= \binom{30}{4} \\ &= \frac{30!}{(30-4)!\,4!} = \frac{30!}{26!\,4!} \\ &= \frac{30 \cdot 29 \cdot 28 \cdot 27 \cdot 26!}{26!\,(4 \cdot 3 \cdot 2 \cdot 1)} \\ &= \boxed{27,405} \end{aligned}$$

There are 27,405 possible committees.

41. 3 out of 6:

$$\begin{aligned} C(6,3) &= 6\,{}_nC_r\,3 = \binom{6}{3} \\ &= \frac{6!}{3!\,3!} = \frac{6 \cdot 5 \cdot 4 \cdot 3!}{3 \cdot 2 \cdot 1 \cdot 3!} = \boxed{20} \end{aligned}$$

20 different kinds of hamburgers can be made.

43. 2 cards out of 5:

$$\begin{aligned} C(5,2) &= 5\,{}_nC_r\,2 = \binom{5}{2} \\ &= \frac{5!}{3!\,2!} = \frac{5 \cdot 4 \cdot 3!}{3!\,(2 \cdot 1)} = \boxed{10} \end{aligned}$$

There are 10 different 2–card combinations.

45. 2 out of 8:

$$C(8,2) = 8\,{}_nC_r\,2 = \binom{8}{2} = \frac{8!}{6!\,2!} = \frac{8\cdot 7\cdot 6!}{6!\cdot 2\cdot 1} = \boxed{28}$$

28 samples of 2 marbles can be drawn in which both marbles are blue.

47. 5 Liberals, 4 Conservatives:

(a) Delegations of 3:

$$C(9,3) = 9\,{}_nC_r\,3 = \binom{9}{3} = \frac{9!}{6!\,3!} = \boxed{84}$$

(b) Delegations of all liberals:

$$C(5,3) = 5\,{}_nC_r\,3 = \binom{5}{3} = \frac{5!}{2!\,3!} = \boxed{10}$$

(c) Delegations of 3; 1 C, 2 L:

$$C(4,1)\cdot C(5,2) = 4\,{}_nC_r\,1\cdot 5\,{}_nC_r\,2 = \binom{4}{1}\cdot\binom{5}{2} = 4\cdot 10 = \boxed{40}$$

(d) Delegations of 3; 1 supervisor:

$$C(1,1)\cdot C(8,2) = 1\,{}_nC_r\,1\cdot 8\,{}_nC_r\,2 = \binom{1}{1}\cdot\binom{8}{2} = \boxed{28}$$

49. The problem asks how many ways can Matthew arrange his schedule. Therefore, order is important, and this is a permutation problem.

There are $P(8,\ 4) = \dfrac{8!}{(8\ -\ 4)!} = \dfrac{8!}{4!} = \boxed{1680}$

ways to arrange his schedule.

51. 4 out of 6:

$$C(6,4) = 6\,{}_nC_r\,4 = \binom{6}{4} = \frac{6!}{2!\,4!} = \boxed{15}$$

There are 15 different soups she can make.

53. 11 out of 12, order matters:

$$P(12,11) = 12\,{}_nP_r\,11 = \frac{12!}{(12-11)!} = \frac{12!}{1!} = \boxed{479{,}001{,}600}$$

55. 8 men, 11 women → 5 member groups:

(a) 5 men:

$$C(8,5) = 8\,{}_nC_r\,5 = \binom{8}{5} = \frac{8!}{3!\,5!} = \boxed{56}$$

(b) 5 women:

$$C(11,5) = 11\,{}_nC_r\,5 = \binom{11}{5} = \frac{11!}{6!\,5!} = \boxed{462}$$

(c) 3 men and 2 women:

$$C(8,3)\cdot C(11,2) = 8\,{}_nC_r\,3\cdot 11\,{}_nC_r\,2 = \binom{8}{3}\cdot\binom{11}{2} = \frac{8!}{5!\,3!}\cdot\frac{11!}{9!\,2!} = 56\cdot 55 = \boxed{3080}$$

(d) 0 women:

$$C(8,5) = 8\,{}_nC_r\,5 = \mathbf{56}$$

1 woman:

$$C(11,1)\cdot C(8,4) = 11\,{}_nC_r\,1\cdot 8\,{}_nC_r\,4 = 11\cdot 70 = \mathbf{770}$$

2 women:

$$C(11,2)\cdot C(8,3) = 11\,{}_nC_r\,2\cdot 8\,{}_nC_r\,3 = 55\cdot 56 = \mathbf{3080}$$

3 women:

$$C(11,3)\cdot C(8,2) = 11\,{}_nC_r\,3\cdot 8\,{}_nC_r\,2 = 165\cdot 28 = \mathbf{4620}$$

$$\mathbf{56 + 770 + 3080 + 4620} = \boxed{8526}$$

57. Each combination can vary between 000 and 999. Thus, there are 1000 possibilities for each lock. That is, there are

$$1000 \cdot 1000 = 1,000,000$$

different combinations in all.

59. There are 2 settings (on/off) for each of the 12 switches. There are

$$2 \cdot 2 \cdot 2 \cdot 2 \cdot 2 \cdot 2 \cdot 2 \cdot 2 \cdot 2 \cdot 2 \cdot 2 \cdot 2 = 2^{12} = 4096$$

different codes for the garage door opener.

61. Initially any key can be put on the ring. Then the remaining 3 keys can be put on in $3! = 6$ different ways. If the keys are numbered 1 to 4, it is important to remember it is not possible to distinguish between the arrangements 1234, 2341, 3412, and 4123 on the key ring. There are only 6 ways to arrange the keys.

63. Prove: $P(n,\ n-1) = P(n,\ n)$

$$\text{(i) } P(n,\ n-1) = \frac{n!}{[n-(n-1)]!} = \frac{n!}{1!} = n!$$

$$\text{(ii) } P(n,\ n) = \frac{n!}{(n-n)!} = \frac{n!}{0!} = n!$$

(i) = (ii)

65. Prove $P(n,\ 0) = 1$

$$P(n,\ 0) = \frac{n!}{(n-0)!} = \frac{n!}{n!} = 1$$

67. Prove $\binom{n}{0} = 1$

$$\binom{n}{0} = \frac{n!}{(n-0)!\,0!} = \frac{n!}{n!\,1} = 1$$

69. Prove: $\binom{n}{n-r} = \binom{n}{r}$

$$\binom{n}{n-r} = \frac{n!}{(n-(n-r))!\,(n-r)!} = \frac{n!}{r!\cdot(n-r)!} = \binom{n}{r}$$

Relating Concepts (Exercises 71 and 72)

71. $\log 50! = \log 1 + \log 2 + \ldots + \log 50$

Using a sum and sequence utility on a calculator, we obtain

$$\log 50! \approx 64.48307487$$
$$50! \approx 10^{64.48307487}$$
$$50! \approx 10^{.48307487} \times 10^{64}$$
$$50! \approx \boxed{3.04140932 \times 10^{64}}$$

Computing the value directly, we obtain $50! \approx 3.041409632 \times 10^{64}$.

(b) $\log 60! = \log 1 + \log 2 + \ldots + \log 60$

Using a sum and sequence utility on a calculator, we obtain

$$\log 60! \approx 81.92017485$$
$$60! \approx 10^{81.92017485}$$
$$60! \approx 10^{.92017485} \times 10^{81}$$
$$60! \approx \boxed{8.320987113 \times 10^{81}}$$

Computing the value directly, we obtain $60! \approx 8.320987113 \times 10^{81}$.

(c) $\log 65! = \log 1 + \log 2 + \ldots + \log 65$

Using a sum and sequence utility on a calculator, we obtain

$$\log 65! \approx 90.91633025$$
$$65! \approx 10^{90.91633025}$$
$$65! \approx 10^{.91633025} \times 10^{90}$$
$$65! \approx \boxed{8.247650592 \times 10^{90}}$$

Computing the value directly, we obtain $65! \approx 8.247650592 \times 10^{90}$.

72. (a) $P(47,\ 13) = \dfrac{47!}{34!}$

$P(47,\ 13) = 47 \cdot 46 \cdot 45 \cdot\ \ldots\ \cdot 35$

$\log P(47,\ 13) = \log 47 + \log 46 + \log 45 + \ \ldots\ + \log 35$

$\log P(47,\ 13) \approx 20.94250295$

$P(47,\ 13) \approx 10^{20.94250295}$

$P(47,\ 13) \approx 10^{.94250295} \times 10^{20}$

$P(47,\ 13) \approx \boxed{8.759976613 \times 10^{20}}$

The calculator sequence for $P(47,\ 13)$ is:

$\dfrac{10\hat{\ }\mathrm{sum}(\mathrm{seq}(\log(x), x, 1, 47, 1))}{10\hat{\ }\mathrm{sum}(\mathrm{seq}(\log(x), x, 1, 34, 1))}$

(b) $P(50,\ 4) = \dfrac{50!}{46!}$

$P(50,\ 4) = 50 \cdot 49 \cdot 48 \cdot 47$

$\log P(50,\ 4) = \log 50 + \log 49 + \log 48 + \log 47$

$\log P(50,\ 4) \approx 6.74250518$

$P(50,\ 4) \approx 10^{6.74250518}$

$P(50,\ 4) \approx 10^{.74250518} \times 10^{6}$

$P(50,\ 4) \approx \boxed{5,527,200}$

The calculator sequence for $P(50,\ 4)$ is:

$\dfrac{10\hat{\ }\mathrm{sum}(\mathrm{seq}(\log(x), x, 1, 50, 1))}{10\hat{\ }\mathrm{sum}(\mathrm{seq}(\log(x), x, 1, 46, 1))}$

(c) $P(29,\ 21) = \dfrac{29!}{8!}$

$P(29,\ 21) = 29 \cdot 28 \cdot 27 \cdot\ \ldots\ \cdot 9$

$\log P(29,\ 21) = \log 29 + \log 28 + \log 27 + \ \ldots\ + \log 9$

$\log P(29,\ 21) \approx 26.34101830$

$P(29,\ 21) \approx 10^{26.34101830}$

$P(29,\ 21) \approx 10^{.34101830} \times 10^{26}$

$P(29,\ 21) \approx \boxed{2.19289732 \times 10^{26}}$

The calculator sequence for $P(29,\ 21)$ is:

$\dfrac{10\hat{\ }\mathrm{sum}(\mathrm{seq}(\log(x), x, 1, 29, 1))}{10\hat{\ }\mathrm{sum}(\mathrm{seq}(\log(x), x, 1, 8, 1))}$

Section 11.7

1. Let h = heads, t = tails: $S = \{h\}$

3. Let h = heads, t = tails:

$S = \{(h,h,h), (h,h,t), (h,t,t), (h,t,h), (t,t,t), (t,t,h), (t,h,t), (t,h,h)\}$

5. $S = \{(1,1), (1,2), (1,3), (2,1), (2,2), (2,3), (3,1), (3,2), (3,3)\}$

7. (a) $S = \{h\}$

$P(E) = \dfrac{n(E)}{n(S)}$

$P = \mathbf{1}$

(b) $S : \emptyset$

$P(E) = \dfrac{n(E)}{n(S)}$

$P = \mathbf{0}$

9. $S = \{(1,1), (1,2), (1,3), (2,1), (2,2), (2,3), (3,1), (3,2), (3,3)\}$

(a) $E = \{(1,1), (2,2), (3,3)\}$

$P(E) = \dfrac{n(E)}{n(S)}$

$P = \dfrac{3}{9} = \mathbf{\dfrac{1}{3}}$

(b) $E = \{(1,1), (1,3), (2,1), (2,3), (3,1), (3,3)\}$

$P(E) = \dfrac{n(E)}{n(S)}$

$P = \dfrac{6}{9} = \mathbf{\dfrac{2}{3}}$

(c) $E = \{(2,1), (2,3)\}$

$P(E) = \dfrac{n(E)}{n(S)}$

$P = \mathbf{\dfrac{2}{9}}$

11. A probability will always be a number between 0 and 1, inclusive of both. $\frac{6}{5} > 1$, and so this answer must be incorrect.

13. 3 yellow, 4 white, 8 blue

$$P(E) = \frac{n(E)}{n(S)}; \quad n(S) = 15$$

(a) $P(\text{yellow}) = \frac{3}{15} = \mathbf{\frac{1}{5}}$

(b) $P(\text{black}) = \frac{0}{15} = \mathbf{0}$

(c) $P(\text{yellow or white}) = \mathbf{\frac{7}{15}}$

(d) odds of yellow $= \frac{1/5}{4/5} = \frac{1}{4}$: **1 to 4**

(e) odds of no blue $= \frac{7/15}{8/15} = \frac{7}{8}$: **7 to 8**

15. $\{1, 2, 3, 4, 5\}$; Sum $= 5$:

$1 + 4$ and $2 + 3$:

$P(E): 2,\ P(E'): 8$

Odds: $\frac{P(E)}{P(E')} = \frac{2}{8}$ $\boxed{\text{Odds: } 1 \text{ to } 4}$

17. Odds of winning: 3 to 2
win: 3, loss: 2, total: 5

$$P(\text{loss}) = \frac{n(E)}{n(S)} = \boxed{\frac{2}{5} = .4}$$

19. Total 10; mother: 1, uncles: 2 brothers: 3, cousins: 4

(a) $P(\text{uncle or brother})$

$$= \frac{n(E)}{n(S)} = \frac{5}{10} = \boxed{\frac{1}{2}}$$

(b) $P(\text{brother or cousin})$

$$= \frac{n(E)}{n(S)} = \boxed{\frac{7}{10}}$$

(c) $P(\text{brother or mother})$

$$= \frac{n(E)}{n(S)} = \frac{4}{10} = \boxed{\frac{2}{5}}$$

21. **(a) vi** **(b) iv** **(c) i** **(d) vi**
(e) iii **(f) ii** **(g) v**

23. Let E be the event "English is spoken." This corresponds to $100\% - 20.4\% = 79.6\%$ of the households, so $P(E) = .796$. Thus, $P(E') = 1 - P(E) = 1 - .796 = .204$.
The odds that English is spoken is

$$\frac{P(E)}{P(E')} = \frac{.796}{.204} = \frac{199}{51} = 199 \text{ to } 51.$$

25. (a) $P(< \$20)$

$= P(\$5 - \$19.99 \text{ or below } \$5)$

$= P(\$5 - \$19.99) + P(\text{below } \$5)$

$= .37 + .25 = \boxed{.62}$

(b) $P(\geq \$40)$

$= P(\$40 - \$69.99 \text{ or } \$70 - \$99.99 \text{ or } \$100 - \$149.99 \text{ or } \geq \$150)$

$= P(\$40 - \$69.99) + P(\$70 - \$99.99) + P(\$100 - \$149.99) + P(\geq \$150)$

$= .09 + .07 + .08 + .03 = \boxed{.27}$

(c) $P(> \$99.99)$

$= P(\$100 - \$149.99 \text{ or } \geq \$150)$

$= P(\$100 - \$149.99) + P(\geq \$150)$

$= .08 + .03 = \boxed{.11}$

(d) $P(< \$100)$

$= P(< \$5 \text{ or } \$5 - \$19.99 \text{ or } \$20 - \$39.99 \text{ or } \$40 - \$69.99 \text{ or } \$70 - \$99.99)$

$= P(< \$5) + P(\$20 - \$39.99) + (\$40 - \$69.99) + P(\$70 - \$99.99)$

$= .25 + .37 + .11 + .09 + .07 = \boxed{.89}$

27. Each suit has 13 cards, and the probability of choosing the correct card in that suit is $\frac{1}{13}$. The probability of picking the incorrect card in a given suit is $\frac{12}{13}$ and there are 4 suits to pick from.

$P(3 \text{ correct choices})$

$$= \frac{1}{13} \times \frac{1}{13} \times \frac{1}{13} \times \left(4 \cdot \frac{12}{13}\right)$$

$$= \boxed{\frac{48}{28{,}561} \approx .001681}$$

29. More than 5 years: 30%,
Female: 28%, Retirement: 65%,
Female retire.: 14%

(a) Male worker:

$$P(E) = \frac{n(E)}{n(S)} = \frac{1 - .28}{1} = \boxed{.72}$$

(b) < 5 years:

$$P(E) = \frac{n(E)}{n(S)} = \frac{1 - .3}{1} = \boxed{.70}$$

(c) Retirement fund or female:

$$\begin{aligned} P(E \cup F) &= P(E) + P(F) - P(E \cap F) \\ &= .65 + .28 - .14 = \boxed{.79} \end{aligned}$$

31. P (2 girls and 3 boys)

$$\begin{aligned} &= P(2G) \qquad [\text{or } P(3B)] \\ &= \binom{5}{2}\left(\frac{1}{2}\right)^2\left(\frac{1}{2}\right)^3 \\ &= \boxed{.3125} \end{aligned}$$

33. $$P(0 \text{ girls}) = \binom{5}{0}\left(\frac{1}{2}\right)^0\left(\frac{1}{2}\right)^5 = \boxed{.03125}$$

35. P (at least 3 boys)

$$\begin{aligned} &= P(3B) + P(4B) + P(5B) \\ &= \binom{5}{3}\left(\frac{1}{2}\right)^3\left(\frac{1}{2}\right)^2 \\ &\quad + \binom{5}{4}\left(\frac{1}{2}\right)^4\left(\frac{1}{2}\right)^1 \\ &\quad + \binom{5}{5}\left(\frac{1}{2}\right)^5\left(\frac{1}{2}\right)^0 \\ &= .3125 + .15625 + .03125 \\ &= \boxed{.5 \text{ or } \frac{1}{2}} \end{aligned}$$

37. If $P(1 \text{ on } 1 \text{ roll}) = \frac{1}{6}$, on 12 rolls of a die, find P (exactly twelve 1's) :

$$\begin{aligned} &= \binom{12}{12}\left(\frac{1}{6}\right)^{12}\left(1 - \frac{1}{6}\right)^0 = 1\left(\frac{1}{6}\right)^{12} \\ &\approx 4.5939 \times 10^{-10} \approx \boxed{4.6 \times 10^{-10}} \end{aligned}$$

39. 12 rolls of a die; find
P (no more than three 1's) :

$$\begin{aligned} &= P(0 \text{ ones}) + P(1 \text{ one}) + P(2 \text{ ones}) + P(3 \text{ ones}) \\ &= \binom{12}{0}\left(\frac{1}{6}\right)^0\left(\frac{5}{6}\right)^{12} + \binom{12}{1}\left(\frac{1}{6}\right)^1\left(\frac{5}{6}\right)^{11} \\ &\quad + \binom{12}{2}\left(\frac{1}{6}\right)^2\left(\frac{5}{6}\right)^{10} + \binom{12}{3}\left(\frac{1}{6}\right)^3\left(\frac{5}{6}\right)^9 \\ &\approx .11215665 + .26917597 + .29609357 + .1973957 \\ &\approx .8748219 \approx \boxed{.875} \end{aligned}$$

41. In this binomial experiment, we call "smoked less than 10" a success. Then $n = 10$, $r = 4$, and $p = .45 + .24 = .69$.

$$\begin{aligned} P(4 \text{ smoked} < 10) &= \binom{10}{4}(.69)^4(1 - .69)^6 \\ &= 210(.69)^4(.31)^6 \approx .042246 \end{aligned}$$

43. In this binomial experiment, we call "smoked between 1 and 19" a success. Then,

$$\begin{aligned} &P(\text{smoked between 1 and 19}) \\ &\quad = P(\text{smoked 1 to 9}) + P(\text{smoked 10 to 19}) \\ &\quad = .24 + .20 = .44 \end{aligned}$$

Also, "fewer than 2" corresponds to "0 or 1." Thus, $n = 10$, $r = 0$ or 1, and $p = .44$.

$$\begin{aligned} &P(\text{fewer than 2 smoked between 1 and 19}) \\ &\quad = P(0 \text{ smoked between 1 and 19}) \\ &\qquad + P(1 \text{ smoked between 1 and 19}) \\ &= \binom{10}{0}(.44)^0(1 - .44)^{10} + \binom{10}{1}(.44)^1(1 - .44)^9 \\ &= (1)(.44)^0(.56)^{10} + (10)(.44)^1(.56)^9 \\ &\approx .003033 + .023831 \approx .026864 \end{aligned}$$

45. In these binomial experiments, we call "the man is color-blind" a success.

(a) For "exactly 5 are color-blind," we have $n = 53$, $r = 5$, and $p = .042$.

$$P\,(\text{exactly 5 are color-blind}) = \binom{53}{5}(.042)^5(1-.042)^{48} = 2,869,685\,(.042)^5(.958)^{48} \approx .047822$$

(b) "No more than 5" $\Rightarrow$ 0, 1, 2, 3, 4, or 5

Then $n = 53$, $r = 0, 1, 2, 3, 4$, or 5, and $p = .042$.

P (no more than 5 are color-blind)

$= P$ (0 is color-blind) $+ P$ (1 is color-blind) $+ P$ (2 are color-blind) $+ P$ (3 are color-blind) $+ P$ (4 are color-blind) $+ P$ (5 are color-blind)

$$= \binom{53}{0}(.042)^0(1-.042)^{53} + \binom{53}{1}(.042)^1(1-.042)^{52} + \binom{53}{2}(.042)^2(1-.042)^{51} + \binom{53}{3}(.042)^3(1-.042)^{50} + \binom{53}{4}(.042)^4(1-.042)^{49} + \binom{53}{5}(.042)^5(1-.042)^{48}$$

$$= 1\,(.042)^0(.958)^{53} + 53\,(.042)^1(.958)^{52} + 1378\,(.042)^2(.958)^{51} + 23,426\,(.042)^3(.958)^{50} + 292,852\,(.042)^4(.958)^{49} + 2,869,685\,(.042)^5(.958)^{48}$$

$$\approx .102890 + .239074 + .272514 + .203105 + .111315 + .047822$$

$\approx .976710$ (Answers may vary due to rounding)

continued

45. continued

(c) Let E be "at least 1 is color-blind." Consider the complementary event E' : "0 are color-blind." Then

$$P(E) = 1 - P(E') = 1 - \binom{53}{0}(.042)^0(1-.042)^{53}$$

$= 1 - .102890$ from part **(b)**

$= .897110$

47. (a) First compute q :

$$q = (1-p)^I = (1-.1)^2 = .81$$

Then, with $S = 4$, $k = 3$, $q = .81$,

$$P = \binom{S}{k} q^k (1-q)^{s-k} = \binom{4}{3} .81^3 (1-.81)^{4-3} = 4 \times .81^3 \times .19^1 \approx .404$$

There is about a **40.4%** chance of exactly 3 people not becoming infected.

(b) $q = (1-p)^I = (1-.5)^2 = .25$

Then, with $S = 4$, $k = 3$, $q = .25$,

$$P = \binom{S}{k} q^k (1-q)^{s-k} = \binom{4}{3} .25^3 (1-.25)^{4-3} = 4 \times .25^3 \times .75^1 \approx .047$$

There is about a **4.7%** chance of this occurring when the disease is highly infectious.

(c) $q = (1-p)^I = (1-.5)^1 = .5$

Then, with $S = 9$, $k = 0$, $q = .5$,

$$P = \binom{S}{k} q^k (1-q)^{s-k} = \binom{9}{0} .5^0 (1-.5)^{9-0} = 1 \times 1 \times .5^9 \approx .002$$

There is about a **.2%** chance of everyone becoming infected. In a large family or group of people, it is unlikely that everyone will become sick even though the disease is highly infectious.

Relating Concepts (Exercises 49 and 50)

49. **(a)** $P_{i,j} = \dfrac{\binom{2i}{j}\binom{4-2i}{2-j}}{\binom{4}{2}}$

$P_{0,0} = \dfrac{\binom{0}{0}\binom{4}{2}}{\binom{4}{2}} = 1$

$P_{0,1} = \dfrac{\binom{0}{1}\binom{4}{1}}{\binom{4}{2}} = 0$

$P_{0,2} = \dfrac{\binom{0}{2}\binom{4}{0}}{\binom{4}{2}} = 0$

$P_{1,0} = \dfrac{\binom{2}{0}\binom{2}{2}}{\binom{4}{2}} = \dfrac{1}{6}$

$P_{1,1} = \dfrac{\binom{2}{1}\binom{2}{1}}{\binom{4}{2}} = \dfrac{2}{3}$

$P_{1,2} = \dfrac{\binom{2}{2}\binom{2}{0}}{\binom{4}{2}} = \dfrac{1}{6}$

$P_{2,0} = \dfrac{\binom{4}{0}\binom{0}{2}}{\binom{4}{2}} = 0$

$P_{2,1} = \dfrac{\binom{4}{1}\binom{0}{1}}{\binom{4}{2}} = 0$

$P_{2,2} = \dfrac{\binom{4}{2}\binom{0}{0}}{\binom{4}{2}} = 1$

(b) $P = \begin{bmatrix} P_{0,0} & P_{0,1} & P_{0,2} \\ P_{1,0} & P_{1,1} & P_{1,2} \\ P_{2,0} & P_{2,1} & P_{2,2} \end{bmatrix}$

$= \begin{bmatrix} 1 & 0 & 0 \\ \frac{1}{6} & \frac{2}{3} & \frac{1}{6} \\ 0 & 0 & 1 \end{bmatrix}$

(c) The matrix exhibits symmetry. The sum of the probabilities in each row is equal to 1. The greatest probabilities lie along the diagonal. This means that a mother cell is most likely to produce a daughter cell like itself.

50. **(a)** It is reasonable to conjecture that since all the cells are drug resistant to both antibiotics, future generations will continue to be resistant to both. Another conjecture is that there will be $\frac{1}{3}$ of the cells in each of the three possible categories, that is,

$$A = \begin{bmatrix} \frac{1}{3} & \frac{1}{3} & \frac{1}{3} \end{bmatrix}$$

(b) $A_{i+i} = A_i P$

$= [a_1 \ a_2 \ a_3] \cdot \begin{bmatrix} P_{0,0} & P_{0,1} & P_{0,2} \\ P_{1,0} & P_{1,1} & P_{1,2} \\ P_{2,0} & P_{2,1} & P_{2,2} \end{bmatrix}$

$A_2 = A_1 P$

$= [0 \ 1 \ 0] \cdot \begin{bmatrix} 1 & 0 & 0 \\ \frac{1}{6} & \frac{2}{3} & \frac{1}{6} \\ 0 & 0 & 1 \end{bmatrix}$

$= \begin{bmatrix} \frac{1}{6} & \frac{2}{3} & \frac{1}{6} \end{bmatrix}$

Similarly

$A_3 = \begin{bmatrix} \frac{5}{18} & \frac{4}{9} & \frac{5}{18} \end{bmatrix}$

$A_4 = \begin{bmatrix} \frac{19}{54} & \frac{8}{27} & \frac{19}{54} \end{bmatrix}$

$A_5 = \begin{bmatrix} \frac{65}{162} & \frac{16}{81} & \frac{65}{162} \end{bmatrix}$

$A_6 = \begin{bmatrix} \frac{211}{486} & \frac{32}{243} & \frac{211}{486} \end{bmatrix}$

$A_7 \approx [.456 \quad .0878 \quad .456]$

$A_8 \approx [.47 \quad .06 \quad .47]$

$A_9 \approx [.48 \quad .04 \quad .48]$

$A_{10} \approx [.487 \quad .026 \quad .487]$

$A_{11} \approx [.491 \quad .017 \quad .491]$

$A_{12} \approx [.494 \quad .012 \quad .494]$

The values appear to be approaching $[.5 \quad 0 \quad .5]$. This means that as time progresses, half the bacteria are resistant to one antibiotic, half are resistant to the other antibiotic, and *none of the bacteria are resistant to both.*

Reviewing Basic Concepts (Sections 8.6 and 8.7)

1. There are $4! = 24$ arrangements.

2. $$P(7,\ 3) = \frac{7!}{(7-3)!} = \frac{7!}{4!} = \frac{7 \cdot 6 \cdot 5 \cdot 4!}{4!} = \boxed{210}$$

3. $$\binom{6}{2} \cdot \binom{5}{2} \cdot \binom{3}{1} = \frac{6!}{(6-2)!\,2!} \cdot \frac{5!}{(5-2)!\,2!} \cdot \frac{3!}{(3-1)!\,1!} = \frac{6!}{4!\,2!} \cdot \frac{5!}{3!\,2!} \cdot \frac{3!}{2!\,1!} = \frac{6 \cdot 5 \cdot 4!}{4!\,2 \cdot 1} \cdot \frac{5 \cdot 4 \cdot 3!}{3!\,2 \cdot 1} \cdot \frac{3 \cdot 2!}{2!\,1} = 15 \cdot 10 \cdot 3 = \boxed{450}$$

4. $$\binom{10}{4} = \frac{10!}{(10-4)!\,4!} = \frac{10!}{6!\,4!} = \frac{10 \cdot 9 \cdot 8 \cdot 7 \cdot 6!}{6!\,4 \cdot 3 \cdot 2 \cdot 1} = \boxed{210}$$

5. Use the multiplication principal of counting:
$9 \cdot 4 \cdot 2 = \boxed{72}$

6. $S = \{\text{hh, ht, th, tt}\}$

7. The sample space for rolling 2 dice:
$n(S) = 6 \cdot 6 = 36$

The event in which the sum is 11 is
$E = \{(5,6),\ (6,5)\}$, so $n(E) = 2$

$$P(E) = \frac{n(E)}{n(S)} = \frac{2}{36} = \boxed{\frac{1}{18}}$$

8. There are $\binom{4}{4}$ ways to draw 4 aces and there are $\binom{4}{1}$ ways to draw 1 queen.

There are $\binom{52}{5}$ different poker hands.

The probability of drawing 4 aces and 1 queen:

$$P(E) = \frac{n(E)}{n(S)} = \frac{\binom{4}{4}\binom{4}{1}}{\binom{52}{5}} = \frac{1 \cdot 4}{2,598,960} = \boxed{\frac{4}{2,598,960} \approx .0000015}$$

9. Odds of rain: 3 to 7
rain: 3, no rain: 7, total: 10

$$P(\text{rain}) = \frac{n(E)}{n(S)} = \boxed{\frac{3}{10} = .3}$$

10. Females: $2.81 - 1.45 = 1.36$ million

$$P(E) = \frac{n(E)}{n(S)} = \frac{1.36}{2.81} \approx \boxed{.484}$$

Chapter 11 Review

1. $a_n = \dfrac{n}{n+1}$

$a_1 = \dfrac{\mathbf{1}}{\mathbf{2}},\quad a_2 = \dfrac{\mathbf{2}}{\mathbf{3}},$

$a_3 = \dfrac{\mathbf{3}}{\mathbf{4}},\quad a_4 = \dfrac{\mathbf{4}}{\mathbf{5}},$

$a_5 = \dfrac{\mathbf{5}}{\mathbf{6}}$

Neither

2. $a_n = (-2)^n$

$a_1 = (-2)^1 = \mathbf{-2}$

$a_2 = (-2)^2 = \mathbf{4}$

$a_3 = (-2)^3 = \mathbf{-8}$

$a_4 = (-2)^4 = \mathbf{16}$

$a_5 = (-2)^5 = \mathbf{-32}$

Geometric

3. $a_n = 2(n+3)$

$a_1 = 2 \cdot 4 = \mathbf{8}$

$a_2 = 2 \cdot 5 = \mathbf{10}$

$a_3 = 2 \cdot 6 = \mathbf{12}$

$a_4 = 2 \cdot 7 = \mathbf{14}$

$a_5 = 2 \cdot 8 = \mathbf{16}$

Arithmetic

4. $a_n = n(n+1)$

$a_1 = 1 \cdot 2 = \mathbf{2}$

$a_2 = 2 \cdot 3 = \mathbf{6}$

$a_3 = 3 \cdot 4 = \mathbf{12}$

$a_4 = 4 \cdot 5 = \mathbf{20}$

$a_5 = 5 \cdot 6 = \mathbf{30}$

Neither

5. $n \geq 2,\ a_n = a_{n-1} - 3$

$$\begin{aligned} a_1 &= & &\mathbf{5} \\ a_2 &= & 5-3 = &\mathbf{2} \\ a_3 &= & 2-3 = &\mathbf{-1} \\ a_4 &= & -1-3 = &\mathbf{-4} \\ a_5 &= & -4-3 = &\mathbf{-7} \end{aligned}$$

Arithmetic

6. Arithmetic,

$a_2 = 10,\quad d = -2$

$a_1 = a_2 - d = 10 + 2 = 12$

$a_n = a_{n-1} + d$

$$\begin{aligned} a_1 &= & &\mathbf{12} \\ a_2 &= & &\mathbf{10} \\ a_3 &= & 10-2 = &\mathbf{8} \\ a_4 &= & 8-2 = &\mathbf{6} \\ a_5 &= & 6-2 = &\mathbf{4} \end{aligned}$$

7. Arithmetic,

$a_3 = \pi,\quad a_4 = 1$

$d = a_4 - a_3 = 1 - \pi$

$a_n = a_1 + (n-1)d,\quad n = 3$

$\pi = a_1 + 2(1-\pi)$

$a_1 = \pi - (2 - 2\pi) = 3\pi - 2$

$a_2 = (3\pi - 2) + (1 - \pi) = 2\pi - 1$

$a_5 = 1 + (1 - \pi) = -\pi + 2$

$\boxed{3\pi - 2,\ 2\pi - 1,\ \pi,\ 1,\ -\pi + 2}$

8. Geometric

$a_1 = \mathbf{6}\quad r = 2,\ a_n = a_1 r^{n-1}$

$a_2 = 6(2) = \mathbf{12}$

$a_3 = 6(2)^2 = \mathbf{24}$

$a_4 = 6(2)^3 = \mathbf{48}$

$a_5 = 6(2)^4 = \mathbf{96}$

9. Geometric

$a_1 = -5\quad a_2 = -1,$

$r = \dfrac{a_2}{a_1} = \dfrac{-1}{-5} = \dfrac{1}{5}$

$a_n = a_1 r^{n-1}$

$a_1 = \mathbf{-5}$

$a_2 = \mathbf{-1}$

$a_3 = -5\left(\dfrac{1}{5}\right)^2 = -\dfrac{\mathbf{1}}{\mathbf{5}}$

$a_4 = -5\left(\dfrac{1}{5}\right)^3 = -\dfrac{\mathbf{1}}{\mathbf{25}}$

$a_5 = -5\left(\dfrac{1}{5}\right)^4 = -\dfrac{\mathbf{1}}{\mathbf{125}}$

10. Arithmetic; $a_5 = -3,\ a_{15} = 17$

$$a_n = a_1 + (n-1)\,d$$

$$n = 5 : -3 = a_1 + 4d \quad [1]$$
$$n = 15 : 17 = a_1 + 14d \quad [2]$$

$[2] - [1]$:

$$\begin{aligned} 17 &= a_1 + 14d \\ 3 &= -a_1 - 4d \\ \hline 20 &= 10d \longrightarrow d = 2 \end{aligned}$$

$d = 2 \Rightarrow [1]$:

$$-3 = a_1 + 4\,(2) \qquad \boxed{a_1 = -11}$$

$$a_n = -11 + (n-1)\,2$$
$$a_n = -11 + 2n - 2$$
$$\boxed{a_n = 2n - 13}$$

[Note: Other methods are possible.]

11. $a_1 = -8,\ a_7 = -\dfrac{1}{8},\quad a_n = a_1\,r^{n-1}$

$$n = 7 : \ -\frac{1}{8} = -8r^6$$
$$\frac{1}{64} = r^6 \Longrightarrow r = \pm\frac{1}{2}$$

Find a_4 if $r = \dfrac{1}{2}$: $a_4 = a_1\,r^3$

$$a_4 = -8\left(\frac{1}{2}\right)^3$$
$$\boxed{a_4 = -1}$$

Find a_4 if $r = -\dfrac{1}{2}$: $a_4 = a_1\,r^3$

$$a_4 = -8\left(-\tfrac{1}{2}\right)^3$$
$$\boxed{a_4 = 1}$$

Find a_n : $a_n = a_1\,r^{n-1}$

$$\boxed{a_n = -8\left(\frac{1}{2}\right)^{n-1} \quad \text{or} \quad a_n = -8\left(-\frac{1}{2}\right)^{n-1}}$$

There are other ways to express a_n.

12. $a_1 = 6,\ d = 2$

$$a_n = a_1 + (n-1)\,d$$
$$a_8 = 6 + 7\,(2) \qquad \boxed{a_8 = 20}$$

13. $a_1 = 6x - 9,\ a_2 = 5x + 1$

$$\begin{aligned} d &= a_2 - a_1 \\ &= 5x + 1 - (6x - 9) \\ &= -x + 10 \end{aligned}$$

$$a_n = a_1 + (n-1)\,d$$
$$\begin{aligned} a_8 &= 6x - 9 + 7\,(-x + 10) \\ &= 6x - 9 - 7x + 70 \end{aligned}$$
$$\boxed{a_8 = -x + 61}$$

14. $a_1 = 2,\ d = 3$

$$S_n = \frac{n}{2}\,[2a_1 + (n-1)\,d]$$
$$\begin{aligned} S_{12} &= \frac{12}{2}\,[2\cdot 2 + 11\cdot 3] \\ &= 6\,[4 + 33] = 6\,(37) \end{aligned}$$
$$\boxed{S_{12} = 222}$$

15. $a_2 = 6,\ d = 10$

$$a_1 = a_2 - d = 6 - 10 = -4$$
$$S_n = \frac{n}{2}\,[2a_1 + (n-1)\,d]$$
$$\begin{aligned} S_{12} &= \frac{12}{2}\,[2\,(-4) + 11\cdot 10] \\ &= 6\,[-8 + 110] = 6\,(102) \end{aligned}$$
$$\boxed{S_{12} = 612}$$

16. $a_1 = -2 \quad r = 3$

$$a_n = a_1\,r^{n-1}$$
$$a_5 = -2\cdot 3^4 \qquad \boxed{a_5 = -162}$$

17. $a_3 = 4 \quad r = \dfrac{1}{5}$

$$a_n = a_1 r^{n-1}$$

$$n = 3: \quad a_3 = a_1 r^2$$

$$4 = a_1 \left(\frac{1}{5}\right)^2 \Longrightarrow a_1 = 100$$

$$a_5 = a_1 \left(\frac{1}{5}\right)^4 = (100)\left(\frac{1}{5}\right)^4$$

$$= (100)\left(\frac{1}{625}\right) \quad \boxed{a_5 = \frac{4}{25}}$$

$$\left[\text{Or } a_5 = a_3 r^2 = 4\left(\frac{1}{5}\right)^2 = \frac{4}{25}\right]$$

18. $a_1 = 3,\ r = 2$

$$S_n = \frac{a_1(1 - r^n)}{1 - r}$$

$$S_4 = \frac{3(1 - 2^4)}{1 - 2} = \frac{3(1 - 16)}{-1} = \frac{-45}{-1}$$

$$\boxed{S_4 = 45}$$

19. $a_1 = -1,\ r = 3$

$$S_n = \frac{a_1(1 - r^n)}{1 - r}$$

$$S_4 = \frac{-1(1 - 3^4)}{1 - 3}$$

$$= \frac{-1(1 - 81)}{-2} = \frac{-1(-80)}{-2} \quad \boxed{S_4 = -40}$$

20. $\dfrac{3}{4},\ -\dfrac{1}{2},\ \dfrac{1}{3},\ \ldots$

$$a_1 = \frac{3}{4},\ r = \frac{a_2}{a_1} = \frac{-1/2}{3/4} = -\frac{2}{3}$$

$$S_n = \frac{a_1(1 - r^n)}{1 - r}$$

$$S_4 = \frac{\frac{3}{4}\left(1 - \left(-\frac{2}{3}\right)^4\right)}{1 - \left(-\frac{2}{3}\right)} = \frac{\frac{3}{4}\left(1 - \frac{16}{81}\right)}{\frac{5}{3}}$$

$$= \frac{\frac{3}{4}\left(\frac{65}{81}\right)}{\frac{5}{3}} \times \frac{108}{108} = \frac{65}{180} \quad \boxed{S_4 = \frac{13}{36}}$$

21. $\displaystyle\sum_{i=1}^{7} (-1)^{i-1}$; $a_1 = 1; \quad r = -1$

$$S_n = \frac{a_1(1 - r^n)}{1 - r}$$

$$S_7 = \frac{1\left(1 - (-1)^7\right)}{1 - (-1)} = \frac{1(2)}{2} = \boxed{1}$$

22. $\displaystyle\sum_{i=1}^{5} (i^2 + i)$

$$= \sum_{i=1}^{5} i^2 + \sum_{i=1}^{5} i$$

$$= \frac{5(5+1)(2 \cdot 5 + 1)}{6} + \frac{5(5+1)}{2}$$

$$= \frac{5(6)(11)}{6} + \frac{5(6)}{2}$$

$$= 55 + 15 = \boxed{70}$$

23. $\displaystyle\sum_{i=1}^{4} \tfrac{i+1}{i}$

$$= \frac{2}{1} + \frac{3}{2} + \frac{4}{3} + \frac{5}{4}$$

$$= \frac{24 + 18 + 16 + 15}{12} = \boxed{\frac{73}{12}}$$

24. $\displaystyle\sum_{j=1}^{10} (3j - 4)$

$$= \sum_{j=1}^{10} 3j - \sum_{j=1}^{10} 4$$

$$= 3\sum_{j=1}^{10} j - \sum_{j=1}^{10} 4$$

$$= 3 \cdot \frac{10(10+1)}{2} - 10(4)$$

$$= 165 - 40 = \boxed{125}$$

25. $\displaystyle\sum_{j=1}^{2500} j = \frac{2500(2500+1)}{2}$

$$= \frac{2500}{2}(2501) = \boxed{3{,}126{,}250}$$

26. $\sum_{i=1}^{5} 4 \cdot 2^i$; $a_1 = 8$; $r = 2$

$$S_n = \frac{a_1(1-r^n)}{1-r}$$

$$S_5 = \frac{8(1-2^5)}{1-2} = \frac{8(-31)}{-1} = \boxed{248}$$

27. $\sum_{i=1}^{\infty} \left(\frac{4}{7}\right)^i$; $a_1 = \frac{4}{7}$; $r = \frac{4}{7}$

$$S_\infty = \frac{a_1}{1-r} = \frac{\frac{4}{7}}{1-\frac{4}{7}} \cdot \frac{7}{7}$$

$$= \frac{4}{7-4} = \boxed{\frac{4}{3}}$$

28. $\sum_{i=1}^{\infty} (-2)\left(\frac{6}{5}\right)^i$

This is the sum of an infinite geometric sequence with $r = \frac{6}{5}$. Since $\frac{6}{5} > 1$, the sum does not exist.

29. $24 + 8 + \frac{8}{3} + \frac{8}{9} + \ldots$

$$r = \frac{a_2}{a_1} = \frac{8}{24} = \frac{1}{3}$$

$$S_\infty = \frac{a_1}{1-r} = \frac{24}{1-\frac{1}{3}} = \frac{24}{\frac{2}{3}}$$

$$= \frac{24}{1} \cdot \frac{3}{2} = \boxed{36}$$

30. $-\frac{3}{4} + \frac{1}{2} - \frac{1}{3} + \frac{2}{9} \cdots$

$$r = \frac{a_2}{a_1} = \frac{\frac{1}{2}}{-\frac{3}{4}} = \frac{1}{2} \cdot \left(-\frac{4}{3}\right) = -\frac{2}{3}$$

$$S_\infty = \frac{a_1}{1-r} = \frac{-\frac{3}{4}}{1-\left(-\frac{2}{3}\right)}$$

$$= -\frac{3}{4} \cdot \frac{3}{5} = \boxed{-\frac{9}{20}}$$

31. $\frac{1}{12} + \frac{1}{6} + \frac{1}{3} + \frac{2}{3} \cdots$

$$r = \frac{a_2}{a_1} = \frac{\frac{1}{6}}{\frac{1}{12}} = 2$$

Since $|r| > 1$, the series diverges.

32. $.9 + .09 + .009 + .0009 + \ldots$

$$r = \frac{a_2}{a_1} = \frac{.09}{.9} = .1$$

$$S_\infty = \frac{a_1}{1-r} = \frac{.9}{1-.1} = \frac{.9}{.9} = \boxed{1}$$

33. $\sum_{i=1}^{4} (x_i{}^2 - 6)$; $x_1 = 0,\ x_2 = 1,\ x_3 = 2,\ x_4 = 3$

$$\sum_{i=1}^{4} (x_i{}^2 - 6)$$

$$= (0^2 - 6) + (1^2 - 6) + (2^2 - 6) + (3^2 - 6)$$

$$= (-6) + (1-6) + (4-6) + (9-6)$$

$$= -6 - 5 - 2 + 3 = \boxed{-10}$$

34. $f(x) = (x-2)^3,\ \Delta x = .1$

$x_1 = 0,\ x_2 = 1,\ x_3 = 2,$
$x_4 = 3,\ x_5 = 4,\ x_6 = 5$

$$\sum_{i=1}^{6} f(x_i)\,\Delta x$$

$$= f(x_1)\,\Delta x + f(x_2)\,\Delta x + f(x_3)\,\Delta x$$
$$+ f(x_4)\,\Delta x + f(x_5)\,\Delta x + + f(x_6)\,\Delta x$$
$$= (x_1-2)^3(.1) + (x_2-2)^3(.1)$$
$$+ (x_3-2)^3(.1) + (x_4-2)^3(.1)$$
$$+ (x_5-2)^3(.1) + (x_6-2)^3(.1)$$
$$= (0-2)^3(.1) + (1-2)^3(.1)$$
$$+ (2-2)^3(.1) + (3-2)^3(.1)$$
$$+ (4-2)^3(.1) + (5-2)^3(.1)$$
$$= (-2)^3(.1) + (-1)^3(.1) + (0)^3(.1)$$
$$+ (1)^3(.1) + (2)^3(.1) + (3)^3(.1)$$
$$= -.8 - .1 + 0 + .1 + .8 + 2.7$$
$$= \boxed{2.7}$$

35. $4 - 1 - 6 - \ldots - 66$

This series is the sum of an arithmetic sequence with $a_1 = 4$ and $d = -1 - 4 = -5$. Therefore, the ith term is

$$\begin{aligned} a_i &= a_1 + (i-1)\,d \\ &= 4 + (i-1)(-5) \\ &= 4 - 5i + 5 \\ &= -5i + 9 \end{aligned}$$

The last term of the series is -66, so

$$\begin{aligned} -66 &= -5i + 9 \\ -75 &= -5i \\ i &= 15 \end{aligned}$$

This indicates that the series consists of 15 terms and is equivalent to:

$$\boxed{\sum_{i=1}^{15} (-5i + 9)}$$

36. $10 + 14 + 18 + \ldots + 86$

This series is the sum of an arithmetic sequence with $a_1 = 10$ and $d = 14 - 10 = 4$. Therefore, the ith term is

$$\begin{aligned} a_i &= a_1 + (i-1)\,d \\ &= 10 + (i-1)(4) \\ &= 10 + 4i - 4 \\ &= 4i + 6 \end{aligned}$$

The last term of the series is 86, so

$$\begin{aligned} 86 &= 4i + 6 \\ 80 &= 4i \\ i &= 20 \end{aligned}$$

This indicates that the series consists of 20 terms and is equivalent to:

$$\boxed{\sum_{i=1}^{20} (4i + 6)}$$

37. $4 + 12 + 36 + \ldots + 972$

This series is the sum of a geometric sequence with $a_1 = 4$ and $r = \dfrac{12}{4} = 3$.

Find the ith term:

$$a_i = a_1 r^{i-1} = 4(3)^{i-1}$$

Find the number of terms:

$$972 = 4(3)^{i-1}$$
$$243 = (3)^{i-1}$$
$$3^5 = (3)^{i-1}$$
$$5 = i - 1 \rightarrow i = 6$$

This indicates that the series consists of 6 terms and is equivalent to:

$$\boxed{\sum_{i=1}^{6} 4(3)^{i-1}}$$

38. $\frac{5}{6}+\frac{6}{7}+\frac{7}{8}+\ldots+\frac{12}{13}$

This series is neither arithmetic nor geometic, but notice that each denominator is 1 larger than its corresponding numerator. This means

$$a_i = \frac{i}{i+1}$$

Find the number of terms:

The numerators of the terms of the series begin at $i = 5$, and increase by 1 until $i = 12$. We conclude that

$$\boxed{\sum_{i=5}^{12} \frac{i}{i+1}}$$

39. $(x+2y)^4$

$$= x^4 + \binom{4}{1} x^3 (2y) + \binom{4}{2} x^2 (2y)^2 + \binom{4}{3} x (2y)^3 + (2y)^4$$

$$= x^4 + 4x^3 (2y) + 6x^2 (4y^2) + 4x (8y^3) + 16y^4$$

$$= \boxed{x^4 + 8x^3y + 24x^2y^2 + 32xy^3 + 16y^4}$$

40. $(3z-5w)^3$

$$= (3z)^3 + \binom{3}{1} (3z)^2 (-5w) + \binom{3}{2} (3z)(-5w)^2 + (-5w)^3$$

$$= 27z^3 + 3(9z^2)(-5w) + 3(3z)(25w^2) - 125w^3$$

$$= \boxed{27z^3 - 135z^2w + 225zw^2 - 125w^3}$$

41. $\left(3\sqrt{x} - \frac{1}{\sqrt{x}}\right)^5$

$$= \left(3x^{1/2}\right)^5 + \binom{5}{1} \left(3x^{1/2}\right)^4 \left(-x^{-1/2}\right) + \binom{5}{2} \left(3x^{1/2}\right)^3 \left(-x^{-1/2}\right)^2 + \binom{5}{3} \left(3x^{1/2}\right)^2 \left(-x^{-1/2}\right)^3 + \binom{5}{4} \left(3x^{1/2}\right) \left(-x^{-1/2}\right)^4 + \left(-x^{-1/2}\right)^5$$

$$= 243x^{5/2} + (5)\left(81x^2\right)^4 \left(-x^{-1/2}\right) + (10)\left(27x^{3/2}\right)\left(x^{-1}\right) + (10)(9x)\left(-x^{-3/2}\right) + (5)\left(3x^{1/2}\right)\left(x^{-2}\right) + \left(-x^{-5/2}\right)$$

$$= \boxed{243x^{5/2} - 405x^{3/2} + 270x^{1/2} - 90x^{-1/2} + 15x^{-3/2} - x^{-5/2}}$$

42. $\left(m^3 - m^{-2}\right)^4$

$$= \left(m^3\right)^4 + \binom{4}{1}\left(m^3\right)^3\left(-m^{-2}\right) + \binom{4}{2}\left(m^3\right)^2\left(-m^{-2}\right)^2 + \binom{4}{3}\left(m^3\right)\left(-m^{-2}\right)^3 + \left(-m^{-2}\right)^4$$

$$= m^{12} + 4m^9\left(-m^{-2}\right) + 6m^6\left(m^{-4}\right) + 4m^3\left(-m^{-6}\right) + \left(m^{-8}\right)$$

$$= \boxed{m^{12} - 4m^7 + 6m^2 - 4m^{-3} + m^{-8}}$$

43. $\binom{n}{r-1} \mathbf{x}^{n-(r-1)}\mathbf{y}^{r-1}$

6th term of $(4x-y)^8$:

$n = 8,\ r = 6,\ \mathbf{x} = 4x,\ \mathbf{y} = -y$

$$\binom{8}{5}(4x)^3(-y)^5 = 56\left(64x^3\right)\left(-y^5\right)$$

$$= \boxed{-3584x^3y^5}$$

44. $\binom{n}{r-1} \mathbf{x}^{n-(r-1)}\mathbf{y}^{r-1}$

7th term of $(m-3n)^{14}$:

$n = 14,\ r = 7,\ \mathbf{x} = m,\ \mathbf{y} = -3n$

$$\binom{14}{6}(m)^8(-3n)^6 = 3003\,(m^8)\,(-3)^6\,n^6$$

$$= \boxed{3003\,(-3)^6\,m^8n^6}$$

$$= 2189187m^8n^6$$

45. 1st 4 terms: $(x+2)^{12}$

$$= x^{12} + \binom{12}{1}x^{11}(2) + \binom{12}{2}x^{10}(2)^2 + \binom{12}{3}x^9(2)^3$$

$$= x^{12} + 12\,(2x^{11}) + 66\,(4x^{10}) + 220\,(8x^9)$$

$$= \boxed{x^{12} + 24x^{11} + 264x^{10} + 1760x^9}$$

46. Last 3 terms: $(2a+5b)^{16}$

$$= \binom{16}{14}(2a)^2(5b)^{14} + \binom{16}{15}(2a)(5b)^{15} + (5b)^{16}$$

$$= 120\,(4a^2)\,(5)^{14}\,b^{14} + 16\,(2a)\,(5)^{15}\,b^{15} + (5)^{16}\,b^{16}$$

$$= \boxed{480\,(5)^{14}\,a^2b^{14} + 32\,(5)^{15}\,ab^{15} + (5)^{16}\,b^{16}}$$

47. Statements containing the set of natural numbers as their domains are proved by mathematical induction. For example, if n is a natural number, then

$1+2+3+\ldots+n = \frac{n\,(n+1)}{2}$ and

$1+3+5+\ldots+(2n-1) = n^2.$

48. For a proof by mathematical induction, first, show that the statement is true for $n = 1$. Then show that if it is true for $n = k$, it must also be true for $n = k+1$.

49. $1+3+5+7+\ldots+(2n-1) = n^2$

(i) Show true for $n = 1$:

$1 = (1)^2 = 1$

(ii) Assume S_k is true:

$1+3+5+7+\ldots+(2k-1) = k^2$

Show S_{k+1} is true:

$1+3+5+7+\ldots+(2(k+1)-1) = (k+1)^2$

Add $2(k+1)-1 = 2k+1$ to both sides of S_k :

$$1+3+5+\ldots+(2k-1)+2k+1 = k^2+2k+1 = (k+1)^2$$

Since S_k implies S_{k+1}, the statement is true for every positive value of n.

50. $2+6+10+14+\ldots+(4n-2) = 2n^2$

(i) Show true for $n = 1$:

$2 = 2\,(1)^2 = 2$

(ii) Assume S_k is true:

$2+6+\ldots+(4k-2) = 2k^2$

Show S_{k+1} is true:

$2+6+\ldots+(4(k+1)-2) = 2(k+1)^2$

Add $4(k+1)-2 = 4k+2$ to both sides of S_k :

$$2+6+\ldots+(4k-2)+4k+2 = 2k^2+4k+2 = 2\,(k^2+2k+1) = 2\,(k+1)^2$$

Since S_k implies S_{k+1}, the statement is true for every positive value of n.

51. $2 + 2^2 + 2^3 + ... + 2^n = 2(2^n - 1)$

(i) Show true for $n = 1$:

$$2^1 = 2(2^1 - 1)$$
$$2 = 2(1)$$
$$2 = 2$$

(ii) Assume S_k is true:

$$2 + 2^2 + ... + 2^k = 2(2^k - 1)$$

Show S_{k+1} is true:

$$2 + 2^2 + ... + 2^{k+1} = 2(2^{k+1} - 1)$$

Add 2^{k+1} to both sides of S_k:

$$\begin{aligned} 2 + 2^2 + ... + 2^k + 2^{k+1} &= 2(2^k - 1) + 2^{k+1} \\ &= 2^{k+1} - 2 + 2^{k+1} \\ &= 2(2^{k+1}) - 2 \\ &= 2(2^{k+1} - 1) \end{aligned}$$

Since S_k implies S_{k+1}, the statement is true for every positive value of n.

52. $1^3 + 3^3 + 5^3 + ... + (2n-1)^3 = n^2(2n^2 - 1)$

(i) Show true for $n = 1$:

$$1^3 = 1^2(2(1^2) - 1)$$
$$1 = 1(2 - 1)$$
$$1 = 1$$

(ii) Assume S_k is true:

$$1^3 + 3^3 + 5^3 + ... + (2k-1)^3 = k^2(2k^2 - 1)$$

Show S_{k+1} is true:

$$1^3 + 3^3 + 5^3 + ... + (2(k+1)-1)^3 = (k+1)^2\left(2(k+1)^2 - 1\right)$$

Add $(2(k+1)-1)^3$ to both sides of S_k:

$$\begin{aligned} 1^3 + 3^3 + 5^3 + ... + (2k-1)^3 &+ (2(k+1)-1)^3 \\ &= k^2(2k^2 - 1) + (2(k+1)-1)^3 \\ &= 2k^4 - k^2 + (2k+1)^3 \\ &= 2k^4 - k^2 + 8k^3 + 12k^2 + 6k + 1 \\ &= 2k^4 + 8k^3 + 11k^2 + 6k + 1 \\ &= (k^2 + 2k + 1)(2k^2 + 4k + 1) \\ &= (k+1)^2\left(2(k+1)^2 - 1\right) \end{aligned}$$

Since S_k implies S_{k+1}, the statement is true for every positive value of n.

53. A student identification number is an example of a permutation.

54. $P(9, 2) = \dfrac{9!}{(9-2)!} = \dfrac{9!}{7!} = \dfrac{9 \cdot 8 \cdot 7!}{7!} = \boxed{72}$

[Calculator: $9 \; {}_nP_r \; 2$]

55. $P(6, 0) = \dfrac{6!}{(6-0)!} = \dfrac{6!}{6!} = \boxed{1}$

[Calculator: $6 \; {}_nP_r \; 0$]

56. $\dbinom{8}{3} = \dfrac{8!}{5!\,3!} = \dfrac{8 \cdot 7 \cdot 6 \cdot 5!}{5!\,(3 \cdot 2 \cdot 1)} = \boxed{56}$

[Calculator: $8 \; {}_nC_r \; 3$]

57. $9! = 9 \cdot 8 \cdot 7 \cdot 6 \cdot 5 \cdot 4 \cdot 3 \cdot 2 \cdot 1 = \boxed{362{,}880}$

58. $C(10, 5) = \dfrac{10!}{5!\,5!} = \dfrac{10 \cdot 9 \cdot 8 \cdot 7 \cdot 6 \cdot 5!}{5!\,(5 \cdot 4 \cdot 3 \cdot 2 \cdot 1)} = \boxed{252}$

[Calculator: $10 \; {}_nC_r \; 5$]

59. 2 chapels, 4 soloists, 3 organists, 2 ministers:

$2 \cdot 4 \cdot 3 \cdot 2 = 48$

$\boxed{\text{48 possible arrangements}}$

60. 5 styles, 3 fabrics, 6 colors:

$5 \cdot 3 \cdot 6 = 90$

$\boxed{\text{90 different couches}}$

61. $4! = \boxed{\text{24 ways}}$

62. P, VP, ST, R, R, R:

(a) 3 members: ${}_6C_3 = \boxed{20}$

(b) President + 2 members:

${}_5C_2 = \boxed{10}$

63. 9 teams, 3 places:

${}_9P_3 = \boxed{504}$

64. 1 letter + 3 digits + 3 letters:

$26 \cdot 10 \cdot 10 \cdot 10 \cdot 26 \cdot 26 \cdot 26$

$\boxed{456{,}976{,}000}$

No repetitions:

$26 \cdot 10 \cdot 9 \cdot 8 \cdot 25 \cdot 24 \cdot 23$

$\boxed{258{,}336{,}000}$

65. 4 green, 5 black, 6 white:

$P(E) = \dfrac{n(E)}{n(S)}$

(a) $P(\text{green}) = \mathbf{\dfrac{4}{15}}$

(b) $P(\text{black}') = \dfrac{10}{15} = \mathbf{\dfrac{2}{3}}$

(c) $P(\text{blue}) = \mathbf{0}$

66. 4 green, 5 black, 6 white:

(a) odds of green:

$\dfrac{P(\text{G})}{P(\text{G}')} = \dfrac{4/15}{11/15} = \dfrac{4}{11}$ $\boxed{\text{4 to 11}}$

(b) odds of 0 white:

$\dfrac{P(\text{W}')}{P(\text{W})} = \dfrac{9/11}{6/11} = \dfrac{9}{6}$ $\boxed{\text{3 to 2}}$

(c) The odds in favor of drawing a marble that is not white are the same as the odds against drawing a white marble. Therefore, the answer is the same as the answer to part (b): $\boxed{\text{3 to 2}}$

67. 52 cards, draw a black king:

$P(E) = \dfrac{n(E)}{n(S)} = \dfrac{2}{52} = \boxed{\dfrac{1}{26}}$

68. 52 cards, draw a face card or ace:

$P(E) + P(F) - P(E \cap F)$

$\dfrac{12}{52} + \dfrac{4}{52} - 0 = \dfrac{16}{52} = \boxed{\dfrac{4}{13}}$

69. 52 cards, draw an ace or a diamond:

$P(E) + P(F) - P(E \cap F)$

$\dfrac{4}{52} + \dfrac{13}{52} - \dfrac{1}{52} = \dfrac{16}{52} = \boxed{\dfrac{4}{13}}$

70. 52 cards, do not draw a diamond:

$P(E) = \dfrac{n(E)}{n(S)} = \dfrac{39}{52} = \boxed{\dfrac{3}{4}}$

71. No more than three defective filters:

$.31 + .25 + .18 + .12 = \boxed{.86}$

72. At least 2 defective filters:

$.18 + .12 + .08 + .06 = \boxed{.44}$

73. More than 5 defective filters: $= \boxed{0}$

74. If $P(5 \text{ on } 1 \text{ roll}) = \frac{1}{6}$, on 12 rolls of a die, find $P(\text{exactly two } 5\text{'s})$:

$$= \binom{12}{2}\left(\frac{1}{6}\right)^2\left(1-\frac{1}{6}\right)^{10}$$

$$= 66\left(\frac{1}{36}\right)\left(\frac{5}{6}\right)^{10}$$

$$= .29609357 \approx \boxed{.296}$$

75. If $P(\text{h on } 1 \text{ roll}) = \frac{1}{2}$, on 10 tosses of a coin, find $P(\text{exactly four h's})$:

$$= \binom{10}{4}\left(\frac{1}{2}\right)^4\left(\frac{1}{2}\right)^6$$

$$= 210\left(\frac{1}{2}\right)^{10}$$

$$\approx .205078 \approx \boxed{.205}$$

76. The total number of students polled is $n(S) = 1640$.

(a) Let E be "selected student is conservative."

$$P(E) = \frac{n(E)}{n(S)} = \frac{303.4}{1640} = .185$$

(b) Let E be "selected student is on the far left or the far right."

$$P(E) = \frac{n(E)}{n(S)}$$

$$= P(\text{student on far left}) + P(\text{student on far right})$$

$$= \frac{n(\text{far left})}{n(S)} + \frac{n(\text{far right})}{n(S)}$$

$$= \frac{44.28}{1640} + \frac{24.6}{1640}$$

$$= \frac{68.88}{1640} = .042$$

(c) Let E be "selected student is middle of the road."

Then $n(E) = 926.6$

$n(E') = 1640 - 926.6 = 713.4$

Thus, the odds against E are

$$\frac{n(E')}{n(E)} = \frac{713.4}{926.6} = \frac{87}{113}$$

87 to 113

Chapter 11 Test

1. (a) $a_n = (-1)^n(n+2)$

$a_1 = (-1)^1 \cdot 3 = \mathbf{-3}$

$a_2 = (-1)^2 \cdot 4 = \mathbf{4}$

$a_3 = (-1)^3 \cdot 5 = \mathbf{-5}$

$a_4 = (-1)^4 \cdot 6 = \mathbf{6}$

$a_5 = (-1)^5 \cdot 7 = \mathbf{-7}$

Neither

(b) $a_n = -3\left(\frac{1}{2}\right)^n$

$a_1 = -3\left(\frac{1}{2}\right)^1 = -\mathbf{\frac{3}{2}}$

$a_2 = -3\left(\frac{1}{2}\right)^2 = -\mathbf{\frac{3}{4}}$

$a_3 = -3\left(\frac{1}{2}\right)^3 = -\mathbf{\frac{3}{8}}$

$a_4 = -3\left(\frac{1}{2}\right)^4 = -\mathbf{\frac{3}{16}}$

$a_5 = -3\left(\frac{1}{2}\right)^5 = -\mathbf{\frac{3}{32}}$

Geometric

(c) $a_1 = 2,\ a_2 = 3,$
$n \geq 3,\ a_n = a_{n-1} + 2a_{n-2}$

$a_1 = \mathbf{2}$

$a_2 = \mathbf{3}$

$a_3 = 3 + 2(2) = \mathbf{7}$

$a_4 = 7 + 2(3) = \mathbf{13}$

$a_5 = 13 + 2(7) = \mathbf{27}$

Neither

2. (a) Arithmetic

$a_1 = 1,\ a_3 = 25$

$$d = \frac{1}{2}(a_3 - a_1) = \frac{1}{2}(25 - 1) = 12$$

$$a_n = a_1 + (n-1)\,d$$

$$a_5 = 1 + 4\,(12)$$

$$\boxed{a_5 = 49}$$

(b) Geometric

$a_1 = 81 \quad r = -\frac{2}{3}$

$$a_n = a_1\,r^{n-1}$$

$$a_5 = 81 \cdot \left(-\frac{2}{3}\right)^4$$

$$\boxed{a_5 = 16}$$

3. (a) Arithmetic

$a_1 = -43,\ d = 12$

$$S_n = \frac{n}{2}\,[2a_1 + (n-1)\,d]$$

$$S_{10} = \frac{10}{2}\,[2\,(-43) + 9 \cdot 12] = 5\,[-86 + 108] = 5\,(22)$$

$$\boxed{S_{10} = 110}$$

(b) Geometric

$a_1 = 5,\ r = -2$

$$S_n = \frac{a_1\,(1 - r^n)}{1 - r}$$

$$S_{10} = \frac{5\left(1 - (-2)^{10}\right)}{1 - (-2)} = \frac{5\,(1 - 1024)}{3} = \frac{-5115}{3}$$

$$\boxed{S_{10} = -1705}$$

4. (a) $\sum_{i=1}^{30} (5i + 2)$

$$= \sum_{i=1}^{30} 5i + \sum_{i=1}^{30} 2 = 5\sum_{i=1}^{30} i + \sum_{i=1}^{30} 2 = 5 \cdot \frac{30\,(30+1)}{2} + 30\,(2) = 2325 + 60 = \boxed{2385}$$

(b) $\sum_{i=1}^{5} (-3 \cdot 2^i)$; $a_1 = -6; \quad r = 2$

$$S_n = \frac{a_1\,(1 - r^n)}{1 - r}$$

$$S_5 = \frac{-6\,(1 - 2^5)}{1 - 2} = \frac{-6\,(-31)}{-1} = \boxed{-186}$$

(c) $\sum_{i=1}^{\infty} (2^i) \cdot 4$; $a_1 = 8; \quad r = 2$

This is the sum of an infinite geometric sequence with $r = 2$. Since $2 > 1$, the sum does not exist.

(d) $\sum_{i=1}^{\infty} 54\left(\frac{2}{9}\right)^i$; $a_1 = 12; \quad r = \frac{2}{9}$

$$S_\infty = \frac{a_1}{1 - r} = \frac{12}{1 - \frac{2}{9}} \cdot \frac{9}{9} = \frac{108}{9 - 2} = \boxed{\frac{108}{7}}$$

5. **(a)** $(2x-3y)^4$

$$= (2x)^4 + \binom{4}{1}(2x)^3(-3y) + \binom{4}{2}(2x)^2(-3y)^2 + \binom{4}{3}(2x)(-3y)^3 + (-3y)^4$$

$$= 16x^4 + 4(8x^3)(-3y) + 6(4x^2)(9y^2) + 4(2x)(-27y^3) + 81y^4$$

$$= \boxed{16x^4 - 96x^3y + 216x^2y^2 - 216xy^3 + 81y^4}$$

(b) $\binom{n}{r-1}\mathbf{x}^{n-(r-1)}\mathbf{y}^{r-1}$

3rd term of $(w-2y)^6$:

$n=6,\ r=3,\ \mathbf{x}=w,\ \mathbf{y}=-2y$

$$\binom{6}{2}(w)^4(-2y)^2 = 15(w^4)(4y^2) = \boxed{60w^4y^2}$$

6. **(a)** $10\,{}_nC_r\,2 = \dfrac{10!}{8!\,2!} = \dfrac{10\cdot 9\cdot 8!}{8!\,(2\cdot 1)} = \boxed{45}$

(b) $\binom{7}{3} = \dfrac{7!}{4!\,3!} = \dfrac{7\cdot 6\cdot 5\cdot 4!}{4!\,(3\cdot 2\cdot 1)} = \boxed{35}$

[Calculator: $7\,{}_nC_r\,3$]

(c) $7! = 7\cdot 6\cdot 5\cdot 4\cdot 3\cdot 2\cdot 1 = \boxed{5040}$

(d) $P(11,\,3) = \dfrac{11!}{(11-3)!} = \dfrac{11!}{8!} = \dfrac{11\cdot 10\cdot 9\cdot 8!}{8!} = \boxed{990}$

[Calculator: $11\,{}_nP_r\,3$]

7. $8+14+20+26+...+(6n+2) = 3n^2+5n$

(i) Show true for $n=1$:

$$8 = 3(1)^2 + 5(1) = 8$$

(ii) Assume S_k is true:

$$8+14+...+(6k+2) = 3k^2+5k$$

Show S_{k+1} is true:

$$8+14+...+(6(k+1)+2) = 3(k+1)^2 + 5(k+1)$$

Add $6(k+1)+2 = 6k+8$ to both sides of S_k :

$$\begin{aligned} 8+14+...+(6k+2)+6k+8 &= 3k^2+5k+6k+8 \\ &= (3k^2+6k+3)+5k+5 \\ &= 3(k+1)^2+5(k+1) \end{aligned}$$

Since S_k implies S_{k+1}, the statement is true for every positive value of n.

8. 6 styles, 3 colors, 2 shades:

$$4\cdot 3\cdot 2 = 24$$

$\boxed{\text{24 types of shoes}}$

9. 20 members, 3 offices:

$$P(20,3) = {}_{20}P_3 = \frac{20!}{17!} = \boxed{6840}$$

10. The number of ways to select the men is $\binom{8}{2}$. The number of ways to select the women is $\binom{12}{3}$. The number of ways to choose the 5 people to attend the conference is given by

$$\binom{8}{2}\cdot\binom{12}{3} = 28\cdot 220 = \boxed{6160}$$

11. **(a)** 52 cards, draw a red three:

$$P(E) = \frac{n(E)}{n(S)} = \frac{2}{52} = \boxed{\frac{1}{26}}$$

(b) 52 cards, do not draw a face card:

$$P(E) = \frac{n(E)}{n(S)} = \frac{40}{52} = \boxed{\frac{10}{13}}$$

(c) 52 cards, draw a king or a spade:

$$P(E) + P(F) - P(E \cap F)$$

$$\frac{4}{52} + \frac{13}{52} - \frac{1}{52} = \frac{16}{52} = \boxed{\frac{4}{13}}$$

(d) 52 cards, odds of drawing a face card:

$$P(E) = \frac{3}{13}, \quad P(E') = \frac{10}{13}$$

$$\frac{P(E)}{P(E')} = \frac{\frac{3}{13}}{\frac{10}{13}} = \frac{3}{10}$$

$$\boxed{3 \text{ to } 10}$$

12. **(a)** If $P(4 \text{ on } 1 \text{ roll}) = \frac{1}{6}$, on 8 rolls of a die find $P(\text{exactly three } 4\text{'s})$:

$$= \binom{8}{3}\left(\frac{1}{6}\right)^3\left(1 - \frac{1}{6}\right)^5$$

$$= 56\left(\frac{1}{6^3}\right)\left(\frac{5^5}{6^5}\right)$$

$$\approx .1041905$$

$$\approx \boxed{.104}$$

(b) If $P(6 \text{ on } 1 \text{ roll}) = \frac{1}{6}$, on 8 rolls of a die find $P(\text{exactly eight } 6\text{'s})$:

$$= \binom{8}{8}\left(\frac{1}{6}\right)^8\left(1 - \frac{1}{6}\right)^0$$

$$= 1\left(\frac{1}{6^8}\right)(1)$$

$$\approx \boxed{.000000595}$$

Chapter 11 Project

Using Experimental Probabilities to Simulate Family Makeup.

Note: **RF** will denote Relative Frequency; **EP** will denote Experimental probability (as a decimal) in the tables.

1. Answers will vary. The results here were actually obtained by author Hornsby tossing two United States quarters on January 27, 1998.

Event	RF	EP
2 heads, 0 tails	$\frac{2}{20}$	.10
1 head, 1 tail	$\frac{10}{20}$	.50
0 heads, 2 tails	$\frac{8}{20}$	.40

(a) The sample space is

$$\{(h,h), (h,t), (t,h), (t,t)\}.$$

The theoretical probabilities are

$$P(2 \text{ heads, } 0 \text{ tails}) = \frac{1}{4} = .25$$

$$P(1 \text{ head, } 1 \text{ tail}) = \frac{2}{4} = .5$$

$$P(0 \text{ heads, } 2 \text{ tails}) = \frac{1}{4} = .25$$

(b) The outcome are not equally likely, as seen in the sample space. Of the possible outcomes, 2 heads and 0 tails appear once, 1 head and 1 tail appear twice, and 0 heads and 2 tails appear once.

(c) It is not reasonable to expect the experimental probabilities to match the theoretical probabilities in a small number of trials (twenty). To obtain better results, we must increase the number of trials. By the law of large numbers, as we increase the number of trials, the experimental probabilities become closer to the theoretical probabilities.

2. Answers will vary. The results here were actually obtained by author Hornsby tossing four United States quarters on January 27, 1998.

Event	RF	EP
4 heads, 0 tails	$\frac{3}{20}$	.15
3 heads, 1 tail	$\frac{5}{20}$	.25
2 heads, 2 tails	$\frac{8}{20}$	.40
1 head, 3 tails	$\frac{4}{20}$	.20
0 heads, 4 tails	$\frac{0}{20}$	.00

(a) See the tree diagram that follows for the sample space for this experiment.

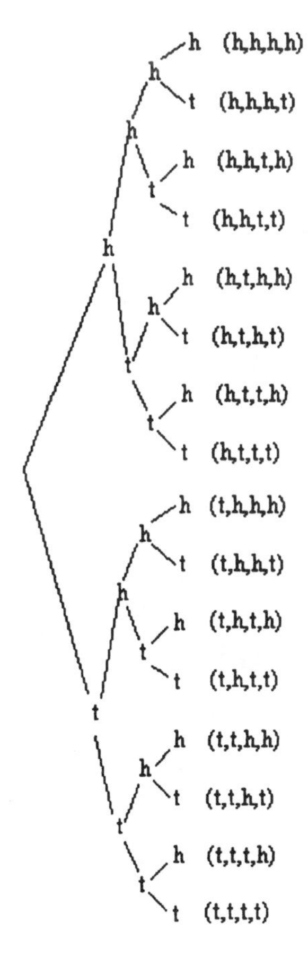

continued

2. continued

The theoretical probabilities are

$$P(4 \text{ heads, } 0 \text{ tails}) = \frac{1}{16} = .0625$$

$$P(3 \text{ heads, } 1 \text{ tails}) = \frac{4}{16} = .25$$

$$P(2 \text{ heads, } 2 \text{ tails}) = \frac{6}{16} = .375$$

$$P(1 \text{ heads, } 3 \text{ tails}) = \frac{4}{16} = .25$$

$$P(0 \text{ heads, } 4 \text{ tails}) = \frac{1}{16} = .0625$$

(b) The outcomes are not equally likely, as seen in the tree diagram. Of the sixteen possible outcomes, 4 heads appear once, 3 heads and 1 tail appear four times, 2 heads and 2 tails appear six times, 1 head and 3 tails appear four times, and 0 heads and 4 tails appear once.

(c) See the answer to Activity **1(c)**.

3. Answers will vary. Using the program shown in the text, we obtained the following histogram, indicating 73 outcomes of 2 (representing 2 girls and 2 boys). Thus, the experimental probability is $\frac{73}{200} = .365$, which is extremely close to the theoretical probability of .375. This is closer than the result found in Activity 2, since the number of trials is much greater.

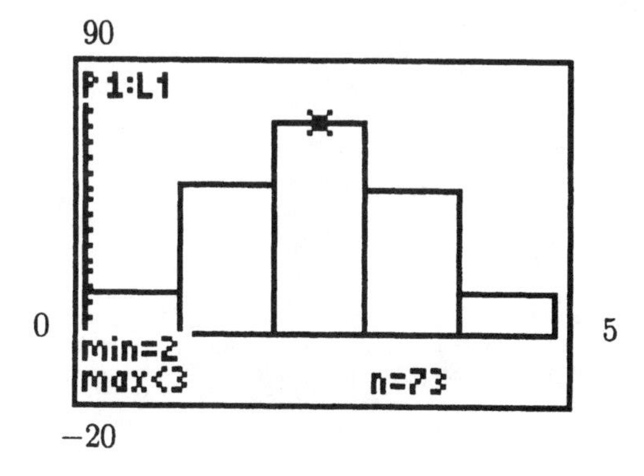

Section 12.1

1. False; the function must be continuous. Consider

$$\lim_{x\to 0} f(x), \text{ where } f(x) = \begin{cases} x & \text{if } x \neq 0 \\ 2 & \text{if } x = 0 \end{cases}.$$

3. False; $\lim_{x\to 1} f(x) = 5$ need only be defined near 1, not at 1.

5. True

7. $\lim_{x\to 2^-} F(x) = 4$, $\lim_{x\to 2^+} F(x) = 4$ $\Longrightarrow$ $\lim_{x\to 2} F(x) = 4$

9. The limit does not exist since as $x \to 3$, y gets arbitrarily large in the negative direction.

11. $\lim_{x\to 1^-} h(x) = 1$, $\lim_{x\to 1^+} h(x) = 1$ $\Longrightarrow$ $\lim_{x\to 1} h(x) = 1$

13. $\lim_{x\to 2^-} f(x) = \infty$, $\lim_{x\to 2^+} f(x) = 2$ $\Longrightarrow$ the limit does not exist

15. $\lim_{x\to 2^-} g(x) = -1$, $\lim_{x\to 2^+} g(x) = -\frac{1}{2}$ $\Longrightarrow$ the limit does not exist

17. $\lim_{x\to 2^-} f(x) = 1.5$, $\lim_{x\to 2^+} f(x) = 1.5$ $\Longrightarrow$ $\lim_{x\to 2} f(x) = 1.5$

19. $\lim_{x\to 2^-} f(x) = -1$, $\lim_{x\to 2^+} f(x) = -1$ $\Longrightarrow$ $\lim_{x\to 2} f(x) = -1$

21. $k(x) = \dfrac{x^3 - 2x - 4}{x - 2}$

x	$k(x)$
1.9	9.41
1.99	9.9401
1.999	9.994001
2.001	10.006001
2.01	10.0601
2.1	10.61

$\lim_{x\to 2^-} k(x) = 10$, $\lim_{x\to 2^+} k(x) = 10$ $\Longrightarrow$ $\lim_{x\to 2} k(x) = 10$

23. $h(x) = \dfrac{\sqrt{x} - 2}{x - 1}$

x	$h(x)$
.9	10.5132
.99	100.501
.999	1000.5
1.001	−999.5
1.01	−99.5012
1.1	−9.51191

$\lim_{x\to 1^-} h(x) = \infty$, $\lim_{x\to 1^+} h(x) = -\infty$ $\Longrightarrow$ $\lim_{x\to 1} h(x)$ does not exist

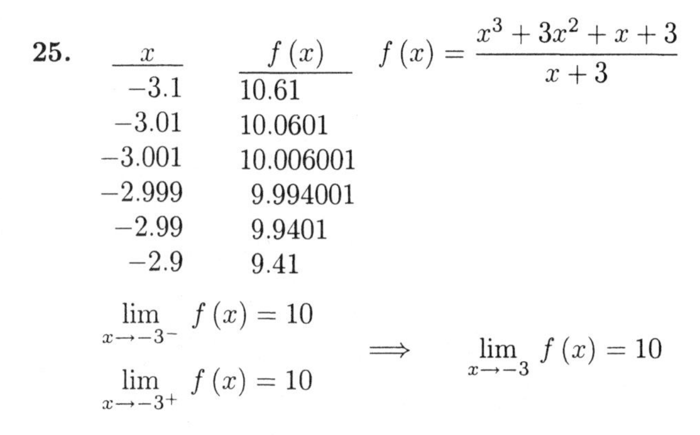

25. $f(x) = \dfrac{x^3 + 3x^2 + x + 3}{x + 3}$

x	$f(x)$
−3.1	10.61
−3.01	10.0601
−3.001	10.006001
−2.999	9.994001
−2.99	9.9401
−2.9	9.41

$\lim_{x\to -3^-} f(x) = 10$, $\lim_{x\to -3^+} f(x) = 10$ $\Longrightarrow$ $\lim_{x\to -3} f(x) = 10$

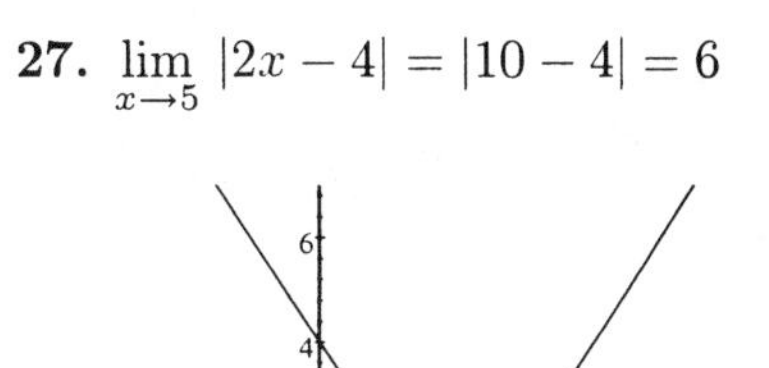

27. $\lim_{x\to 5} |2x - 4| = |10 - 4| = 6$

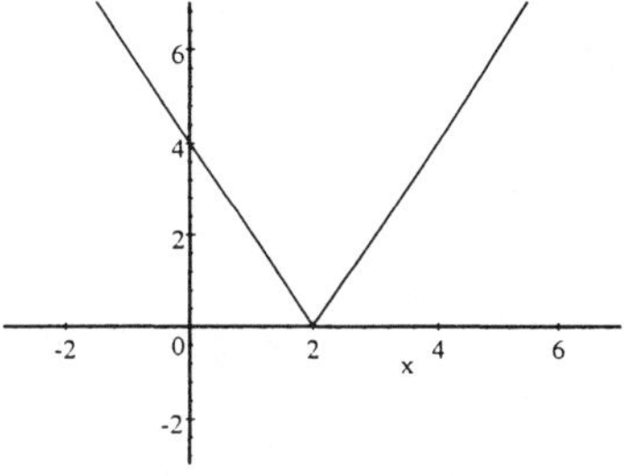

29. $\lim_{x\to5} \frac{x^2-3x-10}{x-5} = \lim_{x\to5} \frac{(x-5)(x+2)}{x-5}$

$= \lim_{x\to5} (x+2) = 5+2 = 7$

A hole is at $(5, 7)$.

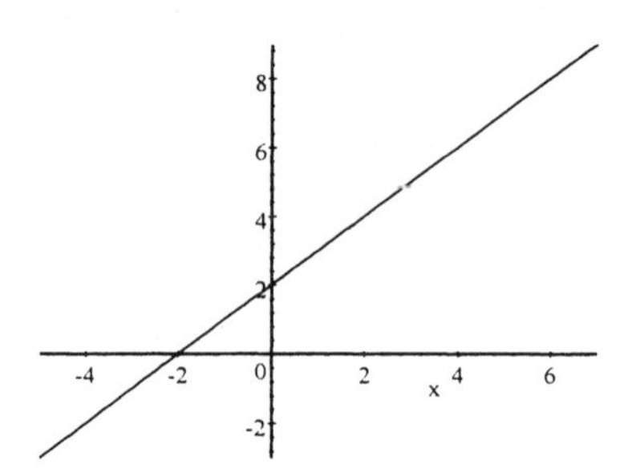

31. $\lim_{x\to-2} \frac{x^2+2}{x+2}$ does not exist.

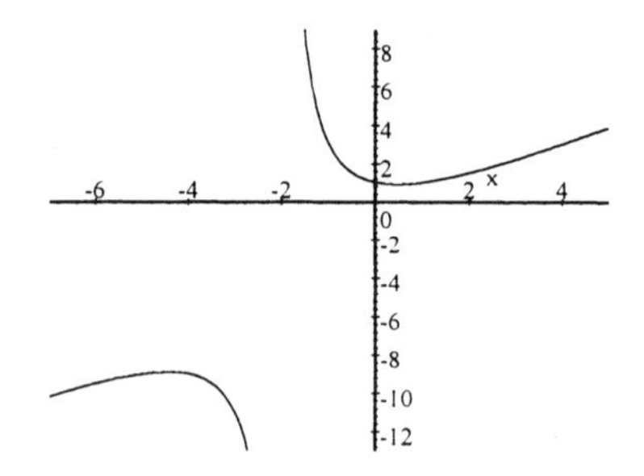

33. $\lim_{x\to2} \frac{x^2-x-2}{x-2} = \lim_{x\to2} \frac{(x-2)(x+1)}{x-2}$

$= \lim_{x\to2} (x+1) = 3$

A hole is at $(2, 3)$.

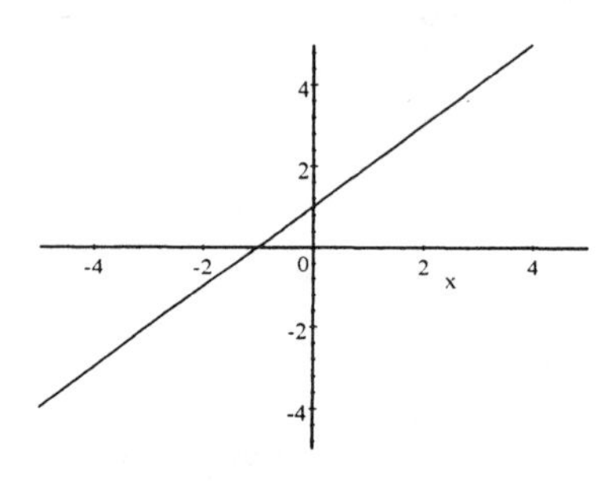

35. $f(x) = \begin{cases} x+7 & \text{if } x \le 3 \\ 5x-5 & \text{if } x > 3 \end{cases}$

$\lim_{x\to3} f(x) = 10$

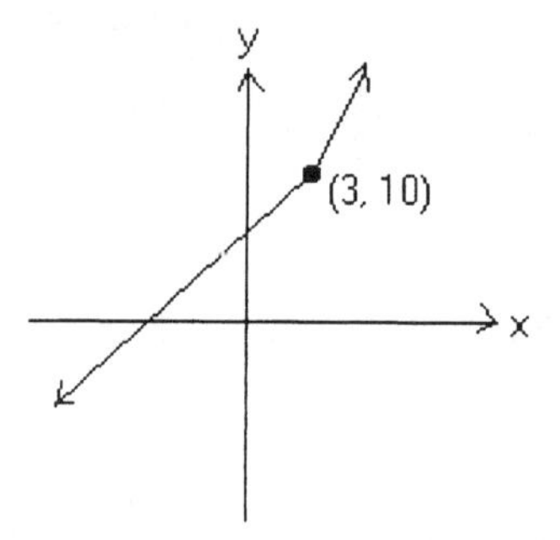

37. $\lim_{x\to1} \frac{\sqrt{x}-1}{x-1} = \frac{1}{2}$

x	y
0	1
.9	.51317
.99	.50126
1.01	.49876
1.1	.48809

39. $\lim_{x\to0} \frac{\sin x}{\sin 2x} = \lim_{x\to0} \frac{\sin x}{2\sin x\cos x}$

$= \lim_{x\to0} \frac{1}{2\cos x} = \frac{1}{2}$

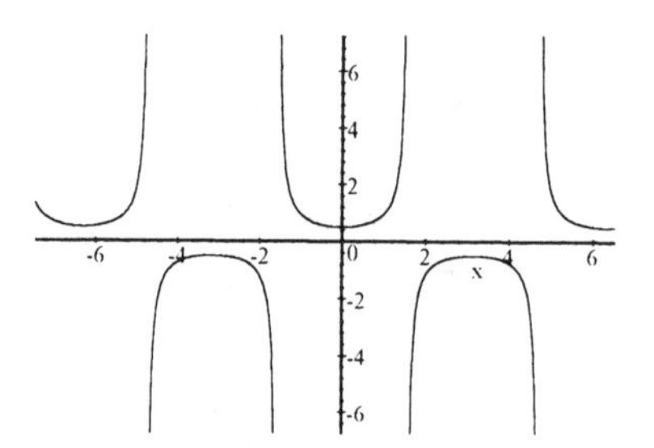

41. $\lim_{x\to\pi} \frac{\tan^2 x}{1+\sec x} = -2$

x	y
3	-2.01
3.1	-2.001
3.14	-2
3.2	-2.002

43. $\lim_{x\to 1} \frac{\ln x}{x-1} = 1$

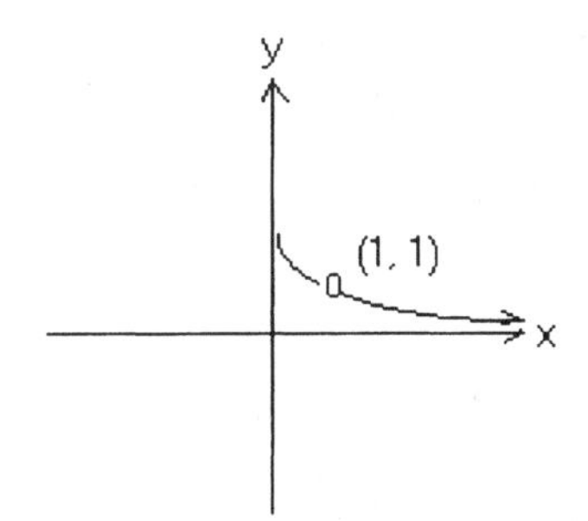

45. $\lim_{x\to 0} \frac{e^{-x}-1}{x} = -1$

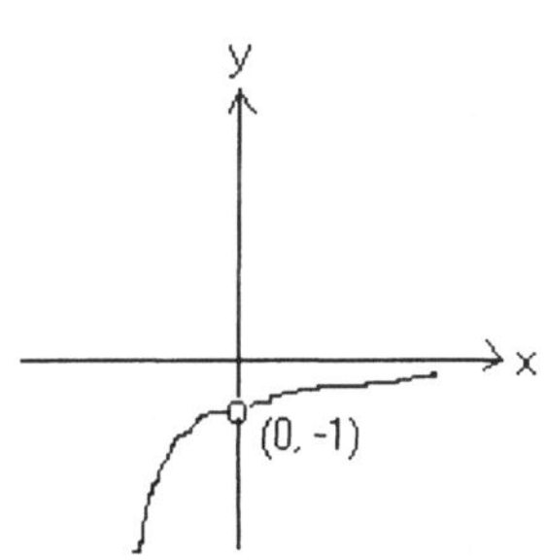

47. $\lim_{x\to 0} \cos\frac{1}{x}$ does not exist

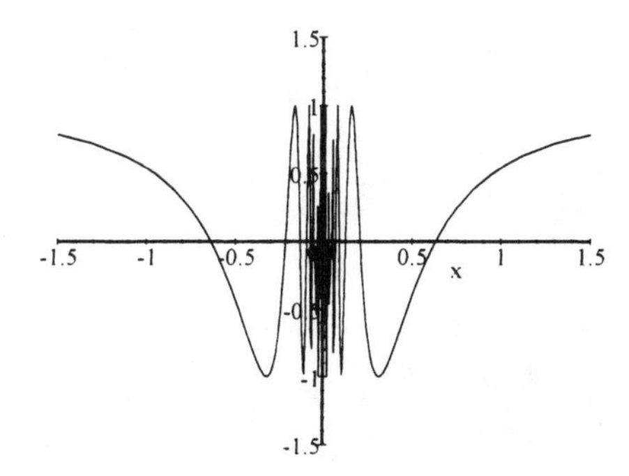

Section 12.2

For Exercises 1 - 10, $\lim_{x\to 4} f(x) = 16,$

$\lim_{x\to 4} g(x) = 8$

1. $\lim_{x\to 4} [f(x) - g(x)] = \lim_{x\to 4} f(x) - \lim_{x\to 4} g(x)$

$= 16 - 8 = \boxed{8}$

[Rule 3]

3. $\lim_{x\to 4} \left[\frac{f(x)}{g(x)}\right] = \frac{\lim_{x\to 4} f(x)}{\lim_{x\to 4} g(x)}$

$= \frac{16}{8} = \boxed{2}$

[Rule 5]

5. $\lim_{x\to 4} \sqrt{f(x)} = \lim_{x\to 4} f(x)^{1/2}$

$= \left[\lim_{x\to 4} f(x)\right]^{1/2} = 16^{1/2} = \boxed{4}$

[Rule 9]

7. $\lim_{x\to 4} 2^{g(x)} = 2^{\lim_{x\to 4} g(x)} = 2^8 = \boxed{256}$

[Rule 10]

9. $\lim_{x\to 4} \frac{f(x)+g(x)}{2g(x)} = \frac{\lim_{x\to 4} f(x) + \lim_{x\to 4} g(x)}{2\lim_{x\to 4} g(x)}$

$= \frac{16+8}{2(8)} = \frac{24}{16} = \boxed{\frac{3}{2}}$

[Rules 3, 5]

11. $\lim_{x\to -3} 7 = \boxed{7}$ [Rule 1]

13. $\lim_{x\to \pi} x = \boxed{\pi}$ [Rule 2]

15. $\lim_{x\to 1} (5x^8 - 3x^2 + 2) = 5(1)^8 - 3(1)^2 + 2$

$= 5 - 3 + 2 = \boxed{4}$

[Rule 6]

17. $$\lim_{x\to 3}\frac{x^3-1}{x^2+1}=\frac{3^3-1}{3^2+1}=\frac{27-1}{9+1}=\frac{26}{10}=\boxed{\frac{13}{5}}$$

[Rule 7]

19. $$\lim_{x\to 0}\frac{x^2+2x}{x}=\lim_{x\to 0}\frac{x(x+2)}{x}=\lim_{x\to 0}(x+2)=0+2=\boxed{2}$$

[Rules 6, 8]

21. $$\lim_{x\to -2}\frac{x^2-4}{x+2}=\lim_{x\to -2}\frac{(x-2)(x+2)}{x+2}=\lim_{x\to -2}(x-2)=-2-2=\boxed{-4}$$

[Rules 6, 8]

23. $$\lim_{x\to 5}\frac{x^2-3x-10}{x-5}=\lim_{x\to 5}\frac{(x-5)(x+2)}{x-5}=\lim_{x\to 5}(x+2)=5+2=\boxed{7}$$

[Rules 6, 8]

25. $$\lim_{x\to 5}\frac{x^2-7x+10}{x^2-25}=\lim_{x\to 5}\frac{(x-5)(x-2)}{(x-5)(x+5)}=\lim_{x\to 5}\frac{(x-2)}{x+5}=\frac{5-2}{5+5}=\boxed{\frac{3}{10}}$$

[Rules 5, 8]

27. $$\lim_{x\to -4}(1-6x)^{3/2}=\left[\lim_{x\to -4}(1-6x)\right]^{3/2}=(1-6(-4))^{3/2}=25^{3/2}=\boxed{125}$$

[Rules 3, 9]

29. $$\lim_{x\to 3}5^{\sqrt{x+1}}=5^{\lim_{x\to 3}\sqrt{x+1}}=5^{\sqrt{4}}=5^2=\boxed{25}$$

[Rule 10]

31. $$\lim_{x\to 4}\left[\log_2(14+\sqrt{x})\right]=\log_2\left[\lim_{x\to 4}(14+\sqrt{x})\right]=\log_2(14+\sqrt{4})=\log_2 16=\boxed{4}$$

[Rule 11]

33. $$\lim_{x\to 0}\left[2^{3x}-\ln(x+1)\right]=\lim_{x\to 0}2^{3x}-\lim_{x\to 0}[\ln(x+1)]=2^{\lim_{x\to 0}3x}-\ln[\lim_{x\to 0}(x+1)]=2^0-\ln 1=1-0=\boxed{1}$$

[Rules 3, 10, 11]

35. $$\lim_{x\to 0}\frac{\sin x-3x}{x}=\lim_{x\to 0}\left[\frac{\sin x}{x}-\frac{3x}{x}\right]=\lim_{x\to 0}\frac{\sin x}{x}-\lim_{x\to 0}3=1-3=\boxed{-2}$$

37. $$\lim_{x\to 0}(x\cot x)=\lim_{x\to 0}\left[x\cdot\frac{\cos x}{\sin x}\right]=\lim_{x\to 0}\frac{x}{\sin x}\cdot\lim_{x\to 0}\frac{\cos x}{1}=\frac{1}{\lim_{x\to 0}\frac{\sin x}{x}}\cdot\lim_{x\to 0}\cos x=\frac{1}{1}\cdot 1=\boxed{1}$$

39. $$\lim_{x\to 0}\frac{\cos x-1}{3x}=\lim_{x\to 0}\frac{\cos x}{3x}-\lim_{x\to 0}\frac{1}{3x}=\lim_{x\to 0}\frac{1}{3x}-\lim_{x\to 0}\frac{1}{3x}=\boxed{0}$$

Relating Concepts (Exercises 41 - 46)

41. $$\lim_{x\to 0}\sin x=\lim_{x\to 0}AB=0$$

$$\lim_{x\to 0}\cos x=\lim_{x\to 0}OB=1$$

42. $\triangle OAB$: Area $= \frac{1}{2}(OB)(AB)$
$= \frac{1}{2}\cos x \sin x = \frac{1}{2}\sin x \cos x$

43. $\triangle OCD$: Area $= \frac{1}{2}(OD)(CD)$
$= \frac{1}{2}(1)\tan x = \frac{1}{2}\tan x$

44. Area of sector $OAD = \frac{1}{2}r^2\theta$
$= \frac{1}{2}(1)^2 x = \frac{x}{2}$

$\text{Area}_{OAB} < \text{Area of sector } OAD < \text{Area}_{OCD}$

$$\frac{1}{2}\sin x \cos x < \frac{x}{2} < \frac{1}{2}\tan x$$

$$\sin x \cos x < x < \frac{\sin x}{\cos x}$$

$$\cos x < \frac{x}{\sin x} < \frac{1}{\cos x}$$

$$\cos x < \frac{\sin x}{x} < \frac{1}{\cos x}$$

Since $0 < \sin x < 1$, and $0 < x < \frac{\pi}{2}$

$$\frac{\sin x}{x} < \frac{x}{\sin x}$$

45. As $x \to 0$, $\frac{\sin x}{x} \to 1$

x	1	.5	.1	.01
$\frac{\sin x}{x}$	.8415	.9589	.9983	.99998

46.
$$\lim_{x\to 0}\frac{1-\cos x}{x} = \lim_{x\to 0}\frac{1-\cos x}{x}\cdot\frac{1+\cos x}{1+\cos x}$$
$$= \lim_{x\to 0}\frac{1-\cos^2 x}{x(1+\cos x)} = \lim_{x\to 0}\frac{\sin^2 x}{x(1+\cos x)}$$
$$= \lim_{x\to 0}\left[\frac{\sin x}{x}\cdot\frac{\sin x}{1+\cos x}\right]$$
$$= \lim_{x\to 0}\frac{\sin x}{x}\cdot\lim_{x\to 0}\frac{\sin x}{1+\cos x}$$
$$= 1\cdot\frac{0}{1+1} = 1\cdot 0 = 0$$

Section 12.3

1. $f(x) = \begin{cases} x & \text{if } x < 2 \\ 3 & \text{if } x = 2 \\ 4 & \text{if } x > 2 \end{cases}$

(a) $\lim_{x\to 2^+} f(x) = 4$

(b) $\lim_{x\to 2^-} f(x) = 2$

3. $f(x) = \frac{x}{5(3-x)^3}$

(a) $\lim_{x\to 3^+} f(x) = -\infty$

(b) $\lim_{x\to 3^-} f(x) = \infty$

5. $f(x) = \frac{x}{(x+1)^2}$

(a) $\lim_{x\to -1^+} f(x) = -\infty$

(b) $\lim_{x\to -1^-} f(x) = -\infty$

7. $\lim_{x\to\infty}\frac{6x^2+1}{2x^2+3} = 3$

9. $\lim_{x\to\infty}(x\sin x)$ does not exist

11. $\lim_{x\to\infty}\left(x\sin\frac{5}{x}\right) = 5$

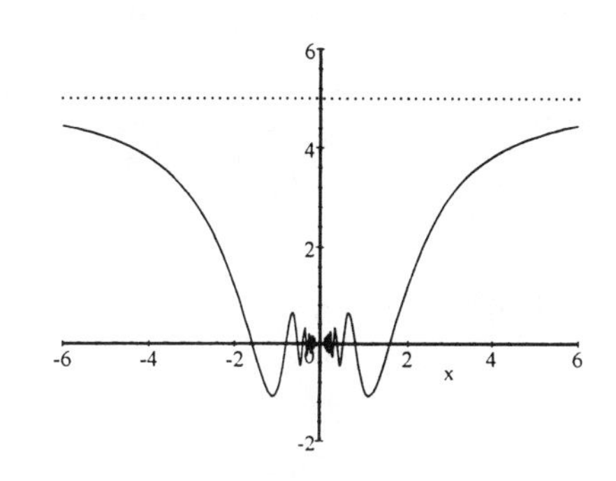

13. $\lim\limits_{x\to\infty}\left(\sqrt{x^2+x}-x\right)=\dfrac{1}{2}$ or .5

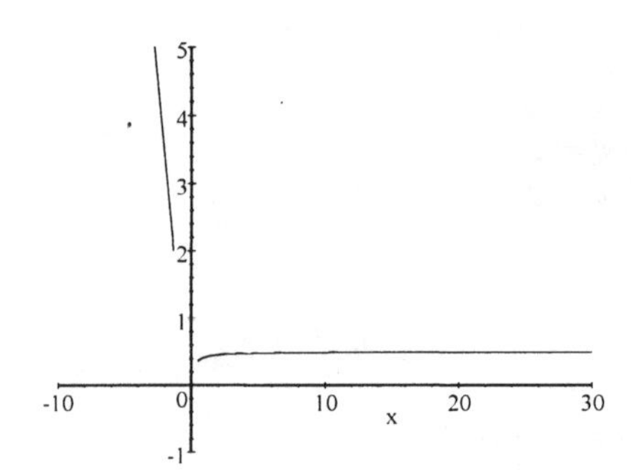

15. $f(x)=\dfrac{e^x}{e^x-1}$

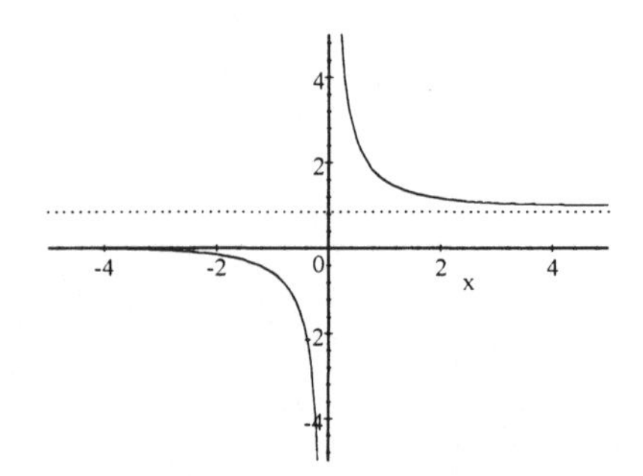

Asymptotes: $y=0,\ y=1,\ x=0$

17. $f(x)=\dfrac{x-\cos x}{x+\sin x}$

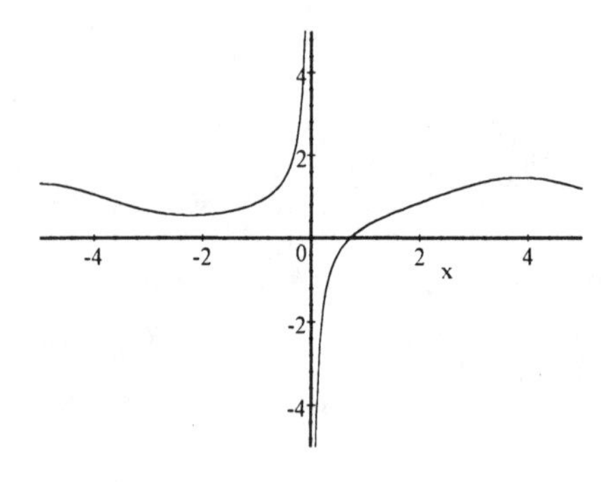

Asymptotes: $y=1,\ x=0$

19. $f(x)=\begin{cases}2x+3 & \text{if } x<1\\ 4 & \text{if } x=1\\ x^2 & \text{if } x>1\end{cases}$

(a) $\lim\limits_{x\to1^+}f(x)=\lim\limits_{x\to1^+}x^2=1$

(b) $\lim\limits_{x\to1^-}f(x)=\lim\limits_{x\to1^-}(2x+3)=5$

21. $f(x)=\dfrac{1}{(1+x)^3}$

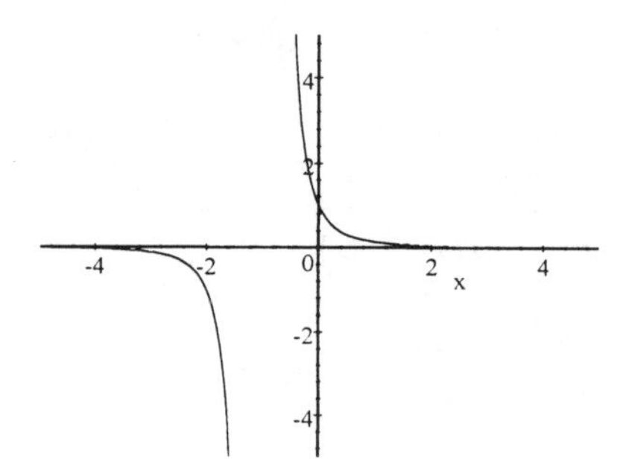

(a) $\lim\limits_{x\to-1^+}f(x)=\infty$

(b) $\lim\limits_{x\to-1^-}f(x)=-\infty$

23. $f(x)=\dfrac{1}{(x-3)^2}$

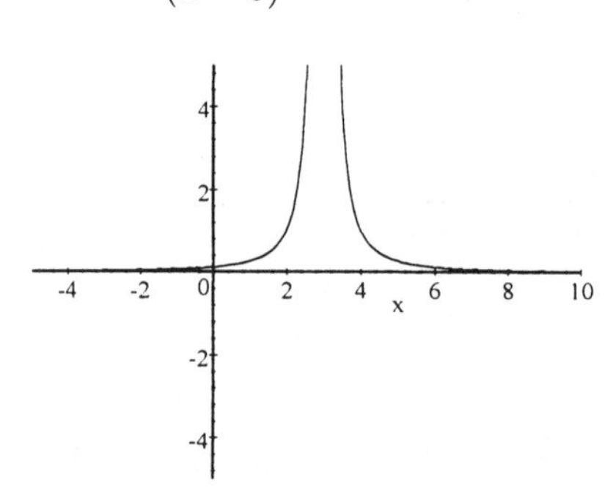

(a) $\lim\limits_{x\to3^+}f(x)=\infty$

(b) $\lim\limits_{x\to3^-}f(x)=\infty$

25. $\lim\limits_{x\to\infty}\dfrac{3x}{5x-1}=\dfrac{3}{5}$

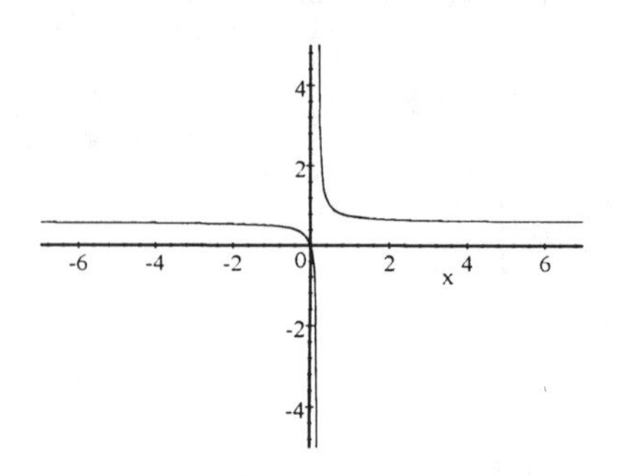

27. $\lim_{x\to-\infty} \dfrac{2x+3}{4x-7} = \dfrac{1}{2}$

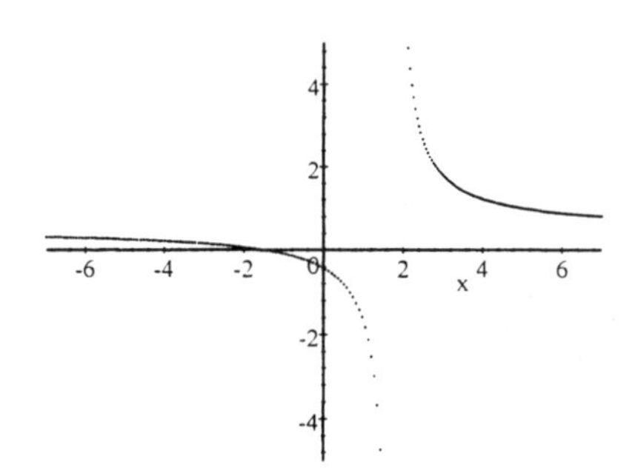

29. $\lim_{x\to\infty} \dfrac{x^2+2x}{2x^3-2x+1} = \lim_{x\to\infty} \dfrac{\dfrac{x^2}{x^3}+\dfrac{2x}{x^3}}{\dfrac{2x^3}{x^3}-\dfrac{2x}{x^3}+\dfrac{1}{x^3}}$

$= \lim_{x\to\infty} \dfrac{\dfrac{1}{x}+\dfrac{2}{x^2}}{2-\dfrac{2}{x^2}+\dfrac{1}{x^3}} = 0$

31. $\lim_{x\to\infty} \dfrac{3x^3+2x-1}{2x^4-3x^3-2} = \lim_{x\to\infty} \dfrac{\dfrac{3x^3}{x^4}+\dfrac{2x}{x^4}-\dfrac{1}{x^4}}{\dfrac{2x^4}{x^4}-\dfrac{3x^3}{x^4}-\dfrac{2}{x^4}}$

$= \lim_{x\to\infty} \dfrac{\dfrac{3}{x}+\dfrac{2}{x^3}-\dfrac{1}{x^4}}{2-\dfrac{3}{x}-\dfrac{2}{x^4}} = 0$

33. $\lim_{x\to\infty} \dfrac{2x^3-x-3}{6x^2-x-1} = \infty$

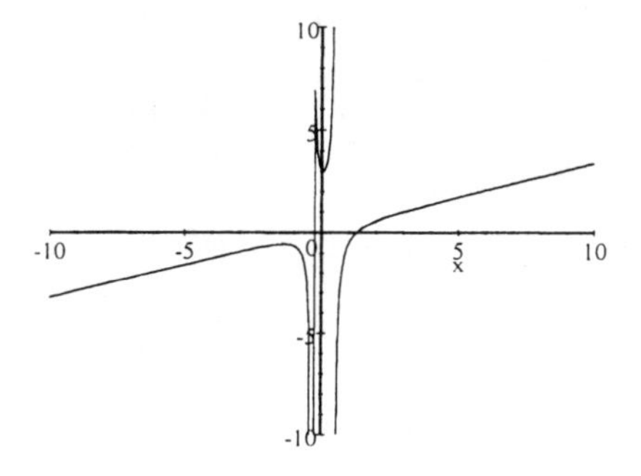

Note: In Exercises 35 - 41, other answers are possible.

35. If $\lim_{x\to5^+} f(x) = \infty$, and $\lim_{x\to5^-} f(x) = -\infty$, $x-5$ must be a factor in the denominator; for example, $f(x) = \dfrac{1}{x-5}$.

37. If $f(x)$ is a polynomial, $\lim_{x\to\infty} f(x) = -\infty$. This will be true when $y \to -\infty$ as $x \to \infty$; for example, $f(x) = -x^3$.

39. $\lim_{x\to\infty} f(x) = \infty$, then $\lim_{x\to\infty} g(x) = \infty$, and $\lim_{x\to\infty} [f(x) - g(x)] = \infty$. Two lines with positive slopes would satisfy these conditions. For example, $f(x) = 2x$, $g(x) = x$.

41. $\lim_{x\to\infty} f(x) = 0$, then $\lim_{x\to\infty} g(x) = \infty$, and $\lim_{x\to\infty} [f(x) \cdot g(x)] = \infty$.
$f(x) = \dfrac{1}{x}$, $g(x) = x^2$ would satisfy these conditions.

43. $y = f(x)$; asymptote: $y = \dfrac{1}{2}x + 3$.
The $\lim_{x\to\infty} f(x)$ will approach the asymptote, whose limit is ∞; so $\lim_{x\to\infty} f(x) = \infty$.

45. **(a)** $y = xe^{-x}$

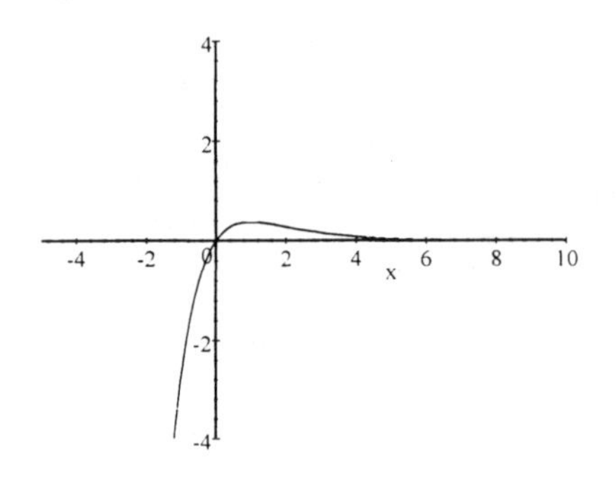

$\lim_{x\to\infty} (xe^{-x}) = 0$

$f(100) \approx 3.72 \times 10^{-42}$

$f(1000) \approx 0$

(b) $y = x^2e^{-x}$

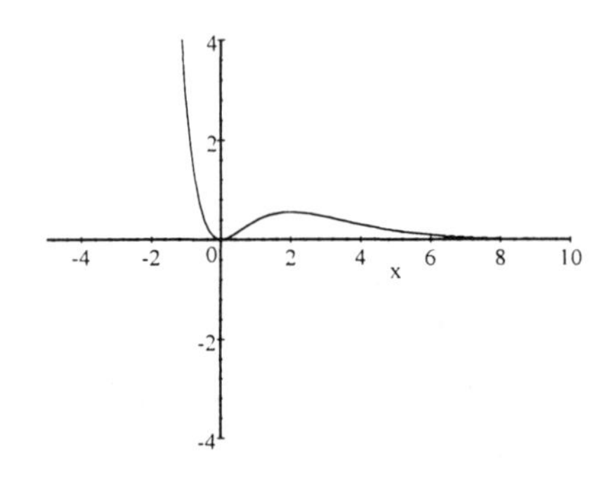

$\lim_{x\to\infty} (x^2e^{-x}) = 0$

$f(100) \approx 3.72 \times 10^{-40}$

$f(1000) \approx 0$

(c) $\lim_{x\to\infty} (x^ne^{-x}) = 0$

47. $\lim_{t\to 1^-} f(t) \approx .03; \quad \lim_{t\to 1^+} f(t) \approx .08$

$\lim_{t\to 1^-} f(t)$ describes the amount of drug in the body just before the second injection is given, while $\lim_{t\to 1^+} f(t)$ describes the amount of drug in the body following that injection.

49. $p(t) = 12 - 4e^{-.5t}$

$$\lim_{t\to\infty} \left[12 - 4e^{-.5t}\right]$$
$$= \lim_{t\to\infty} 12 - 4\lim_{t\to\infty} \left(e^{-.5t}\right)$$
$$= 12 - 4(0) = 12$$

In the long run the price will level off near \$12.

Relating Concepts (Exercises 51 - 54)

51.
$$\begin{aligned} f(x) &= a_nx^n + a_{n-1}x^{n-1} + \ldots + a_1x + a_0 \\ &= a_nx^n\left(1 + \frac{a_{n-1}}{a_n}x^{-1} + \ldots + \frac{a_1}{a_n}x^{1-n} + \frac{a_0}{a_n}\right) \\ &= a_nx^n\left(1 + \frac{a_{n-1}}{a_nx} + \ldots + \frac{a_1}{a_nx^{n-1}} + \frac{a_0}{a_n}\right) \end{aligned}$$

52. $\lim_{x\to\infty} f(x)$

$$= \lim_{x\to\infty} \left[a_nx^n\left(1 + \frac{a_{n-1}}{a_nx} + \ldots + \frac{a_1}{a_nx^{n-1}} + \frac{a_0}{a_n}\right)\right]$$
$$= \lim_{x\to\infty} (a_nx^n) \cdot \lim_{x\to\infty}\left(1 + \frac{a_{n-1}}{a_nx} + \ldots + \frac{a_1}{a_nx^{n-1}} + \frac{a_0}{a_n}\right)$$
$$= a_n \lim_{x\to\infty} x^n \cdot (1)$$
$$= a_n \lim_{x\to\infty} x^n$$

In the same manner, $\lim_{x\to -\infty} f(x) = a_n \lim_{x\to -\infty} x^n$.

53. **(a)** $a_n > 0$, n is even (for example, $f(x) = 3x^4$) :

$\lim_{x\to\infty} f(x) = \infty, \quad \lim_{x\to -\infty} f(x) = \infty$

(b) $a_n < 0$, n is even (for example, $f(x) = -3x^4$) :

$\lim_{x\to\infty} f(x) = -\infty, \quad \lim_{x\to -\infty} f(x) = -\infty$

(c) $a_n > 0$, n is odd (for example, $f(x) = 3x^5$) :

$\lim_{x\to\infty} f(x) = \infty, \quad \lim_{x\to -\infty} f(x) = -\infty$

(d) $a_n < 0$, n is odd (for example, $f(x) = -3x^5$) :

$\lim_{x\to\infty} f(x) = -\infty, \quad \lim_{x\to -\infty} f(x) = \infty$

54. **(a)** ↖ ↗ **(b)** ↙ ↘
(c) ↙ ↗ **(d)** ↖ ↘

Reviewing Basic Concepts (Sections 12.1 - 12.3)

1. **(a)** $\lim_{x\to 3} f(x) = 3$ **(b)** $\lim_{x\to 2} F(x) = 4$
(c) $\lim_{x\to 0} f(x) = 0$ **(d)** $\lim_{x\to 3} g(x) = 2$

2. **(a)** $\lim_{x \to -2^-} f(x) = -1, \quad \lim_{x \to -2^+} f(x) = -\frac{1}{2}$

$\lim_{x \to -2} f(x)$ does not exist

(b) $\lim_{x \to -1^-} f(x) = -\frac{1}{2}, \quad \lim_{x \to -1^+} f(x) = -\frac{1}{2}$

$\lim_{x \to -1} f(x) = -\frac{1}{2}$

3. **(a)** $\lim_{x \to 1^-} f(x) = 1, \quad \lim_{x \to 1^+} f(x) = 1$

$\lim_{x \to 1} f(x) = 1$

(b) $\lim_{x \to 2^-} f(x) = 0, \quad \lim_{x \to 2^+} f(x) = 0$

$\lim_{x \to 2} f(x) = 0$

4. **(a)** $\lim_{x \to \infty} f(x) = 3$

(b) $\lim_{x \to -\infty} g(x) = \infty$

5. $\lim_{x \to 4} \frac{\sqrt{x} - 2}{x - 4} = .25 = \frac{1}{4}$

x	$\frac{\sqrt{x}-2}{x-4}$
4.01	.24984
4.001	.24998
3.99	.25016
3.999	.25002

6. $\lim_{x \to 8} f(x) = 32; \quad \lim_{x \to 8} g(x) = 4$

(a) $\lim_{x \to 8} [f(x) - g(x)] = 32 - 4 = 28$

(b) $\lim_{x \to 8} [g(x) \cdot f(x)] = 4 \cdot 32 = 128$

(c) $\lim_{x \to 8} \frac{f(x)}{g(x)} = \frac{32}{4} = 8$

(d) $\lim_{x \to 8} \log_2 f(x) = \log_2 32 = 5$

(e) $\lim_{x \to 8} \sqrt{f(x)} = \left(\lim_{x \to 8} f(x) \right)^{1/2} = 32^{1/2} = 4\sqrt{2}$

(f) $\lim_{x \to 8} \sqrt[3]{g(x)} = \left(\lim_{x \to 8} g(x) \right)^{1/3} = 4^{1/3} = \sqrt[3]{4}$

(g) $\lim_{x \to 8} 2^{g(x)} = 2^{\lim_{x \to 8} g(x)} = 2^4 = 16$

continued

6. continued

(h) $\lim_{x \to 8} [1 + f(x)]^2 = \left(\lim_{x \to 8} [1 + f(x)] \right)^2$

$= \left(\lim_{x \to 8} 1 + \lim_{x \to 8} f(x) \right)^2$

$= (1 + 32)^2 = 1089$

(i) $\lim_{x \to 8} \frac{f(x) - g(x)}{4g(x)} = \frac{32 - 4}{4(4)} = \frac{28}{16} = \frac{7}{4}$

(j) $\lim_{x \to 8} \frac{2g(x) + 3}{1 + f(x)} = \frac{2(4) + 3}{1 + 32} = \frac{11}{33} = \frac{1}{3}$

7. $\lim_{x \to 3} \frac{x^2 - 9}{x - 3} = \lim_{x \to 3} \frac{(x - 3)(x + 3)}{x - 3}$

$= \lim_{x \to 3} (x + 3) = 6$

8. $\lim_{x \to \infty} \frac{2x^2 - 1}{3x^4 + 5} = 0$

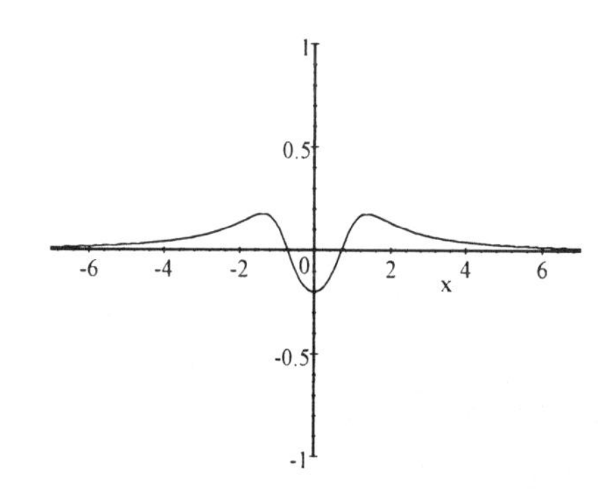

9. $\lim_{x \to \infty} \frac{2x^3 - x + 3}{6x^3 + 4x - 9} = \lim_{x \to -\infty} \frac{\frac{2x^3}{x^3} - \frac{x}{x^3} + \frac{3}{x^3}}{\frac{6x^3}{x^3} + \frac{4x}{x^3} - \frac{9}{x^3}}$

$= \lim_{x \to -\infty} \frac{2 - \frac{1}{x^2} + \frac{3}{x^3}}{6 + \frac{4}{x^2} - \frac{9}{x^3}}$

$= \frac{2}{6} = \frac{1}{3}$

10. $f(x) = \begin{cases} x + 2 & \text{if } x < 4 \\ x^2 - 6 & \text{if } x \geq 4 \end{cases}$

$\lim_{x \to 4^+} f(x) = 10; \quad \lim_{x \to 4^-} f(x) = 6$

$\lim_{x \to 4} f(x)$ does not exist

Section 12.4

1. $(6,\ 5),\ (5,\ 3)$

$$m = \frac{5-3}{6-5} = \boxed{2}$$

3. $(-2,\ 2),\ (3,\ 3)$

$$m = \frac{2-3}{-2-3} = \frac{-1}{-5} = \boxed{\frac{1}{5}}$$

5. $f(x) = x^2$

$$\begin{aligned} m &= \lim_{x\to 4} \frac{x^2-4^2}{x-4} \\ &= \lim_{x\to 4} \frac{(x-4)(x+4)}{x-4} \\ &= \lim_{x\to 4} (x+4) = 4+4 = \boxed{8} \end{aligned}$$

7. $f(x) = -4x^2 + 11x$

$$\begin{aligned} m &= \lim_{x\to -2} \frac{(-4x^2+11x) - \left(-4(-2)^2 + 11(-2)\right)}{x+2} \\ &= \lim_{x\to -2} \frac{-4x^2+11x+38}{x+2} \\ &= \lim_{x\to -2} \frac{(x+2)(-4x+19)}{x+2} \\ &= \lim_{x\to -2} (-4x+19) = -4(-2)+19 = \boxed{27} \end{aligned}$$

9. $f(x) = -\frac{2}{x}$

$$\begin{aligned} m &= \lim_{x\to 4} \frac{\left(-\frac{2}{x}\right) - \left(-\frac{2}{4}\right)}{x-4} \\ &= \lim_{x\to 4} \frac{-\frac{2}{x}+\frac{1}{2}}{x-4} \cdot \frac{2x}{2x} \\ &= \lim_{x\to 4} \frac{-4+x}{2x(x-4)} \\ &= \lim_{x\to 4} \frac{1}{2x} = \boxed{\frac{1}{8}} \end{aligned}$$

11. $f(x) = -3\sqrt{x}$

$$\begin{aligned} m &= \lim_{x\to 1} \frac{-3\sqrt{x} - (-3\sqrt{1})}{x-1} \\ &= \lim_{x\to 1} \frac{-3\sqrt{x}+3}{x-1} \\ &= \lim_{x\to 1} \frac{-3(\sqrt{x}-1)}{(\sqrt{x}+1)(\sqrt{x}-1)} \\ &= \lim_{x\to 1} \frac{-3}{\sqrt{x}+1} = \boxed{-\frac{3}{2}} \end{aligned}$$

13. $f(x) = x^2 + 2x$ at $x = 3$

$$\begin{aligned} m &= \lim_{x\to 3} \frac{(x^2+2x) - (3^2+2\cdot 3)}{x-3} \\ &= \lim_{x\to 3} \frac{x^2+2x-15}{x-3} \\ &= \lim_{x\to 3} \frac{(x+5)(x-3)}{x-3} \\ &= \lim_{x\to 3} (x+5) = 8 \end{aligned}$$

$f(3) = 9+6 = 15 \longrightarrow (3,\ 15)$

$$\begin{aligned} y - y_1 &= m(x-x_1) \\ y - 15 &= 8(x-3) \\ y - 15 &= 8x - 24 \qquad \boxed{y = 8x-9} \end{aligned}$$

15. $f(x) = \frac{5}{x}$ at $x = 2$

$$\begin{aligned} m &= \lim_{x\to 2} \frac{\left(\frac{5}{x}\right) - \left(\frac{5}{2}\right)}{x-2} \cdot \frac{2x}{2x} \\ &= \lim_{x\to 2} \frac{10-5x}{2x(x-2)} \\ &= \lim_{x\to 2} \frac{-5(-2+x)}{2x(x-2)} \\ &= \lim_{x\to 2} \frac{-5}{2x} = \boxed{-\frac{5}{4}} \end{aligned}$$

$f(2) = \frac{5}{2} \longrightarrow \left(2,\ \frac{5}{2}\right)$

$$\begin{aligned} y - y_1 &= m(x-x_1) \\ y - \frac{5}{2} &= -\frac{5}{4}(x-2) \\ y - \frac{5}{2} &= -\frac{5}{4}x + \frac{5}{2} \qquad \boxed{y = -\frac{5}{4}x + 5} \end{aligned}$$

17. $f(x) = 4\sqrt{x}$ at $x = 9$

$$\begin{aligned} m &= \lim_{x\to 9} \frac{4\sqrt{x} - 4\sqrt{9}}{x-9} \\ &= \lim_{x\to 9} \frac{4\sqrt{x} - 12}{x-9} \\ &= \lim_{x\to 9} \frac{4(\sqrt{x}-3)}{(\sqrt{x}+3)(\sqrt{x}-3)} \\ &= \lim_{x\to 9} \frac{4}{(\sqrt{x}+3)} = \frac{2}{3} \end{aligned}$$

$f(9) = 4\sqrt{9} = 12 \longrightarrow (9,\ 12)$

$$\begin{aligned} y - y_1 &= m(x - x_1) \\ y - 12 &= \frac{2}{3}(x-9) \\ y - 12 &= \frac{2}{3}x - 6 \end{aligned}$$

$$\boxed{y = \frac{2}{3}x + 6}$$

19. $f(x) = 5$, the slope is zero
$f'(x) = 0$, so $f'(2) = 0$

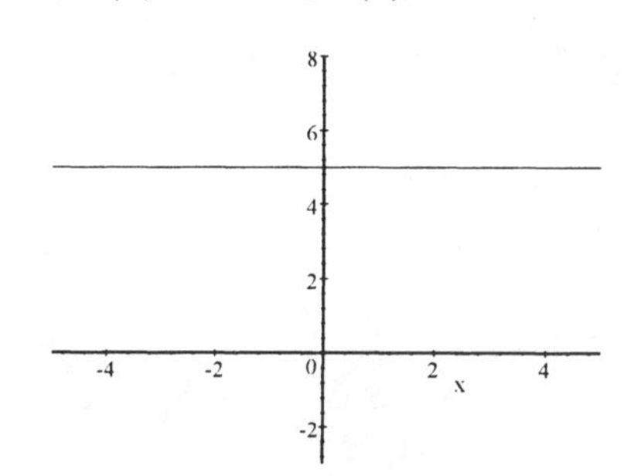

21. $f(x) = -x$, the slope is -1
$f'(x) = -1$, so $f'(2) = -1$

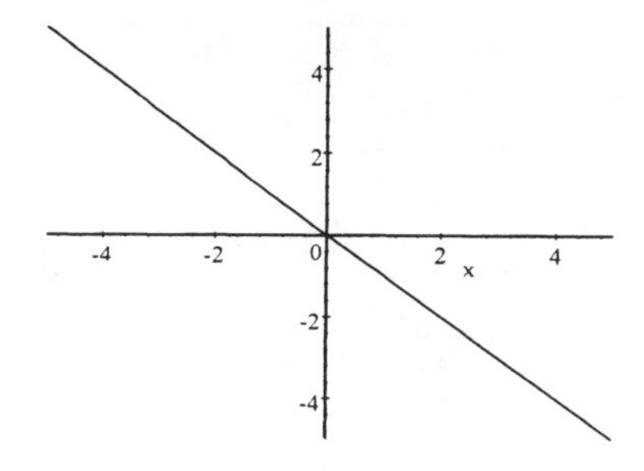

Note: Using the calculator function nDeriv may lead to rounding errors.

23. $f(x) = e^x; \quad a = 0$

Calculator: nDeriv(e^x, x, 0) = $\boxed{1}$

25. $f(x) = \dfrac{10x}{1 + .25x^2}; \quad a = 2$

Calculator: nDeriv$((10x)/(1 + .25x^2),\ x,\ 2) = \boxed{0}$

27. $f(x) = x\cos x; \quad a = \dfrac{\pi}{4}$

Calculator: nDeriv$\left(x\cos x\ ,\ x,\ \dfrac{\pi}{4}\right) \approx \boxed{.1517}$

29. $f'(0) = \dfrac{1}{3}$, the slope at $x = 0$.

31. For $x: 2.03x - .53 = 2.01x - .51$

$.02x = .02 \quad \boxed{x = 1}$

For $y: x = 1 \Longrightarrow 2.01x - .51$

$2.01 - .51 \quad \boxed{y = 1.5}$

$f'(a) = \text{slope} \approx \boxed{2}$

33. $(12,\ 10),\ (6,\ 5): m = \dfrac{10-5}{12-6} = \dfrac{5}{6}$

Interest rates were rising at $\frac{5}{6}\%$ per year.

35. $w(t) = .1t^2$

$$\begin{aligned}
w'(4) = &= \lim_{t\to 4} \frac{w(t) - w(4)}{t-4} \\
&= \lim_{t\to 4} \frac{.1t^2 - 1.6}{t-4} \\
&= \lim_{t\to 4} \frac{.1\left(t^2 - 16\right)}{t-4} \\
&= \lim_{t\to 4} \frac{.1(t-4)(t+4)}{t-4} \\
&= \lim_{t\to 4} [.1(t+4)] = .1(8) = .8
\end{aligned}$$

The tumor is growing at .8 grams per week.

37. $s(t) = t^2 + t$

(a) $20 = t^2 + t$
$0 = t^2 + t - 20$
$0 = (t+5)(t-4)$
$t = -5$ (extraneous)
$t =$ 4 seconds

(b)
$$\begin{aligned}
s'(4) = &= \lim_{t\to 4} \frac{s(t) - s(4)}{t-4} \\
&= \lim_{t\to 4} \frac{t^2 + t - 20}{t-4} \\
&= \lim_{t\to 4} \frac{(t-4)(t+5)}{t-4} \\
&= \lim_{t\to 4} (t+5) = 4 + 5 = 9
\end{aligned}$$

9 feet per second

39. $R(x) = 10x - .002x^2$

$$\begin{aligned}
R'(1000) = &= \lim_{x\to 1000} \frac{R(x) - R(1000)}{x - 1000} \\
&= \lim_{x\to 1000} \frac{10x - .002x^2 - 8000}{x - 1000} \\
&= \lim_{x\to 1000} \frac{-.002x^2 + 10x - 8000}{x - 1000} \\
&= \lim_{x\to 1000} \frac{(x-1000)(-.002x + 8)}{x - 1000} \\
&= \lim_{x\to 1000} (-.002x + 8) = 6 \text{ (thousand)}
\end{aligned}$$

Marginal revenue is \$6000/unit.

41. $f(t) = \dfrac{100,000}{1 + 9.134(.8)^t}$

Calculator:

nDeriv$(100000/(1 + 9.134(.8)\,\hat{}\,x), x, 8)$

$f'(100,000) \approx 5331.9987$

About 5332 people per day.

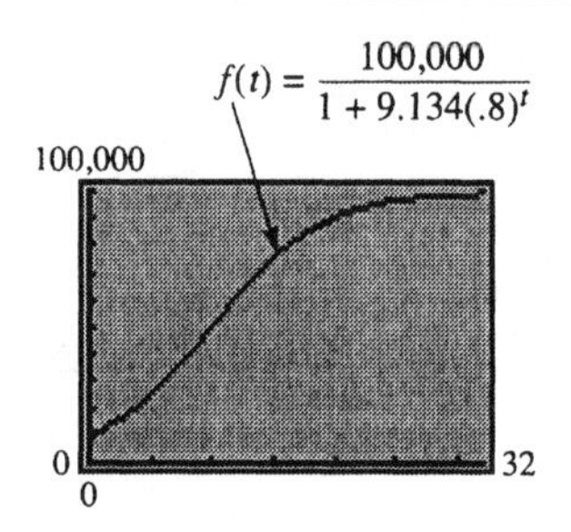

43. The monthly payment on a 30-year mortgage of \$100,000 at 8% interest is

$$f(8) = \frac{100,000\left(\frac{8}{1200}\right)\left(1 + \frac{8}{1200}\right)^{360}}{\left(1 + \frac{8}{1200}\right)^{360} - 1}$$

$f(8) \approx$ \$733.76

45. We want the equation of the line passing through the point (8, 733.76) with slope $m = 69.71$:

$y - 733.76 = 69.71(x - 8)$

$y = 69.71x + 176.08$

Section 12.5

1. $f(x) = 2x + 1$

$x_1 = 0,\ x_2 = 2,\ x_3 = 4,\ x_4 = 6,\ \triangle x = 2$

(a) $$\sum_{i=1}^{4} f(x_i)\triangle x = \sum_{i=1}^{4}(2x_i + 1)\cdot 2$$
$$= 2[(2(0)+1) + (2(2)+1) + (2(4)+1) + (2(6)+1)]$$
$$= 2[1 + 5 + 9 + 13]$$
$$= 2[28] = \boxed{56}$$

(b) Starting at 0, using 4 rectangles, with $\triangle x = 2$, the definite integral would end at 8. So,

$$\int_0^8 (2x+1)\,dx$$

3. $f(x) = 3x + 2$ from $x = 1$ to $x = 5$

$$\triangle x = \frac{5-1}{4} = 1$$

(a) Left endpoints:

$$A = 1[f(1) + f(2) + f(3) + f(4)]$$
$$= 5 + 8 + 11 + 14 = \mathbf{38}$$

(b) Right endpoints:

$$A = 1[f(2) + f(3) + f(4) + f(5)]$$
$$= 8 + 11 + 14 + 17 = \mathbf{50}$$

(c) Average: $A = \dfrac{38 + 50}{2} = \mathbf{44}$

(d) Midpoint:

$$A = 1[f(1.5) + f(2.5) + f(3.5) + f(4.5)]$$
$$= 6.5 + 9.5 + 12.5 + 15.5 = \mathbf{44}$$

5. $f(x) = x + 5$ from $x = 2$ to $x = 4$

$$\triangle x = \frac{4-2}{4} = \frac{1}{2} = .5$$

(a) Left endpoints:

$$A = .5[f(2) + f(2.5) + f(3) + f(3.5)]$$
$$= .5[31] = \mathbf{15.5}$$

(b) Right endpoints:

$$A = .5[f(2.5) + f(3) + f(3.5) + f(4)]$$
$$= .5[33] = \mathbf{16.5}$$

(c) Average: $A = \dfrac{15.5 + 16.5}{2} = \mathbf{16}$

(d) Midpoint:

$$A = .5[f(2.25) + f(2.75) + f(3.25) + f(3.75)]$$
$$= .5[7.25 + 7.75 + 8.25 + 8.75] = \mathbf{16}$$

7. $f(x) = x^2$ from $x = 1$ to $x = 5$

$$\triangle x = \frac{5-1}{4} = 1$$

(a) Left endpoints:

$$A = 1[f(1) + f(2) + f(3) + f(4)]$$
$$= 1[1 + 4 + 9 + 16] = \mathbf{30}$$

(b) Right endpoints:

$$A = 1[f(2) + f(3) + f(4) + f(5)]$$
$$= 1[4 + 9 + 46 + 25] = \mathbf{54}$$

(c) Average: $A = \dfrac{30 + 54}{2} = \mathbf{42}$

(d) Midpoint:

$$A = 1[f(1.5) + f(2.5) + f(3.5) + f(4.5)]$$
$$= 1[2.25 + 6.25 + 12.25 + 20.25] = \mathbf{41}$$

9. $f(x) = e^x - 1$ from $x = 0$ to $x = 4$

$$\triangle x = \frac{4-0}{4} = 1$$

(a) Left endpoints:

$$A = 1[f(0) + f(1) + f(2) + f(3)] \approx \mathbf{27.19}$$

(b) Right endpoints:

$$A = 1[f(1) + f(2) + f(3) + f(4)] \approx \mathbf{80.79}$$

(c) Average: $A = \dfrac{27.19 + 80.79}{2} \approx \mathbf{53.99}$

(d) Midpoint:

$$A = 1[f(.5) + f(1.5) + f(2.5) + f(3.5)]$$
$$\approx \mathbf{47.43}$$

11. $f(x) = \frac{1}{x}$ from $x = 1$ to $x = 5$

$\triangle x = \frac{5-1}{4} = 1$

(a) Left endpoints:

$A = 1[f(1) + f(2) + f(3) + f(4)]$

$\approx \mathbf{\frac{25}{12}} \approx \mathbf{2.08}$

(b) Right endpoints:

$A = 1[f(2) + f(3) + f(4) + f(5)]$

$\approx \mathbf{\frac{77}{60}} \approx \mathbf{1.28}$

(c) Average: $A = \frac{\frac{25}{12} + \frac{77}{60}}{2} = \mathbf{\frac{101}{60}} \approx \mathbf{1.68}$

(d) Midpoint:

$A = 1[f(1.5) + f(2.5) + f(3.5) + f(4.5)]$

$\approx \mathbf{\frac{496}{315}} \approx \mathbf{1.57}$

13. $f(x) = \frac{x}{2}$ from $x = 0$ to $x = 4$

$\triangle x = \frac{4-0}{4} = 1$

(a) $A = 1[f(.5) + f(1.5) + f(2.5) + f(3.5)]$

$= \frac{1}{4} + \frac{3}{4} + \frac{5}{4} + \frac{7}{4} = 4$

(b) $\int_0^4 f(x)\,dx : \; A = \frac{1}{2}bh = \frac{1}{2}(4)(2) = 4$

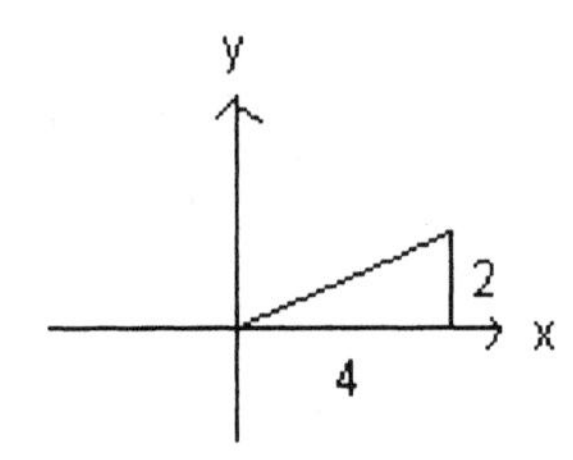

15. $\int_0^4 f(x)\,dx$

(a) $A = \frac{1}{2}bh = \frac{1}{2}(4)(2) = 4$

(b) $A = \frac{1}{2}b_1h_1 + \frac{1}{2}b_2h_2$

$= \frac{1}{2}(3)(3) + \frac{1}{2}(1)(1)$

$= \frac{9}{2} + \frac{1}{2} = 5$

17. $\int_{-4}^0 \sqrt{16 - x^2}\,dx$ (graph includes point $(-4, 0)$)

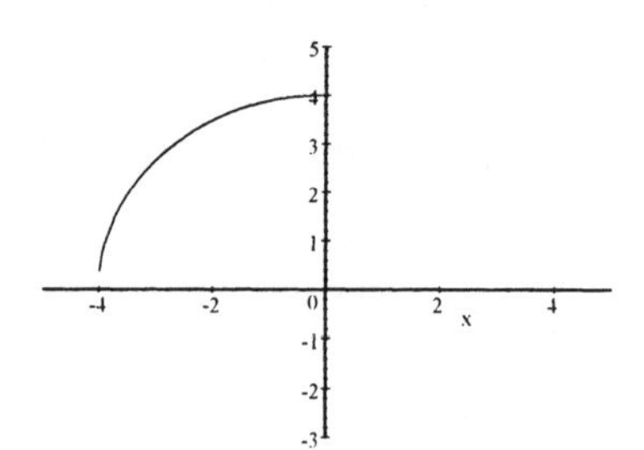

$A = \frac{1}{4}\pi r^2$ where $r = 4$

$= \frac{1}{4}\pi(16) = 4\pi$

19. $\int_2^5 (1 + 2x)\,dx$

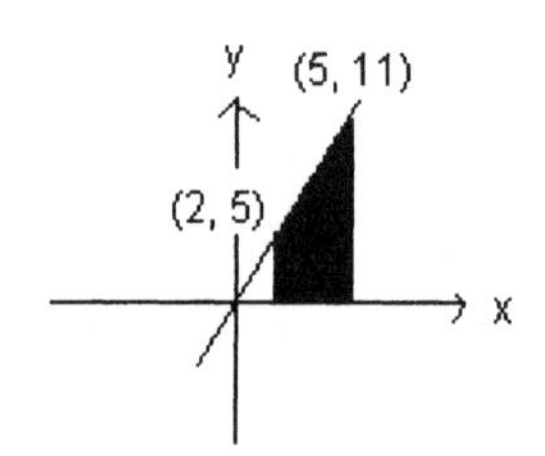

$A = \frac{1}{2}(b_1 + b_2)h = \frac{1}{2}(5 + 11)3 = 24$

Reviewing Basic Concepts (Sections 12.4 and 12.5)

1. $(-3,\ 4)$ to $(3,\ -1)$

$m = \frac{4+1}{-3-3} = -\frac{5}{6}$

2. $(0,\ 2)$ to $(1,\ 2) : m = 0$

3. $(4,\ 0)$ to $(4,\ 2) : m$ does not exist

4. $f(x) = k$ is the graph of a horizontal line, whose slope is 0; therefore a tangent line to $f(x)$ will also have slope 0, so $f'(x) = 0$.

5. $R(x) = -.0012x^2 + 3x$

$R'(1800)$

$$= \lim_{x \to 1800} \frac{(-.0012x^2 + 3x) - (-3888 + 5400)}{x - 1800}$$

$$= \lim_{x \to 1800} \frac{-.0012x^2 + 3x - 1512}{x - 1800}$$

$$= \lim_{x \to 1800} \frac{-.0012(x^2 - 2500x + 1,260,000)}{x - 1800}$$

$$= \lim_{x \to 1800} \frac{-.0012(x - 1800)(x - 700)}{x - 1800}$$

$$= \lim_{x \to 1800} [-.0012(x - 700)] = -1.32$$

The marginal revenue is −$1.32 per unit (decreasing).

6. $f(x) = 6 - x^2$ at $x = -1$

$$f'(-1) = \lim_{x \to -1} \frac{(6 - x^2) - 5}{x + 1}$$

$$= \lim_{x \to -1} \frac{-x^2 + 1}{x + 1}$$

$$= \lim_{x \to -1} \frac{-(x^2 - 1)}{x + 1}$$

$$= \lim_{x \to -1} \frac{-(x - 1)(x + 1)}{x + 1}$$

$$= \lim_{x \to -1} [-1(x - 1)] = 2$$

$f(-1) = 6 - (-1)^2 = 5$

$\Rightarrow (-1,\ 5),\ m = 2:$

$y - 5 = 2(x + 1)$

$y - 5 = 2x + 2$

$\boxed{y = 2x + 7}$

7. $f(x) = \sqrt{x} + 2$

$$f'(9) = \lim_{x \to 9} \frac{(\sqrt{x} + 2) - 5}{x - 9}$$

$$= \lim_{x \to 9} \frac{\sqrt{x} - 3}{(\sqrt{x} - 3)(\sqrt{x} + 3)}$$

$$= \lim_{x \to 9} \frac{1}{\sqrt{x} + 3} = \boxed{\frac{1}{6}}$$

8. $f(x) = x^2$ from $x = 0$ to $x = 5$

$\Delta x = \frac{5 - 0}{5} = 1$

Midpoint:

$$A = 1\left[f\left(\tfrac{1}{2}\right) + f\left(\tfrac{3}{2}\right) + f\left(\tfrac{5}{2}\right) + f\left(\tfrac{7}{2}\right) + f\left(\tfrac{9}{2}\right)\right]$$

$$= \tfrac{1}{4} + \tfrac{9}{4} + \tfrac{25}{4} + \tfrac{49}{4} + \tfrac{81}{4} = \tfrac{165}{4} = \boxed{41.25}$$

9. $y = 8 - x:\ A = \frac{1}{2}bh = \frac{1}{2}(8)(8) = 32$

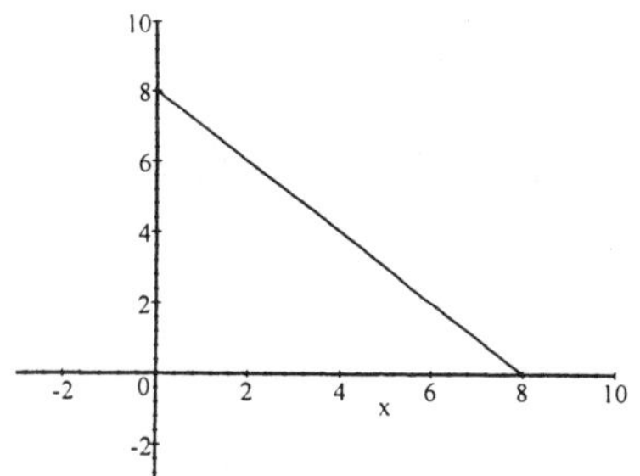

10. $\int_0^3 \sqrt{9 - x^2}\,dx$ (graph includes point $(3, 0)$)

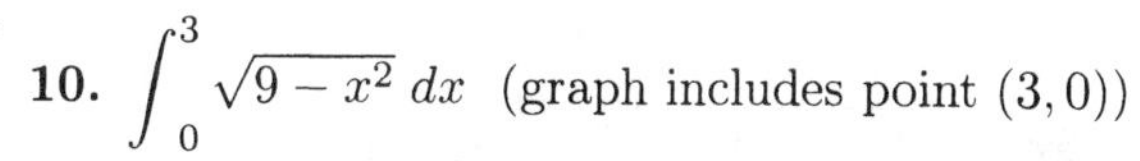

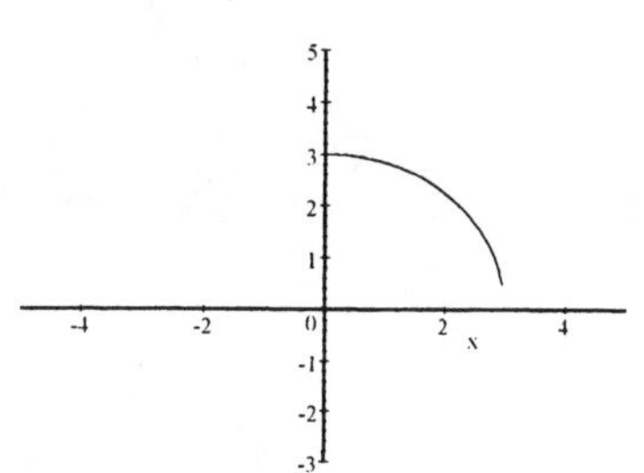

$$A = \frac{1}{4}\pi r^2 \text{ where } r = 3$$

$$= \frac{1}{4}\pi(9) = \frac{9}{4}\pi$$

Chapter 12 Review

1. (a) $\lim_{x\to 1^-} f(x) = 2$

(b) $\lim_{x\to 1^+} f(x) = 2$

(c) $\lim_{x\to 1} f(x) = 2$

2. (a) $\lim_{x\to -1^-} f(x) = -2$

(b) $\lim_{x\to -1^+} f(x) = 2$

(c) $\lim_{x\to -1} f(x)$ does not exist

3. (a) $\lim_{x\to 4^-} f(x) = \infty$

(b) $\lim_{x\to 4^+} f(x) = -\infty$

(c) $\lim_{x\to 4} f(x)$ does not exist

4. $\lim_{x\to\infty} f(x) = -3$

5. $\lim_{x\to 1} (2x^2 - 3x) = 2(1)^2 - 3(1) = -1$

6. $\lim_{x\to -1} (x - x^3) = (-1) - (-1)^3 = 0$

7. $\lim_{x\to 2} \dfrac{3x+4}{x+3} = \dfrac{3(2)+4}{2+3} = \dfrac{10}{5} = 2$

8. $\lim_{x\to 4} \dfrac{x-4}{x^3+5x} = \dfrac{4-4}{4^3+5(4)} = \dfrac{0}{84} = 0$

9. $\lim_{x\to -1} \sqrt{5x+21} = \sqrt{5(-1)+21} = \sqrt{16} = 4$

10. $\lim_{x\to 2} 27^{\frac{3-x}{1+x}} = 27^{\frac{3-2}{1+2}} = 27^{1/3} = 3$

11. $\lim_{x\to 1} [\log_2(5x+3)] = \log_2[5(1)+3]$
$= \log_2 8 = 3$

12. $\lim_{x\to -3} (5-9x)^{2/5} = [5-9(-3)]^{2/5}$
$= 32^{2/5} = 4$

13. $\lim_{x\to 5} \dfrac{2x-10}{5-x} = \lim_{x\to 5} \dfrac{-2(-x+5)}{5-x}$
$= \lim_{x\to 5} (-2) = -2$

14. $\lim_{x\to 6} \dfrac{x^2-36}{x-6} = \lim_{x\to 6} \dfrac{(x-6)(x+6)}{x-6}$
$= \lim_{x\to 6} (x+6) = 12$

15. $\lim_{x\to -3} \dfrac{x^2+2x-3}{x+3} = \lim_{x\to -3} \dfrac{(x-1)(x+3)}{x+3}$
$= \lim_{x\to -3} (x-1) = -4$

16. $\lim_{x\to 1} \dfrac{x^2-2x+1}{x-1} = \lim_{x\to 1} \dfrac{(x-1)(x-1)}{x-1}$
$= \lim_{x\to 1} (x-1) = 0$

17. $\lim_{x\to 2} \dfrac{x^2-x-2}{x^2-5x+6} = \lim_{x\to 2} \dfrac{(x-2)(x+1)}{(x-2)(x-3)}$
$= \lim_{x\to 2} \dfrac{(x+1)}{(x-3)} = \dfrac{3}{-1} = -3$

18. $\lim_{x\to -2} \dfrac{x^3+3x^2+2x}{x^2+2x} = \lim_{x\to -2} \dfrac{x(x+2)(x+1)}{x(x+2)}$
$= \lim_{x\to -2} (x+1)$
$= -2+1 = -1$

19. $\lim_{x\to 1} \dfrac{x^2+x}{x-1}$ does not exist

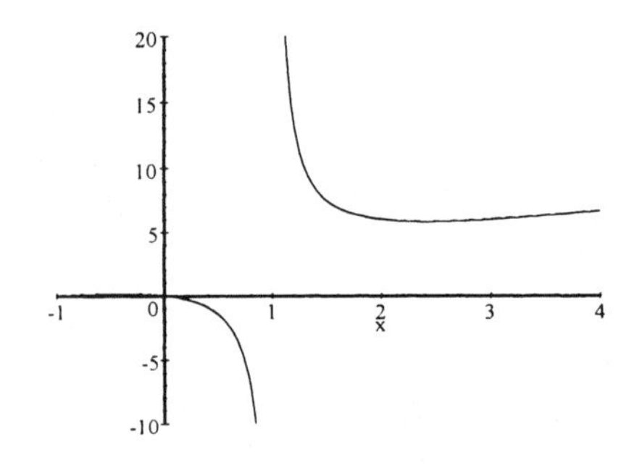

20. $\lim_{x\to 0} \frac{x^2+1}{x}$ does not exist

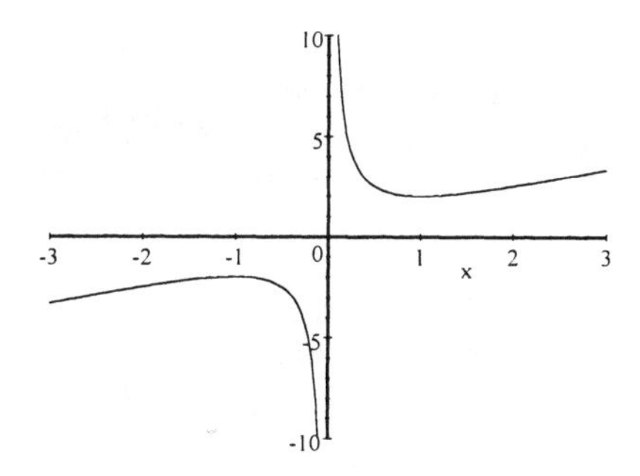

21. $\lim_{x\to 0} \frac{\sin x}{3x} = \frac{\lim_{x\to 0} \frac{\sin x}{x}}{\lim_{x\to 0} 3} = \frac{1}{3}$

22. $\lim_{x\to 0} \frac{\sin \frac{1}{x}}{x}$ does not exist

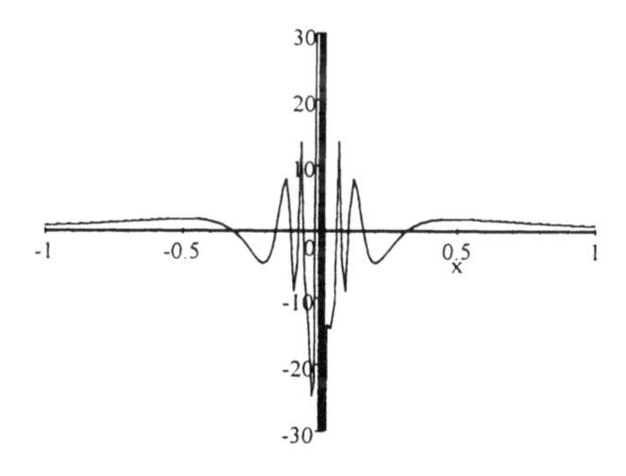

23. $\lim_{x\to 0} \frac{x\cos x - 1}{x^2} = -\infty$

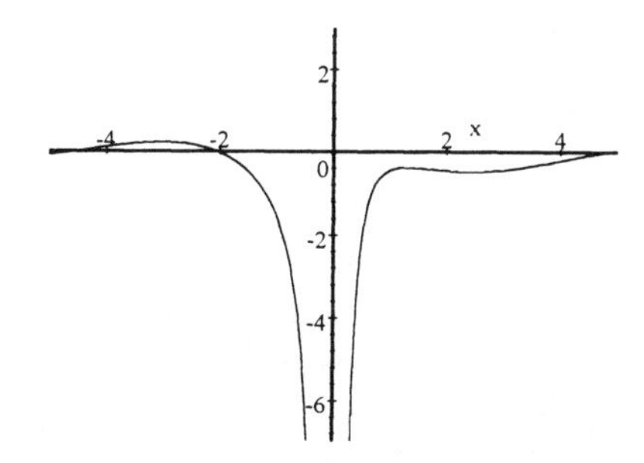

24.
$$\begin{aligned}\lim_{x\to 0} \frac{\tan x}{2x} &= \lim_{x\to 0} \frac{\sin x}{2x\cos x} \\ &= \frac{1}{2}\left[\lim_{x\to 0} \frac{\sin x}{x}\right]\cdot\left[\lim_{x\to 0} \frac{1}{\cos x}\right] \\ &= \frac{1}{2}(1)\left(\frac{1}{\cos 0}\right) \\ &= \frac{1}{2}(1)(1) = \frac{1}{2}\end{aligned}$$

25. $\lim_{x\to 2^-} f(x) = \lim_{x\to 2^-} (3x-1) = 3(2) - 1 = 5$

$\lim_{x\to 2^+} f(x) = \lim_{x\to 2^+} (x+3) = 2+3 = 5$

Therefore, $\lim_{x\to 2} f(x) = 5$.

26. $\lim_{x\to 1^-} f(x) = \lim_{x\to 1^-} x^2 = 1^2 = 1$

$\lim_{x\to 1^+} f(x) = \lim_{x\to 1^+} (3x-2) = 3(1) - 2 = 1$

Therefore, $\lim_{x\to 1} f(x) = 1$.

27. $\lim_{x\to 2^-} f(x) = \lim_{x\to 2^-} \left(x^2-1\right) = (2)^2 - 1 = 3$

28. $\lim_{x\to 0^+} f(x) = \lim_{x\to 0^+} \sqrt{x} = \sqrt{0} = 0$

29. $\lim_{x\to 0^-} f(x) = -\infty$

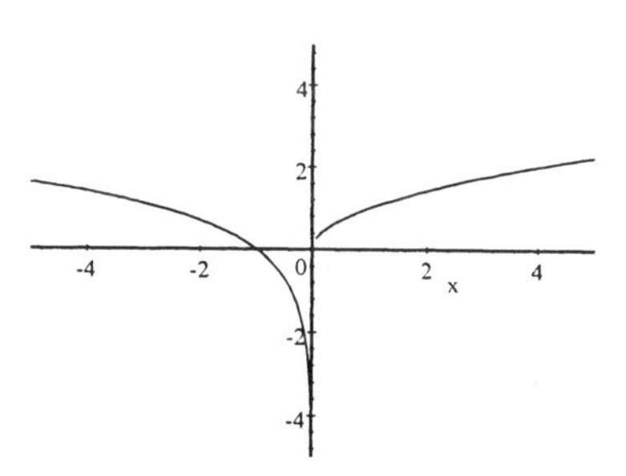

30. $\lim_{x\to 0^-} f(x) = \lim_{x\to 0^-} \frac{\sin x}{x} = 1$

31. $f(x) = \frac{x}{(x-4)^2}$

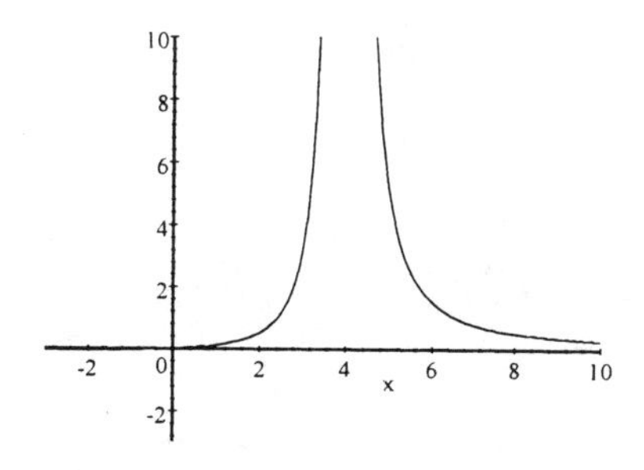

(a) $\lim_{x\to 4^+} \frac{x}{(x-4)^2} = \infty$

(b) $\lim_{x\to 4^-} \frac{x}{(x-4)^2} = \infty$

32. $f(x) = \dfrac{x^2}{(x-1)^3}$

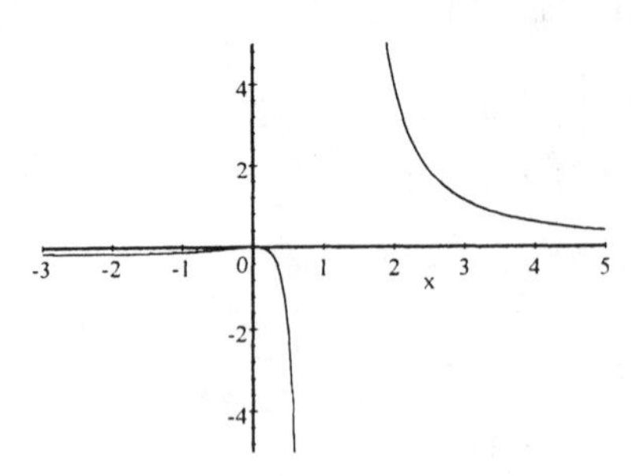

(a) $\lim\limits_{x\to 1^+} \dfrac{x^2}{(x-1)^3} = \infty$

(b) $\lim\limits_{x\to 1^-} \dfrac{x^2}{(x-1)^3} = -\infty$

33.
$$\lim_{x\to\infty} \frac{5x+1}{2x-7} = \lim_{x\to\infty} \frac{\frac{5x}{x}+\frac{1}{x}}{\frac{2x}{x}-\frac{7}{x}}$$
$$= \lim_{x\to\infty} \frac{5+\frac{1}{x}}{2-\frac{7}{x}} = \frac{5}{2}$$

34.
$$\lim_{x\to\infty} \frac{4x^2-5x}{2x^2} = \lim_{x\to\infty} \frac{\frac{4x^2}{x^2}-\frac{5x}{x^2}}{\frac{2x^2}{x^2}}$$
$$= \lim_{x\to\infty} \frac{4-\frac{5}{x}}{2}$$
$$= \frac{4}{2} = 2$$

35. $\lim\limits_{x\to-\infty} \dfrac{x^3+1}{x^2-1} = -\infty$

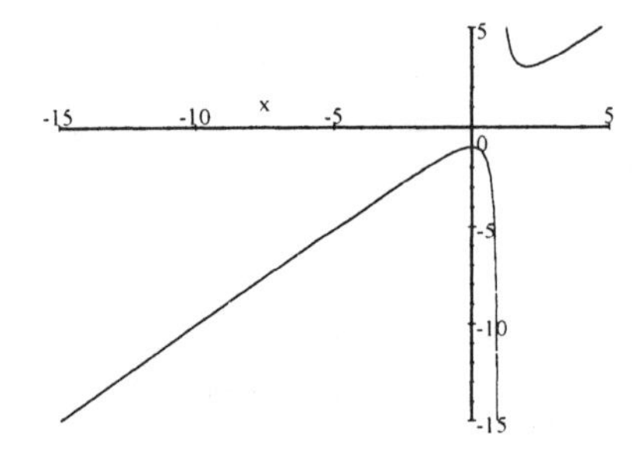

36.
$$\lim_{x\to-\infty} \frac{x^4+x+1}{x^5-2} = \lim_{x\to\infty} \frac{\frac{x^4}{x^5}+\frac{x}{x^5}+\frac{1}{x^5}}{\frac{x^5}{x^5}-\frac{2}{x^5}}$$
$$= \lim_{x\to\infty} \frac{\frac{1}{x}+\frac{1}{x^4}+\frac{1}{x^5}}{1-\frac{2}{x^5}}$$
$$= \frac{0}{1} = 0$$

37.
$$\lim_{x\to\infty} \left(5+\frac{x}{1+x^2}\right) = \lim_{x\to\infty} 5 + \lim_{x\to\infty} \frac{\frac{x}{x^2}}{\frac{1}{x^2}+\frac{x^2}{x^2}}$$
$$= 5 + \lim_{x\to\infty} \frac{\frac{1}{x}}{\frac{1}{x^2}+1}$$
$$= 5 + 0 = 5$$

38. $\lim\limits_{x\to\infty} \left(e^{-2x}+7\right) = 7$

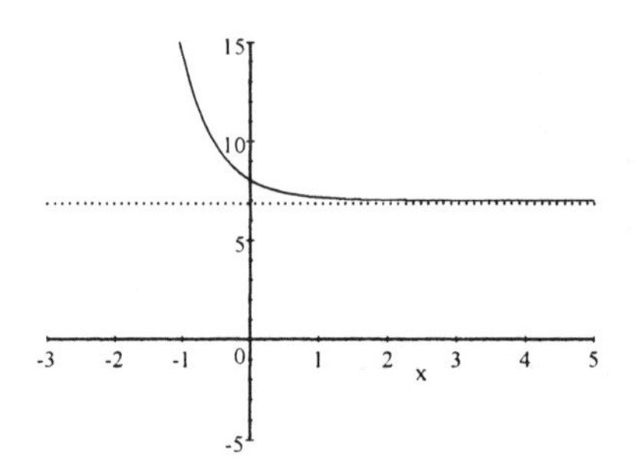

39. $\lim\limits_{x\to\infty} \left(2-\dfrac{3}{1-e^{-x}}\right) = -1$

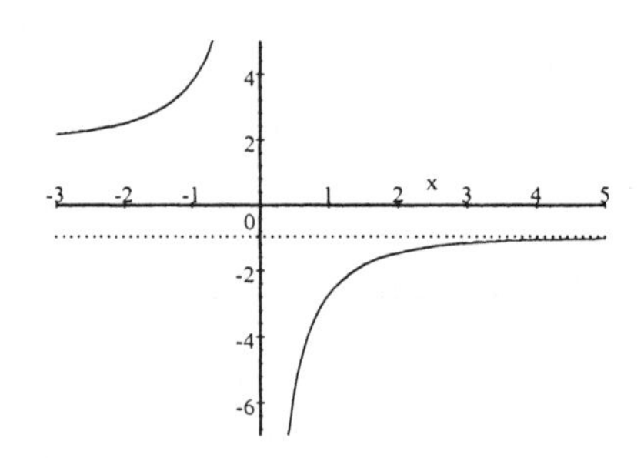

40. $\lim\limits_{x\to\infty} \left(\dfrac{5}{1+2^{-x}}-3\right) = 2$

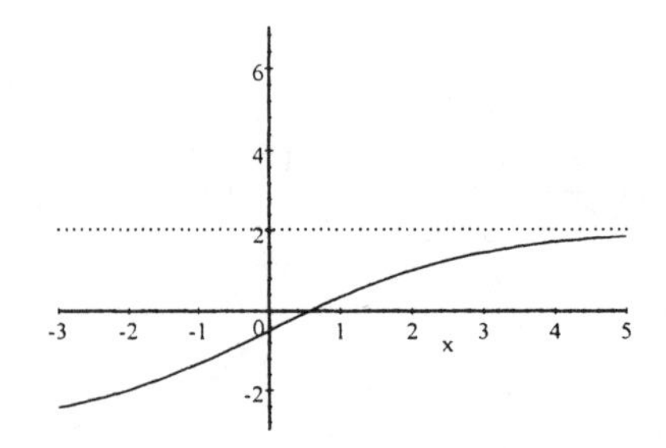

Note: In Exercises 41 and 42, other correct answers are possible.

41. $f(x) = \dfrac{1}{x-2}, \quad g(x) = \dfrac{-1}{x-2}$

42. $f(x) = x^2, \quad g(x) = x$

43. $\lim\limits_{t\to\infty} \left(70 + 110e^{-.25t}\right) = 70$

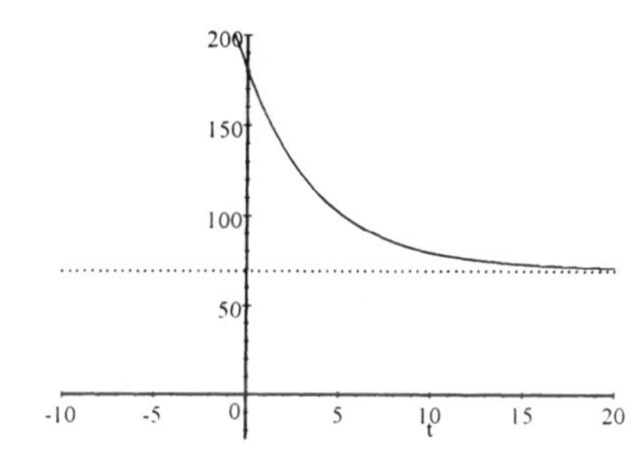

The coffee cools to 70° F.

44. $f(x) = 1 + x^2$

$$\begin{aligned} m &= \lim_{x\to3} \frac{(1+x^2)-(1+3^2)}{x-3} \\ &= \lim_{x\to3} \frac{x^2-9}{x-3} \\ &= \lim_{x\to3} \frac{(x-3)(x+3)}{x-3} \\ &= \lim_{x\to3} (x+3) = 3+3 = \boxed{6} \end{aligned}$$

45. $f(x) = \dfrac{3}{x}$

$$\begin{aligned} m &= \lim_{x\to1} \frac{\left(\frac{3}{x}\right)-\left(\frac{3}{1}\right)}{x-1} \\ &= \lim_{x\to1} \frac{\frac{3}{x}-3}{x-1} \cdot \frac{x}{x} \\ &= \lim_{x\to1} \frac{3-3x}{x(x-1)} \\ &= \lim_{x\to1} \frac{-3(x-1)}{x(x-1)} \\ &= \lim_{x\to1} \frac{-3}{x} = \boxed{-3} \end{aligned}$$

46. $f(x) = \dfrac{4}{x}$

$$\begin{aligned} m &= \lim_{x\to2} \frac{\left(\frac{4}{x}\right)-\left(\frac{4}{2}\right)}{x-2} \\ &= \lim_{x\to2} \frac{\frac{4}{x}-2}{x-2} \cdot \frac{x}{x} \\ &= \lim_{x\to2} \frac{4-2x}{x(x-2)} \\ &= \lim_{x\to2} \frac{-2(x-2)}{x(x-2)} \\ &= \lim_{x\to2} \frac{-2}{x} = \boxed{-1} \end{aligned}$$

$$f(2) = \frac{4}{2} = 2 \longrightarrow (2,\ 2)$$

$$\begin{aligned} y - y_1 &= m(x-x_1) \\ y - 2 &= -1(x-2) \\ y - 2 &= -x+2 \\ y &= -x+4 \end{aligned}$$

47. $f(x) = x^2 - x$

$$\begin{aligned} m &= \lim_{x\to2} \frac{x^2-x-(2^2-2)}{x-2} \\ &= \lim_{x\to2} \frac{x^2-x-2}{x-2} \\ &= \lim_{x\to2} \frac{(x-2)(x+1)}{(x-2)} \\ &= \lim_{x\to2} (x+1) = 2+1 = 3 \end{aligned}$$

$$f(2) = 2^2 - 2 = 2 \longrightarrow (2,\ 2)$$

$$\begin{aligned} y - y_1 &= m(x-x_1) \\ y - 2 &= 3(x-2) \\ y - 2 &= 3x-6 \\ y &= 3x-4 \end{aligned}$$

48. $\lim\limits_{x\to4} \dfrac{\sqrt{x}-2}{x-4}$ is $f'(4)$, which is the slope of the tangent line to the graph of $f(x) = \sqrt{x}$ at the point $(4,\ 2)$. Thus,

$$\lim_{x\to4} \frac{\sqrt{x}-2}{x-4} = .25.$$

49. Since $f'\left(\frac{\pi}{2}\right)$ is the slope of the tangent line to the graph of $f(x) = \sin x$ at the point $\left(\frac{\pi}{2}, 1\right)$ and the tangent line is horizontal (slope = 0), $f'\left(\frac{\pi}{2}\right) = 0$.

50. $f(x) = x \sin x; \quad a = 0$

Calculator:

$\text{nDeriv}(x \sin x, x, \pi) \approx -3.14159$

A reasonable estimate of $f'(\pi)$ is $-\pi$.

51. $f(x) = e^x; \quad a = 0$

Calculator:

$\text{nDeriv}(e\hat{\ }x, x, 0) \approx 1.00000017$

A reasonable estimate of $f'(0)$ is 1.

52. The tangent line appears to pass through the points (40, 150) and (80, 400). Thus, its slope is

$$m = \frac{400 - 150}{80 - 40} = \frac{250}{40} = 6.25.$$

On January 1, 1950, the average farm was growing at a rate of 6.25 acres per year.

53.
$$\begin{aligned} s(t) &= -16t^2 + v_o t + h_o \\ &= -16t^2 + 100t + 4 \end{aligned}$$

$$\begin{aligned} s'(3) = &= \lim_{t \to 3} \frac{s(t) - s(3)}{t - 3} \\ &= \lim_{t \to 3} \frac{-16t^2 + 100t + 4 - 160}{t - 3} \\ &= \lim_{t \to 3} \frac{-16t^2 + 100t - 156}{t - 3} \\ &= \lim_{t \to 3} \frac{(-16t + 52)(t - 3)}{t - 3} \\ &= \lim_{t \to 3} [-16t + 52] \\ &= -16(3) + 52 = 4 \end{aligned}$$

After 3 seconds, the ball is moving upward at a rate of 4 feet per second.

54. $f(x) = .006x^3 - .7x^2 + 32x + 250; \quad x = 72$

Calculator:

$\text{nDeriv}(.006x\hat{\ }3 - .7x^2 + 32x + 250, x, 72)$
≈ 24.512

When 72 units are produced, the marginal cost is \$2451.20.

55. $f(x) = 3x + 1$

$x_1 = -1,\ x_2 = 0,\ x_3 = 1,\ x_4 = 2,\ x_5 = 3$

$$\begin{aligned} \sum_{i=1}^{5} f(x_i) &= \sum_{i=1}^{5} (3x_i + 1) \\ &= (3(-1) + 1) + (3(0) + 1) + (3(1) + 1) \\ &\quad + (3(2) + 1) + (3(3) + 1)] \\ &= -2 + 1 + 4 + 7 + 10 = 20 \end{aligned}$$

56. $\int_0^4 f(x)\,dx =$ Area of lower rectangle + Area of upper triangle.

$$\begin{aligned} \int_0^4 f(x)\,dx &= (3 \cdot 1) + \frac{1}{2}(3 \cdot 1) \\ &= 3 + 1.5 = 4.5 \end{aligned}$$

Note: $\int_0^3 f(x)\,dx$ gives the same answer.

57. $f(x) = 2x + 3$ from $x = 0$ to $x = 4$

$$\Delta x = \frac{4 - 0}{4} = 1$$

$$\begin{aligned} A &= 1[f(0) + f(1) + f(2) + f(3)] \\ &= 3 + 5 + 7 + 9 = 24 \end{aligned}$$

58. $\int_0^4 (2x+3)\,dx$

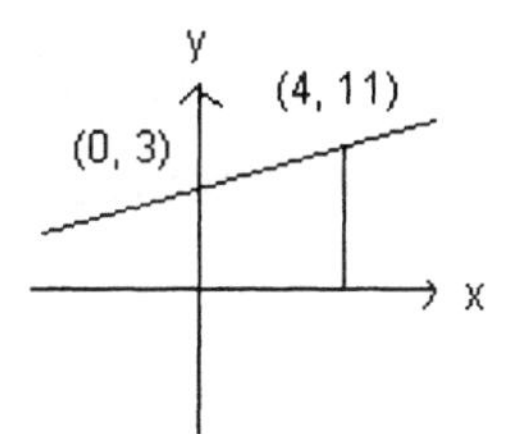

$$A = \frac{1}{2}(b_1+b_2)h$$
$$= \frac{1}{2}(3+11)4 = 28$$

It is 4 units greater than the approximation.

59. $\int_0^5 (5-x)\,dx$

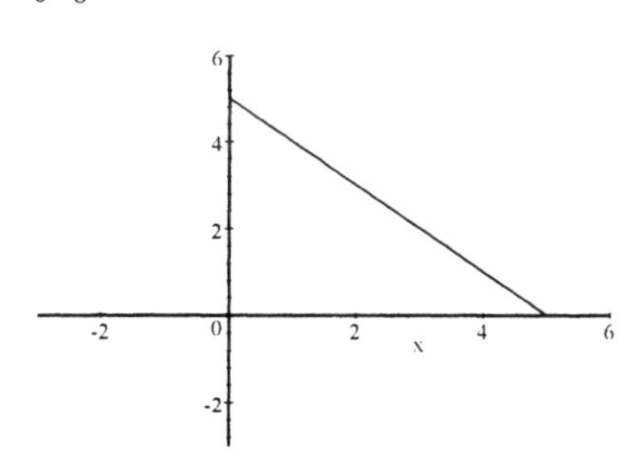

$$A = \frac{1}{2}bh = \frac{1}{2}(5)(5) = 12.5$$

60. $\int_0^1 \sqrt{1-x^2}\,dx$ (graph includes the point $(1,0)$)

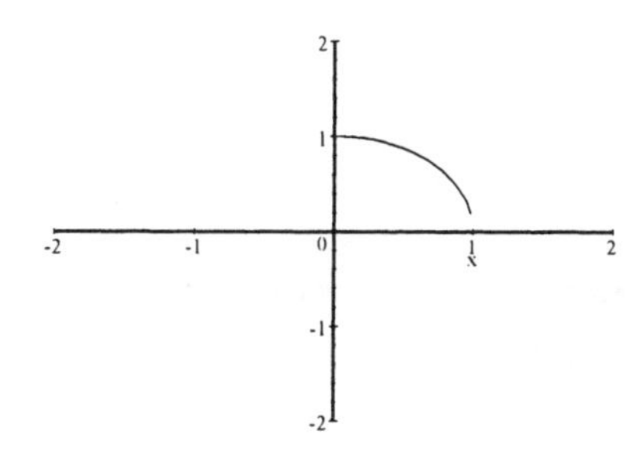

$$A = \frac{1}{4}\pi r^2 \text{ where } r=1$$
$$= \frac{1}{4}\pi(1) = \frac{\pi}{4}$$

Chapter 12 Test

1. $\lim_{x\to 4^-} f(x) = 5$

2. $\lim_{x\to\infty} f(x) = 0$

3. (a) $\lim_{x\to 3^+} f(x) = \infty$

   (b) $\lim_{x\to 3^-} f(x) = 4$

4. $\lim_{x\to 1} \frac{x^2+x+1}{x^2+1} = \frac{1^2+1+1}{1^2+1} = \frac{3}{2}$

5. $$\lim_{x\to -2} \frac{x^2+2x}{x+2} = \lim_{x\to -2} \frac{x(x+2)}{x+2} = \lim_{x\to -2} x = -2$$

6. $$\lim_{x\to 3} \frac{x^2-6x+9}{x-3} = \lim_{x\to 3} \frac{(x-3)(x-3)}{x-3} = \lim_{x\to 3}(x-3) = 3-3 = 0$$

7. $$\lim_{x\to 2} \frac{x^2+x-6}{x^2-4} = \lim_{x\to 2} \frac{(x-2)(x+3)}{(x-2)(x+2)} = \lim_{x\to 2} \frac{x+3}{x+2} = \frac{2+3}{2+2} = \frac{5}{4}$$

8. $\lim_{x\to 3} \sqrt{x^2+7} = \sqrt{3^2+7} = \sqrt{16} = 4$

9. $$\lim_{x\to\infty} \frac{2x^2-3}{5x^2+x+1} = \lim_{x\to\infty} \frac{\frac{2x^2}{x^2}-\frac{3}{x^2}}{\frac{5x^2}{x^2}+\frac{x}{x^2}+\frac{1}{x^2}} = \lim_{x\to\infty} \frac{2-\frac{3}{x^2}}{5+\frac{1}{x}+\frac{1}{x^2}} = \frac{2}{5}$$

10. $$\lim_{x\to -\infty} \frac{3x-4}{4x-3} = \lim_{x\to\infty} \frac{\frac{3x}{x}-\frac{4}{x}}{\frac{4x}{x}-\frac{3}{x}} = \lim_{x\to\infty} \frac{3-\frac{4}{x}}{4-\frac{3}{x}} = \frac{3}{4}$$

11. $\lim_{x\to 0} \dfrac{\sin x}{x\cos x} = \dfrac{\lim_{x\to 0} \frac{\sin x}{x}}{\lim_{x\to 0} \cos x} = \dfrac{1}{1} = 1$

12. $\lim_{x\to 0} \dfrac{x^2-10}{x^3+1} = \dfrac{0^2-10}{0^3+1} = -10$

13. $\lim_{x\to 1^+} f(x) = \lim_{x\to 1^+} (1-2x^2) = 1-2(1)^2 = -1$

14. $f(x) = 2x^2 - 1$

$$\begin{aligned} m &= \lim_{x\to 1} \frac{(2x^2-1)-(2\cdot 1^2-1)}{x-1} \\ &= \lim_{x\to 1} \frac{2x^2-2}{x-1} \\ &= \lim_{x\to 1} \frac{2(x+1)(x-1)}{x-1} \\ &= \lim_{x\to 1} 2(x+1) = 2(1+1) = 4 \end{aligned}$$

15. $f(x) = \dfrac{-3}{x}$

$$\begin{aligned} m &= \lim_{x\to 1} \frac{\left(\frac{-3}{x}\right)-(-3)}{x-1} \\ &= \lim_{x\to 1} \frac{\frac{-3}{x}+3}{x-1} \cdot \frac{x}{x} \\ &= \lim_{x\to 1} \frac{-3+3x}{x(x-1)} \\ &= \lim_{x\to 1} \frac{3(x-1)}{x(x-1)} \\ &= \lim_{x\to 1} \frac{3}{x} = \frac{3}{1} = 3 \end{aligned}$$

$f(1) = \dfrac{-3}{1} = -3 \longrightarrow (1, -3)$

$y - y_1 = m(x - x_1)$

$y + 3 = 3(x-1)$

$y + 3 = 3x - 3$

$y = -3x - 6$

16. $f(x) = \dfrac{1+e^x}{x}$; $a = 4$

Calculator:

nDeriv$((1+e\hat{\ }(x))/x, x, 4) \approx 10.17465434$

A reasonable estimate of $f'(4)$ is 10.1747.

17. (6, 1.6) and (14, .8)

$m = \dfrac{1.6-.8}{6-14} = \dfrac{.8}{-8} = -\dfrac{1}{10}$

Decreasing $\dfrac{1}{10}$ gram/year.

18. $f(x) = 40,000\left(1-e^{-.25x}\right)$; $a = 3$

Calculator:

nDeriv$(40000(1+e\hat{\ }(-.25x)), x, 3) \approx 4723.6656$

A reasonable estimate of $f'(3)$ is 4723.67

After 3 days the information is spreading at the rate of about 4724 people per day.

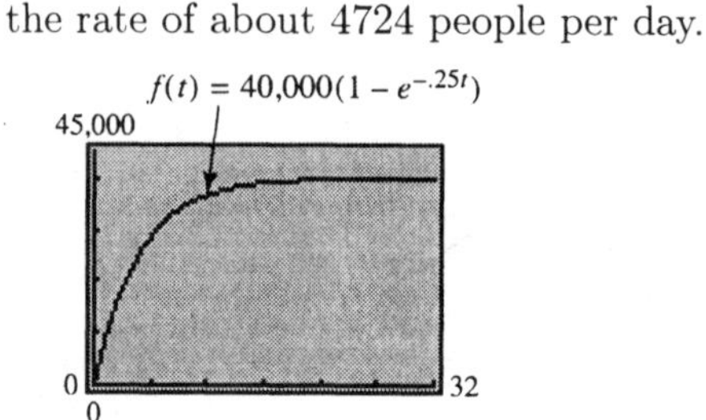

19. $f(x) = x^2$ from $x = 0$ to $x = 4$

$\Delta x = \dfrac{4-0}{4} = 1$

$$\begin{aligned} A &= 1[f(.5)+f(1.5)+f(2.5)+f(3.5)] \\ &= .25+2.25+6.25+12.25 = 21 \end{aligned}$$

20. $\int_0^6 (6-x)\,dx$

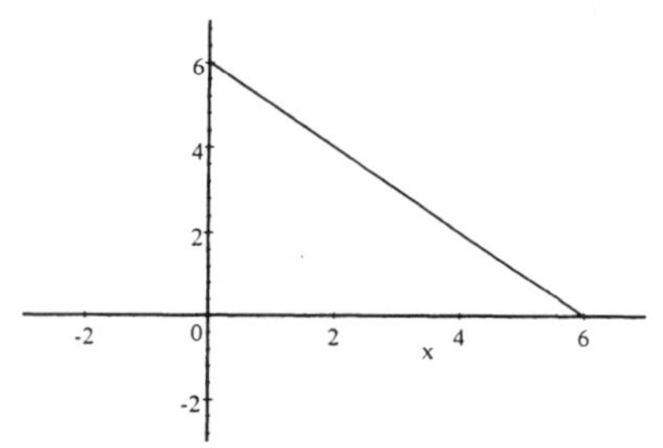

$A = \dfrac{1}{2}bh = \dfrac{1}{2}(6)(6) = 18$

Chapter 12 Project

Instantaneous Rate of Change

1. $g(x) = .0004985x^2 + .04527x + 4.054$; Answers will vary.

Graph $y_1 = g(x)$. For 1949, use dy/dx at $x = 49$; the value is $\approx .094$, so the instantaneous rate of change is $\approx 9.4\%$.

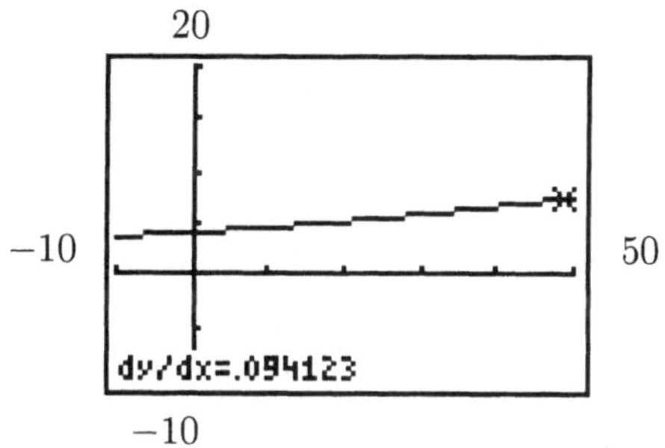

2. Graph $y_1 = 3.706x^3 + 42.48x^2 - 142.7x + 170$. For 1998, use dy/dx at $x = 8$; the value is ≈ 1248.53, so the instantaneous rate of change is ≈ 1249.

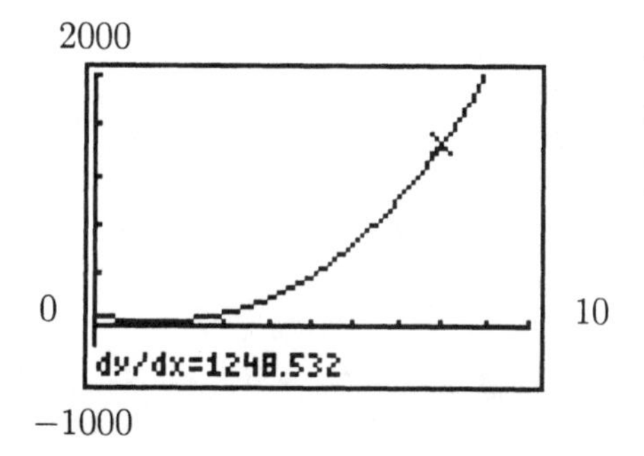

Graph $y_2 = -1.619x^4 + 36.09x^3 - 155.5x^2 + 218.1x + 127$. For 1998, use dy/dx at $x = 8$; the value is ≈ 1343.668, so the instantaneous rate of change is ≈ 1344.

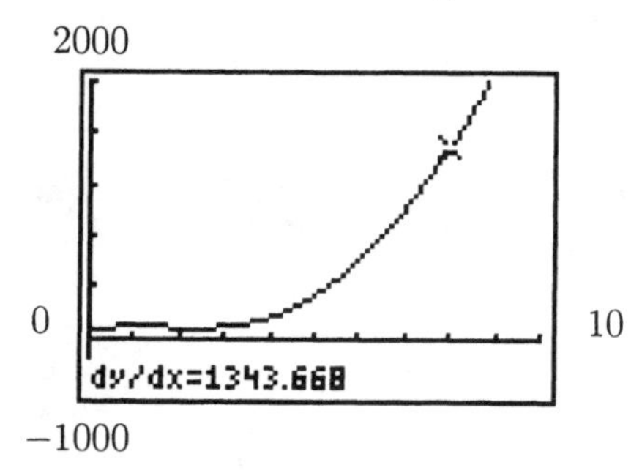

3. Answers will vary.

Graph $y_1 = 364\,(1.005)^x$. For 1949, $(100 + 1949) - 2000$, use dy/dx at $x = 49$; the value is ≈ 2.31806, so the instantaneous rate of change is ≈ 2.32.

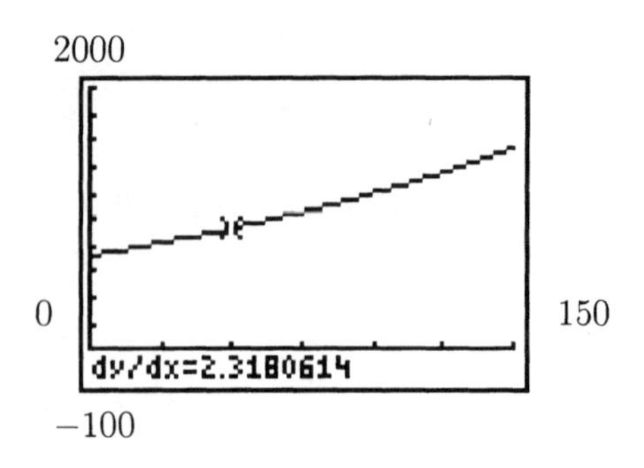

4. Answers will vary.

Graph $y_1 = 9.02 + 1.03 \ln x$. If the current year is 2004, $2004 - 1996 = 8$, use dy/dx at $x = 8$; the value is $\approx .12875$, so the instantaneous rate of change is $\approx .13$.

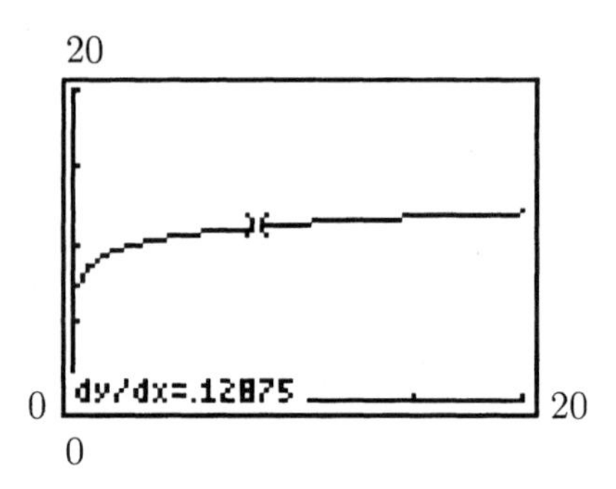

R.1

1. $(-4)^3 \cdot (-4)^2 = (-4)^{3+2} = (-4)^5$

3. $2^0 = 1$

5. $(5m)^0 = 1, \quad \text{if } m \neq 0$

7. $(2^2)^5 = 2^{2 \cdot 5} = 2^{10}$

9. $(2x^5y^4)^3$
$= 2^3 (x^5)^3 (y^4)^3 = 2^3x^{15}y^{12} \quad \text{or} \quad 8x^{15}y^{12}$

11. $-\left(\frac{p^4}{q}\right)^2 = -\left(\frac{p^{4 \cdot 2}}{q^2}\right) = -\frac{p^8}{q^2}$

13. $-5x^{11}$ is a polynomial. It is a monomial since it has one term. It has degree 11 since 11 is the highest exponent.

15. $18p^5q + 6pq$ is a polynomial. It is a binomial since it has two terms. It has degree 6 since 6 is the sum of the exponents in the term $18p^5q$. (The term $6pq$ has degree 2.)

17. $\sqrt{2}\,x^2 + \sqrt{3}\,x^6$ is a polynomial. It is a binomial since it has two terms. It has degree 6 since 6 is the highest exponent

19. $\frac{1}{3}r^2s^2 - \frac{3}{5}r^4s^2 + rs^3$ is a polynomial. It is a trinomial since it has three terms. It has degree 6 since 6 is the sum of the exponents in the term $-\frac{3}{5}r^4s^2$. (The other terms have degree 4.)

21. $-5\sqrt{z} + 2\sqrt{z^3} - 5\sqrt{z^5}$
$= -5z^{1/2} + 2z^{3/2} - 5z^{5/2}$
is not a polynomial since the exponents are not integers

23. $(4m^3 - 3m^2 + 5) + (-3m^3 - m^2 + 5)$
$= (4m^3 - 3m^3) + (-3m^2 - m^2) + (5 + 5)$
$= m^3 - 4m^2 + 10$

25. $(8p^2 - 5p) - (3p^2 - 2p + 4)$
$= 8p^2 - 5p - 3p^2 + 2p - 4$
$= 5p^2 - 3p - 4$

27. $-(8x^3 + x - 3) + (2x^3 + x^2) - (4x^2 + 3x - 1)$
$= -8x^3 - x + 3 + 2x^3 + x^2 - 4x^2 - 3x + 1$
$= -6x^3 - 3x^2 - 4x + 4$

29. $(5m - 6)(3m + 4)$
$= 5m \cdot 3m + 5m \cdot 4 - 6 \cdot 3m - 6 \cdot 4$
$= 15m^2 + 20m - 18m - 24$
$= 15m^2 + 2m - 24$

31. $\left(2m - \frac{1}{4}\right)\left(3m + \frac{1}{2}\right)$
$= 2m \cdot 3m + 2m \cdot \frac{1}{2} - \frac{1}{4} \cdot 3m - \frac{1}{4} \cdot \frac{1}{2}$
$= 6m^2 + m - \frac{3}{4}m - \frac{1}{8}$
$= 6m^2 + \frac{1}{4}m - \frac{1}{8}$

33. $2b^3(b^2 - 4b + 3)$
$= 2b^3 \cdot b^2 + 2b^3(-4b) + 2b^3 \cdot 3$
$= 2b^5 - 8b^4 + 6b^3$

35. $(m - n + k)(m + 2n - 3k)$
$= m^2 + 2mn - 3km - mn - 2n^2$
$+ 3kn + km + 2kn - 3k^2$
$= m^2 + mn - 2km - 2n^2 + 5kn - 3k^2$

37. To find the square of a binomial, find the sum of the square of the first term, twice the product of the two terms, and the square of the last term.

39. $(2m + 3)(2m - 3)$
$= (2m)^2 - (3)^2 = 4m^2 - 9$

41. $(4m + 2n)^2 = (4m)^2 + 2(4m)(2n) + (2n)^2$
$= 16m^2 + 16mn + 4n^2$

43. $(5r + 3t^2)^2 = (5r)^2 + 2(5r)(3t^2) + (3t^2)^2$
$= 25r^2 + 30rt^2 + 9t^4$

45. $[(2p-3)+q]^2$
$= (2p-3)^2 + 2(2p-3)(q) + (q)^2$
$= \left[(2p)^2 - 2(2p)(3) + 3^2\right] + 4pq - 6q + q^2$
$= 4p^2 - 12p + 9 + 4pq - 6q + q^2$

47. $[(3q+5)-p][(3q+5)+p]$
$= (3q+5)^2 - (p)^2$
$= \left[(3q)^2 + 2(3q)(5) + (5)^2\right] - p^2$
$= 9q^2 + 30q + 25 - p^2$

49. $[(3a+b)-1]^2$
$= (3a+b)^2 - 2(3a+b)(1) + (-1)^2$
$= \left[(3a)^2 + 2(3a)(b) + (b)^2\right] - 6a - 2b + 1$
$= 9a^2 + 6ab + b^2 - 6a - 2b + 1$

51. $(6p+5q)(3p-7q)$
$= (6p)(3p) + (6p)(-7q)$
$+ (5q)(3p) + (5q)(-7q)$
$= 18p^2 - 42pq + 15pq - 35q^2$
$= 18p^2 - 27pq - 35q^2$

53. $(p^3 - 4p^2 + p) - (3p^2 + 2p + 7)$
$= p^3 - 4p^2 + p - 3p^2 - 2p - 7$
$= p^3 - 7p^2 - p - 7$

55. $(4x+3y)(4x-3y) = (4x)^2 - (3y)^2$
$= 16x^2 - 9y^2$

57. $(2z+y)(3z-4y)$
$= (2z)(3z) + (2z)(-4y) + (y)(3z) + (y)(-4y)$
$= 6z^2 - 8yz + 3yz - 4y^2$
$= 6z^2 - 5yz - 4y^2$

59. $(3p+5)^2 = (3p)^2 + 2(3p)(5) + 5^2$
$= 9p^2 + 30p + 25$

61. $p(4p-6) + 2(3p-8)$
$= 4p^2 - 6p + 6p - 16$
$= 4p^2 - 16$

63. $-y(y^2-4) + 6y^2(2y-3)$
$= -y^3 + 4y + 12y^3 - 18y^2$
$= 11y^3 - 18y^2 + 4y$

R.2

1. **(a)** $(x+5y)^2 = x^2 + 10xy + 25y^2$: **B**
(b) $(x-5y)^2 = x^2 - 10xy + 25y^2$: **C**
(c) $(x+5y)(x-5y) = x^2 - 25y^2$: **A**
(d) $(5y+x)(5y-x) = 25y^2 - x^2$: **D**

3. $4k^2m^3 + 8k^4m^3 - 12k^2m^4$
$= 4k^2m^3(1 + 2k^2 - 3m)$

5. $2(a+b) + 4m(a+b)$
$= (a+b)(2+4m)$
$= 2(a+b)(1+2m)$

7. $(2y-3)(y+2) + (y+5)(y+2)$
$= (y+2)(2y-3+y+5)$
$= (y+2)(3y+2)$

9. $(5r-6)(r+3) - (2r-1)(r+3)$
$= (r+3)(5r-6-(2r-1))$
$= (r+3)(3r-5)$

11. $2(m-1) - 3(m-1)^2 + 2(m-1)^3$
$= (m-1)\left(2 - 3(m-1) + 2(m-1)^2\right)$
$= (m-1)\left(2 - 3m + 3 + 2(m^2 - 2m + 1)\right)$
$= (m-1)(5 - 3m + 2m^2 - 4m + 2)$
$= (m-1)(2m^2 - 7m + 7)$

13. $6st + 9t - 10s - 15$
$= 3t(2s+3) - 5(2s+3)$
$= (2s+3)(3t-5)$

15. $10x^2 - 12y + 15x - 8xy$
$= 10x^2 + 15x - 8xy - 12y$
$= 5x(2x+3) - 4y(2x+3)$
$= (2x+3)(5x-4y)$

17. $t^3 + 2t^2 - 3t - 6$
$= t^2(t+2) - 3(t+2)$
$= (t+2)(t^2-3)$

19. $(8a-3)(2a-5)$
$= [-1(-8a+3)][-1(-2a+5)]$
$= (3-8a)(5-2a)$

Both are correct.

21. $8h^2 - 24h - 320$
$= 8\left(h^2 - 3h - 40\right)$
$= 8(h-8)(h+5)$

23. $9y^4 - 54y^3 + 45y^2$
$= 9y^2\left(y^2 - 6y + 5\right)$
$= 9y^2(y-1)(y-5)$

25. $14m^2 + 11mr - 15r^2$
$= (7m - 5r)(2m + 3r)$

27. $12s^2 + 11st - 5t^2$
$= (3s - t)(4s + 5t)$

29. $30a^2 + am - m^2$
$= (5a + m)(6a - m)$

31. $18x^5 + 15x^4z - 75x^3z^2$
$= 3x^3\left(6x^2 + 5xz - 25z^2\right)$
$= 3x^3(2x + 5z)(3x - 5z)$

33. $16p^2 - 40p + 25 = (4p-5)^2$

35. $20p^2 - 100pq + 125q^2$
$= 5\left(4p^2 - 20pq + 25q^2\right)$
$= 5(2p - 5q)^2$

37. $9m^2n^2 - 12mn + 4 = (3mn - 2)^2$

39. $(2p+q)^2 - 10(2p+q) + 25$
$= ((2p+q)-5)((2p+q)-5)$
$= (2p+q-5)^2$

41. $9a^2 - 16 = (3a+4)(3a-4)$

43. $25s^4 - 9t^2 = \left(5s^2 + 3t\right)\left(5s^2 - 3t\right)$

45. $(a+b)^2 - 16$
$= (a+b+4)(a+b-4)$

47. $p^4 - 625 = \left(p^2 + 25\right)\left(p^2 - 25\right)$
$= \left(p^2 + 25\right)(p+5)(p-5)$

49. $x^4 - 1$
$= \left(x^2 + 1\right)\left(x^2 - 1\right)$
$= \left(x^2 + 1\right)(x+1)(x-1)$: **b**

51. $8 - a^3 = 2^3 - a^3$
$= (2-a)\left(4 + 2a + a^2\right)$

53. $125x^3 - 27 = (5x)^3 - 3^3$
$= (5x - 3)\left(25x^2 + 15x + 9\right)$

55. $27y^9 + 125z^6$
$= \left(3y^3\right)^3 + \left(5z^2\right)^3$
$= \left(3y^3 + 5z^2\right)\left(9y^6 - 15y^3z^2 + 25z^4\right)$

57. $(r+6)^3 - 216$
$= (r+6)^3 - 6^3$
$= (r+6-6)\left((r+6)^2 + 6(r+6) + 36\right)$
$= r\left(r^2 + 12r + 36 + 6r + 36 + 36\right)$
$= r\left(r^2 + 18r + 108\right)$

59. $27 - (m+2n)^3$
$= 3^3 - (m+2n)^3$
$= (3 - (m+2n))$
$\times \left(9 + 3(m+2n) + (m+2n)^2\right)$
$= (3 - m - 2n)$
$\times \left(9 + 3m + 6n + m^2 + 4mn + 4n^2\right)$

61. $3a^4 + 14a^2 - 5$
Let $u = a^2$:
$3u^2 + 14u - 5 = (3u-1)(u+5)$
$a^2 = u$: $\left(3a^2 - 1\right)\left(a^2 + 5\right)$
a^2 was not substituted back in for u.

63. $a^4 - 2a^2 - 48$
Let $u = a^2$:
$u^2 - 2u - 48 = (u-8)(u+6)$
$a^2 = u$: $\left(a^2 - 8\right)\left(a^2 + 6\right)$

65. $6(4z-3)^2 + 7(4z-3) - 3$
Let $u = 4z - 3$:
$6u^2 + 7u - 3 = (3u-1)(2u+3)$
$4z - 3 = u$:
$= (3(4z-3) - 1)(2(4z-3) + 3)$
$= (12z - 10)(8z - 3)$
$= 2(6z - 5)(8z - 3)$

67. $20(4-p)^2 - 3(4-p) - 2$
Let $u = 4 - p$:
$20u^2 - 3u - 2 = (5u-2)(4u+1)$
$4 - p = u$:
$= (5(4-p)-2)(4(4-p)+1)$
$= (18-5p)(17-4p)$

69. $4b^2 + 4bc + c^2 - 16$
$= (4b^2 + 4b + c^2) - 16$
$= (2b+c)^2 - 16$
$= (2b+c+4)(2b+c-4)$

71. $x^2 + xy - 5x - 5y$
$= x(x+y) - 5(x+y)$
$= (x+y)(x-5)$

73. $p^4(m-2n) + q(m-2n)$
$= (m-2n)(p^4+q)$

75. $4z^2 + 28z + 49 = (2z+7)^2$

77. $1000x^3 + 343y^3$
$= (10x)^3 + (7y)^3$
$= (10x+7y)(100x^2 - 70xy + 49y^2)$

79. $125m^6 - 216$
$= (5m^2)^3 - 6^3$
$= (5m^2-6)(25m^4 + 30m^2 + 36)$

81. $12m^2 + 16mn - 35n^2$
$= (6m-7n)(2m+5n)$

83. $4p^2 + 3p - 1 = (4p-1)(p+1)$

85. $144z^2 + 121$
does not factor; it is prime.

87. $(x+y)^2 - (x-y)^2$
$= (x+y+(x-y))(x+y-(x-y))$
$= (2x)(2y) = 4xy$

R.3

1. $\dfrac{x-2}{x+6}$; $x+6=0$, $\{x|x \neq -6\}$

3. $\dfrac{2x}{5x-3}$; $5x-3=0$, $\left\{x|x \neq \dfrac{3}{5}\right\}$

5. $\dfrac{-8}{x^2+1}$; No restrictions since $x^2+1>0$.
Domain: $(-\infty, \infty)$

7. $\dfrac{3x+7}{(4x+2)(x-1)}$

$4x+2=0$ $\quad x-1=0$
$4x=-2$ $\quad x=1$
$x = -\dfrac{2}{4}$

$\left\{x|x \neq -\dfrac{1}{2}, 1\right\}$

9. $\dfrac{25p^3}{10p^2} = \dfrac{5p^2(5p)}{5p^2(2)} = \dfrac{5p}{2}$

11. $\dfrac{8k+16}{9k+18} = \dfrac{8(k+2)}{9(k+2)} = \dfrac{8}{9}$

13. $\dfrac{3(t+5)}{(t+5)(t-3)} = \dfrac{3}{t-3}$

15. $\dfrac{8x^2+16x}{4x^2} = \dfrac{8x(x+2)}{4x^2}$
$= \dfrac{2(x+2)}{x} = \dfrac{2x+4}{x}$

17. $\dfrac{m^2-4m+4}{m^2+m-6} = \dfrac{(m-2)^2}{(m-2)(m+3)}$
$= \dfrac{m-2}{m+3}$

19. $\dfrac{8m^2+6m-9}{16m^2-9} = \dfrac{(4m-3)(2m+3)}{(4m-3)(4m+3)}$
$= \dfrac{2m+3}{4m+3}$

21. $\dfrac{15p^3}{9p^2} \div \dfrac{6p}{10p^2} = \dfrac{15p^3}{9p^2} \times \dfrac{10p^2}{6p} = \dfrac{25p^2}{9}$

23. $\dfrac{2k+8}{6} \div \dfrac{3k+12}{2}$

$$= \frac{2\,(k+4)}{6} \times \frac{2}{3\,(k+4)} = \frac{2}{9}$$

25. $\dfrac{x^2+x}{5} \cdot \dfrac{25}{xy+y}$

$$= \frac{x\,(x+1)}{5} \cdot \frac{25}{y\,(x+1)} = \frac{5x}{y}$$

27. $\dfrac{4a+12}{2a-10} \div \dfrac{a^2-9}{a^2-a-20}$

$$= \frac{4\,(a+3)}{2\,(a-5)} \cdot \frac{(a+4)\,(a-5)}{(a-3)\,(a+3)}$$

$$= \frac{2\,(a+4)}{a-3}$$

29. $\dfrac{p^2-p-12}{p^2-2p-15} \cdot \dfrac{p^2-9p+20}{p^2-8p+16}$

$$= \frac{(p-4)\,(p+3)}{(p+3)\,(p-5)} \cdot \frac{(p-5)\,(p-4)}{(p-4)\,(p-4)} = 1$$

31. $\dfrac{m^2+3m+2}{m^2+5m+4} \div \dfrac{m^2+5m+6}{m^2+10m+24}$

$$= \frac{(m+2)\,(m+1)}{(m+4)\,(m+1)} \cdot \frac{(m+4)\,(m+6)}{(m+3)\,(m+2)}$$

$$= \frac{m+6}{m+3}$$

33. $\dfrac{2m^2-5m-12}{m^2-10m+24} \div \dfrac{4m^2-9}{m^2-9m+18}$

$$= \frac{(2m+3)\,(m-4)}{(m-4)\,(m-6)} \cdot \frac{(m-3)\,(m-6)}{(2m+3)\,(2m-3)}$$

$$= \frac{m-3}{2m-3}$$

35. $\dfrac{x^3+y^3}{x^2-y^2} \cdot \dfrac{x+y}{x^2-xy+y^2}$

$$= \frac{(x+y)\,(x^2-xy+y^2)}{(x+y)\,(x-y)} \cdot \frac{x+y}{x^2-xy+y^2}$$

$$= \frac{x+y}{x-y}$$

37. $\dfrac{x^3+y^3}{x^3-y^3} \cdot \dfrac{x^2-y^2}{x^2+2xy+y^2}$

$$= \frac{(x+y)\,(x^2-xy+y^2)}{(x-y)\,(x^2+xy+y^2)} \cdot \frac{(x-y)\,(x+y)}{(x+y)\,(x+y)}$$

$$= \frac{x^2-xy+y^2}{x^2+xy+y^2}$$

39. **A:** $\dfrac{x-4}{x+4} \neq -1$

B: $\dfrac{-x-4}{x+4} = \dfrac{-1\,(x+4)}{x+4} = -1$ ✓

C: $\dfrac{x-4}{4-x} = \dfrac{x-4}{-1\,(x-4)} = -1$ ✓

D: $\dfrac{x-4}{-x-4} = \dfrac{x-4}{-1\,(x+4)} \neq -1$

41. $\dfrac{3}{2k} + \dfrac{5}{3k} = \dfrac{9}{6k} + \dfrac{10}{6k} = \dfrac{19}{6k}$

43. $\dfrac{a+1}{2} - \dfrac{a-1}{2}$

$$= \frac{a+1-(a-1)}{2} = \frac{2}{2} = 1$$

45. $\dfrac{3}{p} + \dfrac{1}{2} = \dfrac{6}{2p} + \dfrac{p}{2p} = \dfrac{6+p}{2p}$

47. $\dfrac{1}{6m} + \dfrac{2}{5m} + \dfrac{4}{m}$

$$= \frac{5\,(1)+6\,(2)+30\,(4)}{30m}$$

$$= \frac{5+12+120}{30m} = \frac{137}{30m}$$

49. $\dfrac{1}{a+1} - \dfrac{1}{a-1} = \dfrac{a-1-(a+1)}{(a+1)\,(a-1)}$

$$= \frac{-2}{(a+1)\,(a-1)}$$

51. $$\frac{m+1}{m-1}+\frac{m-1}{m+1}=\frac{(m+1)^2+(m-1)^2}{(m-1)(m+1)}$$
$$=\frac{m^2+2m+1+m^2-2m+1}{(m-1)(m+1)}$$
$$=\frac{2m^2+2}{(m-1)(m+1)}$$

53. $$\frac{3}{a-2}-\frac{1}{2-a}=\frac{3}{a-2}-\frac{-1}{a-2}$$
$$=\frac{3+1}{a-2}=\frac{4}{a-2}\quad\left(\text{Or }\frac{-4}{2-a}\right)$$

55. $$\frac{x+y}{2x-y}-\frac{2x}{y-2x}$$
$$=\frac{x+y}{2x-y}-\frac{-2x}{2x-y}$$
$$=\frac{x+y+2x}{2x-y}=\frac{3x+y}{2x-y}\quad\left(\text{Or }\frac{-3x-y}{y-2x}\right)$$

57. $$\frac{1}{a^2-5a+6}-\frac{1}{a^2-4}$$
$$=\frac{1}{(a-2)(a-3)}-\frac{1}{(a-2)(a+2)}$$
$$=\frac{1(a+2)-1(a-3)}{(a-2)(a-3)(a+2)}$$
$$=\frac{a+2-a+3}{(a-2)(a-3)(a+2)}$$
$$=\frac{5}{(a-2)(a-3)(a+2)}$$

59. $$\frac{1}{x^2+x-12}-\frac{1}{x^2-7x+12}+\frac{1}{x^2-16}$$
$$=\frac{1}{(x+4)(x-3)}-\frac{1}{(x-3)(x-4)}+\frac{1}{(x-4)(x+4)}$$
$$=\frac{1(x-4)-1(x+4)+1(x-3)}{(x+4)(x-3)(x-4)}$$
$$=\frac{x-4-x-4+x-3}{(x+4)(x-3)(x-4)}$$
$$=\frac{x-11}{(x+4)(x-3)(x-4)}$$

61. $$\frac{3a}{a^2+5a-6}-\frac{2a}{a^2+7a+6}$$
$$=\frac{3a}{(a+6)(a-1)}-\frac{2a}{(a+6)(a+1)}$$
$$=\frac{3a(a+1)-2a(a-1)}{(a+6)(a-1)(a+1)}$$
$$=\frac{3a^2+3a-2a^2+2a}{(a+6)(a-1)(a+1)}$$
$$=\frac{a^2+5a}{(a+6)(a-1)(a+1)}$$

63. $$\frac{1+\frac{1}{x}}{1-\frac{1}{x}}=\frac{x\left(1+\frac{1}{x}\right)}{x\left(1-\frac{1}{x}\right)}=\frac{x+1}{x-1}$$

65. $$\frac{\frac{1}{x+1}-\frac{1}{x}}{\frac{1}{x}}=\frac{x(x+1)\left(\frac{1}{x+1}-\frac{1}{x}\right)}{x(x+1)\left(\frac{1}{x}\right)}$$
$$=\frac{x-(x+1)}{x+1}=\frac{-1}{x+1}$$

67. $$\frac{1+\frac{1}{1-b}}{1-\frac{1}{1+b}}=\frac{(1-b)(1+b)\left(1+\frac{1}{1-b}\right)}{(1-b)(1+b)\left(1-\frac{1}{1+b}\right)}$$
$$=\frac{(1-b)(1+b)+(1+b)}{(1-b)(1+b)-(1-b)}$$
$$=\frac{1-b^2+1+b}{1-b^2-1+b}$$
$$=\frac{-b^2+b+2}{-b^2+b}=\frac{(2-b)(1+b)}{b(1-b)}$$

69. $$\frac{m-\frac{1}{m^2-4}}{\frac{1}{m+2}}=\frac{(m^2-4)\left(m-\frac{1}{m^2-4}\right)}{(m^2-4)\left(\frac{1}{m+2}\right)}$$
$$=\frac{m(m^2-4)-1}{m-2}$$
$$=\frac{m^3-4m-1}{m-2}$$

R.4

1. $\left(\frac{4}{9}\right)^{3/2} = \left[\left(\frac{4}{9}\right)^{1/2}\right]^3 = \left(\frac{2}{3}\right)^3 = \frac{2^3}{3^3} = \frac{8}{27}$: **E**

3. $-\left(\frac{9}{4}\right)^{3/2} = -\left[\left(\frac{9}{4}\right)^{1/2}\right]^3 = -\left(\frac{3}{2}\right)^3 = -\frac{3^3}{2^3} = -\frac{27}{8}$: **F**

5. $\left(\frac{8}{27}\right)^{2/3} = \left[\left(\frac{8}{27}\right)^{1/3}\right]^2 = \left(\frac{2}{3}\right)^2 = \frac{2^2}{3^2} = \frac{4}{9}$: **D**

7. $-\left(\frac{27}{8}\right)^{2/3} = -\left[\left(\frac{27}{8}\right)^{1/3}\right]^2 = -\left(\frac{3}{2}\right)^2 = -\frac{3^2}{2^2} = -\frac{9}{4}$: **B**

9. $(-4)^{-3} = \left(-\frac{1}{4}\right)^3 = -\frac{1}{64}$

11. $\left(\frac{1}{2}\right)^{-3} = 2^3 = 8$

13. $-4^{1/2} = -\left(4^{1/2}\right) = -2$

15. $8^{2/3} = \left(8^{1/3}\right)^2 = 2^2 = 4$

17. $27^{-2/3} = \left[\left(\frac{1}{27}\right)^{1/3}\right]^2 = \left(\frac{1}{3}\right)^2 = \frac{1}{9}$

19. $\left(\frac{27}{64}\right)^{-4/3} = \left[\left(\frac{64}{27}\right)^{1/3}\right]^4 = \left(\frac{4}{3}\right)^4 = \frac{256}{81}$

21. $\left(16p^4\right)^{1/2} = 4p^2$

23. $\left(27x^6\right)^{2/3} = \left[\left(27x^6\right)^{1/3}\right]^2 = \left(3x^2\right)^2 = 9x^4$

25. $2^{-3} \cdot 2^{-4} = 2^{-7} = \frac{1}{2^7}$

27. $27^{-2} \cdot 27^{-1} = 27^{-3} = \frac{1}{27^3}$

29. $\frac{4^{-2} \cdot 4^{-1}}{4^{-3}} = 4^{(-2+(-1)-(-3))} = 4^0 = 1$

31. $\left(m^{2/3}\right)\left(m^{5/3}\right) = m^{(2/3+5/3)} = m^{7/3}$

33. $(1+n)^{1/2}(1+n)^{3/4} = (1+n)^{(2/4+3/4)}$
$= (1+n)^{5/4}$

35. $\left(2y^{3/4}z\right)\left(3y^{-2}z^{-1/3}\right) = 6\,y^{(3/4-8/4)}\,z^{(1-1/3)}$
$= 6\,y^{-5/4}\,z^{2/3} = \frac{6\,z^{2/3}}{y^{5/4}}$

37. $\left(4a^{-2}b^7\right)^{1/2} \cdot \left(2a^{1/4}b^3\right)^5$
$= \left(2a^{-1}b^{7/2}\right)\left(2^5a^{5/4}b^{15}\right)$
$= 2^6\,a^{(-1+5/4)}\,b^{(7/2+30/2)} = 2^6\,a^{1/4}\,b^{37/2}$

39. $\left(\frac{r^{-2}}{x^{-5}}\right)^{-3} = \frac{r^6}{s^{15}}$

41. $\left(\frac{-a}{b^{-3}}\right)^{-1} = \left(\frac{b^{-3}}{-a}\right)^1 = \frac{-1}{ab^3}$

43. $\frac{12^{5/4} \cdot y^{-2}}{12^{-1}\,y^{-3}} = 12^{(5/4-(-1))}\,y^{(-2-(-3))}$
$= 12^{9/4}\,y$

45. $\frac{8p^{-3}\left(4p^2\right)^{-2}}{p^{-5}} = \frac{8p^5}{p^3\left(4p^2\right)^2}$
$= \frac{8p^5}{16p^7} = \frac{1}{2p^2}$

47. $\frac{m^{7/3}\,n^{-2/5}\,p^{3/8}}{m^{-2/3}\,n^{3/5}\,p^{-5/8}} = m^{(7/3+2/3)}\,n^{(-2/5-3/5)}\,p^{(3/8+5/8)}$
$= m^3\,n^{-1}\,p^1 = \frac{m^3p}{n}$

49. $\dfrac{-4a^{-1}a^{2/3}}{a^{-2}} = -4a^{(-3/3+2/3+6/3)} = -4a^{5/3}$

51. $\dfrac{(k+5)^{1/2}(k+5)^{-1/4}}{(k+5)^{3/4}} = (k+5)^{(1/2-1/4-3/4)}$
$= (k+5)^{-1/2}$
$= \dfrac{1}{(k+5)^{1/2}}$

53. $y^{5/8}\left(y^{3/8} - 10y^{11/8}\right) = y^{(5/8+3/8)} - 10y^{(5/8+11/8)}$
$= y - 10y^2$

55. $-4k\left(k^{7/3} - 6k^{1/3}\right) = -4k^{(1+7/3)} + 24k^{(1+1/3)}$
$= -4k^{10/3} + 24k^{4/3}$

57. $\left(x + x^{1/2}\right)\left(x - x^{1/2}\right)$
$= x^2 - x^{(1+1/2)} + x^{(1+1/2)} - x^{(1/2+1/2)}$
$= x^2 - x$

59. $\left(r^{1/2} - r^{-1/2}\right)^2$
$= \left(r^{1/2}\right)^2 - 2\left(r^{1/2}\right)\left(r^{-1/2}\right) + \left(r^{-1/2}\right)^2$
$= r - 2r^0 + r^{-1} = r - 2 + \dfrac{1}{r}$

61. $4k^{-1} + k^{-2} = 4k^1 \cdot k^{-2} + k^{-2}$
$= k^{-2}(4k+1)$

63. $9z^{-1/2} + 2z^{1/2} = 9z^{-1/2} + 2z^{-1/2} \cdot z^{2/2}$
$= z^{-1/2}(9+2z)$

65. $p^{-3/4} - 2p^{-7/4} = p^{-7/4} \cdot p^{4/4} - 2p^{-7/4}$
$= p^{-7/4}(p-2)$

67. $(p+4)^{-3/2} + (p+4)^{-1/2} + (p+4)^{1/2}$
$= (p+4)^{-3/2}\left(1 + (p+4) + (p+4)^2\right)$
$= (p+4)^{-3/2}\left(1 + p + 4 + p^2 + 8p + 16\right)$
$= (p+4)^{-3/2}\left(p^2 + 9p + 21\right)$

R.5

1. $(-3x)^{1/3} = \sqrt[3]{-3x}$: **F**

3. $(-3x)^{-1/3} = \dfrac{1}{(-3x)^{1/3}} = \dfrac{1}{\sqrt[3]{-3x}}$: **H**

5. $(3x)^{1/3} = \sqrt[3]{3x}$: **G**

7. $(3x)^{-1/3} = \dfrac{1}{(3x)^{1/3}} = \dfrac{1}{\sqrt[3]{3x}}$: **C**

9. $(-m)^{2/3} = \sqrt[3]{(-m)^2}$ or $\left(\sqrt[3]{(-m)}\right)^2$

11. $(2m+p)^{2/3}$
$= \sqrt[3]{(2m+p)^2}$ or $\left(\sqrt[3]{(2m+p)}\right)^2$

13. $\sqrt[5]{k^2} = k^{2/5}$

15. $-3\sqrt{5p^3} = -3\left(5p^3\right)^{1/2} = -3 \cdot 5^{1/2}p^{3/2}$

17. **A:** $\sqrt{ab} = \sqrt{a} \cdot \sqrt{b}$ is true for $a > 0,\ b > 0$.

19. $\sqrt{9ax^2} = 3x\sqrt{a}$ is true for all $x \geq 0$.

21. $\sqrt[3]{125} = \sqrt[3]{5^3} = 5$

23. $\sqrt[5]{-3125} = \sqrt[5]{(-5)^5} = -5$

25. $\sqrt{50} = \sqrt{25 \cdot 2} = 5\sqrt{2}$

27. $\sqrt[3]{81} = \sqrt[3]{3^4} = 3\sqrt[3]{3}$

29. $-\sqrt[4]{32} = -\sqrt[4]{2^5} = -2\sqrt[4]{2}$

31. $-\sqrt{\dfrac{9}{5}} = -\dfrac{\sqrt{9}}{\sqrt{5}} \cdot \dfrac{\sqrt{5}}{\sqrt{5}} = -\dfrac{3\sqrt{5}}{5}$

33. $-\sqrt[3]{\dfrac{4}{5}} = -\dfrac{\sqrt[3]{4}}{\sqrt[3]{5}} \cdot \dfrac{\sqrt[3]{5^2}}{\sqrt[3]{5^2}} = -\dfrac{\sqrt[3]{100}}{5}$

35.
$$\begin{aligned} \sqrt[3]{16(-2)^4(2)^8} &= \sqrt[3]{(2)^4(-2)^4(2)^8} \\ &= \sqrt[3]{(2)^{12}(-2)^4} \\ &= 2^4(-2)\sqrt[3]{-2} \\ &= -32\sqrt[3]{-2} \text{ or } 32\sqrt[3]{2} \end{aligned}$$

37. $\sqrt{8x^5z^8} = \sqrt{2^3x^5z^8} = 2x^2z^4\sqrt{2x}$

39.
$$\begin{aligned} \sqrt[3]{16z^5x^8y^4} &= \sqrt[3]{2^4z^5x^8y^4} \\ &= 2zx^2y\sqrt[3]{2z^2x^2y} \end{aligned}$$

41. $\sqrt[4]{m^2n^7p^8} = np^2\sqrt[4]{m^2n^3}$

43. $\sqrt[4]{x^4+y^4}$ cannot be simplified.

45. $\sqrt{\dfrac{2}{3x}} = \dfrac{\sqrt{2}}{\sqrt{3x}} \cdot \dfrac{\sqrt{3x}}{\sqrt{3x}} = \dfrac{\sqrt{6x}}{3x}$

47. $\sqrt{\dfrac{x^5y^3}{z^2}} = \dfrac{\sqrt{x^5y^3}}{\sqrt{z^2}} = \dfrac{x^2y\sqrt{xy}}{z}$

49. $\sqrt[3]{\dfrac{8}{x^2}} = \dfrac{\sqrt[3]{2^3}}{\sqrt[3]{x^2}} \cdot \dfrac{\sqrt[3]{x}}{\sqrt[3]{x}} = \dfrac{2\sqrt[3]{x}}{x}$

51.
$$\begin{aligned} \sqrt[4]{\frac{g^3h^5}{9r^6}} &= \frac{\sqrt[4]{g^3h^5}}{\sqrt[4]{3^2r^6}} \\ &= \frac{h\sqrt[4]{g^3h}}{r\sqrt[4]{3^2r^2}} \cdot \frac{\sqrt[4]{3^2r^2}}{\sqrt[4]{3^2r^2}} \\ &= \frac{h\sqrt[4]{9g^3hr^2}}{3r^2} \end{aligned}$$

53.
$$\begin{aligned} \frac{\sqrt[3]{mn}\cdot\sqrt[3]{m^2}}{\sqrt[3]{n^2}} \cdot \frac{\sqrt[3]{n}}{\sqrt[3]{n}} &= \frac{\sqrt[3]{m^3n^2}}{n} \\ &= \frac{m\sqrt[3]{n^2}}{n} \end{aligned}$$

55.
$$\begin{aligned} \frac{\sqrt[4]{32x^5y}\cdot\sqrt[4]{2xy^4}}{\sqrt[4]{4x^3y^2}} &= \frac{\sqrt[4]{2^6x^6y^5}}{\sqrt[4]{2^2x^3y^2}} \cdot \frac{\sqrt[4]{2^2xy^2}}{\sqrt[4]{2^2xy^2}} \\ &= \frac{\sqrt[4]{2^8x^7y^7}}{2xy} \\ &= \frac{4xy\sqrt[4]{x^3y^3}}{2xy} = 2\sqrt[4]{x^3y^3} \end{aligned}$$

57. $\sqrt[3]{\sqrt{4}} = \sqrt[3]{2}$

59. $\sqrt[6]{\sqrt[3]{x}} = \left(x^{1/3}\right)^{1/6} = x^{1/18} = \sqrt[18]{x}$

61.
$$\begin{aligned} &4\sqrt{3} - 5\sqrt{12} + 3\sqrt{75} \\ &= 4\sqrt{3} - 5\left(2\sqrt{3}\right) + 3\left(5\sqrt{3}\right) \\ &= 4\sqrt{3} - 10\sqrt{3} + 15\sqrt{3} \\ &= 9\sqrt{3} \end{aligned}$$

63.
$$\begin{aligned} &3\sqrt{28p} - 4\sqrt{63p} + \sqrt{112p} \\ &= 3\left(2\sqrt{7p}\right) - 4\left(3\sqrt{7p}\right) + 4\sqrt{7p} \\ &= 6\sqrt{7p} - 12\sqrt{7p} + 4\sqrt{7p} \\ &= -2\sqrt{7p} \end{aligned}$$

65.
$$\begin{aligned} &2\sqrt[3]{3} + 4\sqrt[3]{24} - \sqrt[3]{81} \\ &= 2\sqrt[3]{3} + 4\left(2\sqrt[3]{3}\right) - 3\sqrt[3]{3} \\ &= 2\sqrt[3]{3} + 8\sqrt[3]{3} - 3\sqrt[3]{3} \\ &= 7\sqrt[3]{3} \end{aligned}$$

67. $\frac{1}{\sqrt{3}} - \frac{2}{\sqrt{12}} + 2\sqrt{3}$

$$= \frac{1}{\sqrt{3}} \cdot \frac{\sqrt{3}}{\sqrt{3}} - \frac{2}{2\sqrt{3}} \cdot \frac{\sqrt{3}}{\sqrt{3}} + 2\sqrt{3}$$

$$= \frac{\sqrt{3}}{3} - \frac{2\sqrt{3}}{6} + 2\sqrt{3}$$

$$= \frac{2\sqrt{3} - 2\sqrt{3} + 12\sqrt{3}}{6}$$

$$= \frac{12\sqrt{3}}{6} = 2\sqrt{3}$$

69. $\frac{5}{\sqrt[3]{2}} - \frac{2}{\sqrt[3]{16}} + \frac{1}{\sqrt[3]{54}}$

$$= \frac{5}{\sqrt[3]{2}} \cdot \frac{\sqrt[3]{4}}{\sqrt[3]{4}} - \frac{2}{2\sqrt[3]{2}} \cdot \frac{\sqrt[3]{4}}{\sqrt[3]{4}} + \frac{1}{3\sqrt[3]{2}} \cdot \frac{\sqrt[3]{4}}{\sqrt[3]{4}}$$

$$= \frac{5\sqrt[3]{4}}{2} - \frac{\sqrt[3]{4}}{2} + \frac{\sqrt[3]{4}}{6}$$

$$= \frac{15\sqrt[3]{4} - 3\sqrt[3]{4} + \sqrt[3]{4}}{6} = \frac{13\sqrt[3]{4}}{6}$$

71. $(\sqrt{2} + 3)(\sqrt{2} - 3) = (\sqrt{2})^2 - 3^2$

$$= 2 - 9 = -7$$

73. $(\sqrt[3]{11} - 1)\left(\sqrt[3]{11^2} + \sqrt[3]{11} + 1\right)$

$$= \sqrt[3]{11} \cdot \sqrt[3]{11^2} + \sqrt[3]{11} \cdot \sqrt[3]{11} + \sqrt[3]{11} \cdot 1 - \sqrt[3]{11^2} - \sqrt[3]{11} - 1$$

$$= 11 + \sqrt[3]{11^2} + \sqrt[3]{11} - \sqrt[3]{11^2} - \sqrt[3]{11} - 1$$

$$= 11 - 1 = 10$$

75. $(\sqrt{3} + \sqrt{8})^2$

$$= (\sqrt{3})^2 + 2\sqrt{3} \cdot \sqrt{8} + (\sqrt{8})^2$$

$$= 3 + 2\sqrt{24} + 8$$

$$= 11 + 2(2\sqrt{6})$$

$$= 11 + 4\sqrt{6}$$

77. $(3\sqrt{2} + \sqrt{3})(2\sqrt{3} - \sqrt{2})$

$$= 3\sqrt{2} \cdot 2\sqrt{3} + 3\sqrt{2}(-\sqrt{2}) + \sqrt{3} \cdot 2\sqrt{3} + \sqrt{3}(-\sqrt{2})$$

$$= 6\sqrt{6} - 6 + 6 - \sqrt{6} = 5\sqrt{6}$$

79. $\frac{\sqrt{3}}{\sqrt{5} + \sqrt{3}} \cdot \frac{\sqrt{5} - \sqrt{3}}{\sqrt{5} - \sqrt{3}}$

$$= \frac{\sqrt{3} \cdot \sqrt{5} - \sqrt{3} \cdot \sqrt{3}}{(\sqrt{5})^2 - (\sqrt{3})^2}$$

$$= \frac{\sqrt{15} - 3}{5 - 3} = \frac{\sqrt{15} - 3}{2}$$

81. $\frac{1 + \sqrt{3}}{3\sqrt{5} + 2\sqrt{3}} \cdot \frac{3\sqrt{5} - 2\sqrt{3}}{3\sqrt{5} - 2\sqrt{3}}$

$$= \frac{3\sqrt{5} - 2\sqrt{3} + \sqrt{3} \cdot 3\sqrt{5} - 2(\sqrt{3})^2}{(3\sqrt{5})^2 - (2\sqrt{3})^2}$$

$$= \frac{3\sqrt{5} - 2\sqrt{3} + 3\sqrt{15} - 6}{45 - 12}$$

$$= \frac{3\sqrt{5} - 2\sqrt{3} + 3\sqrt{15} - 6}{33}$$

83. $\frac{p}{\sqrt{p} + 2} \cdot \frac{\sqrt{p} - 2}{\sqrt{p} - 2} = \frac{p\sqrt{p} - 2p}{(\sqrt{p})^2 - (2)^2}$

$$= \frac{p(\sqrt{p} - 2)}{p - 4}$$

85. $\frac{a}{\sqrt{a+b} - 1} \cdot \frac{\sqrt{a+b} + 1}{\sqrt{a+b} + 1}$

$$= \frac{a(\sqrt{a+b} + 1)}{(\sqrt{a+b})^2 - (1)^2}$$

$$= \frac{a(\sqrt{a+b} + 1)}{a + b - 1}$$

Appendix A

1. The plane determined by the x-axis and the z-axis is called the $\underline{xz\text{-plane}}$.

3. The component form of the position vector with terminal point $(5,\ 3,\ -2)$ is $\underline{\langle 5,\ 3,\ -2\rangle}$.

5. $P = (0,\ 0,\ 0)\,;\quad Q = (2,\ -2,\ 5)$

$$d = \sqrt{(2-0)^2 + (-2-0)^2 + (5-0)^2}$$
$$= \sqrt{4+4+25} = \boxed{\sqrt{33}}$$

7. $P = (10,\ 15,\ 9)\,;\quad Q = (8,\ 3,\ -4)$

$$d = \sqrt{(8-10)^2 + (3-15)^2 + (-4-9)^2}$$
$$= \sqrt{(-2)^2 + (-12)^2 + (-13)^2}$$
$$= \sqrt{4+144+169} = \boxed{\sqrt{317}}$$

9. $P = (20,\ 25,\ 16)\,;\quad Q = (5,\ 5,\ 6)$

$$d = \sqrt{(5-20)^2 + (5-25)^2 + (6-16)^2}$$
$$= \sqrt{(-15)^2 + (-20)^2 + (-10)^2}$$
$$= \sqrt{225+400+100} = \sqrt{725} = \boxed{5\sqrt{29}}$$

11. $P = (0,\ 0,\ 0)\,;\quad Q = (2,\ -2,\ 5)$

$\mathbf{PQ} = \langle 2,\ -2,\ 5\rangle$

$\mathbf{v} = (2-0)\,\mathbf{i} + (-2-0)\,\mathbf{j} + (5-0)\,\mathbf{k}$
$\mathbf{v} = 2\mathbf{i} - 2\mathbf{j} + 5\mathbf{k}$

13. $P = (10,\ 15,\ 0)\,;\quad Q = (8,\ 3,\ -4)$

$\mathbf{PQ} = \langle (8-10)\,,\ (3-15)\,,\ (-4-0)\rangle$

$\mathbf{PQ} = \langle -2,\ -12,\ -4\rangle$

$\mathbf{v} = -2\mathbf{i} - 12\mathbf{j} - 4\mathbf{k}$

15. $P = (20,\ 25,\ 6)\,;\quad Q = (5,\ 5,\ 16)$

$\mathbf{PQ} = \langle (5-20)\,,\ (5-25)\,,\ (16-6)\rangle$

$\mathbf{PQ} = \langle -15,\ -20,\ 10\rangle$

$\mathbf{v} = -15\mathbf{i} - 20\mathbf{j} + 10\mathbf{k}$

17. $P = (0,\ 0,\ 0)\,;\quad Q = (2,\ -2,\ 5)$

$\mathbf{QP} = \langle (0-2)\,,\ (0+2)\,,\ (0-5)\rangle$

$\mathbf{QP} = \langle -2,\ 2,\ -5\rangle$

$\mathbf{QP} = -\mathbf{PQ}$

For Exercises 19 - 29:

$\mathbf{u} = 2\mathbf{i} + 4\mathbf{j} + 7\mathbf{k}$
$\mathbf{v} = -3\mathbf{i} + 5\mathbf{j} + 2\mathbf{k}$
$\mathbf{w} = 4\mathbf{i} - 3\mathbf{j} - 6\mathbf{k}$

19. $\mathbf{u} - \mathbf{w} = (2-4)\,\mathbf{i} + (4+3)\,\mathbf{j} + (7+6)\,\mathbf{k}$
$= -2\mathbf{i} + 7\mathbf{j} + 13\mathbf{k}$

21. $4\mathbf{u} = 4\,(2\mathbf{i} + 4\mathbf{j} + 7\mathbf{k}) = 8\mathbf{i} + 16\mathbf{j} + 28\mathbf{k}$
$5\mathbf{v} = 5\,(-3\mathbf{i} + 5\mathbf{j} + 2\mathbf{k}) = -15\mathbf{i} + 25\mathbf{j} + 10\mathbf{k}$
$4\mathbf{u} + 5\mathbf{v} = (8-15)\,\mathbf{i} + (16+25)\,\mathbf{j} + (28+10)\,\mathbf{k}$
$= -7\mathbf{i} + 41\mathbf{j} + 38\mathbf{k}$

23. $|\mathbf{u}| = \sqrt{2^2+4^2+7^2} = \sqrt{4+16+49} = \sqrt{69}$

25. $|\mathbf{w}+\mathbf{u}| = \sqrt{(4+2)^2 + (-3+4)^2 + (-6+7)^2}$
$= \sqrt{6^2+1^2+1^2} = \sqrt{36+1+1} = \sqrt{38}$

27. $\mathbf{v}\cdot\mathbf{w} = (-3\cdot 4) + (5\cdot -3) + (2\cdot -6)$
$= -12 - 15 - 12 = -39$

29. $\mathbf{v}\cdot\mathbf{v} = (-3\cdot -3) + (5\cdot 5) + (2\cdot 2)$
$= 9 + 25 + 4 = 38$

31. Let $\mathbf{v} = \langle 2,\ -2,\ 0\rangle$, $\mathbf{w} = \langle 5,\ -2,\ -1\rangle$

$$\begin{aligned}\mathbf{v}\cdot\mathbf{w} &= (2\cdot 5) + (-2\cdot -2) + (0\cdot -1)\\ &= 10 + 4 + 0 = 14\end{aligned}$$

$$\begin{aligned}|\mathbf{v}| &= \sqrt{2^2 + (-2)^2 + 0^2}\\ &= \sqrt{4+4} = 2\sqrt{2}\end{aligned}$$

$$\begin{aligned}|\mathbf{w}| &= \sqrt{5^2 + (-2)^2 + (-1)^2}\\ &= \sqrt{25+4+1} = \sqrt{30}\end{aligned}$$

$$\cos\theta = \frac{\mathbf{v}\cdot\mathbf{w}}{|\mathbf{v}|\cdot|\mathbf{w}|} = \frac{14}{2\sqrt{2}\cdot\sqrt{30}} \approx .903696$$

$$\theta \approx 25.352^\circ \qquad \boxed{\theta \approx 25.4^\circ}$$

33. Let $\mathbf{v} = \langle 6,\ 0,\ 0\rangle$, $\mathbf{w} = \langle 8,\ 3,\ -4\rangle$

$$\mathbf{v}\cdot\mathbf{w} = (6\cdot 8) + (0\cdot 3) + (0\cdot -4) = 48$$

$$\begin{aligned}|\mathbf{v}| &= \sqrt{6^2 + 0^2 + 0^2}\\ &= \sqrt{36} = 6\end{aligned}$$

$$\begin{aligned}|\mathbf{w}| &= \sqrt{8^2 + 3^2 + (-4)^2}\\ &= \sqrt{64+9+16} = \sqrt{89}\end{aligned}$$

$$\cos\theta = \frac{\mathbf{v}\cdot\mathbf{w}}{|\mathbf{v}|\cdot|\mathbf{w}|} = \frac{48}{6\cdot\sqrt{89}} \approx .8479983$$

$$\theta \approx 32.00538^\circ \qquad \boxed{\theta \approx 32.0^\circ}$$

35. Let $\mathbf{v} = \langle 1,\ 0,\ 0\rangle$, $\mathbf{w} = \langle 0,\ 1,\ 0\rangle$

$$\mathbf{v}\cdot\mathbf{w} = (1\cdot 0) + (0\cdot 1) + (0\cdot 0) = 0$$

$$|\mathbf{v}| = \sqrt{1^2} = 1 \qquad |\mathbf{w}| = \sqrt{1^2} = 1$$

$$\cos\theta = \frac{\mathbf{v}\cdot\mathbf{w}}{|\mathbf{v}|\cdot|\mathbf{w}|} = \frac{0}{1\cdot 1} = 0$$

$$\boxed{\theta = 90^\circ}$$

37. $\mathbf{u} = 2\mathbf{i} + 4\mathbf{j} + 7\mathbf{k}$

$$\begin{aligned}|\mathbf{u}| &= \sqrt{2^2 + 4^2 + 7^2}\\ &= \sqrt{4+16+49} = \sqrt{69}\end{aligned}$$

$$\cos\alpha = \frac{2}{\sqrt{69}} \approx .2407717$$

$$\cos\beta = \frac{4}{\sqrt{69}} \approx .4815434$$

$$\cos\gamma = \frac{7}{\sqrt{69}} \approx .84270097$$

$$\boxed{\alpha \approx 76.1^\circ,\quad \beta \approx 61.2^\circ,\quad \gamma \approx 32.6^\circ}$$

$$\text{Ck: } \left(\tfrac{2}{\sqrt{69}}\right)^2 + \left(\tfrac{4}{\sqrt{69}}\right)^2 + \left(\tfrac{7}{\sqrt{69}}\right)^2 = \tfrac{4}{69} + \tfrac{16}{69} + \tfrac{49}{69} = \tfrac{69}{69} = 1$$

39. $\mathbf{w} = 4\mathbf{i} - 3\mathbf{j} - 6\mathbf{k}$

$$\begin{aligned}|\mathbf{w}| &= \sqrt{4^2 + (-3)^2 + (-6)^2}\\ &= \sqrt{16+9+36} = \sqrt{61}\end{aligned}$$

$$\cos\alpha = \frac{4}{\sqrt{61}} \approx .5121475$$

$$\cos\beta = \frac{-3}{\sqrt{61}} \approx -.3841106$$

$$\cos\gamma = \frac{-6}{\sqrt{61}} \approx -.7682213$$

$$\boxed{\alpha \approx 59.2^\circ,\quad \beta \approx 112.6^\circ,\quad \gamma \approx 140.2^\circ}$$

$$\text{Ck: } \left(\tfrac{4}{\sqrt{61}}\right)^2 + \left(\tfrac{-3}{\sqrt{61}}\right)^2 + \left(\tfrac{-6}{\sqrt{61}}\right)^2 = \tfrac{16}{61} + \tfrac{9}{61} + \tfrac{36}{61} = \tfrac{61}{61} = 1$$

41.

$$\begin{aligned}\cos^2\gamma &= 1 - \cos^2\alpha - \cos^2\beta\\ &= 1 - (\cos 45^\circ)^2 - (\cos 120^\circ)^2\\ &= 1 - \left(\frac{\sqrt{2}}{2}\right)^2 - \left(-\frac{1}{2}\right)^2\\ &= 1 - \frac{2}{4} - \frac{1}{4}\\ &= \frac{1}{4}\end{aligned}$$

$$\cos\gamma = \frac{1}{2} \qquad \boxed{\gamma = 60^\circ}$$

43. For two vectors in space to be parallel, they must have the same position vector.

45. $\mathbf{F} = \langle 2,\ 0,\ 5 \rangle$, $\mathbf{P} = \langle 0,\ 0,\ 0 \rangle$, $\mathbf{Q} = \langle 1,\ 3,\ 2 \rangle$

$$\mathbf{PQ} = \langle (1-0),\ (3-0),\ (2-0) \rangle = \langle 1,\ 3,\ 2 \rangle$$

$$\begin{aligned}\mathbf{F} \cdot \mathbf{PQ} &= \langle 2,\ 0,\ 5 \rangle \cdot \langle 1,\ 3,\ 2 \rangle \\ &= (2 \cdot 1) + (0 \cdot 3) + (5 \cdot 2) \\ &= 2 + 0 + 10 \\ &= \boxed{12 \text{ work units}}\end{aligned}$$

47. $\mathbf{F} = \mathbf{i} + 2\mathbf{j} - \mathbf{k} = \langle 1,\ 2,\ -1 \rangle$,

$\mathbf{P} = \langle 2,\ -1,\ 2 \rangle$, $\mathbf{Q} = \langle 5,\ 7,\ 8 \rangle$

$$\mathbf{PQ} = \langle (5-2),\ (7+1),\ (8-2) \rangle = \langle 3,\ 8,\ 6 \rangle$$

$$\begin{aligned}\mathbf{F} \cdot \mathbf{PQ} &= \langle 1,\ 2,\ -1 \rangle \cdot \langle 3,\ 8,\ 6 \rangle \\ &= (1 \cdot 3) + (2 \cdot 8) + (-1 \cdot 6) \\ &= 3 + 16 - 6 \\ &= \boxed{13 \text{ work units}}\end{aligned}$$

49. $\cos^2 \alpha + \cos^2 \beta + \cos^2 \gamma$

$$\begin{aligned}&= \left(\frac{a}{|\mathbf{v}|}\right)^2 + \left(\frac{b}{|\mathbf{v}|}\right)^2 + \left(\frac{c}{|\mathbf{v}|}\right)^2 \\ &= \frac{a^2 + b^2 + c^2}{|\mathbf{v}|^2} \\ &= \frac{a^2 + b^2 + c^2}{\left(\sqrt{a^2 + b^2 + c^2}\right)^2} \\ &= \frac{a^2 + b^2 + c^2}{a^2 + b^2 + c^2} = 1\end{aligned}$$

Appendix B

1. $r = \dfrac{6}{3 + 3\sin\theta} = \dfrac{2}{1 + \sin\theta}$

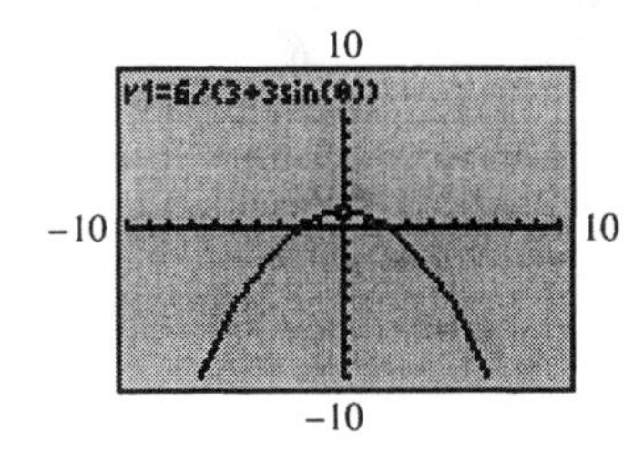

3. $r = \dfrac{-4}{6 + 2\cos\theta} = \dfrac{-2}{3 + \cos\theta}$

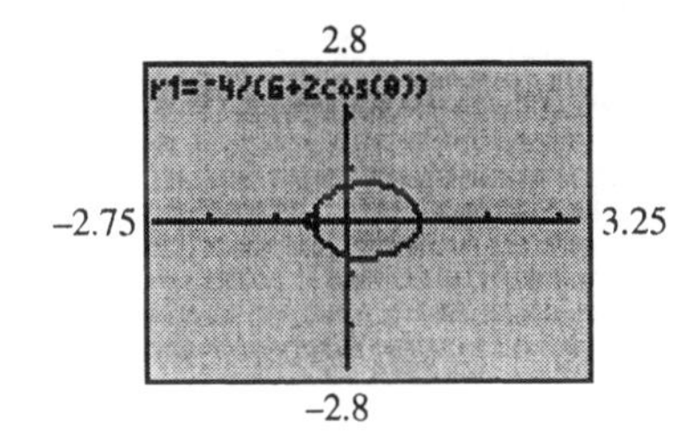

5. $r = \dfrac{2}{2 - 4\sin\theta} = \dfrac{1}{1 - 2\sin\theta}$

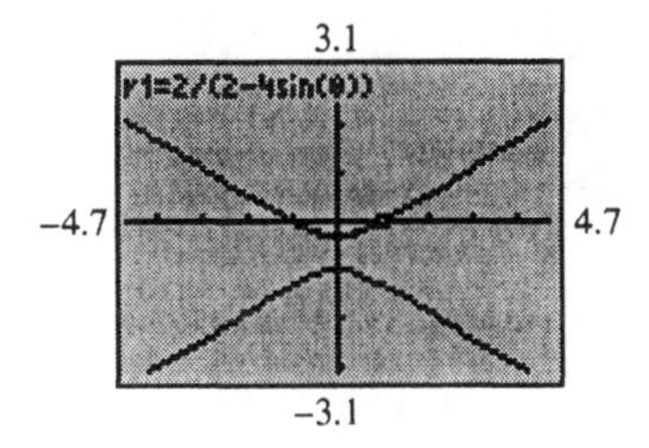

7. $r = \dfrac{4}{2 - 4\cos\theta} = \dfrac{2}{1 - 2\cos\theta}$

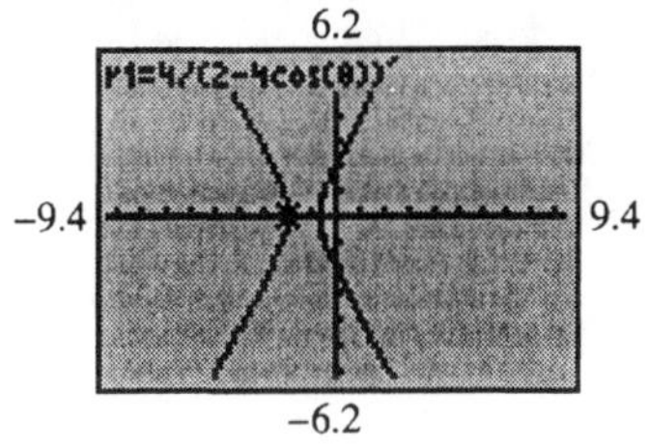

6.2
r1=4/(2−4cos(θ))
−9.4
9.4
−6.2

9. $r = \dfrac{-1}{1 + 2\sin\theta}$

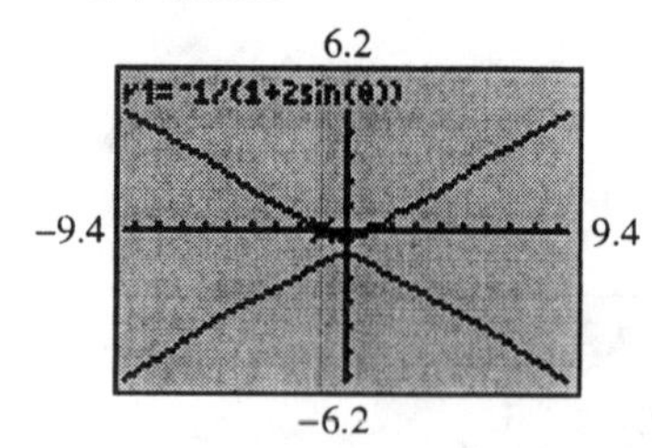

11. $r = \dfrac{-1}{2 + \cos\theta}$

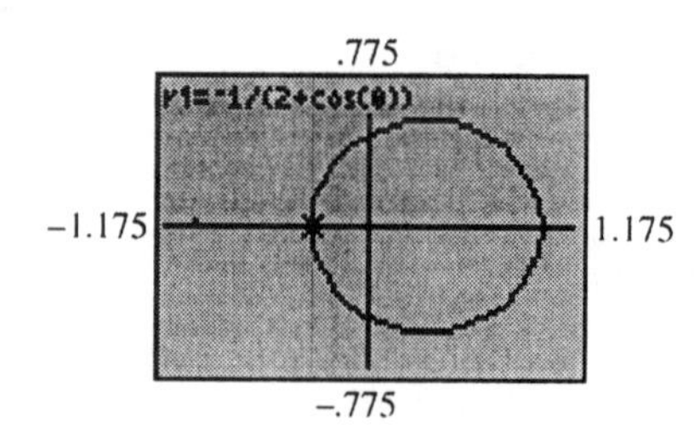

13. Parabola, vertical directrix 3 units to the right of the pole.

$$r = \frac{e\,(3)}{1 + e\cos\theta}, \quad e = 1 \text{ (parabola)}$$

$$\boxed{r = \frac{3}{1 + \cos\theta}}$$

15. Parabola, horizontal directrix 5 units below the pole.

$$r = \frac{e\,(5)}{1 - e\sin\theta}, \quad e = 1 \text{ (parabola)}$$

$$\boxed{r = \frac{5}{1 - \sin\theta}}$$

17. $e = \frac{4}{5}$; vertical directrix 5 units to the right of the pole. Since $0 \le e \le 1$, the conic is an **ellipse.**

$$r = \frac{\frac{4}{5}\,(5)}{1 + \frac{4}{5}\cos\theta} = \frac{4}{1 + \frac{4}{5}\cos\theta} \cdot \frac{5}{5}$$

$$\boxed{r = \frac{20}{5 + 4\cos\theta}}$$

19. $e = \frac{5}{4}$; horizontal directrix 8 units below the pole. Since $\frac{5}{4} > 1$, the conic is a **hyperbola.**

$$r = \frac{\frac{5}{4}\,(8)}{1 - \frac{5}{4}\sin\theta} = \frac{10}{1 - \frac{5}{4}\sin\theta} \cdot \frac{4}{4}$$

$$\boxed{r = \frac{40}{4 - 5\sin\theta}}$$

21. $r = \dfrac{6}{3 - \cos\theta} \cdot \dfrac{1/3}{1/3} = \dfrac{2}{1 - \frac{1}{3}\cos\theta}$

$$e = \frac{1}{3} \longrightarrow \textbf{ellipse}$$

$$r = \frac{6}{3 - \cos\theta}$$

$$3r - r\cos\theta = 6$$

$$3r = r\cos\theta + 6$$

$$(3r)^2 = (r\cos\theta + 6)^2$$

$$9r^2 = (x + 6)^2$$

$$9\left(x^2 + y^2\right) = x^2 + 12x + 36$$

$$9x^2 + 9y^2 = x^2 + 12x + 36$$

$$\boxed{8x^2 + 9y^2 - 12x - 36 = 0}$$

23. $r = \dfrac{-2}{1+2\cos\theta}$; $e = 2 \longrightarrow$ **hyperbola**

$$r + 2r\cos\theta = -2$$
$$r = -2r\cos\theta - 2$$
$$(r)^2 = (-2r\cos\theta - 2)^2$$
$$r^2 = (-2x-2)^2$$
$$x^2 + y^2 = 4x^2 + 8x + 4$$
$$-3x^2 + y^2 - 8x - 4 = 0$$
$$\boxed{3x^2 - y^2 + 8x + 4 = 0}$$

25. $r = \dfrac{-6}{4+2\sin\theta} \cdot \dfrac{1/4}{1/4} = \dfrac{-3/2}{1+\frac{1}{2}\sin\theta}$

$e = \dfrac{1}{2} \longrightarrow$ **ellipse**

$$r = \frac{-6}{4+2\sin\theta}$$
$$4r + 2r\sin\theta = -6$$
$$4r = -2r\sin\theta - 6$$
$$(4r)^2 = (-2r\sin\theta - 6)^2$$
$$16r^2 = (-2y-6)^2$$
$$16\left(x^2+y^2\right) = 4y^2 + 24y + 36$$
$$16x^2 + 16y^2 = 4y^2 + 24y + 36$$
$$16x^2 + 12y^2 - 24y - 36 = 0$$
$$\boxed{4x^2 + 3y^2 - 6y - 9 = 0}$$

27. $r = \dfrac{10}{2-2\sin\theta} \cdot \dfrac{1/2}{1/2} = \dfrac{5}{1-\sin\theta}$

$e = 1 \longrightarrow$ **parabola**

$$r = \frac{10}{2-2\sin\theta}$$
$$2r - 2r\sin\theta = 10$$
$$2r = 2r\sin\theta + 10$$
$$r = r\sin\theta + 5$$
$$r^2 = (r\sin\theta + 5)^2$$
$$r^2 = (y+5)^2$$
$$x^2 + y^2 = y^2 + 10y + 25$$
$$\boxed{x^2 - 10y^2 - 25 = 0}$$

Appendix C

1. $4x^2 + 3y^2 + 2xy - 5x = 8$

$$B^2 - 4AC = 2^2 - 4(4)(3) = 4 - 48 < 0$$

Circle, ellipse, or a point.

3. $2x^2 + 3xy - 4y^2 = 0$

$$B^2 - 4AC = 3^2 - 4(2)(-4) = 9 + 32 > 0$$

Hyperbola or 2 intersecting lines.

5. $4x^2 + 4xy + y^2 + 15 = 0$

$$B^2 - 4AC = 4^2 - 4(4)(1) = 16 - 16 = 0$$

Parabola, one line, or 2 parallel lines.

7. $2x^2 + \sqrt{3}xy + y^2 + x = 5$

$$\cot 2\theta = \frac{A-C}{B} = \frac{2-1}{\sqrt{3}}$$
$$\cot 2\theta = \frac{1}{\sqrt{3}} \Longrightarrow 2\theta = 60^\circ$$
$$\boxed{\theta = 30^\circ}$$

9. $3x^2 + \sqrt{3}xy + 4y^2 + 2x - 3y = 12$

$$\cot 2\theta = \frac{A-C}{B} = \frac{3-4}{\sqrt{3}}$$
$$\cot 2\theta = -\frac{1}{\sqrt{3}} \Longrightarrow 2\theta = 120^\circ$$
$$\boxed{\theta = 60^\circ}$$

11. $x^2 - 4xy + 5y^2 = 18$

$$\cot 2\theta = \frac{A-C}{B} = \frac{1-5}{-4} = 1$$
$$\cot 2\theta = 1 \Longrightarrow 2\theta = 45^\circ$$
$$\boxed{\theta = 22.5^\circ}$$

13. $x^2 - xy + y^2 = 6 \quad [\mathbf{1}]; \quad \theta = 45°$

$$x = x'\cos\theta - y'\sin\theta = \frac{\sqrt{2}}{2}x' - \frac{\sqrt{2}}{2}y' \quad [\mathbf{2}]; \qquad y = x'\sin\theta + y'\cos\theta = \frac{\sqrt{2}}{2}x' + \frac{\sqrt{2}}{2}y' \quad [\mathbf{3}]$$

$[\mathbf{2}], [\mathbf{3}] \longrightarrow [\mathbf{1}]$:

$$\left(\frac{\sqrt{2}}{2}x' - \frac{\sqrt{2}}{2}y'\right)^2 - \left(\frac{\sqrt{2}}{2}x' - \frac{\sqrt{2}}{2}y'\right)\left(\frac{\sqrt{2}}{2}x' + \frac{\sqrt{2}}{2}y'\right) + \left(\frac{\sqrt{2}}{2}x' + \frac{\sqrt{2}}{2}y'\right)^2 = 6$$

$$\frac{1}{2}x'^2 - x'y' + \frac{1}{2}y'^2 - \frac{1}{2}x'^2 + \frac{1}{2}y'^2 + \frac{1}{2}x'^2 + x'y' + \frac{1}{2}y'^2 = 6$$

$$\frac{1}{2}x'^2 + \frac{3}{2}y'^2 = 6 \longrightarrow \boxed{\frac{x'^2}{12} + \frac{y'^2}{4} = 1}$$

Graph for 13:

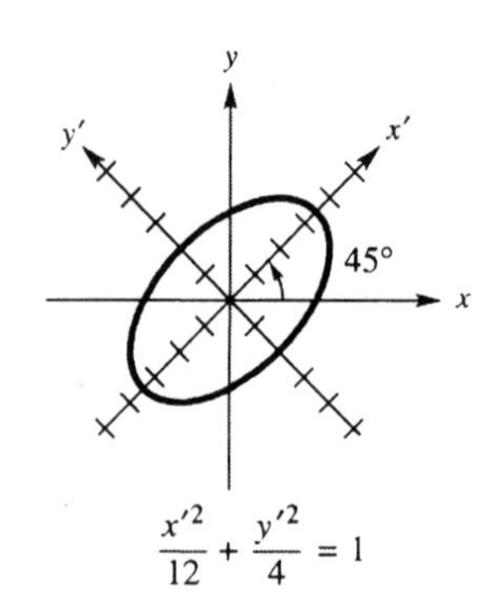

Graph for 15:

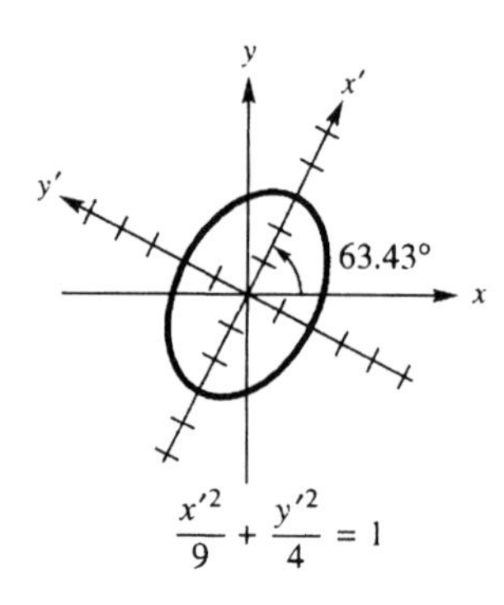

15. $8x^2 - 4xy + 5y^2 = 36 \quad [\mathbf{1}];$

$$\sin\theta = \frac{2}{\sqrt{5}}, \quad y = 2,\ r = \sqrt{5}, \quad x = \sqrt{5-4} = 1 \quad \longrightarrow \quad \cos\theta = \frac{1}{\sqrt{5}}$$

$$x = x'\cos\theta - y'\sin\theta = \frac{1}{\sqrt{5}}x' - \frac{2}{\sqrt{5}}y' \quad [\mathbf{2}]; \qquad y = x'\sin\theta + y'\cos\theta = \frac{2}{\sqrt{5}}x' + \frac{1}{\sqrt{5}}y' \quad [\mathbf{3}]$$

$[\mathbf{2}], [\mathbf{3}] \longrightarrow [\mathbf{1}]$:

$$8\left(\frac{1}{\sqrt{5}}x' - \frac{2}{\sqrt{5}}y'\right)^2 - 4\left(\frac{1}{\sqrt{5}}x' - \frac{2}{\sqrt{5}}y'\right)\left(\frac{2}{\sqrt{5}}x' + \frac{1}{\sqrt{5}}y'\right) + 5\left(\frac{2}{\sqrt{5}}x' + \frac{1}{\sqrt{5}}y'\right)^2 = 36$$

$$8\left(\frac{1}{5}x'^2 - \frac{4}{5}x'y' + \frac{4}{5}y'^2\right) - 4\left(\frac{2}{5}x'^2 - \frac{3}{5}x'y' - \frac{2}{5}y'^2\right) + 5\left(\frac{4}{5}x'^2 + \frac{4}{5}x'y' + \frac{1}{5}y'^2\right) = 36$$

$$\frac{8}{5}x'^2 - \frac{32}{5}x'y' + \frac{32}{5}y'^2 - \frac{8}{5}x'^2 + \frac{12}{5}x'y' + \frac{8}{5}y'^2 + 4x'^2 + 4x'y' + y'^2 = 36$$

$$4x'^2 + 9y'^2 = 36 \longrightarrow \boxed{\frac{x'^2}{9} + \frac{y'^2}{4} = 1}$$

17. $3x^2 - 2xy + 3y^2 = 8 \quad [\mathbf{1}]$

$$\cot 2\theta = \frac{A - C}{B} = \frac{3 - 3}{-2} = 0 \longrightarrow 2\theta = 90^\circ \longrightarrow \theta = 45^\circ$$

$$x = x' \cos\theta - y' \sin\theta = \frac{\sqrt{2}}{2}x' - \frac{\sqrt{2}}{2}y' \quad [\mathbf{2}]; \qquad y = x' \sin\theta + y' \cos\theta = \frac{\sqrt{2}}{2}x' + \frac{\sqrt{2}}{2}y' \quad [\mathbf{3}]$$

$[\mathbf{2}]$, $[\mathbf{3}] \longrightarrow [\mathbf{1}]$:

$$3\left(\frac{\sqrt{2}}{2}x' - \frac{\sqrt{2}}{2}y'\right)^2 - 2\left(\frac{\sqrt{2}}{2}x' - \frac{\sqrt{2}}{2}y'\right)\left(\frac{\sqrt{2}}{2}x' + \frac{\sqrt{2}}{2}y'\right) + 3\left(\frac{\sqrt{2}}{2}x' + \frac{\sqrt{2}}{2}y'\right)^2 = 8$$

$$3\left(\frac{1}{2}x'^2 - x'y' + \frac{1}{2}y'^2\right) - 2\left(\frac{1}{2}x'^2 - \frac{1}{2}y'^2\right) + 3\left(\frac{1}{2}x'^2 + x'y' + \frac{1}{2}y'^2\right) = 8$$

$$\frac{3}{2}x'^2 - 3x'y' + \frac{3}{2}y'^2 - x'^2 + y'^2 + \frac{3}{2}x'^2 + 3x'y' + \frac{3}{2}y'^2 = 8$$

$$2x'^2 + 4y'^2 = 8 \quad \longrightarrow \quad \boxed{\frac{x'^2}{4} + \frac{y'^2}{2} = 1}$$

Graph for 17:

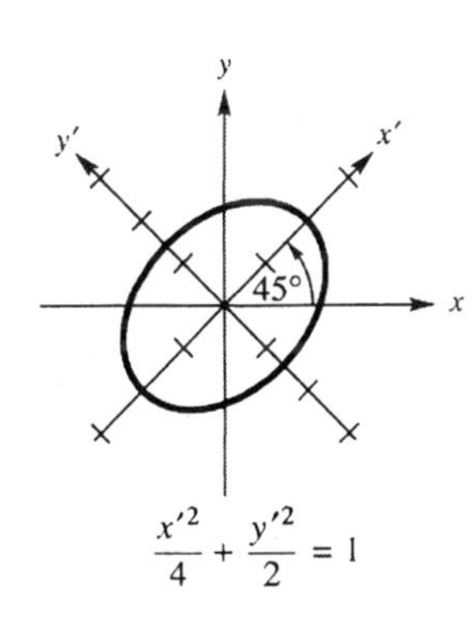

Graph for 19:

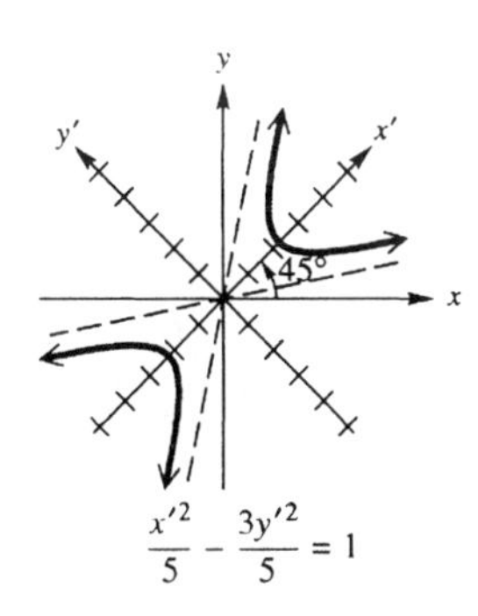

19. $x^2 - 4xy + y^2 = -5 \quad [\mathbf{1}]$

$$\cot 2\theta = \frac{A - C}{B} = \frac{1 - 1}{-4} = 0 \longrightarrow 2\theta = 90^\circ \longrightarrow \theta = 45^\circ$$

$$x = x' \cos\theta - y' \sin\theta = \frac{\sqrt{2}}{2}x' - \frac{\sqrt{2}}{2}y' \quad [\mathbf{2}]; \qquad y = x' \sin\theta + y' \cos\theta = \frac{\sqrt{2}}{2}x' + \frac{\sqrt{2}}{2}y' \quad [\mathbf{3}]$$

$[\mathbf{2}]$, $[\mathbf{3}] \longrightarrow [\mathbf{1}]$:

$$\left(\frac{\sqrt{2}}{2}x' - \frac{\sqrt{2}}{2}y'\right)^2 - 4\left(\frac{\sqrt{2}}{2}x' - \frac{\sqrt{2}}{2}y'\right)\left(\frac{\sqrt{2}}{2}x' + \frac{\sqrt{2}}{2}y'\right) + \left(\frac{\sqrt{2}}{2}x' + \frac{\sqrt{2}}{2}y'\right)^2 = -5$$

$$\frac{1}{2}x'^2 - x'y' + \frac{1}{2}y'^2 - 4\left(\frac{1}{2}x'^2 - \frac{1}{2}y'^2\right) + \frac{1}{2}x'^2 + x'y' + \frac{1}{2}y'^2 = -5$$

$$\frac{1}{2}x'^2 - x'y' + \frac{1}{2}y'^2 - 2x'^2 + 2y'^2 + \frac{1}{2}x'^2 + x'y' + \frac{1}{2}y'^2 = -5$$

$$-x'^2 + 3y'^2 = -5 \quad \longrightarrow \quad \boxed{\frac{x'^2}{5} - \frac{3y'^2}{5} = 1}$$

21. $7x^2 + 6\sqrt{3}xy + 13y^2 = 64 \quad [\mathbf{1}]$

$$\cot 2\theta = \frac{A-C}{B} = \frac{7-13}{6\sqrt{3}} = \frac{-6}{6\sqrt{3}} = -\frac{1}{\sqrt{3}} \longrightarrow 2\theta = 120^\circ \longrightarrow \theta = 60^\circ$$

$$x = x'\cos\theta - y'\sin\theta = \frac{1}{2}x' - \frac{\sqrt{3}}{2}y' \quad [\mathbf{2}]; \qquad y = x'\sin\theta + y'\cos\theta = \frac{\sqrt{3}}{2}x' + \frac{1}{2}y' \quad [\mathbf{3}]$$

$[\mathbf{2}], [\mathbf{3}] \longrightarrow [\mathbf{1}]:$

$$7\left(\frac{1}{2}x' - \frac{\sqrt{3}}{2}y'\right)^2 + 6\sqrt{3}\left(\frac{1}{2}x' - \frac{\sqrt{3}}{2}y'\right)\left(\frac{\sqrt{3}}{2}x' + \frac{1}{2}y'\right) + 13\left(\frac{\sqrt{3}}{2}x' + \frac{1}{2}y'\right)^2 = 64$$

$$7\left(\frac{1}{4}x'^2 - \frac{\sqrt{3}}{2}x'y' + \frac{3}{4}y'^2\right) + 6\sqrt{3}\left(\frac{\sqrt{3}}{4}x'^2 - \frac{1}{2}x'y' - \frac{\sqrt{3}}{4}y'^2\right) + 13\left(\frac{3}{4}x'^2 + \frac{\sqrt{3}}{2}x'y' + \frac{1}{4}y'^2\right) = 64$$

$$\frac{7}{4}x'^2 - \frac{7\sqrt{3}}{2}x'y' + \frac{21}{4}y'^2 + \frac{18}{4}x'^2 - \frac{6\sqrt{3}}{2}x'y' - \frac{18}{4}y'^2 + \frac{39}{4}x'^2 + \frac{13\sqrt{3}}{2}x'y' + \frac{13}{4}y'^2 = 64$$

$$16x'^2 + 4y'^2 = 64 \longrightarrow \boxed{\frac{x'^2}{4} + \frac{y'^2}{16} = 1}$$

Graph for 21:

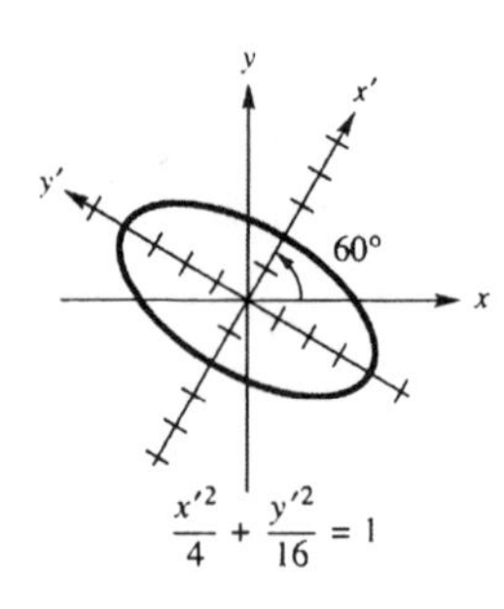

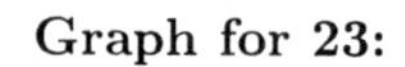
Graph for 23:

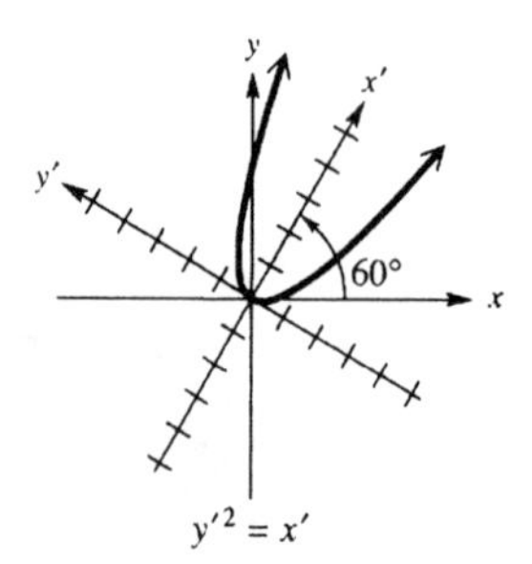

23. $3x^2 - 2\sqrt{3}xy + y^2 - 2x - 2\sqrt{3}y = 0 \quad [\mathbf{1}]$

$$\cot 2\theta = \frac{A-C}{B} = \frac{3-1}{-2\sqrt{3}} = -\frac{1}{\sqrt{3}} \longrightarrow 2\theta = 120^\circ \longrightarrow \theta = 60^\circ$$

$$x = x'\cos\theta - y'\sin\theta = \frac{1}{2}x' - \frac{\sqrt{3}}{2}y' \quad [\mathbf{2}]; \qquad y = x'\sin\theta + y'\cos\theta = \frac{\sqrt{3}}{2}x' + \frac{1}{2}y' \quad [\mathbf{3}]$$

$[\mathbf{2}], [\mathbf{3}] \longrightarrow [\mathbf{1}]:$

$$3\left(\frac{1}{2}x' - \frac{\sqrt{3}}{2}y'\right)^2 - 2\sqrt{3}\left(\frac{1}{2}x' - \frac{\sqrt{3}}{2}y'\right)\left(\frac{\sqrt{3}}{2}x' + \frac{1}{2}y'\right) + \left(\frac{\sqrt{3}}{2}x' + \frac{1}{2}y'\right)^2$$

$$-2\left(\frac{1}{2}x' - \frac{\sqrt{3}}{2}y'\right) - 2\sqrt{3}\left(\frac{\sqrt{3}}{2}x' + \frac{1}{2}y'\right) = 0$$

$$3\left(\frac{1}{4}x'^2 - \frac{\sqrt{3}}{2}x'y' + \frac{3}{4}y'^2\right) - 2\sqrt{3}\left(\frac{\sqrt{3}}{4}x'^2 - \frac{1}{2}x'y' - \frac{\sqrt{3}}{4}y'^2\right) + \left(\frac{3}{4}x'^2 + \frac{\sqrt{3}}{2}x'y' + \frac{1}{4}y'^2\right)$$

$$-x' + \sqrt{3}y' - 3x' - \sqrt{3}y' = 0$$

$$\frac{3}{4}x'^2 - \frac{3\sqrt{3}}{2}x'y' + \frac{9}{4}y'^2 - \frac{3}{2}x'^2 + \sqrt{3}x'y' + \frac{3}{2}y'^2 + \frac{3}{4}x'^2 + \frac{\sqrt{3}}{2}x'y' + \frac{1}{4}y'^2 - 4x' = 0$$

$$4y'^2 - 4x' = 0 \longrightarrow 4y'^2 = 4x' \longrightarrow \boxed{y'^2 = x'}$$

25. $x^2 + 3xy + y^2 - 5\sqrt{2}y = 15$ **[1]**

$$\cot 2\theta = \frac{A - C}{B} = \frac{1-1}{3} = 0 \longrightarrow 2\theta = 90^\circ \longrightarrow \theta = 45^\circ$$

$$x = x'\cos\theta - y'\sin\theta = \frac{\sqrt{2}}{2}x' - \frac{\sqrt{2}}{2}y' \quad [\mathbf{2}]; \qquad y = x'\sin\theta + y'\cos\theta = \frac{\sqrt{2}}{2}x' + \frac{\sqrt{2}}{2}y' \quad [\mathbf{3}]$$

[2], **[3]** $\longrightarrow$ **[1]** :

$$\left(\frac{\sqrt{2}}{2}x' - \frac{\sqrt{2}}{2}y'\right)^2 + 3\left(\frac{\sqrt{2}}{2}x' - \frac{\sqrt{2}}{2}y'\right)\left(\frac{\sqrt{2}}{2}x' + \frac{\sqrt{2}}{2}y'\right) + \left(\frac{\sqrt{2}}{2}x' + \frac{\sqrt{2}}{2}y'\right)^2 - 5\sqrt{2}\left(\frac{\sqrt{2}}{2}x' + \frac{\sqrt{2}}{2}y'\right) = 15$$

$$\frac{1}{2}x'^2 - x'y' + \frac{1}{2}y'^2 + 3\left(\frac{1}{2}x'^2 - \frac{1}{2}y'^2\right) + \frac{1}{2}x'^2 + x'y' + \frac{1}{2}y'^2 - 5x' - 5y' = 15$$

$$\frac{1}{2}x'^2 - x'y' + \frac{1}{2}y'^2 + \frac{3}{2}x'^2 - \frac{3}{2}y'^2 + \frac{1}{2}x'^2 + x'y' + \frac{1}{2}y'^2 - 5x' - 5y' = 15$$

$$\frac{5}{2}x'^2 - \frac{1}{2}y'^2 - 5x' - 5y' = 15$$

$$5x'^2 - 10x' - y'^2 - 10y' = 30$$

$$5\left(x'^2 - 2x' + 1\right) - \left(y'^2 + 10y' + 25\right) = 30 + 5 - 25$$

$$5(x'-1)^2 - (y'+5)^2 = 10 \longrightarrow \boxed{\frac{(x'-1)^2}{2} - \frac{(y'+5)^2}{10} = 1}$$

The graph of the equation is a hyperbola with its center at $(1, -5)$. By translating the axes of the $x'y'$-system down 5 units and right 1 unit, we get an $x''y''$-coordinate system, in which the hyperbola is centered at the origin. Thus $\dfrac{x''^2}{2} - \dfrac{y''^2}{10} = 1$.

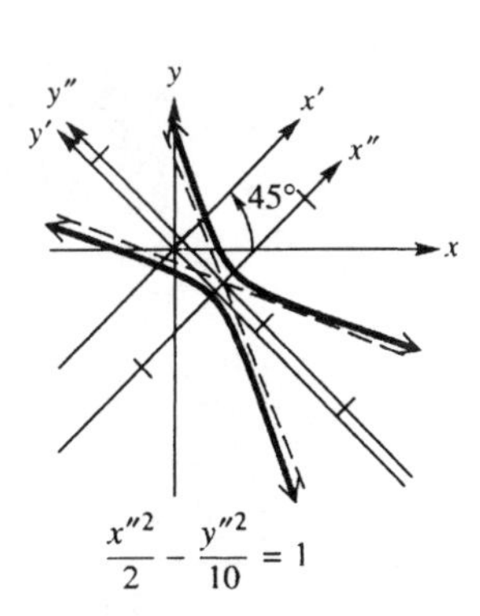

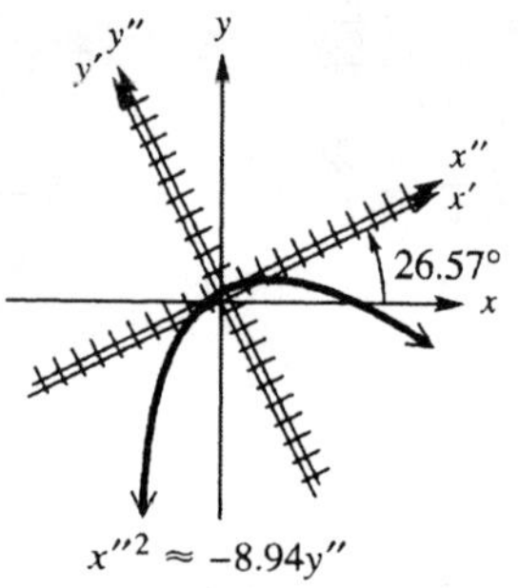

27. $4x^2 + 4xy + y^2 - 24x + 38y - 19 = 0 \quad [1]$

$$\cot 2\theta = \frac{A-C}{B} = \frac{4-1}{4} = \frac{3}{4} \longrightarrow 2\theta \approx 53.13^\circ \longrightarrow \theta \approx 26.57^\circ$$

For $2\theta : x = 3,\ y = 4,\ r = \sqrt{3^2+4^2} = 5 \quad \longrightarrow \quad \cos 2\theta = \frac{3}{5}$

$$\sin\theta = \sqrt{\frac{1-\cos 2\theta}{2}} = \sqrt{\frac{1-\frac{3}{5}}{2}} = \sqrt{\frac{2}{10}} = \frac{\sqrt{5}}{5}$$

$$\cos\theta = \sqrt{\frac{1+\cos 2\theta}{2}} = \sqrt{\frac{1+\frac{3}{5}}{2}} = \sqrt{\frac{8}{10}} = \frac{2\sqrt{5}}{5}$$

x''

x'

y'', y'

y

x

26.57°

$x''^2 \approx -8.94y''$

$$x = x'\cos\theta - y'\sin\theta = \frac{2\sqrt{5}}{5}x' - \frac{\sqrt{5}}{5}y' \quad [2]; \qquad y = x'\sin\theta + y'\cos\theta = \frac{\sqrt{5}}{5}x' + \frac{2\sqrt{5}}{5}y' \quad [3]$$

$[2],\ [3] \longrightarrow [1]:$

$$4\left(\frac{2\sqrt{5}}{5}x' - \frac{\sqrt{5}}{5}y'\right)^2 + 4\left(\frac{2\sqrt{5}}{5}x' - \frac{\sqrt{5}}{5}y'\right)\left(\frac{\sqrt{5}}{5}x' + \frac{2\sqrt{5}}{5}y'\right) - \left(\frac{\sqrt{5}}{5}x' + \frac{2\sqrt{5}}{5}y'\right)^2$$

$$-24\left(\frac{2\sqrt{5}}{5}x' - \frac{\sqrt{5}}{5}y'\right) + 38\left(\frac{\sqrt{5}}{5}x' + \frac{2\sqrt{5}}{5}y'\right) = 19$$

$$4\left(\frac{4}{5}x'^2 - \frac{4}{5}x'y' + \frac{1}{5}y'^2\right) + 4\left(\frac{2}{5}x'^2 + \frac{3}{5}x'y' - \frac{2}{5}y'^2\right) + \left(\frac{1}{5}x'^2 + \frac{4}{5}x'y' + \frac{4}{5}y'^2\right)$$

$$-\frac{48\sqrt{5}}{5}x' + \frac{24\sqrt{5}}{5}y' + \frac{38\sqrt{5}}{5}x' + \frac{76\sqrt{5}}{5}y' = 19$$

$$\frac{16}{5}x'^2 - \frac{16}{5}x'y' + \frac{4}{5}y'^2 + \frac{8}{5}x'^2 + \frac{12}{5}x'y' - \frac{8}{5}y'^2 + \frac{1}{5}x'^2 + \frac{4}{5}x'y' + \frac{4}{5}y'^2 - \frac{48\sqrt{5}}{5}x'$$

$$+\frac{24\sqrt{5}}{5}y' + \frac{38\sqrt{5}}{5}x' + \frac{76\sqrt{5}}{5}y' = 19$$

$$5x'^2 - 2\sqrt{5}x' + 20\sqrt{5}y' = 19$$

$$5\left(x'^2 - \frac{2\sqrt{5}}{5}x' + \frac{1}{5}\right) + 20\sqrt{5}y' = 19 + 1$$

$$5\left(x' - \frac{\sqrt{5}}{5}\right)^2 = 20 - 20\sqrt{5}y$$

$$\left(x' - \frac{\sqrt{5}}{5}\right)^2 = 4 - 4\sqrt{5}y' \quad \longrightarrow \quad \boxed{\left(x' - \frac{\sqrt{5}}{5}\right)^2 = -4\sqrt{5}\left(y' - \frac{\sqrt{5}}{5}\right)}$$

The graph of the equation is a parabola with its vertex at $\left(\frac{\sqrt{5}}{5}, \frac{\sqrt{5}}{5}\right)$. By translating the axes of the $x'y'$-system up $\frac{\sqrt{5}}{5}$ units and left $\frac{\sqrt{5}}{5}$ unit, we get an $x''y''$-coordinate system, in which the parabola is centered at the origin. Thus $x''^2 = -4\sqrt{5}y''$

29. $16x^2+24xy+9y^2-130x+90y=0$ [**1**]

$$\cot 2\theta=\frac{A-C}{B}=\frac{16-9}{24}=\frac{7}{24}\longrightarrow 2\theta\approx 73.74^\circ\longrightarrow\theta\approx 36.87^\circ$$

For $2\theta: x=7,\ y=24,\ r=\sqrt{7^2+24^2}=25 \longrightarrow \cos 2\theta=\frac{7}{25}$

$$\sin\theta=\sqrt{\frac{1-\cos 2\theta}{2}}=\sqrt{\frac{1-\frac{7}{25}}{2}}=\sqrt{\frac{18}{50}}=\frac{3}{5}$$

$$\cos\theta=\sqrt{\frac{1+\cos 2\theta}{2}}=\sqrt{\frac{1+\frac{7}{25}}{2}}=\sqrt{\frac{32}{50}}=\frac{4}{5}$$

$x=x'\cos\theta-y'\sin\theta=\frac{4}{5}x'-\frac{3}{5}y'$ [**2**]; $\quad y=x'\sin\theta+y'\cos\theta=\frac{3}{5}x'+\frac{4}{5}y'$ [**3**]

[**2**], [**3**] $\longrightarrow$ [**1**] :

$$16\left(\frac{4}{5}x'-\frac{3}{5}y'\right)^2+24\left(\frac{4}{5}x'-\frac{3}{5}y'\right)\left(\frac{3}{5}x'+\frac{4}{5}y'\right)+9\left(\frac{3}{5}x'+\frac{4}{5}y'\right)^2-130\left(\frac{4}{5}x'-\frac{3}{5}y'\right)+90\left(\frac{3}{5}x'+\frac{4}{5}y'\right)=0$$

$$16\left(\frac{16}{25}x'^2-\frac{24}{25}x'y'+\frac{9}{25}y'^2\right)+24\left(\frac{12}{25}x'^2+\frac{7}{25}x'y'-\frac{12}{25}y'^2\right)+9\left(\frac{9}{25}x'^2+\frac{24}{25}x'y'+\frac{16}{25}y'^2\right)$$

$$-104x'+78y'+54x'+72y'=0$$

$$\frac{256}{25}x'^2-\frac{384}{25}x'y'+\frac{144}{25}y'^2+\frac{288}{25}x'^2+\frac{168}{25}x'y'-\frac{288}{25}y'^2+\frac{81}{25}x'^2+\frac{216}{25}x'y'+\frac{144}{25}y'^2-50x'+150y'=0$$

$$25x'^2-50x'+150y'=0$$

$$25\left(x'^2-2x'+1\right)=-150y'+25$$

$$25(x'-1)^2=-150\left(y'-\frac{1}{6}\right)\longrightarrow\boxed{(x'-1)^2=-6\left(y'-\frac{1}{6}\right)}$$

The graph of the equation is a parabola with its vertex at $\left(1,\ \frac{1}{6}\right)$. By translating the axes of the $x'y'$-system up $\frac{1}{6}$ units and right 1 unit, we get an $x''y''$-coordinate system, in which the parabola is centered at the origin. Thus $x''^2=-6y''$

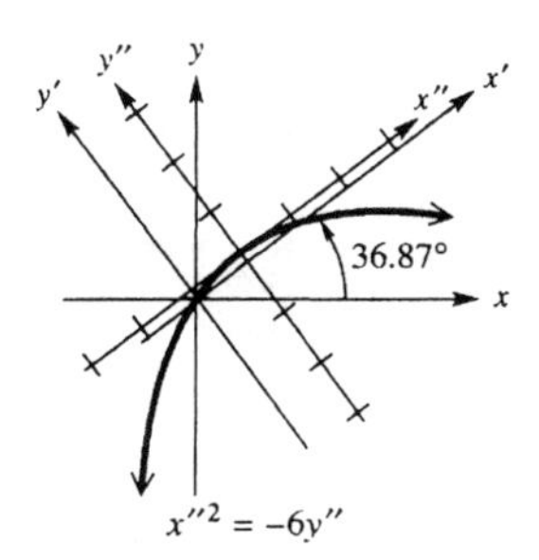

31. The purpose of rotation is to remove the xy-term. If the coefficient of this term is 0, $\cot 2\theta$ is undefined, which implies that $2\theta=0^\circ$, $\theta=0^\circ$.